Precalculus Mathematics

Precalculus Mathematics

FIFTH EDITION

Max A. Sobel
Montclair State University

Norbert Lerner
State University of New York at Cortland

Prentice Hall

Upper Saddle River, New Jersey 07458

Library of Congress Cataloging-in-Publication Data

Sobel, Max A.
  Precalculus mathematics/Max Sobel, Norbert Lerner.—5th ed.
        p.   cm.
  Includes index.
  ISBN 0-13-112095-6
  1. Algebra.   2. Trigonometry.   3. Functions.   I. Lerner, Norbert.   II. Title.
QA154.2.S63   1995   94-11433
512'.1—dc20   CIP

Acquisitions Editor: Sally Denlow
Editorial Production/Supervision: Rachel J. Witty, Letter Perfect, Inc.
Senior Production Editor: Elaine Wetterau
Buyer: Trudy Pisciotti
Marketing Manager: Karie Jabe
Art Director: Amy Rosen
Creative Director: Paula Maylahn
Interior Design and Layout: Judith A. Matz-Coniglio
Page Layout: Meryl Poweski
Cover Designer: De Franco Design, Inc.
Cover Art: Patrick Gorman
Photo Editor: Lorinda Morris-Nantz
Photo Researcher: Rhoda Sidney
Copy Editor: Donna Mulder
Proofreaders: Debby Goldman, Tamar M. Solomon, Reva Goldman
Editorial Assistant: Joanne Wendelkin

©2000 by PRENTICE-HALL, INC.
A Pearson Education Company
Upper Saddle River, New Jersey 07458

Printed in the United States of America
10 9 8 7 6 5 4

ISBN 0-13-112095-6

Prentice-Hall International (UK) Limited,London
Prentice-Hall of Australia Pty. Limited, Sydney
Prentice-Hall Canada Inc., Toronto
Prentice-Hall Hispanoamericana, S.A., Mexico
Prentice-Hall of India Private Limited, New Delhi
Prentice-Hall of Japan, Inc., Tokyo
Pearson Education Asia Pte. Ltd., Singapore
Editora Prentice-Hall do Brasil, Ltda., Rio de Janeiro

Contents

v

4 Circles, Additional Curves, and the Algebra of Functions   *247*

5 Exponential and Logarithmic Functions   *305*

6 The Trigonometric Functions   *361*

7 Right Triangle Trigonometry, Identities, and Equations   *425*

8 Additional Applications of Trigonometry *489*

9 Linear Systems, Matrices, and Determinants *549*

10 The Conic Sections *633*

Preface

This fifth edition of *Precalculus Mathematics* has been written to provide the essential concepts and skills of algebra, trigonometry, and the study of functions that are needed for further study in mathematics. Since calculus is a subject numerous students study after this course, a special emphasis is given to the preparation for the study of calculus. Thus, one of the *major objectives* of the book is *to help you make a comfortable transition from elementary mathematics to calculus.* (However, the objectives of your particular course may not require this special pre-calculus emphasis. You will find that out from your instructor or from the course syllabus.)

A major difficulty that students have in future mathematics courses, especially in calculus, involves the lack of adequate algebraic skills. To help you overcome such deficiencies, an extensive review of the fundamentals of algebra has been included in Chapter 1. You are encouraged to refer to this chapter throughout the course if you encounter algebraic difficulties.

Of special interest are the numerous features in this book, which have been designed to assist you in learning the subject matter of the course. These features are listed below, with descriptions of their purpose and suggested use.

Margin Notes

Throughout the text, notes appear in the margin to enhance the exposition, raise questions, point out interesting facts, explain why some things are done as they are, show alternate procedures, provide historical items of interest, give references and reminders, and caution you to avoid certain common errors.

Test Your Understanding

These are short sets of exercises (in addition to the end-of-section exercises) that are found within almost every section of the text. These encourage you to test your knowledge of new material just developed, prior to attempting to solve the exercises at the end of each section. Answers to all of these are given at the end of each chapter, and thus provide an excellent means of self-study.

Caution Items

Where appropriate, these items appear in the margin notes or in the text and alert you to the typical kinds of errors that you should avoid. Sometimes caution items will show the errors and the corrections side by side as demonstrated in the following CAUTION, which appears in Section 5.3:

| CAUTION: Learn to Avoid These Mistakes ||
WRONG	**RIGHT**
$\log_b A + \log_b B = \log_b(A + B)$	$\log_b A + \log_b B = \log_b AB$
$\log_b(x^2 - 4) = \log_b x^2 - \log_b 4$	$\log_b(x^2 - 4)$ $= \log_b(x + 2)(x - 2)$ $= \log_b(x + 2) + \log_b(x - 2)$
$(\log_b x)^2 = 2 \log_b x$	$(\log_b x)^2 = (\log_b x)(\log_b x)$
$\log_b A - \log_b B = \dfrac{\log_b A}{\log_b B}$	$\log_b A - \log_b B = \log_b \dfrac{A}{B}$
If $2 \log_b x = \log_b(3x + 4)$, then $2x = 3x + 4$.	If $2 \log_b x = \log_b(3x + 4)$, then $\log_b x^2 = \log_b(3x + 4)$ and $x^2 = 3x + 4$.
$\log_b \dfrac{x}{2} = \dfrac{\log_b x}{2}$	$\log_b \dfrac{x}{2} = \log_b x - \log_b 2$
$\log_b(x^2 + 2) = 2 \log_b(x + 2)$	$\log_b(x^2 + 2)$ cannot be simplified further.

Illustrative Examples and Exercises

The text contains numerous illustrative examples with detailed solutions, designed to help you to understand new concepts and learn new skills. You should study these examples, and carefully follow each step with paper and pencil in hand. If this is done, you will be well prepared for the exercises at the end of each section.

Written Assignments

It is generally agreed that students need to be able to practice writing skills in their mathematics courses. Therefore, throughout the book, written assignments will be found that ask you to write an explanation or description rather than just to solve a problem. These are designated in the text by this symbol: ▭⟹

Challenges

Near the end of almost every exercise set, you will be challenged to solve a problem that is more difficult than the typical exercises, or that has an unusual twist to it. Here is a Challenge that appears in Section 1.2 that can be solved using basic arithmetic and logic. Try it!

⁇ **Challenge**   Two boats begin their journeys back and forth across a river at the same time, but from opposite sides of the river. The first time that they pass each other, they are 700 feet from one of the shores of the river. After they each make one turn, they pass each other once again at a distance of 400 feet from the other shore. Assuming that each boat goes at a constant speed and that there is no loss of time making the turn, how wide is the river?

To solve this problem, you might consider the problem-solving strategy of drawing a diagram and then using it to find the solution as follows:

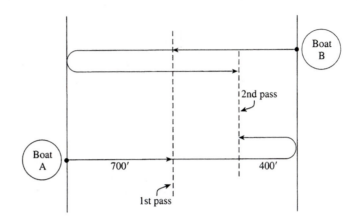

Critical Thinking

Within each chapter there are brief sets of questions that encourage you to think critically. These require higher-order thinking skills and do not depend alone on the routine application of basic skills. Critical thinking has become an important educational objective.

Critical Thinking exercises, the Challenges described above, and numerous other more difficult questions found throughout the exercise sets are of particular value in developing your creative thinking skills. The following illustration of Critical Thinking exercises is presented at the end of Section 1.9.

Critical Thinking

1. Provide a convincing argument to explain why $3\left(\dfrac{x+1}{x-1}\right) \neq \dfrac{3x+1}{3x-1}$.

2. Note that $\dfrac{12}{\sqrt{150}} = \dfrac{2}{5}\sqrt{6}$. Is either form preferable to the other? Explain your answer.

3. Translate the following statement into symbolic form:

   The square of the sum of two numbers is at least as large as four times the product.

   Prove the statement is true. *Hint:* Assume the inequality is true and work backward.

4. Three numbers can be arranged vertically in fraction form to produce these cases:

$$\dfrac{\dfrac{a}{b}}{c} = \dfrac{a}{bc} \qquad \dfrac{a}{\dfrac{b}{c}} = \dfrac{ac}{b}$$

   Find all possible distinct cases using the four numbers a, b, c, and d arranged vertically in fractional form.

Boxed Displays

Boxed displays that highlight important results, definitions, formulas, and summaries appear throughout the text and serve to alert you to major concepts and results.

Review Exercises

At the end of each chapter, a set of exercises, arranged by section, can be used as a review of the chapter prior to taking the chapter tests. (Answers are given at the back of the book.)

Chapter Tests

Each chapter concludes with two forms of a chapter test: standard answer and multiple choice. You should use these to test your knowledge of the work of the chapter, and check your answers with those provided at the back of the book.

Graphing Calculator Exercises

At the end of almost all exercise sets (after Chapter 1), there is an additional set of exercises that require the use of a graphing calculator. Since these are optional, your instructor will let you know whether or not they are applicable for your class. There is also a *Graphing Calculator Appendix* in the back of the book to provide you with some assistance in using a graphing calculator. Even if a graphing calculator is not required, you will still need a scientific calculator throughout the course of study. As a minimum, your calculator should include keys such as the following:

We thank Prof. Kenneth Kalmanson of Montclair State University for contributing the many graphing calculator exercises and the Graphing Calculator Appendix that are new in this edition.

Inside Covers

The inside of the covers contain summaries of useful information. The inside front cover contains algebraic and geometric formulas; the back contains a collection of basic graphs and a summary of trigonometric formulas.

Supplements for the Student

• Solutions Manual
There is a Solutions Manual available for this text that includes worked-out solutions to the odd-numbered exercises in every section, and to all problems in the Chapter Review Exercises and the Chapter Tests. Answers to the Graphing Calculator Exercises are also included in the Solutions Manual. Note that answers to the odd-numbered exercises in each section, and to all Chapter Review Exercises and Chapter Test questions are given at the back of this text.

• Visual Precalculus
This software is available through college bookstores at a minimal cost. It contains over 20 routines that provide additional insight into topics discussed in the text.

How does one succeed in a mathematics course? Unfortunately, there is no universal prescription guaranteed to work. However, our experience with many students suggests that the most important thing to do is to *get involved in the mathematical process*. Don't use this text only as a source for exercises. Rather, read the book, attend class regularly, study your class notes, and make use of the special features described in this Preface. Furthermore, keep up to date. Don't let yourself fall behind; it often leads to poor results. If difficulties arise and you begin to fall behind, then don't hesitate to get additional assistance from your instructor or from a classmate. Working together with a friend can often be beneficial, as long as individual efforts are also made.

Be positive! Don't give up! We are convinced that even if you have an initial negative attitude toward mathematics, an honest attempt to learn it properly will result not only in greater success, but will lead to a self-awareness that you have more ability and talent than you ever gave yourself credit for!

We hope that you will find this book enjoyable, and that you will learn the skills and concepts for future study of mathematics. We encourage you to write to us; your comments, criticisms, and suggestions might be useful for future editions. Also, despite all of our efforts, errors may occasionally creep into a book. It would be appreciated if you would call any of these to our attention. We promise to respond with our letter of thanks.

MAX A. SOBEL
NORBERT LERNER

Supplements for the Instructor
• **Annotated Instructor's Edition**
Contains teaching notes in the margins and suggested homework assignments from the authors, as well as answers to every exercise.

• **Instructor's Solutions Manual**
Contains worked solutions to all even-numbered problems in the book (worked solutions to all odd-numbered exercises are in the Student Solutions Manual, which is available to the instructor). Also included are the solutions to the Challenges, Written Assignments, Critical Thinking exercises, and Graphing Calculator Exercises.

• **Transparencies**
An assortment of masters with helpful illustrations and examples from the text.

• **Test Item File**
A printed version of our algorithmically driven testing software with multiple choice and free-answer exams for each chapter.

• **TestPro**
This versatile testing system allows the instructor to easily create up to 99 versions of a customized text. Users may add their own test items and create existing items in Lotus WYSIWYG format. Each objective in the text has at least one multiple choice and free-response algorithm. IBM format only.

• **Graphing Calculator Videos**
Tutorial videos that cover basic principles and operations of the TI-81 and Casio 7700-G graphing calculators for precalculus. To receive your free copy (upon adoption) of the videos, please contact your local Prentice Hall representative. To find out who your local representative is, call Prentice Hall Faculty Services at 1-800-526-0485.

Precalculus Mathematics

CHAPTER

1

Fundamentals of Algebra

Mathematician Pierre de Fermat, 1601–1665.
Source: Art Resource/Giraudon.

Mathematics is often viewed as a subject that never changes. This is far from the truth! Dramatic new discoveries are not uncommon. For example, in 1995 Dr. Andrew Wiles, a British mathematician working at Princeton University, discovered a proof of **Fermat's Last Theorem,** which had eluded some of the best mathematical scholars for more than 300 years.

Fermat's theorem states that it is not possible to find positive integers x, y, and z such that $x^n + y^n = z^n$ if n is any integer greater than 2. That is, it is not possible to find positive integers such that

$$x^3 + y^3 = z^3 \quad \text{or} \quad x^4 + y^4 = z^4, \text{ etc.}$$

(Of course, there is no problem when $n = 2$. Why not?) Fermat claimed to have found a proof of this theorem but said that he was unable to reproduce it in the narrow margins of his book.

Dr. Wiles, using techniques unknown in Fermat's time, discovered a proof that covered about 200 pages! Did Fermat really have a simpler proof? Will anyone ever find such a proof? The challenge remains.

1.1
REAL NUMBERS AND THEIR PROPERTIES

Although mathematics is constantly growing and changing, the fundamentals are seldom affected and are needed as the foundation for future study. Thus, in Chapter 1 you will find a review of many of the essentials of algebra that are necessary for success in this and other courses. It might be helpful to refer back to this chapter as you discover areas where you need such help.

Throughout this course we will be working with the **set of real numbers.** Here are some examples of such numbers:

$$-5 \qquad -3.25 \qquad 0 \qquad \frac{3}{4} \qquad \sqrt{2} \qquad \pi \qquad 6$$

We can show these numbers on a **number line,** where each real number is the **coordinate** of some point on the line.

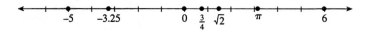

*WATCH THE MARGINS!
We will use these for special notes, added explanations, and hints throughout the book.*

At times we will use a **subset,** or part, of the real numbers in a discussion. For example, we will deal with such subsets of numbers as the following:

The set N of **natural** or **counting numbers:**	$\{1, 2, 3, \dots\}$
The set W of **whole numbers:**	$\{0, 1, 2, 3, \dots\}$
The set I of **integers:**	$\{\dots, -3, -2, -1, 0, 1, 2, 3, \dots\}$

*The set of natural numbers is also called the **set of positive integers,** and the set of whole numbers is often referred to as the **set of nonnegative integers.***

Note that every integer can be written in fractional form. For example:

$$3 = \frac{3}{1} \qquad -2 = \frac{-2}{1} \qquad 0 = \frac{0}{1}$$

However, there are fractions that cannot be written as integers, such as $\frac{2}{3}$ and $-\frac{3}{4}$. The collection of all such fractions and integers is called the set of **rational numbers.**

For every integer a we have $a = \dfrac{a}{1}$. Therefore, every integer is a rational number.

> A **rational number** is one that can be written in the form $\dfrac{a}{b}$ where a and b are integers, $b \neq 0$.

In Exercises 57 and 58 you will learn how to convert repeating decimals into the form a/b.

Every rational number a/b can be converted into decimal form by dividing b into a. The decimal form will either *terminate* as in $11/4 = 2.75$, or it will *repeat endlessly.* Use a calculator to confirm each of the following for the first three repeating cycles:

$$\frac{2}{3} = .666\dots \qquad \frac{4}{11} = .363636\dots \qquad \frac{38}{111} = .342342342\dots$$

The irrational number Pi (π) is the ratio of the circumference of a circle to its diameter. (See Exercise 62.) Using a supercomputer, the decimal form of π has been calculated to more than two billion decimal places.

Decimals that neither terminate nor repeat are called **irrational numbers** such as

$$\sqrt{3} = 1.73205\dots \qquad \pi = 3.14159\dots$$

Irrational numbers *cannot* be expressed as the ratio of two integers. Some other examples of irrational numbers are

$$\sqrt{5} \qquad \sqrt{12} \qquad \sqrt[3]{4} \qquad -\sqrt{17} \qquad \sqrt[4]{\frac{16}{9}} \qquad \frac{3\pi}{4}$$

The collection of rational numbers and irrational numbers comprises the set of **real numbers.** The relationships between these various sets of numbers can be shown by means of this diagram:

Reading upward, each set of numbers is a subset of the set of numbers listed above it.

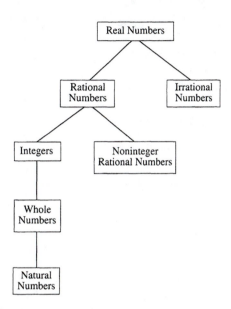

EXAMPLE 1   To which subsets of the real numbers does each of the following numbers belong?

(a) 5 **(b)** $\frac{2}{3}$ **(c)** $\sqrt{7}$ **(d)** -14

Solution
(a) 5 is a natural number, a whole number, an integer, a rational number, and a real number.

(b) $\frac{2}{3}$ is a rational number and a real number.

(c) $\sqrt{7}$ is an irrational number and a real number.

(d) -14 is an integer, a rational number, and a real number. ■

EXAMPLE 2   Classify as true or false: Every whole number is a natural number.

Solution   In order for such a statement to be true, it must be true for all possible cases; otherwise it is false. Since 0 is a whole number but is *not* a natural number, the statement is false; 0 is said to be a *counterexample* of the given statement. As long as we can find a single case where the statement is not true, we must conclude that the statement is false. ■

Throughout this course we shall be using various properties of the set of real numbers, most of which you have encountered before. Here is a summary of some of these important properties.

PROPERTIES OF THE REAL NUMBERS		
For All Real Numbers a, b, and c:	**Addition**	**Multiplication**
Closure Properties	$a + b$ is a real number	$a \cdot b$ is a real number
Commutative Properties	$a + b = b + a$	$a \cdot b = b \cdot a$
Associative Properties	$(a + b) + c = a + (b + c)$	$(a \cdot b) \cdot c = a \cdot (b \cdot c)$
Distributive Property	$a(b + c) = ab + ac$	
Identity Properties	$0 + a = a + 0 = a$ (0 is the **identity element of addition.**)	$1 \cdot a = a \cdot 1 = a$ (1 is the **identity element for multiplication.**)
Inverse Properties	$a + (-a) = (-a) + a = 0$	$a \cdot \dfrac{1}{a} = \dfrac{1}{a} \cdot a = 1, a \neq 0$
Multiplication Property of Zero		$0 \cdot a = a \cdot 0 = 0$
Zero–Product Property	If $ab = 0$, then $a = 0$ or $b = 0$, or both $a = 0$ and $b = 0$.	

TEST YOUR UNDERSTANDING

Throughout this book we shall occasionally pause for you to test your understanding of the ideas just presented. If you have difficulty with these brief sets of exercises, you should reread the material in the section before going ahead. Answers are given at the end of the chapter.

Name the property of real numbers being illustrated.

1. $3 + \left(\frac{1}{2} + 5\right) = \left(3 + \frac{1}{2}\right) + 5$

2. $3 + \left(\frac{1}{2} + 5\right) = \left(\frac{1}{2} + 5\right) + 3$

3. $6 + 4(2) = 4(2) + 6$

4. $8\left(-6 + \frac{2}{3}\right) = 8(-6) + 8\left(\frac{2}{3}\right)$

5. $(17 \cdot 23)59 = (23 \cdot 17)59$

6. $(17 \cdot 23)59 = 17(23 \cdot 59)$

7. Does $3 - 5 = 5 - 3$? Is there a commutative property for subtraction?

8. Give a *counterexample* to show that the set of real numbers is not commutative with respect to division. (That is, use a specific example to show that $a \div b \neq b \div a$.)

9. Does $(8 - 5) - 2 = 8 - (5 - 2)$? Is there an associative property for subtraction?

10. Give a counterexample to show that the set of real numbers is not associative with respect to division.

11. Are $3 + \sqrt{7}$ and $3\sqrt{7}$ real numbers? Explain.

(Answers: Page 83)

The preceding properties have been described primarily in terms of addition and multiplication. The basic operations of subtraction and division can now be defined in terms of addition and multiplication, respectively.

Subtraction

The difference $a - b$ of two real numbers a and b is defined as

Use this fact to show that multiplication distributes over subtraction:
$$a(b - c) = ab - ac.$$
Begin with $a[b + (-c)]$.

$$a - b = a + (-b) \qquad \text{For example, } 5 - 8 = 5 + (-8) = -3.$$

Alternatively, we say

$$a - b = c \text{ if and only if } c + b = a$$

Thus $5 - 8 = -3$ because $-3 + (8) = 5$.

When we say "statement p *if and only if* statement q" it means that if either statement is true, then so is the other. Thus the statement $a - b = c$ *if and only if* $c + b = a$ means:

If $a - b = c$, then $c + b = a$ and if $c + b = a$, then $a - b = c$.

Division

The quotient $a \div b$ of two real numbers a and b is defined as

$$a \div b = a \cdot \frac{1}{b} \qquad b \neq 0 \qquad \text{For example, } 8 \div 2 = 8 \times \tfrac{1}{2} = 4.$$

Alternatively, we say

$$a \div b = c \text{ if and only if } c \times b = a \qquad b \neq 0$$

Thus $8 \div 2 = 4$ because $4 \times 2 = 8$.

The statement $a \div b = c$ *if and only if* $c \times b = a$ means:

If $a \div b = c$, then $c \times b = a$ and if $c \times b = a$, then $a \div b = c$.

*This is very important: DIVISION BY ZERO IS NOT POSSIBLE. We demonstrate this here by using an **indirect proof**, that is, by assuming the opposite to be true and arriving at a false or contradictory result.*

Using this definition of division we can see why division by zero is not possible. Suppose we assume that division by zero is possible. *Assume,* for example, that $2 \div 0 = x$, where x is some real number. Then, by the definition of division, $0 \cdot x = 2$. But $0 \cdot x = 0$, leading to the false statement $2 = 0$. This argument can be duplicated where 2 is replaced by any nonzero number. Can you explain why $0 \div 0$ is *indeterminate?* (See Exercise 59.)

Note that the set of real numbers is closed with respect to subtraction since $a - b = a + (-b)$ and the real numbers have the closure property for addition. That is, the difference of any two real numbers is always another real number. Similarly, except for division by 0, the quotient of any two real numbers is a real number. However, some subsets of the real numbers do not have these properties. For example, the set of whole numbers is *not* closed with respect to subtraction, and the set of integers is *not* closed with respect to division.

EXERCISES 1.1

Classify each number as a member of one or more of these sets: (a) natural numbers; (b) whole numbers; (c) integers; (d) rational numbers; (e) irrational numbers; (f) real numbers.

1. -15 **2.** 72 **3.** $\sqrt{51}$ **4.** $-\frac{3}{4}$ **5.** $\frac{16}{2}$

6. 0.01 **7.** $0.$ **8.** 2π **9.** $\sqrt{12}$ **10.** $-\sqrt{2}$

List or describe the elements in each set.

11. The set of natural numbers less than 5.

12. The set of whole numbers greater than 100.

13. The set of whole numbers between 2 and 7.

14. The set of integers greater than -3.

15. The set of negative integers greater than -3.

16. The set of positive integers less than 5.

17. The set of integers less than 1.

18. The set of integers that are not whole numbers.

19. The set of whole numbers that are not integers.

20. The set of integers that are also rational numbers.

Answer true or false to each statement. If false, give a specific counterexample to justify your answer. (Recall that a true statement must be true for all possible cases; otherwise it is false.)

21. The set of whole numbers is closed with respect to multiplication.

22. The set of natural numbers is closed with respect to subtraction.

23. The set of integers is closed with respect to division.

24. Except for 0, the set of rational numbers is closed with respect to division.

25. The set of integers is commutative with respect to subtraction.

26. The set of rational numbers is associative with respect to multiplication.

27. The set of rational numbers contains the additive inverse for each of its members.

28. Except for 0, the set of rational numbers contains the multiplicative inverse for each of its members.

29. The product of any two real numbers is a real number.

30. The quotient of any two real numbers is a real number.

Name the property illustrated by each of the following.

31. $\sqrt[3]{5} + 7$ is a real number.   **32.** $8 + \sqrt{7} = \sqrt{7} + 8$ **33.** $(-5) + 5 = 0$

34. $9 + (7 + 6) = (9 + 7) + 6$ **35.** $(5 \times 7) \times 8 = (7 \times 5) \times 8$ **36.** $(5 \times 7) \times 8 = 5 \times (7 \times 8)$

37. $\frac{1}{4} + \frac{1}{2} = \frac{1}{2} + \frac{1}{4}$ **38.** $(4 \times 5) + (4 \times 8) = 4(5 + 8)$ **39.** $-13 + 0 = -13$

40. $1 \times \frac{1}{9} = \frac{1}{9}$ **41.** $\frac{1}{2} + \left(-\frac{1}{2}\right) = 0$ **42.** $3 - 7 = 3 + (-7)$

43. $0\left(\sqrt{2} + \sqrt{3}\right) = 0$ **44.** $\sqrt{2} \times \pi$ is a real number.   **45.** $(3 + 9)(7) = (3)(7) + (9)(7)$

46. $\frac{1}{\sqrt{2}} \cdot \sqrt{2} = 1$

Replace the variable n by a real number to make each statement true.

47. $7 + n = 3 + 7$

48. $6 \times n = \sqrt{5} \times 6$

49. $(3 + 7) + n = 3 + (7 + 5)$

50. $6 \times (5 \times 4) = (6 \times n) \times 4$

51. $5(8 + n) = (5 \times 8) + (5 \times 7)$

52. $(3 \times 7) + (3 \times n) = 3(7 + 5)$

53. If $5(x - 2) = 0$, then $x = 2$. Explain this statement using an appropriate property of zero.

54. Give an example showing that addition does *not* distribute over multiplication; your example should show that $a + (b \cdot c) \neq (a + b) \cdot (a + c)$.

55. Note that $2^4 = 4^2$. Is the operation of raising a number to a power a commutative operation? Justify your answer.

56. Explain why the real numbers have the closure property for both subtraction and division, excluding division by 0.

57. Every repeating decimal can be expressed as a rational number in the form a/b. Consider, for example, the decimal $0.727272\ldots$ and the following process:

$$\text{Let } n = 0.727272\ldots \text{. Then } 100n = 72.727272\ldots$$

$$100n = 72.727272\ldots$$

$$\text{Subtract:} \quad n = 0.727272\ldots$$

$$99n = 72$$

$$\text{Solve for } n: \quad n = \frac{72}{99} = \frac{8}{11}$$

Use this method to express each decimal as the quotient of integers and check your answer by division.

(a) $0.454545\ldots$ **(b)** $0.373737\ldots$ **(c)** $0.234234\ldots$

(*Hint:* In part (c) let $n = 0.234234\ldots$; multiply by 1000.)

58. Study the illustration given below for $n = 0.2737373\ldots$ and then convert each repeating decimal into a quotient of integers. Often a bar is placed over the set of digits that repeat. Thus we may write $0.2737373\ldots$ as $0.2\overline{73}$.

Multiply n by 1000: $1000n = 273.\overline{73}$ The decimal point is *behind* the first cycle.

Multiply n by 10: $10n = 2.\overline{73}$ The decimal point is in *front* of the first cycle.

Subtract: $990n = 271$

Divide: $n = \dfrac{271}{990}$

(a) $0.4585858\ldots$ **(b)** $3.21444\ldots$ **(c)** $2.0\overline{146}$ **(d)** $0.00\overline{123}$

59. Explain why $\frac{0}{0}$ is indeterminate. (*Hint:* If $\frac{0}{0}$ is to be some value, then it must be a unique value.) What happens if you try to find the value of $0 \div 0$ on a calculator?

60. Throughout history mathematicians have made attempts to provide approximations for π. In the thirteenth century Fibonacci gave π as $\frac{864}{274}$. Later, in the sixteenth century, Tycho Brahe used the value of $\frac{88}{\sqrt{785}}$ for π. Use a calculator to determine which of these is the better approximation.

61. Begin with any fraction $\frac{a}{b}$ and write a new fraction $\dfrac{(a + b) + b}{a + b}$. Continue in this manner until you have written at least six fractions by this rule. For example, starting with $\frac{2}{3}$, you will form this sequence of fractions: $\frac{2}{3}$, $\frac{8}{5}$, $\frac{18}{13}$, $\frac{44}{31}$, $\frac{106}{75}$, $\frac{256}{181}$. Use a calculator to change each fraction to a decimal, correct to three decimal places. Repeat this for at least three other fractions of your choice, and comment on the results. What do you expect to happen if you continue to form more fractions in such a sequence?

62. Take a circular object (like a tin can) and let the diameter be 1 unit. Use this as a unit on a number line. Mark a point on the circular object with a dot and place the circle so that the dot coincides with zero on the number line. Roll the circle to the right and mark the point where the dot coincides with the number line after one revolution. What number corresponds to this position? Why? Estimate the number of times that the diameter fits into the unrolled circle and compare your results with those of your classmates.

63. We can locate a point with irrational coordinate $\sqrt{2}$ on the number line. At the point with coordinate 1 construct a 1-unit segment perpendicular to the number line. Connect the endpoint of this segment to the point labeled 0. This becomes the hypotenuse of a right triangle, and by the Pythagorean theorem is equal to the square root of 2, written as $\sqrt{2}$. Using a compass, this length can then be transferred to the number line, thus locating a point with coordinate $\sqrt{2}$. This demonstrates that irrational numbers correspond to points on the number line.

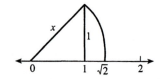

$$x^2 = 1^2 + 1^2$$
$$x^2 = 2$$
$$x = \sqrt{2}$$

Show how to locate a point on the number line with coordinate $\sqrt{5}$.

Written Assignment   From a mathematical point of view, it is common practice to introduce the integers after the counting numbers have been discussed and then develop the rational numbers. Historically, however, the positive rational numbers were developed before the negative integers.

1. Describe some everyday situations where the negative integers are not usually used but could easily be introduced.

2. Speculate on the historical necessity for the development of the positive rational numbers.

1.2 INTRODUCTION TO EQUATIONS AND PROBLEM SOLVING

A statement such as $3(x + 3) = x + 5$ is an example of a **linear equation** because the variable x appears only to the first power. It is also said to be a **conditional equation;** it is true for some replacements of the variable x, but not true for others. For example, it is a true statement for $x = -2$, but it is false for $x = 1$. On the other hand, an equation such as $3(x + 2) = 3x + 6$ is called an **identity** because it is true for *all* real numbers x.

To *solve* an equation means to find the real numbers x for which the given equation is true; these are called the **solutions** or **roots** of the given equation. Let us solve the equation $3(x + 3) = x + 5$, showing the important steps. Note that the strategy here is to collect all terms with the variable on one side of the equation, and the constants on the other side.

The first step is to eliminate the parentheses by applying the distributive property.

$$3(x + 3) = x + 5$$

$$3x + 9 = x + 5$$

$$3x + 9 + (-9) = x + 5 + (-9)$$ Add -9 to each side of the equation.

$$3x = x - 4$$

$$3x + (-x) = x - 4 + (-x)$$ Add $-x$ to each side of the equation.

$$2x = -4$$

$$\tfrac{1}{2}(2x) = \tfrac{1}{2}(-4)$$ Multiply each side by $\tfrac{1}{2}$.

$$x = -2$$

Note: We should not call -2 a solution until it has been checked.

Show that $x = -2$ is the solution by checking in the original equation. Does $3[(-2) + 3] = (-2) + 5$?

In the preceding solution, we made use of the following two basic *properties of equality*.

*The strength of these two properties is that they produce **equivalent equations**, equations having the same roots. Thus the addition property converts the equation $2x - 3 = 7$ into the equivalent form $2x = 10$.*

ADDITION PROPERTY OF EQUALITY

For all real numbers a, b, c, if $a = b$, then $a + c = b + c$.

MULTIPLICATION PROPERTY OF EQUALITY

For all real numbers a, b, c, if $a = b$, then $ac = bc$.

TEST YOUR UNDERSTANDING

(Answers: Page 83)

Solve each linear equation for x.

1. $x + 3 = 9$
2. $x - 5 = 12$
3. $x - 3 = -7$
4. $2x + 5 = x + 11$
5. $3x - 7 = 2x + 6$
6. $3(x - 1) = 2x + 7$
7. $3x + 2 = 5$
8. $5x - 3 = 3x + 1$
9. $x + 3 = 13 - x$
10. $2(x + 2) = x - 5$
11. $2x - 7 = 5x + 2$
12. $4(x + 2) = 3(x - 1)$

The properties of equality can be used to solve equations having more than one variable. The next example shows the use of properties of equality to solve a *formula* for one of the variables in terms of the others.

EXAMPLE 1   The formula relating degrees Fahrenheit and degrees Celsius is $\dfrac{5F - 160}{9} = C$. Solve for F in terms of C.

Solution Try to explain each step shown.

Celsius	Fahrenheit	
100	212	water boils
⋮	⋮	
37	98.6	normal body temperature
⋮	⋮	
0	32	water freezes

$$\frac{5F - 160}{9} = C$$

$$5F - 160 = 9C$$

$$5F = 9C + 160$$

$$F = \left(\tfrac{1}{5}\right)(9C + 160)$$

$$F = \tfrac{9}{5}C + 32 \qquad \blacksquare$$

Let us now explore the solution of problems that are expressed in words. The task will be to translate the English sentences of a problem into suitable mathematical language and to develop an equation that we can solve.

In the following problem, guidelines are suggested that are important to develop critical thinking skills. Study the solution carefully, as well as those in the examples that follow.

The lack of a strategy for solving word problems often creates difficulties for many students. To be a good problem solver you need some patience and much practice.

Problem: The length of a rectangle is 1 centimeter less than twice the width. The perimeter is 28 centimeters. Find the dimensions of the rectangle.

1. *Reread the problem and try to picture the situation given. Make note of all the information stated in the problem.*

   The length is one less than twice the width.

   The perimeter is 28.

2. *Determine what it is you are asked to find. Introduce a suitable variable, usually to represent the quantity to be found. When appropriate, draw a figure.*

   Let w represent the width.

   Then $2w - 1$ represents the length.

w ... $2w - 1$

3. *Use the available information to compose an equation that involves the variable.*

   The perimeter is the distance around the rectangle. This provides the necessary information to write an equation.

   $$w + (2w - 1) + w + (2w - 1) = 28$$

4. *Solve the equation.*

   $$w + (2w - 1) + w + (2w - 1) = 28$$
   $$6w - 2 = 28$$
   $$6w = 30$$
   $$w = 5$$

5. *Return to the original problem to see whether the answer obtained makes sense. Does it appear to be a reasonable solution? Have you answered the question posed in the problem?*

   The original problem asked for both dimensions. If the width w is 5 centimeters, then the length $2w - 1$ must be 9 centimeters.

6. *Check the solution by direct substitution of the answer into the original statement of the problem.*

   As a check, note that the length of the rectangle, 9 centimeters, is 1 centimeter less than twice the width, 5, as given in the problem. Also, the perimeter is 28 centimeters.

7. *Finally, state the solution in terms of the appropriate units of measure.*

   The dimensions are 5 centimeters by 9 centimeters.

Although some of these problems may not seem to be very practical to you, they will help you develop the basic skills of problem solving that will be helpful later when you encounter more realistic applications.

GUIDELINES FOR PROBLEM SOLVING

Read the problem. List the given information.

↓

What is to be found? Introduce a variable and state what the variable represents. Draw a figure or use a table if appropriate.

↓

Write an equation.

↓

Solve the equation.

↓

Does the answer seem reasonable? Have you answered the question stated in the problem?

↓

Check your answer to the equation in the original problem.

↓

State the solution to the problem.

EXAMPLE 2 A car leaves a certain town at noon, traveling due east at 40 miles per hour. At 1:00 P.M. a second car leaves the town traveling in the same direction at a rate of 50 miles per hour. In how many hours will the second car overtake the first car?

Solution Problems of motion of this type often prove difficult to students of algebra, but need not be. The basic relationship to remember is that *rate multiplied by time equals distance* ($r \times t = d$). For example, a car traveling at a rate of 60 miles per hour for 5 hours will travel 60×5 or 300 miles.

Now we need to reread the problem to see what part of the information given will help form an equation. The two cars travel at different rates, and for different amounts of time, but both travel the same distance from the point of departure until they meet. This is the clue: *Represent the distance each travels and equate these quantities.*

Let us use x to represent the number of hours it will take the second car to overtake the first. Then the first car, having started an hour earlier, travels $x + 1$ hours before they meet. You may find it helpful to summarize this information in tabular form.

	Rate	Time	Distance
First car	40	$x + 1$	$40(x + 1)$
Second car	50	x	$50x$

Equating the distances, we have an equation that can be solved for x:

$$50x = 40(x + 1)$$
$$50x = 40x + 40$$
$$10x = 40$$
$$x = 4$$

The second car overtakes the first in 4 hours. Does this answer seem reasonable? Let us check the result. The first car travels 5 hours at 40 miles per hour for a

Guidelines such as the preceding are useful in organizing your work, which, in turn, will assist you to think more critically and ultimately lead to more success in problem solving.

total of 200 miles. The second car travels 4 hours at 50 miles per hour for the same total of 200 miles.

Answer The second car takes 4 hours to overtake the first. ■

EXAMPLE 3 Mrs. Dougherty invested $5000 in a certificate of deposit (CD) that paid an annual interest rate of 4%. Her financial advisor invested another $3000 for her that also paid interest at an annual rate. If her total interest earned for the year was $395, what was the annual rate on this second investment?

Solution We need to make use of the formula $I = Prt$, where I is the interest earned on a principal of P dollars invested at the rate r (in decimal form) for t years. In this case, $t = 1$. Let r represent the interest rate earned on her $3000 investment.

$$
\begin{array}{ccccc}
\underset{\downarrow}{\substack{\text{Interest} \\ \text{at } 4\%}} & + & \underset{\downarrow}{\substack{\text{Interest} \\ \text{at } r\%}} & = & \underset{\downarrow}{\substack{\text{Total} \\ \text{Interest}}} \\
(5000)(0.04)(1) & + & (3000)(r)(1) & = & 395 \\
200 & + & 3000r & = & 395 \\
& & 3000r & = & 195 \\
& & r & = & \dfrac{195}{3000} = 0.065
\end{array}
$$

Always check by returning to the original statement of the problem. If you just check in the constructed equation, and the equation you formed was incorrect, you will not detect the error.

Answer The rate on her investment of $3000 was 6.5%. ■

EXAMPLE 4 Marcella has a base salary of $250 per week. In addition, she receives a commission of 12% of her sales. Last week her total earnings were $520. What were her total sales for the week?

Solution Let x represent her total sales for the week. Then we can make use of this relationship:

$$
\begin{array}{ccccc}
\text{Salary} & + & \text{Commission} & = & \substack{\text{Total} \\ \text{Earnings}} \\
\downarrow & & \downarrow & & \downarrow \\
250 & + & 0.12x & = & 520 \qquad (\text{Recall: } 12\% = 0.12.)
\end{array}
$$

$$0.12x = 270$$

$$x = \frac{270}{0.12}$$

$$x = 2250$$

Answer Marcella's total sales for the week were $2250.

Check: According to the statement of the problem, show that $250 plus 12% of $2250 is equal to $520. Show that $250 + 0.12(2250) = 520$. ■

Now it's your turn! Use the guidelines suggested earlier in this section and try to solve as many of the problems as you can. Don't become discouraged if you have difficulty; be assured that most mathematics students have trouble with word problems. Time and practice will most certainly help you develop your critical thinking skills.

EXERCISES 1.2

Solve for x and check each result.

1. $3x - 2 = 10$ **2.** $5x + 1 = 21$ **3.** $-2x + 1 = 9$

4. $-3x - 2 = 10$ **5.** $-3x - 5 = 7$ **6.** $3x + 2 = -13$

7. $2x - 1 = -17$ **8.** $-2x + 3 = -12$ **9.** $2(x + 1) = 11$

10. $3(x - 2) = 15$ **11.** $3x + 7 = 2x - 2$ **12.** $2.5x - 8 = x + 3$

13. $\frac{1}{2}x + 7 = 2x - 3$ **14.** $\frac{5}{2}x - 5 = 3x + 7$ **15.** $\frac{4}{3}x - 7 = \frac{1}{3}x + 8$

16. $5x - 1 = 5x + 1$ **17.** $\frac{3}{5}(x - 5) = x + 1$ **18.** $5(x + 4) = \frac{5}{2}x - 5$

19. $\frac{7}{2}x + 5 + \frac{1}{2}x = \frac{5}{2}x - 6$ **20.** $2(x + 3) - x = 2x + 8$ **21.** $-3(x + 2) + 1 = x - 25$

22. $\frac{4}{3}(x + 8) = \frac{3}{4}(2x + 12)$ **23.** $1 - 12x = 7(1 - 2x)$ **24.** $2(3x - 7) - 4x = -2$

25. $x + 2\left(\frac{1}{6}x + 2\right) = \frac{6}{5}x + 16$

Solve for the indicated variable for each formula.

26. Perimeter of a rectangle:

27. Area of a trapezoid:

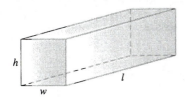

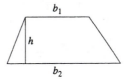

$P = 2\ell + 2w$
Solve for ℓ.

$A = \frac{1}{2}h(b_1 + b_2)$
Solve for h.

28. Surface area of a rectangular box:

29. Volume of a cylinder:

$A = 2\ell w + 2\ell h + 2wh$
Solve for h.

$V = \pi r^2 h$
Solve for h.

30. The formula relating degrees Fahrenheit and degrees Celsius is $F = \frac{9}{5}C + 32$. Solve for C in terms of F. Find C if $F = 77°$.

31. The formula $I = Prt$ gives the interest I earned on P dollars invested at the rate r (in decimal form) per year for t years. Solve for P in terms of the other variables. Find P if $I = \$220$, $r = 5.5\%$, and $t = 2$.

32. Solve for b: $c = \dfrac{2ab}{a + b}$ **33.** Solve for R: $V = \dfrac{1}{3}\pi h^2(3R - h)$

34. Find a number such that two-thirds of the number increased by one is 13.

35. Find the dimensions of a rectangle whose perimeter is 56 inches if the length is 4 inches greater than the width.

36. Each of the two equal sides of an isosceles triangle is 3 inches longer than the base of the triangle. The perimeter is 21 inches. Find the length of each side.

37. Carlos spent $6.15 on stamps, in denominations of 10¢, 25¢, and 30¢. He bought one-half as many 25¢ stamps as 10¢ stamps, and three more 30¢ stamps than 10¢ stamps. How many of each type did he buy? (*Hint:* Whenever you have a certain number of a particular stamp, the *total value* is the number of stamps times the value of that stamp.)

38. Maria has $169 in ones, fives, and tens. She has twice as many one-dollar bills as she has five-dollar bills, and five more ten-dollar bills than five-dollar bills. How many of each type bill does she have? (See the hint in Exercise 37.)

39. Two cars leave a town at the same time and travel in opposite directions. One car travels at the rate of 45 miles per hour, and the other at 55 miles per hour. In how many hours will the two cars be 350 miles apart?

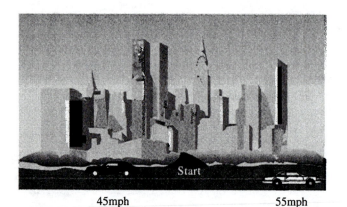

45mph   55mph

40. Robert goes for a walk at a speed of 3 miles per hour. Two hours later Roger attempts to overtake him by jogging at the rate of 7 miles per hour. How long will it take him to reach Robert?

41. Prove that the measures of the angles of a triangle cannot be represented by consecutive odd integers. (*Hint:* The sum of the measures is 180.)

42. The width of a painting is 4 inches less than the length. The frame that surrounds the painting is 2 inches wide and has an area of 240 square inches. What are the dimensions of the painting? (*Hint:* The total area minus the area of the painting alone is equal to the area of the frame.)

43. The length of a rectangle is 1 inch less than three times the width. If the length is increased by 6 inches and the width is increased by 5 inches, then the length will be twice the width. Find the dimensions of the rectangle.

44. The units' digit of a two-digit number is three more than the tens' digit. The number is equal to four times the sum of the digits. Find the number. (*Hint:* We can represent a two-digit number as $10t + u$.)

45. Find three consecutive odd integers such that their sum is 237. (*Hint:* The three integers can be represented as $x, x + 2$, and $x + 4$.)

46. The length of a rectangle is 1 inch less than twice the width. If the length is increased by 11 inches and the width is increased by 5 inches, then the length will be twice the width. What can you conclude about the data for this problem?

47. Amy travels 27.5 miles to get to work by car. The first part of her trip is along a country road on which she averages 35 miles per hour, and the second part is on a highway where she averages 48 miles per hour. If the time she travels on the highway is 5 times the amount of time she travels on the country road, what is the total time for the trip?

48. A taxi charges 80¢ for the first $\frac{1}{6}$ mile and 20¢ for each additional $\frac{1}{6}$ mile. If a passenger paid $6.00, how far did the taxi travel?

49. A financial advisor invested an amount of money at an annual rate of interest of 9%. She invested $2700 more than this amount at 12% annually. The total yearly income from these investments was $1794. How much did she invest at each rate? (*Note:* Use the formula $I = Prt$, where I is the interest earned on the principal of P dollars invested at the rate r (in decimal form) per year. In this case the time, t, is 1 year.)

50. Luis earns a monthly salary of $2250, plus a commission of 4% of his total sales for the month. Last month his total earnings were $2512. What were his total sales?

51. Leslie paid $9010 for a used car, which included a 6% sales tax on the base cost of the car. What was the cost of the car, without the sales tax?

52. The total cost of two certificates of deposit is $12,800. The annual interest rates of these certificates are 3.5% and 4%. The yearly interest on the 4% certificate is $92 more than that on the 3.5% certificate. What is the cost of each certificate?

Challenge   Two boats begin their journeys back and forth across a river at the same time, but from opposite sides of the river. The first time that they pass each other they are 700 feet from one of the shores of the river. After they each make one turn, they pass each other once again at a distance of 400 feet from the other shore. How wide is the river? Assume each boat travels at a constant speed and that there is no loss of time in making a turn.

Written Assignment   Following is a set of directions for a mathematical trick. First try it. Then use algebraic representations for each phrase (direction) and explain why the trick works. Create a similar puzzle of your own.

> Think of a number.
> Add 2.
> Multiply by 3.
> Add 9.
> Multiply by 2.
> Divide by 6.
> Subtract the number with which you started.
> The result is 5.

1.3 STATEMENTS OF INEQUALITY AND THEIR GRAPHS

As you continue your study of mathematics you will find a great deal of attention given to *inequalities*. We begin our discussion of this topic by considering the ordering of the real numbers on the number line. In the following figure we say that *a is less than b* because *a* lies to the left of *b*. In symbols, we write $a < b$.

Also note that *b* lies to the right of *a*. That is, $b > a$; this is read "*b is greater than a.*" Two inequalities, one using the symbol $<$ and the other $>$, are said to have the *opposite sense*.

Here are two examples of the use of these *symbols of inequality*.

$$3 < 7 \qquad 3 \text{ is less than } 7$$
$$5 > -2 \qquad 5 \text{ is greater than } -2$$

Since positive numbers are to the right of the origin on a number line, $a > 0$ means that *a* is positive. Similarly, $a < 0$ means that *a* is negative.

Algebraically, we define $a < b$ as follows:

It may help to note that the inequality symbol "points" to the smaller of the two numbers and "opens wide" to the larger.

DEFINITION OF $a < b$

For any two real numbers *a* and *b*, $a < b$ (or $b > a$) if and only if $b - a$ is a positive number; that is, if and only if $b - a > 0$.

For example, $3 < 7$ because $7 - 3 = 4$, a positive number; that is, $7 - 3 > 0$. Also, since $5 - (-2) > 0$, then $5 > -2$.

A fundamental property of inequalities states that for any two real numbers *a* and *b*, either *a* is less than *b*, or *a* equals *b*, or *a* is greater than *b*. In symbols, we have the following property:

TRICHOTOMY PROPERTY

For any real numbers *a* and *b*, exactly one of the following is true:

$$a < b \qquad a = b \qquad a > b$$

The same number may be added to, or subtracted from, each side of an inequality and the *sense* of the new inequality will be the same. For example:

Note: $5 < 10$ and $8 < 13$ have the same sense.

Since $5 < 10$, then $5 + 3 < 10 + 3$; that is, $8 < 13$.

Since $9 > 5$, then $9 - 2 > 5 - 2$; that is, $7 > 3$.

This gives rise to the following important property of inequalities:

ADDITION PROPERTY OF ORDER

For all real numbers a, b, and c:

$$\text{If } a < b, \text{ then } a + c < b + c.$$

$$\text{If } a > b, \text{ then } a + c > b + c.$$

Notice how this property is used to solve the following inequality. We begin by simplifying the left side.

As with equations, the strategy is to isolate the variable on one side of the inequality, and the constants on the other side.

$$-4x - (3 - 5x) > 8$$
$$-4x - 3 + 5x > 8$$
$$x - 3 > 8$$
$$x - 3 + 3 > 8 + 3 \qquad \text{addition property of order}$$
$$x > 11$$

When nothing is said to the contrary, it will always be assumed that we are using the set of real numbers. Therefore, the *solution set* here consists of all real numbers that are greater than 11. Rather than use a verbal description, we can use braces and write this solution set in **set-builder notation** as

$$\{x \mid x > 11\}$$

This is read as "the set of all x such that x is greater than 11."

The addition property of order also applies for inequalities using the symbols $\leq$ and $\geq$. For example, if $a \leq b$, then $a + c \leq b + c$.

The additional symbols $\leq$ and $\geq$ of inequality are also used quite often:

$a \leq b$ means *a is less than or is equal to b;* that is, $a < b$ or $a = b$.

$a \geq b$ means *a is greater than or is equal to b;* that is, $a > b$ or $a = b$.

EXAMPLE 1   Find the solution set: $3x + 7 \leq 2x - 1$

Solution Apply the addition property of order twice.

$$3x + 7 \leq 2x - 1$$
$$3x + 7 + (-7) \leq 2x - 1 + (-7)$$
$$3x \leq 2x - 8$$
$$3x + (-2x) \leq 2x - 8 + (-2x)$$
$$x \leq -8$$

The solution set consists of all real numbers that are less than or equal to -8, that is, $\{x \mid x \leq -8\}$. ∎

When the addition property of order is applied, **equivalent inequalities** are produced. That is, the new inequality has the same solution set as the original one. Now let us see what happens when we multiply (or divide) each side of an inequality by the same number. Here are several illustrations:

$$8 < 12 \rightarrow 2(8) < 2(12); \text{ that is, } 16 < 24$$
$$20 > -15 \rightarrow \tfrac{1}{5}(20) > \tfrac{1}{5}(-15); \text{ that is, } 4 > -3$$

The sense is preserved.

$$5 < 6 \rightarrow -2(5) > -2(6); \text{ that is, } -10 > -12$$
$$6 > -4 \rightarrow -\tfrac{1}{2}(6) < -\tfrac{1}{2}(-4); \text{ that is, } -3 < 2$$

The sense is reversed.

The preceding are illustrations of the following property:

Note: Similar properties hold for the inequality when $a > b$. Also, note that when $c = 0$, then $ac = bc$; that is, both sides will be equal to 0.

MULTIPLICATION PROPERTY OF ORDER

For all real numbers a, b, and c:

If $a < b$ and c is positive, then $ac < bc$.

If $a < b$ and c is negative, then $ac > bc$.

EXAMPLE 2   Solve for x: $5(3 - 2x) \geq 10$

Solution Multiply each side by $\tfrac{1}{5}$ (or divide by 5).

$$\tfrac{1}{5} \cdot 5(3 - 2x) \geq \tfrac{1}{5}(10) \qquad \text{multiplication property of order}$$
$$3 - 2x \geq 2$$

Add -3 to each side.

$$3 - 2x + (-3) \geq 2 + (-3) \qquad \text{addition property of order}$$
$$-2x \geq -1$$

Multiply by $-\tfrac{1}{2}$ (or divide by -2).

$$-\tfrac{1}{2}(-2x) \leq -\tfrac{1}{2}(-1) \qquad \text{multiplication property of order}$$

$$x \leq \tfrac{1}{2}$$

The solution set is $\{x \mid x \leq \tfrac{1}{2}\}$. ∎

At times a statement of inequality can be solved by our knowledge of the properties of numbers. Thus, in the following example, the solution is obtained because we know that the given fraction will be negative if the numerator and denominator are of opposite signs.

EXAMPLE 3   Solve: $\dfrac{2}{x+4} < 0$

Solution   Since the numerator of the fraction is positive, the fraction will be less than zero if the denominator is negative. Thus

$$x + 4 < 0$$

$$x < -4$$

The solution set is $\{x \mid x < -4\}$.   ∎

TEST YOUR UNDERSTANDING

(Answers: Page 83)

Find the solution set.

1. $x + 3 < 12$

2. $x - 5 < 13$

3. $x - 1 > 8$

4. $x + 7 > 2$

5. $x + (-5) \le 9$

6. $x + (-3) \ge -5$

7. $3x + 8 < 2x + 12$

8. $3x - 6 \ge x + 8$

9. $5(x + 7) \le 3x - 7$

10. $2(x - 1) \le 5x + 1$

11. $\dfrac{5}{3-x} < 0$

12. $\frac{1}{2}x + 3 < \frac{1}{3}x - 2$

The two inequalities $a < b$ *and* $b < c$ may be written as $a < b < c$. Similar forms apply when one or both of the inequality symbols $<$ are replaced by $\le$. Thus, $a \le b < c$ means $a \le b$ *and* $b < c$. Also, forms such as $a \ge b \ge c$ are obtained when the sense is reversed. Such inequalities can be used to define *bounded intervals* on the number line. This is shown in the following figure together with a specific example of each type.

The solid dot means that the boundary point is included in the interval, and the open circle means it is not included.

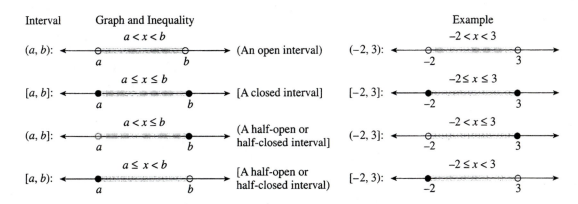

Interval	Graph and Inequality		Example
(a, b):	$a < x < b$	(An open interval)	$(-2, 3)$: $-2 < x < 3$
$[a, b]$:	$a \le x \le b$	[A closed interval]	$[-2, 3]$: $-2 \le x \le 3$
$(a, b]$:	$a < x \le b$	(A half-open or half-closed interval]	$(-2, 3]$: $-2 < x \le 3$
$[a, b)$:	$a \le x < b$	[A half-open or half-closed interval)	$[-2, 3)$: $-2 \le x < 3$

There are also *unbounded intervals*. For example, the set of all $x > 5$ is denoted by $(5, \infty)$. Similarly, $(-\infty, 5)$ represents all $x < 5$. The symbols ∞ and $-\infty$ are read "plus infinity" and "minus infinity" but do *not* represent numbers. They are symbolic devices used to indicate that *all x* in a given direction, without end, are included, as in the following figure.

Note, for example, that $(-\infty, b)$ means $\{x \mid x < b\}$.

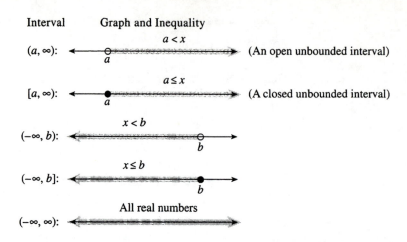

Interval   Graph and Inequality

The number line can be used to show the graphs of solution sets of inequalities, as in the following example.

EXAMPLE 4   Graph the solution set: $-2(x - 1) \geq 4$

Solution

Because we are multiplying (or dividing) by a negative number, the sense is reversed.

$$-2(x - 1) \geq 4$$
$$x - 1 \leq -2$$
$$x \leq -1$$

The solution set is $\{x \mid x \leq -1\}$ and has the following graph:

$(-\infty, -1]$:

The heavily shaded arrow is used to show that all points in the indicated direction are included. The solid dot indicates that -1 is also included in the solution. When the solid dot is replaced by an open circle, we have the graph of $-2(x - 1) > 4$:

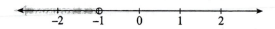

Can you describe the graph of $-2(x - 1) < 4$?   ■

Inequalities can also be used to solve applied problems. This is illustrated in Example 5, which makes use of inequalities of the form $a \leq b \leq c$.

EXAMPLE 5   A small family business employs two part-time workers per week. The total amount of wages they pay to these employees ranges from $128 to $146 per week. If one employee will earn $18 more than the other, what are the possible amounts earned by each per week?

Solution Let x = wages paid to the employee who earns the smaller amount. Then $x + 18$ = wages for the other employee. Since the sum of the wages is at least \$128 but no more than \$146, the sum $x + (x + 18)$ satisfies this *compound inequality:*

$$128 \le x + (x + 18) \le 146$$

Now simplify to get the possibilities for x.

$$128 \le 2x + 18 \le 146$$
$$110 \le 2x \le 128$$
$$55 \le x \le 64$$

Note: -18 is added to each part of the compound inequality, and then each part is divided by 2.

To get the result for the other employee, add 18 to the preceding inequality and simplify.

$$55 + 18 \le x + 18 \le 64 + 18$$
$$73 \le x + 18 \le 82$$

One part-time employee earns from \$55 to \$64 per week and the other earns from \$73 to \$82 per week. ■

There are a number of additional properties of order that are fundamental for later work. We list them in terms of the following rules.

RULE 1.   **If $a < b$ and $b < c$, then $a < c$.**
This is known as the **transitive property of order.** Geometrically, it says that if a is to the left of b, and b is the left of c, then a must be to the left of c on a number line.

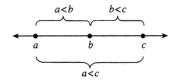

RULE 2.   **If $a < b$ and $c < d$, then $a + c < b + d$.**
Illustration: Since $5 < 10$ and $-15 < -4$, then $5 + (-15) < 10 + (-4)$; that is, $-10 < 6$.

These rules can also be restated by reversing the sense. For instance, Rule 2 would then read as: If $a > b$ and $c > d$, then $a + c > b + d$.

RULE 3.   **If $0 < a < b$ and $0 < c < d$, then $ac < bd$.**
Illustration: Since $3 < 7$ and $5 < 9$, then $(3)(5) < (7)(9)$; that is, $15 < 63$.

RULE 4.   **If $a < b$ and $ab > 0$, then $\dfrac{1}{a} > \dfrac{1}{b}$.**

Illustrations: Since $5 < 10$, then $\frac{1}{5} > \frac{1}{10}$.

Since $-3 < -2$, then $-\frac{1}{3} > -\frac{1}{2}$.

EXERCISES 1.3

Classify each statement as true or false. If it is false, give a specific counterexample to explain your answer.

1. If $x < 2$, then x is negative.

2. If $x > 1$ and $y > 2$, then $x + y > 3$.

3. If $0 < x$, then $-x < 0$.

4. If $x < 5$ and $y < 6$, then $xy < 30$.

5. If $0 < x$, then $x < x^2$.

6. If $x < y < -2$, then $\dfrac{1}{x} > \dfrac{1}{y}$.

7. If $x \le -5$, then $x - 2 \le -7$.

8. If $x \le y$ and $y < z$, then $x < z$.

Express each inequality in interval notation.

9. $-5 \le x \le 2$

10. $0 < x < 7$

11. $-6 \le x < 0$

12. $-2 < x \le 4$

13. $-10 < x < 10$

14. $3 \le x \le 7$

15. $x < 5$

16. $x \le -2$

17. $-2 \le x$

18. $2 < x$

19. $x \le -1$

20. $x < 3$

Show each of the following intervals as a graph on a number line.

21. $(-3, -1)$

22. $(-3, -1]$

23. $[-3, -1)$

24. $[-3, -1]$

25. $[0, 5]$

26. $(-1, 3)$

27. $(-\infty, 0]$

28. $[2, \infty)$

Write an inequality for each of the following graphs. Also express each as an interval of real numbers.

For example: 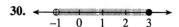 $-1 \le x \le 3$; $[-1, 3]$

29.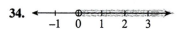

30.

31.

32.

33.

34.

Find the solution set.

35. $x + 5 > 17$

36. $x - 8 < 5$

37. $x - 7 \ge -3$

38. $x + 6 \le -7$

39. $5x - 4 < 6 + 4x$

40. $3x + 12 > 2x - 5$

41. $3x > -21$

42. $9x \le -45$

43. $-5x < 50$

44. $3x + 5 \ge 17$

45. $5x - 3 \le 22$

46. $-2x + 1 \le 19$

47. $2x + 7 \le 5 - 6x$

48. $3x - 2 > x + 5$

49. $-5x + 5 < -3x + 1$

50. $3(x - 1) \ge 2(x - 1)$

51. $2(x + 1) < x - 1$

52. $3x + 5 + x > 2(x - 1)$

53. $\frac{1}{2}x - 5 > \frac{1}{4}x + 3$

54. $\frac{3}{4}x + 2 < \frac{5}{8}x - 3$

55. $-\frac{3}{5}x - 6 < -\frac{2}{5}x + 7$

56. $\dfrac{2 - x}{5} \ge 0$

57. $\dfrac{1}{x} < 0$

58. $-\dfrac{2}{x + 1} > 0$

Solve for x.

59. $3x + 5 \neq 8$ (The symbol $\neq$ is read "is not equal to.")

60. $2x + 1 \not> 5$ (The symbol $\not>$ is read "is not greater than.")

61. $3x - 2 \not< 1$ (The symbol $\not<$ is read "is not less than.")

62. $12x + 9 \neq 15(x - 2)$ **63.** $3(x - 1) \not> 5(x + 2)$ **64.** $2(x + 3) \not< 3(x + 1)$

65. $\frac{5}{3}x \not< 2x - 1$ **66.** $\frac{2}{9}(3x + 7) \not\geq 1 - \frac{4}{3}x$

Solve and graph each inequality.

67. $2x + 3 < 11$ **68.** $-3x + 1 > -2$ **69.** $\frac{1}{2}x + 2 \leq 1$

70. $-3(x + 1) \geq 3$ **71.** $2(x + 1) < 3(x + 2)$ **72.** $-2(x - 2) > 3x - 3$

73. The sum of an integer and 5 less than three times this integer is between 34 and 54. Find all possible pairs of such integers.

74. In order to get a grade of B⁺ in an algebra course, a student must have a test average of at least 86% but less than 90%. If the student's grades on the first three tests were 85%, 86%, and 93%, what grades on the fourth test would guarantee a grade of B⁺?

75. In Exercise 74, what grades on the fourth test would guarantee a B⁺ if the fourth test counts twice as much as each of the other tests?

76. If x satisfies $\frac{7}{4} < x < \frac{9}{4}$, then what are the possible values of y where $y = 4x - 8$?

(*Hint:* Apply the multiplication and addition properties of inequality to the given inequality to obtain $4x - 8$ as the middle part.)

77. For a given time period, the temperature in degrees Celsius varied between 25° and 30°. What was the range in degrees Fahrenheit for this time period? (*Hint:* Begin with $25 < C < 30$ and apply the idea in Exercise 76 using $F = \frac{9}{5}C + 32$.)

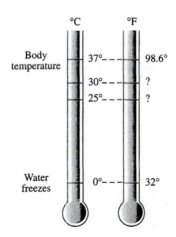

78. Suppose that a machine is programmed to produce rectangular metal plates such that the length of the plate will be 1 more than twice the width w. When the entry $w = 2$ cm is made, the design of the machine will guarantee only that the width is within a one-tenth tolerance of 2. That is, $2 - 0.1 < w < 2 + 0.1$.

(a) Within what tolerance of 5 centimeters is the length?

(b) Find the range of values for the area.

79. A delivery service will accept a package only if the sum of its length ℓ and its girth g is no more than 110 inches. It also requires that each of the three dimensions be at least 2 inches.

    (a) If $\ell = 42$ inches, what are the allowable values for the girth g?

    (b) If $\ell = 42$ inches and $w = 18$ inches, what are the allowable values of h?

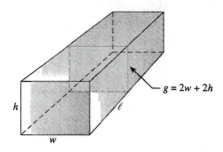

$g = 2w + 2h$

80. A store has two part-time employees who together are paid a weekly total from $150 to $180. If one of the employees earns $15 more than the other, what are the possible amounts earned by each per week?

81. A store has three part-time employees who together are paid a weekly total from $210 to $252. Two of them earn the same amount and the third earns $12 less than the others. Find the possible amounts earned by each per week.

82. A supermarket has twenty part-time employees who together are paid a weekly total from $1544 to $1984. Twelve of them earn the same amount and the remaining eight each earn $22 less. Find the possible amounts earned by each employee per week.

Written Assignment   In your own words, state the addition and multiplication properties of equality and of order without using any mathematical symbolism. For example, your statement for the addition property of order could begin with: "Adding the same number to both sides of . . . ".

Graphing Calculator Exercises   Some graphing calculators allow the use of *relational (logical) operators*. If yours does, you can obtain a graphical solution to inequalities such as $x^2 - 1 > 0$ by graphing $y = (x^2 - 1 > 0)(1)$. In other words, place the statement of the inequality in parentheses followed by the number 1 in parentheses in the formula for y. The result is the solution interval(s) $\{x \mid x^2 - 1 > 0\}$ raised one unit above the x-axis on your calculator's "window."

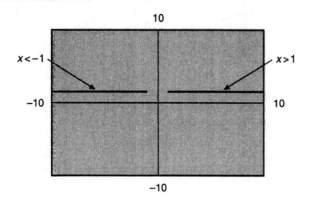

Raising the graph of the solution above the x-axis makes the result clearer. Otherwise, the graph would not be visible on the axis.

If your calculator permits this technique, use it to solve the following inequalities, using the RANGE settings $-10 \leq x \leq 10$ by $-10 \leq y \leq 10$. Check your results using the algebraic techniques of this section.

1. $3x + 1 \leq 2x + 5$

2. $4x - 2 > 6x + 7$

3. $3 < 2x + 4 < 15$ (*Hint:* Multiply two inequality statements.)

1.4 ABSOLUTE VALUE

What do the numbers -5 and $+5$ have in common? Obviously, they are different numbers and are the coordinates of two distinct points on the number line. However, they are both the same distance from 0, the **origin,** on the number line.

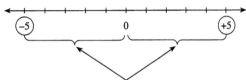

Same distance from the origin

In other words, -5 is as far to the left of 0 as $+5$ is to the right of 0. We show this fact by using **absolute value notation**:

$$|-5| = 5 \text{ read as “The } absolute\ value \text{ of } -5 \text{ is 5.”}$$

$$|+5| = 5 \text{ read as “The } absolute\ value \text{ of } +5 \text{ is 5.”}$$

Geometrically, for any real number x, $|x|$ is the distance (without regard to direction) that x is from the origin. Note that for a positive number, $|x| = x$; $|+5| = 5$. For a negative number, $|x| = -x$; that is, $|-5| = -(-5) = 5$. Also, since 0 is the origin, it is natural to have $|0| = 0$.

We can summarize this with the following definition.

In words, if a is positive or zero, the absolute value of a is a. If a is negative, the absolute value is the opposite of a. For example:

$$|+3| = 3$$
$$|-3| = -(-3) = 3$$
$$|0| = 0$$

DEFINITION OF ABSOLUTE VALUE

For any real number a,

$$|a| = \begin{cases} a & \text{if } a \geq 0 \\ -a & \text{if } a < 0 \end{cases}$$

EXAMPLE 1   Solve for x: $\dfrac{x - 3}{|x - 3|} = 1$

The solution set for Example 1 is $\{x \mid x > 3\}$. However, for the sake of simplicity, we will often omit the set-builder notation. Thus, when we write $x > 3$ as the solution, it is understood that the solution consists of all x such that $x > 3$.

Solution   This fraction will be equal to 1 if the numerator and denominator have the same value. By definition, $|x - 3| = x - 3$ only if $x - 3 \geq 0$ or, $x \geq 3$. However, *to avoid division by 0, $x \neq 3$.* Thus the solution is $x > 3$. Test several values of x for $x < 3$ and $x > 3$ to verify this solution. (You may find it easier to think of this problem in the form $\dfrac{a}{|a|} = 1$, where $a = x - 3$.)   ∎

Some important properties of absolute value follow, presented with illustrations but without formal proof.

PROPERTY 1. For $k > 0$, $|a| = k$ if and only if $a = k$ or $-a = k$.

This property follows immediately from the definition of absolute value. For example, an equation such as $|x| = 2$ is just another way of saying that $x = 2$ or $x = -2$. The graph consists of the two points with coordinates 2 and -2; each of these points is 2 units from the origin.

EXAMPLE 2   Solve: $|5 - x| = 7$

We use Property 1 for this solution, noting that $5 - x$ plays the role of a.

Solution

$$5 - x = 7 \quad \text{or} \quad -(5 - x) = 7$$
$$-x = 2 \qquad\qquad -5 + x = 7$$
$$x = -2 \qquad\qquad\quad x = 12$$

Check: $|5 - (-2)| = |7| = 7$; $|5 - 12| = |-7| = 7$ ■

Now consider the inequality $|x| < 2$. Here we are considering all the real numbers whose absolute value *is less than* 2. On the number line, these are the points whose distance from the origin is less than 2 units, that is, all of the real numbers between -2 and 2, or $-2 < x < 2$. The graph of $|x| < 2$ is the interval $(-2, 2)$.

Describe the graph of $|x| \le 2$.

$|x| < 2$:

Conversely, the graph also shows that if $-2 < x < 2$, then $|x| < 2$.
We can now generalize by means of the following property.

Note: $-k < a < k$ is the same as saying that $a < k$ and $a > -k$. Also, $|a| \le k$ if and only if $-k \le a \le k$.

PROPERTY 2. For $k > 0$, $|a| < k$ if and only if $-k < a < k$.

EXAMPLE 3   Solve: $|x - 2| \le 3$

Solution Let $x - 2$ play the role of a in Property 2. Consequently, $|x - 2| \le 3$ is equivalent to $-3 \le x - 2 \le 3$. Now add 2 to each part to isolate x in the middle.

$$-3 \le x - 2 \le 3$$
$$-3 + 2 \le x - 2 + 2 \le 3 + 2$$

Thus the solution consists of all x such that $-1 \le x \le 5$.   ■

Absolute value can be used to find the distance between two points on the number line. For example, the distance between 5 and -3 can be found as follows:

$$|5 - (-3)| = |8| = 8 \quad \text{or} \quad |(-3) - 5| = |-8| = 8$$

This leads to the following definition:

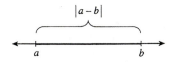

> ### DEFINITION OF DISTANCE ON THE NUMBER LINE
>
> $|a - b|$ represents the distance between points a and b on the number line.

The idea of distance between points on a number line can be used to give an alternative way to solve Example 3. Think of the expression $|x - 2|$ as the distance between x and 2 on the number line, and then consider all the points x whose distance from 2 is less than or equal to 3 units.

Think: $|x - 2| < 3$ means that x is within 3 units of 2. Thus $-1 < x < 5$.

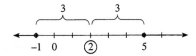

Observe that the *midpoint* or *center* of the interval has coordinate 2. This value can be found by taking the average of the numbers -1 and 5; $\dfrac{-1 + 5}{2} = 2$. In general, we have the following result.

> ### MIDPOINT COORDINATE OF A LINE SEGMENT
>
> If x_1 and x_2 are the endpoints of an interval on the number line, then the coordinate of the midpoint is
>
> $$\frac{x_1 + x_2}{2}$$

EXAMPLE 4 Find the coordinate of the midpoint M of each line segment AB.

(a)

```
 -4      0      6
  ├──────┼──────┤
  A      M      B
```

(b)

```
 -10           -2  0
  ├──────┼──────┤──┤
  A      M      B
```

Solution **(a)** $\dfrac{-4 + 6}{2} = 1$ **(b)** $\dfrac{-10 + (-2)}{2} = -6$ ∎

Next, consider the inequality $|x| > 2$. We know that $|x| < 2$ means that $-2 < x < 2$. Therefore, $|x| > 2$ consists of those values of x for which $x < -2$ or $x > 2$.

$|x| > 2$:

```
  ──┼──○──┼──┼──○──→
   -2     0     2
```

This graph shows the set of points whose distance from 0 is more than 2 units. In general, we have:

Also $|a| \geq k$ if and only if $a \leq -k$ or $a \geq k$.

PROPERTY 3. For $k \geq 0$, $|a| > k$ if and only if $a < -k$ or $a > k$.

EXAMPLE 5   Graph: $|x + 1| > 2$

Solution   Use Property 3 as follows:

$$|x + 1| > 2 \quad \text{means} \quad x + 1 < -2 \quad \text{or} \quad x + 1 > 2$$

Thus $x < -3$ or $x > 1$. The word "or" indicates that we are to consider the values for x that satisfy either one of these two conditions; that is, x may be less than -3 or x may be greater than 1.

Alternatively, $|x + 1|$ can be written as $|x - (-1)|$ to obtain the form $|a - b|$, which represents the distance between x and -1 on the number line. Now locate all of the points that are *more* than 2 units away from the point -1 as follows.

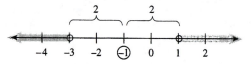

■

> The statement
> $|x + y| \le |x| + |y|$ is
> called the **triangle
> inequality**.

Here are some other useful properties of absolute value:

$$|xy| = |x||y| \qquad \left|\frac{x}{y}\right| = \frac{|x|}{|y|} \qquad |x + y| \le |x| + |y|$$

(See Exercises 46 and 47.)

> Throughout the text you will
> find CAUTION items. These
> illustrate errors that are often
> made by students. Study these
> carefully so that you under-
> stand the errors shown and
> can avoid such mistakes.

CAUTION: Learn to Avoid These Mistakes	
WRONG	**RIGHT**
$\left|\frac{3}{4} - 2\right| = \frac{3}{4} - 2 = -\frac{5}{4}$	$\left|\frac{3}{4} - 2\right| = \left|-\frac{5}{4}\right| = \frac{5}{4}$
$|5 - 7| = |5| - |7| = -2$	$|5 - 7| = |-2| = 2$
$|x| = -2$ has the solution $x = 2$.	There is no solution; the absolute value of a number can never be negative.
$|x - 1| < 3$ if and only if $x < 4$.	$|x - 1| < 3$ if and only if $-3 < x - 1 < 3$; that is, $-2 < x < 4$.
$|2x| > 1$ if and only if $x > \frac{1}{2}$.	$|2x| > 1$ if and only if $x < -\frac{1}{2}$ or $x > \frac{1}{2}$.

EXERCISES 1.4

Classify each statement as true or false.

1. $-\left|-\frac{1}{3}\right| = \frac{1}{3}$

2. $|-1000| < 0$

3. $\left|-\frac{1}{2}\right| = 2$

4. $|\sqrt{2} - 5| = 5 - \sqrt{2}$

5. $\left|\frac{x}{y}\right| = |x| \cdot \frac{1}{|y|}$

6. $2 \cdot |0| = 0$

7. $||x|| = |x|$

8. $|-(-1)| = -1$

9. $|a| - |b| = a - b$ **10.** $|a - b| = b - a$

11. Find the coordinate of the midpoint of a line segment with endpoints as given:

    (a) 3 and 9   **(b)** -8 and -2 **(c)** -12 and 0   **(d)** -5 and 8

12. One endpoint of a line segment is located at -7. The midpoint is located at 3. What is the coordinate of the other endpoint of the line segment?

Solve for x.

13. $|x| = \frac{1}{2}$ **14.** $|3x| = 3$ **15.** $|x - 1| = 3$

16. $|3x - 4| = 0$ **17.** $|2x - 3| = 7$ **18.** $|6 - 2x| = 4$

19. $|4 - x| = 3$ **20.** $|3x - 2| = 1$ **21.** $|3x + 4| = 16$

22. $\left|\dfrac{1}{x - 1}\right| = 2$ **23.** $\dfrac{|x|}{x} = 1$ **24.** $\dfrac{x + 2}{|x + 2|} = -1$

Solve for x and graph.

25. $|x + 1| = 3$ **26.** $|x - 1| \le 3$ **27.** $|x - 1| \ge 3$

28. $|x + 2| = 3$ **29.** $|x + 2| \le 3$ **30.** $|x + 2| \ge 3$

31. $|-x| = 5$ **32.** $|x| \le 5$ **33.** $|x| \ge 5$

34. $|x - 5| \ne 3$ **35.** $|x - 5| \le 3$ **36.** $|x - 5| \ge 3$

37. $|x - 3| < 0.1$ **38.** $|2 - x| < 3$ **39.** $|2x - 1| < 7$

40. $|3x - 6| < 9$ **41.** $|4 - x| < 2$ **42.** $|1 + 5x| < 1$

43. $|x - 4| \not> 1$ *(Note: $\not>$ means $\le$.)*   **44.** $|x - 2| \not\ge 0$ *(Note: $\not\ge$ means $<$.)*   **45.** $\dfrac{1}{|x - 3|} > 0$

46. (a) Prove the product rule $|xy| = |x| \cdot |y|$ for the case $x < 0$ and $y > 0$.

    (b) Prove the quotient rule $\left|\dfrac{x}{y}\right| = \dfrac{|x|}{|y|}$ for the case $x < 0$ and $y < 0$.

47. Cite four different examples to confirm each inequality, using the cases
    (i) $x > 0$, $y > 0$; (ii) $x > 0$, $y < 0$; (iii) $x < 0$, $y > 0$; (iv) $x < 0$, $y < 0$.
    (a) $|x - y| \ge \big||x| - |y|\big|$
    (b) $|x + y| \le |x| + |y|$

 Written Assignment

1. Explain why $x^2 = |x^2| = |x|^2$ for any real number x.

2. Explain the meaning of the inequality $|x - 4| < 1$ in terms of distance on the number line. Repeat for $|x + 4| > 1$, noting that $x + 4 = x - (-4)$.

3. Complete each statement without using mathematical symbolism, so that together they become equivalent to the definition of absolute value:

    For positive x, . . .

    For negative x, . . .

Critical Thinking

1. How can you show that the set of real numbers is *not* associative with respect to division?

2. Show that the set of real numbers does not contain the multiplicative inverse for each of its members.

3. Explain the use of a specific counterexample to disprove a statement claimed to be true in general. Give several examples.

4. What are the possible values for the quotient $\dfrac{|2x|}{2x}$?

5. What inequality represents the set of real numbers that are less than three units from the point -2 on the number line?

6. In what sense does the concept of subtraction depend on addition?

7. If $a < b$ and $c < d$, is it true that $a - c < b - d$? Explain your answer.

1.5 INTEGRAL EXPONENTS

Much of mathematical notation can be viewed as efficient abbreviations of lengthier statements. For example:

$$4^9 = 4 \times 4 \times 4 \times 4 \times 4 \times 4 \times 4 \times 4 \times 4$$

This illustration makes use of a *positive integral exponent*. In this section we shall explore the use of integers as exponents.

> ### DEFINITION OF POSITIVE INTEGRAL EXPONENT
>
> If n is a positive integer and b is any real number, then
>
> $$b^n = \underbrace{b \cdot b \cdot \cdots \cdot b}_{n \text{ factors}}$$
>
> The number b is called the **base** and n is called the **exponent**.

The most common ways of referring to b^n are "b to the nth power," "b to the nth," or "the nth power of b."

Here are some illustrations of the definition:

$$b^1 = b$$
$$(a + b)^2 = (a + b)(a + b)$$
$$(-2)^3 = (-2)(-2)(-2) = -8$$
$$10^6 = 10 \cdot 10 \cdot 10 \cdot 10 \cdot 10 \cdot 10 = 1{,}000{,}000$$

CAUTION
Be careful when you work with parentheses. Note that $(-3)^2 \neq -3^2$.
$(-3)^2 = (-3)(-3) = 9$
$-3^2 = -(3 \times 3) = -9$

You can use a calculator to evaluate the power of a number. Study the manual that accompanies your calculator to determine how to do so. Here are typical key sequences to evaluate 10^6:

A number of important rules concerning positive integral exponents are easily established on the basis of the preceding definition. Here is a list of these rules in which m and n are positive integers, a and b are any real numbers, and denominators cannot be zero.

When multiplying powers with a common base, add the powers and use the same base.

RULE 1. $b^m b^n = b^{m+n}$

Illustrations: $2^3 \cdot 2^4 = 2^{3+4} = 2^7$ $x^3 \cdot x^4 = x^7$

RULE 2. $\dfrac{b^m}{b^n} = \begin{cases} b^{m-n} & \text{if } m > n \\ 1 & \text{if } m = n \\ \dfrac{1}{b^{n-m}} & \text{if } m < n \end{cases}$

Illustrations: $(m > n)$ $(m = n)$ $(m < n)$

$$\frac{2^5}{2^2} = 2^{5-2} = 2^3 \qquad \frac{5^2}{5^2} = 1 \qquad \frac{2^2}{2^5} = \frac{1}{2^{5-2}} = \frac{1}{2^3}$$

$$\frac{x^5}{x^2} = x^{5-2} = x^3 \qquad \frac{x^2}{x^2} = 1 \qquad \frac{x^2}{x^5} = \frac{1}{x^{5-2}} = \frac{1}{x^3}$$

The power of a power is the product of the powers with the same base.

RULE 3. $(b^m)^n = b^{mn}$

Illustrations: $(2^3)^2 = 2^{3 \cdot 2} = 2^6$ $(x^3)^2 = x^{3 \cdot 2} = x^6$

The power of a product is the product of the powers.

RULE 4. $(ab)^m = a^m b^m$

Illustrations: $(2 \cdot 3)^5 = 2^5 \cdot 3^5$ $(xy)^5 = x^5 y^5$

The power of a quotient is the quotient of the powers.

RULE 5. $\left(\dfrac{a}{b}\right)^m = \dfrac{a^m}{b^m}$

Illustrations: $\left(\dfrac{3}{2}\right)^5 = \dfrac{3^5}{2^5}$ $\left(\dfrac{x}{y}\right)^5 = \dfrac{x^5}{y^5}$

Eventually, the rules for exponents will be extended to include all the real numbers (not just positive integers) as exponents. We will find that all the rules stated here will still apply.

Here is a proof of Rule 4. You can try proving the other rules in a similar manner.

$$(ab)^m = \underbrace{(ab)(ab) \cdots (ab)}_{m \text{ times}} \qquad \text{by definition}$$

$$= \underbrace{(a \cdot a \cdot \cdots \cdot a)}_{m \text{ times}} \underbrace{(b \cdot b \cdot \cdots \cdot b)}_{m \text{ times}} \qquad \left\{ \begin{array}{l} \text{by repeated use of the} \\ \text{commutative and associative} \\ \text{laws for multiplication} \end{array} \right.$$

$$= a^m b^m \qquad \text{by definition}$$

The proper use of these rules can simplify computations, as in the following example.

In this chapter occasional use is made of some basic properties of fractions with which you should be familiar. These properties are studied in detail in Section 1.9.

EXAMPLE 1 Evaluate in two ways: $12^3(\frac{1}{6})^3$

Solution

(a) $12^3(\frac{1}{6})^3 = 1728(\frac{1}{216})$

$$= \frac{1728}{216}$$

$$= 8$$

(b) $12^3(\frac{1}{6})^3 = (12 \cdot \frac{1}{6})^3$

$$= 2^3$$

$$= 8$$

EXAMPLE 2 Simplify: **(a)** $2x^3 \cdot x^4$ **(b)** $\dfrac{(x^3y)^2 y^3}{x^4 y^6}$

Solution

(a) $2x^3 \cdot x^4 = 2x^{3+4} = 2x^7$ **(b)** $\dfrac{(x^3y)^2 y^3}{x^4 y^6} = \dfrac{(x^3)^2 y^2 y^3}{x^4 y^6} = \dfrac{x^6 y^5}{x^4 y^6} = \dfrac{x^2}{y}$

Care must be taken with both the multiplication and the division rules when the bases are not the same, as in the following example.

CAUTION

$$\frac{4^5}{9^3} \neq \left(\frac{4}{9}\right)^2$$

Rule 2 does not apply here because the bases are different, and Rule 5 does not apply since the exponents are different.

EXAMPLE 3 Simplify: $\dfrac{4^5}{8^3}$

Solution Since the bases are not the same, Rule 2 does not apply. However, rather than finding 4^5 and 8^3, the problem can be simplified in this way:

$$\frac{4^5}{8^3} = \frac{(2^2)^5}{(2^3)^3} = \frac{2^{10}}{2^9} = 2$$

TEST YOUR UNDERSTANDING

(Answers: Page 83)

Evaluate (simplify) each of the following.

1. 5^3

2. $(-\frac{1}{2})^5$

3. $(-\frac{2}{3})^3 + \frac{8}{27}$

4. $(10^3)^2$

5. $2^3(-2)^3$

6. $(\frac{1}{2})^3 8^3$

7. $\dfrac{17^8}{17^9}$

8. $\dfrac{(-2)^3 + 3^2}{3^3 - 2^2}$

9. $\dfrac{(-12)^4}{4^4}$

10. $(ab^2)^3(a^2b)^4$

11. $\dfrac{2^2 \cdot 16^3}{(-2)^8}$

12. $\dfrac{(2x^3)^2(3x)^2}{6x^4}$

This discussion provides meaning for the use of 0 as an exponent. That is, an expression like 5^0 will now be defined.

Our discussion of exponents has been restricted to the use of positive integers only. Now let us consider the meaning of 0 as an exponent. In particular, what is the meaning of 5^0? We know that 5^3 means that 5 is to be used three times as a factor. But certainly it makes no sense to use 5 zero times. The rules of exponents will help to resolve this dilemma.

We would like these laws for exponents to hold even if one of the exponents happens to be zero. That is, we would *like* Rule 2 to give

$$\frac{5^2}{5^2} = 5^{2-2} = 5^0$$

But it is already known that

$$\frac{5^2}{5^2} = \frac{25}{25} = 1$$

Thus 5^0 ought to be assigned the value 1. Consequently, in order to *preserve* the rules of exponents, it is sensible to let $5^0 = 1$. That is, from now on, we agree to the following:

Notice that the definition calls for b to be a real number different from 0. That is, we do not define an expression such as 0^0; this is said to be undefined.

DEFINITION OF ZERO EXPONENT

If b is a real number different from 0, then

$$b^0 = 1$$

Our next objective is to give meaning to negative integer exponents. For example, we want to decide the meaning of x^{-3}. Our guideline in making this decision will be that *the preceding rules of exponents are to apply for all kinds of integer exponents*. That is, we want to *preserve* the structure of the basic rules. With this in mind, observe the effect of Rule 1 when a negative exponent is involved.

$$(x^3)(x^{-3}) = x^{3 + (-3)} = x^0 = 1$$

Dividing both sides of $(x^3)(x^{-3}) = 1$ first by x^3 and then by x^{-3} produces these two statements

$$x^{-3} = \frac{1}{x^3} \quad \text{and} \quad x^3 = \frac{1}{x^{-3}}$$

We are now ready to make the following definition:

Note that the exponents $-n$ and n are opposites, and that $-n$ may be negative or positive.

DEFINITION OF b^{-n}

If n is an integer and $b \neq 0$, then

$$b^{-n} = \frac{1}{b^n}$$

Can you show that $\left(\frac{a}{b}\right)^{-2} = \left(\frac{b}{a}\right)^2$?

It follows from this definition that $\left(\frac{a}{b}\right)^{-1} = \frac{b}{a}$, since $\left(\frac{a}{b}\right)^{-1} = \frac{1}{\frac{a}{b}} = \frac{b}{a}$. In other words, *a fraction to the -1 power is the reciprocal of the fraction.*

As in the first illustration more than one correct procedure is often possible. Finding the most efficient procedure depends largely on experience.

Illustrations: $\left(\frac{1}{7}\right)^{-2} = \frac{1}{\left(\frac{1}{7}\right)^2} = \frac{1}{\frac{1}{49}} = 49$ or $\left(\frac{1}{7}\right)^{-2} = (7^{-1})^{-2} = 7^2 = 49$

$$\frac{5^{-2}}{15^{-2}} = \left(\frac{5}{15}\right)^{-2} = \left(\frac{1}{3}\right)^{-2} = (3^{-1})^{-2} = 3^2 = 9$$

In view of the definitions for b^0 and b^{-n}, the three cases of Rule 2 can now be condensed into this single form.

CAUTION
Be careful when applying Rule 2, especially when an exponent is negative. Thus
$$\frac{b^3}{b^{-2}} = b^{3-(-2)} = b^5;$$
$$\frac{b^3}{b^{-2}} \neq b^{3-2}.$$

RULE 2. (revised) $\dfrac{b^m}{b^n} = b^{m-n}$

Illustrations: $\dfrac{3^8}{3^2} = 3^6$ $\dfrac{3^2}{3^8} = 3^{-6}$ $\dfrac{3^2}{3^2} = 3^0 = 1$

$\dfrac{x^8}{x^2} = x^6$ $\dfrac{x^2}{x^8} = x^{-6}$ $\dfrac{x^2}{x^2} = x^0 = 1$

EXAMPLE 4 Simplify $\left(\dfrac{a^{-2}b^3}{a^3b^{-2}}\right)^5$ and express the answer using only positive exponents.

Solution There are several ways to proceed; here are two.

(a) $\left(\dfrac{a^{-2}b^3}{a^3b^{-2}}\right)^5 = (a^{-5}b^5)^5 = (a^{-5})^5(b^5)^5 = a^{-25}b^{25} = \dfrac{b^{25}}{a^{25}}$

(b) $\left(\dfrac{a^{-2}b^3}{a^3b^{-2}}\right)^5 = \left(\dfrac{b^5}{a^5}\right)^5 = \dfrac{(b^5)^5}{(a^5)^5} = \dfrac{b^{25}}{a^{25}}$ ■

Exponential notation is used in a variety of situations. Example 5 shows the use of exponents to analyze a situation in which a substance is decreasing exponentially.

EXAMPLE 5 Suppose that a radioactive substance decays so that $\frac{1}{2}$ the amount remains after each hour. If at a certain time there are 320 grams of the substance, how much will remain after 8 hours? How much after n hours?

Solution Since the amount remaining after each hour is $\frac{1}{2}$ of the grams at the end of the preceding hour, we find the remaining amount by multiplying the preceding number of grams by $\frac{1}{2}$.

	Grams Remaining
Start: 0 hours	$320\left(\frac{1}{2}\right)^0 = 320$
After 1 hour	$320\left(\frac{1}{2}\right)^1 = 160$
After 2 hours	$320\left(\frac{1}{2}\right)^2 = 80$
After 3 hours	$320\left(\frac{1}{2}\right)^3 = 40$
$\vdots$	$\vdots$
After 8 hours	$320\left(\frac{1}{2}\right)^8 = 1.25$

Observe that the power of $\frac{1}{2}$ is the same as the number of hours that the substance has been decaying. Assuming that the same pattern will continue, we conclude that there are $320 \left(\frac{1}{2}\right)^n = \dfrac{320}{2^n}$ grams remaining after n hours.   ∎

Exponents can be used to represent very large and very small numbers in compact form. For example, the distance from the earth to the sun is approximately 93,000,000 miles. This can be written in **scientific notation** as follows:

$$93,000,000 = 9.3 \times 10^7$$

Also, a recently developed optical microscope can resolve images down to approximately 12 *nanometers* or billionths of a meter. That is, in scientific notation,

$$0.0000000012 = 1.2 \times 10^{-9}$$

The preceding illustrations indicate that *a number N has been put into scientific notation when it has been expressed as the product of a number between 1 and 10 and an integral power of 10.* Thus:

$$N = x(10^c)$$

where $1 \le x < 10$ and c is an integer.

If the number is greater than 1, then the power of 10 is positive; if the number is less than 1, the power of 10 is negative.

WRITING A NUMBER IN SCIENTIFIC NOTATION

Place the decimal point behind the first nonzero digit. (This produces the number between 1 and 10.) Then determine the power of 10 by counting the number of places you moved the decimal point. If you moved the decimal point to the left, then the power is positive; and if you moved it to the right, it is negative.

Illustrations:

$$2,070,000. = 2.07 \times 10^6$$

six places left

$$0.00000084 = 8.4 \times 10^{-7}$$

seven places right

To convert a number given in scientific notation back into standard notation, all you need do is move the decimal point as many places as indicated by the exponent of 10. Move the decimal point to the right if the exponent is positive and to the left if it is negative.

When necessary, calculators will display numbers in scientific notation. For example, the answer to the product $3,500,000 \times 450,000$ might be shown in one of these ways:

$$1.575E+12 \qquad 1.575E12 \qquad 1.575 \ 12$$

Each of these means

$$1.575 \times 10^{12} = 1,575,000,000,000$$

Many errors are made when working with exponents because of misuses of the basic rules and definitions. This list shows some of the common errors that you should try to avoid.

| CAUTION: Learn to Avoid These Mistakes ||
WRONG	RIGHT
$5^2 \cdot 5^4 = 5^8$ (Do not multiply exponents.) $5^2 \cdot 5^4 = 25^6$ (Do not multiply the base numbers.)	$5^2 \cdot 5^4 = 5^6$ (Rule 1)
$\dfrac{5^6}{5^2} = 5^3$ (Do not divide the exponents.) $\dfrac{5^6}{5^2} = 1^4$ (Do not divide the base numbers.)	$\dfrac{5^6}{5^2} = 5^4$ (Rule 2)
$(5^2)^6 = 5^8$ (Do not add the exponents.)	$(5^2)^6 = 5^{12}$ (Rule 3)
$(-2)^4 = -2^4$ (Misreading the parentheses)	$(-2)^4 = (-1)^4 2^4 = 2^4$ (Rule 4)
$(-5)^0 = -1$ (Misreading definition of b^0)	$(-5)^0 = 1$ (Definition of b^0)
$2^{-3} = -\dfrac{1}{2^3}$ (Misreading definition of b^{-n})	$2^{-3} = \dfrac{1}{2^3}$ (Definition of b^{-n})

EXERCISES 1.5

Classify each statement as true or false. If it is false, correct the right side of the equality to obtain a true statement.

1. $3^4 \cdot 3^2 = 3^8$ **2.** $(2^2)^3 = 2^8$ **3.** $2^5 \cdot 2^2 = 4^7$ **4.** $\dfrac{9^3}{9^3} = 1$

5. $\dfrac{10^4}{5^4} = 2^4$ **6.** $\left(\dfrac{2}{3}\right)^4 = \dfrac{2^4}{3}$ **7.** $(-27)^0 = 1$ **8.** $(2^0)^3 = 2^3$

9. $3^4 + 3^4 = 3^8$ **10.** $(a^2b)^3 = a^2b^3$ **11.** $(a+b)^0 = a+1$ **12.** $a^2 + a^2 = 2a^2$

13. $\dfrac{1}{2^{-3}} = -2^3$ **14.** $(2+\pi)^{-2} = \dfrac{1}{4} + \dfrac{1}{\pi^2}$ **15.** $\dfrac{2^{-5}}{2^3} = 2^{-2}$

Evaluate.

16. -10^5 **17.** $2^0 + 2^1 + 2^2$ **18.** $(-3)^2(-2)^3$ **19.** $\left(\dfrac{2}{3}\right)^0 + \left(\dfrac{2}{3}\right)^1$

20. $\left[\left(\dfrac{1}{2}\right)^3\right]^2$ **21.** $\left(\dfrac{1}{2}\right)^4(-2)^4$ **22.** $\dfrac{3^2}{3^0}$ **23.** $\dfrac{(-2)^5}{(-2)^3}$

24. $\left(-\dfrac{3}{4}\right)^3$ **25.** $\dfrac{2^3 \cdot 3^4 \cdot 4^5}{2^2 \cdot 3^3 \cdot 4^4}$ **26.** $\dfrac{8^3}{16^2}$ **27.** $\dfrac{2^{10}}{4^3}$

28. $\dfrac{3^{-3}}{4^{-3}}$ **29.** $\left(\dfrac{2}{3}\right)^{-2} + \left(\dfrac{2}{3}\right)^{-1}$ **30.** $[(-7)^2(-3)^2]^{-1}$

Write each number in scientific notation.

31. 200,000,000   **32.** 32,100   **33.** 0.00037   **34.** 0.57721   **35.** 0.0000000555

Write each number in standard form.

36. 7.89×10^4 **37.** 7.89×10^{-4} **38.** 1.75×10^{-1} **39.** 2.25×10^5 **40.** 1.11×10

Simplify and express each answer using positive exponents only.

41. $(x^{-3})^2$ **42.** $(x^3)^{-2}$ **43.** $x^3 \cdot x^9$ **44.** $\dfrac{x^9}{x^3}$

45. $(2a)^3(3a)^2$ **46.** $(-2x^3y)^2(-3x^2y^2)^3$ **47.** $(-2a^2b^0)^4$ **48.** $(2x^3y^2)^0$

49. $\dfrac{(x^2y)^4}{(xy)^2}$ **50.** $\left(\dfrac{3a^2}{b^3}\right)^2\left(\dfrac{-2a}{3b}\right)^2$ **51.** $\left(\dfrac{x^3}{y^2}\right)^4\left(\dfrac{-y}{x^2}\right)^2$ **52.** $\dfrac{(x-2y)^6}{(x-2y)^2}$

53. $\dfrac{x^{-2}y^3}{x^3y^{-4}}$ **54.** $\dfrac{(x^{-2}y^2)^3}{(x^3y^{-2})^2}$ **55.** $\dfrac{5x^0y^{-2}}{x^{-1}y^{-2}}$ **56.** $\dfrac{(2x^3y^{-2})^2}{8x^{-3}y^2}$

57. $\dfrac{(-3a)^{-2}}{a^{-2}b^{-2}}$ **58.** $\dfrac{3a^{-3}b^2}{2^{-1}c^2d^{-4}}$ **59.** $\dfrac{(a+b)^{-2}}{(a+b)^{-8}}$ **60.** $\dfrac{8x^{-8}y^{-12}}{2x^{-2}y^{-6}}$

61. $\dfrac{-12x^{-9}y^{10}}{4x^{-12}y^7}$ **62.** $\dfrac{(2x^2y^{-1})^6}{(4x^{-6}y^{-5})^2}$ **63.** $\dfrac{(a+3b)^{-12}}{(a+3b)^{10}}$ **64.** $\dfrac{(-a^{-5}b^6)^3}{(a^8b^4)^2}$

65. $x^{-2} + y^{-2}$ **66.** $(a^{-2}b^3)^{-1}$ **67.** $\left(\dfrac{x^{-2}}{y^3}\right)^{-1}$ **68.** $-2(1+x^2)^{-3}(2x)$

69. $-2(4-5x)^{-3}(-5)$ **70.** $-7(x^2-3x)^{-8}(2x-3)$

Find a value of x to make each statement true.

71. $2^x \cdot 2^3 = 2^{12}$ **72.** $2^{-3} \cdot 2^x = 2^6$ **73.** $2^x \cdot 2^x = 2^{16}$

74. $2^x \cdot 2^{x-1} = 2^7$ **75.** $\dfrac{2^x}{2^2} = 2^{-5}$ **76.** $\dfrac{2^{-3}}{2^x} = 2^4$

77. Assume that a substance decays such that $\frac{1}{2}$ the amount remains after each hour. If there were 640 grams at the start, how much remains after 7 hours? How much after n hours?

78. If a rope is 243 feet long and you successively cut off $\frac{2}{3}$ of the rope, how much remains after five cuts? How much after n cuts?

79. For the rope in Exercise 78, how much remains after five cuts if each time you cut off $\frac{1}{3}$? How much is left after n cuts?

80. A company has a 4-year plan to increase its work force by $\frac{1}{4}$ for each of the 4 years. If the current work force is 2560, how big will it be at the end of the 4-year plan? Write an exponential expression that gives the work force after n years.

81. When an investment of P dollars earns $i\%$ interest per year and the interest is compounded annually, the formula for the final amount A after n years is $A = P(1 + i)^n$, where i is in decimal form. Find the amount A if $1000 is invested at 10% compounded annually for 3 years.

82. Use the formula in Exercise 81 to approximate the number of years it would take for a $1000 investment to double when it is invested at 10% interest compounded annually.

83. "Raising to a power" is neither a commutative nor an associative property. Verify this by showing a counterexample for each property.

84. Use the definition of a^{-n} to prove that $\dfrac{1}{a^{-n}} = a^n$.

85. Light travels at a rate of about 186,000 miles per second. The average distance from the sun to the earth is 93,000,000 miles. Use scientific notation to find how long it takes light to reach the earth from the sun.

86. Based on information given in Exercise 85, use scientific notation to show that 1 light-year (the distance light travels in 1 year) is approximately $5.87 \times 10^{12} = 5,870,000,000,000$ miles.

?

Challenge Suppose you snap your fingers and then wait 1 minute before you snap them again. Next, wait 2 minutes and snap your fingers, then 4 minutes, 8 minutes, 16 minutes, etc. That is, double the interval of time between successive snaps. If you were to keep this up for 1 year, how many times would you snap your fingers in that interval? First guess before doing any computation.

Written Assignment It has been stated that 0^0 is undefined. The following shows why it was not defined to be equal to 1.

$$\textit{Suppose that } 0^0 = 1. \text{ Then } 1 = \frac{1}{1} = \frac{1^0}{0^0} = \left(\frac{1}{0}\right)^0.$$

1. What rule for exponents is being used in the last step?
2. What went wrong?
3. Describe what happens if you attempt to calculate 0^0 on a calculator.

1.6 RADICALS AND RATIONAL EXPONENTS

What is the value of $\sqrt{25}$? Did you give the answer as ± 5? If so, you are making a very common mistake! Note that $\sqrt{25}$ is called a **radical** and stands only for the positive square root of 25; that is, $\sqrt{25} = +5$. To express the negative square root of 25, we write $-\sqrt{25} = -5$. In summary:

> If $a > 0$, then $\sqrt{a} = x$ where $x > 0$ and $x^2 = a$.

The positive square root $\sqrt{a}$ is called the *principal square root* of a.

In general, the **principal nth root** of a real number a is denoted by $\sqrt[n]{a}$, but the expression $\sqrt[n]{a}$ does not always have meaning. For example, let us try to evaluate $\sqrt[4]{-16}$:

It is a fundamental property of real numbers that every positive real number a has exactly one positive nth root. Furthermore, every negative real number has a negative nth root provided that n is odd.

$$2^4 = 16 \qquad (-2)^4 = 16$$

It appears that there is no real number x such that $x^4 = -16$. In general, *there is no real number that is the even root of a negative number.*

1. $\sqrt[3]{64} = 4$ *because*
   $4^3 = 64$
2. $\sqrt[3]{0} = 0$ *because* $0^3 = 0.$
3. $\sqrt[3]{-8} = -2$ *because*
   $(-2)^3 = -8$
4. $\sqrt[4]{-16}$ *is not a real*
   number

DEFINITION OF $\sqrt[n]{a}$: THE PRINCIPAL nTH ROOT OF a

Let a be a real number and n a positive integer, $n \geq 2.$

1. If $a > 0$, then $\sqrt[n]{a}$ is the positive number x such that $x^n = a.$
2. $\sqrt[n]{0} = 0.$
3. If $a < 0$ and n is odd, then $\sqrt[n]{a}$ is the negative number x such that $x^n = a.$
4. If $a < 0$ and n is even, then $\sqrt[n]{a}$ is not a real number.

The symbol $\sqrt[n]{a}$ is also said to be a **radical;** n is the **index** or **root,** and a is called the **radicand.**

EXAMPLE 1   Evaluate those radicals that are real numbers and check. If an expression is not a real number, give a reason.

(a) $\sqrt[3]{-125}$ **(b)** $\sqrt{-9}$ **(c)** $\left(\sqrt[3]{8}\right)^3$

Solution

(a) $\sqrt[3]{-125} = -5$

   Check: $(-5)^3 = -125$

(b) $\sqrt{-9}$ is not a real number since it is the even root of a negative number.

In general, $\left(\sqrt[n]{a}\right)^n = a.$ **(c)** $\left(\sqrt[3]{8}\right)^3 = 8$

   Check: $\sqrt[3]{8} = 2$ and $2^3 = 8$ ∎

In order to multiply or divide radicals, the index must be the same, as shown below.

Illustrations:

$$\left.\begin{array}{l} \sqrt[3]{8} \cdot \sqrt[3]{-27} = (2)(-3) = -6 \\[2mm] \sqrt[3]{(8)(-27)} = \sqrt[3]{-216} = -6 \end{array}\right\} \quad \sqrt[3]{8} \cdot \sqrt[3]{-27} = \sqrt[3]{(8)(-27)}$$

$$\left.\begin{array}{l} \dfrac{\sqrt{36}}{\sqrt{4}} = \dfrac{6}{2} = 3 \\[4mm] \sqrt{\dfrac{36}{4}} = \sqrt{9} = 3 \end{array}\right\} \quad \dfrac{\sqrt{36}}{\sqrt{4}} = \sqrt{\dfrac{36}{4}}$$

In the following rules it is assumed that all radicals exist according to the definition of $\sqrt[n]{a}$ and, as usual, denominators are not zero.

The proofs of these rules are called for in Exercises 72–74.

RULES FOR RADICALS

If all the indicated radicals are real numbers, then

1. $\sqrt[n]{a} \cdot \sqrt[n]{b} = \sqrt[n]{ab}$ multiplication of radicals

2. $\dfrac{\sqrt[n]{a}}{\sqrt[n]{b}} = \sqrt[n]{\dfrac{a}{b}}$ division of radicals, $b \neq 0$

3. $\sqrt[m]{\sqrt[n]{a}} = \sqrt[mn]{a}$

It is understood that $\sqrt{a}$ means $\sqrt[2]{a}$.

Illustrations: (Assume that all variables represent positive numbers.)

$$\sqrt{6x} \cdot \sqrt{7y} = \sqrt{6x \cdot 7y} = \sqrt{42xy}$$

$$\frac{\sqrt[3]{81x^7}}{\sqrt[3]{-3x}} = \sqrt[3]{\frac{81x^7}{-3x}} = \sqrt[3]{-27x^6} = -3x^2 \qquad \textit{Note: } (-3x^2)^3 = -27x^6$$

$$\sqrt[2]{\sqrt[3]{64}} = \sqrt[6]{64} = 2 \qquad\qquad \textit{Note: } 2^6 = 64$$

We are now ready to make the extension of the exponential concept to include fractional exponents. Once again our guideline will be to *preserve* the earlier rules for integer exponents. First, consider the expression $b^{1/5}$. If Rule 3 for exponents is to apply, (page 31), then $(b^{1/5})^5 = b^{(1/5)(5)} = b$. Thus $b^{1/5}$ is the fifth root of b; $b^{1/5} = \sqrt[5]{b}$. This motivates the following definition for $b^{1/n}$.

DEFINITION OF $b^{1/n}$

For a real number b and a positive integer n ($n \geq 2$),

$$b^{1/n} = \sqrt[n]{b}$$

provided that $\sqrt[n]{b}$ exists.

Since $\sqrt{-1}$ is not a real number, $(-1)^{1/2}$ is not defined. In general, $b^{1/n}$ is not defined within the set of real numbers when $b < 0$ and n is even.

Illustrations:

$$(-27)^{1/3} = \sqrt[3]{-27} = -3 \qquad 9^{-1/2} = \frac{1}{9^{1/2}} = \frac{1}{\sqrt{9}} = \frac{1}{3}$$

$$(-16)^{1/4} \text{ is not defined since } \sqrt[4]{-16} \text{ is not a real number}$$

Now that $b^{1/n}$ has been defined, we are able to define $b^{m/n}$, where $\dfrac{m}{n}$ is any rational number. Once again we want the earlier rules of exponents to apply. Observe, for example, the two ways that $8^{2/3}$ can be evaluated on the *assumption* that the rules for integral exponents apply.

$$8^{2/3} = 8^{(1/3) \cdot 2} = (8^{1/3})^2 = \left(\sqrt[3]{8}\right)^2 = 2^2 = 4$$

$$\uparrow$$
$$\text{Rule 3}$$
$$\downarrow$$

$$8^{2/3} = 8^{2 \cdot (1/3)} = (8^2)^{1/3} = \sqrt[3]{8^2} = \sqrt[3]{64} = 4$$

These observations lead to this definition:

> **DEFINITION OF $b^{m/n}$**
>
> Let $\dfrac{m}{n}$ be a rational number with $n \geq 2$. If b is a real number such that $\sqrt[n]{b}$ is defined, then
>
> $$b^{m/n} = \left(\sqrt[n]{b}\right)^m = \sqrt[n]{b^m} \qquad \text{or} \qquad b^{m/n} = (b^{1/n})^m = (b^m)^{1/n}$$

Note that a rational number can always be expressed with a positive denominator; for example, $\dfrac{2}{-3} = \dfrac{-2}{3}$.

Illustrations:

For most such problems it is easier to take the nth root first and then raise to the mth power, rather than the reverse.

or

$$(-64)^{2/3} = \left(\sqrt[3]{-64}\right)^2 = (-4)^2 = 16 \qquad \text{using } b^{m/n} = \left(\sqrt[n]{b}\right)^m$$

$$(-64)^{2/3} = \sqrt[3]{(-64)^2} = \sqrt[3]{4096} = 16 \qquad \text{using } b^{m/n} = \sqrt[n]{b^m}$$

Observe that the earlier definition $b^{-n} = \dfrac{1}{b^n}$ extends to the case $b^{-(m/n)}$ as follows:

$$b^{-(m/n)} = b^{-m/n} = (b^{1/n})^{-m} = \frac{1}{(b^{1/n})^m} = \frac{1}{b^{m/n}}$$

EXAMPLE 2   Evaluate: $8^{-2/3} + (-32)^{-2/5}$

Solution   First rewrite each part using positive exponents. Then apply the definition of $b^{m/n}$ and add:

Here is another procedure for a negative fractional exponent:

$$8^{-2/3} = (8^{1/3})^{-2} = 2^{-2} = \frac{1}{4}$$

$$8^{-2/3} + (-32)^{-2/5} = \frac{1}{8^{2/3}} + \frac{1}{(-32)^{2/5}} = \frac{1}{\left(\sqrt[3]{8}\right)^2} + \frac{1}{\left(\sqrt[5]{-32}\right)^2}$$

$$= \frac{1}{(2)^2} + \frac{1}{(-2)^2} = \frac{1}{4} + \frac{1}{4} = \frac{1}{2} \qquad\blacksquare$$

TEST YOUR UNDERSTANDING	*Write each of the following using fractional exponents.*

1. $\sqrt{7}$ **2.** $\sqrt[3]{-10}$ **3.** $\sqrt[4]{7}$ **4.** $\sqrt[3]{7^2}$ **5.** $(\sqrt[4]{5})^3$

Evaluate.

6. $25^{1/2}$ **7.** $64^{1/3}$ **8.** $\left(\frac{1}{36}\right)^{1/2}$ **9.** $49^{-1/2}$ **10.** $\left(-\frac{1}{27}\right)^{-1/3}$

11. $4^{3/2}$ **12.** $4^{-3/2}$ **13.** $\left(\frac{81}{16}\right)^{3/4}$ **14.** $(-8)^{2/3}$ **15.** $(-8)^{-2/3}$

Simplify. Assume that all variables represent positive numbers.

16. $\sqrt[3]{-8x} \cdot \sqrt[3]{-27x^2}$ **17.** $\dfrac{\sqrt[5]{-4x^3}}{\sqrt[5]{128x^8}}$ **18.** $\sqrt[3]{\dfrac{1}{24}} \cdot \sqrt[3]{-81x^6}$

(Answers: Page 83)

The basic rules for integral exponents apply to rational exponents as well. Try to explain each step in the following example.

EXAMPLE 3   Simplify: $\left(\dfrac{-8a^3}{b^{-6}}\right)^{2/3}$

Solution

$$\left(\frac{-8a^3}{b^{-6}}\right)^{2/3} = \frac{(-8a^3)^{2/3}}{(b^{-6})^{2/3}} \qquad \text{Rule 5}$$

$$= \frac{(-8)^{2/3}(a^3)^{2/3}}{(b^{-6})^{2/3}} \qquad \text{Rule 4}$$

$$= \frac{\sqrt[3]{(-8)^2}\,a^2}{b^{-4}}$$

$$= 4a^2b^4 \qquad\qquad \sqrt[3]{(-8)^2} = \sqrt[3]{64} = 4 \qquad \blacksquare$$

When the index of a radical is n, and the radicand is a perfect nth power, then there is usually no difficulty in computing with the radical. For example:

$$\sqrt{36} + \sqrt[3]{-27} = 6 + (-3) = 3$$

Note: The fundamental idea used here is that $\sqrt[n]{ab} = \sqrt[n]{a} \cdot \sqrt[n]{b}$ where $a \geq 0$ and $b \geq 0$.

When the radicand is not a perfect nth power, as in $\sqrt{24}$, we can simplify the radical so that no perfect square appears as a factor under the radical sign. Thus

$$\sqrt{24} = \sqrt{4 \cdot 6} = \sqrt{4} \cdot \sqrt{6} = 2\sqrt{6}$$

We say that $2\sqrt{6}$ is the *simplified form* of $\sqrt{24}$.

EXAMPLE 4   Simplify: **(a)** $\sqrt{50}$ **(b)** $\sqrt[3]{16}$ **(c)** $\sqrt[3]{-54}$

Solution

(a) $\sqrt{50} = \sqrt{25 \cdot 2} = \sqrt{25} \cdot \sqrt{2} = 5\sqrt{2}$

In parts (b) and (c) we search for a perfect cube as a factor under the radical sign.

(b) $\sqrt[3]{16} = \sqrt[3]{8 \cdot 2} = \sqrt[3]{8} \cdot \sqrt[3]{2} = 2\sqrt[3]{2}$

Recall: In order to be able to multiply two radicals, they must have the same index.

(c) $\sqrt[3]{-54} = \sqrt[3]{(-27)(2)} = \sqrt[3]{-27} \cdot \sqrt[3]{2} = -3\sqrt[3]{2} \qquad \blacksquare$

The rule for multiplication of radicals provides a way to find the product of any two radicals having the same index. Does a similar pattern work for the addition of radicals? That is, is the sum of the square roots of two numbers equal to the square root of their sum? Does $\sqrt{4} + \sqrt{9} = \sqrt{4 + 9}$? This can easily be checked as follows:

$$\sqrt{4} + \sqrt{9} = 2 + 3 = 5$$

But $\sqrt{4 + 9} = \sqrt{13}$. Therefore $\sqrt{4} + \sqrt{9} \neq \sqrt{4 + 9}$.

In general, $\sqrt{a} \pm \sqrt{b} \neq \sqrt{a \pm b}$.

In order to be able to add or subtract radicals, they must have the same index and the same radicand.

For example, we may use the distributive property to add radicals as follows:

$$3\sqrt{5} + 4\sqrt{5} = (3 + 4)\sqrt{5} = 7\sqrt{5}$$

You may find the following list of squares and cubes helpful when simplifying radicals.

EXAMPLE 5   Simplify: $\sqrt[3]{-24x^3} + 2\sqrt[3]{375x^3} - \sqrt[4]{162x^4}$ $(x \geq 0)$

Solution   Simplify each term by searching for perfect powers as factors.

$$\sqrt[3]{-24x^3} = \sqrt[3]{(-8x^3)(3)} = -2x\sqrt[3]{3} \qquad (-2x)^3 = -8x^3$$

$$2\sqrt[3]{375x^3} = 2\sqrt[3]{(125x^3)(3)} = 2(5x)\sqrt[3]{3} = 10x\sqrt[3]{3} \qquad (5x)^3 = 125x^3$$

$$\sqrt[4]{162x^4} = \sqrt[4]{(81x^4)(2)} = 3x\sqrt[4]{2} \qquad (3x)^4 = 81x^4$$

	Perfect Squares	Perfect Cubes
Integer		
1	1	1
2	4	8
3	9	27
4	16	64
5	25	125
6	36	216
7	49	343
8	64	512
9	81	729
10	100	1000

Now combine the terms that have the same index and radicand.

$$-2x\sqrt[3]{3} + 10x\sqrt[3]{3} - 3x\sqrt[4]{2} = 8x\sqrt[3]{3} - 3x\sqrt[4]{2} \qquad \blacksquare$$

The major reason for rationalizing denominators is to obtain a form for radical expressions in which they are more easily combined or compared.

At times a fraction can be changed by a process known as **rationalizing the denominator**. This consists of eliminating a radical from the denominator of a fraction. For example, consider the fraction $4/\sqrt{2}$. To rationalize the denominator, multiply the numerator and denominator by $\sqrt{2}$.

$$\frac{4}{\sqrt{2}} = \frac{4 \cdot \sqrt{2}}{\sqrt{2} \cdot \sqrt{2}} = \frac{4\sqrt{2}}{2} = 2\sqrt{2}$$

EXAMPLE 6   Rationalize the denominator: **(a)** $\dfrac{6}{\sqrt{8}}$ **(b)** $\dfrac{5}{\sqrt[3]{2}}$

Solution

(a) $\dfrac{6}{\sqrt{8}} = \dfrac{6 \cdot \sqrt{2}}{\sqrt{8} \cdot \sqrt{2}} = \dfrac{6\sqrt{2}}{\sqrt{16}} = \dfrac{6\sqrt{2}}{4} = \dfrac{3\sqrt{2}}{2}$

Here is another way to simplify part (a):

$$\frac{6}{\sqrt{8}} = \frac{6}{2\sqrt{2}} = \frac{3}{\sqrt{2}} = \frac{3\sqrt{2}}{2}$$

(b) Multiply numerator and denominator by $\sqrt[3]{4}$ in order to have the cube root of a perfect cube in the denominator.

$$\frac{5}{\sqrt[3]{2}} = \frac{5 \cdot \sqrt[3]{4}}{\sqrt[3]{2} \cdot \sqrt[3]{4}} = \frac{5\sqrt[3]{4}}{\sqrt[3]{8}} = \frac{5\sqrt[3]{4}}{2} \qquad \blacksquare$$

EXAMPLE 7   Combine: $\dfrac{6}{\sqrt{3}} + 2\sqrt{75} - \sqrt{3}$

Solution   Rationalize the denominator in the first term, simplify the second term, and combine:

$$\frac{6}{\sqrt{3}} + 2\sqrt{75} - \sqrt{3} = \frac{6 \cdot \sqrt{3}}{\sqrt{3} \cdot \sqrt{3}} + 2\sqrt{25 \cdot 3} - \sqrt{3}$$

$$= 2\sqrt{3} + 10\sqrt{3} - 1\sqrt{3} = 11\sqrt{3} \qquad \blacksquare$$

TEST YOUR UNDERSTANDING	*Simplify each expression, if possible.*

Simplify each expression, if possible.

1. $\sqrt{8} + \sqrt{32}$ **2.** $\sqrt{12} + \sqrt{48}$ **3.** $\sqrt{45} - \sqrt{20}$

4. $\sqrt[3]{-16} + \sqrt[3]{54}$ **5.** $\sqrt[3]{128} + \sqrt[3]{125}$ **6.** $\sqrt[3]{-81} - \sqrt[3]{-24}$

7. $\frac{8}{\sqrt{2}} + \sqrt{98}$ **8.** $\frac{9}{\sqrt{3}} + \sqrt{300}$ **9.** $2\sqrt{20} - \frac{5}{\sqrt{5}}$

10. $3\sqrt{63} - \frac{14}{\sqrt{7}}$ **11.** $\frac{8}{\sqrt[3]{4}} + \sqrt[3]{16}$ **12.** $\sqrt[3]{81} - \frac{3}{\sqrt[3]{9}}$

(Answers: Page 83)

Do you think that $\sqrt{x^2} = x$? If this were true, then for $x = -5$, we would have $\sqrt{(-5)^2} = -5$. However, as stated on page 38, the radical sign, $\sqrt{}$, means the *positive square root*. Therefore, $\sqrt{(-5)^2} = \sqrt{5^2} = 5$. This leads to the following important result:

> For all real numbers a, $\sqrt{a^2} = |a|$

This result can be extended as follows:

Recall that $\left(\sqrt[n]{a}\right)^n = a$ for n even or odd as long as $\sqrt[n]{a}$ is a real number.

> $\sqrt[n]{a^n} = |a|$,   if n is even.
>
> $\sqrt[n]{a^n} = a$,   if n is odd.

Illustrations:

$$\sqrt{75x^2} = \sqrt{25 \cdot 3}\sqrt{x^2} = 5\sqrt{3}|x|$$

$$\sqrt[7]{(-3)^7} = -3$$

$$\sqrt[8]{\left(-\tfrac{1}{2}\right)^8} = \left|-\tfrac{1}{2}\right| = \tfrac{1}{2}$$

EXAMPLE 8   Simplify: $2\sqrt{8x^3} + 3x\sqrt{32x} - x\sqrt{18x}$

Note: The expressions under the radicals would be negative for $x < 0$. Since the index is even, we must assume that $x \geq 0$ in order for these to have meaning.

Solution   For this problem, $x \geq 0$. Thus we need not make use of absolute-value notation.

$$2\sqrt{8x^3} = 2\sqrt{4 \cdot 2 \cdot x^2 \cdot x} = 4x\sqrt{2x}$$

$$3x\sqrt{32x} = 3x\sqrt{16 \cdot 2x} = 12x\sqrt{2x}$$

$$x\sqrt{18x} = x\sqrt{9 \cdot 2x} = 3x\sqrt{2x}$$

In each case the radicand and the index are the same, so the distributive property can be used to simplify.

$$4x\sqrt{2x} + 12x\sqrt{2x} - 3x\sqrt{2x} = (4x + 12x - 3x)\sqrt{2x} = 13x\sqrt{2x} \quad \blacksquare$$

CAUTION: Learn to Avoid These Mistakes	
WRONG	**RIGHT**
$\sqrt{9 + 16} = \sqrt{9} + \sqrt{16}$	$\sqrt{9 + 16} = \sqrt{25}$
$(a + b)^{1/3} = a^{1/3} + b^{1/3}$	$(a + b)^{1/3} = \sqrt[3]{a + b}$
$\sqrt[3]{8} \cdot \sqrt[2]{8} = \sqrt[6]{64}$	$\sqrt[3]{8} \cdot \sqrt[2]{8} = 2 \cdot 2\sqrt{2} = 4\sqrt{2}$
$2\sqrt{x + 1} = \sqrt{2x + 2}$	$2\sqrt{x + 1} = \sqrt{4(x + 1)}$
	$= \sqrt{4x + 4}$
$\sqrt{(x - 1)^2} = x - 1$	$\sqrt{(x - 1)^2} = \lvert x - 1 \rvert$
$\sqrt{x^9} = x^3$	$\sqrt{x^9} = \sqrt{x^8 \cdot x} = x^4\sqrt{x}$
$a^{-1/2} + b^{-1/2} = \dfrac{1}{\sqrt{a + b}}$	$a^{-1/2} + b^{-1/2} = \dfrac{1}{\sqrt{a}} + \dfrac{1}{\sqrt{b}}$

EXERCISES 1.6

Evaluate.

1. $81^{-1/2}$

2. $\sqrt[3]{-64}$

3. $(64)^{-2/3}$

4. $(-64)^{1/3}$

5. $(-125)^{2/3}$

6. $(-125)^{-2/3}$

7. $\sqrt[3]{9} \cdot \sqrt[3]{-3}$

8. $\sqrt{5} \cdot \sqrt{20}$

9. $\dfrac{\sqrt[3]{-3}}{\sqrt[3]{-24}}$

10. $\dfrac{\sqrt{75}}{\sqrt{3}}$

11. $\dfrac{\sqrt{9}}{27^{-1/3}}$

12. $\dfrac{9^{1/2}}{\sqrt[3]{27}}$

13. $\sqrt[3]{(-125)(-1000)}$

14. $\sqrt{\sqrt{625}}$

15. $\sqrt[3]{\sqrt[3]{-512}}$

16. $\sqrt{\sqrt[3]{729}}$

17. $\sqrt{144 + 25}$

18. $\sqrt[5]{(-243)^2} \cdot (49)^{-1/2}$

19. $\left(\dfrac{1}{8} + \dfrac{1}{27}\right)^{1/3}$

20. $\sqrt{144} + \sqrt{25}$

21. $\left(\dfrac{16}{81}\right)^{3/4} + \left(\dfrac{256}{625}\right)^{1/4}$

22. $\left(\dfrac{1}{8}\right)^{1/3} + \left(\dfrac{1}{27}\right)^{-1/3}$

23. $\left(-\dfrac{125}{8}\right)^{1/3} - \left(\dfrac{1}{64}\right)^{1/3}$

24. $\left(\dfrac{8}{27}\right)^{-2/3} + \left(-\dfrac{32}{243}\right)^{2/5}$

Simplify.

25. $\sqrt{2} + \sqrt{18}$

26. $\sqrt{32} + \sqrt{72}$

27. $\sqrt{6} \cdot \sqrt{12}$

28. $\sqrt[3]{4} \cdot \sqrt[3]{12}$

29. $2\sqrt{5} + 3\sqrt{125}$

30. $-5\sqrt{24} - 2\sqrt{54}$

31. $2\sqrt{200} - 5\sqrt{8}$

32. $3\sqrt{45} - 2\sqrt{20}$

33. $\sqrt[3]{128} + \sqrt[3]{16}$

34. $\sqrt[3]{-24} + \sqrt[3]{81}$

35. $\dfrac{8}{\sqrt{2}} + 2\sqrt{50}$

36. $\sqrt[3]{-54x} + \sqrt[3]{250x}$

37. $3\sqrt{8x^2} - \sqrt{50x^2}$

38. $5\sqrt{75x^2} - 2\sqrt{12x^2}$

39. $3\sqrt{10} + 4\sqrt{90} - 5\sqrt{40}$

40. $\dfrac{2}{\sqrt{3}} + 10\sqrt{3} - 2\sqrt{12}$

41. $\dfrac{10}{\sqrt{5}} + 3\sqrt{45} - 2\sqrt{20}$

42. $10\sqrt{3x} - 2\sqrt{75x} + 3\sqrt{243x}$

43. $3\sqrt{9x^2} + 2\sqrt{16x^2} - \sqrt{25x^2}$

44. $\sqrt{2x^2} + 5\sqrt{32x^2} - 2\sqrt{98x^2}$

45. $\sqrt{x^2y} + \sqrt{8x^2y} + \sqrt{200x^2y}$

Rationalize the denominators and simplify.

46. $\dfrac{24}{\sqrt{6}}$

47. $\dfrac{8x}{\sqrt{2}}$

48. $\dfrac{10}{\sqrt{5}} + \dfrac{8}{\sqrt{4}}$

49. $\dfrac{1}{\sqrt{18}}$

50. $\dfrac{1}{\sqrt{27}}$

51. $\dfrac{24}{\sqrt{3x^2}}$

52. $\dfrac{20x}{\sqrt{5x^3}}$

53. $\dfrac{8}{\sqrt[3]{2}}$

Simplify, and express all answers with positive exponents. (Assume that all letters represent positive numbers.)

54. $(a^{-4}b^{-8})^{3/4}$

55. $(a^{-1/2}b^{1/3})(a^{1/2}b^{-1/3})$

56. $\dfrac{a^2b^{-1/2}c^{1/3}}{a^{-3}b^0c^{-1/3}}$

57. $\left(\dfrac{64a^6}{b^{-9}}\right)^{2/3}$

58. $\dfrac{(49a^{-4})^{-1/2}}{(81b^6)^{-1/2}}$

59. $\left(\dfrac{a^{-2}b^3}{a^4b^{-3}}\right)^{-1/2}\left(\dfrac{a^4b^{-5}}{ab}\right)^{-1/3}$

60. $\frac{2}{3}(3x - 1)^{-1/3} \cdot 3$

61. $\frac{1}{2}(3x^2 + 2)^{-1/2} \cdot 6x$

62. $\frac{1}{3}(x^3 + 2)^{-2/3} \cdot 3x^2$

63. $\frac{1}{2}(x^2 + 4x)^{-1/2}(2x + 4)$

64. $\frac{2}{3}(x^3 - 6x^2)^{-1/3}(3x^2 - 12x)$

65. The diagonal d of a rectangle is given by the formula $d = \sqrt{\ell^2 + w^2}$, where ℓ is the length and w is the width.

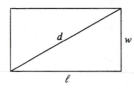

(a) Find d if $\ell = 20$ centimeters and $w = 15$ centimeters.

(b) Find d if $\ell = 16$ centimeters and $w = 10$ centimeters. Give the answer to one decimal place.

66. The formula $A = \sqrt{s(s - a)(s - b)(s - c)}$ is known as **Heron's formula**. It gives the area A of a triangle with sides of length a, b, c and semiperimeter $s = \frac{1}{2}(a + b + c)$. Show that for an equilateral triangle each of whose sides is of length a, Heron's formula gives $A = \dfrac{\sqrt{3}}{4}a^2$.

HISTORICAL NOTE
Heron (see Exercise 66) lived about the second half of the first century A.D. and was also known as Hero of Alexandria. He wrote in the area of applied mathematics and proved, for example, that the angles of incidence and reflection in a mirror are equal. For more information on Heron see Eves, An Introduction to the History of Mathematics, 5th ed. (Philadelphia: Saunders, 1983).

Combine and simplify.

67. $\sqrt[3]{\dfrac{32}{x^2}} - \dfrac{2\sqrt[3]{x}}{\sqrt[3]{2x^3}}$

68. $\dfrac{\sqrt{72a^3}}{3b} - \dfrac{a\sqrt{50a}}{2b} + \dfrac{12a^2}{b\sqrt{2a}}$

Simplify, and express the answers without radicals, using only positive exponents. (Assume that n is a positive integer and that all other letters represent positive numbers.)

69. $\sqrt[3]{\dfrac{x^{3n+1}y^n}{x^{3n+4}y^{4n}}}$

70. $\left(\dfrac{x^n}{x^{n-2}}\right)^{-1/2}$

71. $\sqrt{\dfrac{x^n}{x^{n-2}}}$

Prove the following assuming all radicals are real numbers.

72. $\sqrt[n]{a} \cdot \sqrt[n]{b} = \sqrt[n]{ab}$ (*Hint:* Let $\sqrt[n]{a} = x$ and $\sqrt[n]{b} = y$. Then $x^n = a$ and $y^n = b$.)

73. $\dfrac{\sqrt[n]{a}}{\sqrt[n]{b}} = \sqrt[n]{\dfrac{a}{b}}$

74. $\sqrt[m]{\sqrt[n]{a}} = \sqrt[mn]{a}$ (*Hint:* Let $x = \sqrt[mn]{a}$. Then $a = x^{mn}$.)

75. Use the result $\sqrt{a^2} = |a|$ and the rules for radicals to prove that $|xy| = |x| \cdot |y|$, where x and y are real numbers.

 Written Assignment

1. Use the definitions of $\sqrt{a}$ and $|a|$ to explain why $\sqrt{x^2} = |x|$ for all real numbers x.

2. Consider the following "proof" that $4 = -4$. Explain the error in this development.

$$4 = \sqrt{16} = \sqrt{(-4)(-4)} = \sqrt{-4} \cdot \sqrt{-4} = \left(\sqrt{-4}\right)^2 = -4$$

1.7
FUNDAMENTAL OPERATIONS WITH POLYNOMIALS

Note: All the exponents of the variable of a polynomial must be nonnegative integers. Therefore, $x^3 + x^{1/2}$ and $x^{-2} + 3x + 1$ are not polynomials because of the fractional and negative exponents.

Some of these polynomials have "missing" terms. For example, $x^3 - 3x + 12$ has no x^2 term, but it is still a third-degree polynomial.

The expression $5x^3 - 7x^2 + 4x - 12$ is called a **polynomial in the variable x.** Its *degree* is 3 because 3 is the largest power of the variable x. The *terms* of this polynomial are $5x^3$, $-7x^2$, $4x$, and -12. The *coefficients* are 5, -7, 4, and -12.

A nonzero constant, like 9, is classified as a polynomial of degree zero, since $9 = 9x^0$. The number zero is also referred to as a constant polynomial, but it is not assigned any degree.

A polynomial is in *standard form* if its terms are arranged so that the powers of the variable are in descending or ascending order. Here are some illustrations.

Polynomial	Degree	Standard Form
$x^3 - 3x + 12$	3	Yes
$\frac{2}{3}x^{10} - 4x^2 + \sqrt{2}\,x^4$	10	No
$32 - y^5 + 2y^2$	5	No
$6 + 2x - x^2 + x^3$	3	Yes

Polynomials having one, two, or three terms have special names:

Number of Terms	Name of Polynomial	Illustration
One	Monomial	$17x^4$
Two	Binomial	$\frac{1}{2}x^3 - 6x$
Three	Trinomial	$x^5 - x^2 + 2$

In general, an nth-degree polynomial in the variable x may be written in either one of these standard forms:

$$a_n x^n + a_{n-1}x^{n-1} + \cdots + a_2 x^2 + a_1 x + a_0$$

$$a_0 + a_1 x + a_2 x^2 + \cdots + a_{n-1}x^{n-1} + a_n x^n$$

The coefficients $a_n, a_{n-1}, \ldots, a_0$ are real numbers and the exponents are nonnegative integers. The *leading coefficient* is $a_n \neq 0$, and a_0 is called the *constant term*. (a_0 may also be considered as the coefficient of the term $a_0 x^0$.)

In a polynomial like $3x^2 - x + 4$ the variable x represents a real number. Therefore, when a specific real value is substituted for x, the result will be a real number. For instance, using $x = -3$ in this polynomial gives

$$3(-3)^2 - (-3) + 4 = 34$$

Note: Computations with polynomials are based on the fundamental properties for real numbers since the variables represent real numbers.

Adding or subtracting polynomials involves the combining of **like terms** (those having the same exponent on the variable). This can be accomplished by first rearranging and regrouping the terms (associative and commutative properties) and then combining by using the distributive property.

EXAMPLE 1   Add: $(4x^3 - 10x^2 + 5x + 8) + (12x^2 - 9x - 1)$

Solution
(a) $(4x^3 - 10x^2 + 5x + 8) + (12x^2 - 9x - 1)$
$$= 4x^3 + (12x^2 - 10x^2) + (5x - 9x) + (8 - 1)$$
$$= 4x^3 + 2x^2 - 4x + 7$$

In method (b) we list the polynomials in column form, putting like terms in the same column.

(b)
$$\begin{array}{r} 4x^3 - 10x^2 + 5x + 8 \\ (+)\underline{12x^2 - 9x - 1} \\ 4x^3 + 2x^2 - 4x + 7 \end{array}$$

■

Polynomials may contain more than one variable, as in Example 2. Again, the subtraction can be completed in two different ways, as shown.

Example 2 is of the form $a - b$. Think of this as $a - 1 \cdot b$ and use the distributive property to simplify in method (a).

EXAMPLE 2   Subtract: $(4a^3 - 10a^2 b + 5b + 8) - (12a^2 b - 9b - 1)$

Solution
(a) $(4a^3 - 10a^2 b + 5b + 8) - (12a^2 b - 9b - 1)$
$$= 4a^3 - 10a^2 b + 5b + 8 - 12a^2 b + 9b + 1$$
$$= 4a^3 - 10a^2 b - 12a^2 b + 5b + 9b + 8 + 1$$
$$= 4a^3 - 22a^2 b + 14b + 9$$

(b)
$$
\begin{array}{r}
4a^3 - 10a^2b + 5b + 8 \\
(-)\ \underline{\hspace{1em} 12a^2b - 9b - 1} \\
4a^3 - 22a^2b + 14b + 9
\end{array}
$$

■

The use of the distributive property is fundamental when multiplying polynomials. Perhaps the simplest situation calls for the product of a **monomial** (a polynomial having only one term) times a polynomial of two or more terms, as follows:

$$3x^2(4x^7 - 3x^4 - x^2 + 15) = 3x^2(4x^7) - 3x^2(3x^4) - 3x^2(x^2) + 3x^2(15)$$

$$= 12x^9 - 9x^6 - 3x^4 + 45x^2$$

In the first line we used an extended version of the distributive property, namely,

$$a(b - c - d + e) = ab - ac - ad + ae$$

Next, observe how the distributive property is used to multiply two **binomials** (polynomials having two terms).

$$(2x + 3)(4x + 5) = (2x + 3)4x + (2x + 3)5$$

$$= (2x)(4x) + (3)(4x) + (2x)(5) + (3)(5)$$

$$= 8x^2 + 12x + 10x + 15$$

$$= 8x^2 + 22x + 15$$

Here is a shortcut that can be used to multiply two binomials.

$8x^2$ is the product of the *first* terms in the binomials. $10x$ and $12x$ are the products of the *outer* and *inner* terms. 15 is the product of the *last* terms in the binomials.

$$(2x + 3)(4x + 5) = 8x^2 + 22x + 15$$

In general, we may write the product $(a + b)(c + d)$ in this way:

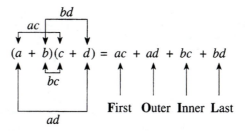

$$(a + b)(c + d) = ac + ad + bc + bd$$

First Outer Inner Last

Keep this diagram in mind as an aid to finding the product of two binomials mentally. Some students find it helpful to remember the first letter of each step, FOIL.

EXAMPLE 3　Find the product of $ax + b$ and $cx + d$ by using the visual inspection method (FOIL).

Solution

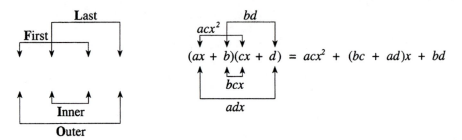

■

The distributive property can be extended to multiply polynomials with three terms, **trinomials**, as in Example 4.

EXAMPLE 4　Multiply $3x^3 - 8x + 4$ by $2x^2 + 5x - 1$.

Solution

Note: The column method is a convenient way to organize your work. Be certain to keep like terms in the same column. Let $x = 2$ and check the solution.

$$
\begin{array}{r}
3x^3 - 8x + 4 \\
2x^2 + 5x - 1 \\
\hline
\end{array}
$$

$$(add) \begin{cases} \quad\;\; -3x^3 \qquad\quad + 8x - 4 \\ \qquad 15x^4 \qquad - 40x^2 + 20x \\ 6x^5 \qquad\quad - 16x^3 + 8x^2 \end{cases}$$

-1 times $3x^3 - 8x + 4$
$5x$ times $3x^3 - 8x + 4$
$2x^2$ times $3x^3 - 8x + 4$

$$\overline{6x^5 + 15x^4 - 19x^3 - 32x^2 + 28x - 4}$$

■

TEST YOUR UNDERSTANDING

Combine.

1. $(x^2 + 2x - 6) + (-2x + 7)$　　**2.** $(x^2 + 2x - 6) - (-2x + 7)$

3. $(5x^4 - 4x^3 + 3x^2 - 2x + 1) + (-5x^4 + 6x^3 + 10x)$

4. $(x^2y + 3xy + xy^2) + (3x^2y - 6xy) - (2xy - 5xy^2)$

Find the products.

5. $(-3x)(x^3 + 2x^2 - 1)$　　　　**6.** $(2x + 3)(3x + 2)$

7. $(6x - 1)(2x + 5)$　　　　　　**8.** $(4x + 7)(4x - 7)$

9. $(2x - 3y)(2x - 3y)$　　　　**10.** $(x^2 - 3x + 5)(2x^3 + x^2 - 3x)$

(Answers: Page 83)

Sometimes more than one operation is involved, as shown in the next example.

EXAMPLE 5　Simplify by performing the indicated operations:

$$(x^2 - 5x)(3x^2) + (x^3 - 1)(2x - 5)$$

Solution Multiply first and then combine like terms.

$$(x^2 - 5x)(3x^2) + (x^3 - 1)(2x - 5) = (3x^4 - 15x^3) + (2x^4 - 5x^3 - 2x + 5)$$
$$= 5x^4 - 20x^3 - 2x + 5 \qquad \blacksquare$$

EXAMPLE 6 Expand: $(a + b)^3$

CAUTION
$(a + b)^2 \neq a^2 + b^2.$

Solution First write $(a + b)^3 = (a + b)(a + b)^2$. Then use the **expanded form** of $(a + b)^2 = a^2 + 2ab + b^2$. Thus:

*The result of Example 6 provides a **formula for the cube of a binomial**, that is, for all expansions of the form $(a + b)^3$.*

$$(a + b)^3 = (a + b)(a^2 + 2ab + b^2)$$
$$= a^3 + 2a^2b + ab^2 + a^2b + 2ab^2 + b^3$$
$$= a^3 + 3a^2b + 3ab^2 + b^3 \qquad \blacksquare$$

The following example shows an application of the product of binomials; it makes use of the product $(a - b)(a + b) = a^2 - b^2$ to rationalize denominators. Each of the two factors $a - b$ and $a + b$ is called the *conjugate* of the other.

EXAMPLE 7 Rationalize the denominator: $\dfrac{5}{\sqrt{10} - \sqrt{3}}$

Solution

$\sqrt{10} + \sqrt{3}$ *is the conjugate of* $\sqrt{10} - \sqrt{3}$.

$$\frac{5}{\sqrt{10} - \sqrt{3}} = \frac{5}{\sqrt{10} - \sqrt{3}} \cdot \frac{\sqrt{10} + \sqrt{3}}{\sqrt{10} + \sqrt{3}}$$

$$= \frac{5(\sqrt{10} + \sqrt{3})}{10 - 3}$$

$$= \frac{5(\sqrt{10} + \sqrt{3})}{7} \qquad \blacksquare$$

EXERCISES 1.7

Add.

1. $3x^2 + 5x - 2$
    $\underline{5x^2 - 7x + 9}$

2. $5x^2 - 9x - 1$
    $\underline{2x^2 + 2x + 7}$

3. $3x^3 - 7x^2 + 8x + 12$
    $\underline{x^3 - 2x^2 + 8x - 9}$

4. $x^3 - 3x^2 + 2x - 5$
    $\underline{5x^2 - x + 9}$

5. $4x^2 + 9x - 17$
    $\underline{2x^3 - 3x^2 + 2x - 11}$

6. $2x^3 + x^2 - 7x + 1$
    $\underline{ - 2x^2 - x + 8}$

Subtract.

7. $3x^3 - 2x^2 - 8x + 9$
    $\underline{2x^3 + 5x^2 + 2x + 1}$

8. $x^3 - 2x^2 + 6x + 1$
    $\underline{ - x^2 - 6x - 1}$

9. $4x^3 + x^2 - 2x - 13$
    $\underline{ 2x^2 + 3x + 9}$

Simplify by performing the indicated operations.

10. $(3x + 5) + (3x - 2)$

11. $5x + (1 - 2x)$

12. $(7x + 5) - (2x + 3)$

13. $(y + 2) + (2y + 1) + (3y + 3)$

14. $h - (h + 2)$

15. $(x^3 + 3x^2 + 3x + 1) - (x^2 + 2x + 1)$

16. $7x - (3 - x) - 2x$

17. $5y - [y - (3y + 8)]$

18. $(2x^3y^2 - 5xy + x^2y^3) + (3xy - x^2y^3)$

19. $(5x - 2xy + x^2y^2) - (2x + xy - x^2y^2)$

20. $x(3x^2 - 2x + 5)$

21. $2x^2(2x + 1 - 10x^2)$

22. $-4t\left(t^4 - \frac{1}{4}t^3 + 4t^2 - \frac{1}{16}t + 1\right)$

23. $(x + 1)(x + 1)$

24. $(2x + 1)(2x - 1)$

25. $(4x - 2)(x + 7)$

26. $(12x - 8)(7x + 4)$

27. $(-2x + 3)(3x + 6)$

28. $(-2x - 3)(3x + 6)$

29. $(-2x - 3)(3x - 6)$

30. $\left(\frac{1}{2}x + 4\right)\left(\frac{1}{2}x - 4\right)$

31. $\left(\frac{2}{3}x + 6\right)\left(\frac{2}{3}x + 6\right)$

32. $(7 + 3x)(9 - 4x)$

33. $(7 - 3x)(4x - 9)$

34. $\left(x + \frac{3}{4}\right)\left(x + \frac{3}{4}\right)$

35. $\left(\frac{1}{5}x - \frac{1}{4}\right)\left(\frac{1}{5}x - \frac{1}{4}\right)$

36. $(x - \sqrt{3})(x + \sqrt{3})$

37. $(\sqrt{x} - 10)(\sqrt{x} + 10)$

38. $\left(\frac{1}{10}x - \frac{1}{100}\right)\left(\frac{1}{10}x - \frac{1}{100}\right)$

39. $(\sqrt{x} + \sqrt{2})(\sqrt{x} - \sqrt{2})$

40. $(x^2 - 3)(x^2 + 3)$

41. $(x^2 + x + 9)(x^2 - 3x - 4)$

42. $(x^{1/3} - 2)(x^{2/3} + 2x^{1/3} + 4)$

43. $(x - 2)(x^2 + 2x + 4)$

44. $(x - 2)(x^3 + 2x^2 + 4x + 8)$

45. $(x - 2)(x^4 + 2x^3 + 4x^2 + 8x + 16)$

46. $(x^n - 4)(x^n + 4)$

47. $(x^{2n} + 1)(x^{2n} - 2)$

48. $5(x + 5)(x - 5)$

49. $3x(1 - x)(1 - x)$

50. $(x + 3)(x + 1)(x - 4)$

51. $(2x + 1)(3x - 2)(3 - x)$

52. $(2x^2 + 3)(9x^2) + (3x^3 - 2)(4x)$

53. $(x^3 - 2x + 1)(2x) + (x^2 - 2)(3x^2 - 2)$

54. $(2x^3 - x^2)(6x - 5) + (3x^2 - 5x)(6x^2 - 2x)$

55. $(x^4 - 3x^2 + 5)(2x + 3) + (x^2 + 3x)(4x^3 - 6x)$

Expand each of the following and combine like terms.

56. $(a - b)^2$

57. $(x - 1)^3$

58. $(x + 1)^4$

59. $(a + b)^4$

60. $(a - b)^4$

61. $(2x + 3)^3$

62. $\left(\frac{1}{2}x - 4\right)^2$

63. $\left(\frac{1}{3}x + 3\right)^3$

64. $\left(\frac{1}{2}x - 1\right)^3$

Rationalize the denominator and simplify.

65. $\dfrac{12}{\sqrt{5} - \sqrt{3}}$

66. $\dfrac{20}{3 - \sqrt{2}}$

67. $\dfrac{14}{\sqrt{2} - 3}$

68. $\dfrac{\sqrt{x}}{\sqrt{x} + \sqrt{y}}$

69. $\dfrac{\sqrt{x} + \sqrt{y}}{\sqrt{x} - \sqrt{y}}$

70. $\dfrac{1}{\sqrt{x} + 2}$

Rationalize the numerator.

71. $\dfrac{\sqrt{5} + 3}{\sqrt{5}}$

72. $\dfrac{\sqrt{5} + \sqrt{3}}{\sqrt{2}}$

73. $\dfrac{\sqrt{x} + \sqrt{y}}{\sqrt{x} - \sqrt{y}}$

Show how to convert the first fraction given into the form of the second one.

74. $\dfrac{\sqrt{x}-2}{x-4}$; $\dfrac{1}{\sqrt{x}+2}$ **75.** $\dfrac{\sqrt{4+h}-2}{h}$; $\dfrac{1}{\sqrt{4+h}+2}$

 Challenge Without doing all of the indicated arithmetic, find the value of P, if $P^3 = 2^{18} + 2^{12} \cdot 3^5 + 2^6 \cdot 3^9 + 3^{12}$. (*Hint:* Consider the expansion of $(a+b)^3$.)

 Written Assignment In your own words, accompanied by minimal algebraic symbolism, explain how the following figures provide a geometric interpretation for the expansion of $(a+b)^2$. Then find a geometric interpretation for the formula $(a-b)^2 = a^2 - 2ab + b^2$.

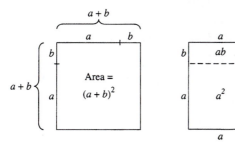

1.8 FACTORING POLYNOMIALS

In a sense, factoring is "unmultiplying."

It is not difficult to multiply three binomials, such as $x+2$, $x-2$, and $x-3$, to obtain the polynomial $x^3 - 3x^2 - 4x + 12$. However, it is more challenging to *begin* with $x^3 - 3x^2 - 4x + 12$ and **factor** (*unmultiply*) it into the form $(x+2)(x-2)(x-3)$.

One of the basic methods of factoring is the reverse of multiplying by a monomial. Consider this multiplication problem:

$$6x^2(4x^7 - 3x^4 - x^2 + 15) = 24x^9 - 18x^6 - 6x^4 + 90x^2$$

As you read this equation from left to right it involves multiplication by $6x^2$. Reading from right to left, it involves "factoring out" the *common monomial factor* $6x^2$. These are the details of this factoring process:

Note the use of the distributive property in this illustration.

$$24x^9 - 18x^6 - 6x^4 + 90x^2 = 6x^2(4x^7) - 6x^2(3x^4) - 6x^2(x^2) + 6x^2(15)$$
$$= 6x^2(4x^7 - 3x^4 - x^2 + 15)$$

Suppose, instead, we were to use $3x$ as the common factor:

$$24x^9 - 18x^6 - 6x^4 + 90x^2 = 3x(8x^8 - 6x^5 - 2x^3 + 30x)$$

The instruction "to factor" will always mean that we are to find the complete factored form for the given expression.

This is a correct factorization but it is not considered the *complete factored form* because $2x$ can still be factored out of the expression in the parentheses.

Factoring techniques extend to polynomials with more than one variable as in the example that follows.

EXAMPLE 1   Factor:

(a) $21x^4y - 14x^5y^2 - 63x^8y^3 + 91x^{11}y^4$

(b) $8x^2(x - 1) + 4x(x - 1) + 2(x - 1)$

Solution

(a) $21x^4y - 14x^5y^2 - 63x^8y^3 + 91x^{11}y^4 = 7x^4y(3 - 2xy - 9x^4y^2 + 13x^7y^3)$

(b) $8x^2(x - 1) + 4x(x - 1) + 2(x - 1) = 2(x - 1)(4x^2 + 2x + 1)$ ■

We will now consider several basic procedures for factoring polynomials. You should check each of these by multiplication. These forms are very useful and you need to learn to recognize and apply each one.

THE DIFFERENCE OF TWO SQUARES

$$a^2 - b^2 = (a - b)(a + b)$$

Illustrations:

$$x^2 - 9 = x^2 - 3^2 = (x - 3)(x + 3)$$

$$3x^2 - 75 = 3(x^2 - 25) = 3(x - 5)(x + 5)$$

Be sure to check each of these facts by multiplication.

THE DIFFERENCE (SUM) OF TWO CUBES

$$a^3 - b^3 = (a - b)(a^2 + ab + b^2)$$

$$a^3 + b^3 = (a + b)(a^2 - ab + b^2)$$

Illustrations:

Remember: First search for a common monomial factor.

$$8x^3 - 27 = (2x)^3 - 3^3 = (2x - 3)(4x^2 + 6x + 9)$$

$$2x^3 + 128y^3 = 2(x^3 + 64y^3) = 2[x^3 + (4y)^3] = 2(x + 4y)(x^2 - 4xy + 16y^2)$$

TEST YOUR UNDERSTANDING

Factor out the common monomial.

1. $3x - 9$ **2.** $-5x + 15$ **3.** $5xy + 25y^2 + 10y^5$

Factor as the difference of squares.

4. $x^2 - 36$ **5.** $4x^2 - 49$ **6.** $(a + 2)^2 - 25b^2$

Factor as the difference (sum) of two cubes.

7. $x^3 - 27$ **8.** $x^3 + 27$ **9.** $8a^3 - 125$

Factor the following by first considering common monomial factors.

10. $3x^2 - 48$ **11.** $ax^3 + ay^3$ **12.** $2hx^2 - 8h^3$

(Answers: Page 83)

It is not possible to factor $x^2 - 5$ or $x - 8$ as the difference of two squares by using polynomial factors with integer coefficients. There are times, however, when it is desirable to allow other factorizations, such as the following:

$$x^2 - 5 = x^2 - \left(\sqrt{5}\right)^2 = \left(x + \sqrt{5}\right)\left(x - \sqrt{5}\right)$$

$$x - 8 = \left(\sqrt{x}\right)^2 - \left(\sqrt{8}\right)^2 = \left(\sqrt{x} + \sqrt{8}\right)\left(\sqrt{x} - \sqrt{8}\right)$$

$$x - 8 = (x^{1/3})^3 - 2^3 = (x^{1/3} - 2)(x^{2/3} + 2x^{1/3} + 4)$$

In general, we follow this rule when factoring:

> Unless otherwise indicated, use the same type of numerical coefficients and exponents in the factors as appear in the given unfactored form.

Just as we can factor the difference of two squares or cubes, so can we also factor the difference of two fourth powers, two fifth powers, and so on. All these situations can be collected in the following single general form that gives the factorization of the difference of two *n*th powers, where *n* is an integer greater than 1.

The factorization of $a^n - b^n$ is one of the most useful in mathematics; it will be needed in calculus.

THE DIFFERENCE OF TWO *n*TH POWERS

$$a^n - b^n = (a - b)(a^{n-1} + a^{n-2}b + a^{n-3}b^2 + \cdots + ab^{n-2} + b^{n-1})$$

To help remember the second factor, we may rewrite it like this:

$$a^{n-1}b^0 + a^{n-2}b^1 + a^{n-3}b^2 + \cdots + a^1 b^{n-2} + a^0 b^{n-1}$$

You can see that the exponents for *a* begin with $n - 1$ and decrease to 0, whereas for *b* they begin with 0 and increase to $n - 1$. Note also that the sum of the exponents for *a* and *b*, for each term, is $n - 1$.

EXAMPLE 2 Factor: $3y^5 - 96$

Solution First factor out 3; then use the form $a^5 - b^5$.

$$3y^5 - 96 = 3(y^5 - 32) = 3(y^5 - 2^5)$$

$$= 3(y - 2)(y^4 + 2y^3 + 4y^2 + 8y + 16) \qquad \blacksquare$$

EXAMPLE 3 Factor: $x^6 - 1$

Solution Several different approaches are possible.
(a) Using the formula for the difference of two *n*th powers:

$$x^6 - 1 = (x - 1)(x^5 + x^4 + x^3 + x^2 + x + 1)$$

(b) Using the formula for the difference of two cubes:

$$x^6 - 1 = (x^2)^3 - 1^3 = (x^2 - 1)(x^4 + x^2 + 1)$$

$$= (x - 1)(x + 1)(x^4 + x^2 + 1)$$

Notice that in (c) we continued to factor the difference and sum of two cubes. This gives the complete factorization for $x^6 - 1$ since neither of the quadratic factors is factorable.

(c) Using the formula for the difference of two squares:

$$x^6 - 1 = (x^3)^2 - 1^2 = (x^3 - 1)(x^3 + 1)$$
$$= (x - 1)(x^2 + x + 1)(x + 1)(x^2 - x + 1) \qquad \blacksquare$$

Some polynomials of four terms that do not appear to be factorable, because there is no common monomial factor in all the terms, can be factored by a method known as **grouping.** In the following, we group the given polynomial into two binomials, each of which is factorable.

CAUTION
$x^2(x + 3) + 9(x + 3)$ is not a factored form of the given expression. Why not?

$$x^3 + 3x^2 + 9x + 27 = (x^3 + 3x^2) + (9x + 27)$$
$$= x^2(x + 3) + 9(x + 3)$$
$$= (x^2 + 9)(x + 3)$$

Here is an alternative grouping that leads to the same answer.

$$x^3 + 3x^2 + 9x + 27 = (x^3 + 27) + (3x^2 + 9x)$$
$$= (x + 3)(x^2 - 3x + 9) + (x + 3)(3x)$$
$$= (x + 3)(x^2 - 3x + 9 + 3x)$$
$$= (x + 3)(x^2 + 9)$$

Not all groupings are productive. Thus in Example 4 the grouping $(ax^2 + 15) + (-3x - 5ax)$ does not lead to a solution.

EXAMPLE 4 Factor $ax^2 + 15 - 5ax - 3x$ by grouping.

Solution

$$ax^2 + 15 - 5ax - 3x = (ax^2 - 5ax) + (15 - 3x)$$
$$= ax(x - 5) + 3(5 - x)$$
$$= ax(x - 5) - 3(x - 5)$$
$$= (ax - 3)(x - 5) \qquad \blacksquare$$

By multiplying $a + b$ by $a + b$ and $a - b$ by $a - b$, the following formulas can be established:

Observe that in $a^2 \pm 2ab + b^2$ the middle term is plus or minus twice the product of a with b. Hence the factored form is the square of the sum $a + b$, or the difference $a - b$.

PERFECT SQUARE TRINOMIALS

$$a^2 + 2ab + b^2 = (a + b)^2$$
$$a^2 - 2ab + b^2 = (a - b)^2$$

EXAMPLE 5 Factor: **(a)** $x^2 + 10x + 25$ **(b)** $9x^3 - 42x^2 + 49x$

Solution

In Example 5(b), let $a = 3x$ and $b = 7$. Then $2ab = 2(3x)(7) = 42x$.

(a) $x^2 + 10x + 25 = x^2 + 2(x \cdot 5) + 5^2 = (x + 5)^2$

(b) $9x^3 - 42x^2 + 49x = x(9x^2 - 42x + 49) = x(3x - 7)^2 \qquad \blacksquare$

Another factoring technique that we will consider deals with trinomials that are not necessarily perfect squares, such as $x^2 + 7x + 12$. From our experience with multiplying binomials we can anticipate that the factors of $x^2 + 7x + 12$ will be of this form:

$$(x + \underline{\ ?\ })(x + \underline{\ ?\ })$$

We need to fill in the blanks with two integers whose product is 12. Furthermore, the middle term of the product must be $+7x$. The possible choices for the two integers are

$$12 \text{ and } 1 \qquad 6 \text{ and } 2 \qquad 4 \text{ and } 3$$

To find the correct pair is now a matter of trial and error. These are the possible factorizations:

$$(x + 12)(x + 1) \qquad (x + 6)(x + 2) \qquad (x + 4)(x + 3)$$

Only the last form gives the correct middle term of $7x$. Therefore, we conclude that $x^2 + 7x + 12 = (x + 4)(x + 3)$.

EXAMPLE 6 Factor: $x^2 - 10x + 24$

Solution The final term, $+24$, must be the product of two positive numbers or two negative numbers. (Why?) Since the middle term, $-10x$, has a minus sign, the factorization must be of this form:

$$(x - \underline{\ ?\ })(x - \underline{\ ?\ })$$

Now try all the pairs of integers whose product is 24: 24 and 1, 12 and 2, 8 and 3, 6 and 4. Only the last of these gives $-10x$ as the middle term. Thus

$$x^2 - 10x + 24 = (x - 6)(x - 4) \qquad\blacksquare$$

Let us now consider a more complicated factoring problem. For example, to factor the trinomial $15x^2 + 43x + 8$, we need to consider possible factors both of 15 and 8:

$$15 = 15 \cdot 1 \qquad 15 = 5 \cdot 3 \qquad 8 = 8 \cdot 1 \qquad 8 = 4 \cdot 2$$

The plus signs in the trial factors are due to the two plus signs in the given trinomial.

Using $15 \cdot 1$, write the form

$$(15x + \underline{\ ?\ })(x + \underline{\ ?\ })$$

*With a little luck, and much more experience, you can often avoid exhausting **all** the possibilities before finding the correct factors. You will then find that such work can often be shortened significantly.*

Try 8 and 1 in the blanks, *both ways*, namely:

$$(15x + 8)(x + 1) \qquad (15x + 1)(x + 8)$$

Neither gives a middle term of $43x$; so now try 4 and 2 in the blanks both ways. Again you will find that neither case works. Next consider the form

$$(5x + \underline{\ ?\ })(3x + \underline{\ ?\ })$$

Once again try 4 and 2 both ways, and 8 with 1 both ways. The correct answer is:

$$15x^2 + 43x + 8 = (5x + 1)(3x + 8)$$

Example 7 shows us that not every trinomial can be factored.

EXAMPLE 7   Factor: $12x^2 - 9x + 2$

Solution Consider the forms

$$(12x - \underline{\ ?\ })(x - \underline{\ ?\ }) \qquad (6x - \underline{\ ?\ })(2x - \underline{\ ?\ }) \qquad (4x - \underline{\ ?\ })(3x - \underline{\ ?\ })$$

In each form try 2 with 1 both ways. None of these produces a middle term of $-9x$; hence we say that $12x^2 - 9x + 2$ is *not factorable* with integral coefficients.   ∎

TEST YOUR UNDERSTANDING

Factor the following trinomial squares.

1. $a^2 + 6a + 9$ **2.** $x^2 - 10x + 25$ **3.** $4x^2 + 12xy + 9y^2$

Factor each trinomial if possible.

4. $x^2 + 8x + 15$ **5.** $a^2 - 12a + 20$ **6.** $x^2 + 3x + 4$

7. $10x^2 - 39x + 14$ **8.** $6x^2 - 11x - 10$ **9.** $6x^2 + 6x - 5$

(Answers: Page 83)

EXAMPLE 8   Factor $15x^2 + 7x - 8$

You may find it easier to leave out the sign in the two binomial forms as you try various cases. Because of the term -8, the signs must be opposites.

Solution Try these forms:

$$(5x \quad \underline{\ ?\ })(3x \quad \underline{\ ?\ }) \qquad (15x \quad \underline{\ ?\ })(x \quad \underline{\ ?\ })$$

In each form try the factors of 8 (2 and 4, 1 and 8) and search for a middle term of $7x$. Then insert the appropriate signs to obtain $(15x - 8)(x + 1)$.   ∎

All the trinomials we have considered were of degree 2. The same methods can be modified to factor certain trinomials of higher degree, as in Example 9.

EXAMPLE 9   Factor: **(a)** $x^4 - x^2 - 12$ **(b)** $a^6 - 3a^3b - 18b^2$

Solution
(a) Note that $x^4 = (x^2)^2$ and let $u = x^2$. Then

The substitution step in Example 9(a) changes the factoring of a fourth degree trinomial into the factoring of a second degree trinomial. Letting $u = a^3$ in part (b) will have the same effect.

$$
\begin{aligned}
x^4 - x^2 - 12 &= u^2 - u - 12 \\
&= (u - 4)(u + 3) \\
&= (x^2 - 4)(x^2 + 3) \\
&= (x - 2)(x + 2)(x^2 + 3)
\end{aligned}
$$

(b) $a^6 - 3a^3b - 18b^2 = (a^3 - 6b)(a^3 + 3b)$ ∎

CAULION: Learn to Avoid These Mistakes	
WRONG	**RIGHT**
$(x + 2)3 + (x + 2)y = (x + 2)3y$	$(x + 2)3 + (x + 2)y$ $= (x + 2)(3 + y)$
$3x + 1 = 3(x + 1)$	$3x + 1$ is not factorable by using integers.
$x^3 - y^3 = (x - y)(x^2 + y^2)$	$x^3 - y^3 = (x - y)(x^2 + xy + y^2)$
$x^3 + 8$ is not factorable.	$x^3 + 8 = (x + 2)(x^2 - 2x + 4)$
$x^2 + y^2 = (x + y)(x + y)$	$x^2 + y^2$ is not factorable by using real numbers.
$4x^2 - 6xy + 9y^2 = (2x - 3y)^2$	$4x^2 - 6xy + 9y^2$ is not a perfect square trinomial and cannot be factored using integers.

EXERCISES 1.8

Factor as the difference of two squares.

1. $4x^2 - 9$ **2.** $25x^2 - 144y^2$ **3.** $a^2 - 121b^2$

Factor as the difference (sum) of two cubes.

4. $x^3 - 8$ **5.** $x^3 + 64$ **6.** $8x^3 + 1$

7. $125x^3 - 64$ **8.** $8 - 27a^3$ **9.** $8x^3 + 343y^3$

Factor as the difference of two squares, allowing irrational numbers as well as radical expressions. (All letters represent positive numbers.)

10. $a^2 - 15$ **11.** $3 - 4x^2$ **12.** $x - 1$

13. $x - 36$ **14.** $2x - 9$ **15.** $8 - 3x$

Factor as the difference (sum) of two cubes, allowing irrational numbers as well as radical expressions.

16. $x^3 - 2$ **17.** $7 + a^3$ **18.** $1 - h$

19. $27x + 1$ **20.** $27x - 64$ **21.** $3x - 4$

Factor by grouping.

22. $a^2 - 2b + 2a - ab$ **23.** $x^2 - y - x + xy$ **24.** $x + 1 + y + xy$

25. $-y - x + 1 + xy$ **26.** $ax + by + ay + bx$ **27.** $2 - y^2 + 2x - xy^2$

Factor completely.

28. $8a^2 - 2b^2$ **29.** $7x^3 + 7h^3$ **30.** $81x^4 - 256y^4$

31. $a^8 - b^8$ **32.** $40ab^3 - 5a^4$ **33.** $a^5 - 32$

34. $x^6 + x^2y^4 - x^4y^2 - y^6$ **35.** $a^3x - b^3y + b^3x - a^3y$ **36.** $7a^2 - 35b + 35a - 7ab$

37. $x^5 - 16xy^4 - 2x^4y + 32y^5$

Factor each trinomial.

38. $12a^2 - 13a + 1$ **39.** $20a^2 - 9a + 1$ **40.** $4 - 5b + b^2$ **41.** $9x^2 + 6x + 1$

42. $5x^2 + 31x + 6$ **43.** $14x^2 + 37x + 5$ **44.** $9x^2 - 18x + 5$ **45.** $8x^2 - 9x + 1$

46. $6x^2 + 12x + 6$ **47.** $8x^2 - 16x + 6$ **48.** $12x^2 + 92x + 15$ **49.** $12a^2 - 25a + 12$

50. $6a^2 - 5a - 21$ **51.** $4x^2 + 4x - 3$ **52.** $15x^2 + 19x - 56$ **53.** $24a^2 + 25ab + 6b^2$

Factor each trinomial when possible. When appropriate, first factor out the common monomial.

54. $x^2 + x + 1$ **55.** $a^2 - 2a + 2$ **56.** $49r^2s - 42rs + 9s$

57. $6x^2 + 2x - 20$ **58.** $15 + 5y - 10y^2$ **59.** $2b^2 + 12b + 16$

60. $4a^2x^2 - 4abx^2 + b^2x^2$ **61.** $a^3b - 2a^2b^2 + ab^3$ **62.** $12x^2y + 22xy^2 - 60y^3$

63. $16x^2 - 24x + 8$ **64.** $16x^2 - 24x - 8$ **65.** $25a^2 + 50ab + 25b^2$

66. $x^4 - 2x^2 + 1$ **67.** $a^6 - 2a^3 + 1$ **68.** $2x^4 + 8x^2 - 42$

69. $6x^5y - 3x^3y^2 - 30xy^3$

Simplify by factoring.

70. $(1 + x)^2(-1) + (1 - x)(2)(1 + x)$ **71.** $(x + 2)^3(2) + (2x + 1)(3)(x + 2)^2$

72. $(x^2 + 2)^2(3) + (3x - 1)(2)(x^2 + 2)(2x)$ **73.** $(x^3 + 1)^3(2x) + (x^2 - 1)(3)(x^3 + 1)^2(3x^2)$

74. Find these products
   (a) $(a + b)(a^2 - ab + b^2)$
   (b) $(a + b)(a^4 - a^3b + a^2b^2 - ab^3 + b^4)$
   (c) $(a + b)(a^6 - a^5b + a^4b^2 - a^3b^3 + a^2b^4 - ab^5 + b^6)$

75. Use the results of Exercise 74 to factor the following:
   (a) $x^5 + 32$ **(b)** $128x^8 + xy^7$

76. Write the factored form of $a^n + b^n$, where n is an odd positive integer greater than 1.

77. Compare the results in parts (b) and (c) of Example 3, page 55, to obtain the factorization of $x^4 + x^2 + 1$.

78. The result in Exercise 77 can also be found using this procedure:

$$x^4 + x^2 + 1 = (x^4 + 2x^2 + 1) - x^2 = (x^2 + 1)^2 - x^2$$

$$= (x^2 + 1 - x)(x^2 + 1 + x)$$

$$= (x^2 - x + 1)(x^2 + x + 1)$$

Use this procedure to factor the following:
   (a) $x^4 + 3x^2 + 4$ **(b)** $9x^4 + 3x^2 + 4$ **(c)** $a^4 + 2a^2b^2 + 9b^4$ **(d)** $a^4 - 3a^2 + 1$

79. Factor completely: $5(4x^2 + 4x + 1)^4(8x + 4)$

80. Factor completely: $3(x^4 - 2x^2 + 1)^2(4x^3 - 4x)$

81. If $y = x\sqrt{x^2 + 2}$, show that $\sqrt{1 + y^2} = x^2 + 1$

82. If $y = 2x(x^2 + 1)^{1/2}$, show that $\sqrt{1 + y^2} = 2x^2 + 1$

83. The large circle has radius R and each small circle has radius a. Write the area of the shaded portion in factored form and use the result to find this area when radius $R = 15.7$ and radius $a = 3.1$.

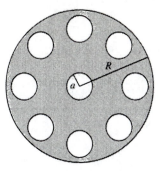

84. (a) If the four congruent square corners are cut from the large square, write the area of the resulting figure in factored form. Use this result to find this area when $y = 12.8$ and $x = 2.4$.

(b) Explain why the expression $(y - 2x)^2 + 4x(y - 2x)$ is also the area of the remaining part and show that it is equivalent to the result in part (a).

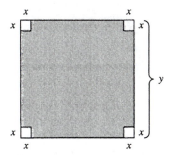

 Challenge   Factor: $x^3 - y^3 + xy^2 - x^2y - x + y$

Written Assignment   It is common to hear someone state that an expression such as $x^2 - 10$ cannot be factored as the difference of two squares. Explain what is meant by such a statement, and discuss its accuracy.

1.9 FUNDAMENTAL OPERATIONS WITH RATIONAL EXPRESSIONS

A **rational expression** is a ratio of polynomials. Rational expressions are the "algebraic extensions" of rational numbers, so the fundamental rules for operating with rational numbers extend to rational expressions.

The important rules for operating with rational expressions, also referred to as *algebraic fractions*, will now be considered. In each case we shall give an example of the rule under discussion in terms of arithmetic fractions so that you can compare the procedures being used. Also, it is assumed that values of the variable for which the denominator is equal to zero are excluded.

Negative of a Fraction

RULE 1. $\quad -\dfrac{a}{b} = \dfrac{-a}{b} = \dfrac{a}{-b} \qquad \left[-\dfrac{2}{3} = \dfrac{-2}{3} = \dfrac{2}{-3} \right]$

Reducing Fractions

RULE 2. $\quad \dfrac{ac}{bc} = \dfrac{a}{b} \qquad \left[\dfrac{2 \cdot 3}{5 \cdot 3} = \dfrac{2}{5} \right]$

Multiplication of Fractions

RULE 3. $\quad \dfrac{a}{b} \cdot \dfrac{c}{d} = \dfrac{ac}{bd} \qquad \left[\dfrac{2}{3} \cdot \dfrac{4}{5} = \dfrac{2 \cdot 4}{3 \cdot 5} = \dfrac{8}{15} \right]$

Division of Fractions

RULE 4. $\quad \dfrac{a}{b} \div \dfrac{c}{d} = \dfrac{a}{b} \cdot \dfrac{d}{c} = \dfrac{ad}{bc} \qquad \left[\dfrac{3}{5} \div \dfrac{2}{3} = \dfrac{3}{5} \cdot \dfrac{3}{2} = \dfrac{3 \cdot 3}{5 \cdot 2} = \dfrac{9}{10} \right]$

Addition and Subtraction of Fractions—Same Denominators

RULE 5. $\quad \dfrac{a}{d} + \dfrac{c}{d} = \dfrac{a + c}{d} \qquad \left[\dfrac{3}{7} + \dfrac{2}{7} = \dfrac{3 + 2}{7} = \dfrac{5}{7} \right]$

RULE 6. $\quad \dfrac{a}{d} - \dfrac{c}{d} = \dfrac{a - c}{d} \qquad \left[\dfrac{7}{9} - \dfrac{2}{9} = \dfrac{7 - 2}{9} = \dfrac{5}{9} \right]$

Addition and Subtraction of Fractions—Different Denominators

RULE 7. $\quad \dfrac{a}{b} + \dfrac{c}{d} = \dfrac{ad + bc}{bd} \qquad \left[\dfrac{2}{3} + \dfrac{3}{4} = \dfrac{2 \cdot 4 + 3 \cdot 3}{3 \cdot 4} = \dfrac{8 + 9}{12} = \dfrac{17}{12} \right]$

RULE 8. $\quad \dfrac{a}{b} - \dfrac{c}{d} = \dfrac{ad - bc}{bd} \qquad \left[\dfrac{4}{5} - \dfrac{2}{3} = \dfrac{4 \cdot 3 - 5 \cdot 2}{5 \cdot 3} = \dfrac{12 - 10}{15} = \dfrac{2}{15} \right]$

Following are a number of examples that illustrate these rules. Be certain that you understand each of the steps shown.

EXAMPLE 1 Simplify: **(a)** $\dfrac{x^2 + 5x - 6}{x^2 + 6x}$ **(b)** $\dfrac{5a - 3b}{3b - 5a}$

Solution

(a) $\dfrac{x^2 + 5x - 6}{x^2 + 6x} = \dfrac{(x - 1)(x + 6)}{x(x + 6)} = \dfrac{x - 1}{x} \qquad$ Rule 2

(b) $\dfrac{5a - 3b}{3b - 5a} = \dfrac{(-1)(-5a + 3b)}{(1)(3b - 5a)} = \dfrac{(-1)(3b - 5a)}{(1)(3b - 5a)} = -1$ ∎

The work in the preceding example can be shortened by dividing the numerator and denominator by $3b - 5a$:

$$\frac{5a - 3b}{3b - 5a} = \frac{\overset{-1}{\cancel{5a - 3b}}}{\underset{1}{\cancel{3b - 5a}}} = \frac{-1}{1} = -1$$

Any nonzero number divided by its opposite is equal to -1:
$$\frac{x - y}{y - x} = -1$$

EXAMPLE 2   Find the product: $\dfrac{x + 1}{x - 1} \cdot \dfrac{2 - x - x^2}{5x}$

Solution

$$\frac{x + 1}{x - 1} \cdot \frac{2 - x - x^2}{5x} = \frac{(x + 1)(2 - x - x^2)}{(x - 1)5x} \qquad \text{Rule 3}$$

Whenever we are working with rational expressions it is taken for granted that final answers have been reduced to lowest terms, using Rule 2.

$$= \frac{(x + 1)(\overset{-1}{\cancel{1 - x}})(2 + x)}{\underset{1}{\cancel{(x - 1)}}5x}$$

$$= -\frac{(x + 1)(x + 2)}{5x} \qquad \text{or} \qquad -\frac{x^2 + 3x + 2}{5x} \qquad \blacksquare$$

EXAMPLE 3   Find the quotient: $\dfrac{(x + 1)^2}{x^2 - 6x + 9} \div \dfrac{3x + 3}{x - 3}$

Solution

$$\frac{(x + 1)^2}{x^2 - 6x + 9} \div \frac{3x + 3}{x - 3} = \frac{(x + 1)^2}{(x - 3)^2} \cdot \frac{x - 3}{3(x + 1)} \qquad \text{Rule 4}$$

$$= \frac{x + 1}{3(x - 3)} \qquad \blacksquare$$

EXAMPLE 4   Find the sum: $\dfrac{3}{x^2 + x} + \dfrac{2}{x^2 - 1}$

Solution

$$\frac{3}{x^2 + x} + \frac{2}{x^2 - 1} = \frac{3(x^2 - 1) + 2(x^2 + x)}{(x^2 + x)(x^2 - 1)} \qquad \text{Rule 7}$$

$$= \frac{5x^2 + 2x - 3}{(x^2 + x)(x^2 - 1)} \qquad \text{combining terms}$$

$$= \frac{(5x - 3)(x + 1)}{x(x + 1)(x^2 - 1)} \qquad \text{factoring}$$

$$= \frac{5x - 3}{x(x^2 - 1)} \qquad \text{Rule 2} \qquad \blacksquare$$

Here is an alternative method for the preceding illustration that often is more convenient. It makes use of the **least common denominator (LCD)** of the two fractions.

The LCD of the two fractions is $x(x + 1)(x - 1)$. We express each fraction using this common denominator and then add the numerators.

$$\frac{3}{x^2 + x} + \frac{2}{x^2 - 1} = \frac{3}{x(x + 1)} + \frac{2}{(x + 1)(x - 1)}$$

$$= \frac{3(x - 1)}{x(x + 1)(x - 1)} + \frac{2x}{(x + 1)(x - 1)x} \qquad \text{Rule 2}$$

$$= \frac{3(x - 1) + 2x}{x(x + 1)(x - 1)} \qquad \text{Rule 5}$$

$$= \frac{5x - 3}{x(x^2 - 1)}$$

EXAMPLE 5 Combine and simplify: $\dfrac{3}{2} - \dfrac{4}{3x(x + 1)} - \dfrac{x - 5}{3x^2}$

Solution The least common denominator of the fractions is $6x^2(x + 1)$.

To find the LCD, use each factor that appears in the factored forms of the denominators the maximum number of times it appears in any one of the factored forms.

$$\frac{3}{2} - \frac{4}{3x(x + 1)} - \frac{x - 5}{3x^2} = \frac{3 \cdot 3x^2(x + 1)}{2 \cdot 3x^2(x + 1)} - \frac{4 \cdot 2x}{3x(x + 1) \cdot 2x} - \frac{(x - 5) \cdot 2(x + 1)}{3x^2 \cdot 2(x + 1)}$$

$$= \frac{(9x^3 + 9x^2) - 8x - (2x^2 - 8x - 10)}{6x^2(x + 1)}$$

$$= \frac{9x^3 + 7x^2 + 10}{6x^2(x + 1)} \qquad \blacksquare$$

Rational expressions are also involved when dividing a polynomial by a monomial or by another polynomial. First, consider division by a monomial in Example 6.

EXAMPLE 6 Simplify: $(4a^2b^2 + 12a^4b^3 - 8a^3b^2) \div 4a^2b$

Solution

An alternative method is to factor $4a^2b$ out of the numerator and then use Rule 2.

$$\frac{4a^2b^2 + 12a^4b^3 - 8a^3b^2}{4a^2b} = \frac{4a^2b^2}{4a^2b} + \frac{12a^4b^3}{4a^2b} - \frac{8a^3b^2}{4a^2b}$$

$$= b + 3a^2b^2 - 2ab \qquad \blacksquare$$

To divide a polynomial by another polynomial, we make use of a long division process as shown in the following example.

EXAMPLE 7 Divide: $(2x^3 + 3x^2 - x + 16) \div (x^2 + 2x - 3)$

Note the steps in the process. Divide $2x^3$ by x^2 to get $2x$. Multiply $2x$ by the divisor and subtract. Then divide $-x^2$ by x^2 again, and so forth. Stop when the remainder has degree less than that of the divisor.

Solution

$$
\begin{array}{r}
2x - 1 \\
x^2 + 2x - 3 \overline{\smash{\big)}\, 2x^3 + 3x^2 - x + 16} \\
\underline{2x^3 + 4x^2 - 6x } \\
-x^2 + 5x + 16 \\
\underline{-x^2 - 2x + 3} \\
7x + 13
\end{array}
$$

Multiply $x^2 + 2x - 3$ by $2x$.
Subtract.
Multiply $x^2 + 2x - 3$ by -1.
Subtract.

Stop when the remainder has degree *less* than degree of divisor.

The result of this division example can also be stated using rational expressions:

$$\frac{2x^3 + 3x^2 - x + 16}{x^2 + 2x - 3} = (2x - 1) + \frac{7x + 13}{x^2 + 2x - 3}$$

To check the solution, show that the following is correct:

$$\underbrace{2x^3 + 3x^2 - x + 16}_{\text{Dividend}} = \underbrace{(2x - 1)}_{\text{= quotient}} \underbrace{(x^2 + 2x - 3)}_{\times \text{ divisor}} + \underbrace{(7x + 13)}_{+ \text{ remainder}}$$ ∎

TEST YOUR UNDERSTANDING

Simplify each expression by reducing to lowest terms.

1. $\dfrac{x^2}{x^2 + 2x}$ 2. $\dfrac{4b^2 - 4ab}{3a^2 - 3ab}$ 3. $\dfrac{3x^2 + x - 10}{5x - 3x^2}$

Perform the indicated operation and simplify.

4. $\dfrac{4 - 2x}{2} \cdot \dfrac{x + 2}{x^2 - 4}$ 5. $\dfrac{x + y}{x - y} \cdot \dfrac{x^2 - 2xy + y^2}{x^2 - y^2}$

Simplify.

6. $\dfrac{2}{3x^2} - \dfrac{1}{2x}$ 7. $\dfrac{3}{2x} + \dfrac{5}{3x} + \dfrac{1}{x}$ 8. $\dfrac{7}{x - 2} + \dfrac{3}{x + 2}$

9. $\dfrac{5}{(x - 1)(x + 2)} - \dfrac{8}{4 - x^2}$ 10. $\dfrac{1 - 4x}{2x + 5} + \dfrac{8x^2 - 16x}{4x^2 - 25} - \dfrac{1}{2x - 5}$

11. $\dfrac{15x^4y^6 - 10x^3y^3 + 20x^6y^4}{5x^2y^2}$ 12. $\dfrac{2x^3 + x^2 - x + 3}{x + 2}$

(Answers: Page 84)

The fundamental properties of fractions can be used to simplify rational expressions whose numerators and denominators may themselves also contain fractions. These are sometimes referred to as *complex fractions*.

The expression in Example 8 is a type that you will encounter in calculus. Be certain that you understand each step in the solution.

EXAMPLE 8 Simplify: $\dfrac{\dfrac{1}{5+h}-\dfrac{1}{5}}{h}$

Solution Combine the fractions in the numerator and then divide.

$$\frac{\dfrac{1}{5+h}-\dfrac{1}{5}}{h}=\frac{\dfrac{5-(5+h)}{5(5+h)}}{h}=\frac{\dfrac{-h}{5(5+h)}}{h}$$

$$=\frac{-h}{5(5+h)}\div\frac{h}{1}=\frac{-h}{5(5+h)}\cdot\frac{1}{h}=-\frac{1}{5(5+h)}\qquad\blacksquare$$

Working with fractions often creates difficulties for many students. Study this list; it may help you avoid some common pitfalls.

CAUTION: Learn to Avoid These Mistakes	
WRONG	**RIGHT**
$\dfrac{2}{3}+\dfrac{x}{5}=\dfrac{2+x}{3+5}$	$\dfrac{2}{3}+\dfrac{x}{5}=\dfrac{2\cdot5+3\cdot x}{3\cdot5}=\dfrac{10+3x}{15}$
$\dfrac{1}{a}+\dfrac{1}{b}=\dfrac{1}{a+b}$	$\dfrac{1}{a}+\dfrac{1}{b}=\dfrac{b+a}{ab}$
$\dfrac{2x+5}{4}=\dfrac{x+5}{2}$	$\dfrac{2x+5}{4}=\dfrac{2x}{4}+\dfrac{5}{4}=\dfrac{x}{2}+\dfrac{5}{4}$
$2+\dfrac{x}{y}=\dfrac{2+x}{y}$	$2+\dfrac{x}{y}=\dfrac{2y+x}{y}$
$3\left(\dfrac{x+1}{x-1}\right)=\dfrac{3(x+1)}{3(x-1)}$	$3\left(\dfrac{x+1}{x-1}\right)=\dfrac{3(x+1)}{x-1}$
$a\div\dfrac{b}{c}=\dfrac{1}{a}\cdot\dfrac{b}{c}$	$a\div\dfrac{b}{c}=a\cdot\dfrac{c}{b}=\dfrac{ac}{b}$
$\dfrac{1}{a^{-1}+b^{-1}}=a+b$	$\dfrac{1}{a^{-1}+b^{-1}}=\dfrac{1}{\dfrac{1}{a}+\dfrac{1}{b}}=\dfrac{ab}{b+a}$
$\dfrac{x^2+4x+6}{x+2}=\dfrac{x^2+4x+\overset{3}{\cancel{6}}}{\underset{1}{\cancel{x+2}}}$ $=\dfrac{x^2+4x+3}{x+1}$	$\dfrac{x^2+4x+6}{x+2}$ is in simplest form.

Example 9 calls for the simplification of a complex fraction involving negative exponents.

EXAMPLE 9   Simplify: $\dfrac{x^{-2} - y^{-2}}{\dfrac{1}{x} - \dfrac{1}{y}}$

In this method both the numerator and denominator are multiplied by x^2y^2 to simplify.

Solution

$$\frac{x^{-2} - y^{-2}}{\dfrac{1}{x} - \dfrac{1}{y}} = \frac{\dfrac{1}{x^2} - \dfrac{1}{y^2}}{\dfrac{1}{x} - \dfrac{1}{y}} = \frac{\left(\dfrac{1}{x^2} - \dfrac{1}{y^2}\right)(x^2y^2)}{\left(\dfrac{1}{x} - \dfrac{1}{y}\right)(x^2y^2)} \qquad \text{Rule 2}$$

This problem can also be solved by using the procedure shown in Example 8.

$$= \frac{y^2 - x^2}{xy^2 - x^2y}$$

$$= \frac{(y - x)(y + x)}{xy(y - x)}$$

$$= \frac{y + x}{xy}$$ ∎

EXERCISES 1.9

Classify each statement as true or false. If it is false, correct the right side to get a correct equality.

1. $\dfrac{5}{7} - \dfrac{2}{3} = \dfrac{3}{4}$

2. $\dfrac{2x + y}{y - 2x} = -2\left(\dfrac{x + y}{x - y}\right)$

3. $\dfrac{3ax - 5b}{6} = \dfrac{ax - 5b}{2}$

4. $\dfrac{x + x^{-1}}{xy} = \dfrac{x + 1}{x^2y}$

5. $x^{-1} + y^{-1} = \dfrac{y + x}{xy}$

6. $\dfrac{2}{\frac{3}{4}} = \dfrac{8}{3}$

Simplify, if possible.

7. $\dfrac{8xy}{12yz}$

8. $\dfrac{24abc^2}{36bc^2d}$

9. $\dfrac{45x^3 + 15x^2}{15x^2}$

10. $\dfrac{9y^2 + 12y^8 - 15y^6}{3y^2}$

11. $\dfrac{12x^3 + 8x^2 + 4x}{4x}$

12. $\dfrac{5a^2 - 10a^3 + 15a^4}{5a^2}$

13. $\dfrac{a^2b^2 + ab^2 - a^2b^3}{ab^2}$

14. $\dfrac{-6a^3 + 9a^6 - 12a^9}{-3a^3}$

15. $\dfrac{6a^2x^2 - 8a^4x^6}{2a^2x^2}$

16. $\dfrac{-8a^3x^3 + 4ax^3 - 12a^2x^6}{-4ax^3}$

17. $\dfrac{x^2 - 5x}{5 - x}$

18. $\dfrac{n - 1}{n^2 - 1}$

19. $\dfrac{n + 1}{n^2 + 1}$

20. $\dfrac{(x + 1)^2}{1 - x^2}$

21. $\dfrac{3x^2 + 3x - 6}{2x^2 + 6x + 4}$

22. $\dfrac{x^3 - x}{x^3 - 2x^2 + x}$

23. $\dfrac{4x^2 + 12x + 9}{4x^2 - 9}$

24. $\dfrac{x^2 + 2x + xy + 2y}{x^2 + 4x + 4}$

25. $\dfrac{a^2 - 16b^2}{a^3 + 64b^3}$

26. $\dfrac{a^2 - b^2}{a^2 - 6b - ab + 6a}$

Perform the indicated operations and simplify.

27. $\dfrac{2x^2}{y} \cdot \dfrac{y^2}{x^3}$

28. $\dfrac{3x^2}{2y^2} \div \dfrac{3x^3}{y}$

29. $\dfrac{2a}{3} \cdot \dfrac{3}{a^2} \cdot \dfrac{1}{a}$

30. $\left(\dfrac{a^2}{b^2} \cdot \dfrac{b}{c^2}\right) \div a$

31. $\dfrac{3x}{2y} - \dfrac{x}{2y}$

32. $\dfrac{a + 2b}{a} + \dfrac{3a + b}{a}$

33. $\dfrac{a - 2b}{2} - \dfrac{3a + b}{3}$

34. $\dfrac{7}{5x} - \dfrac{2}{x} + \dfrac{1}{2x}$

35. $\dfrac{x - 1}{3} \cdot \dfrac{x^2 + 1}{x^2 - 1}$

36. $\dfrac{x^2 - x - 6}{x^2 - 3x} \cdot \dfrac{x^3 + x^2}{x + 2}$

37. $\dfrac{1 - x}{2 + x} \div \dfrac{x^2 - x}{x^2 + 2x}$

38. $\dfrac{x^2 + 3x}{x^2 + 4x + 3} \div \dfrac{x^2 - 2x}{x + 1}$

39. $\dfrac{2}{x} - y$

40. $\dfrac{x^2}{x - 1} - \dfrac{1}{1 - x}$

41. $\dfrac{3y}{y + 1} + \dfrac{2y}{y - 1}$

42. $\dfrac{2a}{a^2 - 1} - \dfrac{a}{a + 1}$

43. $\dfrac{2x^2}{x^2 + x} + \dfrac{x}{x + 1}$

44. $\dfrac{3x + 3}{2x^2 - x - 1} + \dfrac{1}{2x + 1}$

45. $\dfrac{5}{x^2 - 4} - \dfrac{3 - x}{4 - x^2}$

46. $\dfrac{1}{a^2 - 4} + \dfrac{3}{a - 2} - \dfrac{2}{a + 2}$

47. $\dfrac{2x}{x^2 - 9} + \dfrac{x}{x^2 + 6x + 9} - \dfrac{3}{x + 3}$

48. $\dfrac{x}{x - 1} + \dfrac{x + 7}{x^2 - 1} - \dfrac{x - 2}{x + 1}$

49. $\dfrac{x + 3}{5 - x} - \dfrac{x - 5}{x + 5} + \dfrac{2x^2 + 30}{x^2 - 25}$

50. $\dfrac{a^2 + 2ab + b^2}{a^2 - b^2} \div \dfrac{a^2 + 3ab + 2b^2}{a^2 - 3ab + 2b^2}$

51. $\dfrac{x^3 + x^2 - 12x}{x^2 - 3x} \cdot \dfrac{3x^2 - 10x + 3}{3x^2 + 11x - 4}$

52. $\dfrac{n^2 + n}{2n^2 + 7n - 4} \cdot \dfrac{4n^2 - 4n + 1}{2n^2 - n - 3} \cdot \dfrac{2n^2 + 5n - 12}{2n^3 - n^2}$

53. $\dfrac{n^3 - 8}{n + 2} \cdot \dfrac{2n^2 + 8}{n^3 - 4n} \cdot \dfrac{n^3 + 2n^2}{n^3 + 2n^2 + 4n}$

54. $\dfrac{a^3 - 27}{a^2 - 9} \div \left(\dfrac{a^2 + 2ab + b^2}{a^3 + b^3} \cdot \dfrac{a^3 - a^2b + ab^2}{a^2 + ab}\right)$

As in Example 7, page 65, use the division algorithm to find the quotient and remainder. Check each result.

55. $(x^3 - 2x^2 - 13x + 6) \div (x + 3)$

56. $(x^3 + 4x^2 + 3x - 2) \div (x + 2)$

57. $(x^3 - x^2 + 7) \div (x - 1)$

58. $(5x + 2x^3 - 3) \div (x + 2)$

59. $(5x^2 - 7x + x^3 + 8) \div (x - 2)$

60. $(2x^3 + 9x^2 - 3x - 1) \div (2x - 1)$

61. $(4x^3 - 5x^2 + x - 7) \div (x^2 - 2x)$

62. $(8x^4 - 8x^2 + 6x + 6) \div (2x^2 - x)$

63. $(x^3 - x^2 - x + 10) \div (x^2 - 3x + 5)$

64. $(3x^3 + 4x^2 - 13x + 6) \div (x^2 + 2x - 3)$

Simplify.

65. $\dfrac{\dfrac{5}{x^2 - 4}}{\dfrac{10}{x - 2}}$

66. $\dfrac{\dfrac{1}{x} - \dfrac{1}{4}}{x - 4}$

67. $\dfrac{\dfrac{1}{4 + h} - \dfrac{1}{4}}{h}$

68. $\dfrac{\dfrac{1}{x^2} - \dfrac{1}{9}}{x - 3}$

69. $\dfrac{\dfrac{1}{x + 3} - \dfrac{1}{3}}{x}$

70. $\dfrac{\dfrac{1}{4} - \dfrac{1}{x^2}}{x - 2}$

71. $\dfrac{\dfrac{1}{x^2} - \dfrac{1}{16}}{x + 4}$

72. $\dfrac{x^{-2} - \dfrac{1}{4}}{\dfrac{1}{x} - \dfrac{1}{2}}$

73. $\dfrac{x^{-1} - y^{-1}}{\dfrac{1}{x^2} - \dfrac{1}{y^2}}$

74. $\dfrac{\dfrac{4}{x^2} - \dfrac{1}{y^2}}{\dfrac{2}{x} - \dfrac{1}{y}}$

75. $\dfrac{(1 + x^2)(-2x) - (1 - x^2)(2x)}{(1 + x^2)^2}$

76. $\dfrac{(x^2 - 9)(2x) - x^2(2x)}{(x^2 - 9)^2}$

77. $\dfrac{x^2(4 - 2x) - (4x - x^2)(2x)}{x^4}$

78. $\dfrac{(x + 1)^2(2x) - (x^2 - 1)(2)(x + 1)}{(x + 1)^4}$

Simplify, and express as a single fraction without negative exponents.

79. $\dfrac{a^{-1} - b^{-1}}{a - b}$ **80.** $\dfrac{(a + b)^{-1}}{a^{-1} + b^{-1}}$ **81.** $\dfrac{x^{-2} - y^{-2}}{xy}$

82. There are three tests and a final examination given in a mathematics course. Let a, b, and c be the numerical grades of the tests, and let d represent the examination grade.

    (a) If the final grade is computed by allowing the exam to count the same as the average of the three tests, show that the final average is given by the expression $\dfrac{a + b + c + 3d}{6}$.

    (b) Assume that the average of the three tests accounts for 60% of the final grade and that the examination accounts for 40%. Show that the final average is given by the expression $\dfrac{a + b + c + 2d}{5}$.

83. Some calculators require that certain calculations be performed in a different manner to accommodate the machine. Show that in each case the expression on the left can be computed by using the equivalent expression on the right.

    (a) $\dfrac{A}{B} + \dfrac{C}{D} = \dfrac{\dfrac{A \cdot D}{B} + C}{D}$ **(b)** $(A \cdot B) + (C \cdot D) + (E \cdot F) = \left[\dfrac{\left(\dfrac{A \cdot B}{D} + C\right) \cdot D}{F} + E\right] \cdot F$

84. Archimedes, 287–212 B.C., discovered an interesting relationship between a cylinder and an inscribed sphere. Find, in simplest form, the ratio of the volume of the cylinder to the volume of the sphere.

85. The distance between town A and town B is 120 miles. If you drive one way at an average speed of 60 mph, and return at 40 mph, what is your average speed for the round trip? (Contrary to intuition, it is *not* 50 mph.) Find the answer by using the formula for the average speed for a round trip with average speeds in each direction of s_1 and s_2:

$$\dfrac{2}{\dfrac{1}{s_1} + \dfrac{1}{s_2}}$$

Find a simplified form for this complex fraction and check your answer in this form.

86. If $x^2 + y^2 = 4$, show that $-\dfrac{y - x\left(-\dfrac{x}{y}\right)}{y^2} = -\dfrac{4}{y^3}$.

87. If $y^3 - x^3 = 8$, show that $\dfrac{2xy^2 - 2x^2y\left(\dfrac{x^2}{y^2}\right)}{y^4} = \dfrac{16x}{y^5}$.

88. If $y = x^2 - \dfrac{1}{4x^2}$, show that $\sqrt{1 + y^2} = x^2 + \dfrac{1}{4x^2}$.

89. If $y = \dfrac{x^2}{8} - \dfrac{2}{x^2}$, show that $\sqrt{1 + y^2} = \dfrac{x^2}{8} + \dfrac{2}{x^2}$.

?

Challenge A man left 17 horses to his three children. He left one-half to the oldest, one-third to the middle child, and one-ninth to the youngest. Since 17 is not divisible by 2, 3, or 9, the children borrowed a horse from a neighbor in order to have a total of 18 horses. Then the oldest child received $\frac{1}{2} \times 18 = 9$ horses, the middle child received $\frac{1}{3} \times 18 = 6$ horses, the youngest child received $\frac{1}{9} \times 18 = 2$ horses. Since $9 + 6 + 2 = 17$, the number of horses left to the three children, it was possible to return the extra horse to the neighbor! What is wrong with this story?

Critical Thinking

1. Provide a convincing argument to explain why $3\left(\dfrac{x+1}{x-1}\right) \neq \dfrac{3x+1}{3x-1}$.

2. Note that $\dfrac{12}{\sqrt{150}} = \dfrac{2}{5}\sqrt{6}$. Is either form preferable to the other? Explain your answer.

3. Translate the following statement into symbolic form:

> The square of the sum of two numbers is *at least as large* as four times the product.

Prove the statement is true. *Hint:* Assume the inequality is true and work backward.

4. Three numbers can be arranged vertically in fraction form to produce these cases:

$$\dfrac{\dfrac{a}{b}}{c} = \dfrac{a}{bc} \qquad \dfrac{a}{\dfrac{b}{c}} = \dfrac{ac}{b}$$

Find all possible distinct cases using the four numbers a, b, c, and d arranged vertically in fractional form.

**1.10
INTRODUCTION
TO COMPLEX
NUMBERS**

In the definition of a radical, care was taken to avoid the even root of a negative number, such as $\sqrt{-4}$. This was necessary because there is no real number x whose square is -4. Consequently, there can be no real number that satisfies the equation $x^2 + 4 = 0$. Suppose, for the moment, that we could solve $x^2 + 4 = 0$ using algebraic methods:

$$x^2 + 4 = 0$$
$$x^2 = -4$$
$$x = \pm\sqrt{-4}$$
$$= \pm\sqrt{4(-1)}$$
$$= \pm\sqrt{4}\sqrt{-1}$$
$$= \pm 2\sqrt{-1}$$

Although it could be claimed that $2\sqrt{-1}$ is a solution of $x^2 + 4 = 0$, it is certainly not a *real number* solution. Therefore we introduce $\sqrt{-1}$ as a new kind of number; it will be the unit for a new set of numbers, the *imaginary numbers*. The symbol i is used to stand for this number and is defined as follows:

DEFINITION OF i

$$i = \sqrt{-1} \quad \text{and} \quad i^2 = -1$$

Using i, the square root of a negative real number is now defined:

For $x > 0$, $\sqrt{-x} = \sqrt{-1}\sqrt{x} = i\sqrt{x}$

EXAMPLE 1   Simplify: **(a)** $\sqrt{-16} + \sqrt{-25}$ **(b)** $\sqrt{-16} \cdot \sqrt{-25}$

Solution

In the example, 4i and 5i are combined by using the usual rules of algebra. It will be shown that such procedures apply for this new kind of number.

(a) $\sqrt{-16} + \sqrt{-25} = \sqrt{-1} \cdot \sqrt{16} + \sqrt{-1} \cdot \sqrt{25}$

$$= i \cdot 4 + i \cdot 5 = 4i + 5i = 9i$$

(b) $\sqrt{-16} \cdot \sqrt{-25} = (4i)(5i) = 20i^2 = 20(-1) = -20$ ∎

TEST YOUR UNDERSTANDING	*Express as the product of a real number and i.*

1. $\sqrt{-9}$ **2.** $\sqrt{-49}$ **3.** $\sqrt{-5}$ **4.** $-2\sqrt{-1}$ **5.** $\sqrt{-\dfrac{4}{9}}$

Simplify.

6. $\sqrt{-64} \cdot \sqrt{-225}$ **7.** $\sqrt{9} \cdot \sqrt{-49}$

(Answers: Page 84)   **8.** $\sqrt{-50} + \sqrt{-32} - \sqrt{-8}$ **9.** $3\sqrt{-20} + 2\sqrt{-45}$

An indicated product of a real number times the imaginary unit i, such as $7i$ or $\sqrt{2}\,i$, is called a **pure imaginary number.** The sum of a real number and a pure imaginary number is called a **complex number.**

A complex number has the form $a + bi$, where a and b are real numbers and $i = \sqrt{-1}$.

We say that the real number a is the **real part** of $a + bi$ and the real number b is called the **imaginary part** of $a + bi$. In general, two complex numbers are equal only when both their real parts and their imaginary parts are equal. Thus

$$a + bi = c + di \quad \text{if and only if} \quad a = c \quad \text{and} \quad b = d$$

Note, for example, that $\sqrt{\frac{4}{9}}$ is complex, real, and rational; $\sqrt{-10}$ is complex and pure imaginary.

The collection of complex numbers contains all the real numbers, since any real number a can also be written as $a = a + 0i$. Similarly, if b is real, $bi = 0 + bi$, so that the complex numbers also contain the pure imaginaries.

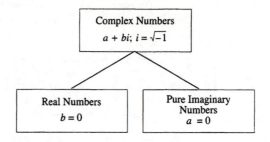

How should complex numbers be added, subtracted, multiplied, or divided? In answering this question, it must be kept in mind that the real numbers are included in the collection of complex numbers. Thus the definitions we construct for the complex numbers must preserve the established operations for the reals.

We add and subtract complex numbers by combining their real and their imaginary parts separately, according to these definitions.

SUM AND DIFFERENCE OF COMPLEX NUMBERS

$$(a + bi) + (c + di) = (a + c) + (b + d)i$$

$$(a + bi) - (c + di) = (a - c) + (b - d)i$$

These procedures are quite similar to those used for combining polynomials. For example:

$$(2 + 3i) + (5 + 7i) = (2 + 5) + (3 + 7)i = 7 + 10i$$

$$(8 + 5i) - (3 + 2i) = (8 - 3) + (5 - 2)i = 5 + 3i$$

Multiplication of two complex numbers is similar to the multiplication of two binomials. For example:

$$(3 + 2i)(5 + 3i) = 15 + 10i + 9i + 6i^2$$

This can be simplified by noting that $10i + 9i = 19i$ and $6i^2 = 6(-1) = -6$. The result is $9 + 19i$.

In general:

$$(a + bi)(c + di) = ac + adi + bci + bdi^2$$

$$= ac + (ad + bc)i + bd(-1)$$

$$= (ac - bd) + (ad + bc)i$$

PRODUCT OF COMPLEX NUMBERS

$$(a + bi)(c + di) = (ac - bd) + (ad + bc)i$$

Note: In practice, it is easier to find the product by using the procedure for multiplying binomials rather than by memorizing the formal definition.

Now consider the quotient of two complex numbers such as

$$\frac{2 + 3i}{3 + i}$$

Our objective is to express this quotient in the form $a + bi$. To do so, we use a method similar to rationalizing the denominator. Note what happens when $3 + i$ is multiplied by its **conjugate**, $3 - i$:

$$(3 + i)(3 - i) = 9 + 3i - 3i - i^2 = 9 - i^2 = 9 + 1 = 10$$

In general, $(a + bi)(a - bi) = a^2 - b^2 i^2 = a^2 + b^2$.

We are now ready to complete the division problem.

$$\frac{2 + 3i}{3 + i} = \frac{2 + 3i}{3 + i} \cdot \frac{3 - i}{3 - i}$$

$$= \frac{6 + 9i - 2i - 3i^2}{9 - i^2}$$

$$= \frac{9 + 7i}{10} \qquad -3i^2 = -3(-1) = 3$$

$$= \frac{9}{10} + \frac{7}{10}i \qquad \leftarrow \text{This is the quotient in the form } a + bi.$$

When the numerator and denominator of $\dfrac{a + bi}{c + di}$ is multiplied by the conjugate of $c + di$, we arrive at the following definition for division (see Exercise 59).

Rather than memorize this definition, simply find quotients as in the preceding illustration.

> ## QUOTIENT OF COMPLEX NUMBERS
> $$\frac{a + bi}{c + di} = \frac{ac + bd}{c^2 + d^2} + \frac{bc - ad}{c^2 + d^2}i \qquad c + di \neq 0$$

Although we will not go into the details here, it can be shown that some of the basic rules for real numbers apply for the complex numbers. For example, the commutative, associative, and distributive laws hold, whereas the rules of order do not apply.

It is also true that the rules for integer exponents apply for complex numbers. For example, $(2 - 3i)^0 = 1$ and $(2 - 3i)^{-1} = \dfrac{1}{2 - 3i}$. In particular, the positive integral powers of i are easily evaluated. For example:

$$i = \sqrt{-1} \qquad\qquad i^5 = i^4 \cdot i = 1 \cdot i = i$$
$$i^2 = -1 \qquad\qquad i^6 = i^4 \cdot i^2 = 1 \cdot i^2 = -1$$
$$i^3 = i^2 \cdot i = -1 \cdot i = -i \qquad i^7 = i^4 \cdot i^3 = 1 \cdot i^3 = -i$$
$$i^4 = i^2 \cdot i^2 = (-1)(-1) = 1 \qquad i^8 = i^4 \cdot i^4 = 1 \cdot i^4 = 1$$

Observe that the first four powers of i at the left repeat for the next four powers shown at the right. This cycle of i, -1, $-i$, and 1 continues endlessly. Therefore, to simplify i^n when $n > 4$, we search for the largest multiple of 4 in the integer n as in the next example.

EXAMPLE 2   Simplify: **(a)** i^{22} **(b)** i^{39}

Solution

(a) $i^{22} = i^{20} \cdot i^2 = (i^4)^5 \cdot i^2 = 1^5 \cdot i^2 = i^2 = -1$

Since $20 = 5(4)$, 5 is the largest multiple of 4 in 22.
Since $36 = 9(4)$, 9 is the largest multiple of 4 in 39.

(b) $i^{39} = i^{36} \cdot i^3 = (i^4)^9 \cdot i^3 = 1^9 \cdot i^3 = i^3 = -i$ ∎

The next example illustrates how a negative power of i can be simplified.

EXAMPLE 3   Express $2i^{-3}$ as the indicated product of a real number and i.

Solution   First note that $2i^{-3} = \dfrac{2}{i^3}$. Next multiply numerator and denominator by i to obtain a real number in the denominator.

As an alternative solution, write $\dfrac{2}{i^3} = \dfrac{2}{-i}$. Then multiply numerator and denominator by i.

$$2i^{-3} = \frac{2}{i^3} \cdot \frac{i}{i} = \frac{2i}{i^4} = \frac{2i}{1} = 2i$$ ∎

EXERCISES 1.10

Classify each statement as true or false.

1. Every real number is a complex number.

2. Every complex number is a real number.

3. Every irrational number is a complex number.

4. Every integer can be written in the form $a + bi$.

5. Every complex number may be expressed as an irrational number.

6. Every negative integer may be written as a pure imaginary number.

Express each of the following numbers in the form $a + bi$.

7. $5 + \sqrt{-4}$ **8.** $7 - \sqrt{-7}$ **9.** -5 **10.** $\sqrt{25}$

Express in the form bi.

11. $\sqrt{-16}$ **12.** $\sqrt{-81}$ **13.** $\sqrt{-144}$ **14.** $-\sqrt{-9}$

15. $\sqrt{-\frac{9}{16}}$ **16.** $\sqrt{-3}$ **17.** $-\sqrt{-5}$ **18.** $-\sqrt{-8}$

Simplify:

19. $\sqrt{-9} \cdot \sqrt{-81}$ **20.** $\sqrt{4} \cdot \sqrt{-25}$ **21.** $\sqrt{-3} \cdot \sqrt{-2}$ **22.** $(2i)(3i)$

23. $(-3i^2)(5i)$ **24.** $(i^2)(i^2)$ **25.** $\sqrt{-9} + \sqrt{-81}$ **26.** $\sqrt{-12} + \sqrt{-75}$

27. $\sqrt{-8} + \sqrt{-18}$ **28.** $2\sqrt{-72} - 3\sqrt{-32}$ **29.** $\sqrt{-9} - \sqrt{-3}$ **30.** $3\sqrt{-80} - 2\sqrt{-20}$

Complete the indicated operation. Express all answers in the form $a + bi$.

31. $(7 + 5i) + (3 + 2i)$ **32.** $(8 + 7i) + (9 - i)$ **33.** $(8 + 2i) - (3 + 5i)$

34. $(7 + 2i) - (4 - 3i)$ **35.** $(7 + \sqrt{-16}) + (3 - \sqrt{-4})$ **36.** $(8 + \sqrt{-49}) - (2 - \sqrt{-25})$

37. $2i(3 + 5i)$ **38.** $3i(5i - 2)$ **39.** $(3 + 2i)(2 + 3i)$

40. $(\sqrt{5} + 3i)(\sqrt{5} - 3i)$ **41.** $(5 - 2i)(3 + 4i)$ **42.** $(\sqrt{3} + 2i)^2$

43. $\dfrac{3 + 5i}{i}$ **44.** $\dfrac{5 - i}{i}$ **45.** $\dfrac{5 + 3i}{2 + i}$

46. $\dfrac{7 - 2i}{2 - i}$ **47.** $\dfrac{3 - i}{3 + i}$ **48.** $\dfrac{8 + 3i}{3 - 2i}$

Simplify.

49. $3i^3$ **50.** $-5i^5$ **51.** $2i^7$ **52.** $3i^{-3}$ **53.** $-4i^{18}$ **54.** i^{-32}

Simplify and express each answer in the form $a + bi$.

55. $(3 + 2i)^{-1}$ **56.** $(3 + 2i)^{-2}$

57. One of the basic rules for operating with radicals is that $\sqrt{ab} = \sqrt{a} \cdot \sqrt{b}$, where a and b are nonnegative real numbers. Prove that this rule does not work when both a and b are negative by showing that $\sqrt{(-4)(-9)} \neq \sqrt{-4} \cdot \sqrt{-9}$.

58. Use the definition $\sqrt{-x} = i\sqrt{x}\,(x > 0)$ to prove that $\sqrt{ab} = \sqrt{a} \cdot \sqrt{b}$ when $a < 0$ and $b \geq 0$.

59. Write $\dfrac{a + bi}{c + di}$ in the form $x + yi$. (*Hint:* Multiply the numerator and denominator by the conjugate of $c + di$.)

60. The set of complex numbers satisfies the associative property for addition. Verify this by completing this problem in two different ways.

$$(3 + 5i) + (2 + 3i) + (7 + 4i)$$

61. Repeat Exercise 60 for multiplication, using $(3 + i)(3 - i)(4 + 3i)$.

62. We say that $0 = 0 + 0i$ is the additive identity for the complex numbers since $0 + z = z$ for any $z = a + bi$. Find the additive inverse (negative) of z.

Perform the indicated operations and express the answers in the form $a + bi$.

63. $(5 + 4i) + 2(2 - 3i) - i(1 - 5i)$ **64.** $2i(3 - 4i)(3 - 6i) - 7i$

65. $\dfrac{(2 + i)^2(3 - i)}{2 + 3i}$ **66.** $\dfrac{1 - 2i}{3 + 4i} - \dfrac{2i - 3}{4 - 2i}$

Use complex numbers to factor each polynomial.

Example: $x^2 + 9 = x^2 - (-9) = x^2 - (3i)^2 = (x + 3i)(x - 3i)$

67. $x^2 + 1$ **68.** $9x^2 + 4$ **69.** $3x^2 + 75$ **70.** $x^2 + 4ix - 3$

71. Find the product: $i \cdot i^2 \cdot i^3 \cdot i^4 \cdot i^5 \cdot i^6 \cdot \ldots \cdot i^{100}$

 Written Assignment Find the possible values of the sum

$$i + i^2 + i^3 + \cdots + i^{n-1} + i^n$$

where n is any positive integer. Comment on your answers.

CHAPTER REVIEW EXERCISES

Section 1.1 Real Numbers and Their Properties (Page 2)

1. State the definition of a rational number.

2. Explain how to classify real numbers in terms of their decimal representations.

3. To which subsets of the real numbers do each of the following belong?

    (a) -13 (b) $\sqrt{13}$ (c) $\frac{3}{5}$ (d) 7   (e) $\sqrt{36}$

4. Name the property illustrated by each of the following:

    (a) $3(7 + 8) = (3)(7) + (3)(8)$ (b) $(3 \times 9) \times 5 = (9 \times 3) \times 5$

    (c) $8 + (6 + 9) = (8 + 6) + 9$ (d) $12 + (-12) = 0$

    (e) $4 + 2\sqrt{5}$ is a real number.

5. Give counterexamples to show that the set of real numbers is neither commutative nor associative with respect to subtraction.

6. State, in symbols, the zero-product property for real numbers.

7. What is the identity element for addition? For multiplication?

8. Explain why division by zero is not possible.

Replace the variable n by a real number to make each equation true.

9. $15 + (8 + n) = (15 + 8) + 9$

10. $(12 \times n) \times 7 = 12 \times (n \times 7)$

11. $2.5\,(8 + n) = (2.5 \times 8) + (2.5 \times 10)$

12. $(7 \times 13) + (7 \times n) = 7(13 + 8)$

Section 1.2 Introduction to Equations and Problem Solving (Page 8)

13. State the addition property of equality.

14. State the multiplication property of equality.

15. Solve for x: (a) $3(x - 1) = x + 2$ (b) $5(x + 1) = 2(x - 2)$

16. The formula for the circumference of a circle, C, in terms of its radius, r, is $C = 2\pi r$. Solve for r in terms of C. Find r if $C = 88$ cm. (Use $\frac{22}{7}$ as an approximation for π.)

17. The length of a rectangle is one centimeter more than twice the width. The perimeter is 32 centimeters. Find the dimensions of the rectangle.

18. Tom earns a salary of $475 per week, plus a commission of 4% of his sales. If his total earnings one week were $520, what were his sales for that week?

19. Lisa invested $4500 for 2 years at simple interest and earned $405. At what rate of interest was her investment?

20. Car A leaves a certain town and travels due east at 50 miles per hour. One hour later, car B leaves the same town and travels due west at 55 miles per hour. How many hours after car A leaves will they be 260 miles apart?

Section 1.3 Statements of Inequality and Their Graphs (Page 16)

21. State the algebraic definition for the inequality $a < b$.

22. State the addition property of order.

23. State the multiplication property of order.

24. Solve the inequality: $-2n - (1 - 3n) < 5$

25. Find the solution set: $5x - 2 \le 3x + 6$

26. Solve for x: **(a)** $3(2 - x) > 9$ **(b)** $\dfrac{5}{x - 2} > 0$

27. Show each of the following intervals as a graph on the number line:

  (a) $(-5, 2)$ **(b)** $[-5, 2]$ **(c)** $(-5, 2]$ **(d)** $[-5, 2)$

28. Graph the solution set: **(a)** $-3(x - 2) \leq x + 2$ **(b)** $2x + 5 > 5x - 1$

29. State the transitive property of order for real numbers a, b, and c.

30. If $a < b$ and $ab > 0$, what can be said of the reciprocals of the two numbers? Give an example where both numbers are positive, and one where they are both negative.

31. State the trichotomy property in terms of two real numbers a and b.

32. The sum of an integer and 2 more than three times this integer is between 12 and 31. Find all possible pairs of such integers.

Section 1.4 Absolute Value (Page 25)

33. State the definition of the absolute value of any real number a.

34. State, in symbols, another way of writing $|a| > k$ $(k > 0)$ that does not use the absolute value notation.

35. Repeat Exercise 34 for $|a| < k$.

36. What is the coordinate of the midpoint of a line segment with endpoints x_1 and x_2?

37. Solve for x: $\dfrac{|x - 2|}{x - 2} = 1$

38. Solve for x: **(a)** $|2 - x| = 5$ **(b)** $|2x - 1| \leq 3$

39. Graph: **(a)** $|x| < 5$ **(b)** $|x + 2| > 1$

40. What geometric property does $|a - b|$ represent with respect to points a and b on a number line? Illustrate with two specific examples.

41. Under what conditions does $|x + y| = |x| + |y|$? When is the equality not true?

42. Is the absolute value of the product of two numbers the same as the product of their absolute values? Explain your answer.

Section 1.5 Integral Exponents (Page 30)

43. Explain the difference, if any, between -5^2 and $(-5)^2$.

44. Describe the motivation for the definition of a zero exponent, such as b^0.

45. Simplify: **(a)** $3x^2 \cdot x^4$ **(b)** $\dfrac{-8^3}{(-4)^2}$ **(c)** $10^3(\tfrac{1}{5})^3$ **(d)** $\dfrac{x^{12}y^0}{(2x)^2(2y)^{-1}}$

46. Simplify and write without negative exponents: $\left(\dfrac{a^3b^{-2}}{a^{-2}b^3}\right)^3$

Find a value of x to make each statement true.

47. $3^{-2} \cdot 3^x = 3^5$ **48.** $3^x \cdot 3^{x+2} = 3^{10}$ **49.** $\dfrac{3^x}{3^2} = 3^5$

50. Write in scientific notation: **(a)** 420,000,000   **(b)** 0.000000023

51. Write in standard form: **(a)** 3.25×10^5 **(b)** 2.5×10^{-6}

52. Assume that a substance decays such that $\frac{1}{2}$ the amount remains after each hour. If there were 960 grams at the start, how much remains after 6 hours? How much after n hours?

Section 1.6 Radicals and Rational Exponents (Page 38)

53. State the definition of $\sqrt[n]{a}$ for $a > 0$.

Classify as true or false.

54. (a) $\sqrt{9} + \sqrt{25} = \sqrt{9 + 25}$ **(b)** $\sqrt{36} = \pm 6$

55. (a) $(x + y)^{1/5} = \sqrt[5]{x + y}$ **(b)** $\sqrt{x^{16}} = x^4$

56. Evaluate: **(a)** $(\sqrt[3]{5})^3$ **(b)** $\sqrt[3]{64}$ **(c)** $\sqrt[3]{-\dfrac{1000}{125}}$

57. Simplify: **(a)** $\sqrt{3a} \cdot \sqrt{5b}$ **(b)** $\dfrac{\sqrt[3]{-54x^5}}{\sqrt[3]{2x^2}}$ **(c)** $\sqrt[5]{16x} \cdot \sqrt[5]{-2x^4}$

58. Evaluate: **(a)** $\left(\dfrac{16}{81}\right)^{-3/4}$ **(b)** $64^{-2/3} + (-32)^{-3/5}$

59. Simplify: $\left(\dfrac{-64a^{-3}}{b^6}\right)^{2/3}$

60. Simplify: **(a)** $\sqrt{75}$ **(b)** $\sqrt[3]{24}$ **(c)** $\sqrt[3]{-250}$

61. Simplify: $\sqrt[3]{16x^3} + 2\sqrt[3]{-54x^3}$

62. Rationalize the denominator: **(a)** $\dfrac{125}{\sqrt{125}}$ **(b)** $\dfrac{6}{\sqrt[3]{2}}$

63. Simplify: $\dfrac{6}{\sqrt{12}} + 2\sqrt{3} - 3\sqrt{75}$

64. What is the value of $\sqrt{a^2}$ for real numbers a?

65. What is the rule for evaluating $\sqrt[m]{\sqrt[n]{a}}$? Use the rule to evaluate $\sqrt[2]{\sqrt[3]{64}}$.

66. Is the square root of the sum of two numbers the same as the sum of their square roots? Explain your answer with a specific example.

Section 1.7 Fundamental Operations with Polynomials (Page 47)

67. Write the general form of an nth degree polynomial in the variable x.

68. When is a polynomial in one variable said to be in standard form?

69. Add: $(3x^3 - 8x^2 + 2x - 5) + (x^2 - 7x + 1)$

70. Combine: $(5a^3 - 6a^2b + 2b + 1) - (3a^2b - 2b + 1)$

71. Multiply: $2x^2(3x^3 - 2x^2 + x - 5)$

72. Find the product: $(ax + b)(cx + d)$

Find each product.

73. $(2x + 1)(3x - 5)$ **74.** $(4x + 5)(4x - 5)$

75. Multiply $2x^3 - 5x + 1$ by $x^2 - 3x + 2$.

76. Expand: **(a)** $(2a - b)^2$ **(b)** $(a - 2b)^3$

77. Rationalize the denominator: $\dfrac{6}{\sqrt{5} - \sqrt{2}}$

Section 1.8 Factoring Polynomials (Page 53)

78. State the rules for factoring the sum and the difference of two cubes.

Factor completely.

79. $5x^6 + 25x^4 - 15x^2$ **80.** $3x^2 - 75$ **81.** $27 - 8x^3$ **82.** $(x - 1)^2 - 9$

83. $a^4 - b^4$ **84.** $2a^5 - 64$ **85.** $15 + ax^2 - 3x - 5ax$ **86.** $x^2 + 2xy + y^2$

87. $x^2 - 2xy + y^2$ **88.** $8x^2 + 29x - 12$ **89.** $6x^2 + x - 1$ **90.** $4x^3 + 4x^2 + x$

91. Factor as the difference of squares using irrational numbers: $x^2 - 3$

92. Factor as the difference of cubes using radical expressions: $x - 8$

93. Write the factored form of $a^n - b^n$ for any integer $n \geq 2$.

94. Give the form for a perfect square trinomial, and state the factored form.

Section 1.9 Fundamental Operations with Rational Expressions (Page 61)

95. What is meant by the term "rational expression"? Give an example.

Reduce to lowest terms, if possible.

96. $\dfrac{x^2 + x - 6}{x^2 + 3x}$ **97.** $\dfrac{2a - 3b}{3b - 2a}$ **98.** $\dfrac{x^3 + 3}{x + 3}$

99. Multiply: $\dfrac{x^2 + 2x}{x - 3} \cdot \dfrac{3 + 2x - x^2}{x}$ **100.** Divide: $\dfrac{8x^2 - 10x - 3}{x^2 - 4x + 4} \div \dfrac{3 - 2x}{2 - x}$

101. Find the sum and simplify: $\dfrac{4}{x^2 - x} + \dfrac{3}{1 - x^2}$

102. Combine and simplify: $\dfrac{2}{3} - \dfrac{3}{6x(x - 1)} + \dfrac{3 - x}{2x^2}$

Divide and check each result.

103. $(2x^3 + x^2 - 5x + 2) \div (2x - 1)$ **104.** $(3x^2 - 8x + 2x^3 + 3) \div (x + 3)$

105. Simplify: $\dfrac{\dfrac{1}{2 + h} - \dfrac{1}{2}}{h}$ **106.** Simplify: $\dfrac{\dfrac{1}{x} + \dfrac{1}{y}}{x^{-2} - y^{-2}}$

Section 1.10 Introduction to Complex Numbers (Page 70)

107. State the definition of the imaginary number i.

108. Explain what is meant by the cycle of powers of i.

109. Simplify: **(a)** $\sqrt{-9} + \sqrt{-36}$ **(b)** $\sqrt{-9} \cdot \sqrt{-36}$

110. Multiply: $(2 + 3i)(3 + 5i)$ **111.** Divide: $\dfrac{3 + 2i}{2 + i}$

Perform the indicated operations.

112. $8i[(3 + 2i) + (7 + 5i)]$ **113.** $(9 - 5i) - (2 + 3i)$

114. Simplify: **(a)** i^{32} **(b)** i^{18} **(c)** i^9 **(d)** i^{27}

115. Express $5i^{-3}$ as the indicated product of a real number and i.

116. Find the real and imaginary parts of $\dfrac{1}{1 + 2i}$.

117. Find and simplify the product of the two complex numbers $a + bi$ and $c + di$.

118. Express the quotient $\dfrac{3 + 2i}{2 + i}$ in the form $a + bi$.

CHAPTER 1: STANDARD ANSWER TEST

Use these questions to test your knowledge of the basic skills and concepts of Chapter 1. Then check your answers with those given at the back of the book.

1. Classify each statement as true or false.

   (a) Negative irrational numbers are not real numbers.

   (b) Every integer is a rational number.

   (c) Some irrational numbers are integers.

   (d) Zero is a rational number.

   (e) If $x < y$, then $x - 5 > y - 5$.

   (f) The absolute value of a sum equals the sum of the absolute values.

   (g) If $-5x < -5y$, then $x > y$.

   (h) $|x - 2| < 3$ means that x is within 2 units of 3 on the number line.

2. One endpoint of a line segment is located at -7. The midpoint is located at -2. What is the coordinate of the other endpoint of the line segment?

3. Solve for x: $\frac{2}{3}(x - 3) + 1 = 2x + 3$

4. Find the dimensions of a rectangle whose perimeter is 52 inches if the length is 5 inches more than twice the width.

5. A car leaves from point B at noon traveling at the rate of 55 miles an hour. One hour later a second car leaves from the same point, traveling in the opposite direction at 45 miles per hour. At what time will they be 200 miles apart?

6. Write in scientific notation:

   (a) 375,000,000   (b) 0.0000318

Solve for x.

7. $\dfrac{|x + 2|}{x + 2} = -1$ 8. $|x + 2| < 1$

9. Solve and graph each inequality: (a) $2(5x - 1) < x$ (b) $|2x - 1| \geq 3$

10. Classify each statement as true or false:

    (a) $\dfrac{x^3(-x)^2}{x^5} = x$ (b) $\left(\dfrac{3}{2 + a}\right)^{-1} = \dfrac{2}{3} + \dfrac{a}{3}$ (c) $(-27)^{-1/3} = 3$

    (d) $(x + y)^{3/5} = \left(\sqrt[3]{x + y}\right)^5$ (e) $\sqrt{9x^2} = 3\,|x|$ (f) $(8 + a^3)^{1/3} = 2 + a$

11. Simplify. Express the answers using positive exponents: (a) $\dfrac{(2x^3y^{-2})^2}{x^{-2}y^3}$ (b) $\dfrac{(3x^2y^{-3})^{-1}}{(2x^{-2}y^2)^{-2}}$

Perform the indicated operations and simplify.

12. (a) $\sqrt{8} \cdot \sqrt{6}$ (b) $\dfrac{\sqrt{360}}{2\sqrt{2}}$ (c) $\dfrac{\sqrt[3]{-243x^8}}{\sqrt[3]{3x^2}}$

13. (a) $\sqrt{50} + 3\sqrt{18} - 2\sqrt{8}$ (b) $\dfrac{12}{\sqrt{3}} + 2\sqrt{3}$

14. $5\sqrt{3x^2} + 2\sqrt{27x^2} - 3\sqrt{48x^2}$

15. $(x^2 + 3x)(3x^2) + (x^3 - 1)(2x + 3)$

16. $(6x^3 + x^2 + 3x + 2) \div (2x + 1)$

Factor completely.

17. $64 - 27b^3$ **18.** $6x^2 - 7x - 3$ **19.** $2x^2 - 6xy - 3y^3 + xy^2$

Perform the indicated operations and simplify.

20. $\dfrac{x^2 - 9}{x^3 + 4x^2 + 4x} \cdot \dfrac{2x^2 + 4x}{x^2 + 2x - 15}$ **21.** $\dfrac{x^3 + 8}{x^2 - 4x - 12} \div \dfrac{x^3 - 2x^2 + 4x}{x^3 - 6x^2}$

22. $\dfrac{\dfrac{1}{x^2} - \dfrac{1}{49}}{x - 7}$ **23.** $\dfrac{1}{x + 3} - \dfrac{2}{x^2 - 9} + \dfrac{x}{2x^2 + x - 15}$

24. Multiply the complex numbers $3 + 7i$ and $5 - 4i$ and express the result in the form $a + bi$.

25. Divide $3 + 7i$ by $5 - 4i$ and express the result in the form $a + bi$.

CHAPTER 1: MULTIPLE CHOICE TEST

1. Which of these statements are true?

   I. Every integer is the coordinate of some point on the number line.

  II. Every rational number is a real number.

 III. Every point on the number line can be named by a rational number.

   (a) Only I   **(b)** Only II   **(c)** Only I and II   **(d)** I, II, and III   **(e)** None of the preceding

2. Express the inequality $-5 \le x$ in interval notation.

   (a) $(\infty, -5)$ **(b)** $(-\infty, -5)$ **(c)** $[-5, \infty)$ **(d)** $(\infty, -5]$ **(e)** None of the preceding

3. If $a + b$ is negative, then $|a + b| =$

   (a) 0 **(b)** $a + b$ **(c)** $-a + b$ **(d)** $-(a + b)$ **(e)** None of the preceding

4. Which of the following is *false*?

   (a) $|8 - 2| = 8 - 2$ **(b)** $|2 - 7| = 2 - 7$ **(c)** $|6 + 8| = |6| + |8|$ **(d)** $|3 - 5| = -(3 - 5)$

   (e) None of the preceding

5. The inequality $|x| \ge k \ (k \ge 0)$ is true if and only if

   (a) $x \le -k$ or $x \ge k$ **(b)** $x \le k$ or $x \ge -k$ **(c)** $-k \le x \le k$ **(d)** $x \ge k$ **(e)** None of the preceding

6. Which of these statements are true for all real numbers a, b, and c?

   I. If $a < b$, then $a + c < b + c$.

  II. If $a < b$, then $ac < bc$ for $c > 0$.

 III. If $a < b$, then $ac > bc$ for $c < 0$.

   (a) Only I   **(b)** Only II   **(c)** Only III   **(d)** I, II, and III   **(e)** None of the preceding

7. In interval notation, the solution of the inequality $-3(x + 1) < 2x + 2$ is

   (a) $(-\infty, -1)$ **(b)** $(-1, \infty)$ **(c)** $(1, \infty)$ **(d)** $(-\infty, 1)$ **(e)** None of the preceding

8. The length of a rectangle is 2 inches less than three times the width. If the length is increased by 5 inches and the width decreased by 1 inch, the length will be five times the width. Using x to denote the original width, which of these equations can be used to solve for x?

   (a) $x - 1 = 5(3x + 3)$ **(b)** $3x - 2 = 5x - 1$ **(c)** $3x + 3 = 5(x - 1)$ **(d)** $3x + 3 = 5x - 1$

   (e) None of the preceding

9. Which of the following is equivalent to $(a + b)^{-1}$?

   (a) $a^{-1} + b^{-1}$ (b) $\dfrac{1}{a + b}$ (c) $(-a) + (-b)$ (d) $\dfrac{1}{a} + \dfrac{1}{b}$ (e) None of the preceding

10. Which of the following are true?

    I. $a^{-1/2} + b^{-1/2} = \dfrac{1}{\sqrt{a}} + \dfrac{1}{\sqrt{b}}$ II. $(-x)^{-1/3} = x^{1/3}$ III. $x^{3/4} = \left(\sqrt[3]{x}\right)^4$

    (a) Only I   (b) Only II   (c) Only III   (d) I, II, and III   (e) None of the preceding

11. Rationalize the denominator: $\dfrac{8}{\sqrt{5} - 1}$.

    (a) $8(\sqrt{5} + 1)$ (b) $2(\sqrt{5} - 1)$ (c) $2(\sqrt{5} + 1)$ (d) $2\sqrt{5} + 1$ (e) None of the preceding

12. Which of the following are true?

    I. $2\sqrt{x + 1} = \sqrt{2x + 1}$ II. $\sqrt{(x - 1)^2} = |x - 1|$ III. $(x + y)^{1/3} = x^{1/3} + y^{1/3}$

    (a) Only I   (b) Only II   (c) Only III   (d) I, II, and III   (e) None of the preceding

13. What is the complete factored form for the expression $x^2(x - 1) - 2x(x - 1) + (x - 1)$?

    (a) $x(x - 2)(x - 1)$ (b) $(x - 1)(x^2 - 2x + 1)$ (c) $(x - 2x)(3x - 3)$ (d) $(x - 1)^3$

    (e) None of the preceding

14. Which of the following are true?

    I. $x^3 - y^3 = (x - y)(x^2 + y^2)$ II. $x^2 + y^2 = (x + y)(x + y)$ III. $4x^2 - 6xy + 9y^2 = (2x - 3y)^2$

    (a) Only I   (b) Only II   (c) Only III   (d) Only II and III   (e) None of the preceding

15. Which of the following are true?

    I. Every integer can be written in the form $a + bi$, where $i = \sqrt{-1}$.

    II. Every real number is a complex number.

    III. The sum, difference, and product of two complex numbers are complex numbers.

    (a) Only I   (b) Only II   (c) Only III   (d) I, II, and III   (e) None of the preceding

16. Which of the following can be used to show that the real number x is within 5 units of 2 on a number line?

    (a) $|x - 2| < 5$ (b) $|x + 2| < 5$ (c) $|x - 5| < 2$ (d) $|x + 5| < 2$ (e) None of the preceding

17. In scientific notation, $325{,}000{,}000 =$

    (a) 3.25×10^{-8} (b) 3.25×10^8 (c) 325×10^6 (d) 0.325×10^9 (e) None of the preceding

18. Which of the following represents the graph shown?

    (a) $|x + 4| \le 2$ (b) $|x - 4| \le 2$ (c) $|x - 2| \le 4$ (d) $|x + 2| \le 4$

    (e) None of the preceding

19. Which of the following is *false*?

    (a) $(x + y)^{1/3} = \sqrt[3]{x + y}$ (b) $x^{-1} + y^{-1} = \dfrac{x + y}{xy}$ (c) $\sqrt[3]{x^9} = x^{1/3}$ (d) $\sqrt{(x - 1)^2} = |x - 1|$

    (e) None of the preceding

20. Which of the following are true?

    I. $\dfrac{1}{i} = -i$ II. $(3 - 2i)^0 = 1$ III. $\dfrac{1 + i}{1 - i} = i$

    (a) Only I   (b) Only II   (c) Only III   (d) I, II, and III   (e) None of the preceding

ANSWERS TO THE TEST YOUR UNDERSTANDING EXERCISES

Page 4
1. Associative property for addition.
2. Commutative property for addition.
3. Commutative property for addition.
4. Distributive property.
5. Commutative property for multiplication.
6. Associative property for multiplication.
7. No; no.
8. $12 \div 3 \neq 3 \div 12$
9. No; no.
10. $(8 \div 4) \div 2 \neq 8 \div (4 \div 2)$
11. Yes; by the closure properties for addition and multiplication of real numbers, respectively.

Page 9
1. $x = 6$
2. $x = 17$
3. $x = -4$
4. $x = 6$
5. $x = 13$
6. $x = 10$
7. $x = 1$
8. $x = 2$
9. $x = 5$
10. $x = -9$
11. $x = -3$
12. $x = -11$

Page 19
1. $\{x \mid x < 9\}$
2. $\{x \mid x < 18\}$
3. $\{x \mid x > 9\}$
4. $\{x \mid x > -5\}$
5. $\{x \mid x \leq 14\}$
6. $\{x \mid x \geq -2\}$
7. $\{x \mid x < 4\}$
8. $\{x \mid x \geq 7\}$
9. $\{x \mid x \leq -21\}$
10. $\{x \mid x \geq -1\}$
11. $\{x \mid x > 3\}$
12. $\{x \mid x < -30\}$

Page 32
1. 125
2. $-\frac{1}{32}$
3. 0
4. 1,000,000
5. -64
6. 64
7. $\frac{1}{17}$
8. $\frac{1}{23}$
9. 81
10. $a^{11}b^{10}$
11. 64
12. $6x^4$

Page 41
1. $7^{1/2}$
2. $(-10)^{1/3}$
3. $7^{1/4}$
4. $7^{2/3}$
5. $5^{3/4}$
6. 5
7. 4
8. $\frac{1}{6}$
9. $\frac{1}{7}$
10. -3
11. 8
12. $\frac{1}{8}$
13. $\frac{27}{8}$
14. 4
15. $\frac{1}{4}$
16. $6x$
17. $-\dfrac{1}{2x}$
18. $-\frac{3}{2}x^2$

Page 44
1. $6\sqrt{2}$
2. $6\sqrt{3}$
3. $\sqrt{5}$
4. $\sqrt[3]{2}$
5. $4\sqrt[3]{2} + 5$
6. $-\sqrt[3]{3}$
7. $11\sqrt{2}$
8. $13\sqrt{3}$
9. $3\sqrt{5}$
10. $7\sqrt{7}$
11. $6\sqrt[3]{2}$
12. $2\sqrt[3]{3}$

Page 50
1. $x^2 + 1$
2. $x^2 + 4x - 13$
3. $2x^3 + 3x^2 + 8x + 1$
4. $4x^2y - 5xy + 6xy^2$
5. $-3x^4 - 6x^3 + 3x$
6. $6x^2 + 13x + 6$
7. $12x^2 + 28x - 5$
8. $16x^2 - 49$
9. $4x^2 - 12xy + 9y^2$
10. $2x^5 - 5x^4 + 4x^3 + 14x^2 - 15x$

Page 54
1. $3(x - 3)$
2. $-5(x - 3)$
3. $5y(x + 5y + 2y^4)$
4. $(x + 6)(x - 6)$
5. $(2x + 7)(2x - 7)$
6. $(a + 2 + 5b)(a + 2 - 5b)$
7. $(x - 3)(x^2 + 3x + 9)$
8. $(x + 3)(x^2 - 3x + 9)$
9. $(2a - 5)(4a^2 + 10a + 25)$
10. $3(x + 4)(x - 4)$
11. $a(x + y)(x^2 - xy + y^2)$
12. $2h(x + 2h)(x - 2h)$

Page 58
1. $(a + 3)^2$
2. $(x - 5)^2$
3. $(2x + 3y)^2$
4. $(x + 3)(x + 5)$
5. $(a - 10)(a - 2)$
6. Not factorable
7. $(5x - 2)(2x - 7)$
8. $(3x + 2)(2x - 5)$
9. Not factorable.

Page 65

1. $\dfrac{x}{x+2}$ **2.** $-\dfrac{4b}{3a}$ **3.** $-\dfrac{x+2}{x}$ **4.** -1 **5.** 1 **6.** $\dfrac{4-3x}{6x^2}$ **7.** $\dfrac{25}{6x}$ **8.** $\dfrac{2(5x+4)}{(x-2)(x+2)}$

9. $\dfrac{13x-18}{(x-1)(x^2-4)}$ **10.** $\dfrac{2}{2x+5}$ **11.** $3x^2y^4 - 2xy + 4x^4y^2$ **12.** $2x^2 - 3x + 5$; rem. $= -7$.

Page 71

1. $3i$ **2.** $7i$ **3.** $i\sqrt{5}$ **4.** $-2i$ **5.** $\frac{2}{3}i$ **6.** -120 **7.** $21i$ **8.** $7i\sqrt{2}$ **9.** $12i\sqrt{5}$

Linear and Quadratic Functions with Applications

Does the function $y = R(x) = 60,000 + 2000\,x - 250x^2$ have a largest or smallest value? If so, how can it be found? Consider this problem:

> The marketing department of the TENRAQ Tennis Company found that, on the average, 600 tennis racquets will be sold monthly at the unit price of $100. The department store also observed that for each $5 reduction in price, an extra 50 racquets will be sold monthly. What price will bring the largest monthly income?

Let x represent the number of $5 reductions. Then it can be shown that the monthly income R depends on x and has the preceding form, that is:

$$y = R(x) = 60,000 + 2000\,x - 250\,x^2$$

The graph of this equation turns out to be a parabola that opens downward so that the vertex of the curve is the highest point. To locate this point, a procedure will be developed in this chapter that will make it possible to rewrite the equation in an equivalent form in which the vertex becomes evident.

In Section 2.8 the details of this problem are presented. You will learn how to solve this and other interesting applied problems, such as finding the maximum height reached by a ball thrown straight upward from ground level.

2.1
INTRODUCTION TO THE FUNCTION CONCEPT

Suppose that you are riding in a car that is averaging 40 miles per hour. Then the distance traveled is determined by the time traveled.

$$\text{distance} = \text{rate} \times \text{time}$$

Symbolically, this relationship can be expressed by the equation

$$s = 40t$$

where s is the distance traveled in time t (measured in hours). For $t = 2$ hours, the distance traveled is

$$s = 40(2) = 80 \text{ miles}$$

Similarly, for each specific value of $t \geq 0$ the equation produces *exactly one* value for s. This correspondence between the distance s and the time t is an example of a *functional relationship*. More specifically, we say that the equation $s = 40t$ defines s as a *function* of t because for *each* choice of t there corresponds *exactly one* value for s. We first choose a value of t. Then there is a corresponding value of s that depends on t; s is the *dependent variable* and t is the *independent variable* of the function defined by $s = 40t$.

Because the variable t represents time in the equation $s = 40t$, it is reasonable to say that $t \geq 0$. This set of allowable values for the independent variable is called the **domain** of the function. The set of corresponding values for the dependent variable is called the **range** of the function. This leads to the following important definition.

*Note, for example, that the equation $y^2 = 12x$ does **not** define y as a function of x; for a given value of $x > 0$ there is **more than one** corresponding value for y. If $x = 3$, for example, then $y^2 = 36$ and $y = 6$ or $y = -6$.*

DEFINITION OF FUNCTION

A **function** is a correspondence between two sets, the domain and the range, such that for each value in the domain there corresponds exactly one value in the range.

The specific letters used for the independent and dependent variables are of no consequence. Usually, we will use x for the independent variable and y for the dependent variable.

Most of the expressions encountered earlier in this text can be used to define functions. Here are some illustrations. Note that in each case *the domain of the function is taken as the largest set of real numbers for which the defining expression in x leads to a real value,* as in the following illustrations.

Note that we must restrict the domain of a function so that the denominator of a fraction is not zero, and so that there is not an even root of a negative number. These two cases produce results that are not real numbers.

Function Given By	Domain
$y = 6x^4 - 3x^2 + 7x + 1$	All real numbers
$y = \dfrac{2x}{x^2 - 4}$	All reals except 2 and -2
$y = |x|$	All real numbers
$y = \sqrt{x}$	All real $x \geq 0$

EXAMPLE 1 Explain why the following equation defines y as a function of x, and find the domain: $y = \dfrac{1}{\sqrt{x-1}}$.

Solution For each *allowable* x the expression $\dfrac{1}{\sqrt{x-1}}$ produces just one y-value. Therefore, the given equation defines a function. To find the domain, note that $\sqrt{x-1}$ is only defined if $x - 1 \geq 0$, or $x \geq 1$. However, since $\sqrt{x-1}$ is in the denominator, $x = 1$ produces division by 0 and so must also be excluded. Thus the domain consists of all $x > 1$. ∎

TEST YOUR UNDERSTANDING	*Decide whether the given equation defines y to be a function of x. For each function, find the domain.*

1. $y = (x + 2)^2$ **2.** $y = \dfrac{1}{(x+2)^2}$ **3.** $y = \dfrac{1}{x^2 + 2}$

4. $y = \pm 3x$ **5.** $y = \dfrac{1}{\sqrt{x^2 + 2x + 1}}$ **6.** $y^2 = x^2$

7. $y = \dfrac{x}{|x|}$ **8.** $y = \sqrt{2 - x}$

(Answers: Page 172)

Sometimes we say that an equation, such as $y = 40x$, is a function. Such informal language is commonly used and should not cause difficulty.

Thus far we have used equations to define functions. However, since a function is a correspondence between two sets, it can be given in other ways as well. Here, for example, is a *table of values*. The table defines y to be a function of x because for each domain value x there corresponds *exactly one* value for y.

In the table, note that for each x there is exactly one value for y. However, it is permissible to have a value in the range (such as 6) associated with more than one value in the domain.

x	1	2	5	-7	23	$\sqrt{2}$
y	6	-6	6	-4	0	5

Instead of using a single equation to define a function, there will be times when a function is defined in terms of more than one equation. For instance, the following three equations define a function whose domain is the set of all real numbers. The function is said to be a **piecewise-defined function.**

$$y = \begin{cases} 1 & \text{if } x \leq -6 \\ x^2 & \text{if } -6 < x < 0 \\ 2x + 1 & \text{if } x \geq 0 \end{cases}$$

Here are several illustrations of y-values found for specific replacements of x in this function:

For $x = -7$, $y = 1$ since $-7 \leq -6$.
For $x = -5$, $y = x^2 = (-5)^2 = 25$ since $-6 < -5 < 0$.
For $x = 5$, $y = 2x + 1 = 2(5) + 1 = 11$ since $5 \geq 0$.

EXAMPLE 2 Decide whether the following defines y to be a function of x.

$$y = \begin{cases} 3x - 1 & \text{if } x \le 1 \\ 2x + 1 & \text{if } x \ge 1 \end{cases}$$

Solution If $x = 1$, the first equation gives $y = 3(1) - 1 = 2$. However, for $x = 1$ the second equation gives $y = 2(1) + 1 = 3$. Since we have two different y-values for the same x-value, the two equations do *not* define y as a function of x. ■

A useful way to refer to a function is to name it by using a specific letter, such as f, g, F, and the like. For example, the function defined by the equation $y = x^2 + 3x$ may be referred to as f and written in the form

$$f(x) = x^2 + 3x$$

CAUTION
*Note that $f(x)$ does **not** mean that we are to multiply f by x; f does not stand for a number.*

The symbol $f(x)$ is read as "f of x" and represents the range value of the function for the value of x that is shown within the parentheses. That is, for each x in the domain of the function, $f(x)$ represents the corresponding range value. For example:

$$f(2) = 2^2 + 3(2) = 10 \qquad \text{When } x = 2, y = f(2) = 10.$$

$$f(-1) = (-1)^2 + 3(-1) = -2 \qquad \text{When } x = -1, y = f(-1) = -2.$$

Observe that each of the forms $y = x^2 + 3x$, $f(x) = x^2 + 3x$, and $y = f(x) = x^2 + 3x$ represents the same function f.

A function may also be viewed as a *rule* that shows how to take an *input* (a domain value) and produce the corresponding *output* (the range value), as in this diagram:

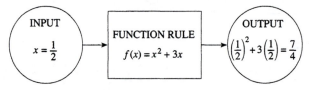

Keep in mind that the variable x in $g(x) = x^2$ is only a placeholder. Any letter could serve the same purpose. For example, $g(t) = t^2$ and $g(z) = z^2$ both define the same function with domain all real numbers.

Let us explore the function notation with another example. If g is the function defined by $y = g(x) = x^2$, then

$$g(1) = 1^2 = 1 \qquad g(2) = 2^2 = 4 \qquad g(3) = 3^2 = 9$$

Note that $g(1) + g(2) \ne g(3)$. To write $g(1) + g(2) = g(1 + 2)$ would be to assume, *incorrectly,* that the distributive property holds for the functional notation. This is not true in general, which comes as no great surprise since g is not a number.

EXAMPLE 3 Let $f(x) = -x^2 + 5x$ and find $f(x - 2)$.

CAUTION
Note that $-x^2 \ne (-x)^2$; $-x^2$ means $-(x^2)$. For example, let $x = 5$: $-x^2 = -5^2 = -(5^2) = -25$, $(-x)^2 = (-5)^2 = 25$.

Solution Whenever there is an x in the expression that defines the given function, replace it by $x - 2$. In a sense, x is a placeholder, and to emphasize this we can use a box instead of x to obtain this form:

$$f(\boxed{}) = -\boxed{}^2 + 5\boxed{}$$

Now enter $x - 2$ into each box. Thus

$$f(x - 2) = -(x - 2)^2 + 5(x - 2)$$
$$= -(x^2 - 4x + 4) + 5(x - 2)$$
$$= -x^2 + 4x - 4 + 5x - 10$$
$$= -x^2 + 9x - 14 \qquad\blacksquare$$

EXAMPLE 4 For the function f defined by $f(x) = \dfrac{1}{x - 3}$, find:

(a) $4f(x)$ **(b)** $f(4x)$ **(c)** $f(4 + x)$ **(d)** $f(4) + f(x)$

Solution

(a) $4f(x) = 4 \cdot \dfrac{1}{x - 3} = \dfrac{4}{x - 3}$

(b) $f(4x) = \dfrac{1}{4x - 3}$

(c) $f(4 + x) = \dfrac{1}{(4 + x) - 3} = \dfrac{1}{x + 1}$

(d) $f(4) + f(x) = \dfrac{1}{4 - 3} + \dfrac{1}{x - 3} = 1 + \dfrac{1}{x - 3} \qquad\blacksquare$

TEST YOUR UNDERSTANDING

Let $f(x) = x^2 - 3x$ and find each of the following.

1. $f(-3)$ **2.** $f(5)$ **3.** $f(0)$ **4.** $f\left(\tfrac{1}{2}\right)$ **5.** $f\left(-\tfrac{1}{2}\right)$

6. $f(a)$ **7.** $f(2x)$ **8.** $2f(x)$ **9.** $f(x + 3)$ **10.** $f(x) + f(3)$

11. $f\left(\dfrac{1}{x}\right)$ **12.** $\dfrac{1}{f(x)}$

(Answers: Page 172)

EXAMPLE 5 Let $g(x) = x^2$. Evaluate and simplify the *difference quotient:*

Difference quotients will be used in the study of calculus.

$$\frac{g(x) - g(4)}{x - 4}, \qquad x \neq 4$$

Solution $g(x) = x^2$ and $g(4) = 16$

$$\frac{g(x) - g(4)}{x - 4} = \frac{x^2 - 16}{x - 4}$$
$$= \frac{(x - 4)(x + 4)}{x - 4}$$
$$= x + 4 \qquad\blacksquare$$

EXAMPLE 6   Let $g(x) = \dfrac{1}{x}$. Evaluate and simplify the *difference quotient:*

$$\frac{g(4 + h) - g(4)}{h}, \qquad h \neq 0$$

Solution   Three steps are involved in finding this difference quotient:

1. Find $g(4 + h)$: $\qquad g(4 + h) = \dfrac{1}{4 + h}$

After some practice you should be able to do this work without these three separate steps. All that is needed is the work shown in step 3.

2. Subtract $g(4)$: $\qquad g(4 + h) - g(4) = \dfrac{1}{4 + h} - \dfrac{1}{4}$

3. Divide by h and simplify:

$$\frac{g(4 + h) - g(4)}{h} = \frac{\dfrac{1}{4 + h} - \dfrac{1}{4}}{h} = \frac{\dfrac{4 - (4 + h)}{4(4 + h)}}{h}$$

$$= \frac{-h}{4h(4 + h)}$$

$$= -\frac{1}{4(4 + h)} \qquad \blacksquare$$

CAUTION: Learn to Avoid These Mistakes	
In each of the following, the function f is defined by $f(x) = 3x^2 - 4$.	
WRONG	**RIGHT**
$f(0) = 0$	$f(0) = 3(0)^2 - 4 = -4$
$f(-2) = -f(2)$	$f(-2) = 3(-2)^2 - 4 = 8$ $-f(2) = -[3(2)^2 - 4] = -8$
$f\left(\dfrac{1}{2}\right) = \dfrac{1}{f(2)}$	$f\left(\dfrac{1}{2}\right) = 3\left(\dfrac{1}{2}\right)^2 - 4 = -\dfrac{13}{4}$ $\dfrac{1}{f(2)} = \dfrac{1}{8}$
$[f(2)]^2 = f(4)$	$[f(2)]^2 = 8^2 = 64$ $f(4) = 3(4)^2 - 4 = 44$
$2 \cdot f(5) = f(10)$	$2 \cdot f(5) = 2[3(5)^2 - 4] = 142$ $f(10) = 3(10)^2 - 4 = 296$
$f(5) + f(2) = f(7)$	$f(5) + f(2) =$ $[3(5)^2 - 4] + [3(2)^2 - 4] = 79$ $f(7) = 3(7)^2 - 4 = 143$

Functions arise and are used throughout mathematics and its applications. The following examples demonstrate how geometric situations give rise to algebraic functions.

A summary of some useful geometric formulas is given on the inside front cover of the text.

EXAMPLE 7 In the figure, right triangle ABE is similar to triangle ACD; $CD = 8$ and $BC = 10$; h and x are the measures of the altitude and base of triangle ABE. Express h as a function of x.

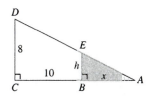

Solution Since corresponding sides of similar triangles are proportional, $\dfrac{BE}{AB} = \dfrac{CD}{AC}$. Substitute as follows:

$$\frac{h}{x} = \frac{8}{x + 10} \qquad AC = AB + BC = x + 10$$

$$h = \frac{8x}{x + 10} \qquad \text{Multiply by } x.$$

To emphasize that h is a function of x, use the functional notation to write the answer in this form:

$$h(x) = \frac{8x}{x + 10} \qquad \blacksquare$$

EXAMPLE 8 A water tank is in the shape of a right circular cone with altitude 30 feet and radius 8 feet. The tank is filled to a depth of h feet. Let x be the radius of the circle at the top of the water level. Solve for h in terms of x and use this to express the volume of water as a function of x.

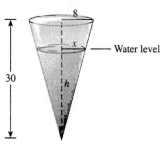

Solution The shaded right triangle is similar to the larger right triangle having base 8 and altitude 30. Therefore

$$\frac{h}{x} = \frac{30}{8} \qquad \text{Solve for } h.$$

$$h = \frac{15}{4}x$$

Now substitute for h in the formula for the volume of a right circular cone.

$$V = \tfrac{1}{3}\pi r^2 h \qquad \text{volume of a right circular cone}$$

$$V(x) = \tfrac{1}{3}\pi x^2 \left(\tfrac{15}{4}x\right) \qquad \text{Substitute for } h \text{ and } r.$$

$$= \tfrac{5}{4}\pi x^3$$

EXERCISES 2.1

Classify each statement as true or false. If it is false, correct the right side to get a correct equation. For each of these statements, use $f(x) = -x^2 + 3$.

1. $f(3) = -6$

2. $f(2)f(3) = -33$

3. $3f(2) = -33$

4. $f(3) + f(-2) = 2$

5. $f(3) - f(2) = -5$

6. $f(2) - f(3) = 11$

7. $f(x) - f(4) = -(x - 4)^2 + 3$

8. $f(x - 4) = -x^2 + 16$

9. $f(4 + h) = -h^2 - 8h - 13$

10. $f(4) + f(h) = -h^2 - 10$

Decide whether the given equation defines y to be a function of x. For each function, find the domain.

11. $y = x^3$

12. $y = \sqrt[3]{x}$

13. $y = \dfrac{1}{\sqrt{x}}$

14. $y = |2x|$

15. $y^2 = 2x$

16. $y = x \pm 3$

17. $y = \dfrac{1}{x + 1}$

18. $y = \dfrac{x - 2}{x^2 + 1}$

19. $y = \dfrac{1}{1 \pm x}$

20. $y = \dfrac{1}{\sqrt[3]{x^2 - 4}}$

Find **(a)** $f(-1)$, **(b)** $f(0)$, *and* **(c)** $f\left(\tfrac{1}{2}\right)$, *if they exist.*

21. $f(x) = 2x - 1$

22. $f(x) = -5x + 6$

23. $f(x) = x^2$

24. $f(x) = x^2 - 5x + 6$

25. $f(x) = x^3 - 1$

26. $f(x) = (x - 1)^2$

27. $f(x) = x^4 + x^2$

28. $f(x) = -3x^3 + \tfrac{1}{2}x^2 - 4x$

29. $f(x) = \dfrac{1}{x - 1}$

30. $f(x) = \sqrt{x}$

31. $f(x) = \dfrac{1}{\sqrt[3]{x}}$

32. $f(x) = \dfrac{1}{3|x|}$

33. For $g(x) = x^2 - 2x + 1$, find:

(a) $g(10)$ **(b)** $5g(2)$ **(c)** $g\left(\tfrac{1}{2}\right) + g\left(\tfrac{1}{3}\right)$ **(d)** $g\left(\tfrac{1}{2} + \tfrac{1}{3}\right)$

34. Let h be given by $h(x) = x^2 + 2x$. Find $h(3)$ and $h(1) + h(2)$ and compare.

35. Let h be given by $h(x) = x^2 + 2x$. Find $3h(2)$ and $h(6)$ and compare.

36. Let h be given by $h(x) = x^2 + 2x$. Find:

(a) $h(2x)$ **(b)** $h(2 + x)$ **(c)** $h\left(\dfrac{1}{x}\right)$ **(d)** $h(x^2)$

Find the value for y for these values of x: **(a)** -5 **(b)** -2 **(c)** 0 **(d)** 2 **(e)** 5

37. $y = \begin{cases} 2x - 1 & \text{if } x \le -2 \\ 1 - 2x & \text{if } x > 2 \end{cases}$

38. $y = \begin{cases} |1 - x| & \text{if } x < -2 \\ 2x - 3 & \text{if } -2 \le x \le 2 \\ x^2 - 2 & \text{if } x > 2 \end{cases}$

Find the difference quotient $\dfrac{f(x) - f(3)}{x - 3}$ *and simplify for the given function f. Rationalize the numerator in Exercise* 42.

39. $f(x) = x^2$ **40.** $f(x) = x^2 - 1$ **41.** $f(x) = \dfrac{1}{x}$

42. $f(x) = \sqrt{x}$ **43.** $f(x) = 2x + 1$ **44.** $f(x) = -x^3 + 1$

Find the difference quotient $\dfrac{f(2 + h) - f(2)}{h}$ *and simplify for the given function f. Rationalize the numerator in Exercise* 48.

45. $f(x) = x$ **46.** $f(x) = -x + 3$ **47.** $f(x) = -x^2$

48. $f(x) = \sqrt{x + 2}$ **49.** $f(x) = \dfrac{1}{x^2}$ **50.** $f(x) = \dfrac{1}{x - 1}$

51. The height, h, of an object above the ground after t seconds that is thrown upward with an initial velocity v_0 is given by $h(t) = -16t^2 + v_0 t$. Express h as a function of t for an initial velocity of 128 ft/sec. Find $h(0)$ and $h(8)$ and interpret your answers.

52. The volume of a sphere can be written as a function of its radius, r, as $V = \dfrac{4}{3}\pi r^3$. Express the radius as a function of the volume, V. Find r when $V = 108\,\pi$ cm^3.

Exercises 53–60 call for the extraction of functions from geometric situations.

(Note: Geometric formulas needed for Exercises 53–60 can be found on the inside front cover of the text.)

53. In the figure, the shaded right triangle, with altitude x, is similar to the larger triangle that has altitude h.

  (a) Express h as a function of x.

  (b) Use the result in **(a)** to express the area A of the larger triangle as a function of x.

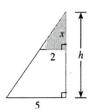

54. In the figure, the shaded triangle is similar to triangle ABC.

  (a) If $BC = 20$ and the altitude of triangle $ABC = 9$, express w as a function of the altitude h of the shaded triangle.

  (b) Express the area of the shaded triangle as a function of w.

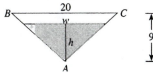

55. In the figure, s is the length of the shadow cast by a 6-foot person standing x feet from a light source that is 24 feet above the level ground. Express s as a function of x.

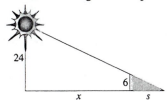

56. Triangle ABC is an isosceles right triangle with right angle at C. h is the measure of the perpendicular from C to side AB. Express the area of the triangle ABC as a function of h.

57. A square piece of tin is 50 centimeters on a side. Congruent squares are cut from the four corners of the square so that when the sides are folded up a rectangular box (without a top) is formed. If the four congruent squares are x centimeters on a side, what is the volume of the box?

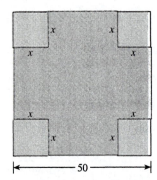
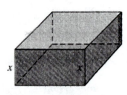

58. Replace the square piece of tin in Exercise 57 by a rectangular piece of tin with dimensions 30 centimeters by 60 centimeters and find the volume of the resulting box in terms of x.

59. A closed rectangular-shaped box is x units wide and it is twice as long. Let h be the altitude of the box. If the total surface area of the box is 120 square units, express the volume of the box as a function of x. (*Hint:* First solve for h in terms of x.)

60. A closed tin can with height h and radius r has volume 5 cubic centimeters. Solve for h in terms of r, and express the surface area of the tin can as a function of r.

   Challenge   Find the domain of the following functions:

1. $f(x) = \sqrt{x^2 + x - 6}$ **2.** $f(x) = \dfrac{1}{\sqrt{4 - x^2}}$

   Written Assignment

1. Explain the meaning of a function. Illustrate with a specific example of your own.

2. Explain why the equation $y = |x|$ defines a function whose domain consists of all real numbers. What is the range of this function?

3. Give at least two different nonmathematical examples of the use of the word *function* that might be found in everyday life situations. Describe the domain and the range of your functions.

   Graphing Calculator Exercises   In order to evaluate a function such as $f(x) = (x^2 + x)/\sqrt{x + 1}$ where $x = 3$ with a graphing calculator, you can enter the expression as $(3^2 + 3)/\sqrt{}(3 + 1)$ on your screen and press ENTER or EXE. (See your

Note: On some calculators the screen will show $\div$ for division rather than /. We will also use $\sqrt{}$ to resemble the calculator key, but note that $\sqrt{}(3 + 1)$ means $\sqrt{(3 + 1)}$.

manual for the correct way to enter and evaluate expressions. Make sure you place parentheses around the "3 + 1" or the root sign will only apply to the 3.) The number 6 that appears on the next line is the value of the function.

If your calculator has a *replay key,* you can evaluate the same function for many values of x without rewriting the function's expression each time. Each time you want to evaluate an expression just replay it, and then substitute a different number for the last value of x by writing over it. (See your manual on how to replay and edit expressions.)

1. Evaluate the function $f(x) = 2x^2 + 3x - 1$ at the given values of x by entering the expression for $f(x)$ on your calculator with the value of x substituted, as explained previously.
   (a) 1   **(b)** 2   **(c)** −2   **(d)** 0   **(e)** 4.73

2. Evaluate the function $f(x) = (x^2 + x)/(\sqrt{x + 1})$ for the given values of x.
   (a) 0   **(b)** 8   **(c)** 1   **(d)** 2   **(e)** 2.15

3. What does your calculator do when you try to evaluate the function in exercise 2 at the following values of x? Why? What is the domain of the function?   **(a)** –1   **(b)** –2

2.2 GRAPHING LINES IN THE RECTANGULAR COORDINATE SYSTEM

*The union of algebra and geometry, credited to French mathematician René Descartes (1596–1650), led to the development of analytic geometry. In his honor, we often refer to the rectangular coordinate system as the **Cartesian coordinate system,** or simply the **Cartesian plane.***

A great deal of information can be learned about a functional relationship by studying its graph. A fundamental objective of this course is to acquaint you with the graphs of some important functions, as well as to develop basic graphing procedures. First we need to review the structure of a **rectangular coordinate system.**

In a plane take any two lines that intersect at right angles and call their point of intersection the **origin.** Let each of these lines be a number line with the origin corresponding to zero for each line. Unless otherwise specified, the unit length is the same on both lines. On the horizontal line the positive direction is taken to be to the right of the origin, and on the vertical line it is taken to be above the origin. Each of these two lines will be referred to as an **axis** of the system (plural: **axes**).

The horizontal line is usually called the **x-axis,** and the vertical line the **y-axis.** The axes divide the plane into four regions called **quadrants.** The quadrants are numbered in a counterclockwise direction as shown in the following figure.

French stamp of René Descartes.
Source: Dr. Marvin Lang.

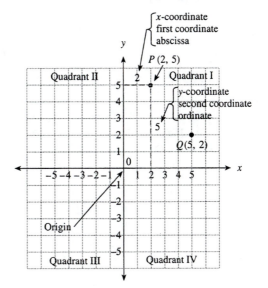

The points in the plane (denoted by the capital letters) are matched with pairs of numbers, referred to as the **coordinates** of these points. For example, starting at the origin, P can be reached by moving 2 units to the right, along the x-axis; then 5 units up, parallel to the y-axis. Thus the first coordinate, 2, of P is called the **x-coordinate** (another name is **abscissa**) and the second coordinate, 5, is the **y-coordinate** (also called **ordinate**). We say that the *ordered pair of numbers* (2, 5) are the coordinates of point P.

All points in the first quadrant are to the right and above the origin and therefore have positive coordinates. Any point in quadrant II is to the left and above the origin and therefore has a negative x-coordinate and a positive y-coordinate. In quadrant III both coordinates are negative, and in the fourth quadrant they are positive and negative, respectively.

Note that the ordered pair (2, 5) is not the same as the pair (5, 2). Each gives the coordinates of a different point on the plane.

Usually, the points on the axes are labeled with just a single number. In that case it is understood that the missing coordinate is 0. Thus point M has coordinates (0, 2) and the coordinates of point N are (3, 0). The coordinates of the origin are (0, 0).

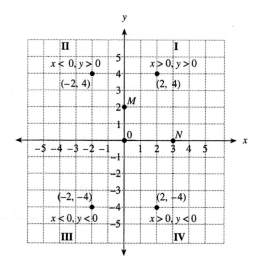

The equality $y = x + 2$ is an equation in two variables. When a specific value for x is substituted into this equation, we get a corresponding y-value. For example, substituting 3 for x gives $y = 3 + 2 = 5$. We therefore say that the ordered pair (3, 5) *satisfies* the equation $y = x + 2$.

DEFINITION OF A GRAPH OF AN EQUATION

The graph of an equation in the variables x and y consists of all the points in a rectangular system whose coordinates satisfy the given equation.

An infinite number of ordered pairs satisfy the equation $y = x + 2$, and all are located on the same straight line. The following table of values shows some ordered pairs of numbers that satisfy the equation $y = x + 2$. These have been *plotted* (located) in a rectangular system and connected by a straight line. The arrowheads in the figure suggest that the line continues endlessly in both directions.

x	y = x + 2
−3	−1
−2	0
−1	1
0	2
1	3
2	4

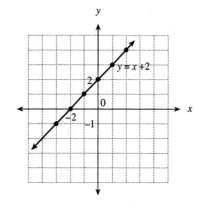

The straight line contains exactly those points whose ordered pairs (x, y) satisfy the equation y = x + 2. Any point not on the line has an ordered pair (x, y) where y ≠ x + 2. Thus (1, 5) is not on the line, since 5 ≠ 1 + 2.

The graph of $y = x + 2$ is a straight line and the equation is called a **linear equation.** Since two points determine a line, a convenient way to graph a line is to locate its two **intercepts.** The **x-intercept** for $y = x + 2$ is -2, the abscissa of the point where the line crosses the x-axis. The **y-intercept** is 2, the ordinate of the point where the line crosses the y-axis.

EXAMPLE 1   Graph the linear equation $y = 2x - 1$ using the intercepts.

Solution   To find the x-intercept, let $y = 0$.

When the line crosses the x-axis, the y-value is 0.

$$2x - 1 = 0$$
$$2x = 1$$
$$x = \tfrac{1}{2} \quad \longleftarrow \text{ the } x\text{-intercept}$$

When the line crosses the y-axis, the x-value is 0.

To find the y-intercept, let $x = 0$.

$$y = 2(0) - 1$$
$$y = -1 \quad \longleftarrow \text{ the } y\text{-intercept}$$

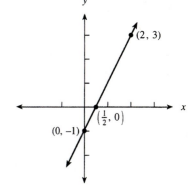

It is generally wise to locate a third point to verify your work. Thus for x = 2, y = 3, and the line passes through the point (2, 3).

Plot the points $\left(\tfrac{1}{2}, 0\right)$ and $(0, -1)$ and draw the line through them to determine the graph.   ∎

Note that an equation such as $y = 2x - 1$ defines y as a **linear function** of x and may be written in the form $f(x) = 2x - 1$. Unless otherwise indicated, the domain consists of all real numbers. The following figure shows the graph of the linear function $y = f(x) = 2x - 4$, including the coordinates of four specific points. From the diagram you will see that the y-value increases 2 units each time that the x-value increases by 1 unit. The ratio of this change in y compared to the corresponding change in x is $\tfrac{2}{1} = 2$. Using the coordinates of the points $(3, 2)$ and $(2, 0)$, and the symbols Δx and Δy for the change in x and the change in y respectively, we have:

$$\frac{\Delta y}{\Delta x} = \frac{\text{change in } y-\text{values}}{\text{change in } x-\text{values}} = \frac{2 - 0}{3 - 2} = \frac{2}{1} = 2$$

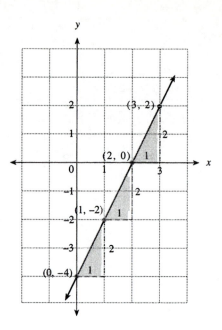

Show that the same ratio is obtained by using any other two points on the line, such as $(1, -2)$ and $(0, -4)$.

We call this ratio the **slope** of the line, defined as follows.

Notice that in the definition $x_2 - x_1$ cannot be zero; that is, $x_2 \neq x_1$. The only time that $x_2 = x_1$ is when the line is vertical, in which case the slope is undefined.

DEFINITION OF SLOPE

If two points (x_1, y_1) and (x_2, y_2) are on a line ℓ, then the slope m of line ℓ is defined by

$$m = \frac{y_2 - y_1}{x_2 - x_1}, \qquad x_2 \neq x_1$$

This discussion shows that there can be only one slope for a given line.

In the following figure the coordinates of two points A and B have been labeled (x_1, y_1) and (x_2, y_2). The change in the y direction from A to B is given by the difference $y_2 - y_1$; the change in the x direction is $x_2 - x_1$. If a different pair of points is chosen, such as P and Q, then the ratio of these differences is still the same because the resulting triangles (ABC and PQR) are similar. Thus, since corresponding sides of similar triangles are proportional, we have

$$m = \frac{y_2 - y_1}{x_2 - x_1} = \frac{AC}{CB} = \frac{PR}{RQ}$$

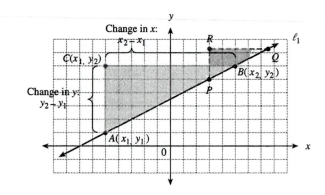

Another descriptive language for slope is $m = \dfrac{rise}{run}$, where **rise** *is the vertical change and* **run** *is the horizontal change.*

It may be helpful to think of the slope of a line in any of these ways:

$$m = \frac{\Delta y}{\Delta x} = \frac{y_2 - y_1}{x_2 - x_1} = \frac{\text{change in } y}{\text{change in } x} = \frac{\text{vertical change}}{\text{horizontal change}}$$

EXAMPLE 2 Find the slope of line ℓ determined by the points $(-3, 4)$ and $(1, -6)$.

Solution Use $(x_1, y_1) = (-3, 4)$; $(x_2, y_2) = (1, -6)$. Then

$$m = \frac{-6 - 4}{1 - (-3)} = -\frac{10}{4} = -\frac{5}{2}$$

CAUTION
Do not mix up the coordinates like this:
$$\frac{y_2 - y_1}{x_1 - x_2} = \frac{4 - (-6)}{1 - (-3)} = \frac{5}{2}$$
This is the negative of the slope.

Note: It makes no difference which of the two points is called (x_1, y_1) or (x_2, y_2) since the ratio will still be the same. If $(x_1, y_1) = (1, -6)$ and $(x_2, y_2) = (-3, 4)$, for example, then $\dfrac{y_2 - y_1}{x_2 - x_1} = \dfrac{4 - (-6)}{-3 - 1} = -\dfrac{5}{2} = m.$ ∎

EXAMPLE 3 Graph the line with slope $\frac{3}{2}$ that passes through the point $(-2, -2)$.

Reading from left to right, a rising line has a positive slope and a falling line has a negative slope.

Solution Think of $\dfrac{3}{2}$ as $\dfrac{\text{change in } y}{\text{change in } x}$. Now start at $(-2, -2)$ and move 3 units up and 2 units to the right. This locates the point $(0, 1)$. Draw the straight line through these two points.

Alternately, start at $(-2, -2)$ and move 3 units down and 2 units to the left to locate $(-4, -5)$.

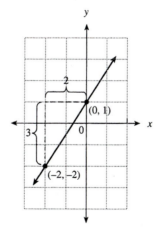

TEST YOUR UNDERSTANDING

Find the slopes of the lines determined by the given pairs of points.

1. $(1, 5)$; $(4, 6)$ **2.** $(3, -5)$; $(-3, 3)$

3. $(-2, -3)$; $(-1, 1)$ **4.** $(-1, 0)$; $(0, 1)$

Draw the line through the given point and with the given slope.

(Answers: Page 172)

5. $(0, 0)$; $m = 2$ **6.** $(-3, 4)$; $m = -\dfrac{3}{2}$

Next consider lines that are parallel to the axes, that is, horizontal and vertical lines.

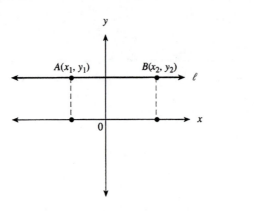 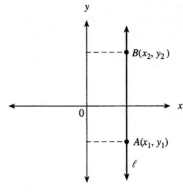

Since ℓ is parallel to the x-axis, $y_1 = y_2$ and $y_2 - y_1 = 0$. Thus *the slope of a horizontal line is 0.*

$$m = \frac{y_2 - y_1}{x_2 - x_1} = \frac{0}{x_2 - x_1} = 0$$

Since ℓ is parallel to the y-axis, $x_1 = x_2$ and $x_2 - x_1 = 0$. However, since division by 0 is undefined, we say that *the slope of a vertical line is undefined;* that is, vertical lines do not have slope.

Two nonvertical lines are parallel if and only if they have the same slope (see Exercise 44). The slope property for perpendicular lines is not as obvious. The figure below suggests the following (see Exercise 45).

Two lines not parallel to the coordinate axes are perpendicular if and only if their slopes are negative reciprocals of one another.

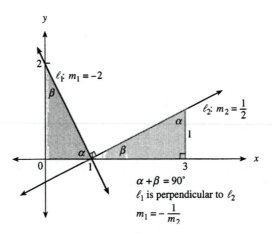

EXAMPLE 4 In the figure, line ℓ_1 has slope $\frac{2}{3}$ and is perpendicular to ℓ_2. If the lines intersect at $P(-1, 4)$, use the slope of ℓ_2 to find the coordinates of another point on ℓ_2.

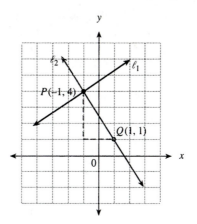

Solution Since the lines are perpendicular, the slope of ℓ_2 is $-\frac{3}{2}$. Now start at P and count 3 units downward and 2 units to the right to reach point $(1, 1)$ on ℓ_2. Other solutions are possible. Can you locate a point on ℓ_2 in the fourth quadrant? ■

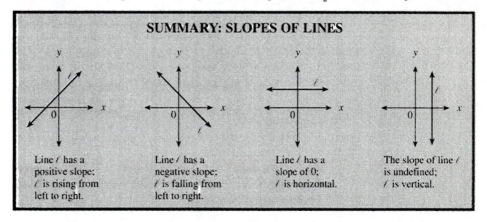

Linear functions play an important role in a wide variety of applications of mathematics. We begin with an illustration related to economics. A business that produces and sells a product needs to understand how the *selling price* and the *demand* for the product are related. That is, usually the higher the price, the less is the demand. Sometimes this relationship can be expressed as a linear function having a *negative* slope. For example, suppose that for a certain calculator it is determined that the weekly demand, D, is related to the price x (in dollars) by the following function:

$$y = D(x) = 500 - 20x$$

$$\text{where } 0 < x \le 25$$

From the graph we see that the demand is high when the price is close to zero dollars! The demand is zero when the price is \$25.

Note that the graph uses a different scale on each axis in order to accommodate the data shown.

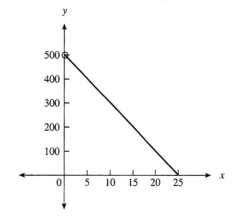

Now suppose that the weekly supply of calculators, S, is also a linear function of the price, x, and is given by $y = S(x) = 10x + 200$. When the graph of this supply function is drawn in the same coordinate system as the demand function, then the **equilibrium point** is the point where the two lines intersect. At this point, the demand = supply for the same price x. To find this x-value, set $D = S$ and solve for the common value x.

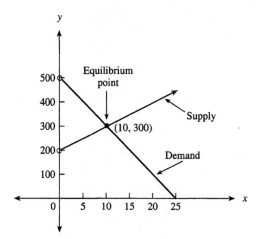

$$500 - 20x = 10x + 200$$

Note that the point of intersection is (10, 300), the equilibrium point. At this point the coordinates are the same for both lines.

$$300 = 30x$$

$$10 = x$$

At $10 per calculator, 300 calculators will be supplied and sold per week. This result, $10, represents the price at which supply equals demand; the demand will be met, and no surplus will exist.

EXERCISES 2.2

Copy and complete each table of values. Then graph the line given by the equation.

1. $y = x - 2$

x	−3	−2	−1	0	1	2
y						

2. $y = -x + 1$

x	−3	−2	−1	0	1	2
y						

3. $y = 2x - 4$

x	−2	−1	0	1	2
y					

4. $y = -2x + 3$

x	−2	−1	0	1	2
y					

Find the x- and y-intercepts and use these to graph each of the following lines.

5. $x + 2y = 4$ **6.** $2x + y = 4$ **7.** $x - 2y = 4$ **8.** $2x - y = 4$

9. $2x - 3y = 6$ **10.** $3x + y = 6$ **11.** $y = -3x - 9$ **12.** $y + 2x = -5$

13. $y = 2x - 1$

14. Sketch the following on the same set of axes.

  (a) $y = x$

  (b) By adding 1 to each y-value (ordinate) in part **(a)**, graph $y = x + 1$. In other words, *shift* each point of $y = x$ one unit up.

  (c) By subtracting 1 from each y-value in part **(a)**, graph $y = x - 1$. That is, shift each point of $y = x$ one unit down.

15. Repeat Exercise 14 for:

  (a) $y = -x$ **(b)** $y = -x + 1$ **(c)** $y = -x - 1$

16. Sketch the following on the same set of axes.

  (a) $y = x$

  (b) By multiplying each y-value in part **(a)** by 2, graph $y = 2x$. In other words, *stretch* each y-value of $y = x$ to twice its size.

  (c) Graph $y = 2x + 3$ by shifting $y = 2x$ three units upward.

Graph the points that satisfy each equation for the given values of x.

17. $y = \frac{1}{2}x;\ \ -6 \le x \le 6$ **18.** $y = -2x + 1;\ \ \ -2 \le x \le 2$

19. $y = 3x - 5;\ \ \ 1 \le x \le 4$ **20.** $y = -\frac{1}{2}x + 2;\ \ \ -2 \le x \le 2$

21. Use the coordinates of each of the following pairs of points to find the slope of ℓ.

  (a) A, C **(b)** B, D **(c)** C, D **(d)** A, E **(e)** B, E **(f)** C, E

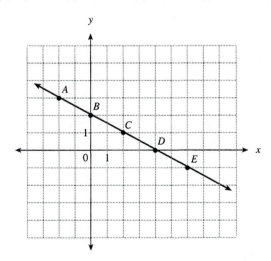

Compute the slope, if it exists, for the line determined by the given pair of points.

22. $(3, 4);\ (2, -5)$ **23.** $(4, 3);\ (-5, 2)$ **24.** $(-7, 6);\ (-7, 106)$

25. $(6, -7);\ (106, -7)$ **26.** $\left(-9, \frac{1}{2}\right);\ \left(2, \frac{1}{2}\right)$ **27.** $\left(2, -\frac{3}{4}\right);\ \left(-\frac{1}{3}, \frac{2}{3}\right)$

Draw the line through the given point having slope m.

28. $(-\frac{1}{2}, 0)$; $m = -1$ **29.** $(0, 2)$; $m = \frac{3}{4}$ **30.** $(3, -4)$; $m = 4$

31. $(-3, 4)$; $m = -\frac{1}{4}$ **32.** $(1, 1)$; $m = 2$ **33.** $\left(-2, \frac{3}{2}\right)$; $m = 0$

34. Graph each of the lines with the following slopes through the point $(5, -3)$:
$$m = -2 \qquad m = -1 \qquad m = 0 \qquad m = 1 \qquad m = 2$$

35. In the same coordinate system, draw the line:

  (a) Through $(1, 0)$ with $m = -1$ **(b)** Through $(0, 1)$ with $m = 1$

  (c) Through $(-1, 0)$ with $m = -1$ **(d)** Through $(0, -1)$ with $m = 1$

36. Line ℓ passes through $(-4, 5)$ and $(8, -2)$.

  (a) Draw the line through $(0, 0)$ perpendicular to ℓ.

  (b) What is the slope of any line perpendicular to line ℓ?

37. Why is the line determined by the points $(6, -5)$ and $(8, -8)$ parallel to the line through $(-3, 12)$ and $(1, 6)$?

38. Verify that the points $A(1, 2)$, $B(4, -1)$, $C(2, -2)$, and $D(-1, 1)$ are the vertices of a parallelogram. Sketch the figure.

39. Consider the four points $P(5, 11)$, $Q(-7, 16)$, $R(-12, 4)$, and $S(0, -1)$. Show that the four angles of the quadrilateral $PQRS$ are right angles. Also show that the diagonals are perpendicular.

40. Lines ℓ_1 and ℓ_2 are perpendicular and intersect at point $(-2, -6)$. ℓ_1 has slope $-\frac{2}{5}$. Use the slope of ℓ_2 to find the y-intercept of ℓ_2.

41. Any horizontal line is perpendicular to any vertical line. Why were such lines excluded from the result, which states that lines are perpendicular if and only if their slopes are negative reciprocals?

42. Find t if the line through $(-1, 1)$ and $(3, 2)$ is parallel to the line through $(0, 6)$ and $(-8, t)$.

43. Find t if the line through $(-1, 1)$ and $\left(1, \frac{1}{2}\right)$ is perpendicular to the line through $\left(1, \frac{1}{2}\right)$ and $(7, t)$. Use the fact that two perpendicular lines have slopes that are negative reciprocals of one another.

44. **(a)** Prove that nonvertical parallel lines have equal slopes by considering two parallel lines ℓ_1, ℓ_2 as in the figure. On ℓ_1 select points A and B, and choose A' and B' on ℓ_2. Now form the appropriate right triangles ABC and $A'B'C'$ using points C and C' on the x-axis. Prove they are similar and write a proportion to show the slopes of ℓ_1 and ℓ_2 are equal.

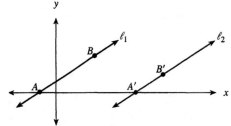

  (b) Why is the converse of the fact given in part **(a)** also true?

45. This exercise gives a proof that if two lines are perpendicular, then they have slopes that are negative reciprocals of one another.

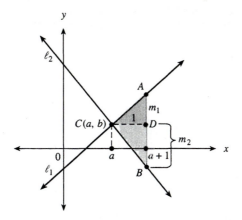

Let line ℓ_1 be perpendicular to line ℓ_2 at the point $C(a, b)$. Use m_1 for the slope of ℓ_1, and m_2 for ℓ_2. We want to show the following:

$$m_1 m_2 = -1 \quad \text{or} \quad m_1 = -\frac{1}{m_2}$$

Add 1 to the x-coordinate a of point C and draw the vertical line through $a + 1$ on the x-axis. This vertical line will meet ℓ_1 at some point A and ℓ_2 at some point B, forming right triangle ABC with right angle at C. Draw the perpendicular from C to AB meeting AB at D. Then CD has length 1.

(a) Using the right triangle CDA, show that $m_1 = DA$.

(b) Show that $m_2 = DB$. Is m_2 positive or negative?

(c) For right triangle ABC, CD is the mean proportional between segments BD and DA on the hypotenuse. Use this fact to conclude that $\dfrac{m_1}{1} = \dfrac{1}{-m_2}$, or $m_1 m_2 = -1$.

46. For a certain type of toy, the weekly demand, D, is related to the price x, in dollars, by the function $y = D = 1000 - 40x; \ 0 < x \le 25$. The weekly supply is given by the function $y = S = 15x + 340$. Find the price at which the supply equals the demand. Graph both equations on the same set of axes and show the equilibrium point.

47. The monthly demand, D, for a certain portable radio is related to the price x by the function $D = 5000 - 25x; \ 0 < x \le 200$. The monthly supply is given by the function $S = 10x + 275$. At what price will the supply equal the demand? How many radios will be sold at this price?

48. Express the area of rectangle $PQRS$ as a function of $x = OP$.

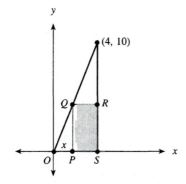

49. Express the area of triangle OPQ in Exercise 48 as a function of x.

50. A window is in the shape of a rectangle with a semicircular top as shown. The perimeter of the window is 15 feet. Use r as the radius of the semicircle and express the area of the window as a function of r.

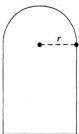

(*Hint:* The length of the semicircle is $\frac{1}{2}(2\pi r) = \pi r$. Now subtract this from 15 to get $15 - \pi r$ as the total length of the three sides. Also note that the base of the rectangle has length $2r$.)

51. A 50-inch piece of wire is to be cut into two parts AP and PB as shown. If part AP is to be used to form a square and PB is used to form a circle, express the total area enclosed by these figures as a function of x.

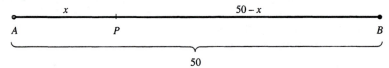

(*Hint:* Since the perimeter of the square is x, one of its sides is $\frac{x}{4}$. Also the circumference of the circle is $50 - x$, which can be used to find the radius of the circle.)

Graphing Calculator Exercises   For each of Exercises 1–6, use a *standard window* (RANGE: $-10 \le x \le 10$ and $-10 \le y \le 10$) unless noted otherwise. Check that your MODE setting graphs functions *sequentially* so that the first function you list is graphed first, the second next, and so on.

1. Graph all of the following equations of straight lines on the same set of axes, and describe their similarities and their differences in slopes. How does the equation seem to be related to the slope of the line?
   (a) $y = x$ **(b)** $y = 2x$ **(c)** $y = 3x$ **(d)** $y = 4x$

2. Graph all of the following equations of straight lines on the same set of axes and describe their similarities and differences in slopes. How do you think each equation is related to the slope of the line?
   (a) $y = -x + 1$ **(b)** $y = -2x + 1$ **(c)** $y = -3x + 1$ **(d)** $y = -4x + 1$

3. Graph each of the following equations of straight lines on the same set of axes, as in Exercises 1 and 2, and describe their difference in y-intercept (which is the point where the graph crosses the y-axis). How is the equation related to this point?
   (a) $y = -2x$ **(b)** $y = -2x + 3$ **(c)** $y = -2x + 6$ **(d)** $y = -2x - 3$

4. How are the slopes of the lines in Exercise 3 related? How does this affect the positions of the lines in your graphs? (See Exercise 44.)

5. Graph the equation $y = 12 - x$. Change range settings so that the graph shows the x- and y-intercepts. Try to adjust the range settings and scale markings so that you can read these intercepts. Check by substituting the coordinates of the intercepts in the equation.

6. **(a)** Graph the equations of straight lines $y = 2x$ and $y = -.5x$ on the same set of axes. According to Exercise 45, they should be perpendicular to one another.
   (b) Do they look perpendicular?   **(c)** If not, why not?
   (d) If necessary, try to adjust the range settings so that they will look perpendicular.

2.3 ALGEBRAIC FORMS OF LINEAR FUNCTIONS

Pictured below is a line with slope m and y-intercept b. To find the equation of ℓ we begin by considering any point $P(x, y)$ on ℓ other than $(0, b)$.

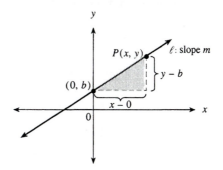

Since the slope of ℓ is given by any two of its points, we may use $(0, b)$ and (x, y) to write

$$m = \frac{y - b}{x - 0} = \frac{y - b}{x}$$

If both sides are multiplied by x, then

$$y - b = mx \quad \text{or} \quad y = mx + b$$

*$y = mx + b$ is sometimes referred to as the **y-form of the equation of a line**.*

SLOPE-INTERCEPT FORM OF A LINE

$$y = mx + b$$

where m is the slope and b is the y-intercept.

A point (x, y) is on this line if and only if the coordinates satisfy this equation.

The equation $y = mx + b$ also defines a function; thus we consider $y = f(x) = mx + b$ as a *linear function* with the domain consisting of all the real numbers.

EXAMPLE 1 Graph the linear function f defined by $y = f(x) = 2x - 1$ by using the slope and y-intercept. Display $f(2) = 3$ geometrically; that is, show the point P corresponding to $f(2) = 3$.

Solution The y-intercept is -1. Locate $(0, -1)$ and use $m = 2 = \frac{2}{1}$ to reach $(1, 1)$, another point on the line. At point P, $x = 2$ and $f(2) = 3$.

Both the domain and range of f consist of all real numbers.

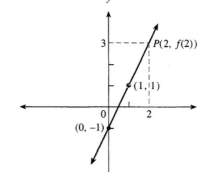

■

EXAMPLE 2   Write, in slope-intercept form, the equation of the line with slope $m = \frac{2}{3}$ passing through the point $(0, -5)$.

Solution   Since $m = \frac{2}{3}$ and $b = -5$, the slope-intercept form $y = mx + b$ gives

$$y = \frac{2}{3}x + (-5)$$

$$= \frac{2}{3}x - 5 \qquad\blacksquare$$

A special case of $y = f(x) = mx + b$ is obtained when $m = 0$. Then

$$y = 0(x) + b \quad \text{or} \quad y = b$$

This says that for *each* input x, the output $f(x)$ is always the same value b.

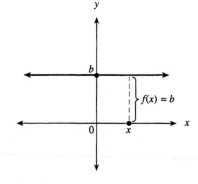

Domain: all reals
Range: the single
         value b

Since $f(x) = b$ is constant for all x, this linear function is also referred to as a **constant function.** Its graph is a horizontal line.

EXAMPLE 3   Write and graph the equation of the line through the point $(2, 3)$ that is:
(a) Parallel to the x-axis   **(b)** Parallel to the y-axis

Solution
(a) The y-intercept is 3 and the slope is 0. Thus the equation is

$$y = 0(x) + 3 \text{ or } y = 3.$$

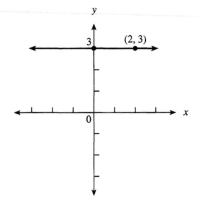

Algebraically, a horizontal line through the point (h, k) has the form $y = k$ for all x.

Can a vertical line be the graph of a function where y depends on x? Explain.

(b) The line has no y-intercept. Also, the slope is undefined. From the figure we note that y can be any value but that x is always 2. Thus the equation of the line is $x = 2$.

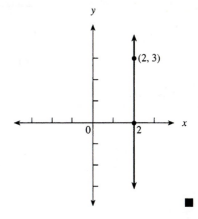

Algebraically, a vertical line through the point (h, k) has the form $x = h$ for all y.   ■

Now let ℓ be a line with slope m that passes through a specific point $Q(x_1, y_1)$. We wish to determine the conditions on the coordinates of any other point $P(x, y)$ that is on the line ℓ.

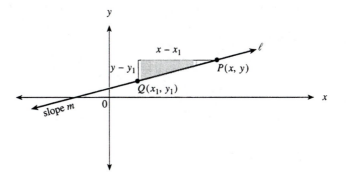

From the figure you can see that $P(x, y)$ will be on ℓ if and only if the ratio $\dfrac{y - y_1}{x - x_1}$ is the same as m. That is, P is on ℓ if and only if

$$m = \frac{y - y_1}{x - x_1}$$

Multiply both sides of this equation by $x - x_1$ to obtain

$$m(x - x_1) = y - y_1$$

which also holds for the coordinates of Q. This leads to another form for the equation of a straight line.

This is the form of a line that is most frequently used in calculus. A point (x, y) is on this line if and only if the coordinates satisfy this equation.

POINT-SLOPE FORM OF A LINE

$$y - y_1 = m(x - x_1)$$

where m is the slope and (x_1, y_1) is a point on the line.

EXAMPLE 4 Write the point-slope form of the line ℓ with slope $m = 3$ that passes through the point $(-1, 1)$. Verify that $(-2, -2)$ is on the line.

Solution Since $m = 3$, any (x, y) on ℓ satisfies this equation:

$$y - 1 = 3[x - (-1)]$$
$$y - 1 = 3(x + 1)$$

CAUTION
Pay attention to the minus signs on the coordinates when used in the point-slope form. Note the substitution of $x_1 = -1$ in this example.

Let $x = -2$:

$$y - 1 = 3(-2 + 1) = -3$$
$$y = -2$$

Thus $(-2, -2)$ is on the line. ∎

TEST YOUR UNDERSTANDING

Write the slope-intercept form of the line with the given slope and y-intercept.

1. $m = 2;\ b = -2$ **2.** $m = -\frac{1}{2};\ b = 0$ **3.** $m = \sqrt{2};\ b = 1$

Write the point-slope form of the line through the given point with slope m.

4. $(2, 6);\ m = -3$ **5.** $(-1, 4);\ m = \frac{1}{2}$ **6.** $\left(5, -\frac{2}{3}\right);\ m = 1$

7. $(0, 0);\ m = -\frac{1}{4}$ **8.** $(-3, -5);\ m = 0$ **9.** $(1, -1);\ m = -1$

(Answers: Page 172)

10. Which, if any, of the preceding produce a constant linear function? State its range.

EXAMPLE 5 Write the slope-intercept form of the line through the two points $(6, -4)$ and $(-3, 8)$.

Solution First compute the slope.

$$m = \frac{-4 - 8}{6 - (-3)} = \frac{-12}{9} = -\frac{4}{3}$$

Use either point to write the point-slope form, and then convert to the slope-intercept form. Thus, using the point $(6, -4)$,

$$y - (-4) = -\frac{4}{3}(x - 6)$$

Verify that the point $(9, -7)$ is not on this line by showing that the coordinates do not satisfy the equation.

$$y + 4 = -\frac{4}{3}x + 8$$

$$y = -\frac{4}{3}x + 4$$

Show that the same final form is obtained using the point $(-3, 8)$. ∎

EXAMPLE 6   Write an equation of the line that is perpendicular to the line $5x - 2y = 2$ and that passes through the point $(-2, -6)$.

Solution   First find the slope of the given line by writing it in slope-intercept form.

$$5x - 2y = 2$$

$$-2y = -5x + 2$$

$$y = \frac{5}{2}x - 1 \qquad \text{The slope is } \frac{5}{2}.$$

Recall that two perpendicular lines have slopes that are negative reciprocals of one another.

The perpendicular line has slope $= -\frac{2}{5}$. Since this line also goes through $P(-2, -6)$, the point-slope form gives

$$y + 6 = -\frac{2}{5}(x + 2) \qquad \text{or} \qquad y = -\frac{2}{5}x - \frac{34}{5} \qquad \blacksquare$$

A linear equation such as $y = -\frac{2}{3}x + 4$ can be converted to other equivalent forms. In particular, when this equation is multiplied by 3, we obtain the form $2x + 3y = 12$. This is an illustration of the *general linear equation*.

GENERAL LINEAR EQUATION

$$Ax + By = C$$

where A, B, C are constants and A, B are not both 0.

The general linear equation $Ax + By = C$ is said to define y as a function of x *implicitly* if $B \neq 0$. In other words, we have these equivalent forms:

Note here that the slope is $-\dfrac{A}{B}$ and the y-intercept is $\dfrac{C}{B}$.

$$Ax + By = C \qquad \longleftarrow \text{\textit{implicit} form of the linear function}$$

$$By = -Ax + C$$

$$y = -\frac{A}{B}x + \frac{C}{B} \qquad \longleftarrow \text{\textit{explicit} form of the linear function; slope-intercept form}$$

As an illustration, we can find the slope and y-intercept of the line with equation $-6x + 2y - 5 = 0$ by converting to the explicit form $y = f(x)$:

$$-6x + 2y - 5 = 0$$

$$2y = 6x + 5$$

$$y = 3x + \tfrac{5}{2} \qquad \longleftarrow \text{slope-intercept form}$$

Thus $m = 3$ and $b = \frac{5}{2}$.

EXAMPLE 7 Find the equation of the line through the points $(2, -3)$ and $(3, -1)$. Write the equation:

(a) In point-slope form
(b) In slope-intercept form
(c) In the general form

Solution

(a) The slope m is $\dfrac{-1 - (-3)}{3 - 2} = 2$. Use this slope and either point, such as $(2, -3)$.

$$y - y_1 = m(x - x_1): \qquad y - (-3) = 2(x - 2) \qquad \text{point-slope form}$$
$$y + 3 = 2(x - 2)$$

(b) Use the solution for part (a) and solve for y.

$$y = mx + b: \qquad y + 3 = 2(x - 2)$$
$$y + 3 = 2x - 4$$
$$y = 2x - 7 \qquad \text{slope-intercept form}$$

(c) Rewrite the solution for part (b).

$$Ax + By = C: \qquad y = 2x - 7$$
$$-2x + y = -7$$
$$\text{or} \qquad 2x - y = 7$$

Note that the form $Ax + By = C$ is not unique. Thus, by multiplying by 2 we obtain the equivalent form $4x - 2y = 14$.

Note that all three forms are different ways of expressing the same equation for the given line through $(2, -3)$ and $(3, -1)$. Show that $x = 3$ and $y = -1$ satisfies each form. ∎

SUMMARY OF THE ALGEBRAIC FORMS OF A LINE		
Slope-Intercept Form	**Point-Slope Form**	**General Linear Equation**
$y = mx + b$	$y - y_1 = m(x - x_1)$	$Ax + By = C$
Line with slope m and y-intercept b.	Line with slope m and through the point (x_1, y_1).	Line with slope $-\dfrac{A}{B}$ and y-intercept $\dfrac{C}{B}$, if $B \neq 0$.

For the general linear equation $Ax + By = C$, note the following:

If $B = 0$ and $A \neq 0$, then $x = \dfrac{C}{A}$, the equation of a vertical line.

If $A = 0$ and $B \neq 0$, then $y = \dfrac{C}{B}$, the equation of a horizontal line.

CAUTION: Learn to Avoid These Mistakes	
WRONG	**RIGHT**
The slope of the line through $(2, 3)$ and $(5, 7)$ is $$m = \frac{7 - 3}{2 - 5} = -\frac{4}{3}$$	The slope of the line through $(2, 3)$ and $(5, 7)$ is $$m = \frac{7 - 3}{5 - 2} = \frac{4}{3}$$
The slope of the line $2x - 3y = 7$ is $-\frac{2}{3}$.	The slope is $\frac{2}{3}$.
The line through $(-4, -3)$ with slope 2 is $y - 3 = 2(x - 4)$.	The equation is $y - (-3) = 2[x - (-4)]$ or $y + 3 = 2(x + 4)$.

EXERCISES 2.3

Write the equation of the line with the given slope m and y-intercept b.

1. $m = 2$, $b = 3$ **2.** $m = -2$, $b = 1$ **3.** $m = 1$, $b = 1$ **4.** $m = -1$, $b = 2$

5. $m = 0$, $b = 5$ **6.** $m = 0$, $b = -5$ **7.** $m = \frac{1}{2}$, $b = 3$ **8.** $m = -\frac{1}{2}$, $b = 2$

9. $m = \frac{1}{4}$, $b = -2$

Write the point-slope form of the line through the given point with the indicated slope.

10. $(3, 4)$; $m = 2$ **11.** $(2, 3)$; $m = 1$ **12.** $(1, -2)$; $m = 0$

13. $(-2, 3)$; $m = 4$ **14.** $(-3, 5)$; $m = -2$ **15.** $(-3, 5)$; $m = 0$

16. $(8, 0)$; $m = -\frac{2}{3}$ **17.** $(2, 1)$; $m = \frac{1}{2}$ **18.** $(-6, -3)$; $m = \frac{4}{3}$

19. $(0, 0)$; $m = 5$ **20.** $\left(-\frac{3}{4}, \frac{2}{5}\right)$; $m = 1$ **21.** $(\sqrt{2}, -\sqrt{2})$; $m = 1$

22. **(a)** Find the slope of the line determined by the points $A(-3, 5)$ and $B(1, 7)$, and write its equation in point-slope form, using the coordinates of A.

     (b) Do the same in part **(a)** using the coordinates of B.

     (c) Verify that the equations obtained in parts **(a)** and **(b)** give the same slope-intercept form.

Write each equation in slope-intercept form; give the slope and y-intercept.

23. $3x + y = 4$ **24.** $2x - y = 5$ **25.** $6x - 3y = 1$ **26.** $4x + 2y = 10$

27. $3y - 5 = 0$ **28.** $x = \frac{3}{2}y + 3$ **29.** $4x - 3y - 7 = 0$ **30.** $5x - 2y + 10 = 0$

31. $\frac{1}{4}x - \frac{1}{2}y = 1$

Write the equation of the line through the two given points in the form Ax + By = C.

32. $(-1, 2)$, $(2, -1)$ **33.** $(2, 3)$, $(3, 2)$ **34.** $(1, 1)$, $(-1, -1)$ **35.** $(3, 0)$, $(0, -3)$

36. $(3, -4)$, $(0, 0)$ **37.** $(-1, -13)$, $(-8, 1)$ **38.** $\left(\frac{1}{2}, 7\right), \left(-4, -\frac{3}{2}\right)$ **39.** $(10, 27)$, $(12, 27)$

40. $(\sqrt{2}, 4\sqrt{2})$, $(-3\sqrt{2}, -10\sqrt{2})$

41. Two lines, parallel to the coordinate axes, intersect at the point $(5, -7)$. What are their equations?

42. Write the equation of the line that is parallel to $y = -3x - 6$ and with y-intercept 6.

43. Write the equation of the line parallel to $2x + 3y = 6$ that passes through the point $(1, -1)$.

Write the equation in slope-intercept form of the line that is perpendicular to the given line and passes through the indicated point.

44. $y = -10x$; $(0, 0)$ **45.** $y = 3x - 1$; $(4, 7)$ **46.** $3x + 2y = 6$; $(6, 7)$ **47.** $y - 2x = 5$; $(-5, 1)$

48. The vertices of a triangle are located at $(-1, -1)$, $(1, 3)$, and $(4, 2)$. Write the equations for the sides of the triangle.

49. In Exercise 48, write the equations of the three altitudes of the triangle.

50. The vertices of a rectangle are located at $(2, 2)$, $(6, 2)$, $(6, -3)$, and $(2, -3)$. What is the relationship between the slopes of the diagonals?

51. The vertices of a square are located at $(2, 2)$, $(5, 2)$, $(5, -1)$, and $(2, -1)$. What is the relationship between the slopes of the diagonals?

Graph the linear function f by using the slope and y-intercept. Display the point corresponding to $y = f(-2)$ on the graph.

52. $y = f(x) = -2x + 1$ **53.** $y = f(x) = x + 3$ **54.** $y = f(x) = 3x - \frac{1}{2}$ **55.** $y = f(x) = \frac{1}{2}x - 3$

56. The sides of the parallelogram with vertices $(-1, 1)$, $(0, 3)$, $(2, 1)$, and $(3, 3)$ are the graphs of four different functions. In each case find the equation that defines the function and state the domain.

57. Any line having a nonzero slope that does not pass through the origin always has both an x- and y-intercept. Let ℓ be such a line having equation $ax + by = c$.

(a) Why is $c \neq 0$?

(b) What are the x- and y-intercepts?

(c) Derive the equation $\dfrac{x}{q} + \dfrac{y}{p} = 1$ where q and p are the x- and y-intercepts, respectively. This is known as the *intercept form* of a line.

(d) Use the intercept form to write the equation of the line passing through $\left(\frac{3}{2}, 0\right)$ and $(0, -5)$.

(e) Use the two points in part (d) to find the slope, write the slope-intercept form, and compare with the result in part (d).

58. Replace m in the point-slope form of a line through points (x_1, y_1) and (x_2, y_2) by $\dfrac{y_2 - y_1}{x_2 - x_1}$. Show that this gives the *two-point form* for the equation of a line:

$$\frac{y - y_1}{y_2 - y_1} = \frac{x - x_1}{x_2 - x_1}$$

Use this result to find the equation of the line through $(-2, 3)$ and $(5, -2)$. Write the equation in point-slope form.

Written Assignment Select a point and a nonzero slope of your choice. Then describe the various algebraic forms that can be used to identify the line through your given point and with your given slope.

Graphing Calculator Exercises In order to graph straight lines on your graphing calculator, you may have to change the form of the equation into its y-form, that is, to slope-intercept form. For example, the point-slope form $y - 1 = 2(x + 3)$ must be changed to the form $y = 2x + 7$. Use a standard window ($-10 \le x \le 10$, $-10 \le y \le 10$) for Exercises 1–5.

Change to the slope-intercept form and graph the resulting equation.

1. $y + 1 = 3(x - 4)$ **2.** $y - 3 = 2(x + 1)$

3. $3y + 6x = 9$ **4.** $2y - 4x = 12$

5. Graph the line passing through the points $(0, 2)$ and $(5, 0)$.

   Your graphing calculator has a TRACE facility that lets you move along the graph. The coordinates of points on the curve are shown at the bottom of the graphics screen. (Consult your manual for instructions appropriate for your calculator.) However, if you just use a standard window ($-10 \le x \le 10$, $-10 \le y \le 10$), you may not be able to TRACE many points on the curve having integer or even decimal coefficients to one or two places.

6. Using the standard window described previously and your TRACE facility, try to locate the point $(4, -1)$ on the graph of Exercise 1. How close do you get?

7. Using the standard window described previously and your TRACE facility, try to locate both points $(0, 2)$ and $(5, 0)$ on the graph of Exercise 5. How close do you get?

   You can TRACE points on your graphs having integer coefficients if you use a *friendly window*. For example, $-4.8 \le x \le 4.7$, $-3.2 \le y \le 3.1$, or integer multiples of these values, is "friendly" on the TI-81. The *default window* $-4.7 \le x \le 4.7$, $-3.1 \le y \le 3.1$ is friendly on the Casio fx-7700G. (Check your user's manual, and see the Graphing Calculator Appendix.)

8. Repeat Exercise 6, this time using a friendly window.

9. Repeat Exercise 7, this time using a friendly window.

Critical Thinking

1. Suppose $f(x) = 2x + 3$. Find a meaning for the expression $f(f(x))$.

2. How can you prove that a point $P(x, y)$ is *not* located on a line with a given equation?

3. Consider the line defined by the equation $y = mx + b$. Under what conditions on m and b will the line pass through quadrants I, III, and IV?

4. Consider the graphs of the two equations $y = ax + c$ and $y = bx + d$. Under what conditions are the x-intercepts the same? Under what conditions will the two lines intersect?

**2.4
PIECEWISE
LINEAR
FUNCTIONS**

When a function is defined by an equation, such as $y = f(x) = 2x - 1$ it is assumed that the domain is the largest set of real numbers for which the expression $2x - 1$ is a real number. In this case it is the set of real numbers. However, there are times when we may wish to restrict the allowable values of x. For example, the graph of

$y = 2x - 1$ for $-2 \le x \le 1$ is a *line segment* with endpoints at $(-2, -5)$ and $(1, 1)$, as in the following figure.

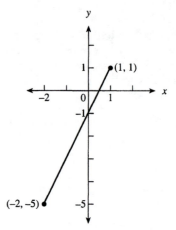

$$y = 2x - 1$$

Domain: $-2 \le x \le 1$

Range: $\quad -5 \le y \le 1$

We can describe this function as a piece or part of the function $y = 2x - 1$ restricted to $-2 \le x \le 1$.

As the preceding suggests, linear functions can be used to define other functions which, in themselves, are not linear but may be described as being "partly" or "piecewise" linear. An important example is the **absolute-value function.**

$$y = f(x) = |x| = \begin{cases} x & \text{if } x \ge 0 \\ -x & \text{if } x < 0 \end{cases}$$

See Section 1.4 to review the meaning of absolute value.

To graph this function, first draw the line $y = -x$ and eliminate all those points on it for which x is positive. Then draw the line $y = x$ and eliminate the part for which x is negative. Now join these two parts to get the graph of $y = |x|$:

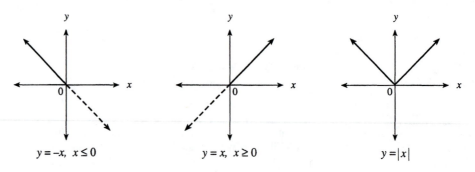

$y = -x, \; x \le 0$ $\qquad\qquad$ $y = x, \; x \ge 0$ $\qquad\qquad$ $y = |x|$

The graph of $y = |x|$ consists of two perpendicular rays intersecting at the origin. Now $y = |x|$ is not a linear function, but it is linear in parts; the two halves $y = x$, $y = -x$ are linear.

Note that the graph is symmetric about the y-axis. (If the paper were folded along the y-axis, the two parts would coincide.) This symmetry can be observed by noting that the y-values for x and $-x$ are the same. That is,

$$|x| = |-x| \qquad \text{for all } x$$

EXAMPLE 1   Graph: $y = |x - 2|$

Solution For $x \geq 2$, we find that $x - 2 \geq 0$, which implies that $y = |x - 2| = x - 2$. This is the ray through (and to the right of) the point $(2, 0)$, with slope 1. For $x < 2$, we get $x - 2 < 0$, which implies that $y = |x - 2| = -(x - 2) = -x + 2$. This gives the left half of the graph shown.

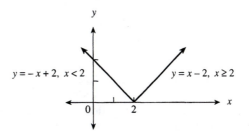

See Exercise 39 for an alternative way to draw this graph.

Just as $y = f(x) = |x|$ was defined in two parts, so can other functions be defined in several parts, as in Example 2.

EXAMPLE 2   Graph: $y = f(x) = \begin{cases} 2x & \text{if } 0 \leq x \leq 1 \\ -x + 2 & \text{if } 1 < x \end{cases}$

What are the domain and range of f?

Solution The domain of f is all $x \geq 0$. From the graph we see that the range consists of all $y \leq 2$.

Note: The open dot at $(1, 1)$ means that this point is not part of the graph; but the point $(1, 2)$ is on the graph since $f(1) = 2 \cdot 1 = 2$.

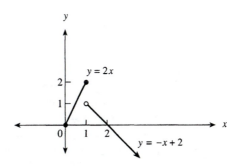

EXAMPLE 3   Graph the function f given by $y = f(x) = \dfrac{|x|}{x}$. What is the domain of f? What is the range of f?

Solution The domain consists of all $x \neq 0$. When $x > 0$, $|x| = x$ and

$$f(x) = \frac{|x|}{x} = \frac{x}{x} = 1$$

Thus $f(x)$ is the constant 1, for all positive x. Similarly, for $x < 0$, $|x| = -x$ and

$$f(x) = \frac{|x|}{x} = \frac{-x}{x} = -1.$$

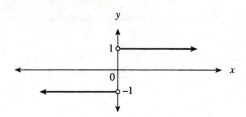

Since -1 and 1 are the only y-values possible, the range consists of just these two numbers. ∎

Example 3 is an illustration of a **step function.** Such a function may be described as a function whose graph consists of parts of horizontal lines. Here is another step function defined for $-2 \leq x < 4$. Note that each step is the graph of one of the six equations used to define f.

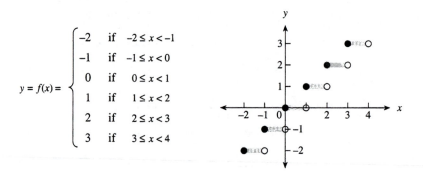

$$y = f(x) = \begin{cases} -2 & \text{if} \quad -2 \leq x < -1 \\ -1 & \text{if} \quad -1 \leq x < 0 \\ 0 & \text{if} \quad 0 \leq x < 1 \\ 1 & \text{if} \quad 1 \leq x < 2 \\ 2 & \text{if} \quad 2 \leq x < 3 \\ 3 & \text{if} \quad 3 \leq x < 4 \end{cases}$$

Observe that in each case, say $2 \leq x < 3$, the integer 2 at the left of the inequality is also the corresponding y-value for each x within this inequality. Putting it another way, we say that the y-value 2 is *the greatest integer less than or equal to x.*

For each number x there is an integer n such that $n \leq x < n + 1$. Therefore, the greatest integer less than or equal to x equals n. In other words, the preceding step function may be extended to a step function with domain of *all* real numbers; its graph would consist of an infinite number of steps.

[x] is the greatest integer not exceeding x itself.

We use the symbol **[x]** to mean *the greatest integer less than or equal to x.* Thus, the **greatest integer function** is given by

$$y = [x] = \text{greatest integer} \leq x$$

Here are a few illustrations:

$$[2\tfrac{1}{2}] = \quad 2 \text{ because 2 is the greatest integer} \leq 2\tfrac{1}{2}$$

$$[0.64] = \quad 0 \text{ because 0 is the greatest integer} \leq 0.64$$

$$[-2.8] = -3 \text{ because } -3 \text{ is the greatest integer} \leq -2.8$$

$$[-5] = -5 \text{ because } -5 \text{ is the greatest integer} \leq -5$$

TEST YOUR UNDERSTANDING	*Evaluate:*			
	1. [12.3]	**2.** [12.5]	**3.** [12.9]	**4.** [13]
	5. $\left[-3\frac{3}{4}\right]$	**6.** $\left[-3\frac{1}{4}\right]$	**7.** [−3]	**8.** [0]
(Answers: Page 172)	**9.** [−0.25]	**10.** [0.25]	**11.** [−0.75]	**12.** [0.75]

Postage charges for mailing packages depend on both weight and destination, and this leads to an application of step functions. For example, the rates for a certain destination shown in the table at the left defines a step function whose graph is at the right.

x = weight (pounds)	y = postage (cost)
Under 1	$0.80
1 or more but under 2	$0.90
2 or more but under 3	$1.00
3 or more but under 4	$1.10
4 or more but under 5	$1.20
5 or more but under 6	$1.30
6 or more but under 7	$1.40
7 or more but under 8	$1.50
8 or more but under 9	$1.60
9 or more but under 10	$1.70

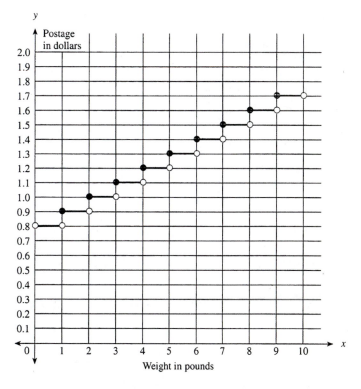

This table defines y to be a function of x. If P is used for postage, then the equation $y = P(x)$ gives the cost C in dollars for mailing x pounds. For example, reading from the table or from the graph, we find that $P = 1.20$ for $4 \leq x < 5$. Can you find a formula for $P(x)$? (See Exercise 41.)

EXERCISES 2.4

Evaluate.

1. [99.1] **2.** [100.1] **3.** [−99.1] **4.** [−100.1]

5. $\left[-\frac{7}{2}\right]$ **6.** $[10^3]$ **7.** $\left[\sqrt{2}\right]$ **8.** $[\pi]$

Graph each function and state the domain and range.

9. $y = |x - 1|$ **10.** $y = |x + 3|$ **11.** $y = |2x|$

12. $y = |2x - 1|$ **13.** $y = |3 - 2x|$ **14.** $y = \left|\frac{1}{2}x + 4\right|$

15. $y = \begin{cases} 3x & \text{if } -1 \le x \le 1 \\ -x & \text{if } 1 < x \end{cases}$ **16.** $y = \begin{cases} -2x + 3 & \text{if } x < 2 \\ x + 1 & \text{if } x > 2 \end{cases}$

17. $y = \begin{cases} x & \text{if } -2 < x \le 0 \\ 2x & \text{if } 0 < x \le 2 \\ -x + 3 & \text{if } 2 < x \le 3 \end{cases}$ **18.** $y = \begin{cases} x & \text{if } 0 \le x < 1 \\ x-1 & \text{if } 1 \le x < 2 \\ x-2 & \text{if } 2 \le x < 3 \\ x-3 & \text{if } 3 \le x \le 4 \end{cases}$

Graph each step function for its given domain.

19. $y = \dfrac{x}{|x|}$; all $x \ne 0$ **20.** $y = \dfrac{|x - 2|}{x - 2}$; all $x \ne 2$ **21.** $y = \dfrac{x + 3}{|x + 3|}$; all $x \ne -3$

22. $y = \begin{cases} -1 & \text{if } -1 \le x \le 0 \\ 0 & \text{if } 0 < x \le 1 \\ 1 & \text{if } 1 < x \le 2 \\ 2 & \text{if } 2 < x \le 3 \end{cases}$ **23.** $y = [x]$; $-3 \le x \le 3$ **24.** $y = [2x]$; $0 \le x \le 2$

25. $y = 2[x]$; $0 \le x \le 2$ **26.** $y = [3x]$; $-1 \le x \le 1$ **27.** $y = [x - 1]$; $-2 \le x \le 3$

The domain is given for the function defined by the equation $y = 3x - 7$. Graph each function.

28. All $x \le 4$.   **29.** All $x \ge 0$.   **30.** All x where $-1 \le x \le 3$.   **31.** $x = -1, 0, 1, 2, 3$.

Write the equation for each graph. State the domain and range if it is the graph of a function.

32.

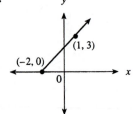

33.

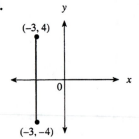

34.

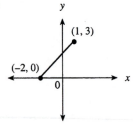

35.

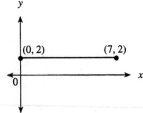

36.

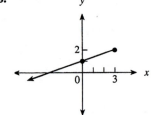

37.

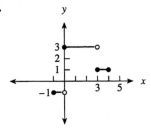

38. (a) The following information shows the first class letter rates in a recent year for pieces not exceeding 6 ounces.

For Pieces Not Exceeding x Ounces	The Rate $R(x)$ Is
1	0.29
2	0.52
3	0.75
4	0.98
5	1.21
6	1.44

Draw the graph of $y = R(x)$.

(b) Find a formula for the rate $R(x)$ that makes use of the greatest integer function. (*Hint:* For $0 < x \leq 6$, consider $[-x]$.)

39. The graph of $y = |x - 2|$ was found in Example 1. Here is an alternative procedure. First graph $y = x - 2$ using a dashed line for the part below the x-axis. Now reflect the negative part (the dashed part) through the x-axis to get the final graph of $y = |x - 2|$.

40. Follow the procedure in Exercise 39 to graph $y = |2x + 5|$.

41. A formula for $P(x)$ on page 119 can be given in terms of the greatest integer function. Find such a formula. (*Hint:* Note for all x, where $3 \leq x < 4$, $P(x) = 1.1 = 0.8 + 0.3 = \frac{1}{10}(8 + 3)$.)

Challenge

1. Graph $y = |[x]|$ and $y = [|x|]$. For what values of x are the two equal to each other?

2. Graph: $y = |x| + [x]$

Written Assignment

1. Describe at least one example, other than the postage function, of a step function that can be found in everyday life. Explain your answer.

2. Does the step function of a sum equal the sum of the step functions? That is, for real numbers a and b, does $[a + b] = [a] + [b]$? Explain your answer with specific examples.

Graphing Calculator Exercises   Your graphing calculator may be able to graph piecewise-defined functions if it allows the use of *relational operators* in the definitions. For example, to graph

$$f(x) = \begin{cases} x & \text{if } x < 0 \\ 3x & \text{if } 0 \leq x < 5 \end{cases}$$

graph the function written like this:

$$(x) * (x < 0) + (3x) * (x \geq 0) * (x < 5)$$

in DOT, or disconnected, mode. (What looks like multiplication works like the word *and*, and what looks like addition works like the word *or*. Look up the use of logical operators or inequality graphs in your user's manual.) Using this technique, graph the piecewise-defined linear functions in Exercises 1–3.

1. $f(x) = \begin{cases} x & \text{if } x < 0 \\ 3x & \text{if } 0 \le x < 2 \end{cases}$

2. $f(x) = \begin{cases} x & \text{if } x < 0 \\ x + 1 & \text{if } x \ge 0 \end{cases}$

3. $f(x) = \begin{cases} x & \text{if } x < 0 \\ 3x & \text{if } 0 \le x < 2 \\ 7 & \text{if } x \ge 2 \end{cases}$

4. What happens if you fail to use DOT mode in Exercise 3?

The technique just described can also be applied to *graphing one-dimensional inequalities*. For example, to graph the set of all x that satisfy $-3 < 2x + 1 \le 4$ we can graph the function

$$f(x) = (1) * (2x + 1 > -3) * (2x + 1 \le 4)$$

on a friendly window such as $-4.8 \le x \le 4.7$, $-3.2 \le y \le 3.1$ on the TI-81 or the default settings on the Casio fx-7700G. The result is the line segment of $y = 1$ such that $-2 < x \le 1.5$. You can check with your TRACE facility, as shown in the following figure. The corresponding x interval on the x-axis solves the given inequality.

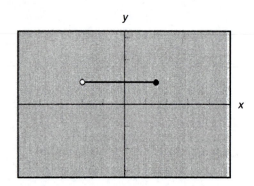

5. Solve the following inequalities by graphing a suitable function with relational operators:
(a) $2x + 1 > 0$ (b) $3x - 2 \le 1$ (c) $2 < 4x + 5 \le 6$

2.5
INTRODUCTION
TO SYSTEMS
OF EQUATIONS

Any two nonparallel lines in a plane intersect in exactly one point. Our objective here is to find the coordinates of this point using the equations of the lines. Here, for example, is a *system* of two linear equations in two variables and their graphs, drawn in the same coordinate system.

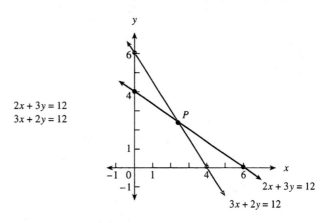

The coordinates of the point of intersection P could be estimated by careful inspection of the graph. However, to find the exact coordinates requires algebraic procedures. These coordinates are called the solution of the system of equations.

Two procedures will now be described to solve such systems. The underlying idea for each of them is that *the coordinates of the point of intersection must satisfy each equation.* The **substitution method** will be taken up first.

EXAMPLE 1 Solve the system by substitution.

$$2x + 3y = 12$$
$$3x + 2y = 12$$

Solution We begin by letting (x, y) be the coordinates of the point of intersection of the preceding system. Then these x- and y-values fit both equations. Hence either equation may be solved for x or y and then substituted into the other equation. For example, solve the second equation for y:

$$y = -\frac{3}{2}x + 6$$

Substitute this expression into the first equation and solve for x.

$$2x + 3y = 12$$
$$2x + 3\left(-\frac{3}{2}x + 6\right) = 12$$
$$2x - \frac{9}{2}x + 18 = 12$$
$$-\frac{5}{2}x = -6$$
$$x = \left(-\frac{2}{5}\right)(-6) = \frac{12}{5}$$

To find the y-value, substitute $x = \frac{12}{5}$ into either of the given equations. It is easiest to use the equation that was written in y-form:

Check the ordered pair by substituting these values into each of the given equations for this system. Recall that we really cannot call this a solution until it has been checked.

$$y = -\frac{3}{2}x + 6$$

$$y = -\frac{3}{2}\left(\frac{12}{5}\right) + 6 = \frac{12}{5}$$

The solution is $\left(\frac{12}{5}, \frac{12}{5}\right)$. ∎

Now we will solve the same system as before, this time using the **multiplication-addition** (or **multiplication-subtraction**) **method**. The idea here is to alter the equations so that the coefficients of one of the variables are either negatives of one another or equal to each other. We may, for example, multiply the first equation by 3 and multiply the second by -2:

Note that all these methods have as their objective the elimination of one variable.

Original System	Resulting System
$2x + 3y = 12$	$6x + 9y = 36$
$3x + 2y = 12$	$-6x - 4y = -24$

Keep in mind that we are looking for the pair (x, y) that fits both equations. Thus, for these x and y values, we may *add equals to equals* in the resulting system to eliminate the variable x.

$$5y = 12$$

$$y = \frac{12}{5}$$

As before, x is now found by substituting $y = \frac{12}{5}$ into one of the given equations.

If in the preceding solution the second equation is multiplied by 2 instead of -2, the system becomes

$$6x + 9y = 36$$

$$6x + 4y = 24$$

Now x can be eliminated by *subtracting* the equations.

As illustrated in the next example, this method can be condensed into a compact procedure.

EXAMPLE 2 Solve the system by the multiplication-addition method.

$$\frac{1}{3}x - \frac{2}{5}y = 4$$

$$7x + 3y = 27$$

Solution

Multiply both sides of the first equation by 15, and both sides of the second equation by 2. Observe that
$2(7x + 3y = 27)$
is an abbreviation for
$2(7x + 3y) = 2(27)$.

$$
\begin{aligned}
15\left(\tfrac{1}{3}x - \tfrac{2}{5}y = 4\right) &\Rightarrow & 5x - 6y &= 60 \\
2(7x + 3y = 27) &\Rightarrow & \underline{14x + 6y} &= \underline{54} \\
& & \text{Add: } 19x &= 114 \\
& & x &= 6
\end{aligned}
$$

Substitute to solve for y.

$$7x + 3y = 27 \Rightarrow 7(6) + 3y = 27$$
$$3y = -15$$
$$y = -5$$

Check: $\frac{1}{3}(6) - \frac{2}{5}(-5) = 4$;
$7(6) + 3(-5) = 27$

The solution is the ordered pair $(6, -5)$. ∎

TEST YOUR UNDERSTANDING

Use the substitution method to solve each linear system.

1. $y = 3x - 1$
$\quad\; y = -5x + 7$

2. $y = 4x + 16$
$\quad\; y = -\frac{2}{5}x + \frac{14}{5}$

3. $4x - 3y = 11$
$\quad\; y = 6x - 13$

4. $2x + 2y = \frac{4}{5}$
$\quad\; -7x + 2y = -1$

5. $x + 7y = 3$
$\quad\; 5x + 12y = -8$

6. $4x - 2y = 40$
$\quad\; -3x + 3y = 45$

Use the multiplication-addition (or subtraction) method to solve each linear system.

7. $3x + 4y = 5$
$\quad\; 5x + 6y = 7$

8. $-8x + 5y = -19$
$\quad\; 4x + 2y = -4$

9. $\frac{1}{7}x + \frac{5}{2}y = 2$
$\quad\; \frac{1}{2}x - 7y = -\frac{17}{4}$

(Answers: Page 172)

When a linear system has a unique solution, as in the preceding examples, we say that the system is **consistent**. Graphically, this means that the two lines intersect. There are two other possibilities, as demonstrated next.

An Inconsistent System

$$39x - 91y = -28$$
$$6x - 14y = 7$$

A Dependent System

$$y = -\frac{2}{3}x + 5$$
$$2x + 3y = 15$$

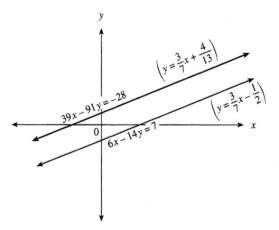

A dependent system can be identified algebraically when a statement that is always true (an identity) is obtained when trying to solve the system.

The two lines have the same slope, $\frac{3}{7}$, but different y-intercepts. There is no point of intersection. *An inconsistent system has no solution.*

The graph is a single line. Each ordered pair (x, y) that satisfies one equation must also satisfy the other. *A dependent system has an infinite number of solutions, that is, all points on the line.*

An inconsistent system can also be identified algebraically by obtaining a *false result* when trying to solve such a system. Thus for the preceding system, if the second equation is solved for *y*, we obtain

$$y = \tfrac{3}{7}x - \tfrac{1}{2}$$

*Arriving at a false result, as shown here, is not a wasted effort. As long as there are no computational errors, such a false conclusion tells us that the given system has no solution. That is, **the system has been solved by learning that it has no solution.***

Now substitute into the first equation.

$$39x - 91(\tfrac{3}{7}x - \tfrac{1}{2}) = -28$$

$$39x - 39x + \tfrac{91}{2} = -28$$

$$\tfrac{91}{2} = -28 \quad \text{False}$$

This false result tells us that the system is inconsistent.

There are many word problems that can be solved by using linear systems. When solving such a problem you must first use the given statement to obtain a system of equations that can then be used to find the answer. You will find that the translation of the verbal form into the mathematical form is usually the most difficult part of the solution. Unfortunately, because of the variety of problems, as well as the numerous ways in which a problem can be stated, there are no fixed methods of translation that apply to all situations. However, the general guidelines given in Section 1.2 for solving word problems in one variable can be adjusted and used here.

3x   4y

4x   3y

EXAMPLE 3   For her participation in a recent "walk-for-poverty," Ellen collected $2 per mile for a total of $52. She recorded that for a certain time she walked at the rate of 3 miles per hour (mph) and the rest at 4 mph. Afterward she mentioned that it was too bad that she did not have the energy to reverse the rates. For if she could have walked 4 mph for the same time that she actually walked 3 mph, and vice versa, she would have collected a total of $60. How long did her walk take?

Solution   The problem asks for the total time of the walk. This time is broken into two parts: the time she walked at 3 mph and the time at 4 mph.

$$\text{Let} \quad x = \text{time at 3 mph}$$
$$\text{and} \quad y = \text{time at 4 mph}$$

We want to find $x + y$.

Since *distance = rate × time,* $3x$ is the distance at the 3-mph rate and $4y$ is the distance at 4 mph; the total distance is the sum $3x + 4y$. This total must equal 26 because she earned $52 at $2 per mile. Thus

$$3x + 4y = 26$$

The $60 she could have earned would have required walking 30 miles. And this 30 miles, she said, would have been possible by reversing the rates for the actual times that she did walk. This says that $4x + 3y = 30$. We now have the system

Notice that the answer in Example 3 was checked by returning to the language of the original problem rather than to any of the equations. It is essential to do this for word problems because the equations themselves may be incorrect translations of the problem.

$$3x + 4y = 26$$

$$4x + 3y = 30$$

The common solution is (6, 2), and therefore the total time for the walk is $6 + 2$, or 8 hours.

Check: She walked $3 \times 6 = 18$ miles at 3 mph, and $4 \times 2 = 8$ miles at 4 mph, for a total of 26 miles. At \$2 per mile, she collected \$52. Reversing the rates gives $(4 \times 6) + (3 \times 2) = 30$ miles, for which she would have earned \$60. ■

EXAMPLE 4 A grocer sells Brazilian coffee at \$5.00 per pound and Colombian coffee at \$8.50 per pound. How many pounds of each should he mix in order to have a blend of 50 pounds that he can sell at \$7.10 per pound?

Solution

Let x = the number of pounds of Brazilian coffee to be used.

Let y = the number of pounds of Colombian coffee to be used.

It is frequently helpful, although not necessary, to summarize the given information in a table such as the following.

	Number of Pounds	**Cost per Pound**	**Total Cost**
Brazilian	x	5.00	5.00x
Colombian	y	8.50	8.50y
Blend	50	7.10	50(7.10)

Now use the information in the table to obtain this system of equations:

$$x + \quad y = 50$$

$$5.00x + 8.50y = 355 \qquad 50(7.10) = 355$$

To check this solution show that 20 pounds of the \$5.00 coffee and 30 pounds of \$8.50 coffee will give the same income as 50 pounds of a \$7.10 blend.

Show that the solution for the system is $x = 20$ and $y = 30$. Thus the grocer should mix 20 pounds of the Brazilian coffee with 30 pounds of the Colombian coffee. This will give him 50 pounds of a blend to sell at \$7.10 per pound. ■

The next example demonstrates how a company can find the **break-even point** for a product that they produce and sell. This is the point at which the company's expenses equal its revenues. Beyond this point the company makes a profit, and below it the company has a loss.

EXAMPLE 5 It costs the Roller King Company \$8 to produce one skateboard. In addition, there is a \$200 daily fixed cost for building maintenance.
(a) Find the total cost for producing x skateboards per day.
(b) Find the total revenue if the company sells x skateboards per day for \$16 each.
(c) Find the daily break-even point. That is, find the coordinates of the point at which the cost equals the revenue.

Solution

(a)

$$\text{Total Daily Cost} = \left(\text{Cost per Item}\right) \cdot \left(\text{Number of Items}\right) + \text{Fixed Cost}$$

$$y = 8 \cdot x + 200$$

(b)

$$\text{Total Daily Revenue} = \left(\text{Sale Price per Item}\right) \cdot \left(\text{Number of Items}\right)$$

$$y = 16 \cdot x$$

Dollars

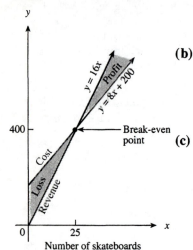

For x > 25, the revenue is greater than the cost, resulting in a profit. For x < 25, the cost is more than the revenue, resulting in a loss.

(c) The daily profit or loss can be observed from the graphs of the cost and revenue lines drawn in the same coordinate system. The point where the lines intersect is the break-even point. At this point the cost for producing x skateboards is the same as the revenue for selling x skateboards on a daily basis. To find the break-even point, we solve this system.

$$\text{Cost:} \qquad y = 8x + 200$$
$$\text{Revenue:} \qquad y = 16x$$

Thus, $16x = 8x + 200$, which gives $x = 25$ and $y = 16(25) = 400$. The break-even point is (25, 400). When the company produces and sells 25 skateboards per day, their cost and revenue are $400 each. ∎

EXERCISES 2.5

Solve each system by the substitution method.

1. $4x - y = 6$
$2x + 3y = 10$

2. $x + 5y = -9$
$4x - 3y = -13$

3. $s + 2t = 5$
$-3s + 10t = -7$

Solve each system by the multiplication-addition or subtraction method.

4. $2u - 6v = -16$
$5u - 3v = 8$

5. $4x - 5y = 3$
$16x + 2y = 3$

6. $\frac{1}{2}x + 3y = 6$
$-x - 8y = 18$

Solve each system.

7. $x - 2y = 3$
$y - 3x = -14$

8. $2x + y = 6$
$3x - 4y = 12$

9. $-3x + 8y = 16$
$16x - 5y = 103$

10. $3x - 8y = -16$
$7x + 19y = -188$

11. $16x - 5y = 103$
$7x + 19y = -188$

12. $4x = 7y - 6$
$9y = -12x + 12$

13. $3s + t - 3 = 0$
$2s - 3t - 2 = 0$

14. $\frac{1}{2}x - \frac{1}{3}y = 2$
$\frac{3}{4}x + \frac{2}{5}y = -1$

15. $\frac{1}{4}x + \frac{1}{3}y = \frac{5}{12}$
$\frac{1}{2}x + y = 1$

16. $0.1x + 0.2y = 0.7$
$0.01x - 0.01y = 0.04$

17. $\dfrac{x}{2} + \dfrac{y}{6} = \dfrac{1}{2}$
$0.2x - 0.3y = 0.2$

18. $2(x + y) = 4 - 3y$
$\frac{1}{2}x + y = \frac{1}{2}$

19. $2(x - y - 1) = 1 - 2x$
$6(x - y) = 4 - 3(3y - x)$

20. $\dfrac{x - 2}{5} + \dfrac{y + 1}{10} = 1$
$\dfrac{x + 2}{3} - \dfrac{y + 3}{2} = 4$

Decide whether the given systems are consistent, inconsistent, or dependent.

21. $4x - 12y = 3$
$x + \frac{1}{3}y = 3$

22. $3x - y = 7$
$-9x + 3y = -21$

23. $2x + 5y = -20$
$x = -\frac{5}{2}y - 10$

24. $x - y = 3$
$-\frac{1}{3}x + \frac{1}{3}y = 1$

25. $x - 5y = 15$
$0.01x - 0.05y = 0.5$

26. $4y = 3x + 2$
$2x = 3y - 3$

Solve the systems for the common pair (x, y) in terms of c.

27. $-x + y = 6c$
$x + y = 3c$

28. $2x - 10c = -y$
$7x - 2y = 2c$

29. Points $(-8, -16)$, $(0, 10)$, and $(12, 14)$ are three vertices of a parallelogram. Find the coordinates of the fourth vertex if it is located in the third quadrant.

30. Find the point of intersection for the diagonals of the parallelogram given in Exercise 29.

31. A line with equation $ax + by = 3$ passes through $(6, 3)$ and $(-1, -1)$. Find a and b without finding the slope.

32. A line with equation $y = mx + b$ passes through the points $(-\frac{1}{3}, -6)$ and $(2, 1)$. Find m and b by substituting the coordinates into the equation and solving the resulting system.

Use a system of two linear equations in two variables to solve the following problems.

33. The total points that a basketball team scored was 96. If there were two-and-a-half times as many field goals as free throws, how many of each were there? (Field goals count 2 points; free throws count 1 point. There are no 3-point field goals.)

34. During a game, a golfer scored only fours and fives per hole. If he played 18 holes and his total score was 80, how many holes did he play in four strokes and how many in five?

35. The perimeter of a rectangle is 60 centimeters. If the length is three more than twice the width, find the dimensions.

36. Three times the larger of two numbers is 10 more than twice the smaller. Five times the smaller is 11 less than four times the larger. What are the numbers?

37. A shopper pays \$3.82 for $9\frac{1}{2}$ pounds of vegetables consisting of potatoes and string beans. If potatoes cost 35¢ per pound and string beans cost 68¢ per pound, how much of each was purchased?

38. A football team scored 54 points. They scored touchdowns at 6 points each, some points-after-touchdown at 1 point each, and some field goals at 3 points each. If there were two-and-one-third as many touchdowns as field goals, and just as many points-after-touchdown as field goals, how many of each were there?

39. The tuition fee at a college plus the room and board comes to \$8400 per year. The room and board is \$600 more than half the tuition. How much is the tuition, and what does the room and board cost?

40. A college student had a work-study scholarship that paid $4.20 per hour. He also made $2 per hour babysitting. His income one week was $80 for a total of $23\frac{1}{2}$ hours of employment. How many hours did he spend on each of the two jobs?

41. There are only nickels and quarters in a child's bank. There are 22 coins in the bank having a total value of $3.90. How many nickels and how many quarters are there?

42. The perimeter of a rectangle is 72 inches. The length is three-and-one-half times as large as the width. Find the dimensions.

43. The treasurer of the student body reported that the receipts for the last concert totaled $916 and that 560 people attended. If students paid $1.25 per ticket and nonstudents paid $2.25 per ticket, how many students attended the concert?

44. The cost of 10 pounds of potatoes and 4 pounds of apples is $6.16, while 4 pounds of potatoes and 8 pounds of apples cost $6.88. What is the cost per pound of potatoes and of apples?

45. David walked 19 kilometers. The first part of the walk was at 5 kph (kilometers per hour) and the rest at 3 kph. He would have covered 2 kilometers less if he had reversed the rates, that is, if he had walked at 3 kph and at 5 kph for the same times that he actually walked at 5 kph and 3 kph, respectively. How long did it take him to walk the 19 kilometers?

46. A swimmer going downstream takes 1 hour, 20 minutes to travel a certain distance. It takes the swimmer 4 hours to make the return trip against the current. If the river flows at the rate of $1\frac{1}{2}$ miles per hour, find the rate of the swimmer in still water and the distance traveled one way. (*Hint:* Let x = rate of swimmer in still water.)

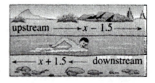

47. A store paid $299.50 for a recent mailing. Some of the letters cost 25¢ postage and the rest needed 45¢ postage. How many letters at each rate were mailed if the total number sent out was 910?

48. Ellen and Robert went to the store to buy some presents. They had a total of $22.80 to spend and came home with $6.20. If Ellen spent two-thirds of her money and Robert spent four-fifths of his money, how much did they each have to begin with?

49. The annual return on two investments totals $464. One investment gives 8% interest and the other $7\frac{1}{2}$%. How much is invested at each rate if the total investment is $6000?

50. A wholesaler has two grades of oil that ordinarily sell for $1.14 per quart and 94¢ per quart. She wants a blend of the two oils to sell at $1.05 per quart. If she anticipates selling 400 quarts of the new blend, how much of each grade should she use? (One of the equations makes use of the fact that the total income will be 400(1.05), or $420.)

51. A student in a chemistry laboratory wants to form a 32-milliliter mixture of two solutions to contain 30% acid. Solution *A* contains 42% acid and solution *B* contains 18% acid. How many milliliters of each solution must be used? (*Hint:* Use the fact that the final mixture will have 0.30(32) = 9.6 milliliters of acid.)

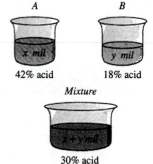

52. A purchase of 6 dozen oranges and 10 pounds of peaches costs $10.60. If the price per dozen oranges increases 10%, and if peaches increase 5% per pound, the same order would cost $11.46. What is the initial cost of a pound of peaches and of a dozen oranges?

53. To go to work Juanita first averages 36 miles per hour driving her car to the train station and then rides the train, which averages 60 miles per hour. The entire trip takes 1 hour and 22 minutes. It costs her 15¢ per mile to drive the car and 6¢ per mile to ride the train. If the total cost is $5.22, find the distances traveled by car and by train.

54. Suppose that someone asked you to find the two numbers in the following puzzle:

The larger of two numbers is 16 more than twice the smaller. The difference between $\frac{1}{4}$ of the larger and $\frac{1}{2}$ the smaller is 2.

55. How many answers are there for the following puzzle?

The difference between two numbers is 3. The larger number decreased by 1 is the same as $\frac{1}{3}$ the sum of the smaller plus twice the larger.

56. How many answers are there to the puzzle in Exercise 55 if the difference between the two numbers is 2?

57. A bag containing a mixture of 6 oranges and 12 tangerines sold for $2.34. A smaller bag containing 2 oranges and 4 tangerines sold for 77¢. An alert shopper asked the salesclerk if it was a better buy to purchase the larger bag. The clerk was not sure, but said that it really made no difference because the price of each package was based on the same unit price for each kind of fruit. Why was the clerk wrong?

58. If Sue gave Sam one of her dollars, then Sam would have half as many dollars as Sue. If Sam gave Sue one of his dollars, then Sue would have five times as many dollars as Sam. How many dollars did each of them have?

59. Refer to Example 5 of this section and determine the daily profit when 40 skateboards are produced and sold per day.

60. Refer to Example 5 of this section and determine the loss when 20 skateboards are produced and sold per day.

61. Refer to Example 5 of this section and determine how many skateboards must be made and sold per day to obtain a $520 profit.

62. Find the break-even point for each pair of cost and revenue equations. (Assume the equations represent dollars.)

	Cost	**Revenue**
(a)	$y = 15x + 450$	$y = 30x$
(b)	$y = 240x + 1600$	$y = 400x$

63. The Electro Calculator Company can produce a calculator at a cost of $12. Their daily fixed cost is $720, and they plan to sell each calculator for $20. Find the break-even point.

64. What is the profit or loss for 200 calculators? For 300 calculators? (Refer to Exercise 63.)

65. How many calculators must be sold for there to be a $200 daily profit? (See Exercise 63.)

66. A bakery makes extra large chocolate chip cookies at a cost of $0.12 per cookie and sells them for $0.60 apiece. If the daily overhead expenses are $60, find the break-even point.

Challenge   Solve the system for the common pair (x, y).

$$ax + by = c$$
$$dx + ey = f$$

Assume a, b, c, d, e, and f are constants and $ae - bd \neq 0$. Explain how these results can be used to solve any such pair of equations, and illustrate with a specific system.

Written Assignment   In your own words, explain the rationale behind the substitution and the multiplication-addition methods for solving a system of equations.

Graphing Calculator Exercises   For the following exercises, use a friendly window (such as $-4.8 \leq x \leq 4.7$, $-3.2 \leq y \leq 3.1$ on the TI-81 or $-4.7 \leq x \leq 4.7$, $-3.1 \leq y \leq 3.1$ on the Casio fx-7700G or the TI-82).

1. Solve the following system of equations by graphing each equation of the system on the same set of axes and finding the point of intersection of the graphs:

$$3x - 4y = 2$$
$$2x + y = 5$$

2. Does the following system of equations have a common solution? Answer by graphing each equation of the system on the same set of axes and trying to find a common point of intersection:

$$3x - 4y = 2$$
$$2x + y = 5$$
$$x + y = 1$$

3. Solve the following system of equations graphically to the nearest hundredth by ZOOMing in repeatedly on the apparent point of intersection until the thousandths place no longer changes. It might help to use the directional cursor keys to toggle between the lines to see if what appears to be a common point really is one. Then compare your answer with the exact answer in fractional form and convert that form to a decimal with your calculator:

$$3x - 4y = 5$$
$$6x + 7y = 8$$

2.6 GRAPHING QUADRATIC FUNCTIONS

A function defined by a polynomial expression of degree 2 is referred to as a **quadratic function** in x. Thus the following are all examples of quadratic functions in x:

$$f(x) = -3x^2 + 4x + 1 \qquad g(x) = 7x^2 - 4 \qquad h(x) = x^2$$

The most general form of such a quadratic function is

$$f(x) = ax^2 + bx + c$$

where a, b, and c represent constants, with $a \neq 0$.

If $a = 0$, then the resulting polynomial no longer represents a quadratic function; $f(x) = bx + c$ is a linear function.

The simplest quadratic function is given by $f(x) = x^2$. The graph of this quadratic function will serve as the basis for drawing the graph of any quadratic function $f(x) = ax^2 + bx + c$. First, observe the *symmetry* that exists. For example,

$$f(-3) = f(3) = 9 \qquad f(-1) = f(1) = 1$$

This is true for all x. That is:

$$f(-x) = (-x)^2 = x^2 = f(x)$$

In general:

> When a function f has the property $f(-x) = f(x)$ for all domain values, then its graph is symmetric about the y-axis and it is called an **even function.**

The following table of values gives several ordered pairs of numbers that are coordinates of points on the graph of $y = x^2$. When these points are located on a rectangular system and connected by a smooth curve, the graph of $y = f(x) = x^2$ is obtained. The curve is called a **parabola,** and every quadratic function $y = ax^2 + bx + c$ has a parabola as its graph. The domain of the function is the set of all real numbers. The range of the function depends on the constants a, b, and c. For the function $y = f(x) = x^2$, the range consists of all $y \geq 0$.

Greater accuracy can be obtained by using more points. But since we can never locate an infinite number of points, we must admit that there is a certain amount of faith involved in connecting the points as we did.

x	$y = x^2$
-3	9
-2	4
-1	1
0	0
1	1
2	4
3	9

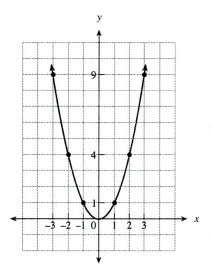

An important feature of such a parabola is that it is symmetric about a vertical line called its **axis of symmetry.** The graph of $y = x^2$ is symmetric with respect to the y-axis. This symmetry is due to the fact that $(-x)^2 = x^2$. Consequently, this is an even function.

The parabola has a *turning point,* called the **vertex,** which is located at the intersection of the parabola with its axis of symmetry. For the preceding graph the coordinates of the vertex are $(0, 0)$, and 0 is the *minimum value* of the function.

From the graph you can see that, reading from left to right, the curve is "falling" down to the origin and then is "rising." These features are technically described as f **decreasing** and f **increasing.**

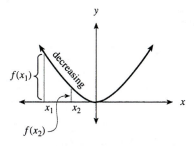

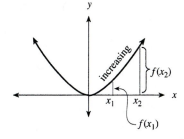

$f(x) = x^2$ is decreasing on $(-\infty, 0]$ because for *each* pair x_1, x_2 in this interval, if $x_1 < x_2$ then $f(x_1) > f(x_2)$.

$f(x) = x^2$ is increasing on $[0, \infty)$ because for *each* pair x_1, x_2 in this interval, if $x_1 < x_2$ then $f(x_1) < f(x_2)$.

The graph of $y = x^2$ can be used as a guide to draw the graphs of other quadratic functions. In the following illustrations, the graph of $y = x^2$ is shown as a dashed curve. The other graphs show examples of *vertical translations* of $y = x^2$.

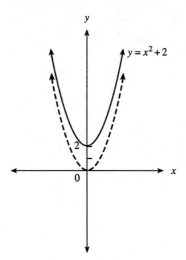

The graph of $y = x^2 + 2$
is congruent to that of
$y = x^2$, but is shifted
2 units up.

The graph of $y = x^2 - 2$
is congruent to that of
$y = x^2$, but is shifted
2 units down.

Next consider the graph of $y = (x + 2)^2$. In this case first add 2 to x, and then square. For $x = -2$, $y = 0$ and the vertex of the curve is at $(-2, 0)$. The graph is congruent to that for $y = x^2$, but is shifted 2 units to the left. In a similar fashion, note that the graph of $y = (x - 2)^2$ is shifted 2 units to the right. Both of these are examples of *horizontal translations* of $y = x^2$.

Both of these parabolas are congruent to the basic parabola $y = x^2$. Each may be graphed by translating (shifting) the parabola $y = x^2$ by 2 units, to the right for $y = (x - 2)^2$ and to the left for $y = (x + 2)^2$.

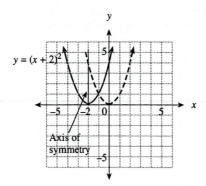

The axis of symmetry is the line $x = -2$.
The vertex is at $(-2, 0)$.

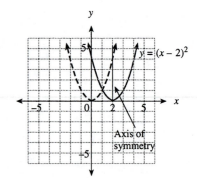

The axis of symmetry is the line $x = 2$.
The vertex is at $(2, 0)$.

In each of the graphs drawn thus far, the coefficient of x^2 has been 1. If the coefficient is -1, it has the effect of reflecting the graph through the x-axis. The domain

*The graph of $y = -x^2$ may also be obtained by multiplying each of the ordinates of $y = x^2$ by -1. This step has the effect of "flipping" the parabola $y = x^2$ downward, a **reflection** in the x-axis.*

of the function is still the set of real numbers, but the range is the set of nonpositive real numbers. The vertex is at (0, 0) and 0 is the *maximum value* of the function. Again, the graph of $y = x^2$ is shown as a dashed curve. Since the graph of $y = x^2$ bends "upward," we say that the curve is **concave up.** Also, since $y = -x^2$ bends "downward," we say that the curve is **concave down.**

$y = -x^2$

x	-3	-2	-1	0	1	2	3
y	-9	-4	-1	0	-1	-4	-9

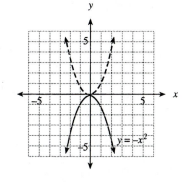

When the coefficient a in $y = ax^2$ is other than 1, then the graph of $y = ax^2$ can be obtained by multiplying each ordinate of $y = x^2$ by the number a, as in the following example.

EXAMPLE 1 Graph: **(a)** $y = \frac{1}{2}x^2$ **(b)** $y = 2x^2$

Solution

(a) $y = \frac{1}{2}x^2$ **(b)** $y = 2x^2$

Note that the graph of $y = 2x^2$ is "steeper" than that of $y = x^2$; the graph of $y = \frac{1}{2}x^2$ is not as steep as that of $y = x^2$.

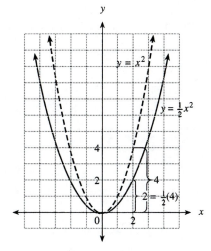

Each ordinate (y-value) is one-half that of the graph for $y = x^2$, shown as a dashed curve.

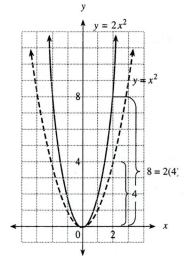

Each ordinate is twice that of the graph for $y = x^2$, shown as a dashed curve. ∎

EXAMPLE 2 Graph $y = f(x) = -x^2 + 2$. State where the function is increasing or decreasing. What is the concavity?

Observe that to say "f is increasing for all $x \leq 0$" means the same as saying "f is increasing on $(-\infty, 0]$. Similarly, we say that f is decreasing on $[0, \infty)$.

Solution Consider the graph of $y = -x^2$ and shift it up 2 units. The function f is increasing for all $x \leq 0$, and it is decreasing for all $x \geq 0$. The curve is concave down.

$y = -x^2 + 2$

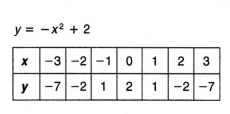

x	−3	−2	−1	0	1	2	3
y	−7	−2	1	2	1	−2	−7

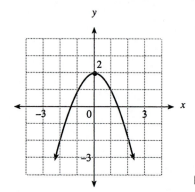

■

TEST YOUR UNDERSTANDING	*Match each graph with one of the given quadratic equations.*

(a) $y = (x + 1)^2$ **(b)** $y = x^2 + 1$ **(c)** $y = (x - 1)^2$

(d) $y = x^2 - 1$ **(e)** $y = -(x - 1)^2$ **(f)** $y = -x^2 + 1$

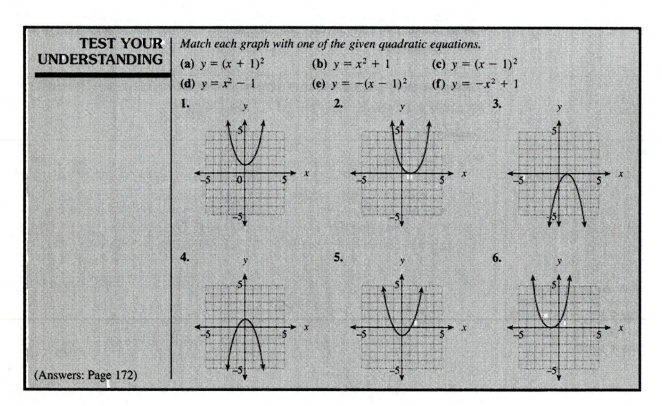

(Answers: Page 172)

Let us now put several ideas together and draw the graph of this function:

$$y = f(x) = (x + 2)^2 - 3$$

An effective way to do this is to begin with the graph of $y = x^2$ and shift the graph

2 units to the left for $y = (x + 2)^2$ and then 3 units down for the graph of $f(x) = (x + 2)^2 - 3$, as in the following figures.

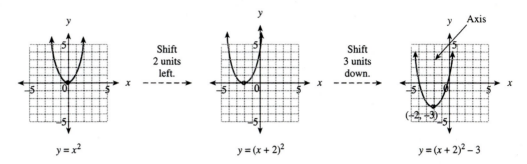

$y = x^2$ $\qquad$ $y = (x + 2)^2$ $\qquad$ $y = (x + 2)^2 - 3$

Note that the graph of $y = (x + 2)^2 - 3$ is congruent to the graph of $y = x^2$. The vertex of the curve is at $(-2, -3)$, and the axis of symmetry is the line $x = -2$. The minimum value of the function, -3, occurs at the vertex. Also observe that the domain consists of all numbers x, and the range consists of all numbers $y \geq -3$.

EXAMPLE 3 Graph $y = f(x) = -(x - 2)^2 + 1$, give the coordinates of the vertex, the equation of the axis of symmetry, and state the domain and range of f.

Solution Consider the graph of $y = -x^2$, and shift this 2 units to the right and 1 unit up. The vertex is at $(2, 1)$, the highest point of the curve, and the axis of symmetry is the line $x = 2$. Also, the domain of f is the set of all real numbers and the range consists of all $y \leq 1$.

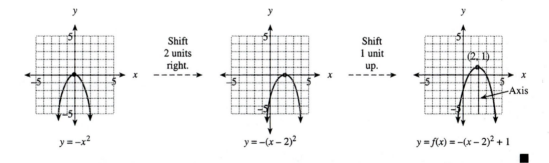

$y = -x^2$ $\qquad$ $y = -(x - 2)^2$ $\qquad$ $y = f(x) = -(x - 2)^2 + 1$

$\blacksquare$

All such parabolas may be described as being vertical since their axis of symmetry is vertical. Vertical parabolas open either upward or downward.

SUMMARY: THE GRAPH OF $y = a(x - h)^2 + k$

The graph of $y = a(x - h)^2 + k$ is congruent to the graph of $y = ax^2$ but is shifted h units horizontally and k units vertically.

(a) The horizontal shift is to the right if $h > 0$ and to the left if $h < 0$.

(b) The vertical shift is upward if $k > 0$ and downward if $k < 0$.

The axis of symmetry is $x = h$; the vertex (h, k) is the highest point if $a < 0$, and the lowest point if $a > 0$.

EXAMPLE 4   Graph the parabola $y = f(x) = -2(x - 3)^2 + 4$.

Solution   The graph will be a parabola congruent to $y = -2x^2$, with vertex at $(3, 4)$ and with $x = 3$ as axis of symmetry. A brief table of values, together with the graph, is shown.

The function is increasing on $(-\infty, 3]$, decreasing on $[3, \infty)$, and the curve is concave down.

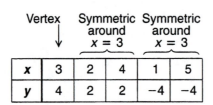

	Vertex	Symmetric around $x = 3$		Symmetric around $x = 3$	
x	3	2	4	1	5
y	4	2	2	-4	-4

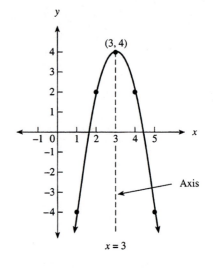

Interchanging the variables x and y in the equation $y = x^2$ produces $x = y^2$, whose graph is a *horizontal parabola* that opens to the right.

x	4	1	0	1	4
y	-2	-1	0	1	2

The axis of symmetry is the x-axis.
The vertex is at $(0, 0)$.

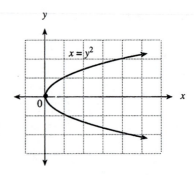

The graph of any quadratic equation of the form $x = ay^2 + by + c$ is a horizontal parabola that opens to the right when $a > 0$, or the left when $a < 0$. These parabolas can be graphed using procedures similar to those used for the vertical parabolas. For example, the parabola $x = y^2 + 3$ can be obtained by shifting the basic horizontal parabola 3 units to the right. The axis of symmetry is the x-axis.

EXAMPLE 5   Graph $x = (y + 2)^2 - 4$ and identify the vertex and axis of symmetry.

Solution   Begin with the parabola $x = y^2$; shift 2 units downward and 4 units to the left as shown. The vertex is $(-4, -2)$ and $y = -2$ is the axis of symmetry.

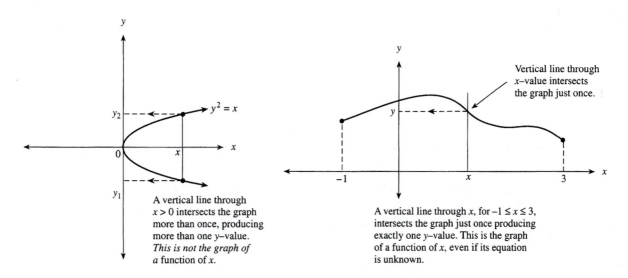

Observe that a horizontal parabola is *not* the graph of a function of x, since there are two y-values for a given x. This can also be observed using the following geometric test.

Vertical Line Test for Functions

The vertical line test applied to a horizontal parabola shows that such a parabola cannot be the graph of a function of x. Thus the equations $x = y^2$ and $x = (y + 2)^2 - 4$ do *not* define functions of x. However, any equation in two variables is said to define a **relation.** Thus all functions are relations, but many relations are not functions.

A vertical line through $x > 0$ intersects the graph more than once, producing more than one y-value. *This is not the graph of a function of x.*

Vertical line through x–value intersects the graph just once.

A vertical line through x, for $-1 \leq x \leq 3$, intersects the graph just once producing exactly one y-value. This is the graph of a function of x, even if its equation is unknown.

EXERCISES 2.6

Draw each set of graphs on the same axes.

1. (a) $y = x^2$ (b) $y = (x - 1)^2$ (c) $y = (x - 1)^2 + 3$
2. (a) $y = x^2$ (b) $y = (x + 1)^2$ (c) $y = (x + 1)^2 - 3$
3. (a) $y = -x^2$ (b) $y = -(x - 1)^2$ (c) $y = -(x - 1)^2 + 3$
4. (a) $y = -x^2$ (b) $y = -(x + 1)^2$ (c) $y = -(x + 1)^2 - 3$

5. (a) $y = x^2$ (b) $y = 2x^2$ (c) $y = 3x^2$

6. (a) $y = -x^2$ (b) $y = -\frac{1}{2}x^2$ (c) $y = -\frac{1}{2}x^2 + 1$

Draw the graph of each function.

7. $y = x^2 + 3$ 8. $y = (x + 3)^2$ 9. $y = -x^2 + 3$ 10. $y = -(x + 3)^2$

11. $y = 3x^2$ 12. $y = 3x^2 + 1$ 13. $y = \frac{1}{4}x^2$ 14. $y = \frac{1}{4}x^2 - 1$

15. $y = \frac{1}{4}x^2 + 1$ 16. $y = -2x^2$ 17. $y = -2x^2 + 2$ 18. $y = -2x^2 - 2$

Graph each of the following functions. Where is the function increasing and decreasing? What is the concavity?

19. $f(x) = (x - 1)^2 + 2$ 20. $f(x) = (x + 1)^2 - 2$ 21. $f(x) = -(x + 1)^2 + 2$

22. $f(x) = -(x + 1)^2 - 2$ 23. $f(x) = 2(x - 3)^2 - 1$ 24. $f(x) = 2\left(x + \frac{5}{4}\right)^2 + \frac{5}{4}$

State (a) the coordinates of the vertex, (b) the equation of the axis of symmetry, (c) the domain, and (d) the range for each of the following functions.

25. $y = f(x) = (x - 3)^2 + 5$ 26. $y = f(x) = (x + 3)^2 - 5$ 27. $y = f(x) = -(x - 3)^2 + 5$

28. $y = f(x) = -(x + 3)^2 - 5$ 29. $y = f(x) = 2(x + 1)^2 - 3$ 30. $y = f(x) = \frac{1}{2}(x - 4)^2 + 1$

31. $y = f(x) = -2(x - 1)^2 + 2$ 32. $y = f(x) = -\frac{1}{2}(x + 2)^2 - 3$ 33. $y = f(x) = \frac{1}{4}(x + 2)^2 - 4$

34. $y = f(x) = 3\left(x - \frac{3}{4}\right)^2 + \frac{4}{5}$

35. Graph the function f where

$$f(x) = \begin{cases} x^2 & \text{if } -2 \le x \le 1 \\ x & \text{if } 1 < x \le 3 \end{cases}$$

36. Graph f where

$$f(x) = \begin{cases} x^2 - 9 & \text{if } -2 \le x < 4 \\ -3x + 15 & \text{if } 4 \le x < 6 \\ 3 & \text{if } x = 6 \end{cases}$$

37. Consider the equation $x = y^2$. Why is y not a function of x?

38. Compare the graphs of $y = x^2$ and $y = |x|$. In what ways are they alike?

39. What is the relationship between the graph of $y = x^2 - 4$ on a plane and of $x^2 - 4 > 0$ on a line?

40. Repeat Exercise 39 for $y = x^2 - 9$ and for $x^2 - 9 < 0$.

41. The graph of $y = ax^2$ passes through the point $(1, -2)$. Find a.

42. The graph of $y = ax^2 + c$ has its vertex at $(0, 4)$ and passes through the point $(3, -5)$. Find the values for a and c.

43. Find the value of k so that the graph of $y = (x - 2)^2 + k$ will pass through the point $(5, 12)$.

44. Find the value for h so that the graph of $y = (x - h)^2 + 5$ will pass through the point $(3, 6)$.

Graph each horizontal parabola.

45. $x = y^2 - 4$ 46. $x = -y^2 + 3$ 47. $x = (y - 3)^2$

48. $x = (y + 1)^2$ 49. $x = 2y^2$ 50. $x = -\frac{1}{2}y^2$

Draw each set of parabolas on the same axes.

51. (a) $x = y^2$ **(b)** $x = (y - 1)^2$ **(c)** $x = (y - 1)^2 - 5$

52. (a) $x = -y^2$ **(b)** $x = -(y + 3)^2$ **(c)** $x = -(y + 3)^2 + 4$

Write the equation of the parabola labeled P, which is obtained from the dashed curve by shifting it horizontally and vertically.

53.

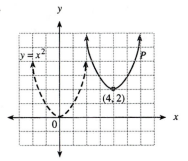

54.

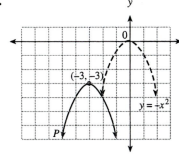

55.

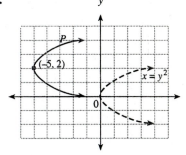

56.

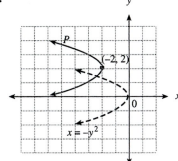

57.

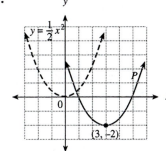

58.

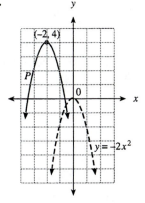

Use the vertical line test to decide if the given graph can be the graph of a function having the indicated set A as the domain.

59.

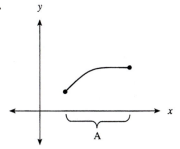

60.

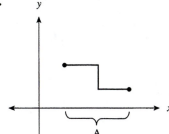

61.

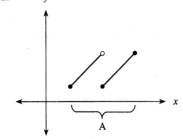

62.

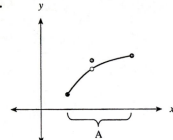

63.

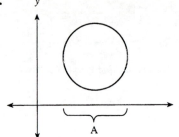

64.

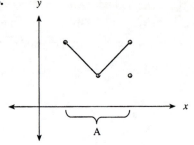

65.

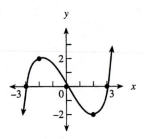

66.

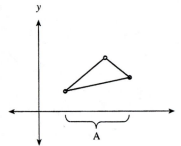

Copy the graph shown on graph paper. Then sketch the graph for parts **(a)** *and* **(b)** *on the same coordinate axes.*

67.

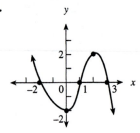

(a) Graph $y = f(x - 2)$.

(b) Graph $y = f(x - 2) + 1$.

68.

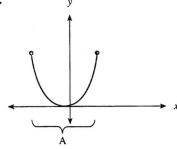

(a) Graph $y = f(x + 2)$.

(b) Graph $y = f(x + 2) - 1$.

69. Rectangle $ABCD$ is inscribed in the arc of the parabola $y = 9 - x^2$ so that the vertices C and D are on the parabola. Write the area of the rectangle as a function of x. What is the domain of this function?

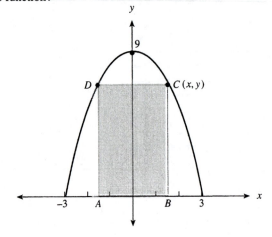

70. The polygon in the figure consists of four rectangles. The bases of the rectangles are of equal length. The altitude of each rectangle is drawn at the right endpoint of its base.

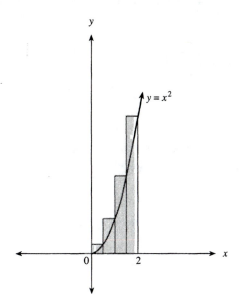

(a) Compute the area of the polygon; that is, the sum of the areas of the four rectangles.

(b) Modify the figure so that the polygon consists of eight rectangles with equal bases. In each case, draw the altitudes at the right endpoint of each base and terminating on the curve $y = x^2$, similar to the figure shown. Compute the area of this polygon and note that this area is less than the area found in part (a).

Challenge   Give an algebraic proof that the function $f(x) = (x - 3)^2$ is symmetric about the axis $x = 3$. (*Hint:* Any two x-values on opposite sides of the line $x = 3$, and equidistant from it, must have the same y-values.)

Written Assignment   Refer to Exercise 70. Without doing any further computation, consider polygons consisting of 16 rectangles, then 32 rectangles, and so forth. Thus consider doubling the number of rectangles in each successive step. As the number of rectangles keeps on increasing, discuss what happens to the areas of the resulting polygons.

Graphing Calculator Exercises   Graph each of the following sets of equations on the same viewing screen. In each case, explain how the graph in part (a) is changed by each of the subsequent equations.

1. (a) $y = x^2$ (b) $y = x^2 + 2$ (c) $y = x^2 - 3$ (d) $y = x^2 + 5$
2. (a) $y = x^2$ (b) $y = 2x^2$ (c) $y = -2x^2$ (d) $y = -3x^2$
3. (a) $y = x^2$ (b) $y = (x - 3)^2$ (c) $y = (x + 3)^2$
4. (a) $y = x^2$ (b) $y = (x - 2)^2$ (c) $y = 0.5(x - 2)^2$ (d) $y = 0.5(x - 2)^2 + 3$
5. Graph the horizontal parabola $x = y^2 + 2$ by solving for y as $y = \pm\sqrt{x - 2}$. Then graph each of the functions $y = \sqrt{x - 2}$ and $y = -\sqrt{x - 2}$ on the same axes.

2.7 THE QUADRATIC FORMULA

When a quadratic function is given in the form $f(x) = ax^2 + bx + c$, some of the properties of the graph are not evident. However, if this function is converted to the **standard form** $f(x) = a(x - h)^2 + k$, then the methods of Section 2.6 can be used to sketch the parabola.

Let us begin with the quadratic function given by $y = x^2 + 4x + 3$. First rewrite the equation in this way:

$$y = (x^2 + 4x + \underline{\ ?\ }) + 3$$

Note that if the question mark is replaced by 4, then we will have a *perfect square trinomial* within the parentheses. However, since this changes the original equation, we must also subtract 4. The completed work follows:

$$y = x^2 + 4x + 3$$
$$= (x^2 + 4x + \mathbf{4}) + 3 - \mathbf{4}$$
$$= (x + 2)^2 - 1$$

The graph is a parabola with vertex at $(-2, -1)$, and $x = -2$ as axis of symmetry.

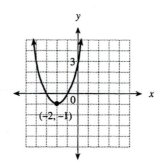

The technique that we have just used is called **completing the square.** This process can be extended to the case where the coefficient of x^2 is a number different from 1, such as $y = -2x^2 + 12x - 11$. The first step is to factor the coefficient of x^2 from the two variable terms only.

$$y = -2x^2 + 12x - 11 = -2(x^2 - 6x + \underline{\ ?\ }) - 11$$

Next, add 9 within the parentheses to form the perfect square $x^2 - 6x + 9 = (x - 3)^2$. However, because of the coefficient in front of the parentheses, we are really adding $-2 \times 9 = -18$; thus 18 must also be added outside the parentheses.

Observe that the 9 added inside the parentheses is the square of one-half of 6;
$$\left(\frac{6}{2}\right)^2 = 9.$$

$$y = -2(x^2 - 6x + 9) - 11 + 18 = -2(x - 3)^2 + 7$$

This is the graph of a parabola that opens downward with vertex at $(3, 7)$; $x = 3$ is the equation of the axis of symmetry.

The process of completing the square may be summarized as follows.

> **COMPLETING THE SQUARE**
>
> To complete the square of quadratic expressions, such as $x^2 + bx$, add the square of one-half of b, the coefficient of x: $\left(\dfrac{b}{2}\right)^2 = \dfrac{b^2}{4}$
>
> Thus: $x^2 + bx + \left(\dfrac{b}{2}\right)^2 = \left(x + \dfrac{b}{2}\right)^2$

By completing the square any quadratic equation $y = ax^2 + bx + c$ may be written in the **standard form** $y = a(x - h)^2 + k$. From this form we can identify the vertex, (h, k), the axis of symmetry, $x = h$, and other information to help us graph the parabola, as illustrated in the following example.

EXAMPLE 1 Write $y = -2 - 4x + 3x^2$ in the form $y = a(x - h)^2 + k$. Find the vertex, axis of symmetry, and graph. On which interval is the function increasing or decreasing? What is the concavity?

Solution First write the given equation in the form $y = 3x^2 - 4x - 2$. Then complete the square.

$$y = 3x^2 - 4x - 2$$

$$= 3\left(x^2 - \tfrac{4}{3}\, x\right) - 2$$

$$= 3\left(x^2 - \tfrac{4}{3}\, x + \tfrac{4}{9}\right) - 2 - \tfrac{4}{3}$$

$$= 3\left(x - \tfrac{2}{3}\right)^2 - \tfrac{10}{3}$$

*Note: Since the original equation is in the form $y = ax^2 + bx + c$, it is very easy to find the y-intercept by letting $x = 0$. This gives the point $(0, -2)$. Since this point is $\tfrac{2}{3}$ unit to the left of the axis of symmetry $x = \tfrac{2}{3}$, we can find the **symmetric point** $\left(\tfrac{4}{3}, -2\right)$.*

Vertex: $\left(\tfrac{2}{3}, -\tfrac{10}{3}\right)$

Axis of symmetry: $x = \tfrac{2}{3}$

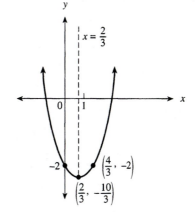

The function is decreasing on $\left(-\infty, \tfrac{2}{3}\right]$ and increasing on $\left[\tfrac{2}{3}, \infty\right)$, and the curve is concave up. ∎

The graph of a quadratic function may or may not intersect the x-axis. Here are some typical cases:

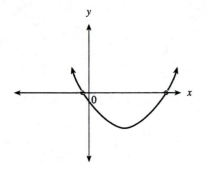

Two x–intercepts;
two solutions for
$y = ax^2 + bx + c = 0$

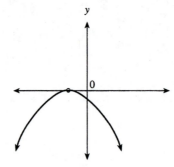

One x–intercept;
one solution for
$y = ax^2 + bx + c = 0$

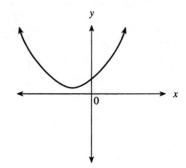

No x–intercepts;
no solutions for
$y = ax^2 + bx + c = 0$

It is clear from these figures that if there are x-intercepts, then these values of x are the solutions to the equation $ax^2 + bx + c = 0$. If there are no x-intercepts, then the equation will not have any real solutions.

As an example let us find the x-intercepts of the parabola $y = f(x) = x^2 - x - 6$. This calls for those values of x for which $y = 0$. That is, we need to solve the equation $y = f(x) = 0$ for x. This can be done by factoring:

$$x^2 - x - 6 = 0$$

$$(x + 2)(x - 3) = 0$$

To solve a quadratic equation by factoring, we make use of this fact: If $A \cdot B = 0$, then $A = 0$ or $B = 0$ or both $A = 0$ and $B = 0$.

Since the product of two factors is zero only when one or both of them are zero, it follows that

$$x + 2 = 0 \quad \text{or} \quad x - 3 = 0$$

$$x = -2 \quad \text{or} \quad x = 3$$

Observe that the x-intercepts are symmetrically spaced about the axis of symmetry,

$x = \frac{1}{2}$. *That is, -2 and 3 are*

each $2\frac{1}{2}$ units from the axis.

The x-intercepts are -2 and 3. The x-intercepts of the parabola are also called the *roots* of the equation $f(x) = 0$, or the *zeros* of the function $f(x)$.

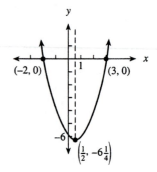

Having found the graph of $y = x^2 - x - 6$, it is relatively easy to draw the graph of $y = |x^2 - x - 6|$. Thus take all of the points below the x-axis, those for which $y < 0$, and reflect them through the x-axis, as shown.

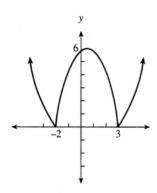

*Note that the term "x-inter-
cepts of a function" is used
to mean the x-intercepts of
the graph of the function.*

EXAMPLE 2   Find the x-intercepts of the function $f(x) = 25x^2 + 30x + 9$.

Solution Let $f(x) = 0$:

$$25x^2 + 30x + 9 = 0$$

$$(5x + 3)(5x + 3) = 0$$

$$5x + 3 = 0$$

$$x = -\tfrac{3}{5}$$

*The number $-\tfrac{3}{5}$ is referred
to as a **double root** of
$25x^2 + 30x + 9 = 0$.*

Check: $f\left(-\tfrac{3}{5}\right) = 25\left(\tfrac{9}{25}\right) + 30\left(-\tfrac{3}{5}\right) + 9 = 9 - 18 + 9 = 0.$

Note: Since there is only *one* answer to this quadratic equation, it follows that the parabola $y = 25x^2 + 30x + 9$ has only one x-intercept; the x-axis is *tangent* to the parabola at its vertex $\left(-\tfrac{3}{5}, 0\right)$.   ∎

The x-intercepts of a quadratic function, if there are any, can be found by completing the square, as in the following example.

EXAMPLE 3   Find the x-intercepts of $f(x) = 2x^2 - 9x - 18$.

Solution Set $f(x) = 0$ and solve for x.

$$2x^2 - 9x - 18 = 0$$

$$2x^2 - 9x = 18$$

$$x^2 - \tfrac{9}{2}x = 9$$

$$x^2 - \tfrac{9}{2}x + \tfrac{81}{16} = 9 + \tfrac{81}{16}$$

$$\left(x - \tfrac{9}{4}\right)^2 = \tfrac{225}{16}$$

$$x - \tfrac{9}{4} = \pm\tfrac{15}{4} \qquad \text{Take the square root of each side.}$$

$$x = \tfrac{9}{4} \pm \tfrac{15}{4}$$

$$x = 6 \quad \text{or} \quad x = -\tfrac{3}{2}$$

As another check, solve the given quadratic equation by factoring.

Check: $2(6)^2 - 9(6) - 18 = 72 - 54 - 18 = 0$

$2\left(-\frac{3}{2}\right)^2 - 9\left(-\frac{3}{2}\right) - 18 = \frac{9}{2} + \frac{27}{2} - 18 = 0$ ∎

In the solution for Example 3, we took the square root of each side. The basis for doing this is the following property:

> ## SQUARE-ROOT PROPERTY FOR EQUATIONS
>
> If $n^2 = k$, then $n = \sqrt{k}$ or $n = -\sqrt{k}$.

A proof of this property can be based on the *zero-product property*. Thus $n^2 = k$ (for $k > 0$), implies $n^2 - k = 0$ or $(n - \sqrt{k})(n + \sqrt{k}) = 0$. Then $n = \sqrt{k}$ or $n = -\sqrt{k}$; that is, $n = \pm\sqrt{k}$.

The method of completing the square can always be used to solve quadratic equations of the form $ax^2 + bx + c = 0$. However, as you have seen in Example 3, this method soon becomes tedious; therefore, we will now develop an efficient formula that can be used in all cases. This will be done by solving the general quadratic equation $ax^2 + bx + c = 0$ for x in terms of the constants, a, b, and c.

We begin with the general quadratic equation.

$$ax^2 + bx + c = 0 \qquad a \neq 0$$

Add $-c$ to each side:

$$ax^2 + bx = -c$$

Divide each side by a ($a \neq 0$).

$$x^2 + \frac{b}{a}x = -\frac{c}{a}$$

Add $\left[\frac{1}{2}\left(\frac{b}{a}\right)\right]^2 = \frac{b^2}{4a^2}$ to each side:

$$x^2 + \frac{b}{a}x + \frac{b^2}{4a^2} = \frac{b^2}{4a^2} - \frac{c}{a}$$

Factor on the left and combine on the right:

$$\left(x + \frac{b}{2a}\right)^2 = \frac{b^2 - 4ac}{4a^2}$$

Take the square root of each side and solve for x.

$$x + \frac{b}{2a} = \pm\sqrt{\frac{b^2 - 4a}{4a^2}} \qquad \text{If } t^2 = k, \text{ then } t = \pm\sqrt{k}.$$

$$x + \frac{b}{2a} = \pm \frac{\sqrt{b^2 - 4ac}}{2a}$$

$$x = -\frac{b}{2a} \pm \frac{\sqrt{b^2 - 4ac}}{2a}$$

Combine terms to obtain the **quadratic formula.**

QUADRATIC FORMULA

If $ax^2 + bx + c = 0$, $a \neq 0$,

then $x = \dfrac{-b \pm \sqrt{b^2 - 4ac}}{2a}$

The values $x = \dfrac{-b + \sqrt{b^2 - 4ac}}{2a}$ and $x = \dfrac{-b - \sqrt{b^2 - 4ac}}{2a}$ are the **roots** of the quadratic equation. These values are also the x-intercepts of the parabola $y = ax^2 + bx + c$, provided they are real numbers.

This formula now allows you to solve any quadratic equation in terms of the constants used. Let us apply it to the equation $2x^2 - 5x + 1 = 0$:

$$2x^2 - 5x + 1 = 0 \qquad x = \frac{-b \pm \sqrt{b^2 - 4ac}}{2a}$$

$$a = 2$$

$$b = -5 \qquad = \frac{-(-5) \pm \sqrt{(-5)^2 - 4(2)(1)}}{2(2)}$$

$$c = 1$$

$$= \frac{5 \pm \sqrt{17}}{4}$$

Thus

$$x = \frac{5 + \sqrt{17}}{4} \quad \text{or} \quad x = \frac{5 - \sqrt{17}}{4}$$

When the radicand $b^2 - 4ac$ in the quadratic formula is negative, then the roots will be imaginary numbers, since the square root of a negative number is imaginary. Recall that $\sqrt{-7} = i\sqrt{7}$ where $i = \sqrt{-1}$. Example 4 is an illustration of a quadratic equation whose roots are imaginary numbers.

EXAMPLE 4 Solve for x: $2x^2 = x - 1$.

Solution First rewrite the equation in the general form $ax^2 + bx + c = 0$.

$$2x^2 - x + 1 = 0$$

Use the quadratic formula with $a = 2$, $b = -1$, and $c = 1$.

$$x = \frac{-(-1) \pm \sqrt{(-1)^2 - 4(2)(1)}}{2(2)}$$

$$= \frac{1 \pm \sqrt{-7}}{4} \qquad\qquad b^2 - 4ac = -7$$

$$= \frac{1 \pm i\sqrt{7}}{4} \quad \text{or} \quad \frac{1}{4} \pm \frac{\sqrt{7}}{4}i \qquad\qquad\blacksquare$$

TEST YOUR UNDERSTANDING

Write in standard form: $y = a(x - h)^2 + k$.

1. $y = x^2 + 4x - 3$ **2.** $y = x^2 - 6x + 7$ **3.** $y = x^2 - 2x + 9$

4. $y = 2x^2 + 8x - 1$ **5.** $y = -x^2 + x - 2$ **6.** $y = \frac{1}{2}x^2 - 3x + 2$

7. Find the x-intercepts of $y = 2x^2 - 2x - 4$ by completing the square.

Use the quadratic formula to solve for x.

8. $x^2 + 3x - 10 = 0$ **9.** $x^2 - x - 12 = 0$ **10.** $x^2 - 2x - 4 = 0$

11. $x^2 + 6x + 6 = 0$ **12.** $x^2 + 6x + 12 = 0$ **13.** $2x^2 + x - 4 = 0$

(Answers: Page 172

As in Example 4, if $b^2 - 4ac < 0$, then no real square roots are possible. Geometrically, this means that the parabola $y = ax^2 + bx + c$ does not meet the x-axis; there are no real solutions for $ax^2 + bx + c = 0$.

When $b^2 - 4ac = 0$, only the solution $x = -\dfrac{b}{2a}$ is possible; the x-axis is tangent to the parabola. Finally, when $b^2 - 4ac > 0$, we have two solutions that are the x-intercepts of the parabola. Since $b^2 - 4ac$ tells us how many (if any) x-intercepts the graph of $y = ax^2 + bx + c$ has, it is called the **discriminant.**

The following equations and graphs illustrate the three possible cases for the value of the discriminant.

Note that for parabolas opening downward, we have the same three possibilities for the x-intercepts.

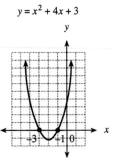

$y = x^2 + 4x + 3$

$b^2 - 4ac = 4 > 0$
Two x-intercepts.
The curve crosses
the x-axis at two
points.

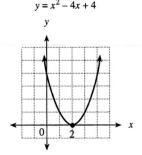

$y = x^2 - 4x + 4$

$b^2 - 4ac = 0$
One x-intercept.
The curve touches
the x-axis at
one point.

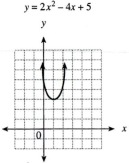

$y = 2x^2 - 4x + 5$

$b^2 - 4ac = -24 < 0$
No x-intercept.
The curve does
not cross or
touch the x-axis.

In summary:

USING THE DISCRIMINANT WHEN a, b, c, ARE REAL NUMBERS

1. **If $b^2 - 4ac > 0$,** then $ax^2 + bx + c = 0$ has two real solutions and the graph of $y = ax^2 + bx + c$ crosses the x-axis at two points. If a, b, c are rational numbers and the discriminant is a perfect square, these roots will be rational numbers; if not, they will be irrational.
2. **If $b^2 - 4ac = 0$,** the solution for $ax^2 + bx + c = 0$ is only one real number (a double root), and the graph of $y = ax^2 + bx + c$ touches the x-axis at one point.
3. **If $b^2 - 4ac < 0$,** $ax^2 + bx + c = 0$ has no real solutions, and the graph of $y = ax^2 + bx + c$ does not cross the x-axis. (The roots are two imaginary numbers.)

Note that if $b^2 - 4ac$ is a perfect square, then the given quadratic will be factorable.

A **quadratic inequality in one variable** can be solved by making use of the graph of the related quadratic function, as demonstrated in Example 5.

EXAMPLE 5   Solve the inequality $x^2 - x - 6 < 0$.

Solution Let $y = f(x) = x^2 - x - 6$. Since this parabola opens upward and has x-intercepts at -2 and 3 (see page 146), the solution of the given inequality can be found as follows.

$y = x^2 - x - 6$ is negative when the parabola is *below* the x-axis; that is, when x is between -2 and 3.

The solution consists of all real numbers x such that $-2 < x < 3$.

These results can be obtained without constructing the graph. All you really need to know are the x-intercepts and whether the parabola opens up or down.

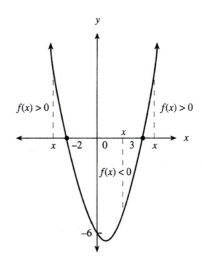

From the figure in Example 5 we can also conclude that $x^2 - x - 6 > 0$ for $x < -2$ or $x > 3$. In general, if $f(x) = ax^2 + bx + c$ has two x-intercepts and $a > 0$, then $f(x) < 0$ for values of x between the intercepts and $f(x) > 0$ for values of x "outside" the intercepts. Can you state similar results for the case $a < 0$? Also, can you decide where $f(x) < 0$ or $f(x) > 0$ when the parabola has only one x-intercept? or no x-intercepts? (See the Written Assignment.)

EXERCISES 2.7

Write in the standard form $y = a(x - h)^2 + k$.

1. $y = x^2 + 2x - 5$ **2.** $y = x^2 - 2x + 5$ **3.** $y = -x^2 - 6x + 2$

4. $y = x^2 - 3x + 4$ **5.** $y = -x^2 + 3x - 4$ **6.** $y = x^2 - 5x - 2$

7. $y = x^2 + 5x - 2$ **8.** $y = 2x^2 - 4x + 3$ **9.** $y = 5 - 6x + 3x^2$

10. $y = -5 + 6x + 3x^2$ **11.** $y = -3x^2 - 6x + 5$ **12.** $y = x^2 - \frac{1}{2}x + 1$

Write each of the following in standard form. Identify the coordinates of the vertex, the equation of the axis of symmetry, the y-intercept, and check by graphing.

13. $y = x^2 + 2x - 1$ **14.** $y = x^2 - 4x + 7$ **15.** $y = -x^2 + 4x - 1$

16. $y = x^2 - 6x + 5$ **17.** $y = 3x^2 + 6x - 3$ **18.** $y = 2x^2 - 4x - 4$

Solve for x by factoring.

19. $x^2 - 5x + 6 = 0$ **20.** $x^2 - 4x + 4 = 0$ **21.** $x^2 - 10x = 0$

22. $2x^2 = x$ **23.** $10x^2 - 13x - 3 = 0$ **24.** $4x^2 - 15x = -9$

25. $4x^2 = 32x - 64$ **26.** $(x + 1)(x + 2) = 30$ **27.** $2x(x + 6) = 22x$

Solve for x by completing the square.

28. $x^2 - 2x - 15 = 0$ **29.** $x^2 - 4x + 1 = 0$ **30.** $2x^2 - 2x = 15$

Use the quadratic formula to solve for x. When the roots are irrational numbers, give their radical forms and also use a calculator to find approximations to two decimal places.

31. $x^2 - 3x - 10 = 0$ **32.** $2x^2 + 3x - 2 = 0$ **33.** $x^2 - 2x - 4 = 0$

34. $2x^2 - 3x - 1 = 0$ **35.** $3x^2 + 7x + 2 = 0$ **36.** $x^2 + 4x + 1 = 0$

37. $x^2 - 6x + 6 = 0$ **38.** $x^2 - 2x + 2 = 0$ **39.** $-x^2 + 6x - 14 = 0$

40. $2 - 4x - x^2 = 0$ **41.** $3x + 1 = 2x^2$ **42.** $6x^2 + 2x = -3$

Find the x-intercepts.

43. $y = 2x^2 - 5x - 3$ **44.** $y = x^2 - 10x + 25$ **45.** $y = x^2 - x + 3$

46. $y = x^2 + 4x + 1$ **47.** $y = 3x^2 + x - 1$ **48.** $y = -x^2 + 4x - 7$

Use the discriminant to describe the solutions as (a) a single real number, (b) two rational numbers, (c) two irrational numbers, or (d) two imaginary numbers.

49. $x^2 - 8x + 16 = 0$ **50.** $2x^2 - x + 1 = 0$ **51.** $-x^2 + 2x + 15 = 0$

52. $4x^2 + x = 3$ **53.** $x^2 + 3x - 1 = 0$ **54.** $2x + 3 = -4x^2$

Use the discriminant to predict how many times, if any, the parabola will be tangent to or cross the x-axis. Then find (a) the vertex, (b) the y-intercept, and (c) the x-intercepts.

55. $y = x^2 - 4x + 4$ **56.** $y = x^2 - 6x + 13$ **57.** $y = 9x^2 - 6x + 1$

58. $y = x^2 - 9x$ **59.** $y = 2x^2 - 4x + 3$ **60.** $y = -x^2 - 4x + 3$

Find the values of b so that the x-axis will be tangent to the parabola. (Hint: Let $b^2 - 4ac = 0$.)

61. $f(x) = x^2 + bx + 9$ **62.** $f(x) = 4x^2 + bx + 25$

63. $f(x) = x^2 - bx + 7$ **64.** $f(x) = 9x^2 + bx + 14$

Find the values of k so that the parabola will intersect the x-axis at two points. (Hint: Let $b^2 - 4ac > 0$.)

65. $f(x) = -x^2 + 4x + k$ **66.** $f(x) = 2x^2 - 3x + k$

67. $f(x) = kx^2 - x - 1$ **68.** $f(x) = kx^2 + 3x - 2$

Find the values of t so that the parabola will not cross the x-axis. (Hint: Let $b^2 - 4ac < 0$.)

69. $f(x) = x^2 - 6x + t$ **70.** $f(x) = 2x^2 + tx + 8$

71. $f(x) = tx^2 - x - 1$ **72.** $f(x) = tx^2 + 3x + 7$

73. Show that the sum of the two roots of $ax^2 + bx + c = 0$ is $-\dfrac{b}{a}$ and that their product is $\dfrac{c}{a}$.

Solve each quadratic inequality.

74. $8 + 2x - x^2 < 0$ **75.** $x^2 - 2x - 3 < 0$ **76.** $2x^2 + 3x - 2 \leq 0$

77. $x^2 + 3x - 10 \geq 0$ **78.** $3 - -x^2 > 0$ **79.** $x^2 + 6x + 9 < 0$

Graph. Show all intercepts.

80. $f(x) = \left(x - \frac{1}{2}\right)(5 - x)$ **81.** $f(x) = (x - 3)(x + 1)$ **82.** $f(x) = x^2 + 1$

83. $f(x) = 4 - 4x - x^2$ **84.** $f(x) = -x^2 + 1$ **85.** $f(x) = 3x^2 - 4x + 1$

Solve for x.

86. $x^4 - 5x^2 + 4 = 0$ *(Hint: Use $u = x^2$ and solve $u^2 - 5u + 4 = 0$.)*

87. $2x^4 - 13x^2 - 7 = 0$ *(Hint: As in Exercise 86 use $u = x^2$.)*

88. $\left(x - \dfrac{2}{x}\right)^2 - 3\left(x - \dfrac{2}{x}\right) + 2 = 0$ $\left(\text{Hint: Let } u = x - \dfrac{2}{x}.\right)$

89. $x^3 + 3x^2 - 4x - 12 = 0$ *(Hint: Factor by grouping.)*

90. $a^3x^2 - 2ax - 1 = 0$ $(a > 0)$

91. $\dfrac{4x}{2x - 1} + \dfrac{3}{2x + 1} = 0$ **92.** $x^2 + 2\sqrt{3}x + 2 = 0$

Graph the function f and decide where f is increasing or decreasing. Where is the curve concave up or down? What is the range?

93. $f(x) = |9 - x^2|$ **94.** $f(x) = |x^2 - 1|$ **95.** $f(x) = |x^2 - x - 6|$

96. Graph the function f where
$$f(x) = \begin{cases} -x^2 + 4 & \text{if } -2 \leq x < 2 \\ x^2 - 10x + 21 & \text{if } 2 \leq x \leq 7 \end{cases}$$

97. Graph the function f where
$$f(x) = \begin{cases} 1 & \text{if } -3 \leq x < 0 \\ x^2 - 4x + 1 & \text{if } 0 \leq x < 5 \\ -2x + 16 & \text{if } 5 \leq x < 9 \end{cases}$$

State the conditions on the values a and k so that the parabola $y = a(x - h)^2 + k$ has the properties indicated.

98. Concave down and has range $y \leq 0$.   **99.** Concave up and has vertex at $(h, 0)$.

100. Concave up and does not intersect the x-axis.   **101.** Concave up and has range $y \geq 2$.

Challenge

1. Assume that a, b, and c in the quadratic equation $ax^2 + bx + c = 0$ are integers. Explain why the discriminant $b^2 - 4ac$ cannot have the value 23. (*Hint:* Either b is even or b is odd. Let $b = 2n$ or $b = 2n + 1$.)

2. Solve for x: $\left(\dfrac{2x}{x + 6}\right)^{2x^2 - 9x + 4} = 1$

Written Assignment

1. State the conditions on the values a and k so that the parabola $y = a(x - h)^2 + k$ opens downward and intersects the x-axis in two points. Explain how you arrived at your choices. What are the domain and range of this function?

2. Suppose the graph of $f(x) = ax^2 + bx + c$ has two x-intercepts and that $a < 0$. In terms of the intercepts, explain the conditions under which $f(x) < 0$, and under which $f(x) > 0$.

3. Suppose the parabola $f(x) = ax^2 + bx + c$ has only one x-intercept. Decide where $f(x) < 0$ or $f(x) > 0$ when $a > 0$ and when $a < 0$. Answer the same questions when there are no x-intercepts.

Graphing Calculator Exercises

1. Check your answers to Exercises 93–95 by graphing the functions with your graphing calculator. Notice that without the absolute value you just get a parabola in each case. But with the absolute value, part of the parabola is flipped over the x-axis on some interval $a \leq x \leq b$. What are a and b for each graph?

2. Check your piecewise-defined graphs in Exercise 96 with your graphing calculator by having it graph $(-x^2 + 4) * (-2 \leq x) * (x < 2) + (x^2 - 10x + 21) * (2 \leq x) * (x \leq 7)$. (Look up the use of relational operators in your user's manual.)

3. Solve the quadratic inequality $x^2 - 10x + 21 \leq 15$ by graphing a function defined with relational operators.

4. Use the quadratic formula to solve for y in the following equation, where x is a quadratic function of y:

$$x = 0.5y^2 + 4y - 2$$

(*Hint:* Think of x as a constant in $0 = 0.5y^2 + 4y - (2 + x)$.) This gives you two functions of x, one for the plus sign and one for the minus sign. Graph both functions on the same viewing screen. Where is the vertex of the resulting horizontal parabola? Check with your TRACE facility.

Critical Thinking

1. How can you determine the maximum or minimum value of a function given by $f(x) = ax^2 + bx + c$?

2. Suppose that a function is increasing on one side of a point P with coordinates (a, b) and decreasing on the other side of this point. What can be said about the point P? Explain.

3. Consider the graph of $y = (x + 2)^2 - 1$. How can you use this to determine the graph of $x = (y + 2)^2 - 1$?

4. Describe the graph of $y = a(x - h)^2 + k$ for various values of a, h, and k. For example, consider $a > 0$, $h < 0$, and $k > 0$ as one case.

5. Sketch the graph of a function such that any horizontal line drawn through a range value will intersect the curve exactly once. Such functions are called *one-to-one functions*. If f is a one-to-one function, and if $f(x_1) = f(x_2)$, what can you say about x_1 and x_2?

2.8 APPLICATIONS OF QUADRATIC FUNCTIONS

The algebraic and graphical techniques developed in this chapter can be used to solve a variety of problems that involve quadratic equations. The first two examples illustrate situations that call for the solution of a quadratic equation of the form $ax^2 + bx + c = 0$.

EXAMPLE 1 The length of a rectangular piece of cardboard is 2 inches more than its width. As in the following figure, an open box is formed by cutting out 4-inch squares from each corner and folding up the sides. If the volume of the box is 672 cubic inches, find the dimensions of the original cardboard.

Consider the steps used to solve this problem:
(a) Select the necessary information from the statement of the problem.
(b) Represent the information algebraically.
(c) Draw a figure.
(d) Write and solve an equation.
(e) Interpret the solution to meet the given conditions of the problem.

Solution Let x be the width of the cardboard. Then $x + 2$ is the length.

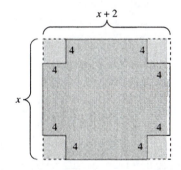

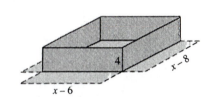

After the squares are cut off and the sides are folded up, the dimensions of the box are:

$$\text{length} = \ell = x + 2 - 8 = x - 6$$

$$\text{width} = w = x - 8$$

$$\text{height} = h = 4$$

Since the volume is to be 672 cubic inches, and $V = \ell wh$, we have

$$(x - 6)(x - 8)4 = 672$$

$$(x - 6)(x - 8) = 168$$

$$x^2 - 14x + 48 = 168$$

$$x^2 - 14x - 120 = 0$$

Note: We reject the root $x = -6$ since the dimensions must be positive numbers.

$$(x - 20)(x + 6) = 0$$

$$x = 20 \quad \text{or} \quad x = -6$$

The dimensions of the original cardboard are: $w = 20$ inches, $\ell = 20 + 2 = 22$ inches. Check this solution. ∎

EXAMPLE 2 Find two consecutive positive integers such that the sum of their squares is 113.

Solution Let $x =$ the first integer. Then $x + 1 =$ the next consecutive integer. The sum of their squares can be represented as $x^2 + (x + 1)^2$. Thus

$$x^2 + (x + 1)^2 = 113$$

$$x^2 + x^2 + 2x + 1 = 113$$

$$2x^2 + 2x + 1 = 113$$

$$2x^2 + 2x - 112 = 0$$

$$x^2 + x - 56 = 0$$

$$(x - 7)(x + 8) = 0$$

$$x = 7 \quad \text{or} \quad x = -8 \qquad \text{If } A \cdot B = 0, \text{ then } A = 0 \text{ or } B = 0.$$

If it were not required that the integers be positive, then another solution consists of the consecutive integers $-8, -7$.

Since the integers were said to be positive, we must reject the value $x = -8$. Therefore, $x = 7$ and $x + 1 = 8$. The two consecutive integers are 7 and 8. ∎

Some problems require the minimum or maximum value of a quadratic function. Such problems can be solved by using the standard form

$$y = f(x) = a(x - h)^2 + k$$

which shows the vertex of the parabola to be (h, k). As observed in Section 2.6, this point is the lowest point or the highest point on the parabola, depending on the sign of a. When $a > 0$, the vertex is the lowest point on the parabola; when $a < 0$, it is the highest point. These special points will be useful in solving certain applied problems.

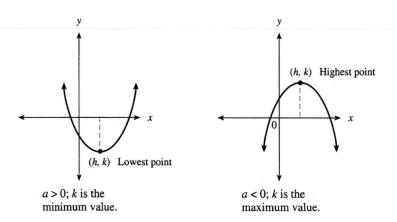

$a > 0$; k is the minimum value.

$a < 0$; k is the maximum value.

In summary, we have the following:

MINIMUM AND MAXIMUM VALUES OF QUADRATIC FUNCTIONS

$$y = f(x) = a(x - h)^2 + k \qquad (a \neq 0)$$

If $a > 0$, the graph is a parabola that opens upward and f has a **minimum value** $f(h) = k$.

If $a < 0$, the graph is a parabola that opens downward and f has a **maximum value** $f(h) = k$.

The examples that follow involve the concepts of minimum and maximum values of quadratic functions.

EXAMPLE 3   Find the maximum value or minimum value of the function $f(x) = 2(x + 3)^2 + 5$.

Solution   Since $2(x + 3)^2 + 5 = 2[x - (-3)]^2 + 5$, we note that $(-3, 5)$ is the turning point. Also since $a = 2 > 0$, the parabola opens upward and $f(-3) = 5$ is the minimum value.   ■

EXAMPLE 4   Find the maximum value of the quadratic function $f(x) = -\frac{1}{3}x^2 + x + 2$. At which value x does f achieve this maximum?

The purpose of converting to the form $a(x - h)^2 + k$ is to find the vertex (h, k): $h = \frac{3}{2}$ and $k = \frac{11}{4}$.

Solution   Convert to the form $a(x - h)^2 + k$:

$$y = f(x) = -\tfrac{1}{3}(x^2 - 3x) + 2$$

$$= -\tfrac{1}{3}(x^2 - 3x + \tfrac{9}{4}) + 2 + \tfrac{3}{4}$$

$$= -\tfrac{1}{3}(x - \tfrac{3}{2})^2 + \tfrac{11}{4}$$

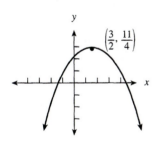

From this form we have $a = -\frac{1}{3}$. Since $a < 0$, $\left(\frac{3}{2}, \frac{11}{4}\right)$ is the highest point of the parabola. Thus f has a maximum value of $\frac{11}{4}$ when $x = \frac{3}{2}$.   ■

TEST YOUR UNDERSTANDING

(Answers: Page 172)

Find the maximum or minimum value of each quadratic function and state the x-value at which this occurs.

1. $f(x) = x^2 - 10x + 21$ **2.** $f(x) = x^2 + \frac{4}{3}x - \frac{7}{18}$

3. $f(x) = 10x^2 - 20x + \frac{21}{2}$ **4.** $f(x) = -8x^2 - 64x + 3$

5. $f(x) = -2x^2 - 1$ **6.** $f(x) = x^2 - 6x + 9$

7. $f(x) = 25x^2 + 70x + 49$ **8.** $f(x) = (x - 3)(x + 4)$

For each x between 0 and 30, such a rectangle is possible. Here are a few.

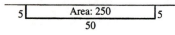

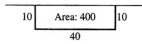

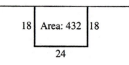

EXAMPLE 5   Suppose that 60 meters of fencing is available to enclose a rectangular garden, one side of which will be against the side of a house. What dimensions of the garden will guarantee a maximum area?

Solution   Use the given information to write a quadratic function. Then write the function in standard form so as to determine the vertex of the corresponding parabola. The maximum or minimum value always occurs at the vertex.

From the sketch you can see that the 60 meters need only be used for three sides, two of which are of the same length x.

Example 5 shows how to select the rectangle of maximum area from such a vast collection of possibilities. Can you explain why the domain of A(x) is $0 < x < 30$?

Let x represent the width of the garden. Then $60 - 2x$ represents the length, and the area A is given by

$$A(x) = x(60 - 2x)$$
$$= 60x - 2x^2$$

To "maximize" A, convert to the form $a(x - h)^2 + k$. Thus

$$A(x) = -2(x^2 - 30x)$$
$$= -2(x^2 - 30x + 225) + 450$$
$$= -2(x - 15)^2 + 450$$

Therefore, the maximum area of 450 square meters is obtained when the dimensions are $x = 15$ meters by $60 - 2x = 30$ meters.   ∎

EXAMPLE 6   The sum of two numbers is 24. Find the two numbers if their product is to be a maximum.

Solution   Let x represent one of the numbers. Since the sum is 24, the other number is $24 - x$. Now let p represent the product of these numbers.

$$p = x(24 - x)$$
$$= -x^2 + 24x$$
$$= -(x^2 - 24x)$$
$$= -(x^2 - 24x + 144) + 144$$
$$= -(x - 12)^2 + 144$$

Try to solve Example 6 if the product is to be a minimum.

The product has a maximum value of 144 when $x = 12$. Hence the numbers are 12 and 12. ∎

EXAMPLE 7 A ball is thrown straight upward from ground level with an initial velocity of 32 feet per second. The formula $s = 32t - 16t^2$ gives its height in feet, s, after t seconds.
(a) What is the maximum height reached by the ball?
(b) When does the ball return to the ground?

Solution
(a) First complete the square in t.

$$s = 32t - 16t^2$$
$$= -16t^2 + 32t$$
$$= -16(t^2 - 2t)$$
$$= -16(t^2 - 2t + 1) + 16$$
$$= -16(t - 1)^2 + 16$$

You should now recognize this as a parabola with vertex at (1, 16). Because the coefficient of t^2 is negative, the curve opens downward. The maximum height, 16 feet, is reached after 1 second.

The motion of the ball is straight up and down.

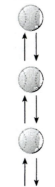

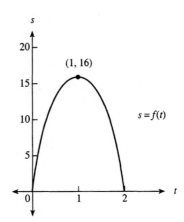

This parabolic arc is the graph of the relation between time t and distance s. It is *not* the path of the ball.

(b) The ball hits the ground when the distance $s = 0$ feet. Thus

$$s = 32t - 16t^2 = 0$$
$$16t(2 - t) = 0$$
$$t = 0 \text{ or } t = 2$$

Since time $t = 0$ is the starting time, the ball returns to the ground 2 seconds later, when $t = 2$. ∎

EXAMPLE 8 The marketing department of the TENRAQ Tennis Company found that, on the average, 600 tennis racquets will be sold monthly at the unit price of $100. The department also observed that for each $5 reduction in price, an extra 50 racquets will be sold monthly. What price will bring the largest monthly income?

Solution Let x be the number of $5 reductions in price for the racquet. Then $5x$ is the total reduction and

$$100 - 5x = \text{reduced unit price}$$

Also, $50x$ is the increase in sales per month and

$$600 + 50x = \text{number of racquets sold monthly}$$

For example, if $x = 2$ the reduction is $10, the sales would be 700, and the income would be ($90)(700) = $63,000.

The monthly income, R, will be the unit price times the number of units sold. Then

$$R = (\text{unit price})(\text{units sold})$$
$$R(x) = (100 - 5x)(600 + 50x)$$
$$= 60{,}000 + 2000x - 250x^2$$
$$= -250(x^2 - 8x) + 60{,}000$$
$$= -250(x^2 - 8x + 16) + 60{,}000 + 4000$$
$$= -250(x - 4)^2 + 64{,}000 \qquad \text{This is in the form } a(x - h)^2 + k.$$

Since $a = -250 < 0$, the maximum monthly income of $64,000 is obtained when $x = 4$. The unit price should be set at $100 - 5(4)$ or $80. ∎

EXAMPLE 9 Each point Q between the endpoints of the segment OT determines a rectangle $PQRS$, as shown. Using $x = OP$, write the area of such a rectangle as $A(x)$, a function of x. Find the coordinates of Q that give the rectangle of maximum area.

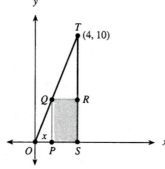

Since the coordinates of T are $(4, 10)$, note that $OS = 4$ and $ST = 10$.

Solution Triangles *OPQ* and *OST* are similar. Thus

$$\frac{PQ}{x} = \frac{ST}{OS} = \frac{10}{4} \quad \text{or} \quad \frac{PQ}{x} = \frac{5}{2} \quad \text{and} \quad PQ = \frac{5}{2}x$$

Then the area of *PQRS* is given by

$$A(x) = (PQ)(PS)$$

$$= \frac{5}{2}x\,(4 - x)$$

Now convert $A(x)$ to standard form:

$$A(x) = \frac{5}{2}x\,(4 - x)$$

$$= -\frac{5}{2}(x^2 - 4x)$$

$$= -\frac{5}{2}(x^2 - 4x + 4) + 10$$

$$= -\frac{5}{2}(x - 2)^2 + 10$$

The graph of this function is a parabola that opens downward and has a maximum of 10 for $x = 2$. Since $PQ = \frac{5}{2}x$, the rectangle of maximum area occurs when $x = 2$ and $y = \frac{5}{2}(2) = 5$. That is, the coordinates of *Q* will be (2, 5) at this point.

EXERCISES 2.8

Write and solve a quadratic equation for each exercise.

1. Find two consecutive positive integers whose product is 210.

2. The sum of a number and its square is 56. Find the number. (There are two possible answers.)

3. One positive integer is 3 greater than another. The sum of the squares of the two integers is 89. Find the integers.

4. Find two integers whose sum is 26 and whose product is 165.
   (*Hint:* Let the two integers be represented by x and $26 - x$.)

5. How wide a border of uniform width should be added to a rectangle that is 8 feet by 12 feet in order to double the area?

6. The length of a rectangle is 3 centimeters greater than its width. The area is 70 square centimeters. Find the dimensions of the rectangle.

7. The area of a rectangle is 15 square centimeters and the perimeter is 16 centimeters. What are the dimensions of the rectangle? (*Hint:* If x represents the width, then $\dfrac{16 - 2x}{2} = 8 - x$ represents the length.)

8. The altitude of a triangle is 5 centimeters less than the base to which it is drawn. The area of the triangle is 21 square centimeters. Find the length of the base.

9. A backyard swimming pool is rectangular in shape, 10 meters wide and 18 meters long. It is surrounded by a walk of uniform width whose area is 52 square meters. How wide is the walk?

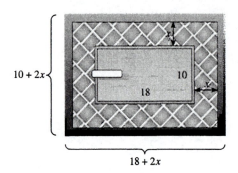

10. A square piece of tin is to be used to form a box without a top by cutting off a 2-inch square from each corner and then folding up the sides. The volume of the box will be 128 cubic inches. Find the length of a side of the original square.

11. The sum S of the first n consecutive positive integers, $1, 2, 3, \ldots, n$, is given by the formula $S = \frac{1}{2}n(n + 1)$. Find n when $S = 120$ and check your answer by addition.

12. The sum S of the first n consecutive even positive integers, $2, 4, 6, \ldots, 2n$, is given by the formula $S = n(n + 1)$. How many such consecutive even positive integers must be added to get a sum of 342?

13. A boat travels downstream (with the current) for 36 miles and then makes the return trip upstream (against the current). The trip downstream took $\frac{3}{4}$ of an hour less than the trip upstream. If the rate of the current is 4 mph, find the rate of the boat in still water and the time for each part of the trip.

14. Two motorcycles each make the same 220-mile trip, but one of them travels 5 mph faster than the other. Find the rate of each motorcycle if the slower one takes 24 minutes longer to complete the trip than the faster one.

Find the maximum or minimum value of the quadratic function and state the x-value at which this occurs.

15. $f(x) = -x^2 + 10x - 18$

16. $f(x) = x^2 + 18x + 49$

17. $f(x) = 16x^2 - 64x + 100$

18. $f(x) = -\frac{1}{2}x^2 + 3x - 6$

19. $f(x) = 49 - 28x + 4x^2$

20. $f(x) = x(x - 10)$

21. $f(x) = -x\left(\frac{2}{3} + x\right)$

22. $f(x) = (x - 4)(2x - 7)$

23. A manufacturer is in the business of producing small models of the Statue of Liberty. He finds that the daily cost in dollars, C, of producing n statues is given by the quadratic formula $C = n^2 - 120n + 4200$. How many statues should be produced per day so that the cost will be minimum? What is the minimal daily cost?

24. A company's daily profit, P, in dollars, is given by $P = -2x^2 + 120x - 800$, where x is the number of articles produced per day. Find x so that the daily profit is a maximum.

25. The sum of two numbers is 12. Find the two numbers if their product is to be a maximum. (*Hint:* Find the maximum value for $y = x(12 - x)$.)

26. The sum of two numbers is n. Find the two numbers such that their product is a maximum. Are there two numbers that will give a minimum product? Explain.

27. The difference of two numbers is 22. Find the numbers if their product is to be a minimum and also find this product.

28. A homeowner has 40 feet of wire and wishes to use it to enclose a rectangular garden. What should be the dimensions of the garden so as to enclose the largest possible area?

29. A gardener has 300 feet of fencing to enclose three adjacent rectangular growing areas. If all three rectangles are to have the same dimensions, and if one side is to be against a building as shown, what dimensions should be used so that the maximum growing area will be enclosed?

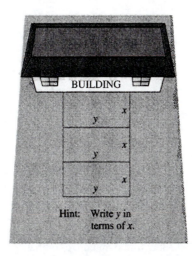

Hint: Write y in terms of x.

The formula $h = 128t - 16t^2$ gives the distance in feet above the ground, h, reached by an object in t seconds. Use this formula for Exercises 30 through 33. (Assume the object is thrown straight upward from ground level.)

30. What is the maximum height reached by the object?

31. How long does it take for the object to reach its maximum height?

32. How long does it take for the object to return to the ground from the time it is first thrown?

33. In how many seconds will the object be at a height of 192 feet? Why are there two possible answers?

34. Suppose it is known that if 65 apple trees are planted in a certain orchard, the average yield per tree will be 1500 apples per year. For each additional tree planted in the same orchard, the annual yield per tree drops by 20 apples. How many trees should be planted in order to produce the maximum crop of apples per year? (*Hint:* If n trees are added to the 65 trees, then the yield per tree is $1500 - 20n$.)

35. It is estimated that 14,000 people will attend a basketball game when the admission price is $7.00. For each 25¢ added to the price, the attendance will decrease by 280. What admission price will produce the largest gate receipts? (*Hint:* If x quarters are added, the attendance will be $14,000 - 280x$.)

36. The sum of the lengths of the two perpendicular sides of a right triangle is 30 centimeters. What are the lengths if the square of the hypotenuse is a minimum?

37. Each point P on the line segment between endpoints $(0, 4)$ and $(2, 0)$ determines a rectangle with dimensions x by y as shown in the figure. Find the coordinates of P that give the rectangle of maximum area.

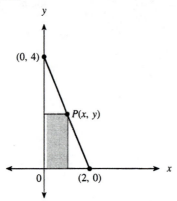

38. Each point $P(x, y)$ on the line segment between $(0, 0)$ and $(5, 7)$ determines a rectangle, as shown. Express the area of the rectangle as a function of x.

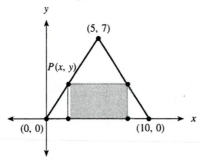

39. For Exercise 38, find the coordinates of P that give the rectangle of maximum area. What is the maximum area?

40. Verify that the coordinates of the vertex of the parabola $y = ax^2 + bx + c$ are

$$\left(-\frac{b}{2a}, \quad \frac{4ac - b^2}{4a}\right)$$

Use the result in Exercise 40 to find the coordinates of the vertex of each parabola, and decide whether $\dfrac{4ac - b^2}{4a}$ is a maximum or minimum value.

41. $y = 2x^2 - 6x + 9$ **42.** $y = -3x^2 + 24x - 41$ **43.** $y = -\frac{1}{2}x^2 - \frac{1}{3}x + 1$

44. $y - x^2 + 5x = 0$ **45.** $y + \frac{2}{3}x^2 = 9$ **46.** $y = 10x^2 + 100x + 1000$

47. The following quadratic formula can be used to approximate the distance in feet, d, that it takes to stop a car after the brakes are applied for a car traveling on a dry road at the rate of r miles per hour: $d = 0.045r^2 + 1.1r$. To the nearest foot, how far will a car travel once the brakes are applied at (a) 40 mph, (b) 55 mph, and (c) 65 mph?

48. The police measured the skid marks made by a car that crashed into a tree. If the measurement gave a braking distance of 250 feet, was the driver exceeding the legal speed limit of 55 miles per hour? To the nearest mph, what was the speed of the car before the brakes were applied? (Use the formula in Exercise 47.)

49. When a department store sold a certain style of shirt for $20, the average number of shirts sold per week was 100. The store observed that with each $1 decrease in price, 10 more shirts were sold weekly. What unit price should be set for the shirts in order to realize the maximum weekly revenue?

 Graphing Calculator Exercises

1. Check your answer to Exercise 41 by graphing the quadratic function. Use a friendly window (for example, $-4.8 \le x \le 4.7$, $-9.6 \le y \le 9.3$ on the TI-81; $-4.7 \le x \le 4.7$, $-9.3 \le y \le 9.3$ on the Casio fx-7700G) and your TRACE facility to locate the vertex of the parabola.

2. Graph the function $y = x^4 - 2x^2$ on a friendly window. Does the function have a maximum value for all x? A minimum value? ZOOM out to see more of the graph. Does this change your mind about the maximum or minimum?

3. Graph the function $y = x^3 - 4x$ on a friendly window. Does this function have a maximum value for all x? A minimum value?

CHAPTER REVIEW EXERCISES

Section 2.1   Introduction to the Function Concept   (Page 86)

1. State the definition of a function.

Decide whether the given equation defines y to be a function of x. For each function, find the domain.

2. $y = (x - 3)^2$ **3.** $y = \dfrac{1}{x^2 + 1}$ **4.** $y = \sqrt{x - 5}$

5. $y^2 = x + 1$ **6.** $y = \dfrac{1}{\sqrt{1 - x}}$ **7.** $y = \pm x$

8. y is a function of x defined by

$$y = \begin{cases} x^2 & x < -2 \\ 2 - x & -2 \le x < 1 \\ 3x + 2 & x \ge 1 \end{cases}$$

Find the values of y for **(a)** $x = -3$, **(b)** $x = -1$, and **(c)** $x = 1$.

9. For $f(x) = x^2 + 3x - 2$, find $2f(3)$ and $f(6)$, and compare.

10. Let $f(x) = 2x^2 - 5x$ and find: **(a)** $f(-2)$, **(b)** $f(0)$, **(c)** $2f(x)$, and **(d)** $f(2x)$.

11. For $g(x) = x^2 + 1$, evaluate and simplify the difference quotient $\dfrac{g(x) - g(2)}{x - 2}$.

12. For $g(x) = -x^2$, evaluate and simplify the difference quotient $\dfrac{g(1 + h) - g(1)}{h}$.

13. For the function $f(x) = \dfrac{2}{x - 1}$, find **(a)** $f(3x)$ and **(b)** $3f(x)$.

14. In the figure, right triangle ABC is similar to triangle ADE. $BC = 8$, $EC = 5$, and h and x are the measures of the altitude and base of triangle ADE. Express h as a function of x. Also express the area of $\triangle ADE$ as a function of x.

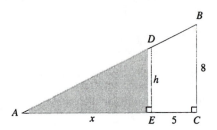

Section 2.2 Graphing Lines in the Rectangular Coordinate System (Page 95)

15. State the definition of the slope of a line between two points (x_1, y_1) and (x_2, y_2).

16. What is the slope of a horizontal line? Of a vertical line?

17. What is the relationship between the slopes of two perpendicular lines?

18. Graph the equation $y = x - 3$.

19. Find the x- and y-intercepts for $x - 3y = 6$. Use these to graph the line.

20. Find the slope of the line determined by the points $(3, -4)$ and $(-1, 6)$.

21. Graph the line with slope $\frac{3}{2}$ that passes through the point $(2, -3)$.

22. Describe the nature of a line that has a positive slope as compared with one that has a negative slope.

23. The weekly demand, D, for a certain calculator is related to the price, x, in dollars by the function $y = D = 2000 - 50x$, $0 < x \le 40$. The weekly supply is given by the function $y = S = 20x + 600$. Find the equilibrium point and explain what it means.

Section 2.3 Algebraic Forms of Linear Equations (Page 107)

24. Write the slope-intercept form of a line where m is the slope and b is the y-intercept.

25. Write the point-slope form of a line where m is the slope and (x_1, y_1) is a point on the line.

26. Describe the graph of a constant function. State the domain and range of the function.

27. Graph the linear function f defined by $y = f(x) = 2x + 1$ by using the slope and y-intercept.

28. Find the equation of the line with slope $-\frac{3}{4}$ and passing through the point $(0, 2)$.

29. Write and graph an equation of the line through the point $(-3, 2)$ that is **(a)** parallel to the x-axis, **(b)** parallel to the y-axis.

30. Write the point-slope form of a line with slope $m = -2$ that passes through the point $(1, -3)$.

31. Write the equation of the line that passes through the points $(-3, 2)$ and $(-2, 5)$:

   (a) in point-slope   **(b)** in slope-intercept form   **(c)** in the general form $Ax + By = C$.

32. Find the slope and y-intercept of the line $-3x + 2y - 6 = 0$.

33. Write the equation of the line parallel to $2x + 3y = 4$ that passes through the point $(-3, 1)$. What is the equation of the line through $(-3, 1)$ that is perpendicular to the given line?

Section 2.4 Piecewise Linear Functions (Page 115)

34. Graph: $y = 1 - 2x$ for $-1 \leq x \leq 2$.

35. Graph: $y = |x + 3|$ **36.** Graph: $y = |1 - x|$

37. Graph: $y = f(x) = \begin{cases} x & \text{if } -2 \leq x < 1 \\ x + 2 & \text{if } x \geq 1 \end{cases}$

38. Evaluate: **(a)** $[4.7]$ **(b)** $[-4.7]$ **(c)** $[\pi]$ **(d)** $[-0.45]$

39. Graph the step function $y = f(x) = [x]$ for $-3 < x \leq 1$. What is the domain of f? What is the range of f?

40. Repeat Exercise 39 for the function $f(x) = -[x]$.

Section 2.5 Introduction to Systems of Equations (Page 122)

Solve each system by the substitution method.

41. $3x + y = 9$
   $x + 2y = 8$

42. $-2x + 3y = 17$
   $3x - y = -8$

43. $2s - 5t = 15$
   $-3s + 2t = -17$

Solve each system.

44. $-2x + 2y = 5$
   $4x + 8y = -4$

45. $2x + 4y = -2$
   $3x - 2y = -5$

46. $x + 2y = -8$
   $\frac{1}{2}x - \frac{2}{3}y = 6$

State whether the given systems are consistent, inconsistent, or dependent.

47. $3x + 2y = 6$
   $y = -\frac{3}{2}x + 3$

48. $2x - 3y = -12$
   $10x - 15y = -3$

49. $3x - y = 8$
   $2x + 3y = 4$

Use a system of two linear equations in two variables to solve each problem.

50. The sum of two numbers is 20. The larger number is four less than twice the smaller. What are the numbers?

51. It costs a manufacturer $6 to produce one calculator. In addition, the company has a daily fixed cost of $240.
   (a) Find the total cost for producing x calculators per day.
   (b) Find the total daily revenue if the company sells x calculators per day for $10 each.
   (c) Find the daily break-even point.

52. A merchant said that it did not matter whether one pair of shoes was sold for $31 or two pairs for $49, because the profit was the same for each sale. How much does one pair of shoes cost the merchant and what is the profit?

Section 2.6 Graphing Quadratic Equations (Page 132)

53. What is meant by a quadratic function?

54. Under what conditions is the graph of a function $f(x)$ said to be symmetric with respect to the y-axis?

55. Draw each graph on the same set of axes:
   (a) $y = x^2$ **(b)** $y = (x + 2)^2$ **(c)** $y = (x + 2)^2 - 3$

Draw the graph of each function.

56. $y = x^2 - 2$ **57.** $y = -x^2 - 2$ **58.** $y = -(x - 2)^2$

59. $y = 2x^2 - 1$ **60.** $y = \frac{1}{2}x^2 + 3$ **61.** $y = 2(x - 1)^2$

For each of the following functions, tell where the function is increasing and decreasing. What is the concavity? Give the coordinates of the vertex and the equation of the axis of symmetry.

62. $f(x) = 4(x - 2)^2 + 1$ **63.** $f(x) = -3(x + 2)^2 + 3$ **64.** $f(x) = 5(x - 1)^2 - 3$

65. Consider the graph of $y = a(x - h)^2 + k$.

     (a) What are the coordinates of the vertex?

     (b) What is the equation of the axis of symmetry?

66. Using a suitable diagram, describe the vertical line test for functions.

67. Graph the parabola $x = (y - 2)^2 + 3$. Identify the vertex and the axis of symmetry.

Section 2.7   The Quadratic Formula   (Page 144)

Write in standard form.

68. $y = x^2 + 4x - 3$ **69.** $y = -2x^2 - 12x + 11$ **70.** $y = \frac{1}{2}x^2 + 3x - 3$

71. State the quadratic formula.

72. Solve for x by completing the square: $3x^2 + 4x - 2 = 0$

Use the quadratic formula to solve for x.

73. $x^2 + 7x + 10 = 0$ **74.** $x^2 - 10x + 25 = 0$ **75.** $x^2 + 4x - 7 = 0$

76. $x^2 - 4x + 7 = 0$ **77.** $2x^2 - x - 3 = 0$ **78.** $3x^2 + x + 3 = 0$

79. State the conditions on the values a and k so that the parabola $y = a(x - h)^2 + k$ opens downward and intersects the x-axis in two points. What are the domain and range of this function?

80. Describe how the discriminant can be used to tell the nature of the roots of a quadratic equation of the form $ax^2 + bx + c = 0$.

81. Find the x-intercepts of the function $f(x) = 2x^2 + x - 6$ by **(a)** factoring and **(b)** completing the square.

Use the discriminant to describe the solutions of each equation.

82. $x^2 - 2x + 5 = 0$ **83.** $4x^2 - 4x = -1$ **84.** $2x^2 + x - 5 = 0$

85. Graph: $y = |x^2 + 3x - 4|$

86. Solve the quadratic inequality $6x^2 + 7x - 5 > 0$.

Section 2.8   Applications of Quadratic Functions   (Page 155)

State the maximum or minimum value of each quadratic function and name the x-value at which this occurs.

87. $f(x) = 2(x - 1)^2$ **88.** $f(x) = -2(x + 1)^2 - 5$ **89.** $f(x) = 3(x + 2)^2 - 1$

90. $f(x) = x^2 - 6x + 8$ **91.** $f(x) = x^2 + 4x + 4$ **92.** $f(x) = -2x^2 - 6x + 5$

93. The sum of two numbers is 40. Find the two numbers if their product is a maximum.

94. Suppose that 120 meters of fencing is available to enclose a rectangular garden. What dimensions should be used in order to have a garden with maximum area if one side of the garden is to be alongside a house?

95. The formula $h = -16t^2 + 32t + 80$ gives the distance in feet above the ground, h, reached by an object in t seconds if thrown straight upward from the top of an 80-foot building.

     (a) What is the maximum height reached by the object?

(b) How long does it take to reach its maximum height?

(c) How long does it take for the object to hit the ground?

96. Describe the minimum or maximum value of the quadratic function $y = f(x) = a(x - h)^2 + k$ for **(a)** $a < 0$ and **(b)** $a > 0$.

CHAPTER 2: STANDARD ANSWER TEST

Use these equations to test your knowledge of the basic skills and concepts of Chapter 2. Then check your answers with those given at the back of the book.

1. Find the domain of the function given by $y = \dfrac{1}{\sqrt[3]{x^3 + 8}}$.

2. Let $g(x) = \dfrac{3}{x}$. Find **(a)** $g(2 + x)$ **(b)** $g(2) + g(x)$.

3. Let $g(x) = x^2 - x$. Evaluate and simplify: $\dfrac{g(x) - g(9)}{x - 9}$.

4. Find the x- and y-intercepts and use these to graph $3x - 2y = 6$.

5. Find the slope of the line determined by these points:

   (a) $(2, -3)$ and $(-1, 4)$ **(b)** $(-3, 2)$ and $(4, 2)$

6. Graph. State the domain and range: $y = 2x + 1$

7. Write the equation of the line through the point $(3, -2)$ that is:

   (a) Parallel to the y-axis   **(b)** Parallel to the x-axis

8. Write the equation of the line with slope $m = \frac{1}{2}$ and y-intercept $b = -3$.

9. Write the equation $2x - 3y = 5$ in slope-intercept form. Give the slope and y-intercept.

10. Write the slope-intercept form of the line through the points $(3, -5)$ and $(-2, 4)$.

11. Write the point-slope form of the line through the point $(2, 8)$ that is perpendicular to the line $y = -\frac{2}{5}x + 3$.

12. Graph: $y = |x + 2|$ **13.** Graph the step function $y = \dfrac{|x + 1|}{x + 1}$.

14. Graph $y = 2 - x$ for $-1 \leq x \leq 2$.   **15.** Graph $y = (x - 2)^2 + 3$.

16. Graph the function $y = f(x) = x^2 - 9$.

17. Let $y = -5x^2 + 20x - 1$.

   (a) Write the quadratic in the standard form $y = a(x - h)^2 + k$.

   (b) Give the coordinates of the vertex.

   (c) Write the equation of the axis of symmetry.

   (d) State the domain and range of the quadratic function.

18. Solve for x: $3x^2 - 8x - 3 = 0$.

19. Find the x-intercepts of $y = -x^2 + 4x + 7$.

20. Find two positive consecutive odd integers such that the sum of their squares is 202.

21. Give the value of the discriminant and use this result to describe the x-intercepts, if any:

   (a) $y = x^2 + 3x + 1$ **(b)** $y = 6x^2 + 5x - 6$

22. Solve the system: $3x + 4y = 7$
$$2x - 3y = 16$$

23. Find the maximum or minimum value of the quadratic function and state the x-value at which this occurs: $f(x) = -\frac{1}{2}x^2 - 6x + 2$.

24. In the figure the altitude BC of triangle ABC is 4 feet. The part of the perimeter $PQRCB$ is to be a total of 28 feet. How long should x be so that the area of rectangle $PCRQ$ is a maximum?

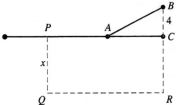

25. The formula $h = 64t - 16t^2$ gives the distance in feet above ground, h, reached by an object in t seconds. What is the maximum height reached by the object?

CHAPTER 2: MULTIPLE CHOICE TEST

1. The line parallel to the line $2x - 3y = 5$ that passes through the point $(-8, 4)$ has equation:

   (a) $y - 4 = -\frac{2}{3}(x + 8)$ **(b)** $y - 4 = \frac{2}{3}(x + 8)$ **(c)** $y - 4 = \frac{3}{2}(x + 8)$

   (d) $y - 4 = -\frac{3}{2}(x + 8)$ **(e)** None of the preceding

2. Let $f(x) = x - 2$. Then $f(x - 2) =$

   (a) $x - 2$ **(b)** $2x - 2$ **(c)** $2x - 4$ **(d)** $x - 4$ **(e)** None of the preceding

3. Let $g(x) = (x - 2)^2$. Then $\dfrac{g(x) - g(7)}{x - 7} =$

   (a) $x - 7$ **(b)** $x + 7$ **(c)** $x - 3$ **(d)** $x + 3$ **(e)** None of the preceding

4. Which of the following is the equation of the line that passes through the point $(2, -3)$ and is parallel to the y-axis?

   (a) $x = 2$ **(b)** $x = -2$ **(c)** $y = 3$ **(d)** $y = -3$ **(e)** None of the preceding

5. Which of the following statements are correct?

   I. The slope of a horizontal line is undefined.

   II. The slope of a vertical line is 0.

   III. The slopes of two perpendicular lines (not parallel to the coordinate axes) are negative reciprocals of one another.

   (a) Only I   **(b)** Only II   **(c)** Only III   **(d)** Only I and II   **(e)** None of the preceding

6. Which of the following is the equation of a line perpendicular to $2x - 3y = 6$?

   (a) $2x + 3y = 6$ **(b)** $3x + 2y = 6$ **(c)** $3x - 2y = 6$ **(d)** $3y - 2x = 6$ **(e)** None of the preceding

7. What is the slope-intercept form of the line through $(2, -3)$ and $(-1, 6)$?

   (a) $y = 3x + 3$ **(b)** $y = -3x + 3$ **(c)** $y = -\frac{1}{3}x + \frac{11}{3}$ **(d)** $y = -3x - 3$ **(e)** None of the preceding

8. The vertices of a right triangle are at $(0, 0)$, $(x, 0)$, and (x, y), with (x, y) in the first quadrant. If $(5, 2)$ is on the hypotenuse of the triangle, then the area of the triangle as a function of x is given by:

   (a) $A(x) = \frac{5}{2}x^2$ **(b)** $A(x) = \frac{5}{4}x^2$ **(c)** $A(x) = \frac{2}{5}x^2$ **(d)** $A(x) = \frac{1}{5}x^2$ **(e)** None of the preceding

9. Which of the following is the equation of a line with intercepts at $(0, -4)$ and $(2, 0)$?

   (a) $2x + 4y = 8$ **(b)** $4x + 2y = 8$ **(c)** $4x + 2y = -8$ **(d)** $2x - 4y = 8$ **(e)** None of the preceding

10. Consider the function defined below and find $f(-3) + f(1)$.

$$f(x) = \begin{cases} x - 2 & \text{if } -3 \le x \le 1 \\ -x + 1 & \text{if } 1 < x \end{cases}$$

(a) -6 (b) 5   (c) 4   (d) -5 (e) None of the preceding

11. Consider the function defined as $y = f(x) = \dfrac{|x - 1|}{x - 1}$. What are the domain and range of this function?

(a) Domain: all x; range: all y (b) Domain: all $x \ne 1$; range: $y = 1$ or $y = -1$

(c) Domain: all $x \ne 1$; range: all y (d) Domain: all x; range: $y = 1$ or $y = -1$ (e) None of the preceding

12. Given: $y = [\,|x|\,]$. For $-2 < x < -1$, $y =$

(a) -2 (b) 2   (c) -1 (d) 1   (e) None of the preceding

13. The coordinates of the vertex and the equation of the axis of symmetry for the graph of $y = (x - 2)^2 - 5$ are

(a) $(2, 5); x = -2$ (b) $(-2, -5); x = 2$ (c) $(2, -5); x = 2$

(d) $(-2, 5); x = -2$ (e) None of the preceding

14. The range of the function given by $y = -2(x + 1)^2 - 3$ is

(a) all real numbers   (b) all $x \ge 2$ (c) all $y \le 3$ (d) all $y \ge -3$ (e) None of the preceding

15. The graph of $y = 3(x + 1)^2 - 2$ is congruent to that for $y = 3x^2$ but is

(a) shifted 1 unit to the right and 2 units down   (b) shifted 1 unit to the left and 2 units down

(c) shifted 1 unit to the right and 2 units up   (d) shifted 1 unit to the left and 2 units up

(e) None of the preceding

16. The function $f(x) = x^2 - 8x + 10$ has

(a) the minimum value -6 when $x = 4$ (b) the minimum value 6 when $x = -4$

(c) the minimum value 26 when $x = 4$ (d) the minimum value 10 when $x = 8$

(e) None of the preceding

17. When a border of uniform width is put around a 6 foot by 10 foot rectangle, the area is enlarged by 80 square feet. Which equation can be used to find the uniform width x?

(a) $x(10 + 2x) + x(6 + 2x) = 80$ (b) $x^2 + 16x - 20 = 0$ (c) $x^2 + 16x - 80 = 0$

(d) $x^2 + 8x - 20 = 0$ (e) None of the preceding

18. Which of the following is true for the parabola $y = -4x^2 + 20x - 25$?

(a) It opens downward and has two x-intercepts.   (b) It opens downward and has no x-intercepts.

(c) It opens downward and has one x-intercept.   (d) It opens to the left and has two y-intercepts.

(e) None of the preceding

19. The solutions of $3x^2 + 6x + 2 = 0$ are:

(a) $\dfrac{-3 \pm \sqrt{3}}{3}$ (b) $\dfrac{-18 \pm \sqrt{3}}{3}$ (c) $-1 \pm \sqrt{2}$ (d) $-6 \pm 2\sqrt{3}i$ (e) None of the preceding

20. Which of the following statements is true for the system shown at the right? $2x - y = 4$

$x + 3y = 7$

(a) The graph consists of two parallel lines.   (b) The graph consists of a single line.

(c) The graph consists of two lines that intersect in the first quadrant.

(d) The graph consists of two lines that intersect in the fourth quadrant.

(e) None of the preceding

ANSWERS TO THE TEST YOUR UNDERSTANDING EXERCISES

Page 87

1. Function; all reals. **2.** Function; all real $x \neq -2$. **3.** Function; all reals. **4.** Not a function.

5. Function; all real $x \neq -1$. **6.** Not a function **7.** Function; all $x \neq 0$. **8.** Function; $x \leq 2$

Page 89

1. 18 **2.** 10 **3.** 0 **4.** $-\frac{5}{4}$ **5.** $\frac{7}{4}$ **6.** $a^2 - 3a$ **7.** $4x^2 - 6x$

8. $2x^2 - 6x$ **9.** $x^2 + 3x$ **10.** $x^2 - 3x$ **11.** $\frac{1}{x^2} - \frac{3}{x}$ **12.** $\frac{1}{x^2 - 3x}$

Page 99

1. $\dfrac{6 - 5}{4 - 1} = \dfrac{1}{3}$ **2.** $\dfrac{3 - (-5)}{-3 - 3} = \dfrac{8}{-6} = -\dfrac{4}{3}$ **3.** $\dfrac{1 - (-3)}{-1 - (-2)} = \dfrac{4}{1} = 4$ **4.** $\dfrac{1 - 0}{0 - (-1)} = \dfrac{1}{1} = 1$

5.

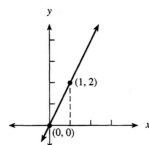

6.

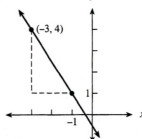

Page 110

1. $y = 2x - 2$ **2.** $y = -\frac{1}{2}x$ **3.** $y = \sqrt{2}\,x + 1$ **4.** $y - 6 = -3(x - 2)$

5. $y - 4 = \frac{1}{2}(x + 1)$ **6.** $y + \frac{2}{3} = 1(x - 5)$ **7.** $y = -\frac{1}{4}x$ **8.** $y + 5 = 0(x + 3)$

9. $y + 1 = -(x - 1)$ **10.** Exercise 8; $y = -5$ and the range is -5.

Page 119

1. 12 **2.** 12 **3.** 12 **4.** 13 **5.** -4 **6.** -4

7. -3 **8.** 0 **9.** -1 **10.** 0 **11.** -1 **12.** 0

Page 125

1. $(1, 2)$ **2.** $(-3, 4)$ **3.** $(2, -1)$ **4.** $(\frac{1}{5}, \frac{1}{5})$ **5.** $(-4, 1)$ **6.** $(35, 50)$

7. $(-1, 2)$ **8.** $\left(\frac{1}{2}, -3\right)$ **9.** $\left(\frac{3}{2}, \frac{5}{7}\right)$

Page 136

1. (b) **2.** (c) **3.** (e) **4.** (f) **5.** (d) **6.** (a)

Page 150

1. $y = (x + 2)^2 - 7$ **2.** $y = (x - 3)^2 - 2$ **3.** $y = (x - 1)^2 + 8$ **4.** $y = 2(x + 2)^2 - 9$

5. $y = -\left(x - \frac{1}{2}\right)^2 - \frac{7}{4}$ **6.** $y = \frac{1}{2}(x - 3)^2 - \frac{5}{2}$ **7.** $x = -1, x = 2$ **8.** $x = -5, x = 2$

9. $x = -3, x = 4$ **10.** $x = 1 \pm \sqrt{5}$ **11.** $x = -3 \pm \sqrt{3}$ **12.** $-3 \pm \sqrt{3}i$ **13.** $x = \dfrac{-1 \pm \sqrt{33}}{4}$

Page 157

1. Minimum value $= -4$ at $x = 5$. **2.** Minimum value $= -\frac{5}{6}$ at $x = -\frac{2}{3}$. **3.** Minimum value $= \frac{1}{2}$ at $x = 1$.

4. Maximum value $= 131$ at $x = -4$. **5.** Maximum value $= -1$ at $x = 0$. **6.** Minimum value $= 0$ at $x = 3$.

7. Minimum value $= 0$ at $x = -\frac{7}{5}$. **8.** Minimum value $= -\frac{49}{4}$ at $x = -\frac{1}{2}$.

Polynomial and Rational Functions

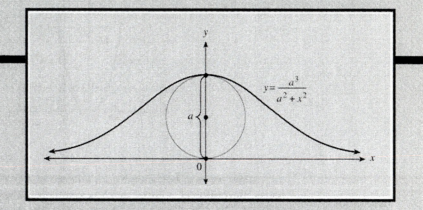

$$y = \frac{a^3}{a^2 + x^2}$$

Of the many functions and their graphs that we will explore in this chapter, perhaps none has as rich a history as the curve shown above, which has been referred to as the "Witch of Agnesi." The number a is the diameter of the circle that is used in generating the points of this curve (see the Challenge near the end of Section 3.3).

Although the French mathematician Pierre de Fermat (see page 1) wrote about this curve earlier, it was the Italian mathematician Marie Gaetana Agnesi (1718–1799) who first presented an extensive discussion of it in her 1748 work entitled *Analytical Institutions*. This was a two-volume treatise that presented an integrated approach to algebra, analytic geometry, and calculus, as it was known in her time. The culmination of ten years of study, this publication brought her instant acclaim throughout the scholarly world of Europe — despite the fact that, in those days, women were discouraged from engaging in mathematical or scientific endeavors. It has been regarded as the most significant mathematical treatise by a woman until that time.

Marie Agnesi, the oldest of twenty-one children, was recognized as a child prodigy. Aside from her exceptional mathematical talents, she was known to be fluent in Latin, Greek, Hebrew, French, and Italian by the age of thirteen!

Her active work in mathematics spanned only about twenty years. Nonetheless, she gained sufficient respect through her work to have been offered her father's chair in mathematics at the University of Bologna upon his death. The last half of her life, about forty years, was devoted to religious contemplation and helping the poor and needy. Ironically, the curve associated with her name, the "Witch of Agnesi," was probably caused through an error in translation of her writings in Latin.

3.1
HINTS
FOR
GRAPHING

The concept of symmetry was used in Chapter 2 when graphing parabolas. Recall that the graph of the function given by $y = f(x) = x^2$ is said to be *symmetric about the y-axis*. (The y-axis is the axis of symmetry.) Observe that points such as $(-a, a^2)$ and (a, a^2) are symmetric points about the axis of symmetry.

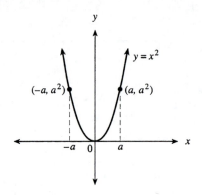

Now we turn our attention to a curve that is *symmetric with respect to a point*. As an illustration of a curve that has symmetry through a point we consider the function given by $y = f(x) = x^3$. This function may be referred to as the *cubing function* because for each domain value x, the corresponding range value is the cube of x.

A table of values is a helpful aid for drawing the graph of this function. Several specific points are located and a smooth curve is then drawn through them to show the graph of $y = x^3$.

$y = f(x) = x^3$

$f(x) = x^3$ is increasing for all x. The curve is concave down on $(-\infty, 0)$ and concave up on $(0, \infty)$.

x	y
-2	-8
-1	-1
$-\frac{1}{2}$	$-\frac{1}{8}$
0	0
$\frac{1}{2}$	$\frac{1}{8}$
1	1
2	8

Domain: all real numbers x
Range: all real numbers y

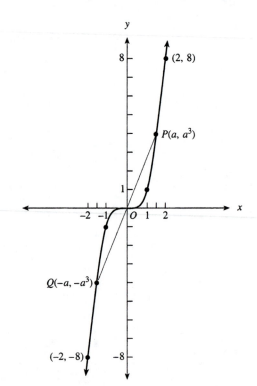

For the function $f(x) = x^3$, we have $f(-x) = (-x)^3 = -x^3 = -f(x)$. Thus $f(-x) = -f(x)$ and we have symmetry with respect to the origin.

The table and the graph reveal that the curve is symmetric through the origin. Geometrically, this means that whenever a line through the origin intersects the curve at a point P, this line will also intersect the curve in another point Q (on the opposite side of the origin) so that the lengths of OP and OQ are equal. This means that both points (a, a^3) and $(-a, -a^3)$ are on the curve for each value $x = a$. These are said to be **symmetric points** through the origin. In particular, since $(2, 8)$ is on the curve, then $(-2, -8)$ is also on the curve. A function that has this symmetric property is called an **odd function.**

In general, the graph of a function $y = f(x)$ is said to be *symmetric through the origin* if for all x in the domain of f, we have

$$f(-x) = -f(x)$$

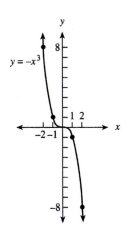

The techniques used for graphing quadratic functions that were discussed in Chapter 2 can be used for other functions as well. For example, the graph of $y = 2x^3$ can be obtained from the graph of $y = x^3$ by multiplying each of its ordinates by 2. Also, the graph of $y = -x^3$ can be obtained by reflecting the graph of $y = x^3$ through the x-axis (or by multiplying the ordinates of $y = x^3$ by -1), as shown in the figure in the margin.

Translations (shifting), as done in Chapter 2, can be applied to the graph of $y = x^3$ as well as to the graph of other functions. This is illustrated in the examples that follow.

The graph of $y = (x + 3)^3$ can be obtained from the graph of $y = x^3$ by a horizontal shift of 3 units to the left.

EXAMPLE 1 Graph $y = g(x) = (x - 3)^3$.

Solution The graph of g is obtained by translating (shifting) the graph of $y = x^3$ three units to the right.

The graph of g is also symmetric through a point. What are its coordinates?

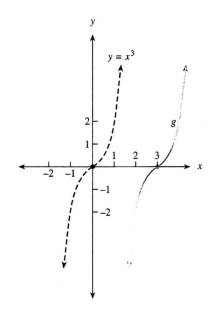

EXAMPLE 2   Graph $y = f(x) = |(x - 3)^3|$.

Solution First graph $y = (x - 3)^3$ as in Example 1. Then take the part of this curve that is below the x-axis $[(x - 3)^3 < 0]$ and reflect it through the x-axis.

The domain of $f(x)$ is all real numbers, and the range is all $y \geq 0$.

Explain how you can use the graph of f to draw the graph of $y = |(x - 3)^3| + 2$.

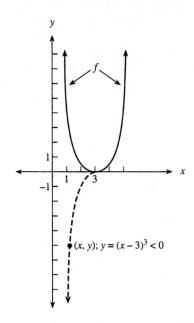

Example 3 illustrates both a horizontal and a vertical translation for the graph of the function $f(x) = x^3$.

EXAMPLE 3   In the figure the curve C_1 is obtained by shifting the curve C with equation $y = x^3$ horizontally and C_2 is obtained by shifting C_1 vertically. What are the equations of C_1 and C_2?

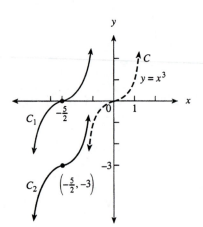

Solution

$$C_1: \qquad y = \left(x + \tfrac{5}{2}\right)^3 \qquad \longleftarrow C \text{ is shifted } \tfrac{5}{2} \text{ units left.}$$

$$C_2: \qquad y = \left(x + \tfrac{5}{2}\right)^3 - 3 \qquad \longleftarrow C_1 \text{ is shifted 3 units down.} \qquad \blacksquare$$

For a given function $f(x)$, the graph of $af(x)$ can be found by means of a **vertical expansion** or a **vertical contraction,** as shown in Example 4. When $a > 1$ we have an expansion, and when $0 < a < 1$ we have a contraction.

EXAMPLE 4 Graph: **(a)** $y = f(x) = x^4$; **(b)** $y = 2x^4$; **(c)** $y = \dfrac{1}{2}x^4$

Solution
(a) Since $f(-x) = (-x)^4 = x^4 = f(x)$, the graph is symmetric about the y-axis. Use the table of values to locate the right half of the curve; the symmetry gives the rest as shown.
(b) Multiply each of the ordinates of $y = x^4$ by 2 to obtain the expansion $y = 2x^4$.
(c) Multiply each of the ordinates of $y = x^4$ by $\dfrac{1}{2}$ to obtain the contraction $y = \dfrac{1}{2}x^4$.

Observe that all three functions are decreasing on $(-\infty, 0]$ and increasing on $[0, \infty)$. Also, the curves are concave up for all x.

x	$y = x^4$	$y = 2x^4$	$y = \tfrac{1}{2}x^4$
0	0	0	0
$\tfrac{1}{2}$	$\tfrac{1}{16}$	$\tfrac{1}{8}$	$\tfrac{1}{32}$
1	1	2	$\tfrac{1}{2}$
$\tfrac{3}{2}$	$\tfrac{81}{16}$	$\tfrac{81}{8}$	$\tfrac{81}{32}$
2	16	32	8

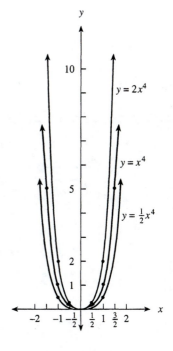

The following table summarizes many of the strategies for graphing functions that were introduced in Chapter 2 and further developed in this section.

GUIDELINES FOR GRAPHING

1. If $f(x) = f(-x)$, the curve is symmetric about the y-axis; it is an even function.

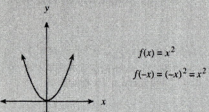

$$f(x) = x^2$$
$$f(-x) = (-x)^2 = x^2$$

2. If $f(-x) = -f(x)$, the curve is symmetric through the origin; it is an odd function.

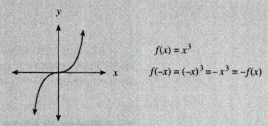

$$f(x) = x^3$$
$$f(-x) = (-x)^3 = -x^3 = -f(x)$$

3. The graph of $y = af(x)$ can be obtained by multiplying the ordinates of the curve $y = f(x)$ by the value a. If $a = -1$, then $y = -f(x)$ is the reflection of $y = f(x)$ through the x-axis.

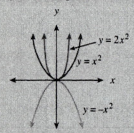

$$y = 2x^2$$
$$y = x^2$$
$$y = -x^2$$

4. The graph of $y = |f(x)|$ can be obtained from the graph of $y = f(x)$ by taking the part of $y = f(x)$ that is below the x-axis and reflecting it through the x-axis.

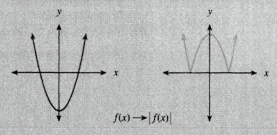

$$f(x) \rightarrow |f(x)|$$

5. The graph of $y = f(x - h)$ can be obtained by shifting $y = f(x)$, h units to the right. $(h > 0)$

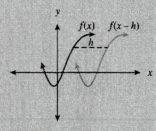

$$f(x) \quad f(x - h)$$

6. The graph of $y = f(x + h)$ can be obtained by shifting $y = f(x)$, h units to the left. $(h > 0)$

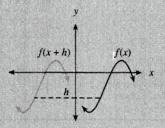

$$f(x + h) \quad f(x)$$

7. The graph of $y = f(x) + k$ can be obtained by shifting $y = f(x)$, k units upward. $(k > 0)$

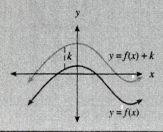

$$y = f(x) + k$$
$$y = f(x)$$

8. The graph of $y = f(x) - k$ can be obtained by shifting $y = f(x)$, k units downward. $(k > 0)$

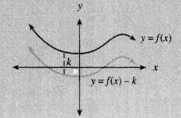

$$y = f(x)$$
$$y = f(x) - k$$

EXERCISES 3.1

For Exercises 1–8, graph each set of curves in the same coordinate system. For each exercise use a dashed curve for the first equation and a solid curve for each of the others. For the last function given, state the domain and range, find where it is increasing and decreasing, and describe the concavity.

1. $y = x^2$, $y = (x - 3)^2$

2. $y = x^3$, $y = (x + 2)^3$

3. $y = x^3$, $y = -x^3$

4. $y = x^4$, $y = (x - 4)^4$

5. $f(x) = x^3$, $g(x) = \frac{1}{2}x^3$, $h(x) = \frac{1}{4}x^3$

6. $f(x) = x^3$, $g(x) = (x - 3)^3 - 3$, $h(x) = (x + 3)^3 + 3$

7. $f(x) = x^4$, $g(x) = (x - 1)^4 - 1$, $h(x) = (x - 2)^4 - 2$

8. $f(x) = x^4$, $g(x) = -\frac{1}{8}x^4$, $h(x) = -x^4 - 4$

9. Graph $y = |(x + 1)^3|$.

10. Graph $y = |[x]|$ for $-3 \le x < 4$.

11. Graph $y = f(x) = |x|$, $y = g(x) = |x - 3|$, and $y = h(x) = |x - 3| + 2$, on the same axes.

Graph each of the following by using translations and reflections.

12. $y = f(x) = (x + 1)^3 - 2$

13. $y = f(x) = (x - 1)^3 + 2$

14. $y = f(x) = 2(x - 3)^3 + 3$

15. $y = f(x) = 2(x + 3)^3 - 3$

16. $y = f(x) = -(x - 1)^3 + 1$

17. $y = f(x) = -(x + 1)^3 - 1$

Find the equation of the curve C which is obtained from the dashed curve by a horizontal or vertical shift, or by a combination of the two.

18.

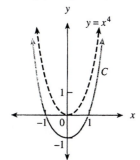

19.

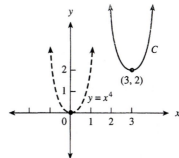

20.

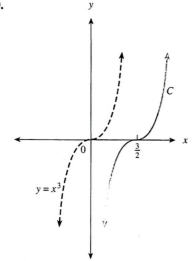

21.

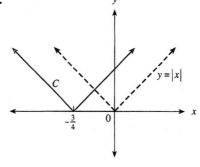

Find the equation of the curve C which is obtained from the dashed curve by using the indicated graphing procedures.

22. Translation *followed* by absolute value.

23. Translation *followed* by absolute value.

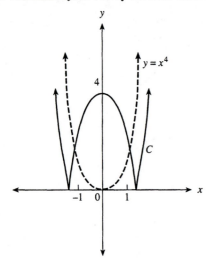

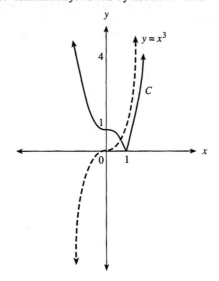

Graph each of the following.

24. $y = |x^4 - 16|$ **25.** $y = |x^3 - 1|$ **26.** $y = |-1 - x^3|$

Graph each of the following. (Hint: Consider the expansion $(a \pm b)^n$ for appropriate values of n; see page 51.)

27. $y = x^3 + 3x^2 + 3x + 1$ **28.** $y = x^3 - 6x^2 + 12x - 8$

29. $y = -x^3 + 3x^2 - 3x + 1$ **30.** $y = x^4 - 4x^3 + 6x^2 - 4x + 1$

Evaluate and simplify the difference quotients.

31. $\dfrac{f(x) - f(3)}{x - 3}$ where $f(x) = x^3$ **32.** $\dfrac{f(-2 + h) - f(-2)}{h}$ where $f(x) = x^3$

33. $\dfrac{f(1 + h) - f(1)}{h}$ where $f(x) = x^4$ **34.** $\dfrac{f(x) - f(-1)}{x + 1}$ where $f(x) = x^4 + 1$

 Challenge   Let $y = f(x) = mx + b$ represent any linear function with positive slope.

1. If the graph of f is translated h units horizontally to the right, which downward vertical translation will have the same result?

2. If the graph of f is translated k units vertically upward, which horizontal translation to the left will have the same result?

 Written Assignment   Consider the "Guidelines for Graphing" on page 178. Supply a specific example of your own to illustrate each of the items listed and explain your choice.

 Graphing Calculator Exercises   The "Guidelines for Graphing" chart can be helpful in deciding on the window in which to graph an unfamiliar function. One can also do a bit of systematic exploration with a graphing calculator, starting by graphing your function on a "standard window" ($-10 \le x \le 10$, $-10 \le y \le 10$). From there, you can use your TRACE facility to reveal points on the graph not shown in your starting window. For example, the graph of $y = |(x - 3)^3|$, discussed in Example 2, is shown in the first quadrant only. TRACING from $x = -10$ to $x = 10$, we see at the bottom of the screen that there are points in the second quadrant as well (shown in the figure as $x = -2$, and $y = 125$). In fact, the y-values range from 0 to 2197 if x is between -10 and 10.

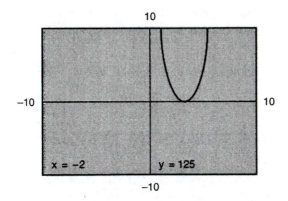

Use the technique described previously to graph the functions in Exercises 1–3. Start with a standard window; TRACE from $x = -10$ to $x = 10$. Then select another window with expanded y-values that includes the points you found with TRACE.

1. $y = -|(x - 4)^3|$

2. $y = (x + 10)(x - 2)^2$

3. $y = (x + 20)^2 - 40$

Another useful exploratory technique is to again begin with a standard window, but now ZOOM out until you include what seems to be a reasonable graph. For example, graphing $y = x^2 + 12$ on a standard window gives a blank screen. But ZOOMing out (roughly doubling or quadrupling all the RANGE settings—see your user's manual on how to adjust ZOOM factors) gives the expected parabola, that is, $y = x^2$, translated upward by 12 units.

Graph the following functions by starting with a standard window. Then ZOOM out once or twice to try to include more of the graph. How many doublings of RANGE values do you need? (Factor = 2)

4. $y = -2(x + 10)^2 - 10$

5. $y = (x^2 - 9)(x - 1)$

6. $y = (x + 10)^4 - 4(x + 10)^2 - 10$

7. Try the ZOOMing out technique with the function in Exercise 1. Does it show you the points on the graph in the second quadrant?

8. Try the TRACE technique on the function in Exercise 4. Does it give you the parabolic shape you expect?

9. After you have used the ZOOM technique on Exercise 6, home in on the "interesting" part of the graph, that is, where its different concavities appear, where its x-intercepts are, and so that its symmetry is revealed. (Your calculator's BOX facility is especially useful here. Look it up in your user's manual.)

3.2 GRAPHING SOME SPECIAL RATIONAL FUNCTIONS

A *rational expression* is a ratio of polynomials. Such expressions may be used to define rational functions and are considered in more detail in later sections. At this point we introduce the topic by exploring several special rational functions.

Consider the function $y = f(x) = \dfrac{1}{x}$, a rational function whose domain consists of all numbers except zero. The denominator x is a polynomial of degree 1, and the numerator $1 = 1x^0$ is a (constant) polynomial of degree zero.

To draw the graph of this function, first observe that it is symmetric through the origin because for all $x \neq 0$

The function $f(x) = \frac{1}{x}$ is decreasing on $(-\infty, 0)$ and on $(0, \infty)$. The curve is concave down on $(-\infty, 0)$ and concave up on $(0, \infty)$.

$$f(-x) = \frac{1}{-x} = -\frac{1}{x} = -f(x)$$

Furthermore, since $y = \dfrac{1}{x}$, it follows that $xy = 1$, a positive number. Therefore, both variables must have the same sign; that is, x and y must both be positive or both negative. Thus, the graph will appear only in quadrants I and III. Select several points for the curve in the first quadrant, and use the symmetry with respect to the origin to obtain the remaining portion of the graph in the third quadrant.

x	$y = \dfrac{1}{x}$
10	.1
100	.01
1000	.001
10,000	.0001
⋮	⋮
↓	↓

Getting very large	Getting close to 0

Horizontal asymptote $y = 0$

x	$y = \dfrac{1}{x}$
$\frac{1}{2}$	2
$\frac{1}{10}$	10
$\frac{1}{100}$	100
$\frac{1}{1000}$	1000
⋮	⋮
↓	↓

Getting close to 0	Getting very large

Vertical asymptote $x = 0$

$y = f(x) = \frac{1}{x}$

x	y
$\frac{1}{5}$	5
$\frac{1}{2}$	2
1	1
2	$\frac{1}{2}$
3	$\frac{1}{3}$
4	$\frac{1}{4}$

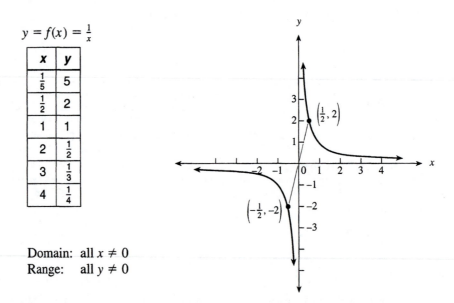

Domain: all $x \neq 0$
Range: all $y \neq 0$

Observe that the curve approaches the x-axis in quadrant I. That is, as the values for x become large, the values for y approach zero. Also, as the values for x approach zero in the first quadrant, the y-values become very large. Similar observations can be made about the curve in the third quadrant. We say that the axes are **asymptotes** for the curve; in particular, the x-axis is a **horizontal asymptote** and the y-axis is a **vertical asymptote.** The tables of values shown in the margin should help to make this clear.

In symbols, using the arrow notation, the discussion on asymptotes in the first quadrant can be summarized as follows:

$f(x) \longrightarrow 0$ as $x \longrightarrow \infty$ As values of x become very large, the corresponding values of $f(x)$ get close to 0. The x-axis is a horizontal asymptote.

$f(x) \longrightarrow \infty$ as $x \longrightarrow 0$ As values of x get very close to 0, the corresponding values of $f(x)$ increase without bound. The y-axis is a vertical asymptote.

In the third quadrant we have the following:

$$f(x) \longrightarrow 0 \text{ as } x \longrightarrow -\infty \quad \text{and} \quad f(x) \longrightarrow -\infty \text{ as } x \longrightarrow 0$$

Recall that the symbols ∞ and $-\infty$ do not represent numbers but are used to indicate that the values increase or decrease without bound.

EXAMPLE 1   Sketch the graph of $g(x) = \dfrac{1}{x-3}$. Find the asymptotes.

Solution   Using $f(x) = \dfrac{1}{x}$ we have $f(x - 3) = \dfrac{1}{x-3} = g(x)$. Therefore, the graph of g can be drawn by shifting the graph of $f(x) = \dfrac{1}{x}$ by 3 units to the right. The x-axis ($y = 0$) is the horizontal asymptote and the line $x = 3$ is the vertical asymptote. The domain is all $x \neq 3$, and the range is all $y \neq 0$.

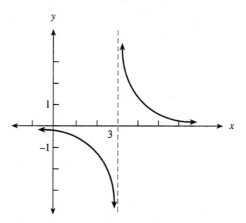

Note that $f(x) \longrightarrow \infty$ as $x \longrightarrow 3^{+}$; that is, as x approaches 3 from the right. Also, $f(x) \longrightarrow -\infty$ as $x \longrightarrow 3^{-}$; that is, as x approaches 3 from the left.   ∎

EXAMPLE 2   Graph $y = f(x) = \dfrac{1}{x^2}$ and find the asymptotes.

Solution   First note that $x \neq 0$. For all other values of x, $x^2 > 0$, so that the curve will appear in quadrants I and II only. Note that the curve is symmetric about the y-axis; that is, f is an even function since

$$f(x) = \frac{1}{x^2} = \frac{1}{(-x)^2} = f(-x)$$

Furthermore, since $f(x) \longrightarrow \infty$ as $x \longrightarrow 0$, the y-axis ($x = 0$) is a vertical asymptote. Also, since $f(x) \longrightarrow 0$ as $x \longrightarrow \pm\infty$, the x-axis ($y = 0$) is a horizontal asymptote. That is, $f(x)$ approaches 0 as x increases or decreases without bound.

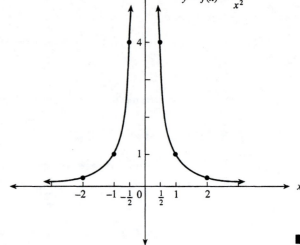

Domain: all $x \neq 0$
Range: all $y > 0$

Select several values for x that get successively closer to 0 and use a calculator to observe the values of $\frac{1}{x^2}$.

■

EXAMPLE 3 Graph $y = \dfrac{1}{(x + 2)^2} - 3$. What are the asymptotes? Find the domain and the range, describe the concavity of the curve, and state where it is increasing or decreasing.

Solution Shift the graph of $y = \dfrac{1}{x^2}$ in Example 2 by 2 units left and then 3 units down.

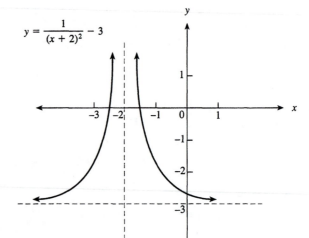

Find the x-intercepts by solving the equation

$$\frac{1}{(x + 2)^2} - 3 = 0$$

Use a calculator to show that, to one decimal place, the solutions are $x = -1.4$ and $x = -2.6$.

Domain: all $x \neq -2$
Range: all $y > -3$

The curve is concave up and increasing on $(-\infty, -2)$ and concave up and decreasing on $(-2, \infty)$.

Vertical asymptote: $x = -2$ $f(x) \longrightarrow \infty$ as $x \longrightarrow -2$

Horizontal asymptote: $y = -3$ $f(x) \longrightarrow -3$ as $x \longrightarrow \pm\infty$ ■

The preceding examples suggest that *a vertical asymptote occurs when the denominator of the rational expression is 0.* Thus, for $g(x) = \dfrac{1}{x - 3}$, the line $x = 3$ is a vertical asymptote. However, this observation does not apply to $f(x) = \dfrac{x + 3}{x^2 - 9}$. The denominator of $f(x)$ is 0 when $x = \pm 3$, but we can reduce as follows:

$$f(x) = \frac{x + 3}{x^2 - 9} = \frac{x + 3}{(x - 3)(x + 3)} = \frac{1}{x - 3} \qquad (x \neq 3)$$

The point $(-3, -\frac{1}{6})$ is on the graph of $y = \dfrac{1}{x - 3}$, but not on the graph of $f(x)$.

This shows that there is no vertical asymptote when x is -3. In fact, the graph of f is the same as the graph in Example 1, except that there would be an open circle at point $(-3, -\frac{1}{6})$ to indicate that this point is *not* on the graph of f. In general

> A graph of a rational function has a vertical asymptote at each value a where the denominator is 0, and the numerator is not 0.

Some rational functions do not have asymptotes, as illustrated in Example 4.

EXAMPLE 4 Graph $y = f(x) = \dfrac{x^2 + x - 6}{x - 2}$.

Solution Factor the numerator and simplify the fraction.

This simplified form indicates that the original rational function is the same as $y = x + 3$, except that $x \neq 2$ since division by 0 is not possible.

$$f(x) = \frac{x^2 + x - 6}{x - 2} = \frac{(x + 3)(x - 2)}{x - 2}$$
$$= x + 3 \quad (x \neq 2)$$

Thus, the graph of the function is the line $y = x + 3$, with an open circle at $(2, 5)$ to show that this point is *not* part of the graph.

Observe that the reduced form $y = x + 3, x \neq 2$, makes it easy to see that the graph of this function has no asymptotes. This was not obvious from the given form.

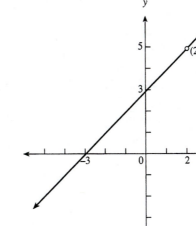

See the inside back cover for these and other useful curves.

The curves studied in this chapter, as well as the parabolas and straight lines discussed in Chapter 2, are very useful in the study of calculus. Having an almost instant recall of the graphs of the following functions will be helpful in future work. Not only should you know what these curves look like, but just as important, you should be able to obtain other curves from them by appropriate translations and reflections and by using the other guidelines listed on page 178.

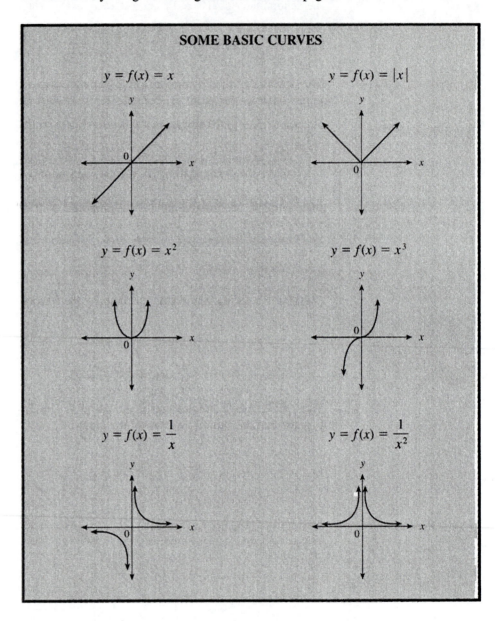

SOME BASIC CURVES

$y = f(x) = x$

$y = f(x) = |x|$

$y = f(x) = x^2$

$y = f(x) = x^3$

$y = f(x) = \dfrac{1}{x}$

$y = f(x) = \dfrac{1}{x^2}$

In addition to understanding how to apply the guidelines on page 178, recognizing the situations when a function has asymptotes and being able to make use of them in graphing the function are valuable skills. A summary of these situations follows.

VERTICAL ASYMPTOTES

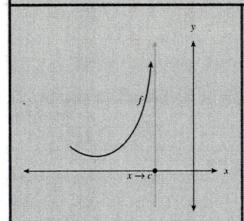

$$f(x) \longrightarrow \infty \text{ as } x \longrightarrow c^-$$

As x gets close to c from the left, $f(x)$ increases without bound.

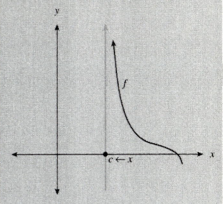

$$f(x) \longrightarrow \infty \text{ as } x \longrightarrow c^+$$

As x gets close to c from the right, $f(x)$ increases without bound.

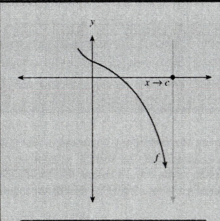

$$f(x) \longrightarrow -\infty \text{ as } x \longrightarrow c^-$$

As x gets close to c from the left, $f(x)$ decreases without bound.

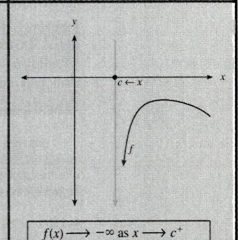

$$f(x) \longrightarrow -\infty \text{ as } x \longrightarrow c^+$$

As x gets close to c from the right, $f(x)$ decreases without bound.

DEFINITION OF A VERTICAL ASYMPTOTE

The line $x = c$ is a vertical asymptote for the graph of a function f if

$$f(x) \longrightarrow \infty \quad \text{or} \quad f(x) \longrightarrow -\infty$$

as x approaches c from either the left or the right.

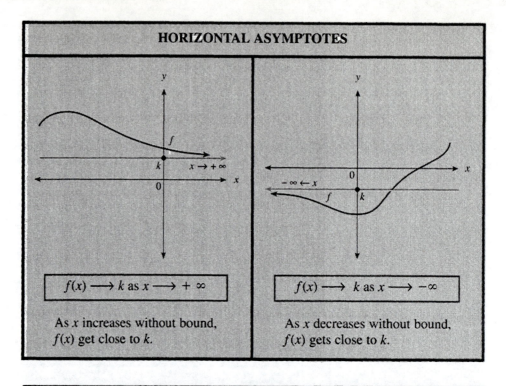

HORIZONTAL ASYMPTOTES

$$f(x) \longrightarrow k \text{ as } x \longrightarrow +\infty$$

As x increases without bound, $f(x)$ get close to k.

$$f(x) \longrightarrow k \text{ as } x \longrightarrow -\infty$$

As x decreases without bound, $f(x)$ gets close to k.

DEFINITION OF A HORIZONTAL ASYMPTOTE

The line $y = k$ is a horizontal asymptote for the graph of a function if

$$f(x) \longrightarrow k \text{ as } x \longrightarrow \infty \text{ or } x \longrightarrow -\infty.$$

Observe from the preceding illustrations that a curve may cross its horizontal asymptote, but it cannot cross its vertical asymptotes. Also note that when the graph of a function has an asymptote, it means that the distance between the curve and the asymptote becomes very small; that is, this distance approaches 0.

EXERCISES 3.2

Graph each set of curves in the same coordinate system. For each exercise use a dashed curve for the first equation and a solid curve for the other.

1. $y = \dfrac{1}{x}$, $y = \dfrac{2}{x}$

2. $y = \dfrac{1}{x}$, $y = -\dfrac{1}{x}$

3. $y = \dfrac{1}{x}$, $y = \dfrac{1}{x + 2}$

4. $y = \dfrac{1}{x}$, $y = \dfrac{1}{x} + 5$

5. $y = \dfrac{1}{x}$, $y = \dfrac{1}{2x}$

6. $y = \dfrac{1}{x}$, $y = \dfrac{1}{x} - 5$

7. $y = \dfrac{1}{x^2}$, $y = \dfrac{1}{(x - 3)^2}$

8. $y = \dfrac{1}{x^2}$, $y = -\dfrac{1}{x^2}$

For Exercises 9–17, graph each of the following. Find all asymptotes, if any. State the domain and range of the function, describe the concavity, and say where it is increasing or decreasing.

9. $y = -\dfrac{1}{x} + 2$

10. $y = -\dfrac{1}{x} - 2$

11. $y = \dfrac{1}{x+4} - 2$

12. $y = \dfrac{1}{x-4} + 2$

13. $y = -\dfrac{1}{x-2} + 1$

14. $y = -\dfrac{1}{x+1} - 2$

15. $y = \dfrac{1}{(x+1)^2} - 2$

16. $y = \dfrac{1}{(x-2)^2} + 3$

17. $y = \dfrac{1}{|x-2|}$

For Exercises 18–30, graph and find the asymptotes, if any.

18. $y = \dfrac{1}{x^3}$

19. $xy = 3$

20. $xy = -2$

21. $xy - y = 1$

22. $xy - 2x = 1$

23. $y = f(x) = \dfrac{x^2 - 9}{x - 3}$

24. $y = f(x) = \dfrac{x^2 - 9}{x + 3}$

25. $y = f(x) = \dfrac{x^2 - x - 6}{x - 3}$

26. $y = f(x) = \dfrac{1 + x - 2x^2}{x - 1}$

27. $y = f(x) = \dfrac{x + 1}{x^2 - 1}$

28. $y = f(x) = \dfrac{x - 1}{x^2 + x - 2}$

29. $y = f(x) = \dfrac{x^3 - 8}{x - 2}$

30. $y = f(x) = \dfrac{x^3 - 2x^2 - 3x + 6}{2 - x}$

31. Graph $y = \left| \dfrac{1}{x} - 1 \right|$.

32. Graph $x = \dfrac{1}{y^2}$. Why is y not a function of x?

Find the difference quotients and simplify.

33. $\dfrac{f(x) - f(3)}{x - 3}$ where $f(x) = \dfrac{1}{x}$

34. $\dfrac{f(1 + h) - f(1)}{h}$ where $f(x) = \dfrac{1}{x^2}$

Written Assignment Explain why the graph of a function that has a vertical asymptote cannot intersect this asymptote, whereas it can intersect a horizontal asymptote. Illustrate your discussion with specific examples for each.

Graphing Calculator Exercises When you graph functions that have vertical asymptotes, it might seem that your graphing calculator is graphing the asymptotes as well as the graph. If yours does, it is only because it is connecting the last point to the left of the asymptote with the first point to the right of it. You can detect these points by slowly tracing the graph in the vicinity of the asymptote. One way to avoid this is to use a window that would normally plot the point with x-value equal to b when the asymptote is $x = b$. (See the Graphing Calculator Appendix for a discussion of *friendly windows*.)

1. Graph the function $y = 1/(x - 3)$ using (**a**) a standard window and (**b**) a friendly window. (**c**) What points are traced just before and just after the asymptote in part (**a**)? (**d**) What does your calculator show with TRACE when $x = 3$ in part (**b**)?
2. Repeat the instructions for Exercise 1 with the function $y = 6x/(x^2 + x - 12)$ and values of x corresponding to the asymptotes $x = a$ and $x = b$.
3. Graph the function $y = (x^2 + x - 6)/(x + 3)$ using (**a**) a standard window and (**b**) a friendly window. (**c**) Does this graph have a vertical asymptote? (**d**) Which window gives a more accurate representation of the graph? Why?

3.3
POLYNOMIAL
AND RATIONAL
FUNCTIONS

In this section we will study procedures for graphing more complicated polynomial and rational functions. Knowing when the graph of a function is above or below the x-axis is a useful aid to graphing and can be summarized by a *table of signs*. To illustrate this procedure, consider the function $f(x) = x^2 - 7x + 10$. Since $(x - 2)(x - 5) = 0$ if and only if $x = 2$ or $x = 5$, we need to consider the other possible values of x. First observe that the numbers 2 and 5 determine three intervals.

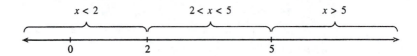

To find the signs of $f(x)$, we select a convenient *test value* within each interval, as follows:

$(-\infty, 2)$:   Test value: $x = 0$ Then $x - 2 < 0$, $x - 5 < 0$, and $(x - 2)(x - 5) > 0$.

$(2, 5)$:   Test value: $x = 3$ Then $x - 2 > 0$, $x - 5 < 0$, and $(x - 2)(x - 5) < 0$.

$(5, \infty)$:   Test value: $x = 6$ Then $x - 2 > 0$, $x - 5 > 0$, and $(x - 2)(x - 5) > 0$.

These values can be summarized, as in the following **table of signs.**

Interval	$(-\infty, 2)$	$(2, 5)$	$(5, \infty)$
Sign of $x - 2$	−	+	+
Sign of $x - 5$	−	−	+
Sign of $(x - 2)(x - 5)$	+	−	+

The test value procedure can be justified algebraically. For example, the sign of $x - 2$ for the interval $(-\infty, 2)$ can be found by taking any value of x in this interval. Then note that $x < 2$ and add -2 to get $x - 2 < 0$. Thus, $x - 2$ is negative for all x in $(-\infty, 2)$.

In the following example a table of signs is used to help find the graph of a third-degree polynomial.

EXAMPLE 1   Graph the polynomial function $f(x) = x^3 + 4x^2 - x - 4$.

Solution   First obtain the factored form of $f(x)$ using factoring by grouping.

$$f(x) = x^3 + 4x^2 - x - 4$$
$$= (x + 4)x^2 - (x + 4)$$
$$= (x + 4)(x^2 - 1)$$
$$= (x + 4)(x + 1)(x - 1)$$

STEP 1 for graphing polynomial functions. Factor the polynomial, if possible, and find the x-intercepts.

Set each factor equal to 0 to get the roots $-4, -1, 1$ of $f(x) = 0$, which are also the x-intercepts of $y = f(x)$. Since there are no other x-intercepts, the curve must stay above or below the x-axis for each of the intervals determined by $-4, -1, 1$.

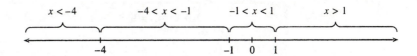

When $f(x) > 0$ the curve is above the x-axis, and it is below for $f(x) < 0$. The following table of signs contains this information and is used to sketch the graph.

STEP 2. *Form a table of signs for* $f(x)$.

Choose convenient test values from each interval. For example, use -5 *in* $(-\infty, -4)$, -2 *in* $(-4, -1)$, 0 *in* $(-1, 1)$, *and* 2 *in* $(1, \infty)$.

Interval	$(-\infty, -4)$	$(-4, -1)$	$(-1, 1)$	$(1, \infty)$
Sign of $x + 4$	−	+	+	+
Sign of $x + 1$	−	−	+	+
Sign of $x - 1$	−	−	−	+
Sign of $f(x)$	−	+	−	+
Location of curve relative to x-axis	below	above	below	above

$$f(x) = x^3 + 4x^2 - x - 4$$

STEP 3. *Sketch the graph using the* x*-intercepts, the signs of* $f(x)$*, and some additional points including the* y*-intercept.*

Note: We can only approximate the coordinates of the turning points. In the study of calculus you will learn how to determine the coordinates of such points accurately.

x	y
-4	0
-3	8
-2	6
-1	0
0	-4
1	0

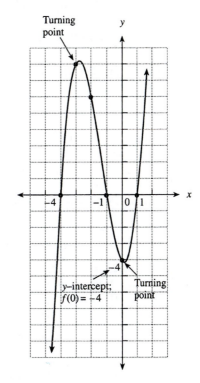

In general, a polynomial of degree n *has at most* $n - 1$ *turning points and at most* n x*-intercepts. (See Exercise 55.)*

Notice that in Example 1 there were two turning points and three x-intercepts. It turns out that for a third-degree polynomial there are at most two turning points and at most three x-intercepts. There may be less of either; for example, $y = x^3$ has no turning points and only one x-intercept (see page 174).

The method for finding the asymptotes of a rational function can be summarized by referring to the *general form of a rational function.*

RATIONAL FUNCTIONS AND ASYMPTOTES

$$\text{Let } f(x) = \frac{p(x)}{q(x)} = \frac{a_n x^n + a_{n-1} x^{n-1} + \cdots + a_0}{b_m x^m + b_{m-1} x^{m-1} + \cdots + b_0}, \qquad a_n \neq 0 \neq b_m$$

Assume that $p(x)$ and $q(x)$ have no common factors. Then:

1. If $n < m$, then $y = 0$ (the x-axis) is a horizontal asymptote.

2. If $n = m$, then $y = \dfrac{a_n}{b_m}$ is a horizontal asymptote.

3. If $n > m$, there are no horizontal asymptotes.

4. If $q(r) = 0$ and $p(r) \neq 0$, then $x = r$ is a vertical asymptote.

A rational function may have numerous vertical asymptotes, but at most one horizontal asymptote.

EXAMPLE 2   Find the asymptotes for each rational function.

(a) $f(x) = \dfrac{x}{x^2 - 2x}$ **(b)** $g(x) = \dfrac{3x^2}{2x^2 + 1}$ **(c)** $h(x) = \dfrac{x^2}{x + 1}$

Solution

(a) $f(x) = \dfrac{x}{x^2 - 2x} = \dfrac{x}{x(x - 2)} = \dfrac{1}{x - 2} \qquad (x \neq 0)$

Then $x = 2$ is a vertical asymptote since the denominator is 0 when x is 2 but the numerator is not 0.

$y = 0$ (the x-axis) is the horizontal asymptote since

degree of numerator $= 1 < 2 =$ degree of denominator

(b) Since the denominator $2x^2 + 1 \neq 0$ for all x, there is no vertical asymptote. Also, $y = \dfrac{3}{2}$ is the horizontal asymptote since

degree of numerator $= 2 =$ degree of denominator

(c) $x = -1$ is a vertical asymptote. There is no horizontal asymptote since

degree of numerator $= 2 > 1 =$ degree of denominator   ■

It is possible to provide a mathematical proof for each of the results shown in Example 2. For example, in part (**b**) we would begin by dividing the numerator and denominator by the largest power of x, namely x^2:

$$g(x) = \frac{3x^2}{2x^2 + 1} = \frac{\dfrac{3x^2}{x^2}}{\dfrac{2x^2}{x^2} + \dfrac{1}{x^2}} = \frac{3}{2 + \dfrac{1}{x^2}}$$

Now as x increases or decreases without bound (that is, $x \to \pm\infty$) the fraction $\dfrac{1}{x^2}$ becomes very small (that is, $\dfrac{1}{x^2} \to 0$). Thus, $g(x) \to \dfrac{3}{2}$ and $y = \dfrac{3}{2}$ is the horizontal asymptote.

TEST YOUR UNDERSTANDING	*Find the vertical and horizontal asymptotes for each rational function.*

1. $y = \dfrac{5}{(x-2)(x-3)}$ **2.** $y = \dfrac{x-2}{x+3}$ **3.** $y = \dfrac{x^2-1}{x-2}$

4. $y = \dfrac{x}{(x-4)^2}$ **5.** $y = \dfrac{2-x}{x^2-x-12}$ **6.** $y = \dfrac{2x^3+x}{x^2-x}$

(Answers: Page 246)

When graphing a rational function of the form $f(x) = \dfrac{p(x)}{q(x)}$, where $p(x)$ and $q(x)$ are polynomials, it will be useful to find the x- and y-intercepts, if there are any. To find the x-intercepts, let $y = 0$ and solve for x. This is done by setting the numerator $p(x) = 0$, since the only way that a fraction can be 0 is when the numerator is 0. To find the y-intercept, let $x = 0$ and solve for y.

Guidelines for Graphing Rational Functions:

STEP 1. *Factor the numerator and denominator, if possible.*

STEP 2. *Find the x- and y-intercepts, if any.*

STEP 3. *Find the vertical asymptotes.*

STEP 4. *Find the horizontal asymptotes.*

STEP 5. *Form a table of signs for f in the intervals determined by the numbers for which the numerator or denominator equals 0. Use a specific test value in each interval.*

EXAMPLE 3   Graph $y = f(x) = \dfrac{x-2}{x^2-3x-4}$.

Solution

$$f(x) = \frac{x-2}{(x+1)(x-4)}$$

$f(x) = 0$ when the numerator $x - 2 = 0$, so the x-intercept is 2. When $x = 0$,
$y = \dfrac{0-2}{(0+1)(0-4)} = \dfrac{1}{2}$, so the y-intercept is $\dfrac{1}{2}$.

Setting the denominator $(x+1)(x-4)$ equal to 0 gives the vertical asymptotes $x = -1$ and $x = 4$.

The horizontal asymptote is $y = 0$ since the degree of the numerator is 1, which is less than 2, the degree of the denominator.

The numbers for which the numerator or denominator equals 0 are -1, 2, and 4. They determine the four intervals in the table of signs for f.

Interval	$(-\infty, -1)$	$(-1, 2)$	$(2, 4)$	$(4, \infty)$
Sign of $x + 1$	$-$	$+$	$+$	$+$
Sign of $x - 2$	$-$	$-$	$+$	$+$
Sign of $x - 4$	$-$	$-$	$-$	$+$
Sign of $f(x)$	$-$	$+$	$-$	$+$
Location of curve relative to x-axis	below	above	below	above

STEP 6. *Graph the curve. First draw the asymptotes and locate the intercepts. Use the table of signs and some selected points to complete the graph.*

When graphing, remember that the closer the curve is to a vertical asymptote, the steeper it gets. And the larger $|x|$ gets, the closer the curve is to the horizontal asymptote; the curve gets "flatter." Using a table of values to locate a few selected points helps to draw the correct shape.

You might find it easier to focus initially on the graph of $f(x)$ within each of the intervals shown in the table of signs. For example, for the interval $(4, \infty)$, first plot the points $(4.2, 2.1)$ and $(5, 0.5)$. Then consider the asymptotes $x = 4$ and $y = 0$ (the x-axis), and draw part of the curve, as shown in the margin. Finally, proceed in this manner for each of the remaining intervals to produce the following completed graph.

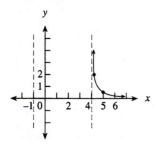

Use a calculator to round off decimals to the nearest tenth. Note that it is easier to substitute into the form
$$\frac{x - 2}{(x + 1)(x - 4)}$$ *than into*
$$\frac{x - 2}{x^2 - 3x - 4}.$$

x	y
−3	−0.4
−2	−0.7
−1.2	−3.1
0	0.5
2	0
3	−0.3
3.9	−3.9
4.2	2.1
5	0.5
6	0.3

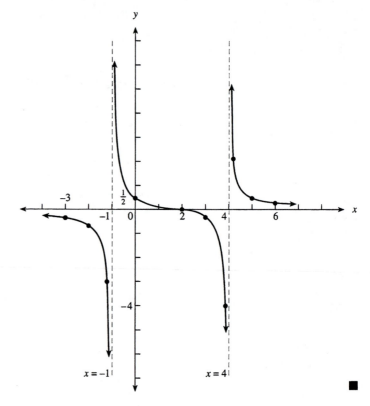

Recall that the degree of the constant 2 is zero. (See page 47.)

EXAMPLE 4   Graph: $y = f(x) = \dfrac{2}{x^2 + 1}$.

Solution There are no vertical asymptotes since the denominator $x^2 + 1 \ne 0$. The x-axis, $y = 0$, is the horizontal asymptote since the degree of the numerator is less than the degree of the denominator.

There is symmetry through the y-axis since

$$f(-x) = \frac{2}{(-x)^2 + 1} = \frac{2}{x^2 + 1} = f(x)$$

The graph is above the x-axis since $y = \dfrac{2}{x^2 + 1} > 0$ for all x.

The y-intercept is 2 since $\dfrac{2}{0^2 + 1} = 2$.

The highest point on the graph is (0, 2) because the denominator is the smallest when $x = 0$; $2 = \dfrac{2}{0^2 + 1} \geq \dfrac{2}{x^2 + 1}$.

The preceding observations are used to obtain the following graph.

$$y = f(x) = \frac{2}{x^2 + 1}$$

x	y
0	2
±0.5	1.6
±1	1
±2	0.4
±3	0.2

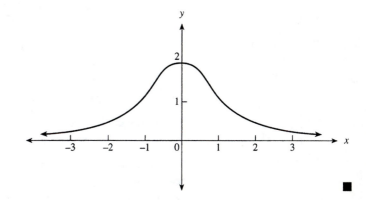

Occasionally we may encounter a rational function whose graph has an **oblique asymptote** (or **slant asymptote**). Consider, for example, the function

$$f(x) = \frac{x^2 - 2x + 1}{x} \qquad \text{or} \qquad f(x) = x - 2 + \frac{1}{x}$$

Note that as $|x|$ becomes very large, then $\dfrac{1}{x}$ approaches 0 and the value of the given function approaches the value of $x - 2$. Thus the line $y = x - 2$ is an oblique asymptote, as in the following graph. Note that there is also a vertical asymptote at $x = 0$.

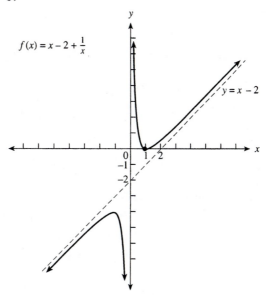

In general, consider a rational function $f(x) = \dfrac{p(x)}{q(x)}$ where the degree of $p(x)$ is *one greater* than the degree of $q(x)$. If $f(x)$ is written in the form

$$f(x) = ax + b + \frac{r(x)}{q(x)}$$

then the line $y = ax + b$ is an oblique asymptote for the graph of $f(x)$.

For example, the function $f(x) = \dfrac{3x^2 - 2x - 1}{x + 1}$ can be written in the form

$$
\begin{array}{r}
3x - 5 \\
x + 1\overline{\smash{)}3x^2 - 2x - 1} \\
\underline{3x^2 + 3x} \\
-5x - 1 \\
\underline{-5x - 5} \\
4
\end{array}
$$

$$f(x) = 3x - 5 + \frac{4}{x + 1}$$

and the line $y = 3x - 5$ is an oblique asymptote for the graph of $f(x)$. (Note that as $|x|$ becomes very large, $\dfrac{4}{x + 1}$ approaches 0 and $f(x) \rightarrow 3x - 5$.)

A considerable effort has been put into the development of graphing procedures, and there is more to come. However, no matter how careful we are, part of our curve sketching is being done on "faith." After all, what is the shape of a curve connecting two points? Following are some possibilities. Locating more points between P and Q will not really answer the question, no matter how close the chosen points are, since the question as to how to connect them still remains. More advanced methods studied in calculus will help answer such questions, and curve sketching can then be done with less ambiguity.

EXERCISES 3.3

Use a table of signs to determine the signs of f.

1. $f(x) = (x - 1)(x - 2)(x - 3)$

2. $f(x) = x(x + 2)(x - 2)$

3. $f(x) = \dfrac{(3x - 1)(x + 4)}{x^2(x - 2)}$

4. $f(x) = \dfrac{x(2x + 3)}{(x + 2)(x - 5)}$

5. $f(x) = (x^2 + 2)(x - 4)(x + 1)$

6. $f(x) = x(x + 1)(x + 2)$

7. $f(x) = \dfrac{x - 10}{3(x + 1)(5x - 1)}$

8. $f(x) = \dfrac{3}{(x + 2)(4x + 3)}$

Sketch the graph of each polynomial function. Indicate all intercepts.

9. $f(x) = (x + 3)(x + 1)(x - 2)$

10. $f(x) = (x - 1)(x - 3)(x - 5)$

11. $f(x) = (x + 3)(x + 1)(x - 1)(x - 3)$

12. $f(x) = -x(x + 4)(x^2 - 4)$

13. $f(x) = x^3 - 4x$

14. $f(x) = 9x - x^3$

15. $f(x) = x^3 + 3x$

16. $f(x) = x^3 + x^2 - 6x$

17. $f(x) = -x^3 - x^2 + 6x$

18. $f(x) = x^3 + 2x^2 - 9x - 18$

19. $f(x) = x^4 - 4x^2$

20. $f(x) = (x - 1)^2(3 + 2x - x^2)$

21. $f(x) = x^4 - 6x^3 + 8x^2$

22. $f(x) = x(x^2 - 1)(x^2 - 4)$

Sketch the graph of each rational function. Indicate the asymptotes and the intercepts, if any.

23. $y = f(x) = \dfrac{1}{(x - 1)(x + 2)}$

24. $y = f(x) = \dfrac{1}{x^2 - 4}$

25. $y = f(x) = \dfrac{1}{4 - x^2}$

26. $y = f(x) = \dfrac{2}{x^2 - x - 6}$

27. $y = f(x) = \dfrac{x}{x^2 - 1}$

28. $y = f(x) = \dfrac{x + 1}{x^2 + x - 2}$

29. $y = f(x) = \dfrac{3}{x^2 + 1}$

30. $y = f(x) = \dfrac{2}{x + x^2}$

31. $y = f(x) = \dfrac{x + 2}{x - 2}$

32. $y = f(x) = \dfrac{x}{1 - x}$

33. $y = f(x) = \dfrac{x - 1}{x + 3}$

34. $y = f(x) = \dfrac{x^2 + x}{x^2 - 4}$

35. $y = f(x) = \dfrac{x^2 + x - 2}{x^2 + x - 12}$

36. $y = f(x) = \dfrac{1 - x^2}{x^2 - 9}$

37. $y = f(x) = \dfrac{9}{3x^2 + 6}$

38. $y = f(x) = -\dfrac{3}{x^2 + 2}$

39. $y = f(x) = \dfrac{2x^2}{x^2 + 1}$

40. $y = f(x) = \dfrac{2x^2}{x^2 + 1} - 1$

Give the equation of the oblique asymptote for each of the following.

41. $f(x) = \dfrac{x^2 + 3}{x}$

42. $f(x) = \dfrac{x^2 + 3x + 1}{x}$

43. $f(x) = \dfrac{x^2 - 3x + 1}{x - 1}$

44. $f(x) = \dfrac{x^2 - x + 3}{x - 2}$

45. $f(x) = \dfrac{x^3 - x^2 - x - 2}{x^2 - 1}$

46. $f(x) = \dfrac{x^3 + x^2 + x + 3}{x^2 + 1}$

Simplify each function, obtain factored forms, and determine the signs of g. (The combination of such skills will be needed in the study of calculus.)

47. $g(x) = 2(2x - 1)(x - 2) + 2(x - 2)^2$

48. $g(x) = 2(x - 3)(x + 2)^2 + 2(x + 2)(x - 3)^2$

49. $g(x) = \dfrac{x^2 - 2x(x + 2)}{x^4}$

50. $g(x) = \dfrac{(x - 3)(2x) - (x^2 - 5)}{(x - 3)^2}$

51. $g(x) = 2\left(\dfrac{x - 5}{x + 2}\right)\dfrac{(x + 2) - (x - 5)}{(x + 2)^2}$

52. $g(x) = \dfrac{5(x + 10)^2 - 10x(x + 10)}{(x + 10)^4}$

53. Use the method shown at the bottom of page 192 to verify that the horizontal asymptote
for the function $y = f(x) = \dfrac{2x^2}{3 - 5x^2}$ is $y = -\dfrac{2}{5}$.

54. Repeat Exercise 53 to show that the horizontal asymptote for $y = g(x) = \dfrac{3x^3 - x^2 + 1}{x^3 + 2}$
is $y = 3$.

55. The graph of an nth-degree polynomial has at most n x-intercepts, and at most $n - 1$
turning points. When n is even, the number of x-intercepts will range from 0 to n; when n
is odd, there must be at least one x-intercept so that the number of x-intercepts will range
from 1 to n. Graph each of the following and show that the graphs confirm the preceding
statements:

(a) $y = x^2 + 1$ **(b)** $y = x^4 - 5x^2 + 4$ **(c)** $y = x^5 + 1$ **(d)** $y = x^3 - x^2 - 4x + 4$

Written Assignment

1. Use specific examples of your own to describe the conditions for the graph of a rational function to have vertical, horizontal, and oblique asymptotes.

2. Explain why a rational function can have at most one horizontal asymptote.

? **Challenge** The figure below shows the same curve discussed in the introduction to this chapter. Note that the line ℓ and the x-axis are parallel and tangent to the circle with center $\left(0, \dfrac{a}{2}\right)$ at points O and T. The following procedure shows how to locate point P on the curve. In the figure, a line through O is drawn that intersects the circle at Q, and line ℓ at A. AB is perpendicular to the x-axis, and the horizontal line through Q intersects the y-axis at R, and AB at P. Let P have coordinates (x, y). Then Q has the same second coordinate as P, and let its first coordinate be x'. Observe that ΔOTA is similar to ΔORQ so that $OT/OR = TA/RQ$ and this gives $\dfrac{a}{y} = \dfrac{x}{x'}$. Square this equation to get $\dfrac{a^2}{y^2} = \dfrac{x^2}{(x')^2}$. Now show that $(x')^2 = (a - y)y$. (*Hint:* Use the equation of the circle.) Then show that $y = \dfrac{a^3}{a^2 + x^2}$.

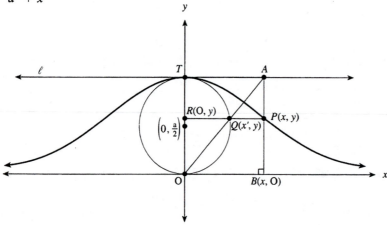

Graphing Calculator Exercises An alternative to making a table of test values to know if the graph of a function is above or below the x-axis is to include the x-axis as well as all the x-intercepts in the x-RANGE values in the initial graph by taking the y-RANGE values a little above and below zero. Then trace the graph to the right and left of the x-intercepts and vertical asymptotes in order to approximate the y-values of the turning points, if any. Finally, adjust the y-RANGE values to include these. Use this technique to graph the functions in Exercises 1–4. Show any x-intercepts, vertical asymptotes, and turning points.

1. $y = (x - 1)(x + 2)(x - 3)(x + 4)$
2. $y = x^5 - x$
3. $y = (x^3 + x - 2)/x$
4. $y = (x^3 + x - 2)/x^2$
5. Find any oblique asymptotes in the graphs of Exercises 1–4 and graph them with the originals.

Critical Thinking

1. Sketch the graph of a function having the following properties: $x = 1$ is a vertical asymptote, $y = -1$ is a horizontal asymptote, the domain is all $x \neq 1$, the range is all $y \neq -1$, the origin is the only x-intercept, f is increasing on both $(-\infty, 1)$ and $(1, \infty)$, and the graph is concave up on $(-\infty, 1)$ and concave down on $(1, \infty)$.

2. Compose a rational function that has $y = x + 1$ as an oblique asymptote, $x = -2$ as a vertical asymptote, and that passes through the origin. Explain your justification for each condition, and then check your result through the use of a graphing calculator.

3. Explain, with a specific example, why the following statement is *not* true:

   A graph of a rational function has a vertical asymptote at each value a where the denominator is 0.

4. Explain the difference between the graph of $f(x) = \dfrac{(x - a)(x - b)}{x - a}$ and that of $g(x) = x - b$.

3.4 EQUATIONS AND INEQUALITIES WITH FRACTIONS

To find the x-intercepts of the graph of a rational function such as

$$y = f(x) = \frac{3x}{x + 7} - \frac{8}{5}$$

we need to let $y = 0$ and solve for x, as shown in Example 1.

EXAMPLE 1 Find the x-intercepts of $y = f(x) = \dfrac{3x}{x + 7} - \dfrac{8}{5}$.

Solution First set $f(x) = 0$ and then multiply both sides of the equation by $5(x + 7)$, the least common denominator (LCD); then solve for x.

$$5(x + 7)\left(\frac{3x}{x + 7} - \frac{8}{5}\right) = 5(x + 7) \cdot 0$$

$$5(x + 7) \cdot \frac{3x}{x + 7} - 5(x + 7) \cdot \frac{8}{5} = 5(x + 7) \cdot 0 \qquad \text{distributive property}$$

$$5 \cdot 3x - (x + 7) \cdot 8 = 0$$

$$15x - 8x - 56 = 0$$

$$7x = 56$$

$$x = 8$$

Notice that multiplication by the LCD transforms the original equation into one that does not involve any fractions.

The x-intercept is 8.

$$Check: \frac{3(8)}{8 + 7} - \frac{8}{5} = \frac{24}{15} - \frac{8}{5} = \frac{8}{5} - \frac{8}{5} = 0$$ ■

EXAMPLE 2 Solve x: $\dfrac{2x - 5}{x + 1} - \dfrac{3}{x^2 + x} = 0$

Solution Factor the denominator in the second fraction.

$$\frac{2x - 5}{x + 1} - \frac{3}{x(x + 1)} = 0$$

Multiply by the LCD, which is $x(x + 1)$.

At times, multiplication by the LCD can give rise to a quadratic equation, as in Example 2. Try to explain each step in the solution.

$$x(x + 1) \cdot \frac{2x - 5}{x + 1} - x(x + 1) \cdot \frac{3}{x(x + 1)} = x(x + 1) \cdot 0$$

$$x(2x - 5) - 3 = 0$$

$$2x^2 - 5x - 3 = 0$$

$$(2x + 1)(x - 3) = 0$$

You should check these answers in the original equation.

$$2x + 1 = 0 \quad \text{or} \quad x - 3 = 0$$

$$x = -\tfrac{1}{2} \quad \text{or} \quad x = 3 \qquad \blacksquare$$

It is especially important to check each solution of a fractional equation. The reason for this can be explained through the use of another example.

EXAMPLE 3 Solve for x: $\dfrac{24}{x^2 - 16} - \dfrac{5}{x + 4} = \dfrac{3}{x - 4}$

Solution Begin by multiplying each side by $(x + 4)(x - 4)$.

$$(x + 4)(x - 4) \cdot \frac{24}{x^2 - 16} - (x + 4)(x - 4) \cdot \frac{5}{x + 4} = (x + 4)(x - 4) \cdot \frac{3}{x - 4}$$

$$24 - (x - 4)5 = (x + 4)3$$

$$24 - 5x + 20 = 3x + 12$$

$$-8x = -32$$

$$x = 4$$

In this example we could have noticed at the outset that $x - 4 \neq 0$, and thus $x \neq 4$. In other words, it is wise to notice such restrictions on the variable before starting the solution. These are the values of the variable that would cause division by zero.

Note that we began with the *assumption* that there was a value x for which the equation was true. This led to the value $x = 4$; that is, we argued that *if* there is a solution, then it must be 4. But if x is replaced by 4 in the given equation, we obtain

$$\frac{24}{0} - \frac{5}{8} = \frac{3}{0}$$

Since division by 0 is meaningless, we have an impossible equation and, therefore, 4 cannot be a solution. We conclude that there is no replacement of x for which the equation is true. Therefore, the solution set is the **empty set**, the set that contains no elements. $\blacksquare$

The equation in Example 1 can also be written as the **proportion** $\dfrac{3x}{x+7} = \dfrac{8}{5}$.

A proportion is a statement that two ratios are equal, such as the following:

$$\frac{a}{b} = \frac{c}{d}$$

The proportion is in the form of a fractional equation and can be simplified by multiplying both sides by bd.

$$\frac{a}{b} = \frac{c}{d}$$

$$(bd)\,\frac{a}{b} = (bd)\,\frac{c}{d}$$

$$ad = bc$$

This is often read "a is to b as c is to d" and may be written in this form:

$$a : b = c : d$$

*Sometimes we say that $ad = bc$ is obtained by **cross-multiplying**. This description is used because ad and bc can be obtained by multiplying the numerators times the denominators along the lines in this cross.*

$$ad = bc$$

An equation such as $\dfrac{x}{3} + \dfrac{x}{2} = 10$ cannot be solved through use of the proportion property as given. Why not? Can you rewrite this equation so that this property can be used?

PROPORTION PROPERTY

If $\dfrac{a}{b} = \dfrac{c}{d}$, then $ad = bc$.

Observe how the proportion property can be used after the equation in Example 1 is written as a proportion.

$$3x \cdot 5 = 8 \cdot (x + 7) \qquad \text{proportion property}$$

$$15x = 8x + 56$$

$$7x = 56$$

$$x = 8$$

TEST YOUR UNDERSTANDING

Solve for x and check your results. Use the proportion property where appropriate.

1. $\dfrac{x}{3} + \dfrac{x}{2} = 10$

2. $\dfrac{3}{x} + \dfrac{2}{x} = 10$

3. $\dfrac{x-3}{8} = 4$

4. $\dfrac{8}{x-3} = 4$

5. $\dfrac{x+3}{x} = \dfrac{2}{3}$

6. $\dfrac{2}{x+3} = \dfrac{3}{x+3}$

7. $\dfrac{2}{3x} + \dfrac{3}{2x} = 4$

8. $\dfrac{x}{x-6} + \dfrac{2}{x} = \dfrac{1}{x-6}$

9. $\dfrac{x}{x-1} - \dfrac{3}{4x} = \dfrac{3}{4} - \dfrac{1}{x-1}$

10. $\dfrac{2x+10}{x+18} = \dfrac{x+3}{4-x}$

(Answers: Page 246)

The general procedures for solving problems outlined in Section 1.2 apply to problems that involve fractions as well. The reader is advised to review that material at this time. The first illustrative problem makes use of the proportion property.

***

Similar triangles have the same shape, but not necessarily the same size. If two triangles are similar, their corresponding angles are equal and their corresponding sides are proportional.

EXAMPLE 4   How high is a tree that casts an 18-foot shadow at the same time that a 3-foot stick casts a 2-foot shadow?

Solution From the diagram we use similar triangles to write a proportion.

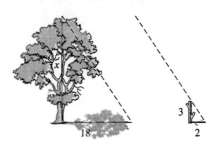

Let x represent the height of the tree. Then, since the triangles are similar,

$$\frac{x}{18} = \frac{3}{2}$$

$$2x = 54 \qquad \text{proportion property}$$

$$x = 27$$

The tree is 27 feet high.   ■

EXAMPLE 5   A wildlife conservation team wants to determine the deer population in a state park. They capture a sample of 80 deer, tag each one, and then release them. After a sufficient amount of time has elapsed for the tagged deer to mix thoroughly with the others in the park, the team captures another sample of 100 deer, five of which had previously been tagged. Estimate the number of deer in the park.

Solution Let x be the total deer population. Assume that the 80 tagged deer were thoroughly mixed with the others. Thus it is reasonable to expect the ratio of 80 tagged deer to the total deer population to be the same as the ratio of the tagged deer, captured a second time, to the total number in the second sample.

$$\frac{\text{total tagged deer}}{\text{total deer population}} = \frac{\text{tagged deer in second sample}}{\text{total deer in second sample}}$$

$$\frac{80}{x} = \frac{5}{100}$$

$$5x = 8000$$

$$x = 1600$$

Source: Bettman/J. D. Taylor.

There are approximately 1600 deer in the park.   ■

Suppose that you are able to work at a steady rate and can paint a room in 6 hours. This means that in 1 hour you have painted $\frac{1}{6}$ of the room. In general, for a job that takes you n hours, the part of the job done in 1 hour is $\frac{1}{n}$. This idea is used to solve certain types of work problems that call for the solution of fractional equations, as in Example 6.

*Working together should enable the boys to do the job in less time than it would take either of them to do the job alone. If Elliot can do the job alone in 2 hours, together the two boys should take less than this time. This kind of reasoning can be **critical**, since an answer of 2 or more indicates that an error has been made in the solution.*

EXAMPLE 6 Working alone, Harry can mow a lawn in 3 hours. Elliot can complete the same job in 2 hours. How long will it take them working together, assuming they both start at the same time?

Solution To solve work problems of this type, we first consider the part of the job that can be done in 1 hour.

$$\text{Let } x = \text{time (in hours) to do the job together}$$

$$\text{Then } \frac{1}{x} = \text{portion of job done in 1 hour working together}$$

$$\text{Also: } \quad \frac{1}{2} = \text{portion of job done by Elliot in 1 hour}$$

$$\frac{1}{3} = \text{portion of job done by Harry in 1 hour}$$

	Time for Whole Job in Hours	Work Done in 1 Hour
Harry	3	$\frac{1}{3}$
Elliot	2	$\frac{1}{2}$
Together	x	$\frac{1}{x}$

Now we have that *each* of the expressions $\frac{1}{2} + \frac{1}{3}$ and $\frac{1}{x}$ represents the part of the job done in 1 hour. Therefore, we may set these expressions equal to each other to form the equation needed to solve this problem.

$$\frac{1}{2} + \frac{1}{3} = \frac{1}{x}$$

$$3x + 2x = 6$$

$$5x = 6$$

$$x = 1\frac{1}{5}$$

Together they need $1\frac{1}{5}$ hours, or 1 hour and 12 minutes, to complete the job. ■

Numerous formulas involving fractions are found in mathematics as well as in areas such as science, business, and industry. When working with such formulas it is sometimes useful to solve for one of the variables in terms of the others. This is illustrated in the next example.

EXAMPLE 7 $r = \dfrac{ab}{a + b + c}$ is the formula for the radius of the inscribed circle of a right triangle in terms of the sides of the triangle. Solve for side b of the adjacent figure.

Solution Begin by clearing fractions:

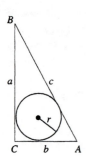

$$r = \frac{ab}{a + b + c}$$

$$r(a + b + c) = ab$$

$$ra + rb + rc = ab$$

$$ra + rc = ab - rb \qquad \text{Bring all terms involving } b \text{ to one side.}$$

$$r(a + c) = (a - r)b \qquad \text{Factor.}$$

$$\frac{r(a + c)}{a - r} = b \qquad \text{Divide by } a - r.$$ ∎

For solving equations with fractions the LCD was used to eliminate the denominators, resulting in a simpler equation. This idea is also used to solve inequalities with fractions.

EXAMPLE 8   Solve: $\dfrac{x}{3} - \dfrac{2x + 1}{4} > 1$

Solution

$$12\left(\frac{x}{3}\right) - 12\left(\frac{2x + 1}{4}\right) > 12(1)$$

$$4x - 3(2x + 1) > 12$$

$$4x - 6x - 3 > 12$$

$$-2x > 15$$

$$x < -\frac{15}{2} \qquad \text{Why?}$$ ∎

EXAMPLE 9   Graph on a number line: $\dfrac{1 - 2x}{x - 3} \le 0$

Solution First find the values of x for which the numerator or the denominator is zero.

$$1 - 2x = 0 \qquad x - 3 = 0$$

$$-2x = -1 \qquad x = 3$$

$$x = \tfrac{1}{2}$$

Note here that $x \ne 3$. For $x = 3$, $x - 3 = 0$ and division by 0 is not possible.

Now locate the three intervals determined by these two points.

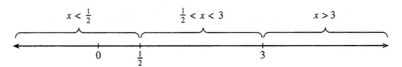

*This solution **cannot** be described using a single inequality. For example, it makes no sense to write $3 < x \leq \frac{1}{2}$ since this implies the false result $3 \leq \frac{1}{2}$.*

Next, select a specific value to test in each interval. You should find that the fraction will be negative when $x < \frac{1}{2}$ or when $x > 3$. Also note that $x = \frac{1}{2}$ satisfies the given inequality but that $x = 3$ is excluded from the solution. Can you explain why this is so? The solution for the given inequality consists of all x such that $x \leq \frac{1}{2}$ or $x > 3$, that is, $\{x \mid x \leq \frac{1}{2}$ or $x > 3\}$.

When the fraction in Example 9 is multiplied by the square of the denominator, we have

$$\frac{1 - 2x}{x - 3} \cdot (x - 3)^2 = (1 - 2x)(x - 3)$$

Since $(x - 3)^2 > 0$ for all $x \neq 3$, the product $(1 - 2x)(x - 3)$ has the same signs as the given fraction. Consequently, the solution for Example 9 can also be found using the method demonstrated in Example 5, page 151.

In the next example, the substitution method studied in Section 2.5 is applied to solving a nonlinear system of equations.

EXAMPLE 10   Solve the system

$$y = \frac{2}{x}$$

$$y + 2x = 5$$

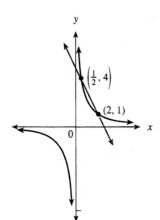

Solution Substitute the y-value, in terms of x, from the first equation into the second.

$$\frac{2}{x} + 2x = 5$$

$$2 + 2x^2 = 5x \qquad \text{Multiply by } x.$$

$$2x^2 - 5x + 2 = 0$$

$$(2x - 1)(x - 2) = 0$$

$$x = \tfrac{1}{2} \quad \text{or} \quad x = 2$$

For $x = \frac{1}{2}$, $y = -2\left(\frac{1}{2}\right) + 5 = 4$. For $x = 2$, $y = -2(2) + 5 = 1$. The points of intersection are $\left(\frac{1}{2}, 4\right)$ and $(2, 1)$. Check both of these points by substituting in the original system. ∎

EXERCISES 3.4

Solve for x and check your results.

1. $\dfrac{x}{2} - \dfrac{x}{5} = 6$

2. $\dfrac{2}{x} - \dfrac{5}{x} = 6$

3. $\dfrac{x-1}{2} = \dfrac{x+2}{4}$

4. $\dfrac{2}{x-1} = \dfrac{4}{x+2}$

5. $\dfrac{5}{x} - \dfrac{3}{4x} = 1$

6. $\dfrac{3x}{4} - \dfrac{3}{2} = \dfrac{x}{6} + \dfrac{4x}{3}$

7. $\dfrac{2x+1}{5} - \dfrac{x-2}{3} = 1$

8. $\dfrac{3x-2}{4} - \dfrac{x-1}{3} = \dfrac{1}{2}$

9. $\dfrac{x+4}{2x-10} = \dfrac{8}{7}$

10. $\dfrac{10}{x} - \dfrac{1}{2} = \dfrac{15}{2x}$

11. $\dfrac{2}{x+6} - \dfrac{2}{x-6} = 0$

12. $\dfrac{1}{3x-1} + \dfrac{1}{3x+1} = 0$

13. $\dfrac{x+1}{x+10} = \dfrac{1}{2x}$

14. $\dfrac{3}{x+1} = \dfrac{9}{x^2-3x-4}$

15. $\dfrac{x^2}{2} - \dfrac{3x}{2} + 1 = 0$

16. $\dfrac{x+1}{x-1} - \dfrac{2}{x(x-1)} = \dfrac{4}{x}$

17. $\dfrac{5}{x^2-9} = \dfrac{3}{x+3} - \dfrac{2}{x-3}$

18. $\dfrac{3}{x^2-25} = \dfrac{5}{x-5} - \dfrac{5}{x+5}$

19. $\dfrac{1}{x^2+4} + \dfrac{1}{x^2-4} = \dfrac{18}{x^4-16}$

20. $\dfrac{3}{x^3-8} = \dfrac{1}{x-2}$

Find the x-intercepts of y = f(x) and state the domain.

21. $f(x) = \dfrac{2x-5}{x+1} - \dfrac{3}{x^2+x}$

22. $f(x) = \dfrac{2}{x} + 2x - 5$

23. $f(x) = \dfrac{10-5x}{3x} - \dfrac{2}{x+5} - \dfrac{8-4x}{x+5}$

24. $f(x) = \dfrac{2-x}{x+1} + \dfrac{x+8}{x-2} - \dfrac{4-x}{x^2-x-2}$

25. $f(x) = \dfrac{3}{2x^2-3x-2} - \dfrac{x+2}{2x+1} - \dfrac{2x}{10-5x}$

26. $f(x) = \dfrac{2x-2}{x^2+2x} - \dfrac{5x-6}{6x} - \dfrac{1-x}{x+5}$

Solve for the indicated variable.

27. $\dfrac{v^2}{K} = \dfrac{2g}{m}$, for m

28. $A = \dfrac{h}{2}(b + B)$, for B

29. $S = \pi(r_1 + r_2)s$, for r_1

30. $V = \dfrac{1}{3}\pi r^2 h$, for h

31. $d = \dfrac{s-a}{n-1}$, for s

32. $S = \dfrac{n}{2}[2a + (n-1)d]$, for d

33. $\dfrac{1}{f} = \dfrac{1}{m} + \dfrac{1}{p}$, for m

34. $c = \dfrac{2ab}{a+b}$, for b

Solve each inequality.

35. $\dfrac{x}{2} - \dfrac{x}{3} \le 5$

36. $\dfrac{x}{3} - \dfrac{x}{2} \le 5$

37. $\dfrac{x+3}{4} - \dfrac{x}{2} > 1$

38. $\dfrac{x}{3} - \dfrac{x-1}{2} < 2$

39. $\frac{1}{2}(x+1) - \frac{2}{3}(x-2) < \frac{1}{6}$

40. $\frac{1}{2}(x-2) - \frac{1}{5}(x-1) > 1$

41. $\dfrac{x-2}{x+3} \geq 0$ **42.** $\dfrac{x+5}{2-x} < 0$ **43.** $\dfrac{(6-x)(3+x)}{x+1} \leq 0$

Solve each inequality. (Hint: Convert each inequality to obtain a fraction of the form $\dfrac{p}{q}$ on one side, and 0 on the other side.

44. $\dfrac{2x+3}{x} < 1$ **45.** $\dfrac{x}{x-1} > 2$ **46.** $\dfrac{x+4}{x+2} > \dfrac{1}{3}$

47. What number must be subtracted from both the numerator and denominator of the fraction $\frac{11}{23}$ to give a fraction whose value is $\frac{2}{5}$?

48. The denominator of a fraction is 3 more than the numerator. If 5 is added to the numerator and 4 is subtracted from the denominator, the value of the new fraction is 2. Find the original fraction.

49. One pipe can empty a tank in 3 hours. A second pipe takes 4 hours to complete the same job. How long will it take to empty the tank if both pipes are used?

50. Working together, Amy and Julie can paint their room in 3 hours. If it takes Amy 5 hours to do the job alone, how long would it take Julie to paint the room working by herself?

51. Find two fractions whose sum is $\frac{2}{3}$ if the smaller fraction is $\frac{1}{2}$ the larger one. (*Hint:* Let x represent the larger fraction.)

52. A rope that is 20 feet long is cut into two pieces. The ratio of the smaller piece to the larger piece is $\frac{3}{5}$. Find the length of the shorter piece.

53. The shadow of a tree is 20 feet long at the same time that a 1-foot-high flower casts a 4-inch shadow. How high is the tree?

54. The area A of a triangle is given by $A = \dfrac{bh}{2}$, where b is the length of the base and h is the length of the altitude.

(a) Solve for h in terms of A and b.   **(b)** Find h when $A = 100$ and $b = 25$.

55. If 561 is divided by a certain number, the quotient is 29 and the remainder is 10. Find the number.

(*Hint:* For $N \div D$ we have $\dfrac{N}{D} = Q + \dfrac{R}{D}$, where Q is the quotient and where R is the remainder.)

56. A student received grades of 72, 75, and 78 on three tests. What must her score on the next test be for her to have an average grade of 80 for all four tests?

57. The denominator of a certain fraction is 1 more than the numerator. If the numerator is increased by $2\frac{1}{2}$, the value of the new fraction will be equal to the reciprocal of the original fraction. Find the original fraction.

58. Dan takes twice as long as George to complete a certain job. Working together, they can complete the job in 6 hours. How long will it take Dan to complete the job by himself?

59. Prove:

If $\dfrac{a}{b} = \dfrac{c}{d}$,   then $\dfrac{a+b}{b} = \dfrac{c+d}{d}$

60. Prove:

If $\dfrac{a}{b} = \dfrac{c}{d}$,   then $\dfrac{a}{a+b} = \dfrac{c}{c+d}$

61. A bookstore has a stock of 30 paperback copies of *Algebra* and 50 hardcover copies of the same book. They wish to increase their stock for the new semester. Based on past experience, they want their final numbers of paperback and hardcover copies to be in the ratio 4 to 3. However, the publisher stipulates that they will sell the store only 2 copies of the paperback edition for each copy of the hardcover edition ordered. Under these conditions, how many of each edition should the store order to achieve the 4 : 3 ratio?

62. To find out how wide a certain river is, a pole 20 feet high is set straight up on one of the banks. Another pole 4 feet long is also set straight up, on the same side, some distance away from the embankment. The observer waits until the shadow of the 20-foot pole just reaches the other side of the river. At that time he measures the shadow of the 4-foot pole and finds it to be 34 feet. Use this information to determine the width of the river.

63. Two hundred fish are caught, tagged, and returned to the lake from which they were caught. Some time later, when the 200 tagged fish had thoroughly mixed with the others, another 160 fish were caught and contained 4 that were tagged previously. Estimate the number of fish in the lake.

64. Estimate the total number of fish in the lake in Exercise 63 if the first catch contained 150 fish and the second catch had 120 fish, 3 of which were tagged previously.

65. A certain college gives 4 points per credit for a grade of *A*, 3 points per credit for a *B*, 2 points for a *C*, 1 point per credit for a *D*, and 0 for an *F*. A student is taking 15 credits for the semester. She expects *A*'s in a 4-credit course and in a 3-credit course. She also expects a *B* in another 3-credit course and a *D* in a 2-credit course. What grade must she get in the fifth course in order to earn a 3.4 grade point average for the term?

$$\left(\text{Hint: Grade point average} = \frac{\text{total points}}{\text{total credits}}.\right)$$

66. For electric circuits, when two resistances R_1 and R_2 are wired in parallel, the single equivalent resistance R can be found by using this relationship: $\dfrac{1}{R} = \dfrac{1}{R_1} + \dfrac{1}{R_2}$. Find the single equivalent resistance for two resistances wired in parallel if these measure 2000 and 3000 ohms, respectively.

67. What resistance, wired in parallel with a 20,000-ohm resistance, will give a single equivalent resistance of 4000 ohms? (See Exercise 66.)

68. A formula used in optics relates the focal length of a lens, f, the distance of an object from the lens, p, and the distance from the lens to the object, q, in this way: $\dfrac{1}{f} = \dfrac{1}{p} + \dfrac{1}{q}$. If the focal length of a lens is 15 centimeters and the distance from the lens to the image is 20 centimeters, find the distance of the object from the lens.

69. Find the focal length of a lens if the distance of an object from the lens is 30 centimeters and the distance of the image from the lens is 15 centimeters. (See Exercise 68.)

70. Solve $f(x) < 0$, where $f(x) = \dfrac{1}{x^3 + 6x^2 + 12x + 8}$, and graph $y = f(x)$.
   (*Hint:* Consider the expansion of $(a + b)^3$.)

Solve each system.

71. $2x + 3y = 7$
   $y = \dfrac{1}{x}$

72. $y = x^2 - 2x - 4$
   $y = -\dfrac{8}{x}$

73. The vertices of the right triangle are $(0, y)$, $(0, 0)$, and $(x, 0)$. The hypotenuse passes through $(3, 1)$. Express the area of the triangle as a function of x, and find the area when $x = 3.1$.

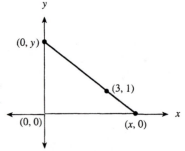

? Challenge The ancient Greeks and some modern artists felt that the rectangular shape most pleasing to the eye is one whose ratio of length to width is approximately 1.62. This ratio is known as the **golden ratio.** The Parthenon in Greece, as well as many famous paintings, makes use of these dimensions.

The Parthenon Acropolis in Athens.
Source: Superstock.

The following *continued fraction* gives rise to the golden ratio x. Solve for x using the hint below.

$$x = 1 + \cfrac{1}{1 + \cfrac{1}{1 + \cfrac{1}{1 + \cfrac{1}{1 + \cdots}}}}$$

Hint: Observe that the circled part shown above equals x and solve the resulting equation.

Also find the reciprocal of your solution, $\dfrac{1}{x}$, and comment on the result.

 Graphing Calculator Exercises For Exercises 1 and 2, solve the nonlinear system of equations by **(a)** solving each equation for y as a function of x, **(b)** graphing the two functions on the same set of axes, and **(c)** using TRACE and ZOOM to find the points of intersection. A friendly window is advised in each case.

1. $xy = 2$ and $y + 2x - 5 = 0$.

2. $x = 9/y$ and $5x - y = -12$.

For Exercises 3 and 4, solve the given inequality by **(a)** substituting y for 0, **(b)** graphing the corresponding equation as a function of y, and **(c)** determining from the graph the intervals where it is above or below the x-axis, depending on whether the inequality sign indicates greater than or less than zero.

3. $(x - 2)/(x + 2) \geq 0$.

4. $(6 - x)(3 + x)/(x + 1) \leq 0$.

5. If your calculator permits functions defined with relational operators (see your user's manual and the Graphing Calculator Appendix), solve the inequality in Exercise 3 by graphing $y = ((x - 2)/(x + 2) \geq 0) * (2)$.

6. Solve the inequality in Exercise 6 using the method of Exercise 5.

3.5
VARIATION

If a car is traveling at a constant rate of 40 miles per hour, then the distance d traveled in t hours is given by $d = 40t$. The change in the distance is "directly" affected by the change in the time; as t increases, so does d. We say that d *is directly proportional to t*. This is because $d = 40t$ converts to the proportion $\dfrac{d}{t} = 40$. We also say that d *varies directly as t* and that 40 is the *constant of variation, or the constant of proportionality*.

Direct variation is a functional relationship in the sense that $y = kx$ *defines* y *to be a function of* x.

DIRECT VARIATION

y varies directly as x if $y = kx$ for some constant of variation k.

EXAMPLE 1

(a) Write the equation that expresses this direct variation: y varies directly as x, and y is 8 when x is 12.

(b) Find y for $x = 30$.

Solution

(a) Since y varies directly as x, we have $y = kx$ for some constant k. To find k, substitute the given values for the variables and solve.

$$8 = k(12)$$

$$\frac{2}{3} = k$$

Thus, $y = \dfrac{2}{3}x$.

(b) For $x = 30$, $y = \dfrac{2}{3}(30) = 20$. ∎

When y varies directly as x, the equation $y = kx$ is equivalent to $\dfrac{y}{x} = k \ (x \neq 0)$. Therefore, if each of the pairs x_1, y_1 and x_2, y_2 satisfies $y = kx$, then each of the ratios y_1 to x_1 and y_2 to x_2 is equal to k. Thus

$$\frac{y_1}{x_1} = \frac{y_2}{x_2}$$

This proportion can be used in some direct variation problems if the constant of proportionality is not required. Thus part (b) of Example 1 can be answered without first writing the equation of variation. In particular, using $x_1 = 12$ with $y_1 = 8$, and $x_2 = 30$, we can solve for y_2 as follows:

$$\frac{y_2}{x_2} = \frac{y_1}{x_1}$$

$$\frac{y_2}{30} = \frac{8}{12}$$

$$y_2 = 30\left(\frac{8}{12}\right) = 20$$

Numerous examples of direct variation can be found in geometry. Here are some illustrations.

Circumference of a circle of radius r:

$C = 2\pi r$; C varies directly as the radius r;
$\quad\quad\quad\quad$ 2π is the constant of variation.

Area of a circle of radius r:

$A = \pi r^2$; A varies directly as the square of r;
$\quad\quad\quad\quad$ π is the constant of variation.

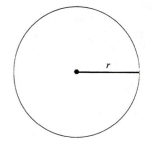

Area of an equilateral triangle of side s:

$A = \dfrac{\sqrt{3}}{4}s^2$; A varies directly as s^2;
$\quad\quad\quad\quad$ $\dfrac{\sqrt{3}}{4}$ is the constant of proportionality.

Volume of a cube of side e:

$V = e^3$; V varies directly as the cube of e (as e^3);
$\quad\quad\quad\quad$ 1 is the constant of variation.

The same principle applies to the force required to hold a spring when compressed x units within its natural length.

EXAMPLE 2   According to Hooke's law, the force F required to hold a spring stretched x units beyond its natural length is directly proportional to x. If a force of 20 pounds is needed to hold a certain spring stretched 3 inches beyond its natural length, how far will 60 pounds of force hold the spring stretched beyond its natural length?

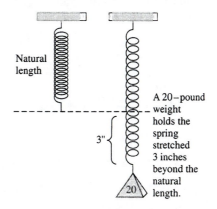

Natural length

A 20–pound weight holds the spring stretched 3 inches beyond the natural length.

3"

20

Note: The force F increases as x increases. This will always be the case in a direct variation situation provided that the constant of variation is positive.

Solution Since F varies directly as x, we have $F = kx$. Solve for k by substituting the known values for F and x.

$$20 = k(3)$$

$$\frac{20}{3} = k$$

Thus $F = \frac{20}{3}x$. Now let $F = 60$ and solve for x.

$$60 = \frac{20}{3}x$$

$$60\left(\frac{3}{20}\right) = x$$

$$9 = x$$

As noted after Example 1, the solution also can be found by solving the proportion
$$\frac{x}{60} = \frac{3}{20}.$$

Thus 60 pounds is needed to hold the spring stretched 9 inches beyond its natural length. ∎

When y varies directly as x and the constant of variation is positive, the variables x and y increase or decrease together. There are other situations where one variable increases as the other decreases. We refer to this as *inverse variation*.

INVERSE VARIATION

y varies inversely as x if $y = \dfrac{k}{x}$ for some constant of variation k.

EXAMPLE 3 According to Boyle's law, the pressure P of a compressed gas is inversely proportional to the volume V. Suppose that there is a pressure of 25 pounds per square inch when the volume of gas is 400 cubic inches. Find the pressure when the gas is compressed to 200 cubic inches.

Solution Since P varies inversely as V, we have

$$P = \frac{k}{V}$$

Substitute the known values for P and V and solve for k.

$$25 = \frac{k}{400}$$

$$10,000 = k$$

Thus $P = \dfrac{10,000}{V}$, and when $V = 200$ we have

Note that the pressure increases as the volume decreases.

$$P = \frac{10,000}{200} = 50$$

The pressure is 50 pounds per square inch. ∎

When y varies inversely as x the equation $y = \dfrac{k}{x}$ is equivalent to $xy = k$.

Therefore, when each of the pairs x_1, y_1 and x_2, y_2 satisfies $y = \dfrac{k}{x}$, we get each of the products equal to k. Thus

$$x_1 y_1 = x_2 y_2$$

This equation provides an alternative way to solve some inverse variation problems, if the equation of the variation is not required. Thus in Example 3, using $V_1 = 400$ with $P_1 = 25$ and $V_2 = 200$, we can solve for P_2 as follows:

$$V_2 P_2 = V_1 P_1$$

$$200\, P_2 = (400)(25)$$

$$P_2 = \frac{(400)(25)}{200} = 50$$

TEST YOUR UNDERSTANDING

(Answers: Page 246)

y varies directly as x.

1. $y = 5$ when $x = 4$.
Find y for $x = 20$.

2. $y = 2$ when $x = 7$.
Find y for $x = 21$.

y varies inversely as x.

3. $y = 3$ when $x = 5$.
Find y when $x = 3$.

4. $y = 4$ when $x = 8$.
Find y when $x = 2$.

The variation of a variable may depend on more than one other variable. Here are some illustrations:

$z = kxy$ z varies *jointly* as x and y.

$z = kx^2 y$ z varies *jointly* as x^2 and y.

$z = \dfrac{k}{xy}$ z varies *inversely* as x and y.

$z = \dfrac{kx}{y}$ z varies *directly* as x and *inversely* as y.

$w = \dfrac{kxy^3}{z}$ w varies *jointly* as x and y^3 and *inversely* as z.

Observe that the word "jointly" is used to imply that z varies directly as the product of the two factors.

EXAMPLE 4 Describe the variation given by these equations:

(a) $z = kx^2 y^3$ **(b)** $z = \dfrac{kx^2}{y}$ **(c)** $V = \pi r^2 h$

Solution
(a) z varies jointly as x^2 and y^3.
(b) z varies directly as x^2 and inversely as y.
(c) V varies jointly as r^2 and h.

■

EXAMPLE 5   Suppose that z varies directly as x and inversely as the square of y. If $z = \frac{1}{3}$ when $x = 4$ and $y = 6$, find z when $x = 12$ and $y = 4$.

Solution

$$z = \frac{kx}{y^2}$$

$$\frac{1}{3} = \frac{k(4)}{6^2}$$

$$\frac{1}{3} = \frac{k}{9}$$

$$3 = k$$

Thus $z = \dfrac{3x}{y^2}$. When $x = 12$ and $y = 4$ we have

$$z = \frac{3(12)}{4^2} = \frac{9}{4} \qquad\blacksquare$$

EXAMPLE 6   According to Newton's Law of Gravitation, the force of attraction between two objects varies jointly as their masses and inversely as the square of the distance between them.
(a) Write an equation to show this relationship.
(b) How does the force change when the distance between the two objects is decreased from 6 feet to 3 feet?
(c) How does the force change if each of their masses is doubled and the distance between them is also doubled?

Solution

(a) Denote the masses of the objects by m_1 and m_2, and the distance between them by d. Then the force F is given by $F = \dfrac{km_1 m_2}{d^2}$, where k is a constant.

(b) At $d = 6$, $F = \dfrac{km_1 m_2}{36} = \dfrac{1}{36} km_1 m_2$. At $d = 3$, $F = \dfrac{km_1 m_2}{9} = \dfrac{1}{9} km_1 m_2$.
Since $\dfrac{1}{9} = 4 \times \dfrac{1}{36}$, the force when $d = 3$ is four times as strong as when $d = 6$. That is,

$$4\left(\frac{1}{36} km_1 m_2\right) = \frac{1}{9} km_1 m_2$$

(c) In the original formula, replace m_1, m_2, and d by $2m_1$, $2m_2$, and $2d$. Call the new force F'. Then

$$F' = \frac{k(2m_1)(2m_2)}{(2d)^2} = \frac{4km_1 m_2}{4d^2} = \frac{km_1 m_2}{d^2}$$

Thus the force is unchanged. $\qquad\blacksquare$

EXERCISES 3.5

Write the equation for the given variation and identify the constant of variation.

1. The perimeter P of a square varies directly as the side s.

2. The area of a circle varies directly as the square of the radius.

3. The area of a rectangle 5 centimeters wide varies directly as its length.

4. The volume of a rectangular-shaped box 10 centimeters high varies jointly as the length and width.

Write the equation for the given variation using k as the constant of variation.

5. z varies jointly as x and y^3.

6. z varies inversely as x and y^3.

7. z varies directly as x and inversely as y^3.

8. w varies jointly as x and y and z.

9. w varies directly as x^2 and inversely as y and z.

Find the constant of variation for Exercises 10–14.

10. y varies directly as x; $y = 4$ when $x = \frac{2}{3}$.

11. s varies directly as t^2; $s = 50$ when $t = 10$.

12. y varies inversely as x; $y = 15$ when $x = \frac{1}{3}$.

13. u varies jointly as v and w; $u = 2$ when $v = 15$ and $w = \frac{2}{3}$.

14. z varies directly as x and inversely as the square of y; $z = \frac{7}{2}$ when $x = 14$ and $y = 6$.

15. a varies inversely as the square of b, and $a = 10$ when $b = 5$. Find a when $b = 25$.

16. z varies jointly as x and y; $z = \frac{3}{2}$ when $x = \frac{5}{6}$ and $y = \frac{9}{20}$. Find z when $x = 2$ and $y = 7$.

17. s varies jointly as ℓ and the square of w; $s = \frac{10}{3}$ when $\ell = 12$ and $w = \frac{5}{6}$. Find s when $\ell = 15$ and $w = \frac{9}{4}$.

18. The cost C of producing x number of articles varies directly as x. If it costs \$560 to produce 70 articles, what is C when $x = 400$?

19. If a ball rolls down an inclined plane, the distance traveled varies directly as the square of the time. If the ball rolls 12 feet in 2 seconds, how far will it roll in 3 seconds?

20. The volume V of a right circular cone varies jointly as the square of the radius r of the base, and the altitude h. If $V = 8\pi$ cubic centimeters when $r = 2$ centimeters and $h = 6$ centimeters, find the formula for the volume V.

21. A force of 2.4 pounds is needed to hold a spring stretched 1.8 inches beyond its natural length. Use Hooke's law to determine the force required to hold the spring stretched 3 inches beyond its natural length.

22. The force required to hold a metal spring compressed from its natural length is directly proportional to the change in the length of the spring. If 235 pounds is required to hold the spring compressed within its natural length of 18 inches to a length of 15 inches, how much force is required to hold it compressed to a length of 12 inches?

23. Fifty pounds per square inch is the pressure exerted by 150 cubic inches of a gas. Use Boyle's law to find the pressure if the gas is compressed to 100 cubic inches.

24. The gas in Exercise 23 expands to 500 cubic inches. What is the pressure?

25. If we neglect air resistance, the distance that an object will fall from a height near the surface of the earth is directly proportional to the square of the time it falls. If the object falls 256 feet in 4 seconds, how far will it fall in 7 seconds?

26. The volume of a right circular cylinder varies jointly as its height and the square of the radius of the base. The volume is 360π cubic centimeters when the height is 10 centimeters and the radius is 6 centimeters. Find V when $h = 18$ centimeters and $r = 5$ centimeters.

27. If the volume of a sphere varies directly as the cube of its radius and $V = 288\pi$ cubic inches when $r = 6$ inches, find V when $r = 2$ inches.

28. The resistance to the flow of electricity through a wire depends on the length and thickness of the wire. The resistance R is measured in *ohms* and varies directly as the length l and inversely as the square of the diameter d. If a wire 200 feet long with diameter 0.16 inch has a resistance of 64 ohms, how much resistance will there be if only 50 feet of wire is used?

29. A wire made of the same material as the wire in Exercise 28 is 100 feet long. Find R if $d = 0.4$ inch. Find R if $d = 0.04$ inch.

30. If z varies jointly as x and y, how does x vary with respect to y and z?

31. A rectangular-shaped beam is to be cut from a round log with a 2.5-foot diameter. The strength s of the beam varies jointly as its height y and the square of its width x.

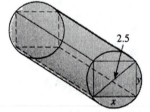

   (a) Write s as a function of y.

   (b) Write s as a function of x.

   (c) Find s to two decimal places when $x = y$. Express the answer in terms of the constant of proportionality.

32. The weight of an object that is above the earth varies inversely as the square of its distance from the center of the earth. Assume that an object weighs 150 pounds on earth. How much would that object weigh at a height of 200 miles above the earth? Assume the radius of the earth to be 4000 miles and give your answer to the nearest pound.

33. The weight of an object that is within the earth varies directly as its distance from the center of the earth. How much would the object in Exercise 32 weigh at a distance of 100 miles below the earth's surface?

34. Suppose that an astronaut weighs 120 pounds on earth. How far above the earth would she have to be in order to weigh just half as much? (See Exercise 32.)

35. Suppose that both x and y vary directly as z. Show that the sum $x + y$ also varies directly as z.

36. Repeat Exercise 35 for inverse variation.

37. In Exercise 35, how does the product xy vary in relation to z?

38. According to Newton's Law of Gravitation, how does the force of attraction between two objects change if each of their masses is cut in half but the distance between them is doubled?

39. The illumination I from a source of light varies inversely as the square of the distance from the source. If $I = 20$ foot-candles when $d = 4$ feet, find the illumination when $d = 8$ feet.

40. In Exercise 39, at what distance from the source will the illumination be double the amount that it is at 4 feet?

41. In the figure, show that the area of $\triangle AOB$ varies jointly as the square of x and the slope of the line $y = mx$.

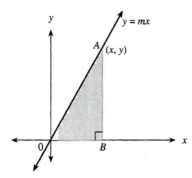

 Written Assignment   Consider the two situations when y varies directly as x for $k > 0$ and for $k < 0$. Sketch the graph of each case for $x > 0$, and determine the behavior of y as x increases. Then do the same for the two cases of inverse variation.

3.6 SYNTHETIC DIVISION, THE REMAINDER AND FACTOR THEOREMS

Any linear equation $ax + b = 0$ is easy to solve, and quadratic equations can always be solved by the quadratic formula. To solve higher-degree polynomial equations, the ability to factor the polynomials will be very helpful. Extending our knowledge of factoring is a major theme of this section, and will be used in the section that follows to expand our abilities to solve polynomial equations of higher degrees.

We begin with a process called **synthetic division,** which is a special case of the long-division method for polynomials (see page 65). To develop this process, we will use the following illustration in which the divisor has the form $x - c$.

Check this result by showing that $f(x) = q(x) \cdot g(x) + r$, where $f(x) = x^3 + x^2 - 11x + 12$, $q(x)$ is the quotient, $g(x)$ is the divisor, and r is the remainder.

$$
\begin{array}{r}
x^2 + 3x - 5 \quad \longleftarrow \text{Quotient} \\
\text{Divisor} \longrightarrow x - 2 \overline{\big)\, x^3 + x^2 - 11x + 12} \quad \longleftarrow \text{Dividend} \\
\underline{x^3 - 2x^2} \qquad\qquad\qquad \\
+\,3x^2 - 11x \qquad\quad \\
\underline{+\,3x^2 - 6x} \qquad\quad \\
-\,5x + 12 \\
\underline{-\,5x + 10} \\
+\,2 \quad \longleftarrow \text{Remainder}
\end{array}
$$

Now it should be clear that all of the work done involved the coefficients of the variables and the constants. Thus we could just as easily complete the division by omitting the variables, as long as we write the coefficients in the proper places. The division problem would then look like this:

$$
\begin{array}{r}
1 + 3 - 5 \\
1 - 2 \overline{\big)\, 1 + 1 - 11 + 12} \\
\underline{① - 2} \\
+\,3 - 11 \\
\underline{+③ - 6} \\
-\,5 + 12 \\
\underline{-⑤ + 10} \\
+\,2
\end{array}
$$

Since the circled numerals are repetitions of those immediately above them, this process can be further shortened by deleting them. Moreover, since these circled numbers are the products of the numbers in the quotient by the 1 in the divisor, we may also eliminate this 1. Thus we have the following:

$$
\begin{array}{r}
1 + 3 - 5 \\
-2\,\overline{\smash{)}\,1 + 1 - 11 + 12} \\
\underline{-\,2} \\
+\,3 - 11 \\
\underline{-\,6} \\
-\,5 + 12 \\
\underline{+\,10} \\
+\,2
\end{array}
\longrightarrow
\begin{array}{l}
\text{It is not} \\
\text{necessary} \\
\text{to bring} \\
\text{down the } -11 \\
\text{and 12.}
\end{array}
$$

$$
\begin{array}{r}
1 + 3 - 5 \\
-2\,\overline{\smash{)}\,1 + 1 - 11 + 12} \\
\underline{-\,2} \\
+\,3 \\
\underline{-\,6} \\
-\,5 \\
\underline{+\,10} \\
+\,2
\end{array}
\longrightarrow
\begin{array}{l}
\text{Move the} \\
\text{numerals} \\
\text{upward.}
\end{array}
$$

$$
\begin{array}{r}
1 + 3 - 5 \\
-2\,\overline{\smash{)}\,1 + 1 - 11 + 12} \\
\underline{-\,2 - 6 + 10} \\
+\,3 - 5 + 2
\end{array}
$$

When the top numeral 1 is brought down, the last line contains the coefficients of the quotient and the remainder. So eliminate the line above the dividend.

$$
\begin{array}{r}
-2\,\overline{\smash{)}\,1 + 1 \;- 11 + 12} \\
\underline{-\,2 \;- 6 + 10} \\
1 + 3 \;- 5\,\boxed{+\,2}
\end{array}
$$
Remainder

Quotient: $\quad x^2 + 3x - 5$

We can further simplify this process by changing the sign of the divisor, making it $+2$ instead of -2. This change allows us to add throughout rather than subtract, as follows.

$$
\begin{array}{r}
+2\,\overline{\smash{)}\,1 + 1 - 11 + 12} \\
\underline{+\,2 + 6 - 10} \\
1 + 3 - 5\,\boxed{+\,2}
\end{array}
$$
Coefficients of dividend

Coefficients of quotient: $\qquad$ Remainder

Quotient: $\quad x^2 + 3x - 5$

The long-division process has now been condensed to this short form. Doing a division problem by this short form is called *synthetic division*, as illustrated in the examples that follow.

EXAMPLE 1 Use synthetic division to find the quotient and the remainder.

$$(2x^3 - 9x^2 + 10x - 7) \div (x - 3)$$

Solution Write the coefficients of the dividend in descending order. Change the sign of the divisor (change -3 to $+3$).

$$
\begin{array}{r}
+3\,\overline{\smash{)}\,2 - 9 + 10 - 7} \\

\end{array}
\qquad
\begin{array}{l}
\text{Now bring down the} \\
\text{first term, 2, and} \\
\text{multiply by } +3.
\end{array}
\qquad
\begin{array}{r}
+3\,\overline{\smash{)}\,2 - 9 + 10 - 7} \\
\underline{+\,6} \\
2
\end{array}
$$

Add -9 and $+6$ to obtain the sum -3. Multiply this sum by $+3$ to obtain -9, and repeat the process to the end. The completed example should look like this:

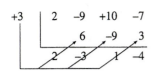

$$+3 \,\big|\, 2 - 9 + 10 - 7$$
$$\underline{ + 6 - 9 + 3}$$
Coefficients of quotient: $2 - 3 + 1\,\big|\,\underline{-4}$ Remainder

Since the original dividend began with x^3 (third degree), the quotient will begin with x^2 (second degree). Thus we read the last line as implying a quotient of $2x^2 - 3x + 1$ and a remainder of -4. (See below for a check of this result.)   ■

The synthetic division process has been developed for divisors of the form $x - c$. (Thus, in Example 1, $c = 3$.) A minor adjustment also permits divisors by polynomials of the form $x + c$. For example, a divisor of $x + 2$ may be written as $x - (-2)$; $c = -2$.

Note: The quotient in a synthetic division problem is always a polynomial of degree one less than that of the dividend. This is so because the divisor has degree 1. The bottom line in the synthetic division process, except for the last entry on the right, gives the coefficients of the quotient: a polynomial in standard form.

EXAMPLE 2   Use synthetic division to find the quotient and the remainder.

$$\left(-\tfrac{1}{3}x^4 + \tfrac{1}{6}x^2 - 7x - 4\right) \div (x + 3)$$

Solution Think of $x + 3$ as $x - (-3)$. Since there is no x^3 term in the dividend, use $0x^3$.

$$-3 \,\Big|\, -\tfrac{1}{3} + 0 + \tfrac{1}{6} - 7 - 4$$
$$\underline{\phantom{-\tfrac{1}{3}} + 1 - 3 + \tfrac{17}{2} - \tfrac{9}{2}}$$
$$-\tfrac{1}{3} + 1 - \tfrac{17}{6} + \tfrac{3}{2}\,\Big|\,-\tfrac{17}{2} = \text{Remainder}$$

Quotient: $-\tfrac{1}{3}x^3 + x^2 - \tfrac{17}{6}x + \tfrac{3}{2}$ Remainder: $-\tfrac{17}{2}$

Check this result.   ■

TEST YOUR UNDERSTANDING

Use synthetic division, as indicated, to find each quotient and remainder.

1. $(x^3 - x^2 - 5x + 6) \div (x - 2)$
   $$+2 \,\big|\, 1 - 1 - 5 + 6$$

2. $(2x^3 + 7x^2 + 5x - 2) \div (x + 2)$
   $$-2 \,\big|\, 2 + 7 + 5 - 2$$

3. $(2x^3 - 3x^2 + 7) \div (x - 3)$
   $$+3 \,\big|\, 2 - 3 + 0 + 7$$

4. $(3x^3 - 7x + 12) \div (x + 1)$
   $$-1 \,\big|\, 3 + 0 - 7 + 12$$

(Answers: Page 246)

In Example 1 we found that when the polynomial $p(x) = 2x^3 - 9x^2 + 10x - 7$ was divided by $x - 3$, the quotient was $q(x) = 2x^2 - 3x + 1$ and the remainder was $r = -4$. As a check we see that

$$\underbrace{2x^3 - 9x^2 + 10x - 7}_{p(x)} = \underbrace{(2x^2 - 3x + 1)}_{q(x)}\underbrace{(x - 3)}_{\cdot\ (x-3)} + \underbrace{(-4)}_{+\ r}$$

In general, whenever a polynomial $p(x)$ is divided by $x - c$ we have

Another form of this is
$$\frac{p(x)}{x - c} = q(x) + \frac{r}{x - c}.$$

$$p(x) = q(x)(x - c) + r$$

where $q(x)$ is the quotient and r is the (constant) remainder. Since this equation holds for all x, we may let $x = c$ and obtain

$$p(c) = q(c)(c - c) + r$$
$$= q(c) \cdot 0 + r$$
$$= r$$

This result may be summarized as follows:

REMAINDER THEOREM

If a polynomial $p(x)$ is divided by $x - c$, the remainder is equal to $p(c)$.

EXAMPLE 3   Find the remainder when $p(x) = 3x^3 - 5x^2 + 7x + 5$ is divided by $x - 2$.

Solution By the remainder theorem, the answer is $p(2)$.

$$p(x) = 3x^3 - 5x^2 + 7x + 5$$
$$p(2) = 3(2)^3 - 5(2)^2 + 7(2) + 5$$

Check this result by dividing $p(x)$ by $x - 2$.

$$= 23$$ ∎

EXAMPLE 4   Let $p(x) = x^3 - 2x^2 + 3x - 1$. Use synthetic division and the remainder theorem to find $p(3)$.

Solution According to the remainder theorem, $p(3)$ is equal to the remainder when $p(x)$ is divided by $x - 3$.

Check this result by substituting $x = 3$ in $p(x)$.

$$
\begin{array}{r|rrrr}
3 & 1 & -2 & +3 & -1 \\
  &   & +3 & +3 & +18 \\
\hline
  & 1 & +1 & +6 & \boxed{+17} = \text{Remainder} = p(3)
\end{array}
$$

∎

TEST YOUR UNDERSTANDING	*Use synthetic division to find the remainder r when $p(x)$ is divided by $x - c$. Verify that $r = p(c)$ by substituting $x = c$ into $p(x)$.*
	1. $p(x) = x^5 - 7x^4 + 4x^3 + 10x^2 - x - 5$; $x - 1$
	2. $p(x) = x^4 + 11x^3 + 11x^2 + 11x + 10$; $x + 10$
	3. $p(x) = x^4 + 11x^3 + 11x^2 + 11x + 10$; $x - 10$
(Answers: Page 246)	**4.** $p(x) = 6x^3 - 40x^2 + 25$; $x - 6$

Once again, we are going to consider the division of a polynomial $p(x)$ by a divisor of the form $x - c$ and make use of the relationship

$$p(x) = q(x)(x - c) + r$$

Now suppose that the remainder $r = 0$. Then the remainder theorem gives $p(c) = r = 0$, and the preceding equation becomes

*If $u = vw$, then v and w are said to be **factors** of u.*

$$p(x) = q(x)(x - c)$$

It follows that $x - c$ is a *factor* of $p(x)$. Conversely, suppose that $x - c$ is a factor of $p(x)$. This means there is another polynomial, say $q(x)$, so that

$$p(x) = q(x)(x - c)$$

or

$$p(x) = q(x)(x - c) + 0$$

*If $p(c) = 0$, then c is said to be a **zero of the polynomial** $p(x)$, and it is also a root of $p(x) = 0$.*

which tells us that when $p(x)$ is divided by $x - c$ the remainder is zero. These observations comprise the following result:

FACTOR THEOREM

A polynomial $p(x)$ has a factor $x - c$ if and only if $p(c) = 0$.

EXAMPLE 5   Show that $x - 2$ is a factor of $p(x) = x^3 - 3x^2 + 7x - 10$.

Solution By the factor theorem we can state that $x - 2$ is a factor of $p(x)$ if $p(2) = 0$.

$$p(2) = 2^3 - 3(2)^2 + 7(2) - 10$$
$$= 0$$

∎

EXAMPLE 6
(a) Use the factor theorem to show that $x + 3$ is a factor of
   $p(x) = x^3 - x^2 - 8x + 12$.
(b) Factor $p(x)$ completely.
(c) Find the zeros of $p(x)$.

Solution
(a) First write $x + 3 = x - (-3)$, so that $c = -3$. Then use synthetic division.

Note: $p(-3) = 0$. We use synthetic division here in order to be able to factor $p(x)$ as shown in part (b).

$$-3 \begin{array}{|rrrr} 1 & -1 & -8 & +12 \\ & -3 & +12 & -12 \\ \hline 1 & -4 & +4 & +0 \end{array}$$

Since $p(-3) = 0$, the factor theorem tells us that $x + 3$ is a factor of $p(x)$.

(**b**) Synthetic division has produced the quotient $x^2 - 4x + 4$. Since $x + 3$ is a factor of $p(x)$, we may write

$$x^3 - x^2 - 8x + 12 = (x^2 - 4x + 4)(x + 3)$$

To find the complete factored form, observe that $x^2 - 4x + 4 = (x - 2)^2$. Thus:

$$x^3 - x^2 - 8x + 12 = (x - 2)^2(x + 3)$$

(**c**) Since $p(x) = (x - 2)^2(x + 3) = 0$, then $x - 2 = 0$ or $x + 3 = 0$. Thus 2 and -3 are the roots of $p(x) = 0$ and are the zeros of $p(x)$. ■

When the degree and the zeros of a polynomial are given, then the polynomial can be determined except for its leading coefficient. That is, if $p(x)$ has degree 3 and the roots are r_1, r_2, and r_3, then $p(x) = a(x - r_1)(x - r_2)(x - r_3)$. To find a, some additional information will be needed, as shown in Example 7.

EXAMPLE 7 A polynomial $p(x)$ of degree 4 has the zeros -2, 0, 1, and $\frac{1}{2}$. If $p(-1) = -6$, find $p(x)$ and write it in standard form.

Solution Each zero r contributes the factor $x - r$. Then

$$p(x) = a(x + 2)(x - 0)(x - 1)(x - \tfrac{1}{2})$$
$$= ax(x + 2)(x - 1)(x - \tfrac{1}{2})$$

To find a, use $p(-1) = -6$ as follows

$$p(-1) = -6 = a(-1)(-1 + 2)(-1 - 1)(-1 - \tfrac{1}{2}) = -3a$$

Therefore, $a = 2$ and

$$p(x) = 2x(x + 2)(x - 1)(x - \tfrac{1}{2})$$
$$= 2x^4 + x^3 - 5x^2 + 2x$$ ■

EXERCISES 3.6

Use synthetic division to find the quotient and remainder. Check each result.

1. $(x^3 - 2x^2 - 5x + 6) \div (x - 3)$

2. $(x^3 - x^2 - 5x + 2) \div (x + 2)$

3. $(2x^3 + x^2 - 3x + 7) \div (x + 1)$

4. $(3x^3 - 2x^2 + x - 1) \div (x - 1)$

5. $(x^3 + 5x^2 - 7x + 8) \div (x - 2)$

6. $(x^3 - 3x^2 + x - 5) \div (x + 3)$

7. $(x^4 - 3x^3 + 7x^2 - 2x + 1) \div (x + 2)$

8. $(x^4 + x^3 - 2x^2 + 3x - 1) \div (x - 2)$

9. $(2x^4 - 3x^2 + 4x - 2) \div (x - 1)$

10. $(3x^4 + x^3 - 2x + 3) \div (x + 1)$

Divide.

11. $(x^3 - 27) \div (x - 3)$

12. $(x^3 - 27) \div (x + 3)$

13. $(x^3 + 27) \div (x + 3)$

14. $(x^3 + 27) \div (x - 3)$

15. $(x^4 - 16) \div (x - 2)$

16. $(x^4 - 16) \div (x + 2)$

17. $(x^4 + 16) \div (x + 2)$

18. $(x^4 + 16) \div (x - 2)$

19. $(x^4 - \frac{1}{2}x^3 + \frac{1}{3}x^2 - \frac{1}{4}x + \frac{1}{5}) \div (x - 1)$

20. $(4x^5 - x^3 + 5x^2 + \frac{3}{2}x - \frac{1}{2}) \div (x + \frac{1}{2})$

Use synthetic division to find each quotient by following the procedure shown in this example:

$$\frac{2x^3 - 7x^2 + 8x + 6}{2x - 3} = \frac{2x^3 - 7x^2 + 8x + 6}{2\left(x - \frac{3}{2}\right)}$$

$$= \frac{1}{2}\left(\frac{2x^3 - 7x^2 + 8x + 6}{x - \frac{3}{2}}\right) \qquad \text{Divisor within the parentheses has the form } x - c.$$

$$= \frac{1}{2}\left(2x^2 - 4x + 2 + \frac{9}{x - \frac{3}{2}}\right) \qquad \text{By synthetic division}$$

$$= x^2 - 2x + 1 + \frac{9}{2x - 3} \qquad \text{Multiplying by } \frac{1}{2}$$

21. $(6x^3 - 5x^2 - 3x + 4) \div (2x - 1)$

22. $(10x^3 - 3x^2 + 4x + 7) \div (2x + 1)$

23. $(6x^3 + 7x^2 + x + 8) \div (2x + 3)$

24. $(9x^4 + 6x^3 - 4x + 5) \div (3x - 1)$

Use synthetic division and the remainder theorem.

25. $f(x) = x^3 - x^2 + 3x - 2$; find $f(2)$.

26. $f(x) = 2x^3 + 3x^2 - x - 5$; find $f(-1)$.

27. $f(x) = x^4 - 3x^2 + x + 2$; find $f(3)$.

28. $f(x) = x^4 + 2x^3 - 3x - 1$; find $f(-2)$.

29. $f(x) = x^5 - x^3 + 2x^2 + x - 3$; find $f(1)$.

30. $f(x) = 3x^4 + 2x^3 - 3x^2 - x + 7$; find $f(-3)$.

Find the remainder for each division by substitution, using the remainder theorem. That is, in Exercise 31 (for example) let $f(x) = x^3 - 2x^2 + 3x - 5$ and find $f(2) = r$.

31. $(x^3 - 2x^2 + 3x - 5) \div (x - 2)$

32. $(x^3 - 2x^2 + 3x - 5) \div (x + 2)$

33. $(2x^3 + 3x^2 - 5x + 1) \div (x - 3)$

34. $(3x^4 - x^3 + 2x^2 - x + 1) \div (x + 3)$

35. $(4x^5 - x^3 - 3x^2 + 2) \div (x + 1)$

36. $(3x^5 - 2x^4 + x^3 - 7x + 1) \div (x - 1)$

Show that the given binomial $x - c$ is a factor of $p(x)$, factor $p(x)$ completely, and solve $p(x) = 0$.

37. $p(x) = x^3 + 6x^2 + 11x + 6$; $x + 1$

38. $p(x) = x^3 - 6x^2 + 11x - 6$; $x - 1$

39. $p(x) = x^3 + 5x^2 - 2x - 24$; $x - 2$

40. $p(x) = -x^3 + 11x^2 - 23x - 35$; $x - 7$

41. $p(x) = -x^3 + 7x + 6$; $x + 2$

42. $p(x) = x^3 + 2x^2 - 13x + 10$; $x + 5$

43. $p(x) = 6x^3 - 25x^2 - 29x + 20$; $x - 5$

44. $p(x) = 12x^3 - 22x^2 - 100x - 16$; $x + 2$

45. $p(x) = x^4 + 4x^3 + 3x^2 - 4x - 4$; $x + 2$

46. $p(x) = x^4 - 8x^3 + 7x^2 + 72x - 144$; $x - 4$

47. $p(x) = x^6 + 6x^5 + 8x^4 - 6x^3 - 9x^2$; $x + 3$

48. $p(x) = 8x^6 - 52x^5 + 6x^4 + 117x^3 - 54x^2$; $2x - 1$

49. Find $p(x)$, a polynomial of degree 3 with zeros of -2, 2, and 3 if $p(1) = 18$.

50. Find $p(x)$, a polynomial of degree 4 with zeros of -1, 1, 3, and 3 if $p(0) = -18$.

Use synthetic division to answer the following.

51. When $x^2 + 5x - 2$ is divided by $x + n$, the reminder is -8. Find all possible values of n and check by division.

52. Find d so that $x + 6$ is a factor of $x^4 + 4x^3 - 21x^2 + dx + 108$.

53. Find b so that $x - 2$ is a factor of $x^3 + bx^2 - 13x + 10$.

54. Find a so that $x - 10$ is a factor of $ax^3 - 25x^2 + 47x + 30$.

55. Find the complete factored form of $p(x) = x^5 + x^4 + 5x^2 - x - 6$ if $p(-2) = p(-1) = p(1) = 0$.

56. $p(x)$ is a fifth-degree polynomial in which 1 is the coefficient of x^5. Write the polynomial in standard form, given that $p(0) = p(1) = p(2) = p(3) = p(4) = 0$.

57. Prove that $p(x) = 2x^4 + 5x^2 + 20$ has no factor of the form $x - c$ for real numbers c.

58. Repeat Exercise 57 for $p(x) = -2x^4 - 5x^2 - 20$.

59. Prove that $x + 1$ is a factor of $x^n + 1$ for any positive odd integer n.

60. Prove that $x - a$ is a factor of $x^n - a^n$ for all positive integers n.

Challenge If the polynomial $p(x)$ contains the factor $(x - r)^2$, then r is called a double root of $p(x) = 0$. Assume that the equation $x^3 + bx - 54 = 0$ has a double root and find the value of b. (*Hint:* Use synthetic division.)

Written Assignment Use your own words and your own specific examples to explain both the remainder and the factor theorem.

Graphing Calculator Exercises Synthetic division can be accomplished on your calculator if a given polynomial, $a_n x^n + a_{n-1} x^{n-1} + \cdots + a_1 x + a_0$ is first written in *nested form*:

$$(\cdots ((a_n x + a_{n-1}) x + a_{n-2}) x + \cdots a_1) x + a_0$$

For example, consider the polynomial $x^4 + x^2 - 11x + 12$. Begin by writing this with terms in descending order, including terms with zero coefficients. Then the polynomial and its nested form appear as follows:

$$1x^4 + 0x^3 + 1x^2 - 11x + 12$$

$$(((1x + 0) x + 1) x - 11) x + 12$$

To divide this polynomial by $x - 2$, for example, the work on a calculator can be organized as follows:

The first coefficient 1 is entered.
Then press $* 2 + 0$ ENTER and 2 appears.
Then press $* 2 + 1$ ENTER and 5 appears.
Then press $* 2 - 11$ ENTER and -1 appears.
Then press $* 2 + 12$ ENTER and 10 appears.

Hence, the quotient is $x^3 + 2x^2 + 5x - 1$, and the remainder is 10. Note that the process is similar to the steps completed for synthetic division.

Complete the division indicated in Exercises 1 and 2 by first expressing the numerator in nested form, and then proceeding as shown previously to find the quotient and remainder.

1. $(x^3 - 3x^2 + 4x + 1)/(x - 5)$

2. $(x^4 + 2x - 7)/(x + 6)$

3. Use the graph of $f(x)$ to approximate all values of k such that $f(x) = x^3 + k^3x^2 + kx - k^4$ is divisible by the linear function $x - 2$. (Approximate k to the nearest tenth.)

3.7 SOLVING POLYNOMIAL EQUATIONS

The factor theorem is useful when solving a polynomial equation $p(x) = 0$, provided that one of the roots of the given equation or a linear factor of $p(x)$ is given. (See Example 6, page 221.) But suppose no such information is given, and it is not clear how $p(x)$ can be factored. It turns out that in many cases the roots can still be found, as we shall now see. To begin, let us first consider this polynomial equation:

$$(3x + 2)(5x - 4)(2x - 3) = 0$$

To find the roots, set each factor equal to zero.

$$3x + 2 = 0 \qquad 5x - 4 = 0 \qquad 2x - 3 = 0$$

$$x = -\frac{2}{3} \qquad\qquad x = \frac{4}{5} \qquad\qquad x = \frac{3}{2}$$

Now multiply the original three factors and keep careful note of the details of this multiplication. Your result should be

$$(3x + 2)(5x - 4)(2x - 3) = 30x^3 - 49x^2 - 10x + 24 = 0$$

which must have the same three rational roots.

As you analyze this multiplication it becomes clear that the constant 24 is the product of the three constants in the binomials, 2, -4, and -3. Also, the leading coefficient, 30, is the product of the three original coefficients of x in the binomials, namely 3, 5, and 2.

Furthermore, 3, 5, and 2 are also the denominators of the roots $-\frac{2}{3}, \frac{4}{5},$ and $\frac{3}{2}$. Therefore, the denominators of the rational roots are all factors of 30, and their numerators are all factors of 24.

These results are not accidental. It turns out that we have been discussing the following general result:

RATIONAL ROOT THEOREM

Let $f(x) = a_n x^n + a_{n-1} x^{n-1} + \cdots + a_1 x + a_0$ $(a_0 \neq 0)$ be an nth-degree polynomial with integer coefficients. If $\frac{p}{q}$ is a rational root of $f(x) = 0$, where $\frac{p}{q}$ is in lowest terms, then p is a factor of a_0 and q is a factor of a_n.

Let us see how this theorem can be applied to find the rational roots of

$$f(x) = 4x^3 - 16x^2 + 11x + 10 = 0$$

Note: The theorem allows only these 16 numbers as possible rational roots: no other rational numbers can be roots.

Begin by listing all factors of the constant 10 and of the leading coefficient 4.

Possible numerators (factors of 10): ± 1, ± 2, ± 5, ± 10

Possible denominators (factors of 4): ± 1, ± 2, ± 4

Possible rational roots (take each number in the first row and divide by each number in the second row):

$$\pm 1, \quad \pm\tfrac{1}{2}, \quad \pm\tfrac{1}{4}, \quad \pm 2, \quad \pm 5, \quad \pm\tfrac{5}{2}, \quad \pm\tfrac{5}{4}, \quad \pm 10$$

To decide which (if any) of these are roots of $f(x) = 0$, we could substitute the values directly into $f(x)$. However, it is easier to use synthetic division because in most cases it leads to easier computations and also makes quotients available. Therefore, we proceed by using synthetic division with divisors c, where c is a possible rational root.

If $f(c) = 0$, then c is a root; if $f(c) \neq 0$, then c is not a root.

Try $c = 1$:

$$1 \, | \, 4 - 16 + 11 + 10$$
$$\underline{+ \;\; 4 - 12 - \;\; 1}$$
$$4 - 12 - \;\; 1 \, | \, + 9$$

Since $f(1) = 9 \neq 0$, 1 is *not* a root.

Try $c = -1$:

$$-1 \, | \, 4 - 16 + 11 + 10$$
$$\underline{- \;\; 4 + 20 - 31}$$
$$4 - 20 + 31 \, | \, - 21$$

Since $f(-1) = -21 \neq 0$, -1 is *not* a root.

Try $c = \tfrac{1}{2}$:

$$\tfrac{1}{2} \, | \, 4 - 16 + 11 \, + 10$$
$$\underline{+ \;\; 2 - \;\; 7 + \;\; 2}$$
$$4 - 14 + \;\; 4 \, | \, + 12$$

Since $f\left(\tfrac{1}{2}\right) = 12 \neq 0$,

$\tfrac{1}{2}$ is *not* a root.

Try $c = -\tfrac{1}{2}$:

$$-\tfrac{1}{2} \, | \, 4 - 16 + 11 \, + 10$$
$$\underline{- \;\; 2 + \;\; 9 - 10}$$
$$4 - 18 + 20 \, | \, + \;\; 0$$

Since $f\left(-\tfrac{1}{2}\right) = 0$,

$-\tfrac{1}{2}$ *is* a root.

By the factor theorem it follows that $x - \left(-\tfrac{1}{2}\right) = x + \tfrac{1}{2}$ is a factor of $f(x)$, and synthetic division gives the other factor, $4x^2 - 18x + 20$.

$$f(x) = \left(x + \tfrac{1}{2}\right)\left(4x^2 - 18x + 20\right)$$

To find other roots of $f(x) = 0$ we could proceed by using the rational root theorem for $4x^2 - 18x + 20 = 0$. But this is unnecessary because the quadratic expression is factorable.

$$f(x) = \left(x + \tfrac{1}{2}\right)\left(4x^2 - 18x + 20\right)$$

$$= \left(x + \tfrac{1}{2}\right)(2)\left(2x^2 - 9x + 10\right)$$

$$= 2\left(x + \tfrac{1}{2}\right)(x - 2)(2x - 5)$$

$$= (2x + 1)(x - 2)(2x - 5)$$

The solution of $f(x) = 0$ can now be found by setting each factor equal to zero. The roots are $x = -\frac{1}{2}$, $x = 2$, and $x = \frac{5}{2}$; these are the zeros of $f(x)$.

Whenever a polynomial has ± 1 as the leading coefficient, then any rational root will be an integer that is a factor of the constant term of the polynomial.

EXAMPLE 1 Factor $f(x) = x^3 + 6x^2 + 11x + 6$.

Solution Since the leading coefficient is 1, whose only factors are ± 1, the possible denominators of a rational root of $f(x) = 0$ can only be ± 1. Hence the possible rational roots must all be factors of ± 6, namely ± 1, ± 2, ± 3, and ± 6. Use synthetic division to test these cases.

$$
\begin{array}{r}
1\,\underline{|\,1 + 6 + 11 + 6} \\
+ 1 + 7 + 18 \\
\hline
1 + 7 + 18\,\boxed{+ 24} = r
\end{array}
$$

Since $r = f(1) \neq 0$, $x - 1$ is *not* a factor of $f(x)$.

$$
\begin{array}{r}
-1\,\underline{|\,1 + 6 + 11 + 6} \\
- 1 - 5 - 6 \\
\hline
1 + 5 + 6\,\boxed{+ 0} = r
\end{array}
$$

Since $r = f(-1) = 0$, $x - (-1) = x + 1$ *is* a factor of $f(x)$.

$$x^3 + 6x^2 + 11x + 6 = (x + 1)(x^2 + 5x + 6)$$

Now factor the trinomial:

$$x^3 + 6x^2 + 11x + 6 = (x + 1)(x + 2)(x + 3)$$ ∎

TEST YOUR UNDERSTANDING	*For each $p(x)$, find **(a)** the possible rational roots of $p(x) = 0$, **(b)** the factored form of $p(x)$, and **(c)** the roots of $p(x) = 0$.*

1. $p(x) = x^3 - 3x^2 - 10x + 24$

2. $p(x) = x^4 + 6x^3 + x^2 - 24x + 16$

3. $p(x) = 4x^3 + 20x^2 - 23x + 6$

4. $p(x) = 3x^4 - 13x^3 + 7x^2 - 13x + 4$

(Answers: Page 246)

EXAMPLE 2 Find all real roots of $p(x) = 2x^5 + 7x^4 - 18x^2 - 8x + 8 = 0$.

Solution Begin by searching for rational roots. The possible rational roots are ± 1, $\pm \frac{1}{2}$, ± 2, ± 4, and ± 8. Testing these possibilities (left to right), the first root we find is $\frac{1}{2}$ as shown.

$$
\begin{array}{r}
\frac{1}{2}\,\underline{|\,2 + 7 + 0 - 18 - 8 + 8} \\
\phantom{\frac{1}{2}\,|\,2}+ 1 + 4 + 2 - 8 - 8 \\
\hline
2 + 8 + 4 - 16 - 16\,\boxed{+ 0}
\end{array}
$$

Therefore, $x - \frac{1}{2}$ is a factor of $p(x)$

$$p(x) = \left(x - \tfrac{1}{2}\right)(2x^4 + 8x^3 + 4x^2 - 16x - 16)$$

$$= 2\left(x - \tfrac{1}{2}\right)(x^4 + 4x^3 + 2x^2 - 8x - 8)$$

To find other roots of $p(x) = 0$ it now becomes necessary to solve the **depressed equation**

$$x^4 + 4x^3 + 2x^2 - 8x - 8 = 0$$

The possible rational roots for this equation are ± 1, ± 2, ± 4, and ± 8. However, values like ± 1 that were tried before, and produced nonzero remainders, need not be tried again. Why? We find that $x = -2$ is a root:

$$
\begin{array}{r}
-2\,|\,1 + 4 + 2 - 8 - 8 \\
\underline{-2 - 4 + 4 + 8} \\
1 + 2 - 2 - 4 \,|\,+0
\end{array}
$$

$$x^4 + 4x^3 + 2x^2 - 8x - 8 = (x + 2)(x^3 + 2x^2 - 2x - 4)$$
$$= (x + 2)[x^2(x + 2) - 2(x + 2)]$$
$$= (x + 2)(x^2 - 2)(x + 2)$$
$$= (x + 2)^2(x^2 - 2)$$

Thus we may now write $p(x)$ as follows:

$$p(x) = 2\left(x - \tfrac{1}{2}\right)(x^4 + 4x^3 + 2x^2 - 8x - 8)$$

This result can also be written as $(2x - 1)(x + 2)^2(x^2 - 2)$.

$$= 2\left(x - \tfrac{1}{2}\right)(x + 2)^2(x^2 - 2)$$

$$= 2\left(x - \tfrac{1}{2}\right)(x + 2)^2\left(x + \sqrt{2}\right)\left(x - \sqrt{2}\right)$$

Setting each factor equal to zero produces the real roots of $p(x) = 0$, namely two rational and two irrational roots.

These four roots of $p(x) = 0$ are also the x-intercepts of the graph of the polynomial function $p(x)$.

$$x = \tfrac{1}{2}, \quad x = -2, \quad x = -\sqrt{2}, \quad x = \sqrt{2} \qquad \blacksquare$$

The roots of the equation $p(x) = 0$ in Example 2 are $\frac{1}{2}$, -2, $-\sqrt{2}$, and $\sqrt{2}$. Since the factored form of $p(x)$ contains the factor $(x + 2)^2$, we say that -2 is a *double root*. Therefore, counting the double root -2 as two roots, there are five roots for $p(x) = 0$, which is the same as the degree of $p(x)$. This is no coincidence. Such a result is true for any polynomial equation and is based upon the following theorems.

HISTORICAL NOTE
This theorem was proved by
Carl Freidrich Gauss in
1797 *when he was 20 years*
old. He is recognized as
one of the greatest mathe-
maticians of all time.

Source: Dr. Marvin Lang.

FUNDAMENTAL THEOREM OF ALGEBRA

If $p(x)$ is a polynomial of degree $n \geq 1$, then $p(x)$ has at least one complex zero. (That is, $p(x) = 0$ has at least one root.)

The proof of this theorem is beyond the level of this course; however, it leads to the next result that tells us that a polynomial equation of degree n cannot have more than n distinct roots.

THE *N*-ROOTS THEOREM

If $p(x)$ is a polynomial of degree $n \geq 1$, then $p(x) = 0$ has exactly n roots, provided that a root of multiplicity k is counted k times.

For example, the fifth-degree equation in Example 2 has five roots, counting -2 twice because it is a double root.

Even though it can be very difficult to find all the roots of a polynomial equation, the preceding theorem does help us to the extent that we know that there can be no more than n roots when $p(x)$ has degree n. However, for illustrative purposes, we will consider polynomial equations whose roots can be found using the methods that have been studied.

EXAMPLE 3   Find all real and imaginary roots: $x^4 - 6x^2 - 8x + 24 = 0$.

Solution   The possible rational roots are

$$\pm 1, \ \pm 2, \ \pm 3, \ \pm 4, \ \pm 6, \ \pm 8, \ \pm 12, \ \pm 24$$

Trying these possibilities, left to right, shows that 2 is a root:

$$
\begin{array}{r}
2\,\big|\,1 + 0 - 6 - \ \ 8 + 24 \\
\underline{+\,2 + 4 - \ \ 4 - 24} \\
1 + 2 - 2 - 12\,\big|\!\underline{+\ \ 0}
\end{array}
$$

Now, by the factor theorem, we have

$$p(x) = (x - 2)(x^3 + 2x^2 - 2x - 12)$$

and any remaining rational root of the given equation must be a root of the *depressed equation*

$$x^3 + 2x^2 - 2x - 12 = 0$$

This work can be condensed as follows:

$$
\begin{array}{r}
2\,\big|\,1 + 0 - 6 - \ \ 8 + 24 \\
\underline{+\,2 + 4 - \ \ 4 - 24} \\
2\,\big|\,1 + 2 - 2 - 12\,\big|\!\underline{+\ \ 0} \\
\underline{+\,2 + 8 + 12} \\
1 + 4 + 6\,\big|\!\underline{+\ \ 0}
\end{array}
$$

Since there is the possibility of a double root, try 2 again.

$$
\begin{array}{r}
2\,\big|\,1 + 2 - 2 - 12 \\
\underline{+\,2 + 8 + 12} \\
1 + 4 + 6\,\big|\!\underline{+\ \ 0}
\end{array}
$$

Then $x^3 + 2x^2 - 2x - 12 = (x - 2)(x^2 + 4x + 6)$

and $p(x) = (x - 2)(x - 2)(x^2 + 4x + 6)$

$$= (x - 2)^2(x^2 + 4x + 6)$$

Using the quadratic formula, the roots of $x^2 + 4x + 6 = 0$ are found to be $-2 \pm \sqrt{2}i$. Thus $p(x) = 0$ has two imaginary roots and the double root 2, a total of four roots. Since the degree of $p(x)$ is 4, we know from the n-roots theorem that there can be no other roots; we have found them all. ∎

Observe that the pair of conjugates $-2 + \sqrt{2}i$ and $-2 - \sqrt{2}i$ are both roots of $p(x) = 0$ in Example 3. This is a special case of the result that states that if a polynomial equation with real coefficients has nonreal complex roots, then they occur in conjugate pairs.

In Example 3, such numbers as $-2 + \sqrt{2}i$ and $-2 - \sqrt{2}i$ are said to be **conjugates** of one another. In general, the conjugate of a complex number $z = a + bi$ is sometimes denoted as $\bar{z}$, so that $\bar{z} = \overline{a + bi} = a - bi$. In the exercises you will be asked to prove that the conjugate of a sum is equal to the sum of the conjugates, $\overline{z + w} = \bar{z} + \bar{w}$, and similarly for differences, products, and quotients. Example 4 demonstrates such results for specific complex numbers.

EXAMPLE 4   For $z = -2 + 5i$ and $w = 4 - 7i$, verify the following:
(a) $\overline{z + w} = \bar{z} + \bar{w}$ and **(b)** $\overline{zw} = \bar{z} \cdot \bar{w}$.

Solution

(a) $\overline{z + w} = \overline{(-2 + 5i) + (4 - 7i)}$ $\qquad$ $\bar{z} + \bar{w} = \overline{-2 + 5i} + \overline{4 - 7i}$

$\qquad\qquad = \overline{2 - 2i}$ $\qquad\qquad\qquad\qquad = (-2 - 5i) + (4 + 7i)$

$\qquad\qquad = 2 + 2i$ $\qquad\qquad\qquad\qquad\qquad = 2 + 2i$

(b) $\overline{zw} = \overline{(-2 + 5i)(4 - 7i)}$ $\qquad$ $\bar{z} \cdot \bar{w} = \overline{(-2 + 5i)}\ \overline{(4 - 7i)}$

$\qquad\qquad = \overline{27 + 34i}$ $\qquad\qquad\qquad\qquad = (-2 - 5i)(4 + 7i)$

$\qquad\qquad = 27 - 34i$ $\qquad\qquad\qquad\qquad\qquad = 27 - 34i$ ∎

See page 95 for a note about René Descartes.

We have seen that the rational root theorem sometimes provides us with a rather lengthy list of possible roots. One way to eliminate some of these possibilities is to make use of a theorem discovered by Descartes that provides information about the number of real roots of a polynomial equation. It is based on the number of *variations in sign* of a polynomial; that is, the number of times that there is a change in sign between terms. Here are several examples to clarify this idea:

$p(x) = x^4 + 3x^3 - 5x^2 - 3x + 7$ $p(x)$ has two variations in sign.

$\qquad\qquad\qquad + \text{ to } - \qquad - \text{ to } +$

$r(x) = 2x^3 + x^2 + 5x + 1$ $r(x)$ has no variations in sign.

$s(x) = x^5 - 2x^4 + x^3 + 2x - 3$ $s(x)$ has three variations in sign.

$\qquad + \text{ to } - \quad - \text{ to } + \quad + \text{ to } -$

DESCARTES' RULE OF SIGNS

Consider a polynomial $p(x)$ with real coefficients, terms in descending powers of x, and with a nonzero constant term. Then:

1. The number of positive zeros of $p(x)$ is either equal to
   (a) the number of variations of sign in $p(x)$, or
   (b) is less than that number by an even integer.
2. The number of negative zeros of $p(x)$ is either equal to
   (a) the number of variations of sign in $p(-x)$, or
   (b) is less than that number by an even integer.

Note: When there is a missing term, that is, a term with a zero coefficient, it is not considered in applying Descartes' rule.

Using Descartes' Rule of Signs in conjunction with the *n*-roots theorem enables us to analyze the composition of the roots of a polynomial equation as to the number of positive and negative real roots and the number of imaginary roots.

EXAMPLE 5 Describe all of the possibilities for the number of roots of $p(x) = x^5 - x^4 - 3x^3 + 3x^2 - 4x + 4 = 0$.

Solution First note that $p(x)$ is a fifth-degree polynomial and therefore will have five roots. (Why?) Then confirm that $p(x)$ has four variations in sign. Therefore, by Descartes' Rule of Signs, $p(x)$ has 4, 2, or 0 positive zeros. Now consider $p(-x)$:

$$p(-x) = (-x)^5 - (-x)^4 - 3(-x)^3 + 3(-x)^2 - 4(-x) + 4$$
$$= -x^5 - x^4 + 3x^3 + 3x^2 + 4x + 4$$

Since $p(-x)$ has only one variation in sign, Descartes' rule implies that $p(x)$ has exactly one negative zero.

This table summarizes the possibilities that we have found:

Observe that in the case where the number of positive roots is 2 and the number of negative roots is 1, the n-roots theorem is used to conclude that there must be 2 imaginary roots.

Number of Positive Zeros	Number of Negative Zeros	Number of Imaginary Zeros	Total Number of Zeros
4	1	0	5
2	1	2	5
0	1	4	5

With the help of the rational root theorem it can be shown that the zeros of $p(x)$ are $-2, 1, 2$, and $\pm i$. The factored form of $p(x)$ is

$$p(x) = (x + 2)(x - 1)(x - 2)(x^2 + 1)$$ ∎

EXAMPLE 6   Explain why there are no real zeros for $f(x) = 2x^4 + 3x^2 + 8$.

Solution   Observe that $f(-x) = 2(-x)^4 + 3(-x)^2 + 8 = 2x^4 + 3x^3 + 8 = f(x)$. Neither $f(x)$ nor $f(-x)$ has any variations of sign so that the number of positive or negative zeros is 0. All four roots of $f(x) = 0$ are imaginary.   ∎

EXERCISES 3.7

Find all real roots.

1. $x^3 + x^2 - 21x - 45 = 0$
2. $x^3 + 2x^2 - 29x + 42 = 0$
3. $3x^3 + 2x^2 - 75x - 50 = 0$
4. $2x^3 - 15x^2 + 24x + 16 = 0$
5. $x^4 + 3x^3 + 3x^2 + x = 0$
6. $x^3 + 3x^2 + 3x + 1 = 0$
7. $x^4 + 6x^3 + 2x^2 - 18x - 15 = 0$
8. $x^4 + 6x^3 + 7x^2 - 12x - 18 = 0$
9. $x^4 + 2x^3 - 7x^2 - 18x - 18 = 0$
10. $x^4 - x^3 - 5x^2 - x - 6 = 0$
11. $-x^5 + 5x^4 - 3x^3 - 15x^2 + 18x = 0$
12. $x^4 - 5x^3 + 3x^2 + 15x - 18 = 0$
13. $2x^3 - 5x - 3 = 0$
14. $x^4 + 4x^3 - 7x^2 - 36x - 18 = 0$
15. $3x^4 - 11x^3 - 3x^2 - 6x + 8 = 0$
16. $6x^3 - 25x^2 + 21x + 10 = 0$

Factor.

17. $-x^3 - 3x^2 + 24x + 80$
18. $x^3 - 3x^2 - 10x + 24$
19. $6x^4 + 9x^3 + 9x - 6$
20. $x^3 - 28x - 48$

Use the rational root theorem to show that $f(x) = 0$ has no rational roots.

21. $f(x) = 2x^3 - 5x^2 - x + 8$
22. $f(x) = x^3 + 3x - 3$

Find all real and imaginary roots.

23. $p(x) = x^4 - 3x^3 + 5x^2 - x - 10 = 0$
24. $p(x) = x^5 - 3x^4 - 3x^3 + 9x^2 - 10x + 30 = 0$
25. $p(x) = 3x^3 - 5x^2 + 2x - 8 = 0$
26. $p(x) = 2x^4 - 5x^3 + x^2 + 4x - 4 = 0$
27. $p(x) = x^5 - 9x^4 + 31x^3 - 49x^2 + 36x - 10 = 0$
28. $p(x) = 2x^6 + 5x^5 + x^4 + 10x^3 - 4x^2 + 5x - 3 = 0$

Use Descartes' Rule of Signs to describe all possibilities for the number of positive, negative, and imaginary zeros of $p(x)$.

29. $p(x) = 4x^3 - 3x^2 - 7x + 9$
30. $p(x) = -x^3 - 2x^2 - 11$
31. $p(x) = x^5 + 4x^4 - 3x^3 - x^2 + x - 1$
32. $p(x) = x^6 + 5x^5 - x^3 + 2x^2 - x - 8$

Solve the system.

33. $y = x^3 - 3x^2 + 3x - 1$
    $y = 7x - 13$

34. $y = -x^3$
    $y = -3x^2 + 4$

35. $y = 4x^3 - 7x^2 + 10$
    $y = x^3 + 43x - 5$

36. Solve the system and graph:

$$y = 2x - 5$$

$$y = \frac{-1}{x^2 - 2x + 1}$$

Use $z = -6 + 8i$ and $w = \frac{1}{2} + \frac{1}{2}i$ to verify the following.

37. $\overline{z + w} = \overline{z} + \overline{w}$ **38.** $\overline{z - w} = \overline{z} - \overline{w}$ **39.** $\overline{zw} = \overline{z} \cdot \overline{w}$ **40.** $\overline{\left(\dfrac{z}{w}\right)} = \dfrac{\overline{z}}{\overline{w}}$

Use $z = a + bi$ and $w = c + di$ to prove the following:

41. $\overline{z + w} = \overline{z} + \overline{w}$ **42.** $\overline{z - w} = \overline{z} - \overline{w}$ **43.** $\overline{\left(\dfrac{z}{w}\right)} = \dfrac{\overline{z}}{\overline{w}}$ **44.** $\overline{zw} = \overline{z} \cdot \overline{w}$

45. Let $p(x) = a_n x^n + a_{n-1} x^{n-1} + \cdots + a_1 x + a_0$, where the a_i are real numbers. Prove that if the complex (imaginary) number $a + bi$ is a root of the polynomial equation $p(x) = 0$, then so is $\overline{z}$; $p(\overline{z}) = 0$.

46. Use the result in Exercise 45 and the *n*-roots theorem to list the possible combinations of real and imaginary roots of each equation. For example, the combinations for $x^4 - 2x^3 + x + 7 = 0$ are four complex (imaginary), or two complex (imaginary) and two real, or four real.

    (a) $x^3 - 7x^2 + 4x - 9 = 0$ **(b)** $2x^5 - 6x^4 + x^2 - 3 = 0$

    (c) $5x^6 - 2x^5 + x^4 - x^3 - 4x + 1 = 0$

47. The *intermediate value theorem* for a polynomial function $f(x)$ states that for $a < b$, if $f(a) \neq f(b)$, then f takes on every value between $f(a)$ and $f(b)$ in the interval $[a, b]$. Use this theorem to show that $f(x) = x^5 - 3x^2 + 2x - 3$ has a zero between 1 and 2.

48. Use the theorem stated in Exercise 47 to approximate the zero between 1 and 2 to the nearest tenth. (*Hint:* Consider the signs of $f(1.1), f(1.2)$, and so on.

49. Use the rational root theorem to prove that the real number $\sqrt{5}$ is an irrational number. (*Hint:* Use an equation that has $\sqrt{5}$ as a root.)

?

Challenge For a polynomial $p(x)$, a real number b is an *upper bound* for the real zeros of $p(x)$, if b is greater than or equal to all the real zeros. If b is less than or equal to all the real zeros, then b is a *lower bound* for the real zeros of $p(x)$. The following theorem can be used to find such upper and lower bounds.

Assume that all coefficients of the polynomial function $p(x) = a_n x^n + a_{n-1} x^{n-1} + \cdots + a_0$ are real numbers and that the leading coefficient, a_n, is positive. If $p(x)$ is divided by $x - b$ by synthetic division, then:

    i. If $b > 0$ and all entries in the last row of the division process are nonnegative, than b is an upper bound for all real zeros of $p(x)$.

    ii. If $b < 0$ and the entries in the last row of the division process alternate in sign, then b is a lower bound for all real zeros of $p(x)$. (The number 0 in the last line can be considered positive or negative, but not both.)

1. Let $p(x) = 2x^3 + 5x^2 - 6x - 2$ and find the following:

    (a) the smallest integer that is an upper bound for all the real zeros of $p(x)$.

    (b) the largest integer that is a lower bound for all the real zeros of $p(x)$.

    (*Hint:* Try positive and negative integer values of b, such as $1, 2, \cdots$ and $-1, -2, \cdots$.)

2. Use a graphing calculator to graph p as a check on your work.

Graphing Calculator Exercises

1. In Exercise 47, you were asked to use the intermediate value theorem to prove that $f(x) = x^5 - 3x^2 + 2x - 3$ has a root between 1 and 2. Using TRACE and ZOOM, try to approximate this root to the nearest tenth, starting with a standard window. What happens? How could you speed up the process?

2. Graph the function $p(x)$ in Example 5 (a friendly window is suggested) and confirm the real roots stated there. How does one confirm the imaginary roots?

3. Graph the function $p(x)$ in Example 6 and confirm that it has no real roots. In particular, how do you know that the graph will always be above the x-axis?

4. If a function has a real root of even multiplicity, its graph will touch the x-axis there, like x^2 or $-x^2$; if the root has odd multiplicity, the graph will cross the x-axis there, like x^3 or $-x^3$. Use this idea to use your calculator's graph of $p(x) = x^5 + x^4 - x - 1$ on a friendly window to guess at the multiplicity of its roots. Then follow up your guesses by synthetic division to factor the polynomial.

 Critical Thinking

1. To solve an equation with fractions, we multiply through by the LCD. What happens if, instead, we just multiply by a common denominator? Thus, in Example 3 on page 200, is it possible to solve the given equation by multiplying by the common denominator $(x^2 - 16)(x + 4)(x - 4)$? Explain your answer.

2. Suppose that y varies directly as x and that x varies inversely as z. How does y vary in relation to z?

3. Show that $x + y$ is a factor of $x^n - y^n$, where n is a positive even integer.

4. A sample of 100 deer is captured, tagged, and released. Later, a second sample is captured and the number previously tagged is counted. This count is used as the basis for estimating the total number of deer in the park. Under what conditions will this be a good estimate? When might it not be a valid basis for estimating the total number in the park?

5. In *base 5*, the numeral 23423_{five} means $2(5^4) + 3(5^3) + 4(5^2) + 2(5) + 3$ and can be thought of as the polynomial $2x^4 + 3x^3 + 4x^2 + 2x + 3$ where $x = 5$. Use synthetic division to find the value of this number in base ten.

6. Assume that $a_n x^n + a_{n-1} x^{n-1} + \cdots + a_1 x + a_0 = 0$ is an nth-degree polynomial equation with integer coefficients, having the root $\frac{6}{8}$. What can you say about the factors of a_0 and a_n? Explain.

3.8 DECOMPOSING RATIONAL FUNCTIONS

In Chapter 1 we learned how to combine rational expressions. For example, combining the fractions in

$$(1) \quad \frac{6}{x - 4} + \frac{3}{x - 2} \qquad \text{produces} \qquad (2) \quad \frac{9x - 24}{(x - 4)(x - 2)}$$

It is now our goal to start with a rational expression such as (2) and *decompose* it into the form (1). When this is accomplished we say that $\dfrac{9x - 24}{(x - 4)(x - 2)}$ has been decomposed into (simpler) **partial fractions**.

$$\frac{9x - 24}{(x - 4)(x - 2)} = \frac{6}{x - 4} + \frac{3}{x - 2}$$

We will only consider examples that involve linear factors in the denominator. Examples involving nonfactorable quadratic factors are considered in the Exercises.

First observe that each factor in the denominator on the left serves as a denominator of a partial fraction on the right. Let us assume, for the moment, that the numerators 6 and 3 are not known. Then it is reasonable to begin by writing

$$\frac{9x - 24}{(x - 4)(x - 2)} = \frac{A}{x - 4} + \frac{B}{x - 2}$$

where A and B are the constants to be found. To find these values, first clear fractions by multiplying both sides by $(x - 4)(x - 2)$.

$$(x - 4)(x - 2) \cdot \frac{9x - 24}{(x - 4)(x - 2)} = (x - 4)(x - 2)\left[\frac{A}{x - 4} + \frac{B}{x - 2}\right]$$

$$9x - 24 = A(x - 2) + B(x - 4)$$

Since we want this equation to hold for all values of x, we may select specific values for x that will produce the constants A and B. Observe that when $x = 4$ the term $B(x - 4)$ will become zero.

$$9(4) - 24 = A(4 - 2) + B(4 - 4)$$

$$12 = 2A + 0$$

$$6 = A$$

Similarly, B can be found by letting $x = 2$.

$$9(2) - 24 = A(2 - 2) + B(2 - 4)$$

$$-6 = 0 - 2B$$

$$3 = B$$

EXAMPLE 1 Decompose $\dfrac{6x^2 + x - 37}{(x - 3)(x + 2)(x - 1)}$ into partial fractions.

Solution Since there are three linear factors in the denominator, we begin with the form

$$\frac{6x^2 + x - 37}{(x - 3)(x + 2)(x - 1)} = \frac{A}{x - 3} + \frac{B}{x + 2} + \frac{C}{x - 1}$$

Multiply by $(x - 3)(x + 2)(x - 1)$ to clear fractions.

$$6x^2 + x - 37 = A(x + 2)(x - 1) + B(x - 3)(x - 1) + C(x - 3)(x + 2)$$

Since the second and third terms on the right have the factor $x - 3$, the value $x = 3$ will make these two terms zero, as shown on the following page.

$$6(3)^2 + 3 - 37 = A(3 + 2)(3 - 1) + B(3 - 3)(3 - 1) + C(3 - 3)(3 + 2)$$
$$54 + 3 - 37 = A(5)(2) + 0 + 0$$
$$20 = 10A$$
$$2 = A$$

To find B, use $x = -2$.

$$6(-2)^2 - 2 - 37 = 0 + B(-5)(-3) + 0$$
$$-15 = 15B$$
$$-1 = B$$

To find C, let $x = 1$.

$$6(1)^2 + 1 - 37 = 0 + 0 + C(1 - 3)(1 + 2)$$
$$-30 = -6C$$
$$5 = C$$

Substituting the values for A, B, and C into the original form produces the desired decomposition.

Check this result by combining the fractions on the right side.

$$\frac{6x^2 + x - 37}{(x - 3)(x + 2)(x - 1)} = \frac{2}{x - 3} + \frac{-1}{x + 2} + \frac{5}{x - 1}$$

$$= \frac{2}{x - 3} - \frac{1}{x + 2} + \frac{5}{x - 1}$$ ∎

TEST YOUR UNDERSTANDING

Decompose into partial fractions.

1. $\dfrac{8x - 19}{(x - 2)(x - 3)}$ **2.** $\dfrac{1}{(x + 2)(x - 4)}$ **3.** $\dfrac{6x^2 - 22x + 18}{(x - 1)(x - 2)(x - 3)}$

Factor the denominator and decompose into partial fractions.

4. $\dfrac{4x + 6}{x^2 + 5x + 6}$ **5.** $\dfrac{23x - 1}{6x^2 + x - 1}$

(Answers: Page 246)

Let us look at a somewhat different example.

Note that the least common denominator is the highest power of the linear factor in either denominator.

$$\frac{7}{x + 3} - \frac{4}{(x + 3)^2} = \frac{7x + 17}{(x + 3)^2}$$

Now assume that the specific numerators are not known, and begin the decomposition process in this way:

$$\frac{7x + 17}{(x + 3)^2} = \frac{A}{x + 3} + \frac{B}{(x + 3)^2}$$

Clear fractions.

$$(1) \quad 7x + 17 = A(x + 3) + B$$

To find B, let $x = -3$.

$$7(-3) + 17 = A(0) + B$$
$$-4 = B$$

Substitute this value for B into Equation (1).

$$(2) \quad 7x + 17 = A(x + 3) - 4$$

Now find A by substituting some easy value for x, say $x = 0$, into (2).

$$7(0) + 17 = A(0 + 3) - 4$$
$$17 = 3A - 4$$
$$7 = A$$

Substituting these values for A and B into our original form produces the decomposition.

Note: If the original denominator had been $(x + 3)^3$, then we could have used the additional fraction $\dfrac{C}{(x + 3)^3}$ to start with.

$$\frac{7x + 17}{(x + 3)^2} = \frac{7}{x + 3} + \frac{-4}{(x + 3)^2} = \frac{7}{x + 3} - \frac{4}{(x + 3)^2}$$

EXAMPLE 2   Decompose $\dfrac{6 + 26x - x^2}{(2x - 1)(x + 2)^2}$ into partial fractions.

Solution Begin with this form:

$$\frac{6 + 26x - x^2}{(2x - 1)(x + 2)^2} = \frac{A}{2x - 1} + \frac{B}{x + 2} + \frac{C}{(x + 2)^2}$$

Clear fractions.

$$(1) \quad 6 + 26x - x^2 = A(x + 2)^2 + B(2x - 1)(x + 2) + C(2x - 1)$$

Find A by substituting $x = \frac{1}{2}$.

$$6 + 13 - \tfrac{1}{4} = A\left(\tfrac{5}{2}\right)^2 + 0 + 0$$
$$\tfrac{75}{4} = \tfrac{25}{4}A$$
$$3 = A$$

Find C by letting $x = -2$.

$$6 - 52 - 4 = 0 + 0 + C(-5)$$
$$-50 = -5C$$
$$10 = C$$

Substitute $A = 3$ and $C = 10$ into Equation (1).

(2) $6 + 26x - x^2 = 3(x + 2)^2 + B(2x - 1)(x + 2) + 10(2x - 1)$

To find B, substitute a simple value like $x = 1$ into (2).

$$6 + 26 - 1 = 3(9) + B(1)(3) + 10(1)$$
$$-6 = 3B$$
$$-2 = B$$

Then the decomposition is

$$\frac{6 + 26x - x^2}{(2x - 1)(x + 2)^2} = \frac{3}{2x - 1} - \frac{2}{x + 2} + \frac{10}{(x + 2)^2}$$

You can check this by combining the right side.   ■

Thus far, in every decomposition problem the degree of the polynomial in the numerator has been less than the degree in the denominator. Here is an example where this is not the case.

$$\frac{2x^3 + 3x^2 - x + 16}{x^2 + 2x - 3}$$

In such cases the *first* step is to divide.

$$\frac{2x^3 + 3x^2 - x + 16}{x^2 + 2x - 3} = \text{quotient} + \frac{\text{remainder}}{\text{divisor}}$$

$$= 2x - 1 + \frac{7x + 13}{x^2 + 2x - 3}$$

The problem will be completed by decomposing $\dfrac{7x + 13}{x^2 + 2x - 3}$. Verify that

$$\frac{7x + 13}{x^2 + 2x - 3} = \frac{7x + 13}{(x - 1)(x + 3)} = \frac{5}{x - 1} + \frac{2}{x + 3}$$

Therefore, the final decomposition is

$$\frac{2x^3 + 3x^2 - x + 16}{x^2 + 2x - 3} = 2x - 1 + \frac{5}{x - 1} + \frac{2}{x + 3}$$

> **CAUTION**
>
> When the degree of the numerator is greater than or equal to the degree in the denominator, you *must* divide first. If this step is ignored, the resulting decomposition will be wrong. For example, suppose that you started *incorrectly* in this way:
>
> $$\frac{2x^3 + 3x^2 - x + 16}{(x - 1)(x + 3)} = \frac{A}{x - 1} + \frac{B}{x + 3}$$
>
> This approach will produce the following *incorrect* answer:
>
> $$\frac{2x^3 + 3x^2 - x + 16}{(x - 1)(x + 3)} = \frac{5}{x - 1} + \frac{2}{x + 3}$$

EXERCISES 3.8

Decompose into partial fractions.

1. $\dfrac{2x}{(x + 1)(x - 1)}$
2. $\dfrac{x}{x^2 - 4}$
3. $\dfrac{x + 7}{x^2 - x - 6}$
4. $\dfrac{4x^2 + 16x + 4}{(x + 3)(x^2 - 1)}$

5. $\dfrac{5x^2 + 9x - 56}{(x - 4)(x - 2)(x + 1)}$
6. $\dfrac{x}{(x - 3)^2}$
7. $\dfrac{3x - 3}{(x - 2)^2}$
8. $\dfrac{2 - 3x}{x^2 + x}$

9. $\dfrac{3x - 30}{15x^2 - 14x - 8}$
10. $\dfrac{2x + 1}{(2x + 3)^2}$
11. $\dfrac{x^2 - x - 4}{x(x + 2)^2}$
12. $\dfrac{x^2 + 5x + 8}{(x - 3)(x + 1)^2}$

First divide and then complete the decomposition into partial fractions.

13. $\dfrac{x^3 - x + 2}{x^2 - 1}$
14. $\dfrac{4x^2 - 14x + 2}{4x^2 - 1}$
15. $\dfrac{12x^4 - 12x^3 + 7x^2 - 2x - 3}{4x^2 - 4x + 1}$

Decompose into partial fractions.

16. $\dfrac{10x^2 - 16}{x^4 - 5x^2 + 4}$
17. $\dfrac{10x^3 - 15x^2 - 35x}{x^2 - x - 6}$
18. $\dfrac{25x^3 + 10x^2 + 31x + 5}{25x^2 + 10x + 1}$

19. $\dfrac{5x^2 - 24x - 173}{x^3 + 4x^2 - 31x - 70}$

The method of this section can be adjusted to fractions involving nonfactorable quadratics in the denominator. Study this example and apply it in Exercises 20–23 to decompose the fractions:

$$\frac{2x}{(x - 1)(x^2 + 1)} = \frac{A}{x - 1} + \frac{Bx + C}{x^2 + 1} \qquad \longleftarrow \text{ Allow for a linear numerator over the quadratic factor}$$

$$2x = A(x^2 + 1) + (x - 1)(Bx + C)$$

$$2 = 2A \qquad \longleftarrow \text{ Use } x = 1.$$

$$1 = A$$

$$0 = 1 + (-1)C \qquad \longleftarrow \text{Let } A = 1 \text{ and use } x = 0.$$

$$1 = C$$

$$-2 = 2 + (-2)(-B + 1) \qquad \longleftarrow \text{Let } A = 1, C = 1, \text{ and use } x = -1.$$

$$-1 = B$$

$$\frac{2x}{(x-1)(x^2+1)} = \frac{1}{x-1} + \frac{-x+1}{x^2+1}$$

20. $\dfrac{x-3}{(x+2)(x^2+1)}$ **21.** $\dfrac{-4x}{(x^2+3)(x-1)}$ **22.** $\dfrac{x^2-5x+19}{(x-3)(x^2+4)}$ **23.** $\dfrac{4x^2+5}{(x-1)(x^2+x+1)}$

CHAPTER REVIEW EXERCISES

Section 3.1 Hints for Graphing (Page 174)

Graph each of the following. State the domain and range for each.

1. $y = 2x^3$ **2.** $y = -2x^3$

3. $y = (x-1)^3$ **4.** $y = -(x+1)^3$

5. $y = -2x^4$ **6.** $y = -\dfrac{1}{2}x^4$

7. Under what conditions is the graph of a function $y = f(x)$ said to be symmetric through the origin?

8. Under what conditions is the graph of $y = f(x)$ symmetric about the y-axis?

9. Explain how the graph of $y = f(x + h)$ can be obtained from the graph of $y = f(x)$ when $h > 0$, and also when $h < 0$.

10. Explain how the graph of $y = f(x) + k$ can be obtained from the graph of $y = f(x)$ when $k > 0$, and also when $k < 0$.

Section 3.2 Graphing Some Special Rational Functions (Page 182)

11. What is meant by a *rational expression*?

12. Explain, in words, the meaning of the symbols: $f(x) \longrightarrow 0$ as $x \longrightarrow \infty$.

Graph each of the following. Find all asymptotes, if any. State the domain and range, and describe the concavity.

13. $y = \dfrac{1}{x+1}$ **14.** $y = \dfrac{1}{x} + 1$

15. $y = -\dfrac{1}{x^2}$ **16.** $y = \dfrac{1}{x-1}$

17. $xy = -1$ **18.** $y = \dfrac{x^2-x-2}{x+1}$

19. What are the conditions needed for the graph of a rational function to have a vertical asymptote?

20. What are the conditions needed for the graph of a rational function to have a horizontal asymptote?

Section 3.3   Polynomial and Rational Functions   (Page 190)

Construct a table of signs to determine the signs of f.

21. $f(x) = x^2 + 3x - 10$ **22.** $f(x) = x^3 - x^2 - 4x + 4$

Find the horizontal and vertical asymptotes for each function.

23. $g(x) = \dfrac{2x^2}{3x^2 - 1}$ **24.** $h(x) = \dfrac{x^2}{x - 2}$ **25.** $r(x) = \dfrac{x}{2x - x^2}$ **26.** $t(x) = \dfrac{x - 1}{x^2 - 1}$

27. Graph and show intercepts and asymptotes, if any: $y = \dfrac{x + 1}{x^2 + x - 2}$

28. Graph: $y = \dfrac{4}{x^2 + 2}$. Discuss the symmetry and find the equations of the asymptotes, if any.

Find the equation of the oblique asymptotes.

29. $f(x) = \dfrac{x^2 - 2x + 3}{x}$ **30.** $f(x) = \dfrac{x^2 - 5x + 3}{x + 2}$

Section 3.4   Equations and Inequalities with Fractions   (Page 199)

Solve for x.

31. $\dfrac{x}{x - 1} - \dfrac{2}{x(x - 1)} = \dfrac{4}{x} - \dfrac{1}{x - 1}$ **32.** $\dfrac{8}{x + 1} + \dfrac{6}{x - 1} = 5$

33. State the proportion property.

34. Use the proportion property to solve for x: $\dfrac{x - 2}{x} = \dfrac{3}{4}$

Find the x-intercepts and state the domain.

35. $f(x) = \dfrac{8}{x} - 2x - 6$ **36.** $f(x) = \dfrac{3x}{x + 5} - \dfrac{3}{2}$

37. Solve for c: $r = \dfrac{ab}{a + b + c}$

38. Solve for x and graph on a number line: $\dfrac{2x + 1}{x - 2} \geq 0$

39. Solve for x: $\dfrac{3x - 2}{2 - x} \leq 0$

40. Solve the system: $y = -\dfrac{6}{x}$

$y - 2x = 7$

41. Suppose that a 3-foot stick casts a $\frac{1}{2}$-foot shadow. How high is a tree that casts a 15-foot shadow at the same time?

42. One pipe can empty a tank in 4 hours. If it takes a second pipe 5 hours to complete the same job, how long will it take to empty the tank if both pipes are used at the same time?

Section 3.5   Variation   (Page 210)

43. Explain the meaning of each statement:

   (a) y varies directly as x.   **(b)** y varies inversely as x.

Write the equation for the given variation and identify the constant of variation.

44. The surface area of a cube varies directly as the square of the length of an edge, e.

45. The area of a triangle varies directly as the altitude, h, and the base, b.

46. Suppose y varies directly as x and $y = 9$ when $x = 15$. Find y for $x = 40$.

47. Suppose y varies inversely as x and $y = 4$ when $x = 3$. Find y for $x = 4$.

48. Suppose z varies directly as x and inversely as the square of y. If $z = \frac{1}{4}$ when $x = 6$ and $y = 8$, find z when $x = 9$ and $y = 2$.

49. According to Boyle's law, the pressure P of a compressed gas is inversely proportional to the volume V. Suppose there is a pressure of 40 pounds per square inch when the volume of the gas is 300 cubic inches. Find the pressure when the gas is compressed to 200 cubic inches.

50. According to Newton's Law of Gravitation, the force of attraction between two objects varies jointly as their masses and inversely as the square of the distance between them. How does the force change if each of their masses is tripled but the distance between them is cut to one-third?

Section 3.6   Synthetic Division, the Remainder and Factor Theorems   (Page 217)

Use synthetic division to find the quotient and remainder. Check each result.

51. $(x^3 + x^2 - 7x + 5) \div (x - 1)$ **52.** $(x^3 + 4x^2 + x - 2) \div (x + 1)$

53. $(2x^3 - x^2 + 5x + 7) \div (x + 2)$ **54.** $(3x^3 + x^2 + x - 2) \div (x - 2)$

55. $(x^4 - 3x^3 + 2x - 8) \div (x - 3)$ **56.** $(2x^4 + 5x^2 - 7x + 8) \div (x + 3)$

57. State the remainder theorem.

58. State the factor theorem.

59. Find the remainder if $p(x) = 2x^3 - 5x^2 + 9x - 3$ is divided by $x - 1$.

60. Show that $x - 3$ is a factor of $x^3 - 3x^2 - x + 3$.

61. Use the factor theorem to show that $x + 5$ is a factor of $p(x) = x^3 + 2x^2 - 13x + 10$. Then factor $p(x)$ completely.

Section 3.7   Solving Polynomial Equations   (Page 225)

62. State the rational root theorem.

63. List all the possible rational roots of $f(x) = 6x^3 + 23x^2 - 6x - 8$.

64. Factor $f(x)$, as given in Exercise 63, completely.

65. Find all real roots of $p(x) = x^4 + 2x^3 - x^2 - 4x - 2$.

66. Find all real and imaginary roots: $x^4 + 3x^3 + x^2 + 4 = 0$

67. State the fundamental theorem of algebra.

68. For $z = 2 - 3i$ and $w = -3 + 4i$, verify that $\overline{z + w} = \overline{z} + \overline{w}$.

69. Repeat Exercise 68 to show that $\overline{zw} = \overline{z} \cdot \overline{w}$.

70. Use Descartes' Rule of Signs to determine all possibilities for the number of roots of $p(x) = x^5 + x^4 - 7x^3 - 7x^2 - 18x - 18$.

Section 3.8   Decomposing Rational Functions (Page 234)

Decompose into partial fractions.

71. $\dfrac{5x + 7}{x^2 + 2x - 3}$

72. $\dfrac{3x + 2}{x^2 + 5x + 4}$

73. $\dfrac{3x^2 + 15x + 8}{(x + 1)(x + 2)(x - 3)}$

74. $\dfrac{2x + 3}{(x - 1)^2}$

75. $\dfrac{2x^3 + x^2 - 3x + 5}{x^2 - 2x - 3}$

CHAPTER 3: STANDARD ANSWER TEST

Use these questions to test your knowledge of the basic skills and concepts of Chapter 3. Then check your answers with those given at the back of the book.

Graph each function and write the equations of the asymptotes if there are any.

1. $f(x) = (x + 2)^3 - \frac{3}{2}$

2. $y = f(x) = x^3$

3. $y = f(x) = -\dfrac{1}{x - 2}$

4. $y = f(x) = \dfrac{x - 1}{x^2 - x - 2}$

5. Graph: $y = x^3 - x^2 - 4x + 4$

6. Construct a table of signs for $f(x) = \dfrac{x^2 - 2x}{x + 3}$.

7. Graph: $y = f(x) = \dfrac{2}{x^2 + 2}$

8. Graph on a number line: $\dfrac{x + 2}{x - 1} \le 0$

9. Solve for a: $S = \dfrac{n}{2}[2a + (n - 1)d]$

Solve for x.

10. $\dfrac{6}{x} = 2 + \dfrac{3}{x + 1}$

11. $\dfrac{x}{2} - \dfrac{3x + 1}{3} > 2$

12. $\dfrac{6}{x^2 - 9} - \dfrac{2}{x - 3} = \dfrac{1}{x + 3}$

13. A piece of wire that is 10 feet long is cut into two pieces. The ratio of the smaller piece to the larger piece is $\frac{3}{4}$. Find the length of the larger piece.

14. Working alone, Dave can wash his car in 45 minutes. If Ellen helps him, they can do the job together in 30 minutes. How long would it take Ellen to wash the car by herself?

15. z varies directly as x and inversely as y. If $z = \frac{2}{3}$ when $x = 2$ and $y = 15$, find z when $x = 4$ and $y = 10$.

16. If the volume of a sphere varies directly as the cube of its radius and $V = 36\pi$ cubic inches when $r = 3$ inches, find V when $r = 6$ inches.

17. Use synthetic division to divide $2x^5 + 5x^4 - x^2 - 21x + 7$ by $x + 3$.

18. Let $p(x) = 27x^4 - 36x^3 + 18x^2 - 4x + 1$. Use the remainder theorem to evaluate $p\left(\frac{1}{3}\right)$.

19. Use the result of Exercise 18 and the factor theorem to determine whether or not $x - \frac{1}{3}$ is a factor of $p(x)$.

20. Show that $x - 2$ is a factor of $p(x) = x^4 - 4x^3 + 7x^2 - 12x + 12$, and factor $p(x)$ completely.

21. Make use of the rational root theorem to factor $f(x) = x^4 + 5x^3 + 4x^2 - 3x + 9$.

22. Find the roots of $p(x) = 0$ for $p(x) = x^4 + 3x^3 - 3x^2 - 11x - 6$.

23. Find all real and imaginary roots for $p(x) = x^4 - 2x^3 - 2x^2 - 2x - 3$.

Decompose into partial fractions.

24. $\dfrac{x - 15}{x^2 - 25}$ **25.** $\dfrac{6x^2 - 2x + 2}{x^3 - 2x^2 - 5x + 6}$

CHAPTER 3: MULTIPLE CHOICE TEST

1. Which of the following are true?

   I. The graph of $f(x) = x^2$ is symmetric about the y-axis.

   II. The graph of $f(x) = x^3$ is symmetric through the origin.

   III. The graph of $f(x) = \dfrac{1}{x}$ is symmetric about the x-axis.

   (a) Only I   **(b)** Only II   **(c)** Only III   **(d)** Only I and II   **(e)** None of the preceding

2. The equation of the horizontal asymptote for $f(x) = \dfrac{1}{x - 3}$ is

   (a) $x = 0$ **(b)** $y = 0$ **(c)** $x = 3$ **(d)** $x = -3$ **(e)** None of the preceding

3. For $h > 0$, the graph of $y = f(x - h)$ can be obtained by shifting the graph of $y = f(x)$

   (a) h units to the right   **(b)** h units to the left   **(c)** h units upward   **(d)** h units downward

   (e) None of the preceding

4. Which of the following are true for the graph of $y = -\dfrac{1}{x^2}$?

   I. The horizontal asymptote is $x = 0$.

   II. The vertical asymptote is $y = 0$.

   III. The graph is symmetric about the y-axis.

   (a) Only I   **(b)** Only II   **(c)** Only III   **(d)** I, II, and III   **(e)** None of the preceding

5. The table below can be used to determine the position of the curve $f(x) = (x + 3)(x + 1)(x - 2)$ relative to the x-axis. Reading left to right, these positions for the given intervals, are

   (a) above, below, above, below

   (b) below, above, below, above

   (c) below, below, above, above

   (d) above, below, below, above

   (e) None of the preceding

Interval	$(-\infty, -3)$	$(-3, -1)$	$(-1, 2)$	$(2, \infty)$
Sign of $f(x)$				

6. What is the equation of the vertical asymptote for $f(x) = \dfrac{3x^2}{x^3 + x}$?

   (a) $x = 0$ **(b)** $x = -1$ **(c)** $x = 3$ **(d)** $y = 3$ **(e)** None of the preceding

7. Working alone, Amy can complete a job in 5 hours. Julie can complete the same job in 4 hours. Which of the following equations can be used to find out how long it will take to complete the job together? (Use x for the time it takes them to complete the job working together.)

   (a) $\dfrac{1}{4} + \dfrac{1}{x} = \dfrac{1}{5}$ **(b)** $\dfrac{1}{4} - \dfrac{1}{5} = \dfrac{1}{x}$ **(c)** $\dfrac{x}{4} - \dfrac{x}{5} = 1$ **(d)** $\dfrac{1}{4} + \dfrac{1}{5} = \dfrac{1}{x}$ **(e)** None of the preceding

8. If $c = \dfrac{a + 2b}{ab}$, then $b =$

   (a) $\dfrac{a}{ac - 2}$ (b) $\dfrac{1}{c} - \dfrac{a}{2}$ (c) $a(c - a)$ (d) $c - 1$ (e) None of the preceding

9. The equation $\dfrac{9x + 14}{2x^2 - x - 6} - \dfrac{1}{x - 2} = \dfrac{2}{2x + 3}$ has

   (a) No solution   (b) Just one solution   (c) Exactly two solutions

   (d) More than two solutions   (e) None of the preceding

10. Neglecting air resistance, the distance an object falls from rest is directly proportional to the square of the time it has fallen. If an object falls 64 feet in the first 2 seconds, how far will it fall in the first 6 seconds?

   (a) 576 ft.   (b) 512 ft.   (c) 384 ft.   (d) 192 ft.   (e) None of the preceding

11. z varies directly as x and inversely as y^2. If $z = 60$ when $x = 3$ and $y = \frac{1}{2}$, then the value of z when $x = 6$ and $y = 4$ is:

   (a) 1920   (b) 30   (c) $\frac{45}{2}$ (d) $\frac{15}{8}$ (e) None of the preceding

12. When the synthetic division at the right is completed, the remainder is

$$-2\,\big|\,1 + 3 - 5 + 7$$

   (a) 0   (b) $+7$ (c) $+21$ (d) -7 (e) None of the preceding

13. Which of the following is a set of possible rational roots of $f(x) = 2x^3 - 8x^2 + 7x - 10$?

   (a) $\pm1, \pm2, \pm5, \pm10$ (b) $\pm1, \pm\frac{1}{2}, \pm2, \pm\frac{5}{2}, \pm5, \pm10$

   (c) $\pm2, \pm10$ (d) $\pm1, \pm2, \pm10$ (e) None of the preceding

14. Consider the decomposition of $\dfrac{6x^2 + x - 37}{(x - 3)(x + 2)(x - 1)}$ into the partial fractions

   $\dfrac{A}{x - 3} + \dfrac{B}{x + 2} + \dfrac{C}{x - 1}$. Then $A =$

   (a) -2 (b) 1   (c) 2   (d) 3   (e) None of the preceding

15. The equation $x^4 - 6x^3 + 14x^2 - 16x + 8 = 0$ has 2 as a double root. The remaining roots are

   (a) $1 \pm i$ (b) $1 \pm \sqrt{3}$ (c) $5 \pm 5i$ (d) ±8 (e) None of the preceding

16. For a given function, $f(x) \longrightarrow \infty$ as $x \longrightarrow c^-$. From this we can conclude

   (a) $x = c$ is a vertical asymptote.   (b) $x = -c$ is a vertical asymptote.

   (c) $y = c$ is a horizontal asymptote.   (d) $y = -c$ is a horizontal asymptote.   (e) None of the preceding

17. For a given function, $f(x) \longrightarrow k$ as $x \longrightarrow \infty$. From this we can conclude

   (a) $x = k$ is a vertical asymptote.   (b) $x = -k$ is a vertical asymptote.

   (c) $y = k$ is a horizontal asymptote.   (d) $y = -k$ is a horizontal asymptote.   (e) None of the preceding

18. What is the oblique asymptote for $f(x) = \dfrac{x^2 + 3x - 1}{x - 2}$?

   (a) $x = 2$ (b) $y = x - 2$ (c) $y = x^2 + 3x - 1$ (d) $y = x + 5$ (e) None of the preceding

19. According to Descartes' Rule of Signs, find the number of possible positive zeros of $p(x) = x^4 - 3x^3 - x^2 + 2x + 1$.

   (a) 2   (b) 2 or 0   (c) 4, 2, or 0   (d) 4   (e) None of the preceding

20. The graph of $f(x) = x^3 + 2x + 1$ has how many x-intercepts?

   (a) 0   (b) 1   (c) 2   (d) 3   (e) None of the preceding

ANSWERS TO THE TEST YOUR UNDERSTANDING EXERCISES

Page 193

1. Vertical asymptotes: $x = 2$, $x = 3$; horizontal asymptote: $y = 0$
2. Vertical asymptote: $x = -3$; horizontal asymptote: $y = 1$
3. Vertical asymptote: $x = 2$; no horizontal asymptote
4. Vertical asymptote: $x = 4$; horizontal asymptote: $y = 0$
5. Vertical asymptotes: $x = -3$, $x = 4$; horizontal asymptote: $y = 0$
6. Vertical asymptotes: $x = 1$; no horizontal asymptote

Page 201

1. 12   2. $\frac{1}{2}$ 3. 35   4. 5   5. -9 6. No solution   7. $\frac{13}{24}$

8. $-4, 3$ 9. $-3, -1$ 10. $-7, -\frac{2}{3}$

Page 213

1. 25   2. 6   3. 5   4. 16

Page 219

1. $x^2 + x - 3$; $r = 0$ 2. $2x^2 + 3x - 1$; $r = 0$ 3. $2x^2 + 3x + 9$; $r = 34$
4. $3x^2 - 3x - 4$; $r = 16$

Page 220

1.
$$
\begin{array}{r}
1 \underline{\big|\, 1 - 7 + 4 + 10 - 1 - 5} \\
+ 1 - 6 - 2 + 8 + 7 \\
\hline
1 - 6 - 2 + 8 + 7 \boxed{+ 2} = r
\end{array}
$$
$p(1) = 1 - 7 + 4 + 10 - 1 - 5 = 2$

2.
$$
\begin{array}{r}
-10 \underline{\big|\, 1 + 11 + 11 + 11 + 10} \\
- 10 - 10 - 10 - 10 \\
\hline
1 + 1 + 1 + 1 \boxed{\;\; 0} = r
\end{array}
$$
$p(-10) = 10{,}000 - 11{,}000 - 1100 - 110 + 10 = 0$

3.
$$
\begin{array}{r}
10 \underline{\big|\, 1 + 11 + 11 + 11 + 10} \\
+ 10 + 210 + 2210 + 22210 \\
\hline
1 + 21 + 221 + 2221 \boxed{+ 22220} = r
\end{array}
$$
$p(10) = 10{,}000 + 11{,}000 + 1100 + 110 + 10$
$\quad\quad = 22{,}220$

4.
$$
\begin{array}{r}
6 \underline{\big|\, 6 - 40 + 0 + 25} \\
+ 36 - 24 - 144 \\
\hline
6 - 4 - 24 \boxed{- 119} = r
\end{array}
$$
$p(6) = 6(6)^3 - 40(6)^2 + 25 = -119$

Page 227

1. (a) $\pm 1, \pm 2, \pm 3, \pm 4, \pm 6, \pm 8, \pm 12, \pm 24$ (b) $(x + 3)(x - 2)(x - 4)$ (c) $-3, 2, 4$
2. (a) $\pm 1, \pm 2, \pm 4, \pm 8, \pm 16$ (b) $(x + 4)^2(x - 1)^2$ (c) $-4, 1$
3. (a) $\pm 1, \pm\frac{1}{2}, \pm\frac{1}{4}, \pm 2, \pm 3, \pm\frac{3}{2}, \pm\frac{3}{4}, \pm 6$ (b) $(x + 6)(2x - 1)^2$ (c) $-6, \frac{1}{2}$
4. (a) $\pm 1, \pm\frac{1}{3}, \pm 2, \pm\frac{2}{3}, \pm 4, \pm\frac{4}{3}$ (b) $(3x - 1)(x - 4)(x^2 + 1)$ (c) $\frac{1}{3}, 4$

Page 236

1. $\dfrac{3}{x - 2} + \dfrac{5}{x - 3}$ 2. $-\dfrac{\frac{1}{6}}{x + 2} + \dfrac{\frac{1}{6}}{x - 4}$ 3. $\dfrac{1}{x - 1} + \dfrac{2}{x - 2} + \dfrac{3}{x - 3}$

4. $\dfrac{6}{x + 3} - \dfrac{2}{x + 2}$ 5. $\dfrac{4}{3x - 1} + \dfrac{5}{2x + 1}$

Circles, Additional Curves, and the Algebra of Functions

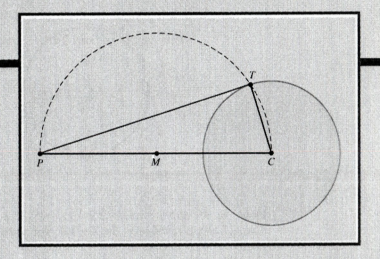

Circles have been important in the study of mathematics dating back to the ancient Greeks, who discovered many of their properties, all from a pure geometric point of view. The figure above shows how to construct a tangent to a circle from a point outside of the circle. This construction is based on the fact that an angle inscribed in a semicircle is a right angle. (See Exercise 48, page 257.)

Draw the line connecting point P with the center of the circle C, and bisect PC. Using the midpoint M of PC as center and MC as radius, draw the semicircle intersecting the given circle at point T. Then, since $\angle PTC$ is inscribed in a semicircle, it is a right angle. Now draw line PT, which is the desired tangent since PT is perpendicular to the radius at T.

Not until the seventeenth century, with the development of the Cartesian plane (a rectangular coordinate system) by René Descartes, did the algebraic formulation of curves take hold. If, for example, the preceding configuration were placed into a rectangular coordinate system, the equation of the circle becomes available and the coordinates of the point of tangency could be found. (See Exercise 49, page 257).

Circles and their equations are studied in the first section of this chapter, and in Section 4.2 a circle is divided into two semicircles that introduce radical functions. Such functions are the focus of study in Sections 4.2 and 4.3, which are then used for demonstrative purposes in the remainder of the chapter.

4.1 CIRCLES

You have learned to recognize the graphs of many equations, some of which are shown on the back inside cover of this text. One curve that undoubtedly is familiar to most students is that of a *circle*. As you can see, the vertical line test (page 139) shows that a circle cannot be the graph of a function.

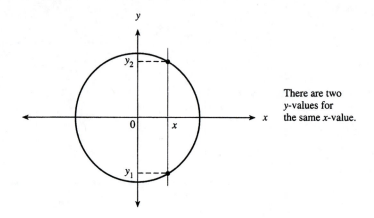

There are two y-values for the same x-value.

However, the *upper semicircle* is the graph of a function, as is the *lower semicircle*. Through any allowable x, the vertical line shown intersects each semicircle exactly once.

The circle also belongs to the group of curves called the conic sections. (See the chapter on conic sections.)

The curves previously studied were essentially defined by their equations. For example, the equation $y = x^3$ *defines* a function of x that is called the cubic function. For a circle, we will *not* begin with an equation. Rather, a more fundamental approach is taken by *starting* with a geometric definition that leads to the equation. In order to do this, we first need to know how to find the distance AB between two points $A(x_1, y_1)$ and $B(x_2, y_2)$ in a rectangular system. This is given by the **distance formula:**

DISTANCE FORMULA

The distance AB between points $A = (x_1, y_1)$ and $B = (x_2, y_2)$ is given by

$$AB = \sqrt{(x_1 - x_2)^2 + (y_1 - y_2)^2}$$

The proof that follows is based on this typical figure:

Note that the diagram was set up so that $x_2 - x_1 > 0$ and $y_2 - y_1 > 0$. Other situations may have negative values, but it doesn't make any difference because $(x_2 - x_1)^2 = (x_1 - x_2)^2$ and $(y_2 - y_1)^2 = (y_1 - y_2)^2$.

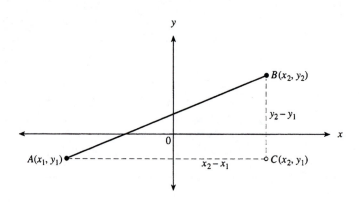

Since AB is the hypotenuse of the right triangle ABC, the Pythagorean theorem gives

$$AB^2 = AC^2 + CB^2$$

But $AC = x_2 - x_1$ and $CB = y_2 - y_1$. Thus

$$AB^2 = (x_2 - x_1)^2 + (y_2 - y_1)^2$$

If A and B are on the same horizontal line, then $y_1 = y_2$ and $AB = \sqrt{(x_1 - x_2)^2} = |x_1 - x_2|$.

Taking the positive square root gives the stated result.

EXAMPLE 1 Find the length of the line segment determined by points $A(-2, 2)$ and $B(6, -4)$.

Solution

$$\begin{aligned} AB &= \sqrt{(-2 - 6)^2 + [2 - (-4)]^2} \\ &= \sqrt{(-8)^2 + 6^2} \\ &= \sqrt{64 + 36} \\ &= \sqrt{100} \\ &= 10 \end{aligned}$$ ∎

Now we are ready to study the circle and its properties and will be able to use the distance formula to derive the equation of a circle.

DEFINITION OF A CIRCLE

A **circle** is the set of all points in the plane, each of which is at a fixed distance r from a given point called the **center** of the circle; r is the **radius** of the circle.

The figure below is a circle with center at the origin and radius r. A point will be on this circle *if and only if* its distance from the origin is equal to r. That is, $P(x, y)$ is on this circle if and only if $OP = r$. Since the origin has coordinates $(0, 0)$, the distance formula gives

$$\begin{aligned} r &= \sqrt{(x - 0)^2 + (y - 0)^2} \\ &= \sqrt{x^2 + y^2} \end{aligned}$$

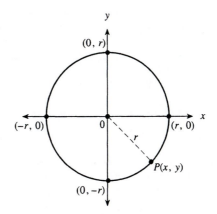

Squaring produces this result:

$$r^2 = x^2 + y^2$$

The words "if and only if" here mean that if P is on the circle, then OP = r, and if OP = r, then P is on the circle.

We conclude that $P(x, y)$ is on the circle with center at $(0, 0)$ and radius r if and only if the coordinates of P satisfy the preceding equation.

STANDARD FORM FOR THE EQUATION OF A CIRCLE WITH CENTER AT THE ORIGIN AND RADIUS r

$$x^2 + y^2 = r^2$$

Now consider any circle of radius r, not necessarily one with the origin as center. Let the center C have coordinates (h, k). Then, using the distance formula, a point $P(x, y)$ is on this circle if and only if

$$CP = r = \sqrt{(x - h)^2 + (y - k)^2}$$

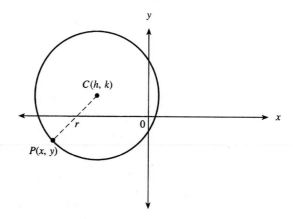

Observe that this form includes the earlier case where $(h, k) = (0, 0)$.

By squaring CP, we obtain the following:

STANDARD FORM FOR THE EQUATION OF A CIRCLE WITH CENTER AT (h, k) AND RADIUS r

$$(x - h)^2 + (y - k)^2 = r^2$$

Another way to obtain the graph of a circle with center $(h, k) \neq (0, 0)$, and radius r, is to translate the circle $x^2 + y^2 = r^2$ horizontally h units and vertically k units. Thus, in Example 2, translate $x^2 + y^2 = 4$, 2 units to the right and 3 units down.

EXAMPLE 2 Find the center and the radius of the circle with this equation: $(x - 2)^2 + (y + 3)^2 = 4$.

Solution Using $y + 3 = y - (-3)$, rewrite the equation in this form:

$$(x - 2)^2 + [(y - (-3)]^2 = 2^2$$

Thus the radius $r = 2$ and the center is at $(2, -3)$. ∎

EXAMPLE 3 Write the equation of the circle with center $(-3, 5)$ and radius $\sqrt{2}$.

Solution Use $h = -3$, $k = 5$, and $r = \sqrt{2}$ in the standard form to obtain:

$$[x - (-3)]^2 + (y - 5)^2 = (\sqrt{2})^2$$

or

$$(x + 3)^2 + (y - 5)^2 = 2 \qquad \blacksquare$$

TEST YOUR UNDERSTANDING

Find the length of the line segment determined by the two points.

1. $(4, 0)$; $(-8, -5)$ **2.** $(9, -1)$; $(2, 3)$ **3.** $(-7, -5)$; $(3, -13)$

Find the center and radius of each circle.

4. $x^2 + y^2 = 100$ **5.** $x^2 + y^2 = 10$

6. $(x - 1)^2 + (y + 1)^2 = 25$ **7.** $\left(x + \frac{1}{2}\right)^2 + y^2 = 256$

8. $(x + 4)^2 + (y + 4)^2 = 50$

Write the equation of the circle with the given center and radius in standard form.

(Answers: Page 303)   **9.** Center at $(0, 4)$; $r = 5$ **10.** Center at $(1, -2)$; $r = \sqrt{3}$

The equation in Example 2 can be written in another form.

$$(x - 2)^2 + (y + 3)^2 = 4$$
$$x^2 - 4x + 4 + y^2 + 6y + 9 = 4$$
$$x^2 - 4x + y^2 + 6y = -9$$

Note: The major reason for writing the equation of a circle in standard form is that this form enables us to identify the center and the radius of the circle. This information is sufficient to allow us to draw the circle.

This last equation no longer looks like the equation of a circle. Starting with such an equation, we can convert it back into the standard form of a circle by completing the square in both variables, if necessary. For example, let us begin with

$$x^2 - 4x + y^2 + 6y = -9$$

To complete the squares, $\left(\frac{1}{2} \cdot 4\right)^2 = 4$ and $\left(\frac{1}{2} \cdot 6\right)^2 = 9$ are added to each side.

Then complete the squares in x and y:

$$(x^2 - 4x + 4) + (y^2 + 6y + 9) = -9 + 4 + 9$$
$$(x - 2)^2 + (y + 3)^2 = 4$$

EXAMPLE 4 Find the center and the radius of the circle with equation $9x^2 + 12x + 9y^2 = 77$.

Solution First divide by 9 so that the x^2 and y^2 terms each have a coefficient of 1.

$$x^2 + \frac{4}{3}x + y^2 = \frac{77}{9}$$

Observe that in these expanded forms of circles the coefficients of the x^2 and y^2 terms are equal.

Complete the square in x; add $\frac{4}{9}$ to both sides of the equation.

$$\left(x^2 + \tfrac{4}{3}x + \tfrac{4}{9}\right) + y^2 = \tfrac{77}{9} + \tfrac{4}{9}$$

$$\left(x + \tfrac{2}{3}\right)^2 + y^2 = 9$$

In standard form:

$$[x - \left(-\tfrac{2}{3}\right)]^2 + (y - 0)^2 = 3^2$$

The center is at $\left(-\frac{2}{3}, 0\right)$ and $r = 3$.   ∎

TEST YOUR UNDERSTANDING

(Answers: Page 303)

Find the center and radius of each circle.

1. $x^2 - 6x + y^2 - 10y = 2$ **2.** $x^2 + y^2 + y = \frac{19}{4}$

3. $x^2 - x + y^2 + 2y = \frac{23}{4}$ **4.** $16x^2 + 16y^2 - 8x + 32y = 127$

A line that is tangent to a circle touches the circle at only one point and is perpendicular to the radius drawn to the point of tangency.

When the equation of a circle is given and the coordinates of a point P on the circle are known, then the equation of the tangent line to the circle at P can be found. For example, the circle $(x + 3)^2 + (y + 1)^2 = 25$ has center $(-3, -1)$ and $r = 5$. The point $P(1, 2)$ is on this circle because its coordinates satisfy the equation of the circle.

$$(1 + 3)^2 + (2 + 1)^2 = 4^2 + 3^2 = 25$$

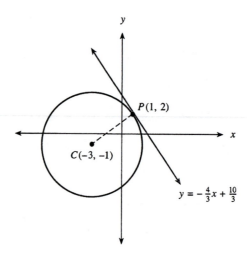

Recall that perpendicular lines have slopes that are negative reciprocals of one another.

The slope of radius CP is $\dfrac{2 - (-1)}{1 - (-3)} = \dfrac{3}{4}$. Then since the tangent at P is perpendicular to the radius CP, its slope is the negative reciprocal of $\frac{3}{4}$, namely $-\frac{4}{3}$. Now using the point-slope form, we get this equation of the tangent at P:

$$y - 2 = -\frac{4}{3}(x - 1)$$

In slope-intercept form this becomes

$$y = -\frac{4}{3}x + \frac{10}{3}$$

The midpoint of a segment on a number line is discussed on page 27.

The center of a circle is the midpoint of each of its diameters. So if the endpoints of a diameter are given, the coordinates of the center can be found using the *midpoint formula,* as shown below. In the figure, PQ is a line segment with midpoint M having the coordinates (x', y'). Since x_1, x', and x_2 are on the x-axis, $x' = \dfrac{x_1 + x_2}{2}$. Similarly, $y' = \dfrac{y_1 + y_2}{2}$.

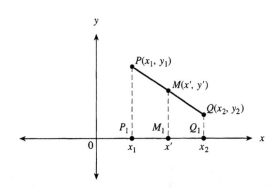

MIDPOINT FORMULA

If (x_1, y_1) and (x_2, y_2) are the endpoints of a line segment, the midpoint of the segment has coordinates

$$\left(\frac{x_1 + x_2}{2}, \frac{y_1 + y_2}{2}\right)$$

EXAMPLE 5   Points $P(2, 5)$ and $Q(-4, -3)$ are the endpoints of a diameter of a circle. Find the center, radius, and equation of the circle.

Solution   The center is the midpoint of PQ whose coordinates (x', y') are given by

$$x' = \frac{2 + (-4)}{2} = -1, \qquad y' = \frac{5 + (-3)}{2} = 1$$

The center is located at $C(-1, 1)$. To find the radius, use the distance formula between $C(-1, 1)$ and $P(2, 5)$.

$$r = \sqrt{(-1 - 2)^2 + (1 - 5)^2} = \sqrt{25} = 5$$

The equation of the circle is

$$(x + 1)^2 + (y - 1)^2 = 25 \qquad \blacksquare$$

CAUTION: Learn to Avoid These Mistakes	
WRONG	**RIGHT**
The circle $(x + 3)^2 + (y - 2)^2 = 7$ has center $(3, -2)$ and radius 7.	The circle has center $(-3, 2)$ and radius $\sqrt{7}$.
The equation of the circle with center $(-1, 0)$ and the radius 5 has equation $x^2 + (y + 1)^2 = 5$.	The circle has equation $(x + 1)^2 + y^2 = 25$.

EXERCISES 4.1

1. Graph these circles in the same coordinate system.

   (a) $x^2 + y^2 = 25$ (b) $x^2 + y^2 = 16$ (c) $x^2 + y^2 = 4$ (d) $x^2 + y^2 = 1$

2. Graph these circles in the same coordinate system.

   (a) $(x - 3)^2 + (y - 3)^2 = 9$ (b) $(x + 3)^2 + (y - 3)^2 = 9$

   (c) $(x + 3)^2 + (y + 3)^2 = 9$ (d) $(x - 3)^2 + (y + 3)^2 = 9$

Write the equation of each circle in the standard form $(x - h)^2 + (y - k)^2 = r^2$. Find the center and radius for each.

3. $x^2 - 4x + y^2 = 21$ 4. $x^2 + y^2 + 8y = -12$

5. $x^2 - 2x + y^2 - 6y = -9$ 6. $x^2 + 4x + y^2 + 10y + 20 = 0$

7. $x^2 - 4x + y^2 - 10y = -28$ 8. $x^2 - 10x + y^2 - 14y = -25$

9. $x^2 - 8x + y^2 = -14$ 10. $x^2 + y^2 + 2y = 7$

11. $x^2 - 20x + y^2 + 20y = -100$ 12. $4x^2 - 4x + 4y^2 = 15$

13. $16x^2 + 24x + 16y^2 - 32y = 119$ 14. $36x^2 - 48x + 36y^2 + 180y = -160$

Write the equation of each circle in standard form.

15. Center at $(2, 0)$; $r = 2$ 16. Center at $\left(\frac{1}{2}, 1\right)$; $r = 10$

17. Center at $(-3, 3)$; $r = \sqrt{7}$ 18. Center at $(-1, -4)$; $r = 2\sqrt{2}$

19. Draw the circle $x^2 + y^2 = 25$ and the tangent lines at the points $(3, 4)$, $(-3, 4)$, $(3, -4)$, and $(-3, -4)$. Write the equations of these tangent lines.

20. Where are the tangents to the circle $x^2 + y^2 = 4$ whose slopes equal 0? Write their equations.

21. Write the equations of the tangents to the circle $x^2 + y^2 = 4$ whose slopes are undefined.

Draw the given circle and the tangent line at the indicated point for each of the following. Write the equation of the tangent line.

22. $x^2 + y^2 = 80$; $(-8, 4)$

23. $x^2 - 2x + y^2 - 2y = 8$; $(4, 2)$

24. $(x - 4)^2 + (y + 5)^2 = 45$; $(1, 1)$

25. $x^2 + 4x + y^2 - 6y = 60$; $(6, 0)$

26. $x^2 + 14x + y^2 + 18y = 39$; $(5, -4)$

27. $x^2 + y^2 = 9$; $(-2, \sqrt{5})$

Find the coordinates of the midpoint of a line segment with endpoints as given.

28. $P(3, 2)$ and $Q(-2, 1)$

29. $P(-2, 4)$ and $Q(3, -8)$

30. $P(-1, 0)$ and $Q(0, 5)$

31. $P(-8, 7)$ and $Q(3, -6)$

32. Points $P(3, -5)$ and $Q(-1, 3)$ are the endpoints of a diameter of a circle. Find the center, radius, and equation of the circle.

33. Write the equation of the tangent line to the circle $x^2 + y^2 = 80$ at the point in the first quadrant where $x = 4$.

34. Write the equation of the tangent line to the circle $x^2 + y^2 = 9$ at the point in the third quadrant where $y = -1$.

35. Write the equation of the tangent line to the circle $x^2 + 14x + y^2 + 18y = 39$ at the point in the second quadrant where $x = -2$.

Exercises 36–44 are directly preparatory for topics in calculus that call for the extraction of functions from geometric figures.

36. Suppose a kite is flying at a height of 300 feet above a point P on the ground. The kite string is anchored in the ground x feet from P.

    (a) If s is the length of the string (assume the string forms a straight line), write s as a function of x.

    (b) Find s when $x = 400$ feet.

37. A 13-foot-long board is leaning against the wall of a house so that its base is 5 feet from the wall. When the base of the board is pulled y feet further from the wall, the top of the board drops x feet.

    (a) Express y as a function of x.   (b) Find y when $x = 7$.

38. Let s be the square of the distance from the origin to point $P(x, y)$ on the line through points $(0, 4)$ and $(2, 0)$. Find the coordinates of P such that s is a minimum value.

39. Let s be the square of the distance from point $(6, 6)$ to point $P(x, y)$ on the line $y = -2x + 4$. Find the coordinates of P such that s is a minimum value.

40. The figure represents a solid with a circular base having radius 2. A plane cutting the solid perpendicular to the xy-plane and to the x-axis, between -2 and 2, forms a cross section in the shape of a square. Express the area of the cross section in terms of the variable x. (*Hint:* The length of a side of the square is $2y$.)

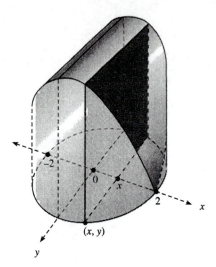

41. A rectangle is inscribed in a semicircle of radius 12 as shown. Express the area of the rectangle as a function of x.

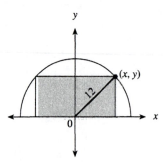

42. Triangle ABC is inscribed in a semicircle of radius 8 so that one of its sides coincides with a diameter. Express the area of the triangle as a function of $x = AC$.

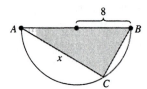

43. $ABCD$ is an isosceles trapezoid in which sides AB and DC are parallel. Express the area of the trapezoid as a function of altitude h.

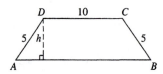

44. The figure shows a right circular cone in which r is the radius of the base, and the slant height is 10. Express the volume of the cone as a function of r.

45. In the figure, each side of triangle OAB is the diameter of a semicircle as shown. Show that the area of the semicircle on AB is equal to the sum of the areas of the semicircles on the other two sides.

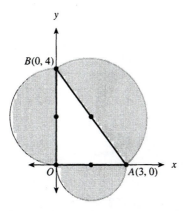

46. Generalize the result in Exercise 45 for any right triangle. (*Hint:* Use c as the length of the hypotenuse and a, b as the lengths of the other sides, and place the triangle in a coordinate system as in Exercise 45.)

47. An architect's design of a decorative window consists of a rectangle surmounted by a semicircle of radius 2 feet. She wants to place a circular stained glass window within the semicircular part with its center on the vertical radius of the semicircle. The distance from the top of the circle to the semicircle as well as the distance from the bottom of the circle to the top of the rectangle is 4 inches. Using the center of the semicircle as the origin, find an equation of the circular stained glass window.

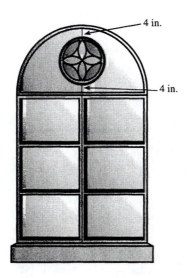

48. $A(3, 0)$ and $B(-3, 0)$ are the endpoints of a diameter of the circle $x^2 + y^2 = 9$. Let $C(x, y)$ be any other point on the circle and use the distance formula to prove that triangle ABC is a right triangle.

49. Draw a coordinate system in which one unit on each axis is 0.5 inch. Now use a compass to draw the circle of radius 2 units, centered at the origin, and locate point $P(8, 0)$. Construct the two tangents from P to the circle as described in the introduction of this chapter, and algebraically find the exact coordinates of the points of tangency.

50. This exercise outlines a proof of the Pythagorean theorem.

   (a) Begin with a right triangle as shown at the left. Construct a square as on the right, where each side has length $a + b$.

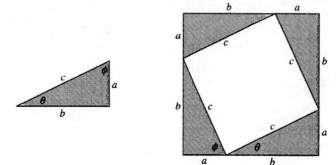

   Why is the inner quadrilateral a square whose sides are of length c?

   (b) Express the area of the outer square as the sum of the areas of the five parts.

   (c) Since $(a + b)^2$ is also the area of the outer square, how can we conclude that $a^2 + b^2 = c^2$?

? Challenge   Part of an ancient wheel was found. It had been a solid wheel without spokes. To find out how large the original wheel was, the archaeologist placed the part onto a coordinate system using a unit length of 1 inch on each axis. Three points on the rim of the wheel part are located at $A(-3, -3)$, $B(-1, 11)$, and $C(5, 13)$. What was the archeologist's conclusion as to the length of the radius of the original wheel?

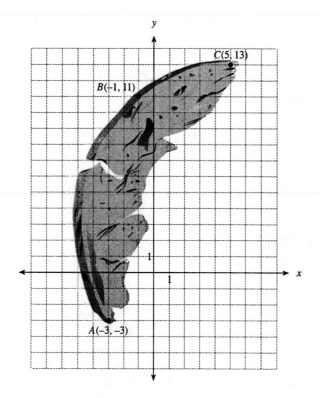

Written Assignment   Take a circle of radius r and cut it into two semicircles. Divide each semicircle into 8 congruent sectors and fit them together as shown.

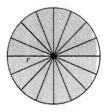

The area of this figure is the same as the area of the original circle. Now imagine a similar configuration after each semicircle is divided into $n = 16$ congruent sectors, then again when $n = 32$, and so on. Describe what these configurations tend toward as n gets large, and explain why it is reasonable to conclude that the area of the circle is πr^2.

 Graphing Calculator Exercises

1. **(a)** Graph the equation $x^2 + y^2 = 9$ by solving for y to obtain two functions of x and graphing both on the same set of axes using a standard window ($-10 \leq x \leq 10$, $-10 \leq y \leq 10$). Does your graph look like a circle?

   (b) Graph the circle in part **(a)** using a window in which the x- and y-scales have equal length.

2. Graph the equation $(x - 1)^2 + (y + 2)^2 = 2$, following the instructions for Exercise 1.

3. Using the circle $x^2 + y^2 = 9$ for the head, find circles and semicircles that can be used to draw the face in the following figure. Graph the result with your graphing calculator.

(*Hint:* The Casios will draw as many circles as you like in the same window. For the TI's, look up the DRAW facility in your manual.)

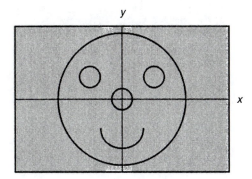

**4.2
GRAPHING
RADICAL
FUNCTIONS**

*When the defining expression
of a function involves the
independent variable within
a radical sign, it is referred
to as a **radical function**.*

When the graph of the circle $x^2 + y^2 = 9$ is separated into the upper and lower semi-circles, two *radical functions* are produced as follows:

$$x^2 + y^2 = 9$$
$$y^2 = 9 - x^2$$
$$y = \pm\sqrt{9 - x^2}$$

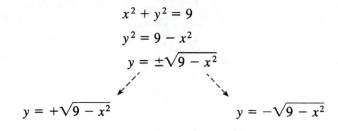

$$y = +\sqrt{9 - x^2} \qquad\qquad y = -\sqrt{9 - x^2}$$

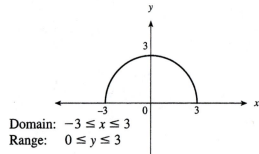

Domain: $-3 \le x \le 3$
Range: $0 \le y \le 3$

Domain: $-3 \le x \le 3$
Range: $-3 \le y \le 0$

The horizontal parabola $x = y^2$ (see page 138) is not the graph of a function of x. However, when $x = y^2$ is solved for y, two radical functions are also produced as shown.

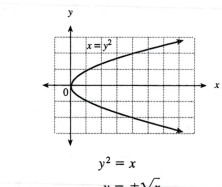

$$y^2 = x$$
$$y = \pm\sqrt{x}$$

$$y = +\sqrt{x} \qquad\qquad y = -\sqrt{x}$$

*Another way to draw the
graph of $y = \sqrt{x}$ is to plot
specific points such as $(0, 0)$,
$(1, 1)$, and $(4, 2)$, and connect them with a smooth
curve. Accuracy can be
gained by using a calculator
for points such as $(2, 1.41)$
and $(7, 2.65)$.*

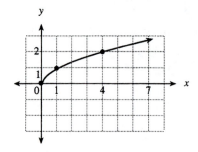

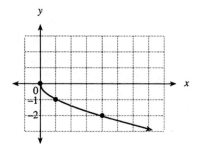

The function at the left, $y = \sqrt{x}$, is known as the **square root function,** whose domain consists of all real numbers $x \geq 0$ (the square root of a negative number is not a real number), and the range is all $y \geq 0$. Observe that the function is increasing and the graph is concave down.

EXAMPLE 1   Find the domain and range of $y = g(x) = \sqrt{x - 2}$ and graph.

Solution Since the square root of a negative number is not a real number, the expression $x - 2$ must be nonnegative; therefore, the domain of g consists of all $x \geq 2$. The graph of g may be found by shifting the graph of $y = \sqrt{x}$ by 2 units to the right.

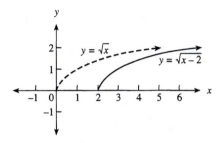

The range of $y = \sqrt{x}$ is all $y \geq 0$, and since the graph of g was obtained from $y = \sqrt{x}$ by a horizontal translation, the range of g is also all $y \geq 0$. ∎

EXAMPLE 2   Find the domain of $y = f(x) = \sqrt{|x|}$ and graph.

Solution Since $|x| \geq 0$ for all x, the domain of f consists of all real numbers. To graph f, first note that the graph is symmetric about the y-axis:

It is also helpful to locate a few specific points on the curve as an aid to graphing.

$$f(-x) = \sqrt{|-x|} = \sqrt{|x|} = f(x)$$

f is decreasing and concave down for $x < 0$, and it is increasing and concave down for $x > 0$.

x	y
0	0
1	1
4	2
9	3
−1	1
−4	2
−9	3

$y = \sqrt{|x|} = \sqrt{-x}, x < 0$ $y = \sqrt{|x|} = \sqrt{x}, x \geq 0$

In drawing the graph of $f(x)$, we first found the graph for $x \geq 0$ and used symmetry to obtain the rest. For $x \geq 0$, we get $|x| = x$ and $f(x) = \sqrt{|x|} = \sqrt{x}$. ∎

EXAMPLE 3 Find the domain of $y = h(x) = x^{-1/2}$ and graph.

Solution Note that $h(x) = x^{-1/2} = \dfrac{1}{x^{1/2}} = \dfrac{1}{\sqrt{x}}$. Thus the domain consists of all $x > 0$. Furthermore, $\dfrac{1}{\sqrt{x}} > 0$ for all x, so we know that the graph must be in the first quadrant only. Plot a few points and connect them with a smooth curve.

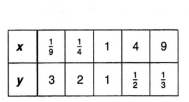

x	$\frac{1}{9}$	$\frac{1}{4}$	1	4	9
y	3	2	1	$\frac{1}{2}$	$\frac{1}{3}$

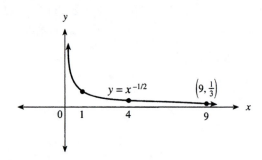

The equations of the two asymptotes are $x = 0$ and $y = 0$.

Observe that the closer x is to zero, the larger are the corresponding y-values. Also, as the values of x get larger, the corresponding y-values get closer to 0. That is, $h(x) \longrightarrow \infty$ as $x \longrightarrow 0^+$ and $h(x) \longrightarrow 0$ as $x \longrightarrow \infty$. Thus the coordinate axes are asymptotes to the curve $y = x^{-1/2}$. ∎

EXAMPLE 4 Use the graph of $y = \dfrac{1}{\sqrt{x}}$ to graph $\dfrac{1}{\sqrt{x}} + 1$ and $y = \dfrac{1}{\sqrt{x+1}}$ and find the asymptotes.

Solution The graphing procedures developed earlier can be used here. Thus the graph of $y = \dfrac{1}{\sqrt{x}} + 1$ can be obtained from the graph for Example 3 by shifting up one unit. Then the asymptotes will be the y-axis and the line $y = 1$. The graph of $y = \dfrac{1}{\sqrt{x+1}}$ is found by shifting the graph for $y = \dfrac{1}{\sqrt{x}}$ to the left one unit. The equations of the asymptotes are $x = -1$ and $y = 0$.

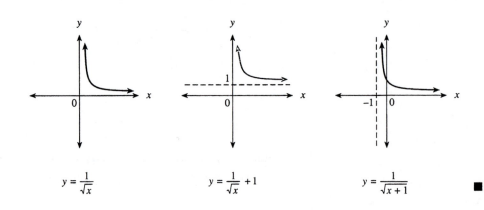

$$y = \frac{1}{\sqrt{x}} \qquad\qquad y = \frac{1}{\sqrt{x}} + 1 \qquad\qquad y = \frac{1}{\sqrt{x+1}}$$

∎

TEST YOUR UNDERSTANDING	*Graph each pair of functions on the same set of axes and state the domain of g.*

1. $f(x) = \sqrt{x}$, $g(x) = \sqrt{x} - 2$ **2.** $f(x) = -\sqrt{x}$, $g(x) = -\sqrt{x - 1}$

3. $f(x) = \sqrt{x}$, $g(x) = 2\sqrt{x}$ **4.** $f(x) = \dfrac{1}{\sqrt{x}}$, $g(x) = \dfrac{1}{\sqrt{x + 2}}$

(Answers: Page 303)

Another radical function that you should learn to recognize is the *cube root function,* $y = \sqrt[3]{x}$. An interesting way to graph this function is to first take $y = \sqrt[3]{x}$ and cube both sides to get $y^3 = x$. Now recall the graph of $y = x^3$ and obtain the graph of $x = y^3$ by reversing the role of the variables.

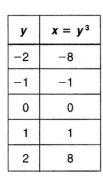

y	$x = y^3$
-2	-8
-1	-1
0	0
1	1
2	8

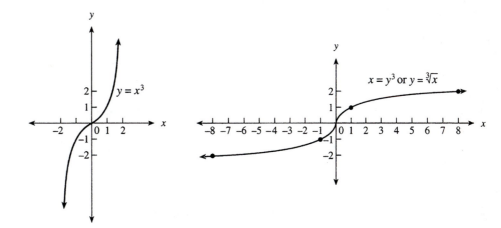

The next example illustrates a function that is defined piecewise by using both a polynomial and a radical expression.

EXAMPLE 5   Graph f defined on the domain $-2 \le x < 5$ as follows:

$$f(x) = \begin{cases} x^2 - 1 & \text{for } -2 \le x < 2 \\ \sqrt{x - 2} & \text{for } 2 \le x < 5 \end{cases}$$

When $x = 2$, the radical part of f is used;
$f(2) = \sqrt{2 - 2} = 0$. *So there is a solid dot at $(2, 0)$ and an open dot at $(2, 3)$. Also, there is an open dot for $x = 5$ since 5 is not a domain value of f.*

Solution   The first part of f is given by $f(x) = x^2 - 1$ for $-2 \le x < 2$. This is an arc of a parabola obtained by shifting the graph of $y = x^2$ downward one unit.

The second part of f is given by $f(x) = \sqrt{x - 2}$ for $2 \le x < 5$. This is an arc of the square root curve obtained by shifting the graph of $y = \sqrt{x}$ two units to the right.

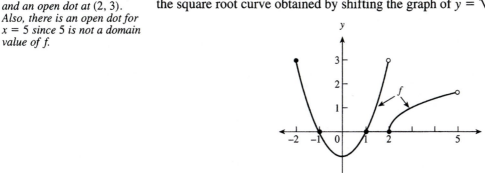

■

EXERCISES 4.2

Graph each set of curves on the same coordinate system. Use a dashed curve for the first equation and a solid curve for the second.

1. $f(x) = \sqrt{x}$, $g(x) = \sqrt{x - 1}$ **2.** $f(x) = -\sqrt{x}$, $g(x) = -2\sqrt{x}$

3. $f(x) = \sqrt{x}$, $g(x) = \sqrt{x + 1}$ **4.** $f(x) = -\sqrt{x}$, $g(x) = -\sqrt{x - 2}$

5. $f(x) = \sqrt[3]{x}$, $g(x) = \sqrt[3]{x + 2}$ **6.** $f(x) = -\sqrt[3]{x}$, $g(x) = -\sqrt[3]{x - 3}$

7. $f(x) = \sqrt[3]{x}$, $g(x) = \sqrt[3]{x - 1}$ **8.** $f(x) = \sqrt[3]{x}$, $g(x) = -\sqrt[3]{x + 1}$

Find the domain of f, sketch the graph, and give the equations of the asymptotic lines if there are any. Also state where f is increasing or decreasing and describe the concavity.

9. $f(x) = \sqrt{x + 2}$ **10.** $f(x) = x^{1/2} + 2$ **11.** $f(x) = \sqrt{x - 3} - 1$ **12.** $f(x) = -\sqrt{x} + 3$

13. $f(x) = \sqrt{-x}$ **14.** $f(x) = \sqrt{(x - 2)^2}$ **15.** $f(x) = 2\sqrt[3]{x}$ **16.** $f(x) = |\sqrt[3]{x}|$

17. $f(x) = -x^{1/3}$ **18.** $f(x) = \sqrt[3]{-x}$ **19.** $f(x) = \dfrac{1}{\sqrt{x}} - 1$ **20.** $f(x) = \dfrac{1}{\sqrt{x - 2}}$

21. (a) Explain why the graph of $f(x) = \dfrac{1}{\sqrt[3]{x}}$ is symmetric through the origin.

     (b) What is the domain of f?

     (c) Use a table of values to graph f.

     (d) What are the equations of the asymptotes?

22. Find the domain of $f(x) = \dfrac{1}{\sqrt[3]{x + 1}}$, sketch the graph, and give the equations of the asymptotes.

23. Find the graph of the function $y = \sqrt[4]{x}$ by raising both sides of the equation to the fourth power and comparing to the graph of $y = x^4$.

24. Reflect the graph of $y = x^2$, for $x \geq 0$, through the line $y = x$. Obtain the equation of this new curve by interchanging variables in $y = x^2$ and solving for y.

25. Follow the instruction of Exercise 24 with $y = x^3$ for all values x.

In Exercises 26 and 27, the function f is defined by using more than one expression. Graph f on its given domain.

26. $f(x) = \begin{cases} \sqrt{x} & \text{for } 0 \leq x \leq 4 \\ 10 - \frac{1}{2}x^2 & \text{for } 4 < x < 6 \end{cases}$

27. $f(x) = \begin{cases} -2x - 1 & \text{for } -3 \leq x < 0 \\ \sqrt[3]{x - 1} & \text{for } 0 \leq x \leq 2 \end{cases}$

Verify the equation involving the difference quotient for the given radical function (see section 2.1).

28. $f(x) = \sqrt{x}$; $\dfrac{f(x) - f(25)}{x - 25} = \dfrac{1}{\sqrt{x} + 5}$ Factor $x - 25$ as the difference of squares.

29. $f(x) = \sqrt{x}$; $\dfrac{f(4 + h) - f(4)}{h} = \dfrac{1}{\sqrt{4 + h} + 2}$ Rationalize the numerator.

30. $f(x) = -\sqrt{x}$; $\dfrac{f(x) - f(9)}{x - 9} = -\dfrac{1}{\sqrt{x} + 3}$

Exercises 31–34 *call for the extraction of radical functions from geometric situations (see pages* 91, 93, 94).

31. A runner starts at point A, goes to point P that is x miles from B, and then runs to D. (Angles at B and C are right angles.)

   (a) Write the total distance d traveled as a function of x.

   (b) The runner averages 12 miles per hour from A to P and 10 miles per hour from P to D. Write the time t for the trip as a function of x. (*Hint:* Use time = distance/rate.)

   (c) Approximate, to the nearest tenth of an hour, the time for the trip when $x = 5$ miles.

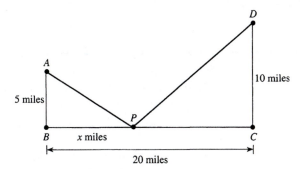

32. In the figure, AC is along the shoreline of a lake, and the distance from A to C is 12 miles. P represents the starting point of a swimmer who swims at 3 miles per hour along the hypotenuse PB. P is 5 miles from point A on the shoreline. After reaching B, he walks at 6 miles per hour to C.

   (a) Express the total time t of the trip as a function of x, where x is the distance from A to B. (*Hint:* Use time = distance/rate and assume angle A is a right angle.

   (b) Approximate, to the nearest tenth of an hour, the time for the trip when $x = 4$ miles.

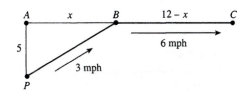

33. Express the distance d from the origin to a point (x, y) on the curve $y = \dfrac{1}{\sqrt{x}}$ as a function of x.

34. Express the distance d from the point $(2, 5)$ to a point (x, y) on the line $3x + y = 6$ as a function of x.

 Challenge In Exercise 34, find the shortest distance from point $(2, 5)$ to the line $3x + y = 6$. (*Hint:* Consider d^2 and observe that $y = \sqrt{x}$ is an increasing function.)

Written Assignment Discuss how the graph of $y = f(x) = \sqrt{2 - x}$ can be obtained from the graph of $y = g(x) = \sqrt{x - 2}$ in Example 1.

Graphing Calculator Exercises

1. A student tried to graph $y = x^{1/2}$, entering the function as $y = x \wedge 1/2$. What will be obtained instead of the graph expected? How should the function have been entered?

2. Graph the semicircles $y = \sqrt{4 - (x + 3)^2} + 3$ and $y = -\sqrt{4 - (x + 3)^2} + 3$ on the same set of axes, using a standard window of $-10 \le x \le 10$, $-10 \le y \le 10$. (a) What is the equation of the corresponding circle? (b) Why are there gaps between the semicircles? (c) Why doesn't the graph look circular? (d) Find a suitable window that will not show gaps in the graphs or distort the circle.

4.3 RADICAL EQUATIONS AND GRAPHS

Additional graphs of radical functions will be studied here, although the main focus will be on the solution of *radical equations*. These two developments are connected in the sense that to find the x-intercepts for a function such as $f(x) = \sqrt{x + 4} + 2 - x$, it is necessary to solve the radical equation $\sqrt{x + 4} + 2 - x = 0$. (See Example 1.)

An equation in which a variable occurs in a radicand, as illustrated by $\sqrt{x + 2} + 2 = 5$, is called a **radical equation.** To solve this equation we first isolate the radical on one side and then square both sides to eliminate the radical.

$$\sqrt{x + 2} + 2 = 5$$
$$\sqrt{x + 2} = 3$$
$$\left(\sqrt{x + 2}\right)^2 = 3^2 \qquad \text{Square both sides.}$$
$$x + 2 = 9$$
$$x = 7$$

You can check this result by substituting 7 for x in the original equation.

In general, when solving radical equations, we will be making use of the following principle:

If $a = b$, then $a^n = b^n$

This statement says that every solution of $a = b$ will also be a solution of $a^n = b^n$.

Sometimes the method of raising both sides of an equation to the same power can produce a new equation that has more solutions than the original equation, as in the following example.

EXAMPLE 1 Solve for x: $\sqrt{x + 4} + 2 = x$

Solution

$$\sqrt{x + 4} + 2 = x$$
$$\sqrt{x + 4} = x - 2$$
$$\left(\sqrt{x + 4}\right)^2 = (x - 2)^2 \qquad \text{Square both sides.}$$
$$x + 4 = x^2 - 4x + 4$$
$$0 = x^2 - 5x$$
$$0 = x(x - 5)$$

Note: If the radical in $\sqrt{x + 4} + 2 = x$ is not first isolated, it is still possible to solve the equation, but the work will be more involved. Try it. Also observe that $x = 5$ is the x-intercept of the curve $y = \sqrt{x + 4} + 2 - x$.

From the last step we conclude that $x = 0$ or $x = 5$. (This is based on the *zero-product property;* see page 4). Now check each result.

Check:

$$\text{Let } x = 0: \qquad\qquad \text{Let } x = 5:$$

$$\sqrt{0 + 4} + 2 = 0 \qquad \sqrt{5 + 4} + 2 = 5$$

$$2 + 2 = 0 \quad \text{No} \qquad 3 + 2 = 5 \quad \text{Yes}$$

We conclude that the only solution for the given equation is 5. The number 0 is called an *extraneous solution* or *extraneous root.* ∎

How did the extraneous solution in Example 1 arise? In going from

$$\sqrt{x + 4} = x - 2 \quad \text{to} \quad \left(\sqrt{x + 4}\right)^2 = (x - 2)^2$$

we used the basic principle: If $a = b$, then $a^n = b^n$. Therefore, every solution of the first equation is also a solution for the second. But *this principle is not always reversible.* In particular, both 0 and 5 are solutions of the second equation, but only 5 is a solution of the first. In summary:

> When raising both sides of an equation to the same power, **extraneous solutions** might be introduced. Therefore, whenever this method is used, all solutions must be checked in the original equation.

We have been looking at the solutions of radical equations in one variable. Keep in mind, however, that the solutions of $f(x) = 0$ can be regarded as the x-intercepts of the graph of the related equation in two variables, $y = f(x)$. This is illustrated in the following example.

EXAMPLE 2 Draw the graph of the function $y = f(x) = \sqrt{x^2 - 4}$.

Solution The domain of f is the set of all $x \leq -2$ or $x \geq 2$. This is found by solving the inequality $x^2 - 4 \geq 0$.

The x-intercepts are -2 and 2, found by solving $\sqrt{x^2 - 4} = 0$.

$f(x) > 0$ for all domain values x except the x-intercepts. Therefore, the graph is above the x-axis on the intervals $(-\infty, -2)$ and $(2, \infty)$.

The graph is symmetric about the y-axis since $f(-x) = \sqrt{(-x)^2 - 4} = \sqrt{x^2 - 4} = f(x)$.

Use the preceding information and some selected points to draw the graph on $(2, \infty)$. Complete the graph by using the symmetry through the y-axis.

x	y
2	0
4	$\sqrt{12} = 3.5$
6	$\sqrt{32} = 5.7$
8	$\sqrt{60} = 7.7$

y-values are rounded to tenths

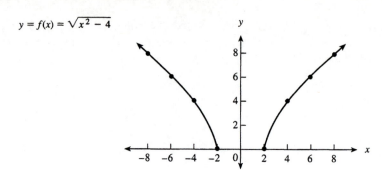

$$y = f(x) = \sqrt{x^2 - 4}$$

EXAMPLE 3   Graph $y = f(x) = \dfrac{8}{\sqrt{16 - x^2}}$.

Solution f is symmetric around the y-axis since

$$f(-x) = \frac{8}{\sqrt{16 - (-x)^2}} = \frac{8}{\sqrt{16 - x^2}} = f(x)$$

The inequality $16 - x^2 > 0$ can also be solved by using the procedure shown in Example 5, page 151, or by using a table of signs as on page 190.

The domain of f is the solution of the inequality $16 - x^2 > 0$, or $x^2 < 16$ which consists of all x where $-4 < x < 4$. (Any number larger than 4 has a square larger than 16, and the same is true for numbers less than -4.)

Since $f(x) = \dfrac{8}{\sqrt{16 - x^2}} = 0$ has no solution, there are no x-intercepts. Letting $x = 0$ gives the y-intercept $f(0) = \dfrac{8}{\sqrt{16}} = 2$.

There are asymptotes because the numerator of f is constant and $16 - x^2$ gets close to 0. In particular, $f(x) \longrightarrow \infty$ as $x \longrightarrow 4^-$ and as $x \longrightarrow -4^+$ so that $x = 4$ and $x = -4$ are vertical asymptotes. To graph the given function, first draw the asymptotes. Plot several convenient points selected from the table. Then sketch the right half through the points so that the curve is asymptotic to the line $x = 4$. The left half is obtained by reflecting the right half through the y-axis.

x	y
0	2
1	2.1
2	2.3
3	3.0
3.5	4.1
3.7	5.3
3.9	9.0
3.99	28.3
3.999	89.4
$\downarrow$	$\downarrow$
4	∞

The y-values are rounded to tenths.

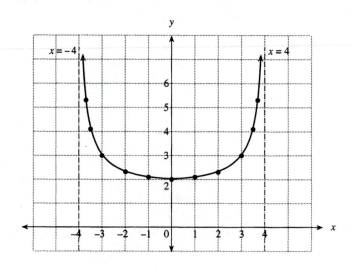

Radical equations may contain more than one radical. For such cases it is usually best to transform the equation first into one with as few radicals on each side as possible. Consider, for example, the following equation:

CAUTION
$(\sqrt{x} + 1)^2 \neq x + 1.$ *Use*
$(a + b)^2 = a^2 + 2ab + b^2$
with $a = \sqrt{x}$ *and* $b = 1.$

$$\sqrt{x - 7} - \sqrt{x} = 1$$

$$\sqrt{x - 7} = \sqrt{x} + 1$$

$$\left(\sqrt{x - 7}\right)^2 = \left(\sqrt{x} + 1\right)^2 \qquad \text{Square both sides.}$$

$$x - 7 = x + 2\sqrt{x} + 1$$

$$-8 = 2\sqrt{x}$$

$$-4 = \sqrt{x}$$

$$16 = x \qquad \text{Square again.}$$

Check: $\sqrt{16 - 7} - \sqrt{16} = \sqrt{9} - \sqrt{16} = 3 - 4 = -1 \neq 1.$ Therefore, this equation has no solution, which could have been observed at an earlier stage as well. For example, $-4 = \sqrt{x}$ has no solution (Why not?)

In the preceding illustration, do you see why the given equation cannot have a solution without having to do any algebraic work?

TEST YOUR UNDERSTANDING	*Find the x-intercepts and the domain of f.*

Find the x-intercepts and the domain of f.

1. $f(x) = \sqrt{x} - 5$ **2.** $f(x) = \sqrt{x^2 - 3x}$

Solve for x.

3. $\sqrt{x + 1} = 3$ **4.** $\sqrt{x - 2} = 3$

5. $\sqrt{x + 2} = \sqrt{2x - 5}$ **6.** $\sqrt{x^2 + 9} = -5$

7. $\dfrac{1}{\sqrt{x}} = 3$ **8.** $\dfrac{4}{\sqrt{x - 1}} = 2$

9. $\sqrt{x^2 - 5} = 2$ **10.** $\sqrt{x} + \sqrt{x - 5} = 5$

(Answers: Page 303)

11. $\sqrt{x + 16} - x = 4$ **12.** $2x = 1 + \sqrt{1 - 2x}$

The next example shows how to change the algebraic form of a radical function so as to make it easier to explore the properties of that function. That is, in the given form, it is not clear where the function is positive, negative, or zero. However, after changing into the form shown, it becomes easy to determine the signs of $f(x)$.

EXAMPLE 4 Find the domain of f, the roots of $f(x) = 0$, and determine the signs of f where

$$f(x) = \frac{x^2 - 1}{2\sqrt{x - 1}} + 2x\sqrt{x - 1}$$

Solution The radical in the denominator calls for $x - 1 > 0$. Therefore, the domain consists of all $x > 1$. Now simplify as follows:

$$f(x) = \frac{x^2 - 1}{2\sqrt{x - 1}} + 2x\sqrt{x - 1}$$

$$= \frac{x^2 - 1}{2\sqrt{x - 1}} + \frac{(2x\sqrt{x - 1})(2\sqrt{x - 1})}{2\sqrt{x - 1}}$$

$$= \frac{x^2 - 1 + 4x(x - 1)}{2\sqrt{x - 1}}$$

$$= \frac{5x^2 - 4x - 1}{2\sqrt{x - 1}}$$

$$= \frac{(5x + 1)(x - 1)}{2\sqrt{x - 1}}$$

$$= \frac{5\left(x + \frac{1}{5}\right)(x - 1)}{2\sqrt{x - 1}}$$

Any of the last three forms is an acceptable simplification, but the last form is more advantageous for determining the signs of $f(x)$.

The numerator is zero when $x = -\frac{1}{5}$ or $x = 1$. But these values are *not* in the domain of f. Since a fraction can be zero only when the numerator is zero, it follows that $f(x) = 0$ has no solutions and the graph of f has no x-intercepts. The signs of $f(x)$ are given in this brief table:

The ability to determine the signs of a function f, that is, finding where $f(x)$ is positive or negative, is useful when graphing the function. Furthermore, this approach will prove to be helpful in the study of calculus.

$$f(x) = \frac{5\left(x + \frac{1}{5}\right)(x - 1)}{2\sqrt{x - 1}}$$

Interval	$(1, \infty)$
Sign of $x + \frac{1}{5}$	+
Sign of $x - 1$	+
Sign of $\sqrt{x - 1}$	+
Sign of $f(x)$	+

Thus $f(x) > 0$ on its domain $(1, \infty)$.   ∎

All the examples thus far have involved square roots. However, the method of raising both sides of an equation to the same power can also be applied to radical equations involving cube roots or fourth roots, and so on.

EXAMPLE 5   Solve: $\sqrt[3]{x + 3} = 2$

Solution In this case, cube each side of the equation.

$$(\sqrt[3]{x + 3})^3 = (2)^3$$

$$x + 3 = 8 \qquad [(\sqrt[3]{a})^3 = a]$$

$$x = 5$$

Check: $\sqrt[3]{5 + 3} = \sqrt[3]{8} = 2$ ∎

The equation in Example 5 as well as any other radical equation can also be written using fractional exponents. Thus, for Example 5 we have the equation $(x + 3)^{1/3} = 2$, and to solve it we use the same steps as before.

$$(x + 3)^{1/3} = 2$$

$$[(x + 3)^{1/3}]^3 = 2^3 \qquad \text{Cube both sides.}$$

$$x + 3 = 8$$

$$x = 5$$

EXAMPLE 6   Solve for x:

$$\frac{3(2x - 1)^{1/2}}{x - 3} - \frac{(2x - 1)^{3/2}}{(x - 3)^2} = 0$$

Solution Multiply through by $(x - 3)^2$ to clear fractions, as shown next.

$$(x - 3)^2 \left[\frac{3(2x - 1)^{1/2}}{x - 3} \right] - (x - 3)^2 \left[\frac{(2x - 1)^{3/2}}{(x - 3)^2} \right] = (x - 3)^2(0)$$

$$3(x - 3)(2x - 1)^{1/2} - (2x - 1)^{3/2} = 0$$

$$(2x - 1)^{1/2}[3(x - 3) - (2x - 1)] = 0$$

$$(2x - 1)^{1/2}(x - 8) = 0$$

Factor out $(2x - 1)^{1/2}$. This is easier to see if we let $a = (2x - 1)^{1/2}$ and factor a out of $3(x - 3)a - a^3$.

Set each factor equal to zero.

$$(2x - 1)^{1/2} = 0 \qquad \text{or} \qquad x - 8 = 0$$

$$2x - 1 = 0 \qquad \text{or} \qquad x = 8$$

$$x = \tfrac{1}{2} \qquad \text{or} \qquad x = 8$$

Check in the original equation to show that both $x = \tfrac{1}{2}$ and $x = 8$ are solutions. ∎

As you may have noticed, the algebraic techniques developed earlier involving other kinds of expressions often carry over into this work. Following is an example that uses our knowledge of quadratics in conjunction with radicals.

EXAMPLE 7   Solve for x: $\sqrt[3]{x^2} + \sqrt[3]{x} - 20 = 0$

Solution First rewrite the equation by using rational exponents.

$$x^{2/3} + x^{1/3} - 20 = 0$$

Then think of $x^{2/3}$ as the square of $x^{1/3}$, $x^{2/3} = (x^{1/3})^2$, and use the substitution $u = x^{1/3}$ as follows:

$$x^{2/3} + x^{1/3} - 20 = 0$$

$$(x^{1/3})^2 + x^{1/3} - 20 = 0$$

Alternatively, we may keep the radical sign and proceed with $u = \sqrt[3]{x}$.

$$u^2 + u - 20 = 0$$

$$(u + 5)(u - 4) = 0 \qquad \text{factoring the quadratic}$$

$$u + 5 = 0 \qquad \text{or} \qquad u - 4 = 0$$

$$u = -5 \qquad \text{or} \qquad u = 4$$

Now replace u by $x^{1/3}$.

$$x^{1/3} = -5 \qquad \text{or} \qquad x^{1/3} = 4$$

Thus

$$x = -125 \qquad \text{or} \qquad x = 64$$

Check to show that both values are solutions of the given equation. ■

The final example calls for the solution of a system of equations, one of which is a radical function.

EXAMPLE 8   Solve the system and graph:

$$y = \sqrt[3]{x}$$

$$y = \tfrac{1}{4}x$$

Solution For the points of intersection the x- and y-values are the same in both equations. Thus, for such points, we may set the y-values equal to one another.

CAUTION
A common error is to take $x^3 - 64x = 0$ and divide through by x to get $x^2 - 64 = 0$. This step produces the roots ± 8. The root 0 has been lost because we divided by x, and 0 is the number for which the factor x in $x(x^2 - 64)$ is zero. You may always divide by a nonzero expression and get an equivalent form of the equation. But when you divide by a variable quantity there is the danger of losing some roots, those for which the divisor is 0.

$$\tfrac{1}{4}x = \sqrt[3]{x} \quad \longleftarrow \quad x^{1/3} \text{ can be used in place of } \sqrt[3]{x}$$

Cube both sides and solve for x.

$$\tfrac{1}{64}x^3 = x$$

$$x^3 = 64x$$

$$x^3 - 64x = 0$$

$$x(x^2 - 64) = 0$$

$$x(x + 8)(x - 8) = 0$$

$$x = 0 \quad \text{or} \quad x = -8 \quad \text{or} \quad x = 8$$

Substitute these values into either of the given equations to obtain the corresponding y-values. The remaining equation can be used for checking. The solutions are $(-8, -2)$, $(0, 0)$, and $(8, 2)$.

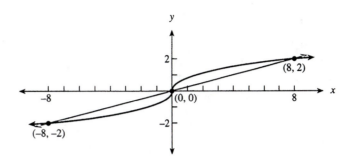

CAUTION: Learn to Avoid These Mistakes	
WRONG	**RIGHT**
$(9 - x^2)^{1/2}(9 - x^2)^{3/2}$ $= (9 - x^2)^{3/4}$	$(9 - x^2)^{1/2}(9 - x^2)^{3/2}$ $= (9 - x^2)^{1/2 + (3/2)}$ $= (9 - x^2)^2$
$\dfrac{2}{\sqrt{9 - x^2}} + 3\sqrt{9 - x^2}$ $= \dfrac{2 + 3\sqrt{9 - x^2}}{\sqrt{9 - x^2}}$	$\dfrac{2}{\sqrt{9 - x^2}} + 3\sqrt{9 - x^2}$ $= \dfrac{2}{\sqrt{9 - x^2}} + \dfrac{3(9 - x^2)}{\sqrt{9 - x^2}}$ $= \dfrac{2 + 3(9 - x^2)}{\sqrt{9 - x^2}}$
$x^{-1/3} + x^{2/3} = x^{-1/3}(1 + x^{1/3})$	$x^{-1/3} + x^{2/3} = x^{-1/3}(1 + x)$
$x^2\sqrt{1 + x} = \sqrt{x^2 + x^3}$	$x^2\sqrt{1 + x} = \sqrt{x^4}\sqrt{1 + x}$ $= \sqrt{x^4 + x^5}$

EXERCISES 4.3

Find the x-intercepts of each curve and find the domain of f. (Hint: In Exercise 5, for example, note that $x^2 - 5x - 6$ cannot be negative. Thus solve $x^2 - 5x - 6 \geq 0$.)

1. $f(x) = \sqrt{x - 1} - 4$ **2.** $f(x) = \sqrt{3x - 2} - 5$ **3.** $f(x) = \sqrt{x^2 + 2x}$

4. $f(x) = \sqrt{x - x^2}$ **5.** $f(x) = \sqrt{x^2 - 5x - 6}$ **6.** $f(x) = \sqrt{2x^2 - x - 2} - 1$

Solve each equation.

7. $\sqrt{4x + 9} - 7 = 0$

8. $2 + \sqrt{7x - 3} = 7$

9. $(3x + 1)^{1/2} = (2x + 6)^{1/2}$

10. $\sqrt{x - 1} - \sqrt{2x - 11} = 0$

11. $\sqrt{x^2 - 36} = 8$

12. $3x = \sqrt{3 - 5x - 3x^2}$

13. $\sqrt{x^2 + \dfrac{1}{2}} = \dfrac{1}{\sqrt{3}}$

14. $3\sqrt{x} = 2\sqrt{3}$

15. $\dfrac{8}{\sqrt{x + 2}} = 4$

16. $\left(\dfrac{5x + 4}{2}\right)^{1/3} = 3$

17. $\dfrac{1}{\sqrt{2x - 1}} = \dfrac{3}{\sqrt{5 - 3x}}$

18. $\sqrt[3]{2x + 7} = 3$

19. $\sqrt[4]{1 - 3x} = \frac{1}{2}$

20. $2 + \sqrt[5]{7x - 4} = 0$

21. $\sqrt{x} + \sqrt{x - 5} = 5$

22. $\sqrt{x - 7} = 7 - \sqrt{x}$

23. $\sqrt{x - 1} + \sqrt{3x - 2} = 3$

24. $\sqrt{10 - x} - \sqrt{x + 3} = 1$

25. $\sqrt{4x + 1} + \sqrt{x + 7} = 6$

26. $\sqrt{x + 4} + \sqrt{3x + 1} = 7$

27. $x\sqrt{4 - x} - \sqrt{9x - 36} = 0$

28. $2x - 5\sqrt{x} - 3 = 0$

29. $x = 8 - 2\sqrt{x}$

30. $\dfrac{(2x + 2)^{3/2}}{(x + 9)^2} - \dfrac{(2x + 2)^{1/2}}{x + 9} = 0$

31. $\sqrt{x^2 - 6x} = x - \sqrt{2x}$

32. $x^{1/3} - 3x^{1/6} + 2 = 0$

33. $4x^{2/3} - 12x^{1/3} + 9 = 0$

34. $\sqrt[4]{3x^2 + 4} = x$

35. $(5x - 6)^{1/5} + \dfrac{x}{(5x - 6)^{4/5}} = 0$

36. $\frac{3}{2}x^{1/2} - \frac{3}{2}x^{-1/2} = 0$

37. $x^{-3/2} - \frac{1}{9}x^{-1/2} = 0$

38. $x^{1/2} + \frac{1}{2}x^{-1/2}(x - 9) = 0$

39. $x^{2/3} + \frac{2}{3}x^{-1/3}(x - 10) = 0$

40. $\frac{1}{2}x^2(x + 5)^{-1/2} + 2x(x + 5)^{1/2} = 0$

*In the following exercises **(a)** find the domain of f, **(b)** determine the signs of f, and **(c)** find the roots of $f(x) = 0$.*

41. $f(x) = (x + 4)\sqrt{x - 2}$

42. $f(x) = (x + 4)\sqrt[3]{x - 2}$

43. $f(x) = (x - 2)^{2/3}$

44. $f(x) = \frac{2}{3}(x - 2)^{-1/3}$

45. $f(x) = \dfrac{9 + x}{9\sqrt{x}}$

46. $f(x) = \dfrac{9 - x}{9x^{3/2}}$

47. $f(x) = \dfrac{(x + 4)\sqrt[3]{x - 2}}{3\sqrt[3]{x}}$

48. $f(x) = \dfrac{-5x(x - 4)}{2\sqrt{5 - x}}$

Take the expression at the left and change it into the equivalent form given at the right.

EXAMPLE

Change $x^{2/3} + \frac{2}{3}x^{-1/3}(x - 10)$ to the form $\dfrac{5(x - 2)}{3\sqrt[3]{x}}$.

$$x^{2/3} + \frac{2}{3}x^{-1/3}(x - 10) = x^{2/3} + \dfrac{2(x - 10)}{3x^{1/3}} = \dfrac{3x^{2/3}x^{1/3} + 2(x - 10)}{3x^{1/3}}$$

$$= \dfrac{3x + 2x - 10}{3\sqrt[3]{x}} = \dfrac{5(x - 2)}{3\sqrt[3]{x}}$$

49. $x^{1/2} + (x - 4)\frac{1}{2}x^{-1/2}$; $\dfrac{3x - 4}{2\sqrt{x}}$

50. $x^{1/3} + (x - 1)\frac{1}{3}x^{-2/3}$; $\dfrac{4x - 1}{3\sqrt[3]{x^2}}$

51. $\frac{1}{2}(4 - x^2)^{-1/2}(-2x)$; $-\dfrac{x}{\sqrt{4 - x^2}}$

52. $\dfrac{-\frac{1}{2}(25 - x^2)^{-1/2}(-2x)}{25 - x^2}$; $\dfrac{x}{\sqrt{(25 - x^2)^3}}$

53. $\dfrac{x}{3(x-1)^{2/3}} + (x-1)^{1/3}; \dfrac{4x-3}{3(\sqrt[3]{x-1})^2}$

54. $(x+1)^{1/2}(2) + (2x+1)\frac{1}{2}(x+1)^{-1/2}; \dfrac{6x+5}{2\sqrt{x+1}}$

Simplify and determine the signs of f.

55. $f(x) = \sqrt{x} - \dfrac{1}{\sqrt{x}}$

56. $f(x) = \frac{3}{2}x^{1/2} - \frac{3}{2}x^{-1/2}$

57. $f(x) = \dfrac{x^2}{2}(x-2)^{-1/2} + 2x(x-2)^{1/2}$

58. $f(x) = \dfrac{\sqrt{x-1} - \dfrac{x}{2\sqrt{x-1}}}{x-1}$

Find the domain and graph the function.

59. $y = f(x) = \sqrt{x^2 - 9}$

60. $y = f(x) = \dfrac{1}{\sqrt{x^2 - 9}}$

61. $y = f(x) = \sqrt{9 - x^2}$

62. $y = f(x) = \dfrac{1}{\sqrt{9 - x^2}}$

63. $y = f(x) = (x-3)\sqrt{x}$

64. $y = (x+3)\sqrt{x}$

65. The radius of a sphere with surface area A is given by the formula $r = \dfrac{1}{2}\sqrt{\dfrac{A}{\pi}}$. Use this formula to solve for A in terms of π and r. Then find A when $r = 2$ centimeters.

66. The radius of a right circular cylinder with height h and volume V is given by this formula: $r = \sqrt{\dfrac{V}{\pi h}}$. Use this formula to solve for V in terms of r and h. Then find V when $r = 2$ centimeters and $h = 3$ centimeters.

67. The slant height s of a right circular cone is given by the formula $s = \sqrt{r^2 + h^2}$, where r is the radius of the base and h is the altitude. Solve for h in terms of r and s, and then find h when $s = 17.23$ cm and $r = 8.96$ cm, to two decimal places.

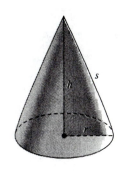

68. In the figure, rectangle $ABCD$ is inscribed in a circle of radius 10. Express the area of the rectangle as a function of x. (*Hint:* solve for y in terms of x.)

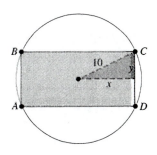

69. The volume of a right circular cylinder with altitude h and radius r is 5 cm³. Solve for r in terms of h and express the surface area of the cylinder as a function of h.

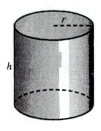

70. A right circular cone with height h and radius r is inscribed in a sphere of radius 1 as shown. Solve for r in terms of h and express the volume of the cone as a function of h.

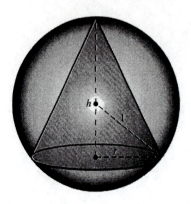

71. The distance from the point $(3, 0)$ to a point $P(x, y)$ on the curve $y = \sqrt{x}$ is $3\sqrt{5}$. Find the coordinates of P. (*Hint:* Use the formula for the distance between two points.)

Solve the system and then sketch each graph.

72. $y = \dfrac{2}{\sqrt{x}}$ **73.** $y = \sqrt{x}$

   $x + 3y = 7$ $y = \tfrac{1}{2}x$

Solve for the indicated variable in terms of others.

74. $y = \sqrt{16 - x^2}$, for x, $0 < x < 4$

75. $x = \tfrac{3}{5}\sqrt{25 - y^2}$, for y, $0 < y < 5$

76. $\dfrac{1}{2\sqrt{x}} + \dfrac{f}{2\sqrt{y}} = 0$, for f

77. $\dfrac{1}{\sqrt{xy}}(xf + y) = f$, for f

 Challenge   The table of values in Example 2, page 268, shows that as x gets larger, the y-values get close to the x-values. Here are two more points: $(20, 19.90)$, $(50, 49.96)$. This suggests that $y = x$ could be an oblique asymptote. Prove that this is the case by showing the difference in the values $y_1 = x$ and $y_2 = \sqrt{x^2 - 4}$ approaches 0 as $x \longrightarrow \infty$. (*Hint:* Multiply $y_1 - y_2 = x - \sqrt{x^2 - 4}$ by the conjugate of the difference divided by itself.)

Written Assignment   Explain what is meant by an extraneous solution to an equation and how such a solution can be generated.

Graphing Calculator Exercise   Verify that the function in Example 4 has no roots by graphing the function.

4.4
COMBINING
AND
DECOMPOSING
FUNCTIONS

Just as we can add, subtract, multiply, and divide real numbers, so can we combine functions. These operations turn out to be relatively simple. For example, consider two functions f and g:

$$f(x) = x^2 + 2x \qquad g(x) = 3x - 2$$

The *sum* of these two functions is designated as $f + g$ and is defined to be $(f + g)(x) = f(x) + g(x)$. That is,

$$(f + g)(x) = f(x) + g(x) = (x^2 + 2x) + (3x - 2) = x^2 + 5x - 2$$

Specifically, for $x = 4$ we have

$$(f + g)(4) = f(4) + g(4) = (16 + 8) + (12 - 2) = 24 + 10 = 34$$

Also, since $(f + g)(x) = x^2 + 5x - 2$, then $(f + g)(4) = 16 + 20 - 2 = 34$.

The *difference* $f - g$, *product* $f \cdot g$, and *quotient* $\dfrac{f}{g}$ are found in a similar manner. Using f and g as given, we have

$$(f - g)(x) = f(x) - g(x) = (x^2 + 2x) - (3x - 2) = x^2 - x + 2$$

$$(f \cdot g)(x) = f(x)g(x) = (x^2 + 2x)(3x - 2) = 3x^3 + 4x^2 - 4x$$

$$\frac{f}{g}(x) = \frac{f(x)}{g(x)} = \frac{x^2 + 2x}{3x - 2} \qquad \left(x \neq \frac{2}{3}\right)$$

For each of f and g, the domain is the set of all real numbers. The same domain applies for $f + g$, $f - g$, and $f \cdot g$. However, for $\dfrac{f}{g}$, the denominator cannot equal 0, so that the domain is the set of all real numbers except $\dfrac{2}{3}$.

EXAMPLE 1   Let $f(x) = \dfrac{1}{x^3 - 1}$ and $g(x) = \sqrt{x}$.

(a) Evaluate $f(4)$, and $g(4)$, and compute $f(4) \cdot g(4)$.

(b) Use the expression for $(f \cdot g)(x)$ to evaluate $(f \cdot g)(4)$. Also state the domain of $f \cdot g$.

(c) Find $\dfrac{f}{g}(x)$ and the domain of $\dfrac{f}{g}$.

Solution

(a) $f(4) = \dfrac{1}{4^3 - 1} = \dfrac{1}{63}$, $g(4) = \sqrt{4} = 2$

$$f(4) \cdot g(4) = \frac{1}{63} \cdot 2 = \frac{2}{63}$$

(b) $(f \cdot g)(x) = f(x)g(x) = \dfrac{1}{x^3 - 1} \cdot \sqrt{x} = \dfrac{\sqrt{x}}{x^3 - 1}$

$(f \cdot g)(4) = \dfrac{\sqrt{4}}{4^3 - 1} = \dfrac{2}{63}$

The domain of f consists of all $x \neq 1$, and g has domain all $x \geq 0$. Thus the domain of $f \cdot g$ consists of all values x that are common to the domains of f and g; that is, $f \cdot g$ has domain all $x \geq 0$ and $x \neq 1$.

(c) $\dfrac{f}{g}(x) = \dfrac{f(x)}{g(x)} = \dfrac{\dfrac{1}{x^3 - 1}}{\sqrt{x}} = \dfrac{1}{(x^3 - 1)\sqrt{x}}$

Domain: All $x > 0$ and $x \neq 1$ ∎

A summary of the general definitions for forming the sum, difference, product, and quotient of two functions follows.

COMBINING FUNCTIONS

For functions f and g:

Sum:	$(f + g)(x) = f(x) + g(x)$
Difference:	$(f - g)(x) = f(x) - g(x)$
Product:	$(f \cdot g)(x) = f(x)g(x)$
Quotient:	$\left(\dfrac{f}{g}\right)(x) = \dfrac{f(x)}{g(x)}$

Each of the domains of $f + g$, $f - g$, and $f \cdot g$ consists of all x common to the domains of f and g. The domain of $\dfrac{f}{g}$ has all x common to the domains of f and g except for those x where $g(x) = 0$.

There is another way to combine functions. Let f and g be given by

$$f(x) = x^2 + 2x \quad \text{and} \quad g(x) = 3x - 2$$

Now take a specific domain value of x, such as $x = 3$. Then $f(3) = 9 + 6 = 15$. Use 15 as a domain value of g to get $g(15) = 45 - 2 = 43$. This work may be condensed as follows:

$$g(f(3)) = g(15) = 43 \quad \text{Read this as "g of f of 3."}$$

The output of f becomes the input of g: $f(3) = 15$ comes out of f and goes into g.

That is, the domain value of f, $x = 3$, gives a corresponding range value of 15. This value, 15, then becomes a domain value for g.

The roles of f and g may be interchanged. Thus

$$f(g(3)) = f(7) = 63 \qquad \text{Read this as "f of g of 3."}$$

EXAMPLE 2   Let $f(x) = \dfrac{1}{x-2}$ and $g(x) = \sqrt{x}$. Find **(a)** $g(f(6))$ and **(b)** $f(g(6))$.

Solution

(a) $g(f(6)) = g\left(\frac{1}{4}\right) = \sqrt{\frac{1}{4}} = \frac{1}{2}$ **(b)** $f(g(6)) = f\left(\sqrt{6}\right) = \dfrac{1}{\sqrt{6}-2}$ ■

The preceding procedure that we have been using is referred to as the **composition of functions.** Note that the composition of functions does not always produce a real number. For instance, in Example 2, if $x = -3$, then $f(-3) = -\frac{1}{5}$; $g\left(-\frac{1}{5}\right) = \sqrt{-\frac{1}{5}}$ is not a real number. We therefore say that $g(f(-3))$ is undefined or that it does not exist.

TEST YOUR UNDERSTANDING	*For each pair of functions f and g, evaluate (if possible) each of the following:*
	(a) $g(f(1))$ **(b)** $f(g(1))$ **(c)** $f(g(0))$ **(d)** $g(f(-2))$
	1. $f(x) = 3x - 1$; $g(x) = x^2 + 4$ **2.** $f(x) = \sqrt{x}$; $g(x) = x^2$
(Answers: Page 303)	**3.** $f(x) = \sqrt[3]{3x - 1}$; $g(x) = 5x$ **4.** $f(x) = \dfrac{x+2}{x-1}$; $g(x) = x^3$

Now consider the composition of functions for *any* allowable x. As in Example 2, use $f(x) = \dfrac{1}{x-2}$ and $g(x) = \sqrt{x}$. Then

$$g(f(x)) = g\left(\frac{1}{x-2}\right) = \sqrt{\frac{1}{x-2}} = \frac{1}{\sqrt{x-2}}$$

This new correspondence between a domain value x and the range value $\dfrac{1}{\sqrt{x-2}}$ is referred to as the **composite function of g by f** (or the *composition of g by f*). This composite function is denoted by $g \circ f$. That is:

$g(f(x))$ is read as "g of f of x."

$$(g \circ f)(x) = g(f(x)) = \frac{1}{\sqrt{x-2}}$$

The domain of $g \circ f$ will consist of all values x in the domain of f such that $f(x)$ is in the domain of g. Since $f(x) = \dfrac{1}{x-2}$ has domain all $x \neq 2$ and the domain of $g(x) = \sqrt{x}$ is all $x \geq 0$, the domain of $g \circ f$ is all $x \neq 2$ for which $\dfrac{1}{x-2}$ is positive, that is, all $x > 2$.

Reversing the roles of the two functions gives the composite of f by g, where

$$(f \circ g)(x) = f(g(x)) = f(\sqrt{x}) = \frac{1}{\sqrt{x} - 2}$$

$f(g(x))$ *is read as "f of g of x."*

The domain of $f \circ g$ consists of all $x \geq 0$ and $x \neq 4$.

Following is the definition of composite functions:

DEFINITION OF COMPOSITE FUNCTION

For functions f and g the composite function of g by f, denoted $g \circ f$, has range values defined by

$$(g \circ f)(x) = g(f(x))$$

and domain consisting of all x in the domain of f for which $f(x)$ is in the domain of g.

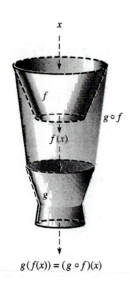

$g(f(x)) = (g \circ f)(x)$

It may help you to remember the construction of composites by looking at the following schematic diagram, as well as the figure in the margin.

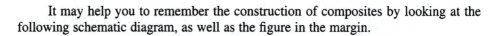

It is also helpful to view the composition $(g \circ f)(x) = g(f(x))$ as consisting of an "inner" function f and an "outer" function g.

EXAMPLE 3 Form the composite functions $f \circ g$ and $g \circ f$ and give their domains, where $f(x) = \dfrac{1}{x^2 - 1}$ and $g(x) = \sqrt{x}$.

Solution We find that $f \circ g$ is given by

$$(f \circ g)(x) = f(g(x)) = f(\sqrt{x}) = \frac{1}{(\sqrt{x})^2 - 1} = \frac{1}{x - 1}$$

The domain of $f \circ g$ excludes $x < 0$ since such x are not in the domain of g. And it excludes $x = 1$ since $g(1) = 1$, which is not in the domain of f. Therefore, the domain of $f \circ g$ is $x \geq 0$ and $x \neq 1$. Also, $g \circ f$ is given by

From example 3 it follows that $(f \circ g)(x) \neq (g \circ f)(x)$. We conclude that, in general, the composite of f by g is not equal to the composite of g by f; that is, $f(g(x)) \neq g(f(x))$. Composition is not commutative.

$$(g \circ f)(x) = g(f(x)) = g\left(\frac{1}{x^2 - 1}\right) = \sqrt{\frac{1}{x^2 - 1}} = \frac{1}{\sqrt{x^2 - 1}}$$

The domain of $g \circ f$ excludes $x = \pm 1$ since these are not in the domain of f. And it excludes $-1 < x < 1$ since negative values of $f(x)$ are not in the domain of g. Therefore, the domain of $g \circ f$ is seen to be $x < -1$ or $x > 1$. ∎

The composition of functions may be extended to include more than two functions.

EXAMPLE 4 Let $f(x) = \sqrt{x}$, $g(x) = x^2 + 1$, and $h(x) = \dfrac{1}{x}$. Find the composition of f by g by h, denoted $f \circ g \circ h$.

Solution

$$(f \circ g \circ h)(x) = f(g(h(x))) \qquad h \text{ is the "inner" function.}$$

$$= f\left(g\left(\frac{1}{x}\right)\right) \qquad g \text{ is the "middle" function.}$$

$$= f\left(\frac{1}{x^2} + 1\right) \qquad f \text{ is the "outer" function.}$$

$$= \sqrt{\frac{1}{x^2} + 1} \qquad\qquad\qquad \blacksquare$$

One of the most useful skills needed in the study of calculus is the ability to recognize that a given function may be viewed as the **composition** of two or more functions. For instance, let h be given by

$$h(x) = \sqrt{x^2 + 2x + 2}$$

If we let $f(x) = x^2 + 2x + 2$ and $g(x) = \sqrt{x}$, then the composite g by f is

$$(g \circ f)(x) = g(f(x))$$
$$= g(x^2 + 2x + 2)$$
$$= \sqrt{x^2 + 2x + 2} = h(x)$$

Thus the given function h has been **decomposed** into the composition of the two functions f and g. Such decompositions are not unique. More than one decomposition is possible. For h we may also use

$$t(x) = x^2 + 2x \qquad s(x) = \sqrt{x + 2}$$

Then

You would most likely agree that the first of these decompositions is more "natural." Just which decomposition one is to choose will, in later work, depend on the situation.

$$(s \circ t)(x) = s(t(x))$$
$$= s(x^2 + 2x)$$
$$= \sqrt{x^2 + 2x + 2} = h(x)$$

EXAMPLE 5 Find f and g such that $h = f \circ g$, where $h(x) = \left(\dfrac{1}{3x-1}\right)^5$ and the inner function g is rational.

Solution Let $g(x) = \dfrac{1}{3x-1}$ and $f(x) = x^5$.

$$(f \circ g)(x) = f(g(x))$$

$$= f\left(\frac{1}{3x-1}\right)$$

$$= \left(\frac{1}{3x-1}\right)^5 = h(x) \qquad \blacksquare$$

As another example of a decomposition where the inner function is a binomial, let $g(x) = 3x - 1$ and $f(x) = \dfrac{1}{x^5}$.

EXAMPLE 6 Decompose $h(x) = \sqrt{(x^2 - 3x)^5}$ into two functions so that the outer function is a monomial.

Solution Write $h(x) = (\sqrt{x^2 - 3x})^5$. Let $f(x) = \sqrt{x^2 - 3x}$ and $g(x) = x^5$.

$$(g \circ f)(x) = g(f(x))$$

$$= g(\sqrt{x^2 - 3x})$$

$$= (\sqrt{x^2 - 3x})^5$$

$$= h(x) \qquad \blacksquare$$

TEST YOUR UNDERSTANDING

(Answers: Page 304)

For each function h, find functions f and g so that $h = g \circ f$.

1. $h(x) = (8x - 3)^5$ **2.** $h(x) = \sqrt[5]{8x - 3}$ **3.** $h(x) = \sqrt{\dfrac{1}{8x - 3}}$

4. $h(x) = \left(\dfrac{5}{7 + 4x^2}\right)^3$ **5.** $h(x) = \dfrac{(2x+1)^4}{(2x-1)^4}$ **6.** $h(x) = \sqrt{(x^4 - 2x^2 + 1)^3}$

Example 6 showed a way of decomposing $h(x) = \sqrt{(x^2 - 3x)^5}$ into the composition of two functions. It is also possible to express h as the composition of three functions. For example, we may use these functions.

$$f(x) = x^2 - 3x \qquad g(x) = x^5 \qquad t(x) = \sqrt{x}$$

Then

$$(t \circ g \circ f)(x) = t(g(f(x))) \qquad \text{f is the "inner" function.}$$

$$= t(g(x^2 - 3x)) \qquad \text{g is the "middle" function.}$$

$$= t((x^2 - 3x)^5) \qquad \text{t is the "outer" function.}$$

$$= \sqrt{(x^2 - 3x)^5}$$

$$= h(x)$$

The final example for this section is an application of the composition of functions.

EXAMPLE 7 The figure shows a rocket that has been launched from a point P on the ground. The path of the rocket is being tracked by a camera at a point C on the ground that is 1500 feet from P. Let h be the height of the rocket after t seconds and assume that it is rising at a rate of 600 feet per second. Use d for the distance from the camera to the rocket and write d as a function of h, h as a function of t, and interpret $(d \circ h)(t)$. Also, approximate the height of the rocket after 5 seconds.

Space Shuttle climbs from Pad 39A toward orbit.
Source: NASA.

Solution By the Pythagorean theorem, $d(h) = \sqrt{1500^2 + h^2}$. Since h is increasing at the rate of 600 ft/sec, $h(t) = 600t$. Thus

$$(d \circ h)(t) = d(h(t))$$
$$= d(600t)$$
$$= \sqrt{1500^2 + (600t)^2}$$

The expression $(d \circ h)(t)$ is the distance in feet that the rocket is from point C after t seconds. Then, letting $t = 5$, the distance is given by

$$(d \circ h)(5) = \sqrt{1500^2 + (600 \cdot 5)^2}$$

which is approximately equal to 3350 ft. ∎

EXERCISES 4.4

In Exercises 1–6, let $f(x) = 2x - 3$ *and* $g(x) = 3x + 2$.

1. **(a)** Find $f(1)$, $g(1)$, and $f(1) + g(1)$.
   (b) Find $(f + g)(x)$ and state the domain of $f + g$.
   (c) Use the result in part (b) to evaluate $(f + g)(1)$.

2. **(a)** Find $g(2)$, $f(2)$, and $g(2) - f(2)$.
   (b) Find $(g - f)(x)$ and state the domain of $g - f$.
   (c) Use the result in part (b) to evaluate $(g - f)(2)$.

3. **(a)** Find $f\left(\frac{1}{2}\right)$, $g\left(\frac{1}{2}\right)$, and $f\left(\frac{1}{2}\right) \cdot g\left(\frac{1}{2}\right)$.
   (b) Find $(f \cdot g)(x)$ and state the domain of $f \cdot g$.
   (c) Use the result in part (b) to evaluate $(f \cdot g)\left(\frac{1}{2}\right)$.

4. **(a)** Find $g(-2)$, $f(-2)$, and $\dfrac{g(-2)}{f(-2)}$.
   (b) Find $\dfrac{g}{f}(x)$ and state the domain of $\dfrac{g}{f}$.
   (c) Use the result in part (b) to evaluate $\dfrac{g}{f}(-2)$.

5. **(a)** Find $g(0)$ and $f(g(0))$.
   (b) Find $(f \circ g)(x)$ and state the domain of $f \circ g$.
   (c) Use the result in part (b) to evaluate $(f \circ g)(0)$.

6. **(a)** Find $f(0)$ and $g(f(0))$.
   (b) Find $(g \circ f)(x)$ and state the domain of $g \circ f$.
   (c) Use the result in part (b) to evaluate $(g \circ f)(0)$.

For each pair of functions, find the following:

   (a) $(f + g)(x)$; domain of $f + g$.
   (b) $\left(\dfrac{f}{g}\right)(x)$; domain $\dfrac{f}{g}$.
   (c) $(f \circ g)(x)$; domain of $f \circ g$.

7. $f(x) = x^2$, $g(x) = \sqrt{x}$

8. $f(x) = 5x - 1$, $g(x) = \dfrac{5}{1 + 3x}$

9. $f(x) = x^3 - 1$, $g(x) = \dfrac{1}{x}$

10. $f(x) = 3x - 1$, $g(x) = \frac{1}{3}x + \frac{1}{3}$

11. $f(x) = x^2 + 6x + 8$, $g(x) = \sqrt{x - 2}$

12. $f(x) = \sqrt[3]{x}$, $g(x) = x^2$

For each pair of functions, find the following:

   (a) $(g - f)(x)$; domain of $g - f$.
   (b) $(g \cdot f)(x)$; domain of $g \cdot f$.
   (c) $(g \circ f)(x)$; domain of $g \circ f$.

13. $f(x) = -2x + 5, g(x) = 4x - 1$

14. $f(x) = |x|, g(x) = 3|x|$

15. $f(x) = 2x^2 - 1, g(x) = \dfrac{1}{2x}$

16. $f(x) = \sqrt{2x + 3}, g(x) = x^2 - 1$

Find $(f \circ g)(x)$ and $(g \circ f)(x)$.

17. $f(x) = x^2, g(x) = x - 1$

18. $f(x) = |x - 3|, g(x) = 2x + 3$

19. $f(x) = \dfrac{x}{x - 2}, \ g(x) = \dfrac{x + 3}{x}$

20. $f(x) = x^3 - 1, \ g(x) = \dfrac{1}{x^3 + 1}$

21. $f(x) = \sqrt{x + 1}, g(x) = x^4 - 1$

22. $f(x) = 2x^3 - 1, g(x) = \sqrt[3]{\dfrac{x + 1}{2}}$

23. $f(x) = \sqrt{x}, \ g(x) = 4$

24. $f(x) = \sqrt[3]{1 - x}, \ g(x) = 1 - x^3$

25. Let $f(x) = \dfrac{1}{x}, g(x) = 2x - 1$, and $h(x) = x^{1/3}$. Find the following:

  (a) $(f \circ g \circ h)(x)$ **(b)** $(g \circ f \circ h)(x)$ **(c)** $(h \circ f \circ g)(x)$

26. Let $f(x) = x + 2, g(x) = \sqrt{x}$ and $h(x) = x^3$. Find the following:

  (a) $(f \circ g \circ h)(x)$ **(b)** $(f \circ h \circ g)(x)$ **(c)** $(g \circ f \circ h)(x)$

  (d) $(g \circ h \circ f)(x)$ **(e)** $(h \circ f \circ g)(x)$ **(f)** $(h \circ g \circ f)(x)$

27. Let $f(x) = \dfrac{1}{x}$. Find $(f \circ f)(x)$ and $(f \circ f \circ f)(x)$.

28. Let $f(x) = x^2, g(x) = \dfrac{1}{x - 1}$, and $h(x) = 1 + \dfrac{1}{x}$. Find the following:

  (a) $(f \circ h \circ g)(x)$ **(b)** $(g \circ h \circ f)(x)$ **(c)** $(h \circ g \circ f)(x)$

Find functions f and g so that $h(x) = (f \circ g)(x)$. In each case let the inner function g be a polynomial or a rational function.

29. $h(x) = (3x + 1)^2$

30. $h(x) = (x^2 - 2x)^3$

31. $h(x) = \sqrt{1 - 4x}$

32. $h(x) = \sqrt[3]{x^2 - 1}$

33. $h(x) = \left(\dfrac{x + 1}{x - 1}\right)^2$

34. $h(x) = \left(\dfrac{1 - 2x}{1 + 2x}\right)^3$

35. $h(x) = (3x^2 - 1)^{-3}$

36. $h(x) = \left(1 + \dfrac{1}{x}\right)^{-2}$

37. $h(x) = \sqrt{\dfrac{x}{x - 1}}$

38. $h(x) = \sqrt[3]{\dfrac{x - 1}{x}}$

39. $h(x) = \sqrt{(x^2 - x - 1)^3}$

40. $h(x) = \sqrt[3]{(1 - x^4)^2}$

41. $h(x) = \dfrac{2}{\sqrt{4 - x^2}}$

42. $h(x) = -\left(\dfrac{3}{x - 1}\right)^5$

Find three functions, f, g, and h such that $k(x) = (h \circ g \circ f)(x)$.

43. $k(x) = \left(\sqrt{2x + 1}\right)^3$

44. $k(x) = \sqrt[3]{(2x - 1)^2}$

45. $k(x) = \sqrt{\left(\dfrac{x}{x + 1}\right)^5}$

46. $k(x) = \left(\sqrt[7]{\dfrac{x^2 - 1}{x^2 + 1}}\right)^4$

47. $k(x) = (x^2 - 9)^{2/3}$

48. $k(x) = (5 - 3x)^{5/2}$

49. $k(x) = -\sqrt{(x^2 - 4x + 7)^3}$

50. $k(x) = -\left(\dfrac{2x}{1 - x}\right)^{2/5}$

51. $k(x) = \left(1 + \sqrt{2x - 11}\right)^2$

52. $k(x) = \sqrt[3]{(x^2 - 4)^5 - 1}$

53. Find f so that $((x + 1)^2 + 1)^2 = (f \circ f)(x)$.

54. Find f so that $\sqrt{1 + \sqrt{1 + \sqrt{1 + x}}} = (f \circ f \circ f)(x)$.

55. Let $f(x) = x^3$. Find a function g so that $(f \circ g)(x) = x$ and $(g \circ f)(x) = x$.

56. Let $f(x) = x$. Find $(f \circ g)(x)$ and $(g \circ f)(x)$ for any function g.

57. If $f(x) = 2x + 1$, find $g(x)$ so that $(f \circ g)(x) = 2x^2 - 4x + 1$.

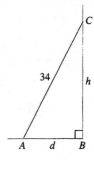

58. A 34-foot ladder is leaning against a wall. The foot of the ladder is moving away from point B at the base of the wall at the rate of 2 ft/sec. Express the height $h = BC$ as a function of the distance $d = BA$, and express d as a function of the time t. Interpret $(h \circ d)(t)$ and find $(h \circ d)(8)$.

59. Gas is being pumped into a spherical balloon at the rate of 50 cubic feet per minute. Express the radius r as a function of the volume V, and V as a function of t. Then interpret $(r \circ V)(t)$ where t is the time in minutes. Approximate $(r \circ V)(t)$ when $t = 10$ minutes.

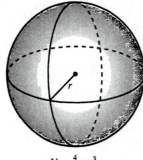

$$V = \tfrac{4}{3}\pi r^3$$

 Challenge   Let $f(x) = \dfrac{1}{x^2 - 4x + 4}$. Find a function g such that $(f \circ g)(x)$ is a polynomial.

 Written Assignment   Explain, with specific examples, the distinction between the composition and the decomposition of functions.

 Graphing Calculator Exercises   For Exercises 1 and 2, let $f(x) = x^2 - 1$ and $g(x) = x + 2$.

1. (a) Graph $y = f(x)g(x)$ on an interval including the zeros of $f(x)$ and those of $g(x)$.
  (b) What do you notice about these zeros and the graph of the product?

2. (a) Graph $y = f(x)/g(x)$ on an interval including the zeros of $f(x)$ and those of $g(x)$.
  (b) What do you notice about these zeros and the graph of the quotient?

3. Let $f(x) = |x|$ and $g(x) = x - 2$. Graph the given composition and describe its relation to $y = |x|$:
  (a) $y = f(g(x))$ **(b)** $y = g(f(x))$

4. The graph of a function is symmetric about the vertical line $x = c$ provided that $f(c + t) = f(c - t)$ for all t such that $c + t$ is in the domain of f. Verify by graphing that $g(x) = \sqrt{|x - 2|}$ is symmetric about the line $x = 2$.

Critical Thinking

1. Certain operations on an equation can produce extraneous roots. Others can have the effect of losing a root. Explain, with examples, how each of these situations can occur.
2. If $f(x) = x^2$, what is $f(f(x))$? Find a formula for $f(f(\cdots f(x))\cdots)$ where f appears n times.
3. In general, for two functions $f(x)$ and $g(x)$, $f(g(x)) \neq g(f(x))$. Show that for the functions $f(x) = 3x - 2$ and $g(x) = \frac{x+2}{3}$ the equality does hold. Explain what that means in terms of the graphs of the two functions.
4. Sketch graphs of two functions $f(x)$ and $g(x)$ such that the domain of $f(x)$ is equal to the range of $g(x)$ and the domain of $g(x)$ is equal to the range of $f(x)$. Furthermore, for each function, the domain and range are not to be the same set of numbers.
5. Is it true that $\sqrt[3]{|x|} = |\sqrt[3]{x}|$ for all real numbers x? Justify your answer.

4.5 INVERSE FUNCTIONS

A real number a and its negative $-a$ are said to be *inverse elements* with respect to the operation of addition because $a + (-a) = 0$. Similarly, a nonzero real number a and its reciprocal are inverse elements with respect to multiplication since $a \cdot \frac{1}{a} = 1$. Now we are going to extend this concept and learn that there are *inverse functions* with respect to the operation of composition.

First recall that, by definition, a function has each domain value x corresponding to exactly one range value y. Some (but not all) functions have the additional property that to every range value y there corresponds exactly one domain value x. Such functions are said to be **one-to-one functions.** To understand this concept, consider the functions $y = x^2$ and $y = x^3$.

For a one-to-one function you can start at a range value y and trace it back to exactly one domain value x.

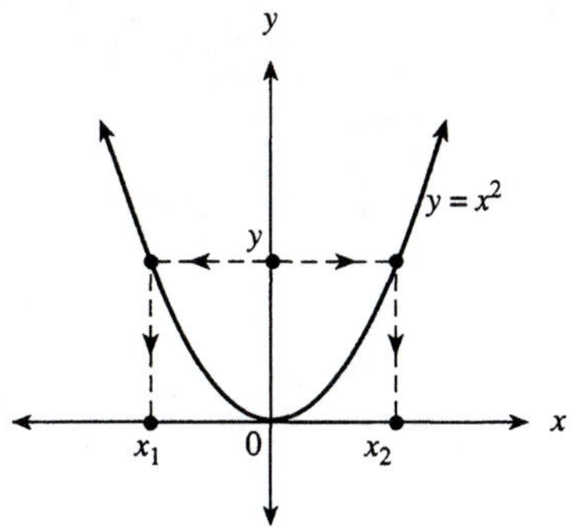

$y = x^2$ *is not* a one-to-one function. There are two domain values for a range value $y > 0$.

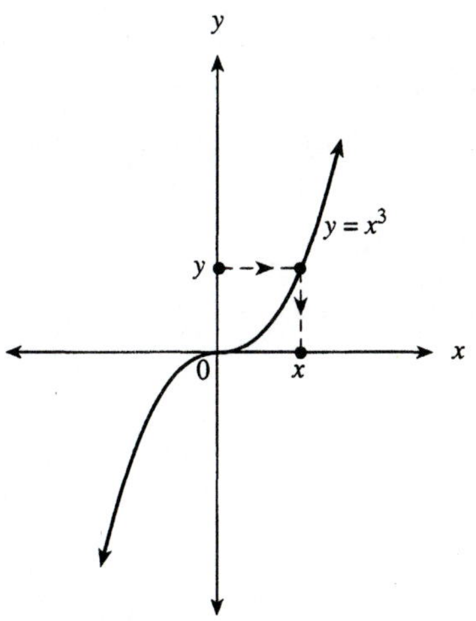

$y = x^3$ *is* a one-to-one function. There is exactly one domain value x for each range value y.

DEFINITION OF A ONE-TO-ONE FUNCTION

A function f is a one-to-one function if and only if for each range value there corresponds exactly one domain value.

Once a graph of a function is known, there is a simple **horizontal line test** for determining the one-to-one property. Consider a horizontal line through each range value y, as in the following figures. If the line meets the curve exactly once, then we have a one-to-one function; otherwise, it is not one-to-one.

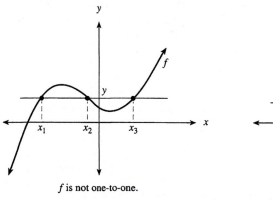

f is not one-to-one.

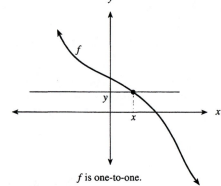

f is one-to-one.

HORIZONTAL LINE TEST FOR ONE-TO-ONE FUNCTIONS

A function f is one-to-one if and only if the horizontal lines through the range values intersect the graph of f in exactly one point.

TEST YOUR UNDERSTANDING

Use the horizontal line test to determine which of the following are one-to-one functions.

1. $y = x^2 - 2x + 1$ **2.** $y = \sqrt{x}$ **3.** $y = \dfrac{1}{x}$

4. $y = |x|$ **5.** $y = 2x + 1$ **6.** $y = \sqrt[3]{x}$

7. $y = [x]$ **8.** $y = \dfrac{1}{x^2}$ **9.** $y = \begin{cases} x + 1 \text{ if } -1 \le x < 2 \\ -\dfrac{1}{2}x + 3 \text{ if } 2 \le x < 4 \end{cases}$

(Answers: Page 304)

The one-to-one property of a function f can also be stated algebraically in the following equivalent forms:

1. If $x_1 \ne x_2$, then $f(x_1) \ne f(x_2)$.

(If two domain values are unequal, then the corresponding range values are unequal.)

2. If $f(x_1) = f(x_2)$, then $x_1 = x_2$.

(If two range values are equal, then the corresponding domain values are equal.)

EXAMPLE 1   Prove algebraically that $f(x) = -5x + 7$ is a one-to-one function.

Solution   Using condition 2, assume that $f(x_1) = f(x_2)$, then

$$-5x_1 + 7 = -5x_2 + 7$$

$$-5x_1 = -5x_2$$

$$x_1 = x_2$$

Therefore, f is one-to-one.   ■

The graphs of $y = x^3$ and $x = y^3$ were used on page 263 to introduce the radical function $y = \sqrt[3]{x}$. Now they will be used to introduce the inverse function concept. We begin with the curve $y = x^3$ and interchange variables to obtain $x = y^3$. In the following figures the two curves are shown separately and then brought together on the same set of axes in figure (c). Because of the interchange of the variables, the two curves are *reflections* of each other in the line $y = x$. Another way to describe this relationship is to say that they are *mirror images* of one another through the "mirror line" $y = x$. For example, the point (2, 8) on $y = x^3$ and the point (8, 2) on $x = y^3$ are reflections of each other.

If the paper were folded along the line $y = x$, the two curves would coincide.

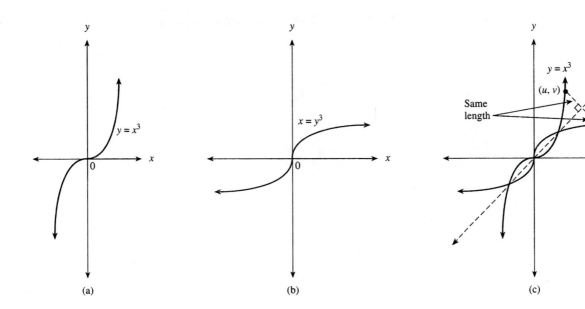

(a)   (b)   (c)

We began with the one-to-one function $y = x^3$ and by interchanging variables obtained $x = y^3$. This equation can be solved for y by taking the cube root of both sides:

$$x = y^3 \longrightarrow y = \sqrt[3]{x}$$

These two equations are equivalent and therefore define the same function of x. However, since $y = \sqrt[3]{x}$ shows *explicitly* how y depends on x, it is the preferred form.

Now let $f(x) = x^3$ and $g(x) = \sqrt[3]{x}$. Then, if the composites $f \circ g$ and $g \circ f$ are formed, something surprising happens.

Function f cubes and function g "uncubes"; f and g are inverse functions.

$$(f \circ g)(x) = f(g(x)) = f(\sqrt[3]{x}) = (\sqrt[3]{x})^3 = x$$
$$(g \circ f)(x) = g(f(x)) = g(x^3) = \sqrt[3]{x^3} = x$$

In each case we obtained the same value x that we started with; whatever one of the functions does to a value x, the other function undoes. Whenever two functions act on each other in such a manner, we say that they are **inverse functions** or that either function is the inverse of the other. Consequently, the two functions

$$y = f(x) = x^3 \quad \text{and} \quad y = g(x) = \sqrt[3]{x}$$

are inverse functions; each is the inverse of the other. Note that a function has to be a one-to-one function in order to have an inverse function. For example, $y = x^2$ is *not* one-to-one. Its reflection in the line $y = x$ has the equation $x = y^2$, or $y = \pm\sqrt{x}$, which is *not* a function of x. (Why not?)

The preceding work leads to the formal definition of inverse functions.

DEFINITION OF INVERSE FUNCTIONS

Two functions f and g are said to be *inverse functions* if and only if:

1. For each x in the domain of g, $g(x)$ is in the domain of f and

$$(f \circ g)(x) = f(g(x)) = x$$

2. For each x in the domain of f, $f(x)$ is in the domain of g and

$$(g \circ f)(x) = g(f(x)) = x$$

The notation f^{-1} is often used to represent the inverse of f. For example, if $f(x) = x^3$, then $f^{-1}(x) = \sqrt[3]{x}$. We read f^{-1} as "the inverse of f" or as "f inverse." Note that if $f(a) = b$, then $f^{-1}(b) = a$. For example, let $a = 2$:

$$f(x) = x^3: \qquad f(2) = 2^3 = 8$$
$$f^{-1}(x) = \sqrt[3]{x}: \qquad f^{-1}(8) = \sqrt[3]{8} = 2$$

Also, using this notation and the definition of inverse functions, we have

$$f(f^{-1}(x)) = x \quad \text{and} \quad f^{-1}(f(x)) = x$$

Thus

$$f(f^{-1}(8)) = 8 \quad \text{and} \quad f^{-1}(f(2)) = 2$$

It turns out (as suggested by our work with $y = x^3$) that *every one-to-one function f has an inverse g and that their graphs are reflections of each other through the line $y = x$.* The technique of interchanging variables, used to obtain $y = \sqrt[3]{x}$ from $y = x^3$, can be applied to many situations by following this procedure:

To find the inverse g of a function f:

1. Begin with $y = f(x)$.
2. Interchange variables to obtain $x = f(y)$.
3. Solve for y in terms of x to obtain $y = g(x) = f^{-1}(x)$.

EXAMPLE 2 Find the inverse g of $y = f(x) = 2x + 3$. Then show that $(f \circ g)(x) = x$ and $(g \circ f)(x) = x$ and graph both functions on the same axes.

Solution To find the inverse of f, begin with $y = 2x + 3$. Then interchange variables and solve for y in terms of x.

$$y = 2x + 3$$

$$x = 2y + 3$$

$$2y = x - 3$$

$$y = \tfrac{1}{2}x - \tfrac{3}{2} \quad \text{or} \quad y = g(x) = \tfrac{1}{2}x - \tfrac{3}{2}$$

Using $y = g(x) = \tfrac{1}{2}x - \tfrac{3}{2}$, we have

$$(f \circ g)(x) = f(g(x)) = f\left(\tfrac{1}{2}x - \tfrac{3}{2}\right)$$

$$= 2\left(\tfrac{1}{2}x - \tfrac{3}{2}\right) + 3$$

$$= x - 3 + 3$$

$$= x$$

$$(g \circ f)(x) = g(f(x)) = g(2x + 3)$$

$$= \tfrac{1}{2}(2x + 3) - \tfrac{3}{2}$$

$$= x + \tfrac{3}{2} - \tfrac{3}{2}$$

$$= x$$

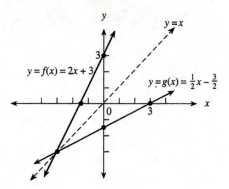

■

EXAMPLE 3   Follow the instructions in Example 2 for $y = f(x) = \sqrt{x}$.

Solution Interchange variables in $y = \sqrt{x}$ to get $x = \sqrt{y}$ and note that x cannot be negative: $x \geq 0$. Solving for y by squaring produces $y = x^2$. Therefore, the inverse function is $y = g(x) = x^2$ with domain $x \geq 0$.

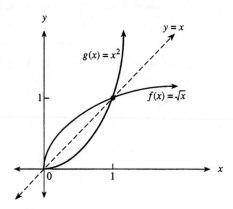

Using $f(x) = \sqrt{x}$ and $g(x) = x^2$, we have:

$$(f \circ g)(x) = f(g(x)) = f(x^2) = \sqrt{x^2} = |x| = x \qquad \text{since } x \geq 0$$

$$(g \circ f)(x) = g(f(x)) = g\left(\sqrt{x}\right) = \left(\sqrt{x}\right)^2 = x$$

■

The technique of interchanging the variables in a one-to-one function f to obtain the inverse $y = f^{-1}(x)$ causes the domains and ranges to be interchanged. That is,

$$\text{Domain of } f = \text{Range of } f^{-1}$$

$$\text{Range of } f = \text{Domain of } f^{-1}$$

This observation is used in the next example.

EXAMPLE 4   Find $f^{-1}(x)$ and its domain and range for:

(a) $f(x) = \dfrac{1}{x-3}$ (b) $f(x) = x^{3/2} - 3$

Solution

(a)
$$y = \frac{1}{x-3}$$

$$x = \frac{1}{y-3} \qquad \text{Interchange variables.}$$

$$xy - 3x = 1 \qquad \text{Multiply by } y - 3.$$

$$xy = 3x + 1$$

$$y = 3 + \frac{1}{x} \qquad \text{Divide by } x.$$

$$\text{or } f^{-1}(x) = 3 + \frac{1}{x}$$

The graph and the domain and range of $y = \dfrac{1}{x-3}$ *is given on page* 183.

Domain of f = all numbers $\neq 3$ = Range of f^{-1}

Range of f = all numbers $\neq 0$ = Domain of f^{-1}

(b)
$$y = x^{3/2} - 3$$

$$x = y^{3/2} - 3$$

$$y^{3/2} = x + 3$$

$$y = (x + 3)^{2/3} \qquad \text{Raise both sides to the } \tfrac{2}{3} \text{ power.}$$

$$\text{or } f^{-1}(x) = (x + 3)^{2/3}$$

Domain of f = all nonnegative numbers = Range of f^{-1}

Range of f = all numbers ≥ -3 = Domain of f^{-1}

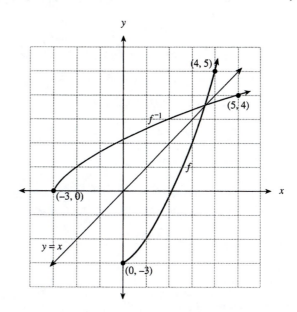

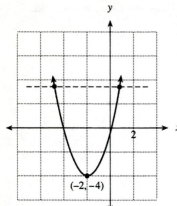

It is important to note that not every function has an inverse. For example, the parabola $y = x^2 + 4x$ is not a one-to-one function and consequently does not have an inverse function. However, if x is restricted so that $x \geq -2$, then $y = f(x) = x^2 + 4x$ is one-to-one and has an inverse. The graph of f^{-1} can be found by reflecting the graph of f through the line $y = x$ as shown.

Domain of f: $x \geq -2$ Range of f^{-1}: $y \geq -2$

Range of f: $y \geq -4$ Domain of f^{-1}: $x \geq -4$

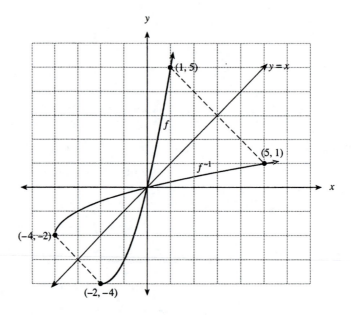

To find an algebraic form of the inverse, proceed as in the earlier examples. In this case, the method of completing the square is needed.

$$y = x^2 + 4x$$

$$x = y^2 + 4y \qquad \text{Interchange variables.}$$

$$x + 4 = y^2 + 4y + 4 \qquad \text{Complete the square in } y.$$

$$x + 4 = (y + 2)^2$$

$$\pm\sqrt{x + 4} = y + 2 \qquad \text{Square root property for equations; page 148.}$$

$$y = -2 \pm \sqrt{x + 4}$$

Since the range of f^{-1} is $y \geq -2$ and since $\sqrt{x + 4} \geq 0$, we must use the plus sign to obtain the equation for the inverse:

$$y = f^{-1}(x) = -2 + \sqrt{x + 4}$$

EXERCISES 4.5

Determine if f is one-to-one. If so, sketch its inverse by reflecting through the line y = x.

1.
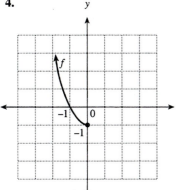

2.
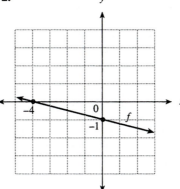

3.
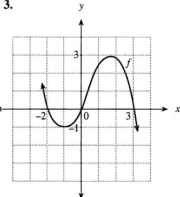

4.
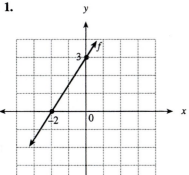

5.
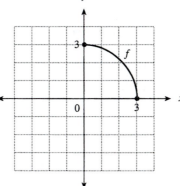

6.
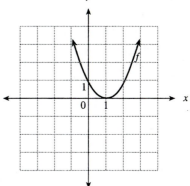

Use the horizontal line test to decide if the function is one-to-one.

7. $f(x) = (x - 1)^2$ **8.** $f(x) = x^3 - 3x^2$ **9.** $f(x) = \dfrac{1 - x}{x}$

10. $f(x) = \dfrac{1}{x - 2}$ **11.** $f(x) = \sqrt{x} + 3$ **12.** $f(x) = \dfrac{1}{x^2 - x - 2}$

Prove algebraically that the given function is one-to-one.

13. $f(x) = \frac{3}{5}(x - 4) + 6$ **14.** $f(x) = \dfrac{3x - 8}{x - 3}$

Show that f and g are inverse functions according to the criteria $(f \circ g)(x) = x$ and $(g \circ f)(x) = x$. Then graph both functions and the line $y = x$ on the same axes.

15. $f(x) = \frac{1}{3}x - 3$; $g(x) = 3x + 9$ **16.** $f(x) = 2x - 6$; $g(x) = \frac{1}{2}x + 3$

17. $f(x) = (x + 1)^3$; $g(x) = \sqrt[3]{x} - 1$ **18.** $f(x) = -(x + 2)^3$; $g(x) = -\sqrt[3]{x} - 2$

19. $f(x) = \dfrac{1}{x - 1}$; $g(x) = \dfrac{1}{x} + 1$ **20.** $f(x) = x^2 + 2$ for $x \geq 0$; $g(x) = \sqrt{x - 2}$

Find the inverse function g of the given function f.

21. $y = f(x) = (x - 5)^3$ **22.** $y = f(x) = x^{1/3} - 3$ **23.** $y = f(x) = \frac{2}{3}x - 1$

24. $y = f(x) = -4x + \frac{2}{5}$ **25.** $y = f(x) = (x - 1)^5$ **26.** $y = f(x) = -x^5$

27. $y = f(x) = x^{3/5}$ **28.** $y = f(x) = x^{5/3} + 1$

Find $f^{-1}(x)$ and show that $f(f^{-1}(x)) = x$ and $f^{-1}(f(x)) = x$.

29. $f(x) = \dfrac{2}{x - 2}$ **30.** $f(x) = -\dfrac{1}{x} - 1$ **31.** $f(x) = \dfrac{3}{x + 2}$

32. $f(x) = \dfrac{x}{x + 1}$ **33.** $f(x) = x^{-5}$ **34.** $f(x) = \dfrac{1}{\sqrt[3]{x - 2}}$

Verify that the function is its own inverse by showing that $(f \circ f)(x) = x$

35. $f(x) = \dfrac{1}{x}$ **36.** $f(x) = \sqrt{4 - x^2}; 0 \le x \le 2$

37. $f(x) = \dfrac{x}{x - 1}$ **38.** $f(x) = \dfrac{3x - 8}{x - 3}$

Find the inverse g of the given function f, graph both in the same coordinate system, and state the domain and range of g.

39. $y = f(x) = (x + 1)^2; x \ge -1$ **40.** $y = f(x) = x^2 - 4x + 4; x \ge 2$

41. $y = f(x) = \dfrac{1}{\sqrt{x}}$ **42.** $y = f(x) = -\sqrt{x}$

43. $y = f(x) = x^2 - 4$ for $x \ge 0$ **44.** $y = f(x) = 4 - x^2$ for $0 \le x \le 2$

45. $y = f(x) = x^2 - 4x$ for $x \ge 2$ **46.** $y = f(x) = 4x - x^2$ for $x \ge 2$

47. Aside from the linear function $y = x$, what other linear functions are their own inverses? (*Hint:* Inverse functions are reflections of one another through the line $y = x$.)

48. In the figure, curve C_1 is the graph of a one-to-one function $y = f(x)$ and curve C_2 is the graph of $y = g(x)$. The points on C_2 have been obtained by interchanging the first and second coordinates of the points on curve C_1. For a specific value x in the domain of f, the point $P(x, f(x))$ on C_1 produces $Q(f(x), x)$ on C_2. How does this figure demonstrate that $(g \circ f)(x) = x$?

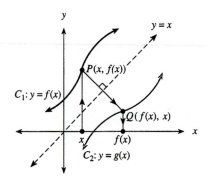

Challenge Suppose that a function f is increasing on its domain D. Use the algebraic condition for one-to-one functions to decide if f is one-to-one.

 Written Assignment Let $f(x) = \dfrac{3}{4x + 5}$ and let g be the inverse of f. Make direct use of the property $f(g(x)) = x$ to find g, without the use of the variable y. Explain why the process used in this section to find an inverse is consistent with the work done here.

Graphing Calculator Exercises In Exercises 1–2, use the graph of f to determine whether the function is one-to-one.

1. $f(x) = x^3 - 3x^2 + 3x - 1$
2. $f(x) = (x^2 - 1)/\sqrt{x + 4}$
3. Estimate the largest closed interval $[a, b]$, (within the window), on which $f(x) = x^3 - 3x^2 - x + 3$ is one-to-one, and decreasing, from the graph of $y = f(x)$. Then use the graph to estimate the domain of $f^{-1}(x)$ having this interval as its range (rounding a and b to the nearest tenth so that you stay in the interval), and graph $y = f^{-1}(x)$ and $y = f(x)$ on the same set of axes.

CHAPTER REVIEW EXERCISES

Section 4.1   Circles (Page 248)

Consider two points $A(x_1, y_1)$ and $B(x_2, y_2)$.

1. State the formula for finding the distance AB.
2. What are the coordinates of the midpoint of segment AB?
3. Write the standard form for the equation of a circle
   (a) with the center at the origin and radius r.
   (b) with center at the point (h, k) and radius r.
4. Find the length of the segment determined by the points $(-4, 3)$ and $(6, -1)$.
5. Find the midpoint of the segment determined by the points in Exercise 4.
6. What is the equation of the circle with center at $(2, -5)$ and radius $\sqrt{3}$?

Find the center and radius for each circle.

7. $(x + 3)^2 + (y - 1)^2 = 9$ 8. $x^2 - 4x + y^2 + 2y = -1$
9. Find the equation of the tangent to the circle $(x - 3)^2 + (y - 2)^2 = 2$ at the point $(2, 1)$.
10. Points $(-3, 5)$ and $(1, -3)$ are the endpoints of a diameter of a circle. Find the center, radius, and equation of the circle.

Section 4.2   Graphing Radical Functions (Page 260)

Find the domain and graph.

11. $y = -\sqrt{x}$ 12. $y = \sqrt{4 - x^2}$ 13. $y = \sqrt{x + 2}$

14. $y = -\sqrt{|x|}$ 15. $y = -x^3$ 16. $y = \dfrac{1}{\sqrt{x - 1}}$

17. Find the equations of the asymptotes for the graph in Exercise 16.

18. Graph f defined on the domain $-1 < x \leq 4$ as follows:

$$f(x) = \begin{cases} x^2 + 1 & \text{for } -1 < x \leq 1 \\ \sqrt{x - 1} & \text{for } 1 < x \leq 4 \end{cases}$$

Section 4.3   Radical Equations and Graphs (Page 266)

Solve for x.

19. $\sqrt{x - 4} + 3 = 5$

20. $\sqrt{x + 7} + 5 = x$

21. $\sqrt{x - 5} + \sqrt{x} = 5$

22. $\sqrt[3]{x - 4} = 2$

Graph; state the domain for each function.

23. $y = f(x) = \sqrt{x^2 - 9}$

24. $y = f(x) = -\sqrt{x^2 - 4}$

25. $y = f(x) = \sqrt{x^2 - 2x - 8}$

26. $y = f(x) = -\dfrac{1}{\sqrt{4 - x^2}}$

27. Find the domain of f, the roots of $f(x) = 0$, and determine the signs of f where

$$f(x) = \frac{x^2 - 1}{\sqrt{x + 1}} + \sqrt{x + 1}$$

28. Solve the system and graph:

$$y = \sqrt{x}$$

$$y = \frac{1}{2}x$$

Solve for x.

29. $\sqrt[3]{x^2} + \sqrt[3]{x} - 6 = 0$

30. $\dfrac{(2x + 1)^{3/2}}{(x - 1)^2} + \dfrac{(2x + 1)^{1/2}}{(x - 1)} = 0$

31. Change $\dfrac{x}{(5x - 6)^{4/5}} + (5x - 6)^{1/5}$ to $\dfrac{6(x - 1)}{(5x - 6)^{4/5}}$

32. Simplify and determine the signs of f: $f(x) = x^{1/2} + \frac{1}{2}x^{-1/2}(x - 9)$

Section 4.4   Combining and Decomposing Functions (Page 277)

Let $f(x) = \sqrt{x + 1}$ and $g(x) = \dfrac{1}{x + 1}$. Find each of the following and state the domain for each.

33. $(f + g)(x)$

34. $(f - g)(x)$

35. $(f \cdot g)(x)$

36. $\dfrac{f}{g}(x)$

Use f and g as defined above and evaluate.

37. $f(g(0))$

38. $g(f(3))$

39. $(f \circ g)(x)$

40. $(g \circ f)(x)$

41. Let $f(x) = \dfrac{1}{x^2}$, $g(x) = x - 1$, and $h(x) = \sqrt{x}$. Find $f \circ g \circ h$.

42. Decompose $h(x) = \sqrt{(x^2 + x)^3}$ into the composition of two functions so that the outer function is a monomial.

43. Express the function in Exercise 42 as the composition of three functions.

44. Find f and g such that $h = f \circ g$ where $h(x) = \left(\dfrac{1}{2x - 3}\right)^3$ and the inner function g is rational.

45. A car is traveling from A toward B at an average speed of 45 miles per hour. Express the distance $d = PC$ as a function of y, and y as a function of the time t. (P represents a police control car.) Interpret the composite $(d \circ y)(t)$, and find d when $t = 3$ minutes, assuming that $t = 0$ at point A.

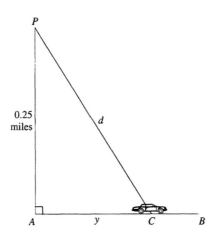

Section 4.5   Inverse Functions (Page 287)

46. State the definition of a one-to-one function.

47. Describe the horizontal line test for one-to-one functions.

48. State the definition of inverse functions.

49. Use the horizontal line test to determine which of the following are one-to-one functions:
    (a) $y = 2x^2 + 5x - 3$ (b) $y = |x - 2|$ (c) $y = x^{-1/2}$

50. Draw the graph of $y = -x^3$ and its reflection in the line $y = x$.

51. Repeat Exercise 50 for the function $f(x) = \dfrac{1}{x - 1}$.

52. Let $f(x) = 2x + 5$ and find (a) $f^{-1}(x)$ and (b) $\dfrac{1}{f(x)}$.

53. Prove that $f(x) = 1 - 2x$ and $g(x) = \dfrac{1}{2} - \dfrac{x}{2}$ are inverse functions.

Find the inverse function g of the given function, and state the domain and range of g.

54. $y = f(x) = (x + 2)^3$ 55. $y = f(x) = x^{2/3} + 1, x \geq 0$

56. Draw the graph of f and f^{-1} on the same set of axes for the function in Exercise 52.

CHAPTER 4: STANDARD ANSWER TEST

Use these questions to test your knowledge of the basic skills and concepts of Chapter 4. Then check your answers with those given at the back of the book.

1. (a) Draw the circle $(x - 3)^2 + (y + 4)^2 = 25$.

   (b) Write the equation of the tangent line to this circle at the point $(6, 0)$.

2. Find the center and radius of the circle $4x^2 + 4x + 4y^2 - 56y = -97$.

Graph each function, state its domain, and then give the equations of the asymptotes if there are any.

3. $f(x) = \sqrt[3]{x + 2}$ 4. $g(x) = \dfrac{1}{\sqrt{x}} + 2$ 5. $f(x) = -\sqrt[3]{x - 2}$ 6. $g(x) = \dfrac{1}{\sqrt{x - 3}}$

7. Graph: $f(x) = \begin{cases} 1 - x^2 & \text{for } -3 \le x < 0 \\ \sqrt{x} + 1 & \text{for } \;\;0 \le x < 2 \end{cases}$

Solve each equation.

8. $\sqrt{18x + 5} - 9x = 1$ 9. $6x^{2/3} + 5x^{1/3} - 4 = 0$ 10. $\sqrt{x - 7} + \sqrt{x + 9} = 8$

11. Convert $(x + 1)^{1/2}(2x) + (x + 1)^{-1/2}\left(\dfrac{x^2}{2}\right)$ into the equivalent form $\dfrac{x(5x + 4)}{2\sqrt{x + 1}}$. Show your work.

Determine the signs of f.

12. $f(x) = \dfrac{\sqrt[3]{x - 4}}{x - 2}$ 13. $f(x) = x^{1/2} - \frac{1}{2}x^{-1/2}(x + 4)$

14. Find the x-intercepts and domain of the function $f(x) = \sqrt{x^2 + x - 2}$.

Let $f(x) = \dfrac{1}{x^2 - 1}$ and $g(x) = \sqrt{x + 2}$. Find and state the domain for each of the following.

15. (a) $(f + g)(x)$ (b) $(f - g)(x)$ 16. (a) $\dfrac{f}{g}(x)$ (b) $(f \cdot g)(x)$

17. Describe how the graph of $g(x) = -\sqrt[3]{x - 4}$ can be obtained from the graph of $f(x) = \sqrt[3]{x}$.

18. Let $f(x) = \dfrac{1}{x^2 + 1}$ and $g(x) = 2\sqrt{x}$.

    (a) Find $f(4)$, $g(4)$, and $(f \cdot g)(4)$.   (b) Evaluate $f(g(9))$ and $g(f(9))$.

19. For $f(x) = \dfrac{1}{1 - x^2}$ and $g(x) = \sqrt{x}$ find the composites $(f \circ g)(x)$ and $(g \circ f)(x)$ and state their domains.

20. Find the functions f and g so that $h(x) = \sqrt[3]{(x - 2)^2} = (f \circ g)(x)$, where g is a binomial.

21. Let $F(x) = \dfrac{1}{(2x - 1)^{3/2}}$. Find functions f, g, and h so that $(f \circ g \circ h)(x) = F(x)$.

22. Complete the following for the function $f(x) = 3x - 2$.

    (a) Find the inverse function, $f^{-1}(x)$.   (b) Graph $f(x)$ and $f^{-1}(x)$ on the same axes.

23. Find the inverse g of $y = f(x) = \sqrt[3]{x} - 1$ and show that $(f \circ g)(x) = x$.

24. State the domain of $f(x) = \sqrt[3]{3x + 1} - x - 1$ and find the x-intercepts.

25. Solve the system: $y = 2\sqrt{x - 1}$

$$y = \tfrac{1}{2}x + 1$$

CHAPTER 4: MULTIPLE CHOICE TEST

1. The circle $x^2 + y^2 + 3y = -\frac{1}{4}$ has

   (a) center $\left(0, \frac{3}{2}\right)$ and radius $r = \frac{1}{2}$

   (b) center $(0, -3)$ and radius $r = 2$

   (c) center $(3, 0)$ and radius $r = \frac{1}{2}$

   (d) center $\left(0, -\frac{3}{2}\right)$ and radius $r = \sqrt{2}$

   (e) None of the preceding

2. The equation of the circle with center at $(2, -3)$ and radius $r = 4$ is

   (a) $(x + 2)^2 + (y - 3)^2 = 16$ (b) $(x - 2)^2 + (y + 3)^2 = 16$

   (c) $(x - 2)^2 + (y + 3)^2 = 4$ (d) $(x + 2)^2 + (y - 3)^2 = 4$ (e) None of the preceding

3. The equation $\sqrt{x^2 - 9x} + \sqrt{3x} = x$ has

   (a) No solution   (b) One solution   (c) Two solutions

   (d) Three solutions   (e) None of the preceding

4. The graph of $y = \dfrac{1}{\sqrt{x - 1}}$ is found by shifting the graph of $y = \dfrac{1}{\sqrt{x}}$

   (a) One unit downward   (b) One unit upward   (c) One unit to the left

   (d) One unit to the right   (e) None of the preceding

5. For which of the following functions is both the domain and the range the set of all real numbers greater than or equal to 0?

   (a) $y = \sqrt{x}$ (b) $y = x^3$ (c) $y = |x|$ (d) $y = \sqrt[3]{x}$ (e) None of the preceding

6. What are the x-intercepts for the graph of $y = \sqrt{2x^2 - 8x - 24}$?

   (a) $2, -6$ (b) $-4, 3$ (c) $-6, 4$ (d) 0 (e) None of the preceding

7. Which of the following are the solutions for the equation $\sqrt[3]{x^2} - \sqrt[3]{x} - 6 = 0$?

   (a) $2, -3$ (b) $-2, 3$ (c) $-8, 27$ (d) $8, -27$ (e) None of the preceding

8. The radius of a right circular cylinder of height h and volume V is given by $r = \sqrt{\dfrac{V}{\pi h}}$.

   Using this formula to solve for h in terms of r and V, $h =$

   (a) $\dfrac{\pi r^2}{V}$ (b) $\dfrac{V}{\pi r^2}$ (c) $\dfrac{(\pi r)^2}{V}$ (d) $\dfrac{V}{\pi r}$ (e) None of the preceding

9. Which of the following are the signs of $f(x) = \dfrac{x - 2}{\sqrt{x^2 + 4}}$?

   (a) For $x \neq \pm 2$, $f(x) > 0$ (b) For $x \neq 2$, $f(x) > 0$

   (c) For $x < 2$, $f(x) < 0$; for $x > 2$, $f(x) > 0$

   (d) For $x < -2$, $f(x) < 0$; for $x > -2$, $f(x) > 0$ (e) None of the preceding

10. Which of the following intervals represents the domain of the function

    $f(x) = \dfrac{x^2}{\sqrt{1 - x^2}}$?

    (a) $(1, \infty)$ (b) $(-\infty, -1)$ (c) $(-\infty, -1)$ and $(1, \infty)$ (d) $(-1, 1)$ (e) None of the preceding

11. Which of the following are correct?

    I. $x\sqrt{x^2 + 1} = \sqrt{x^3 + x}$

    II. $x^{-1/5} + x^{3/5} = x^{-1/5}(1 + x^{2/5})$

    III. $(x^2 - 3)^{1/3}(x^2 - 3)^{2/3} = (x^2 - 3)^{2/9}$

    (a) Only I   **(b)** Only II   **(c)** Only III   **(d)** Only I and II   **(e)** None of the preceding

12. Let $f(x) = \dfrac{1}{x^2 - 1}$ and $g(x) = \sqrt{x}$. Then $\dfrac{f}{g}(4) =$

    (a) $\dfrac{2}{15}$ **(b)** $\dfrac{15}{2}$ **(c)** $\dfrac{1}{30}$ **(d)** 30   **(e)** None of the preceding

13. $x^{1/3} + (x - 8)\dfrac{1}{3}x^{-2/3} =$

    (a) $\dfrac{x^{1/3} + x - 8}{3x^{2/3}}$ **(b)** $\dfrac{3x^{1/3} + x - 8}{x^{2/3}}$ **(c)** $\dfrac{x - 8}{3\sqrt[3]{x}}$ **(d)** $\dfrac{4(x - 2)}{3\sqrt[3]{x^2}}$ **(e)** None of the preceding

14. If $f(x) = \sqrt{x^3 + 1}$ and $g(x) = 2x - 3$, then $f(g(2)) =$

    (a) $\sqrt{2}$ **(b)** 1   **(c)** 3   **(d)** 25   **(e)** None of the preceding

15. Let $f(x) = \sqrt[3]{x}$ and $g(x) = \dfrac{1}{x^3 - 1}$. Then $g(f(x)) =$

    (a) $\dfrac{1}{\sqrt[3]{x^3 - 1}}$ **(b)** $\dfrac{1}{\sqrt[3]{x - 1}}$ **(c)** $\sqrt[3]{x^3 - 1}$ **(d)** $\dfrac{1}{x - 1}$ **(e)** None of the preceding

16. Which of the following is a one-to-one function of x?

    (a) $y = x^4$ **(b)** $y = \dfrac{1}{x}$ **(c)** $y = |x|$ **(d)** $x = y^2$ **(e)** None of the preceding

17. Which of the following is the inverse of $y = f(x) = 3x + 2$?

    (a) $y = 2x + 3$ **(b)** $y = \frac{1}{3}x - \frac{2}{3}$ **(c)** $y = 2 + 3x$ **(d)** $y = \frac{1}{3}x + 2$ **(e)** None of the preceding

18. Which of the functions f and g are such that $h(x) = \sqrt{(4 - x^2)^3} = (f \circ g)(x)$?

    (a) $f(x) = 4 - x^2$; $g(x) = x^{3/2}$ **(b)** $f(x) = x^{2/3}$; $g(x) = 4 - x^2$

    (c) $f(x) = x^{3/2}$; $g(x) = 4 - x^2$ **(d)** $f(x) = x^{1/3}$; $g(x) = (4 - x^2)^2$ **(e)** None of the preceding

19. If $f(x) = \dfrac{2x}{x - 3}$, then $f^{-1}(x) =$

    (a) $\dfrac{x - 3}{2x}$ **(b)** $\dfrac{x - 2}{3x}$ **(c)** $\dfrac{3x}{x - 2}$

    (d) $\dfrac{xy - 3x}{2}$ **(e)** None of the preceding

20. The tangent line to the circle $(x - 3)^2 + (y + 4)^2 = 25$ at the point $(6, 0)$ is which of the following?

    (a) $4x + 3y = 24$ **(b)** $4x - 3y = 24$ **(c)** $3x + 4y = 18$

    (d) $3x - 4y = 18$ **(e)** None of the preceding

ANSWERS TO THE TEST YOUR UNDERSTANDING EXERCISES

Page 251

1. 13   **2.** $\sqrt{65}$ **3.** $2\sqrt{41}$ **4.** (0, 0); 10   **5.** (0, 0); $\sqrt{10}$ **6.** $(1, -1)$; 5

7. $\left(-\frac{1}{2}, 0\right)$; 16   **8.** $(-4, -4)$; $5\sqrt{2}$ **9.** $x^2 + (y - 4)^2 = 25$

10. $(x - 1)^2 + (y + 2)^2 = 3$

Page 252

1. (3, 5); 6   **2.** $\left(0, -\frac{1}{2}\right)$; $\sqrt{5}$ **3.** $\left(\frac{1}{2}, -1\right)$; $\sqrt{7}$ **4.** $\left(\frac{1}{4}, -1\right)$; 3

Page 263

1. Domain of g: $x \geq 0$

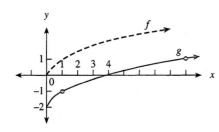

2. Domain of g: $x \geq 1$

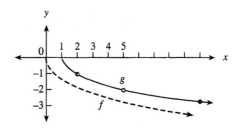

3. Domain of g: $x \geq 0$

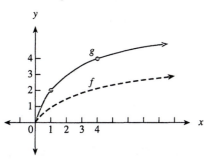

4. Domain of g: $x > -2$

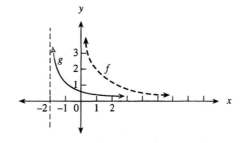

Page 269

1. 25; all $x \geq 0$ **2.** 0, 3; all $x \leq 0$ or all $x \geq 3$ **3.** $x = 8$ **4.** $x = 25$ **5.** $x = 7$

6. No solution   **7.** $x = \frac{1}{9}$ **8.** $x = 5$ **9.** $x = -3$ or $x = 3$ **10.** $x = 9$ **11.** $x = 0$

12. $x = \frac{1}{2}$

Page 279

1. (a) $g(f(1)) = g(2) = 8$

  (b) $f(g(1)) = f(5) = 14$

  (c) $f(g(0)) = f(4) = 11$

  (d) $g(f(-2)) = g(-7) = 53$

2. (a) $g(f(1)) = g(1) = 1$

  (b) $f(g(1)) = f(1) = 1$

  (c) $f(g(0)) = f(0) = 0$

  (d) $g(f(-2))$ is undefined.

3. (a) $g(f(1)) = g(\sqrt[3]{2}) = 5\sqrt[3]{2}$

   (b) $f(g(1)) = f(5) = \sqrt[3]{14}$

   (c) $f(g(0)) = f(0) = -1$

   (d) $g(f(-2)) = g(\sqrt[3]{-7}) = -5\sqrt[3]{7}$

4. (a) $g(f(1))$ is undefined.

   (b) $f(g(1))$ is undefined.

   (c) $f(g(0)) = f(0) = -2$

   (d) $g(f(-2)) = g(0) = 0$

Page 282
(Other answers are possible.)

1. $f(x) = 8x - 3$; $g(x) = x^5$

2. $f(x) = 8x - 3$; $g(x) = \sqrt[5]{x}$

3. $f(x) = \dfrac{1}{8x - 3}$; $g(x) = \sqrt{x}$

4. $f(x) = \dfrac{5}{7 + 4x^2}$; $g(x) = x^3$

5. $f(x) = \dfrac{2x + 1}{2x - 1}$; $g(x) = x^4$

6. $f(x) = x^4 - 2x^2 + 1$; $g(x) = \sqrt{x^3}$

Page 288
The functions given in 2, 3, 5, and 6 are one-to-one. The others are not.

Exponential and Logarithmic Functions

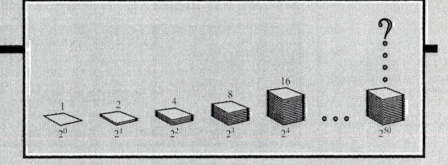

When quantities are repeatedly doubled, their numbers grow dramatically. Here is an illustration of a problem that involves a doubling procedure. Suppose you place a piece of paper on the floor, then place another piece on top of this, then place two pieces on the pile, then four pieces, then eight pieces, and so forth. In other words, at each step you double the number of pieces on the pile. Powers of 2 can be used to keep track of the growth of the pile of paper:

$$\begin{array}{ll}
\text{Begin with one piece} & 1 = 2^0 \\
\text{Step 1: add 1 piece} & 2 = 2^1 \\
\text{Step 2: add 2 pieces} & 4 = 2^2 \\
\text{Step 3: add 4 pieces} & 8 = 2^3 \\
\text{Step 4: add 8 pieces} & 16 = 2^4 \\
\quad \ldots \text{etc.}
\end{array}$$

Now continue this process for 50 doublings. From the pattern in the table you should see that there will be 2^{50} pieces of paper in the pile at that time. Suppose each piece of paper is 0.003 inches thick. How high do you think the pile will be? Will it be over your head? Will it reach the ceiling of your room? Will it be as high as your building?

Take a guess, and then use a calculator to approximate the answer. The height will be $2^{50} \times 0.003$ in inches. Divide this result by 12 to change to feet, and then by 5280 to change to miles (if necessary). Thus, your computation will look like this:

$$(2^{50} \times 0.003 \div 12) \div 5280$$

Are you surprised by the answer?

5.1 EXPONENTIAL FUNCTIONS AND EQUATIONS

The table of powers of 2 shown in the preceding introduction can be expressed by the equation

$$y = f(n) = 2^n$$

This equation defines y to be a function of n, and gives y, the number of pieces of paper at the nth step. (In this case n is a nonnegative integer.) Such a function with the variable as an exponent is known as an *exponential function*.

We use $b > 0$ in order to avoid even roots of negative numbers, such as

$$(-4)^{1/2} = \sqrt{-4}$$

Why do we have $b \neq 1$?

> ### EXPONENTIAL FUNCTION
>
> For any constant $b > 0$, $b \neq 1$ the equation
>
> $$y = b^x$$
>
> defines an **exponential function** with base b and domain all real numbers x.

First consider the specific exponential function $y = f(x) = 2^x$ and note the following:

1. The function is defined for all real values of x. When x is negative, we may apply the definition for negative exponents. Thus, for $x = -2$,

$$2^x = 2^{-2} = \frac{1}{2^2} = \frac{1}{4}$$

The domain of the function is the set of real numbers.

2. For all replacements of x, the function takes on a positive value. That is, 2^x can never represent a negative number, nor can 2^x be equal to 0. The range of the function is the set of all positive real numbers.

Finally, as shown in the table of values, a few specific ordered pairs of numbers can be used as an aid to graphing the function $f(x) = 2^x$.

The function is increasing and the curve is concave up. The x-axis is a horizontal asymptote toward the left.

x	$y = 2^x$
-3	$\frac{1}{8}$
-2	$\frac{1}{4}$
-1	$\frac{1}{2}$
0	1
1	2
2	4
3	8

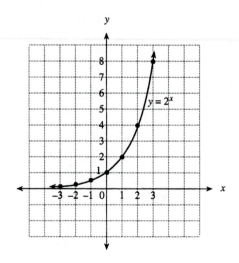

If desired, the accuracy of this graph can be improved by plotting more points. For example, consider such rational values of x as $\frac{1}{2}$ or $\frac{3}{2}$. Rounding to two decimal places, we have:

$$2^{1/2} = \sqrt{2} = 1.41$$
$$2^{3/2} = \left(\sqrt{2}\right)^3 = 2.83$$

Using irrational values for x, such as $\sqrt{2}$ or π, is another matter entirely. (In Section 1.6 our development of exponents stopped with the rationals.) To give a precise meaning of such numbers is beyond the scope of this course. However, using a scientific calculator you can obtain approximations such as $\left(\sqrt{2}, 2^{\sqrt{2}}\right) \approx (1.41, 2.67)$ and $\left(\sqrt{5}, 2^{\sqrt{5}}\right) \approx (2.24, 4.71)$ and then verify that these points "fit" the curve.

When the preceding table of values, used to graph $y = 2^x$, is extended for negative x-values, we obtain additional numerical evidence that the y-values get close to 0.

x	$y = 2^x = \dfrac{1}{2^{-x}}$
-1	$\dfrac{1}{2^1} = 0.5$
-5	$\dfrac{1}{2^5} = \dfrac{1}{32} = 0.03125$
-10	$\dfrac{1}{2^{10}} = \dfrac{1}{1024} = 0.000976562 \approx 9.766 \times 10^{-4}$
-20	$\dfrac{1}{2^{20}} = \dfrac{1}{1,048,576} \approx 9.537 \times 10^{-7}$

Thus $y = 2^x \to 0$ as $x \to -\infty$ so that $y = 0$ (the x-axis) is a horizontal asymptote to the left.

When graphing $y = 2^x$, we made use of the property $2^{-n} = \dfrac{1}{2^n}$. This and other properties listed below have been studied previously for rational exponents r and s.

$$b^r b^s = b^{r+s} \qquad \frac{b^r}{b^s} = b^{r-s} \qquad (b^r)^s = b^{rs}$$

$$a^r b^r = (ab)^r \qquad b^0 = 1 \qquad b^{-r} = \frac{1}{b^r}$$

In more advanced work it can be shown that these rules hold for positive bases a and b and all real numbers r and s.

The graphs of all exponential functions of the form $y = b^x$, for $b > 1$, have the same basic shape as that for $y = 2^x$. This is demonstrated in the following example.

EXAMPLE 1 Graph the curves $y = (1.5)^x$, $y = 2^x$, $y = 3^x$, and $y = 4^x$ on the same coordinate system.

Solution

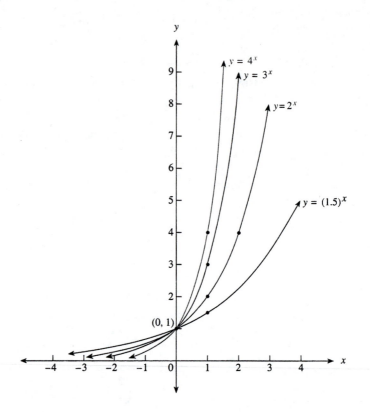

Note that all of the curves are increasing, they are all concave up, they all have the x-axis as a horizontal asymptote, and they all pass through the point $(0, 1)$. Also for $x > 0$, and reading from left to right, the curves grow faster as the base b gets larger. For $x < 0$, but reading from right to left, the curves come down faster with increasing values for b. ■

The next example demonstrates how the graphs of exponential functions can be translated.

EXAMPLE 2 Use the graph of $y = f(x) = 2^x$ to sketch the curves

$$y = g(x) = 2^{x-3} \qquad \text{and} \qquad y = h(x) = 2^x - 1$$

Even though functions g and h are not exactly of the form b^x, they are also classified as being exponential functions because the variable is within the exponent.

Write the equations of the asymptotes.

Solution Since $g(x) = f(x - 3)$, the graph of g can be obtained by translating $y = 2^x$ by 3 units to the right. Moreover, since $h(x) = f(x) - 1$ the graph of h can be found by translating $y = 2^x$ down 1 unit.

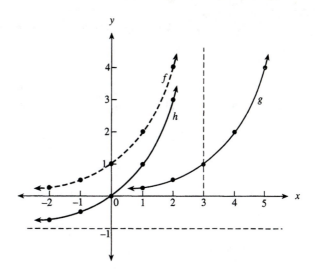

Note that the range of g is the same as for f, all $y > 0$. The range for h consists of all $y > -1$.

The asymptote for g is the same as that for f, namely $y = 0$, the x-axis. The asymptote for h is the line $y = -1$, caused by the downward translation of f by 1 unit to obtain h. ■

Thus far we have restricted our attention to exponential functions of the form $y = f(x) = b^x$, where $b > 1$. All of these have graphs that are of the same general shape as that for $y = 2^x$. For $b = 1$, $y = b^x = 1^x = 1$ for all x. Since this is a constant function, $f(x) = 1$, we do not use the base $b = 1$ in the classification of exponential functions.

Now let us explore exponential functions $y = f(x) = b^x$ for which $0 < b < 1$. In particular, if $b = \dfrac{1}{2}$, we get $y = \left(\dfrac{1}{2}\right)^x = \dfrac{1}{2^x}$ or $y = 2^{-x}$.

All curves $y = b^x$ for $0 < b < 1$ have this same basic shape. The curve is concave up, the function is decreasing, and the x-axis is a horizontal asymptote toward the right as x becomes large.

x	$y = 2^{-x}$
-3	8
-2	4
-1	2
0	1
1	$\dfrac{1}{2}$
2	$\dfrac{1}{4}$
3	$\dfrac{1}{8}$

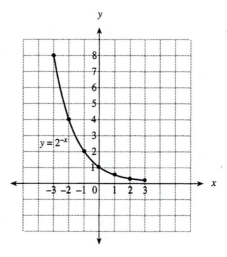

As shown in the following figure, the graph of $y = g(x) = \dfrac{1}{2^x}$ can also be found by comparing it to the graph of $y = f(x) = 2^x$. Since $g(x) = \dfrac{1}{2^x} = 2^{-x} = f(-x)$, the y-values for g are the same as the y-values for f *on the opposite side of the y-axis*. In other words, the graph of g is the *reflection* of the graph of f through the y-axis.

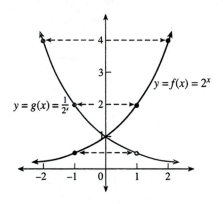

Note that $-3^{-x} = -(3^{-x})$.

EXAMPLE 3 Graph the function $y = f(x) = -3^{-x}$.

Solution The graph of this function will be the reflection of the graph of $y = 3^{-x}$ in the x-axis. If taken step by step, we have the following:

(a) Draw the graph of $y = 3^x$.

(b) Reflect in the y-axis to obtain $y = 3^{-x}$.

(c) Reflect in the x-axis to obtain $y = -3^{-x}$.

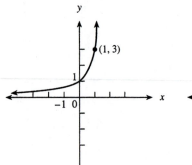 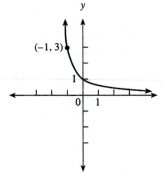 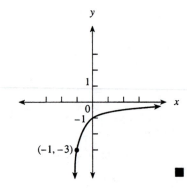

We have explored functions of the form $y = f(x) = b^x$ for specific values of b. In each case the graphs went through the point $(0, 1)$ since $y = b^0 = 1$. Also, each such graph has the x-axis as a one-sided asymptote and there are no x-intercepts. These and other properties of $y = f(x) = b^x$ for $b > 0$ and $b \neq 1$ are summarized as follows.

PROPERTIES OF $y = f(x) = b^x$

1. The domain consists of all real numbers x.
2. The range consists of all positive numbers y.
3. The function is increasing (the curve is rising) when $b > 1$, and it is decreasing (the curve is falling) when $0 < b < 1$.
4. The curve is concave up for $b > 1$ and for $0 < b < 1$.
5. It is a one-to-one function.
6. The point $(0, 1)$ is on the curve. There is no x-intercept.
7. The x-axis is a horizontal asymptote to the curve, toward the left for $b > 1$ and toward the right for $0 < b < 1$.
8. $b^{x_1} b^{x_2} = b^{x_1 + x_2};\ b^{x_1}/b^{x_2} = b^{x_1 - x_2};\ (b^{x_1})^{x_2} = b^{x_1 x_2}$.

The one-to-one property of a function f may be stated in this way: (See page 288.)

$$\text{If } f(x_1) = f(x_2), \text{ then } x_1 = x_2.$$

That is, since $f(x_1)$ and $f(x_2)$ represent the same range value, there can only be one corresponding domain value; consequently, $x_1 = x_2$. Using $f(x) = b^x$, this statement means the following:

$$\text{If } b^{x_1} = b^{x_2}, \text{ then } x_1 = x_2.$$

This property can be used to solve certain **exponential equations,** such as $5^{x^2} = 625$. First note that 625 can be written as 5^4.

$$5^{x^2} = 625$$
$$5^{x^2} = 5^4$$

By the one-to-one property applied to the function $f(t) = 5^t$, we may equate the exponents and solve for x.

CAUTION
a^{bc} means $a^{(b^c)}$, and $(a^b)^c = a^{bc}$. Thus, in general, $a^{bc} = a^{(b^c)} \neq (a^b)^c$. Compare the values of $(2^2)^3$ and $2^{(2^3)}$

$$x^2 = 4$$
$$x = \pm 2 \qquad (x = 2 \quad \text{or} \quad x = -2)$$

To check these solutions, note that $5^{2^2} = 5^4 = 625$ and $5^{(-2)^2} = 5^4 = 625$.

EXAMPLE 4   Solve for x: $\dfrac{1}{3^{x-1}} = 81$

Solution Write 81 as 3^4 and $\dfrac{1}{3^{x-1}}$ as $3^{-(x-1)}$.

Note that the one-to-one property here is being applied to function $f(t) = 3^t$.

$$3^{-(x-1)} = 3^4$$
$$-(x - 1) = 4 \qquad \text{by the one-to-one property}$$
$$-x + 1 = 4$$
$$-x = 3$$
$$x = -3$$

∎

TEST YOUR UNDERSTANDING	*Solve for x.*

Solve for x.

1. $2^{x-1} = 32$ **2.** $2^{x^2} = 16$ **3.** $8^{2x+1} = 64$

4. $\dfrac{1}{2^x} = 64$ **5.** $\dfrac{1}{5^{x+1}} = 125$ **6.** $\dfrac{1}{4^{x-2}} = 64$

7. $27^x = 3$ **8.** $27^x = 9$ **9.** $125^x = 25$

10. $\left(\dfrac{1}{4}\right)^x = 32$ **11.** $\left(\dfrac{3}{5}\right)^x = \dfrac{27}{125}$ **12.** $\left(\dfrac{9}{25}\right)^x = \dfrac{5}{3}$

(Answers: Page 359)

EXERCISES 5.1

Graph the exponential function f by making use of a brief table of values. Then use this curve to sketch the graph of g. Write the equation of the horizontal asymptote of g.

1. $f(x) = 2^x$; $g(x) = 2^{x+3}$ **2.** $f(x) = 3^x$; $g(x) = 3^x - 2$

3. $f(x) = 4^x$; $g(x) = -(4^x)$ **4.** $f(x) = 5^x$; $g(x) = \left(\tfrac{1}{5}\right)^x$

5. $f(x) = \left(\tfrac{3}{2}\right)^x$; $g(x) = \left(\tfrac{3}{2}\right)^{-x}$ **6.** $f(x) = 8^x$; $g(x) = 8^{x-2} + 3$

7. $f(x) = 3^x$; $g(x) = 2(3^x)$ **8.** $f(x) = 3^x$; $g(x) = \tfrac{1}{2}(3^x)$

9. $f(x) = 2^{x/2}$; $g(x) = 2^{x/2} - 3$ **10.** $f(x) = 4^x$; $g(x) = 4^{1-x}$

11. $f(x) = 4^{-x}$; $g(x) = -4^{-x}$ **12.** $f(x) = 2^{-x}$; $g(x) = 2^{-x} + 2$

Sketch the curves on the same axes.

13. $y = \left(\tfrac{3}{2}\right)^x$, $y = 2^x$, $y = \left(\tfrac{5}{2}\right)^x$ **14.** $y = \left(\tfrac{1}{4}\right)^x$, $y = \left(\tfrac{1}{3}\right)^x$, $y = \left(\tfrac{1}{2}\right)^x$

15. $y = 2^{|x|}$, $y = -(2^{|x|})$

16. $y = 2^x$, $y = 2^{-x}$, $y = 2^x - 2^{-x}$ (*Hint:* Subtract ordinates.)

17. Find an equation of the given curve that has been obtained from the curve $y = 2^x$ by the indicated procedures.

(a) One translation   **(b)** Two translations

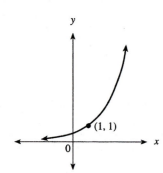

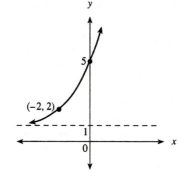

18. Find an equation of the given curve that has been obtained from the curve $y = \left(\frac{1}{2}\right)^x = 2^{-x}$ by the indicated procedures.

(a) Two translations

(b) A reflection followed by a translation

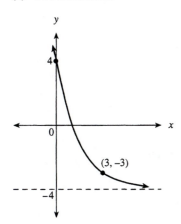
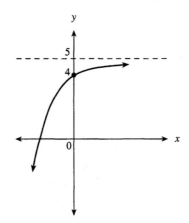

Use the one-to-one property of an appropriate exponential function to solve the equation.

19. $2^x = 64$

20. $3^x = 81$

21. $2^{x^2} = 512$

22. $3^{x-1} = 27$

23. $5^{2x+1} = 125$

24. $2^{x^3} = 256$

25. $7^{x^2+x} = 49$

26. $b^{x^2+x} = 1$

27. $\dfrac{1}{2^x} = 32$

28. $\dfrac{1}{10^x} = 10{,}000$

29. $9^x = 3$

30. $64^x = 8$

31. $9^x = 27$

32. $64^x = 16$

33. $\left(\frac{1}{49}\right)^x = 7$

34. $5^x = \frac{1}{125}$

35. $\left(\frac{27}{8}\right)^x = \frac{9}{4}$

36. $(0.01)^x = 1000$

37. $3^{2x-1} = \dfrac{729}{9^x+1}$

38. $2^x - 2^{-x} = 2^{-x}$

39. $2^{x^2+x} = 4^{1+x}$

40. $3^{2x+2} = 27^{x^2-1}$

41. Graph the function $y = 2^x$ and $y = x^2$ in the same coordinate system for the interval $[0, 5]$. (Use a larger unit on the x-axis than on the y-axis.) What are the points of intersection?

42. Solve for x: $(6^{2x})(4^x) = 1728$.   **43.** Solve for x: $(5^{2x+1})(7^{2x}) = 175$.

44. Use a calculator to verify that $\sqrt{3} = 1.732050.\ldots$ Now fill in the powers of 2 in the table, rounding off each entry to four decimal places. (Note that the numbers across the top of the table are successive decimal approximations of $\sqrt{3}$.)

x	1.7	1.73	1.732	1.7320	1.73205
2^x					

On the basis of the results, what is your estimate of $2^{\sqrt{3}}$ to three decimal places? Now find $2^{\sqrt{3}}$ directly on the calculator and compare.

45. Follow the instructions in Exercise 44 for these numbers:
(a) $3^{\sqrt{2}}$ **(b)** $3^{\sqrt{3}}$ **(c)** $2^{\sqrt{5}}$ **(d)** 4^{π}

46. Begin with the graph of $y = 2^{-x}$ and describe a sequence of three more curves, the last of which is $y = |-2^{-x} + 1|$. (Use translations and reflections as necessary.)

47. Find a such that $y = a2^x$ is an expansion of $y = 2^x$ that gives the same result as a shift of $y = 2^x$ to the left by 3 units.

48. Find a such that $y = a2^x$ is a contraction of $y = 2^x$ that gives the same result as a shift of $y = 2^x$ to the right by 2 units.

Written Assignment Each of the following are *incorrect* attempts to solve the equation $(5^{2x+1})(6^{4x}) = 30^2$. Explain what is wrong in each case.

1. $(5^{2x+1})(6^{4x}) = 30^2$
$30^{6x+1} = 30^2$
$6x + 1 = 2$
$x = \dfrac{1}{6}$

2. $(5^{2x+1})(6^{4x}) = 30^2$
$(2x + 1)(4x) = 2$
$4x^2 + 2x - 1 = 0$
$x = \dfrac{-1 \pm \sqrt{5}}{4}$

Graphing Calculator Exercises

1. Solve the equation $2^x = x^3$ by graphing $y = 2^x$ and $y = x^3$ on the same set of axes. Determine any points of intersection to the nearest hundredth of a unit. (*Hint:* Before you jump to the conclusion that there is only one point of intersection, notice that $2^{10} > 10^3$.)

2. A student wanted a graphing calculator to show the graph of $y = 3^{2x}$. But instead, when graphed, the student got the straight line $y = 6x$. What do you think might have been the trouble with this student's approach?

3. The graph of $y = 2^x + 2^{-x} - 2$ resembles a parabola. Graph it together with $y = x^2$ on the same set of axes. At what positive values of x is the graph of $y = x^2$ below that of $y = 2^x + 2^{-x} - 2$?

4. The function $y = (1+1/x)^x$ is important in describing compound interest. Graph this function for $x > 0$ and discuss its intercepts and asymptotes, if any.

5. A certain population of 1000 bacteria doubles in size every 30 minutes according to the equation $P = 1000*2^{(x/30)}$. Graph this function and determine from your graph how long it will be before the population reaches a million bacteria.

**5.2
LOGARITHMIC
FUNCTIONS**

One of the properties of the exponential function $y = f(x) = b^x$ is that it is a one-to-one function. This means that it has an inverse function whose graph can be obtained by reflecting the graph of $y = b^x$ through the line $y = x$. This is illustrated in the following figures for the case where $b > 1$ and also where $0 < b < 1$.

Recall from Section 4.5 that $f^{-1}(x)$ is a notation used to represent the inverse of a function f.

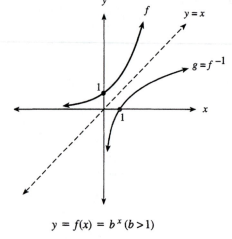

$y = f(x) = b^x \, (b > 1)$

$y = f(x) = b^x (0 < b < 1)$

An equation for the inverse function g can be obtained by interchanging the roles of the variables as follows:

$$\text{Function, } f: \quad y = f(x) = b^x$$

$$\text{Inverse function, } g: \quad x = g(y) = b^y$$

Thus $x = b^y$ is an equation for g. Unfortunately, we have no way of solving $x = b^y$ to get y explicitly in terms of x. To overcome this difficulty, we create some new terminology.

The equation $x = b^y$ tells us that y *is the exponent on b that produces x*. In situations like this, the word **logarithm** is used in place of *exponent*. Now we may say that y *is the logarithm on b that produces x*. This description can be abbreviated to $y = \text{logarithm}_b x$, and abbreviating further we reach the final form

$$y = \log_b x$$

which is read "y equals log x to the base b" or "y equals log x base b."

It is important to realize that we are only defining (not proving) the equation $y = \log_b x$ to have the same meaning as $x = b^y$. In other words, these two forms are equivalent:

$$\text{Exponential form:} \quad x = b^y$$

$$\text{Logarithmic form:} \quad y = \log_b x$$

And since they are equivalent they define the same function g:

$$y = g(x) = \log_b x$$

LOGARITHMIC FUNCTION

For any constant $b > 0$, $b \neq 1$, the equation

$$y = \log_b x$$

defines a **logarithmic function** with base b and domain all $x > 0$.

($y = \log_b x$ is equivalent to $x = b^y$)

Since $y = \log_b x$ and $y = b^x$ are inverse functions, we know that their domains and ranges are interchanged. (See page 292.) In particular,

Domain of $y = \log_b x$ consists of all positive numbers = Range of $y = b^x$

Range of $y = \log_b x$ consists of all real numbers = Domain of $y = b^x$

Furthermore, when the curve $y = b^x$, $b > 1$ is reflected through the line $y = x$ to obtain the inverse curve $y = \log_b x$, the horizontal asymptote to the left for $y = b^x$ converts to a vertical asymptote downward for $y = \log_b x$ as demonstrated in Example 1. A similar observation holds for the case $0 < b < 1$.

Converting a logarithmic function to exponential form can be helpful when graphing, as in the following example.

EXAMPLE 1   Graph: $y = f(x) = \log_2 x$

Solution   Rewrite f in the form $x = 2^y$ and select appropriate values of y, as in the table of values shown. For example, when $y = 3$, $x = 2^3 = 8$ and when $y = -3$, $x = 2^{-3} = \frac{1}{8}$.

x	y
$\frac{1}{8}$	-3
$\frac{1}{4}$	-2
$\frac{1}{2}$	-1
1	0
2	1
4	2
8	3

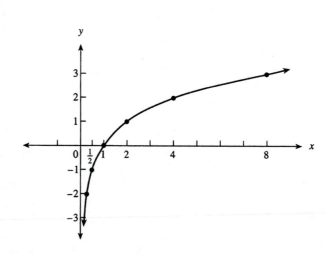

An alternate procedure is to graph the inverse of the given function, $y = 2^x$, and reflect this graph through the line $y = x$, as shown below. Also, as noted in the remarks preceding this example, the reflection of the horizontal asymptote for $y = 2^x$, the x-axis, produces the vertical asymptote for $y = \log_2 x$, the y-axis. That is, in symbols, $y = \log_2 x \longrightarrow -\infty$ as $x \longrightarrow 0^+$.

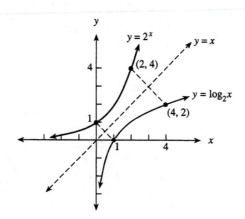

EXAMPLE 2   Find the domain of $y = \log_2(x - 3)$ and graph the function.

Solution   In $y = \log_2(x - 3)$ the quantity $x - 3$ plays the role that x does in $\log_2 x$. Thus $x - 3 > 0$ and the domain consists of all $x > 3$. Thus, to obtain the graph, shift the graph of $y = \log_2 x$ three units to the right. Note that the equation of the vertical asymptote is $x = 3$.

How does this graph differ from that for $y = \log_2 x - 3$?

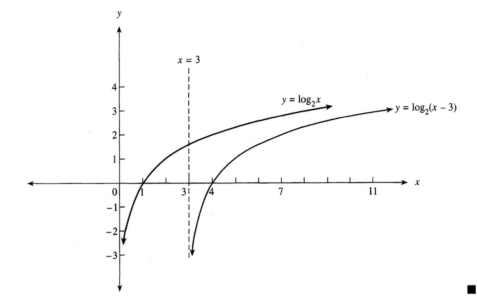

At times it is not clear how to graph a given logarithmic function using translations and reflections of some known curve. In that case, converting to exponential form as in Example 3 will be useful.

EXAMPLE 3   Graph the function $y = f(x) = \log_2(2x + 4)$.

Solution   Since logarithms can be taken only of positive numbers, the domain of the function is the solution of $2x + 4 > 0$; that is, all $x > -2$. Furthermore, the asymptote is $x = -2$. To locate points on the curve, first rewrite the logarithmic function in exponential form:

$$y = \log_2(2x + 4) \longleftrightarrow 2x + 4 = 2^y$$

Find the coordinates of a few points and connect them with a smooth curve. For example:

$$\text{if } y = 1, 2x + 4 = 2 \text{ and } x = -1;$$

$$\text{if } y = -1, 2x + 4 = 2^{-1} = \frac{1}{2} \text{ and } x = -\frac{7}{4}.$$

The completed graph is shown on the following page.

x	y
$-\frac{7}{4}$	-1
$-\frac{3}{2}$	0
-1	1
0	2
2	3
6	4

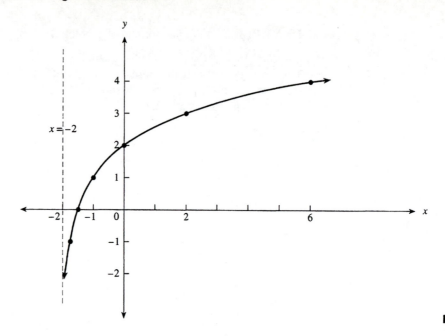

Another way to graph the function in the preceding example becomes possible after the laws of logarithms (Section 5.3) become available. In that setting, $\log_2(2x + 4)$ can be rewritten as $\log_2 2 + \log_2(x + 2) = 1 + \log_2(x + 2)$. Consequently, the curve can be obtained by translating the graph of $\log_2 x$ two units left and one unit up.

TEST YOUR UNDERSTANDING

(Answers: Page 359)

1. Find the equation of the inverse of $y = 3^x$ and graph both on the same axes.

2. Find the equation of the inverse of $y = \left(\frac{1}{3}\right)^x$ and graph both on the same axes.

Let $y = f(x) = \log_5 x$. Describe how the graph of each of the following can be obtained from the graph of f.

3. $g(x) = \log_5(x + 2)$ 4. $g(x) = 2 + \log_5 x$

5. $g(x) = -\log_5 x$ 6. $g(x) = 2 \log_5 x$

Since we know that the exponential function $y = f(x) = b^x$ and the **logarithmic function** $y = g(x) = \log_b x$ are inverse functions, it follows that

$$f(g(x)) = f(\log_b x) = b^{\log_b x} = x \qquad \text{for all } x > 0,$$

Recall from Section 4.5 that for inverse functions, $f(g(x)) = x$ and $g(f(x)) = x$.

and

$$g(f(x)) = g(b^x) = \log_b(b^x) = x \qquad \text{for all } x.$$

The important properties of the function $y = \log_b x$, where $b > 0$ and $b \neq 1$ are summarized in the following list.

PROPERTIES OF $y = f(x) = \log_b x$ **for** $b > 0$ **and** $b \neq 1$

1. The domain consists of all positive numbers x.
2. The range consists of all real numbers y.
3. The function increases (the curve is rising) for $b > 1$, and it decreases (the curve is falling) for $0 < b < 1$.
4. The curve is concave down for $b > 1$, and it is concave up for $0 < b < 1$.
5. It is a one-to-one function; if $\log_b(x_1) = \log_b(x_2)$, then $x_1 = x_2$.
6. The point $(1, 0)$ is on the graph. There is no y-intercept.
7. The y-axis is a vertical asymptote to the curve in the downward direction for $b > 1$ and in the upward direction $0 < b < 1$.
8. $\log_b(b^x) = x$ and $b^{\log_b x} = x$.

Converting from one of the equivalent forms $y = \log_b x$ and $x = b^y$ to the other is a useful skill. (You might find the schematic approach shown in the margin to be helpful.) The following table demonstrates specific cases of such conversions.

The conversion can be remembered by following the arrows from bottom to top in the counterclockwise direction.

$\log_5 25 = 2$

Logarithmic Form $\log_b x = y$	Exponential Form $b^y = x$
$\log_5 25 = 2$	$5^2 = 25$
$\log_{27} 9 = \frac{2}{3}$	$27^{2/3} = 9$
$\log_6 \frac{1}{36} = -2$	$6^{-2} = \frac{1}{36}$

Of the two forms $y = \log_b x$ and $x = b^y$, the exponential form is usually easier to work with. Consequently, when there is a question concerning $y = \log_b x$ it is often useful to convert to the exponential form. For instance, here are two logarithmic results that are easily verified using the exponential form:

$$\log_b 1 = 0 \qquad \text{since } b^0 = 1$$

$$\log_b b = 1 \qquad \text{since } b^1 = b$$

The following examples show how to solve for the variables x, y, or b in the form $\log_b x = y$.

EXAMPLE 4   Evaluate: **(a)** $\log_2 64$ **(b)** $\log_3 \dfrac{1}{243}$

Solution

(a) Let $y = \log_2 64$ and change to exponential form.

Check: $\log_2 64 = 6$ *is correct since* $2^6 = 64$.

$$y = \log_2 64$$

$$2^y = 64 = 2^6$$

$$y = 6 \qquad \text{by the one-to-one property}$$

(b) Let $y = \log_3 \dfrac{1}{243}$

Check: $\log_3 \dfrac{1}{243} = -5$

is correct since

$3^{-5} = \dfrac{1}{3^5} = \dfrac{1}{243}.$

$$3^y = \dfrac{1}{3^5} = 3^{-5}$$

Thus $y = -5$ ∎

EXAMPLE 5   Solve for the indicated variable:

(a) $y = \log_9 27$ **(b)** $\log_b 8 = \tfrac{3}{4}$ **(c)** $\log_{49} x = -\tfrac{1}{2}$

Solution

(a) Convert $y = \log_9 27$ to exponential form.

$$9^y = 27$$
$$(3^2)^y = 3^3 \qquad \text{Express each side as a power of 3.}$$
$$3^{2y} = 3^3$$
$$2y = 3 \qquad \text{by the one-to-one property}$$

Check: $\log_9 27 = \tfrac{3}{2}$ *is correct since* $9^{3/2} = 27$.

$$y = \tfrac{3}{2}$$

(b) Convert $\log_b 8 = \tfrac{3}{4}$ to exponential form.

$$b^{3/4} = 8$$

Check: $\log_{16} 8 = \tfrac{3}{4}$ *since* $16^{3/4} = \left(\sqrt[4]{16}\right)^3 = 8$.

$$(b^{3/4})^{4/3} = 8^{4/3} \qquad \text{Raise both sides to the } \tfrac{4}{3} \text{ power.}$$
$$b = 16 \qquad 8^{4/3} = \left(\sqrt[3]{8}\right)^4 = 2^4$$

(c) Convert $\log_{49} x = -\tfrac{1}{2}$ to exponential form.

Check this solution.

$$x = 49^{-1/2} = \dfrac{1}{\sqrt{49}} = \dfrac{1}{7}$$ ∎

EXERCISES 5.2

Sketch the graph of the function f. Reflect this curve through the line $y = x$ to obtain the graph of the inverse function g, and write the equation for g.

1. $y = f(x) = 4^x$ **2.** $y = f(x) = 5^x$ **3.** $y = f(x) = \left(\tfrac{1}{3}\right)^x$ **4.** $y = f(x) = (0.2)^x$

Describe how the graph of h can be obtained from the graph of g. Find the domain of h, and write the equation of the vertical asymptote.

5. $g(x) = \log_3 x$; $h(x) = \log_3 (x + 2)$ **6.** $g(x) = \log_5 x$; $h(x) = \log_5 (x - 1)$

7. $g(x) = \log_8 x$; $h(x) = 2 + \log_8 x$ **8.** $g(x) = \log_{10} x$; $h(x) = 2 \log_{10} x$

Sketch the graph of f and state its domain.

9. $f(x) = \log_{10} x$ **10.** $f(x) = -\log_{10} x$ **11.** $f(x) = |\log_{10} x|$

12. $f(x) = \log_{10}(-x)$ **13.** $f(x) = \log_{10}|x|$ **14.** $f(x) = \log_{1/10}(x + 1)$

15. Find an equation of the given curve that has been obtained from the curve $y = \log_2 x$ by the indicated procedure.

(a) One translation

(b) One translation

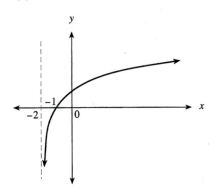
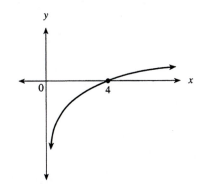

16. Find an equation of the given curve that has been obtained from the curve $y = \log_{1/2} x$ by the indicated procedures.

(a) A translation followed by taking absolute value

(b) Two translations

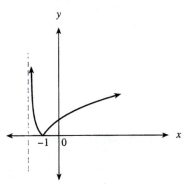
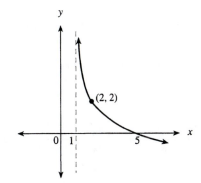

Convert from the exponential to the logarithmic form.

17. $2^8 = 256$ **18.** $5^{-3} = \frac{1}{125}$ **19.** $\left(\frac{1}{3}\right)^{-1} = 3$

20. $81^{3/4} = 27$ **21.** $17^0 = 1$ **22.** $\left(\frac{1}{49}\right)^{-1/2} = 7$

Convert from the logarithmic form to the exponential form.

23. $\log_{10} 0.0001 = -4$ **24.** $\log_{64} 4 = \frac{1}{3}$ **25.** $\log_{\sqrt{2}} 2 = 2$

26. $\log_{13} 13 = 1$ **27.** $\log_{12} \frac{1}{1728} = -3$ **28.** $\log_{27/8} \frac{9}{4} = \frac{2}{3}$

Evaluate:

29. $\log_{10} 10,000$ **30.** $\log_8 8$ **31.** $\log_5 625$ **32.** $\log_8 1$

33. $\log_{2/3} \frac{8}{27}$ **34.** $\log_{1/2} 32$ **35.** $\log_{0.3} \frac{1000}{27}$ **36.** $\log_{3/2} \frac{8}{27}$

Solve for the indicated quantity: y, x, or b.

37. $\log_2 16 = y$

38. $\log_4 32 = y$

39. $\log_{1/3} 27 = y$

40. $\log_7 x = -2$

41. $\log_{1/6} x = 3$

42. $\log_8 x = -\frac{2}{3}$

43. $\log_b 125 = 3$

44. $\log_b 8 = \frac{3}{2}$

45. $\log_b \frac{1}{8} = -\frac{3}{2}$

46. $\log_{100} 10 = y$

47. $\log_{27} 3 = y$

48. $\log_{1/16} x = \frac{1}{4}$

49. $\log_b \frac{16}{81} = 4$

50. $\log_8 x = -3$

51. $\log_b \frac{1}{27} = -\frac{3}{2}$

52. $\log_{\sqrt{3}} x = 2$

53. $\log_{\sqrt{8}}\left(\frac{1}{8}\right) = y$

54. $\log_b \frac{1}{128} = -7$

55. $\log_{0.001} 10 = y$

56. $\log_{0.2} 5 = y$

57. $\log_9 x = 1$

58. $\frac{1}{2}\log_{10} x = 3$

59. $\log_{10} 1000 = \frac{y}{2}$

60. $\log_{2b} 0.1 = -1$

Evaluate each expression.

61. $\log_{3/4}\left(\log_{1/27}\frac{1}{81}\right)$

62. $\log_2(\log_4 256)$

By interchanging the roles of the variables, find the inverse function g. Show that $(f \circ g)(x) = x$ and $(g \circ f)(x) = x$.

63. $y = f(x) = \log_3(x + 3)$

64. $y = f(x) = 2^{x+1}$

Graph the function using the procedure in Example 3. Give the equation of the asymptote.

65. $y = \log_2(2x - 4)$

66. $y = \log_4(3x - 2)$

67. Let $f(x) = \log_2(2x + 4)$ and $g(x) = 2^{x-1} - 2$. Show that these are inverse functions by showing that $f(g(x)) = x$ for all x and $g(f(x)) = x$ for all $x > -2$.

68. In Exercise 67 graph $f(x)$ by drawing the graph of $g(x)$ using translations, and then reflecting this through the line $y = x$.

Challenge Solve the equation $\log_4(2x + 3) = -\frac{5}{2}$ without converting it into exponential form, and explain the steps in your work. (*Hint:* Use the one-to-one property.)

Graphing Calculator Exercises

1. Graph the functions (**a**) $y = \log_{10}|x|$ and (**b**) $y = \log_{10} x^2$. How are they alike, and how do they differ?

2. Graph $y = x \log_{10} x^2$. What is the domain of the function? Find intercepts and asymptotes, if any. Describe the symmetry.

3. (**a**) Graph the function $y = 2 + \log_2 x$ and its inverse on the same set of axes. (*Hint:* Find and graph the inverse first.)

(**b**) Where do the graphs intersect? (*Hint:* Use the line $y = x$.)

4. Graph the function $y = x^{(1/\log x)}$. What happens as x gets closer and closer to zero? What does this tell you about 0^0? In particular, does it make sense to define $0^0 = 1$?

5.3
THE LAWS OF LOGARITHMS

$\log_2 8 = 3; (2^3 = 8)$
$\log_2 16 = 4; (2^4 = 16)$
$\log_2 128 = 7; (2^7 = 128)$

From our knowledge of logarithms we have

$$\log_2 8 + \log_2 16 = 3 + 4$$

$$= 7 = \log_2 128 = \log_2 8 \cdot 16$$

Thus

$$\log_2 8 \cdot 16 = \log_2 8 + \log_2 16$$

It turns out that this equation is a special case of the first law of logarithms.

CAUTION
$\log_b MN \neq (\log_b M)(\log_b N)$

$\log_b \dfrac{M}{N} \neq \dfrac{\log_b M}{\log_b N}$

> ### LAWS OF LOGARITHMS
>
> If M and N are positive, $b > 0$, and $b \neq 1$, then
>
> LAW 1. $\log_b MN = \log_b M + \log_b N$
>
> LAW 2. $\log_b \dfrac{M}{N} = \log_b M - \log_b N$
>
> LAW 3. $\log_b (N^k) = k \log_b N$

Since logarithms are exponents, it is not surprising that these laws can be proved by using the appropriate rules of exponents. Following is a proof of Law 1; the proofs of Laws 2 and 3 are left as exercises.

Let

$$\log_b M = r \quad \text{and} \quad \log_b N = s$$

Convert to exponential form :

Recall: If $\log_b x = y$ then $b^y = x$.

$$M = b^r \quad \text{and} \quad N = b^s$$

Multiply the two equations:

$$MN = b^r b^s = b^{r+s}$$

Then convert to logarithmic form:

$$\log_b MN = r + s$$

Substitute for r and s to get the final result:

$$\log_b MN = \log_b M + \log_b N$$

EXAMPLE 1   For positive numbers A, B, and C, show that

$$\log_b \frac{AB^2}{C} = \log_b A + 2 \log_b B - \log_b C$$

Solution

$$\log_b \frac{AB^2}{C} = \log_b(AB^2) - \log_b C \qquad \text{Law 2}$$

$$= \log_b A + \log_b B^2 - \log_b C \qquad \text{Law 1}$$

$$= \log_b A + 2\log_b B - \log_b C \qquad \text{Law 3} \qquad \blacksquare$$

EXAMPLE 2 Express $\frac{1}{2}\log_b x - 3\log_b(x-1)$ as the logarithm of a single expression in x.

Solution

Identify the laws of logarithms being used in Examples 2 and 3.

$$\frac{1}{2}\log_b x - 3\log_b(x-1) = \log_b x^{1/2} - \log_b(x-1)^3$$

$$= \log_b \frac{x^{1/2}}{(x-1)^3}$$

$$= \log_b \frac{\sqrt{x}}{(x-1)^3} \qquad \blacksquare$$

For some scientific and technical applications, numbers are written in scientific notation and therefore logarithms to base 10 are used. Such logarithms are referred to as **common logarithms**. Furthermore, when a base is not indicated, as in Example 3, we will always consider it to be base 10. That is,

$$\log N \text{ is assumed to mean } \log_{10} N.$$

EXAMPLE 3 Given: $\log 2 = 0.3010$ and $\log 3 = 0.4771$; find $\log\sqrt{12}$.

Solution

Note: $\log 2 = 0.3010$ *means* $\log_{10} 2 = 0.3010$ *which implies* $10^{0.3010} = 2$; $\log 3 = 0.4771$ *implies* $10^{0.4771} = 3$.

$$\log\sqrt{12} = \log 12^{1/2} = \frac{1}{2}\log 12$$

$$= \frac{1}{2}\log(3 \cdot 4) = \frac{1}{2}[\log 3 + \log 4]$$

$$= \frac{1}{2}[\log 3 + \log 2^2]$$

$$= \frac{1}{2}[\log 3 + 2\log 2]$$

$$= \frac{1}{2}\log 3 + \log 2$$

$$= \frac{1}{2}(0.4771) + 0.3010$$

$$= 0.5396 \qquad \blacksquare$$

Example 3 was used to illustrate the use of the laws of logarithms. However, in actual work, to find the logarithm of a number to base 10 we would use a scientific calculator. Thus, to find log $\sqrt{12}$ on a calculator use this key sequence:

$$\boxed{12} \quad \boxed{\sqrt{}} \quad \boxed{\log}$$

On a graphing calculator follow this sequence:

$$\boxed{\log} \quad \boxed{\sqrt{}} \quad \boxed{12} \quad \boxed{\text{ENTER}}$$

The following formula allows us to change bases of a logarithm:

FORMULA FOR CHANGE OF BASE OF LOGARITHMS

For $N > 0$, and a and b positive real numbers $\neq 1$,

$$\log_b N = \frac{\log_a N}{\log_a b}$$

Here is a proof of this formula:

$$\text{Let } y = \log_b N$$

Then

$$b^y = N \qquad \text{Change to exponential form.}$$

$$\log_a b^y = \log_a N \qquad \text{Take the log base } a \text{ of each side.}$$

$$y \log_a b = \log_a N \qquad \text{Law 3}$$

$$y = \frac{\log_a N}{\log_a b}$$

Most logarithms are irrational numbers. When found on a calculator and rounded to four decimal places, the results are reasonably close approximations. However, for the sake of simplicity, we use the equal sign, as done here, for $\log_5 130 = 3.0244$.

As an example of the use of this formula, let us evaluate $\log_5 130$. Change to base 10, and then use a calculator to find the result rounded to four decimal places as follows:

$$\log_5 130 = \frac{\log_{10} 130}{\log_{10} 5} = \frac{\log 130}{\log 5} = 3.0244$$

Logarithms are involved in a variety of formulas that deal with natural phenomena. For example, the loudness of sound is measured in **decibels,** named after Alexander Graham Bell, and can be found by the following formula:

$$D = 10 \log\left(\frac{I}{I_0}\right)$$

D: number of decibels

I: intensity of sound, as measured in watts per square centimeter

I_0:   intensity of the lowest sound that can be heard, universally agreed to be 10^{-16} watts per square centimeter

Thus the formula shows D as a function of I.

EXAMPLE 4   Find the number of decibels in normal conversation if the intensity is 10^{-10} watts per square centimeter.

Solution   Use the formula with $I = 10^{-10}$ and $I_0 = 10^{-16}$.

$$D = 10 \log\left(\frac{10^{-10}}{10^{-16}}\right) = 10(\log 10^6)$$

$$= 10 \cdot 6 = 60 \qquad \log_{10} 10^6 = 6$$

The number of decibels is 60.   ■

EXAMPLE 5   Find the intensity of a sound of 120 decibels.

Solution

$$120 = 10 \log\left(\frac{I}{I_0}\right)$$

$$12 = \log\left(\frac{I}{I_0}\right)$$

$$10^{12} = \left(\frac{I}{I_0}\right) \qquad\qquad \text{Change to exponential form.}$$

Thus $I = (I_0)(10^{12}) = (10^{-16})(10^{12})$

$$= 10^{-4}$$

The intensity is 10^{-4} watts per square centimeter, considered to be the threshold of pain from sound.   ■

TEST YOUR UNDERSTANDING

Convert the given logarithms into expressions involving $\log_b A$, $\log_b B$, and $\log_b C$.

1. $\log_b ABC$
2. $\log_b \dfrac{A}{BC}$
3. $\log_b \dfrac{(AB)^2}{C}$

4. $\log_b AB^2C^3$
5. $\log_b \dfrac{A\sqrt{B}}{C}$
6. $\log_b \dfrac{\sqrt[3]{A}}{(BC)^3}$

Change each expression into the logarithm of a single expression in x.

7. $\log_b x + \log_b x + \log_b 3$
8. $2 \log_b(x - 1) + \frac{1}{2} \log_b x$

9. $\log_b(2x - 1) - 3 \log_b(x^2 + 1)$

10. $\log_b x - \log_b(x - 1) - 2 \log_b(x - 2)$

Use the change of base formula and a calculator to find these logarithms to four decimal places.

11. $\log_3 75$
12. $\log_2 145$

(Answers: Page 359)

The following examples illustrate how the laws of logarithms can be used to solve **logarithmic equations**.

EXAMPLE 6   Solve for x: $\log_8(x - 6) + \log_8(x + 6) = 2$

Solution   First note that in $\log_8(x - 6)$ we must have $x - 6 > 0$, or $x > 6$. Similarly, $\log_8(x + 6)$ calls for $x > -6$. Therefore, the only solutions, if there are any, must satisfy $x > 6$.

CAUTION
$\log_8(x^2 - 36) \neq$
$\log_8 x^2 - \log_8 36$

$$\log_8(x - 6) + \log_8(x + 6) = 2$$

$$\log_8(x - 6)(x + 6) = 2 \qquad \text{Law 1}$$

$$\log_8(x^2 - 36) = 2$$

$$x^2 - 36 = 8^2 \qquad \text{converting to exponential form}$$

$$x^2 - 100 = 0$$

$$(x + 10)(x - 10) = 0$$

$$x = -10 \quad \text{or} \quad x = 10$$

The only possible solutions are -10 and 10. Our initial observation that $x > 6$ automatically eliminates -10. (If that initial observation had not been made, -10 could still have been eliminated by checking in the given equation.) The value $x = 10$ can be checked as follows:

$$\log_8(10 - 6) + \log_8(10 + 6) = \log_8 4 + \log_8 16$$

$$= \tfrac{2}{3} + \tfrac{4}{3} = 2 \qquad \blacksquare$$

EXAMPLE 7   Solve for x: $\log_{10}(x^3 - 1) - \log_{10}(x^2 + x + 1) = 1$

Solution

$$\log_{10}(x^3 - 1) - \log_{10}(x^2 + x + 1) = 1$$

$$\log_{10} \frac{x^3 - 1}{x^2 + x + 1} = 1 \qquad \text{Law 2}$$

See page 54 for the factorization of the difference of two cubes; $a^3 - b^3$.

Then factor as follows:

$$\log_{10} \frac{(x - 1)(x^2 + x + 1)}{x^2 + x + 1} = 1$$

$$\log_{10}(x - 1) = 1$$

$$x - 1 = 10^1 \qquad \text{Why?}$$

$$x = 11$$

Check: $\log_{10}(11^3 - 1) - \log_{10}(11^2 + 11 + 1) = \log_{10} 1330 - \log_{10} 133$

$$= \log_{10} \tfrac{1330}{133}$$

$$= \log_{10} 10 = 1 \qquad \blacksquare$$

Sometimes it is convenient to solve a logarithmic equation using the one-to-one property of logarithmic functions. This property (stated on page 288) says:

$$\text{If } \log_b M = \log_b N, \text{ then } M = N.$$

Without the one-to-one property, the solution begins with

$$\log_3 \frac{2x}{x + 5} = 0$$

$$\frac{2x}{x + 5} = 3^0 = 1$$

EXAMPLE 8   Solve for x: $\log_3 2x - \log_3(x + 5) = 0$

Solution

$$\log_3 2x - \log_3(x + 5) = 0$$

$$\log_3 2x = \log_3(x + 5)$$

$$2x = x + 5 \qquad \text{by the one-to-one property}$$

$$x = 5$$

Check: $\log_3 2(5) - \log_3(5 + 5) = \log_3 10 - \log_3 10 = 0$ ∎

The final example demonstrates how the laws of logarithms can be used when graphing a logarithmic function.

EXAMPLE 9   Find the domain of $y = f(x) = \log_2 \dfrac{4}{x - 2}$ and graph. Identify the asymptote and the intercepts if any.

Solution Since $\dfrac{4}{x - 2}$ must be positive, the domain consists of all $x > 2$. There is no y-intercept since $x = 0$ is not in the domain of f. To find the x-intercepts, set $y = 0$ and solve for x:

$$\log_2 \frac{4}{x - 2} = 0$$

$$\frac{4}{x - 2} = 2^0 = 1 \qquad \text{Convert to exponential form.}$$

$$4 = x - 2$$

$$x = 6 \qquad \text{The } x\text{-intercept is 6.}$$

Using Law 2 of logarithms, we have

$$y = \log_2 \frac{4}{x - 2} = \log_2 4 - \log_2(x - 2)$$

$$= 2 - \log_2(x - 2)$$

This form shows that $x = 2$ is the vertical asymptote. The graph of f can now be obtained in three stages:

1. Graph $y = \log_2(x - 2)$, which is the same curve as $y = \log_2 x$ moved two units to the right. It is not necessary to graph $y = \log_2 x$.

2. Reflect the curve in stage 1 through the x-axis to obtain the curve $y = -\log_2(x - 2)$.

3. Shift the curve in stage 2 two units up to get the graph of f.

Can you find the coordinates of the point P of intersection algebraically?

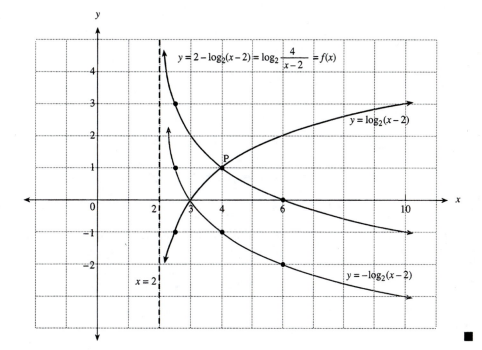

CAUTION: Learn to Avoid These Mistakes	
WRONG	**RIGHT**
$\log_b A + \log_b B = \log_b(A + B)$	$\log_b A + \log_b B = \log_b AB$
$\log_b(x^2 - 4) = \log_b x^2 - \log_b 4$	$\log_b(x^2 - 4)$ $= \log_b(x + 2)(x - 2)$ $= \log_b(x + 2) + \log_b(x - 2)$
$(\log_b x)^2 = 2 \log_b x$	$(\log_b x)^2 = (\log_b x)(\log_b x)$
$\log_b A - \log_b B = \dfrac{\log_b A}{\log_b B}$	$\log_b A - \log_b B = \log_b \dfrac{A}{B}$
If $2 \log_b x = \log_b(3x + 4)$, then $2x = 3x + 4$.	If $2 \log_b x = \log_b(3x + 4)$, then $\log_b x^2 = \log_b(3x + 4)$ and $x^2 = 3x + 4$.
$\log_b \dfrac{x}{2} = \dfrac{\log_b x}{2}$	$\log_b \dfrac{x}{2} = \log_b x - \log_b 2$
$\log_b(x^2 + 2) = 2 \log_b(x + 2)$	$\log_b(x^2 + 2)$ cannot be simplified further.

EXERCISES 5.3

Use the laws of logarithms (as much as possible) to convert the given logarithms into expressions involving sums, differences, and multiples of logarithms.

1. $\log_b \dfrac{3x}{x+1}$ **2.** $\log_b \dfrac{x^2}{x-1}$ **3.** $\log_b \dfrac{\sqrt{x^2-1}}{x}$ **4.** $\log_b \dfrac{1}{x}$

5. $\log_b \dfrac{1}{x^2}$ **6.** $\log_b \sqrt{\dfrac{x+1}{x-1}}$ **7.** $\log_b \left(\dfrac{2x-5}{x^3}\right)$ **8.** $\log_b \dfrac{\sqrt[3]{x+1}}{(x-2)(x+2)^5}$

Convert each expression into the logarithm of a single expression in x in simplified form.

9. $\log_b(x+1) - \log_b(x+2)$ **10.** $\log_b x + 2\log_b(x-1)$ **11.** $\frac{1}{2}\log_b(x^2-1) - \frac{1}{2}\log_b(x^2+1)$

12. $\log_b(x+2) - \log_b(x^2-4)$ **13.** $3\log_b x - \log_b 2 - \log_b(x+5)$ **14.** $\frac{1}{3}\log_b(x-1) + \log_b 3 - \frac{1}{3}\log_b(x+1)$

15. $\log_b(x^2-x-6) - \log_b(x+2)$ **16.** $\frac{1}{2}[\log_b(x^2+4x+4) - \log_b(x+2)]$ **17.** $\frac{1}{3}[\log_b(x^3-8) - 2\log_b(x-2)]$

Use the appropriate laws of logarithms to explain why each statement is correct.

18. $\log_b 16 + \log_b 4 = \log_b 64$ **19.** $\log_b 27 + \log_b 3 = \log_b 243 - \log_b 3$

20. $\frac{1}{2}\log_b 0.0001 = -\log_b 100$ **21.** $-2\log_b \frac{4}{9} = \log_b \frac{81}{16}$

Find the logarithms by using the laws of logarithms and the given information that $\log_b 2 = 0.3010$, $\log_b 3 = 0.4771$, and $\log_b 5 = 0.6990$. Assume that all logs have base b.

22. (a) $\log \sqrt{2}$ **(b)** $\log 9$ **(c)** $\log 12$ **23. (a)** $\log 4$ **(b)** $\log 8$ **(c)** $\log \frac{1}{2}$

24. (a) $\log 50$ **(b)** $\log 10$ **(c)** $\log \frac{25}{6}$ **25. (a)** $\log 48$ **(b)** $\log \frac{2}{3}$ **(c)** $\log 125$

26. (a) $\log 0.2$ **(b)** $\log 0.25$ **(c)** $\log 2.4$ **27. (a)** $\log \sqrt[3]{5}$ **(b)** $\log \sqrt{20^3}$ **(c)** $\log \sqrt{900}$

Use the change of base formula and a calculator to find each logarithm rounded to four decimal places.

28. $\log_3 28$ **29.** $\log_4 120$ **30.** $\log_5 95$ **31.** $\log_8 64$ **32.** $\log_{12} 257$

Simplify. (Hint: Use Property 8, page 319.)

33. $b^{3\log_b 4}$ **34.** $5^{-4\log_5 2}$ **35.** $(6^{\log_6 x})^3$ **36.** $\log_2\left(\dfrac{2^{x+3}}{2^{x-1}}\right)$

Solve for x and check.

37. $\log_{10} x + \log_{10} 5 = 2$

38. $\log_{10} x + \log_{10} 5 = 1$

39. $\log_{10} 5 - \log_{10} x = 2$

40. $\log_{10}(x+21) + \log_{10} x = 2$

41. $\log_{12}(x-5) + \log_{12}(x-5) = 2$

42. $\log_3 x + \log_3(2x+51) = 4$

43. $\log_{16} x + \log_{16}(x-4) = \frac{5}{4}$

44. $\log_2(x^2) - \log_2(x-2) = 3$

45. $\log_{10}(3-x) - \log_{10}(12-x) = -1$

46. $\log_{10}(3x^2-5x-2) - \log_{10}(x-2) = 1$

47. $\log_{1/7} x + \log_{1/7}(5x-28) = -2$

48. $\log_{1/3} 12x^2 - \log_{1/3}(20x-9) = -1$

49. $\log_{10}(x^3-1) - \log_{10}(x^2+x+1) = -2$

50. $2\log_{10}(x-2) = 4$

51. $2\log_{25} x - \log_{25}(25-4x) = \frac{1}{2}$

52. $\log_3(8x^3+1) - \log_3(4x^2-2x+1) = 2$

53. Prove Law 2. $\left(Hint\text{: Follow the proof of Law 1 using } \dfrac{b^r}{b^s} = b^{r-s}.\right)$

54. Prove Law 3. (*Hint:* Use $(b^r)^k = b^{rk}$.)   **55.** Solve for x: $(x + 2) \log_b b^x = x$.

56. Solve for x: $\log_{N^2} N = x$.   **57.** Solve for x: $\log_x (2x)^{3x} = 4x$.

58. (a) Explain why $\log_b b = 1$.

    (b) Show that $(\log_b a)(\log_a b) = 1$. (*Hint:* Use Law 3 and the result $b^{\log_b x} = x$.)

59. Find the intensity of the sound of a train that is measured at 100 decibels.

60. Find the number of decibels for the sound of a jet takeoff if the intensity is measured as 10^{-2} watts per square centimeter.

61. How does a sound that measures 120 decibels compare with one of 100 decibels?

62. The measurement of the strength of earthquakes is done using common logarithms. In particular, the Richter scale measures the magnitude R of an earthquake of intensity I by the formula

$$R = \log \frac{I}{I_0} \quad \text{where } I_0 \text{ is a minimum perceptible intensity.}$$

    (a) Find the magnitude of an earthquake whose intensity is $100\,I_0$.

    (b) One earthquake is 1000 times as intense as another. How do their magnitudes on the Richter scale compare?

Find the domain and graph each function. Identify the asymptotes and intercepts, if any.

63. $y = \log_5 x^3$

64. $y = \log_2 \sqrt[3]{8x}$

65. $y = \log_4 \sqrt{x + 1}$

66. $y = \log_3 (3x - 6)$

67. $y = \log_5 \dfrac{1}{x - 2}$

68. $y = \log_{10} \dfrac{100}{x + 2}$

69. $y = \log_2 (x^2 + x) - \log_2 x$

70. $y = \log_5 \dfrac{1}{x} + \log_5 5x$

  Challenge Prove that $\log_9 16 = \log_3 4$. Do not use log tables or a calculator.

  Written Assignment State Laws 2 and 3 of logarithms in your own words, that is, without symbols.

  Graphing Calculator Exercises

1. The formula $D = 10 \log(I/I_0)$ can also be written in the form $D = 10 \log I + 120$, where $I_0 = 10^{-12}$ and both I_0 and I are measured in watts per square meter. (a) Verify that, given the change in units from centimeters to meters, this formula for D is equivalent to that in Example 4. (b) Using a suitable window, graph D as a function of $\log I$. (This is known as a "semi-log plot.")

2. In the graphing calculator exercises of Section 5.2, you were asked to graph $f(x) = 2 + \log_2 x$ and its inverse on the same set of axes. Since your calculator probably does not have a $\log_2 x$ key, you had to graph the inverse first and then use it to find the graph of $y = f(x)$. In this exercise, graph $y = f(x)$ directly, using the results of this section, and compare your answer to the one you gave there.

3. Kepler's third law of planetary motion states that $Y^2/D^3 = 0.001$ (approximately), where D is the average distance, in millions of miles, of the planet from the sun and Y is the time, in years, of its period of revolution around the sun. Express log Y as a linear function of log D, and graph this function. (This is known as a log-log plot.)

4. Graph $y = \log(x^2 - 4)$. State the domain, and indicate the intercepts and asymptotes, if any.

5. Repeat Exercise 4 for the function $y = \log(x^2 + 4)$.

Critical Thinking

1. Consider the inequalities $2^x < 3^x$, $2^x > 3^x$ and the equation $2^x = 3^x$. Decide where each of them is true and where they are false.

2. A simplified form of $\sqrt{3^{2x} + 2 + 3^{-2x}}$ that does not involve radicals or fractional exponents can be obtained after factoring the radicand. Find this simplification.

3. For the function $f(x) = \log_2 x$, form the composition $f \circ f \circ f$ and determine x for which $f(f(f(x))) = 2$. Also, generalize the preceding by replacing the base 2 by b and solving for x.

4. Make use of your knowledge of logarithmic functions to determine the values of c and d, in $y = f(x) = \log_b(cx + d)$, so that the graph of f has the vertical asymptote $x = -\frac{1}{2}$ and x-intercept 0.

5. The function $f(x) = \log_{10} x$ is increasing on the interval $(0, \infty)$. It is also true that the value of $\log_{10} x$ becomes arbitrarily large with increasing values of x. Find the values of x for which each of the following inequalities is true:

$$\log_{10} x > 3, \quad \log_{10} x > 30, \quad \log_{10} x > 300$$

As x increases, would you describe the growth of $\log_{10} x$ as fast, moderate, or slow?

5.4 THE NATURAL EXPONENTIAL AND LOGARITHMIC FUNCTIONS

The expression $\left(1 + \frac{1}{n}\right)^n$, where n represents a positive integer, plays an important role in the study of calculus and, in the next section, it is involved in the development of compound interest. When this expression is evaluated for n equal to 2, 4, and 5, the resulting values, to two decimal places are 2.25, 2.44, and 2.48, respectively. As you can see, the values are increasing. Will this continue? How large can $\left(1 + \frac{1}{n}\right)^n$ become as n gets larger?

Since $\frac{1}{n} \longrightarrow 0$ as $n \longrightarrow \infty$, it follows that $1 + \frac{1}{n} \longrightarrow 1$, and you might conclude that the nth power of these numbers, $\left(1 + \frac{1}{n}\right)^n$, will also get close to 1. This is a common reaction, but with the help of a calculator we can explore the values of this expression as n gets large and see what does happen.

n	$\left(1 + \dfrac{1}{n}\right)^n$
1	2.0000000
10	2.5937425
100	2.7048138
1,000	2.7169239
10,000	2.7181459
100,000	2.7182682
1,000,000	2.7182805
10,000,000	2.7182817

From the results in the table, it appears that the values of the given expression increase and get close to 2.71828. In fact, it can be shown in calculus that $\left(1 + \dfrac{1}{n}\right)^n$ approaches a specific number that is designated by the letter e. That is

$$\left(1 + \frac{1}{n}\right)^n \longrightarrow e \quad \text{as} \quad n \longrightarrow \infty$$

Like the number π, e is an irrational number and plays an important role in mathematics and its applications. (See Section 5.5.) Since it is an irrational number, its decimal representation is a nonterminating, nonrepeating decimal. Here are the first 15 decimal places for e:

$$e = 2.718281828459045 \ldots$$

e is the most important base for both exponential and logarithmic functions. We begin with the exponential function.

HISTORICAL NOTE
The discovery of e is generally attributed to the Swiss mathematician Leonhard Euler (1707–1783). Euler was one of the most prolific mathematicians of all times, having written one of the first textbooks on calculus, as well as over 60 volumes of original work. He is given credit for a formula that relates some of the most important numbers in mathematics:

$e^{\pi i} + 1 = 0$
where $i = \sqrt{-1}$.

A Swiss stamp honoring Euler.
Source: Dr. Marvin Lang.

DEFINITION OF THE NATURAL EXPONENTIAL FUNCTION

The natural exponential function for all real numbers x is defined by the following equation:

$$y = e^x$$

Since $2 < e < 3$, the graph of the natural exponential function $y = e^x$ has the same shape as the curves $y = b^x$ for $b > 1$, and lies between the graphs of $y = 2^x$ and $y = 3^x$, as shown. You can use the $\boxed{e^x}$ key on your calculator for specific values of x to locate points on the graph.

In the study of calculus, the concept of a tangent to a circle is extended to other curves. There you will learn that the tangent to the curve $y = e^x$ at point $(0, 1)$ has a slope equal to 1. If you carefully draw the line that touches this curve at $(0, 1)$, and at no other point, you should find its slope to be close to 1. Also see Critical Thinking, Exercise 3, page 352.

x	e^x
-3	0.05
-2	0.14
-1	0.37
0	1
1	2.72
2	7.39

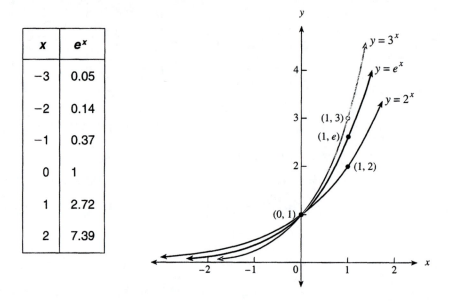

EXAMPLE 1 Graph $y = g(x) = e^{-x}$ and $y = h(x) = -e^x$ on the same axes. Find the range of each function.

Solution Rather than plot points, we can sketch these curves by making use of the graph of $y = f(x) = e^x$, as shown in the following figure. The reflection of the graph of $y = e^x$ through the y-axis gives the graph of $y = e^{-x}$. (Compare this with the graphs of $y = 2^x$ and $y = 2^{-x}$ on page 310.) Then reflect the graph of $y = e^x$ through the x-axis to find the graph of $y = -e^x$.

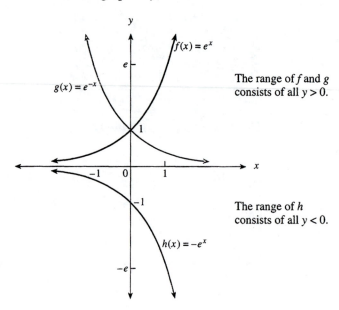

The range of f and g consists of all $y > 0$.

The range of h consists of all $y < 0$.

■

The inverse of $y = e^x$ is given by $y = \log_e x$. In place of $\log_e x$ we usually write **ln x**, which is called the **natural log of x**. Thus $x = e^y$ and $y = \ln x$ are equivalent. As noted when we studied inverse functions, the graph of $y = \ln x$ can be obtained by reflecting the graph of $y = e^x$ through the line $y = x$. The reflections of three key points are also shown.

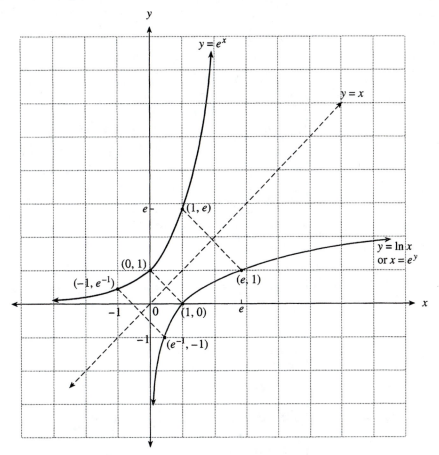

Since $f(x) = e^x$ and $g(x) = \ln x$ are inverse functions, we know that

$$x = f(g(x)) = f(\ln x) = e^{\ln x} \qquad \text{for all } x > 0$$

and

$$x = g(f(x)) = g(e^x) = \ln e^x \qquad \text{for all } x$$

This gives these two important properties:

$$e^{\ln x} = x \qquad \text{for all } x > 0$$

$$\ln e^x = x \qquad \text{for all real } x$$

Illustrations:

$$e^{\ln 9.3} = 9.3; \qquad e^{-\ln 4} = \frac{1}{e^{\ln 4}} = \frac{1}{4}$$

$$\ln e^{12} = 12; \qquad \ln \sqrt{e} = \ln e^{1/2} = \frac{1}{2}$$

These results are included in the lists that follow. Note that since $e > 1$, the properties of $y = b^x$ and $y = \log_b x$ ($b > 1$) carry over to $y = e^x$ and $y = \ln x$.

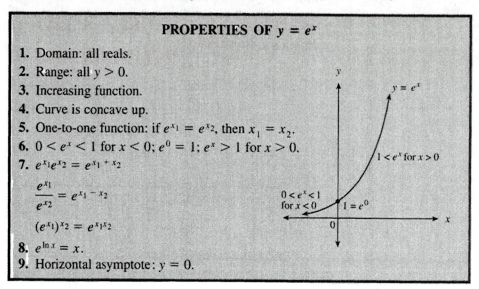

PROPERTIES OF $y = e^x$

1. Domain: all reals.
2. Range: all $y > 0$.
3. Increasing function.
4. Curve is concave up.
5. One-to-one function: if $e^{x_1} = e^{x_2}$, then $x_1 = x_2$.
6. $0 < e^x < 1$ for $x < 0$; $e^0 = 1$; $e^x > 1$ for $x > 0$.
7. $e^{x_1}e^{x_2} = e^{x_1 + x_2}$

$$\frac{e^{x_1}}{e^{x_2}} = e^{x_1 - x_2}$$

$$(e^{x_1})^{x_2} = e^{x_1 x_2}$$

8. $e^{\ln x} = x$.
9. Horizontal asymptote: $y = 0$.

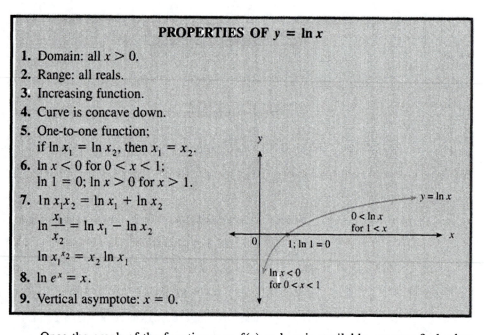

PROPERTIES OF $y = \ln x$

1. Domain: all $x > 0$.
2. Range: all reals.
3. Increasing function.
4. Curve is concave down.
5. One-to-one function: if $\ln x_1 = \ln x_2$, then $x_1 = x_2$.
6. $\ln x < 0$ for $0 < x < 1$; $\ln 1 = 0$; $\ln x > 0$ for $x > 1$.
7. $\ln x_1 x_2 = \ln x_1 + \ln x_2$

$$\ln \frac{x_1}{x_2} = \ln x_1 - \ln x_2$$

$$\ln x_1^{x_2} = x_2 \ln x_1$$

8. $\ln e^x = x$.
9. Vertical asymptote: $x = 0$.

Note that the graph of $y = \ln x - 3$ would be found by shifting the graph of $y = \ln x$ three units downward.

Once the graph of the function $y = f(x) = \ln x$ is available, we can find other graphs in much the same manner as we did earlier for other bases. For example, see page 317 where the graph of $y = \log_2(x - 3)$ was found by shifting the graph of $y = \log_2 x$ three units to the right. In a similar manner, to find the graph of $y = \ln(x - 3)$ we also shift the graph of $y = \ln x$ three units to the right. The domain of $f(x) = \ln(x - 3)$ consists of all x for which $x - 3 > 0$, that is, $x > 3$. The next example calls both for the use of symmetry as well as Law 3 of logarithms.

EXAMPLE 2 Graph $y = f(x) = \ln x^2$.

Solution Since $x^2 \geq 0$, the domain consists of all $x \neq 0$. Also the curve is symmetric about the y-axis since $f(-x) = \ln(-x)^2 = \ln x^2 = f(x)$. (It is an even function.) Also note that by Law 3 for logarithms, $\ln x^2 = 2 \ln x$ for $x > 0$. Thus we can draw the graph of $y = 2 \ln x$ by doubling the ordinates of $y = \ln x$, and then reflect through the y-axis to complete the graph.

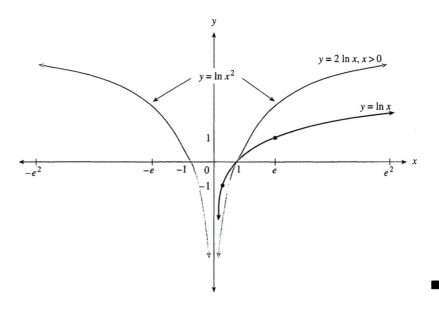

The examples that follow illustrate a variety of algebraic techniques for simplifying expressions and for solving exponential and logarithmic equations.

EXAMPLE 3 Let $f(x) = \dfrac{3x^2}{x^2 + 4}$. Use the laws of logarithms to write $\ln f(x)$ as an expression involving sums, differences, and multiples of natural logarithms.

Whenever M = N it follows from the definition of a function that ln M = ln N. That is, for equal domain values M and N, there can be only one range value.

Solution Since $f(x) = \dfrac{3x^2}{x^2 + 4}$, we can proceed as follows.

$$\ln f(x) = \ln \frac{3x^2}{x^2 + 4}$$

$$= \ln 3x^2 - \ln(x^2 + 4) \qquad \text{by Law 2 of logarithms}$$

$$= \ln 3 + \ln x^2 - \ln(x^2 + 4) \qquad \text{by Law 1 of logarithms}$$

$$= \ln 3 + 2 \ln x - \ln(x^2 + 4) \qquad \text{by Law 3 of logarithms} \qquad \blacksquare$$

EXAMPLE 4 Convert $\dfrac{1}{2} \ln(x^2 + 1) - \ln(x^2 - x - 2) + \ln(x + 1)$ into the natural log of a single expression in x.

Solution

Provide a reason for each step in this solution.

$$\frac{1}{2}\ln(x^2 + 1) - \ln(x^2 - x - 2) + \ln(x + 1)$$

$$= \ln(x^2 + 1)^{1/2} + \ln(x + 1) - \ln(x^2 - x - 2)$$

$$= \ln\sqrt{x^2 + 1} + \ln(x + 1) - \ln(x - 2)(x + 1)$$

$$= \ln\frac{\sqrt{x^2 + 1}(x + 1)}{(x - 2)(x + 1)}$$

$$= \ln\frac{\sqrt{x^2 + 1}}{x - 2}$$

∎

EXAMPLE 5 Solve for t: $e^{\ln(2t - 1)} = 5$

Solution

$$e^{\ln(2t - 1)} = 5$$

$$2t - 1 = 5 \qquad \text{Using } e^{\ln x} = x \text{ with } x = 2t - 1$$

$$2t = 6$$

$$t = 3$$

∎

EXAMPLE 6 Solve for t: $e^{2t - 1} = 5$

Solution Rewrite the exponential expression in logarithmic form.

$$e^{2t - 1} = 5$$

$$2t - 1 = \ln 5 \qquad\qquad \log_e 5 = \ln 5 = 2t - 1$$

$$2t = 1 + \ln 5$$

Using a calculator, to three decimal places

$t = \frac{1}{2}(1 + \ln 5) = 1.305.$

$$t = \frac{1}{2}(1 + \ln 5)$$

Check: $e^{2[1/2(1 + \ln 5)] - 1} = e^{1 + \ln 5 - 1} = e^{\ln 5} = 5$

∎

EXAMPLE 7 Solve for x: $\ln(x + 1) = 1 + \ln x$

Solution

$$\ln(x + 1) - \ln x = 1$$

$$\ln\frac{x + 1}{x} = 1$$

Now convert to exponential form:

$$\frac{x + 1}{x} = e$$

Recall: If $\log_b x = y$, then $b^y = x$.

$\ln\frac{x + 1}{x} = \log_e\frac{x + 1}{x} = 1$

So, $e^1 = \frac{x + 1}{x}$.

$$ex = x + 1$$

$$ex - x = 1$$

$$(e - 1)x = 1$$

$$x = \frac{1}{e - 1}$$

Check: $\ln\left(\dfrac{1}{e-1} + 1\right) = \ln\dfrac{e}{e-1} = \ln e - \ln(e-1)$

$$= 1 + \ln(e-1)^{-1} = 1 + \ln\dfrac{1}{e-1} \qquad\blacksquare$$

EXAMPLE 8 **(a)** Express $h(x) = \ln(x^2 + 5)$ as the composite of two functions.
(b) Express $F(x) = e^{\sqrt{x^2-3x}}$ as the composite of three functions.

Solution
(a) Let $f(x) = \ln x$ and $g(x) = x^2 + 5$. Then

$$(f \circ g)(x) = f(g(x)) = f(x^2 + 5) = \ln(x^2 + 5) = h(x)$$

(b) Let $f(x) = e^x$, $g(x) = \sqrt{x}$, and $h(x) = x^2 - 3x$. Then

$$
\begin{aligned}
(f \circ g \circ h)(x) &= f(g(h(x))) &&\text{\textit{h} is the ``inner'' function.}\\
&= f(g(x^2 - 3x)) &&\text{\textit{g} is the ``middle'' function.}\\
&= f(\sqrt{x^2 - 3x}) &&\text{\textit{f} is the ``outer'' function.}\\
&= e^{\sqrt{x^2-3x}} = F(x)
\end{aligned}
$$

(Other solutions are possible.) $\qquad\blacksquare$

EXAMPLE 9 Find the x-intercepts and domain of $f(x) = x^2 e^x + 2xe^x$.

Solution The x-intercepts can be found by solving the equation

$$x^2 e^x + 2xe^x = 0$$

By factoring, this equation can be rewritten as

$$xe^x(x + 2) = 0$$

Since $e^x > 0$ for all x, the product can only be equal to 0 if $x = 0$ or if $x + 2 = 0$. Thus the solutions to the equation are 0 and -2, which are also the x-intercepts for the graph of f. Also, the domain of f consists of all real numbers since x^2, e^x, and $2x$ are all defined for such x. $\qquad\blacksquare$

The final example illustrates how the quadratic formula is used to solve an exponential equation.

EXAMPLE 10 Solve for x: $\dfrac{e^x - e^{-x}}{2} = 2$

Solution

$$\frac{e^x - e^{-x}}{2} = 2$$

$$e^x - e^{-x} = 4$$

$$(e^x)(e^x) - (e^{-x})(e^x) = 4e^x \qquad \text{Multiply by } e^x.$$

$$(e^x)^2 - 1 = 4e^x \qquad e^{-x} \cdot e^x = e^0 = 1$$

$$(e^x)^2 - 4e^x - 1 = 0$$

Now let $u = e^x$ to give a quadratic equation in the variable u.

*Note: The Challenge on page
342 calls for a check of this
solution.*

$$u^2 - 4u - 1 = 0$$

$$u = 2 \pm \sqrt{5} \qquad \text{by the quadratic formula}$$

$$e^x = 2 \pm \sqrt{5} \qquad \text{since } u = e^x$$

Since $e^x > 0$ for all x, and $2 - \sqrt{5} < 0$, we only use the positive solution for e^x:

$$e^x = 2 + \sqrt{5}$$

To find the solution for x, convert to logarithmic form:

$$x = \ln(2 + \sqrt{5})$$ ■

EXERCISES 5.4

Sketch each pair of functions on the same axes.

1. $y = e^x; \ y = e^{x-2}$ **2.** $y = e^x; \ y = 2e^x$

3. $y = \ln x; \ y = \frac{1}{2} \ln x$ **4.** $y = \ln x; \ y = \ln(x + 2)$

5. $y = \ln x; \ y = \ln(-x)$ **6.** $y = e^x; \ y = e^{|x|}$

7. $y = e^x; \ y = e^x + 2$ **8.** $y = \ln x; \ y = \ln|x|$

9. $f(x) = -e^{-x}; \ g(x) = 1 - e^{-x}$ **10.** $g(x) = 1 - e^{-x}; \ s(x) = 1 - e^{-2x}$

11. $g(x) = 1 - e^{-x}; \ t(x) = 1 - e^{(-1/2)x}$ **12.** $u(x) = 1 - e^{-3x}; \ v(x) = 1 - e^{(-1/3)x}$

Explain how the graph of f can be obtained from the curve $y = \ln x$. (Hint: First apply the appropriate laws of logarithms.)

13. $f(x) = \ln ex$ **14.** $f(x) = \ln \dfrac{x}{e}$ **15.** $f(x) = \ln \sqrt{x}$ **16.** $f(x) = \ln \dfrac{1}{x}$

17. $f(x) = \ln(x^2 - 1) - \ln(x + 1)$ **18.** $f(x) = \ln x^{-3}$

Find the domain and any x-intercepts for the graph of f.

19. $f(x) = \ln(x + 2)$ **20.** $f(x) = \ln|2x + 1|$ **21.** $f(x) = \ln(2x - 1)$

22. $f(x) = \dfrac{1}{\ln x}$ **23.** $f(x) = \dfrac{\ln(x - 1)}{x - 2}$ **24.** $f(x) = \ln(\ln x)$

Use the laws of logarithms (as much as possible) to write $\ln f(x)$ as an expression involving sums, differences, and multiples of natural logarithms.

25. $f(x) = \dfrac{5x}{x^2 - 4}$ **26.** $f(x) = x\sqrt{x^2 + 1}$ **27.** $f(x) = \dfrac{(x - 1)(x + 3)^2}{\sqrt{x^2 + 2}}$

28. $f(x) = \sqrt{\dfrac{x + 7}{x - 7}}$ **29.** $f(x) = \sqrt{x^3(x + 1)}$ **30.** $f(x) = \dfrac{x}{\sqrt[3]{x^2 - 1}}$

Convert each expression into the logarithm of a single expression.

31. $\frac{1}{2} \ln x + \ln(x^2 + 5)$ **32.** $\ln 2 + \ln x - \ln(x - 1)$

33. $3 \ln(x + 1) + 3 \ln(x - 1)$ **34.** $\ln(x^3 - 1) - \ln(x^2 + x + 1)$

35. $\frac{1}{2} \ln x - 2 \ln(x - 1) - \frac{1}{3} \ln(x^2 + 1)$

Simplify.

36. $\ln(e^{3x})$ **37.** $e^{\ln \sqrt{x}}$ **38.** $\ln(x^2 e^3)$ **39.** $e^{-2 \ln x}$ **40.** $(e^{\ln x})^2$ **41.** $\ln\left(\dfrac{e^x}{e^x - 1}\right)$

Solve for x.

42. $e^{3x+5} = 100$ **43.** $e^{-0.01x} = 27$ **44.** $e^{x^2} = e^x e^{3/4}$

45. $e^{\ln(1-x)} = 2x$ **46.** $\ln x + \ln 2 = 1$ **47.** $\ln(x + 1) = 0$

48. $\ln x = -2$ **49.** $\ln e^{\sqrt{x+1}} = 3$ **50.** $e^{\ln(6x^2 - 4)} = 5x$

51. $\ln(x^2 - 4) - \ln(x + 2) = 0$ **52.** $(e^{x+2} - 1)\ln(1 - 2x) = 0$ **53.** $\ln x = \frac{1}{2} \ln 4 + \frac{2}{3} \ln 8$

54. $\frac{1}{2} \ln(x + 4) = \ln(x + 2)$ **55.** $\ln x = 2 + \ln(1 - x)$ **56.** $\ln(x^2 + x - 2) = \ln x + \ln(x - 1)$

Graph each of the following. State the domain, intercepts, and equations of the asymptotes, if any.

57. $f(x) = \ln(x - 4)$ **58.** $f(x) = \ln x - 4$ **59.** $f(x) = \ln(4 - x)$

60. $f(x) = \ln x^2 - 4$ **61.** $f(x) = \ln x^3$

Show that each function is the composite of two functions.

62. $h(x) = e^{2x+3}$ **63.** $h(x) = e^{-x^2+x}$ **64.** $h(x) = \ln(1 - 2x)$

65. $h(x) = \ln \dfrac{x}{x + 1}$ **66.** $h(x) = (e^x + e^{-x})^2$ **67.** $h(x) = \sqrt[3]{\ln x}$

Show that each function is the composite of three functions.

68. $F(x) = e^{\sqrt{x+1}}$ **69.** $F(x) = e^{(3x-1)^2}$

70. $F(x) = [\ln(x^2 + 1)]^3$ **71.** $F(x) = \ln \sqrt{e^x + 1}$

Determine the domain of each function. Also find the roots of $f(x) = 0$.

72. $f(x) = xe^x + e^x$ **73.** $f(x) = e^{2x} - 2xe^{2x}$

74. $f(x) = -3x^2 e^{-3x} + 2xe^{-3x}$ **75.** $f(x) = 1 + \ln x$

76. Show that $(e^x + e^{-x})^2 - (e^x - e^{-x})^2 = 4$.

77. Show that $\ln\left(\dfrac{x}{4} - \dfrac{\sqrt{x^2 - 4}}{4}\right) = -\ln\left(x + \sqrt{x^2 - 4}\right)$.   **78.** Solve for x: $\dfrac{e^x + e^{-x}}{2} = 1$.

79. Solve for x in terms of y if $y = \dfrac{e^x}{2} - \dfrac{1}{2e^x}$. (*Hint:* Let $u = e^x$ and solve the resulting quadratic in u.)

80. For $f(x) = e^x$, show that $\dfrac{f(2 + h) - f(2)}{h} = e^2\left(\dfrac{e^h - 1}{h}\right)$.

81. For $g(x) = \ln x$ show that $\dfrac{g(2 + h) - g(2)}{h} = \dfrac{1}{h} \ln\left(1 + \dfrac{h}{2}\right)$.

82. For $f(x) = e^x$, find and simplify $\dfrac{f\left(\ln\frac{x}{2}\right) - f\left(\ln\frac{1}{2}\right)}{x - 1}$.

83. For $g(x) = \ln x$, find and simplify $\dfrac{g(e^{3x}) - g(e^6)}{x - 2}$.

84. Solve the inequality $\ln x > 3$.

85. Solve the inequality $e^x > 500$.

Challenge

1. Solve for x: $\log_b(\log_b nx) = 1 \ (n > 0)$.

2. Check the solution for Example 10 on page 339.

Graphing Calculator Exercises

1. Draw the graphs of $y = e^x$ and $y = e^{(-x)}$ on the same coordinate axes. By estimating the addition and averaging of ordinates try to sketch the function $y = (0.5)(e^x + e^{(-x)})$. Then use your graphing calculator to show the graph of this function.

2. Solve the equation $e^x = x^3$ by drawing the graphs of $y = e^x$ and $y = x^3$ on the same axes. Find an appropriate window and use the ZOOM and TRACE functions to find the result to two decimal places.

3. Graph the following function that defines the normal distribution curve used extensively in statistics. Estimate the maximum of f to two decimal places.

$$f(x) = \frac{1}{\sqrt{2\pi}} \, e^{-\frac{x^2}{2}}$$

4. The graph of $f(x) = 2/(1 + e^{-2x})$ is an example of a limited growth curve, or S-curve. Graph it on a suitable window, and find its intercepts and asymptotes, if any.

5.5
APPLICATIONS: EXPONENTIAL GROWTH AND DECAY

The number e to three decimal places is 2.718. This can be written as $e^1 = 2.718$. Other powers of e can be found by using a calculator with an $\boxed{e^x}$ key. For example, verify the following on your calculator.

$$e^{2.1} = 8.166 \qquad e^{-2.1} = 0.122$$

Here and throughout we will use the equal sign even though it is evident that the results are approximations that have been rounded (in this case) to three decimal places.

The common logarithms of numbers can be found on a calculator by using the $\boxed{\log}$ key, and the $\boxed{\ln}$ key can be used to find the natural logarithm of numbers. Verify each of the following, rounded to three decimal places:

Recall that $\log x$ *implies* $\log_{10} x$, *and* $\ln x$ *implies* $\log_e x$.

$$\log 2.1 = 0.322 \qquad \log 0.25 = -0.602$$

$$\ln 2.1 = 0.742 \qquad \ln 0.25 = -1.386$$

Although tables are available in the back of the book to find these values, all of the work that follows assumes that a scientific calculator will be used. We begin with the solution of an exponential equation in Example 1.

EXAMPLE 1   Solve the exponential equation $2^x = 35$ for x.

Solution

$$2^x = 35$$

$$\ln 2^x = \ln 35 \qquad \text{If } A = B, \text{ then } \ln A = \ln B.$$

$$x \ln 2 = \ln 35 \qquad \text{Why?}$$

$$x = \frac{\ln 35}{\ln 2}$$

The key sequence to evaluate this quotient on a scientific calculator is as follows:

Note that the key sequence will be different if a graphing calculator is being used.

Then press ENTER or EXE (for "execute"), or some other version of "=" to obtain the result of 5.129, rounded to three decimal places.   ∎

TEST YOUR UNDERSTANDING

(Answers: Page 360)

Solve each equation for x using natural logarithms. Approximate the answer to three decimal places.

1. $4^x = 5$ **2.** $4^{-x} = 5$ **3.** $\left(\frac{1}{2}\right)^x = 12$

4. $2^{3x} = 10$ **5.** $4^x = 15$ **6.** $67^{5x} = 4$

The procedure used in solving the preceding exponential equations will be needed in a variety of applied problems. The first of these is an application of exponential growth. Imagine that a bacterial culture is growing at a rate such that after each hour the number of bacteria has doubled. Thus, if there were 10,000 bacteria when the culture started to grow, then after 1 hour the number would have grown to 20,000, after 2 hours there would be 40,000, and so on. It becomes reasonable to say that

$$20,000 = (10,000)2^1$$
$$40,000 = (10,000)2^2$$
$$80,000 = (10,000)2^3$$

$$y = f(x) = (10,000)2^x$$

gives the number y of bacteria present after x hours of growth. At this rate, how long will it take for this bacterial culture to grow to 100,000? To answer this question we let $y = 100,000$ and solve for x.

$$(10,000)2^x = 100,000$$

$$2^x = 10 \qquad \text{Divide by 10,000.}$$

$$x \ln 2 = \ln 10$$

$$x = \frac{\ln 10}{\ln 2}$$

$$= 3.32$$

It will take about 3.3 hours.

In the preceding illustration the exponential and logarithmic functions were used to solve a problem of *exponential growth*. Many problems involving **exponential growth** or **exponential decay** can be solved by using the general formula

$$y = f(x) = Ae^{kx}$$

which shows how the amount of a substance y depends on the time x. Since $f(0) = A$, A represents the initial amount of the substance and k is a constant. In a given situation $k > 0$ signifies that y is growing (increasing) with time. For $k < 0$ the substance is decreasing. (Compare to the graphs of $y = e^x$ and $y = e^{-x}$ for $x \geq 0$.)

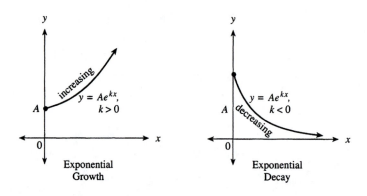

Exponential Growth

Exponential Decay

The preceding bacterial problem also fits this general form. This can be seen by substituting $2 = e^{\ln 2}$ into $y = (10,000)2^x$:

$$y = (10,000)2^x = (10,000)(e^{\ln 2})^x = 10,000e^{(\ln 2)x}$$

in which $A = 10,000$ and $k = \ln 2$.

EXAMPLE 2 A radioactive substance is decaying (it is changing into another element) according to the formula $y = Ae^{-0.2x}$, where y is the amount of the substance remaining after x years.

(a) If the initial amount $A = 80$ grams, how much is left after 3 years?
(b) The **half-life** of a radioactive substance is the time it takes for half of it to decompose. Find the half-life of this substance in which $A = 80$ grams.

Solution
(a) Since $A = 80$, $y = 80e^{-0.2x}$. We need to solve for the amount y when $x = 3$.

$$y = 80e^{-0.2x}$$
$$= 80e^{-0.2(3)}$$
$$= 80e^{-0.6} = 43.905$$

There will be about 43.9 grams after 3 years.

(b) This question calls for the time x at which only half of the initial amount is left. Consequently, the half-life x is the solution to $40 = 80e^{-0.2x}$. Divide each side by 80:

$$\frac{1}{2} = e^{-0.2x}$$

Take the natural log of both sides, or change to logarithmic form, to obtain $-0.2x = \ln \frac{1}{2}$. Since $\ln \frac{1}{2} = \ln 1 - \ln 2 = -\ln 2$, we solve for x as follows.

$$-0.2x = -\ln 2$$

$$x = \frac{\ln 2}{0.2} = 3.466$$

The half-life is approximately 3.47 years. ∎

Observe that the preceding computation could also be done using common logarithms, although it is somewhat more complicated. Begin by taking the common logarithm of $0.5 = e^{-0.2x}$ and solve for x to obtain

This process of finding the age of the remains is referred to as radioactive carbon dating.

$$x = \frac{\log 0.5}{-0.2 \log e} = 3.466$$

Carbon-14, also written as ^{14}C, is a radioactive isotope of carbon with a half-life of about 5750 years. By finding how much ^{14}C is contained in the remains of a formerly living organism, it becomes possible to determine what percentage this is of the original amount of ^{14}C at the time of death. Once this information is given, the formula $y = Ae^{kx}$ will enable us to date the age of the remains. The dating will be done after we solve for the constant k. Since the amount of ^{14}C after 5750 years will be $\frac{A}{2}$, we have the following:

A calculator gives $k = -0.0001205$. This verifies that the amount, y, of carbon-14 is decreasing (decaying).

$$\frac{A}{2} = Ae^{5750k}$$

$$\frac{1}{2} = e^{5750k}$$

$$5750k = \ln \frac{1}{2}$$

$$k = \frac{\ln 0.5}{5750}$$

Substitute this value for k into $y = Ae^{kx}$ to obtain the following formula for the amount of carbon-14 remaining after x years:

$$y = Ae^{(\ln 0.5/5750)x}$$

EXAMPLE 3 An animal skeleton is found to contain one-third of its original amount of ^{14}C. How old is the skeleton?

Solution Let x be the age of the skeleton. Then

$$\frac{1}{3}A = Ae^{(\ln 0.5/5750)x}$$

$$\frac{1}{3} = e^{(\ln 0.5/5750)x}$$

$$\left(\frac{\ln 0.5}{5750}\right)x = \ln\frac{1}{3} = -\ln 3$$

$$x = \frac{(5750)(-\ln 3)}{\ln 0.5}$$

$$= \frac{-5750 \ln 3}{\ln 0.5} = 9113.5$$

The skeleton is about 9000 years old, to the nearest 500 years.   ∎

The formulas used in the evaluation of **compound interest** are also applications of exponential growth. The statement that an investment earns compound interest means that the interest earned over a fixed period of time is added to the initial investment, and then the new total earns interest for the next investment period, and so on. For example, suppose an investment of P dollars earns interest at a yearly rate of r percent, and the interest is compounded annually. Then, after the first year the total value is the sum of the initial investment P plus the interest Pr (r is being used as a decimal fraction). Thus, the total after one year is

$$P + Pr = P(1 + r)$$

After the second year the total amount is $P(1 + r)$ plus the interest this amount earns, which is $P(1 + r)r$. Then the total after two years is

$$P(1 + r) + P(1 + r)r = P(1 + r)(1 + r) = P(1 + r)^2$$

This expression is simplified by factoring out $P(1 + r)$.

Similarly, after three years the total is

$$P(1 + r)^2 + P(1 + r)^2 r = P(1 + r)^2(1 + r) = P(1 + r)^3$$

and after t years the final amount A is given by

$$A = P(1 + r)^t$$

The interest periods for compound interest are usually less than one year. They could be quarterly (4 times per year), monthly, or daily, and so forth. For such cases the interest rate per period is the annual rate r divided by the number of interest periods per year. Thus, if the interest is compounded quarterly, the rate per period is $r/4$. Now, following the reasoning used to obtain $A = P(1 + r)^t$, the final amount A after one year (4 interest periods) is

$$A_1 = P\left(1 + \frac{r}{4}\right)^4$$

If there are n interest periods per year, the rate per period is r/n, and after one year we have

$$A_1 = P\left(1 + \frac{r}{n}\right)^n$$

Likewise, after t years the final amount A_t is given by

This result can be derived from the preceding result. See Exercise 58.

$$A_t = P\left(1 + \frac{r}{n}\right)^{nt}$$

COMPOUND INTEREST FORMULA

$$A_t = P\left(1 + \frac{r}{n}\right)^{nt}$$

P = initial investment

r = annual interest rate (as a decimal)

n = number of annual interest periods

t = number of years of investment

A_t = final value after t years

EXAMPLE 4   A \$5000 investment earns interest at the annual rate of 8.4% compounded monthly. Use a calculator to answer the following:

(a) What is the investment worth after one year?
(b) What is it worth after 10 years?
(c) How much interest was earned in 10 years?

Use the formula $I = pr$ to compute the simple interest that \$5000 would earn in 1 year. Compare this to the compound interest earned in part (a).

Solution

(a) Since the annual rate $r = 8.4\% = 0.084$ and the compounding is done monthly, the interest rate per month is $r/n = 0.084/12 = 0.007$. Substitute this, together with $P = 5000$ and $n = 12$, into $A = P\left(1 + \dfrac{r}{n}\right)^n$.

$$A = 5000(1 + 0.007)^{12} = 5000(1.007)^{12} = 5436.55$$

To the nearest dollar, the amount on deposit after one year is \$5437.

(b) Use $A_t = P\left(1 + \dfrac{r}{n}\right)^{nt}$ where $P = 5000$, $\dfrac{r}{n} = 0.007$, $n = 12$, and $t = 10$.

$$A = 5000(1.007)^{12(10)} = 5000(1.007)^{120} = 11{,}547.99$$

The amount after 10 years is approximately \$11,548.

(c) After 10 years the interest earned is

$$11{,}548 - 5000 = 6548 \text{ dollars}$$ ∎

Recall, from Section 5.4, that the values of $\left(1 + \dfrac{1}{n}\right)^n$ approach the number e as n gets larger and larger. It is also true that $\left(1 + \dfrac{r}{n}\right)^n$ approaches e^r as n gets larger and larger. This can be seen as follows

$$\left(1 + \frac{r}{n}\right)^n = \left(1 + \frac{1}{\frac{n}{r}}\right)^n = \left(1 + \frac{1}{m}\right)^n, \text{ where } m = \frac{n}{r}$$

$$= \left(1 + \frac{1}{m}\right)^{\left(\frac{n}{r}\right)r}$$

$$= \left(1 + \frac{1}{m}\right)^{mr}$$

$$= \left[\left(1 + \frac{1}{m}\right)^m\right]^r$$

As an illustration, let $r = 0.2$ and use a calculator to verify the following computations, rounded to 5 decimal places. They demonstrate that

$\left(1 + \dfrac{0.2}{n}\right)^n$ *approaches* $e^{0.2}$

as n gets very large.

$$\left(1 + \frac{0.2}{10}\right)^{10} = 1.21899$$

$$\left(1 + \frac{0.2}{100}\right)^{100} = 1.22116$$

$$\left(1 + \frac{0.2}{1000}\right)^{1000} = 1.22138$$

Also, $e^{0.2} = 1.22140.$

Then, since r is constant, $m = \dfrac{n}{r} \longrightarrow \infty$ as $n \longrightarrow \infty$. Consequently,

$$\left(1 + \frac{r}{n}\right)^n = \left[\left(1 + \frac{1}{m}\right)^m\right]^r \longrightarrow [e]^r = e^r.$$

Now, applying this to the compound interest formula, we have

$$P\left(1 + \frac{r}{n}\right)^{nt} = P\left[\left(1 + \frac{r}{n}\right)^n\right]^t \longrightarrow P[e^r]^t = Pe^{rt}, \text{ as } n \longrightarrow \infty.$$

In summary, we started with $P\left(1 + \dfrac{r}{n}\right)^{nt}$, in which n is the number of annual interest periods, and obtained Pe^{rt} by letting the number of annual interest periods get arbitrarily large; $n \longrightarrow \infty$. The result, $A = Pe^{rt}$, is called the *continuous compound interest formula*.

CONTINUOUS COMPOUND INTEREST FORMULA

$$A = Pe^{rt}$$

P = initial investment
r = annual interest rate (as a decimal)
t = number of years of the investment
A = final amount after t years

EXAMPLE 5 Find the value of a $1000 investment, compounded continuously at an annual rate of 8% after 10 years.

Solution Using $p = \$1000$, $r = 0.08$, and $t = 10$ years, the final value is given by

$$A = Pe^{rt}$$
$$= 1000\, e^{(0.08)10}$$
$$= 1000\, e^{0.8} = 2225.54$$

The final value is approximately $2225. ∎

In the preceding example, the final value is more than double the initial investment. How long did it take to double? The next example gives the answer.

EXAMPLE 6 Suppose that $1000 is invested at the annual rate of 8% and compounded continuously. How long will it take for this investment to double?

Solution In this question, the final amount $2000 is given, but the time it takes to reach this amount is unknown. Therefore, we substitute $A = 2000$, $P = 1000$, $r = 0.08$ into $A = Pe^{rt}$ and solve for t.

$$2000 = 1000e^{(0.08)t}$$
$$2 = e^{0.08t} \qquad \text{Divide by 1000.}$$
$$\ln 2 = 0.08t \qquad \text{Write in logarithmic form.}$$
$$\frac{\ln 2}{0.08} = t$$
$$8.664 = t$$

It will take approximately $8\frac{2}{3}$ years for the investment to double in value. As a check note that $1000\, e^{(0.08)(8.664)} \approx 1999.95$. ∎

EXERCISES 5.5

Evaluate each of the following, rounded to three decimal places.

1. $\dfrac{\ln 6}{\ln 2}$ 2. $\dfrac{\ln 10}{\ln 5}$ 3. $\dfrac{\ln 8}{\ln 0.2}$ 4. $\dfrac{\ln 0.8}{\ln 4}$

5. $\dfrac{\ln 15}{2 \ln 3}$ 6. $\dfrac{\ln 25}{3 \ln 5}$ 7. $\dfrac{\ln 100}{-4 \ln 10}$ 8. $\dfrac{\ln 80}{-5 \ln 8}$

Solve for x, rounded to three decimal places.

9. $2^x = 45$ 10. $5^x = 70$ 11. $2^{-x} = 125$ 12. $3^{x+1} = 100$

13. $2^{3x} = 80$ 14. $100(2^x) = 2000$ 15. $80e^x = 120$ 16. $45^{4x} = 23$

Find the value of y in $y = Ae^{kx}$ for the given values of A, k, and x rounded to two decimal places.

17. $A = 100, k = 0.75, x = 4$ **18.** $A = 25, k = 0.5, x = 10$

19. $A = 1000, k = -1.8, x = 2$ **20.** $A = 12.5, k = -0.04, x = 50$

Solve for k. Leave the answer in terms of natural logarithms.

21. $5000 = 50e^{2k}$ **22.** $75 = 150e^{10k}$ **23.** $\dfrac{A}{3} = Ae^{4k}$ **24.** $\dfrac{A}{2} = Ae^{100k}$

25. A bacterial culture is growing according to the formula $y = 10,000\,e^{0.6x}$, where x is the time in days. Estimate the number of bacteria after 1 week.

26. Estimate the number of bacteria after the culture in Exercise 25 has grown for 12 hours.

27. How long will it take for the bacterial culture in Exercise 25 to triple in size?

28. How long will it take until the number of bacteria in Exercise 25 reaches 1,000,000?

29. A certain radioactive substance decays according to the exponential formula

$$S = S_0 e^{-0.04t}$$

where S_0 is the initial amount of the substance and S is the amount of the substance left after t years. If there were 50 grams of the radioactive substance to begin with, how long will it take for half of it to decay?

30. Show that when the formula in Exercise 29 is solved for t, the result is $t = -25\ln\dfrac{S}{S_0}$.

31. A radioactive substance is decaying according to the formula $y = Ae^{kx}$, where x is the time in years. The initial amount $A = 10$ grams, and 8 grams remain after 5 years.

(a) Find k. Leave the answer in terms of natural logs.

(b) Estimate the amount remaining after 10 years.

(c) Find the half-life to the nearest tenth of a year.

32. The half-life of radium is approximately 1690 years. A laboratory has 50 milligrams of radium.

(a) Use the half-life to solve for k in $y = Ae^{kx}$. Answer in terms of natural logs.

(b) To the nearest 10 years, how long does it take until there are 40 milligrams left?

33. Suppose that 5 grams of a radioactive substance decreases to 4 grams in 30 seconds. What is its half-life to the nearest tenth of a second?

34. How long does it take for two-thirds of the radioactive material in Exercise 33 to decay? Give your answer to the nearest tenth of a second.

35. When the population growth of a certain city was first studied, the population was 22,000. It was found that the population P grows with respect to time t (in years) by the exponential formula $P = (22,000)(10^{0.0163t})$. How long will it take for the city to double its population?

36. How long will it take for the population of the city described in Exercise 35 to triple?

37. An Egyptian mummy is found to contain 60% of its ^{14}C. To the nearest 100 years, how old is the mummy? (*Hint:* If A is the original amount of ^{14}C, then $\frac{3}{5}A$ is the amount left. Also, see Example 3.)

38. A skeleton contains one-hundredth of its original amount of ^{14}C. To the nearest 1000 years, how old is the skeleton?

39. Answer the question in Exercise 38 if one-millionth of its ^{14}C is left.

40. Suppose that the value of an automobile is given by $V(t) = Ae^{kt}$ where A is the initial cost and t is the number of years after purchase. If a car was bought on May 1, 1995 for $14,000 and its value on May 1, 1996 was $11,600, what will its value be on May 1, 2000?

41. In how many years will the value of the car in Exercise 40 be $8000?

42. Suppose that sugar dissolves in water so that the amount of undissolved sugar is given by $S(t) = Ae^{kt}$, where A is the initial amount of sugar and t is the number of hours after the sugar was added to the water. If 5 pounds of sugar was put into water and 2 pounds dissolved in 2 hours, how much undissolved sugar remains after 6 hours?

43. In Exercise 42, how long will it take for 75% of the sugar to dissolve?

44. Suppose that a $10,000 investment earns compound interest at the annual interest rate of 9%. If the time of deposit of the investment is one year ($t = 1$), find the value of the investment for each of the following types of compounding.

   (a) $n = 4$ (quarterly)   **(b)** $n = 12$ (monthly)   **(c)** $n = 52$ (weekly)

   (d) $n = 365$ (daily)   **(e)** continuously

45. Follow the instructions in Exercise 44, but change the time of deposit to 5 years.

46. Compute the interest earned for each case in Exercise 44.

47. Follow the instructions in Exercise 44, but change the time of deposit to 3.5 years.

48. Suppose that $1500 is invested at the annual interest rate of 8%, compounded continuously. How much will be on deposit after 5 years? After 10 years?

49. Mrs. Kassner deposits $5000 at the annual interest rate of 9%. How long will it take her investment to double in value? How long would it take if the interest rate were 12%? Assume continuous compounding in both cases.

50. How long would it take for a $1000 investment to double if it earns 12% interest annually, compounded continuously? How long would it take to triple?

51. A $1000 investment earns interest at the annual rate of r%, compounded continuously. If the investment doubles in 5 years, what is r?

52. How long does it take a $4000 investment to double if it earns interest at the annual rate of 8%, compounded quarterly?

53. In Exercise 52, how long would it take if the compounding were done monthly?

54. An investment P earns 9% interest per year, compounded continuously. After 3 years, the value of the investment is $5000. Find the initial amount P. (*Hint:* Solve $A = Pe^{rt}$ for P.)

55. Answer the question in Exercise 54, using 6 years as the time of deposit.

56. An investment P earns 8% interest per year, compounded quarterly. After one year the value of the investment is $5000. Find the initial amount P. (*Hint:* Solve $A_1 = P\left(1 + \frac{r}{n}\right)^n$ for P.)

57. How many dollars must be invested at the annual rate of 12%, compounded monthly, in order for the value of the investment to be $20,000 after 5 years? (*Hint:* Solve $A_t = P\left(1 + \frac{r}{n}\right)^{nt}$ for P.)

58. Explain how the result $A_t = P\left(1 + \frac{r}{n}\right)^{nt}$ can be obtained from $A_1 = P\left(1 + \frac{r}{n}\right)^n$ [*Hint:* A_2, the value of the investment after 2 years, is obtained when A_1 has earned interest for one year compounded n times. Thus $A_2 = A_1\left(1 + \frac{r}{n}\right)^n$.]

59. Solve for y: $x = -\dfrac{1}{k}(\ln A - \ln y)$. Use your result to describe the relationship between the variables x and y, assuming that x represents time, A is a positive constant, and k is a nonzero constant.

60. In 1944 a stamp for first-class postage cost 3¢. By the start of 1994 the cost had increased to 29¢. Assuming that the average annual rate of inflation during this period was 5%, was this a reasonable price for a stamp in 1994? If not, what should have the price been?

61. Use the information in Exercise 60 to determine an average annual inflation rate that would make the 1994 cost of 29¢ appropriate.

 Written Assignment   Contact at least two local banks and determine how much interest you can earn on a certificate of deposit of $5000 for five years. Find out what the annual interest rate is and how it is compounded. Use appropriate results developed in this section to compute the interest earned over the five-year period and compare your results with the information obtained from the bank.

Critical Thinking

1. The solution of Example 7, page 327, was accomplished by converting from the logarithmic to the exponential form. Instead of making this conversion, what should be done so that the equation can be solved using the one-to-one property?

2. It is not true, in general, that $(\log_b x)^3$ is equal to $3 \log_b x$ for all $x > 0$. However, there are specific values of x for which they are equal. Find these values.

3. On each of three sheets of graph paper draw a coordinate system using a unit length of at least 1 inch. Use a sharpened pencil to make careful drawings of $y = 2^x$, $y = 3^x$, and $y = e^x$, for $-2 \le x \le 2$, one on each sheet. Do this using the x-values of -2, -1, -0.5, -0.25, 0, 0.25, 0.5, 1, 2, and use a calculator, when necessary, for the y-values. Now use a straightedge to draw the line tangent to each curve at point $P(0, 1)$. This line should touch the curve *only* at P. Use the grid lines to estimate the slope of each tangent and decide which slope is closest to the value 1.

4. What is the half-life of a substance that decays exponentially according to $y = Ae^{kt}$, $k < 0$? Use the result to verify the answer in Exercise 29, page 350. What is the doubling time for a substance that grows exponentially according to $y = Ae^{kt}$, $k > 0$?

CHAPTER REVIEW EXERCISES

Section 5.1 Exponential Functions and Equations (Page 306)

1. List the important properties of the exponential function $f(x) = b^x$ for $b > 1$ and for $0 < b < 1$.

2. Use a table of values to sketch $y = 8^x$ on the interval $[-1, 1]$.

3. Sketch $y = 3^x$ and $y = -3^x$ on the same axes.

4. Explain how to obtain the graphs of $y = 5^{x+3}$ and $y = 5^x + 3$ from the graph of $y = 5^x$.

5. Graph: $y = \left(\frac{1}{3}\right)^{x-2} + 1$

6. If f is a one-to-one function and $f(x_1) = f(x_2)$, what can be said about x_1 and x_2?

Solve for x:

7. $8^{3x-2} = 64$ **8.** $\left(\dfrac{4}{5}\right)^{x^3} = \dfrac{64}{125}$ **9.** $\dfrac{1}{3^{x+2}} = 27$ **10.** $\left(\dfrac{2}{3}\right)^x = \dfrac{27}{8}$

11. Graph $y = f(x) = b^x$ for $b = 1$.

Section 5.2 Logarithmic Functions (Page 314)

12. Which of the following statements are true?

  (a) If $0 < b < 1$, the function $f(x) = b^x$ decreases.

  (b) The point $(0, 1)$ is on the curve $y = \log_b x$.

  (c) $y = \log_b x$, for $b > 1$, increases and the curve is concave down.

  (d) The domain of $y = b^x$ is the same as the range of $y = \log_b x$.

  (e) The x-axis is an asymptote to $y = \log_b x$ and the y-axis is an asymptote to $y = b^x$.

13. Graph $y = f(x) = \log_3 x$ by converting to exponential form and using selected values.

14. Repeat Exercise 13 by graphing the inverse function and reflecting the graph of the inverse in the line $y = x$.

15. Find the domain for $y = \log_2(x + 3)$ and graph the function. Show the asymptote and write its equation.

16. Graph: $y = \log_2(2x - 4)$

17. Write in exponential form: **(a)** $\log_3 81 = 4$ **(b)** $\log_2 32 = 5$

18. Write in logarithmic form: **(a)** $(0.1)^3 = \dfrac{1}{1000}$ **(b)** $\left(\dfrac{1}{64}\right)^{-2/3} = 16$

Solve for the indicated variable:

19. $\log_2 64 = y$ **20.** $\log_{1/2} x = 5$ **21.** $\log_{\sqrt{5}} x = -4$

Section 5.3 The Laws of Logarithms (Page 323)

22. Write the three laws of logarithms.

23. Express $\log_b \dfrac{AB}{C^2}$ in terms of $\log_b A$, $\log_b B$, and $\log_b C$.

24. Express as the logarithm of a single expression in x: $3 \log_b x - \dfrac{1}{2} \log_b(x + 2)$

25. Explain what is meant by a common logarithm.

26. State the formula for change of base of logarithms.

Use the formula in Exercise 26 to evaluate, rounded to four decimal places:

27. $\log_2 75$ **28.** $\log_5 135$ **29.** $\log_{12} 1000$

Solve for x.

30. $\log_4(x + 6) + \log_4(x - 6) = 3$ **31.** $\log_{10}(x^2 - 1) - \log_{10}(x + 1) = 1$

32. $\log_5 3x - \log_5 (x + 2) = 0$ **33.** $3 \log_{10}(x + 3) = 6$

34. Describe how to obtain the graph of $y = f(x) = \log_2 8(x + 2)$ from the graph of $y = \log_2 x$. Identify the asymptote and intercepts, if any.

35. Graph: $y = f(x) = \log(x^2 - 2x) - \log(x - 2)$

Section 5.4 The Natural Exponential and Logarithmic Functions (Page 332)

36. Define a natural exponential function.

37. What is meant by $\ln x$?

38. List the important properties of the function $f(x) = e^x$.

39. List the important properties of the function $f(x) = \ln x$.

40. Explain how to sketch the graph of $y = \ln(x - 2)$ from the graph of $y = \ln x$.

Use the laws of logarithms to write $\ln f(x)$ as an expression involving sums, differences, and multiples of natural logarithms.

41. $f(x) = \dfrac{2x^2}{x - 4}$ **42.** $f(x) = \dfrac{x^2\sqrt{x - 1}}{x^2 - 25}$

Convert each expression into the logarithm of a single expression in x in simplified form.

43. $\ln 3 + \ln(x^2 - 1) - \ln(x - 1)$ **44.** $\ln x + 2\ln(x^2 + 4) - \frac{1}{2}\ln(x + 7)$

45. What is the domain of $f(x) = \ln(3x - 2)$? Write the equation of the asymptote.

46. Simplify: **(a)** $\ln\left(e^{x^2}\right)$ **(b)** $e^{3\ln x}$

47. Solve for x: **(a)** $e^{\ln(3 - 2x)} = 5x$ **(b)** $\ln x - \ln 2 = 1$

48. Express $h(x) = \ln(3x - 2)$ as the composite of two functions.

49. Solve for x: $\ln(x - 1) + 1 = \ln x$

50. Find the x-intercepts and domain of $f(x) = xe^x - 2x^2 e^x$.

51. Match the columns.

(i) $\ln x < 0$	**(a)** $x < 0$
(ii) $\ln x = 0$	**(b)** $x = 0$
(iii) $\ln x > 0$	**(c)** $x > 0$
(iv) $0 < e^x < 1$	**(d)** $0 < x < 1$
(v) $e^x = 1$	**(e)** $x = 1$
(vi) $e^x > 1$	**(f)** $x > 1$

52. Graph: **(a)** $y = \ln x^3$ **(b)** $y = \ln(ex)$

Section 5.5 Applications: Exponential Growth and Decay (Page 342)

Solve for x, rounded to three decimal places.

53. $2^x = 132$ **54.** $2^{-x} = 80$ **55.** $5^{x+1} = 145$

Find the value of $y = Ae^{kx}$ for the given values of A, k, and x.

56. $A = 1000, k = 0.55, x = 2$ **57.** $A = 45, k = 0.06, x = 20$

58. Suppose that 8 grams of a radioactive substance decreases to 6 grams in 30 seconds. What is its half-life to the nearest tenth of a second?

59. How long does it take for three-fourths of the radioactive material in Exercise 58 to decay? Give your answer to the nearest tenth of a second.

60. State the formula for compound interest when interest is compounded n times per year, at an annual rate of $r\%$ for t years.

61. Repeat Exercise 60 for interest compounded continuously.

62. Find the total amount on deposit after 3 years if $5000 is deposited at a rate of 8% in a bank that pays interest semiannually.

63. Repeat Exercise 62 for interest compounded quarterly.

64. Repeat Exercise 62 for interest compounded continuously.

65. In Exercise 62, how long will it take for the money to double if interest is compounded continuously?

66. State the general exponential growth or exponential decay functions, and show the difference between the two graphically.

CHAPTER 5: STANDARD ANSWER TEST

Use these questions to test your knowledge of the basic skills and concepts of Chapter 5. Then test your answers with those given at the back of the book.

1. Match each curve with one of the given equations listed below.

(i)
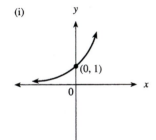

(ii)
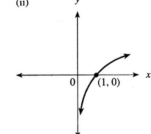

(iii)
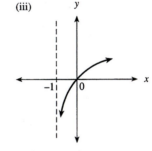

(iv)
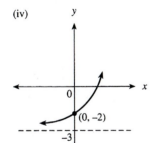

(v)
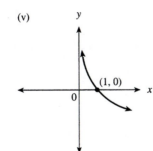

(vi)
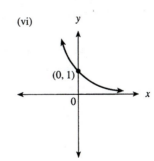

(a) $y = b^x; b > 1$ **(b)** $y = b^x; 0 < b < 1$ **(c)** $y = \log_b x; b > 1$ **(d)** $y = \log_b x; 0 < b < 1$

(e) $y = \log_b(x + 1); b > 1$ **(f)** $y = \log_b(x - 1); b > 1$ **(g)** $y = b^{x+2}; b > 1$ **(h)** $y = b^x - 3; b > 1$

2. (a) Convert $\log_5 125 = 3$ into exponential form.

   (b) Convert $16^{3/4} = 8$ into logarithmic form.

3. Solve for x.

   (a) $81^x = 9$ (b) $e^{\ln(x^2-x)} = 6$

4. Solve for x: $\dfrac{1}{2^{x+1}} = 64$

5. Simplify: $80\,e^{3(\ln 1/2)}$

6. (a) Solve for b: $\log_b \dfrac{27}{8} = -3$ (b) Evaluate: $\log_{10} 0.01$

7. Describe the graph of $y = 3^x$ by answering the following:

   (a) What are the domain and range?

   (b) Where is the function increasing or decreasing?

   (c) Describe the concavity.

   (d) Find the x- and y-intercepts, if any.

   (e) Write the equations of the asymptotes, if any.

8. Sketch the graphs of $y = e^{-x}$ and its inverse on the same axes. Write an equation of the inverse in the form $y = g(x)$.

Find the domain of each function and give the equation of the vertical or horizontal asymptote.

9. $y = f(x) = 2^x - 4$ 10. $y = f(x) = \log_3(x + 4)$

11. Use the laws of logarithms (as much as possible) to write the following as an expression involving sums, differences, and multiples of logarithms:

   (a) $\log_b x(x^2 + 1)^{10}$ (b) $\ln \dfrac{x^3}{(x + 1)\sqrt{x^2 + 2}}$

12. Express as the logarithm of a single expression in x:

   (a) $\log_7 x + \log_7 2x + \log_7 5$ (b) $\frac{1}{2}\ln x - 2\ln(x + 2)$

13. Use the change of base formula to evaluate $\log_2 75$, rounded to four decimal places.

14. Graph the function $f(x) = \log_2(x + 2)$ and state the domain, intercepts, and the equation of the asymptote, if any.

15. Graph the function $f(x) = e^{x+1}$ and state the domain, intercepts, and the equation of the asymptote, if any.

16. Solve for x: $\log_{25} x^2 - \log_{25}(2x - 5) = \frac{1}{2}$.

17. Solve for x: $\log_{10} x + \log_{10}(3x + 20) = 2$.

18. Explain how the graph of $y = \ln e^2(x - 1)$ can be obtained from the graph of $y = \ln x$ by using translations.

19. Solve for x, rounded to four decimal places: $4^{2x} = 5$.

20. A radioactive substance decays according to the formula $y = Ae^{-0.04t}$, where t is the time in years. If the initial amount $A = 50$ grams, find the half-life to the nearest tenth of a year.

21. A \$2000 investment earns interest at the annual rate of 8%, compounded quarterly. Write the expression that equals the value of the investment after 6 years. (Leave the answer in exponential form.)

22. How long will it take the investment in Exercise 21 to double?

23. If $6000 is invested at the annual interest rate of 7%, compounded continuously, what is the investment worth in 10 years to the nearest dollar?

24. Solve for k in $y = 120e^{kt}$ if $y = 180$ when $t = 3$.

25. Solve for x: $\ln 2 - \ln(1 - x) = 1 - \ln(x + 1)$.

CHAPTER 5: MULTIPLE CHOICE TEST

1. Which of the following are true for the function $y = f(x) = b^x$; $b > 0$ and $b \neq 1$?

   I. The domain consists of all real numbers x.

   II. The range consists of all positive numbers y.

   III. It is a one-to-one function.

   (a) Only I   **(b)** Only II   **(c)** Only III   **(d)** I, II, and III   **(e)** None of the preceding

2. Solve for x: $b^{x^2-2x} = 1$

   (a) $x = 0,\ x = 2$ **(b)** $x = 1,\ x = 3$ **(c)** $x = 1$ **(d)** $x = 0$ **(e)** None of the preceding

3. For which of the following is the point $\left(-1, \frac{1}{4}\right)$ on the graph of the function?

   (a) $y = f(x) = -\log_2(x - 1)$ **(b)** $y = f(x) = \log_2(x - 1)$

   (c) $y = f(x) = 2^{x-1}$ **(d)** $y = f(x) = -2^{x-1}$ **(e)** None of the preceding

4. Which of the following is the domain for the function $y = f(x) = \log_2(x + 3)$?

   (a) $x < 3$ **(b)** $x \geq 3$ **(c)** $x < -3$ **(d)** $x \geq -3$ **(e)** None of the preceding

5. Solve for x: $\log_3 x + \log_3(2x + 51) = 4$

   (a) $\frac{3}{2}$ **(b)** -27 **(c)** $\frac{47}{3}$ **(d)** 10 **(e)** None of the preceding

6. Solve for b: $\log_b \frac{1}{64} = -\frac{3}{2}$.

   (a) $\frac{1}{16}$ **(b)** 8 **(c)** 16 **(d)** 512 **(e)** None of the preceding

7. Which of the following are true?

   I. $(\log_b x)^2 = 2\log_b x$

   II. $\log_b A + \log_b B = \log_b(A + B)$

   III. $\log_b A - \log_b B = \dfrac{\log_b A}{\log_b B}$

   (a) Only I   **(b)** Only II   **(c)** Only III   **(d)** I, II, and III   **(e)** None of the preceding

8. Which of the following are true for the function $y = f(x) = e^x$?

   I. The domain consists of all real numbers x.

   II. The curve is concave up.

   III. It is a one-to-one function.

   (a) Only I   **(b)** Only II   **(c)** Only III   **(d)** I, II, and III   **(e)** None of the preceding

9. Which of the following is equal to $\log_5 85$?

   (a) $\dfrac{\log_{10} 5}{\log_{10} 85}$ (b) $\dfrac{\log_{10} 85}{\log_{10} 5}$ (c) $\dfrac{\log_5 10}{\log_5 85}$ (d) $\dfrac{\log_5 85}{\log_5 10}$ (e) None of the preceding

10. Solve for x: $e^{\ln(2-x)} = 2x$

   (a) e^2 (b) $\ln 2$ (c) $\frac{3}{2}$ (d) $\frac{2}{3}$ (e) None of the preceding

11. Which of the following are true for the function $y = f(x) = \ln x$?

   I. The domain consists of all real numbers x.

   II. The range consists of all $y > 0$.

   III. The curve is concave down.

   (a) Only I   (b) Only II   (c) Only III   (d) I, II, and III   (e) None of the preceding

12. Solve for x: $\ln(x + 1) - 1 = \ln x$.

   (a) $e - 1$ (b) $\dfrac{1}{e - 1}$ (c) 1 (d) e (e) None of the preceding

13. Solve for x: $9^{x-1} = 4$

   (a) $\dfrac{\ln 4}{\ln 9} + 1$ (b) $\dfrac{\ln 5}{\ln 9}$ (c) $\ln 4 - \ln 9 + 1$ (d) $\dfrac{\ln 9}{\ln 4} + 1$ (e) None of the preceding

14. As n becomes larger and larger, what value does the expression $\left(1 + \dfrac{r}{n}\right)^n$ approach?

   (a) e^r (b) e^n (c) e (d) 1 (e) None of the preceding

15. Which of the following reflections is appropriate to obtain the graph of $y = \dfrac{1}{2^x}$?

   (a) $y = 2^x$ in the x-axis   (b) $y = 2^{-x}$ in the x-axis   (c) $y = 2^x$ in the y-axis
   (d) $y = 2^{-x}$ in the y-axis   (e) None of the preceding

16. To show the graph of $y = \log_2 4x$, which of the following translations of $y = \log_2 x$ is correct?

   (a) Two units down   (b) Two units up   (c) Multiply the ordinates by 4
   (d) Four units to the right   (e) None of the preceding

17. What is the relationship between the graphs of $y = 8^x$ and $y = \log_8 x$?

   (a) They are reflections of each other in the x-axis.   (b) They are reflections of each other in the y-axis.
   (c) They are reflections of each other in the line $y = x$.   (d) There is no reflective relationship to each other.
   (e) None of the preceding

18. A radioactive substance decays according to the formula $y = Ae^{kx}$, where the initial amount $A = 40$ grams. If after 8 years 30 grams remain, then the amount y remaining after x years is given by

   (a) $y = 30e^{(x \ln 0.75)/8}$ (b) $y = 40e^{(x \ln 0.75)/8}$ (c) $y = 40e^{(x \ln 1.3)/8}$
   (d) $y = 40e^{(x \ln 0.75)/\ln 8}$ (e) None of the preceding

19. A $3000 investment earns interest at the annual rate of 8.4% compounded monthly. After five years the total amount of the investment, A_5, is given by

(a) $A_5 = 3000(1.084)^{60}$ (b) $A_5 = 3000(1.084)^5$ (c) $A_5 = 3000(1.007)^{60}$

(d) $A_5 = 3000(1.007)^{12}$ (e) None of the preceding

20. If $Q = \dfrac{(\sqrt[5]{409})(0.0058)}{7.29}$, then $\log_b Q$ is which of the following?

(a) $\frac{1}{5}(\log_b 409 + \log_b 0.0058) - \log 7.29$ (b) $5\log_b 409 + \log_b 0.0058 - \log_b 7.29$

(c) $\dfrac{\frac{1}{5}\log_b 409 + \log_b 0.0058}{\log_b 7.29}$ (d) $\frac{1}{5}\log_b 409 + \log_b 0.0058 - \log_b 7.29$

(e) None of the preceding

ANSWERS TO THE TEST YOUR UNDERSTANDING EXERCISES

Page 312

1. 6   **2.** 2 or -2 **3.** $\frac{1}{2}$ **4.** -6 **5.** -4 **6.** -1 **7.** $\frac{1}{3}$ **8.** $\frac{2}{3}$ **9.** $\frac{2}{3}$

10. $-\frac{5}{2}$ **11.** 3   **12.** $-\frac{1}{2}$

Page 318

1.

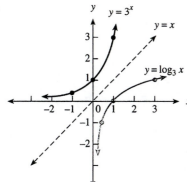

2.

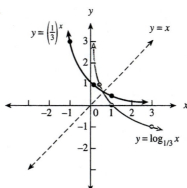

3. Shift 2 units left.   **4.** Shift 2 units up.   **5.** Reflect through x-axis   **6.** Double the size of each ordinate.

Page 326

1. $\log_b A + \log_b B + \log_b C$ **2.** $\log_b A - \log_b B - \log_b C$ **3.** $2\log_b A + 2\log_b B - \log_b C$

4. $\log_b A + 2\log_b B + 3\log_b C$ **5.** $\log_b A + \frac{1}{2}\log_b B - \log_b C$ **6.** $\frac{1}{3}\log_b A - 3\log_b B - 3\log_b C$

7. $\log_b 3x^2$ **8.** $\log_b[\sqrt{x}(x-1)^2]$ **9.** $\log_b \dfrac{2x-1}{(x^2+1)^3}$ **10.** $\log_b \dfrac{x}{(x-1)(x-2)^2}$

11. 3.9299 **12.** 7.1799

Page 343
1. 1.161 **2.** −1.161 **3.** −3.585 **4.** 1.107 **5.** 1.953 **6.** 0.066

The Trigonometric Functions

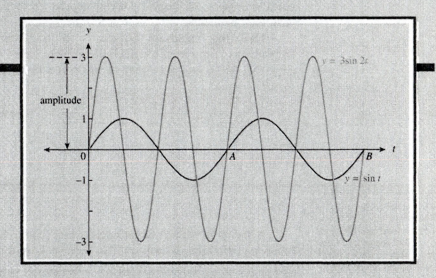

The search for a mathematical description of sound dates back to the classical Greek period. The Pythagoreans discovered that when two strings held with the same tension are plucked, the sounds produced harmonize, provided that the length of the strings had simple numerical ratios such as 2 to 1 or 3 to 2. Over the centuries musicians, scientists, and mathematicians added to the understanding of the relationships between music and mathematics.

Underlying the mathematical description of musical sound is the basic sine curve that has the formula $y = \sin t$, where t represents time in seconds. Its graph (in red) shows two repetitions of the same cycle; one from 0 to A and another from A to B. This pattern continues as t increases. All musical sounds have such periodicities, whether the sound comes from a tuning fork, a musical instrument, or the human voice.

A simple sound, such as that made by a tuning fork, is described by the equation $y = a \sin bt$, a variation of the basic sine curve. The positive number a is called the amplitude and is the maximum distance that the curve is from the horizontal axis. The positive number b is called the frequency, and it tells how many cycles take place in one second. As an example, the graph of $y = 3 \sin 2t$ (in blue) has basically the same shape as the graph of $y = \sin t$, except that the amplitude is 3 times the amplitude of $y = \sin t$, and as shown in the figure, it has two cycles during the same time that $y = \sin t$ has one.

Most musical sounds are much more complicated than the simple sounds described by $y = a \sin bt$. However, the French mathematician Joseph Fourier (1768–1830) proved that any periodic sound is described by the sum of sine terms of the form $a \sin bt$ that are studied in this chapter.

6.1 ANGLE MEASURE

Distance along a straight line can be measured using different kinds of units. For example, inches and centimeters are commonly used **units of measure.** Likewise, there are different ways of measuring angles. The two most common ways make use of *degrees* and *radians.* In this chapter, the primary focus will be on radians, which is the most important angular measurement in mathematics. However, since we will be using degrees in later work, and because radians are better understood when compared to degrees, we begin by reviewing the connections between these kinds of angular measurements.

Measuring angles in degrees is the traditional method inherited from the Babylonians. They used a numeration system based on groups of 60 and divided a circle into 360 equal parts, each of which is called a **degree.**

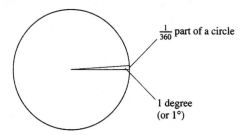

$\frac{1}{360}$ part of a circle

1 degree (or 1°)

Each degree can be divided into 60 equal parts called **minutes,** *and each minute can be divided into 60 equal parts called* **seconds.** *In our work, however, degrees will be used to the nearest tenth of a degree.*

A radian is a much larger angular unit than a degree, as you can see from the following definition. It turns out that 1 radian is approximately equal to 57°.

DEFINITION OF A RADIAN

One radian is the measure of a central angle of a circle that is subtended by an arc whose length is equal to the radius of the circle.

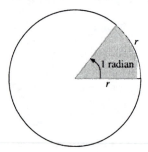

1 radian

In the following figure ∠AOB has been doubled so that ∠AOC has measure 2 radians, and arc length $AC = 2r$. This demonstrates that

$$\text{Arc length} = \begin{pmatrix} \text{angle measure} \\ \text{in radians} \end{pmatrix} \times (\text{radius})$$

$$2r = (2) \times (r)$$

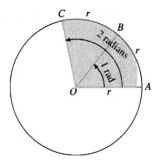

Using s for the arc length, θ for the angle measure, and r for the radius, the preceding can be generalized as follows:

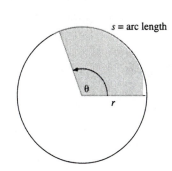

s = arc length

$$s = r\theta \quad \text{or} \quad \theta = \frac{s}{r}$$

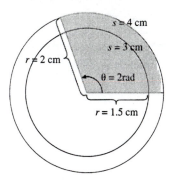

$s = 4$ cm
$s = 3$ cm
$r = 2$ cm
$\theta = 2$ rad
$r = 1.5$ cm

We then say that the angle measure θ is given by the quotient $\dfrac{s}{r} = \dfrac{\text{arc length}}{\text{radius}}$. Note that the size of the circle does not affect the radian measure of the angle; the measure is the *ratio* of intercepted arc length to the radius. This is demonstrated in the figure in the margin, where $\theta = \dfrac{s}{r} = \dfrac{4}{2} = \dfrac{3}{1.5} = 2$ radians. In particular, when $s = r$ then $\theta = \dfrac{s}{r} = 1$, and the angle has a measure of 1 radian.

From geometry we know that the circumference C of a circle is given by the formula $C = 2\pi r$. Thus $\dfrac{C}{r} = 2\pi$, so there are 2π radians in a complete rotation of $360°$ about a point. Therefore

Since $2r = d$, the diameter of the circle, $C = 2\pi r$ gives $\pi = C/d$, showing that π is the ratio of the circumference to the diameter of the circle. This ratio is the same for all circles. (Also, see Ex. 62, page 8.)

$$2\pi \text{ radians} = 360°$$

Dividing by 2 produces this fundamental relationship between radians and degrees:

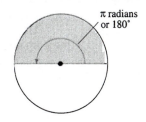

π radians
or 180°

$$\pi \text{ radians} = 180°$$

When both sides of the equation $180° = \pi$ radians are divided by 180, we obtain this conversion formula:

$1°$ contains $\dfrac{\pi}{180}$ radian, which is approximately 0.01745 radian. You can verify this using a calculator.

$$1° = \frac{\pi}{180} \text{ radian}$$

Since each degree equals $\dfrac{\pi}{180}$ radian, degrees can be converted into radians by multiplying the number of degrees by $\dfrac{\pi}{180}$.

We can also convert 30° to radians by using a proportion: $\dfrac{30}{180} = \dfrac{x}{\pi}$.

For simplicity we use the equals sign and write 30° = 0.524 radian, even though it is not an exact equality.

EXAMPLE 1   Convert 30° and 135° into radians. Express the results in terms of π and as a decimal rounded to three places.

Solution

$$30° = \left(30 \cdot \frac{\pi}{180}\right) \text{ radian} = \frac{\pi}{6} \text{ radian} = 0.524$$

$$135° = \left(135 \cdot \frac{\pi}{180}\right) \text{ radians} = \frac{3\pi}{4} \text{ radian} = 2.356 \quad\blacksquare$$

From now on radian measures will (in most instances) be stated without using the word "radian." Thus the angle measure 2 automatically means 2 radians unless degree measure is explicitly stated.

EXAMPLE 2   A central angle in a circle of radius 4 centimeters is 75°. Find the length of the intercepted arc to the nearest tenth of a centimeter.

Solution   First change to radians:

$$75° = \left(75 \cdot \frac{\pi}{180}\right) = \frac{5\pi}{12}$$

Then

$$s = r\theta$$

$$= 4\left(\frac{5\pi}{12}\right)$$

$$= \frac{5\pi}{3} = 5.2 \text{ centimeters} \qquad \text{to one decimal place} \quad\blacksquare$$

The conversion formula to change from radians to degrees is obtained by dividing both sides of the equation π radians = 180° by π.

A radian contains $\left(\dfrac{180}{\pi}\right)°$, *which is approximately 57.296°.*

$$1 \text{ radian} = \frac{180}{\pi} \text{ degrees}$$

Since each radian equals $\dfrac{180}{\pi}$ degrees, radians can be converted to degrees by multiplying the number of radians by $\dfrac{180}{\pi}$.

EXAMPLE 3

(a) Express $\dfrac{\pi}{4}$ and $\dfrac{5\pi}{6}$ in degree measure.

We can also express $\dfrac{\pi}{4}$ in degree measure by using a proportion:

$$\frac{\dfrac{\pi}{4}}{\pi} = \frac{x}{180}$$

Consult your calculator manual for the procedures that convert one kind of angle measure to another.

(b) Convert $\frac{3}{5}$ radian to degrees. State the result in terms of π and also to the nearest tenth of a degree.

Solution

(a) $\dfrac{\pi}{4} = \left(\dfrac{\pi}{4} \cdot \dfrac{180}{\pi}\right)^{\circ} = 45°$

$\dfrac{5\pi}{6} = \left(\dfrac{5\pi}{6} \cdot \dfrac{180}{\pi}\right)^{\circ} = 150°$

(b) $\dfrac{3}{5} = \left(\dfrac{3}{5} \cdot \dfrac{180}{\pi}\right)^{\circ} = \left(\dfrac{108}{\pi}\right)^{\circ} = 34.4°$ ∎

TEST YOUR UNDERSTANDING

Convert the degrees to radians and express the answers in terms of π.

1. 15°   **2.** 36°   **3.** 90°   **4.** 150°

5. 195°   **6.** 315°   **7.** 320°   **8.** 340°

Convert the radians to degrees.

9. $\dfrac{\pi}{15}$ **10.** $\dfrac{4\pi}{9}$ **11.** $\dfrac{7\pi}{6}$ **12.** $\dfrac{7}{2}$

13. $\dfrac{17\pi}{20}$ **14.** $\dfrac{23\pi}{15}$ **15.** $\dfrac{5\pi}{3}$ **16.** $\dfrac{11\pi}{6}$

(Answers: Page 424)

The following discussion shows how angular measure in radians is used to determine the area of a **sector of a circle.**

In the figure notice that the area of the shaded sector depends on the central angle θ. That is, $A = k \cdot \theta$, where k is some constant to be determined. To find k, we take the special case where $\theta = 2\pi$. This means that we have the entire circle whose area is πr^2. Using this information, we may then proceed as follows:

$$A = k \cdot \theta$$

$$\pi r^2 = k \cdot 2\pi$$

$$k = \tfrac{1}{2} r^2$$

Thus the area of a sector of a circle with radius r and central angle θ in radians is given by

$$A = \frac{1}{2} r^2 \theta$$

EXAMPLE 4 Find the area of a sector of a circle of radius 6 centimeters if the central angle is 60°.

Solution First convert 60° to radian measure:

$$60° = 60 \times \frac{\pi}{180} = \frac{\pi}{3}$$

Then $A = \dfrac{1}{2}(6)^2\left(\dfrac{\pi}{3}\right) = 6\pi$ square centimeters. ∎

Angular measure is also used to study motion on a circle. First, recall how speed is determined when the motion is linear. When a car travels 120 miles in 3 hours, its speed is $120 \div 3 = 40$ miles per hour (mph). Or, if an object travels 120 feet in 3 seconds, its speed is 40 feet per second (ft/sec). In general,

The speed is assumed to be constant.

Speed is the distance traveled per unit of time.

This concept also applies to circular motion. Suppose that one end of a level board is tangent to a circle at point P, as shown, and that the board is pushed horizontally at a constant rate, causing the circle to rotate. Assuming that the board remains tangent to the circle and that there is no slipping, you can see that the arc length s that P has moved is the same as the straight line or linear distance the board has moved. Thus, for a point moving on a circle of radius r we define its **linear speed** v to be the arc length s it has traveled per unit of time.

$$\text{linear speed} = v = \frac{\text{distance}}{\text{time}} = \frac{s}{t}$$

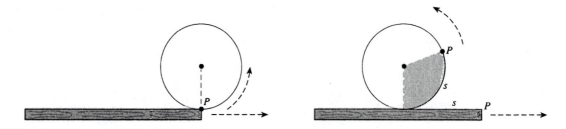

ω is the lowercase Greek letter omega.

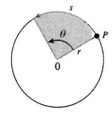

Similarly, the **angular speed** ω is defined as the number of radians turned (angular distance) per unit of time. Therefore, when OP rotates through θ radians, the angular speed is given by

$$\omega = \frac{\text{angular rotation}}{\text{time}} = \frac{\theta}{t}$$

Substituting $s = r\theta$ into $v = s/t$ gives $v = (r\theta)/t = r(\theta/t) = r\omega$, and we have the following relationship between the linear and angular speeds;

$$v = r\omega$$

When applying these results θ is in radians, the same linear units are used for r and s, and the same units of time are used for v and ω.

EXAMPLE 5 Assuming the earth's radius to be 4000 miles, find the linear speed of a point on the equator in mph and also in ft/sec. What is the angular speed in radians per hour (rad/hr)?

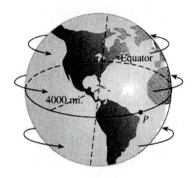

Solution As the earth makes one complete rotation in 24 hours, the angular rotation is 2π radians. Thus

$$v = \frac{s}{t} = \frac{r\theta}{t} = \frac{(4000)(2\pi)}{24} = 1047$$

The linear speed is 1047 mph rounded to the nearest mph. Since 1 mile = 5280 feet and 1 hour = (60)(60) = 3600 seconds.

$$v = 1047 \text{ mph} = 1047(5280) \text{ ft/hr}$$

$$= \frac{1047(5280)}{3600} \text{ ft/sec}$$

$$= 1536 \text{ ft/sec} \qquad \text{to the nearest foot per second}$$

Note:

$$\frac{\pi}{12} \, rad/hr = \frac{\pi}{12}\left(\frac{180}{\pi}\right)$$

$$= 15° \, per \, hour.$$

Also, the angular speed ω is given by

$$\omega = \frac{\theta}{t} = \frac{2\pi}{24} = \frac{\pi}{12} \text{ rad/hr} \qquad \blacksquare$$

EXERCISES 6.1

Convert the degrees to radians. Leave the answers in terms of π.

1. 45°	**2.** 60°	**3.** 90°	**4.** 180°	**5.** 270°
6. 360°	**7.** 150°	**8.** 135°	**9.** 225°	**10.** 240°
11. 210°	**12.** 300°	**13.** 330°	**14.** 345°	**15.** 75°

Convert the degrees to radians and give the results to two decimal places.

16. 10°   **17.** 100°   **18.** 220°   **19.** 340°

Convert the radians to degrees.

20. $\dfrac{\pi}{2}$ **21.** π **22.** $\dfrac{3\pi}{2}$ **23.** 2π **24.** $\dfrac{\pi}{3}$ **25.** $\dfrac{5\pi}{9}$ **26.** $\dfrac{\pi}{6}$ **27.** $\dfrac{2\pi}{3}$

28. $\dfrac{3\pi}{4}$ **29.** $\dfrac{5\pi}{4}$ **30.** $\dfrac{7\pi}{6}$ **31.** $\dfrac{5\pi}{3}$ **32.** $\dfrac{\pi}{12}$ **33.** $\dfrac{5\pi}{18}$ **34.** $\dfrac{11\pi}{36}$

Convert the radians to degrees to give the results to the nearest tenth of a degree.

35. $\dfrac{\pi}{15}$ **36.** $\dfrac{1}{2}$ **37.** 3   **38.** 6

Solve for the indicated part s, r, θ (in radians).

39.

40.

41.

42.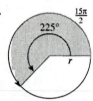

43.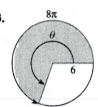

44.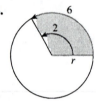

A circle has a radius of 12 centimeters. Find the area of a sector of this circle for each given central angle.

45. 30°   **46.** 90°   **47.** 135°   **48.** 225°   **49.** 315°

50. The area of a sector of a circle with radius 6 centimeters is 15 square centimeters. Find the measure of the central angle of the sector in degrees.

51. Find the area of a sector of a circle whose radius is 2 inches if the length of the intercepted arc is 8 inches.

52. The area of a sector of a circle with radius 4 centimeters is $\dfrac{16\pi}{3}$ square centimeters. Find the measure of the central angle of the sector in degrees.

53. Find the area of a circular sector with central angle 45° if the length of the intercepted arc is $\dfrac{\pi}{2}$ centimeters. Give the exact answer in terms of π, and approximate the area to the nearest tenth of a square centimeter.

54. A flower garden is a 270° circular sector with a 6-foot radius. Find the exact area of the garden in terms of π, and approximate the area to the nearest tenth of a square foot.

55. A curve along a highway is an arc of a circle with a 250-meter radius. If the curve is 50 meters long, by how many degrees does the highway change its direction? Give the answer to the nearest degree.

56. A 40-inch pendulum swings through an angle of 15°. Find the length of the arc through which the tip of the pendulum swings to the nearest tenth of an inch.

57. A cup is in the shape of a right circular cone made from a circular sector with an 8-inch radius and a central angle of 270°. Find the surface area of the cup to the nearest tenth of a square inch.

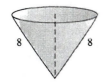

58. Find the area of the sector inscribed in the square *ABCD*.

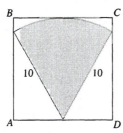

59. The equilateral triangle *ABC* in the following figure represents a wooden platform standing in a lawn. A goat is tied to a corner with a 15-foot rope. Assuming that the goat cannot climb onto the platform, what is the maximum amount of grazing area available to the goat? Assume that the rope does not go over the platform.

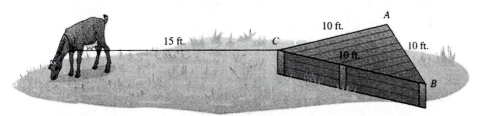

60. The shaded region is called a segment of the circle. Show that its area is given by

$$\tfrac{1}{2}r^2\left(\theta - \frac{\sqrt{3}}{2}\right).$$

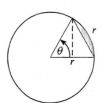

61. (a) Suppose that a satellite has a circular orbit that is 300 miles above the earth's surface and that one revolution around the earth takes 1 hour and 45 minutes. Using $r = 4000$ miles as the radius of the earth, find the satellite's linear speed in mph and also in ft/sec. (Use a calculator and round the answers to the nearest mph and nearest ft/sec.)

(b) Find the angular speed of the satellite in rad/hr.

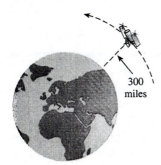

62. Answer the questions in Exercise 61 if the satellite has a circular orbit that is 1000 miles above the earth and if it remains above the same point on the earth throughout its orbit.

63. Suppose that a fly is sitting on the end of the minute hand of a clock. Find the angular speed of the fly in rad/min and in rad/hr.

64. A point on the rim of a wheel is turning with an angular speed of 220 rad/sec. If the wheel's diameter is 5 feet, find the linear speed in ft/sec.

65. (a) The outer diameter of the wheels on a bicycle is 22 inches. If the wheels are turning at a rate of 240 revolutions per minute, find the linear speed of the bicycle.

    (b) Use a calculator to estimate how many minutes it takes the bicycle to travel 1 mile.

22 in.

66. A car is traveling at 50 mph. Find the revolutions per minute of the wheels if the car's tires have a 26-inch diameter.

67. A motorcycle is traveling on a curve along a highway. The curve is an arc of a circle with a radius of $\frac{1}{4}$ mile. If the motorcycle's speed is 42 mph, what is the angle through which the motorcycle will turn in $\frac{1}{2}$ minute?

Challenge In the figure triangle ABC is a $30°-60°-90°$ triangle. M is the midpoint of AB and BC is tangent to the circular sector with center M. If $AB = 1$, find the area of the shaded circular sector.

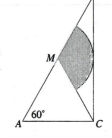

Written Assignment

1. Which is bigger, an angle of 1° or an angle of 1 radian? Discuss and justify your observations.

2. To convert $\pi/8$ radians into degree measure we multiply by $180/\pi$. Why does this make sense? Explain.

**6.2
TRIGONOMETRIC
FUNCTIONS**

The circle having radius 1 with center at the origin is called the **unit circle.** The circle has equation $x^2 + y^2 = 1$, and it will be of fundamental importance in the development of the **trigonometric functions.**

As shown in the following figure, if we start at point $A(1, 0)$ and move counterclockwise on the circle to a point $P(x, y)$, we will have traced the arc AP whose length s and angle measure θ are assumed to be positive. If the movement is clockwise, as in the figure at the right, the arc length s and angle measure θ are assumed to be negative.

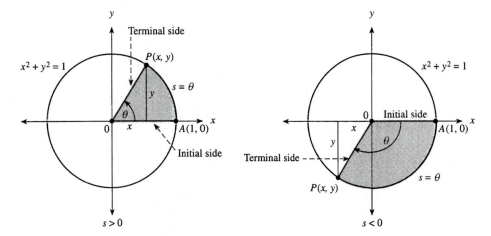

Note: The point $P(x, y)$ is the intersection of the terminal side of $\angle AOP$ and the unit circle.

The angle AOP is said to be in **standard position** because its initial side OA is on the positive part of the x-axis and its vertex is at the origin. Side OP is the terminal side of $\angle AOP$. Using θ for the radian measure of this angle, we have $\theta = s$ because the radius $r = 1$ and

$$s = r\theta = 1 \cdot \theta = \theta$$

For any such value θ we define the trigonometric function **sine of θ**, written as **sin θ**, to be the second coordinate of point $P(x, y)$, and the trigonometric function **cosine of θ**, written **cos θ**, to be the first coordinate. Thus

$$\sin \theta = y \quad \text{and} \quad \cos \theta = x$$

This can be done for any arc length $s = \theta$ on the unit circle, and since arc lengths can be measured using real numbers, we have that *the domain of these two functions is the set of real numbers.*

There are four other trigonometric functions, called the **tangent, cotangent, secant,** and **cosecant,** defined as follows, provided the denominators are not 0.

*The trigonometric functions are also called **circular functions.***

$$\text{tangent of } \theta = \tan \theta = \frac{y}{x} = \frac{\sin \theta}{\cos \theta}$$

$$\text{cotangent of } \theta = \cot \theta = \frac{x}{y} = \frac{\cos \theta}{\sin \theta}$$

$$\text{secant of } \theta = \sec \theta = \frac{1}{x} = \frac{1}{\cos \theta}$$

$$\text{cosecant of } \theta = \csc \theta = \frac{1}{y} = \frac{1}{\sin \theta}$$

Three values of θ will be of special importance in our work: $\dfrac{\pi}{6}$ $(30°)$,

$\frac{\pi}{4}$ (45°), and $\frac{\pi}{3}$ (60°). First consider $\theta = \frac{\pi}{4}$. In the figure we have drawn the **reference triangle** OPQ by inserting the perpendicular from point P to the x-axis.

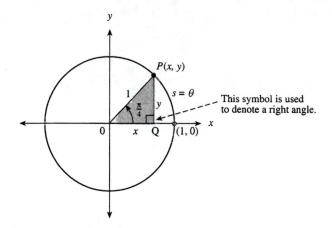

This reference triangle OPQ is an isosceles right triangle with right angle at Q and acute angles each of measure $\frac{\pi}{4}$, sometimes referred to as a 45°–45°–90° triangle. Then $x = y$, and by the Pythagorean theorem

$$x^2 + y^2 = x^2 + x^2 = 1$$
$$2x^2 = 1$$
$$x^2 = \frac{1}{2}$$
$$x = \frac{1}{\sqrt{2}} \qquad \text{The positive root is used since P is in quadrant I.}$$

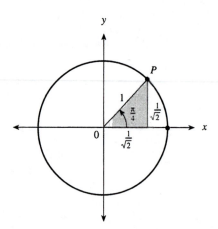

Also $y = \frac{1}{\sqrt{2}}$ and we have the following function values of $\frac{\pi}{4}$ (45°):

TRIGONOMETRIC FUNCTION VALUES OF $\frac{\pi}{4}$

$$\sin \frac{\pi}{4} = y = \frac{1}{\sqrt{2}} \qquad \csc \frac{\pi}{4} = \frac{1}{y} = \sqrt{2}$$

$$\cos \frac{\pi}{4} = x = \frac{1}{\sqrt{2}} \qquad \sec \frac{\pi}{4} = \frac{1}{x} = \sqrt{2}$$

$$\tan \frac{\pi}{4} = \frac{y}{x} = 1 \qquad \cot \frac{\pi}{4} = \frac{x}{y} = 1$$

Angles in standard position are *coterminal* provided they have the same terminal side. For such angles the same point $P(x, y)$ on the unit circle determines the trigonometric function values. For example, in the figure below, the trigonometric function values for $\frac{\pi}{4}$ and $-\frac{7\pi}{4}$ are exactly the same because the two angles having these measures are coterminal.

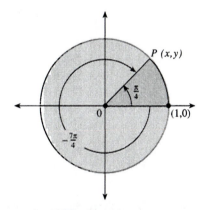

EXAMPLE 1 Find $\tan\left(-\frac{5\pi}{4}\right)$ and $\csc\left(-\frac{5\pi}{4}\right)$.

Solution For $-\frac{5\pi}{4}$ (or $-225°$) we have a $45°–45°–90°$ reference triangle in quadrant II. Then the coordinates of P are $\left(-\frac{1}{\sqrt{2}}, \frac{1}{\sqrt{2}}\right)$. Therefore,

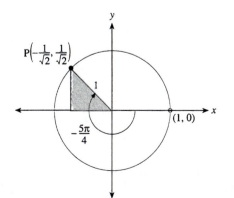

$$\tan\left(-\frac{5\pi}{4}\right) = \frac{\dfrac{1}{\sqrt{2}}}{-\dfrac{1}{\sqrt{2}}} = -1$$

$$\csc\left(-\frac{5\pi}{4}\right) = \frac{1}{\dfrac{1}{\sqrt{2}}} = \sqrt{2} \qquad \blacksquare$$

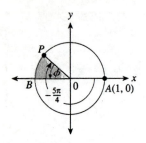

Reference angle BOP has measure $\phi = \dfrac{\pi}{4}$.

Another way to find the values in Example 1 is to make use of the **reference angle.** As shown in the margin, this is the acute angle of positive measure ϕ, between the terminal side of $\angle AOP$ and the x-axis. For Example 1 the reference angle has measure $\dfrac{\pi}{4}$. Then the trigonometric values can be found by first finding the values for $\dfrac{\pi}{4}$ and then prefixing the correct signs by noting the quadrant containing point P. Thus

$$\tan\left(-\frac{5\pi}{4}\right) = -\tan\frac{\pi}{4} = -1$$

$$\csc\left(-\frac{5\pi}{4}\right) = \csc\frac{\pi}{4} = \sqrt{2}$$

In general, the signs of the trigonometric function values depend on the location of point $P(x, y)$ in the coordinate system, as shown in the following figure.

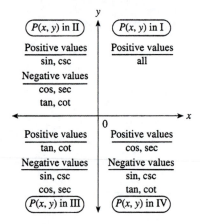

The other special angles of $\dfrac{\pi}{6}$ and $\dfrac{\pi}{3}$ radians are *complementary* since $\dfrac{\pi}{6} + \dfrac{\pi}{3} = \dfrac{\pi}{2}$. Consequently, we can find the sides of their reference triangles simultaneously. In the following figure, the hypotenuse of right $\triangle ABC$ has length 1. Now extend side BC to D so that $DC = CB$ and connect A to D. Then triangles ABC and ADC are congruent (by side-angle-side), so that $\triangle ABD$ is equilateral. Then $DB = 1$ and $CB = \frac{1}{2}$. Also, by the Pythagorean theorem, $(AC)^2 = 1^2 - \left(\frac{1}{2}\right)^2 = \frac{3}{4}$ so that $AC = \dfrac{\sqrt{3}}{2}$.

$\triangle ABC$ is referred to as a $30°$-$60°$-$90°$ triangle.

The trigonometric function values of $\frac{\pi}{6}$ and $\frac{\pi}{3}$ are now obtained using the figures below.

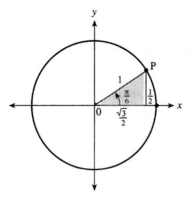 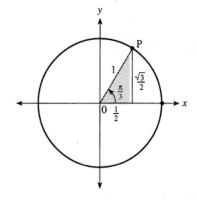

TRIGONOMETRIC FUNCTION VALUES OF $\frac{\pi}{6}$ AND $\frac{\pi}{3}$

$$\sin \frac{\pi}{6} = \frac{1}{2} \qquad \csc \frac{\pi}{6} = 2 \qquad\qquad \sin \frac{\pi}{3} = \frac{\sqrt{3}}{2} \qquad \csc \frac{\pi}{3} = \frac{2}{\sqrt{3}}$$

$$\cos \frac{\pi}{6} = \frac{\sqrt{3}}{2} \qquad \sec \frac{\pi}{6} = \frac{2}{\sqrt{3}} \qquad\qquad \cos \frac{\pi}{3} = \frac{1}{2} \qquad \sec \frac{\pi}{3} = 2$$

$$\tan \frac{\pi}{6} = \frac{1}{\sqrt{3}} \qquad \cot \frac{\pi}{6} = \sqrt{3} \qquad\qquad \tan \frac{\pi}{3} = \sqrt{3} \qquad \cot \frac{\pi}{3} = \frac{1}{\sqrt{3}}$$

EXAMPLE 2 Find the six trigonometric function values for $\theta = \frac{5\pi}{6}$ (150°).

Solution The terminal side of θ is in the second quadrant and the reference angle has measure $\phi = \pi - \frac{5\pi}{6} = \frac{\pi}{6}$. Therefore there is a 30°–60°–90° reference triangle with $x = -\frac{\sqrt{3}}{2}$ and $y = \frac{1}{2}$ as shown.

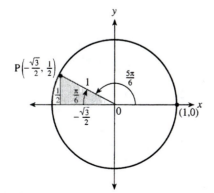

$$\sin \frac{5\pi}{6} = y = \frac{1}{2} \qquad\qquad \csc \frac{5\pi}{6} = \frac{1}{y} = 2$$

$$\cos \frac{5\pi}{6} = x = -\frac{\sqrt{3}}{2} \qquad\qquad \sec \frac{5\pi}{6} = \frac{1}{x} = -\frac{2}{\sqrt{3}}$$

$$\tan \frac{5\pi}{6} = \frac{y}{x} = -\frac{1}{\sqrt{3}} \qquad\qquad \cot \frac{5\pi}{6} = \frac{x}{y} = -\sqrt{3}$$

■

TEST YOUR UNDERSTANDING	*Determine each of the following trigonometric function values.*

TEST YOUR UNDERSTANDING

Determine each of the following trigonometric function values.

1. $\sin \dfrac{3\pi}{4}$ 2. $\csc\left(-\dfrac{3\pi}{4}\right)$ 3. $\tan \dfrac{5\pi}{4}$ 4. $\sec\left(-\dfrac{11\pi}{6}\right)$

5. $\cot \dfrac{10\pi}{3}$ 6. $\csc \dfrac{7\pi}{4}$ 7. $\sin\left(-\dfrac{4\pi}{3}\right)$ 8. $\cos\left(-\dfrac{19\pi}{6}\right)$

(Answers: Page 424)

EXAMPLE 3   Find $\csc \theta$ if $\cos \theta = \dfrac{\sqrt{3}}{2}$ and $\cot \theta = -\sqrt{3}$.

Solution The angle of θ radians has its terminal side in quadrant IV since $\cos \theta > 0$ and $\cot \theta < 0$. Also, since $\cos \theta = \dfrac{\sqrt{3}}{2}$ the reference angle must be $\dfrac{\pi}{6}$ radian, and we may use the following reference triangle.

In the figure $\theta = \dfrac{11\pi}{6}$.

However, any θ with the same terminal side could be used here.

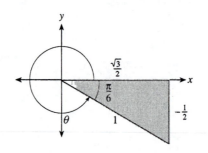

$$\csc \theta = \frac{1}{y} = \frac{1}{-\frac{1}{2}} = -2$$

∎

The trigonometric function values of $\dfrac{\pi}{6}$, $\dfrac{\pi}{4}$ and $\dfrac{\pi}{3}$ that have been found are exact values. It is important to become familiar with these results because they will be very useful for demonstrative purposes as the study of trigonometry continues. However, in some situations decimal approximations will be preferred and an appropriate calculator should be used to obtain such values. For example, most graphing calculators, set in radian mode, use key-stroke sequences as follows.

Verify that when the exact trigonometric values, such as

$$\tan \frac{\pi}{6} = \frac{1}{\sqrt{3}}$$

are converted into decimal form and rounded to 4 decimal places, the results are the same as those obtained on a calculator.

$$\tan \frac{\pi}{6} = \boxed{\text{TAN}}\ \boxed{(}\ \boxed{\pi}\ \boxed{\div}\ \boxed{6}\ \boxed{)}\ \boxed{\text{ENTER}} = 0.5774 \qquad \text{rounded to four decimal places}$$

Also, in degree mode the same result is obtained when rounded to four decimal places.

$$\tan 30° = \boxed{\text{TAN}}\ \boxed{30}\ \boxed{\text{ENTER}} = 0.5774$$

The following sequences can be used on a scientific calculator:

$$\tan \frac{\pi}{6} = \boxed{\pi}\ \boxed{\div}\ \boxed{6}\ \boxed{=}\ \boxed{\text{TAN}}$$

$$\tan 30° = \boxed{30°}\ \boxed{\text{TAN}}$$

Other calculators might use sec 45° = [45] [COS] [1/x].

Consult your calculator manual for appropriate sequences. Note that some calculators use the [=] *key*

or the [EXE] *key in place of* [ENTER].

EXAMPLE 4 Use a calculator to find sec 45°.

Solution Most calculators do not have a [SEC] key. Therefore, since the secant and cosine are reciprocals key-stroke sequences such as these can be used to find sec 45°:

[1] [÷] [COS] [45] [ENTER] = 1.4142

[COS] [45] [ENTER] [x⁻¹] [ENTER] = 1.4142   rounded to four decimal places

[(] [COS] [45] [)] [x⁻¹] [ENTER] = 1.4142   ∎

When the reference angle is not one of the special angles, then a calculator can be used to obtain the trigonometric function values. For example, to find cos 200° note that the terminal side is in quadrant III and the reference angle is 20°. Then

$$\cos 200° = -\cos 20° = -[COS]\ [20]\ [ENTER] = -0.9397$$

The preceding can be done more efficiently. Calculators are constructed to give the ratios directly, without having to enter the reference angle or the quadrant location. Thus

$$\cos 200° = [COS]\ [200]\ [ENTER] = -0.9397$$

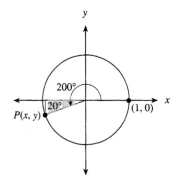

EXAMPLE 5 Use a calculator to evaluate the trigonometric function values rounded to four decimal places: **(a)** $\cos(-383°)$ **(b)** $\sec 932°$ **(c)** $\sin \dfrac{9\pi}{5}$

Solution Use degree mode for parts **(a)** and **(b)**

(a) $\cos(-383°) = [COS]\ [-383]\ [ENTER] = 0.9205$

(b) $\sec 932° = (\cos 932°)^{-1}$

$$= [COS]\ [932]\ [ENTER]\ [x⁻¹]\ [ENTER] = -1.1792$$

Use radian mode for part (c)

(c) $\sin \dfrac{9\pi}{5} = [SIN]\ [(]\ [9]\ [\pi]\ [÷]\ [5]\ [)]\ [ENTER] = -0.5878$ ∎

On a scientific calculator cos (−383°) *may be found as follows:*

[383] [+/−] [COS]

Verify these results by making use of the reference angles.

The procedure used in Example 5 is recommended for future work and will automatically be used, unless exact values are required, and the calculator key sequence may not always be shown. However, it is important for your understanding of trigonometry that you know how to make use of reference angles and triangles that are directly related to the definitions of the trigonometric function values.

An angle that has its terminal side on one of the coordinate axes is called a **quadrantal angle.** Such angles do not have reference triangles or reference angles. However, the trigonometric function values for such angles are easily determined by using the following unit circle diagram. For example, to find the values of the quadrantal angle with measure $\theta = \pi$ use the point $P(-1, 0)$ where $x = -1$ and $y = 0$, to

obtain

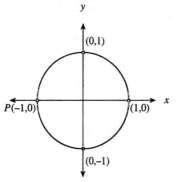

$\sin \pi = y = 0$ $\csc \pi$ is undefined

$\cos \pi = x = -1$ $\sec \pi = \dfrac{1}{-1} = -1$

$\tan \pi = \dfrac{0}{-1} = 0$ $\cot \pi$ is undefined

As demonstrated here, when determining the trigonometric ratios of a quadrantal angle, two of the ratios will be undefined.

Why are the trigonometric values for $\theta = -3\pi$ the same as for $\theta = \pi$?

EXAMPLE 6   Evaluate the trigonometric function values for $\theta = \dfrac{\pi}{2}$.

Solution   The terminal side OP of $\angle AOP$ is on the y-axis because $\theta = \dfrac{\pi}{2}$ is the radian measure of a right angle. Then, since P is also on the unit circle, its coordinates are (0, 1). Thus

$\sin \dfrac{\pi}{2} = y = 1$ $\csc \dfrac{\pi}{2} = \dfrac{1}{y} = \dfrac{1}{1} = 1$

$\cos \dfrac{\pi}{2} = x = 0$ $\sec \dfrac{\pi}{2} = \dfrac{1}{x}$ is undefined since $x = 0$

$\tan \dfrac{\pi}{2} = \dfrac{y}{x}$ is undefined since $x = 0$ $\cot \dfrac{\pi}{2} = \dfrac{x}{y} = \dfrac{0}{1} = 0$ ∎

CAUTION: Learn to Avoid These Mistakes	
WRONG	**RIGHT**
$\cos\left(-\dfrac{\pi}{4}\right) = -\cos\dfrac{\pi}{4}$	$\cos\left(-\dfrac{\pi}{4}\right) = \cos\dfrac{\pi}{4}$
$\cos\dfrac{7\pi}{8} = \cos\dfrac{\pi}{8}$	$\cos\dfrac{7\pi}{8} = -\cos\dfrac{\pi}{8}$
$\sin\left(-\dfrac{\pi}{3}\right) = \dfrac{\sqrt{3}}{2}$	$\sin\left(-\dfrac{\pi}{3}\right) = -\sin\dfrac{\pi}{3} = -\dfrac{\sqrt{3}}{2}$
$\sec\dfrac{\pi}{2} = \dfrac{1}{\cos\dfrac{\pi}{2}}$	$\cos\dfrac{\pi}{2} = 0;\ \sec\dfrac{\pi}{2}$ is undefined.
$\tan\left(-\dfrac{9\pi}{4}\right) = \tan\dfrac{\pi}{4}$	$\tan\left(-\dfrac{9\pi}{4}\right) = -\tan\dfrac{\pi}{4}$

EXERCISES 6.2

For Exercises 1–28 do not use a calculator.

Find the trigonometric function values.

1. $\sin \dfrac{\pi}{4}$ **2.** $\sec \dfrac{\pi}{4}$ **3.** $\cos \dfrac{\pi}{3}$ **4.** $\csc \dfrac{\pi}{3}$ **5.** $\tan \dfrac{\pi}{6}$ **6.** $\sec \dfrac{\pi}{6}$

Assume that θ is the measure of an angle in standard position. Find the quadrant containing the terminal side of the angle and the measure of the reference angle φ.

7. $\theta = \dfrac{2\pi}{3}$ **8.** $\theta = \dfrac{5\pi}{3}$ **9.** $\theta = \dfrac{5\pi}{4}$ **10.** $\theta = -\dfrac{9\pi}{4}$ **11.** $\theta = -\dfrac{13\pi}{3}$ **12.** $\theta = \dfrac{25\pi}{6}$

Assume that θ is the measure of an angle in standard position. Find the quadrant containing the terminal side of the angle and the measure of the coterminal angle between 0 and 2π radians.

13. $\theta = -\dfrac{7\pi}{6}$ **14.** $\theta = -\dfrac{5\pi}{36}$ **15.** $\theta = \dfrac{11\pi}{3}$ **16.** $\theta = \dfrac{25\pi}{3}$ **17.** $\theta = -\dfrac{17\pi}{4}$ **18.** $\theta = -\dfrac{25\pi}{6}$

Locate an angle of θ radians in a coordinate system, include the reference triangle, and find the measure of the reference angle. Then find the coordinates of P(x, y) on the terminal side and on the unit circle, and write the six trigonometric function values.

19. $\theta = \dfrac{2\pi}{3}$ **20.** $\theta = \dfrac{29\pi}{6}$ **21.** $\theta = -\dfrac{7\pi}{4}$ **22.** $\theta = -\dfrac{\pi}{6}$ **23.** $\theta = -\dfrac{7\pi}{6}$ **24.** $\theta = -\dfrac{31\pi}{6}$

Find the coordinates of P(x, y) on the unit circle and on the terminal side of the angle of measure θ. Write the trigonometric function values.

25. $\theta = -\dfrac{\pi}{2}$ **26.** $\theta = 3\pi$ **27.** $\theta = -\dfrac{7\pi}{2}$

28. Complete this table of trigonometric values that includes all quadrantal angles.

θ coterminal with	sin θ	cos θ	tan θ	cot θ	sec θ	csc θ
0						
$\dfrac{\pi}{2}$						
π						
$\dfrac{3\pi}{2}$						

Find the trigonometric function value (if it exists). Use a calculator (in radian mode) only when necessary.

29. $\tan \dfrac{3\pi}{4}$ **30.** $\sec \left(-\dfrac{\pi}{3}\right)$ **31.** $\sin \dfrac{4\pi}{3}$ **32.** $\sec (-\pi)$ **33.** $\csc (-\pi)$

34. $\tan \dfrac{7\pi}{2}$ **35.** $\cot \left(\dfrac{7\pi}{2}\right)$ **36.** $\tan 0$ **37.** $\cot 0$ **38.** $\cot \left(-\dfrac{5\pi}{2}\right)$

39. $\tan 8\pi$ **40.** $\sin \dfrac{9\pi}{2}$ **41.** $\cos 0.23$ **42.** $\csc 0.23$ **43.** $\cot 0.95$

44. $\sin 0.95$ **45.** $\cos \dfrac{19\pi}{4}$ **46.** $\tan \left(-\dfrac{23\pi}{6}\right)$ **47.** $\tan 1.48$ **48.** $\sec 1.48$

Use a calculator (in degree mode) to find the trigonometric function value rounded to four decimal places.

49. tan 220°   **50.** sec (−72°)   **51.** sin 261°   **52.** cos (−275°)

53. cot 1200°   **54.** csc (−480°)   **55.** cos (−792.5°)   **56.** sin 242.4°

Find the six trigonometric function values of θ, rounded to four decimal places.

57. θ = 29.3°   **58.** θ = −40.7°   **59.** θ = 152.8°   **60.** θ = −96.1°

Find the coordinates of P_1 through P_6 on the unit circle.

61.

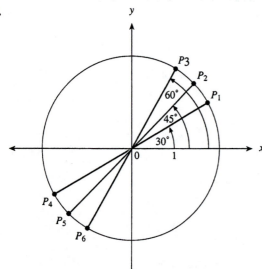

62.

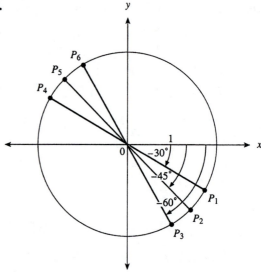

63. (a) Verify that $P\left(\dfrac{2}{3}, \dfrac{\sqrt{5}}{3}\right)$ is on the unit circle.

  (b) Draw a diagram including θ, where $0 < θ < 2\pi$, so that its terminal side intersects the unit circle at P.

  (c) Write the six trigonometric function values of θ.

64. Follow the instructions in Exercise 63 for $P\left(-\dfrac{3}{4}, \dfrac{\sqrt{7}}{4}\right)$.

65. Find y so that $P\left(\dfrac{\sqrt{3}}{4}, y\right)$ is on the unit circle in the fourth quadrant and evaluate tan θ where P is on the terminal side of the angle of θ radians.

66. The terminal side of an angle of θ radians coincides with the line $y = -x$ and lies in the second quadrant. Find cos θ.

Find all possible values of θ where $0 \le θ < 2\pi$. (Hint: For quadrantal angles use the appropriate points on the unit circle. Otherwise use reference triangles in the appropriate quadrants.)

67. sin θ = 1   **68.** cos θ = −1   **69.** sin θ = 0   **70.** cos θ = 0

71. sin θ = $\dfrac{1}{2}$ **72.** cos θ = $-\dfrac{1}{2}$ **73.** cos θ = $-\dfrac{\sqrt{2}}{2}$ **74.** sin θ = $-\dfrac{\sqrt{3}}{2}$

Use the given information to determine the quadrant containing the terminal side of the angle of θ radians and find the remaining trigonometric values.

75. $\tan \theta = 1$, $\sin \theta = -\dfrac{\sqrt{2}}{2}$

76. $\cos \theta = -\dfrac{1}{2}$, $\csc \theta = \dfrac{2}{\sqrt{3}}$

77. $\cot \theta = -\sqrt{3}$, $\sin \theta = -\dfrac{1}{2}$

78. $\sec \theta = \sqrt{2}$, $\cot \theta = 1$

79. Find $\cos \theta$ if $\dfrac{\pi}{2} < \theta < \pi$ and $\sin \theta = \dfrac{1}{3}$.

80. Find $\tan \theta$ if $-\dfrac{\pi}{2} < \theta < 0$ and $\sec \theta = \dfrac{5}{3}$.

81. Find $\sec \theta$ if the terminal side of the angle of θ radians is in quadrant I and $\csc \theta = 4$.

82. Explain why $\sin^2 \theta + \cos^2 \theta = 1$, where θ is any angle measure.

83. In the figure, AB is tangent to the circle at A and meets the terminal side of the angle of θ radians at B. Why is AB equal to $\tan \theta$? Which segment has measure equal to $\sec \theta$?

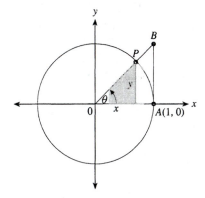

84. The terminal side of an angle of θ radians coincides with the line $y = \frac{1}{2}x$ and lies in the third quadrant. Find $\sin \theta$.

 Challenge   Construct a figure, somewhat similar to the one in Exercise 83, that includes two line segments, one of which has the length of $\cot \theta$ and the other $\csc \theta$.

 Written Assignment   From the definition of the sine function, we have that the domain is the set of all real numbers. This definition can also be used to find the range of this function. Find the range and explain how you arrived at your decision.

 Graphing Calculator Exercises

1. Put your calculator in Parametric and Radian mode. Set your Range variables so that $0 \le T \le 2\pi$, $-1.5 \le X \le 1.5$, and $-1 \le Y \le 1$. Also set your T step (or "pitch") at 0.1. Then **(a)** graph $X_1(T) = \cos T$ and $Y_1(T) = \sin T$, and explain what you get. **(b)** What is the meaning of the variables X, Y, and T in part **(a)**? Suggest a better T step if necessary.

2. Repeat Exercise 1, changing the mode to Degrees and the T range to $0° \le T \le 360°$, with T step $= 5°$.

3. How could you graph the circle with radius equal to 2 and center at the origin in Parametric and Degree mode? Test your solution on your calculator.

4. Set your calculator as in Exercise 1. Then graph the unit circle as in part (a), adding the straight line $X_2(T) = 1$ and $Y_2(T) = \tan T$. Then slowly Trace the circle, starting at $T = 0$, and every 0.1 radian (visible on the bottom of the screen) switch between the two graphs with the down cursor key. (To be able to do this on some calculators, you have to graph the two graphs with a multi-line statement. See your user's manual.) How does this illustrate Exercise 83?

5. Try this approach to Exercise 84. Graph the unit circle as in Exercise 1 together with $X_1(T) = -T$ and $Y_1(T) = -(0.5)T$. **(a)** Does this give the whole straight line? Why? **(b)** Use Trace to find the intersection of the circle and the curve in the third quadrant. The values of T, X and Y are given at the bottom of the screen. How does this agree with the exact answer to Exercise 84?

6. Use the approach in Exercise 5 to find the angle T, $\sin T$, and $\cos T$ of the intersection of the unit circle in the third quadrant with the lines **(a)** $y = 2x$ and **(b)** $y = 3x$.

Critical Thinking

1. The measure of the central angle in a circle subtended by an arc that is $\frac{1}{400}$ part of a circle is called a gradient. (This is another way of measuring angles, and you will find that your calculator can be set in the gradient or GRAD mode.) Find the conversion factors for changing degrees into gradients and vice versa, and do the same for radians and gradients.

2. Explain why the linear speed on a circle increases as the radius increases for a fixed angular speed $\omega > 0$.

3. Study the figure in Exercise 83 and give a geometric explanation of why there is no tangent value when $\theta = \frac{\pi}{2}$.

4. For any value $x = \cos \theta$, $0 \leq \theta \leq \pi$, what is the value of $\sin \theta$ in terms of x?

5. Why are the tangent values for $\pi < \theta < \frac{3\pi}{2}$ the same as the tangent values for $0 < \theta < \frac{\pi}{2}$?

**6.3
GRAPHING
THE SINE
AND COSINE
FUNCTIONS**

The domain of the sine function is the set of all real numbers. The domain values may be interpreted as being either radians θ or arc lengths s; it makes no difference because $s = \theta$ on the unit circle. Also, since $y = \sin \theta$ is the second coordinate of point P, which is the intersection of the unit circle with the terminal side of the angle of θ radians, the range values of the sine function are all such y-values. From the unit circle diagram below, you can see that the range consists of all numbers y where $-1 \leq y \leq 1$.

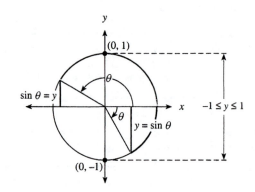

You may think of the circle as being "unrolled" to get the θ values on the horizontal.

One way to graph $y = \sin \theta$ on a rectangular system is to use the θ values on the horizontal axis. Then the corresponding values $y = \sin \theta$ become the ordinates.

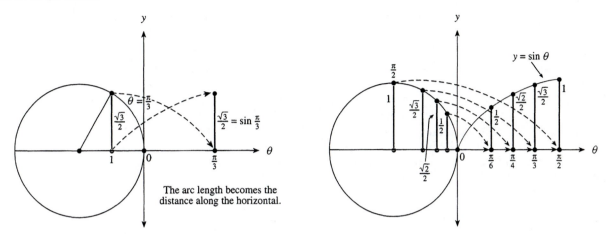

The arc length becomes the distance along the horizontal.

Now that you see the connection between $y = \sin \theta$, defined on the unit circle, and its graph in a rectangular system, we will obtain a more accurate graph using a table of values. First recall that $\sin \theta$ is positive for θ in quadrants I and II; $0 < \theta < \pi$. Also, $\sin \theta$ is negative for $\pi < \theta < 2\pi$. Let us form a table of values by using intervals of $\dfrac{\pi}{6}$ for $0 \le \theta \le 2\pi$.

θ	0	$\frac{\pi}{6}$	$\frac{\pi}{3}$	$\frac{\pi}{2}$	$\frac{2\pi}{3}$	$\frac{5\pi}{6}$	π	$\frac{7\pi}{6}$	$\frac{4\pi}{3}$	$\frac{3\pi}{2}$	$\frac{5\pi}{3}$	$\frac{11\pi}{6}$	2π
$y = \sin \theta$	0	$\frac{1}{2}$	$\frac{\sqrt{3}}{2}$	1	$\frac{\sqrt{3}}{2}$	$\frac{1}{2}$	0	$-\frac{1}{2}$	$-\frac{\sqrt{3}}{2}$	-1	$-\frac{\sqrt{3}}{2}$	$-\frac{1}{2}$	0

We plot these points $\left(\text{using the approximation } \dfrac{\sqrt{3}}{2} = 0.87\right)$ and connect the points with a smooth curve to obtain the graph of $y = \sin \theta$ for $0 \le \theta \le 2\pi$. The segment from zero to 2π on the horizontal axis may be viewed as the unit circle after it has been "unrolled."

The sine function increases on the intervals $\left[0, \dfrac{\pi}{2}\right]$ and $\left[\dfrac{3\pi}{2}, 2\pi\right]$ and decreases on $\left[\dfrac{\pi}{2}, \dfrac{3\pi}{2}\right]$. The curve is concave down on $(0, \pi)$ and concave up on $(\pi, 2\pi)$.

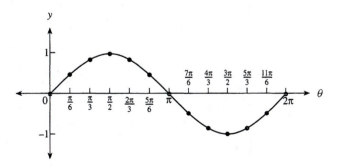

Additional sine values can be found using a calculator set in radian mode. For example,

$$\sin \frac{\pi}{5} = \boxed{\text{SIN}} \ \boxed{(} \ \boxed{\pi} \ \boxed{\div} \ \boxed{5} \ \boxed{)} \ \boxed{\text{ENTER}} = 0.59 \qquad \text{rounded to two} \atop \text{decimal places}$$

On a scientific calculator this sequence can be used:

$$\sin \frac{\pi}{5} = \boxed{\pi} \ \boxed{\div} \ \boxed{5} \ \boxed{=} \ \boxed{\text{SIN}}$$

For angles of measure θ, where $2\pi \le \theta \le 4\pi$, we know that the terminal sides (in the unit circle) are the same as for those angles from 0 to 2π. Hence everything repeats. Similarly for $-2\pi \le \theta \le 0$, and so on. Thus we say that the sine function is *periodic*, with **period** 2π; $\sin(\theta + 2\pi) = \sin \theta$. That is, the sine curve repeats itself to the right and left as shown in the figure below.

Similar unit circle diagrams can also be drawn to demonstrate that
$\sin(\theta - 2\pi) = \sin \theta$.

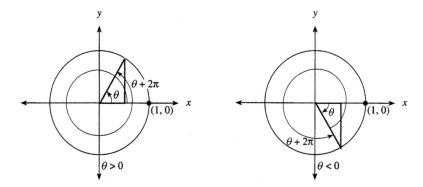

These two unit circle diagrams show coterminal angles having measures θ and $\theta + 2\pi$. At the left θ is positive and at the right θ is negative. In either case, $\sin \theta = \sin(\theta + 2\pi)$.

Even though $\sin(\theta + 4\pi) = \sin \theta$, the period is not 4π because 4π is not the smallest positive number having this property.

A function f is said to be periodic with period p provided that p is the smallest positive constant, if any, such that $f(x + p) = f(x)$ for all x in the domain of f.

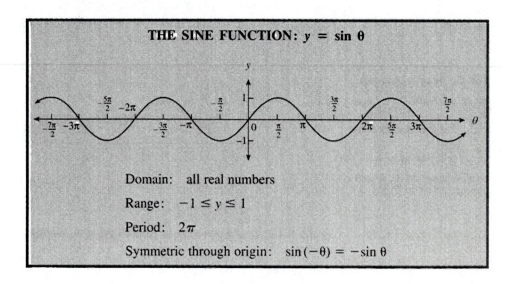

THE SINE FUNCTION: $y = \sin \theta$

Domain: all real numbers

Range: $-1 \le y \le 1$

Period: 2π

Symmetric through origin: $\sin(-\theta) = -\sin \theta$

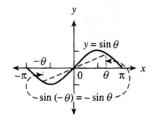

The graph of $y = \sin \theta$ indicates that the sine function is symmetric through the origin; this is an odd function. This symmetry can be observed in the figure in the margin and it can be verified by returning to the unit circle below.

A similar unit circle diagram can be drawn when the terminal sides of the angles are in quadrants II and III.

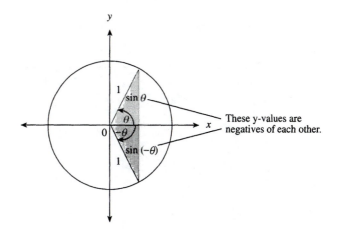

These y-values are negatives of each other.

Letting $f(\theta) = \sin \theta$, we have

$$f(-\theta) = \sin(-\theta) = -\sin(\theta) = -f(\theta)$$
$$f(-\theta) = -f(\theta)$$

The graph of the cosine function can be obtained from the graph of $y = \sin \theta$ by observing that $\cos \theta = \sin\left(\theta + \frac{\pi}{2}\right)$. To see why this is true, consider these two typical situations:

Similar diagrams can be drawn for the cases where the terminal side is in the 3rd or 4th quadrant, as well as for the cases where $\theta < 0$.

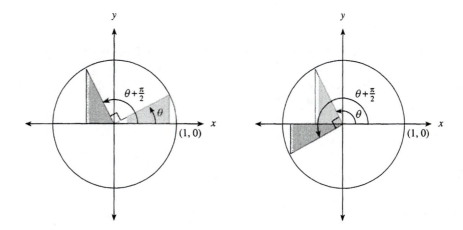

In each case the reference triangles for θ and $\theta + \frac{\pi}{2}$ are congruent. Consequently, the side adjacent to the reference angle for θ radians has the same length as the side opposite the reference angle for $\theta + \frac{\pi}{2}$ radians. It follows that $\cos \theta = \sin\left(\theta + \frac{\pi}{2}\right)$.

The graph of $y = \sin\left(\theta + \dfrac{\pi}{2}\right)$ can be obtained by shifting the graph of $y = \sin\theta$ by $\dfrac{\pi}{2}$ units to the left. Therefore, the cosine curve can be obtained by shifting the sine curve $\dfrac{\pi}{2}$ units to the left.

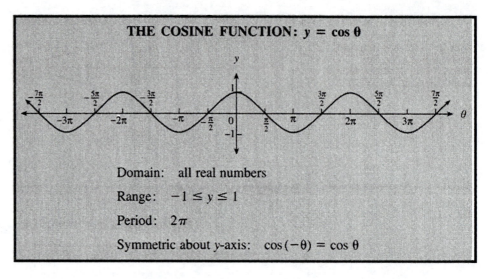

THE COSINE FUNCTION: $y = \cos\theta$

Domain: all real numbers

Range: $-1 \le y \le 1$

Period: 2π

Symmetric about y-axis: $\cos(-\theta) = \cos\theta$

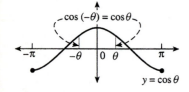

The cosine is an even function.

In Section 6.2 we used the unit circle to arrive at $\cos\theta = x$. However, to be consistent with the usual labeling of the vertical axis in a rectangular system, we have written the equation in the form $y = \cos\theta$. Furthermore, from now on we will use x instead of θ for the horizontal axis. We use the letter θ when making direct reference to the unit circle.

The cosine curve may be regarded as being $\dfrac{\pi}{2}$ units ahead of (or behind) the sine curve, and vice versa. Both functions have the same period, 2π. The symmetry about the y-axis can be observed in the figure in the margin and it can be verified by studying the following unit circle diagram for an obtuse angle. (Similar diagrams can be drawn for any size angle.)

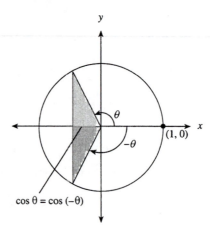

$\cos\theta = \cos(-\theta)$

The sine function has a maximum value of $M = 1$ and a minimum value of $m = -1$. One-half of the difference is called the **amplitude.**

$$\text{Amplitude} = \frac{M - m}{2}$$

Note that $-1 \le \sin x \le 1$ is equivalent to $-2 \le 2 \sin x \le 2$.

For $y = \sin x$, amplitude $= \dfrac{1 - (-1)}{2} = 1$. Thus for $y = 2 \sin x$, $M = 2$, $m = -2$, and the amplitude is $\dfrac{2 - (-2)}{2} = 2$. In general, the amplitude of $y = a \sin x$ or of $y = a \cos x$ is equal to $|a|$.

In the next figure the idea of amplitude is illustrated for a few cases. Notice that each of these functions has period 2π.

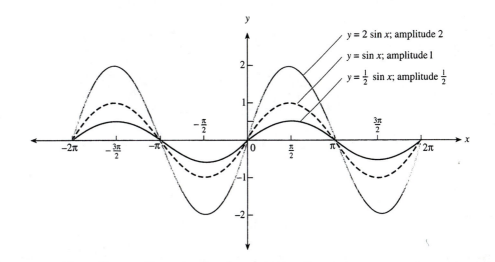

Graph the curve for the values $-2\pi \le x \le 2\pi$.

1. $y = -\sin x$ **2.** $y = 3 \cos x$ **3.** $y = \sin(-x)$

Find the amplitude.

4. $y = 10 \sin x$ **5.** $y = -\frac{2}{3} \cos x$ **6.** $y = \frac{1}{2} - \cos x$

For $y = \sin x$, the coefficient of x is 1. By changing the coefficient, we alter the period of the function. Consider, for example, the graph of $y = \sin 2x$. As x assumes values from 0 through π, $2x$ takes on values from 0 through 2π. That is, $0 \le 2x \le 2\pi$ is equivalent to $0 \le x \le \pi$. Thus the graph goes through a complete cycle for $0 \le x \le \pi$ and has period π. This information is shown in the following table of values and graph. Note that the graph completes two full cycles in the interval $0 \le x \le 2\pi$.

x	0	$\frac{\pi}{4}$	$\frac{\pi}{2}$	$\frac{3\pi}{4}$	π
$2x$	0	$\frac{\pi}{2}$	π	$\frac{3\pi}{2}$	2π
$y = \sin 2x$	0	1	0	-1	0

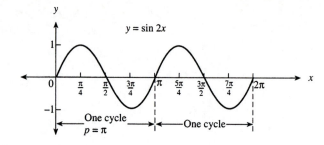

The equivalence of $0 \le 2x \le 2\pi$ and $\frac{0}{2} \le \frac{2x}{2} \le \frac{2\pi}{2}$ gave the period $p = \pi$ for $y = \sin 2x$. In a similar manner you can show that both $y = a \sin bx$ and $y = a \cos bx$ have period $p = \frac{2\pi}{|b|}$.

The preceding observations regarding the amplitude and period are included in the following guidelines for graphing $y = a \sin bx$ and $y = a \cos bx$.

GUIDELINES FOR GRAPHING $y = a \sin bx$, $y = a \cos bx$

1. Find the period $p = \frac{2\pi}{|b|}$ and the amplitude $|a|$.

2. Divide the segment $[0, p]$ into four equal parts:

$$0 \qquad \frac{p}{4} \qquad \frac{p}{2} \qquad \frac{3p}{4} \qquad p$$

3. For $y = a \sin bx$, $y = 0$ at $x = 0$, $\frac{p}{2}$, and p; $y = a$ at $x = \frac{p}{4}$; and $y = -a$ at $x = \frac{3p}{4}$.

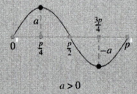
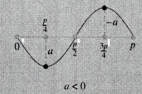

$a > 0$ $\qquad\qquad\qquad$ $a < 0$

4. For $y = a \cos bx$, $y = 0$ at $x = \frac{p}{4}$ and $\frac{3p}{4}$; $y = a$ at $x = 0$, and at $x = p$; $y = -a$ at $x = \frac{p}{2}$.

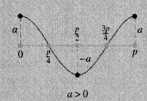
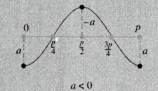

$a > 0$ $\qquad\qquad\qquad$ $a < 0$

5. Connect the five points as shown and repeat the basic cycle as required in either direction.

EXAMPLE 1   Graph $y = 2 \cos \dfrac{x}{2}$ on the interval $[-p, p]$, where p is the period, and compare to the graph of $y = \cos x$.

Solution The amplitude is $|a| = |2| = 2$, and $p = \dfrac{2\pi}{|b|} = \dfrac{2\pi}{\frac{1}{2}} = 4\pi$. Now divide $[0, 4\pi]$ into four equal parts and note that $a = 2$ is positive to get the following:

$$y = 0 \qquad \text{at } x = \frac{4\pi}{4} = \pi \text{ and } \frac{3(4\pi)}{4} = 3\pi$$

$$y = 2 \qquad \text{at } x = 0 \text{ and } 4\pi$$

$$y = -2 \qquad \text{at } x = \frac{4\pi}{2} = 2\pi$$

Connect the five points as shown. Complete the problem by repeating the cycle to -4π at the left and include the graph of $y = \cos x$.

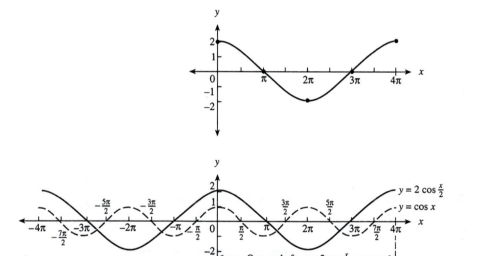

The methods of shifting or translating the graph of a given equation to obtain the graph of a more complicated equation can be applied to the circular functions. This was done earlier (page 386) where the graph of $y = \sin \theta$ was shifted $\dfrac{\pi}{2}$ units to the left to obtain the graph of $y = \sin\left(\theta + \dfrac{\pi}{2}\right) = \sin\left(\theta - \left(-\dfrac{\pi}{2}\right)\right) = \cos x$. The number $-\dfrac{\pi}{2}$ is called the **phase shift.**

In general, the graph of $y = a \sin b(x - h)$ is the same as the graph of $y = a \sin bx$ except that it has been shifted h units to the right if $h > 0$, and h units to the left if $h < 0$. In either case, h is called the phase shift. Similar observations hold for the graph of $y = a \cos b(x - h)$.

In general, for
$y = a \sin(bx + c)$ *or*
$y = a \cos(bx + c)$, *the phase*

shift is $-\dfrac{c}{b}$ *since* $bx + c$ *can*

be written as $b\left(x - \left(-\dfrac{c}{b}\right)\right)$.
Consequently the phase shift

here is not $\dfrac{\pi}{2}$. *The form*

$2x - \dfrac{\pi}{2}$ *must first be*

changed so that x has a
coefficient of 1. Thus, write

$2x - \dfrac{\pi}{2} = 2\left(x - \dfrac{\pi}{4}\right)$

and note that the phase shift

is $\dfrac{\pi}{4}$.

EXAMPLE 2   For $y = 3 \sin\left(2x - \dfrac{\pi}{2}\right)$ determine the amplitude, period, and phase shift, and sketch the graph for one period.

Solution Since $y = 3 \sin\left(2x - \dfrac{\pi}{2}\right) = 3 \sin 2\left(x - \dfrac{\pi}{4}\right)$, the amplitude is 3, the

period is $\dfrac{2\pi}{2} = \pi$, and the phase shift is $\dfrac{\pi}{4}$. Therefore, the graph is obtained by

shifting the graph of $y = 3 \sin 2x$ to the right $\dfrac{\pi}{4}$ units, as shown in the graph.

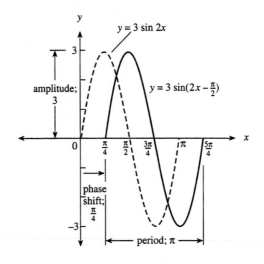

The next example demonstrates how to graph the sum of two circular functions by **adding the ordinates.**

EXAMPLE 3   Graph $y = \sin x + \cos x$ on $[-2\pi, 2\pi]$.

Solution First graph $y = \sin x$ and $y = \cos x$ on the same set of axes as shown by the dashed curves in the figure on the next page.

x	0	$\dfrac{\pi}{4}$	$\dfrac{\pi}{2}$	$\dfrac{3\pi}{4}$	π	$\dfrac{5\pi}{4}$	$\dfrac{3\pi}{2}$	$\dfrac{7\pi}{4}$	2π
$\sin x$	0	$\dfrac{\sqrt{2}}{2}$	1	$\dfrac{\sqrt{2}}{2}$	0	$-\dfrac{\sqrt{2}}{2}$	-1	$-\dfrac{\sqrt{2}}{2}$	0
$\cos x$	1	$\dfrac{\sqrt{2}}{2}$	0	$-\dfrac{\sqrt{2}}{2}$	-1	$-\dfrac{\sqrt{2}}{2}$	0	$\dfrac{\sqrt{2}}{2}$	1
$\sin x + \cos x$	1	$\sqrt{2}$	1	0	-1	$-\sqrt{2}$	-1	0	1

Use a graphing calculator to verify the graph of $y = \sin x + \cos x$ obtained here.

Select specific values of x in $[0, 2\pi]$ for which $\sin x$ and $\cos x$ are easy to find and add these ordinates. The preceding table contains such values, and the resulting points have been indicated by the dots. After connecting the dots by a smooth curve, the graph is completed by repeating the process described on $[-2\pi, 0]$.

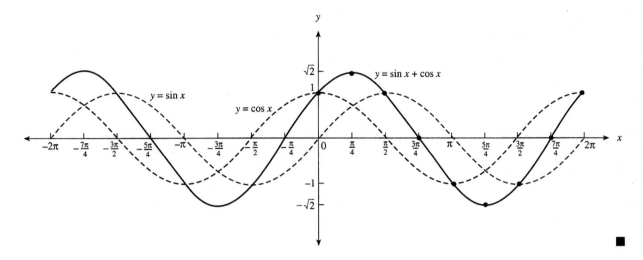

■

EXERCISES 6.3

1. Complete the table and use these points and the symmetry through the origin to graph $y = \sin x$ for $-\pi \le x \le \pi$.

x	$-\pi$	$-\dfrac{5\pi}{6}$	$-\dfrac{2\pi}{3}$	$-\dfrac{\pi}{2}$	$-\dfrac{\pi}{3}$	$-\dfrac{\pi}{6}$	0
$y = \sin x$							

2. (a) Complete the table and use these points and the symmetry about the y-axis to graph $y = \cos x$ for $-\pi \le x \le \pi$.

x	0	$\dfrac{\pi}{4}$	$\dfrac{\pi}{2}$	$\dfrac{3\pi}{4}$	π
$y = \cos x$					

(b) On which intervals is the cosine increasing? decreasing? On which intervals is the graph of the cosine concave up? down?

For Exercises 3–8, explain how the graph of g can be obtained from the graph of f by using an appropriate phase shift. Graph both functions on the same axes for $0 \le x \le 2\pi$.

3. $f(x) = \sin x, \ g(x) = \sin\left(x - \dfrac{\pi}{2}\right)$ **4.** $f(x) = \sin x, \ g(x) = \sin\left(x + \dfrac{\pi}{4}\right)$

5. $f(x) = \cos x, \ g(x) = \cos\left(x - \dfrac{\pi}{3}\right)$ **6.** $f(x) = \cos x, \ g(x) = \cos\left(x + \dfrac{\pi}{3}\right)$

7. $f(x) = \sin x$, $g(x) = 2 \sin(x + \pi)$ **8.** $f(x) = \cos x$, $g(x) = -\cos(x - \pi)$

Graph for $-2\pi \le x \le 2\pi$.

9. $y = |\sin x|$ **10.** $y = -|\cos x|$

11. Graph $y = 3 \sin x$, $y = \frac{1}{3} \sin x$, and $y = -3 \sin x$ on the same axes for $0 \le x \le 2\pi$. Find the amplitudes.

12. Graph $y = 2 \cos x$, $y = \frac{1}{2} \cos x$, and $y = -2 \cos x$ on the same axes for $-\pi \le x \le \pi$. Find the amplitudes.

Sketch the curve on the interval $0 \le x \le 2\pi$. Find the amplitude and period.

13. $y = \cos 2x$ **14.** $y = -\sin 2x$ **15.** $y = -\frac{3}{2} \sin 4x$

16. $y = \cos 4x$ **17.** $y = -\cos \frac{1}{2} x$ **18.** $y = -2 \sin \frac{1}{2} x$

19. Find the period p of $y = \frac{1}{2} \cos \frac{1}{4} x$. Graph this curve and the curve $y = \cos x$ for $-p \le x \le p$ on the same axes.

20. Find the period p of $y = 3 \sin \frac{1}{3} x$. Graph this curve and the curve $y = \sin x$ for $-p \le x \le p$ on the same axes.

21. Find the period p of $y = 2 \sin \pi x$ and sketch the curve for $0 \le x \le p$.

22. Find the period p of $y = -\frac{3}{4} \cos \frac{\pi}{2} x$ and sketch the curve for $0 \le x \le p$.

In the following exercises, the curve has equation of the form $y = a \sin bx$ or $y = a \cos bx$. Find a and b and write the equation.

23.

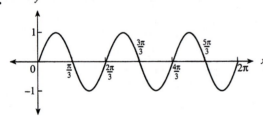

24.

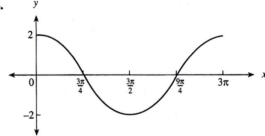

25.

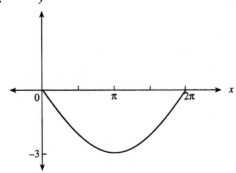

26.

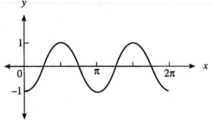

Determine the amplitude, period, and phase shift.

27. $y = 5 \cos 3\left(x - \dfrac{\pi}{6}\right)$ **28.** $y = \dfrac{1}{2} \sin 2(x + \pi)$ **29.** $y = -2 \sin(2x + \pi)$

30. $y = -\cos\left(\dfrac{x}{2} - \dfrac{\pi}{3}\right)$ **31.** $y = \dfrac{3}{2} \cos\left(\dfrac{x}{4} - 1\right)$ **32.** $y = 4 \sin\left(\pi x + \dfrac{\pi}{2}\right)$

Determine the amplitude, period, and phase shift, and sketch the graph for one period.

33. $y = \sin(4x - \pi)$ **34.** $y = \sin\left(2x + \dfrac{\pi}{2}\right)$ **35.** $y = 2 \cos\left(2x - \dfrac{\pi}{2}\right)$

36. $y = \frac{1}{2} \cos(3x + \pi)$ **37.** $y = -\dfrac{5}{2} \sin\left(\dfrac{x}{2} + \dfrac{\pi}{4}\right)$ **38.** $y = -2 \cos\left(\dfrac{\pi x}{2} - \dfrac{\pi}{2}\right)$

Explain how the graph of the first equation can be obtained from the graph of the second equation. Sketch the graph for one period.

39. $y = 3 + \sin\left(\dfrac{x}{2} + \dfrac{\pi}{2}\right)$; $y = \sin\left(\dfrac{x}{2} + \dfrac{\pi}{2}\right)$ **40.** $y = -3 + 2 \cos\left(x - \dfrac{\pi}{6}\right)$; $y = 2 \cos\left(x - \dfrac{\pi}{6}\right)$

Sketch the graphs of the functions for $0 \le x \le 2\pi$.

41. $f(x) = \sin x - \cos x$ **42.** $f(x) = 2 \sin x + \cos x$ **43.** $f(x) = \cos x + \cos 2x$

44. $f(x) = \sin 2x + \frac{1}{2} \cos x$ **45.** $f(x) = |\sin x| + |\cos x|$ **46.** $f(x) = x + \sin x$

47. Let $f(x) = 2x + 5$ and $g(x) = \cos x$. Form the composites $f \circ g$ and $g \circ f$.

48. Let $f(x) = \sqrt{x^2 + 1}$ and $g(x) = \sin x$. Form the composites $f \circ g$ and $g \circ f$.

49. Let $h(x) = \cos(5x^2)$. Find f and g so that $h = f \circ g$, where the inner function g is quadratic.

50. Let $h(x) = \sin(\ln x)$. Find f and g so that $h = f \circ g$, where the outer function is trigonometric.

51. Let $F(x) = \cos \sqrt[3]{1 - 2x}$. Find f, g, and h so that $F = f \circ g \circ h$, where h is linear.

52. Let $F(x) = \ln(\sin(x^2 - 1))$. Find f, g, and h so that $F = f \circ g \circ h$, where h is a binomial.

53. Prove that if p is the period of the function f, then $f(x + 2p) = f(x)$ for x in the domain of f that consists of all real numbers.

Challenge If $f(x) = 3 \sin\left(2x - \dfrac{\pi}{8}\right)$ find $g(x) = a \cos(bx + c)$ so that $f(x) + g(x) = 0$ for all values x.

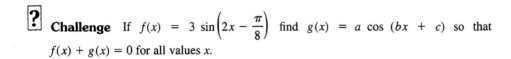

Written Assignment In section 7.3 it will be proved that $\sin(\alpha + \beta) = \sin \alpha \cos \beta + \cos \alpha \sin \beta$, for all values of α and β. Use this result with $\alpha = x$ and $\beta = \dfrac{\pi}{4}$ and find the constant a such that $a \sin\left(x + \dfrac{\pi}{4}\right) = \sin x + \cos x$. Explain how this gives an alternative method to graph the equation in Example 3, page 391.

Graphing Calculator Exercises   For Exercises 1–4, graph the following functions $y = f(x)$ in the window $-2 \le X \le 2$ and $-3 \le Y \le 3$, and describe the behavior of the function as x gets closer and closer to zero. Which of these functions is defined at zero?

1. $f(x) = (\sin x)/x$

2. $f(x) = (1 - \cos x)/x$

3. $f(x) = |1/\sin(x)|$

4. $f(x) = \sin(1/x)$

In calculus courses, it is shown that the difference quotient $q(x) = (f(x + h) - f(x))/h$ is a good approximation to the slope of the tangent line to the graph of $y = f(x)$ at the point $(x, f(x))$, for small values of h.

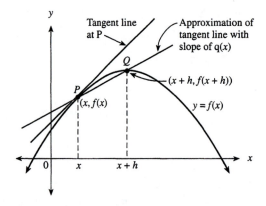

Graph the difference quotient function $y = q(x)$ with $h = 0.5$ and $-2\pi \le x \le 2\pi$ for each of the functions in Exercises 5–6, and tell which trigonmetric function $g(x)$ the graph resembles. ($g(x)$ is called the "derivative of $f(x)$.") Test your guess by graphing $g(x)$ and $q(x)$ together.

5. $f(x) = \sin x$ (Graph $y = (\sin(x + 0.5) - \sin x)/0.5$)

6. $f(x) = \cos x$ (Graph $y = (\cos(x + 0.5) - \cos x)/0.5$)

7. How do your answers in Exercises 5 and 6 change if we set $h = 0.01$?

6.4 GRAPHING OTHER TRIGONOMETRIC FUNCTIONS

The tangent function has been defined by $\tan \theta = \dfrac{y}{x} = \dfrac{\sin \theta}{\cos \theta}$ (see page 371), where x and y are the coordinates of the appropriate point on the unit circle. Since $\tan \theta = \dfrac{\sin \theta}{\cos \theta}$ the properties of the tangent function depend on the sine and cosine functions. First observe that $\cos \theta = 0$ for $\theta = \pm\dfrac{\pi}{2},\ \pm\dfrac{3\pi}{2},\ \pm\dfrac{5\pi}{2}, \ldots$. Therefore, the domain of $\tan \theta = \dfrac{\sin \theta}{\cos \theta}$ consists of all real numbers θ except those of the form $\theta = \dfrac{\pi}{2} + k\pi$, where k is any integer.

The graph of $y = \tan \theta$ is symmetric through the origin, since for any θ in the domain we have

Recall that a function f is symmetric through the origin provided that $f(-t) = -f(t)$.

$$\tan(-\theta) = \frac{\sin(-\theta)}{\cos(-\theta)} = \frac{-\sin \theta}{\cos \theta} = -\frac{\sin \theta}{\cos \theta} = -\tan \theta$$

The period of $y = \tan \theta$ is π, as can be observed by returning to the unit circle. Consider, for example, the case where the terminal side of an angle of θ radians is in quadrant I; in particular, we assume that $0 < \theta < \dfrac{\pi}{2}$. Adding π to θ gives the angle

It is also true that $\tan(\theta + 2\pi) = \tan \theta$. Why, then, is the period not 2π?

of measure $\theta + \pi$, whose terminal side is in quadrant III. Since the two terminal sides are on the same line, the reference triangles are congruent, and the positive ratio $\dfrac{\text{opposite}}{\text{adjacent}}$ is the same in each case. Therefore, $\tan(\theta + \pi) = \tan \theta$. Furthermore, π is

the smallest such positive number since, for any number t, $0 < t < \pi$, a similar diagram would show that $\tan(\theta + t) \neq \tan \theta$.

Similar diagrams can be used when the terminal side of the angle is in any of the other quadrants, as well as for the cases where $\theta < 0$.

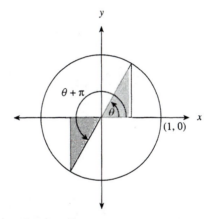

It is possible to construct the graph of the tangent function after considering the geometric interpretation of the tangent. This is suggested in the following figure in which AB is perpendicular to the circle at A and meets the terminal side of $\angle AOP$ at B. (See Exercise 83, page 381)

Since triangles OAB and OQP are similar,

$$\frac{AB}{OA} = \frac{QP}{OQ}$$

$$\frac{AB}{1} = \frac{y}{x}$$

$$AB = \tan \theta$$

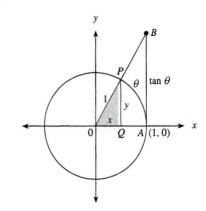

For $0 \leq \theta < \dfrac{\pi}{2}$ put θ on the horizontal axis and place the corresponding tangent segment vertically at the end of θ. As the θ-values "unroll" along the horizontal axis, the tangent line to the circle at 0 is "stretched" and "curved" to become the tangent curve in a rectangular system.

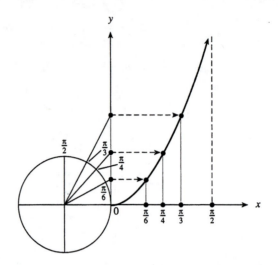

Now reflect this branch of the curve through the origin to get one full cycle on $\left(-\dfrac{\pi}{2}, \dfrac{\pi}{2} \right)$. In doing so, the variable θ has been replaced by x.

The tangent function is increasing on $\left(-\dfrac{\pi}{2}, \dfrac{\pi}{2} \right)$. *It is concave down on* $\left(-\dfrac{\pi}{2}, 0 \right)$ *and concave up on* $\left(0, \dfrac{\pi}{2} \right)$. *There is no amplitude.*

x	$\tan x$
0	0
$\dfrac{\pi}{6}$	$\dfrac{1}{\sqrt{3}} \approx 0.6$
$\dfrac{\pi}{4}$	1
$\dfrac{\pi}{3}$	$\sqrt{3} \approx 1.7$

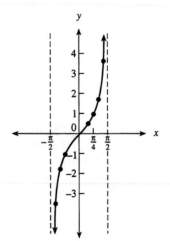

Notice that the vertical lines $x = \pm \dfrac{\pi}{2}$ are asymptotes to the curve. You can observe this "growth" of $y = \tan x$ by noting that as x gets close to $\dfrac{\pi}{2}$, the sine gets close to 1 and the cosine gets close to 0. Hence the fraction $\dfrac{\sin x}{\cos x}$ gets very large. This is also demonstrated by the following table values, to four decimal places obtained by using a calculator set in the radian mode.

θ	1.4	1.5	1.57	1.5704	$\longrightarrow$ **getting close to** $\dfrac{\pi}{2} = \mathbf{1.57079 \cdots}$
$\sin \theta$	.9854	.9974	.9999	.9999	$\longrightarrow$ getting close to 1
$\cos \theta$	.1699	.0707	.0007	.0003	$\longrightarrow$ getting close to 0
$\tan \theta$	5.797	14.10	1255	2523	$\longrightarrow$ getting very large

   Since the period of tan x is π, the preceding figure shows one cycle of the tangent function, which repeats to the left and right. The range of $y = \tan x$ consists of all real numbers.

$\tan x \to \infty$ as $x \to \dfrac{\pi}{2}^{-}$ and $\tan x \to -\infty$ as $x \to -\dfrac{\pi}{2}^{+}$.

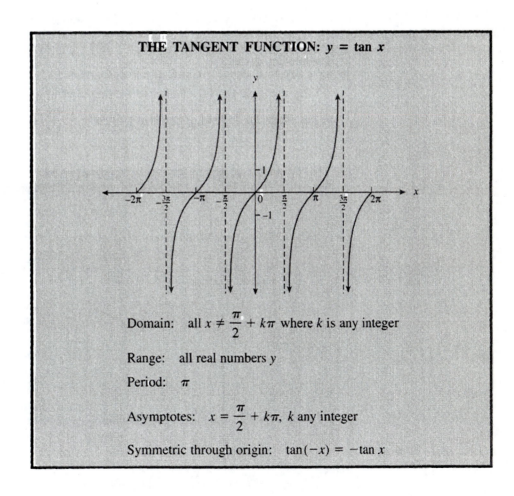

THE TANGENT FUNCTION: $y = \tan x$

Domain: all $x \neq \dfrac{\pi}{2} + k\pi$ where k is any integer

Range: all real numbers y

Period: π

Asymptotes: $x = \dfrac{\pi}{2} + k\pi$, k any integer

Symmetric through origin: $\tan(-x) = -\tan x$

EXAMPLE 1   Graph $y = -\tan 3x$.

Observe that since the period of $y = \tan x$ is π, and the asymptotes occur every π units along the x-axis, the period here is $\dfrac{\pi}{3}$ and the asymptotes occur every $\dfrac{\pi}{3}$ units along the x-axis.

Solution

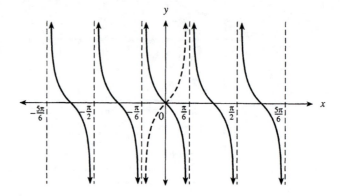

One complete cycle of the tangent curve takes place for x where $-\dfrac{\pi}{2} < x < \dfrac{\pi}{2}$, and the length of this interval, π, is the period of $y = \tan x$. Consequently, since

In general, the period of $y = \tan bx$ or of $y = \tan b(x - h)$ is $\dfrac{\pi}{|b|}$. This is also true for the cotangent graphed below.

$-\dfrac{\pi}{2} < 3x < \dfrac{\pi}{2}$ is equivalent to $-\dfrac{\pi}{6} < x < \dfrac{\pi}{6}$ and $\dfrac{\pi}{6} - \left(-\dfrac{\pi}{6}\right) = \dfrac{\pi}{3}$, the period of $y = \tan 3x$ is $\dfrac{\pi}{3}$. Note that to graph $y = -\tan 3x$, we first graphed $y = \tan 3x$ using a dashed curve, and then reflected this curve through the x-axis, as shown in the preceding graph.   ∎

By using an analysis similar to the one used to obtain the graph of $y = \tan x$, we can obtain the following graph of $y = \cot x = \dfrac{\cos x}{\sin x}$.

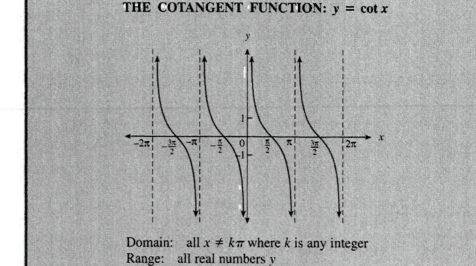

THE COTANGENT FUNCTION: $y = \cot x$

Domain: all $x \neq k\pi$ where k is any integer
Range: all real numbers y
Period: π
Asymptotes: $x = k\pi$, k any integer
Symmetric through origin: $\cot(-x) = -\cot x$

$\cot x \to \infty$ as $x \to 0^+$ and $\cot x \to -\infty$ as $x \to \pi^-$.

The secant function can be graphed by making use of the cosine because $\sec x = \dfrac{1}{\cos x}$ for $\cos x \neq 0$. We need only consider $x \geq 0$ since

$$\sec(-x) = \frac{1}{\cos(-x)} = \frac{1}{\cos x} = \sec x$$

shows that the graph of the secant function is symmetric with respect to the y-axis. Now consider $0 \leq x < \dfrac{\pi}{2}$. For such x take the reciprocal of $\cos x$ to obtain $y = \dfrac{1}{\cos x} = \sec x$. Below are some specific cases to help graph the curve.

Using a calculator in the radian mode
$$\left(\frac{\pi}{2} = 1.570796\ldots \right),$$
we obtain
$$\sec 1.5 = 14$$
$$\sec 1.56 = 93$$
$$\sec 1.57 = 1256$$
$$\sec 1.5707 = 10381$$
As $x \to \dfrac{\pi}{2}^{-}$ *,* $\sec x \to \infty$.

x	$\cos x$	$\sec x = \dfrac{1}{\cos x}$
0	1	1
$\dfrac{\pi}{6}$	$\dfrac{\sqrt{3}}{2}$	$\dfrac{2}{\sqrt{3}} \approx 1.2$
$\dfrac{\pi}{4}$	$\dfrac{1}{\sqrt{2}}$	$\sqrt{2} \approx 1.4$
$\dfrac{\pi}{3}$	$\dfrac{1}{2}$	2

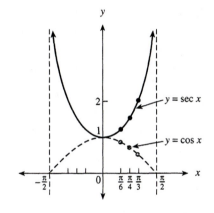

You can see that as x gets close to $\dfrac{\pi}{2}$, $\cos x$ gets close to 0, and therefore the reciprocals get very large. It follows that $x = \dfrac{\pi}{2}$ is a vertical asymptote and, by symmetry, so is $x = -\dfrac{\pi}{2}$. By similar analysis the graph of $y = \sec x$ can be found for $\dfrac{\pi}{2} < x < \dfrac{3\pi}{2}$; the periodicity gives the rest. Note that for all x in the domain of the secant, we have

$$\sec(x + 2\pi) = \frac{1}{\cos(x + 2\pi)} = \frac{1}{\cos x} = \sec x$$

If 2π were replaced by any positive number $a < 2\pi$, then the preceding would no longer be true. Therefore, 2π is the period of $y = \sec x$.

Also, just as the asymptotes above occurred where the cosine was 0, the asymptotes for $y = \sec x$ will occur whenever the $\cos x = 0$, where $x = \dfrac{\pi}{2} + k\pi$.

For what parts of
$\left(-\dfrac{\pi}{2}, \dfrac{3\pi}{2}\right)$ *is the secant increasing or decreasing? Where is the curve concave up or down? Is there an amplitude?*

$\sec x \to \infty$ *as* $x \to -\dfrac{\pi}{2}^{+}$

and as $x \to \dfrac{\pi}{2}^{-}$ *. Also,*

$\sec x \to -\infty$ *as* $x \to \dfrac{\pi}{2}^{+}$

and as $x \to \dfrac{3\pi}{2}^{-}$

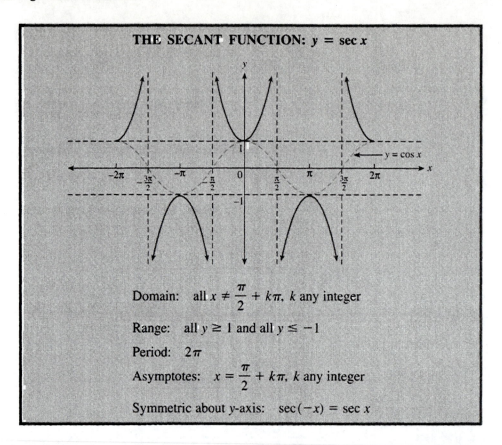

THE SECANT FUNCTION: $y = \sec x$

Domain: all $x \neq \dfrac{\pi}{2} + k\pi$, k any integer

Range: all $y \geq 1$ and all $y \leq -1$

Period: 2π

Asymptotes: $x = \dfrac{\pi}{2} + k\pi$, k any integer

Symmetric about y-axis: $\sec(-x) = \sec x$

The properties of $y =$
$\csc x = \dfrac{1}{\sin x}$ *can be obtained from* $y = \sin x$, *just as the properties of the cosine were used for the secant.*

$\csc x \to \infty$ *as* $x \to 0^{+}$
and as $x \to \pi^{-}$. *Also*
$\csc x \to -\infty$ *as* $x \to \pi^{+}$ *and as* $x \to 2\pi^{-}$.

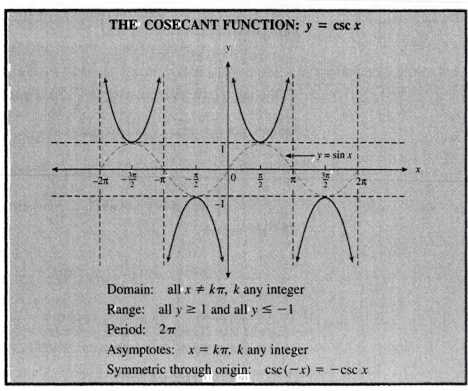

THE COSECANT FUNCTION: $y = \csc x$

Domain: all $x \neq k\pi$, k any integer

Range: all $y \geq 1$ and all $y \leq -1$

Period: 2π

Asymptotes: $x = k\pi$, k any integer

Symmetric through origin: $\csc(-x) = -\csc x$

Phase shifts can be used in any of the circular functions, not just for the sine and cosine. This is illustrated in the next example.

EXAMPLE 2   For $y = \csc\left(2x + \dfrac{\pi}{2}\right)$ determine the period and phase shift, and sketch the graph for one period.

Solution Since $y = \csc\left(2x + \dfrac{\pi}{2}\right) = \csc 2\left(x + \dfrac{\pi}{4}\right)$, the period is $\dfrac{2\pi}{2} = \pi$, and

the phase shift is $-\dfrac{\pi}{4}$. The graph is the same as the graph of $y = \csc 2x$ only

The period p for $y = a \csc b(x - h)$ is the same as for $y = a \csc bx$, which, in turn, is the same as for $y = a \sin bx$; $p = \dfrac{2\pi}{|b|}$.
This is also true for $y = a \sec b(x - h)$.

shifted $\dfrac{\pi}{4}$ units to the left. Just as the asymptotes for $y = \csc x$ are at the endpoints

and midpoint of the interval $[0, 2\pi]$, so are the asymptotes for $y = \csc 2\left(x + \dfrac{\pi}{4}\right)$

at the endpoints and midpoint of the interval $\left[-\dfrac{\pi}{4}, \dfrac{3\pi}{4}\right]$ as shown.

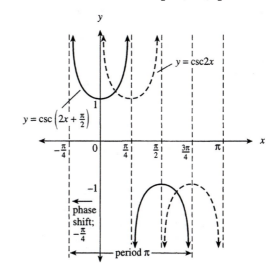

■

EXERCISES 6.4

1. Complete the following table (using a calculator only if necessary) and sketch the curve $y = \tan x$ for $-\dfrac{\pi}{2} < x \le 0$. Then use the symmetry through the origin to obtain the graph for $-\dfrac{\pi}{2} < x < \dfrac{\pi}{2}$.

x	-1.4	-1.3	$-\dfrac{\pi}{3}$	$-\dfrac{\pi}{4}$	$-\dfrac{\pi}{6}$	0
$y = \tan x$						

2. **(a)** Verify that $\cot(\theta + \pi) = \cot\theta$ for typical angles of θ radians in quadrant I or II, using a unit circle diagram.

   (b) Complete the table of values and sketch the graph of $y = \cot x$ for $0 < x < \pi$.

x	$\frac{\pi}{6}$	$\frac{\pi}{4}$	$\frac{\pi}{3}$	$\frac{\pi}{2}$	$\frac{2\pi}{3}$	$\frac{3\pi}{4}$	$\frac{5\pi}{6}$
$y = \cot x$							

   (c) Refer to part (b) and find the subintervals of $(0, \pi)$ for which the cotangent is increasing or decreasing and the subintervals where the curve is concave up or down.

3. **(a)** Show that the cotangent is symmetric through the origin by verifying that $\cot(-x) = -\cot x$.

   (b) Repeat part (a) for the cosecant.   **(c)** Verify that $\csc(x + 2\pi) = \csc x$.

Sketch the graph of f by making an appropriate phase shift of the graph of g.

4. $f(x) = \tan\left(x - \dfrac{\pi}{2}\right)$; $g(x) = \tan x$ 5. $f(x) = \cot\left(x + \dfrac{\pi}{2}\right)$; $g(x) = \cot x$

6. $f(x) = \sec\left(x + \dfrac{\pi}{4}\right)$; $g(x) = \sec x$ 7. $f(x) = \csc\left(x - \dfrac{\pi}{3}\right)$; $g(x) = \csc x$

8. Compare the graph of the tangent and cotangent and decide which of the following equations are true for all allowable x.

   (a) $\cot x = -\tan x$ **(b)** $\tan x = \cot\left(x + \dfrac{\pi}{2}\right)$ **(c)** $\cot x = -\tan\left(x - \dfrac{\pi}{2}\right)$

Sketch the graph of the equation, including at least two cycles. Indicate the period and vertical asymptotes.

9. $y = \cot 3x$ 10. $y = \tan 2x$ 11. $y = -2\cot\dfrac{x}{2}$ 12. $y = \dfrac{1}{2}\tan\dfrac{x}{2}$ 13. $y = \sec 4x$

14. $y = -\sec x$ 15. $y = -\csc\frac{1}{3}x$ 16. $y = \csc 2x$ 17. $y = 2\sec\dfrac{3x}{2}$

Determine the period and phase shift, and sketch the graph for one period.

18. $y = \tan 2\left(x - \dfrac{\pi}{2}\right)$ 19. $y = \dfrac{1}{2}\cot 2\left(x + \dfrac{\pi}{4}\right)$ 20. $y = 2\sec\left(\dfrac{\pi}{2} + \dfrac{\pi}{8}\right)$ 21. $y = -\csc\left(\dfrac{x}{2} - \dfrac{\pi}{4}\right)$

Find the values of the constants a, b, and h for the function with the given graph.

22. $y = a\tan b(x - h)$

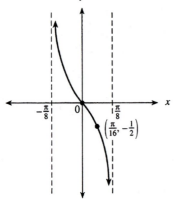

23. $y = a\cot b(x - h)$

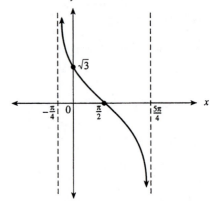

24. $y = a \sec b(x - h)$

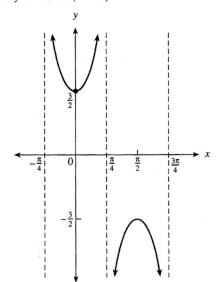

25. $y = a \csc b(x - h)$

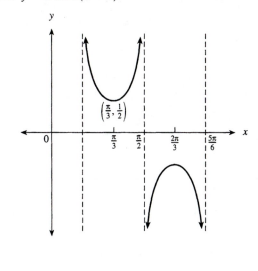

Complete the table of values, using radian mode, giving each value to the nearest unit. Then describe which property of the function is demonstrated by the results in the table.

26.

x	-1.1	-1.5	-1.55	-1.57	-1.5707	-1.57079
$\tan x$						

27.

x	0.5	0.1	0.01	0.001	0.0001	0.00001
$\csc x$						

Sketch the graph of each equation.

28. $y = |\sec x|$ **29.** $y = |\tan x|$ **30.** $y = -|\cot x|$

31. Let $f(x) = x^2$ and $g(x) = \tan x$. Form the composites $f \circ g$ and $g \circ f$.

32. Let $f(x) = \dfrac{x}{x + 1}$ and $g(x) = \sec x$. Form the composites $f \circ g$ and $g \circ f$.

33. Let $f(x) = e^x$, $g(x) = \sqrt{x}$ and $h(x) = \sec x$. Form the composites $f \circ g \circ h$, $g \circ f \circ h$, and $h \circ g \circ f$.

34. Let $h(x) = \cot^3 x$. Find f and g so that $h = f \circ g$, where the outer function f is a polynomial.

35. Let $F(x) = \sqrt{\tan(2x + 1)}$. Find f, g, and h so that $F = f \circ g \circ h$ and g is trigonometric.

36. Let $F(x) = \tan^2 \left(\dfrac{x + 1}{x - 1} \right)$. Find f, g, h so that $F = f \circ g \circ h$, where h is rational and f is quadratic.

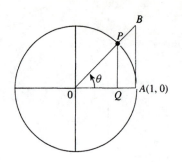

37. Refer to the figure and explain why $0 < \dfrac{\sin \theta}{\theta} < 1$ for $0 < \theta < \dfrac{\pi}{2}$. (*Hint:* In a unit circle the central angle in radians is the same as the length of the intercepted arc.)

38. Refer to the figure in Exercise 37 and prove $\cos \theta < \dfrac{\sin \theta}{\theta} < \dfrac{1}{\cos \theta}$ for $0 < \theta < \dfrac{\pi}{2}$.
(*Hint:* Use the areas of the two right triangles and the sector of the circle.)

39. Use a calculator to complete this table where θ is given in radians.

θ	1	0.5	0.25	0.1	0.01
$\dfrac{\sin \theta}{\theta}$					

As θ approaches 0 in value, what appears to be happening to the ratio $\dfrac{\sin \theta}{\theta}$?

40. Find the subintervals of the given interval on which each function is increasing or decreasing. Also, find the subintervals where the graph is concave up or concave down.

   (a) $y = \tan x$; $\left(-\dfrac{\pi}{2}, \dfrac{\pi}{2}\right)$ **(b)** $y = \cot x$; $(0, \pi)$ **(c)** $y = \sec x$; $\left(-\dfrac{\pi}{2}, \dfrac{3\pi}{2}\right)$ **(d)** $y = \csc x$; $(0, 2\pi)$

 Challenge Construct four unit circle diagrams, one for each of the cases $0 < \theta < \dfrac{\pi}{2}$, $\dfrac{\pi}{2} < \theta < \pi$, $\pi < \theta < \dfrac{3\pi}{2}$, and $\dfrac{3\pi}{2} < \theta < 2\pi$, that demonstrate the equality $\cot \theta = -\tan\left(\theta - \dfrac{\pi}{2}\right)$. Also explain why it is unnecessary to consider the cases $\theta = 0, \dfrac{\pi}{2}, \pi, \dfrac{3\pi}{2}$.

 Written Assignment Assume the result $\cot \theta = -\tan\left(\theta - \dfrac{\pi}{2}\right)$ for all allowable values of θ and explain how the graph of the cotangent function can be obtained from the graph of the tangent function.

 Graphing Calculator Exercises

1. Your calculator has keys for sin, cos, and tan, but not for csc, sec, and cot. However, you can get your calculator to graph the last three functions by graphing the reciprocals of the first three. Use this technique to graph $y = \csc x$, $y = \sec x$, and $y = \cot x$, and compare your graphs with those given in the text. If you get vertical lines as part of your graphs, explain them. (Should they be there?)

   For Exercises 2–4, find approximate solutions to the given equations by finding the intersections of the graphs of $y = $ [the left hand side] and $y = $ [the right hand side], for $0 \le X \le \pi$, to the nearest hundredth. Use Trace and Zoom. (For example, to solve $\sin x = \cos x$, we would graph $y = \sin x$ and $y = \cos x$.)

2. $\sin x = \cos^2 x$

3. $\tan x = \sin x + \cos x$

4. $\cos x + 1 = \sec x$

5. Show that $\cos x < (\sin x)/x < \sec x$ when $0 < x < \pi/2$ by graphing. (Also see Exercise 38.)

Critical Thinking

1. Study the graph of $y = \sin \theta$ on page 383 and explain how the sine curve on the interval $\left[0, \dfrac{\pi}{2}\right]$ could be used to obtain the sine curve on $\left[\dfrac{\pi}{2}, \pi\right]$. Also, explain how the curve on $[0, \pi]$ can be used to obtain the curve on $[\pi, 2\pi]$.

2. Let a, b, and c be any constants where a and b are not zero. Explain why the amplitude of $y = a \sin bx + c$ is $|a|$.

3. Find the equation of a sine curve that has two complete cycles on the interval $\left[0, \dfrac{3\pi}{4}\right]$ with amplitude 1.

4. Compare the graphs of $y = \sec x$ and $y = \csc x$ and describe how either curve can be obtained from the other. For each case write the equation that fits the description that was used.

5. For $-\dfrac{\pi}{2} < x \le 0$, is it true or false that $|\tan x| = \tan |x|$? Explain.

**6.5
INVERSE
TRIGONOMETRIC
FUNCTIONS**

The sine function is not one-to-one. This fact is apparent from its graph, since a horizontal line through a range value y intersects the curve more than once; more than one x corresponds to y.

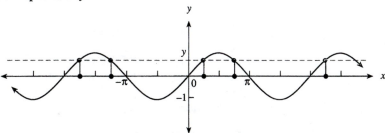

See page 288 for the definition of a one-to-one function.

If we restrict the domain to $\left[-\dfrac{\pi}{2}, \dfrac{\pi}{2}\right]$, then $y = \sin x$ is one-to-one because for each range value there corresponds exactly one domain value. This restricted function will have the same range as the original function: $-1 \le y \le 1$, as shown in this graph.

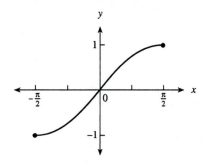

See Section 4.5 to review the inverse function concept.

Since $y = \sin x$ for $-\dfrac{\pi}{2} \le x \le \dfrac{\pi}{2}$ is one-to-one, we know that there is an inverse function. The graph of the inverse is obtained by reflecting the graph of $y = \sin x$ through the line $y = x$, and the equation of the inverse is obtained by interchanging the variables in $y = \sin x$. Also, the domain and range of the restricted function $y = \sin x$ become the range and domain of the inverse, respectively.

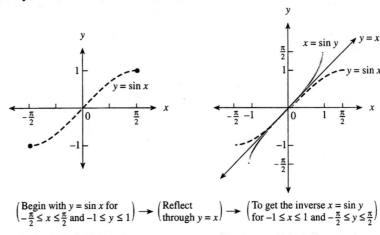

$\left(\begin{array}{c}\text{Begin with } y = \sin x \text{ for}\\ -\frac{\pi}{2} \le x \le \frac{\pi}{2} \text{ and } -1 \le y \le 1\end{array}\right) \rightarrow \left(\begin{array}{c}\text{Reflect}\\ \text{through } y = x\end{array}\right) \rightarrow \left(\begin{array}{c}\text{To get the inverse } x = \sin y\\ \text{for } -1 \le x \le 1 \text{ and } -\frac{\pi}{2} \le y \le \frac{\pi}{2}\end{array}\right)$

The equation of the inverse, $x = \sin y$, does not express y explicitly as a function of x. To do this we create some new terminology. First observe that $x = \sin y$ means that y is the radian whose sine is x, or

y is the *arc* length on the unit circle whose sine is x.

To shorten this, we replace "arc length on the unit circle whose sine is x" by "arc sin of x," and this phrase is further abbreviated to "arcsin x." Thus $y = $ **arcsin x** is, by definition, equivalent to $x = \sin y$. To sum up, here is the basic information about this new function.

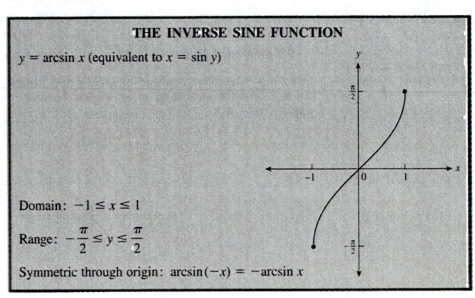

THE INVERSE SINE FUNCTION

$y = \arcsin x$ (equivalent to $x = \sin y$)

Domain: $-1 \le x \le 1$

Range: $-\dfrac{\pi}{2} \le y \le \dfrac{\pi}{2}$

Symmetric through origin: $\arcsin(-x) = -\arcsin x$

The proof of the symmetric property is called for in Exercise 53.

Another common notation for the inverse sine function is $y = \sin^{-1} x$, in which the -1 is not an exponent; $\sin^{-1} x \neq \dfrac{1}{\sin x}$.

Since we are dealing with inverse functions, the following are true:

$$\arcsin(\sin x) = x \qquad \text{if } -\frac{\pi}{2} \leq x \leq \frac{\pi}{2}$$

$$\sin(\arcsin x) = x \qquad \text{if } -1 \leq x \leq 1$$

EXAMPLE 1   Evaluate $\arcsin \frac{1}{2}$.

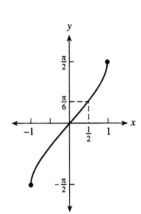

Solution Let $y = \arcsin \frac{1}{2}$. Then y is the radian whose sine is equal to $\frac{1}{2}$. Thus $y = \dfrac{\pi}{6}$. To check this we use the fact that $y = \arcsin x$ is equivalent to $x = \sin y$. Hence

$$\sin y = \sin \frac{\pi}{6} = \frac{1}{2} = x \qquad \blacksquare$$

You will find it helpful in finding values like $\arcsin \frac{1}{2}$ to remember that $y = \arcsin x$ is negative for $-1 \leq x < 0$ and positive for $0 < x \leq 1$. You can see this from its graph.

Caution: Even though $\sin \dfrac{5\pi}{6} = \dfrac{1}{2}$ we do not have $\arcsin \dfrac{1}{2} = \dfrac{5\pi}{6}$ because the range of the arcsin function consists of the numbers in the interval $\left[-\dfrac{\pi}{2}, \dfrac{\pi}{2}\right]$.

When a domain value x of the inverse sine function is not the sine of a special angle, such as $\dfrac{1}{2}$ or $\dfrac{\sqrt{3}}{2}$, a calculator can be used to evaluate this angle, that is, to find $\arcsin x$ (or $\sin^{-1} x$). Note that calculators are constructed so that the $\sin^{-1}$ key gives only the range values of the inverse sine function, so that the type of error demonstrated in the preceding caution cannot occur.

Radian mode is needed here since the range of the inverse sine function consists of radians. Also, note that we use the $\boxed{\text{SIN}^{-1}}$ *key to obtain arcsin values.*

On a scientific calculator, part (a) can be done with this sequence:

$\boxed{0.6123}$ $\boxed{\text{SIN}^{-1}}$

EXAMPLE 2   Evaluate if possible.
(a) $\arcsin(0.6123)$ **(b)** $\arcsin(-0.3981)$
(c) $\sin(\arcsin 0.8)$ **(d)** $\arcsin(1.0475)$

Solution
(a) $\arcsin(0.6123) = \boxed{\text{SIN}^{-1}}\ \boxed{0.6123}\ \boxed{\text{ENTER}} = 0.66$ rounded to two decimal places

(b) $\arcsin(-0.3981) = \boxed{\text{SIN}^{-1}}\ \boxed{-0.3981}\ \boxed{\text{ENTER}} = -0.41$

(c) $\sin(\arcsin 0.8) = 0.8$, since sin and arcsin are inverse functions and 0.8 is in the domain of arcsin.

(d) Undefined since 1.0475 is not in the domain of $\sin^{-1}$. $\blacksquare$

To define the inverse cosine function, known as *arccos*, we begin by restricting $y = \cos x$ to $0 \leq x \leq \pi$ to obtain a one-to-one function. Next, reflect the curve for $0 \leq x \leq \pi$ through the line $y = x$ to obtain the graph of the inverse whose equation is $x = \cos y$.

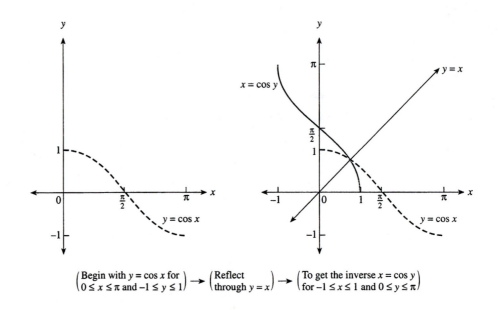

$$\left(\begin{array}{c}\text{Begin with } y = \cos x \text{ for}\\ 0 \leq x \leq \pi \text{ and } -1 \leq y \leq 1\end{array}\right) \rightarrow \left(\begin{array}{c}\text{Reflect}\\ \text{through } y = x\end{array}\right) \rightarrow \left(\begin{array}{c}\text{To get the inverse } x = \cos y\\ \text{for } -1 \leq x \leq 1 \text{ and } 0 \leq y \leq \pi\end{array}\right)$$

Now define $y = \textbf{arccos } x$ to mean $x = \cos y$ and call it the *inverse cosine function*. (This function is also written in the form $y = \cos^{-1} x$.) Since arccos and cos are inverse functions, we have the following:

$$\arccos(\cos x) = x \qquad \text{if } 0 \leq x \leq \pi$$

$$\cos(\arccos x) = x \qquad \text{if } -1 \leq x \leq 1$$

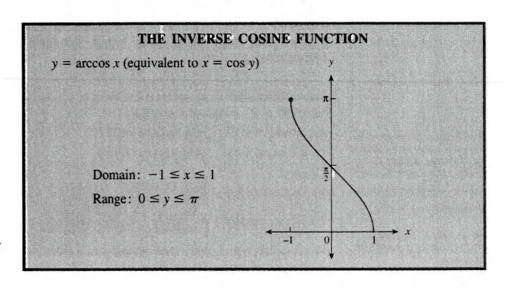

THE INVERSE COSINE FUNCTION

$y = \arccos x$ (equivalent to $x = \cos y$)

Domain: $-1 \leq x \leq 1$

Range: $0 \leq y \leq \pi$

Except for arccos $1 = 0$, *the inverse cosine function is always positive.*

EXAMPLE 3 Evaluate, if possible.

(a) $\arccos(0.2943)$ **(b)** $\arccos(-0.7385)$

(c) $\arccos(-1.6027)$ **(d)** $\cos\left(\arccos\frac{3}{4}\right)$

Solution

(a) $\arccos(0.2943) =$ [COS⁻¹] [0.2943] [ENTER] $= 1.27$

rounded to two decimal places

Using a scientific calculator for (b), we have this sequence:

[0.7385] [+/−] [COS⁻¹]

(b) $\arccos(-0.7385) =$ [COS⁻¹] [−0.7385] [ENTER] $= 2.40$

(c) Undefined: -1.6027 is not in the domain of arccos.

(d) $\cos\left(\arccos\frac{3}{4}\right) = \frac{3}{4}$ since cos and arccos are inverse functions and $\frac{3}{4}$ is in the domain of arccos. ■

EXAMPLE 4 Find the exact value of $\arccos\left(-\frac{1}{2}\right)$.

Solution Since $\cos\frac{2\pi}{3} = -\frac{1}{2}$ and $\frac{2\pi}{3}$ is in the range of the arccos function, we have $\arccos\left(-\frac{1}{2}\right) = \frac{2\pi}{3}$. ■

EXAMPLE 5 Find the exact value of $\sin\left(\arccos\frac{1}{2}\right)$.

CAUTION
Do not confuse $(\sin x)(\arccos x)$ *with* $\sin(\arccos x)$. *The first represents the product of the two numbers* $\sin x$ *and* $\arccos x$. *The second represents the composite of the inner function* arccos *with the outer function* sin.

Solution Since $\arccos\frac{1}{2} = \frac{\pi}{3}$, we have

$$\sin\left(\arccos\frac{1}{2}\right) = \sin\frac{\pi}{3} = \frac{\sqrt{3}}{2}$$ ■

A comparison of the sine and cosine curves (see pages 384 and 386) reveals that $\sin\theta = \cos\left(\theta - \frac{\pi}{2}\right)$. Also, since the cosine function is symmetric through the y-axis, $\cos\left(\theta - \frac{\pi}{2}\right) = \cos\left(-\left(\theta - \frac{\pi}{2}\right)\right) = \cos\left(\frac{\pi}{2} - \theta\right)$ so that

$$\sin\theta = \cos\left(\frac{\pi}{2} - \theta\right)$$

This result can be used to prove the identity connecting the inverse sine and inverse cosine functions in the next example.

EXAMPLE 6 Show that $\arcsin x + \arccos x = \frac{\pi}{2}$ for all x in the common domain $-1 \le x \le 1$.

Solution Let $\arcsin x = y$; $-\dfrac{\pi}{2} \le y \le \dfrac{\pi}{2}$. Then $x = \sin y$. Now for any y value we have $\cos\left(\dfrac{\pi}{2} - y\right) = \sin y$. Therefore,

$$x = \sin y = \cos\left(\dfrac{\pi}{2} - y\right)$$

However, note that $x = \cos\left(\dfrac{\pi}{2} - y\right)$ is equivalent to $\dfrac{\pi}{2} - y = \arccos x$ because $0 \le \dfrac{\pi}{2} - y \le \pi$. Then, adding y yields

Prove that $-\dfrac{\pi}{2} \le y \le \dfrac{\pi}{2}$
implies $0 \le \dfrac{\pi}{2} - y \le \pi$.

$$\dfrac{\pi}{2} = y + \arccos x$$

and substituting for y gives

$$\dfrac{\pi}{2} = \arcsin x + \arccos x \qquad\blacksquare$$

TEST YOUR UNDERSTANDING

Evaluate the following where possible without the use of a calculator.

1. $\arcsin(-1)$ **2.** $\arccos\left(\dfrac{\sqrt{3}}{2}\right)$ **3.** $\arcsin 2$

4. $\arccos\left(-\dfrac{1}{2}\right)$ **5.** $\cos(\arcsin 0)$ **6.** $\cos\left[\arcsin\left(-\dfrac{\sqrt{3}}{2}\right)\right]$

7. $\sin(\arcsin 1)$ **8.** $\arccos(\cos x)$ **9.** $\arccos(-1.5)$

Evaluate using a calculator.

10. $\arcsin(0.7314)$ **11.** $\arcsin(-0.9284)$ **12.** $\arccos(0.2288)$

(Answers: Page 424)   **13.** $\arccos(-0.4447)$ **14.** $\sin(\arccos 0.8419)$ **15.** $\arcsin(\cos 0.89)$

The tangent function has period π and completes a full cycle on the interval $\left(-\dfrac{\pi}{2}, \dfrac{\pi}{2}\right)$. Thus when $y = \tan x$ is restricted to $-\dfrac{\pi}{2} < x < \dfrac{\pi}{2}$ we have a one-to-one function whose range consists of all real numbers.

$y = \tan x$

Domain: $-\dfrac{\pi}{2} < x < \dfrac{\pi}{2}$

Range: all real numbers

Asymptotes: $x = \pm\dfrac{\pi}{2}$

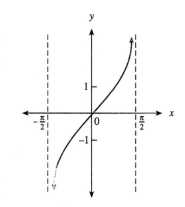

We define the inverse by $y = $ **arctan** x, which means that $x = \tan y$. Its graph is obtained by reflecting the preceding curve through the line $y = x$. (The inverse tangent function is also written in the form $y = \tan^{-1} x$.) Also, since arctan and tan are inverse functions, we have:

$$\arctan(\tan x) = x \qquad \text{if } -\frac{\pi}{2} < x < \frac{\pi}{2}$$

$$\tan(\arctan x) = x \qquad \text{for all } x$$

Observe that $\tan^{-1} x \to \dfrac{\pi}{2}$

as $x \to \infty$ *and*

$\tan^{-1} x \to -\dfrac{\pi}{2}$

as $x \to -\infty$.

The proof of the symmetric property is called for in Exercise 54.

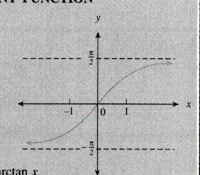

THE INVERSE TANGENT FUNCTION

$y = \arctan x$ (equivalent to $x = \tan y$)

Domain: all real numbers

Range: $-\dfrac{\pi}{2} < y < \dfrac{\pi}{2}$

Asymptotes: $y = \dfrac{\pi}{2}, y = -\dfrac{\pi}{2}$

Symmetric through origin: $\arctan(-x) = -\arctan x$

EXAMPLE 7 Find the exact value of arctan $\sqrt{3}$.

Solution Let $y = \arctan \sqrt{3}$. Then y is the radian whose tangent is $\sqrt{3}$. Therefore, since $\tan \dfrac{\pi}{3} = \sqrt{3}$ and $-\dfrac{\pi}{2} < y < \dfrac{\pi}{2}$, we have $y = \dfrac{\pi}{3}$. ∎

EXAMPLE 8 Evaluate arctan 1000 rounded to four decimal places and use this result to find $\arctan(-1000)$.

Solution arctan $1000 = \boxed{\text{TAN}^{-1}} \ \boxed{1000} \ \boxed{\text{ENTER}} = 1.5698$.

Then, since arctan is symmetric through the origin, we have $\arctan(-1000) = -1.5698$. ∎

EXAMPLE 9 Solve for x: $\arctan\left(\dfrac{x-5}{\sqrt{3}}\right) = -\dfrac{\pi}{3}$.

Solution The given equation is equivalent to

$$\left(\frac{x-5}{\sqrt{3}}\right) = \tan\left(-\frac{\pi}{3}\right)$$

Check:

$\arctan\left(\dfrac{2-5}{\sqrt{3}}\right)$

$= \arctan\left(\dfrac{-3}{\sqrt{3}}\right)$

$= \arctan\left(-\sqrt{3}\right) = -\dfrac{\pi}{3}$

Since $\tan\left(-\dfrac{\pi}{3}\right) = -\sqrt{3}$ we get

$$\frac{x-5}{\sqrt{3}} = -\sqrt{3}$$

$$x - 5 = -\left(\sqrt{3}\right)^2 = -3$$

$$x = 2$$

∎

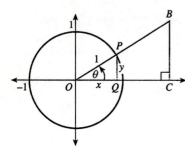

The figure in the margin shows an acute angle with reference triangle OQP. Right triangle OCB is similar to triangle OQP so that

$$\frac{CB}{OB} = \frac{QP}{OP} = \frac{QP}{1} = y = \sin\theta \quad \text{and} \quad \frac{OC}{OB} = \frac{OQ}{OP} = \frac{OQ}{1} = x = \cos\theta$$

Similar results apply to the other trigonometric function values of θ. For example

$$\tan\theta = \frac{y}{x} = \frac{QP}{OQ} = \frac{CB}{OC}$$

The preceding demonstrates that the sides of any right triangle, similar to the unit circle reference triangle for θ, determine the trigonometric values of θ. Furthermore since $\tan\theta = \dfrac{CB}{OC}$ we may write $\theta = \arctan\dfrac{CB}{OC}$. This observation is used in the following application.

EXAMPLE 10   In the figure AB represents an 8-foot-high billboard that subtends the angle of θ radians. $BC = 22$ feet is the distance the billboard is above the ground, and point P is 40 feet from C at ground level. Use the arctan function to solve for θ.

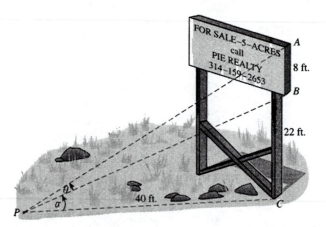

Solution From the figure we have $0 < \alpha < \dfrac{\pi}{2}$ and $0 < \alpha + \theta < \dfrac{\pi}{2}$. Then by the observation preceding this example, it follows that $\tan\alpha = \dfrac{22}{40} = \dfrac{11}{20}$ or $\alpha = \arctan\dfrac{11}{20}$. Also, $\tan(\alpha + \theta) = \dfrac{8 + 22}{40} = \dfrac{3}{4}$ so that

$$\alpha + \theta = \arctan\frac{3}{4}$$

$$\theta = \arctan \frac{3}{4} - \alpha$$

$$\theta = \arctan \frac{3}{4} - \arctan \frac{11}{20}$$

Using a graphing calculator in the radian mode, we may use the following sequence:

$\theta =$ $\boxed{\text{TAN}^{-1}}$ $\boxed{0.75}$ $\boxed{-}$ $\boxed{\text{TAN}^{-1}}$ $\boxed{(}$ $\boxed{11}$ $\boxed{\div}$ $\boxed{20}$ $\boxed{)}$ $\boxed{\text{ENTER}}$ $= 0.14$

rounded to two decimal places

Thus, $\theta = 0.14$ radian. (This is approximately $8°$.)

Using a scientific calculator to find θ, the following sequence can be used:

$\theta =$ $\boxed{0.75}$ $\boxed{\text{TAN}^{-1}}$ $\boxed{-}$ $\boxed{(}$ $\boxed{11}$ $\boxed{\div}$ $\boxed{20}$ $\boxed{)}$ $\boxed{\text{TAN}^{-1}}$ $\boxed{=}$ ∎

The cotangent, secant, and cosecant functions also have inverses. The basic properties and graphs of the inverses are summarized here for reference.

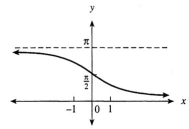

$y = \text{arccot } x$ (equivalent to $x = \cot y$)

Domain: all real numbers

Range: $0 < y < \pi$

Asymptotes: $y = 0, y = \pi$

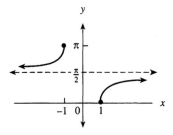

$y = \text{arcsec } x$ (equivalent to $x = \sec y$)

Domain: $x \le -1$ or $x \ge 1$

Range: $0 \le y \le \pi, y \ne \dfrac{\pi}{2}$

Asymptote: $y = \dfrac{\pi}{2}$

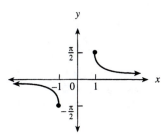

$y = \text{arccsc } x$ (equivalent to $x = \csc y$)

Domain: $x \le -1$ or $x \ge 1$

Range: $-\dfrac{\pi}{2} \le y \le \dfrac{\pi}{2}, y \ne 0$

Asymptote: $y = 0$

EXERCISE 6.5

Find the exact value whenever possible. Otherwise, use a calculator.

1. arcsin 0

2. arccos 0

3. arcsin (−1)

4. arccos (−1)

5. arctan (−1)

6. arcsin $\dfrac{1}{\sqrt{2}}$

7. arccos $\left(-\dfrac{1}{\sqrt{2}}\right)$

8. arctan $\dfrac{1}{\sqrt{3}}$

9. arctan 115

10. arcsin $\left(-\dfrac{3}{2}\right)$

11. arcsin (.5562)

12. arccos 2

13. arccos (−0.6137)

14. arctan (.6128)

15. arcsin (−1.0436)

16. arccos $\dfrac{3}{8}$

17. arctan $\dfrac{7}{11}$

18. arctan (−16.5)

19. $\tan\left(\arctan \dfrac{x}{2}\right)$

20. $\arcsin\left(\tan \dfrac{\pi}{3}\right)$

21. cos (arcsin 0.6208)

22. arctan (cos (−0.7149))

23. $\sin\left(\arctan \frac{1}{12}\right)$

24. $\tan\left(\arcsin \frac{1}{2}\right)$

25. cos [arcsin(−1)]

26. $\tan\left(\arccos \frac{4}{5}\right)$

27. $\sin\left(\arcsin \dfrac{\sqrt{3}}{2}\right)$

28. cos (arctan (−2.35))

29. $\sin\left(\arccos \frac{13}{12}\right)$

30. $\arcsin\left(\cos \dfrac{\pi}{3}\right)$

31. $\arccos\left(\tan \dfrac{\pi}{4}\right)$

32. $\arctan\left(\tan \dfrac{\pi}{6}\right)$

33. Explain why sin (arcsin x) = x for −1 ≤ x ≤ 1.　　**34.** Explain why tan (arctan x) = x for all x.

Graph the curves. State the domain and range.

35. $y = 2 \arcsin x$

36. $y = \arcsin(x - 2)$

37. $y = 2 + \arctan x$

38. $y = -\arcsin x$

39. $y = \frac{1}{2} \arcsin 2x$

40. $y = 2 \arccos \dfrac{x}{2}$

41. $y = \sin(\arcsin x)$

42. $y = \arcsin x + \arccos x$

Evaluate using a calculator.

43. tan (arctan 3 + arctan 4)

44. $\sin\left(\arcsin \frac{2}{3} + \arctan \frac{4}{5}\right)$

45. $\cos\left(\arcsin \frac{8}{17} - \arccos \frac{12}{13}\right)$

46. $\tan\left(2 \arctan \frac{4}{5}\right)$

47. $\cos\left(2 \arcsin \frac{1}{3}\right)$

48. $\sin\left(\frac{1}{2} \arccos \frac{3}{5}\right)$

Solve for x and check.

49. $\arcsin(x + 2) = \dfrac{\pi}{6}$

50. $\arctan \dfrac{x}{3} = \dfrac{\pi}{4}$

51. $\arctan(x^2 + 4x + 3) = -\dfrac{\pi}{4}$

52. $\arccos(2x - 1) = \dfrac{2\pi}{3}$

53. Prove that arcsin (−x) = − arcsin x. (*Hint:* begin with arcsin x = y and multiply by −1.)

54. Prove that arctan (−x) = −arctan x. (*Hint:* begin with arctan x = y and multiply by −1.)

55. In the figure, θ represents the angle subtended by a 5-foot picture when viewed from point P that is 7 feet below the picture and 14 feet away from the wall on which the picture hangs. Solve for θ using the arctan function.

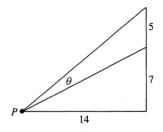

56. (a) When the graph of $y = \cos \theta$ is shifted π units to the right and then reflected through the θ axis, the resulting curve coincides with the original curve so that $\cos \theta = -\cos(\theta - \pi) = -\cos(\pi - \theta)$. Now replace θ by y to get $\cos y = -\cos(y - \pi)$

and let arccos $x = y$, for $0 \le y \le \pi$. Prove that arccos x + arccos$(-x) = \pi$ for $-1 \le x \le 1$. (*Hint:* the proof is similar to the proof of Example 6, page 409.)

(b) Use the result in part (a) to find the exact value of $\arccos\left(-\dfrac{\sqrt{3}}{2}\right)$.

57. Prove that cos (arcsin x) $= \sqrt{1 - x^2}$ for $-1 \le x \le 1$. (*Hint:* Let θ = arcsin x and use the result in Exercise 82, page 381)

58. Let $f(x) = \arctan x$ and $g(x) = x^2$ and form the composites $f \circ g$ and $g \circ f$.

59. Let $f(x) = \arcsin x$ and $g(x) = 3x + 2$ and form the composites $f \circ g$ and $g \circ f$.

60. Let $k(x) = \arctan \sqrt{x^2 - 1}$. Find f, g, and h so that $k = f \circ g \circ h$.

61. Let $h(x) = \ln(\arccos x)$. Find f and g so that $h = f \circ g$.

62. (a) Complete the table rounding each value to six decimal places.

x	1	10	100	1000	10,000
$\tan^{-1} x$					

(b) What property of the inverse tangent function is demonstrated by the results of part **(a)**?

 Written Assignment The inverse sine function was defined after we restricted the domain of the sine function to the interval $\left[-\dfrac{\pi}{2}, \dfrac{\pi}{2}\right]$. Could an inverse sine function also have been defined if we had restricted the sine function to the interval $[0, \pi]$? Justify your answer.

 Graphing Calculator Exercises

1. Are tan (arctan x) and arctan (tan x) the same function? State the domain and range of each function and graph the functions on different sets of axes to justify your answer.

2. Repeat Exercise 1 with sin (arcsin x) and arcsin (sin x).

3. Repeat Exercise 1 with cos (arccos x) and arccos (cos x).

4. Suppose that you are sitting in a classroom next to a wall and you are looking at the blackboard in the front of the room. The blackboard is 15 meters long and starts 2 meters from the wall. Then the viewing angle is given by α = arctan (17/x) − arctan(2/x). How far should you sit from the front of the room to maximize your viewing angle? (*Hint:* Graph α as a function of x. Use Trace and Zoom.)

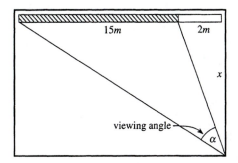

CHAPTER REVIEW EXERCISES

Section 6.1 Angle Measure (Page 362)

1. State the definition of a radian.

2. Convert to radians: **(a)** 115° **(b)** 405° **(c)** 1.5°

3. Convert to degrees: **(a)** $\dfrac{4\pi}{5}$ **(b)** $\dfrac{17\pi}{6}$ **(c)** $\dfrac{1}{5}$

4. A central angle in a circle of radius 12 centimeters is 105°. Find the length of the intercepted arc. Give the answer in terms of π.

5. The length of the arc of a circular sector in a circle of radius 6 centimeters is 10π centimeters. What is the degree measure of the central angle of the sector?

6. State the formula for the area of a circular sector having a central angle of θ radians in a circle of radius r.

7. Find the area of a circular sector in a circle of radius 10 centimeters if the central angle is 120°. Give the answer in terms of π.

8. The area of a sector of a circle with radius 8 centimeters is $\dfrac{64\pi}{3}$ square centimeters. Find the measure of the central angle of the sector in degrees.

9. A 30.5-inch pendulum swings through an angle of 20.5°. Find the length of the arc through which the tip of the pendulum swings to the nearest tenth of an inch.

10. A point on the rim of a wheel is turning with an angular speed of 200 rad/sec. If the wheel's diameter is 3 feet, find the linear speed in ft/sec.

11. **(a)** A satellite revolves around the earth in a circular orbit that is 200 miles above the earth's surface. If one revolution takes 1 hour and 30 minutes, find the satellites linear speed in mph. Give the answer to the nearest mph.

    (b) Find the angular speed of the satellite in rad/hr. Give the answer in terms of π,

12. Triangle ABC is an isosceles right triangle with hypotenuse $AB = 4$ cm. CD is an arc of a circle having radius AC. Find the area of the shaded region.

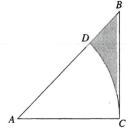

Section 6.2 Trigonometric Functions (Page 370)

Let $P(x, y)$ be the intersection of the terminal side of θ and the unit circle. Determine the coordinates of P and find the exact values of the six trigonometric function values of θ.

13. $\theta = \dfrac{3\pi}{4}$ 14. $\theta = \dfrac{11\pi}{6}$ 15. $\theta = -120°$ 16. $\theta = -420°$

Find the trigonometric function values for θ without using a calculator.

17. $\theta = -2\pi$ 18. $\theta = \dfrac{3\pi}{2}$ 19. $\theta = \dfrac{9\pi}{2}$ 20. $\theta = -3\pi$

Find the six trigonometric function values of the angle to four decimal places.

21. 75°   22. −237°   23. −374°   24. 520°

Find the function value (if it exists). Give the exact value when possible; otherwise use a calculator and round off to four decimal places.

25. $\sin 215°$ 26. $\cos(-85°)$ 27. $\tan 420°$

28. $\cot(-125.8°)$ **29.** $\sec 172.5°$ **30.** $\csc(-261.5°)$ **31.** $\cot\left(\dfrac{7\pi}{5}\right)$

32. $\sin\left(\dfrac{7\pi}{2}\right)$ **33.** $\cos 5\pi$ **34.** $\tan\left(-\dfrac{\pi}{2}\right)$

35. $\cot \pi$ **36.** $\sec\dfrac{7\pi}{4}$ **37.** $\csc\left(-\dfrac{5\pi}{6}\right)$

38. $\cos(-315°)$ **39.** $\sin 200°$ **40.** $\tan(-137.2°)$

Find all exact values for θ where $0 \le \theta < 2\pi$.

41. $\sin \theta = -1$ **42.** $\cos \theta = \dfrac{\sqrt{3}}{2}$ **43.** $\sec \theta = -\sqrt{2}$ **44.** $\tan \theta = -\dfrac{1}{\sqrt{3}}$

Find all exact values for θ where $-360° < \theta \le 0$.

45. $\cot \theta = 0$ **46.** $\sec \theta = -2$ **47.** $\sin \theta = -\dfrac{1}{\sqrt{2}}$

48. Find $\csc \theta$ if the terminal side of θ is in quadrant III and $\cos \theta = -\dfrac{3}{4}$.

49. Find $\tan \theta$ if $\sin \theta = -\dfrac{3}{4}$ and $\cos \theta$ is positive.

50. If $\sec \theta = -\dfrac{2}{\sqrt{3}}$ and $\tan \theta < 0$, find $\sin \theta$.

51. Find $\sin \theta$ if $\cos \theta = \dfrac{3}{8}$ and $-\dfrac{\pi}{2} < \theta < 0$.

Section 6.3 Graphing the Sine and Cosine Functions (Page 382)

52. State the domain and range of the sine function, and describe its symmetry.

53. Complete the equation $\cos \theta = \sin$ _____ so that it can be used to explain how the cosine curve can be obtained from the sine curve, and give the explanation.

54. Graph $y = 3 \cos x$, $y = -3\cos x$, and $y = \dfrac{3}{2} \cos x$ in the same coordinate system for $0 \le x \le 2\pi$.

55. Find the amplitude and period of $y = -\dfrac{1}{2} \sin 2x$ and sketch the graph for $-\pi \le x \le \pi$.

56. Find the amplitude and period p of $y = 2 \cos \dfrac{x}{3}$ and sketch the graph for $0 \le x \le p$.

The given graph has equations of the form $y = a \sin bx$ or $y = a \cos bx$. Find a and b and write the equation.

57.

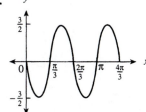

58.

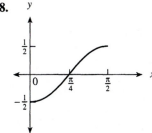

Find the amplitude period, and phase shift and sketch the graph for one period.

59. $y = 4 \sin\left(x + \dfrac{\pi}{4}\right)$ **60.** $y = -\cos\left(2x - \dfrac{\pi}{3}\right)$

Find the amplitude, period, and phase shift.

61. $y = 4 \cos 4\left(x - \dfrac{\pi}{2}\right)$ **62.** $y = 3 \sin \dfrac{2}{3}x$ **63.** $y = \dfrac{2}{3} \sin\left(2x - \dfrac{\pi}{2}\right)$

64. $y = -\dfrac{5}{2} \cos\left(\dfrac{x}{2} + \dfrac{\pi}{4}\right)$ **65.** $y = \dfrac{1}{3} \cos(\pi - 3x)$

Section 6.4 Graphing Other Trigonometric Functions (Page 394)

State the domain and range of the function and describe its symmetry.

66. $y = \tan x$ **67.** $y = \sec x$

68. Complete the equation $\csc x =$ _____ so that it can be used to explain how the graph of the cosecant curve can be obtained from the sine curve, and give the explanation.

Sketch the graph of the function for one period and include the asymptotes.

69. $y = \cot x$ **70.** $y = \csc x$

71. Find the subintervals of the interval $(0, 2\pi)$ for which the cosecant is increasing or decreasing, and where the curve is concave up or down.

72. Prove that the curve $y = \sec x$ is symmetric about the y-axis by establishing that $\sec(-x) = \sec x$ for x in the domain.

73. Find the period of $y = \tan 2x$ and sketch the graph for one period. Include the asymptotes and their equations.

Determine the period and phase shift, and sketch the graph for one period. Include the asymptotes and their equations.

74. $y = \cot 2\left(x - \dfrac{\pi}{4}\right)$ **75.** $y = -2 \sec\left(\dfrac{x}{2} - \dfrac{\pi}{8}\right)$

Find the period and phase shift. Also, write the equation of the first asymptote to the curve to the right of the y-axis.

76. $y = \tan \dfrac{1}{2}\left(x - \dfrac{\pi}{2}\right)$ **77.** $y = 3 \csc\left(4x - \dfrac{\pi}{2}\right)$

78. The given curve is one cycle of $y = a \tan b(x - h)$. Find the constants a, b, and h.

79. Describe how the graph of $y = -|\sec x|$ can be obtained from the graph of $y = \sec x$. (No graph required.)

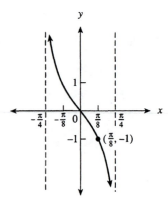

6.5 Inverse Trigonometric Functions (Page 405)

80. State the domain and range of each function:

   (a) $y = \arcsin x$ **(b)** $y = \arccos x$ **(c)** $y = \arctan x$

Find the exact value if possible; otherwise use a calculator.

81. $\arctan(-1)$ **82.** $\arcsin \dfrac{\sqrt{3}}{2}$ **83.** $\arccos \dfrac{3}{2}$

84. $\cos(\arccos 0.3627)$ **85.** $\tan\left(\arcsin\dfrac{5}{13}\right)$ **86.** $\arcsin\left(-\dfrac{2}{5}\right)$

87. $\arctan\dfrac{8}{3}$ **88.** $\cos(\arcsin 1.2)$ **89.** $\arccos\left(\tan\dfrac{\pi}{4}\right)$

90. State the domain and range of $y = 3\arcsin\dfrac{x}{2}$ and sketch the graph.

Evaluate, using a calculator.

91. (a) $\tan(\arctan 2 - \arctan 5)$ **(b)** $\cos\left(\arcsin\dfrac{5}{8} + \arccos\dfrac{2}{5}\right)$

92. (a) $\cos\left(2\arcsin\dfrac{1}{4}\right)$ **(b)** $\tan\left(\dfrac{1}{2}\arccos\dfrac{3}{4}\right)$

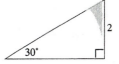

93. Solve for x: $\arccos(5x + 2) = \dfrac{\pi}{3}$

94. Use the arctan function to solve for θ, rounded to two decimal places.

95. Prove that $\sin(\arccos x) = \sqrt{1 - x^2}$ for $-1 \le x \le 1$. (*Hint:* Use $\sin^2\theta + \cos^2\theta = 1$.)

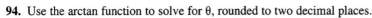

CHAPTER 6: STANDARD ANSWER TEST

Use these questions to test your knowledge of the basic skills and concepts of Chapter 6. Then check your answers with those given at the end of the book.

1. Make the following angle measure conversions without using a calculator or a table.

(a) $\dfrac{5\pi}{12}$ radians to degrees (b) $\tfrac{3}{2}$ radians to degrees

(c) $315°$ to radians (d) $20°$ to radians

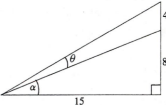

2. A sector is inscribed in a $30°$–$60°$–$90°$ triangle as in the adjoining figure. The side opposite the $30°$ angle is 2 units. Find the area of the shaded part.

3. The wheels on a bicycle have an outer diameter of 26 inches. If the wheels are revolving 246 times per minute, find each of the following. (State the answers in terms of π.)

(a) Find the angular speed in rad/min. (b) Find the linear speed in ft/min.

(c) How many miles will the bicycle travel in 30 minutes?

4. A sector of a circle with radius 10 inches has a central angle of $120°$. Find the arc length of the sector.

5. Find each trigonometric value (if it exists). Do not use a calculator.

(a) $\sin\dfrac{\pi}{3}$ (b) $\cot\dfrac{3\pi}{2}$ (c) $\sin\left(-\dfrac{3\pi}{4}\right)$

6. Assume that the angle of θ radians is in standard position and find the measure of the reference angle.

(a) $\dfrac{19\pi}{4}$ (b) $-\dfrac{25\pi}{6}$

7. Find each trigonometric value (if it exists). Do not use a calculator.

   (a) $\tan \dfrac{17\pi}{6}$ **(b)** $\cos 3\pi$ **(c)** $\csc\left(-\dfrac{7\pi}{6}\right)$

8. Find $\sec \theta$ if $\sin \theta = -\dfrac{\sqrt{3}}{2}$ and $\tan \theta = \sqrt{3}$.

9. Verify that $P\left(-\dfrac{2}{5}, \dfrac{\sqrt{21}}{5}\right)$ is a point on the unit circle and then write the six trigonometric function values of θ, where θ is measure of the angle whose terminal side contains point P.

10. (a) State the domain and range of the sine function.

   (b) Find the subintervals of the interval $[0, 2\pi]$ on which the sine is increasing or decreasing, and those for which the sine curve is concave up or down.

11. Find the amplitude and period of the function $y = 2 \sin 2x$ and graph for x in the interval $0 \le x \le 2\pi$.

12. For $y = -\dfrac{1}{2} \sin\left(3x - \dfrac{\pi}{2}\right)$ find the period, amplitude, and phase shift.

13. Graph $y = 2 \cos(2x + \pi)$ for one period and indicate the period, amplitude, and phase shift.

14. The given curve has equation $y = a \cos bx$ or $y = a \sin bx$. Find a and b and write the equation of the curve.

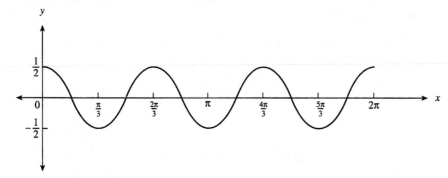

15. (a) Find the subintervals of $\left(-\dfrac{\pi}{2}, \dfrac{\pi}{2}\right)$ where the tangent function is increasing or decreasing, and those for which the curve is concave up or down.

   (b) Explain how the graph of $y = \tan\left(x + \dfrac{\pi}{2}\right)$ can be obtained from the graph of $y = \tan x$.

16. Find the period of $y = \tan 2x$ and write the equation of each asymptote that occurs on the interval $[0, \pi]$.

17. State the domain and range of $y = \sec x$ and graph for x in the interval $-\dfrac{\pi}{2} < x < \dfrac{\pi}{2}$.

18. Make use of the fact that $\csc x = \dfrac{1}{\sin x}$ to show that the cosecant is symmetric through the origin.

19. For $F(x) = \csc^3 2x$ find f, g, and h so that $F = f \circ g \circ h$.

20. State the domain and range of each function:

    (a) $y = \arcsin x$ **(b)** $y = \arccos x$

21. **(a)** State the domain and range of $y = \arctan x$ and graph.

    (b) Describe the symmetry and write the equations of the asymptotes.

22. Evaluate without the use of a calculator:

    (a) $\arcsin \dfrac{1}{\sqrt{2}}$ **(b)** $\arccos(-1)$ **(c)** $\arctan\left(-\dfrac{1}{\sqrt{3}}\right)$

23. Evaluate:

    (a) $\arcsin(0.5398)$ **(b)** $\arccos(-0.1896)$ **(c)** $\arctan(-0.2341)$

24. Evaluate without using a calculator.

    (a) $\sin\left(\arccos -\frac{1}{\sqrt{2}}\right)$ **(b)** $\cos\left(2 \arcsin \frac{1}{2}\right)$ **(c)** $\arctan(\tan x^2)$

25. Solve for x: $\arcsin(4x + 1) = -\dfrac{\pi}{6}$

CHAPTER 6: MULTIPLE CHOICE TEST

1. Express $\dfrac{11\pi}{6}$ radians in degree measure:

    (a) $330°$ **(b)** $660°$ **(c)** $\dfrac{180}{\pi}$ **(d)** $\left(\dfrac{11\pi}{6}\right)\left(\dfrac{\pi}{180}\right)$ **(e)** None of the preceding

2. What is the area in square centimeters of a sector of a circle of radius 6 cm if the central angle is 30°?

    (a) $\dfrac{\pi}{2}$ **(b)** 12π **(c)** 3π **(d)** 540 **(e)** None of the preceding

3. A wheel having a 3-foot diameter is rotating on its axis at a speed of 280 rad/sec. How far has a point on the rim of the wheel traveled after 1 minute?

    (a) 420 ft   **(b)** 840 ft   **(c)** 25,200 ft   **(d)** 50,400 ft   **(e)** None of the preceding

4. Find $\tan\left(-\dfrac{3\pi}{4}\right)$.

    (a) 1 **(b)** -1 **(c)** $\dfrac{1}{\sqrt{2}}$ **(d)** $-\dfrac{1}{\sqrt{2}}$ **(e)** None of the preceding

5. Which of the following are correct?

    I. $\sin\left(-\dfrac{3\pi}{2}\right) = 1$ **II.** $\cos\left(-\dfrac{3\pi}{2}\right) = 0$ **III.** $\tan\left(-\dfrac{3\pi}{2}\right)$ is undefined.

    (a) Only I   **(b)** Only II   **(c)** Only III   **(d)** I, II, and III   **(e)** None of the preceding

6. Find $\sin\left(-\dfrac{7\pi}{3}\right)$.

    (a) $-\dfrac{1}{2}$ **(b)** $-\dfrac{\sqrt{3}}{2}$ **(c)** $\dfrac{\sqrt{3}}{2}$ **(d)** $-\dfrac{2}{\sqrt{3}}$ **(e)** None of the preceding

7. Which of the following are correct?

   I. $\cos\left(-\dfrac{\pi}{6}\right) = \cos\dfrac{\pi}{6}$

   II. $\sin\dfrac{7\pi}{8} = \sin\dfrac{\pi}{8}$

   III. $\tan\left(-\dfrac{\pi}{4}\right) = -\tan\dfrac{\pi}{4}$

   (a) Only I   **(b)** Only II   **(c)** Only III   **(d)** I, II, and III   **(e)** None of the preceding

8. The point $P\left(\dfrac{1}{3}, y\right)$ where $y < 0$ is on the unit circle having center at the origin. Find $\tan\theta$ where θ is an angle in standard position whose terminal side intersects the unit circle at point P.

   (a) $-2\sqrt{2}$ **(b)** -2 **(c)** $\dfrac{-\sqrt{2}}{9}$ **(d)** $2\sqrt{2}$ **(e)** None of the preceding

9. Which of the following are true for the graph of $y = \sin\theta$?

   I. Domain: all real numbers x

   II. Range: $-1 \le y \le 1$

   III. Period: 2π

   (a) Only I   **(b)** Only II   **(c)** Only III   **(d)** I, II, and III   **(e)** None of the preceding

10. The amplitude A and phase shift t for the graph of $y = -\frac{1}{2}\sin(3x - \pi)$ are which of the following?

    (a) $A = \dfrac{1}{2},\ t = \dfrac{\pi}{3}$ **(b)** $A = -\dfrac{1}{2},\ t = \pi$ **(c)** $A = \dfrac{3}{2},\ t = \dfrac{\pi}{3}$ **(d)** $A = \dfrac{1}{2},\ t = 3\pi$

    (e) None of the preceding

11. What is the period of $y = 2\cos 4x$?

    (a) 8π **(b)** 4π **(c)** π **(d)** $\dfrac{\pi}{2}$ **(e)** None of the preceding

12. Which of the following are true for the graph of $y = \tan\left(x - \dfrac{\pi}{4}\right)$?

    I. The domain is the set of all real numbers x.

    II. The vertical lines $x = -\dfrac{\pi}{4}$ and $x = \dfrac{3\pi}{4}$ are asymptotes to the curve.

    III. The curve is symmetric with respect to the x-axis.

    (a) Only I   **(b)** Only II   **(c)** Only III   **(d)** I, II, and III   **(e)** None of the preceding

13. Which of the answers show the correct sequence of true (T) and false (F) for the following statements?

    I. The period of $y = \tan x$ is π.

    II. The period of $y = \cos\dfrac{x}{2}$ is π.

    III. The period of $y = \sec x$ is 2π.

    (a) F, F, F   **(b)** T, T, F   **(c)** F, T, T   **(d)** T, T, T   **(e)** None of the preceding

14. Which of the following are true?

  I. The range of $y = \arcsin x$ is given by $-\dfrac{\pi}{2} \le y \le \dfrac{\pi}{2}$.

  II. The range of $y = \arctan x$ is given by $-\dfrac{\pi}{2} \le y \le \dfrac{\pi}{2}$.

  III. $\arccos\left(-\dfrac{1}{\sqrt{2}}\right) = -\dfrac{\pi}{4}$.

  (a) I, II, and III   **(b)** Only I and II   **(c)** Only I   **(d)** Only II and III   **(e)** None of the preceding

15. $\tan\left(\arccos\left(-\dfrac{\sqrt{3}}{2}\right)\right) =$

  (a) $\dfrac{1}{\sqrt{3}}$ **(b)** $-\dfrac{1}{\sqrt{3}}$ **(c)** $-\sqrt{3}$ **(d)** -1 **(e)** None of the preceding

16. The given curve is the graph of an equation of the form $y = a \csc b(x - h)$. What are the values of a, b, and h?

  (a) $a = \dfrac{2}{3}$, $b = 2$, $h = -\dfrac{\pi}{4}$

  (b) $a = \dfrac{3}{2}$, $b = \dfrac{1}{2}$, $h = \dfrac{\pi}{4}$

  (c) $a = \dfrac{3}{2}$, $b = \dfrac{1}{2}$, $h = \dfrac{\pi}{2}$

  (d) $a = \dfrac{3}{2}$, $b = 2$, $h = \dfrac{\pi}{4}$

  (e) None of the preceding

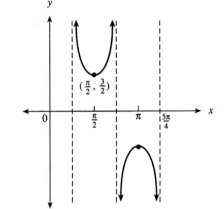

17. The two asymptotes to the curve $y = -\tan \dfrac{x}{3}$ that are closest to the y-axis are

  (a) $x = \pm\pi$ **(b)** $x = \pm\dfrac{3}{2}\pi$ **(c)** $x = \pm\dfrac{2}{3}\pi$ **(d)** $x = \pm\dfrac{\pi}{3}$ **(e)** None of the preceding

18. If $\tan\theta = -\sqrt{3}$ and $\cos\theta = \dfrac{1}{2}$, than $\csc\theta =$

  (a) -2 **(b)** 2 **(c)** $-\dfrac{2}{\sqrt{3}}$ **(d)** $\dfrac{2}{\sqrt{3}}$ **(e)** None of the preceding

19. Which of the following are true?

  I. $\arcsin \dfrac{1}{\sqrt{2}} = \arctan 1$

  II. $\arcsin\left(-\dfrac{1}{\sqrt{2}}\right) = \arccos\left(-\dfrac{1}{\sqrt{2}}\right)$

  III. $\arctan \dfrac{1}{\sqrt{3}} = \arcsin \dfrac{1}{2}$

  (a) I and II only   **(b)** I and III only   **(c)** II and III only   **(d)** I, II, and III   **(e)** None of the preceding

20. The domain of $y = 2 \arccos \dfrac{x}{3}$ is given by

(a) $-\dfrac{1}{3} \le x \le \dfrac{1}{3}$ (b) $-1 \le x \le 1$ (c) $-3 \le x \le 3$ (d) $-2 \le x \le 2$ (e) None of the preceding

ANSWERS TO THE TEST YOUR UNDERSTANDING EXERCISES

Page 365

1. $\dfrac{\pi}{12}$ **2.** $\dfrac{\pi}{5}$ **3.** $\dfrac{\pi}{2}$ **4.** $\dfrac{5\pi}{6}$ **5.** $\dfrac{13\pi}{12}$ **6.** $\dfrac{7\pi}{4}$ **7.** $\dfrac{16\pi}{9}$ **8.** $\dfrac{17\pi}{9}$

9. $12°$ **10.** $80°$ **11.** $210°$ **12.** $\left(\dfrac{630}{\pi}\right)°$ **13.** $153°$ **14.** $276°$ **15.** $300°$ **16.** $330°$

Page 376

1. $\dfrac{1}{\sqrt{2}}$ **2.** $-\sqrt{2}$ **3.** 1 **4.** $\dfrac{2}{\sqrt{3}}$ **5.** $\dfrac{1}{\sqrt{3}}$ **6.** $-\sqrt{2}$ **7.** $\dfrac{\sqrt{3}}{2}$ **8.** $-\dfrac{\sqrt{3}}{2}$

Page 387

1.
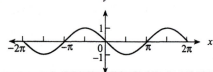

2.
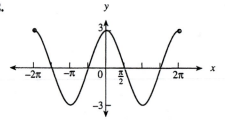

3. Same as in Exercise 1   **4.** 10   **5.** $\dfrac{2}{3}$ **6.** 1

Page 410

1. $-\dfrac{\pi}{2}$ rad   **2.** $\dfrac{\pi}{6}$ rad   **3.** undefined   **4.** $\dfrac{2\pi}{3}$ rad   **5.** 1   **6.** $\dfrac{1}{2}$ **7.** 1   **8.** x

9. undefined   **10.** 0.82 rad   **11.** -1.19 rad   **12.** 1.34 rad   **13.** 2.03 rad   **14.** 0.5396

15. 0.68 rad

Right Triangle Trigonometry, Identities, and Equations

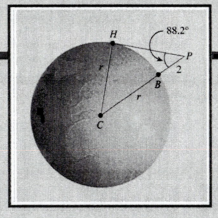

Contrary to common belief, some of the ancient peoples knew that the earth was spherical rather than flat. The ancient (educated) Greeks knew this, and in the second century B.C., the great astronomer Hipparchus (the inventor of trigonometry) was able to approximate the radius of the earth.

Essentially, his method was based on a figure like the one above, in which point P represents the top of a mountain that is 2 miles high. Point B is assumed to be at the base of the mountain directly under point P, so that when line PB is extended, it will go through the center of the earth, point C, as indicated.

From point P, using an appropriate sighting instrument, he located the horizon point H so that the line PH is tangent to the earth's surface at H. Then, since the earth is a sphere, the radius CH is perpendicular to PH, and triangle PCH is a right triangle. Next, he measured the angle at P determined by the lines PH and PC, which was approximately $88.2°$ (depending on the accuracy of the instrument used).

Hipparchus then formed the ratio of the side opposite the angle at P to the hypotenuse PC which is $\dfrac{r}{r+2}$. This ratio is known as the sine ratio of angle P, and Hipparchus already had available the sine ratios of such angles. In particular, the sine ratio of $88.2°$ rounded to five decimal places is 0.99951. Therefore,

$$\frac{r}{r+2} = 0.99951$$

Solving this equation for r gives the approximation $r = 4,080$, miles which is close to the known radius of the earth.

It is believed that the approximation obtained by Hipparchus was in error due to the crudeness of the measuring instruments available in his time.

425

7.1 TRIGONOMETRY OF ACUTE ANGLES

The word trigonometry is based on the Greek words for triangle (trigonon) and measure (metron).

The trigonometric functions, studied in Chapter 6, play an important role in certain branches of modern mathematics and its applications. However, the study of trigonometry dates back more than 3000 years and over the centuries has been instrumental in developing knowledge in areas such as architecture, astronomy, navigation, and surveying. In the first two sections of this chapter, we will study some traditional aspects of trigonometry and its applications. (Additional applications are considered in Chapter 8.)

The trigonometric function values for measures of acute angles, $0 < \theta < \dfrac{\pi}{2}$, have been included in our previous work in Section 6.2. We will now focus on such angles. First, observe that the function values can be found using any triangle that is similar to the reference triangle determined by the unit circle. Since triangles OPQ and OBC are similar,

$$\frac{CB}{OB} = \frac{QP}{OP} = \frac{y}{1} = y = \sin\theta$$

$$\frac{OC}{OB} = \frac{OQ}{OP} = \frac{x}{1} = x = \cos\theta$$

corresponding sides of similar triangles are proportional

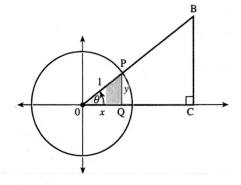

This can also be done for the remaining four trigonometric functions.

The preceding demonstrates that the trigonometric function values *depend only on the size of the angle and not on the size of the triangle.* Therefore, it is no longer essential to have the reference triangle with hypotenuse of length 1, nor is it necessary for such an angle to be in standard position in order to find the trigonometric values. For example,

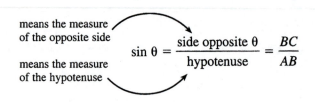

means the measure of the opposite side

$$\sin\theta = \frac{\text{side opposite }\theta}{\text{hypotenuse}} = \frac{BC}{AB}$$

means the measure of the hypotenuse

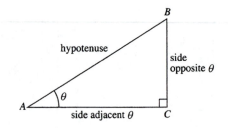

This result is included in the following list, in which **trigonometric ratio** is used in place of *trigonometric function value.* This alternate language emphasizes that the trigonometric function values are equivalent to the ratios of the lengths of the sides of a right triangle.

Trigonometric Ratio	Abbreviation	Definition	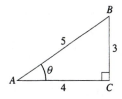
sine of θ	sin θ	$\dfrac{\text{side opposite}}{\text{hypotenuse}}$	$\dfrac{a}{c}$
cosine of θ	cos θ	$\dfrac{\text{side adjacent}}{\text{hypotenuse}}$	$\dfrac{b}{c}$
tangent of θ	tan θ	$\dfrac{\text{side opposite}}{\text{side adjacent}}$	$\dfrac{a}{b}$
cotangent of θ	cot θ	$\dfrac{\text{side adjacent}}{\text{side opposite}}$	$\dfrac{b}{a}$
secant of θ	sec θ	$\dfrac{\text{hypotenuse}}{\text{side adjacent}}$	$\dfrac{c}{b}$
cosecant of θ	csc θ	$\dfrac{\text{hypotenuse}}{\text{side opposite}}$	$\dfrac{c}{a}$

EXAMPLE 1   Find the trigonometric ratios of θ as shown in the figure below.

Solution   Relative to $\angle A$, BC is the side opposite and AC is the side adjacent.

$$\sin \theta = \frac{\text{side opposite}}{\text{hypotenuse}} = \frac{BC}{AB} = \frac{3}{5} \qquad \csc \theta = \frac{\text{hypotenuse}}{\text{side opposite}} = \frac{AB}{BC} = \frac{5}{3}$$

$$\cos \theta = \frac{\text{side adjacent}}{\text{hypotenuse}} = \frac{AC}{AB} = \frac{4}{5} \qquad \sec \theta = \frac{\text{hypotenuse}}{\text{side adjacent}} = \frac{AB}{AC} = \frac{5}{4}$$

$$\tan \theta = \frac{\text{side opposite}}{\text{side adjacent}} = \frac{BC}{AC} = \frac{3}{4} \qquad \cot \theta = \frac{\text{side adjacent}}{\text{side opposite}} = \frac{AC}{BC} = \frac{4}{3} \qquad \blacksquare$$

When it is convenient, the letter for the vertex of an angle will be used in place of the angle measure when writing the ratios. Thus, in the preceding example, we could also write $\sin A = \frac{3}{5}$. In other words, $\sin A = \sin \theta$.

EXAMPLE 2 Use the figure below and write the six trigonometric ratios for angle B.

Solution Note that we need to consider sides opposite and adjacent to $\angle B$.

$$\sin B = \frac{12}{13} \qquad \csc B = \frac{13}{12}$$

$$\cos B = \frac{5}{13} \qquad \sec B = \frac{13}{5}$$

$$\tan B = \frac{12}{5} \qquad \cot B = \frac{5}{12}$$

Observe that the six values are listed so that in each row the pair of values are reciprocals; sin and csc, cos and sec, tan and cot.

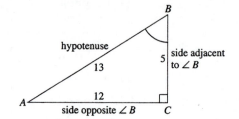

■

TEST YOUR UNDERSTANDING

(Answers: Page 488)

Use the given figure to answer each question.

1. Write the six trigonometric ratios for angle A.
2. Write the six trigonometric ratios for angle B.
3. Find the value of $(\sin A)^2 + (\cos A)^2$.
4. Find the value of $(\sin A)(\csc A)$.
5. Find the value of $(\sec B)^2 - (\tan B)^2$.
6. Find the value of $(\tan B)(\cot B)$.

Note: In later work an expression such as $(\sin A)^2$ is usually written as $\sin^2 A$. Thus $\cos^2 A$ means $(\cos A)^2$, and so on.

When only two sides of a right triangle are given, all the trigonometric ratios can still be found by using the Pythagorean theorem to find the third side. This is illustrated in Example 3.

EXAMPLE 3 Triangle ABC is a right triangle with right angle at C. Find $\sin B$ if $\tan A = \frac{3}{2}$.

Solution Since $\tan A = \dfrac{\text{side opposite}}{\text{side adjacent}} = \dfrac{3}{2}$, label the sides opposite and adjacent to $\angle A$ as shown, and use the Pythagorean theorem to find c.

$$c^2 = a^2 + b^2$$
$$= 9 + 4$$
$$= 13$$

It is common notation to use the same letter, capital and lower case, for an angle and its opposite side such as $\angle A$ and side a.

Then

$$c = \sqrt{13} \qquad \text{and} \qquad \sin B = \frac{AC}{AB} = \frac{2}{\sqrt{13}} \qquad \text{or} \qquad \frac{2\sqrt{13}}{13}$$

■

EXAMPLE 4 In the given figure, use the Pythagorean theorem to solve for AC in terms of x. Then write each of the six trigonometric ratios of $\angle A$ in terms of x.

Solution By the Pythagorean theorem, $(AC)^2 + x^2 = 1^2 = 1$. Thus $(AC)^2 = 1 - x^2$ and $AC = \sqrt{1 - x^2}$.

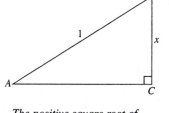

The positive square root of $1 - x^2$ is used since the side of the triangle has a positive measure.

$$\sin A = x \qquad\qquad \csc A = \frac{1}{x}$$

$$\cos A = \sqrt{1 - x^2} \qquad\qquad \sec A = \frac{1}{\sqrt{1 - x^2}}$$

$$\tan A = \frac{x}{\sqrt{1 - x^2}} \qquad\qquad \cot A = \frac{\sqrt{1 - x^2}}{x}$$

■

EXAMPLE 5 Show that $(\sin \theta)(\cos \theta) = \dfrac{x\sqrt{25 - x^2}}{25}$ where θ is an acute angle in a right triangle having one side of measure x.

Solution The difference $25 - x^2$, in conjunction with the Pythagorean theorem, suggests a right triangle with hypotenuse 5 as shown.

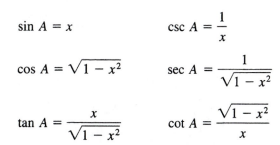

If a given expression in x involved the sum $25 + x^2$, rather than the difference $25 - x^2$, then the two perpendicular sides of the triangle would be labeled with 5 and x, and the hypotenuse would be $\sqrt{25 + x^2}$.

$$(AC)^2 = 5^2 - x^2 = 25 - x^2$$

$$AC = \sqrt{25 - x^2}$$

$$(\sin \theta)(\cos \theta) = \frac{x}{5} \cdot \frac{\sqrt{25 - x^2}}{5} = \frac{x\sqrt{25 - x^2}}{25}$$

(Observe that if $AC = x$, then $CB = \sqrt{25 - x^2}$, and the result would be the same.)

■

The trigonometric ratios of the special acute angles having measures $\dfrac{\pi}{6} = 30°$, $\dfrac{\pi}{4} = 45°$, $\dfrac{\pi}{3} = 60°$ can easily be memorized, or recalled, when associated with right triangles in which the shortest side is 1 unit. Such triangles can be formed by multiplying the sides of the reference triangles used earlier on the unit circle as follows. See the figures below and at the top of page 430.

Multiply the length of the sides of the triangle at the left by $\sqrt{2}$ to obtain the triangle at the right.

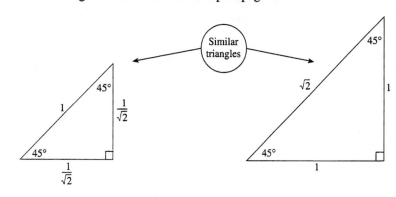

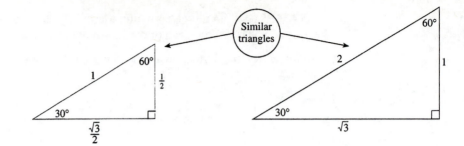

Multiply the length of the sides of the triangle at the left by 2 to obtain the triangle at the right.

Recalling that the ratios are unaffected by the size of the triangle, the ratios found for $\frac{\pi}{4} = 45°$ on page 373, and for $\frac{\pi}{6} = 30°$ and $\frac{\pi}{3} = 60°$ on page 375, can be obtained from the triangles shown at the right in the preceding figures.

EXAMPLE 6 Draw a right triangle whose shortest side is 1 unit, and such that for one of the acute angles cot $\theta = \dfrac{\sqrt{3}}{3}$.

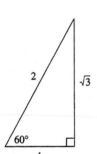

Solution Since cot $\theta = \dfrac{\sqrt{3}}{3} = \dfrac{\sqrt{3}}{3} \cdot \dfrac{\sqrt{3}}{\sqrt{3}} = \dfrac{3}{3\sqrt{3}} = \dfrac{1}{\sqrt{3}}$, and the angle is acute, $\theta = 60°$. Also, since cot $60° = \dfrac{1}{\sqrt{3}} = \dfrac{\text{adjacent}}{\text{opposite}}$, we have the triangle, shown in the margin, whose shortest side is 1 unit. ∎

When the range values of the inverse trigonometric functions are restricted to acute angle measures, there are some useful geometric interpretations involving right triangles.

EXAMPLE 7 Find the exact value of $\cos\left(\arcsin \frac{2}{3}\right)$ and check by calculator.

Solution We know that $y = \arcsin \frac{2}{3}$ is the measure of an acute angle because for $0 < x < 1$, $y = \arcsin x$ satisfies $0 < y < \dfrac{\pi}{2}$ (see page 406). Now construct a right triangle containing y, as shown in the margin, so that $\sin y = \frac{2}{3}$. Then from the triangle

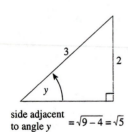

side adjacent to angle y $= \sqrt{9 - 4} = \sqrt{5}$

$$\cos\left(\arcsin \frac{2}{3}\right) = \cos y = \frac{\sqrt{5}}{3} \qquad \text{exact value}$$

Check: (Either radian or degree mode can be used.)

$$\cos\left(\arcsin \frac{2}{3}\right) = \boxed{\text{COS}} \ \boxed{(} \ \boxed{\text{SIN}^{-1}} \ \boxed{(} \ \boxed{2} \ \boxed{\div} \ \boxed{3} \ \boxed{)} \ \boxed{)} \ \boxed{\text{ENTER}}$$

$$= 0.745356 \text{ (approximation)}$$

Also,
$$\frac{\sqrt{5}}{3} = 0.745356.$$

∎

The formula

$$\arcsin x + \arccos x = 90°$$

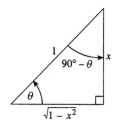

developed on page 410 for $-1 \le x \le 1$, has a useful geometric interpretation suggested by the figure in the margin. Since $\sin \theta = x$ and $\cos(90° - \theta) = x$, the formula says that for acute angles

> An angle whose sine is x and the angle whose cosine is x are complementary.

It can also be shown that

$$\arctan x + \text{arccot } x = 90°$$

and

$$\text{arcsec } x + \text{arccsc } x = 90°$$

In addition to the preceding forms, the reciprocal relationships $\cot \theta = \dfrac{1}{\tan \theta}$, $\sec \theta = \dfrac{1}{\cos \theta}$, and $\csc \theta = \dfrac{1}{\sin \theta}$ lead to the following inverse forms (see Exercise 56).

$$\text{arccot } x = \arctan \frac{1}{x}$$

$$\text{arcsec } x = \arccos \frac{1}{x}$$

$$\text{arccsc } x = \arcsin \frac{1}{x}$$

EXERCISES 7.1

Find the six trigonometric ratios of θ.

1.
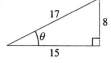

2.

3.
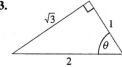

4.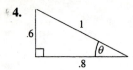

5.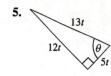

6.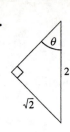

Draw an appropriate right triangle whose shortest side is 1 unit, and find θ, *where* $0 < \theta < \dfrac{\pi}{2}$.

7. $\sin \theta = \dfrac{\sqrt{2}}{2}$ **8.** $\tan \theta = \dfrac{\sqrt{3}}{3}$ **9.** $\tan \theta = \sqrt{3}$

10. $\cos \theta = \frac{1}{2}$ **11.** $\csc \theta = 2$ **12.** $\cot \theta = 1$

Two sides of right $\triangle ABC$ *are given in which* $\angle C$ *is the right angle. Use the Pythagorean theorem to find the third side and then write the six trigonometric ratios of (a) angle A and (b) angle B.*

13. $a = 6$; $b = 8$ **14.** $a = 13$; $b = 5$ **15.** $a = 4$; $b = 5$

16. $a = 1$; $b = 1$ **17.** $a = 1$; $b = 2$ **18.** $a = 4$; $b = 10$

19. $a = 3$; $c = 4$ **20.** $a = 20$; $c = 29$ **21.** $b = 2$; $c = 5$

22. $b = 4$; $c = 7$ **23.** $a = 21$; $c = 29$ **24.** $a = \frac{1}{3}$; $b = \frac{1}{4}$

Use triangle XYZ to find the value for each expression.

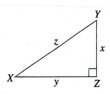

25. $(\tan X)(\cot X)$ **26.** $(\cos Y)(\sec Y)$ **27.** $(\sin X)\left(\dfrac{1}{\csc X}\right)$ **28.** $(\tan Y)\left(\dfrac{1}{\cot Y}\right)$

29. $\sin^2 X + \cos^2 X$ **30.** $\sin^2 Y + \cos^2 Y$ **31.** $\sec^2 X - \tan^2 X$ **32.** $\csc^2 Y - \cot^2 Y$

Angle C is the right angle of right $\triangle ABC$. *Use the given information to find the other five ratios of the indicated angle.*

33. $\sin A = \frac{3}{4}$ **34.** $\cos B = \frac{2}{5}$ **35.** $\tan A = \frac{9}{40}$

36. $\cot A = 1$ **37.** $\sec B = \dfrac{\sqrt{11}}{\sqrt{2}}$ **38.** $\csc B = 10$

Find the exact value of each expression and check by calculator.

39. $\sin\left(\arccos \frac{12}{13}\right)$ **40.** $\sin\left(\arctan \frac{1}{12}\right)$ **41.** $\tan\left(\arcsin \frac{3}{4}\right)$ **42.** $\cot\left(\arcsin \frac{4}{7}\right)$

Use the information given for △ABC and verify the indicated equality.

43. $(\sin A)(\cos A) = x\sqrt{1 - x^2}$

44. $\tan^2 A = \dfrac{x^2}{9 - x^2}$

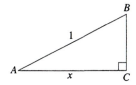

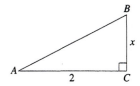

45. $\dfrac{4 \sin^2 A}{\cos A} = \dfrac{x^2}{\sqrt{16 - x^2}}$

46. $(\sin A)(\cos A) = \dfrac{2x}{4 + x^2}$

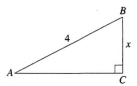

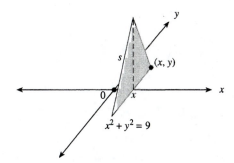

Show that the equation is true for an appropriate acute angle of θ radians in a right triangle, having one side whose measure is x.

47. $(\sin \theta)(\tan \theta) = \dfrac{x^2}{2\sqrt{4 - x^2}}$

48. $\sec^3 \theta = \dfrac{8}{(4 - x^2)^{3/2}}$

49. $(\tan^2 \theta)(\cos^2 \theta) = \dfrac{x^2}{x^2 + 16}$

50. $(\sin^3 \theta)(\cot^3 \theta) = \dfrac{(x^2 - 5)^{3/2}}{x^3}$

51. The side of an equilateral triangle is s centimeters long. Express the area as a function of s.

52. The figure shows an equilateral triangle that is perpendicular to the plane of the circle $x^2 + y^2 = 9$ with one of its sides coinciding with a chord of the circle that is perpendicular to the x-axis. Write the area of the triangle as a function of x and state its domain. (*Hint:* $s = 2y$.)

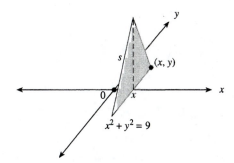

53. Replace the equilateral triangle in Exercise 52 with a right isosceles triangle having one of its equal sides coinciding with the chord that is perpendicular to the x-axis, and find the area of the triangle.

54. Replace the equilateral triangle in Exercise 52 with a right isosceles triangle having its hypotenuse coinciding with the chord that is perpendicular to the x-axis, and find the area of the triangle.

55. The wire AB is 100 inches long and is cut into two pieces at point P as shown. If AP is used to form an equilateral triangle, and PB is used for a circle, express the total area of the two resulting figures as a function of x and state its domain.

56. Prove that $\text{arccot } x = \arctan \dfrac{1}{x}$ for $x > 0.$ $\left(Hint\text{: Let } y = \arctan \dfrac{1}{x} \text{ so that } \tan y = \dfrac{1}{x}. \right)$

57. (a) Let θ be the measure of an acute angle of a right triangle so that $\theta = \arcsin x$. Write the trigonometric ratios of θ in terms of x.

    (b) Follow the instructions in part **(a)** using $\theta = \arctan x$.

 Challenge   In the figure the circle centered at the origin has equation $x^2 + y^2 = 2$. Triangle ABC is a right triangle with $BC = 1$. Use the figure to show that $\tan 22.5° = \sqrt{2} - 1$.

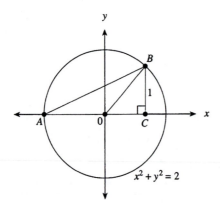

 Written Assignment   Explain why the equation of a linear function f whose graph has positive slope can be written in the form $f(x) = (\tan \theta) x + b$, where θ is the acute angle determined by any horizontal line and the graph of f.

Graphing Calculator Exercises

1. In Example 5 it was shown that $(\sin \theta)(\cos \theta) = \dfrac{x\sqrt{25 - x^2}}{25}$, where θ is an acute angle in a right triangle having one side of measure x. Find the value of x for which $y = \dfrac{x\sqrt{25 - x^2}}{25}$ is a maximum by graphing this function. Use Trace and Zoom to find x to the nearest hundredth. Then find the angle measure of the corresponding θ, to the nearest tenth of a degree, that gives this maximum. $\left(Hint\text{: } \sin \theta = \dfrac{x}{5}, \text{ so } \sin^{-1} \dfrac{x}{5} = \theta. \right)$

2. Using calculus it can be shown that the maximum value of $(\sin \theta)(\cos \theta)$ in Exercise 1 does not depend upon the length of the hypotenuse of the right triangle. Verify that if $(\sin \theta)(\cos \theta) = \dfrac{x\sqrt{16 - x^2}}{16}$, for example, the maximum value of the left-hand side remains the same as in Exercise 1. Show this by graphing the right-hand side as a function of x.

7.2 RIGHT TRIANGLE TRIGONOMETRY

Can you find the distance across the lake between the two cottages at B and C? A surveyor can do this without getting wet. This, and many more problems involving distances that cannot be measured directly, can be solved with the aid of trigonometric ratios.

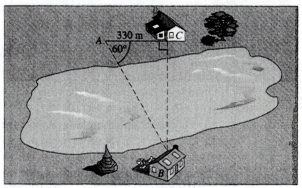

Tan A will be used because this ratio involves the unknown side and the given side. Cot A can also be used.

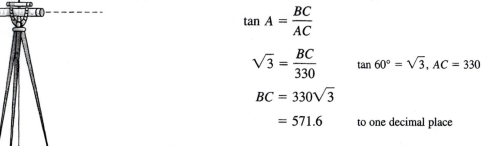

Here is one way the surveyor can find BC. First locate a point A so that the resulting triangle ABC has a right angle at C and a $60°$ angle at A. From A to C is along flat ground and can be measured directly. Assume that $AC = 330$ meters and solve for BC as follows:

$$\tan A = \frac{BC}{AC}$$

$$\sqrt{3} = \frac{BC}{330} \qquad \tan 60° = \sqrt{3},\ AC = 330$$

$$BC = 330\sqrt{3}$$

$$= 571.6 \qquad \text{to one decimal place}$$

Thus the distance across the lake between the cottages is about 572 meters.

A transit is an instrument that surveyors can use to determine such angles. You may have seen a transit being used in highway or building construction.

EXAMPLE 1 Solve for x. Give both the exact value and the approximation to the nearest tenth of a unit.

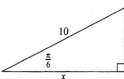

Solution Both the cosine and secant ratios involve the sides labeled x and 10. The cosine is somewhat easier to use.

$$\cos 30° = \frac{x}{10}$$

$$\frac{\sqrt{3}}{2} = \frac{x}{10}$$

$$x = 5\sqrt{3} \qquad \text{exact value}$$

$$= 8.7 \qquad \text{rounded to the nearest tenth} \qquad ■$$

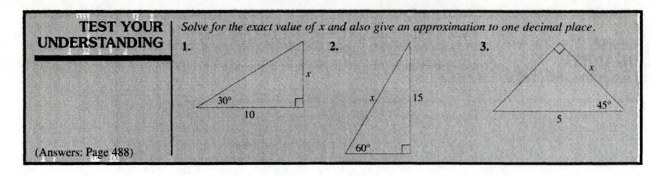

TEST YOUR UNDERSTANDING

Solve for the exact value of x and also give an approximation to one decimal place.

1.

30°
10
x

2.

x
15
60°

3.

x
45°
5

(Answers: Page 488)

The preceding problems called for trigonometric ratios of the special angles 30°, 45°, and 60°, whose ratios had already been determined. However, many problems do not involve such special acute angles. In such a case, the trigonometric ratio, or the angle with a given ratio, can be obtained using a calculator.

Here are some general guidelines for solving trigonometric problems that involve geometric figures:

1. After carefully reading the problem, draw a figure that matches the given information. Try to draw it as close to scale as possible.
2. Record any given values directly on the corresponding parts, and label the required unknown parts with appropriate letters.
3. Determine the trigonometric ratios and geometric formulas that can be used to solve for an unknown part and solve.

The examples that follow make use of these guidelines.

EXAMPLE 2 The **angle of elevation** of an 80-foot ramp leading to a bridge above a highway is 10.5°. Find the height of the bridge above the highway.

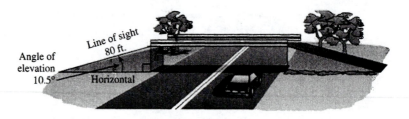

Line of sight
80 ft.

Angle of elevation
10.5°
Horizontal

A sequence such as the following could be used on many scientific calculators:

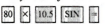

80 × 10.5 SIN =

Solution First note that the angle of elevation is the angle between the horizontal and the line of sight to the top of an object. Since the hypotenuse is given, and the side opposite the 10.5° angle is the unknown, use the sine ratio.

$$\sin 10.5° = \frac{h}{80}$$

$$h = 80 \sin 10.5°$$

$$= \boxed{80} \times \boxed{\text{SIN}} \boxed{10.5} \boxed{\text{ENTER}} = 14.57884204$$

Then $h = 14.6$ feet to the nearest tenth of a foot. ∎

EXAMPLE 3 From the top of a house the **angle of depression** of a point on the ground is 25°. The point is 35 meters from the base of the building. How high is the building?

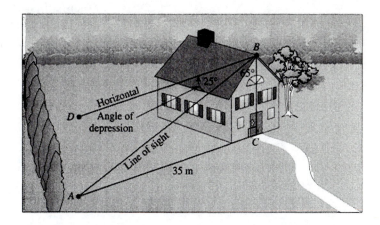

Solution First note that the angle of depression is determined by the horizontal line BD and the line of sight BA as shown. Now observe that the complement of the angle of depression, $\angle ABC = 65°$, and that $\angle CAB = 25°$ since BD and AC are parallel. Consequently, the height BC can be found using either the cot 65° or the tan 25°. It is somewhat simpler to work with the tangent ratio than with the cotangent. Thus,

Solve for BC using the cotangent ratio.

$$\tan 25° = \frac{BC}{35}$$

$$BC = 35 \tan 25° = 16.32076804$$

Then the height is 16 meters to the nearest meter. ∎

In Example 5 the question calls for finding an angle measure. Therefore, an inverse function key, introduced in Section 6.5, will be needed. The next example illustrates the procedure needed.

EXAMPLE 4 Find the acute angle θ to the nearest tenth of a degree for the given ratio.
(a) $\sin \theta = 0.4741$ **(b)** $\cot \theta = 0.4625$

Solution
(a) Since 0.4741 is the sine value of an acute angle θ, and the $\boxed{\text{SIN}^{-1}}$ key gives this angle we have

$$\theta = \boxed{\text{SIN}^{-1}}\ \boxed{0.4741}\ \boxed{\text{ENTER}} = 28.3° \qquad \text{rounded to tenths}$$

(b) Since the tangent and cotangent ratios are reciprocals, we have

$$\tan \theta = \frac{1}{\cot \theta} = \frac{1}{0.4625}$$

When checking these results, use four decimal places for the angle in order to obtain the desired accuracy of the given ratio. Here is a check for (b):

TAN | 65.1795 | ENTER
x^{-1} | ENTER | = 0.4625

Then, the | TAN^{-1} | key will give the corresponding acute angle by one of the following procedures:

$$\theta = \tan^{-1}\frac{1}{0.4625} = \tan^{-1}(0.4625)^{-1}$$

$$= \boxed{TAN^{-1}}\ \boxed{0.4625}\ \boxed{X^{-1}}\ \boxed{ENTER}$$

$$= 65.17945866 \qquad\qquad \text{calculator display}$$

$$= 65.2° \qquad\qquad \text{rounded to tenths}$$

or

$$\theta = \tan^{-1}(1 \div 0.4625)$$

$$= \boxed{TAN^{-1}}\ \boxed{(}\ \boxed{1}\ \boxed{\div}\ \boxed{0.4625}\ \boxed{)}\ \boxed{ENTER} = 65.2°$$

Also, using a scientific calculator we have:

$$\theta = \boxed{0.4625}\ \boxed{1/x}\ \boxed{TAN^{-1}} = 65.2°\qquad\blacksquare$$

TEST YOUR UNDERSTANDING	*Find θ to the nearest tenth of a degree.*

1. $\cos\theta = 0.3907$ **2.** $\tan\theta = 1.4281$ **3.** $\sin\theta = 0.8960$

(Answers: Page 488) **4.** $\sec\theta = 1.0794$ **5.** $\cot\theta = 3.8391$ **6.** $\csc\theta = 10.826$

EXAMPLE 5 The top of a hill is 40 meters higher than a nearby airport, and the horizontal distance from the end of a runway is 325 meters from a point directly below the hilltop. An airplane takes off at the end of the runway in the direction of the hill, at an angle that is to be kept constant until the hill is passed. If the pilot wants to clear the hill by 30 meters, what should be the angle at takeoff in degree measure?

Solution Sketch a figure and let the takeoff angle be θ. Since the side opposite θ is 40 + 30, the tangent ratio gives

$$\tan\theta = \frac{70}{325}$$

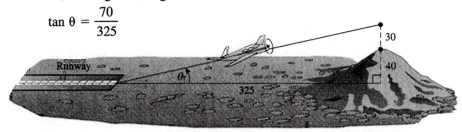

Since θ is the acute angle whose tangent ratio is $\dfrac{70}{325}$, the | TAN^{-1} | key is used to find θ after setting the calculator in degree mode:

This sequence can be used on a scientific calculator:

$$\theta = \tan^{-1} \frac{70}{325}$$

$$= \boxed{\text{TAN}^{-1}} \ \boxed{(} \ \boxed{70} \ \boxed{\div} \ \boxed{325} \ \boxed{)} \ \boxed{\text{ENTER}}$$

$$= 12.1549417$$

Then, to the nearest tenth of a degree, the takeoff angle is 12.2°. ∎

EXERCISES 7.2

Solve for the indicated part of each right triangle without using a calculator. Give the exact values.

1.

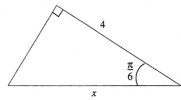

2.

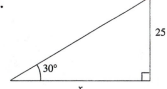

3.

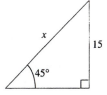

4.

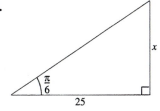

5.

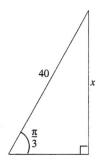

6.

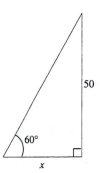

7.

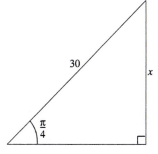

8.

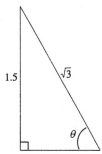

9.

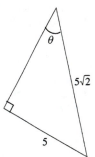

10. For the given $\triangle ABC$ find the lengths of AB and AC. (*Hint:* Draw BD perpendicular to AC.)

Find the measure of the acute angle θ in degrees to the nearest tenth of a degree.

11. $\tan \theta = 6.314$ **12.** $\sin \theta = .7214$ **13.** $\cot \theta = .4592$

14. $\sec \theta = 14.31$ **15.** $\cos \theta = .9940$ **16.** $\csc \theta = 2.763$

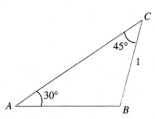

Triangle ABC is a right triangle with C = 90°. Solve for the indicated part. Give the angle measure and the length of the sides to the nearest tenth.

17. $A = 70°$, $a = 35$; find b.
18. $A = 70°$, $a = 35$; find c.
19. $B = 42.3°$, $a = 20$; find b.

20. $B = 42.3°$, $a = 20$; find c.
21. $a = 1$, $b = 3$; find B.
22. $a = 12$, $b = 9.5$; find A.

23. $b = 9$, $c = 25$; find B.
24. $A = 15.5°$, $c = 48$; find a.

25. The angle of elevation to the top of a flagpole is 35° from a point 50 meters from the base of the pole. What is the height of the pole to the nearest meter?

26. How high is a building whose horizontal shadow is 50 meters when the angle of elevation of the sun is 60.4°? Give the answer to the nearest tenth of a meter.

27. At a point 100 feet away from the base of a giant redwood tree a surveyor measures the angle of elevation to the top of the tree to be 70°. How tall is the tree to the nearest foot?

28. From the top of a 172-foot-high water tank, the angle of depression to a house is 13.3°. How far away is the house from the water tank to the nearest foot?

29. An observation post along a shoreline is 225 feet above sea level. If the angle of depression from this post to a ship at sea is 6.7°, how far is the ship from the shore to the nearest foot?

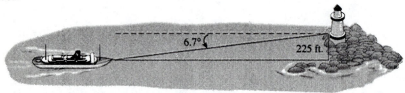

30. A kite string forms a 42.5° angle with the ground when 740 feet of string are out. What is the altitude of the kite to the nearest foot, assuming the string forms a straight line?

31. One of the cables that helps to stabilize a telephone pole is 82 feet long and is anchored into the ground 14.5 feet from the base of the pole. Find the angle that the cable makes with the ground.

32. One of the equal sides of an isosceles triangle is 18.7 units long and the vertex angle is 33°. Find the length of the base to the nearest tenth of a unit.

33. One side of an inscribed angle of a circle is a diameter of the circle, and the other side is a chord of length 10. If the inscribed angle is 66°, what is the length of the radius to the nearest tenth?

34. From an observation point A at the top of a 250-foot cliff, the angles of depression to two points P and Q on opposite sides of a river are 62° and 12°. (Points A, P, and Q determine a vertical plane.) Find the distance between P and Q to the nearest foot.

35. A surveyor finds that from point A on the ground the angle of elevation to the top of a mountain is 23°. When he is at a point B that is $\frac{1}{4}$ mile closer to the base of the mountain, the angle of elevation is 43°. How high is the mountain? One mile = 5280 feet. (Assume that the base of the mountain and the two observation points are on the same line.) Give the answer to the nearest foot.

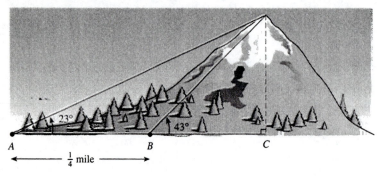

36. A 12-foot flagpole is standing vertically at the edge of the roof of a building. The angle of elevation to the top of the flagpole from a point on the ground that is 64 feet from the building is 78.6°. Find the height of the building to the nearest foot.

37. A circle with a 20-centimeter radius is inscribed in a regular pentagon. Find the perimeter of the pentagon to the nearest centimeter.

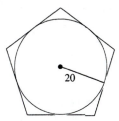

38. In the figure a wedge is formed by a plane cutting into a right circular cylinder of radius 2 at an angle of measure θ. The plane intersects the base of the cylinder in a diameter of the circular base. Express the area of the triangular cross section as a function of x if

(a) θ = 30°   **(b)** θ = 45°   **(c)** θ = 60°

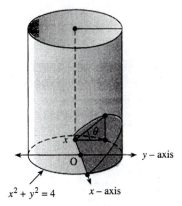

$x^2 + y^2 = 4$ x – axis

Challenge In the figure $DB = 16$, $\angle BAD = 13°$, and $\angle ABD = 22°$. Solve for CD.

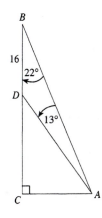

Written Assignment The trigonometric ratios given by the calculator or the tables are found by using more advanced mathematical procedures. However, rough approximations can be made when the definitions of the ratios are applied to carefully drawn right triangles. To do this you will need a ruler on which the inches are subdivided into sixteenths, a protractor, and graph paper. The graph paper should be divided into one-inch squares, each of which is subdivided into 64 squares 1/8 inch on a side.

Construct right triangles ABC so that A is at the origin, the hypotenuse AB is in quadrant I, and C is on the x-axis. Use 6 inches (96 sixteenths) as the unit length, and select the length of AB so that single measurements will give sin A and cos A. Do this for $\angle A = 10°, 20°, \ldots, 80°$. Describe this process, compare your results to the calculator values, and comment on the accuracy obtained.

Graphing Calculator Exercise An isosceles triangle having two equal sides of length 1 and vertex angle 2θ is given. What angle measure for θ will maximize the area of the triangle? (*Hint:* Show that area = sin θ cos θ.) Then graph the function given by $y = \sin x \cos x$ for $0° \leq x \leq 90°$ and $0 \leq y \leq 1$, and find its maximum value with Zoom and Trace.

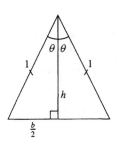

7.3 IDENTITIES

Identities will be useful later in solving equations, and eventually they will play an important role in a wide variety of mathematical situations.

Solving the equation $\dfrac{\sin^2 \theta}{\cos^2 \theta} = \dfrac{1}{3}$ for θ is easier to do if $\dfrac{\sin \theta}{\cos \theta}$ is first replaced by the equivalent expression $\tan \theta$ to get $\tan^2 \theta = \dfrac{1}{3}$. Such equations are discussed in detail in Section 7.6. Our objective here is to learn to recognize and work with **trigonometric identities,** such as $\dfrac{\sin \theta}{\cos \theta} = \tan \theta$. This equation is called an identity because it is a true statement for all values of θ except those when $\cos \theta = 0$.

A trigonometric identity is an equation that is true for all values of the variable for which the expressions in the equation are defined.

The following list of fundamental identities, together with some alternative forms, will be used throughout our work. It is therefore important that you learn to recognize them.

FUNDAMENTAL IDENTITIES		
Fundamental Identity	**Common Alternative Forms**	**Restrictions on θ**
$\csc \theta = \dfrac{1}{\sin \theta}$	$\sin \theta = \dfrac{1}{\csc \theta}$ $\sin \theta \csc \theta = 1$	Not coterminal with $0, \pi$.
$\sec \theta = \dfrac{1}{\cos \theta}$	$\cos \theta = \dfrac{1}{\sec \theta}$ $\cos \theta \sec \theta = 1$	Not coterminal with $\dfrac{\pi}{2}, \dfrac{3\pi}{2}$.
$\cot \theta = \dfrac{1}{\tan \theta}$	$\tan \theta = \dfrac{1}{\cot \theta}$ $\tan \theta \cot \theta = 1$	Not a quadrantal angle.
$\tan \theta = \dfrac{\sin \theta}{\cos \theta}$		Not coterminal with $\dfrac{\pi}{2}, \dfrac{3\pi}{2}$.
$\cot \theta = \dfrac{\cos \theta}{\sin \theta}$		Not coterminal with $0, \pi$.
$\sin^2 \theta + \cos^2 \theta = 1$	$\sin^2 \theta = 1 - \cos^2 \theta$ $\cos^2 \theta = 1 - \sin^2 \theta$	None
$\tan^2 \theta + 1 = \sec^2 \theta$	$\tan^2 \theta = \sec^2 \theta - 1$	Not coterminal with $\dfrac{\pi}{2}, \dfrac{3\pi}{2}$.
$1 + \cot^2 \theta = \csc^2 \theta$	$\cot^2 \theta = \csc^2 \theta - 1$	Not coterminal with $0, \pi$.

*The last three identities are called the **Pythagorean identities** because they are based on the Pythagorean theorem.*

Each of the preceding identities can be established by using the definitions of the trigonometric ratios on the unit circle $x^2 + y^2 = 1$, as shown here.

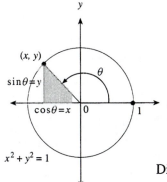

$$\csc \theta = \frac{1}{y} = \frac{1}{\sin \theta} \qquad \sec \theta = \frac{1}{x} = \frac{1}{\cos \theta}$$

$$\tan \theta = \frac{y}{x} = \frac{\sin \theta}{\cos \theta} \qquad \cot \theta = \frac{x}{y} = \frac{\cos \theta}{\sin \theta}$$

$$\tan \theta \cot \theta = \frac{y}{x} \cdot \frac{x}{y} = 1$$

$$\sin^2 \theta + \cos^2 \theta = x^2 + y^2 = 1 \qquad \text{by the Pythagorean theorem}$$

Dividing the last equation by $\cos^2 \theta$ produces

$$\frac{\sin^2 \theta}{\cos^2 \theta} + \frac{\cos^2 \theta}{\cos^2 \theta} = \frac{1}{\cos^2 \theta}$$

$$\left(\frac{\sin \theta}{\cos \theta} \right)^2 + 1 = \left(\frac{1}{\cos \theta} \right)^2$$

$$\tan^2 \theta + 1 = \sec^2 \theta$$

In practice, when working with identities, no special mention is made about the restrictions on the variable. However, you should be able to see what restrictions are necessary. In most cases such restrictions will occur when the denominator of an expression would be equal to 0.

Similarly, when $\sin^2 \theta + \cos^2 \theta = 1$ is divided by $\sin^2 \theta$, we get

$$1 + \cot^2 \theta = \csc^2 \theta$$

EXAMPLE 1 Use fundamental identities to convert $\dfrac{1 - \csc \theta}{\cot \theta}$ into an expression involving only sines and cosines, and simplify.

Solution

Here is another way to simplify the right side.

$$\frac{1 - \dfrac{1}{\sin \theta}}{\dfrac{\cos \theta}{\sin \theta}} = \frac{\sin \theta - 1}{\sin \theta} \cdot \frac{\sin \theta}{\cos \theta}$$

$$= \frac{\sin \theta - 1}{\cos \theta}$$

$$\frac{1 - \csc \theta}{\cot \theta} = \frac{1 - \dfrac{1}{\sin \theta}}{\dfrac{\cos \theta}{\sin \theta}} \qquad \longleftarrow \quad \begin{cases} \csc \theta = \dfrac{1}{\sin \theta} \\[2mm] \cot \theta = \dfrac{\cos \theta}{\sin \theta} \end{cases}$$

$$= \frac{1 - \dfrac{1}{\sin \theta}}{\dfrac{\cos \theta}{\sin \theta}} \cdot \frac{\sin \theta}{\sin \theta}$$

$$= \frac{\sin \theta - 1}{\cos \theta} \qquad \text{multiplying fractions} \qquad \blacksquare$$

The result in Example 1 can be stated as

$$\frac{1 - \csc \theta}{\cot \theta} = \frac{\sin \theta - 1}{\cos \theta}$$

and the solution in Example 1 is the *proof* that this equation is an identity.
The solution to Example 1 also suggests the following procedure.

METHOD I FOR VERIFYING IDENTITIES

Use fundamental or previously proven identities to change one side of the equation into the form of the other side.

"Verify the identity" is another way of saying "Prove that a given equation is an identity."

EXAMPLE 2 Verify this identity:

$$\frac{\tan^2 \theta + 1}{\tan \theta \csc^2 \theta} = \tan \theta$$

Solution In most cases it is easier to work with the more complicated side. Supply a reason for each step of the proof.

It is often helpful to convert the expression into one involving sines and cosines.

$$\frac{\tan^2 \theta + 1}{\tan \theta \csc^2 \theta} = \frac{\sec^2 \theta}{\tan \theta \csc^2 \theta}$$

$$= \frac{\dfrac{1}{\cos^2 \theta}}{\dfrac{\sin \theta}{\cos \theta} \cdot \dfrac{1}{\sin^2 \theta}}$$

$$= \frac{1}{\cos^2 \theta} \cdot \frac{\cos \theta \sin^2 \theta}{\sin \theta}$$

$$= \frac{\sin \theta}{\cos \theta}$$

$$= \tan \theta \qquad \blacksquare$$

Here is another procedure that can be used to verify an identity.

METHOD II FOR VERIFYING IDENTITIES

Convert each side of the given equation until the same form is obtained on each side.

EXAMPLE 3 Verify the identity: $\dfrac{\csc\theta + 1}{\csc\theta - 1} = (\sec\theta + \tan\theta)^2$.

Solution When working separately on each side, a vertical line helps to separate the work.

$$\frac{\csc\theta + 1}{\csc\theta - 1}$$

$$\downarrow$$

$$\frac{\dfrac{1}{\sin\theta} + 1}{\dfrac{1}{\sin\theta} - 1}$$

$$\downarrow$$

$$\frac{\left(\dfrac{1}{\sin\theta} + 1\right)\sin\theta}{\left(\dfrac{1}{\sin\theta} - 1\right)\sin\theta}$$

$$\downarrow$$

$$\frac{1 + \sin\theta}{1 - \sin\theta}$$

$$(\sec\theta + \tan\theta)^2$$

$$\downarrow$$

$$\left(\frac{1}{\cos\theta} + \frac{\sin\theta}{\cos\theta}\right)^2 \qquad \text{Convert to sines and cosines.}$$

$$\downarrow$$

$$\left(\frac{1 + \sin\theta}{\cos\theta}\right)^2 \qquad \text{Combine fractions.}$$

$$\downarrow$$

$$\frac{(1 + \sin\theta)^2}{\cos^2\theta} \qquad \left(\frac{a}{b}\right)^2 = \frac{a^2}{b^2}$$

$$\downarrow$$

$$\frac{(1 + \sin\theta)^2}{1 - \sin^2\theta} \qquad \cos^2\theta = 1 - \sin^2\theta$$

$$\downarrow$$

$$\frac{(1 + \sin\theta)^2}{(1 - \sin\theta)(1 + \sin\theta)} \qquad a^2 - b^2 = (a - b)(a + b)$$

$$\downarrow$$

$$\frac{1 + \sin\theta}{1 - \sin\theta} \qquad \text{Simplify.}$$

Since the same form has been obtained for each side, the proof is complete. ∎

It takes practice to become good at verifying identities. You will have to call on a wide variety of algebraic skills to be successful. Example 4 calls on a process similar to the one used earlier in rationalizing denominators.

EXAMPLE 4 Verify the identity: $\dfrac{\sin \theta}{1 + \cos \theta} = \dfrac{1 - \cos \theta}{\sin \theta}$.

Solution Begin by multiplying the numerator and denominator of the left side by $1 - \cos \theta$.

The motivation for this approach is that it gets $1 - \cos \theta$ into the numerator.

$$\frac{\sin \theta}{1 + \cos \theta} = \frac{(\sin \theta)(1 - \cos \theta)}{(1 + \cos \theta)(1 - \cos \theta)}$$

$$= \frac{(\sin \theta)(1 - \cos \theta)}{1 - \cos^2 \theta}$$

$$= \frac{(\sin \theta)(1 - \cos \theta)}{\sin^2 \theta}$$

$$= \frac{1 - \cos \theta}{\sin \theta} \qquad \blacksquare$$

In the next example an identity is verified that involves trigonometric expressions, natural logarithms, and absolute value.

EXAMPLE 5 Prove the identity:

$$-\ln|\csc x + \cot x| = \ln|\csc x - \cot x|$$

Solution

$$-\ln|\csc x + \cot x| = \ln|\csc x + \cot x|^{-1} \qquad k \ln a = \ln a^k$$

$$= \ln \frac{1}{|\csc x + \cot x|}$$

$$= \ln \left| \frac{1}{\csc x + \cot x} \right| \qquad \frac{1}{|a|} = \frac{|1|}{|a|} = \left| \frac{1}{a} \right|$$

$$= \ln \left| \frac{1}{\csc x + \cot x} \cdot \frac{\csc x - \cot x}{\csc x - \cot x} \right|$$

$$= \ln \left| \frac{\csc x - \cot x}{\csc^2 x - \cot^2 x} \right|$$

$$= \ln \left| \frac{\csc x - \cot x}{1} \right|$$

$$= \ln|\csc x - \cot x| \qquad \blacksquare$$

*To disprove a statement that claims to be true in general, all that is needed is to find one exception; one case for which the statement is false. Such an exception is also called a **counterexample**.*

Notice that in all of the preceding solutions we avoided performing the same operation on each side of the given equality at the same time. Doing so could, at times, lead to false results. For example, $\sin \theta = -|\sin \theta|$ is certainly not an identity. Squaring both sides gives $\sin^2 \theta = \sin^2 \theta$, which might *incorrectly* suggest that the original statement is an identity.

Sometimes a trigonometric equation may have the appearance of an identity but really is not one. A quick way to discover this is to find one value of θ for which the expressions in the equation are defined but for which the equality is false.

A trigonometric equation that is not an identity is false for almost all values of the variable. Therefore, to find such a value, a good strategy is to first try special values such as

$$0, \pm\frac{\pi}{6}, \pm\frac{\pi}{4}, \pm\frac{\pi}{3}, \pm\frac{\pi}{2}.$$

EXAMPLE 6 Prove that $\sin^2 \theta - \cos^2 \theta = 1$ is *not* an identity.

Solution One counterexample is sufficient. Try $\theta = \dfrac{\pi}{4}$.

$$\sin^2 \frac{\pi}{4} - \cos^2 \frac{\pi}{4} = \left(\frac{1}{\sqrt{2}}\right)^2 - \left(\frac{1}{\sqrt{2}}\right)^2 = 0 \neq 1$$

Since the expressions in the equation are defined for $\dfrac{\pi}{4}$, but the equation is false for $\theta = \dfrac{\pi}{4}$, the equation cannot be an identity.

EXERCISES 7.3

Complete by using the fundamental identities.

1. $\dfrac{1}{\csc \theta} =$ **2.** $1 - \sin^2 \theta =$ **3.** $\csc^2 \theta - \cot^2 \theta =$

4. $\tan \theta \cos \theta =$ **5.** $\sin \theta \cot \theta =$ **6.** $-\dfrac{1}{\cos \theta} =$

Without using the trigonometric tables or a calculator, find the value of each of the following.

7. $\sin^2 39° + \cos^2 39°$ **8.** $\sin(-7°) \csc(-7°)$

9. $\tan^2(3.2°) - \sec^2(3.2°)$ **10.** $3 \cos^2 \dfrac{4\pi}{7} \sec^2 \dfrac{4\pi}{7}$

Show that each expression is equal to 1.

11. $(\sin \theta)(\cot \theta)(\sec \theta)$ **12.** $(\cos \theta)(\tan \theta)(\csc \theta)$ **13.** $\cos^2 \theta(\tan^2 \theta + 1)$ **14.** $\tan^2 \theta(\csc^2 \theta - 1)$

Find the values of θ for which the expressions are defined.

15. $\dfrac{1}{\cos \theta}$ **16.** $\tan \theta$ **17.** $\dfrac{1}{\tan \theta}$ **18.** $\cot \theta$ **19.** $\dfrac{\cos \theta}{\cot \theta}$ **20.** $\sec^2 \theta \csc \theta$

Use fundamental identities to convert the expressions into a form involving only sines or cosines of θ, and simplify. Also, find the restrictions on θ, if any.

21. $\dfrac{\tan \theta}{\cot \theta}$ **22.** $\cot \theta \sec^2 \theta$ **23.** $\dfrac{1 - \csc \theta}{\cot \theta}$

24. $\dfrac{1 - \cot^2 \theta}{\cot^2 \theta}$ **25.** $\dfrac{\sec \theta + \csc \theta}{\cos \theta + \sin \theta}$ **26.** $\dfrac{\sec \theta}{\tan \theta + \cot \theta}$

Express in terms of $\sin \theta$ only.

27. $\cos^2 \theta - \sin^2 \theta$ **28.** $\tan^2 \theta \csc^2 \theta$

Express in terms of $\cos \theta$ *only.*

29. $\dfrac{\cot \theta - \sin \theta}{\csc \theta}$ **30.** $\sec \theta - \sin \theta \tan \theta$

Verify each identity using Method I. Find the restrictions on θ, *if any.*

31. $\dfrac{\cos \theta}{\cot \theta} = \sin \theta$ **32.** $\cos^2 \theta - \sin^2 \theta = 1 - 2 \sin^2 \theta$

33. $(\tan \theta - 1)^2 = \sec^2 \theta - 2 \tan \theta$ **34.** $\dfrac{1}{\sec^2 \theta} = 1 - \dfrac{1}{\csc^2 \theta}$

Verify each identity using Method II. Find the restrictions on θ, *if any.*

35. $\sec \theta - \cos \theta = \sin \theta \tan \theta$ **36.** $(1 - \sin \theta)(1 + \sin \theta) = \dfrac{1}{1 + \tan^2 \theta}$

37. $\dfrac{\cot \theta - 1}{1 - \tan \theta} = \dfrac{\csc \theta}{\sec \theta}$ **38.** $\dfrac{1 + \sec \theta}{\csc \theta} = \sin \theta + \tan \theta$

Verify each identity using either method.

39. $\tan \theta + \cot \theta = \dfrac{1}{(\sin \theta)(\cos \theta)}$ **40.** $\cot^2 \theta - \cos^4 \theta \csc^2 \theta = \cos^2 \theta$

41. $(\sec \theta + \tan \theta)(1 - \sin \theta) = \cos \theta$ **42.** $\sec^4 \theta - \tan^4 \theta = 1 + 2 \tan^2 \theta$

43. $(\csc^2 \theta - 1) \sin^2 \theta = \cos^2 \theta$ **44.** $\sin^4 \theta - \cos^4 \theta + \cos^2 \theta = \sin^2 \theta$

45. $\tan^2 \theta - \sin^2 \theta = \sin^2 \theta \tan^2 \theta$ **46.** $\dfrac{\cot^2 \theta + 1}{\tan^2 \theta + 1} = \cot^2 \theta$

47. $\dfrac{1 + \sec \theta}{\csc \theta} = \sin \theta + \tan \theta$ **48.** $\dfrac{\sec^2 \theta}{\sec^2 \theta - 1} = \csc^2 \theta$

49. $\dfrac{1}{1 + \cos \theta} + \dfrac{1}{1 - \cos \theta} = 2 \csc^2 \theta$ **50.** $\dfrac{\tan \theta \sin \theta}{\sec^2 \theta - 1} = \cos \theta$

51. $\dfrac{1 + \tan^2 \theta}{1 + \cot^2 \theta} = \sec^2 \theta - 1$ **52.** $\dfrac{\sin \theta + \tan \theta}{1 + \sec \theta} = \sin \theta$

53. $\dfrac{\sin \theta + \cos \theta}{\sin \theta - \cos \theta} = \dfrac{\sec \theta + \csc \theta}{\sec \theta - \csc \theta}$ **54.** $\tan \theta \sec^2 \theta = \dfrac{\sec \theta}{\csc \theta - \sin \theta}$

55. $\dfrac{1 - \cos \theta}{1 + \cos \theta} = (\csc \theta - \cot \theta)^2$ **56.** $\dfrac{1 + \tan \theta}{1 - \tan \theta} = \dfrac{\cot \theta + 1}{\cot \theta - 1}$

57. $(\sin^2 \theta + \cos^2 \theta)^5 = 1$ **58.** $\dfrac{2 + \cot^2 \theta}{\csc^2 \theta} - 1 = \sin^2 \theta$

59. $\dfrac{1}{\cos^2 \theta} + \dfrac{1}{\sin^2 \theta} = \dfrac{1}{\sin^2 \theta - \sin^4 \theta}$ **60.** $\dfrac{1 + \sin \theta}{\cos \theta} = \dfrac{\cos \theta}{1 - \sin \theta}$

61. $\dfrac{\tan^2 \theta + 1}{\tan^2 \theta} = \csc^2 \theta$ **62.** $\dfrac{\cot \theta}{\csc \theta + 1} = \sec \theta - \tan \theta$

63. $\dfrac{\tan \theta}{\sec \theta - 1} = \dfrac{\sec \theta + 1}{\tan \theta}$ **64.** $\dfrac{\cos \theta}{\csc \theta - 2 \sin \theta} = \dfrac{\tan \theta}{1 - \tan^2 \theta}$

65. $\dfrac{\cos^2 \theta + 2 \sin^2 \theta}{\cos^3 \theta} = \sec^3 \theta + \dfrac{\sin^2 \theta}{\cos^3 \theta}$ **66.** $\dfrac{\cos \theta}{\csc \theta + 2 \sin \theta} = \dfrac{\tan \theta}{1 + 3 \tan^2 \theta}$

67. $\dfrac{1}{(\csc\theta - \sec\theta)^2} = \dfrac{\sin^2\theta}{\sec^2\theta - 2\tan\theta}$

68. $\csc\theta - \cot\theta = \dfrac{1}{\csc\theta + \cot\theta}$

69. $\dfrac{\sec\theta}{\tan\theta - \sin\theta} = \dfrac{1 + \cos\theta}{\sin^3\theta}$

70. $\dfrac{1 - \cos\theta}{\sin\theta} + \dfrac{\sin\theta}{1 - \cos\theta} = 2\csc\theta$

71. $-\ln|\cos x| = \ln|\sec x|$

72. $\ln|\sin x| = -\ln|\csc x|$

73. $-\ln|\sec x - \tan x| = \ln|\sec x + \tan x|$

74. $\ln\left|\dfrac{\sin x}{1 - \cos x}\right| = -\ln\left|\dfrac{\sin x}{1 + \cos x}\right|$

Show that each equation is not an identity by finding one value for which the expressions in the equation are defined but for which the equation is false.

75. $\tan(-x) = \tan x$

76. $\tan\left(x + \dfrac{\pi}{2}\right) = \cot x$

77. $\sin x = -\sqrt{1 - \cos^2 x}$

78. $\tan x = \sqrt{\sec^2 x - 1}$

79. $\dfrac{\cot\theta - 1}{1 - \tan\theta} = \dfrac{\sec\theta}{\csc\theta}$

80. $\dfrac{1 + \sin\theta}{\cos\theta} = \dfrac{\sin\theta}{1 + \cos\theta}$

81. **(a)** Let θ be an acute angle of a right triangle where $\tan\theta = \dfrac{x}{3}$. Show that

$$\ln(\sec\theta + \tan\theta) = \ln\left(\dfrac{\sqrt{x^2 + 9} + x}{3}\right).$$

(b) Let $F(x) = \ln\left(\dfrac{\sqrt{x^2 + 9} + x}{3}\right)$ and evaluate $F(4) - F(0)$.

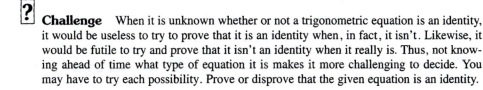

Challenge When it is unknown whether or not a trigonometric equation is an identity, it would be useless to try to prove that it is an identity when, in fact, it isn't. Likewise, it would be futile to try and prove that it isn't an identity when it really is. Thus, not knowing ahead of time what type of equation it is makes it more challenging to decide. You may have to try each possibility. Prove or disprove that the given equation is an identity.

1. $\dfrac{\sin x}{\csc x} - \dfrac{\cos x}{\sec x} = 1$

2. $\dfrac{1 + \csc x}{\sec x} = \cos x - \tan x$

3. $\dfrac{\sec^2 x - 2}{(1 + \tan x)^2} = \dfrac{1 - \cot x}{1 + \cot x}$

 Written Assignment Explain why the value of $\sin(x + y)$ is not necessarily equal to $\sin x + \sin y$.

 Graphing Calculator Exercises

1. Suppose that you are tutoring a student whose answers to certain problems differ from those given at the back of the book. By graphing both forms, check which of the following of the student's answers appear to be identical to the text's answer. Verify the apparent identities by either Method I or Method II.

Student's Answer	Text's Answer
(a) $\sin 2x$	$2 \sin x$
(b) $\dfrac{\sin x}{1 - \cos x}$	$\dfrac{1 + \cos x}{\sin x}$
(c) $\sin(x + 1)$	$\sin x + \sin 1$
(d) $\dfrac{\tan x \csc^2 x}{\tan^2 x + 1}$	$\cot x$

For Exercises 2 and 3, use the graph of $f(x)$ to find a simpler expression, $g(x)$. Then prove, using Method I or Method II, that the statement $f(x) = g(x)$ is an identity where both are defined.

2. $f(x) = \sin^4 x + 2\sin^2 x \cos^2 x + \cos^4 x$

3. $f(x) = \dfrac{\cos x - \cos^3 x}{\sin^4 x + \sin^2 x \cos^2 x}$

？ Critical Thinking

1. The definitions of the trigonometric ratios for acute angles on page 427 can be extended to apply to angles of other sizes by making use of the three sides of the reference triangles. Compare this to the definitions based on the unit circle in Section 6.2. Are these two procedures equivalent for angles of all sizes? Discuss.

2. Let $\triangle ABC$ be a right triangle with right angle of C and base $AC = 1$. Assume that point B is moving vertically upward from C, and let θ represent the increasing angle measures at point A. By comparing appropriate pairs of sides of the resulting triangles ABC, discuss the behavior of $\tan \theta$, $\sin \theta$, and $\cos \theta$ as θ increases toward $90°$.

3. The solution to Exercise 35, page 440, can be generalized after making some substitutions. Replace $23°$ by α, $43°$ by β, $\frac{1}{4}$ mile by d, and let $BC = x$. Now let h be the height of the mountain and show that $h = \dfrac{d \tan \alpha \tan \beta}{\tan \beta - \tan \alpha}$.

4. Some identities have restrictions on the variable θ that involve not only the quadrantal angles. If the expression $\dfrac{\sin \theta - 1}{1 - 3 \tan^2 \theta}$ were part of an identity, what restrictions on θ would the expression call for?

5. Squaring the equation $\sin x = \sqrt{1 - \cos^2 x}$ produces the identity $\sin^2 x = 1 - \cos^2 x$. Does it now follow that the given equation is also an identity? Explain.

**7.4
THE ADDITION
AND
SUBTRACTION
FORMULAS**

Is it true that $\cos(30° + 60°) = \cos 30° + \cos 60°$? Some quick calculations show that it is *not*.

$$\cos(30° + 60°) = \cos 90° = 0 \qquad \text{but} \qquad \cos 30° + \cos 60° = \frac{\sqrt{3}}{2} + \frac{1}{2}$$

To write $\cos(\theta_1 + \theta_2) = \cos \theta_1 + \cos \theta_2$ would be to assume, *incorrectly,* that the

$\cos(30° + 60°)$
$= \cos 30° \cos 60°$
$\qquad - \sin 30° \sin 60°$
$= \dfrac{\sqrt{3}}{2} \cdot \dfrac{1}{2} - \dfrac{1}{2} \cdot \dfrac{\sqrt{3}}{2}$
$= 0$
$= \cos 90°$

cosine function obeys the distributive property. We emphasize that cos is the *name* of a function; it is not a number.

The cosine of the sum of two angles is correctly evaluated (see the computation at the left) by using Formula (3) which is listed below; it is one of several important trigonometric identities in *two* variables that give the trigonometric values of sums and differences of angle measures.

ADDITION AND SUBTRACTION FORMULAS	
(1)	$\sin(\alpha + \beta) = \sin \alpha \cos \beta + \cos \alpha \sin \beta$
(2)	$\sin(\alpha - \beta) = \sin \alpha \cos \beta - \cos \alpha \sin \beta$
(3)	$\cos(\alpha + \beta) = \cos \alpha \cos \beta - \sin \alpha \sin \beta$
(4)	$\cos(\alpha - \beta) = \cos \alpha \cos \beta + \sin \alpha \sin \beta$
(5)	$\tan(\alpha + \beta) = \dfrac{\tan \alpha + \tan \beta}{1 - \tan \alpha \tan \beta}$
(6)	$\tan(\alpha - \beta) = \dfrac{\tan \alpha - \tan \beta}{1 + \tan \alpha \tan \beta}$

These formulas are stated in terms of the variables α and β, which can represent angles measured in either degrees or radians (real numbers). We now prove Formula (4) by making use of the unit circle.

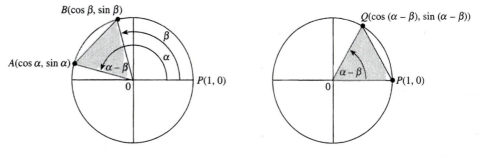

The distance formula is given on page 248.

In the preceding figure the unit circle at the left contains a typical situation in which $\alpha > \beta$. Since point A is on the unit circle, we find that the coordinates of A are $(\cos \alpha, \sin \alpha)$. Similarly, B has coordinates $(\cos \beta, \sin \beta)$. Then, by the distance formula applied to points A and B,

$$(AB)^2 = (\cos \alpha - \cos \beta)^2 + (\sin \alpha - \sin \beta)^2$$

$$= (\cos^2 \alpha - 2 \cos \alpha \cos \beta + \cos^2 \beta) + (\sin^2 \alpha - 2 \sin \alpha \sin \beta + \sin^2 \beta)$$

$$= (\cos^2 \alpha + \sin^2 \alpha) + (\cos^2 \beta + \sin^2 \beta) - 2(\cos \alpha \cos \beta + \sin \alpha \sin \beta)$$

$$= 2 - 2(\cos \alpha \cos \beta + \sin \alpha \sin \beta)$$

In the unit circle at the right in the preceding figure, the central angle $\alpha - \beta$ is in standard position. Therefore, the coordinates of Q are $(\cos(\alpha - \beta), \sin(\alpha - \beta))$. Using the distance formula for points P and Q, we have the following:

Recall that for any θ, $\cos^2\theta + \sin^2\theta = 1$. This is used here for θ = α − β: $\cos^2(\alpha - \beta) + \sin^2(\alpha - \beta) = 1$.

$$(PQ)^2 = [\cos(\alpha - \beta) - 1]^2 + [\sin(\alpha - \beta) - 0]^2$$
$$= \cos^2(\alpha - \beta) - 2\cos(\alpha - \beta) + 1 + \sin^2(\alpha - \beta)$$
$$= 2 - 2\cos(\alpha - \beta)$$

Triangles AOB and QOP are congruent (SAS). Then $AB = PQ$ and this implies $(AB)^2 = (PQ)^2$. We may therefore equate the preceding results and then simplify.

$$2 - 2\cos(\alpha - \beta) = 2 - 2(\cos\alpha\cos\beta + \sin\alpha\sin\beta)$$

(4) $$\cos(\alpha - \beta) = \cos\alpha\cos\beta + \sin\alpha\sin\beta$$

The proofs of Formulas (1) and (2) will be called for in the exercises.

Formula (3) is now easy to prove by using Formula (4). The trick is to write $\alpha + \beta = \alpha - (-\beta)$.

$$\cos(\alpha + \beta) = \cos(\alpha - (-\beta))$$
$$= \cos\alpha\cos(-\beta) + \sin\alpha\sin(-\beta) \quad \text{(by (4))}$$
$$= \cos\alpha\cos\beta + (\sin\alpha)(-\sin\beta) \quad \begin{array}{l}(\cos(-\theta) = \cos\theta, \\ \sin(-\theta) = -\sin\theta)\end{array}$$

(3) $$= \cos\alpha\cos\beta - \sin\alpha\sin\beta$$

EXAMPLE 1 Evaluate $\sin\dfrac{7\pi}{12}$ and $\tan\dfrac{7\pi}{12}$ by using $\dfrac{7\pi}{12} = \dfrac{\pi}{4} + \dfrac{\pi}{3}$. State the answers in radical form, and check by calculator.

Solution Use Formulas (1) and (5) with $\alpha = \dfrac{\pi}{4}$, $\beta = \dfrac{\pi}{3}$.

A graphing calculator can be used to check this result as follows. Check (set in radian mode):

$\sin\dfrac{7\pi}{12} =$

| SIN | (| 7 | π | ÷ | 12 |) | ENTER |

= 0.9659

$\dfrac{1}{4}(\sqrt{2} + \sqrt{6}) =$

| 0.25 | (| √ | 2 | + | √ | 6 |) | ENTER |

= 0.9659

$$\sin\frac{7}{12}\pi = \sin\left(\frac{\pi}{4} + \frac{\pi}{3}\right)$$
$$= \sin\frac{\pi}{4}\cos\frac{\pi}{3} + \cos\frac{\pi}{4}\sin\frac{\pi}{3}$$
$$= \frac{\sqrt{2}}{2}\cdot\frac{1}{2} + \frac{\sqrt{2}}{2}\cdot\frac{\sqrt{3}}{2}$$
$$= \frac{1}{4}(\sqrt{2} + \sqrt{6})$$

$$\tan\frac{7}{12}\pi = \tan\left(\frac{\pi}{4} + \frac{\pi}{3}\right)$$
$$= \frac{\tan\dfrac{\pi}{4} + \tan\dfrac{\pi}{3}}{1 - \tan\dfrac{\pi}{4}\tan\dfrac{\pi}{3}}$$
$$= \frac{1 + \sqrt{3}}{1 - \sqrt{3}}$$

The following sequences can be used on many scientific calculators to check this result:

$$\tan \frac{7\pi}{12} = \boxed{7}\ \boxed{\times}\ \boxed{\pi}\ \boxed{\div}\ \boxed{12}\ \boxed{=}\ \boxed{\text{TAN}} = -3.7321$$

$$\frac{1 + \sqrt{3}}{1 - \sqrt{3}} = \boxed{1}\ \boxed{+}\ \boxed{3}\ \boxed{\sqrt{}}\ \boxed{=}\ \boxed{\div}\ \boxed{(}\ \boxed{1}\ \boxed{-}\ \boxed{3}\ \boxed{\sqrt{}}\ \boxed{)}\ \boxed{=} -3.7321 \quad \blacksquare$$

EXAMPLE 2   Let $\cos \alpha = -\frac{4}{5}$ and $\cos \beta = -\frac{12}{13}$ where α and β have terminal sides in quadrants II and III, respectively. Find the exact values of $\cos(\alpha - \beta)$ and $\sin(\alpha - \beta)$.

Solution Place α and β in standard position and draw reference triangles. Since $\cos \alpha = \dfrac{-4}{5} = \dfrac{\text{adjacent}}{\text{hypotenuse}}$, the triangle for α has hypotenuse 5 and adjacent side -4. The third side is 3 by the Pythagorean theorem. Similarly, the triangle for β has sides 13, -12, -5.

Although a unit circle is not used here, triangles such as these may be used since the ratios do not depend on the size of the triangle.

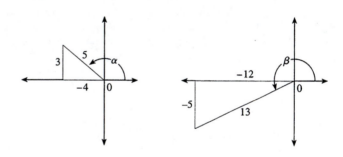

$$\cos(\alpha - \beta) = \cos \alpha \cos \beta + \sin \alpha \sin \beta$$

$$= \left(-\frac{4}{5}\right)\left(-\frac{12}{13}\right) + \left(\frac{3}{5}\right)\left(-\frac{5}{13}\right)$$

$$= \frac{48}{65} - \frac{15}{65}$$

$$= \frac{33}{65}$$

$$\sin(\alpha - \beta) = \sin \alpha \cos \beta - \cos \alpha \sin \beta$$

$$= \left(\frac{3}{5}\right)\left(-\frac{12}{13}\right) - \left(-\frac{4}{5}\right)\left(-\frac{5}{13}\right)$$

$$= -\frac{36}{65} - \frac{20}{65}$$

Can you determine the quadrant containing the terminal side of $\alpha - \beta$ using these two results?

$$= -\frac{56}{65} \quad \blacksquare$$

(Answers: Page 488)

TEST YOUR UNDERSTANDING

Complete each equation to get a specific case of an addition or subtraction formula.

1. $\sin 5° \cos 12° + \cos 5° \sin 12° = ?$

2. $\cos \dfrac{3\pi}{10} \cos \dfrac{\pi}{5} + \sin \dfrac{3\pi}{10} \sin \dfrac{\pi}{5} = ?$

3. $\dfrac{\tan(-15°) - \tan(-20°)}{1 + \tan(-15°)\tan(-20°)} = ?$

4. $\sin(\theta + 5°)\cos(\theta - 5°) - \cos(\theta + 5°)\sin(\theta - 5°) = ?$

State the answers in radical form for Exercises 5–10.

5. Use $15° = 45° - 30°$ and appropriate subtraction formulas to evaluate $\sin 15°$, $\cos 15°$, and $\tan 15°$.

6. Repeat Exercise 5 with $15° = 60° - 45°$.

7. Use $\dfrac{11\pi}{12} = \dfrac{\pi}{6} + \dfrac{3\pi}{4}$ and appropriate addition formulas to evaluate $\sin \dfrac{11\pi}{12}$, $\cos \dfrac{11\pi}{12}$, and $\tan \dfrac{11\pi}{12}$.

8. Repeat Exercise 7 with $\dfrac{11\pi}{12} = \dfrac{7\pi}{6} - \dfrac{\pi}{4}$ using subtraction formulas.

9. Use an addition or subtraction formula to evaluate $\sin 195°$.

10. Use an addition or subtraction formula to evaluate $\cos \dfrac{5\pi}{12}$.

11. $\cos \alpha = \dfrac{3}{5}$ and the terminal side of α is in quadrant IV. $\sin \beta = \dfrac{15}{17}$ and the terminal side of β is in quadrant I. Find $\cos(\alpha - \beta)$.

12. For α and β as in Exercise 11, find $\sin(\alpha - \beta)$.

Formulas (1) and (3) can be used to verify the addition formula for the tangent function.

$$\tan(\alpha + \beta) = \frac{\sin(\alpha + \beta)}{\cos(\alpha + \beta)} = \frac{\sin \alpha \cos \beta + \cos \alpha \sin \beta}{\cos \alpha \cos \beta - \sin \alpha \sin \beta}$$

Divide the numerator and denominator by $\cos \alpha \cos \beta$.

$$\tan(\alpha + \beta) = \frac{\dfrac{\sin \alpha \cos \beta}{\cos \alpha \cos \beta} + \dfrac{\cos \alpha \sin \beta}{\cos \alpha \cos \beta}}{\dfrac{\cos \alpha \cos \beta}{\cos \alpha \cos \beta} - \dfrac{\sin \alpha \sin \beta}{\cos \alpha \cos \beta}}$$

Thus

(5) $$\tan(\alpha + \beta) = \frac{\tan \alpha + \tan \beta}{1 - \tan \alpha \tan \beta}$$

A similar analysis will verify Formula (6).

EXAMPLE 3 Find the exact value of tan(arctan 3 − arctan 2), and check by calculator.

Solution Let $\alpha = $ arctan 3 and $\beta = $ arctan 2. Then apply the following formula for $\tan(\alpha - \beta)$.

$$\tan(\text{arctan } 3 - \text{arctan } 2) = \tan(\alpha - \beta)$$

$$= \frac{\tan \alpha - \tan \beta}{1 + \tan \alpha \tan \beta}$$

Note that $\frac{1}{7}$ is the exact answer and 0.142857 in the check is a close approximation.

$$= \frac{3 - 2}{1 + 3 \cdot 2} \qquad \begin{aligned} &\tan \alpha = \tan(\text{arctan } 3) = 3; \\ &\tan \beta = \tan(\text{arctan } 2) = 2 \end{aligned}$$

$$= \frac{1}{7}$$

Check: (Use radian mode.)

$$\tan(\text{arctan } 3 - \text{arctan } 2)$$

$$= \boxed{\text{TAN}}\ \boxed{(}\ \boxed{\text{TAN}^{-1}}\ \boxed{3}\ \boxed{-}\ \boxed{\text{TAN}^{-1}}\ \boxed{2}\ \boxed{)}\ \boxed{\text{ENTER}} = 0.142857$$

Also, $\dfrac{1}{7} = 0.142857$. ∎

With the aid of the addition and subtraction formulas, we will be able to derive other useful formulas, some of which have already been encountered. For example, the result $\cos \theta = \sin\left(\theta + \dfrac{\pi}{2}\right)$ was derived in Section 6.3 by using the unit circle. This formula can be described as a *reduction formula* in the sense that a trigonometric function of $\theta + \dfrac{\pi}{2}$ is "reduced" to a trigonometric function of just θ. Such reduction formulas can be derived using the addition and subtraction formulas as demonstrated in the next example.

EXAMPLE 4 Use the addition and subtraction formulas to obtain reduction formulas for **(a)** $\sin\left(\dfrac{\pi}{2} - \theta\right)$, **(b)** $\cos\left(\dfrac{\pi}{2} - \theta\right)$, and **(c)** $\cos(\theta + \pi)$.

Solution

*Since $\sin(-x) = -\sin x$ the result in **(a)** can be written as*

$$\sin\left(\theta - \frac{\pi}{2}\right) = -\cos \theta.$$

Also note that all three of these formulas hold for all values of θ.

(a) $\sin\left(\dfrac{\pi}{2} - \theta\right) = \sin\dfrac{\pi}{2}\cos\theta - \cos\dfrac{\pi}{2}\sin\theta$ \qquad Formula (2)

$$= 1(\cos\theta) - 0(\sin\theta) = \cos\theta$$

Therefore,

$$\sin\left(\frac{\pi}{2} - \theta\right) = \cos\theta$$

(b) $\cos\left(\dfrac{\pi}{2} - \theta\right) = \cos\dfrac{\pi}{2}\cos\theta + \sin\dfrac{\pi}{2}\sin\theta = \sin\theta$ Formula (4)

Therefore,

$$\cos\left(\dfrac{\pi}{2} - \theta\right) = \sin\theta$$

(c) $\cos(\theta + \pi) = \cos\theta\cos\pi - \sin\theta\sin\pi = -\cos\theta$ Formula (3)

Therefore,

$$\cos(\theta + \pi) = -\cos\theta$$ ∎

The reduction formula for $\tan\left(\dfrac{\pi}{2} - \theta\right)$ cannot be done directly using Formula (6). (Why not?) However, using the results in Example 3, we get

$$\tan\left(\dfrac{\pi}{2} - \theta\right) = \dfrac{\sin\left(\dfrac{\pi}{2} - \theta\right)}{\cos\left(\dfrac{\pi}{2} - \theta\right)} = \dfrac{\cos\theta}{\sin\theta} = \cot\theta$$

Therefore,

This formula holds for all θ that are not coterminal with 0 or π.

$$\tan\left(\dfrac{\pi}{2} - \theta\right) = \cot\theta$$

There are many reduction formulas possible. For the sine function alone, $\sin\left(\theta + \dfrac{\pi}{2}k\right)$ produces an infinite number of such formulas as k takes on all nonzero integer values. (See Exercise 49.) Of particular interest are the three preceding results involving the sine, cosine, and tangent of $\dfrac{\pi}{2} - \theta$, when θ is an acute angle.

In this case θ and $90° - \theta$ are complementary and the equation $\sin\theta = \cos(90° - \theta)$ shows that the sine of θ equals the cosine of its complement. The preceding equation is an example of a **cofunction identity,** and the sine and cosine are said to be **cofunctions** of one another. Likewise, the tangent and cotangent are cofunctions, as are the secant and cosecant. The following list of cofunction identities shows that *a trigonometric function of an acute angle θ is the same as the cofunction of its complement.*

If we let φ = 90° − θ, then θ = 90° − φ and the first identity can be rewritten in this equivalent form:
$\cos\phi = \sin(90° - \phi)$
Similarly
$\cot\phi = \tan(90° - \phi)$
$\csc\phi = \sec(90° - \phi)$

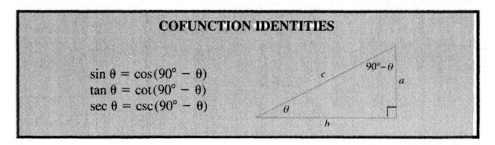

COFUNCTION IDENTITIES

$\sin\theta = \cos(90° - \theta)$
$\tan\theta = \cot(90° - \theta)$
$\sec\theta = \csc(90° - \theta)$

The addition and subtraction formulas can also be used to verify identities other than reduction formulas, as shown next.

EXAMPLE 5 Verify the identity $\sin\left(\theta + \dfrac{\pi}{3}\right) = \dfrac{1}{2}\left(\sin\theta + \sqrt{3}\cos\theta\right)$.

Solution Use Formula (1).

$$\sin\left(\theta + \frac{\pi}{3}\right) = \sin\theta\cos\frac{\pi}{3} + \cos\theta\sin\frac{\pi}{3}$$

$$= \frac{1}{2}\sin\theta + \frac{\sqrt{3}}{2}\cos\theta$$

$$= \frac{1}{2}\left(\sin\theta + \sqrt{3}\cos\theta\right) \qquad\blacksquare$$

Example 3, page 391, shows the graph of $y = \sin x + \cos x$ that was determined by adding the ordinates of the graphs of $y = \sin x$ and $y = \cos x$. This graph can also be found by using this result:

$$y = \sin x + \cos x = \sqrt{2}\sin\left(x + \frac{\pi}{4}\right)$$

Verify this result using Formula (1).

Use a graphing calculator to verify that the graph of $y = \sqrt{2}\sin\left(x + \dfrac{\pi}{4}\right)$ is the same as the curve obtained in Example 3, page 391. An algebraic basis for the equality $\sin x + \cos x = \sqrt{2}\sin\left(x + \dfrac{\pi}{4}\right)$ is given in the following general result, a proof of which is outlined in Exercise 50.

> For nonzero constants a and b, an expression of the form $a\sin x + b\cos x$ is equal to $c\sin(x + \theta)$ where $c = \sqrt{a^2 + b^2}$ and θ is an angle such that $\sin\theta = \dfrac{b}{c}$ and $\cos\theta = \dfrac{a}{c}$.

EXAMPLE 6 Convert $f(x) = \sqrt{3}\sin x - \cos x$ into the form $f(x) = c\sin(x + \theta)$.

Solution Use $a = \sqrt{3}$ and $b = -1$ so that $c = \sqrt{3 + 1} = 2$. Also $\sin\theta = -\dfrac{1}{2}$, $\cos\theta = \dfrac{\sqrt{3}}{2}$ so that we may use $\theta = -\dfrac{\pi}{6}$. Then $f(x) = 2\sin\left(x - \dfrac{\pi}{6}\right)$.

The angle $\dfrac{11\pi}{6}$ could also be used as well as any other coterminal angle. However, it is simpler to use θ where $0 < \theta < 2\pi$ or where $-2\pi < \theta < 0$.

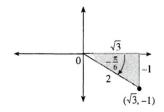

EXERCISES 7.4

Use an addition or subtraction formula to evaluate the expression. Do not use a calculator.

1. $\cos 22° \cos 38° - \sin 22° \sin 38°$

2. $\sin 52° \cos 7° - \cos 52° \sin 7°$

3. $\dfrac{\tan 25° - \tan 55°}{1 + \tan 25° \tan 55°}$

4. $\cos \dfrac{5\pi}{12} \cos \dfrac{7\pi}{12} + \sin \dfrac{5\pi}{12} \sin \dfrac{7\pi}{12}$

Use the addition or subtraction formulas to evaluate sin θ, cos θ, and tan θ for the specified value of θ, and check by calculator.

5. $\theta = 75°$

6. $\theta = 105°$

7. $\theta = \dfrac{\pi}{12}$

8. $\theta = \dfrac{19\pi}{12}$

9. $\theta = 165°$

10. $\theta = 345°$

11. $\theta = \dfrac{17\pi}{12}$

12. $\theta = \dfrac{7\pi}{3}$

Prove the reduction formula by using an appropriate addition or subtraction formula.

13. $\sin(\pi - \theta) = \sin \theta$

14. $\cos\left(\dfrac{\pi}{2} + \theta\right) = -\sin \theta$

15. $\tan(\pi - \theta) = -\tan \theta$

16. $\cos(\theta - 2\pi) = \cos \theta$

17. $\sin\left(\theta + \dfrac{7\pi}{2}\right) = -\cos \theta$

18. $\cos\left(\theta - \dfrac{5\pi}{2}\right) = \sin \theta$

19. If α and β are acute angles having $\sin \alpha = \frac{3}{5}$ and $\cos \beta = \frac{12}{13}$, find the exact values of $\sin(\alpha - \beta)$ and $\tan(\alpha + \beta)$.

20. Suppose that $\sin \alpha$ and $\cos \beta$ are the same as in Exercise 19, but α terminates in quadrant II and β terminates in quadrant IV. Find the exact values of $\sin(\alpha - \beta)$ and $\tan(\alpha + \beta)$.

21. Find the exact values of $\sin(\alpha + \beta)$ and $\cos(\alpha + \beta)$ for α terminating in quadrant III and β terminating in quadrant IV, where $\sin \alpha = -\frac{1}{3}$ and $\cos \beta = \frac{2}{5}$.

22. Let $\cos \alpha = \frac{24}{25}$ with α terminating in quadrant I, and $\tan \beta = -\frac{15}{8}$ with β terminating in quadrant II. Find the exact values of $\cos(\alpha - \beta)$ and $\tan(\alpha - \beta)$.

23. Let $\sin \alpha = \frac{21}{29}$ with α terminating in quadrant II, and $\tan \beta = -\frac{5}{12}$ with β terminating in quadrant IV. Find the exact values of $\sin(\alpha - \beta)$ and $\tan(\alpha + \beta)$.

24. Both α and β terminate in quadrant III so that $\sin \alpha = -\frac{1}{5}$ and $\cos \beta = -\frac{3}{4}$. Find the exact values of $\sin(\alpha + \beta)$, $\cos(\alpha + \beta)$, and $\tan(\alpha + \beta)$.

25. Evaluate the ratio by using a reduction formula that calls for a ratio of an acute angle. For example, $\cos 155° = \cos(180° - 25°) = -\cos 25° = -0.9063$, or, $\cos 155° = \cos(90° + 65°) = -\sin 65° = -0.9063$. Also verify the result by entering the angle directly into the calculator.

(a) $\cos 191°$ (b) $\sin 132°$ (c) $\tan 173°$ (d) $\cos 102°$

Use addition or subtraction formulas to verify the identity.

26. $\tan\left(\dfrac{\pi}{4} + \theta\right) = \dfrac{1 + \tan \theta}{1 - \tan \theta}$

27. $\cos\left(\theta - \dfrac{\pi}{4}\right) = \dfrac{\sqrt{2}}{2}(\cos \theta + \sin \theta)$

28. $\sin\left(\dfrac{\pi}{6} - \theta\right) = \dfrac{1}{2}(\cos \theta - \sqrt{3} \sin \theta)$

29. $\cos(\theta + 30°) + \cos(\theta - 30°) = \sqrt{3} \cos \theta$

30. $\cos\left(\theta + \dfrac{\pi}{6}\right) + \sin\left(\theta - \dfrac{\pi}{3}\right) = 0$

31. $\dfrac{\sin\left(\theta + \dfrac{\pi}{2}\right)}{\cos\left(\theta + \dfrac{\pi}{2}\right)} = -\cot\theta$

32. $\sec\left(\dfrac{\pi}{2} - \theta\right) = \csc\theta$

33. $\csc(\pi - \theta) = \csc\theta$

34. $\cos(\alpha + \beta) + \cos(\alpha - \beta) = 2\cos\alpha\cos\beta$

35. $\sin(\alpha + \beta) - \sin(\alpha - \beta) = 2\cos\alpha\sin\beta$

36. $\dfrac{\sin(\alpha - \beta)}{\cos(\alpha + \beta)} = \dfrac{\cot\beta - \cot\alpha}{\cot\alpha\cot\beta - 1}$

37. $\dfrac{\sin(\alpha + \beta)}{\sin(\alpha - \beta)} = \dfrac{\tan\alpha + \tan\beta}{\tan\alpha - \tan\beta}$

38. $\cos(\alpha + \beta)\cos(\alpha - \beta) = \cos^2\alpha - \sin^2\beta$

39. $2\sin\left(\dfrac{\pi}{4} - \theta\right)\sin\left(\dfrac{\pi}{4} + \theta\right) = \cos^2\theta - \sin^2\theta$

40. $\cos\alpha\cos(\alpha - \beta) + \sin\alpha\sin(\alpha - \beta) = \cos\beta$

41. Let $x = \dfrac{\pi}{2} - \theta$ in the equation $\cos\left(\dfrac{\pi}{2} - x\right) = \sin x$. Then prove that $\sin\left(\dfrac{\pi}{2} - \theta\right) = \cos\theta$.

42. Prove addition Formula (1). $\left(Hint:\ \text{Begin with }\sin(\alpha + \beta) = \cos\left[\dfrac{\pi}{2} - (\alpha + \beta)\right]\text{ and}\right.$

note that $\dfrac{\pi}{2} - (\alpha + \beta) = \left(\dfrac{\pi}{2} - \alpha\right) - \beta.\Big)$

43. Use addition Formula (1) to prove (2). (*Hint:* Use $\alpha - \beta = \alpha + (-\beta)$.)

44. Use addition Formula (5) to prove (6). (*Hint:* Use $\alpha - \beta = \alpha + (-\beta)$.)

45. Derive $\cot(\alpha + \beta) = \dfrac{\cot\alpha\cot\beta - 1}{\cot\alpha + \cot\beta}$ by forming $\dfrac{\cos(\alpha + \beta)}{\sin(\alpha + \beta)}$ and using Formulas (1) and (3).

46. Derive the formula in Exercise 45 by forming $\dfrac{1}{\tan(\alpha + \beta)}$ and using Formula (5).

47. Explain how the graph of the sine can be obtained from the cosine curve by using the reduction formula $\cos\left(\dfrac{\pi}{2} + \theta\right) = -\sin\theta$.

48. Explain how the graph of the sine can be obtained from the cosine curve by using the reduction formula $\cos\left(\dfrac{\pi}{2} - \theta\right) = \sin\theta$.

49. For each nonzero integer k, $\sin\left(\theta + \dfrac{\pi}{2}k\right)$ reduces to the form $\pm f(\theta)$ where f is one of the six basic trigonometric functions. What are the possible forms $\pm f(\theta)$? Explain.

50. Let $y = a\sin x + b\cos x$ where a and b are nonzero constants. In the figure $P(a, b)$ has been placed in quadrant II. (If P were put into any other quadrant, the results would be the same.) Using c and θ as shown, prove that

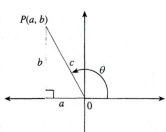

$$a\sin x + b\cos x = c(\cos\theta\sin x + \sin\theta\cos x)$$

Then explain how to obtain $y = c\sin(x + \theta)$.

Convert the equation into the form $y = c\sin(x + \theta)$ as in Exercise 50. Use a calculator to find θ in radians rounded to two decimal places, only if necessary.

51. $y = 3\sin x - 3\cos x$ **52.** $y = -3\sin x + 4\cos x$

Graph the function making use of the result in Exercise 50.

53. $f(x) = 2 \sin x + 2\sqrt{3} \cos x$ **54.** $f(x) = -\sqrt{2} \sin x - \sqrt{2} \cos x$

55. Find $\tan \beta$ for β in the given figure without using a calculator or tables. (*Hint:* Use the formula for $\tan(\alpha + \beta)$.)

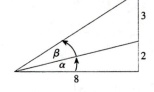

56. Let α, β be acute angles such that $\tan \alpha = \frac{3}{5}$ and $\tan \beta = \frac{1}{4}$. Find $\alpha + \beta$ without using a calculator.

Use appropriate addition or subtraction formulas to find the exact value of each expression and check by calculator.

57. $\tan(\arctan 3 + \arctan 4)$ **58.** $\sin\left(\arcsin \frac{2}{3} + \arctan \frac{4}{5}\right)$ **59.** $\cos\left(\arcsin \frac{8}{17} - \arccos \frac{12}{13}\right)$

60. For θ, $0 < \theta < 90°$, what inverse function identities can be obtained from the three co-function identities on page 456?

61. Let $f(x) = \sin x$ and $g(x) = \cos x$. Prove the following:

(a) $\dfrac{f(x + h) - f(x)}{h} = \left(\dfrac{\cos h - 1}{h}\right) \sin x + \left(\dfrac{\sin h}{h}\right) \cos x$

(b) $\dfrac{g(x + h) - g(x)}{h} = \left(\dfrac{\cos h - 1}{h}\right) \cos x - \left(\dfrac{\sin h}{h}\right) \sin x$

62. In the figure line ℓ_1 has equation $y = -\frac{1}{2}x + \frac{3}{2}$ and ℓ_2 has equation $y = -\frac{4}{5}x + \frac{6}{5}$. The lines intersect at P and the x-axis at A and B.

(a) For angles α and β as shown, explain why $\tan \alpha = $ slope ℓ_1 and $\tan \beta = $ slope ℓ_2.

(b) Explain why $\angle BPA = \alpha - \beta$.

(c) Use the subtraction formula for $\tan(\alpha - \beta)$ to find the angle between the lines, $\alpha - \beta$, to the nearest tenth of a degree.

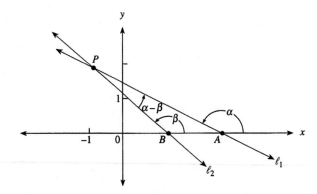

For each pair of lines ℓ_1 and ℓ_2 let P be their point of intersection, let A be the intersection of ℓ_1 with the x-axis, and let B be the intersection of ℓ_2 with the x-axis. Follow the instructions in Exercise 62, using α, β, and $\alpha - \beta$, as in Exercise 62, to find the measure of $\angle BPA$ to the nearest tenth of a degree.

63. $\ell_1: 3x + 2y = 6$
$\ell_2: 6x + y = -6$

64. $\ell_1: y = 5x - 10$
$\ell_2: 5x - 2y = 5$

65. $\ell_1:$ $x + 3y = 17$
$\ell_2: 5x - 6y = -20$
(*Hint:* $\tan(\alpha - \beta) < 0$ implies $\alpha - \beta$ is *obtuse*.)

66. $\ell_1: y - 1 = \sqrt{3}(x - 1)$
$\ell_2: y = x$

Challenge   In the figure α is an acute angle in right triangle OQR, β is an acute angle in right triangle OPQ, and $\alpha + \beta$ is an acute angle in right triangle OPS. Use the information on the figure to prove that $\sin(\alpha + \beta) < \sin \alpha + \sin \beta$. Note that QT is perpendicular to PS, $a = QR$, $b = PQ$, $c = OQ$, $d = OP$, $e = PT$, and $f = ST$.

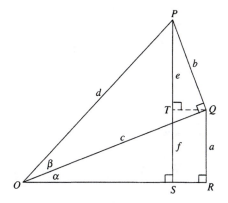

 Graphing Calculator Exercises

1. Verify that $\cos(x + y) \neq \cos x + \cos y$, in general, by graphing both $y = \cos(x + 1)$ and $y = \cos x + \cos 1$.

2. Verify that $\sin(x - y) \neq \sin x - \sin y$, in general, by graphing both $y = \sin(x - 1)$ and $\sin x - \sin 1$.

3. Find a simpler expression $s(x)$ for

$$y = f(x) = \cos x \cos\left(x - \frac{\pi}{3}\right) + \sin x \sin\left(x - \frac{\pi}{3}\right)$$

by graphing $y = f(x)$. Then verify that $s(x) = f(x)$ is an identity, using the subtraction formulas.

7.5
THE DOUBLE-
AND
HALF-ANGLE
FORMULAS

The formula $\sin(\alpha + \beta) = \sin \alpha \cos \beta + \cos \alpha \sin \beta$ is true for all values of α and β. If, in particular, we let $\beta = \alpha$, then

$$\sin 2\alpha = \sin(\alpha + \alpha) = \sin \alpha \cos \alpha + \cos \alpha \sin \alpha$$

Consequently, we have the following *double-angle formula*:

CAUTION
Formula 7 implies that, in general, $\sin 2\alpha \neq 2 \sin \alpha$. *Verify this using* $\alpha = 30°$.

(7) $\sin 2\alpha = 2 \sin \alpha \cos \alpha$

Next use $\alpha = \beta$ in the formula for $\cos(\alpha + \beta)$:

$$\cos(\alpha + \beta) = \cos \alpha \cos \beta - \sin \alpha \sin \beta$$

$$\cos 2\alpha = \cos(\alpha + \alpha) = \cos \alpha \cos \alpha - \sin \alpha \sin \alpha$$

Thus a double-angle formula for the cosine is

(8) $\cos 2\alpha = \cos^2 \alpha - \sin^2 \alpha$

Since $\cos^2 \alpha = 1 - \sin^2 \alpha$, Formula (8) may be written as

(9) $$\cos 2\alpha = 1 - 2 \sin^2 \alpha$$

Similarly, using $\sin^2 \alpha = 1 - \cos^2 \alpha$, we have

(10) $$\cos 2\alpha = 2 \cos^2 \alpha - 1$$

Substituting $\alpha = \beta$ in $\tan(\alpha + \beta) = \dfrac{\tan \alpha + \tan \beta}{1 - \tan \alpha \tan \beta}$ gives

(11) $$\tan 2\alpha = \frac{2 \tan \alpha}{1 - \tan^2 \alpha}$$

Here is a summary of the double-angle formulas stated in terms of the variable θ.

Double-angle formulas can also be derived for other circular functions. (See Exercises 37, 38, and 40.)

DOUBLE-ANGLE FORMULAS

(7) $\sin 2\theta = 2 \sin \theta \cos \theta$

(8) $\cos 2\theta = \cos^2 \theta - \sin^2 \theta$

(9) $\cos 2\theta = 1 - 2 \sin^2 \theta$

(10) $\cos 2\theta = 2 \cos^2 \theta - 1$

(11) $\tan 2\theta = \dfrac{2 \tan \theta}{1 - \tan^2 \theta}$

EXAMPLE 1   Evaluate $\sin 15° \cos 15°$ using a double-angle formula.

Solution   Rewrite $2 \sin \theta \cos \theta = \sin 2\theta$ in the equivalent form

$$\sin \theta \cos \theta = \tfrac{1}{2} \sin 2\theta$$

Then

$$\sin 15° \cos 15° = \tfrac{1}{2} \sin 2(15°)$$

$$= \tfrac{1}{2} \sin 30°$$

$$= \tfrac{1}{2}\left(\tfrac{1}{2}\right)$$

$$= \tfrac{1}{4} \qquad \blacksquare$$

EXAMPLE 2   If θ is obtuse such that $\sin \theta = \frac{5}{13}$, find $\sin 2\theta$, $\cos 2\theta$, and $\tan 2\theta$.

Note: $\cos \theta$ and $\tan \theta$ can also be found without using a reference triangle. Thus

$\cos \theta = -\sqrt{1 - \sin^2 \theta}$

$\qquad = -\sqrt{1 - \frac{25}{169}}$

$\qquad = -\frac{12}{13}$

$\tan \theta = \dfrac{\sin \theta}{\cos \theta} = \dfrac{\frac{5}{13}}{-\frac{12}{13}}$

$\qquad = -\frac{5}{12}$

Solution   Since θ is in quadrant II, and $\sin \theta = \dfrac{5}{13} = \dfrac{\text{opposite}}{\text{hypotenuse}}$, the third side of the reference triangle is found using the Pythagorean theorem; $-\sqrt{13^2 - 5^2} = -12$.

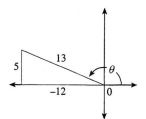

$$\cos \theta = -\frac{12}{13}$$

$$\tan \theta = -\frac{5}{12}$$

Then from the reference triangle,

$$\sin 2\theta = 2 \left(\tfrac{5}{13}\right)\left(-\tfrac{12}{13}\right) = -\frac{120}{169} \qquad \text{Formula (7)}$$

$$\cos 2\theta = 2 \left(-\tfrac{12}{13}\right)^2 - 1 = \frac{119}{169} \qquad \text{Formula (10)}$$

$$\tan 2\theta = \frac{2\left(-\frac{5}{12}\right)}{1 - \left(-\frac{5}{12}\right)^2} = -\frac{120}{119} \qquad \text{Formula (11)} \qquad ■$$

EXAMPLE 3   A photographer wants to take a picture of a 4-foot vase standing on a 3-foot pedestal. She wants to position the camera at a point C on the floor so that the angles subtended by the vase and the pedestal are identical in size. Using the figure below, determine how far away from the foot of the pedestal should the camera be placed?

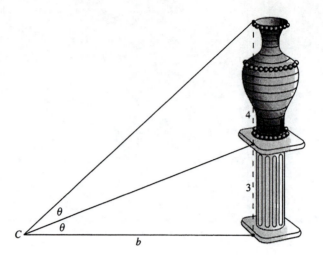

Solution From the figure we have $\tan \theta = \dfrac{3}{b}$ and $\tan 2\theta = \dfrac{4 + 3}{b} = \dfrac{7}{b}$. Now substitute for $\tan \theta$ in the formula for $\tan 2\theta$ as follows.

$$\frac{7}{b} = \tan 2\theta = \frac{2 \tan \theta}{1 - \tan^2 \theta} = \frac{2\left(\dfrac{3}{b}\right)}{1 - \left(\dfrac{3}{b}\right)^2}$$

Then

$$\frac{7}{b} = \frac{\dfrac{6}{b}}{1 - \dfrac{9}{b^2}}$$

$$\frac{7}{b} = \frac{6b}{b^2 - 9}$$

$$7b^2 - 63 = 6b^2$$

$$b^2 = 63$$

$$b = \sqrt{63} = 3\sqrt{7}$$

Since $3\sqrt{7} \approx 7.9$, the camera should be placed approximately 8 feet from the pedestal. ■

EXAMPLE 4 Let θ be an acute angle of a right triangle where $\sin \theta = \dfrac{x}{2}$. Show that $\sin 2\theta = \dfrac{x\sqrt{4 - x^2}}{2}$.

Solution Since $\sin \theta = \dfrac{x}{2}$, use x as the side opposite θ and 2 as the hypotenuse. The Pythagorean theorem gives the third side. Now use Formula (7) to solve for $\sin 2\theta$.

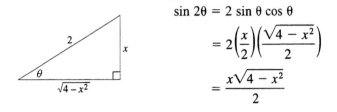

$$\sin 2\theta = 2 \sin \theta \cos \theta$$

$$= 2 \left(\frac{x}{2}\right)\left(\frac{\sqrt{4 - x^2}}{2}\right)$$

$$= \frac{x\sqrt{4 - x^2}}{2} \qquad \blacksquare$$

EXAMPLE 5 Show that $\cos 3\theta = 4 \cos^3 \theta - 3 \cos \theta$.

Solution

$$\cos 3\theta = \cos(2\theta + \theta)$$

$$= \cos 2\theta \cos \theta - \sin 2\theta \sin \theta \qquad \text{by (3)}$$

$$= (2 \cos^2 \theta - 1) \cos \theta - (2 \sin \theta \cos \theta) \sin \theta \qquad \text{by (10) and (7)}$$

$$= 2 \cos^3 \theta - \cos \theta - 2 \cos \theta \sin^2 \theta$$

$$= 2 \cos^3 \theta - \cos \theta - 2 \cos \theta(1 - \cos^2 \theta)$$

$$= 4 \cos^3 \theta - 3 \cos \theta \qquad \blacksquare$$

Formula (9) may be solved for $\sin^2 \alpha$ as follows:

$$\cos 2\alpha = 1 - 2 \sin^2 \alpha$$

$$2 \sin^2 \alpha = 1 - \cos 2\alpha$$

(12) $$\sin^2 \alpha = \frac{1 - \cos 2\alpha}{2}$$

Similarly, Formula (10) produces a formula for $\cos^2 \alpha$ and the result for $\tan^2 \alpha$ is found using $\dfrac{\sin^2 \alpha}{\cos^2 \alpha}$.

SQUARING IDENTITIES	
(12)	$\sin^2 \theta = \dfrac{1 - \cos 2\theta}{2}$
(13)	$\cos^2 \theta = \dfrac{1 + \cos 2\theta}{2}$
(14)	$\tan^2 \theta = \dfrac{1 - \cos 2\theta}{1 + \cos 2\theta}$

Notice that these identities convert the second power of a trigonometric function into an expression involving only the first power. This type of reduction makes these identities useful when it is necessary to reduce the powers within a given trigonometric expression.

EXAMPLE 6   Convert $\sin^4 \theta$ to an expression involving only the first power of the cosine.

Solution

$$\sin^4 \theta = (\sin^2 \theta)^2$$

$$= \left(\frac{1 - \cos 2\theta}{2}\right)^2 \qquad \text{by (12)}$$

$$= \frac{1 - 2\cos 2\theta + \cos^2 2\theta}{4}$$

$$= \tfrac{1}{4} - \tfrac{1}{2}\cos 2\theta + \tfrac{1}{4}\cos^2 2\theta$$

$$= \tfrac{1}{4} - \tfrac{1}{2}\cos 2\theta + \tfrac{1}{4}\left(\frac{1 + \cos 4\theta}{2}\right) \qquad \text{by (13)}$$

$$= \tfrac{1}{4} - \tfrac{1}{2}\cos 2\theta + \tfrac{1}{8} + \tfrac{1}{8}\cos 4\theta$$

$$= \tfrac{3}{8} - \tfrac{1}{2}\cos 2\theta + \tfrac{1}{8}\cos 4\theta \qquad \blacksquare$$

Since (12) is an identity for all values of θ, we may change the form of θ by writing $\theta = \dfrac{\phi}{2}$ and substitute to obtain the **half-angle formula** for the sine function.

If $x^2 = a$, then $x = \pm\sqrt{a}$.

(15)

$$\sin^2 \frac{\phi}{2} = \frac{1 - \cos \phi}{2}$$

$$\sin \frac{\phi}{2} = \pm\sqrt{\frac{1 - \cos \phi}{2}}$$

When using this formula, the appropriate sign will depend on the location of the terminal side of $\dfrac{\phi}{2}$.

EXAMPLE 7   Find the exact value of $\sin 15°$ by using the half-angle formula.

Solution Note that $15° = \dfrac{30°}{2}$ is in quadrant I. Therefore, we use the plus sign in the half-angle formula for $\sin \dfrac{\phi}{2}$:

Check this result using a calculator to find $\sin 15°$ and to evaluate $\dfrac{\sqrt{2 - \sqrt{3}}}{2}$.

$$\sin 15° = \sqrt{\frac{1 - \cos 30°}{2}} = \sqrt{\frac{1 - \dfrac{\sqrt{3}}{2}}{2}} = \frac{\sqrt{2 - \sqrt{3}}}{2} \qquad \blacksquare$$

Following the procedure used to get the formula for $\sin \dfrac{\phi}{2}$ from (12), the half-angle formula for the cosine is obtained from (13). These results are included in the summary below, written in terms of the variable θ. A formula for the tangent can be derived using (7) and (13).

$$\tan \frac{\theta}{2} = \frac{\sin \dfrac{\theta}{2}}{\cos \dfrac{\theta}{2}} = \frac{2 \sin \dfrac{\theta}{2} \cos \dfrac{\theta}{2}}{2 \cos^2 \dfrac{\theta}{2}} \qquad \left\{ \begin{array}{l} \text{multiplying} \\ \text{numerator and} \\ \text{denominator by} \\ 2 \cos \dfrac{\theta}{2} \end{array} \right.$$

Another variation is:
$$\tan \frac{\theta}{2} = \frac{1 - \cos \theta}{\sin \theta}$$
See Example 4, page 446.

$$= \frac{\sin \theta}{2 \left(\dfrac{1 + \cos \theta}{2} \right)} \qquad \text{by (7) and (13)}$$

$$= \frac{\sin \theta}{1 + \cos \theta}$$

HALF-ANGLE FORMULAS

The plus or minus sign depends on the location of the terminal side of $\dfrac{\theta}{2}$.

(15) $\qquad \sin \dfrac{\theta}{2} = \pm \sqrt{\dfrac{1 - \cos \theta}{2}}$

(16) $\qquad \cos \dfrac{\theta}{2} = \pm \sqrt{\dfrac{1 + \cos \theta}{2}}$

(17) $\qquad \tan \dfrac{\theta}{2} = \dfrac{\sin \theta}{1 + \cos \theta} = \dfrac{1 - \cos \theta}{\sin \theta}$

EXAMPLE 8 Solve for the exact value of b.

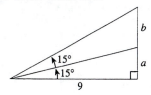

Solution From the preceding figure $\tan 15° = \dfrac{a}{9}$. Also, using (17),

$$\tan 15° = \tan \frac{30°}{2} = \frac{1 - \cos 30°}{\sin 30°}$$

Show that the same result is obtained by using the form
$$\tan 15° = \frac{\sin 30°}{1 + \cos 30°}$$

$$= \frac{1 - \dfrac{\sqrt{3}}{2}}{\dfrac{1}{2}}$$

$$= 2 - \sqrt{3}$$

Now equate this result to $\dfrac{a}{9}$ as follows:

Then $\dfrac{a}{9} = 2 - \sqrt{3}$ or $a = 18 - 9\sqrt{3}$.

Also, since $\tan 30° = \dfrac{\sqrt{3}}{3}$, and since the given figure shows that $\tan 30° = \dfrac{a + b}{9}$, we have

See Critical Thinking 3 on page 472.

$$\dfrac{a + b}{9} = \dfrac{\sqrt{3}}{3}$$

$$a + b = 3\sqrt{3}$$

$$b = 3\sqrt{3} - a$$

$$= 3\sqrt{3} - (18 - 9\sqrt{3}) \qquad (a = 18 - 9\sqrt{3})$$

$$= 12\sqrt{3} - 18 \qquad\qquad\blacksquare$$

CAUTION: Learn to Avoid These Mistakes	
WRONG	**RIGHT**
$\sin 4x = 4 \sin x$	$\sin 4x = \sin 2(2x)$ $\qquad = 2 \sin 2x \cos 2x$
$\cos^4 x = \dfrac{1 + \cos^2 2x}{2}$	$\cos^4 x = \left(\dfrac{1 + \cos 2x}{2}\right)^2$
$\cos \dfrac{\theta}{4} = \pm\sqrt{\dfrac{1 + \cos 4\theta}{2}}$	$\cos \dfrac{\theta}{4} = \pm\sqrt{\dfrac{1 + \cos \dfrac{\theta}{2}}{2}}$

EXERCISES 7.5

Evaluate the expression using an appropriate double-angle formula. Do not use a calculator.

1. $1 - 2 \sin^2 \dfrac{7\pi}{12}$ **2.** $\sin 105° \cos 105°$ **3.** $\dfrac{6 \tan 75°}{1 - \tan^2 75°}$ **4.** $\dfrac{1}{4}\left(\cos^2 \dfrac{3\pi}{8} - \sin^2 \dfrac{3\pi}{8}\right)$

Use half-angle formulas to verify the equation.

5. $\cos 75° = \dfrac{1}{2}\sqrt{2 - \sqrt{3}}$ **6.** $\sin\left(-\dfrac{\pi}{8}\right) = -\dfrac{1}{2}\sqrt{2 - \sqrt{2}}$

7. $\tan \dfrac{3\pi}{8} = \sqrt{2} + 1$ **8.** $\sin \dfrac{\pi}{12} = \dfrac{1}{2}\sqrt{2 - \sqrt{3}}$

9. **(a)** Show that the half-angle formula in (17) gives $\tan 15° = 2 - \sqrt{3}$.

   (b) Use a calculator, to verify that $2 - \sqrt{3}$ and $\tan 15°$ are the same.

10. (a) Use an appropriate formula to show that $\cos^2 \dfrac{\pi}{8} = \dfrac{2 + \sqrt{2}}{4}$.

(b) From part (a) show that $\cos \dfrac{\pi}{8} = \dfrac{1}{2} \sqrt{2 + \sqrt{2}}$.

(c) Use a calculator to verify that $\dfrac{1}{2} \sqrt{2 + \sqrt{2}}$ and $\cos \dfrac{\pi}{8}$ are the same.

11. (a) Use an appropriate formula to show that $\tan(22.5°) = \sqrt{2} - 1$.

(b) Use a calculator, to show that $\sqrt{2} - 1$ and $\tan(22.5°)$ are the same.

12. If $\cos \theta = \frac{12}{13}$ and θ is in the first quadrant, use double-angle formulas to find:

(a) $\sin 2\theta$ **(b)** $\cos 2\theta$ **(c)** $\tan 2\theta$

13. If θ is obtuse and $\tan \theta = -\frac{15}{8}$, use double angle formulas to find:

(a) $\sin 2\theta$ **(b)** $\cos 2\theta$ **(c)** $\tan 2\theta$

14. If $\sin \theta = -\frac{24}{25}$ and $\tan \theta > 0$, find $\cos 2\theta$.

15. If $\tan \theta = -\frac{2}{3}$ and $\cos \theta > 0$, find $\sin 2\theta$.

16. Use the information from the triangle to derive $\cos 2\theta = 1 - \frac{2}{9}x^2$.

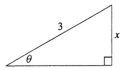

17. Use the information in Exercise 16 to derive $9 \sin 2\theta = 2x \sqrt{9 - x^2}$.

18. Let θ be an acute angle of a right triangle so that $\tan \theta = x$. Find the expressions for $\sin 2\theta$, $\cos 2\theta$, and $\tan 2\theta$ in terms of x.

Solve for the exact value of b.

19.

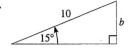

20.

21. At a point A that is 50 meters from the base of a tower, the angle of elevation to the top of the tower is twice as large as is the angle of elevation from a point B that is 150 meters from the tower. Assuming that the base of the tower and the points A and B are in the same line on level ground, find the height of the tower h.

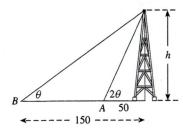

Use a double-angle formula to solve for b without using tables or a calculator.

22.

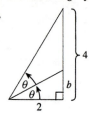

23.

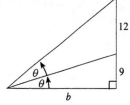

24.

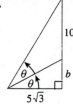

Hint: Use the rational root theorem.

Verify the identities.

25. $\cot x \sin 2x = 2 \cos^2 x$

26. $\dfrac{\sin 2x}{1 + \cos 2x} = \tan x$

27. $\sec^2 \dfrac{x}{2} = \dfrac{2}{1 + \cos x}$

28. $\sin^2 \dfrac{x}{2} = \dfrac{\sec x - 1}{2 \sec x}$

29. $\cos^2 \dfrac{x}{2} = \dfrac{\tan x + \sin x}{2 \tan x}$

30. $\tan^2 \dfrac{x}{2} = \dfrac{\sec x - 1}{\sec x + 1}$

31. $\cot \dfrac{x}{2} = \csc x + \cot x$

32. $\dfrac{1}{2} \tan 2x = \dfrac{1}{\cot x - \tan x}$

33. $(\sin x + \cos x)^2 = 1 + \sin 2x$

34. $\sin 3x = 3 \sin x - 4 \sin^3 x$

35. $\tan 3x = \dfrac{3 \tan x - \tan^3 x}{1 - 3 \tan^2 x}$

36. $\sin 2x = \dfrac{2 \tan x}{\sec^2 x}$

37. $\cot 2x = \dfrac{\cot^2 x - 1}{2 \cot x}$

38. $\csc 2x = \frac{1}{2} \csc x \sec x$

39. $\csc 2x - \cot 2x = \tan x$

40. $\sec 2\theta = \dfrac{\sec^2 \theta}{2 - \sec^2 \theta}$

41. $\cos 2x = \cos^4 x - \sin^4 x$

42. $\sin 4x = 8 \sin x \cos^3 x - 4 \sin x \cos x$

43. $\cos 4x = 8 \cos^4 x - 8 \cos^2 x + 1$

44. $\sin^4 \dfrac{x}{2} = \dfrac{1}{4} - \dfrac{1}{2} \cos x + \dfrac{1}{4} \cos^2 x$

45. $\cos^4 x = \frac{1}{8} \cos 4x + \frac{1}{2} \cos 2x + \frac{3}{8}$

46. $\tan \dfrac{x}{2} + \cot \dfrac{x}{2} = 2 \csc x$

47. $\sin^2 x \cos^2 x = \frac{1}{8}(1 - \cos 4x)$

48. $(1 + 2 \cos x)^2 = 3 + 4 \cos x + 2 \cos 2x$

49. Verify that for nonzero constants $a, b, c,$ and k that $a \sin kx + b \cos kx = c \sin (kx + \theta)$ where $c = \sqrt{a^2 + b^2}$, and θ is an angle with $\sin \theta = \dfrac{b}{c}$ and $\cos \theta = \dfrac{a}{c}$. (See Exercise 50, page 459.)

50. Use the result in Exercise 49 to graph the functions:

(a) $f(x) = \sin 2x - \sqrt{3} \cos 2x$ (b) $f(x) = \sin \dfrac{x}{2} + \cos \dfrac{x}{2}$

51. (a) Add the formulas for $\cos(\alpha + \beta)$ and $\cos(\alpha - \beta)$ to derive this *product formula:*

$$\cos \alpha \cos \beta = \tfrac{1}{2} [\cos(\alpha + \beta) + \cos(\alpha - \beta)]$$

(b) Use a similar analysis to prove these *product formulas:*

$$\sin \alpha \sin \beta = \tfrac{1}{2} [\cos(\alpha - \beta) - \cos(\alpha + \beta)]$$

$$\sin \alpha \cos \beta = \tfrac{1}{2} [\sin(\alpha + \beta) + \sin(\alpha - \beta)]$$

$$\cos \alpha \sin \beta = \tfrac{1}{2} [\sin(\alpha + \beta) - \sin(\alpha - \beta)]$$

52. Use the results of Exercise 51 to express each of the following as a sum or difference:

(a) $\sin 6x \sin 2x$ (b) $2 \cos x \cos 4x$ (c) $3 \cos 5x \sin(-2x)$ (d) $4 \sin x \cos \dfrac{x}{2}$

53. Substitute $\alpha = \dfrac{u + v}{2}$ and $\beta = \dfrac{u - v}{2}$ into the formulas in Exercise 51 and derive these *sum formulas:*

$$\cos u + \cos v = 2 \cos \frac{u + v}{2} \cos \frac{u - v}{2}$$

$$\cos v - \cos u = 2 \sin \frac{u + v}{2} \sin \frac{u - v}{2}$$

$$\sin u + \sin v = 2 \sin \frac{u + v}{2} \cos \frac{u - v}{2}$$

$$\sin u - \sin v = 2 \cos \frac{u + v}{2} \sin \frac{u - v}{2}$$

54. Use the results of Exercise 53 to express each of the following as a product:

(a) $\sin 4x + \sin 2x$ (b) $\cos 6x - \cos 3x$ (c) $2 \sin 5x - 2 \sin x$ (d) $\dfrac{1}{2} \cos \dfrac{x}{2} + \dfrac{1}{2} \cos \dfrac{5x}{2}$

Use double- or half-angle formulas to find the exact value of each expression.

55. $\cos \left(2 \arcsin \frac{1}{3} \right)$ **56.** $\sin \left(\frac{1}{2} \arccos \frac{3}{5} \right)$

57. $\tan \left(2 \arctan \frac{4}{5} \right)$ **58.** $\tan \left(\frac{1}{2} \arccos \frac{3}{4} \right)$

Let θ be the measure of an acute angle so that $\theta = \arctan x$ and verify the identity.

59. $\sin(2 \arctan x) = \dfrac{2x}{1 + x^2}$ **60.** $\tan \left(\dfrac{1}{2} \arctan x \right) = \dfrac{x}{\sqrt{1 + x^2} + 1}$

61. $\tan(2 \arctan x) = \dfrac{2x}{1 - x^2}$ **62.** $\cos(2 \arctan x) = \dfrac{1 - x^2}{1 + x^2}$

Challenge

1. Explain how $\tan \dfrac{\theta}{2} = \dfrac{\sin \theta}{1 + \cos \theta}$ is demonstrated in the figure.

2. Find a geometric interpretation of $\tan \dfrac{\theta}{2} = \dfrac{1 - \cos \theta}{\sin \theta}$.

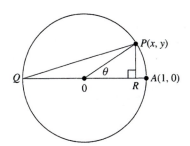

Written Assignment Are the restrictions on θ in the two forms of $\tan \dfrac{\theta}{2}$ in Formula (17) the same? Explain.

Graphing Calculator Exercises For each of the functions in Exercises 1 and 2, use your calculator to find a simpler expression $g(x)$ for the given $f(x)$. Then verify by means of algebra that $f(x) = g(x)$ is indeed an identity.

1. $f(x) = 4 \sin^3 x - 3 \sin x$

2. $f(x) = (\cos^4 x - \sin^4 x)/(1 - 2 \sin^2 x)$

Critical Thinking

1. The addition-subtraction formulas, on page 451, for the sine and cosine apply to all angle measures α and β. What are the restrictions on α and β for Formula (5)?

2. In general, $\tan(\alpha + \beta)$ is not equal to $\tan \alpha + \tan \beta$. However, there are some values of α and β for which they are equal. Find such α and β, and do the same for $\tan(\alpha - \beta)$.

3. A purpose of solving Example 8 on page 467 without the use of a calculator is to become familiar with the use of the half-angle formula for the tangent. Simpler, though approximate, calculator solutions are possible. Write a calculator sequence for finding b.

4. Find a general form for all measures of acute angles whose exact trigonometric ratios can be found using the half-angle formulas, based on the available ratios of 30°, 45°, and 60°.

5. Formulas in Section 7.5 can be verified for specific cases of the variable by using known trigonometric ratios. In this regard, first find $\cos 15°$ using the half-angle formula; then use this in the identity of Exercise 43 on page 470 to verify $\cos 60° = \frac{1}{2}$.

7.6
TRIGONOMETRIC EQUATIONS

Can you find the points of intersection of the sine and cosine curves?

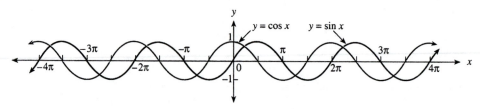

A trigonometric equation that is true for some values of the variables but is not an identity is a conditional equation.

Answering this question amounts to finding all values x for which $\sin x = \cos x$. To solve this **trigonometric equation** for x, first divide by $\cos x$.

$$\sin x = \cos x$$

$$\frac{\sin x}{\cos x} = 1$$

Now use $\dfrac{\sin x}{\cos x} = \tan x$ and solve $\tan x = 1$. For x in the interval $\left(-\dfrac{\pi}{2}, \dfrac{\pi}{2}\right)$,

$\tan x = 1$ has only one solution, $x = \dfrac{\pi}{4}$. Then, since the tangent has period π, all the solutions are obtained by adding on all the multiples of π:

$$x = \frac{\pi}{4} + k\pi, \qquad \text{where } k \text{ is any integer}$$

For $k = -2, -1, 0, 1, 2$, the following specific solutions are produced:

$$-\frac{7\pi}{4}, \ -\frac{3\pi}{4}, \ \frac{\pi}{4}, \ \frac{5\pi}{4}, \ \frac{9\pi}{4}$$

or

$$-315°, \ -135°, \ 45°, \ 225°, \ 405°$$

$x = \dfrac{\pi}{4} + k\pi$, k any integer, is the **general solution** of $\sin x = \cos x$.

EXAMPLE 1 Find the general solution: $\sqrt{3} \csc x - 2 = 0$.

Solution Begin by isolating $\csc x$ on one side.

$$\sqrt{3} \csc x - 2 = 0$$
$$\sqrt{3} \csc x = 2$$
$$\csc x = \frac{2}{\sqrt{3}} \qquad \left(\text{or } \sin x = \frac{\sqrt{3}}{2} \right)$$

Unless otherwise indicated, the solutions for trigonometric equations are assumed to be the exact values, not approximations.

Thus, in the interval $[0, 2\pi)$, $x = \dfrac{\pi}{3}$ or $x = \dfrac{2\pi}{3}$. Since the cosecant has period 2π, we obtain the general solution by adding on the multiples of 2π. Then for any integer k we have

$$x = \begin{cases} \dfrac{\pi}{3} + 2k\pi \\[2mm] \dfrac{2\pi}{3} + 2k\pi \end{cases} \quad \text{or} \quad x = \begin{cases} 60° + k(360°) \\[2mm] 120° + k(360°) \end{cases} \qquad \blacksquare$$

CAUTION
Do not begin by dividing $\cos^2 x = \cos x$ by $\cos x$. This step would produce $\cos x = 1$ and we would have lost all the roots of $\cos x = 0$. (This error is comparable to solving the equation $x^2 = x$ by first dividing each side by x.)

EXAMPLE 2 Find the general solution: $\cos^2 x = \cos x$.

Solution Since the cosine has period 2π, first solve for x in the interval $0 \le x < 2\pi$. Begin by subtracting $\cos x$ from both sides. Thus

$$\cos^2 x - \cos x = 0$$

Factor out $\cos x$.

$$(\cos x)(\cos x - 1) = 0$$

Then

$$\cos x = 0 \quad \text{or} \quad \cos x - 1 = 0$$

Hence $\cos x = 0$ for $x = \dfrac{\pi}{2}$ or $\dfrac{3\pi}{2}$ and $\cos x = 1$ for $x = 0$. Since the period of the cosine is 2π, we obtain all the solutions by taking each solution in $[0, 2\pi)$ and adding all multiples of 2π. The general solution may be presented in these forms, where k is any integer.

$$x = \begin{cases} \dfrac{\pi}{2} + 2k\pi \\ \dfrac{3\pi}{2} + 2k\pi \\ 2k\pi \end{cases} \quad \text{or} \quad x = \begin{cases} 90° + k(360°) \\ 270° + k(360°) \\ k(360°) \end{cases} \quad \blacksquare$$

Observe that in the general solution multiples of π, rather than of 2π, are added. This is due to the fact that $\cos 2x$ has period π.

EXAMPLE 3 Find the general solution: $\cos 2x = \frac{1}{2}$.

Solution Reduce the given equation to a simpler one by substituting $\theta = 2x$ and solve $\cos \theta = \frac{1}{2}$. In $[0, 2\pi)$ the solutions for θ are $\dfrac{\pi}{3}$ and $\dfrac{5\pi}{3}$, so for any integer k

Check for $x = \dfrac{\pi}{6} + k\pi$:

$$\cos 2x = \cos 2\left(\dfrac{\pi}{6} + k\pi\right)$$

$$\theta = \begin{cases} \dfrac{\pi}{3} + 2k\pi \\ \dfrac{5\pi}{3} + 2k\pi \end{cases}$$

$$= \cos\left(\dfrac{\pi}{3} + 2k\pi\right)$$

$$= \cos \dfrac{\pi}{3} = \dfrac{1}{2}$$

Then, since $x = \dfrac{\theta}{2}$, divide the preceding equations by 2 to get the general solution.

The check for $x = \dfrac{5\pi}{6} + k\pi$ is similar.

$$x = \begin{cases} \dfrac{\pi}{6} + k\pi \\ \dfrac{5\pi}{6} + k\pi \end{cases} \quad \blacksquare$$

Example 3 can also be solved using the identity $\cos 2x = 2\cos^2 x - 1$ and noting that the period is π.

$$2\cos^2 x - 1 = \tfrac{1}{2}$$

The given solution for Example 3 and the work that follows it illustrate that there are alternate ways of solving trigonometric equations, as is also true for proving trigonometric identities.

$$\cos^2 x = \tfrac{3}{4}$$

$$\cos x = \pm\dfrac{\sqrt{3}}{2}$$

Then, in $[0, 2\pi)$, $x = \dfrac{\pi}{6}, \dfrac{5\pi}{6}, \dfrac{7\pi}{6}, \dfrac{11\pi}{6}$.

Now, since $\dfrac{\pi}{6} + 1(\pi) = \dfrac{7\pi}{6}$ and $\dfrac{5\pi}{6} + 1(\pi) = \dfrac{11\pi}{6}$, the same general solution is obtained by adding multiples of π to $\dfrac{\pi}{6}$ and $\dfrac{5\pi}{6}$ as before.

Numerous situations in trigonometry only call for the solutions of an equation in the interval $[0, 2\pi)$. This is illustrated in the next example.

EXAMPLE 4   Solve $\sin 2x = \sin x$ for $0 \le x < 2\pi$.

Solution Using the double-angle formula $\sin 2x = 2 \sin x \cos x$, we may write

$$2 \sin x \cos x = \sin x$$

or

$$2 \sin x \cos x - \sin x = 0$$

Factor the left side.

$$(\sin x)(2 \cos x - 1) = 0$$

Then

$$\sin x = 0 \quad \text{or} \quad 2 \cos x = 1$$

$$\sin x = 0 \quad \text{or} \quad \cos x = \tfrac{1}{2}$$

Now $\sin x = 0$ for $x = 0$ and π; $\cos x = \tfrac{1}{2}$ for $x = \dfrac{\pi}{3}$ and $\dfrac{5\pi}{3}$. The solutions in $[0, 2\pi)$ are $0, \dfrac{\pi}{3}, \pi$, and $\dfrac{5\pi}{3}$. ∎

Check:

$x = 0$: $\sin(2 \cdot 0)$
$\quad = 0 = \sin 0$

$x = \dfrac{\pi}{3}$: $\sin \dfrac{2\pi}{3}$

$\quad = \dfrac{\sqrt{3}}{2} = \sin \dfrac{\pi}{3}$

$x = \pi$: $\sin 2\pi = 0 = \sin \pi$

$x = \dfrac{5\pi}{3}$: $\sin \dfrac{10\pi}{3} = \sin \dfrac{4\pi}{3}$

$\quad = -\dfrac{\sqrt{3}}{2} = \sin \dfrac{5\pi}{3}$

EXAMPLE 5   Solve: $\dfrac{\sec x}{\cos x} - \dfrac{1}{2} \sec x = 0$.

Solution

$$\frac{\sec x}{\cos x} - \frac{1}{2} \sec x = 0$$

$$(\sec x)\left(\frac{1}{\cos x} - \frac{1}{2}\right) = 0$$

$$\sec x = 0 \quad \text{or} \quad \frac{1}{\cos x} - \frac{1}{2} = 0$$

$$\sec x = 0 \quad \text{or} \quad \frac{1}{\cos x} = \frac{1}{2}$$

$$\sec x = 0 \quad \text{or} \quad \cos x = 2$$

Some trigonometric equations have no solutions. This is comparable to an algebraic equation such as $x^2 + 1 = 0$, which has no real solutions.

Since $\sec x \ge 1$ or $\sec x \le -1$ for all x, $\sec x = 0$ has no roots. Also, since $|\cos x| \le 1$, $\cos x = 2$ has no roots. Thus the given equation has no solutions. ∎

EXAMPLE 6 Solve: $\cos^2 x + \frac{1}{2}\sin x - \frac{1}{2} = 0$ for $0 \le x < 2\pi$.

Solution Multiply by 2 to clear fractions.

$$2\left(\cos^2 x + \tfrac{1}{2}\sin x - \tfrac{1}{2}\right) = 2 \cdot 0$$

$$2\cos^2 x + \sin x - 1 = 0$$

Convert to an equivalent form involving only $\sin x$ by using $\cos^2 x = 1 - \sin^2 x$.

$$2(1 - \sin^2 x) + \sin x - 1 = 0$$

$$2\sin^2 x - \sin x - 1 = 0$$

The left side is quadratic in $\sin x$, which is factorable.

Think of $u = \sin x$ and factor $2u^2 - u - 1$ as $(2u + 1)(u - 1)$.

$$(2\sin x + 1)(\sin x - 1) = 0$$

$$2\sin x + 1 = 0 \quad \text{or} \quad \sin x - 1 = 0$$

$$\sin x = -\tfrac{1}{2} \quad \text{or} \quad \sin x = 1$$

Now $\sin x = -\dfrac{1}{2}$ for $x = \dfrac{7\pi}{6}$ and $\dfrac{11\pi}{6}$; $\sin x = 1$ for $x = \dfrac{\pi}{2}$. The solutions in $[0, 2\pi)$ are $\dfrac{\pi}{2}, \dfrac{7\pi}{6}$, and $\dfrac{11\pi}{6}$. ∎

EXAMPLE 7 Solve $3\tan\theta - 7 = 0$ for θ to the nearest tenth of a degree, where $0° \le \theta < 360°$.

Solution $3\tan\theta - 7 = 0$ is equivalent to $\tan\theta = \dfrac{7}{3}$. Then, since this tangent value is positive and $0 \le \theta < 360°$, the solutions are in quadrants I and III. After setting the calculator in degree mode, the first quadrant solution is

Here is a keystroke sequence to use with a scientific calculator:

$$\theta = \boxed{\text{TAN}^{-1}}\ \boxed{(}\ \boxed{7}\ \boxed{\div}\ \boxed{3}\ \boxed{)}\ \boxed{\text{ENTER}} = 66.8° \qquad \text{to the nearest tenth}$$

Adding 180° gives the solution in quadrant III as follows:

$$66.8° + 180° = 246.8°$$

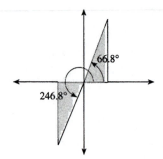

∎

EXAMPLE 8 Solve $2\sin^2 x + 5\sin x - 2 = 0$ for x in the interval $[0°, 360°)$, rounded to the nearest tenth of a degree.

Solution Let $u = \sin x$ and substitute into the equation to obtain $2u^2 + 5u - 2 = 0$. Now use the quadratic formula,

$$u = \frac{-5 \pm \sqrt{25 - 4(2)(-2)}}{2(2)} = \frac{-5 \pm \sqrt{41}}{4}$$

Then

$$u = \sin x = \frac{-5 \pm \sqrt{41}}{4} = \begin{cases} 0.3508 \\ -2.8508 \end{cases}$$

The quadrant I answer can be found directly using

$$\sin^{-1}\left(\frac{-5 + \sqrt{41}}{4}\right).$$

Since $|\sin x| \leq 1$, $\sin x = -2.8504$ has no solutions. For $\sin x = 0.3508$, there will be two solutions, one in quadrant I and one in quadrant II. The quadrant I solution is $x = \boxed{\text{SIN}^{-1}}\ \boxed{0.3508}\ \boxed{\text{ENTER}} = 20.5°$. In quadrant II we have $180° - 20.5° = 159.5°$.

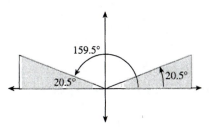

■

EXERCISES 7.6

Find the general solution and check.

1. $\cos x = 1$ **2.** $\sin x = 1$ **3.** $\sin x = \frac{1}{2}$

4. $\cos^2 x = \frac{1}{2}$ **5.** $\sec x = -1$ **6.** $\csc x = 2$

7. $\sin 2x = -\frac{1}{2}$ **8.** $2 \cos \frac{x}{3} = 1$ **9.** $\frac{1}{2} \sin^2 x = 1$

10. $\tan x = \sqrt{3}$ **11.** $\tan 2x = \sqrt{3}$ **12.** $2 \sin(x - 1) = \sqrt{2}$

13. $2 \cos(x + 1) = -2$ **14.** $\dfrac{1}{\sec x} = 2$ **15.** $\cos 3x = 1$

Solve the equation for θ in the interval $[0°, 360°)$ and check.

16. $2 \sin \theta = 1$ **17.** $\sqrt{3} \csc \theta = 2$ **18.** $2 \cos \theta + 3 = 0$

19. $2 \sec \theta - 2\sqrt{2} = 0$ **20.** $\tan^2 \theta - 3 = 0$ **21.** $\sin^2 \theta - \cos^2 \theta = 1$

22. $2 \csc^2 \theta - 1 = 0$ **23.** $2 \sin \frac{\theta}{2} - 1 = 0$ **24.** $1 + \sqrt{2} \sin 2\theta = 0$

25. $-1 + \tan \frac{3\theta}{2} = 0$ **26.** $\sin^2 \theta - \cos^2 \theta = 0$ **27.** $2 \tan \theta \cos^2 \theta = 1$

Solve the equation for x in the interval $0 \le x < 2\pi$.

28. $\sin x (\cos x + 1) = 0$

29. $(\cos x - 1)(2 \sin x + 1) = 0$

30. $\sin x + \cos x = 0$

31. $\sin x - \sqrt{3} \cos x = 0$

32. $\sec x + \tan x = 0$

33. $2 \cos^2 x - \sqrt{3} \cos x = 0$

34. $2 \tan x = \sin x$

35. $\sin^2 x + \sin x - 2 = 0$

36. $2(\cos^2 x - \sin^2 x) = \sqrt{2}$

37. $\sin 2x = \cos x$

38. $\sin^2 x + 2 \cos^2 x = 2$

39. $\sin^2 x + \cos^2 x = 1.5$

40. $\sec^2 x = 2 \tan x$

41. $2 \cos^2 x + 9 \cos x - 5 = 0$

42. $\cos 2x - \cos x = 0$

43. $\cos^2 2x = \cos 2x$

44. $3 \cos^4 x + 4 \cos^2 x = 0$

45. $2 \sin^4 x - 3 \sin^2 x + 1 = 0$

46. $2 \sin^2 x - 1 = \sin x$

47. $2 \tan x - 1 = \tan^2 x$

48. $\sin x \cot^2 x - 3 \sin x = 0$

49. $3 \tan^2 x = 7 \sec x - 5$

50. $\sin^3 2x - \sin 2x = 0$

51. $\cos^2 x - \sin^2 x + \sin x = 1$

52. $2 \cos x - 2 \sec x - 3 = 0$

53. $3 \cos 2x + 2 \sin^2 x = 2$

54. $\cos 4x = \sin 2x$

55. $\sin 2x + \sin 4x = 0$

56. $\sec x + \tan x = 1$ (*Hint*: Square both sides.)

57. $\sin x + \cos x = 1$ (*Hint*: Square both sides.)

Use a calculator to approximate the solutions of each equation to the nearest tenth of a degree for x in the interval $[0°, 360°)$.

58. $3 \sin x = 2$

59. $7 \sec x - 15 = 0$

60. $\cos 2x = .9033$

61. $\sin \dfrac{x}{2} = .8259$

62. $12 \cos^2 x + 5 \cos x - 3 = 0$

63. $4 \cot^2 x - 12 \cot x + 9 = 0$

64. $\tan^2 x + \tan x - 1 = 0$ (*Hint*: Use the quadratic formula.)

65. $\cos^2 x - \sin^2 x = -\dfrac{3}{4}$

66. $4 \sin x - 5 \cos x = 0$

67. In the figure α and β are acute angles.

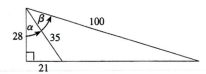

(a) Show that $\alpha < \dfrac{\pi}{4}$ by comparing $\cos \alpha$ and $\cos \dfrac{\pi}{4}$.

(b) Without using tables or a calculator, prove that $\alpha = \beta$. (*Hint*: Apply a double-angle formula to 2α.)

Find the coordinates of the points of intersection of the two curves for the interval $[0, 2\pi]$.

68. $y = \tan x$
$y = \cot x$

69. $y = \sin x$
$y = -\cos x$

70. $y = \sin x$
$y = \cos 2x$

Challenge Observe that in the general solution in Example 2 of this section, the first two forms differ by π for each k. Find a single form that can replace these two forms.

Written Assignment Near the start of Section 7.6, the equation $\sin x = \cos x$ is divided by $\cos x$ to obtain $\dfrac{\sin x}{\cos x} = 1$. Explain why no roots of the first equation are lost even though the divisor $\cos x$ is variable. (Compare to the observation made in the Caution next to Example 2.)

 Graphing Calculator Exercises A graphing calculator may be used to approximate the solutions to trigonometric equations of the form $f(x) = g(x)$ in the following two ways. *Method I:* Graph the functions $y = f(x)$ and $y = g(x)$ on the same set of axes, and use Trace and Zoom to approximate their points of intersection. *Method II:* Graph the function $y = f(x) - g(x)$, and use Trace and Zoom to find where this function intersects the x-axis, that is, where $f(x) - g(x) = 0$.

In Exercises 1–3, use Method I to approximate the solutions of the given equations to the nearest hundredth for $0 \le x \le 2\pi$.

1. $\tan x = 3 \cos x$

2. $\sin^2 x = 3 \sin x + 2$

3. $\sin x = \sin x/2$

In Exercises 4–6, use Method II to approximate the solutions of the given equations to the nearest hundredth for $2\pi \le x \le 4\pi$.

4. Same as Exercise 1.   **5.** Same as Exercise 2.   **6.** Same as Exercise 3.

7. Give the general solutions to each of the equations in Exercises 1–3.

8. Find the general solution to the equation $\sin 2x = x^2 - 1$.

For the inequalities in Exercises 9 and 10, graph $y = $ each side of the inequality, and determine for which values of x the left-hand side is greater than or equal to the right-hand side.

9. $\tan x \ge 3 \cos x$ **10.** $\sin 2x \ge x^2 - 1$

CHAPTER REVIEW EXERCISES

Section 7.1   Trigonometry of Acute Angles   (Page 426)

1. Define each of the six trigonometric ratios for θ in terms of the words *side opposite, side adjacent,* and *hypotenuse* in $\triangle ABC$.

2. In the figure, write the six trigonometric ratios for angle B using the lengths a, b, and c.

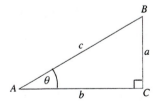

3. Triangle ABC is a right triangle with right angle at C. Find $\cos B$ if $\tan A = \dfrac{2}{3}$.

4. Triangle ABC is a right triangle with right angle at C. If $b = 6$ and $c = 9$, find the six trigonometric ratios of $\angle A$.

5. Write each of the six trigonometric ratios of $\angle B$ in terms of x.

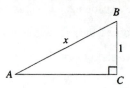

Draw an appropriate right triangle and find θ if:

6. $\cos\theta = \dfrac{\sqrt{3}}{2}$ **7.** $\cos\theta = \dfrac{1}{\sqrt{2}}$ **8.** $\sec\theta = \dfrac{2}{\sqrt{3}}$ **9.** $\cot\theta = \sqrt{3}$

10. Use the figure to show that $\sin^2\theta + \cos^2\theta = 1$.

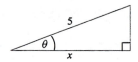

11. Repeat Exercise 10 to show that $\sec^2\theta = \tan^2\theta + 1$.

12. Show that $\cos\theta\,\tan^2\theta = \dfrac{x^2 - 9}{3x}$ where θ is an acute angle in a right triangle having one side of measure x.

Section 7.2   Right Triangle Trigonometry (Page 435)

Find the exact value of the indicated part.

13.   **14.**   **15.** 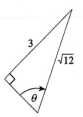   **16.**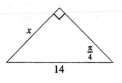

Find each trigonometric ratio to four decimal places.

17. $\sin 73°$ **18.** $\cos 80°$ **19.** $\tan 40.1°$ **20.** $\sec 55.9°$

Find θ to the nearest tenth of a degree.

21. $\sin\theta = 0.2356$ **22.** $\cos\theta = 0.5266$ **23.** $\tan\theta = 1.3452$

24. $\cot\theta = 0.5126$ **25.** $\sec\theta = 4.3997$ **26.** $\csc\theta = 1.6798$

Triangle ABC is a right triangle with $C = 90°$. Solve for the indicated part to the nearest tenth.

27. $A = 52.8°$, $a = 20$; find b.   **28.** $a = 4.2, b = 12$; find A.

29. From the top of a house the angle of depression of a point on the ground is $32°$. If the building is 18 meters high, how far is the point from the base of the building? (Give the answer to the nearest meter.)

30. How high is a building whose horizontal shadow is 45 meters when the angle of elevation of the sun is $55.6°$? (Give the answer to the nearest tenth of a meter.)

31. In the figure $AB = 12$, $\angle C$ is a right angle, $\angle A = 20°$ and $\angle CBD = 65°$. Find h to the nearest tenth.

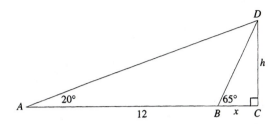

Section 7.3   Identities   (Page 442)

Find the exact value without using a calculator.

32. $\sin^2(-26°) + \cos^2(-26°)$

33. $\cot^2(10°) - \csc^2(10°)$

34. $2 \cot \dfrac{5\pi}{9} \tan \dfrac{5\pi}{9}$

35. $(\sec^2 8° - 1) \cot^2 8°$

Convert the expression into a form involving only sines and cosines, and simplify. Also find the restriction on θ.

36. $\csc \theta \tan^2 \theta$

37. $\dfrac{\sec \theta + \csc \theta}{\tan \theta - \cot \theta}$

38. Use $\sin^2\theta + \cos^2\theta = 1$ to prove **(a)** $\tan^2\theta + 1 = \sec^2\theta$ and **(b)** $1 + \cot^2\theta = \csc^2\theta$.

Verify the identity.

39. $\sin^2\theta - 2\cos^2\theta = 3\sin^2\theta - 2$

40. $(\cot \theta - 1)^2 = \csc^2\theta - 2\cot \theta$

41. $(1 + \cos \theta)(1 - \cos \theta) = \dfrac{1}{1 + \cot^2\theta}$

42. $\dfrac{\tan^2\theta + 1}{\cot^2\theta + 1} = \tan^2\theta$

43. $\dfrac{\cot \theta \cos \theta}{\csc \theta - 1} = 1 + \sin \theta$

44. $\dfrac{\cos \theta}{1 + \sin \theta} = \dfrac{1 - \sin \theta}{\cos \theta}$

45. $\dfrac{\sin \theta}{\sec \theta + 2\cos \theta} = \dfrac{\cot \theta}{1 + 3\cot^2\theta}$

46. $\dfrac{1 + \sin \theta}{\cos \theta} + \dfrac{\cos \theta}{1 + \sin \theta} = 2\sec \theta$

47. $\sec \theta - \tan \theta = \dfrac{1}{\sec \theta + \tan \theta}$

Prove that the equation is not an identity.

48. $\dfrac{1 + \sin \theta}{1 - \sin \theta} = (\sec \theta - \tan \theta)^2$

49. $\sec \theta = \sqrt{1 + \tan^2\theta}$

Section 7.4   The Addition and Subtraction Formulas   (Page 450)

Complete the equation to get a specific case of an addition or subtraction formula.

50. $\sin 10° \cos 4° - \cos 10° \sin 4° =$

51. $\cos (6° - \theta) \cos (6° + \theta) - \sin (6° - \theta) \sin (6° + \theta) =$

52. Use the subtraction formula for the tangent to find $\tan 15°$. Leave the answer in radical form.

53. Use an addition or subtraction formula to evaluate $\cos \dfrac{7\pi}{12}$. Leave the answer in radical form.

54. $\cos \alpha = -\dfrac{4}{5}$ and the terminal side of α is in quadrant II. $\sin \beta = -\dfrac{12}{13}$ and the terminal side of β is in quadrant IV. Find $\sin(\alpha - \beta)$.

55. Use the given information in Exercise 54 to find $\tan(\alpha + \beta)$.

56. Both α and β terminate in quadrant III so that $\sin \alpha = -\dfrac{1}{4}$ and $\cos \beta = -\dfrac{2}{3}$. Find $\cos(\alpha - \beta)$. Leave the answer in radical form.

Verify the identity.

57. $\cos\left(\dfrac{\pi}{6} + \theta\right) = \dfrac{1}{2}\left(\sqrt{3}\cos\theta - \sin\theta\right)$

58. $\dfrac{\cos(\alpha + \beta)}{\sin(\alpha + \beta)} = \dfrac{1 - \tan\alpha\tan\beta}{\tan\alpha + \tan\beta}$

59. $\sin\alpha\cos(\alpha - \beta) - \cos\alpha\sin(\alpha - \beta) = \sin\beta$

60. Convert $y = -4\sin x + 4\cos x$ into the form $y = c\sin(x + \theta)$.

61. Find $\tan\beta$ in the given figure.

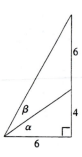

Section 7.5 The Double- and Half-Angle Formulas (Page 461)

Complete the equation using a specific case of a double- or half-angle formula, and evaluate without using a calculator.

62. $1 - 2\sin^2\dfrac{\pi}{12} =$

63. $\dfrac{1}{2}\sin 15° \cos 15° =$

64. $\tan 22.5° =$

65. $\cos\dfrac{\pi}{8} =$

66. $\dfrac{8\tan 105°}{1 - \tan^2 105°} =$

67. If θ is an obtuse angle so that $\tan\theta = -\dfrac{4}{3}$, find $\sin 2\theta$, $\cos 2\theta$, and $\tan 2\theta$.

68. Find the exact value of b.

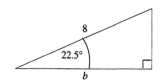

69. Let θ be an acute angle in a right triangle so that $\sin \theta = x$.

    (a) Find $\sin 2\theta$ and $\cos 2\theta$ in terms of x.

    (b) Find $\sin 4\theta$ in terms of x.

70. Use a double-angle formula to solve for the exact value of b.

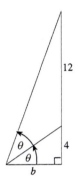

Verify the identity.

71. $\dfrac{1 + \cos 2x}{\sin 2x} = \cot x$

72. $\sin^4 x = \dfrac{1}{4} - \dfrac{1}{2} \cos 2x + \dfrac{1}{4} \cos^2 2x$

73. $\sin 4x = 4 \sin x \cos x - 8 \sin^3 x \cos x$

Section 7.6 Trigonometric Equations (Page 472)

Find the general solution of the equation.

74. $\sin x = \dfrac{1}{2}$ **75.** $\sec x = \dfrac{2}{\sqrt{3}}$ **76.** $\cot^2 x - 3 = 0$ **77.** $\tan 3x = -1$

Solve the equation for θ in the interval $[0°, 360°)$.

78. $\sqrt{3} \sec \theta = -2$

79. $\dfrac{1}{2} \sin \theta - 1 = 0$

80. $\sqrt{2} \cos \dfrac{\theta}{2} - 1 = 0$

81. $2 \sin^2 \theta \cot \theta = -1$

Solve the equation for x in the interval $0 \le x < 2\pi$.

82. $2 \sin^2 x - \sqrt{3} \sin x = 0$

83. $2 \cos^2 x - \cos x - 1 = 0$

84. $2 \sin^2 x + 9 \sin x - 5 = 0$

85. $\sec x \tan^2 x - 3 \sec x = 0$

86. $\cos 2x + \cos 4x + 1 = 0$

Solve the equation for x, rounded to the nearest tenth of a degree, in the interval $[0°, 360°)$.

87. $5 \cos x - 3 = 0$ **88.** $6 \tan^2 x - 7 \tan x - 3 = 0$

CHAPTER 7: STANDARD ANSWER TEST

Use these questions to test your knowledge of the basic skills and concepts of Chapter 7. Then check your answers with those given at the end of the book.

1. Use the given figure to write the following trigonometric ratios:

    (a) tan A **(b)** sin B **(c)** sec A

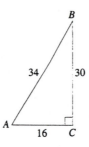

2. If $\angle A$ is an acute angle such that sec $A = \frac{3}{2}$, find sin A, cos A, and tan A.

3. Use the given triangle to show that

$$3(\sin^2 A)(\sec A) = \frac{x^2}{\sqrt{x^2 + 9}}$$

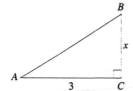

4. Solve for the hypotenuse of an isosceles right triangle, one of whose legs has length 20.

5. A building casts a shadow of 100 feet. The angle of elevation from the tip of the shadow to the top of the building is 75°. Find the height of the building correct to the nearest foot.

6. From the top of a cliff, the angle of depression to a point 120 meters from the base of the cliff (and on the same level as the base) is 47.8°. Find the height of the cliff to the nearest tenth of a meter.

In Questions 7–9 verify the identities.

7. $\cot^2 \theta - \cos^2 \theta = \cos^2 \theta \cot^2 \theta$

8. $\dfrac{1 - \sin \theta}{1 + \sin \theta} = (\sec \theta - \tan \theta)^2$

9. $\dfrac{\tan \theta - \sin \theta}{\csc \theta - \cot \theta} = \sec \theta - \cos \theta$

10. Use 165° = 135° + 30° and an appropriate addition formula to find the exact value of cos 165°.

11. Use a half-angle formula to find the exact value of $\sin \left(-\dfrac{\pi}{12} \right)$.

12. If sin $\theta = \frac{1}{3}$ and θ is acute, find sin 2θ using a double-angle formula.

13. Find the exact value of $\dfrac{\tan 115° - \tan 55°}{1 + \tan 115° \tan 55°}$ without using a calculator.

Verify the identities.

14. $\sin\left(\dfrac{\pi}{4} - \theta\right) = \dfrac{\sqrt{2}}{2}\,(\cos\theta - \sin\theta)$

15. $\csc 4x = \frac{1}{4}\,(\csc x)(\sec x)(\sec 2x)$

16. $\sin^2\dfrac{x}{2} = \dfrac{\tan x - \sin x}{2\tan x}$

Find the exact value of the expression.

17. $\tan\left(\arctan 3 - \arctan \frac{4}{9}\right)$ **18.** $\cos\left(2\arcsin \frac{1}{2}\right)$

19. Solve $\sin^2 x - \cos^2 x + \sin x = 0$ for x in the interval $0 \le x < 2\pi$ without using a calculator.

20. Solve $\cos 2x = 2\sin^2 x - 2$ for $0° \le x < 360°$ without using a calculator.

21. Find the general solution of $\sin 2x = 1$.

22. Use a calculator to solve for x to two decimal places:

$$5\cos^2 x - 17\cos x + 6 = 0 \text{ for } 0 \le x < 2\pi$$

23. Find $\cos(\alpha + \beta)$ if α terminates in quadrant II with $\sin\alpha = \frac{3}{5}$, and β terminates in quadrant III with $\tan\beta = \frac{12}{5}$.

24. $\triangle ABC$ has $\angle C = 90°$, $\angle A = 15°$, and $a = 8$. Find the exact value of b using a half-angle formula.

25. Use a double-angle formula to solve for b.

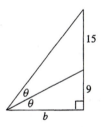

CHAPTER 7:   MULTIPLE CHOICE TEST

1. Triangle ABC is a right triangle with right angle at C. If $\tan A = \frac{2}{3}$, find $\sec B$.

(a) $\dfrac{\sqrt{13}}{3}$ (b) $\dfrac{\sqrt{13}}{2}$ (c) $\dfrac{2\sqrt{13}}{13}$ (d) $\dfrac{3\sqrt{13}}{13}$ (e) None of the preceding

2. Evaluate: $\sin 30° + \cos 60° - \tan 45°$

(a) 1   (b) -1 (c) 0   (d) $1 + \dfrac{\sqrt{2}}{2}$ (e) None of the preceding

3. From the top of a building the angle of depression of a point on the ground is 35°. The point is 50 meters from the base of the building. Which of the following can be used to determine the height x of the building?

(a) $\tan 35° = \dfrac{50}{x}$ (b) $\tan 55° = \dfrac{50}{x}$ (c) $\tan 55° = \dfrac{x}{50}$ (d) $\cot 35° = \dfrac{x}{50}$ (e) None of the preceding

4. Which of the following are true for all acute angles θ?

   I. $\sec^2 \theta - \tan^2 \theta = 1$

   II. $\csc^2 \theta - \cot^2 \theta = 1$

   III. $(\cos \theta)(\tan \theta) = \sin \theta$

   (a) Only I (b) Only II (c) Only III (d) I, II, and III (e) None of the preceding

5. Which of the following values of θ can be used to show that $\sin^2 \theta - \cos^2 \theta = 1$ is *not* an identity?

   (a) $\dfrac{\pi}{2}$ (b) $\dfrac{3\pi}{2}$ (c) $\dfrac{\pi}{4}$ (d) $-\dfrac{\pi}{2}$ (e) None of the preceding

6. When the expression $\sec^2 \theta + 2 \sec \theta \tan \theta + \tan^2 \theta$ is converted into an expression involving only $\sin \theta$, the result is

   (a) $\dfrac{1 + \sin^2 \theta}{1 - \sin^2 \theta}$ (b) $\dfrac{1 + \sin \theta}{1 - \sin \theta}$ (c) $\dfrac{1}{1 - \sin \theta}$ (d) $\dfrac{1 + 2 \sin \theta + \sin^2 \theta}{\sqrt{1 - \sin^2 \theta}}$ (e) None of the preceding

7. Complete: $\cos \dfrac{5\pi}{4} \cos \dfrac{\pi}{2} - \sin \dfrac{5\pi}{4} \sin \dfrac{\pi}{2} = $

   (a) $\cos \dfrac{7\pi}{4}$ (b) $\sin \dfrac{7\pi}{4}$ (c) $\cos \dfrac{3\pi}{4}$ (d) $-\sin \dfrac{3\pi}{4}$ (e) None of the preceding

8. Let $\sin \alpha = \frac{3}{5}$ and $\sin \beta = -\frac{4}{5}$ where α and β have terminal sides in quadrants II and III respectively. Then $\sin (\alpha - \beta) = $

   (a) -1 (b) 1 (c) $-\frac{7}{25}$ (d) $\frac{7}{25}$ (e) None of the preceding

9. Which of the following are true for $\cos 2\theta$?

   I. $\cos 2\theta = \sin^2 \theta - \cos^2 \theta$

   II. $\cos 2\theta = 2 \sin^2 \theta - 1$

   III. $\cos 2\theta = 2 \cos^2 \theta - 1$

   (a) Only I (b) Only II (c) Only III (d) I, II, and III (e) None of the preceding

10. If $\cos \theta = -\dfrac{5}{13}$ and θ is in the second quadrant, then $\sin 2\theta = $

    (a) $\dfrac{119}{169}$ (b) $\dfrac{24}{13}$ (c) $-\dfrac{60}{169}$ (d) $-\dfrac{120}{169}$ (e) None of the preceding

11. Use a half-angle formula to evaluate $\sin \dfrac{\pi}{8}$.

    (a) $\dfrac{\sqrt{2}}{4}$ (b) $\sqrt{2 + \sqrt{2}}$ (c) $\frac{1}{2}\sqrt{2 - \sqrt{2}}$ (d) $\frac{1}{4}\left(2 - \sqrt{2}\right)$ (e) None of the preceding

12. When the double-angle formula for the sine function is used to simplify $\dfrac{\sin 2x}{1 - \cos^2 x}$ the result is

    (a) $-2 \cot x$ (b) $2 \cot x$ (c) $\dfrac{2}{\sin x}$ (d) $\tan 2x$ (e) None of the preceding

13. Which of the following is the general solution of $\sin 2x = 1$, where k represents any integer?

    (a) $\dfrac{\pi}{4} + k\pi$ (b) $\dfrac{\pi}{2} + 2k\pi$ (c) $\dfrac{\pi}{4} + 2k\pi$ (d) $\dfrac{\pi}{2} + k\pi$ (e) None of the preceding

14. Solve for x in the interval $0 \le x < 2\pi$: $\sin^2 x = \sin x$

    (a) $0, \pi$ (b) $\dfrac{\pi}{2}, \dfrac{3\pi}{2}$ (c) $0, \dfrac{\pi}{2}, \dfrac{3\pi}{2}$ (d) $\dfrac{\pi}{2}, \pi$ (e) None of the preceding

15. At a certain time of day a 100-foot tree casts a 23-foot shadow. What is the angle of elevation from the tip of the shadow to the top of the tree when rounded to the nearest degree?

    (a) $13°$ (b) $43°$ (c) $76°$ (d) $77°$ (e) None of the preceding

16. Which of the following are the correct restrictions on θ for the expression $\dfrac{\cot \theta}{\sin \theta - \cos \theta}$?

    (a) θ is not coterminal with any quadrantal angle.

    (b) θ is not coterminal with $0, \pi, \dfrac{\pi}{4},$ and $\dfrac{5\pi}{4}$.

    (c) θ is not coterminal with 0 and π.

    (d) θ is not coterminal with $\dfrac{\pi}{2}$ and $\dfrac{3\pi}{2}$.

    (e) None of the preceding

17. The expression $\cos\left(\theta + \dfrac{\pi}{3}\right) - \cos\left(\theta - \dfrac{\pi}{3}\right)$ is equal to which of the following for all values of θ?

    (a) $\cos \theta$ (b) $\cos 2\theta$ (c) $\sqrt{3} \sin \theta$ (d) $-\sqrt{3} \sin \theta$ (e) None of the preceding

18. When $y = 3 \sin x - 3\sqrt{3} \cos x$ is converted into the form $y = c \sin(x + \theta)$, then θ could be which of these values (radians)?

    (a) $-\dfrac{\pi}{6}$ (b) $\dfrac{\pi}{6}$ (c) $-\dfrac{\pi}{3}$ (d) $-\dfrac{\pi}{4}$ (e) None of the preceding

19. The number of solutions for the equation $5 \sin^2 \theta - 2 \sin \theta = 0$ in the interval $0° \le \theta < 360°$ are

    (a) four   (b) three   (c) two   (d) one   (e) None of the preceding

20. If θ is the measure of an acute angle so that $\theta = \arccos x$, then $\sin(2 \arccos x) =$

    (a) $2\sqrt{1 - x^2}$ (b) $2x\sqrt{1 - x^2}$ (c) $\dfrac{2x}{\sqrt{1 - x^2}}$ (d) $2x^2$ (e) None of the preceding

ANSWERS TO THE TEST YOUR UNDERSTANDING EXERCISES

Page 428

1. and **2.**

$$\sin A = \tfrac{7}{25} = \cos B \qquad \cot A = \tfrac{24}{7} = \tan B$$

$$\cos A = \tfrac{24}{25} = \sin B \qquad \sec A = \tfrac{25}{24} = \csc B$$

$$\tan A = \tfrac{7}{24} = \cot B \qquad \csc A = \tfrac{25}{7} = \sec B$$

3. $\tfrac{49}{625} + \tfrac{576}{625} = \tfrac{625}{625} = 1$ **4.** $\left(\tfrac{7}{25}\right)\left(\tfrac{25}{7}\right) = 1$ **5.** $\tfrac{625}{49} - \tfrac{576}{49} = \tfrac{49}{49} = 1$ **6.** $\left(\tfrac{24}{7}\right)\left(\tfrac{7}{24}\right) = 1$

Page 436

1. $\dfrac{10\sqrt{3}}{3}$; 5.8   **2.** $10\sqrt{3}$; 17.3   **3.** $\dfrac{5\sqrt{2}}{2}$; 3.5

Page 438

1. 67.0   **2.** 55.0   **3.** 63.6   **4.** 22.1   **5.** 14.6°   **6.** 5.3°

Page 454

1. $\sin 17°$ **2.** $\cos \dfrac{\pi}{10}$ **3.** $\tan 5°$ **4.** $\sin 10°$

5. $\tfrac{1}{4}\left(\sqrt{6} - \sqrt{2}\right)$; $\tfrac{1}{4}\left(\sqrt{6} + \sqrt{2}\right)$; $\dfrac{3 - \sqrt{3}}{3 + \sqrt{3}} = 2 - \sqrt{3}$

6. Same as Exercise 5. $\left(Note: \dfrac{\sqrt{3} - 1}{1 + \sqrt{3}} \cdot \dfrac{\sqrt{3}}{\sqrt{3}} = \dfrac{3 - \sqrt{3}}{3 + \sqrt{3}} = 2 - \sqrt{3}\right)$

7. $\tfrac{1}{4}\left(\sqrt{6} - \sqrt{2}\right)$; $-\tfrac{1}{4}\left(\sqrt{6} + \sqrt{2}\right)$; $\dfrac{1 - \sqrt{3}}{1 + \sqrt{3}} = -2 + \sqrt{3}$

8. Same as Exercise 7.   **9.** $\tfrac{1}{4}\left(\sqrt{2} - \sqrt{6}\right)$ **10.** $\tfrac{1}{4}\left(\sqrt{6} - \sqrt{2}\right)$ **11.** $-\tfrac{36}{85}$ **12.** $-\tfrac{77}{85}$

Page 463

1. $\sin 10°$ **2.** $\tan \theta$ **3.** $\cos 6\theta$ **4.** $\cos\dfrac{\alpha}{3}$ **5.** $\dfrac{\sqrt{3}}{2}$ **6.** $\dfrac{\sqrt{2}}{2}$ **7.** 2   **8.** $-\dfrac{\sqrt{3}}{2}$

Additional Applications of Trigonometry

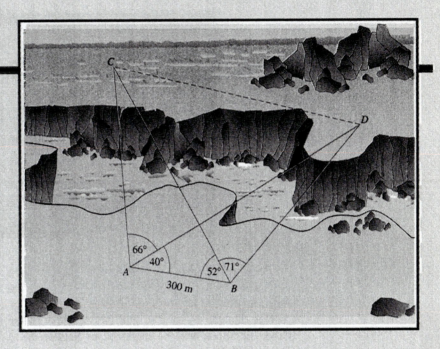

A surveying team, located on one side of a deep canyon, wants to find the distance between points C and D on the other side of the canyon. To accomplish this they first select two points, A and B, that are 300 meters apart on their side of the canyon and then they use *triangulation*. That is, they partition quadrilateral *ABDC* into triangles by drawing the diagonals *AD* and *BC*.

In particular, triangles *ABC* and *BAD* will be used. At points A and B they are able to measure the angles $\angle BAD$, $\angle DAC$, $\angle ABC$, and $\angle CBD$. As the angle measurements shown on the figure indicate, there are no right angles involved. Consequently, the earlier methods used to find inaccessible distances using right triangle trigonometry do not apply here.

However, in this chapter two new procedures, the *law of cosines* and the *law of sines*, will be developed that will make it possible to solve this and many other problems that call for finding unknown parts of triangles that do not involve right angles (see Exercise 31, page 509).

8.1
THE LAW
OF COSINES

Suppose that the distance across a lake, shown as AB in the following figure, cannot be measured directly. It cannot be found by using the methods studied previously in Section 7.2, since that earlier work depended on acute angles within right triangles. However, now that trigonometric ratios are available for obtuse angles, we will be able to find this distance. (See Exercise 17.)

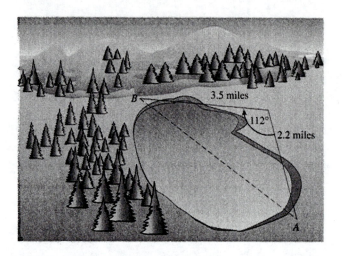

First recall from your work in geometry that a triangle is determined (that is, can be constructed) when you are given the three sides (SSS) or two sides and the included angle (SAS). For both of these cases there is a formula known as the **Law of Cosines** that can be used to find the measures of the remaining parts of the triangle. (Other situations will be explored in Section 8.2.) The derivation of the Law of Cosines follows.

In each of the following figures, vertex A is placed at the origin ($\angle A$ is in standard position) and side AC coincides with the x-axis. The measure of $\angle A$ is denoted by the letter A itself. (This dual use of A, as a vertex and as the angle measure, will not cause confusion since the context will make it clear how A is being used.) The coordinates of C are $(b, 0)$. To find the coordinates of B, first construct altitude BD. From right $\triangle ABD$ we have

Observe that in each case $\triangle ABD$ would be similar to the reference triangle determined by the unit circle. Thus, since the trigonometric ratios are independent of the size of the reference triangle, we use the sides of $\triangle ABD$ to form the ratios.

$$\cos A = \frac{AD}{c} \qquad \text{or} \qquad AD = x = c \cos A \quad \longleftarrow \quad \text{the } x\text{-coordinate of } B$$

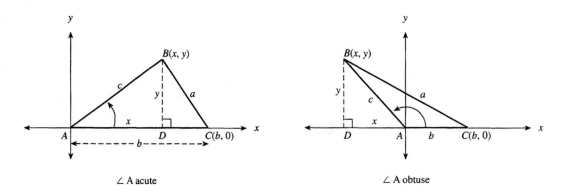

∠ A acute   ∠ A obtuse

Similarly,

$$\sin A = \frac{DB}{c} \quad \text{or} \quad DB = y = c \sin A \quad \longleftarrow \quad \text{the } y\text{-coordinate of } B$$

Apply the distance formula (page 248) to the points $B\,(c \cos A,\ c \sin A)$ and $C\,(b,\ 0)$.

$$a^2 = (c \cos A - b)^2 + (c \sin A - 0)^2$$

$$= c^2\cos^2 A - 2bc \cos A + b^2 + c^2 \sin^2 A$$

$$= b^2 + c^2(\cos^2 A + \sin^2 A) - 2bc \cos A$$

Since $\cos^2 A + \sin^2 A = 1$, we have

$$\mathbf{\mathit{a^2 = b^2 + c^2 - 2bc \cos A}}$$

What happens if angle A is a right angle?

If, in turn, $\triangle ABC$ is oriented so that B and C coincide with the origin, then similar formulas can be derived for b^2 and c^2. These results are summarized in the following.

You may find it easier to remember this law by using this verbalized form: The square of any side of a triangle equals the sum of the squares of the remaining two sides minus twice their product times the cosine of their included angle.

THE LAW OF COSINES

For any $\triangle ABC$ with angle measures A, B, C, and sides of length a, b, c,

$$a^2 = b^2 + c^2 - 2bc \cos A$$

$$b^2 = a^2 + c^2 - 2ac \cos B$$

$$c^2 = a^2 + b^2 - 2ab \cos C$$

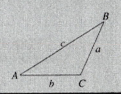

The Law of Cosines can be used to solve for a side of a triangle when the other two sides and their included angle are given. This is known as the case **Side-Angle-Side (SAS)** and is illustrated in Example 1.

EXAMPLE 1   Solve for c in $\triangle ABC$ if $a = 4$ cm, $b = 7$ cm, and $C = 130°$.

Solution   Make a sketch of the triangle showing the given information.

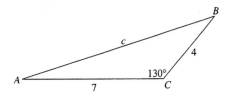

Apply the Law of Cosines:

$$c^2 = a^2 + b^2 - 2ab \cos C$$

$$= 4^2 + 7^2 - 2(4)(7) \cos 130° = 100.9961061$$

Thus c is approximately 10 cm.   ∎

The Law of Cosines can also be used to solve for an angle measure of a triangle when the three sides are given. This is known as the case **Side-Side-Side (SSS).** First, solve $a^2 = b^2 + c^2 - 2bc \cos A$ for $\cos A$ as follows:

$$a^2 = b^2 + c^2 - 2bc \cos A$$

$$2bc \cos A = b^2 + c^2 - a^2$$

$$\cos A = \frac{b^2 + c^2 - a^2}{2bc}$$

Similarly, for $\cos B$ and $\cos C$ the results are:

$$\cos B = \frac{a^2 + c^2 - b^2}{2ac} \quad \text{and} \quad \cos C = \frac{a^2 + b^2 - c^2}{2ab}$$

EXAMPLE 2 Solve for angle measures A and B for $\triangle ABC$ in Example 1.

Solution Apply the Law of Cosines for $\cos A$ using $a = 4$, $b = 7$, and $c = 10$:

$$\cos A = \frac{b^2 + c^2 - a^2}{2bc}$$

$$= \frac{7^2 + 10^2 - 4^2}{2(7)(10)} = 0.95$$

Then $A = \arccos(0.95) = 18°$, rounded to the nearest degree. The measure B can now be found using the fact that the sum of the angle measures of a triangle is $180°$.

$$B = 180° - (A + C)$$

$$= 180° - (18° + 130°)$$

$$= 32° \qquad \blacksquare$$

EXAMPLE 3 Solve for the exact value of a in $\triangle ABC$ if $b = 15$, $c = 10$, and $A = 120°$.

Solution

$$a^2 = b^2 + c^2 - 2bc \cos A$$

$$= 15^2 + 10^2 - 2(15)(10) \cos 120°$$

$$= 225 + 100 - 2(15)(10) \left(-\frac{1}{2} \right)$$

$$= 325 + 150 = 475$$

Then

$$a = \sqrt{475} = \sqrt{25 \cdot 19} = 5\sqrt{19} \qquad \blacksquare$$

Use the Law of Cosines to solve for the indicated part of $\triangle ABC$. Find the exact values without using a calculator.

1. $a = 6$, $b = 4$, $C = 60°$; find c.

2. $b = 15$, $c = 10\sqrt{2}$, $A = \dfrac{\pi}{4}$; find a.

3. $a = 5\sqrt{3}$, $c = 12$, $B = 150°$; find b.

4. $a = \sqrt{67}$, $b = 7$, $c = 9$; find A.

5. $a = 8$, $b = \sqrt{34}$, $c = 3\sqrt{2}$; find B.

6. $a = 4$, $b = 6$, $c = \sqrt{76}$; find C.

(Answers: Page 547)

Example 4 illustrates the application of the Law of Cosines to finding an inaccessible distance between two points.

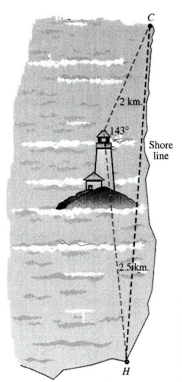

EXAMPLE 4 An offshore lighthouse is 2 kilometers from a Coast Guard station C and 2.5 kilometers from a hospital H near the shoreline. If the angle formed by light beams to C and H is 143°, what is the distance CH (by sea) between the Coast Guard station and the hospital?

Solution Use the Law of Cosines to solve for CH:

$$(CH)^2 = 2^2 + (2.5)^2 - 2(2)(2.5)(\cos 143°) = 18.2363551$$

Taking the square root and rounding off to the nearest tenth, we have

$$CH = 4.3 \text{ kilometers} \qquad \blacksquare$$

The final application shows how the Law of Cosines is used to derive a formula for the area of a triangle known as **Heron's formula.** To derive this formula, it is necessary to first show that the area of a triangle can be found according to this rule:

AREA OF A TRIANGLE

The area K of a triangle is one-half the product of two sides and the sine of the included angle.

$$K = \tfrac{1}{2} bc \sin A$$

$$K = \tfrac{1}{2} ac \sin B$$

$$K = \tfrac{1}{2} ab \sin C$$

A proof for the first of these formulas follows:

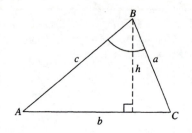

$$\sin A = \frac{h}{c} \quad \text{or} \quad h = c \sin A$$

Geometric formulas are given inside the front cover.

$$K = \tfrac{1}{2} bh$$

$$= \tfrac{1}{2} bc \sin A$$

EXAMPLE 5   Find the area of the triangle shown in the margin:

Solution

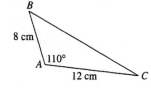

$$K = \tfrac{1}{2} \, bc \sin A$$

$$= \tfrac{1}{2} \, (12)(8) \sin 110° = 45.1052458$$

The area is approximately 45 cm^2.   ■

HISTORICAL NOTE
Heron (or Hero) of Alexandria lived about A.D. 75. Although his name is associated with this formula, it was very likely discovered by Archimedes. This is a frequent event in history. Thus, for example, Pascal's triangle (see Section 12.3) was known by the Chinese at least 300 years prior to Pascal's time.

To derive Heron's formula we begin by squaring $K = \tfrac{1}{2} bc \sin A$ to obtain

$$K^2 = \tfrac{1}{4} b^2 c^2 \sin^2 A$$

$$= \tfrac{1}{4} b^2 c^2 (1 - \cos^2 A) \quad \longleftarrow \quad \sin^2 A = 1 - \cos^2 A$$

$$= \tfrac{1}{4} b^2 c^2 (1 + \cos A)(1 - \cos A)$$

By the Law of Cosines, $\cos A = \dfrac{b^2 + c^2 - a^2}{2bc}$. Then, substituting for cos A into the preceding, you can verify (see Exercise 38) this result:

$$K^2 = \tfrac{1}{16}(a + b + c)(b + c - a)(a + c - b)(a + b - c)$$

Then

Observe that the factor $b + c - a = a + b + c - 2a$; similarly, for the last two factors.

$$K^2 = \tfrac{1}{16}(a + b + c)(a + b + c - 2a)(a + b + c - 2b)(a + b + c - 2c)$$

Now take the **semiperimeter** $s = \dfrac{1}{2}(a + b + c)$ of the triangle and substitute $2s$ for $a + b + c$ in the equation of K^2 to obtain

$$K^2 = \tfrac{1}{16}(2s)(2s - 2a)(2s - 2b)(2s - 2c)$$

$$K^2 = s(s - a)(s - b)(s - c)$$

Finally, take the positive square root to obtain Heron's formula.

HERON'S FORMULA

The area K of a triangle with sides of length a, b, and c is given by

$$K = \sqrt{s(s - a)(s - b)(s - c)}$$

where $s = \frac{1}{2}(a + b + c)$.

EXAMPLE 6   Find the area of the triangle using Heron's formula.

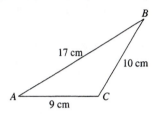

Solution   First find the semiperimeter, s.

$$s = \frac{1}{2}(a + b + c) = \frac{1}{2}(10 + 9 + 17) = 18$$

Then

$$
\begin{aligned}
K &= \sqrt{s(s - a)(s - b)(s - c)} \\
&= \sqrt{18(18 - 10)(18 - 9)(18 - 17)} \\
&= \sqrt{1296} \\
&= 36
\end{aligned}
$$

The area is 36 cm^2.   ∎

EXERCISES 8.1

Use the Law of Cosines to solve for the indicated parts of $\triangle ABC$. Do not use a calculator.

1. Solve for a if $b = 4$, $c = 11$, and $A = 60°$.

2. Solve for b if $a = 20$, $c = 8$, and $B = 45°$.

3. Solve for A if $a = 5\sqrt{3}$, $b = 10\sqrt{3}$, and $c = 15$.

4. Solve for B if $a = 9\sqrt{2}$, $b = \sqrt{337}$, and $c = 7$.

Solve for the indicated parts of $\triangle ABC$. Give the sides to the nearest tenth and the angle measure to the nearest tenth of a degree.

5. $A = 20°$, $b = 8$, $c = 13$; find a and B.

6. $a = 9$, $c = 14$, $B = 110°$; find b and C.

7. $a = 12$, $b = 5$, $c = 13$; find C and A.

8. $a = b = c = 10$; find A, B, and C.

9. $a = 18$, $b = 15$, $c = 4$; find B and C.

10. $A = 65.5°$, $b = 4$, $c = 11$; find a and C.

11. $a = 18$, $b = 9$, $C = 30.2°$; find c and A.

12. $a = 15$, $c = 5$, $B = 157.5°$, find b and A.

13. $b = 2.2$, $c = 6.4$, $A = 42°$; find a and B.

14. $a = 60$, $b = 20$, $c = 75$; find A and B.

15. Two points A and B are on the shoreline of a lake. A surveyor is located at a point C where $AC = 180$ meters and $BC = 120$ meters and finds that $\angle ACB$ has measure $56.3°$. What is the distance between A and B to the nearest meter?

16. Point C is 2.7 kilometers from a house A and 3 kilometers from house B, where A and B are on opposite sides of a valley. If $\angle ACB$ has measure $130.1°$, find the distance between the houses to the nearest kilometer.

17. Find the distance AB, discussed in the opening to this section, to the nearest tenth of a mile.

18. A diagonal of a parallelogram has length 80 and makes an angle of $20°$ with one of the sides. If this side has length 34, find the length of the other side of the parallelogram to the nearest unit.

19. The equal sides of an isosceles trapezoid are 10 cm, and the shorter base is 14 cm. If one pair of equal angles has a measure of $40°$ each, find the length of a diagonal to the nearest tenth of a centimeter.

20. The equal sides of an isosceles triangle are each 30 units long and the vertex angle is $27.7°$. Find the base to the nearest tenth of a unit, and the measure of the base angles to the nearest tenth of a degree.

21. In the accompanying figure, solve for AB to the nearest tenth.

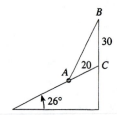

22. The four bases of a baseball diamond form a square 90 feet on a side. The shortstop S is in a position that is 50 feet from second base and forms a $15°$ angle with the base path as shown. Find the distance between the shortstop and first base to the nearest foot.

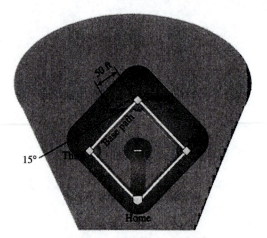

23. In Exercise 22, find the distance between the shortstop and home plate to the nearest foot.

24. Points A and B are the endpoints of a proposed tunnel through a mountain. From a point P, away from the mountain, a surveyor is able to see both points A and B. The surveyor finds that $PA = 620$ meters, $PB = 450$ meters, and $\angle APB$ has measure $83.3°$. In the figure at the top of page 497, find the length AB of the tunnel to the nearest meter.

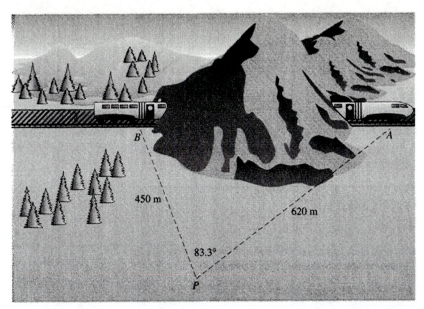

25. Two trains leave the same station at 11 A.M. and travel in straight lines at speeds of 54 miles per hour and 60 miles per hour, respectively. If the difference in their direction is 124°, how far apart are they at 11:20 A.M. to the nearest tenth of a mile?

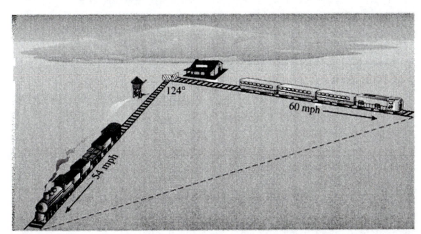

26. Prove that for any △ABC, $c = a \cos B + b \cos A$. (*Hint*: consider the cases where ∠A is acute, obtuse, or a right angle, and put ∠A in standard position.)

Find the area of △ABC having the given parts.

27. $A = 30°$, $b = 25$, $c = 18$

28. $A = 82°$, $b = 14$, $c = 31$

29. $B = 40.5°$, $a = 8.4$, $c = 12.6$

30. Find the area of quadrilateral *ABCD*. (Give the answer to the nearest square unit.)

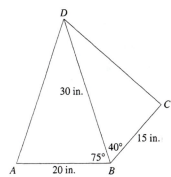

Find the area of the triangle using $K = \frac{1}{2}(base)(height);$ *then verify that Heron's formula gives the same result.*

31.

32.

33.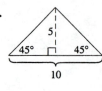

Use Heron's formula to find the areas of the triangles with the given sides.

34. $a = 10$ ft, $b = 13$ ft, $c = 20.1$ ft

35. $a = 10$ ft, $b = 13$ ft, $c = 3.4$ ft

36. $a = 210$ cm, $b = 280$ cm, $c = 425$ cm

37. Use the Law of Cosines to find AD and CD in Exercise 30. Then use Heron's formula to find the area of quadrilateral $ABCD$ and compare to the result in Exercise 30.

38. Show that $K^2 = \frac{1}{4} b^2 c^2 (1 + \cos A)(1 - \cos A)$ is equivalent to

$$K^2 = \tfrac{1}{16}(a + b + c)(b + c - a)(a + c - b)(a + b - c)$$

by following these directions (see page 494):

(a) Substitute $\dfrac{b^2 + c^2 - a^2}{2bc}$ for $\cos A$ and obtain the form

$$K^2 = \tfrac{1}{16}(2bc + b^2 + c^2 - a^2)(2bc - b^2 - c^2 + a^2)$$

(b) Rewrite the factor $2bc + b^2 + c^2 - a^2$ as the difference of two squares $u^2 - a^2$, and factor. Do the same for the last factor $2bc - b^2 - c^2 + a^2$ and obtain the stated result.

39. Use the Law of Cosines to derive each of the following for $\triangle ABC$.

(a) $\dfrac{\cos A}{a} + \dfrac{\cos B}{b} + \dfrac{\cos C}{c} = \dfrac{a^2 + b^2 + c^2}{2abc}$ **(b)** $2(ab \cos C + ac \cos B + bc \cos A) = a^2 + b^2 + c^2$

40. The equations in Exercise 39 can be used to check the results when solving for the parts of a triangle. Find the remaining parts of Example 3, page 492, and check the results using the equation in Exercise 39(b).

?️ Challenge

1. Find the area of the equilateral triangle using $K = \frac{1}{2}(base)(height);$ then verify that Heron's formula gives the same result.

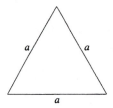

2. Use Heron's formula to verify that the area of the isosceles triangle is $\frac{1}{2}a^2$.

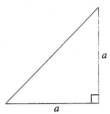

 Written Assignment   The proof of the area formula $K = \frac{1}{2}bc \sin A$ was done on page 494 using a triangle in which $\angle A$ was an acute angle. Explain why such a proof also applies to the case when $\angle A$ is obtuse.

 Graphing Calculator Exercises

1. If we fix the length of two sides of a triangle, then the length of the third side is a function of its opposite angle. Suppose that triangle ABC has sides of length a, b, and c with $a = 3$ and $b = 4$. If the measure of angle C is x radians, **(a)** write c as a function of x, $0 \le x \le \pi$. **(b)** Graph this function. **(c)** Determine for what value of x we get a maximum c, and **(d)** a minimum c.

2. Suppose that the lengths of sides a and b in triangle ABC are both equal to 1. Then Heron's formula makes the area K a function of the length x of the third side c. **(a)** Express K as a function of x, $0 \le x \le 2$, and **(b)** graph this function. **(c)** Approximate the value of x that makes K a maximum, to the nearest thousandth for each.

**8.2
THE LAW
OF SINES**

The Law of Cosines has been used to solve triangles that involve the cases SAS and SSS. However, this law is not adequate for each of the following cases:

> Two angles and a side (**ASA** or **AAS**)
>
> Two sides and an angle *not* included (**SSA**)

For these cases we will need to use another property of triangles called the **Law of Sines.** To establish this new result, first construct altitude DB for $\triangle ABC$, as shown. From right triangles ABD and CBD, we have

$$\sin A = \frac{DB}{c} \qquad \text{and} \qquad \sin C = \frac{DB}{a}$$

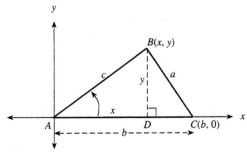

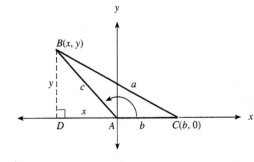

See Exercise 30 for the case
$A = 90°.$

$\angle$ A acute

$\angle$ A obtuse

Then $DB = c \sin A$ and $DB = a \sin C$, which gives

$$a \sin C = c \sin A$$

$$\frac{a}{\sin A} = \frac{c}{\sin C} \qquad \text{Why?}$$

The "double" equality is an abbreviation for these three results:

$$\frac{a}{\sin A} = \frac{b}{\sin B}$$

$$\frac{a}{\sin A} = \frac{c}{\sin C}$$

$$\frac{b}{\sin B} = \frac{c}{\sin C}$$

Also, another version of this law is obtained by taking reciprocals:

$$\frac{\sin A}{a} = \frac{\sin B}{b} = \frac{\sin C}{c}$$

Similar reasoning produces $\dfrac{a}{\sin A} = \dfrac{b}{\sin B}$, which is combined with the first result into the following:

> ### THE LAW OF SINES
>
> For any $\triangle ABC$ with angle measures A, B, C and sides of length a, b, c,
>
> $$\frac{a}{\sin A} = \frac{b}{\sin B} = \frac{c}{\sin C}$$

Our first two examples demonstrate the use of the Law of Sines when two angles and one side are given.

EXAMPLE 1 Solve $\triangle ABC$ if $B = 75°$, $a = 5$, and $C = 41°$.

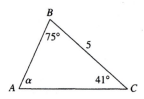

To "solve a triangle" means to solve for each of the unknown parts.

Solution Since two angles are given, the third is easy to find because the sum of the angles of a triangle is $180°$.

$$A = 180° - (75° + 41°) = 64°$$

Now use the Law of Sines to solve for b, and substitute the appropriate values. From $\dfrac{a}{\sin A} = \dfrac{b}{\sin B}$ we obtain $b = \dfrac{a \sin B}{\sin A}$. Then

$$b = \frac{5 \sin 75°}{\sin 64°} = 5.4 \qquad \text{rounded to tenths}$$

If necessary consult your calculator manual for an appropriate key sequence for this computation.

Using a graphing calculator this computation can be performed with this sequence:

On a scientific calculator, the following sequence might be used:

$$\boxed{5}\ \boxed{\times}\ \boxed{75}\ \boxed{\text{SIN}}\ \boxed{\div}\ \boxed{64}\ \boxed{\text{SIN}}\ \boxed{=}$$

Solving for c is done similarly. Thus,

$$c = \frac{a \sin C}{\sin A} = \frac{5 \sin 41°}{\sin 64°} = 3.6$$ ■

EXAMPLE 2   To the nearest meter, find the distance $AB = c$ across the pond if $B = 108°$, $C = 39°$, and $AC = 950$ meters.

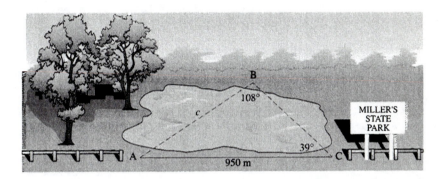

Observe that AAS also fits the case ASA, since two given angles of a triangle determine the third angle.

Solution $\dfrac{c}{\sin C} = \dfrac{b}{\sin B}$ can be used to solve for c since only c is unknown. Thus

$$c = \frac{b \sin C}{\sin B}$$

$$= \frac{950 \sin 39°}{\sin 108°} = 628.62129$$

The distance AB is 629, to the nearest meter. ■

TEST YOUR UNDERSTANDING

(Answers: Page 547)

Use the Law of Sines to answer the following without using a calculator.

1. Find a if $A = 30°$, $B = 45°$, and $b = 8$.
2. Find c if $A = 30°$, $C = 120°$, and $a = 12$.
3. Find b if $B = 60°$, $C = 75°$, and $a = 15$.
4. Find b if $A = 15°$, $B = 135°$, and $c = 4$.

When two sides and an angle opposite one of the sides (SSA) are given, there are a number of possibilities. The following figures illustrate the four possible cases when $\angle A$ is acute. It is assumed that parts a, A, and b are given. Note that the altitude opposite $\angle A$ is given by $h = b \sin A$.

CASE 1. No solution: $a < b \sin A$

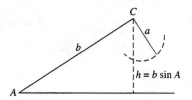

CASE 2. One solution: $a = b \sin A$

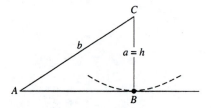

CASE 3. Two solutions: $b \sin A < a < b$

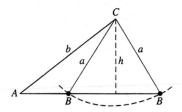

CASE 4. One solution: $a \geq b$

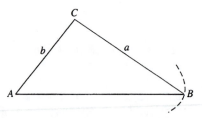

The preceding descriptions are independent of the symbols chosen. Any choice of symbols could be used as long as we have the case SSA.

When $\angle A$ is obtuse, there are three possibilities:

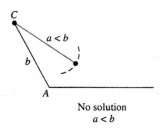

No solution
$a < b$

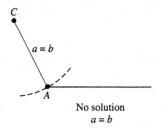

No solution
$a = b$

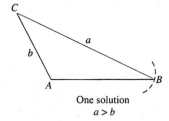

One solution
$a > b$

EXAMPLE 3 Solve $\triangle ABC$ if $a = 50$, $b = 65$, and $A = 57°$.

Solution $b \sin A = 65 \sin 57° = 54.5$. Since $a < 54.5$ we have $a < b \sin A$ and there is no solution possible (case 1). ■

The next example illustrates Case 3, in which $b \sin A < a < b$. This is called the *ambiguous case* because two triangles are possible and, therefore, there is not a unique solution. The two angles obtained are supplementary, and therefore have the same sine value.

EXAMPLE 4 Approximate the remaining parts of $\triangle ABC$ if $a = 10$, $b = 13$, and $A = 25°$.

Solution $b \sin A = 13 \sin 25° = 5.5$. Since $5.5 < 10 < 13$, it follows that $b \sin A < a < b$ (case 3) and there are two solutions for B. From the Law of Sines,

$$\sin B = \frac{b \sin A}{a} = \frac{13(\sin 25°)}{10} = .5494$$

Note that sin B has been rounded to four decimal places, and the measures of the angles and the sides are rounded to tenths.

Then, $B = \arcsin .5494 = 33.3°$ and the solution for the supplementary angle is $180° - 33.3° = 146.7°$.

(a) If $B = 33.3°$, then $C = 180° - (A + B) = 121.7°$ and

$$c = \frac{a \sin C}{\sin A} = \frac{10 \sin(121.7°)}{\sin 25°} = 20.1$$

(b) If $B = 146.7°$, we get $C = 8.3°$ and $c = 3.4$.

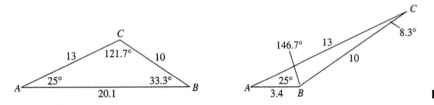

■

The Laws of Sines and Cosines can be helpful in solving applied problems using vectors. A **vector** is a directed line segment such as the following:

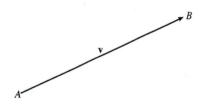

*Boldface is used to distinguish between the geometric vector **v** and the number or variable v. Since it is clumsy to reproduce boldface by hand, you might find it simpler to use the symbol $\vec{v}$ instead.*

A is the *initial point* and B is the *terminal point* of the vector denoted by $\overrightarrow{AB}$. A letter in boldface type, such as **v**, is also used to represent a vector.

Two vectors that have the same length and direction are said to be equal; the

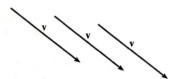

location of the vector in the plane is arbitrary. Thus the same vector **v** in three locations is shown in the margin.

A vector PQ is used in applications to represent a quantity that has both magnitude and direction. The length of $\overrightarrow{PQ}$, denoted by $|\overrightarrow{PQ}|$, is the magnitude, and the direction of $\overrightarrow{PQ}$ is the direction from the initial point P to the terminal point Q. When a vector **v** is used for a quantity with magnitude and direction, we locate **v** in a position that is convenient to the situation. For example:

*In each case **v** has the same length and direction.*

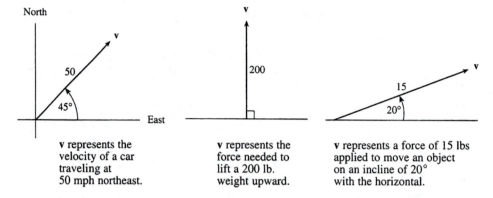

v represents the velocity of a car traveling at 50 mph northeast.

v represents the force needed to lift a 200 lb. weight upward.

v represents a force of 15 lbs applied to move an object on an incline of 20° with the horizontal.

When two vectors are added, the sum is a vector called the **resultant.** It is found using the *parallelogram law* in which the sum is the diagonal of the parallelogram determined by two vectors such as $\overrightarrow{AB}$ and $\overrightarrow{AC}$ as shown.

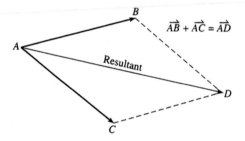

In certain applications involving moving objects, the direction of the motion is stated as a **bearing.** This is the direction given by the acute angle made with the north-south line. Thus, a bearing of North 30° East, abbreviated as N 30° E, means the object is traveling in a direction that is 30° east of the north-south line. A bearing of South 30° West, abbreviated as S 30° W, means the object is traveling in a direction that is 30° west of the south-north line, as shown. (Note that the letter N or S is written first.)

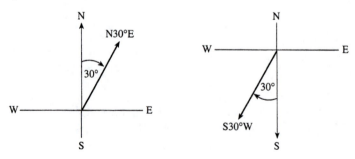

*The **heading** of a moving object is the intended direction.*

EXAMPLE 5 A motorboat starts at the south shore of a river and is *heading* due north to the opposite shore. The boat's speed (in still water) is 15 mph, and the river is flowing due east at 4 mph. What is the actual speed of the boat, and what is the final bearing?

Solution Draw a diagram. Represent the boat's speed by $\overrightarrow{AB}$ and the river's speed by $\overrightarrow{AC}$. The resultant $\overrightarrow{AD} = \overrightarrow{AC} + \overrightarrow{CD}$ gives the final bearing, and its length is the actual speed.

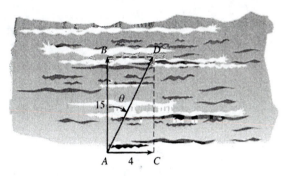

The Pythagorean theorem gives the magnitude.

$$|\overrightarrow{AD}|^2 = 15^2 + 4^2 = 241$$

Thus the actual speed is $\sqrt{241}$ or about 15.5 mph. Also,

$$\tan \theta = \frac{4}{15} = .2667$$

Then $\theta = \arctan \dfrac{4}{15} = 15°$ to the nearest degree, and the direction of the boat is given by the bearing N 15°E. ∎

EXAMPLE 6 An airplane has an airspeed (the speed in still air) of 240 mph with a bearing of S 30°W. If a wind is blowing due west at 40 mph, find the final bearing of the plane and the ground speed (the speed relative to the ground).

Refer to the figure at the top of page 506 as you study this solution.

Solution Draw a diagram in which **u** represents the velocity of the plane ($|\mathbf{u}| = 240$) and **v** represents the velocity of the wind ($|\mathbf{v}| = 40$). Then the magnitude of the resultant $\mathbf{w} = \mathbf{u} + \mathbf{v}$ is the ground speed. If from the tip of $\overline{\mathbf{u}}$ a perpendicular is drawn to the north–south line, a right triangle is formed. Then the angle of 120° shown in the figure is justified, since this is an exterior angle of this triangle. Now use the Law of Cosines to find $|\mathbf{w}|$.

$$|\mathbf{w}|^2 = 240^2 + 40^2 - 2(40)(240) \cos 120$$
$$= 68,800$$

Then the ground speed $|\mathbf{w}|$ is approximately 262 mph. Also, to find θ, we first find α using the Law of Sines as follows:

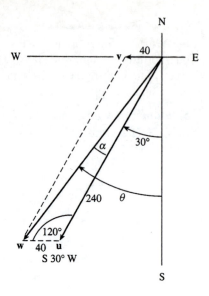

$$\frac{40}{\sin \alpha} = \frac{262}{\sin 120°}$$

$$\sin \alpha = \frac{40 \sin 120°}{262} = 0.1322$$

Then $\alpha = 8°$ to the nearest degree, $\theta = 30° + 8° = 38°$ and the final bearing of the plane is S 38° W. ∎

It is also possible to reverse the addition procedure for vectors. That is, a given vector can be *resolved* into the sum of two *component* vectors, whose sum is the given vector. The figure in the margin demonstrates how vector AB is resolved into the sum of two perpendicular components. This was done by first drawing perpendicular lines through A and then drawing lines through B to complete the parallelogram. The final example makes use of this procedure.

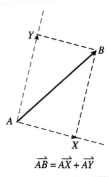

$\vec{AB} = \vec{AX} + \vec{AY}$

$\overline{AB}$

EXAMPLE 7 A 400-pound packing crate is sitting on a ramp that makes a 25° angle with the horizontal. Find the force required to hold the crate from sliding down the ramp. (Assume that there is no friction.)

Solution In the diagram $\vec{PQ}$ is the force vector due to gravity. Thus $|\vec{PQ}| = 400$.

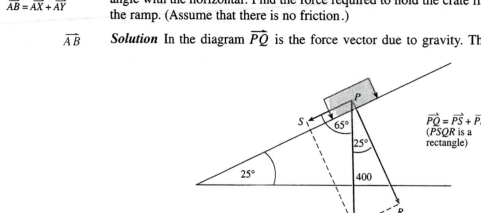

$\vec{PQ} = \vec{PS} + \vec{PR}$
(*PSQR* is a rectangle)

Now resolve $\overrightarrow{PQ}$ into the sum of two *component* vectors $\overrightarrow{PR}$ and $\overrightarrow{PS}$ so that $\overrightarrow{PS}$ is parallel to the ramp, and $\overrightarrow{PR}$ is perpendicular to the ramp. Observe that $\overrightarrow{PS}$ is the-force vector that moves the crate down the ramp. Then a force equal to $|\overrightarrow{PS}|$, acting in the opposite direction, will hold the crate in place. (The magnitude of vector $\overrightarrow{PR}$ is the force that the crate exerts on the ramp, and is counteracted by the ramp itself.) From the diagram,

Observe that a force over 400 pounds is needed to lift the crate vertically, and just over 169 pounds is needed to move it upward on the ramp: an advantage of $400 - 169 = 231$ pounds.

$$|\overrightarrow{PS}| = 400 \cos 65° = 169.04$$

Therefore, a force of 169 pounds (to the nearest pound) is needed to hold the crate in place. ∎

EXERCISES 8.2

Use the Law of Sines to solve for the indicated parts of $\triangle ABC$. Do not use a calculator.

1. Solve for a and c if $A = 30°$, $B = 120°$, and $b = 54$.

2. Solve for c if $C = 30°$, $A = 135°$, and $a = 100$.

3. Solve for C if $a = 12$, $c = 4\sqrt{3}$, and $A = \dfrac{2\pi}{3}$.

4. Solve for B if $a = 5\sqrt{2}$, $b = 10$, and $A = \dfrac{\pi}{6}$.

Use the Law of Sines to solve $\triangle ABC$.

5. $A = 25°$, $C = 55°$, $b = 12$ **6.** $C = 110°$, $B = 28°$, $a = 8$

7. $A = 62.2°$, $B = 50°$, $b = 5$ **8.** $A = 155°$, $B = 15.5°$, $c = 20$

Determine the number of triangles possible with the given parts.

9. $A = 32°$, $a = 5.1$, $b = 10$ **10.** $A = 32°$, $a = 6$, $b = 10$

11. $A = 30°$, $a = 9.5$, $b = 19$ **12.** $A = 50°$, $a = 15$, $b = 14.3$

13. $A = 126°$, $a = 20$, $b = 25$ **14.** $A = 77°$, $a = 49$, $b = 50$

Use the Law of Sines to solve for the indicated parts of $\triangle ABC$ whenever possible.

15. Solve for B and c if $A = 53°$, $a = 12$, and $b = 15$.

16. Solve for C and a if $B = 122°$, $b = 20$, and $c = 8$.

17. Solve for B if $b = 25$, $a = 7$, and $A = 75°$.

18. Solve for B if $A = 44°$, $a = 9$, and $b = 12$.

19. Solve for C if $B = 22.7°$, $b = 25$, and $c = 30$.

20. Solve for C if $B = 22.7°$, $b = 8$, and $c = 30$.

21. Solve for the larger angle of the given parallelogram.

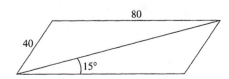

22. Two guy wires support a telephone pole. They are attached to the top of the pole and are anchored into the ground on opposite sides of the pole at points A and B. If $AB = 120$ feet and the angles of elevation at A and B are 72° and 56°, respectively, find the length of the wires to the nearest tenth of a foot.

23. An airplane is flying in a straight line toward an airfield at a fixed altitude. At one point the angle of depression to the airfield is 32°. After flying 2 more miles the angle of depression is 74°. What is the distance between the airplane and the airfield when the angle of depression is 74°? Give your answer to the nearest tenth of a mile.

24. Use the Law of Sines to find AD to the nearest unit.

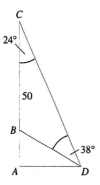

25. From the top of a 250-foot hill, the angles of depression of two cottages A and B on the shore of a lake are 15.5° and 29.2°, respectively. If the cottages are due north of the observation point, find the distance between them to the nearest foot.

26. A 45-foot tower standing vertically on a hillside casts a shadow down the hillside that is 72 feet long. The angle at the tip of the shadow S, subtended by the tower, is 28°. Find the angle of elevation of the sun at S.

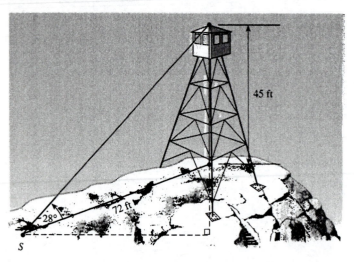

27. In $\triangle ABC$ let D be a point between A and C such that $\angle BDA = 58°$. If $\angle A = 110°$, $\angle C = 43°$, and $DC = 20$ cm, find AD to the nearest tenth of a centimeter.

28. (a) Use the Law of Sines to prove that any two sides of a triangle are in the same ratio as the sines of their opposite angles.

(b) Two of the angles of a triangle are 105° and 45°. The side opposite the smallest angle of the triangle has length 10. Use the result in (a) to find the length of the side opposite the 45° angle without using tables or a calculator.

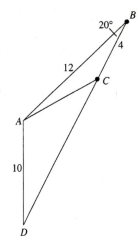

29. In the figure in the margin $AB = 12$, $BC = 4$, $AD = 10$, and the measure of $\angle B$ is 20°. Solve for CD to the nearest unit. (*Hint:* Begin with the Law of Cosines.)

30. Show that if $A = 90°$, the Law of Sines is consistent with the definition of the sine of an acute angle.

31. In the figure, A and B are two points on one side of a canyon with $AB = 300$ meters. Use a calculator to find the distance between the points C and D on the opposite side of the canyon.

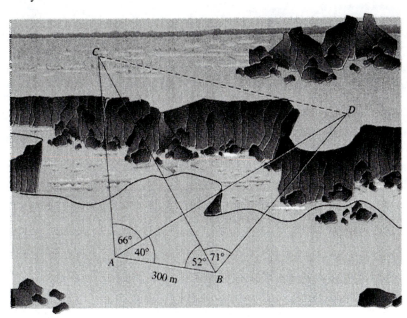

32. Two forces are acting on an object. An 80-pound force is pulling to the right, and a 60-pound force is pulling upward. Find the resultant force and the angle it makes with the 80-pound force. (Give the angle to the nearest degree.)

33. Two forces, one of 24 pounds and the other of 10 pounds, are pulling on an object at right angles to one another. Find the resultant force and the angle it makes with the 24-pound force. (Give the angle to the nearest degree.)

34. Two forces acting on an object form an angle of 60° with each other. One force is 24 pounds and the other is 7 pounds. Find the resultant force and the angle it makes with the smaller force. (Give the force to the nearest pound and the angle to the nearest degree.)

35. Answer the questions in Exercise 34 if the forces are 38 pounds and 12 pounds and the angle between them is 80°.

36. A boat is heading across a river from east to west. The boat's speed (in still water) is 10 mph, and the river is flowing north at 3 mph. What is the actual speed of the boat, and what is its final bearing? (Round the speed to the nearest tenth of an mph and the bearing to the nearest degree.)

37. An airplane has an airspeed of 210 mph with a bearing of N 30° E. A wind is blowing due west at 30 mph. Find the final bearing of the plane, and its ground speed. (Round the speed to the nearest mph and the bearing to the nearest degree.)

38. In Example 7, find the force that the crate exerts on the ramp. (Give the answer to the nearest pound.)

39. Find the force required to keep a vehicle weighing 1800 pounds from rolling down a driveway that makes an angle of 5° with the horizontal. (Round the force to the nearest pound.)

40. Find the force required to push a crate weighing 120 pounds up a ramp that makes an angle of 12° with the horizontal. (Round the force to the nearest pound.)

41. A force of 200 pounds is required to push a vehicle up a driveway that makes a 10° angle with the horizontal. Find the weight of the vehicle to the nearest 10 pounds.

42. A 36-pound force is required to stop a 160-pound skier from sliding down a hill. To the nearest degree, find the angle the hill makes with the horizontal.

43. Use the Law of Sines to show that the area of a triangle ABC may be given as

$$K = \frac{a^2 \sin B \sin C}{2 \sin A}$$

and use the form to verify the result in Example 5 on page 494.

 Written Assignment

1. Make a list of the various cases of solving triangles that can be done with the Law of Cosines and those that can be done with the Law of Sines. (Use abbreviations such as AAS for the given parts angle-angle-side.)

2. Discuss the four possibilities of SSA for an acute $\angle A$ in which parts a, A, and b are given.

 Critical Thinking

1. Prove that the Pythagorean theorem follows from the Law of Cosines. In what sense is this result due to circular reasoning?

2. In $\triangle ABC$, $A = 60°$ and $b = 3c$. Use the Law of Cosines to express a in terms of c. For what value of c does $a = 7$? Is there a value c for which $\triangle ABC$ becomes a right triangle? Explain.

3. $ABCD$ is a parallelogram having diagonals AC and BD with $AC > BD$. Let $AD = a$ and $AB = b$. Observe that the Law of Cosines gives the length of each diagonal in terms of a, b, and the cosine of an appropriate angle. Explain why $AC = \sqrt{a^2 + b^2 + 2ab \cos A}$.

4. As noted in the margin on page 500, the Law of Sines is an abbreviation of three equations. What do these three equations become when the Law of Sines is applied to a right $\triangle ABC$ with right angle at C?

5. In $\triangle ADC$, $\angle C = 90°$ and $\angle A = 23°$. Point B is between A and C so that $AB = .25$ miles and $\angle DBC = 43°$. How can the Law of Sines be used to find CD?

8.3 TRIGONOMETRIC FORM OF COMPLEX NUMBERS

Complex numbers were introduced in Section 1.10 and then used later to solve equations having imaginary roots. Now you will learn how these numbers can be given both a geometric and a trigonometric interpretation that will result in new, and sometimes more efficient, methods for doing some basic computations.

An ordered pair of real numbers (x, y) determines the unique complex number $x + yi$, and vice versa. This correspondence between the complex numbers and the ordered pairs of real numbers allows us to give a geometric interpretation of the complex numbers by using the **complex plane.** This plane is a rectangular coordinate system in which the horizontal axis is called the **real axis** and the vertical axis is the **imaginary axis.** The following figure illustrates how complex numbers are plotted (graphed) in the complex plane.

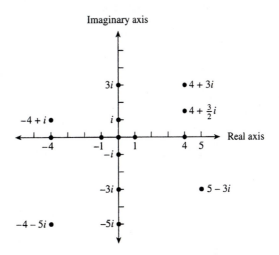

There is a geometric interpretation for the sum of two complex numbers. First locate $z = a + bi$ and $w = c + di$ and use line segments to connect these points to the origin. Then complete the parallelogram as shown. The tip of the diagonal of the parallelogram that passes through the origin will represent the sum $z + w$ (see Exercise 10). This graphic procedure for finding sums is known as the **parallelogram rule.**

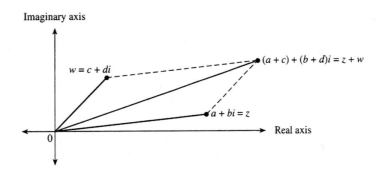

Imaginary

(figure, left margin)

EXAMPLE 1   Find the sum $(3 + 2i) + (2 - 4i)$ using the parallelogram rule, and verify using the definition for addition of complex numbers.

Solution   The parallelogram rule is used in the figure in the margin and computed algebraically as follows:

$$(3 + 2i) + (2 - 4i) = (3 + 2) + (2 - 4)i$$
$$= 5 - 2i \qquad \blacksquare$$

EXAMPLE 2   Describe the diagonal formed when finding the sum of $z = 3 + 2i$ and its conjugate $\bar{z}$ by use of the parallelogram rule.

Solution   The diagonal is six units in length on the x-axis, from $(0, 0)$ to $(6, 0)$, as shown in the next figure. The sum is $6 + 0i = 6$.

Recall that the conjugate of $a + bi$ is $a - bi$.

Imaginary

(figure)

$3 + 2i = z$

0 6 Real

$3 - 2i = \bar{z}$

$\blacksquare$

EXAMPLE 3   Use a parallelogram to find $(3 + 2i) - (2 - 4i)$.

Solution   First locate $3 + 2i$ and $-(2 - 4i) = -2 + 4i$. Then add using the parallelogram rule to get the difference, $1 + 6i$.

Complex numbers can also be subtracted graphically. This is done in Example 3 using a parallelogram after changing the difference $z - w$ into the sum $z + (-w)$.

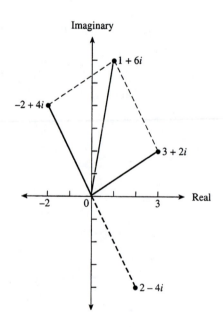

$\blacksquare$

For z not on an axis, the triangle indicates why $r = \sqrt{x^2 + y^2}$. You can verify that this definition also applies for z on an axis. On the x-axis this is consistent with our earlier definition of absolute value.

Any complex number z determines a unique point in the plane. Its distance from the origin, $r \geq 0$, is called the **modulus** or **absolute value** of z and is also denoted by $|z|$.

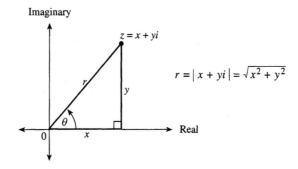

$$r = |x + yi| = \sqrt{x^2 + y^2}$$

The line segment connecting z to the origin and the positive part of the real axis determine an angle of measure θ where $0 \leq \theta < 360°$ (or $0 \leq \theta < 2\pi$). θ is called the **principal argument** of z, which is also referred to simply as the argument of z. (The choice of principal argument can vary. Another interval that is used for this is $-180° \leq \theta < 180°$.) From the preceding figure we have

$$x = r \cos \theta \quad \text{and} \quad y = r \sin \theta$$

You can verify that these formulas also apply when z is on an axis.

Consequently,

$$z = x + iy = r \cos \theta + i\,(r \sin \theta)$$

or, when factored, produces the **trigonometric form**

$$z = r(\cos \theta + i \sin \theta)$$

For $z = 0 + 0i = 0$, $r = 0$ and any θ can be used.

Any angle coterminal with θ will give the same result and may also be referred to as an argument of z. However, we will work primarily with the principal argument.

EXAMPLE 4   Convert **(a)** $\sqrt{3} + i$ and **(b)** $-1 - i$ into trigonometric form.

Solution Graph the points and observe that $\sqrt{3} + i$ determines a $30° - 60° - 90°$ triangle and $-1 - i$ determines a $45° - 45° - 90°$ triangle.

(a) $r = \sqrt{(\sqrt{3})^2 + 1^2} = 2$

r is the modulus and θ is the argument.

$$\theta = 30° = \frac{\pi}{6}$$

$$\sqrt{3} + i = 2\,(\cos 30° + i \sin 30°)$$

$$= 2\left(\cos \frac{\pi}{6} + i \sin \frac{\pi}{6}\right)$$

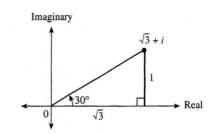

(b) $r = \sqrt{(-1)^2 + (-1)^2} = \sqrt{2}$

$\theta = 225° = \dfrac{5\pi}{4}$

$-1 - i = \sqrt{2}\,(\cos 225° + i \sin 225°)$

$= \sqrt{2}\left(\cos \dfrac{5\pi}{4} + i \sin \dfrac{5\pi}{4}\right)$

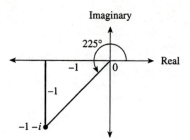

In contrast to the trigonometric form $z = r(\cos \theta + i \sin \theta)$, we call $z = x + yi$ the **rectangular form** of z.

EXAMPLE 5   Convert $3(\cos 50° + i \sin 50°)$ into the rectangular form $x + yi$.

Solution

$$3(\cos 50° + i \sin 50°) = 3(.6428 + .7660i)$$

$$= 1.9284 + 2.2980i \qquad \blacksquare$$

The multiplication and division of complex numbers are sometimes simpler to do using the trigonometric forms. To see how this is done, we begin with

$$z_1 = r_1(\cos \theta_1 + i \sin \theta_1) \qquad \text{and} \qquad z_2 = r_2(\cos \theta_2 + i \sin \theta_2)$$

and multiply.

$$z_1 z_2 = [r_1(\cos \theta_1 + i \sin \theta_1)][r_2(\cos \theta_2 + i \sin \theta_2)]$$

$$= r_1 r_2[(\cos \theta_1 \cos \theta_2 - \sin \theta_1 \sin \theta_2) + i(\sin \theta_1 \cos \theta_2 + \cos \theta_1 \sin \theta_2)]$$

Now use addition formulas for sines and cosines on page 439 to obtain

$$z_1 z_2 = r_1 r_2[\cos(\theta_1 + \theta_2) + i \sin(\theta_1 + \theta_2)]$$

If $\theta_1 + \theta_2$ is not in $[0, 2\pi)$, replace it with the principal argument, which is the smallest nonnegative coterminal angle.

To multiply two complex numbers in trigonometric form, multiply their moduli and add their arguments.

EXAMPLE 6   Let $z_1 = 2 + 2\sqrt{3}i$ and $z_2 = -1 - \sqrt{3}i$.

(a) Evaluate $z_1 z_2$ by using the rectangular forms.

(b) Evaluate $z_1 z_2$ by using trigonometric forms and verify that the result is the same as in part (a).

Solution

(a) $z_1 z_2 = (2 + 2\sqrt{3}i)(-1 - \sqrt{3}i)$

$= (-2 + 6) + (-2\sqrt{3} - 2\sqrt{3})i = 4 - 4\sqrt{3}i$

(b) Converting z_1 and z_2 to trigonometric form, we get the following:

$$z_1 = 4\left(\cos\frac{\pi}{3} + i\sin\frac{\pi}{3}\right) \quad \text{and} \quad z_2 = 2\left(\cos\frac{4\pi}{3} + i\sin\frac{4\pi}{3}\right)$$

Then

$$z_1 z_2 = 4 \cdot 2\left[\cos\left(\frac{\pi}{3} + \frac{4\pi}{3}\right) + i\sin\left(\frac{\pi}{3} + \frac{4\pi}{3}\right)\right]$$

$$= 8\left(\cos\frac{5\pi}{3} + i\sin\frac{5\pi}{3}\right)$$

$$= 8\left[\frac{1}{2} + i\left(-\frac{\sqrt{3}}{2}\right)\right]$$

$$= 4 - 4\sqrt{3}\,i$$

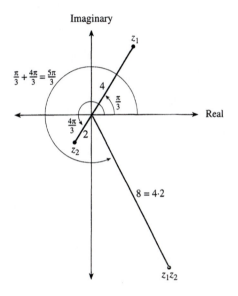

TEST YOUR UNDERSTANDING

(Answers: Page 547)

Let $z_1 = -2 + 2i$ and $z_2 = 3\sqrt{3} - 3i$.
Convert to trigonometric form and evaluate $z_1 z_2$.

By similar reasoning as for the product (see Exercise 46), the result for division is the following:

If $\theta_1 - \theta_2$ is not in $[0, 2\pi)$, replace it with the principal argument, which is the smallest nonnegative coterminal angle.

$$\frac{z_1}{z_2} = \frac{r_1}{r_2}\left[\cos(\theta_1 - \theta_2) + i\sin(\theta_1 - \theta_2)\right] \qquad z_2 \neq 0$$

To divide two complex numbers in trigonometric form, divide their moduli and subtract their arguments.

EXAMPLE 7   For z_1, z_2 as in Example 6, find:

(a) $\dfrac{z_1}{z_2}$ using rectangular forms.

(b) $\dfrac{z_1}{z_2}$ using trigonometric forms and verify that the result is the same as in part (a).

Solution

(a) $\dfrac{2 + 2\sqrt{3}i}{-1 - \sqrt{3}i} = \dfrac{2 + 2\sqrt{3}i}{-1 - \sqrt{3}i} \cdot \dfrac{-1 + \sqrt{3}i}{-1 + \sqrt{3}i}$

$\qquad = \dfrac{(-2 - 6) + (2\sqrt{3} - 2\sqrt{3})i}{1 + 3}$

$\qquad = \dfrac{-8}{4} = -2$

(b) $\dfrac{4\left(\cos \dfrac{\pi}{3} + i \sin \dfrac{\pi}{3}\right)}{2\left(\cos \dfrac{4\pi}{3} + i \sin \dfrac{4\pi}{3}\right)} = 2\left[\cos\left(\dfrac{\pi}{3} - \dfrac{4\pi}{3}\right) + i \sin\left(\dfrac{\pi}{3} - \dfrac{4\pi}{3}\right)\right]$

$\qquad = 2[\cos(-\pi) + i \sin(-\pi)]$

$\qquad = 2(\cos \pi + i \sin \pi)$ π is the principal argument.

$\qquad = 2(-1 + 0)$

$\qquad = -2$ ∎

EXERCISES 8.3

Find z + w using the parallelogram rule and verify using the definition of addition for complex numbers.

1. $z = 3 + 4i$, $w = 4 + 3i$ **2.** $z = 3 - 4i$, $w = 4 - 3i$

3. $z = -3 + 2i$, $w = -2 - 3i$ **4.** $z = 1 + 3i$, $w = -3 - i$

Find z − w using parallelograms and verify using the definition of subtraction for complex numbers.

5. $z = 3 + 4i$, $w = 4 + 3i$ **6.** $z = 3 - 4i$, $w = 4 - 3i$

7. $z = -3 + 2i$, $w = -2 - 3i$ **8.** $z = 1 + 3i$, $w = -3 - i$

9. Find the sum graphically on a complex plane.

$$[(2 + 5i) + (2 - 2i)] + (3 - 5i)$$

10. Verify the parallelogram rule by showing that $x = a + c$ and $y = b + d$.

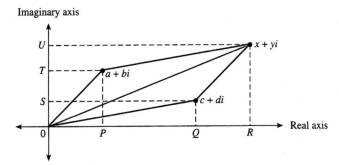

11. Use the parallelogram rule to show that the sum of a complex number and its conjugate is a real number.

12. Let $z = a + bi$ and $w = c + di$. What must be true so that $z + w$ has the form xi where x is a real number?

Convert each of the following into trigonometric form and plot each point in a complex plane.

13. $1 + i$ **14.** $3 + 3i$ **15.** $-1 + \sqrt{3}\,i$ **16.** $-2 + 2\sqrt{3}\,i$ **17.** $\dfrac{\sqrt{3}}{2} - \dfrac{1}{2}i$

18. 5 **19.** -10 **20.** $-4i$ **21.** $-10 - 10i$

Express each of the following complex numbers in the rectangular form $x + yi$. Use a calculator only if necessary.

22. $2(\cos 30° + i \sin 30°)$ **23.** $3(\cos 120° + i \sin 120°)$ **24.** $\dfrac{1}{2}\left(\cos \dfrac{\pi}{2} + i \sin \dfrac{\pi}{2}\right)$

25. $5\left(\cos \dfrac{3\pi}{2} + i \sin \dfrac{3\pi}{2}\right)$ **26.** $9(\cos 0° + i \sin 0°)$ **27.** $\sqrt{2}(\cos 60° + i \sin 60°)$

28. $1(\cos \pi + i \sin \pi)$ **29.** $4(\cos 315° + i \sin 315°)$ **30.** $3(\cos 68° + i \sin 68°)$

31. $\cos 350° + i \sin 350°$

Find zw, $\dfrac{z}{w}$, and $\dfrac{w}{z}$. Give the answers in trigonometric form and convert to rectangular form. Use a calculator only if necessary.

32. $z = 3(\cos 41° + i \sin 41°)$
$w = 2(\cos 20° + i \sin 20°)$

33. $z = \frac{1}{2}(\cos 100° + i \sin 100°)$
$w = 10(\cos 50° + i \sin 50°)$

34. $z = \cos \dfrac{\pi}{3} + i \sin \dfrac{\pi}{3}$
$w = 6\left(\cos \dfrac{\pi}{4} + i \sin \dfrac{\pi}{4}\right)$

35. $z = \sqrt{2}\left(\cos \dfrac{5\pi}{4} + i \sin \dfrac{5\pi}{4}\right)$
$w = 8\left(\cos \dfrac{7\pi}{4} + i \sin \dfrac{7\pi}{4}\right)$

Evaluate zw and $\dfrac{z}{w}$ using trigonometric forms and convert the answers to rectangular form.

36. $z = 1 + i$
$w = -2 - 2i$

37. $z = 7i$
$w = \sqrt{3} + i$

38. $z = -5i$
$w = 7$

39. $z = -2 + 2\sqrt{3}\,i$
$w = -2\sqrt{3} - 2i$

40. $z = 10 - 10i$
$w = -4\sqrt{2} + 4\sqrt{2}\,i$

41. $z = -1 + i$
$w = 1 - i$

Multiply as indicated and express the answer in rectangular form. Use a calculator only if necessary.

42. $[4(\cos 23° + i \sin 23°)][2(\cos 37° + i \sin 37°)]$

43. $\left[\frac{1}{2}(\cos 20° + i \sin 20°)\right]\left[\sqrt{2}(\cos 70° + i \sin 70°)\right]$

44. $[15(\cos 25° + i \sin 25°)][3(\cos 205° + i \sin 205°)]$

45. $\left[\frac{2}{3}(\cos 122° + i \sin 122°)\right]\left[\frac{9}{4}(\cos 77° + i \sin 77°)\right]$

46. Derive the formula for quotients in trigonometric form on page 515. (*Hint:* Use the conjugate of $\cos \theta_2 + i \sin \theta_2$ and the subtraction formulas for sines and cosines.)

Prove the following using $z = a + bi$ and $w = c + di$.

47. $|z| = |\bar{z}|$

48. $z\bar{z} = |z|^2$

49. $|zw| = |z||w|$

50. $\left|\dfrac{z}{w}\right| = \dfrac{|z|}{|w|}$

When the equation $|z| = 2$, for $z = x + yi$ is squared, we obtain $x^2 + y^2 = 4$. Thus $|z| = 2$ is the equation of a circle of radius 2 and center $(0, 0)$. Identify the curve defined by each equation.

51. $|z - 1| = 2$

52. $|z - 2i| = 3$

53. $|z - (2 + 3i)| = 1$

54. $3|z - 1| = |z + i|$

55. $|z + 2| = 8 - |z - 2|$

Challenge Use the result of Exercise 48 and properties of conjugates (see Exercises 41–44, page 233) to prove that $|zw| = |z||w|$, without converting to the $a + bi$ forms.

Written Assignment In Sections 1.1 and 1.3 a number of basic properties of the real numbers are discussed. Review these two sections and list those properties that you think extend to the complex numbers, and make another list of those that do not.

8.4 DE MOIVRE'S THEOREM

HISTORICAL NOTE
Abraham De Moivre made this important contribution to the theory of complex numbers in 1730. He lived from 1667 to 1754 and was responsible for initial work in actuarial mathematics.

The formula for the product of two complex numbers in trigonometric form can be used to derive a special formula for the nth power of a complex number z. We begin with the case $n = 2$. First, let

$$z_1 = z_2 = z = r(\cos \theta + i \sin \theta)$$

Now substitute into

$$z_1 z_2 = r_1 r_2 [\cos (\theta_1 + \theta_2) + i \sin (\theta_1 + \theta_2)]$$

Thus

$$z^2 = r^2 [\cos 2\theta + i \sin 2\theta]$$

For $n = 3$, write $z^3 = z^2 z$ and let $z_1 = z^2$ and $z_2 = z$ in the formula for $z_1 z_2$ to obtain

$$z^3 = z^2 z = [r^2(\cos 2\theta + i \sin 2\theta)][r(\cos \theta + i \sin \theta)]$$

$$= r^3(\cos 3\theta + i \sin 3\theta)$$

In this manner it can be shown that $z^n = r^n(\cos n\theta + i \sin n\theta)$ for each positive integer n. Since $z = r(\cos \theta + i \sin \theta)$, we get the following result:

DE MOIVRE'S THEOREM

For any positive integer n:

$$[r(\cos\theta + i\sin\theta)]^n = r^n(\cos n\theta + i\sin n\theta)$$

To take an nth power of a complex number in trigonometric form, take the nth power of the modulus and n times the argument.

EXAMPLE 1   Use De Moivre's theorem to evaluate $(1 - i)^8$.

Solution Converting $z = 1 - i$ to trigonometric form, we obtain

$$z = 2^{1/2}\left(\cos\frac{7\pi}{4} + i\sin\frac{7\pi}{4}\right)$$

Now use De Moivre's theorem with $r = 2^{1/2}$ and $\theta = \dfrac{7\pi}{4}$.

$$z^8 = (2^{1/2})^8\left[\cos\left(8\cdot\frac{7\pi}{4}\right) + i\sin\left(8\cdot\frac{7\pi}{4}\right)\right]$$

$$= 2^4(\cos 14\pi + i\sin 14\pi)$$

$$= 16(\cos 0 + i\sin 0) \qquad\qquad \text{0 is the principal argument.}$$

$$= 16(1 + 0)$$

$$= 16 \qquad\qquad\qquad\qquad\qquad\qquad\qquad\qquad\qquad ■$$

TEST YOUR UNDERSTANDING

(Answers: Page 547)

Use De Moivre's theorem to evaluate the following powers.

1. $(\cos 15° + i\sin 15°)^3$ **2.** $\left[\frac{1}{2}(\cos 15° + i\sin 15°)\right]^6$

3. $\left(-\frac{1}{2} + \frac{1}{2}i\right)^4$ **4.** $(-\sqrt{3} + i)^{10}$

Example 1 shows that $(1 - i)^8 = 16$. This says that $1 - i$ is an 8th root of 16. There are seven more 8th roots of 16. For the integers $n \geq 2$ every complex number has n distinct nth roots. De Moivre's theorem can be used to derive the following formula, which enables us to find such roots.

The formula can be restated for θ in degree measure. In this case, replace 2π by $360°$.

THE nth-ROOT FORMULA

If n is a positive integer, $n \geq 2$, and if $z = r(\cos\theta + i\sin\theta)$ is a nonzero complex number, the nth roots of z are given by

$$r^{1/n}\left(\cos\frac{\theta + 2k\pi}{n} + i\sin\frac{\theta + 2k\pi}{n}\right) \qquad \text{for } k = 0, 1, 2, \ldots, n - 1$$

When $k = 0, 1, 2, \ldots, n - 1$ the nth-root formula produces the n distinct roots. For larger values of k, say $k = n, n + 1, \ldots, 2n - 1$, the periodicity of sin and cos will produce the same roots. This is discussed in Exercise 33 at the end of this section, and also in Example 2, which is presented after the following verification of the nth-root formula.

One way to justify the nth-root formula is to take an nth root of z, raise it to the nth power, and obtain z. Thus, if $k = 0, 1, 2, \ldots, n - 1$,

$$\left[r^{1/n} \left(\cos \frac{\theta + 2k\pi}{n} + i \sin \frac{\theta + 2k\pi}{n} \right) \right]^n$$

$$= (r^{1/n})^n \left[\cos \left(n \cdot \frac{\theta + 2k\pi}{n} \right) + i \sin \left(n \cdot \frac{\theta + 2k\pi}{n} \right) \right] \qquad \text{by De Moivre's theorem}$$

$$= r[\cos (\theta + 2k\pi) + i \sin (\theta + 2k\pi)] \qquad \text{by the addition formulas for sin and cos}$$

$$= r[\cos \theta + i \sin \theta]$$

$$= z$$

EXAMPLE 2   Find the three cube roots of $z = 8i$.

Solution First convert z into trigonometric form.

$$z = 8i = 0 + 8i = 8 \left(\cos \frac{\pi}{2} + i \sin \frac{\pi}{2} \right)$$

Now use the nth-root formula with $n = 3$, $r = 8$, and $\theta = \dfrac{\pi}{2}$. The values $k = 0$, 1, 2 produce the cube roots as follows:

$$k = 0: \qquad 8^{1/3} \left(\cos \frac{\frac{\pi}{2} + 0}{3} + i \sin \frac{\frac{\pi}{2} + 0}{3} \right) = 2 \left(\cos \frac{\pi}{6} + i \sin \frac{\pi}{6} \right)$$
$$= \sqrt{3} + i$$

Substitute $K = 3, 4, 5$ into the nth-root formula to see that the same three cube roots are obtained.

$$k = 1: \qquad 8^{1/3} \left(\cos \frac{\frac{\pi}{2} + 2\pi}{3} + i \sin \frac{\frac{\pi}{2} + 2\pi}{3} \right) = 2 \left(\cos \frac{5\pi}{6} + i \sin \frac{5\pi}{6} \right)$$
$$= -\sqrt{3} + i$$

$$k = 2: \qquad 8^{1/3} \left(\cos \frac{\frac{\pi}{2} + 4\pi}{3} + i \sin \frac{\frac{\pi}{2} + 4\pi}{3} \right) = 2 \left(\cos \frac{3\pi}{2} + i \sin \frac{3\pi}{2} \right)$$
$$= -2i \qquad \blacksquare$$

The roots found in Example 2 can be checked by showing that their cubes equal $8i$. For example, here is a check for $\sqrt{3} + i$ using rectangular forms.

You can also check this root by applying De Moivre's theorem to the trigonometric form of $\sqrt{3} + i$.

$$(\sqrt{3} + i)^3 = (\sqrt{3} + i)^2(\sqrt{3} + i) = (3 - 1 + 2\sqrt{3}i)(\sqrt{3} + i)$$
$$= (2 + 2\sqrt{3}i)(\sqrt{3} + i) = 2\sqrt{3} - 2\sqrt{3} + 2i + 6i$$
$$= 8i$$

The nth-root formula can also be used to solve equations of the form $z^n - c = 0$, where n is a positive integer and c is a constant.

EXAMPLE 3   Solve the equation $z^4 + 16 = 0$.

Solution $z^4 + 16 = 0$ can be written as $z^4 = -16$. Therefore, the solutions of the given equation are the 4th roots of -16. Writing -16 in trigonometric form, we have

$$-16 = 16(\cos \pi + i \sin \pi)$$

Now use $n = 4$, $r = 16$, and $\theta = \pi$ in the nth-root formula to get this general form of the roots.

$$z = 16^{1/4}\left(\cos \frac{\pi + 2k\pi}{4} + i \sin \frac{\pi + 2k\pi}{4}\right)$$

Specifically, the four roots of the equation are obtained by letting $k = 0, 1, 2, 3$ as follows:

$$k = 0: \quad z = 2\left(\cos \frac{\pi}{4} + i \sin \frac{\pi}{4}\right) \quad = \sqrt{2} + \sqrt{2}i$$

Check for $z = \sqrt{2} - \sqrt{2}i$:
$(\sqrt{2} - \sqrt{2}i)^4$
$= [(\sqrt{2} - \sqrt{2}i)^2]^2$
$= [2 - 4i - 2]^2$
$= [-4i]^2$
$= 16i^2 = -16$

$$k = 1: \quad z = 2\left(\cos \frac{3\pi}{4} + i \sin \frac{3\pi}{4}\right) = -\sqrt{2} + \sqrt{2}i$$

$$k = 2: \quad z = 2\left(\cos \frac{5\pi}{4} + i \sin \frac{5\pi}{4}\right) = -\sqrt{2} - \sqrt{2}i$$

$$k = 3: \quad z = 2\left(\cos \frac{7\pi}{4} + i \sin \frac{7\pi}{4}\right) = \sqrt{2} - \sqrt{2}i \quad \blacksquare$$

The nth-root formula shows that each root has modulus $r^{1/n}$. Therefore, since the modulus of a complex number is its distance from the origin, the nth roots are on a circle with center at the origin having radius $r^{1/n}$. Furthermore, the arguments of a pair of successive roots differ by $\dfrac{2\pi}{n}$. This can be observed by using $\dfrac{\theta + 2k\pi}{n}$, which is the argument of the kth root. For instance, when $k = 2$ and $k = 3$ we have

$$\text{argument for } k = 3: \quad \frac{\theta + 6\pi}{n} = \frac{\theta}{n} + \frac{6\pi}{n}$$
$$\text{argument for } k = 2: \quad \frac{\theta + 4\pi}{n} = \frac{\theta}{n} + \frac{4\pi}{n}$$

$$\text{difference} = \frac{2\pi}{n}$$

Generalize this argument using $k + 1$ and k in place of 3 and 2.

This means that the nth roots are equally spaced around the circle. Here is the diagram for the cube roots of $8i$ (see Example 2).

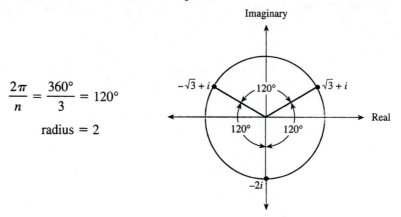

$$\frac{2\pi}{n} = \frac{360°}{3} = 120°$$

$$\text{radius} = 2$$

The trigonometric form of $z = 1$ is $1 = 1(\cos 0 + i \sin 0)$. Then, using $r = 1$ and $\theta = 0$ in the nth-root formula, we obtain the following:

---

nth ROOTS OF UNITY

The nth roots of 1 are given by

$$\cos\frac{2k\pi}{n} + i \sin\frac{2k\pi}{n} \quad \text{for} \quad k = 0, 1, 2, \ldots, n-1$$

These roots are uniformly spaced on the unit circle and determine a regular polygon.

EXAMPLE 4 Find the 6th roots of unity and sketch the regular hexagon.

Solution Use $n = 6$ in the preceding formula.

k	$\cos\dfrac{k\pi}{3} + i \sin\dfrac{k\pi}{3} = x + yi$
0	$\cos 0 + i \sin 0 = 1$
1	$\cos\dfrac{\pi}{3} + i \sin\dfrac{\pi}{3} = \dfrac{1}{2} + \dfrac{\sqrt{3}}{2}i$
2	$\cos\dfrac{2\pi}{3} + i \sin\dfrac{2\pi}{3} = -\dfrac{1}{2} + \dfrac{\sqrt{3}}{2}i$
3	$\cos \pi + i \sin \pi = -1$
4	$\cos\dfrac{4\pi}{3} + i \sin\dfrac{4\pi}{3} = -\dfrac{1}{2} - \dfrac{\sqrt{3}}{2}i$
5	$\cos\dfrac{5\pi}{3} + i \sin\dfrac{5\pi}{3} = \dfrac{1}{2} - \dfrac{\sqrt{3}}{2}i$

Observe that each pair of consecutive roots have arguments that differ by $\dfrac{2\pi}{n} = \dfrac{2\pi}{6}$ radian, or 60°.

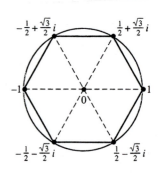

EXERCISES 8.4

Use De Moivre's theorem to evaluate the given powers and express the answers in the rectangular form x + yi. Use a calculator only if necessary.

1. $(\cos 6° + i \sin 6°)^{10}$

2. $\left(\cos \dfrac{\pi}{5} + i \sin \dfrac{\pi}{5}\right)^{15}$

3. $(\cos 40° + i \sin 40°)^8$

4. $\left[2\left(\cos \dfrac{\pi}{9} + i \sin \dfrac{\pi}{9}\right)\right]^6$

5. $\left[\dfrac{1}{2}\left(\cos \dfrac{\pi}{8} + i \sin \dfrac{\pi}{8}\right)\right]^6$

6. $[3(\cos 15° + i \sin 15°)]^5$

7. $(-1 - i)^4$

8. $(1 - \sqrt{3}i)^8$

9. $\left(-\sqrt{3} + \sqrt{3}i\right)^6$

10. $\left(\dfrac{1}{2} + \dfrac{\sqrt{3}}{2}i\right)^6$

11. $\left(\dfrac{\sqrt{2}}{2} - \dfrac{\sqrt{2}}{2}i\right)^{10}$

12. $(2 - 2i)^4$

13. $\left(-\dfrac{\sqrt{3}}{2} + \dfrac{1}{2}i\right)^{12}$

14. $(2^{1/4} + 2^{1/4}i)^{12}$

Find the indicated roots and express them in trigonometric form.

15. The cube roots of $27\left(\cos \dfrac{\pi}{15} + i \sin \dfrac{\pi}{15}\right)$

16. The cube roots of $-4\sqrt{2} - 4\sqrt{2}i$

17. The 5th roots of $32\left(\cos \dfrac{\pi}{8} + i \sin \dfrac{\pi}{8}\right)$

18. The 4th roots of $4 - 4\sqrt{3}i$

Write the roots in rectangular form and graph them on the appropriate circle.

19. The 4th roots of 16   **20.** The cube roots of 27   **21.** The cube roots of -1 **22.** The 6th roots of -1

Find the following nth roots of unity for the given n, and sketch the regular polygon determined by the roots.

23. $n = 3$ **24.** $n = 4$ **25.** $n = 5$ **26.** $n = 8$

Solve the given equation.

27. $z^3 = -27$ **28.** $z^4 + 81 = 0$ **29.** $z^3 + 8i = 0$ **30.** $z^4 - \frac{1}{16}i = 0$

31. $z^3 + i = -1$ **32.** $z^4 + 1 - \sqrt{3}i = 0$

33. (a) Show that $k = 0$ and $k = n$ give the same nth roots in the nth-root formula.

  (b) Repeat part (a) for $k = 1$ and $k = n + 1$.

  (c) Repeat part (a) for k and $k + n$.

34. Prove that $(\cos \theta + i \sin \theta)^{-1} = \cos \theta - i \sin \theta$.

35. (a) Let n be a positive integer. Use the result in Exercise 34 and De Moivre's theorem to prove that $[r(\cos \theta + i \sin \theta)]^{-n} = r^{-n}(\cos n\theta - i \sin n\theta)$.

  (b) Explain why $[r(\cos \theta + i \sin \theta)]^n = r^n(\cos n\theta + i \sin n\theta)$ holds for all integers n.

36. (a) Two complex numbers $a + bi$ and $c + di$ are equal if and only if $a = c$ and $b = d$. Use this criterion of equality in conjunction with De Moivre's theorem to derive the double-angle formulas for the sine and cosine. [*Hint*: Expand the right side of $\cos 2\theta + i \sin 2\theta = (\cos \theta + i \sin \theta)^2$.]

  (b) Follow the procedure that is described in part (a) to derive the formulas $\cos 3\theta = 4 \cos^3 \theta - 3 \cos \theta$ and $\sin 3\theta = 3 \sin \theta - 4 \sin^3 \theta$.

37. $z = \cos \dfrac{2\pi}{3} + i \sin \dfrac{2\pi}{3}$ is one of the cube roots of unity. Verify that $1 + z + z^2 = 0$.

Challenge   Let $z = \cos \dfrac{2\pi}{n} + i \sin \dfrac{2\pi}{n}$ be an nth root of unity for $n \geq 2$.

Prove that $1 + z + z^2 + \cdots + z^{n-1} = 0$. (*Hint:* Consider the fact that $z^n = 1$.)

Written Assignment   The trigonometric form of a complex number given by $z = r(\cos \theta + i \sin \theta)$ can also be written using **Euler's formula,** $e^{i\theta} = \cos \theta + i \sin \theta$, so that $z = re^{i\theta}$. (In Euler's formula, θ is a real number, i is the imaginary unit, and e is the base of the natural logarithms.) Assuming that the fundamental rules for exponents apply, write the rules of multiplication and division of complex numbers, as well as DeMoivre's theorem in terms of the exponential notation. Now use these results to write the solutions of Examples 6(b) and 7(b) of Section 8.3 and Example 1 of Section 8.4 and describe the notational advantages gained using the exponential forms in comparison to using the original notations.

Graphing Calculator Exercises

Set your modes menu to Degrees and Parametric and your Range variables at $T\text{min} = 0$, $T\text{max} = 360$, $T\text{step} = 45$, $-1.5 \leq X \leq 1.5$, and $-1 \leq Y \leq 1$. Set $X_1(T) = \cos T$ and $Y_1(T) = \sin T$. Then the points that are graphed will be on the unit circle, since

$\sin^2 T + \cos^2 T = 1$ for all values of T.

1. Copy the graph on your viewing screen and indicate **(a)** the eighth roots of unity, **(b)** the fourth roots of unity, and **(c)** the square roots of unity.

2. Change the Tstep so that your graph will contain **(a)** the sixth roots of unity and **(b)** the third roots of unity.

3. Find the right settings for graphing the cube roots of -2, and graph these roots.

8.5 POLAR COORDINATES

When a complex number is given in trigonometric form, it can be located in the complex plane using its argument θ and modulus r. A similar method, using **polar coordinates,** can be used for locating points in the plane without using complex numbers. In a **polar coordinate system** a point in the plane is located in reference to a fixed point 0, called the **pole**, and a fixed ray called the **polar axis**. The pole is the endpoint of the polar axis, which is usually horizontal and extends endlessly to the right.

Any point $P \neq 0$ is on some ray that forms an angle of measure θ with the polar axis as its initial side and is some distance r from the pole. The ordered pair (r, θ) are the polar coordinates of P. Consequently, when the polar coordinates (r, θ) of a point P are given, P can be located as follows:

1. Start at the polar axis and rotate through an angle of measure θ to determine the ray θ.

2. On the ray θ move r units from the pole 0 to locate P.

*The terminal side of θ is also referred to as the **ray θ**. The pole 0 is the initial point of ray θ.*

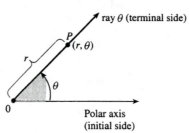

EXAMPLE 1 Plot the points with these polar coordinates:

$$\left(1, \frac{\pi}{6}\right), \qquad \left(2, \frac{2\pi}{3}\right), \qquad \left(\frac{3}{2}, 255°\right), \qquad \left(\frac{2}{3}, 330°\right)$$

Solution

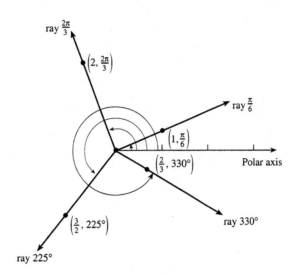

Observe that for positive angles we rotate counter-clockwise.

A ray for an angle θ may be extended through the pole 0 in the opposite or backward direction. Points on this opposite side have polar coordinates (r, θ), where $r < 0$.

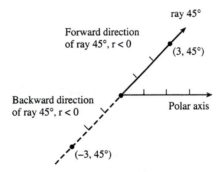

The opposite or backward extension of a ray θ is the same as the ray $\theta + \pi$. Therefore, (r, θ) and $(-r, \theta + \pi)$ are polar coordinates of the same point.

The angle θ may also be negative. In this case we rotate clockwise from the polar axis to determine the ray θ.

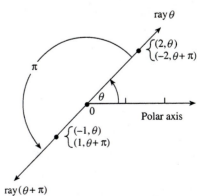

EXAMPLE 2   Plot the points with these polar coordinates:

$$\left(2, -\frac{\pi}{3}\right), \qquad \left(-2, -\frac{\pi}{3}\right)$$

Solution

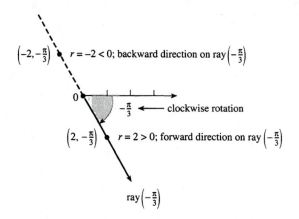

A pair of polar coordinates (r, θ) determine exactly one point. However, a point in the plane has more than one set of polar coordinates. For example, the point P with coordinates $(2, 60°)$ also has these polar coordinates:

$$(2, 420°), \qquad (2, -300°), \qquad (-2, 240°), \qquad (-2, -120°)$$

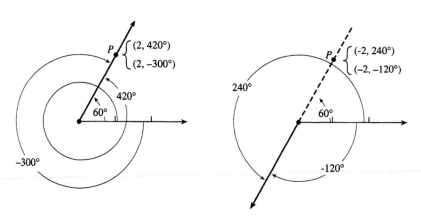

In fact, a point has an infinite number of polar coordinates. If (r, θ) are polar coordinates of point P, then

$$(r, \theta + 2k\pi) \qquad \text{and} \qquad (-r, \theta + (2k + 1)\pi) \qquad \text{for all integers } k$$

give all the polar coordinates of P. The pole 0 is assigned the polar coordinates $(0, \theta)$ for any angle θ.

EXAMPLE 3   Let $\left(3, \dfrac{\pi}{6}\right)$ be polar coordinates of point P. Find polar coordinates (r, θ) of P subject to these conditions:

(a) $r > 0, -2\pi < \theta < 0$ **(b)** $r < 0, 0 < \theta < 2\pi$

Solution

(a)

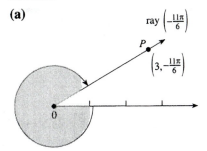

(b)

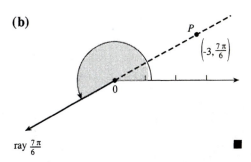

To see the relationship between polar and rectangular coordinates, we superimpose a rectangular system onto a polar system so that the pole coincides with the origin, the polar axis coincides with the positive part of the x-axis, and the ray $\dfrac{\pi}{2}$ coincides with the positive part of the y-axis.

$r^2 = x^2 + y^2$ *is equivalent to*
$r = \pm\sqrt{x^2 + y^2}$. *Also, the signs of x and y together with*
$\tan \dfrac{y}{x}$ *determine* θ.

$$x = r \cos \theta$$
$$y = r \sin \theta$$
$$x^2 + y^2 = r^2$$
$$\tan \theta = \frac{y}{x}$$

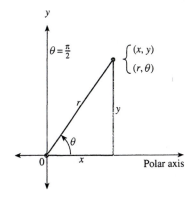

The equations listed next to the figure are easy to derive using the given triangle, which assumes that (x, y) is not on an axis, and $r > 0$. Furthermore, it can be shown that these equations hold when $r \leq 0$ or when (x, y) is on an axis (see Exercises 35, 36, and 37). The preceding equations are used to change from one type of coordinates into the other.

EXAMPLE 4 Convert from polar to rectangular coordinates.

(a) $\left(4, \dfrac{3\pi}{4}\right)$ **(b)** $\left(\dfrac{1}{2}, -\dfrac{\pi}{3}\right)$

Solution

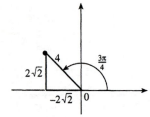

(a) $x = r \cos \theta = 4 \cos\left(\dfrac{3\pi}{4}\right) = 4\left(-\dfrac{\sqrt{2}}{2}\right) = -2\sqrt{2}$

$y = r \sin \theta = 4 \sin\left(\dfrac{3\pi}{4}\right) = 4\left(\dfrac{\sqrt{2}}{2}\right) = 2\sqrt{2}$

$(x, y) = (-2\sqrt{2}, 2\sqrt{2})$

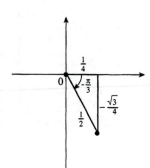

(b) $x = r \cos \theta = \dfrac{1}{2} \cos\left(-\dfrac{\pi}{3}\right) = \dfrac{1}{2} \cos \dfrac{\pi}{3} = \dfrac{1}{2}\left(\dfrac{1}{2}\right) = \dfrac{1}{4}$

$y = r \sin \theta = \dfrac{1}{2} \sin\left(-\dfrac{\pi}{3}\right) = -\dfrac{1}{2} \sin \dfrac{\pi}{3} = -\dfrac{1}{2}\left(\dfrac{\sqrt{3}}{2}\right) = -\dfrac{\sqrt{3}}{4}$

$(x, y) = \left(\dfrac{1}{4}, -\dfrac{\sqrt{3}}{4}\right)$ ∎

EXAMPLE 5 Convert from rectangular to polar coordinate (r, θ) so that $r \geq 0$ and $0 \leq \theta < 2\pi$.

(a) $(3\sqrt{3}, -3)$ **(b)** $(-5, -5)$

Solution

(a) $r = \sqrt{x^2 + y^2} = \sqrt{(3\sqrt{3})^2 + (-3)^2}$

$= \sqrt{27 + 9}$

$= 6$

$\tan \theta = \dfrac{-3}{3\sqrt{3}} = -\dfrac{\sqrt{3}}{3}$. Then $\theta = \dfrac{11\pi}{6}$, since (x, y) is in quadrant IV.

Therefore,

$$(r, \theta) = \left(6, \dfrac{11\pi}{6}\right)$$

(b) $r = \sqrt{(-5)^2 + (-5)^2} = \sqrt{50} = 5\sqrt{2}$

$\tan\theta = \dfrac{-5}{-5} = 1.$ Then $\theta = \dfrac{5\pi}{4}$, since (x, y) is in quadrant III. Therefore,

$$(r, \theta) = \left(5\sqrt{2}, \frac{5\pi}{4}\right)$$

■

EXERCISES 8.5

Plot the points in the same polar coordinate system.

1. $(1, 30°), (-1, 30°), (1, -30°), (-1, -30°)$

2. $(2, 90°), (-2, 90°), (2, -90°), (-2, -90°)$

3. $(4, 0°), (-4, 0°)$

4. $\left(3, \dfrac{\pi}{4}\right), \left(3, -\dfrac{\pi}{4}\right), \left(-3, \dfrac{\pi}{4}\right), \left(-3, -\dfrac{\pi}{4}\right)$

5. $\left(\dfrac{1}{2}, \pi\right), \left(-\dfrac{1}{2}, \pi\right), \left(\dfrac{1}{2}, -\pi\right), \left(-\dfrac{1}{2}, -\pi\right)$

6. $\left(5, \dfrac{4\pi}{3}\right), \left(-5, \dfrac{4\pi}{3}\right), \left(5, -\dfrac{4\pi}{3}\right), \left(-5, -\dfrac{4\pi}{3}\right)$

The polar coordinates of a point P are given. Find the polar coordinates (r, θ) of P subject to each of the following:

(a) $r > 0$ and $2\pi < \theta < 4\pi$ **(b)** $r > 0$ and $-2\pi < \theta < 0$

(c) $r < 0$ and $0 < \theta < 2\pi$ **(d)** $r < 0$ and $-2\pi < \theta < 0$

7. $(1, 45°)$ **8.** $\left(\dfrac{1}{2}, 200°\right)$ **9.** $\left(-3, \dfrac{\pi}{6}\right)$ **10.** $\left(-\dfrac{3}{4}, \dfrac{5\pi}{4}\right)$

Convert the following polar coordinates into rectangular coordinates.

11. $\left(7, -\dfrac{5\pi}{2}\right)$ **12.** $\left(-8, \dfrac{13\pi}{6}\right)$ **13.** $\left(-\dfrac{3}{2}, \dfrac{5\pi}{2}\right)$

14. $(-1, -540°)$ **15.** $\left(\dfrac{4}{5}, 405°\right)$ **16.** $(-10, -420°)$

Convert the following rectangular coordinates into polar coordinates (r, θ) so that $r \geq 0$ and $0 \leq \theta < 2\pi$.

17. $(2, 0)$ **18.** $(0, 2)$ **19.** $(-2, 0)$

20. $(0, -2)$ **21.** $(-6, -6)$ **22.** $\left(3, -3\sqrt{3}\right)$

23. $\left(2\sqrt{2}, 2\sqrt{2}\right)$ **24.** $\left(-2\sqrt{2}, 2\sqrt{2}\right)$ **25.** $\left(\sqrt{3}, -1\right)$

Convert the following rectangular coordinates into polar coordinates (r, θ) so that $r < 0$ and $0 \leq \theta < 2\pi$.

26. $(0, 7)$ **27.** $(7, 0)$ **28.** $(2, 2)$ **29.** $\left(-\sqrt{3}, 1\right)$ **30.** $\left(4, -4\sqrt{3}\right)$ **31.** $\left(\sqrt{3}, \sqrt{3}\right)$

Convert the following rectangular coordinates into polar coordinates (r, θ) so that $r > 0$ and $-180° \leq \theta < 180°$.

32. $(-1, -1)$ **33.** $(0, -10)$ **34.** $\left(4, -4\sqrt{3}\right)$

The following exercises call for the verification of the formulas $x = r \cos \theta$ and $y = r \sin \theta$ in various cases. Assume that they hold when $r > 0$.

35. Verify the formulas for the origin whose polar coordinates are $(0, \theta)$, where θ is any angle measure.

36. Verify the formulas for a point whose rectangular coordinates are $(x, 0)$ with $x < 0$.

37. Verify the formulas for a point P with rectangular coordinates (x, y) and polar coordinates (r, θ) with $r < 0$. (*Hint:* Since P also has polar coordinates $(-r, \theta + \pi)$, you may apply the formulas since $-r > 0$.)

? **Challenge** Graph the set of points whose polar coordinates (r, θ) satisfy the inequalities $\dfrac{\pi}{6} \leq |\theta| \leq \dfrac{5\pi}{6}$ and $1 \leq |r| \leq 3$.

Graphing Calculator Exercises See your calculator's user's manual on how to graph individual points in polar coordinates. Some examples are with the Pt-On ($r \cos T$, $r \sin T$), where $T = \theta$, and Rec (r, θ) commands. (Also see the Graphing Calculator Appendix.)

1. Plot the following points, (r, θ): **(a)** $(4, \pi/3)$; **(b)** $(4, 2\pi/3)$; **(c)** $(4, 3\pi/3)$; **(d)** $(-4, \pi/3)$; **(e)** $(-4, 2\pi/3)$; **(f)** $(-4, 3\pi/3)$.

2. Plot $(r, \theta) = (5, 10°) =$ point A, and then find and plot the remaining seven vertices of a regular octagon $ABCDEFGH$ having the origin as its center.

Critical Thinking

1. Let the real part of $z = x + yi$ be denoted by Re z, and the imaginary part by Im z, so that Re $z = x$ and Im $z = y$. Express Re z and Im z in terms of z and its conjugate $\bar{z}$.

2. Give a geometric justification of the triangle inequality $|z + w| \leq |z| + |w|$ for complex numbers z and w.

3. The cube roots of $8i$ are shown on page 520. Suppose a concentric circle of twice the radius is drawn and the lines through the cube roots of $8i$ are extended to obtain three points on the new circle. What would be the coordinates of these points, and what do they represent?

4. If (x, y) are the rectangular coordinates of the polar point (r, θ), verify that $(r, \theta + 2k\pi)$ and $(-r, \theta + (2k + 1)\pi)$ have the same rectangular coordinates for all integers k.

5. Find the rectangular coordinates of the point $(2, 15°)$ without tables or calculators. Express the coordinates in radical form and verify the results using a calculator.

8.6 GRAPHING POLAR EQUATIONS

The ray $\dfrac{\pi}{4}$ and its opposite ray form a straight line through the pole 0. This line contains all points with polar coordinates $\left(r, \dfrac{\pi}{4}\right)$ for all real numbers r. Thus the graph of the equation $\theta = \dfrac{\pi}{4}$ (for all r) is the line as shown at the top of page 531.

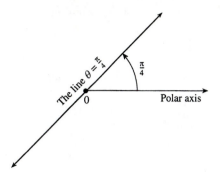

The graph of the equation

θ_0 designates a constant.

$$\theta = \theta_0$$

is a line through the pole forming an angle of measure θ_0 with the polar axis.

EXAMPLE 1   Write a polar equation of the line ℓ through the points with rectangular coordinates $(-2, 2)$ and $(3, -3)$.

Solution ℓ has equation $\theta = 135°$, since ℓ bisects quadrants II and IV.

Some other equations for ℓ are: $\theta = \dfrac{3\pi}{4}$, $\theta = -\dfrac{5\pi}{4}$.

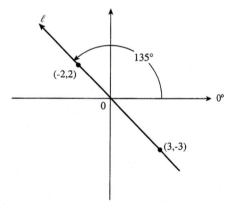

All points of the form $(4, \theta)$ for any value θ are four units from the pole and therefore determine a circle with center 0 and radius 4. Thus the graph of the polar equation $r = 4$ (for all θ) is the circle as shown in the following figure. Note that the points $(-4, \theta)$ are also on this circle. However, these are the same points as before, since $(-4, \theta)$ and $(4, \theta + \pi)$ represent the same point.

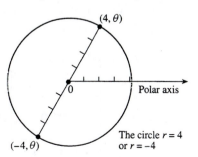

The circle $r = 4$
or $r = -4$

The graph of the equation

$$r = r_0$$

r_0 designates a nonzero constant.

is a circle with center 0 and radius $|r_0|$.

Equations written in terms of polar coordinates are called **polar equations,** and equations using rectangular coordinates may be referred to as **rectangular equations.** When you first learned to graph using rectangular equations, you became accustomed to locating points by moving horizontally for the x coordinate, and vertically for y. With polar coordinates you need to become accustomed to moving in a circular direction for the θ coordinate and forward or backward from the pole along the ray θ for the r coordinate.

EXAMPLE 2   Graph the polar equation $r = 2 \cos \theta$.

Solution As the values of θ increase from 0 to $\dfrac{\pi}{2}$ the corresponding values of r decrease from 2 to 0.

θ	0	$\frac{\pi}{6}$	$\frac{\pi}{4}$	$\frac{\pi}{3}$	$\frac{\pi}{2}$
$\cos \theta$	1	$\frac{\sqrt{3}}{2}$	$\frac{\sqrt{2}}{2}$	$\frac{1}{2}$	0
$r = 2 \cos \theta$	2	$\sqrt{3}$	$\sqrt{2}$	1	0

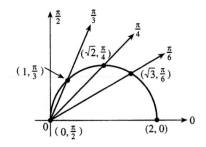

Refer to the graph of the cosine function to see when the cosine is increasing or decreasing.

As the values of θ increase from $\dfrac{\pi}{2}$ to π, the values of r decrease from 0 to -2. Since the rays θ for $\dfrac{\pi}{2} < \theta < \pi$ are in quadrant II, and the r values are negative, the points are in quadrant IV. For values of $\theta = \pi$ to 2π the same points as before are obtained. For example, the pair $\left(-\sqrt{2}, \dfrac{5\pi}{4}\right)$ and $\left(\sqrt{2}, \dfrac{\pi}{4}\right)$ represent one point, as do the pair $\left(1, \dfrac{5\pi}{3}\right)$ and $\left(-1, \dfrac{2\pi}{3}\right)$. Furthermore, the periodicity of

The curve is traced out twice from $\theta = 0$ to $\theta = 2\pi$.

the cosine will produce the same points for $\theta < 0$ or $\theta > 2\pi$. Therefore, the complete graph of $r = 2 \cos \theta$ is obtained for $\theta = 0$ to $\theta = \pi$, as shown at the top of page 533.

θ	$\frac{\pi}{2}$	$\frac{2\pi}{3}$	$\frac{3\pi}{4}$	$\frac{5\pi}{6}$	π
$\cos \theta$	0	$-\frac{1}{2}$	$-\frac{\sqrt{2}}{2}$	$-\frac{\sqrt{3}}{2}$	-1
$r = 2 \cos \theta$	0	-1	$-\sqrt{2}$	$-\sqrt{3}$	-2

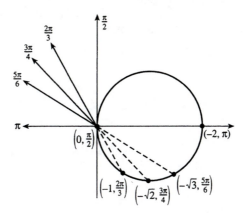

The graph in Example 2 is a circle, as are the graphs of the polar equations $r = 2a \cos \theta$, for $a > 0$. To prove this, consider the circle with radius a, center $(a, 0)$ and let P be a point on this circle with polar coordinates (r, θ). Now recall that an angle inscribed in a semicircle is a right angle. Then, from the right triangle,

Repeat the argument for P on the lower semicircle. $\cos \theta = \dfrac{\text{adjacent}}{\text{hypotenuse}} = \dfrac{r}{2a}$, or $r = 2a \cos \theta$.

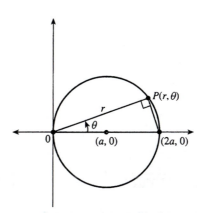

CIRCLE: $r = 2a \cos \theta, \ a > 0$

Similarly, the graph of $r = -2a \cos \theta$, $a > 0$, is the circle with radius a whose center has polar coordinates (a, π) as shown in the figure below. In this case, note that $\cos \theta = -\cos(\pi - \theta) = -\dfrac{r}{2a}$, or $r = -2a \cos \theta$.

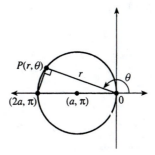

CIRCLE: $r = -2a \cos \theta, a > 0$

In the exercises you will be asked to verify that $r = \pm 2a \sin \theta$, $a > 0$, are polar equations of circles with radius a, as shown here.

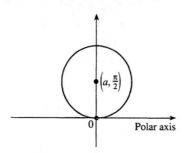

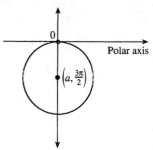

CIRCLE: $r = 2a \sin\theta$, $a > 0$ CIRCLE: $r = -2a \sin\theta$, $a > 0$

EXAMPLE 3 Graph the polar equations.

(a) $r = -4 \cos \theta$ **(b)** $r = \frac{1}{2} \sin \theta$

Solution

(a) $r = -4 \cos \theta$

$\quad = -2(2) \cos \theta$

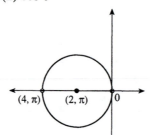

(b) $r = \frac{1}{2} \sin \theta$

$\quad = 2\left(\frac{1}{4}\right) \sin \theta$

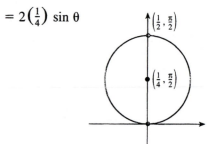

This is the form $-2a \cos \theta$, $a = 2$.   This is the form $2a \sin \theta$, $a = \frac{1}{4}$. ■

(Answers: Page 548)

TEST YOUR UNDERSTANDING

Write a polar equation of the line through the points with the given polar coordinates.

1. $(0, 0°), (2, 30°)$ **2.** $(0, 0°), (-1, 140°)$ **3.** $\left(3, \frac{\pi}{2}\right), \left(-3, \frac{\pi}{2}\right)$

Write a polar equation of the line through the points with the given rectangular coordinates.

4. $(1, 1), (5, 5)$ **5.** $(0, 2), (0, 4)$ **6.** $(0, 0), \left(-\sqrt{3}, 1\right)$

7. Write a polar equation of the circle with center at the pole containing the given point.

(a) $(4, 72°)$ **(b)** $(-5, 5°)$ **(c)** $(1, 1)$; rectangular coordinates

Write the polar equation of the circle with the given radius a and with center in polar coordinates.

8. $a = 2, (2, 0)$ **9.** $a = \frac{3}{2}, \left(\frac{3}{2}, \pi\right)$ **10.** $a = \frac{2}{3}, \left(\frac{2}{3}, \frac{3\pi}{2}\right)$

Another way to show that the graph of $r = 2a \cos \theta$ is a circle is to convert the equation into rectangular form. To do this, begin by multiplying the equation by r.

$$r = 2a \cos \theta$$

$$r^2 = 2a(r \cos \theta) \qquad \text{multiplying by } r$$

The purpose of multiplying by r is to obtain the factor r cos θ on the right and r² on the left, which easily convert into rectangular forms.

Since $r^2 = x^2 + y^2$ and $r \cos \theta = x$, we get

$$x^2 + y^2 = 2ax$$

Now complete the square in x.

$$x^2 - 2ax + y^2 = 0$$

$$x^2 - 2ax + a^2 + y^2 = a^2$$

$$(x - a)^2 + y^2 = a^2 \quad \longleftarrow \quad \text{a circle with radius } a \text{ and center } (a, 0)$$

The preceding demonstrates that converting from polar to rectangular coordinates can be helpful in identifying polar graphs.

EXAMPLE 4   Show that $r = \dfrac{2}{1 - \sin \theta}$ is the equation of a parabola by converting to rectangular coordinates.

Solution

$$r = \frac{2}{1 - \sin \theta}$$

$$r - r \sin \theta = 2$$

$$r = r \sin \theta + 2$$

$$r = y + 2 \qquad\qquad \longleftarrow \quad y = r \sin \theta$$

$$r^2 = y^2 + 4y + 4 \qquad\quad \longleftarrow \quad \text{squaring}$$

$$x^2 + y^2 = y^2 + 4y + 4 \quad \longleftarrow \quad r^2 = x^2 + y^2$$

$$4y = x^2 - 4$$

$$y = \tfrac{1}{4}x^2 - 1 \qquad\qquad \longleftarrow \quad \text{a parabola with vertex } (0, -1),$$
$$\text{opening upward} \qquad\qquad\qquad\qquad\qquad\qquad \blacksquare$$

EXAMPLE 5   Graph the polar curve $r = 2(1 + \sin \theta)$.

Solution   Since the sine has period 2π, it is necessary to consider only values of θ from 0 to 2π. We form a table of selected values for θ (in intervals of 30°), plot the points on a *polar grid system,* and connect the points with a smooth curve.

θ	0°	30°	60°	90°	120°	150°	180°	210°	240°	270°	300°	330°	360°
sin θ	0	.5	.87	1	.87	.5	0	−.5	−.87	−1	−.87	−.5	0
$r = 2(1 + \sin \theta)$	2	3	3.7	4	3.7	3	2	1	.3	0	.3	1	2

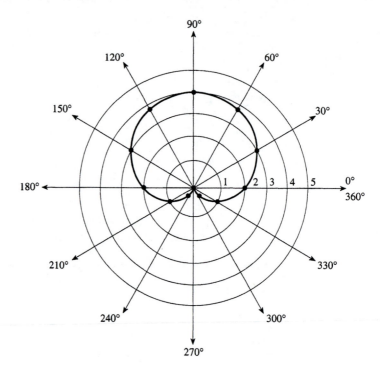

This heart-shaped curve is a **cardioid**. *Cardioids are given by these polar equations:*

$$r = a(1 \pm \sin \theta)$$

$$r = a(1 \pm \cos \theta)$$

If $\sin \theta$ *is involved then* $\theta = \dfrac{\pi}{2}$ *is the axis of symmetry, and if* $\cos \theta$ *is involved, then the symmetry is around the polar axis.*

EXAMPLE 6 Sketch the polar curve $r = \sin 2\theta$.

Solution As θ increases from 0 to $\dfrac{\pi}{4}$, 2θ increases from 0 to $\dfrac{\pi}{2}$, and therefore $r = \sin 2\theta$ increases from 0 to 1. Also, as θ increases from $\dfrac{\pi}{4}$ to $\dfrac{\pi}{2}$, 2θ increases from $\dfrac{\pi}{2}$ to π; therefore r decreases from 1 to 0. This produces the curve for $\theta = 0$ to $\theta = \dfrac{\pi}{2}$.

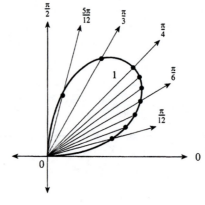

Refer to the graph of the sine function to see when the sine is increasing or decreasing.

r increases from 0 to 1 and then decreases to 0 as θ increases from 0 to $\frac{\pi}{2}$.

As θ increases from $\frac{\pi}{2}$ to $\frac{3\pi}{4}$, 2θ increases from π to $\frac{3\pi}{2}$ and r decreases from 0 to -1. These negative r-values produce points in the fourth quadrant, since the rays are in the second quadrant. Similarly, as θ increases from $\frac{3\pi}{4}$ to π, r increases from -1, to 0. This completes the loop for $\theta = \frac{\pi}{2}$ to $\theta = \pi$.

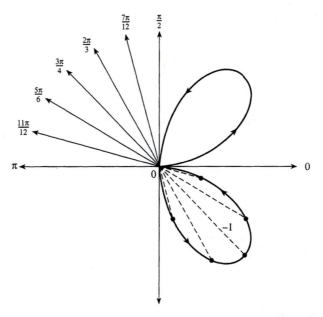

Rose curves are given by
$$r = a \sin n\theta$$
$$r = a \cos n\theta$$
There are n petals if n is odd and 2n petals if n is even.

By similar analysis we find that there is such a loop in each of the remaining quadrants. The complete graph has four loops, as shown in the following figure.

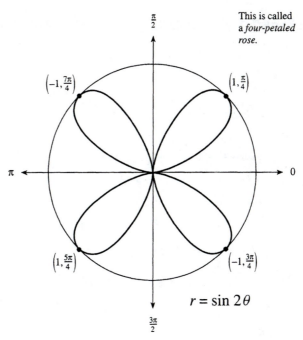

This is called a *four-petaled rose.*

$$r = \sin 2\theta$$

∎

EXAMPLE 7 Graph the polar curve $r = \theta$ for $\theta \geq 0$.

Solution Form a table of values and note that as θ increases, so does r.

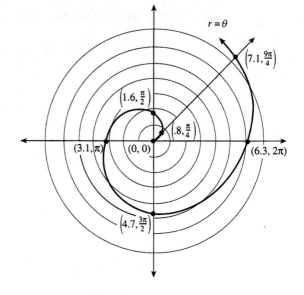

θ	0	$\frac{\pi}{4}$	$\frac{\pi}{2}$	π	$\frac{3\pi}{2}$	2π	$\frac{9\pi}{4}$
$r = \theta$	0	$\frac{\pi}{4}$	$\frac{\pi}{2}$	π	$\frac{3\pi}{2}$	2π	$\frac{9\pi}{4}$
(Approximations) →		.8	1.6	3.1	4.7	6.3	7.1

The curve is an endless, ever widening spiral called the **spiral of Archimedes.** ∎

EXERCISES 8.6

Sketch the graphs of the given polar equations in the same polar system.

1. $\theta = 30°$, $\theta = 210°$, $\theta = -150°$, $\theta = -30°$

2. $\theta = \frac{2\pi}{3}$, $\theta = \frac{5\pi}{3}$, $\theta = -\frac{\pi}{4}$, $\theta = \frac{15\pi}{4}$

3. $\theta = 0$, $\theta = \frac{\pi}{2}$, $\theta = \pi$, $\theta = -\frac{\pi}{2}$

4. $r = 1$, $r = \frac{3}{2}$, $r = 3$

5. $r = -2$, $r = 4$, $r = -4$

Write a polar equation of the line through the given points.

6. $(3, 150°)$, $(2, -30°)$

7. $(0, 0°)$, $(10, -6°)$

8. $(1, -1)$, $(-1, 1)$; (rectangular coordinates)

9. $(\sqrt{3}, 1)$, $(-\sqrt{6}, -\sqrt{2})$; (rectangular coordinates)

Graph the polar equation and convert it into rectangular form.

10. $r = \cos \theta$ **11.** $r = \sin \theta$ **12.** $r = -3 \cos \theta$

13. $r = 5 \sin \theta$ **14.** $r = -5 \sin \theta$ **15.** $r = -\frac{1}{3} \cos \theta$

Convert the polar equations into rectangular equations.

16. (a) $\theta = 45°$ **(b)** $\theta = \frac{\pi}{3}$ **(c)** $\theta = \frac{2\pi}{3}$

17. (a) $r = 2$ **(b)** $r = -2$ **(c)** $r = \sqrt{5}$

Sketch each cardioid.

18. $r = 1 + \cos \theta$ **19.** $r = 3 + 3 \cos \theta$ **20.** $r = 4 - 4 \sin \theta$ **21.** $r = 2(1 - \cos \theta)$

Graph the polar curves.

22. $r \cos \theta = 2$

23. $r \sin \theta = -2$

24. $\tan \theta = 3$

25. $\cot \theta = -3$

26. $r = \frac{1}{2}\theta, \theta \geq 0$

27. $r = \theta, \theta \leq 0$

28. $r = 2 \sin 2\theta$
(four-petal rose)

29. $r = \cos 2\theta$
(four-petal rose)

30. $r = \cos 3\theta$
(three-petal rose)

31. $r = 2 \sin 3\theta$
(three-petal rose)

32. $r^2 = \sin 2\theta$
(lemniscate; two loops)

33. $r^2 = 4 \cos 2\theta$
(lemniscate; two loops)

34. $r = 1 + 2 \sin \theta$
(limaçon; like a cardioid but with an inner loop)

35. $r = 1 - 2 \cos \theta$
(limaçon; like a cardioid but with an inner loop)

Convert the polar equation into rectangular form and identify the curve.

36. $r = \dfrac{1}{1 - \sin \theta}$

37. $r = \dfrac{4}{1 + \cos \theta}$

38. $r = 2 \csc \theta$

39. $r = -5 \sec \theta$

40. $r = 2 \cos \theta + 2 \sin \theta$

41. $r = 6 \cos \theta + 8 \sin \theta$

42. $r^2 = \dfrac{2}{\sin 2\theta}$

43. $r(2 \cos \theta + 3 \sin \theta) = 3$

44. $r(8 \cos \theta - 4 \sin \theta) = 20$

45. $r \cos\left(\theta - \dfrac{\pi}{3}\right) = \sqrt{3}$

46. (a) Show that the polar equation of the given circle is $r = 2a \sin \theta, a > 0$.

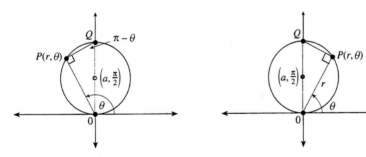

(b) Write the rectangular equation of the circle with center $(0, -a)$ and radius $a > 0$, and convert to the polar form $r = -2a \sin \theta$.

47. Since $x = r \cos \theta$, the equation of the vertical line $x = c$ becomes $r \cos \theta = c$ in polar form. Draw a line $x = c$ and use trigonometry to explain this result.

48. Since $y = r \sin \theta$, the equation of the horizontal line $y = c$ becomes $r \sin \theta = c$ in polar form. Draw a line $y = c$ and use trigonometry to explain this result.

Find the polar coordinates of the points of intersection of the given curves for the indicated interval of θ. (Note: A point of intersection must have the same polar coordinates for each curve.)

49. $r = 3 \cos \theta, r = 1 + \cos \theta; 0 \leq \theta < 2\pi$

50. $r = 1, r = 1 - \sin \theta; 0 \leq \theta \leq 2\pi$

51. $r = \sin 2\theta, r = \cos \theta, 0 \leq \theta \leq \pi$

52. $r = \sin \theta, r = \cos \theta; 0 \leq \theta \leq \dfrac{\pi}{2}$

53. $r \cos \theta = \frac{1}{2}, 2r \sin \theta = -\sqrt{3}; -\dfrac{\pi}{2} \leq \theta \leq 0$

Challenge The figure shows a circle of radius 2 and center with polar coordinates $\left(8, \dfrac{\pi}{2}\right)$. Derive the polar equation $r^2 = 16\, r \sin\theta - 60$ of the circle without using rectangular coordinates. (*Hint:* Use the Law of Cosines.)

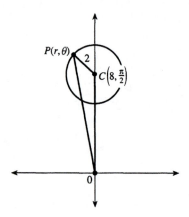

 Written Assignment The maximum distance that the polar curve $r = 4 \cos\theta$ is from the pole 0 is 4. Explain why this is so. Make a chart or list that gives the maximum distance that each of the following curves is from the pole, and include a θ value when this occurs in each case.

1. $r = \pm 2a \cos\theta$ 2. $r = a(1 \pm \cos\theta)$ 3. $r = a \cos n\theta$

   $r = \pm 2a \sin\theta$ $r = a(1 \pm \sin\theta)$ $r = a \sin n\theta$

Graphing Calculator Exercises Some calculators can graph equations of the type $r = f(\theta)$ directly in a polar coordinate mode. (Read your manual to find out if yours can.) Others must first be put into Parametric mode, and the equation is then graphed in the form of points $(X(T), Y(T)) = (f(T) \cos T, f(T) \sin T)$, where $T = \theta$. For example, to graph the polar equation of a four-leaved rose $r = \cos 2\theta$, we graph the points $(X(T), Y(T)) = ((\cos 2T) \cos T, (\cos 2T) \sin T)$, with $0 \leq T \leq 2\pi$, $-1 \leq X \leq 1$, and $-1 \leq Y \leq 1$.

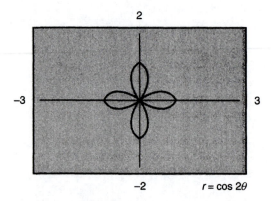

$r = \cos 2\theta$

In Exercises 1–3, graph the polar equation in the indicated window, and use the graph to determine any symmetries. A graph is *symmetric with respect to a given line* if it is its own mirror image with respect to the given line. It is *symmetric with respect to a given point* if all lines through the point intersect the graph at points equidistant from the given point. For example, the circle $r = 1$ is symmetric with respect to the x-axis, the y-axis, and the origin, but the parabola $r = 1/(1 - \cos \theta)$ is symmetric only with respect to the x-axis.

1. $r = 4 \cos 2\theta, 0° \le \theta \le 360°, -5 \le X \le 5, -5 \le Y \le 5$

2. $r = 4 - 4 \cos \theta, 0° \le \theta \le 360°, -12 \le X \le 12, -8 \le Y \le 8$

3. $r = 1/(1 - \sin \theta)$, with window as in Exercise 2

   In Exercises 4–5, graph the polar equations on the same set of axes and estimate the points of intersection of the two graphs.

4. $r = 4 + 4 \cos \theta$ and $r = 6$; window as in Exercise 2

5. $r = 6 \cos 3\theta$ and $r = 8 \sin \theta$; window as in Exercise 2

CHAPTER REVIEW EXERCISES

Section 8.1 The Law of Cosines (Page 490)

1. Write the Law of Cosines for a triangle with sides a, b, c and angle measures A, B, C in terms of a^2.

2. Repeat Exercise 1 and write the Law of Cosines in terms of $\cos A$.

3. Solve for c in $\triangle ABC$ if $a = 5$, $b = 8$, and $C = 110°$. Give your answer to the nearest integer.

4. Solve for angle measures A and B in $\triangle ABC$ as given in Exercise 3. Give your answers to the nearest degree.

5. Two points A and B are on opposite sides of a lake. A surveyor at point C finds that the distance AC is 150 meters, BC is 100 meters, and $\angle ACB$ is 62°. Find the distance across the lake between points A and B, to the nearest meter.

6. Use the Law of Cosines to find c in $\triangle ABC$ where $a = 4$, $b = 5$, and $C = 60°$. Leave your answer in radical form.

7. Write the formula for the area of $\triangle ABC$ in terms of b, c, and the sine of A.

8. Find the area of $\triangle ABC$ where $AB = 10$ centimeters, $AC = 15$ centimeters, and $A = 130°$.

9. State Heron's formula for the area of a triangle with sides of measure a, b, and c.

10. Use Heron's formula to find the exact area of $\triangle ABC$ where $a = 11$ centimeters, $b = 8$ centimeters, and $c = 15$ centimeters.

Section 8.2 The Law of Sines (Page 499)

11. State the Law of Sines for a triangle with sides a, b, c and angle measures A, B, C.

12. Let $\angle A$ in $\triangle ABC$ be an acute angle and prove $\dfrac{a}{\sin A} = \dfrac{b}{\sin B}$.

13. Solve for a in $\triangle ABC$ if $A = 60°$, $C = 45°$, and $c = 10$. Leave your answer in radical form.

14. Solve $\triangle ABC$ if $B = 65°$, $a = 8.0$, and $C = 42°$.

15. Solve for c in $\triangle ABC$, to the nearest meter, if $B = 118°$, $C = 40°$, and $b = 250$ meters.

16. Solve $\triangle ABC$ if $a = 30$, $b = 45$, and $A = 50°$.

17. Approximate the remaining parts of $\triangle ABC$ if $a = 8$, $b = 12$, and $A = 40°$.

18. A motorboat starts at the north shore of a river and heads due south to the opposite shore. The boat's speed (in still water) is 20 mph, and the river is flowing due west at 3 mph. What is the actual speed of the boat, and what is the final bearing?

19. An airplane has an airspeed (the speed in still air) of 280 mph, with a bearing of N 40° W. If a wind is blowing due west at 30 mph, find the final bearing of the plane and the ground speed (the speed relative to the ground).

20. A 350-pound crate is on a ramp that makes a 30° angle with the horizontal. Find the force required to hold the crate from sliding down the ramp, assuming that there is no friction.

Section 8.3   Trigonometric Form of Complex Numbers   (Page 511)

21. Explain what is meant by the complex plane.

22. State the parallelogram rule for the addition of two complex numbers $z = a + bi$ and $w = c + di$.

23. What is the modulus of $z = x + iy$?

24. For $z = 2 + 3i$ and $w = 4 - 2i$, evaluate the following using the parallelogram rule.
   (a) $z + w$ **(b)** $z + \bar{z}$ **(c)** $z - w$

25. What is the trigonometric form of the complex number $z = x + iy$?

26. Convert into trigonometric form: **(a)** $-\sqrt{3} + i$ **(b)** $1 - i$

27. Convert $2(\cos 40° + i \sin 40°)$ into rectangular form.

28. State the rule for multiplying two complex numbers in trigonometric form.

29. Repeat Exercise 28 for the quotient of the two numbers.

30. Let $z_1 = 2 + 2i$ and $z_2 = -1 - i$. Evaluate $z_1 z_2$ by using both rectangular and trigonometric forms.

31. Repeat Exercise 30 for the quotient $\dfrac{z_1}{z_2}$.

Section 8.4   De Moivre's Theorem   (Page 518)

32. Complete this formula for De Moivre's theorem: $[r(\cos \theta + i \sin \theta)]^n = ?$

33. Use De Moivre's theorem to evaluate $(1 + i)^6$.

34. If $z = r(\cos \theta + i \sin \theta)$ is a nonzero complex number, write the form of the nth roots of z for $k = 0, 1, 2, \ldots, n - 1$.

35. What is the angular difference between two consecutive nth roots of a complex number?

36. Find the three cube roots of $27i$.

37. Solve the equation $z^4 + 256 = 0$.

38. Write the form of the nth roots of unity for $k = 0, 1, 2, \ldots, n - 1$.

39. Find the 6th roots of unity.

Section 8.5   Polar Coordinates   (Page 524)

40. Plot the points with these polar coordinates:

$$\left(2, \frac{\pi}{4}\right), \quad \left(1, \frac{3\pi}{4}\right), \quad (3, 210°), \quad (4, 315°)$$

41. Repeat Exercise 40 for the points $\left(3, -\frac{\pi}{6}\right)$, $\left(-3, -\frac{\pi}{6}\right)$, and $\left(-3, \frac{\pi}{3}\right)$.

42. Let P have coordinates $(4, 30°)$. Find r in each case so that the resulting pairs are polar coordinates of P.

(a) $(r, 390°)$ (b) $(r, -330°)$ (c) $(r, 210°)$ (d) $(r, -150°)$

43. Let P have polar coordinates $\left(5, \frac{\pi}{4}\right)$. Find polar coordinates of (r, θ) subject to these conditions:

(a) $r > 0, -2\pi < \theta < 0$ (b) $r < 0, 0 < \theta < 2\pi$

44. Convert $\left(2, \frac{\pi}{4}\right)$ and $\left(\frac{1}{2}, -\frac{\pi}{6}\right)$ into rectangular coordinates.

45. Convert the rectangular coordinates $\left(-\sqrt{3}, -1\right)$ and $(-2, 2)$ into polar coordinates so that $r \geq 0$ and $0 \leq \theta < 2\pi$.

Section 8.6 Graphing Polar Equations (Page 530)

46. Describe the graph with polar equation $\theta = \theta_0$.

47. Repeat Exercise 46 for the polar equation $r = r_0$.

48. Write a polar equation for the line through points with rectangular coordinates $\left(1, -\sqrt{3}\right)$ and $\left(-2, \sqrt{12}\right)$.

49. Write a polar equation of the circle with radius 3 and center having polar coordinates $(3, \pi)$.

Sketch the graph of each polar equation.

50. $r = 4 \cos \theta$ **51.** $r = -2 \cos \theta$ **52.** $r = -3 \sin \theta$

53. How many times is the complete graph of $r = 4 \sin \theta$ traced out as θ varies from $\theta = 0$ to $\theta = 2\pi$?

54. Convert the polar equation $r = 2a \sin \theta$ into a rectangular equation.

55. Show that $r = \dfrac{1}{1 + \cos \theta}$ is the equation of a parabola by converting to rectangular coordinates.

Sketch the graphs of the polar equations.

56. $r = 4(1 + \sin \theta)$ **57.** $r = -\sin 2\theta$ **58.** $r = -\theta$ for $\theta \geq 0$

CHAPTER 8: STANDARD ANSWER TEST

Use these questions to test your knowledge of the basic skills and concepts of Chapter 8. Then check your answers with those given at the end of the book.

1. (a) Solve for a if $A = 60°$, $b = 8$, and $c = 12$. Give A in radians.

(b) Write an expression for $\sin B$ in terms of other known parts of triangle ABC given in part (a).

2. Determine the number of triangles possible with the given parts.

(a) $A = 30°$, $a = 5.8$, $b = 12$ (b) $A = 30°$, $a = 4.1$, $b = 8$

3. Find the exact area of $\triangle ABC$ where $A = 60°$, $b = 20$ cm, and $c = 12$ cm, without using trigonometric tables or a calculator.

4. Solve for a in the figure.

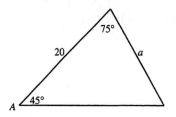

5. In $\triangle ABC$, $A = 18°$, $b = 20$ centimeters, and $c = 34$ centimeters. Find the area of the triangle to the nearest square centimeter.

6. A triangle has sides that measure 7 centimeters, 9 centimeters, and 12 centimeters. Find the area to the nearest square centimeter using Heron's formula.

7. $ABCD$ is a parallelogram. Solve for x and give the exact answer.

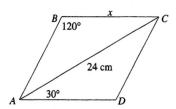

8. Triangle ABC has $a = 9$, $b = 13$, and $c = 6$. Find A and B to the nearest tenth of a degree.

9. Find the remaining parts B, C, and c of all possible triangles ABC if $a = 10$, $b = 16$, and $A = 30°$. (Round the answers to the nearest tenths.)

10. Find the force required to hold a 2,000-pound car from rolling down a ramp that makes a 9° angle with the horizontal. (Round the answer to the nearest pound.)

A boat that can travel at a speed of 18 mph in still water is heading in the direction N 40° W. A current is moving due east at 6 mph.

11. Find the boat's actual speed to the nearest mph.

12. Find the final bearing to the nearest degree.

13. Multiply as indicated and express the answer in rectangular form:

$$\left[5\left(\cos 250° + i \sin 250°\right)\right]\left[\sqrt{2}\left(\cos 245° + i \sin 245°\right)\right]$$

For Exercises 14 and 15, let $z = \dfrac{\sqrt{3}}{4} + \dfrac{1}{4}i$ *and* $w = -1 + \sqrt{3}i$.

14. Convert z and w into trigonometric form and evaluate the quotient $\dfrac{z}{w}$.

15. Use De Moivre's theorem to evaluate z^5 and express the answer in rectangular form.

16. Find the three cube roots of 8 and locate them graphically.

17. Find the four fourth roots of $z = -8\sqrt{2} + 8\sqrt{2}i$ and express them in trigonometric form.

18. Point P has polar coordinates $(-3, 60°)$. Find the polar coordinates (r, θ) of P so that

    (a) $r < 0$ and $-360° < \theta < 0°$

    (b) $r > 0$ and $0 < \theta < 360°$

19. (a) Convert the polar coordinates $\left(\dfrac{1}{3}, -\dfrac{\pi}{6}\right)$ into rectangular coordinates.

    (b) Convert the rectangular coordinates $\left(-7\sqrt{2}, 7\sqrt{2}\right)$ into the polar coordinates.

20. Write the polar equation of the circle with radius 3 and center having polar coordinates $(3, \dfrac{3\pi}{2})$.

Convert the polar equation into a rectangular equation and identify the curve.

21. $r = 2 \sin \theta$ **22.** $r = 4 \sin \theta - 3 \cos \theta$

Sketch the graphs of the polar equations.

23. $2r \sin \theta = 6$ **24.** $r = 2 + 2 \cos \theta$ **25.** $r = 4 \sin 2\theta$

CHAPTER 8: MULTIPLE CHOICE TEST

Do not use tables or a calculator for this test.

1. Which of the following formulas can be used to solve for side c in $\triangle ABC$ given angles A, B, and C and sides a and b?

    I. $c^2 = a^2 + b^2 - 2ac \cos C$

    II. $c^2 = a^2 - b^2 - 2ab \cos C$

    III. $c^2 = a^2 - b^2 + 2ab \cos C$

    (a) Only I   **(b)** Only II   **(c)** Only III   **(d)** I, II, and III   **(e)** None of the preceding

2. Which of the following formulas can be used to solve for angle measure A in $\triangle ABC$?

    I. $\cos A = \dfrac{a^2 + b^2 + c^2}{2bc}$ **II.** $\cos A = \dfrac{b^2 + c^2 - a^2}{2bc}$ **III.** $\cos A = \dfrac{a^2 + b^2 - c^2}{2bc}$

    (a) Only I   **(b)** Only II   **(c)** Only III   **(d)** I, II, and III   **(e)** None of the preceding

3. In $\triangle ABC$, $a = 4$, $b = \sqrt{106}$ and $c = 5\sqrt{2}$. Using the Law of Cosines, the measure of $\angle B$ is found to be

    (a) $-\dfrac{\pi}{4}$ **(b)** $\dfrac{\pi}{4}$ **(c)** $\dfrac{2\pi}{3}$ **(d)** $\dfrac{3\pi}{4}$ **(e)** None of the preceding

4. Find the area of $\triangle ABC$ if $b = 10$ cm, $c = 12$ cm, and $A = 120°$.

    (a) $30\sqrt{3}$ cm^2 **(b)** 30 cm^2 **(c)** $60\sqrt{3}$ cm^2 **(d)** Cannot be determined   **(e)** None of the preceding

5. In $\triangle ABC$, $a = 8$ cm, $b = 8$ cm, and $c = 12$ cm. Using Heron's formula, the area of the triangle is which of the following?

    (a) $280\sqrt{2}$ cm^2 **(b)** $12\sqrt{7}$ cm^2 **(c)** $2\sqrt{7}$ cm^2 **(d)** $\frac{1}{2}(8)(12)(\sin 30°)$ cm^2 **(e)** None of the preceding

6. In $\triangle ABC$, $A = 75°$, $B = 40°$, and $b = 40$. Then side $c =$

    (a) $\dfrac{40 \sin 40°}{\sin 65°}$ **(b)** $\dfrac{40 \sin 65°}{\sin 40°}$ **(c)** $\dfrac{40 \sin 75°}{\sin 40°}$ **(d)** $\dfrac{40 \sin 40°}{\sin 75°}$ **(e)** None of the preceding

7. Determine the number of triangles possible having the parts $a = 4.2$, $b = 7$, and $A = 30°$.

(a) None   (b) One   (c) Two   (d) More than two   (e) None of the preceding

8. For which value of a is it not possible to have $\triangle ABC$ if $A = 30°$ and $b = 10$?

(a) $a = 5$ (b) $a = 12$ (c) $a = 8$ (d) $a = 4$ (e) None of the preceding

9. What is the force required to keep a 300-pound crate from sliding down a ramp that makes an angle of 15° with the horizontal?

(a) $300 \cos 75°$ (b) $300 \cos 15°$ (c) $\dfrac{300}{\cos 15°}$ (d) $\dfrac{\cos 75°}{300}$ (e) None of the preceding

10. Convert $1 - i$ into trigonometric form:

(a) $\sqrt{2}\left(\cos \dfrac{\pi}{4} + i \sin \dfrac{\pi}{4}\right)$ (b) $\sqrt{2} \cos \dfrac{7\pi}{4} + i \sin \dfrac{\pi}{4}$ (c) $\sqrt{2}\left(\cos \dfrac{7\pi}{4} + i \sin \dfrac{7\pi}{4}\right)$

(d) $\sqrt{2} \cos \dfrac{\pi}{4} + i \sin \dfrac{\pi}{4}$ (e) None of the preceding

11. $z = 2(\cos 20° + i \sin 20°)$. Then $z^3 =$

(a) $8(\cos^3 20° + i \sin^3 20°)$ (b) $3 + \dfrac{3\sqrt{3}}{2} i$ (c) $4\sqrt{3} + 4i$ (d) $4 + 4\sqrt{3}i$

(e) None of the preceding

12. Let $z = 8\left(\cos \dfrac{4\pi}{3} + i \sin \dfrac{4\pi}{3}\right)$ and $w = 24\left(\cos \dfrac{11\pi}{6} + i \sin \dfrac{11\pi}{6}\right)$. Then $\dfrac{z}{w} =$

(a) $\dfrac{1}{3}\left(\cos \dfrac{\pi}{2} + i \sin \dfrac{\pi}{2}\right)$ (b) $3\left(\cos \dfrac{19\pi}{6} + i \sin \dfrac{19\pi}{6}\right)$ (c) $\dfrac{1}{3}\left(\cos \dfrac{3\pi}{2} + i \sin \dfrac{3\pi}{2}\right)$

(d) $3\left(\cos \left(-\dfrac{\pi}{2}\right) + i \sin \left(-\dfrac{\pi}{2}\right)\right)$ (e) None of the preceding

13. Convert the polar coordinates $\left(-\dfrac{1}{2}, \dfrac{\pi}{3}\right)$ to rectangular coordinates:

(a) $\left(-\dfrac{1}{4}, -\dfrac{\sqrt{3}}{4}\right)$ (b) $\left(-\dfrac{1}{4}, \dfrac{\sqrt{3}}{4}\right)$ (c) $\left(-\dfrac{\sqrt{3}}{4}, -\dfrac{1}{4}\right)$ (d) $\left(-\dfrac{\sqrt{3}}{4}, \dfrac{1}{4}\right)$ (e) None of the preceding

14. Convert the rectangular coordinates $\left(-3, 3\sqrt{3}\right)$ to polar coordinates (r, θ) so that $r \geq 0$ and $0 \leq \theta < 2\pi$.

(a) $\left(6, \dfrac{5\pi}{3}\right)$ (b) $\left(6, \dfrac{2\pi}{3}\right)$ (c) $\left(-6, \dfrac{2\pi}{3}\right)$ (d) $\left(6, \dfrac{5\pi}{6}\right)$ (e) None of the preceding

15. Which of the following is a cube root of -64?

(a) $-4(\cos \pi + i \sin \pi)$ (b) $2\sqrt{3} + 2i$ (c) $-2 + 2\sqrt{3}i$ (d) $2 - 2\sqrt{3}i$ (e) None of the preceding

16. Which of the following polar equations have circles as graphs?

  I. $r = -5$

  II. $r = -\dfrac{1}{2} \sin \theta$

  III. $r \cos \theta = 6$

(a) Only I   (b) Only II   (c) Only I and II   (d) I, II, and III   (e) None of the preceding

17. The graph of which of the following is a cardioid?

(a) $r = 3 - 3 \cos \theta$ (b) $r = 3 \sin 2\theta$ (c) $r = 2 \cos 3\theta$ (d) $r \cos \theta + r \sin \theta = 1$

(e) None of the preceding

18. A rectangular form of the polar equation $r = -8 \cos \theta$ is

  (a) $x^2 + (y - 4)^2 = 16$ **(b)** $(x + 4)^2 + y^2 = 16$ **(c)** $\sqrt{x^2 + y^2} = -8x$ **(d)** $x^2 + y^2 = 64 \cos^2 \theta$

  (e) None of the preceding

19. Which of the following is equivalent to $[r(\cos \theta + i \sin \theta)]^n$?

  (a) $r^n \cos n\theta + i \sin n\theta$ **(b)** $r^n(\cos \theta + i \sin \theta)$ **(c)** $r^n(\cos n\theta + i \sin n\theta)$ **(d)** $rn(\cos n\theta + i \sin n\theta)$

  (e) None of the preceding

20. Which of the following is a solution of the equation $z^3 + 27i = 0$?

  (a) $\dfrac{3\sqrt{3}}{2} - \dfrac{3}{2}i$ **(b)** $\dfrac{3}{2} - \dfrac{3\sqrt{3}}{2}i$ **(c)** $\dfrac{3\sqrt{3}}{2} + \dfrac{3}{2}i$ **(d)** $27\left(\cos \dfrac{3\pi}{2} + i \sin \dfrac{3\pi}{2}\right)$

  (e) None of the preceding

ANSWERS TO THE TEST YOUR UNDERSTANDING EXERCISES

Page 493

1. $2\sqrt{7}$ **2.** $5\sqrt{5}$ **3.** $\sqrt{399}$ **4.** $\dfrac{\pi}{3}$ **5.** $\dfrac{\pi}{4}$ **6.** $\dfrac{2\pi}{3}$

Page 501

1. $4\sqrt{2}$ **2.** $12\sqrt{3}$ **3.** $\dfrac{15}{2}\sqrt{6}$ **4.** $4\sqrt{2}$

Page 515

$$z_1 = 2\sqrt{2}\left(\cos \frac{3\pi}{4} + i \sin \frac{3\pi}{4}\right) \qquad z_1 z_2 = 12\sqrt{2}\left(\cos \frac{31\pi}{12} + i \sin \frac{31\pi}{12}\right)$$

$$z_2 = 6\left(\cos \frac{11\pi}{6} + i \sin \frac{11\pi}{6}\right) \qquad\qquad = 12\sqrt{2}\left(\cos \frac{7\pi}{12} + i \sin \frac{7\pi}{12}\right)$$

Page 519

1. $\dfrac{\sqrt{2}}{2} + \dfrac{\sqrt{2}}{2}i$ **2.** $\dfrac{1}{64}i$ **3.** $-\dfrac{1}{4}$ **4.** $512 + 512\sqrt{3}i$

Page 527

1.
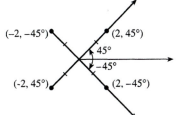

2.
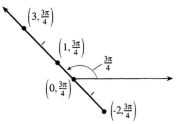

3. **(a)** $\left(5, \dfrac{25\pi}{12}\right)$ **(b)** $\left(-5, \dfrac{13\pi}{12}\right)$ **(c)** $\left(5, -\dfrac{23\pi}{12}\right)$ **(d)** $\left(-5, -\dfrac{11\pi}{12}\right)$

Page 534

1. $\theta = 30°$ **2.** $\theta = 140°$ **3.** $\theta = \dfrac{\pi}{2}$ **4.** $\theta = \dfrac{\pi}{4}$ **5.** $\theta = \dfrac{\pi}{2}$

6. $\theta = \dfrac{5\pi}{6}$ **7. (a)** $r = 4$ **(b)** $r = 5$ **(c)** $r = \sqrt{2}$

8. $r = 4 \cos \theta$ **9.** $r = -3 \cos \theta$ **10.** $r = -\dfrac{4}{3} \sin \theta$

Linear Systems, Matrices, and Determinants

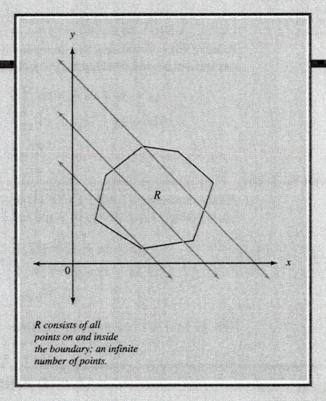

R consists of all points on and inside the boundary; an infinite number of points.

As the chapter title suggests, systems of linear equations and systems of linear inequalities are topics studied in this chapter. The major emphasis will be on systems of equations. New methods of solving such systems, beyond the basic procedures of Section 2.5, will be developed. These procedures are conducive for solving larger systems of linear equations.

The final section of the chapter will deal with linear programming, a mathematical procedure for solving problems related to the logistics of decision making. There you will learn how to select a specific point $P(a, b)$ in a polygonal region R for which a sum like $2x + 3y$ is the largest such sum possible for points in R. That is, $2a + 3b \geq 2x + 3y$ for all (x, y) in R. As you will see, this procedure has a variety of interesting and useful applications. Just for fun, you can also challenge a friend as follows. Each of you selects a point (x, y) in R, and whoever produces the largest sum $2x + 3y$ is the winner. You can't lose if you know the appropriate procedure.

9.1
SOLVING LINEAR SYSTEMS USING MATRICES

The strategy here is to eliminate one of the variables and obtain a system of two equations in two variables, that we already know how to solve. (See Section 2.5.)

The initial work with linear systems was done in Section 2.5. The methods used there for solving linear systems, consisting of two equations in two variables, extend to systems of three linear equations in three variables. For example:

$$\text{(1)} \qquad 2x - 5y + z = -10$$
$$\text{(2)} \qquad x + 2y + 3z = 26$$
$$\text{(3)} \qquad -3x - 4y + 2z = 5$$

To solve this system begin by eliminating the variable x using the first two equations; multiply the second equation by -2 and add to the first:

$$\left.\begin{array}{r} 2x - 5y + z = -10 \\ -2(x + 2y + 3z = 26) \end{array}\right\} \Rightarrow \begin{array}{r} 2x - 5y + z = -10 \\ \underline{-2x - 4y - 6z = -52} \end{array}$$
$$\text{Add:} \qquad\qquad -9y - 5z = -62$$
$$\text{or} \qquad\qquad 9y + 5z = 62$$

Another equation in y and z can be obtained from Equations (2) and (3) by eliminating x; multiply Equation (2) by 3 and add Equation (3):

$$\left.\begin{array}{r} 3(x + 2y + 3z = 26) \\ -3x - 4y + 2z = 5 \end{array}\right\} \Rightarrow \begin{array}{r} 3x + 6y + 9z = 78 \\ \underline{-3x - 4y + 2z = 5} \end{array}$$
$$\text{Add:} \qquad\qquad 2y + 11z = 83$$

Now we have this system in two variables:

$$9y + 5z = 62$$
$$2y + 11z = 83$$

Geometrically, an equation of the form $ax + by + cz = d$ can be interpreted as a plane in space. For this example there are three planes in space that intersect at the unique common point $(-1, 3, 7)$.

Three planes intersecting at a unique common point.

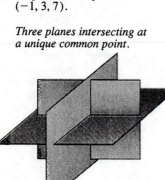

One way to solve this system is to eliminate the variable y by multiplying the first equation by -2, the second by 9, adding the results, and solving for z.

$$\begin{array}{r} -18y - 10z = -124 \\ \underline{18y + 99z = 747} \\ 89z = 623 \\ z = 7 \end{array}$$

Substituting this into $9y + 5z = 62$ will give $y = 3$. Now substitute these values for y and z into an earlier equation, say Equation (2), to find x.

$$x + 2y + 3z = 26$$
$$x + 2(3) + 3(7) = 26$$
$$x = -1$$

The remaining initial equations can be used for checking:

$$(1) \qquad 2(-1) - 5(3) + \quad 7 \; = -10$$

$$(3) \qquad -3(-1) - 4(3) + 2(7) = \quad 5$$

The solution is $x = -1$, $y = 3$, $z = 7$, which may also be written as the *ordered triple* $(-1, 3, 7)$.

The preceding algebraic procedures can become clumsy and tedious, especially when applied to larger linear systems. We are now going to develop another, more efficient procedure, one that easily extends to larger systems. However, for demonstration purposes, systems of three equations in three variables will be used.

We begin by forming the rectangular array of numbers, called a *matrix*, consisting of the coefficients and constants of the system. For example, the following linear system

$$(1) \qquad 2x + 5y + \quad 8z = 11$$

$$(2) \qquad x + 4y + \quad 7z = 10$$

$$(3) \qquad 3x + 6y + 12z = 15$$

is replaced by this corresponding matrix; also called the **augmented matrix** of the system.

The dashed vertical line serves as a reminder that the coefficients of the variables are to the left. The numbers to the right are the constants on the right-hand side of the equal signs in the given system.

Coefficients

$$
\begin{array}{cccc}
x & y & z & \text{constants} \\
\downarrow & \downarrow & \downarrow & \downarrow
\end{array}
$$

$$
\begin{array}{c}
\text{Row 1} \\
\text{Row 2} \\
\text{Row 3}
\end{array}
\left[
\begin{array}{ccc|c}
2 & 5 & 8 & 11 \\
1 & 4 & 7 & 10 \\
3 & 6 & 12 & 15
\end{array}
\right]
$$

Compare the steps in the two columns that follow. Note that the objective is to transform the given linear system into the form reached in step 5.

Working with the Equations	Working with the Matrices
Step 1 Write the system. $$\begin{aligned} 2x + 5y + \;8z &= 11 \\ x + 4y + \;7z &= 10 \\ 3x + 6y + 12z &= 15 \end{aligned}$$	Step 1 Write the augmented matrix. $$\left[\begin{array}{ccc\|c} 2 & 5 & 8 & 11 \\ 1 & 4 & 7 & 10 \\ 3 & 6 & 12 & 15 \end{array}\right]$$
Step 2 Interchange the first two equations. $$\begin{aligned} x + 4y + \;7z &= 10 \\ 2x + 5y + \;8z &= 11 \\ 3x + 6y + 12z &= 15 \end{aligned}$$	Step 2 Interchange the first two rows. $$\left[\begin{array}{ccc\|c} 1 & 4 & 7 & 10 \\ 2 & 5 & 8 & 11 \\ 3 & 6 & 12 & 15 \end{array}\right]$$

Observe that the instructions stated in Steps 2, 3, 4, and 5 refer to what has been done to the preceding system or matrix. (See the next page.)

Working with the Equations	Working with the Matrices
Step 3 Add -2 times the first equation to the second, and -3 times the first to the third. $\begin{aligned} x + 4y + 7z &= 10 \\ -3y - 6z &= -9 \\ -6y - 9z &= -15 \end{aligned}$	**Step 3** Add -2 times the first row to the second, and -3 times the first to the third. $\begin{bmatrix} 1 & 4 & 7 & \vdots & 10 \\ 0 & -3 & -6 & \vdots & -9 \\ 0 & -6 & -9 & \vdots & -15 \end{bmatrix}$
Step 4 Multiply the second equation by $-\frac{1}{3}$. $\begin{aligned} x + 4y + 7z &= 10 \\ y + 2z &= 3 \\ -6y - 9z &= -15 \end{aligned}$	**Step 4** Multiply row 2 by $-\frac{1}{3}$. $\begin{bmatrix} 1 & 4 & 7 & \vdots & 10 \\ 0 & 1 & 2 & \vdots & 3 \\ 0 & -6 & -9 & \vdots & -15 \end{bmatrix}$
Step 5 Add 6 times the second equation to the third. $\begin{aligned} x + 4y + 7z &= 10 \\ y + 2z &= 3 \\ 3z &= 3 \end{aligned}$	**Step 5** Add 6 times row 2 to row 3. $\begin{bmatrix} 1 & 4 & 7 & \vdots & 10 \\ 0 & 1 & 2 & \vdots & 3 \\ 0 & 0 & 3 & \vdots & 3 \end{bmatrix}$

The operations on the equations in the left column produced **equivalent systems** of equations. Equivalent systems have the same solutions. Since each of these systems has its corresponding matrix produced by comparable operations on the rows of the matrices, we say that the matrices are **row-equivalent**.

When a different sequence of operations on the rows is used, it will most likely result in a different, but equivalent, triangular form.

In Step 5, we reached a row-equivalent matrix whose corresponding linear system is said to be in **triangular form.** From this form we solve for the variables using **back-substitution.** That is, we find z from the last equation, then substitute back into the second to find y, and finally substitute back into the first to find x.

$$\begin{aligned} x + 4y + 7z &= 10 \\ y + 2z &= 3 \\ 3z &= 3 \end{aligned}$$

$3z = 3 \;\longrightarrow\; z = 1$

$y + 2(1) = 3 \;\longrightarrow\; y = 1$

$x + 4(1) + 7(1) = 10 \;\longrightarrow\; x = -1$

The solution is the ordered triple $(-1, 1, 1)$.

Our new matrix method for solving a linear system has two major parts:

PART A: Use the following **fundamental row operations** to transform the augmented matrix of the linear system into a row-equivalent matrix that translates into a linear system in triangular form, if possible.

Sometimes, as in Example 4, an incomplete triangular form will be reached.

1. Interchange two rows.
2. Multiply a row by a nonzero constant.
3. Add a multiple of a row to another row.

Part A can be completed by using a variety of row operations. The steps of a solution are therefore not unique, but the final solution will be the same.

PART B: Convert the matrix obtained in Part A into a linear system equivalent to the original system; solve for the variables by back-substitution.

EXAMPLE 1   Solve the linear system using row-equivalent matrices.

$$2x + 14y - 4z = -2$$
$$-4x - 3y + z = 8$$
$$3x - 5y + 6z = 7$$

Solution Begin by writing the augmented matrix of the linear system and apply the fundamental row operations.

$$\left[\begin{array}{ccc|c} 2 & 14 & -4 & -2 \\ -4 & -3 & 1 & 8 \\ 3 & -5 & 6 & 7 \end{array}\right]$$

This is the augmented matrix for the given system.

*The explanations pointing to the rows state that row operations were applied to the rows of the **preceding matrix** to obtain the designated row.*

Getting a 1 in the circled position will make it easier to get zeros *below* this 1 in the next step using row operation (3).

$$\left[\begin{array}{ccc|c} ① & 7 & -2 & -1 \\ -4 & -3 & 1 & 8 \\ 3 & -5 & 6 & 7 \end{array}\right] \longleftarrow \tfrac{1}{2} \times (\text{row 1})$$

$$\left[\begin{array}{ccc|c} 1 & 7 & -2 & -1 \\ 0 & 25 & -7 & 4 \\ 0 & -26 & 12 & 10 \end{array}\right] \begin{array}{l} \\ \longleftarrow 4 \times (\text{row 1}) + \text{row 2} \\ \longleftarrow -3 \times (\text{row 1}) + \text{row 3} \end{array}$$

$$\left[\begin{array}{ccc|c} 1 & 7 & -2 & -1 \\ 0 & 25 & -7 & 4 \\ 0 & -1 & 5 & 14 \end{array}\right] \begin{array}{l} \\ \\ \longleftarrow \text{row 2} + \text{row 3} \end{array}$$

Getting the −1 in the circled position will make it easier to get zero *below* this −1 in the next step.

$$\left[\begin{array}{ccc|c} 1 & 7 & -2 & -1 \\ 0 & ⊖① & 5 & 14 \\ 0 & 25 & -7 & 4 \end{array}\right] \begin{array}{l} \\ \longleftarrow \Big\{ \begin{array}{l} \text{Interchange rows 2} \\ \text{and 3.} \end{array} \end{array}$$

$$\left[\begin{array}{ccc|c} 1 & 7 & -2 & -1 \\ 0 & -1 & 5 & 14 \\ 0 & 0 & 118 & 354 \end{array}\right] \begin{array}{l} \\ \\ \longleftarrow 25 \times (\text{row 2}) + \text{row 3} \end{array}$$

Now, convert to the corresponding linear system, as on the following page, and solve for the variables using back-substitution.

$$
\begin{array}{rcl}
x + 7y - 2z &=& -1 \\
-y + 5z &=& 14 \\
118z &=& 354
\end{array}
$$

$$118z = 354 \longrightarrow z = 3$$

$$-1y + 5(3) = 14 \longrightarrow y = 1$$

$$x + 7(1) - 2(3) = -1 \longrightarrow x = -2$$

Thus the solution is the ordered triple $(-2, 1, 3)$. ∎

TEST YOUR UNDERSTANDING

(Answers: Page 631)

Solve each system. Begin with the augmented matrix, use fundamental row operations to obtain a triangular form, and then use back-substitution.

1. $\begin{array}{r} x + 2y + 3z = -4 \\ 4x + 5y + 6z = -4 \\ 7x - 15y - 9z = 4 \end{array}$

2. $\begin{array}{r} 4x - 2y - z = 1 \\ 2x + y + 2z = 9 \\ x - 3y - z = \frac{3}{2} \end{array}$

In the next example the matrix method is used to solve an applied problem.

EXAMPLE 2 A veterinarian wants to control the diet of an animal so that on a monthly basis the animal consumes (besides hay, grass, and water) 60 pounds of oats, 75 pounds of corn, and 55 pounds of soybeans. The veterinarian has three feeds available, each consisting of oats, corn, and soybeans as shown in the table. How many pounds of each feed should be used to obtain the desired mix?

	Oats	Corn	Soybeans
1 lb of feed A	6 oz	5 oz	5 oz
1 lb of feed B	6 oz	6 oz	4 oz
1 lb of feed C	4 oz	7 oz	5 oz

Solution

$$
\begin{array}{ll}
\text{Let} & x = \text{pounds of feed } A \\
& y = \text{pounds of feed } B \\
& z = \text{pounds of feed } C
\end{array}
$$

Then

$$
\begin{array}{l}
6x = \text{ounces of oats in } x \text{ pounds of feed } A \\
6y = \text{ounces of oats in } y \text{ pounds of feed } B \\
4z = \text{ounces of oats in } z \text{ pounds of feed } C
\end{array}
$$

The total number of pounds of oats required is 60. In ounces this is $60 \times 16 = 960$. Thus:

$$6x + 6y + 4z = 960$$

Analysis for the total ounces of corn and soybeans leads to this linear system:

$$6x + 6y + 4z = 960 \qquad \text{(total ounces of oats)}$$
$$5x + 6y + 7z = 1200 \qquad \text{(total ounces of corn)}$$
$$5x + 4y + 5z = 880 \qquad \text{(total ounces of soybeans)}$$

Now proceed with the matrix method to solve the system.

$$\begin{bmatrix} 6 & 6 & 4 & 960 \\ 5 & 6 & 7 & 1200 \\ 5 & 4 & 5 & 880 \end{bmatrix} \longleftarrow \begin{array}{l}\text{Augmented matrix of}\\ \text{the system}\end{array}$$

$$\begin{bmatrix} 1 & 1 & \frac{2}{3} & 160 \\ 5 & 6 & 7 & 1200 \\ 5 & 4 & 5 & 880 \end{bmatrix} \longleftarrow \tfrac{1}{6} \times (\text{row 1})$$

$$\begin{bmatrix} 1 & 1 & \frac{2}{3} & 160 \\ 0 & 1 & \frac{11}{3} & 400 \\ 0 & -1 & \frac{5}{3} & 80 \end{bmatrix} \begin{array}{l} \\ \longleftarrow -5 \times (\text{row 1}) + \text{row 2} \\ \longleftarrow -5 \times (\text{row 1}) + \text{row 3}\end{array}$$

$$\begin{bmatrix} 1 & 1 & \frac{2}{3} & 160 \\ 0 & 1 & \frac{11}{3} & 400 \\ 0 & 0 & \frac{16}{3} & 480 \end{bmatrix} \begin{array}{l} \\ \\ \longleftarrow 1 \times (\text{row 2}) + \text{row 3}\end{array}$$

Check these results in the original statement of the problem.

Converting into the equivalent triangular system and using back-substitution gives $z = 90$, $y = 70$, and $x = 30$. Therefore, 30 pounds of feed A, 70 pounds of feed B, and 90 pounds of feed C should be combined to obtain the desired mix. ∎

This matrix procedure also reveals when a linear system has no solutions, that is, when it is an *inconsistent system*. In such a case we will obtain a row in a matrix of the form

$$0 \quad 0 \cdots 0 \; \vdots \; p$$

where $p \neq 0$. But when this row is converted to an equation, we get the false statement $0 = p$. This is illustrated in the next example.

EXAMPLE 3 Solve the system.

$$3x - 2y + z = 1$$
$$x - y - z = 2$$
$$6x - 4y + 2z = 3$$

Solution Write the augmented matrix for the linear system and apply the fundamental row operations.

$$\begin{bmatrix} 3 & -2 & 1 & \vdots & 1 \\ 1 & -1 & -1 & \vdots & 2 \\ 6 & -4 & 2 & \vdots & 3 \end{bmatrix}$$

$$\begin{bmatrix} 1 & -1 & -1 & \vdots & 2 \\ 3 & -2 & 1 & \vdots & 1 \\ 6 & -4 & 2 & \vdots & 3 \end{bmatrix} \begin{array}{l} \longleftarrow \\ \longleftarrow \end{array} \left\{ \begin{array}{l} \text{Interchange rows 1} \\ \text{and 2} \end{array} \right.$$

$$\begin{bmatrix} 1 & -1 & -1 & \vdots & 2 \\ 0 & 1 & 4 & \vdots & -5 \\ 0 & 2 & 8 & \vdots & -9 \end{bmatrix} \begin{array}{l} \longleftarrow -3 \times (\text{row 1}) + \text{row 2} \\ \longleftarrow -6 \times (\text{row 1}) + \text{row 3} \end{array}$$

$$\begin{bmatrix} 1 & -1 & -1 & \vdots & 2 \\ 0 & 1 & 4 & \vdots & -5 \\ 0 & 0 & 0 & \vdots & 1 \end{bmatrix} \begin{array}{l} \\ \\ \longleftarrow -2 \times (\text{row 2}) + \text{row 3} \end{array}$$

Three planes, two of which are parallel. No common point of intersection.

The last row gives the false equation $0 = 1$, and this means that the given system has no solutions. Geometrically, in this example, two of the three planes represented by the given equations are parallel as indicated in the adjoining figure. ∎

The next example demonstrates how the matrix method can be used to solve a system that has infinitely many solutions, that is, a *dependent system*.

CAUTION
Note that there is no y-term in the third equation. You should think of this as 0y and record the 0 in the y-position in the third row of the augmented matrix as shown on the next page.

EXAMPLE 4 Solve the system

$$x + 2y - z = 1$$
$$2x - y + 3z = 4$$
$$5x + 5z = 9$$

Solution

$$\begin{bmatrix} 1 & 2 & -1 & \vdots & 1 \\ 2 & -1 & 3 & \vdots & 4 \\ 5 & 0 & 5 & \vdots & 9 \end{bmatrix}$$

$$\begin{bmatrix} 1 & 2 & -1 & \vdots & 1 \\ 0 & -5 & 5 & \vdots & 2 \\ 0 & -10 & 10 & \vdots & 4 \end{bmatrix} \begin{matrix} \leftarrow -2 \times (\text{row } 1) + \text{row } 2 \\ \leftarrow -5 \times (\text{row } 1) + \text{row } 3 \end{matrix}$$

$$\begin{bmatrix} 1 & 2 & -1 & \vdots & 1 \\ 0 & -5 & 5 & \vdots & 2 \\ 0 & 0 & 0 & \vdots & 0 \end{bmatrix} \leftarrow -2 \times (\text{row } 2) + \text{row } 3$$

Since the last row contains *all* zeros, we have an equivalent linear system of two equations in three variables.

$$\begin{aligned} x + 2y - z &= 1 \\ -5y + 5z &= 2 \end{aligned}$$

This system has an *incomplete triangular form*.

Again we use back-substitution. First use the last equation to solve for y in terms of z to get $y = -\frac{2}{5} + z$. Now let $z = c$ represent any number, giving $y = -\frac{2}{5} + c$, and substitute back into the first equation.

$$x + 2y - z = x + 2\left(-\tfrac{2}{5} + c\right) - c = 1 \longrightarrow x = \tfrac{9}{5} - c$$

The solutions are

$$\left(\tfrac{9}{5} - c, \, -\tfrac{2}{5} + c, \, c\right) \qquad \text{for any real number } c$$

Find the specific solutions of the system for the values $c = 0$, $c = 1$, $c = \frac{2}{5}$, and $c = -2$.

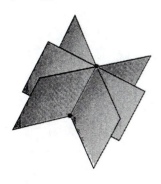

This result can be checked in the original system as follows.

$$x + 2y - \ z = \tfrac{9}{5} - c + 2\left(-\tfrac{2}{5} + c\right) - c = 1$$

$$2x - \ y + \ z = 2\left(\tfrac{9}{5} - c\right) - \left(-\tfrac{2}{5} + c\right) + 3c = 4$$

$$5x \qquad + 5z = 5\left(\tfrac{9}{5} - c\right) + 5c = 9 \qquad\qquad\blacksquare$$

Geometrically, the three planes for the given equations in Example 4 intersect in a common line, as indicated in the adjoining figure. The coordinates of *all* the points on this common line are given by the preceding solutions as c takes on all the real numbers.

You should also be able to obtain the solutions in the form

$$\left(\tfrac{7}{5} - e, e, \tfrac{2}{5} + e\right)$$

for any number e.

The solutions for the system in Example 4 are stated in terms of $z = c$, where c (or z) represents any number. There are other ways in which these solutions can be stated. For example, when c is replaced by z in the given solutions we have

$$x = \tfrac{9}{5} - z \quad \text{or} \quad z = \tfrac{9}{5} - x \quad \text{and} \quad y = -\tfrac{2}{5} + z$$

Now let $x = d$ be any number. Then

$$z = \tfrac{9}{5} - x = \tfrac{9}{5} - d$$

and

$$y = -\tfrac{2}{5} + z = -\tfrac{2}{5} + \left(\tfrac{9}{5} - d\right) = \tfrac{7}{5} - d$$

The solutions can then be stated in this form:

$$\left(d, \tfrac{7}{5} - d, \tfrac{9}{5} - d\right) \qquad \text{for any real number } d$$

In summary, when using the matrix method, keep the following observations in mind.

1. If you reach a linear system in triangular form, as in Example 1, the system has a unique solution that can be found by back-substitution.
2. If you reach a linear system that has an incomplete triangular form, and there are no false equations, as in Example 4, then the system has many solutions that can be found by back-substitution.
3. If you reach a linear system having a false equation, then the system has no solutions.

EXERCISES 9.1

Use matrices and fundamental row operations to solve each system.

1. $x + 5y = -9$
$\quad 4x - 3y = -13$

2. $4x - y = 6$
$\quad 2x + 3y = 10$

3. $3x + 2y = 18$
$\quad 6x + 5y = 45$

4. $2x + 4y = 24$
$\quad -3x + 5y = -25$

5. $4x - 5y = -2$
$\quad 16x + 2y = 3$

6. $x = y - 7$
$\quad 3y = 2x + 16$

7. $2x = -8y + 2$
$\quad 4y = x - 1$

8. $2x - 5y = 4$
$\quad -10x + 25y = -20$

9. $30x + 45y = 60$
$\quad 4x + 6y = 8$

10. $x - y = 3$
$\quad -\tfrac{1}{3}x + \tfrac{1}{3}x = 1$

11. $2x = 8 - 3y$
$\quad 6x + 9y = 14$

12. $-10x + 5y = 8$
$\quad 15x - 10y = -4$

13.
$$x - 2y + 3z = -2$$
$$-4x + 10y + 2z = -2$$
$$3x + y + 10z = 7$$

14.
$$2x + 4y + 8z = 14$$
$$4x - 2y + 2z = 6$$
$$-5x + 3y - z = -4$$

15.
$$-x + 2y + 3z = 11$$
$$2x - 3y = -6$$
$$3x - 3y + 3z = 3$$

16.
$$-2x + y + 2z = 14$$
$$5x + z = -10$$
$$x - 2y - 3z = -14$$

17.
$$x - 2z = 5$$
$$3y + 4z = -2$$
$$-2x + 3y + 8z = 4$$

18.
$$2x + y = 3$$
$$4x + + 5z = 6$$
$$-2y + 5z = -4$$

19.
$$x - 2y + z = 1$$
$$-6x + y + 2z = -2$$
$$-4x - 3y + 4z = 0$$

20.
$$4x - 3y + z = 0$$
$$-3x + y + 2z = 0$$
$$-2x + y + 5z = 0$$

21.
$$w - x + 2y + 2z = 0$$
$$2w - y - 3z = 0$$
$$4x - 3y + z = -2$$
$$-3w + 2x + 4z = 1$$

22.
$$4w - 5x + 2z = 0$$
$$-2w + 10x + y - 3z = 8$$
$$5x - 2y + 4z = -16$$
$$6w + 3z = 0$$

23.
$$v + 2w - x + 3z = 4$$
$$-w + 5x - y - z = 3$$
$$3v + 6x + 2z = 1$$
$$2v + 3w + 3x - y + 5z = 10$$
$$v + 9x - 2y + z = 5$$

24.
$$w + 3y = 10$$
$$-x + 2y + 6z = 2$$
$$-3w - 2x - 4z = 2$$
$$4x - y = 8$$

Write a system of three equations in three variables for each problem, and solve.

25. Daniel has $575 in one-dollar, five-dollar, and ten-dollar bills. Altogether he has 95 bills. The number of one-dollar bills plus the number of ten-dollar bills is five more than twice the number of five-dollar bills. How many of each type of bill does he have?

26. The sum of three numbers is 33. The largest number is one less than twice the smallest number. Three times the smallest number is one less than the sum of the other two numbers. Find the three numbers.

27. The sum of the angles of a triangle is 180°. The largest angle is equal to the sum of the other two angles. Twice the smallest angle is 10° less than the largest angle. Find the measure for each angle.

28. The treasurer of a club invested $5000 of their savings into three different accounts, at annual yields of 8%, 9%, and 10%. The total interest earned for the year was $460. The amount earned by the 10% account was $20 more than that earned by the 9% account. How much was invested at each rate?

29. A grocer sells peanuts at $2.80 per pound, pecans at $4.50 per pound, and Brazil nuts at $5.40 per pound. He wants to make a mixture of 50 pounds of mixed nuts to sell at $4.44 per pound. The mixture is to contain as many pounds of Brazil nuts as the other two types combined. How many pounds of each type must he use in this mixture?

30. Answer the question in Example 2 of this section after making the following changes. The monthly consumption of oats, corn and soybeans is 45, 60, and 45 pounds, respectively. Also, the contents of the feeds are given in this table:

	Oats	Corn	Soybeans
1 lb of feed *A*	4 oz	6 oz	6 oz
1 lb of feed *B*	8 oz	4 oz	4 oz
1 lb of feed *C*	3 oz	8 oz	5 oz

31. A dietician wants to combine three foods so that the resulting mixture contains 900 units of vitamins, 750 units of minerals, and 350 units of fat. The units of vitamins, minerals, and fat contained in each gram of the three foods are shown in the table. How many grams of each food should be combined to obtain the required mixture?

	Vitamins	Minerals	Fat
1 gram of food A	35 units	15 units	10 units
1 gram of food B	10 units	20 units	10 units
1 gram of food C	20 units	15 units	5 units

Written Assignment Continue using row operations after Step 5 on page 552 and reach the form

$$\begin{bmatrix} 1 & 0 & 0 & | & -1 \\ 0 & 1 & 0 & | & 1 \\ 0 & 0 & 1 & | & 1 \end{bmatrix}$$

Describe the steps you have used to obtain this form, and explain why nothing more needs to be done.

Challenge On page 556 the figure shows three planes, two of which are parallel so that the corresponding linear system has no common solutions. A system represented by three parallel planes also has no solutions. Now sketch three planes in space, no two of which are parallel, such that the corresponding system does not have any common solutions.

Graphing Calculator Exercises Some graphing calculators have built-in fundamental row operations which can be of great assistance, especially in large systems of equations. If yours does (see your user's manual), use it to solve the system of equations in Exercise 1.

1. $3x + 4y - z + w + u = 8$
$2x - y + z + 2w - 3u = 1$
$-x + 3y - z - w + 5u = 5$
$4x - 2y + 3w - u = 4$
$9x + y - 3z - w + 5u = 11$

2. (a) Find a quadratic function of x whose graph passes through the points $(2, 1)$, $(3, 2)$, $(4, -1)$. (*Hint:* Set up and solve a system of three equations such that the given points satisfy $y = ax^2 + bx + c$.)

(b) Graph the quadratic function with your graphing calculator, and check that it goes through the three points.

3. (a) Find a cubic function of x whose graph passes through points $(2, 1)$, $(3, 2)$, $(-1, -1)$, and $(-2, -2)$. (*Hint:* The form of the cubic is $y = ax^3 + bx^2 + cx + d$. See Exercise 2.)

(b) Graph the cubic function, and verify that it goes through the given points.

9.2 MATRIX ALGEBRA

The matrix method for solving linear systems of Section 9.1 uses row operations to transform a matrix into a row-equivalent matrix. That method, however, does not combine two or more matrices in any way. In this section we learn how to add and multiply matrices and also learn that with these operations matrices have many, but not all, of the basic properties of the real numbers stated in Section 1.1.

A matrix A with m rows and n columns is said to have dimensions m by n and can be written as

The subscripts on an element a_{ij} identify the row and column that the element is in. The first subscript i is the number of the row, and j is the number of the column; a_{53} is in row 5 and column 3.

$$A = \begin{bmatrix} a_{11} & a_{12} & \cdots & a_{1n} \\ a_{21} & a_{22} & \cdots & a_{2n} \\ \vdots & \vdots & \cdots & \vdots \\ a_{m1} & a_{m2} & \cdots & a_{mn} \end{bmatrix} \begin{matrix} \longleftarrow \text{ 1st row} \\ \longleftarrow \text{ 2nd row} \\ \\ \\ \longleftarrow \text{ mth row} \end{matrix}$$

$$\begin{matrix} \uparrow & \uparrow & & \uparrow \\ \text{1st} & \text{2nd} & & n\text{th} \\ \text{column} & \text{column} & & \text{column} \end{matrix}$$

Instead of the symbol A we sometimes use $A_{m \times n}$ to emphasize the dimensions of A, in which $m \times n$ means m rows by n columns. The preceding rectangular display can be abbreviated by writing

$$A_{m \times n} = [a_{ij}] \longleftarrow a_{ij} \text{ is the element in the } i\text{th row and } j\text{th column}$$

in which a_{ij} represents *each* element in the matrix for $1 \le i \le m$ and $1 \le j \le n$.

Equality of Matrices

Two m by n matrices are equal if and only if they have precisely the same elements in the same positions. In symbols using $A = [a_{ij}]$ and $B = [b_{ij}]$, we have:

$$A = B \text{ if and only if } a_{ij} = b_{ij} \text{ for each } i \text{ and } j$$

To add two m by n matrices, we add *corresponding* elements. For example, here is the computation for the sum of two 3 by 2 matrices.

$$\begin{bmatrix} 2 & -5 \\ 3 & 7 \\ -1 & 0 \end{bmatrix} + \begin{bmatrix} -4 & 1 \\ 0 & 9 \\ 2 & -2 \end{bmatrix} = \begin{bmatrix} 2 + (-4) & -5 + 1 \\ 3 + 0 & 7 + 9 \\ -1 + 2 & 0 + (-2) \end{bmatrix} = \begin{bmatrix} -2 & -4 \\ 3 & 16 \\ 1 & -2 \end{bmatrix}$$

Addition of Matrices

In general, the sum of $A_{m \times n} = [a_{ij}]$ and $B_{m \times n} = [b_{ij}]$ can be expressed as

$$A + B = [a_{ij}] + [b_{ij}] = [a_{ij} + b_{ij}] \text{ for each } i \text{ and } j$$

It is important to note that matrices can be added only when they have the same dimensions. Thus

$$\begin{bmatrix} 1 & 2 & 3 \\ -2 & 0 & 6 \end{bmatrix} \quad \text{and} \quad \begin{bmatrix} 6 & -5 \\ 3 & 3 \end{bmatrix}$$

cannot be added.

Commutative and Associative Properties for Addition

Matrix addition is both commutative and associative. That is, for matrices having the same dimensions:

$$A + B = B + A$$
$$(A + B) + C = A + (B + C)$$

We will not present formal proofs of these and other properties. However, you can see that they make sense by verifying the properties for specific examples.

EXAMPLE 1 Let $A = \begin{bmatrix} 3 & 1 & -2 \\ -1 & 3 & 0 \end{bmatrix}$, $B = \begin{bmatrix} -5 & 5 & 6 \\ -2 & -2 & -2 \end{bmatrix}$, and

$C = \begin{bmatrix} 0 & 1 & 0 \\ 1 & 0 & 1 \end{bmatrix}$. Show that $(A + B) + C = A + (B + C)$.

Solution

$$(A + B) + C = \left(\begin{bmatrix} 3 & 1 & -2 \\ -1 & 3 & 0 \end{bmatrix} + \begin{bmatrix} -5 & 5 & 6 \\ -2 & -2 & -2 \end{bmatrix} \right) + \begin{bmatrix} 0 & 1 & 0 \\ 1 & 0 & 1 \end{bmatrix}$$

$$= \begin{bmatrix} -2 & 6 & 4 \\ -3 & 1 & -2 \end{bmatrix} + \begin{bmatrix} 0 & 1 & 0 \\ 1 & 0 & 1 \end{bmatrix} = \begin{bmatrix} -2 & 7 & 4 \\ -2 & 1 & -1 \end{bmatrix}$$

$$A + (B + C) = \begin{bmatrix} 3 & 1 & -2 \\ -1 & 3 & 0 \end{bmatrix} + \left(\begin{bmatrix} -5 & 5 & 6 \\ -2 & -2 & -2 \end{bmatrix} + \begin{bmatrix} 0 & 1 & 0 \\ 1 & 0 & 1 \end{bmatrix} \right)$$

$$= \begin{bmatrix} 3 & 1 & -2 \\ -1 & 3 & 0 \end{bmatrix} + \begin{bmatrix} -5 & 6 & 6 \\ -1 & -2 & -1 \end{bmatrix} = \begin{bmatrix} -2 & 7 & 4 \\ -2 & 1 & -1 \end{bmatrix}$$

Thus

You should also verify that $A + B = B + A$.

$$(A + B) + C = A + (B + C) \qquad \blacksquare$$

The m by n zero matrix, denoted by O, is the matrix all of whose entries are 0. For example, the 2 by 3 zero matrix is

$$O_{2\times3} = \begin{bmatrix} 0 & 0 & 0 \\ 0 & 0 & 0 \end{bmatrix}$$

Verify the next rule.

Identity Property for Addition

$$A_{m \times n} + O_{m \times n} = A_{m \times n}$$

Negative of a Matrix

The negative of an m by n matrix $A = [a_{ij}]$ is the matrix denoted by $-A$ and is defined in this way:

$$-A = [-a_{ij}] \text{ for each } i \text{ and } j$$

In other words, the negative of a matrix is formed by replacing each number in A by its opposite. Also, since $a_{ij} + (-a_{ij}) = 0$ for each i and j, we have the following:

Additive Inverse Property

$$A + (-A) = O$$

For example, if $A = \begin{bmatrix} 8 & -5 \\ -4 & 9 \\ 0 & 2 \end{bmatrix}$, then $-A = \begin{bmatrix} -8 & 5 \\ 4 & -9 \\ 0 & -2 \end{bmatrix}$ and

$$A + (-A) = \begin{bmatrix} 8 + (-8) & -5 + 5 \\ -4 + 4 & 9 + (-9) \\ 0 + 0 & 2 + (-2) \end{bmatrix} = \begin{bmatrix} 0 & 0 \\ 0 & 0 \\ 0 & 0 \end{bmatrix} = O$$

Subtraction of Matrices

Subtraction of m by n matrices can be defined using the negative of a matrix.

$$A - B = A + (-B)$$

For example,

$$\begin{bmatrix} 2 & 3 & -1 \\ -5 & 7 & 0 \end{bmatrix} - \begin{bmatrix} -4 & 0 & 2 \\ 1 & 9 & -2 \end{bmatrix} = \begin{bmatrix} 2 & 3 & -1 \\ -5 & 7 & 0 \end{bmatrix} + \begin{bmatrix} 4 & 0 & -2 \\ -1 & -9 & 2 \end{bmatrix} = \begin{bmatrix} 6 & 3 & -3 \\ -6 & -2 & 2 \end{bmatrix}$$

Scalar Multiplication

A matrix A can be multiplied by a real number c by multiplying each element in A by c. This is referred to as **scalar multiplication.**

$$cA = c[a_{ij}] = [ca_{ij}] \text{ for each } i \text{ and } j$$

It is customary to write the scalar c to the left of the matrix, thus we do not use the notation Ac for scalar multiplication.

For example,

$$7\begin{bmatrix} 8 & -5 \\ -4 & 9 \\ 0 & 2 \end{bmatrix} = \begin{bmatrix} 56 & -35 \\ -28 & 63 \\ 0 & 14 \end{bmatrix}$$

Using this property, the negative of a matrix A may be written as $-A = -1A$. Here is a list of scalar multiplication properties in which c and d are scalars and matrices A and B have the same dimensions:

Scalar Multiplication Properties

(1)	$(cd)A = c(dA)$	The scalar product involving two scalars and a matrix is associative.
(2)	$c(A + B) = cA + cB$	Scalar multiplication distributes over matrix addition from the left.
(3)	$(c + d)A = cA + dA$	Scalar multiplication distributes over the sum of scalars (from the right).

EXAMPLE 2 Verify $(cd)A = c(dA)$ using

$$c = -2, \ d = \tfrac{1}{3}, \quad \text{and} \quad A = \begin{bmatrix} 6 & -9 \\ 1 & 5 \end{bmatrix}$$

Solution

$$(cd)A = \left(-2 \cdot \tfrac{1}{3}\right)\begin{bmatrix} 6 & -9 \\ 1 & 5 \end{bmatrix} = -\tfrac{2}{3}\begin{bmatrix} 6 & -9 \\ 1 & 5 \end{bmatrix} = \begin{bmatrix} -4 & 6 \\ -\tfrac{2}{3} & -\tfrac{10}{3} \end{bmatrix}$$

$$c(dA) = -2\left(\tfrac{1}{3}\begin{bmatrix} 6 & -9 \\ 1 & 5 \end{bmatrix}\right) = -2\begin{bmatrix} 2 & -3 \\ \tfrac{1}{3} & \tfrac{5}{3} \end{bmatrix} = \begin{bmatrix} -4 & 6 \\ -\tfrac{2}{3} & -\tfrac{10}{3} \end{bmatrix} \quad\blacksquare$$

TEST YOUR UNDERSTANDING

Let $A = \begin{bmatrix} 1 & 4 & -3 \\ 2 & 5 & 0 \\ 0 & 1 & 1 \end{bmatrix}$, $B = \begin{bmatrix} -2 & 2 \\ 3 & -3 \\ 1 & 4 \end{bmatrix}$, $C = \begin{bmatrix} 5 & 1 \\ -1 & 6 \\ 8 & 2 \end{bmatrix}$

Evaluate each of the following, if possible.

1. $A + B$
2. $B + C$
3. $C + (B + C)$
4. $B - C$
5. $C - B$
6. $2A + 0A$
7. $2A + 3C$
8. $(-5)(B + C)$
9. $4\left(-\tfrac{1}{2}\right)C$

(Answers: Page 631)

Multiplication of matrices is a more complicated operation than addition. First, let us consider the product of

$$A_{2\times3} = \begin{bmatrix} -4 & -1 & 2 \\ 3 & 8 & 1 \end{bmatrix} \quad \text{and} \quad B_{3\times2} = \begin{bmatrix} -2 & 1 \\ 7 & 0 \\ 4 & -6 \end{bmatrix}$$

Here is the completed product followed by the explanation of how it was done.

$$\begin{array}{cccc} A & B & C & C \end{array}$$

$$\begin{bmatrix} -4 & -1 & 2 \\ 3 & 8 & 1 \end{bmatrix} \begin{bmatrix} -2 & 1 \\ 7 & 0 \\ 4 & -6 \end{bmatrix} = \begin{bmatrix} 9 & -16 \\ 54 & -3 \end{bmatrix} = \begin{bmatrix} c_{11} & c_{12} \\ c_{21} & c_{22} \end{bmatrix}$$

Row 1 *of A* and *column* 1 *of B* produce 9, the element in row 1, column 1 of C:

$$(-4)(-2) + (-1)(7) + (2)(4) = 9 = c_{11}$$

Row 1 *of A* and *column* 2 *of B* produce -16, the element in row 1, column 2 of C:

$$(-4)(1) + (-1)(0) + (2)(-6) = -16 = c_{12}$$

Row 2 *of A* and *column* 1 *of B* produce 54, the element in row 2, column 1 of C:

$$(3)(-2) + (8)(7) + (1)(4) = 54 = c_{21}$$

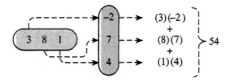

Row 2 *of A* and *column* 2 *of B* produce -3, the element in row 2, column 2 of C:

$$(3)(1) + (8)(0) + (1)(-6) = -3 = c_{22}$$

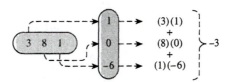

As you can observe in the preceding computations, an element in $AB = C$ is the sum of the products of numbers in a row of A by the numbers in a column B. Furthermore, the ith row of A together with the jth column of B produces the element in row i, column j of C.

Here is a diagram that displays how the row 2, column 3 element in a product is completed.

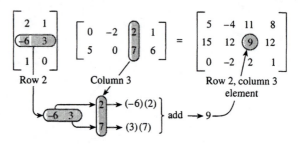

In each of the matrix products that have been found, you can see that it is possible to compute product AB only when *the number of columns of A equals the number of rows of B.* Furthermore, if A is m by n and B is n by p, then AB is m by p. For example,

$$(A_{3 \times 5})(B_{5 \times 7}) = C_{3 \times 7}$$

These must be the same.   number of rows in A   number of columns in B

Using A and D from the preceding set of exercises, we have

$$AD = \begin{bmatrix} 1 & 2 \\ 3 & -5 \end{bmatrix}\begin{bmatrix} 8 & -8 \\ -7 & -7 \end{bmatrix} = \begin{bmatrix} -6 & -22 \\ 59 & 11 \end{bmatrix}$$

and

$$DA = \begin{bmatrix} 8 & -8 \\ -7 & -7 \end{bmatrix}\begin{bmatrix} 1 & 2 \\ 3 & -5 \end{bmatrix} = \begin{bmatrix} -16 & 56 \\ -28 & 21 \end{bmatrix}$$

Recall that a single counterexample is sufficient to disprove a statement that claims to be true in general.

Therefore, $AD \neq DA$, which *proves* that matrix multiplication is not commutative. We do have the following properties, assuming that the operations are possible.

Associative and Distributive Properties

$$(AB)C = A(BC)$$ associative property

$$A(B + C) = AB + AC$$ left distributive property

$$(B + C)A = BA + CA$$ right distributive property

In the exercises you will be asked to verify these properties for specific cases.

Scalar multiplication can also be combined with the product of matrices according to this result, in which c represents any real number.

Properties of Scalar Multiplication of a Matrix Product

$$c(AB) = (cA)B = A(cB)$$

A scalar multiple of a matrix product equals the matrix product of either matrix times the scalar multiple of the other.

EXAMPLE 3
(a) For matrices A and B, for which AB is defined, simplify $(2A)B - A(5B)$.
(b) Evaluate the simplified expression in part (a) for

$$A = \begin{bmatrix} 7 & -5 \\ 0 & 4 \end{bmatrix} \quad \text{and} \quad B = \begin{bmatrix} 0 & 1 & 2 \\ -3 & 2 & -1 \end{bmatrix}$$

Solution
(a) $(2A)B - A(5B) = 2(AB) - 5(AB)$ scalar multiplication of a product

$$= 2(AB) + \left(-5(AB)\right) \quad \text{definition of matrix subtraction}$$

$$= \left(2 + (-5)\right)(AB) \quad \text{distributive}$$

$$= -3(AB)$$

(b) Using the result in part (a) we have

$$-3(AB) = -3\left(\begin{bmatrix} 7 & -5 \\ 0 & 4 \end{bmatrix}\begin{bmatrix} 0 & 1 & 2 \\ -3 & 2 & -1 \end{bmatrix}\right)$$

Evaluating before the algebraic simplification in part (a) would have required two scalar multiplications, two matrix products, and one matrix subtraction. Here all that is needed is one matrix product and one scalar multiplication.

$$= -3\begin{bmatrix} 15 & -3 & 19 \\ -12 & 8 & -4 \end{bmatrix}$$

$$= \begin{bmatrix} -45 & 9 & -57 \\ 36 & -24 & 12 \end{bmatrix} \quad \blacksquare$$

There is also a multiplicative identity for **square matrices,** matrices that are n by n. This is the matrix whose main diagonal consists of 1's and all other entries are 0. We use I_n for this matrix.

In particular, when $n = 3$,

$$I_3 = \begin{bmatrix} 1 & 0 & 0 \\ 0 & 1 & 0 \\ 0 & 0 & 1 \end{bmatrix} \longleftarrow \text{main diagonal}$$

Using $A = \begin{bmatrix} a & b & c \\ d & e & f \\ g & h & i \end{bmatrix}$, we have

$$I_3 A = \begin{bmatrix} 1 & 0 & 0 \\ 0 & 1 & 0 \\ 0 & 0 & 1 \end{bmatrix} \begin{bmatrix} a & b & c \\ d & e & f \\ g & h & i \end{bmatrix} = \begin{bmatrix} a & b & c \\ d & e & f \\ g & h & i \end{bmatrix} = A$$

Similarly, $AI_3 = A$. In general, if A is n by n, then

$$AI_n = I_n A = A$$

Multiplication Identity Property

I_n is the multiplicative identity for the n by n matrices. It is also called the **nth-order identity matrix.**

The following applied example illustrates how matrices can be used to record and organize data, and demonstrates that matrix multiplication can be an efficient method to do computations.

EXAMPLE 4 The Education Toy Company produces three kinds of model airplanes. The major materials needed are metal (m), plastic (p), and wood (w). The 3×3 matrix R shows how much of each material is needed for the models, expressed in terms of convenient units.

Number of Units

$$R = \begin{matrix} & \begin{matrix} m & p & w \end{matrix} & \\ \begin{bmatrix} 2 & 4 & 5 \\ 3 & 5 & 3 \\ 3 & 6 & 4 \end{bmatrix} & \begin{matrix} \text{Model } a \\ \text{Model } b \\ \text{Model } c \end{matrix} \end{matrix}$$

The company sells directly to a chain of department stores that has placed an order for 700 of model a, 800 of model b, and 500 of model c. This purchase order is written as the 1×3 matrix P:

Number of Models

$$\begin{matrix} a & b & c \end{matrix}$$
$$P = [700 \quad 800 \quad 500]$$

(a) Find the matrix that gives the total number of units of each material needed to fill the purchase order.

(b) The toy company buys all the materials from either one of two suppliers. The unit prices of supplier s_1 are \$1.50 for metal, \$0.85 for plastic, and \$1.15 for wood. Supplier s_2 charges \$1.65 for metal, \$0.80 for plastic, and \$1.10 for wood. Write the 3×2 *cost matrix* C whose columns show the unit prices of each supplier, and find the matrix that shows the cost of materials for each model based on the two sets of supplier prices.

(c) Find the 1×2 matrix that shows the total cost of materials to fill the purchase order for each set of supplier prices. Based on cost only, which supplier should the company buy the materials from?

Solution

(a) The total number of metal units needed is

$$(700)(2) + (800)(3) + (500)(3) = 5300$$

Understanding this computation leads to this matrix product:

$$PR = [700 \quad 800 \quad 500] \begin{bmatrix} 2 & 4 & 5 \\ 3 & 5 & 3 \\ 3 & 6 & 4 \end{bmatrix} = [5300 \quad 9800 \quad 7900]$$

(b)

$$C = \begin{bmatrix} 1.50 & 1.65 \\ .85 & .80 \\ 1.15 & 1.10 \end{bmatrix} \begin{matrix} \text{metal} \\ \text{plastic} \\ \text{wood} \end{matrix}$$

$$RC = \begin{bmatrix} 2 & 4 & 5 \\ 3 & 5 & 3 \\ 3 & 6 & 4 \end{bmatrix} \begin{bmatrix} 1.50 & 1.65 \\ .85 & .80 \\ 1.15 & 1.10 \end{bmatrix} = \begin{bmatrix} 12.15 & 12.00 \\ 12.20 & 12.25 \\ 14.20 & 14.15 \end{bmatrix} \begin{matrix} \text{Model } a \\ \text{Model } b \\ \text{Model } c \end{matrix}$$

(c) $P(RC) = [700 \quad 800 \quad 500] \begin{bmatrix} 12.15 & 12.00 \\ 12.20 & 12.25 \\ 14.20 & 14.15 \end{bmatrix} = [25,365 \quad 25,275]$

Verify the result by computing (PR)C. Why must this give the same results?

Since $25,365 - 25,275 = 90$, the company saves \$90 by using supplier s_2. ∎

EXERCISES 9.2

Evaluate the given matrix expression, if possible, using these matrices.

$$A = \begin{bmatrix} 1 \\ -2 \end{bmatrix} \quad B = [3 \quad 5] \quad C = \begin{bmatrix} 5 & -1 \\ -3 & 4 \end{bmatrix} \quad D = \begin{bmatrix} 1 & 2 \\ 3 & 4 \end{bmatrix} \quad E = \begin{bmatrix} 1 & -3 & 2 \\ -5 & 2 & 0 \\ 0 & -1 & 3 \end{bmatrix}$$

$$F = \begin{bmatrix} 1 & 1 & 1 \\ -1 & 1 & 1 \\ -1 & -1 & 1 \end{bmatrix} \quad G = \begin{bmatrix} 1 & -4 & 0 \\ 2 & -3 & 5 \end{bmatrix} \quad H = \begin{bmatrix} 5 & 2 & -3 & 0 \\ 1 & 0 & 3 & -1 \\ 4 & -2 & 0 & 1 \end{bmatrix} \quad J = \begin{bmatrix} 1 & 0 \\ 0 & -1 \\ 1 & 1 \\ 0 & 0 \end{bmatrix}$$

1. $C + D$ **2.** $D - C$ **3.** $A + B$ **4.** $C + G$ **5.** $E + 2F$ **6.** $-2E + 5E$

7. BA **8.** AB **9.** CD **10.** DC **11.** CA **12.** BC

13. GE **14.** EG **15.** GJ **16.** EF **17.** HJ **18.** $(DA)C$

19. GH **20.** $\left(\frac{1}{2}D\right)J$ **21.** $F\begin{bmatrix} 0 \\ 0 \\ 0 \end{bmatrix}$ **22.** $\begin{bmatrix} 0 \\ 0 \\ 0 \end{bmatrix}F$ **23.** I_3H **24.** EI_3

25. $G(-2F)$ **26.** $(-2G)F$ **27.** $(GF)H$ **28.** $(G(-F))H$

Verify the matrix equation using these matrices.

$$A = \begin{bmatrix} 2 & 1 \\ -4 & 1 \end{bmatrix} \quad B = \begin{bmatrix} 0 & 6 \\ 5 & -4 \end{bmatrix} \quad C = \begin{bmatrix} 3 & 7 \\ -1 & -2 \end{bmatrix}$$

29. $B + C = C + B$ **30.** $3(AB) = (3A)B = A(3B)$ **31.** $A + (B + C) = (A + B) + C$

32. $A(B + C) = AB + AC$ **33.** $(A + B)C = AC + BC$ **34.** $(AB)C = A(BC)$

35. Verify that $(AB)C = A(BC)$ using

$$A = \begin{bmatrix} 1 & -2 \\ 3 & 3 \\ 0 & 4 \end{bmatrix} \quad B = \begin{bmatrix} -1 & 6 & 8 \\ 2 & 1 & 3 \end{bmatrix} \quad C = \begin{bmatrix} 5 \\ -2 \\ 1 \end{bmatrix}$$

36. Simplify $B\left(\frac{1}{3}A\right) - \left(\frac{7}{3}B\right)A$ using the matrix properties of this section and evaluate this expression for matrices A and B given in Exercise 35.

37. Convert $A(2B) + (2A)C$ into a form that involves only one matrix product and apply the result to evaluate the given expression for these matrices:

$$A = \begin{bmatrix} 5 & -2 & 1 \end{bmatrix} \quad B = \begin{bmatrix} 5 & -2 \\ 3 & 3 \\ 0 & 4 \end{bmatrix} \quad C = \begin{bmatrix} -1 & 2 \\ 6 & 1 \\ 8 & 3 \end{bmatrix}$$

38. Arrange the four matrices with the indicated dimensions so that a product of all four can be formed. What will be the dimensions of this product?

$$A_{4 \times 1} \quad B_{3 \times 5} \quad C_{1 \times 3} \quad D_{2 \times 4}$$

39. Let $E = \begin{bmatrix} 0 & 1 & 0 \\ 1 & 0 & 0 \\ 0 & 0 & 1 \end{bmatrix}$ and $A = \begin{bmatrix} a & b & c \\ d & e & f \\ g & h & i \end{bmatrix}$

(a) Evaluate EA.

(b) Compare the answer in part (a) to matrix A and find a matrix F so that FA is the same as A except that the first and third rows are interchanged.

40. (a) Use E and A in Exercise 39 and evaluate AE.

(b) Find a matrix F so that AF is the same as A except that the second and third columns are interchanged.

41. Find matrices E and F so that for matrix A in Exercise 39,

$$EA = \begin{bmatrix} a & b & c \\ d & e & f \\ g + 3a & h + 3b & i + 3c \end{bmatrix} \quad \text{and} \quad AF = \begin{bmatrix} a & b - c & c \\ d & e - f & f \\ g & h - i & i \end{bmatrix}$$

42. Find matrices E and F so that for matrix A in Exercise 39,

$$EA = \begin{bmatrix} 2a & 2b & 2c \\ d & e & f \\ g & h & i \end{bmatrix} \quad \text{and} \quad AF = \begin{bmatrix} a & 2b & c \\ d & 2e & f \\ g & 2h & i \end{bmatrix}$$

43. (a) Find x, y so that

$$\begin{bmatrix} 1 & 1 \\ 0 & 0 \end{bmatrix}\begin{bmatrix} 2 & 3 \\ x & y \end{bmatrix} = \begin{bmatrix} 0 & 0 \\ 0 & 0 \end{bmatrix}$$

(b) Explain why it is not possible to find x, y so that

$$\begin{bmatrix} 2 & 3 \\ x & y \end{bmatrix}\begin{bmatrix} 1 & 1 \\ 0 & 0 \end{bmatrix} = \begin{bmatrix} 0 & 0 \\ 0 & 0 \end{bmatrix}$$

44. Find $a, b, c,$ and d so that $\begin{bmatrix} 1 & 1 \\ 0 & 1 \end{bmatrix}\begin{bmatrix} a & b \\ c & d \end{bmatrix} = I_2$ and verify that $\begin{bmatrix} a & b \\ c & d \end{bmatrix}\begin{bmatrix} 1 & 1 \\ 0 & 1 \end{bmatrix} = I_2.$

45. (a) Let $A = \begin{bmatrix} x & 0 & 0 \\ 0 & y & 0 \\ 0 & 0 & z \end{bmatrix}$ and evaluate $A^2 = AA.$

(b) Use the result in part (a) to evaluate $A^3 = AAA.$

(c) If n is a positive integer, find $A^n.$

46. (a) Use $A = \begin{bmatrix} 3 & -1 \\ 2 & 1 \end{bmatrix}$, $B = \begin{bmatrix} 0 & -2 \\ 1 & 4 \end{bmatrix}$

Show that $(A + B)^2 \neq A^2 + 2AB + B^2.$

(b) Give a reason for each numbered step.

$$\begin{aligned} (A + B)^2 &= (A + B)(A + B) \\ &= (A + B)A + (A + B)B & \text{(i)} \\ &= A^2 + BA + AB + B^2 & \text{(ii)} \end{aligned}$$

(c) Verify the result of part (b) using matrices A and B in part (a).

47. The **transpose** of an $m \times n$ matrix A, denoted by A^t, is the $n \times m$ matrix obtained by taking the rows in A, in sequence, and converting each one into a column of A^t. Thus, for

$$A = \begin{bmatrix} 1 & 2 \\ 3 & 4 \\ 5 & 6 \end{bmatrix}, \quad A^t = \begin{bmatrix} 1 & 3 & 5 \\ 2 & 4 & 6 \end{bmatrix}$$

(a) Simplify: $(A^t)^t$

(b) For two $m \times n$ matrices $(A + B)^t = A^t + B^t$.

Prove this for the case using $A = \begin{bmatrix} a_1 & a_2 & a_3 \\ a_4 & a_5 & a_6 \end{bmatrix}$ and $B = \begin{bmatrix} b_1 & b_2 & b_3 \\ b_4 & b_5 & b_6 \end{bmatrix}$

48. For $A = \begin{bmatrix} -3 & 5 & 1 \\ 2 & 0 & 4 \\ -1 & 3 & -2 \end{bmatrix}$ and $B = \begin{bmatrix} 4 & 0 & -1 \\ 1 & 2 & 3 \\ -6 & 0 & 4 \end{bmatrix}$ verify that $(AB)^t = B^tA^t$.

(See Exercise 47.)

49. The Health-Sport Company has two stores identified as Health-Sport I and Health-Sport II. In the month of May, Health-Sport I sold 21 bicycles, 15 rowing machines, and 34 treadmills; the corresponding sales of Health-Sport II were 19, 25, and 28. In June, Store I sold 28 bicycles, 18 rowing machines, and 27 treadmills, and the sales at Store II were 25, 17, and 28, respectively.

(a) Write the 2×3 sales matrix M showing the sales of the three items at each store. Do the same for the month of June, using J for this matrix. Find $M + J$ and interpret the results. Also find $M - J$ and explain what it means when $M - J$ has a positive element, a zero element, or a negative element. (Note: M is the sales matrix for May.)

(b) The combined sales for May, June, and July are given by the sales matrix

$$\begin{array}{cccc} & \text{bic.} & \text{row.} & \text{tread.} \\ S = & \left[\begin{matrix} 67 \\ 59 \end{matrix}\right. & \begin{matrix} 45 \\ 62 \end{matrix} & \left.\begin{matrix} 70 \\ 63 \end{matrix}\right] \end{array} \begin{matrix} \text{Store I} \\ \text{Store II} \end{matrix}$$

Use matrix operations to find the matrix that shows the sales of each item during July at each store.

50. A student is responsible for purchasing the fruit needed for a club picnic; 6 pounds of apples, 10 pounds of grapes, 8 pounds of plums, and 12 pounds of peaches. There are two stores available that have all of this fruit. The Maxi-Mart charges \$0.99 for a pound of apples, \$1.49 for a pound of grapes, \$1.69 for a pound of plums, and \$1.29 for a pound of peaches. At the Super-Duper store, the corresponding prices are \$1.09, \$1.59, \$1.35, and \$1.19.

(a) Write the 1×4 matrix A that gives the number of pounds of each fruit needed and the 4×2 cost matrix C showing the price per pound for each store.

(b) Compute the matrix product AC and explain what the elements of AC represent.

51. The Educational Toy Company (see Example 4) gets another purchase order for 0 of model a, 1200 of model b and 200 of model c.

(a) Write the 1×3 purchase order and use matrix multiplication to find the number of units of each material needed for this purchase order.

(b) Find the total cost of materials needed to fill the order based on each supplier's prices by using matrix multiplication. Which supplier is more economical for the toy company?

Challenge For $A = \begin{bmatrix} 2 & 0 \\ -4 & 2 \end{bmatrix}$ find matrices E_1, E_2 so that $E_2 E_1 A = \begin{bmatrix} 1 & 0 \\ 0 & 1 \end{bmatrix}$.

(*Hint:* Consider the ideas similar to those contained in Exercises 39–42.)

Written Assignment Consider the real number properties for addition and multiplication given in the boxed display on page 4. Some of these properties apply to addition and multiplication of the 2 by 2 matrices. Make lists of those that do apply and those that don't. For those that don't, give counterexamples.

Graphing Calculator Exercises If your graphing calculator can evaluate matrix operations, given a square matrix A, it can be used to find $A^2 = AA$, $A^3 = A^2 A$, and so on. (See your user's manual.)

$$\text{Let } A = \begin{bmatrix} 0 & 1 & 0 & 1 \\ 1 & 0 & 1 & 0 \\ 0 & 1 & 1 & 0 \\ 1 & 0 & 1 & 0 \end{bmatrix}$$

Compute **(a)** A^2 **(b)** A^3 **(c)** A^4 **(d)** A^5

9.3 SOLVING LINEAR SYSTEMS USING INVERSES

You may have wondered why nothing was said in the preceding section about division of matrices. The reason is simple, *division of matrices is undefined.* There is, however, a process possible for some matrices that resembles the division of real numbers when the division is converted to multiplication using reciprocals, also called inverses, as in

$$2 \div 5 = 2 \cdot \frac{1}{5} = 2 \cdot 5^{-1}$$

Similarly, some *square* matrices have inverses, denoted by A^{-1}, which makes it possible to compute products of the form BA^{-1}.

Inverse Matrices

An n by n square matrix B is said to be the **inverse** of an n by n square matrix A if and only if we have the following:

$$AB = BA = I_n, \quad \text{where } I_n \text{ is the identity matrix}$$

In such a case matrix A is also the inverse of B; A and B are inverses of one another. The notation A^{-1} is frequently used for the inverse of a matrix A, so that

This compares to
$x \cdot \frac{1}{x} = \frac{1}{x} \cdot x = 1$ *for real numbers* $x \neq 0$.

$$AA^{-1} = A^{-1}A = I_n$$

When a matrix has an inverse, it has only one inverse. That is, the inverse of a matrix is unique (see Exercise 37).

Does $A = \begin{bmatrix} 1 & -1 \\ 0 & 2 \end{bmatrix}$ have an inverse? If so, then there are numbers v, w, x, and y such that

$$\begin{bmatrix} 1 & -1 \\ 0 & 2 \end{bmatrix} \begin{bmatrix} v & w \\ x & y \end{bmatrix} = \begin{bmatrix} 1 & 0 \\ 0 & 1 \end{bmatrix} = I_2$$

Multiply to get

$$\begin{bmatrix} v - x & w - y \\ 2x & 2y \end{bmatrix} = \begin{bmatrix} 1 & 0 \\ 0 & 1 \end{bmatrix}$$

Since two matrices are equal if and only if their elements in the same positions are equal, we get

$$v - x = 1$$

$$w - y = 0$$

$$2x = 0 \longrightarrow x = 0$$

$$2y = 1 \longrightarrow y = \tfrac{1}{2}$$

$$v - x = v - 0 = 1 \longrightarrow v = 1$$

$$w - y = w - \tfrac{1}{2} = 0 \longrightarrow w = \tfrac{1}{2}$$

Therefore,

$$\begin{bmatrix} v & w \\ x & y \end{bmatrix} = \begin{bmatrix} 1 & \tfrac{1}{2} \\ 0 & \tfrac{1}{2} \end{bmatrix}$$

Now check by multiplying:

This shows that $AA^{-1} = I_2$.

$$\begin{bmatrix} 1 & -1 \\ 0 & 2 \end{bmatrix} \begin{bmatrix} 1 & \tfrac{1}{2} \\ 0 & \tfrac{1}{2} \end{bmatrix} = \begin{bmatrix} 1 + 0 & \tfrac{1}{2} - \tfrac{1}{2} \\ 0 + 0 & 0 + 1 \end{bmatrix} = \begin{bmatrix} 1 & 0 \\ 0 & 1 \end{bmatrix} = I_2$$

Also,

This shows that $A^{-1}A = I_2$.

$$\begin{bmatrix} 1 & \tfrac{1}{2} \\ 0 & \tfrac{1}{2} \end{bmatrix} \begin{bmatrix} 1 & -1 \\ 0 & 2 \end{bmatrix} = \begin{bmatrix} 1 & 0 \\ 0 & 1 \end{bmatrix} = I_2$$

Thus

$$A^{-1} = \begin{bmatrix} 1 & -1 \\ 0 & 2 \end{bmatrix}^{-1} = \begin{bmatrix} 1 & \tfrac{1}{2} \\ 0 & \tfrac{1}{2} \end{bmatrix}$$

Not all square matrices have inverses. For example, the matrix $\begin{bmatrix} 0 & 1 \\ 0 & 1 \end{bmatrix}$ has no inverse because if it did there would have to be v, w, x, and y such that

$$\begin{bmatrix} 0 & 1 \\ 0 & 1 \end{bmatrix} \begin{bmatrix} v & w \\ x & y \end{bmatrix} = \begin{bmatrix} 1 & 0 \\ 0 & 1 \end{bmatrix}$$

$$\begin{bmatrix} x & y \\ x & y \end{bmatrix} = \begin{bmatrix} 1 & 0 \\ 0 & 1 \end{bmatrix}$$

Equating the elements gives

$$x = 1 \quad \text{and} \quad x = 0$$

which is not possible. Consequently, $\begin{bmatrix} 0 & 1 \\ 0 & 1 \end{bmatrix}$ has no inverse.

The method of finding an inverse used in the preceding discussion is not practical for larger matrices. As you have seen, when $n = 2$ we have to solve a system of 4 equations. With $n = 3$, there would be 9 equations, then 16 equations for $n = 4$, and so on.

Fortunately, there are other more efficient methods for finding inverses. We will now present such a procedure and first demonstrate how it works using the matrix $A = \begin{bmatrix} 1 & -1 \\ 0 & 2 \end{bmatrix}$, for which we already know the inverse.

Begin by constructing a 2 by 4 matrix that contains both A and the identity matrix I_2.

$$\begin{array}{cc} A & I_2 \\ \downarrow & \downarrow \end{array}$$

$$\left[\begin{array}{cc:cc} 1 & -1 & 1 & 0 \\ 0 & 2 & 0 & 1 \end{array} \right] \qquad \text{Put } A \text{ to the left and } I_2 \text{ to the right of the dashed line.}$$

Now apply fundamental row operations to the rows of this 2 by 4 matrix until the left half has been transformed into I_2. At this step the right half will be A^{-1}. Here are the details.

$$\begin{array}{cc} A & I_2 \\ \downarrow & \downarrow \end{array}$$

$$\left[\begin{array}{cc:cc} 1 & -1 & 1 & 0 \\ 0 & 2 & 0 & 1 \end{array} \right]$$

$$\tfrac{1}{2} \times (\text{row } 2) \longrightarrow \left[\begin{array}{cc:cc} 1 & -1 & 1 & 0 \\ 0 & 1 & 0 & \tfrac{1}{2} \end{array} \right]$$

$$\text{row } 2 + \text{row } 1 \longrightarrow \left[\begin{array}{cc:cc} 1 & 0 & 1 & \tfrac{1}{2} \\ 0 & 1 & 0 & \tfrac{1}{2} \end{array} \right]$$

$$\begin{array}{cc} \uparrow & \uparrow \\ I_2 & A^{-1} \end{array}$$

From our earlier work we know that $\begin{bmatrix} 1 & \tfrac{1}{2} \\ 0 & \tfrac{1}{2} \end{bmatrix}$ is the inverse of A.

This method can also be used to find out that a matrix does not have an inverse. The fundamental row operations are applied until we discover that it is not possible to reach *I* on the left. When this happens there is no inverse.

EXAMPLE 1 Show that $A = \begin{bmatrix} 2 & -4 \\ -1 & 2 \end{bmatrix}$ has no inverse.

Solution

$$\begin{array}{cc} A & I_2 \\ \downarrow & \downarrow \end{array}$$

$$\begin{bmatrix} 2 & -4 & \vdots & 1 & 0 \\ -1 & 2 & \vdots & 0 & 1 \end{bmatrix}$$

$$\tfrac{1}{2} \times (\text{row 1}) \longrightarrow \begin{bmatrix} 1 & -2 & \vdots & \tfrac{1}{2} & 0 \\ -1 & 2 & \vdots & 0 & 1 \end{bmatrix}$$

$$\text{row 1} + \text{row 2} \longrightarrow \begin{bmatrix} 1 & -2 & \vdots & \tfrac{1}{2} & 0 \\ 0 & 0 & \vdots & \tfrac{1}{2} & 1 \end{bmatrix}$$

Because of the row of zeros to the left it is not possible to use row operations to obtain 1 0 in the first row on the left. Then I_2 cannot be reached on the left, and therefore there is no inverse of *A*. ■

TEST YOUR UNDERSTANDING

Find A^{-1}, if it exists, for the given A. When A^{-1} exists, verify that $AA^{-1} = A^{-1}A = I_2$.

1. $A = \begin{bmatrix} 3 & -6 \\ -2 & 5 \end{bmatrix}$ **2.** $A = \begin{bmatrix} -4 & 8 \\ -3 & 6 \end{bmatrix}$ **3.** $A = \begin{bmatrix} -3 & -2 \\ -1 & 0 \end{bmatrix}$

4. $A = \begin{bmatrix} \tfrac{1}{2} & \tfrac{1}{4} \\ -\tfrac{1}{3} & -\tfrac{1}{3} \end{bmatrix}$ **5.** $A = \begin{bmatrix} 7 & 7 \\ 7 & 7 \end{bmatrix}$ **6.** $A = \begin{bmatrix} 10 & 100 \\ 100 & -100 \end{bmatrix}$

(Answers: Page 631)

The procedure used for finding the inverse of a 2 × 2 matrix also applies to larger square matrices and can be summarized for an $n \times n$ matrix A as follows.

Apply fundamental row operations on the matrix $[A \vdots I_n]$ with the objective of transforming A into I_n.

1. If the form $[I_n \vdots B]$ is obtained, then the $n \times n$ matrix B is the inverse of A.
2. If all zeros are obtained in one or more rows to the left of the dashed vertical line, then A does not have an inverse.

The next example illustrates this procedure for a 3 × 3 matrix. This will be considerably more work than for the 2 × 2 case; however, it is usually more efficient than the method of solving a system of nine linear equations that was mentioned on the preceding page.

EXAMPLE 2 Let $A = \begin{bmatrix} 2 & 0 & -1 \\ -1 & 2 & 1 \\ 3 & -2 & -4 \end{bmatrix}$. Find A^{-1}.

Solution

Our first step will be to interchange the first two rows to get the -1 in the second row into the top left position.

$$\begin{matrix} A & & I_3 \\ \downarrow & & \downarrow \end{matrix}$$

$$\left[\begin{array}{ccc|ccc} 2 & 0 & -1 & 1 & 0 & 0 \\ -1 & 2 & 1 & 0 & 1 & 0 \\ 3 & -2 & -4 & 0 & 0 & 1 \end{array}\right]$$

$$\begin{matrix} \text{Interchange} \longrightarrow \\ R_1 \text{ and } R_2 \longrightarrow \end{matrix} \left[\begin{array}{ccc|ccc} -1 & 2 & 1 & 0 & 1 & 0 \\ 2 & 0 & -1 & 1 & 0 & 0 \\ 3 & -2 & -4 & 0 & 0 & 1 \end{array}\right]$$

$$\begin{matrix} 2 \times R_1 + R_2 \longrightarrow \\ 3 \times R_1 + R_3 \longrightarrow \end{matrix} \left[\begin{array}{ccc|ccc} -1 & 2 & 1 & 0 & 1 & 0 \\ 0 & 4 & 1 & 1 & 2 & 0 \\ 0 & 4 & -1 & 0 & 3 & 1 \end{array}\right]$$

$$R_3 + (-R_2) \longrightarrow \left[\begin{array}{ccc|ccc} -1 & 2 & 1 & 0 & 1 & 0 \\ 0 & 4 & 1 & 1 & 2 & 0 \\ 0 & 0 & -2 & -1 & 1 & 1 \end{array}\right]$$

The explanations of the row operations at the left use the abbreviation R_2 in place of row 2, etc. Recall that these notes describe what was done in the preceding matrix to get the indicated rows.

$$-\tfrac{1}{2} \times R_3 \longrightarrow \left[\begin{array}{ccc|ccc} -1 & 2 & 1 & 0 & 1 & 0 \\ 0 & 4 & 1 & 1 & 2 & 0 \\ 0 & 0 & 1 & \tfrac{1}{2} & -\tfrac{1}{2} & -\tfrac{1}{2} \end{array}\right]$$

$$\begin{matrix} -1 \times R_3 + R_1 \longrightarrow \\ -1 \times R_3 + R_2 \longrightarrow \end{matrix} \left[\begin{array}{ccc|ccc} -1 & 2 & 0 & -\tfrac{1}{2} & \tfrac{3}{2} & \tfrac{1}{2} \\ 0 & 4 & 0 & \tfrac{1}{2} & \tfrac{5}{2} & \tfrac{1}{2} \\ 0 & 0 & 1 & \tfrac{1}{2} & -\tfrac{1}{2} & -\tfrac{1}{2} \end{array}\right]$$

$$\begin{matrix} -1 \times R_1 \longrightarrow \\ \tfrac{1}{4} \times R_2 \longrightarrow \end{matrix} \left[\begin{array}{ccc|ccc} 1 & -2 & 0 & \tfrac{1}{2} & -\tfrac{3}{2} & -\tfrac{1}{2} \\ 0 & 1 & 0 & \tfrac{1}{8} & \tfrac{5}{8} & \tfrac{1}{8} \\ 0 & 0 & 1 & \tfrac{1}{2} & -\tfrac{1}{2} & -\tfrac{1}{2} \end{array}\right]$$

$$2 \times R_2 + R_1 \longrightarrow \left[\begin{array}{ccc|ccc} 1 & 0 & 0 & \tfrac{3}{4} & -\tfrac{1}{4} & -\tfrac{1}{4} \\ 0 & 1 & 0 & \tfrac{1}{8} & \tfrac{5}{8} & \tfrac{1}{8} \\ 0 & 0 & 1 & \tfrac{1}{2} & -\tfrac{1}{2} & -\tfrac{1}{2} \end{array}\right]$$

$$\begin{matrix} \uparrow & & \uparrow \\ I_3 & & A^{-1} \end{matrix}$$

You should now verify that the indicated matrix A^{-1} satisfies

$$AA^{-1} = A^{-1}A = I_3$$

∎

One application of inverses is to the solution of linear systems in which the number of variables is the same as the number of equations. For such a system the coefficients of the variables form a square matrix A. Then, as you will see, if A^{-1} exists the system can be solved using this inverse. For illustrative purposes, we begin with this specific linear system:

$$
\begin{aligned}
2x \quad\;\; - \; z &= 2 \\
-x + 2y + \; z &= 0 \\
3x - 2y - 4z &= 10
\end{aligned}
$$

The first step is to convert the linear system into a matrix equation as follows:

Observe that after the left side has been multiplied we get

$$
\begin{bmatrix} 2x + 0y - z \\ -x + 2y + z \\ 3x - 2y + 4z \end{bmatrix} = \begin{bmatrix} 2 \\ 0 \\ 10 \end{bmatrix}
$$

Now equate the elements to obtain the original system.

$$
\begin{bmatrix} 2 & 0 & -1 \\ -1 & 2 & 1 \\ 3 & -2 & -4 \end{bmatrix} \begin{bmatrix} x \\ y \\ z \end{bmatrix} = \begin{bmatrix} 2 \\ 0 \\ 10 \end{bmatrix}
$$

↑   ↑   ↑

matrix of coefficients   matrix of variables   matrix of constants

Using A for the matrix of coefficients, X for the matrix of variables, and C for the matrix of constants, we have

$$AX = C$$

Now assume that A^{-1} exists, and multiply the preceding equation by A^{-1} on the left. Then

$$A^{-1}(AX) = A^{-1}C$$
$$(A^{-1}A)X = A^{-1}C \qquad \text{Matrix multiplication is associative.}$$
$$I_3X = A^{-1}C \qquad A^{-1}A = I_3$$
$$X = A^{-1}C \qquad I_3X = X$$

The solution can now be found by equating the entries in $X = \begin{bmatrix} x \\ y \\ z \end{bmatrix}$ with the three

$A^{-1}C$ has dimensions 3 by 1.

numbers in $A^{-1}C$. This solution will be completed in Example 4. Note, however, that the preceding result $A^{-1}C$ applies to any linear system as summarized below:

> Let $AX = C$ be the matrix form of a linear system with n equations and n variables, where A is the $n \times n$ matrix of coefficients, X is the $n \times 1$ matrix of variables, and C is the $n \times 1$ matrix of constants. Then if matrix A has an inverse, the solution for the system is given by
>
> $$X = A^{-1}C$$

EXAMPLE 3 Solve the linear system using the inverse of the matrix of coefficients.

$$3x + 4y = 5$$
$$x + 2y = 3$$

Solution The matrix of coefficients is $A = \begin{bmatrix} 3 & 4 \\ 1 & 2 \end{bmatrix}$. Its inverse is found to be

Verify that $AA^{-1} = I_2$ and $A^{-1}A = I_2$.

$A^{-1} = \begin{bmatrix} 1 & -2 \\ -\frac{1}{2} & \frac{3}{2} \end{bmatrix}$. Then, using $X = A^{-1}C$, we have

$$\begin{bmatrix} x \\ y \end{bmatrix} = \begin{bmatrix} 1 & -2 \\ -\frac{1}{2} & \frac{3}{2} \end{bmatrix}\begin{bmatrix} 5 \\ 3 \end{bmatrix} = \begin{bmatrix} -1 \\ 2 \end{bmatrix}$$

Therefore, $x = -1$, $y = 2$. Check this in the original system.   ■

EXAMPLE 4 Solve the linear system using A^{-1}, where A is the matrix of coefficients.

$$2x \quad\;\; - \;\; z = \;\; 2$$
$$-x + 2y + \;\; z = \;\; 0$$
$$3x - 2y - 4z = 10$$

Solution The matrix of coefficients A is given by

$$A = \begin{bmatrix} 2 & 0 & -1 \\ -1 & 2 & 1 \\ 3 & -2 & -4 \end{bmatrix}$$

which is the same matrix as in Example 2. Then, using A^{-1} from Example 2 and the result $X = A^{-1}C$, we get

$$\begin{bmatrix} x \\ y \\ z \end{bmatrix} = \begin{bmatrix} \frac{3}{4} & -\frac{1}{4} & -\frac{1}{4} \\ \frac{1}{8} & \frac{5}{8} & \frac{1}{8} \\ \frac{1}{2} & -\frac{1}{2} & -\frac{1}{2} \end{bmatrix}\begin{bmatrix} 2 \\ 0 \\ 10 \end{bmatrix} = \begin{bmatrix} -1 \\ \frac{3}{2} \\ -4 \end{bmatrix}$$

Therefore, $x = -1$, $y = \frac{3}{2}$, $z = -4$.   ■

The system in Example 4 was solved with little effort because A^{-1} was already available. In fact, any system having the same A as its matrix of coefficients can be solved just as quickly, regardless of the constants. Therefore, this inverse method is particularly useful in solving more than one linear system having the same matrix of coefficients.

EXAMPLE 5 A mathematics class has learned how to solve a linear system of three equations in three variables by using the algebraic procedures used in the example on page 553. The class does not know any other procedures. The algebra teacher gives a quiz to this class of 25 students and each student is asked to solve a system of the form

$$2x + 5y + 3z = a$$
$$x + y + z = b$$
$$-3x + 2y - z = c$$

where no two students are given the same three constants a, b, c.

(a) Solve the systems for those three cases of $C = \begin{bmatrix} a \\ b \\ c \end{bmatrix}$ where

$$C = \begin{bmatrix} 3 \\ 1 \\ 2 \end{bmatrix} \quad C = \begin{bmatrix} 5 \\ 4 \\ -12 \end{bmatrix} \quad C = \begin{bmatrix} -2 \\ -3 \\ 14 \end{bmatrix}$$

(b) What might have been the teacher's purpose in structuring the quiz this way?

Solution

(a) The matrix of coefficients, which is the same for all 25 systems, and its inverse are:

Verify that
$AA^{-1} = I = A^{-1}A.$

$$A = \begin{bmatrix} 2 & 5 & 3 \\ 1 & 1 & 1 \\ -3 & 2 & -1 \end{bmatrix} \quad A^{-1} = \begin{bmatrix} 3 & -11 & -2 \\ 2 & -7 & -1 \\ -5 & 19 & 3 \end{bmatrix}$$

Then, using $X = A^{-1}C$ for each of the three cases, written here in condensed form, the solutions are found as follows.

In this condensed form the three solutions $A^{-1}C$ are in the same color as used for C.

$$\begin{bmatrix} x \\ y \\ z \end{bmatrix} = \begin{bmatrix} 3 & -11 & -2 \\ 2 & -7 & -1 \\ -5 & 19 & 3 \end{bmatrix}$$

(b) Since no two students are given the same three constants, the work in solving their systems must differ to some extent. Consequently, there is an automatic security built into the structure of the quiz. However, the teacher's work in grading will not be excessive since only one matrix inverse needs to be found, and the computations for $A^{-1}C$ are straightforward.

EXERCISES 9.3

Find A^{-1}, if it exists, and verify that $AA^{-1} = A^{-1}A = I_n$.

1. $A = \begin{bmatrix} 4 & -1 \\ 2 & 0 \end{bmatrix}$

2. $A = \begin{bmatrix} -\frac{1}{5} & -\frac{2}{5} \\ 0 & \frac{1}{4} \end{bmatrix}$

3. $A = \begin{bmatrix} \frac{1}{3} & -\frac{4}{3} \\ -2 & 8 \end{bmatrix}$

4. $A = \begin{bmatrix} -\frac{3}{2} & \frac{5}{3} \\ \frac{9}{4} & -\frac{5}{2} \end{bmatrix}$

5. $A = \begin{bmatrix} 2 & -5 \\ -3 & 4 \end{bmatrix}$

6. $A = \begin{bmatrix} 10 & 15 \\ -5 & -1 \end{bmatrix}$

7. $A = \begin{bmatrix} \frac{1}{3} & \frac{2}{3} \\ \frac{2}{3} & \frac{1}{3} \end{bmatrix}$

8. $A = \begin{bmatrix} -3 & 6 \\ 6 & -3 \end{bmatrix}$

9. $A = \begin{bmatrix} 0 & a \\ b & 0 \end{bmatrix}$, $ab \neq 0$

10. $A = \begin{bmatrix} 2 & 1 & 0 \\ 0 & 0 & -2 \\ 4 & 4 & 0 \end{bmatrix}$

11. $A = \begin{bmatrix} 0 & 1 & 0 \\ 1 & 0 & 0 \\ 0 & 0 & 1 \end{bmatrix}$

12. $A = \begin{bmatrix} 1 & 2 & -1 \\ -1 & 3 & 4 \\ 1 & 7 & 2 \end{bmatrix}$

13. $A = \begin{bmatrix} 1 & 0 & 2 \\ 2 & -1 & 0 \\ 0 & 3 & 4 \end{bmatrix}$

14. $A = \begin{bmatrix} 1 & 1 & -1 \\ 1 & -1 & -1 \\ -1 & -1 & -1 \end{bmatrix}$

15. $A = \begin{bmatrix} 4 & -3 & 1 \\ 0 & -1 & 9 \\ -2 & 1 & 4 \end{bmatrix}$

16. $A = \begin{bmatrix} 8 & -13 & 2 \\ -4 & 7 & -1 \\ 3 & -5 & 1 \end{bmatrix}$

17. $A = \begin{bmatrix} -11 & 2 & 2 \\ -4 & 0 & 1 \\ 6 & -1 & -1 \end{bmatrix}$

18. $A = \begin{bmatrix} 1 & 2 & 0 \\ -1 & 1 & 4 \\ 2 & 3 & -1 \end{bmatrix}$

19. $A = \begin{bmatrix} 1 & 1 & 0 & 2 \\ -1 & 0 & 2 & -1 \\ 0 & 2 & 0 & -2 \\ 2 & 0 & 0 & 5 \end{bmatrix}$

20. $A = \begin{bmatrix} 2 & -3 & 0 & 4 \\ -4 & 1 & -1 & 0 \\ -2 & 7 & -2 & 12 \\ 10 & 0 & 3 & -4 \end{bmatrix}$

Write the system of linear equations obtained from the matrix equation $AX = C$ for the given matrices.

21. $A = \begin{bmatrix} 3 & 1 \\ 2 & -2 \end{bmatrix}$, $X = \begin{bmatrix} x \\ y \end{bmatrix}$, $C = \begin{bmatrix} 9 \\ 14 \end{bmatrix}$

22. $A = \begin{bmatrix} 5 & -2 & 3 \\ 0 & 4 & 1 \\ 2 & -1 & 6 \end{bmatrix}$, $X = \begin{bmatrix} x \\ y \\ z \end{bmatrix}$, $C = \begin{bmatrix} -2 \\ 7 \\ 0 \end{bmatrix}$

Solve the linear system using the inverse of the matrix of coefficients. Observe that each matrix of coefficients is one of the matrices given in Exercises 1–20.

23. $\begin{aligned} 2x - 5y &= 7 \\ -3x + 4y &= -14 \end{aligned}$

24. $\begin{aligned} 2x - 5y &= -21 \\ -3x + 4y &= -7 \end{aligned}$

25. $\begin{aligned} \tfrac{1}{3}x + \tfrac{2}{3}y &= -8 \\ \tfrac{2}{3}x + \tfrac{1}{3}y &= 5 \end{aligned}$

26. $\begin{aligned} \tfrac{1}{3}x + \tfrac{2}{3}y &= 0 \\ \tfrac{2}{3}x + \tfrac{1}{3}y &= 1 \end{aligned}$

27. $\begin{aligned} x \quad + 2z &= 4 \\ 2x - y \quad &= -8 \\ 3y + 4z &= 0 \end{aligned}$

28. $\begin{aligned} x \quad + 2z &= -2 \\ 2x - y \quad &= 2 \\ 3y + 4z &= 1 \end{aligned}$

29. $\begin{aligned} x + y - z &= 1 \\ x - y - z &= 2 \\ -x - y - z &= 3 \end{aligned}$

30. $\begin{aligned} x + y - z &= -4 \\ x - y - z &= 6 \\ -x - y - z &= 10 \end{aligned}$

31. $\begin{aligned} 8x - 13y + 2z &= 1 \\ -4x + 7y - z &= 3 \\ 3x - 5y + z &= -2 \end{aligned}$

32. $\begin{aligned} -11x + 2y + 2z &= 0 \\ -4x \qquad + z &= 5 \\ 6x - y - z &= -1 \end{aligned}$

33. $\begin{aligned} w + x \qquad + 2z &= 2 \\ -w \qquad + 2y - z &= -6 \\ 2x \qquad - 2z &= 0 \\ 2w \qquad + 5z &= 8 \end{aligned}$

34. $\begin{aligned} w + x \qquad + 2z &= 1 \\ -w \qquad + 2y - z &= 3 \\ 2x \qquad - 2z &= 2 \\ 2w \qquad + 5z &= -1 \end{aligned}$

35. For $A = \begin{bmatrix} 2 & 1 \\ 0 & -4 \end{bmatrix}$ and $B = \begin{bmatrix} 0 & 2 \\ 1 & -2 \end{bmatrix}$ show that $(AB)^{-1} = B^{-1}A^{-1}$.

36. A shopper went to the market three times per month in September, October, and November, each time purchasing apples, bananas, and grapes. On each first monthly trip, the shopper bought 4 pounds of apples, 3 pounds of bananas, and 2 pounds of grapes. On each second monthly trip the corresponding number of pounds purchased were 3, 4, and 2. And on each final trip per month the corresponding number of pounds were 3, 3, and 2. The total amounts spent by the shopper for each trip are shown in the table.

Monthly Shopping Trips	Totals Spent (in dollars)		
	September	October	November
First	8.5	9.25	8.60
Second	8.25	9.00	8.30
Third	7.75	8.25	7.70

If the market changed its unit prices (price per pound) on a monthly basis, find the unit prices for each month.

37. (a) Assume that the matrix A has an inverse and let B and C be any matrices such that

$$AB = BA = I_n \quad \text{and} \quad AC = CA = I_n$$

Begin with the equation $AB = I_n$ and prove that $B = C$.

(b) What has been proved about the inverse of a matrix in part (a)?

38. (a) Let A and B be n by n matrices having inverses. Simplify the products $(AB)(B^{-1}A^{-1})$ and $(B^{-1}A^{-1})(AB)$.

(b) As a result of part (a), what can you say about the matrix $B^{-1}A^{-1}$?

39. (a) For matrix A in Exercise 35, verify that $(A^{-1})^2 = (A^2)^{-1}$.

(b) Let A be an invertible square matrix. Use the result in Exercise 38(b) to prove that $(A^2)^{-1} = (A^{-1})^2$.

(c) Write a generalization of the result in part (a) for any positive integer power of A.

? **Challenge**

1. Let A be an n by n matrix. Show that if A satisfies the equation $A^2 + 5A - I = 0$, where $I = I_n$ and 0 is the n by n zero matrix, then A has an inverse. Find the inverse.

2. For any square matrices A and B it is true that $(AB)^t = B^t A^t$. (See Exercise 47, page 572 for the definition of the transpose of a matrix.) Use this result to prove that if A is an invertible

$n \times n$ matrix, then the inverse of the transpose is the transpose of the inverse. That is, $(A^t)^{-1} = (A^{-1})^t$. (Verify this using the matrix A in Exercise 17.)

 Written Assignment (a) Study Exercise 38 and state the result as a mathematical theorem about two invertible matrices of the same dimensions. Do not use any symbolism whatever in your statement. (b) Do the same for the result in Exercise 39(c).

 Graphing Calculator Exercises

1. If your calculator has built-in fundamental row operations, use them to find the inverse of the matrix. (See your user's manual for the correct way to use them.)

$$A = \begin{bmatrix} 1 & 0 & 2 & 3 & 1 \\ 1 & 1 & 3 & 4 & 0 \\ 0 & 2 & 1 & 1 & 1 \\ 0 & 0 & 2 & 1 & 0 \\ 2 & 4 & 0 & 0 & 6 \end{bmatrix}$$

2. If your graphing calculator can find the inverse of a matrix with a single operation key, use it to check your answer to Exercise 1. Then check further by multiplying your answer by A. What should the result be?

3. Use your answer to Exercises 1 and 2 to solve the following system of equations:

$$x + 2z + 3w + v = 4$$
$$x + y + 3z + 4w = 5$$
$$2y + z + w + v = 6$$
$$2z + w = 7$$
$$2x + 4y + 6v = 1$$

 Critical Thinking

1. Why is it necessary to verify that both the left and right distributive properties hold for matrices, yet for the real numbers it is only necessary to begin with the left distributive property $a(b + c) = ab + ac$?

2. Let $A = [a_{ij}]$ and $B = [b_{ij}]$ be m by n matrices. The definition of addition states that for all i and j, $A + B = [a_{ij} + b_{ij}]$ and $B + A = [b_{ij} + a_{ij}]$. How does the commutative property for addition of matrices follow from this?

3. Suppose for matrices A, B, and C we have $AB = AC$. Under what conditions does it follow that $B = C$? Justify your answer.

4. If A is an invertible matrix, is it correct to say $(5A)^{-1} = \frac{1}{5}A^{-1}$? Explain.

5. For $A = \begin{bmatrix} a & 0 \\ 0 & b \end{bmatrix}$, where $ab \neq 0$, find $(A^n)^{-1}$ for any positive integer n. Justify your result.

9.4
INTRODUCTION
TO
DETERMINANTS

The general linear equation has the form $ax + by = c$. Since we are now dealing with two linear equations, it is appropriate to use subscripts to distinguish the constants in the first equation from those in the second. Let

$$a_1 x + b_1 y = c_1$$
$$a_2 x + b_2 y = c_2$$

It would be useful to review Section 2.5.

represent any system of linear equations, and refer to the system by the letter S.

Suppose that system S is a consistent system. Then S has a unique solution that can be found by the multiplication-addition method.

Multiply the first equation by b_2 and the second by $-b_1$.

$$a_1 b_2 x + b_1 b_2 y = c_1 b_2$$
$$-a_2 b_1 x - b_1 b_2 y = -c_2 b_1$$

Add to eliminate y.

$$a_1 b_2 x - a_2 b_1 x = c_1 b_2 - c_2 b_1$$

Factor.

$$(a_1 b_2 - a_2 b_1)x = c_1 b_2 - c_2 b_1$$

To solve for x it must be the case that $a_1 b_2 - a_2 b_1 \neq 0$. Then

$$x = \frac{c_1 b_2 - c_2 b_1}{a_1 b_2 - a_2 b_1}$$

Solve
$$8x - 20y = 3$$
$$4x + 10y = \tfrac{3}{2}$$
by substituting into the general results for x and y.

Similarly, multiplying the first and second equations of S by $-a_2$ and a_1, respectively, will produce this solution for y.

$$y = \frac{a_1 c_2 - a_2 c_1}{a_1 b_2 - a_2 b_1}$$

We have found that if $a_1 b_2 - a_2 b_1 \neq 0$, then system S has the common solution given above. The situation $a_1 b_2 - a_2 b_1 = 0$ will be taken up at the end of this section.

It is somewhat clumsy to apply this general solution, since it is necessary to keep track of the position of each constant. (Try this with the example suggested in the margin.) However, this process can be simplified with the introduction of some special symbolism.

First notice the denominator for both x and y is the same value $a_1 b_2 - a_2 b_1$.

This value is called the **determinant** of the matrix $A = \begin{bmatrix} a_1 & b_1 \\ a_2 & b_2 \end{bmatrix}$ and is sym-

bolized by replacing the brackets of the matrix by vertical bars as in the following definition.

$$\begin{vmatrix} a_1 & b_1 \\ a_2 & b_2 \end{vmatrix} = a_1b_2 - a_2b_1$$

You can use this diagram to help remember the definition.

DEFINITION OF A 2 BY 2 DETERMINANT

For a matrix $A = \begin{bmatrix} a_1 & b_1 \\ a_2 & b_2 \end{bmatrix}$ the determinant of A is symbolized by

$|A| = \begin{vmatrix} a_1 & b_1 \\ a_2 & b_2 \end{vmatrix}$ and is defined by $|A| = \begin{vmatrix} a_1 & b_1 \\ a_2 & b_2 \end{vmatrix} = a_1b_2 - a_2b_1.$

Observe that the same four numbers, placed in different positions, can result in unequal determinants, as shown.

Illustrations:

$$\begin{vmatrix} 10 & -20 \\ 8 & 4 \end{vmatrix} = (10)(4) - (8)(-20) = 200$$

$$\begin{vmatrix} -20 & 8 \\ -10 & 4 \end{vmatrix} = (-20)(4) - (10)(8) = -160$$

TEST YOUR UNDERSTANDING

(Answers: Page 632)

Evaluate each of the determinants.

1. $\begin{vmatrix} 1 & 2 \\ 3 & 4 \end{vmatrix}$
2. $\begin{vmatrix} 1 & 3 \\ 2 & 4 \end{vmatrix}$
3. $\begin{vmatrix} 2 & 1 \\ 4 & 3 \end{vmatrix}$
4. $\begin{vmatrix} 1 & 3 \\ 4 & 2 \end{vmatrix}$

5. $\begin{vmatrix} 2 & 3 \\ 1 & 4 \end{vmatrix}$
6. $\begin{vmatrix} 4 & 2 \\ 1 & 3 \end{vmatrix}$
7. $\begin{vmatrix} 10 & -5 \\ 2 & 1 \end{vmatrix}$
8. $\begin{vmatrix} \frac{1}{2} & 6 \\ 0 & 4 \end{vmatrix}$

9. $\begin{vmatrix} -8 & -4 \\ -7 & -3 \end{vmatrix}$
10. $\begin{vmatrix} -1 & 0 \\ 0 & -1 \end{vmatrix}$
11. $\begin{vmatrix} 0 & 2 \\ 2 & 0 \end{vmatrix}$
12. $\begin{vmatrix} \frac{1}{3} & -\frac{1}{4} \\ 8 & -6 \end{vmatrix}$

The determinant notation can be used for the numerator and denominator in the solution for x found previously so that

$$\text{numerator} = \begin{vmatrix} c_1 & c_2 \\ b_1 & b_2 \end{vmatrix} = c_1b_2 - c_2b_1$$

and

$$\text{denominator} = \begin{vmatrix} a_1 & b_1 \\ a_2 & b_2 \end{vmatrix} = a_1b_2 - a_2b_1$$

The same can be done for the numerator and denominator for y, resulting in the following generalized solution of a linear system having two equations in two variables, known as *Cramer's rule*.

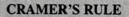

CRAMER'S RULE

The system of equations S

$$a_1 x + b_1 y = c_1$$
$$a_2 x + b_2 y = c_2$$

has the unique solution

$$x = \frac{\begin{vmatrix} c_1 & b_1 \\ c_2 & b_2 \end{vmatrix}}{\begin{vmatrix} a_1 & b_1 \\ a_2 & b_2 \end{vmatrix}} \quad \text{and} \quad y = \frac{\begin{vmatrix} a_1 & c_1 \\ a_2 & c_2 \end{vmatrix}}{\begin{vmatrix} a_1 & b_1 \\ a_2 & b_2 \end{vmatrix}}$$

provided that $a_1 b_2 - a_2 b_1 \neq 0$.

Another way of stating that $a_1 b_2 - a_2 b_1 \neq 0$ is to say that the determinant of the matrix of coefficients is not zero.

Cramer's rule becomes easier to apply after making a few observations. First, the **second-order determinant** of the matrix of coefficients of system S is

$$\begin{vmatrix} a_1 & b_1 \\ a_2 & b_2 \end{vmatrix}$$

which also is the denominator for each fraction. The first column of the determinant consists of the coefficients of the x-terms in S, and the second column contains the coefficients of the y-terms. To write the fraction for x, first record the denominator. Then, to get the determinant in the numerator, simply remove the first column (the coefficients of x) and replace it with constants c_1 and c_2, as shown in the following figure. Similarly, the replacement of the second column by c_1, c_2 gives the numerator in the fraction for y.

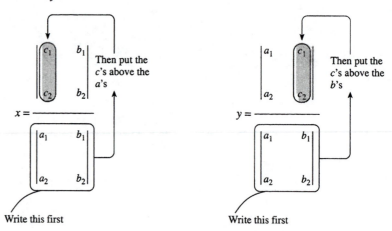

EXAMPLE 1   Solve the following system using Cramer's rule.

$$5x - 9y = 7$$
$$-8x + 10y = 2$$

Solution

$$x = \frac{\begin{vmatrix} 7 & -9 \\ 2 & 10 \end{vmatrix}}{\begin{vmatrix} 5 & -9 \\ -8 & 10 \end{vmatrix}} = \frac{70 - (-18)}{50 - 72} = \frac{88}{-22} = -4$$

Check the ordered pair $(-4, -3)$ in the given system.

$$y = \frac{\begin{vmatrix} 5 & 7 \\ -8 & 2 \end{vmatrix}}{\begin{vmatrix} 5 & -9 \\ -8 & 10 \end{vmatrix}} = \frac{10 - (-56)}{-22} = -3$$

■

EXAMPLE 2   Use determinants to solve the given system.

$$3x = 2y + 22$$

$$2(x + y) = x - 2y - 2$$

Solution   First write the system in standard form.

CAUTION
It is important to have both equations in the standard form

$$ax + by = c$$

since Cramer's rule is based on this form.

$$3x - 2y = 22$$

$$x + 4y = -2$$

$$x = \frac{\begin{vmatrix} 22 & -2 \\ -2 & 4 \end{vmatrix}}{\begin{vmatrix} 3 & -2 \\ 1 & 4 \end{vmatrix}} = \frac{84}{14} = 6 \qquad y = \frac{\begin{vmatrix} 3 & 22 \\ 1 & -2 \end{vmatrix}}{14} = \frac{-28}{14} = -2$$

■

The system

$$a_1 x + b_1 y = c_1$$

$$a_2 x + b_2 y = c_2$$

Recall that a system of two linear equations in two variables is dependent when the two equations are equivalent. Inconsistent means that there are no (common) solutions.

is either dependent or inconsistent when the determinant of the coefficients is zero, that is, when

$$\begin{vmatrix} a_1 & b_1 \\ a_2 & b_2 \end{vmatrix} = 0$$

For example, consider these two systems:

$$\textbf{(a)} \quad 2x - 3y = 5 \qquad \textbf{(b)} \quad 2x - 3y = 8$$

$$-10x + 15y = 8 \qquad \qquad -10x + 15y = -40$$

In each case we have the following:

$$\begin{vmatrix} a_1 & b_1 \\ a_2 & b_2 \end{vmatrix} = \begin{vmatrix} 2 & -3 \\ -10 & 15 \end{vmatrix} = 0$$

Verify these conclusions; see Section 2.5.   System (a) turns out to be inconsistent and (b) is dependent.

EXERCISES 9.4

Evaluate each determinant.

1. $\begin{vmatrix} 5 & -1 \\ -3 & 4 \end{vmatrix}$ **2.** $\begin{vmatrix} 1 & 2 \\ 3 & 4 \end{vmatrix}$ **3.** $\begin{vmatrix} 17 & -3 \\ 20 & 2 \end{vmatrix}$ **4.** $\begin{vmatrix} -7 & 9 \\ -5 & 5 \end{vmatrix}$

5. $\begin{vmatrix} 10 & 5 \\ 6 & -3 \end{vmatrix}$ **6.** $\begin{vmatrix} 6 & 11 \\ 0 & -9 \end{vmatrix}$ **7.** $\begin{vmatrix} 16 & 0 \\ -9 & 0 \end{vmatrix}$ **8.** $\begin{vmatrix} a & b \\ 3a & 3b \end{vmatrix}$

Solve each system using Cramer's rule.

9. $3x + 9y = 15$
$6x + 12y = 18$

10. $x - y = 7$
$-2x + 5y = -8$

11. $-4x + 10y = 8$
$11x - 9y = 15$

12. $7x + 4y = 5$
$-x + 2y = -2$

13. $5x + 2y = 3$
$2x + 3y = -1$

14. $3x + 3y = 6$
$4x - 2y = -1$

15. $\frac{1}{3}x + \frac{3}{8}y = 13$
$x - \frac{9}{4}y = -42$

16. $x - 3y = 7$
$-\frac{1}{2}x + \frac{1}{4}y = 1$

17. $3x + y = 20$
$y = x$

18. $3x + 2y = 5$
$2x + 3y = 0$

19. $\frac{1}{2}x - \frac{2}{7}y = -\frac{1}{2}$
$-\frac{1}{3}x - \frac{1}{2}y = \frac{31}{6}$

20. $-4x + 3y = -20$
$2x + 6y = -15$

21. $9x - 12 = 4y$
$3x + 2y = 3$

22. $\dfrac{x - y}{3} - \dfrac{y}{6} = \dfrac{2}{3}$
$22x + 9(y - 2x) = 8$

Verify that the determinant of the coefficients is zero for each of the following. Then decide whether the system is dependent or inconsistent.

23. $5x - 2y = 3$
$-15x + 6y = -4$

24. $2x - 3y = 5$
$10 - 4x = -6y$

25. $16x - 4y = 20$
$12x - 3y = 15$

26. $2x - 6y = -12$
$-3x + 9y = 18$

27. $3x = 5y - 10$
$6x - 10y = -25$

28. $10y = 2x - 4$
$x - 5y = 2$

When variables are used for some of the entries in the symbolism of a determinant, the determinant itself can be used to state equations. Solve for x.

29. $\begin{vmatrix} x & 2 \\ 5 & 3 \end{vmatrix} = 8$ **30.** $\begin{vmatrix} 7 & 3 \\ 4 & x \end{vmatrix} = 15$ **31.** $\begin{vmatrix} -2 & 4 \\ x & 3 \end{vmatrix} = -1$

Solve each system.

32. $\begin{vmatrix} x & y \\ 3 & 2 \end{vmatrix} = 2$ **33.** $\begin{vmatrix} x & y \\ 2 & 4 \end{vmatrix} = 5$ **34.** $\begin{vmatrix} 3 & x \\ 2 & y \end{vmatrix} = 13$

$\begin{vmatrix} x & -1 \\ y & 3 \end{vmatrix} = 14$ $\begin{vmatrix} 1 & y \\ -1 & x \end{vmatrix} = -\dfrac{1}{2}$ $\begin{vmatrix} 3 & 2 \\ y & x \end{vmatrix} = -12$

35. Show that if the rows and columns of a second-order determinant are interchanged, the value of the determinant remains the same.

36. Show that if one of the rows of $\begin{vmatrix} a_1 & b_1 \\ a_2 & b_2 \end{vmatrix}$ is a multiple of the other, then the determinant is zero. (*Hint:* Let $a_2 = ka_1$ and $b_2 = kb_1$.)

37. Use Exercises 35 and 36 to demonstrate that the determinant is zero if one column is a multiple of the other.

38. If a common factor k is factored out of each element of a row or column of a two-by-two matrix A, resulting in a matrix B, then show that $|A| = k|B|$.

39. Make repeated use of the result in Exercise 38 to show the following:

$$\begin{vmatrix} 27 & 3 \\ 105 & -75 \end{vmatrix} = (45)\begin{vmatrix} 9 & 1 \\ 7 & -5 \end{vmatrix} \quad \text{or} \quad \begin{vmatrix} 27 & 3 \\ 105 & -75 \end{vmatrix} = (45)\begin{vmatrix} 3 & 1 \\ 7 & -15 \end{vmatrix}$$

Then evaluate each side to check.

40. Prove: $\begin{vmatrix} a_1 + t_1 & b_1 \\ a_2 + t_2 & b_2 \end{vmatrix} = \begin{vmatrix} a_1 & b_1 \\ a_2 & b_2 \end{vmatrix} + \begin{vmatrix} t_1 & b_1 \\ t_2 & b_2 \end{vmatrix}$

41. Prove that if to each element of a row (or column) of a second-order determinant we add k times the corresponding element of another row (or column), then the value of the new determinant is the same as that of the original determinant.

42. (a) Evaluate $\begin{vmatrix} 3 & 5 \\ -6 & -1 \end{vmatrix}$ by definition.

   (b) Evaluate the same determinant using the result of Exercise 41 by adding 2 times row one to row two.

   (c) Evaluate the same determinant using the result of Exercise 41 by adding -6 times column two to column one.

Use the results of Exercises 38 and 41 to evaluate each determinant.

43. $\begin{vmatrix} 12 & -42 \\ -6 & 27 \end{vmatrix}$ **44.** $\begin{vmatrix} 45 & 75 \\ 40 & -25 \end{vmatrix}$

Use Cramer's rule to solve each problem.

45. Karin walked at a steady rate for a half-hour and then rode a bicycle for another half-hour, also at a constant rate. The total distance traveled was 7 miles. The next day, going at the same rates, she covered 6 miles, only she walked for two-thirds of an hour and rode for a third of an hour. What were her speeds walking and riding?

46. An airplane, flying with a tail wind, takes 2 hours and 40 minutes to travel 1120 miles. The return trip, against the wind, takes 2 hours and 48 minutes. What is the wind velocity and what is the speed of the plane in still air? (Assume that both velocities are constant; add the velocities for the downwind trip, and subtract them for the return trip.)

**9.5
HIGHER-ORDER
DETERMINANTS
AND THEIR
PROPERTIES**

Just as a system of two linear equations in two variables can be solved using second-order determinants, so can a system of three linear equations in three variables be solved by using *third-order* determinants. In this section, we will study higher-order determinants and their properties, but the focus will be primarily on determinants of order $n = 3$.

A **third-order determinant** can be defined in terms of second-order determinants as follows:

DEFINITION OF A THIRD-ORDER DETERMINANT

$$\begin{vmatrix} a_1 & b_1 & c_1 \\ a_2 & b_2 & c_2 \\ a_3 & b_3 & c_3 \end{vmatrix} = a_1 \begin{vmatrix} b_2 & c_2 \\ b_3 & c_3 \end{vmatrix} - a_2 \begin{vmatrix} b_1 & c_1 \\ b_3 & c_3 \end{vmatrix} + a_3 \begin{vmatrix} b_1 & c_1 \\ b_2 & c_2 \end{vmatrix}$$

Note that the first term on the right is the product of a_1 times a second-order determinant. This determinant, also called the **minor of a_1**, can be found by eliminating the row and column that a_1 is in. Thus

$$\begin{vmatrix} a_1 & b_1 & c_1 \\ a_2 & b_2 & c_2 \\ a_3 & b_3 & c_3 \end{vmatrix} \longrightarrow \begin{vmatrix} b_2 & c_2 \\ b_3 & c_3 \end{vmatrix} \qquad \text{minor of } a_1$$

Similar schemes can be used to obtain the minors for a_2 and a_3.

$$\begin{vmatrix} a_1 & b_1 & c_1 \\ a_2 & b_2 & c_2 \\ a_3 & b_3 & c_3 \end{vmatrix} \longrightarrow \begin{vmatrix} b_1 & c_1 \\ b_3 & c_3 \end{vmatrix} \qquad \text{minor of } a_2$$

$$\begin{vmatrix} a_1 & b_1 & c_1 \\ a_2 & b_2 & c_2 \\ a_3 & b_3 & c_3 \end{vmatrix} \longrightarrow \begin{vmatrix} b_1 & c_1 \\ b_2 & c_2 \end{vmatrix} \qquad \text{minor of } a_3$$

EXAMPLE 1   Evaluate: $\begin{vmatrix} 2 & -2 & 2 \\ 3 & 1 & 0 \\ 2 & -1 & 1 \end{vmatrix}$

Solution

$$\begin{vmatrix} 2 & -2 & 2 \\ 3 & 1 & 0 \\ 2 & -1 & 1 \end{vmatrix} = 2 \begin{vmatrix} 1 & 0 \\ -1 & 1 \end{vmatrix} - 3 \begin{vmatrix} -2 & 2 \\ -1 & 1 \end{vmatrix} + 2 \begin{vmatrix} -2 & 2 \\ 1 & 0 \end{vmatrix}$$

$$= 2(1 - 0) - 3(-2 + 2) + 2(0 - 2)$$

$$= 2 - 0 - 4$$

$$= -2 \qquad \blacksquare$$

A general third-order determinant can be evaluated and simplified as follows:

Simplified Form of a Third-Order Determinant

$$\begin{vmatrix} a_1 & b_1 & c_1 \\ a_2 & b_2 & c_2 \\ a_3 & b_3 & c_3 \end{vmatrix} = a_1(b_2c_3 - b_3c_2) - a_2(b_1c_3 - b_3c_1) + a_3(b_1c_2 - b_2c_1)$$

$$= a_1b_2c_3 + a_2b_3c_1 + a_3b_1c_2 - a_1b_3c_2 - a_2b_1c_3 - a_3b_2c_1$$

The given definition of a third-order determinant can be described as an *expansion by minors* along the first column. It turns out that there are six such expansions, one for each row and column, all giving the same result. Here is the expansion by minors along the first row.

Simplify this expansion to see that it agrees with the preceding simplification of the expansion along the first column.

$$\begin{vmatrix} a_1 & b_1 & c_1 \\ a_2 & b_2 & c_2 \\ a_3 & b_3 & c_3 \end{vmatrix} = + a_1 \begin{vmatrix} b_2 & c_2 \\ b_3 & c_3 \end{vmatrix} - b_1 \begin{vmatrix} a_2 & c_2 \\ a_3 & c_3 \end{vmatrix} + c_1 \begin{vmatrix} a_2 & b_2 \\ a_3 & b_3 \end{vmatrix}$$

When the expansion by minors is done along the second column we have

$$\begin{vmatrix} a_1 & b_1 & c_1 \\ a_2 & b_2 & c_2 \\ a_3 & b_3 & c_3 \end{vmatrix} = - b_1 \begin{vmatrix} a_2 & c_2 \\ a_3 & c_3 \end{vmatrix} + b_2 \begin{vmatrix} a_1 & c_1 \\ a_3 & c_3 \end{vmatrix} - b_3 \begin{vmatrix} a_1 & c_1 \\ a_2 & c_2 \end{vmatrix}$$

$$= -b_1(a_2c_3 - a_3c_2) + b_2(a_1c_3 - a_3c_1) - b_3(a_1c_2 - a_2c_1)$$

$$= a_1b_2c_3 + a_2b_3c_1 + a_3b_1c_2 - a_1b_3c_2 - a_2b_1c_3 - a_3b_2c_1$$

For any expansion along a row or column keep the following display of signs in mind. It gives the signs preceding the row or column elements in the expansion.

Observe how the signs in the second column have been used in the preceding expansion.

$$\begin{vmatrix} + & - & + \\ - & + & - \\ + & - & + \end{vmatrix}$$

EXAMPLE 2 For $A = \begin{bmatrix} 2 & 1 & -3 \\ -4 & 0 & 2 \\ 5 & -1 & 6 \end{bmatrix}$ evaluate $|A|$ using these methods:

(a) Expand by minors along the third column.
(b) Expand by minors along the second row.

Solution

(a) $|A| = \begin{vmatrix} 2 & 1 & -3 \\ -4 & 0 & 2 \\ 5 & -1 & 6 \end{vmatrix} = -3 \begin{vmatrix} -4 & 0 \\ 5 & -1 \end{vmatrix} - 2 \begin{vmatrix} 2 & 1 \\ 5 & -1 \end{vmatrix} + 6 \begin{vmatrix} 2 & 1 \\ -4 & 0 \end{vmatrix}$

$$= -3(4 - 0) - 2(-2 - 5) + 6(0 + 4)$$

$$= -12 + 14 + 24$$

$$= 26$$

CAUTION
Do not confuse the matrix A with its determinant $|A|$. The matrix A is a square array of numbers, whereas the determinant $|A|$ is a single value associated with A.

(b) $|A| = \begin{vmatrix} 2 & 1 & -3 \\ -4 & 0 & 2 \\ 5 & -1 & 6 \end{vmatrix} = -(-4)\begin{vmatrix} 1 & -3 \\ -1 & 6 \end{vmatrix} + 0\begin{vmatrix} 2 & -3 \\ 5 & 6 \end{vmatrix} - 2\begin{vmatrix} 2 & 1 \\ 5 & -1 \end{vmatrix}$

$$= 4(3) + 0 - 2(-7)$$

$$= 26 \qquad \blacksquare$$

Here is another procedure that can be used to evaluate a third-order determinant. Rewrite the first two columns at the right as shown. Follow the arrows pointing downward to get the three products having a plus sign, and the arrows pointing upward give the three products having a negative sign.

Note that this process does not work for higher-order determinants.

$\begin{vmatrix} a_1 & b_1 & c_1 \\ a_2 & b_2 & c_2 \\ a_3 & b_3 & c_3 \end{vmatrix}\begin{matrix} a_1 & b_1 \\ a_2 & b_2 \\ a_3 & b_3 \end{matrix} = a_1 b_2 c_3 + b_1 c_2 a_3 + c_1 a_2 b_3 - a_3 b_2 c_1 - b_3 c_2 a_1 - c_3 a_2 b_1$

EXAMPLE 3 Evaluate by rewriting the first two columns: $\begin{vmatrix} -1 & 3 & 4 \\ 2 & 1 & 2 \\ 5 & 1 & 3 \end{vmatrix}$

Solution

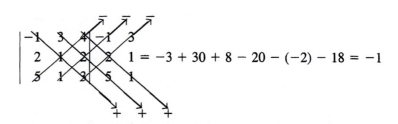

$\begin{vmatrix} -1 & 3 & 4 \\ 2 & 1 & 2 \\ 5 & 1 & 3 \end{vmatrix}\begin{matrix} -1 & 3 \\ 2 & 1 \\ 5 & 1 \end{matrix} = -3 + 30 + 8 - 20 - (-2) - 18 = -1$

$\blacksquare$

TEST YOUR UNDERSTANDING

Evaluate each determinant using each of these methods:
(a) Expand by minors along one of the rows.
(b) Expand by minors along one of the columns.
(c) Rewrite the first two columns.

1. $\begin{vmatrix} 2 & 1 & 0 \\ -1 & 0 & 3 \\ 0 & 4 & -5 \end{vmatrix}$

2. $\begin{vmatrix} 3 & 6 & 9 \\ 0 & 0 & 0 \\ -2 & -4 & -6 \end{vmatrix}$

3. $\begin{vmatrix} 3 & -2 & 7 \\ 0 & 3 & -5 \\ 0 & 0 & -3 \end{vmatrix}$

4. $\begin{vmatrix} -1 & 4 & -2 \\ 6 & -6 & 1 \\ 3 & 3 & 2 \end{vmatrix}$

5. $\begin{vmatrix} 1 & -2 & 3 \\ -4 & 5 & -4 \\ 3 & -2 & 1 \end{vmatrix}$

6. $\begin{vmatrix} 2 & -1 & 9 \\ -7 & 3 & -4 \\ 2 & -1 & 9 \end{vmatrix}$

(Answers: Page 632)

Using the notation for matrices on page 561, the general 3×3 matrix A can be written as

$$A = \begin{bmatrix} a_{11} & a_{12} & a_{13} \\ a_{21} & a_{22} & a_{23} \\ a_{31} & a_{32} & a_{33} \end{bmatrix}$$

When evaluating $|A|$ by expanding along the second column, the minor of a_{12} is the determinant $\begin{vmatrix} a_{21} & a_{23} \\ a_{31} & a_{33} \end{vmatrix}$. This minor is denoted by M_{12}. In general, each element a_{ij} has the minor M_{ij}, which is the determinant of the matrix obtained by deleting the ith row and the jth column of A. Using this minor notation and expanding along the second column of A to find $|A|$, we have

$$|A| = -a_{12}M_{12} + a_{22}M_{22} - a_{32}M_{32}$$

Observe that the signs preceding the terms in this expansion are negative when the sum of the subscripts is odd, and positive when the sum is even. Therefore, the minus signs for the first and last terms may be replaced by $(-1)^{1+2}$ and $(-1)^{3+2}$ respectively, and the plus for the second term by $(-1)^{2+2}$. Then,

$$|A| = a_{12}(-1)^{1+2}M_{12} + a_{22}(-1)^{2+2}M_{22} + a_{32}(-1)^{3+2}M_{32}$$

For each term, the product of the power of -1 and the minor is called a *cofactor* and is denoted by A_{ij}. Thus, the cofactor of a_{32} is $A_{32} = (-1)^{3+2}M_{32}$.

For a square matrix $A = [a_{ij}]$, the cofactor A_{ij} of the element a_{ij} is defined by

$$A_{ij} = (-1)^{i+j}M_{ij}$$

where M_{ij} is the minor of a_{ij}.

Each of the six cofactor expansions is equivalent to the original definition of a third-order determinant.

Using the cofactor notation, the expansion along the second column for the determinant of the 3×3 matrix A becomes

$$|A| = a_{12}A_{12} + a_{22}A_{22} + a_{32}A_{32}$$

EXAMPLE 4   Evaluate $|A|$ using the cofactor expansion along the first row for

$$\begin{bmatrix} 2 & -3 & -1 \\ 1 & 4 & -2 \\ -5 & 0 & 3 \end{bmatrix}$$

Solution

$$|A| = a_{11}A_{11} + a_{12}A_{12} + a_{13}A_{13}$$

$$= 2(-1)^{1+1}\begin{vmatrix} 4 & -2 \\ 0 & 3 \end{vmatrix} + (-3)(-1)^{1+2}\begin{vmatrix} 1 & -2 \\ -5 & 3 \end{vmatrix} + (-1)(-1)^{1+3}\begin{vmatrix} 1 & 4 \\ -5 & 0 \end{vmatrix}$$

$$= 2(1)(12 - 0) + (-3)(-1)(3 - 10) + (-1)(1)(0 + 20)$$

$$= 2(12) + 3(-7) - 20 = -17$$ ∎

The cofactor expansion can be used to define higher-order determinants. In general, for an $n \times n$ matrix A, the determinant of A, using the cofactor expansion along the first row, is

$$|A| = a_{11}A_{11} + a_{12}A_{12} + \cdots + a_{1n}A_{1n}$$

and all such expansions along any row or column are equivalent.

The double subscripts are useful when generalizing the determinant concept to include all orders n. However, when n is small, using a single subscript for the rows and different letters for the columns, as done earlier for n = 3, is less complicated.

Observe that the use of cofactors avoids the dependency on the pattern of signs, as done earlier for third-order determinants. However, the work involved when evaluating determinants of order n, especially when n is relatively small, is about the same whether cofactors or minors are used. When using minor expansions, use the following array of signs as done for third-order determinants.

Sign Pattern for Finding |A| When Using Minors

$$\begin{vmatrix} + & - & + & - & \cdot & \cdot & \cdot \\ - & + & - & + & \cdot & \cdot & \cdot \\ + & - & + & - & \cdot & \cdot & \cdot \\ - & + & - & + & \cdot & \cdot & \cdot \\ \cdot & \cdot & \cdot & \cdot & & & \\ \cdot & \cdot & \cdot & \cdot & & & \\ \cdot & \cdot & \cdot & \cdot & & & \end{vmatrix}$$

Regardless of the order n, there are a number of properties of determinants that can be used to simplify the evaluation of $|A|$. Some of these properties were introduced in the preceding set of exercises. In Exercises 4–9 of this section, the proofs for some special cases of those properties are considered. In particular, the following two fundamental *properties* can be very helpful when evaluating $|A|$.

(1) If a common factor is factored out of each element in a row (or column) of a square matrix A, resulting in a new matrix B, then $|A| = k|B|$. (See Exercise 8.)

(2) If to each element of a row (or column) of a square matrix we add k times the elements of another row (or column), then the resulting matrix has the same determinant as the original matrix. (See Exercise 9.)

EXAMPLE 5   Evaluate $|A|$ using properties (1) and (2) where

$$A = \begin{bmatrix} 4 & 6 & -3 \\ -2 & -18 & 7 \\ 5 & 12 & -8 \end{bmatrix}$$

Solution

Observe that property (2) has been used twice to produce one row containing two zeros. This resulted in only one nonzero term when expanding by minors along that row.

$$|A| = \begin{vmatrix} 4 & 6 & -3 \\ -2 & -18 & 7 \\ 5 & 12 & -8 \end{vmatrix} = 6 \begin{vmatrix} 4 & 1 & -3 \\ -2 & -3 & 7 \\ 5 & 2 & -8 \end{vmatrix} \qquad \text{6 has been factored out of column 2.}$$

$$= 6 \begin{vmatrix} 0 & 1 & 0 \\ 10 & -3 & -2 \\ -3 & 2 & -2 \end{vmatrix} \qquad \begin{array}{l} -4 \times (\text{col. 2}) + \text{col. 1} \\ \text{and } 3 \times (\text{col. 2}) + \text{col. 3} \end{array}$$

$$= 6\,(-1) \begin{vmatrix} 10 & -2 \\ -3 & -2 \end{vmatrix} \qquad \begin{array}{l} \text{expanding by minors along} \\ \text{row 1} \end{array}$$

$$= -6\,(-20 - 6) = 156 \qquad \blacksquare$$

Study the following evaluation of a fourth-order determinant, using properties (1) and (2) and cofactor expansions.

$$\begin{vmatrix} 4 & -3 & 0 & 2 \\ -5 & 3 & 1 & -4 \\ -2 & 6 & 5 & 8 \\ 0 & -9 & 1 & 0 \end{vmatrix} = 3(2) \begin{vmatrix} 4 & -1 & 0 & 1 \\ -5 & 1 & 1 & -2 \\ -2 & 2 & 5 & 4 \\ 0 & -3 & 1 & 0 \end{vmatrix} \qquad \begin{array}{l} \text{property (1) on} \\ \text{columns 2 and 4} \end{array}$$

$$= 6 \begin{vmatrix} 0 & -1 & 0 & 0 \\ -1 & 1 & 1 & -1 \\ 6 & 2 & 5 & 6 \\ -12 & -3 & 1 & -3 \end{vmatrix} \qquad \begin{array}{l} \text{property (2);} \\ 4 \times (\text{col. 2}) + \text{col. 1} \\ \text{and col. 2} + \text{col. 4} \end{array}$$

$$= 6 \left((-1)(-1)^{1+2} \begin{vmatrix} -1 & 1 & -1 \\ 6 & 5 & 6 \\ -12 & 1 & -3 \end{vmatrix} \right) \qquad \begin{array}{l} \text{cofactor expansion} \\ \text{along row 1} \end{array}$$

$$= 6 \begin{vmatrix} 0 & 1 & 0 \\ 11 & 5 & 11 \\ -11 & 1 & -2 \end{vmatrix} \qquad \begin{array}{l} \text{property (2);} \\ \text{col. 2} + \text{col. 1, and} \\ \text{col. 2} + \text{col. 3} \end{array}$$

$$= 6 \left((1)(-1)^{1+2} \begin{vmatrix} 11 & 11 \\ -11 & -2 \end{vmatrix} \right) \qquad \begin{array}{l} \text{cofactor expansion} \\ \text{along row 1} \end{array}$$

$$= -6(11) \begin{vmatrix} 1 & 1 \\ -11 & -2 \end{vmatrix} \qquad \begin{array}{l} \text{property (1) on} \\ \text{row 1} \end{array}$$

$$= -66\,(-2 + 11)$$

$$= -594$$

Third-order determinants can be used to extend Cramer's rule for the solution of three linear equations in three variables. Assume the following general system has a unique solution.

$$a_1x + b_1y + c_1z = d_1$$
$$a_2x + b_2y + c_2z = d_2$$
$$a_3x + b_3y + c_3z = d_3$$

By completing the tedious computations involved, it can be shown that the unique solution for this system is the following. (Also, see Exercises 43 and 44 for a proof of this result that is based on special properties of determinants.)

Note that D is the determinant of coefficients of the variables, in order. Then the numerators for the solutions for x, y, and z consist of the coefficients, but in each case the constants are used to replace the coefficients of the variable under consideration.

$$x = \frac{\begin{vmatrix} d_1 & b_1 & c_1 \\ d_2 & b_2 & c_2 \\ d_3 & b_3 & c_3 \end{vmatrix}}{D} \qquad y = \frac{\begin{vmatrix} a_1 & d_1 & c_1 \\ a_2 & d_2 & c_2 \\ a_3 & d_3 & c_3 \end{vmatrix}}{D} \qquad z = \frac{\begin{vmatrix} a_1 & b_1 & d_1 \\ a_2 & b_2 & d_2 \\ a_3 & b_3 & d_3 \end{vmatrix}}{D}$$

where

$$D = \begin{vmatrix} a_1 & b_1 & c_1 \\ a_2 & b_2 & c_2 \\ a_3 & b_3 & c_3 \end{vmatrix} \quad \text{and} \quad D \neq 0$$

If we let the determinants in the numerators be denoted by D_x, D_y, and D_z so that

$$D_x = \begin{vmatrix} d_1 & b_1 & c_1 \\ d_2 & b_2 & c_2 \\ d_3 & b_3 & c_3 \end{vmatrix} \qquad D_y = \begin{vmatrix} a_1 & d_1 & c_1 \\ a_2 & d_2 & c_2 \\ a_3 & d_3 & c_3 \end{vmatrix} \qquad D_z = \begin{vmatrix} a_1 & b_1 & d_1 \\ a_2 & b_2 & d_2 \\ a_3 & b_3 & d_3 \end{vmatrix}$$

then the solution of the system in Cramer's rule can be stated in the following condensed form.

$$x = \frac{D_x}{D} \qquad y = \frac{D_y}{D} \qquad z = \frac{D_z}{D}, \qquad \text{where } D \neq 0$$

EXAMPLE 6   Use Cramer's rule to solve this system.

$$x + 2y + z = 3$$
$$2x - y - z = 4$$
$$-x - y + 2z = -5$$

Solution First we find D and note that $D \neq 0$.

$$D = \begin{vmatrix} 1 & 2 & 1 \\ 2 & -1 & -1 \\ -1 & -1 & 2 \end{vmatrix} = \begin{vmatrix} 1 & 2 & 1 \\ 0 & -5 & -3 \\ 0 & 1 & 3 \end{vmatrix} \qquad \begin{array}{c} -2 \times (\text{row } 1) + \text{row } 2 \\ \text{and} \\ 1 \times (\text{row } 1) + \text{row } 3 \end{array}$$

$$= (1) \begin{vmatrix} -5 & -3 \\ 1 & 3 \end{vmatrix} \qquad \text{expanding along column 1}$$

$$= -12$$

Verify each of the following computations.

$$x = \frac{D_x}{D} = \frac{\begin{vmatrix} 3 & 2 & 1 \\ 4 & -1 & -1 \\ -5 & -1 & 2 \end{vmatrix}}{D} = \frac{-24}{-12} = 2$$

$$y = \frac{D_y}{D} = \frac{\begin{vmatrix} 1 & 3 & 1 \\ 2 & 4 & -1 \\ -1 & -5 & 2 \end{vmatrix}}{D} = \frac{-12}{-12} = 1$$

$$z = \frac{D_z}{D} = \frac{\begin{vmatrix} 1 & 2 & 3 \\ 2 & -1 & 4 \\ -1 & -1 & -5 \end{vmatrix}}{D} = \frac{12}{-12} = -1$$

Check this solution in the original solution. Thus $x = 2$, $y = 1$, and $z = -1$. ■

EXERCISES 9.5

1. For $A = \begin{bmatrix} 6 & -2 & -1 \\ 0 & -9 & 4 \\ -3 & 5 & 1 \end{bmatrix}$ evaluate $|A|$ using each of these methods.

   (a) Expand by minors along the first column. (b) Expand by minors along the third row.
   (c) Rewrite the first two columns.

2. For $A = \begin{bmatrix} -8 & -1 & 0 \\ 4 & 7 & -5 \\ 3 & 0 & 2 \end{bmatrix}$ evaluate $|A|$ using each of these methods.

   (a) Expand by minors along the second column. (b) Expand by minors along the second row.
   (c) Rewrite the first two columns.

3. Evaluate the determinant using cofactor expansions along

$$\begin{vmatrix} -5 & -2 & 1 \\ -3 & 7 & 4 \\ 1 & -6 & -2 \end{vmatrix}$$

(a) the first row **(b)** the second row **(c)** the third column

*A **fundamental property of determinants**, which holds for all orders n, is stated in each exercise. Prove this property for the indicated special case.*

4. *If a square matrix A contains a row of zeros or a column of zeros, then* $|A| = 0$. Prove this case:

$$\begin{vmatrix} a_1 & b_1 & c_1 \\ 0 & 0 & 0 \\ a_3 & b_3 & c_3 \end{vmatrix} = 0$$

5. *If the rows and columns of a square matrix are interchanged to obtain the matrix A^t (called the transpose of A), then $|A^t| = |A|$.* Prove this case:

$$\begin{vmatrix} a_1 & b_1 & c_1 \\ a_2 & b_2 & c_2 \\ a_3 & b_3 & c_3 \end{vmatrix} = \begin{vmatrix} a_1 & a_2 & a_3 \\ b_1 & b_2 & b_3 \\ c_1 & c_2 & c_3 \end{vmatrix}$$

6. (a) *If one row of a square matrix A is a multiple of another row, then* $|A| = 0$. Prove this case:

$$\begin{vmatrix} a & b & c \\ ka & kb & kc \\ d & e & f \end{vmatrix} = 0$$

(b) Use the property in part (a) and the property in Exercise 5 to prove that *if one column in A is a multiple of another column, then* $|A| = 0$.

7. *Interchanging any two rows or any two columns changes the sign of the determinant.* Prove this case:

$$\begin{vmatrix} c_1 & b_1 & a_1 \\ c_2 & b_2 & a_2 \\ c_3 & b_3 & a_3 \end{vmatrix} = -\begin{vmatrix} a_1 & b_1 & c_1 \\ a_2 & b_2 & c_2 \\ a_3 & b_3 & c_3 \end{vmatrix}$$

8. *If a common factor k is factored out of each element of a row (or column) of a square matrix A, resulting in a matrix B, then $|A| = k|B|$.*

Prove the following case in which the elements in the first row of A may be written as $a_1 = ka_1'$, $b_1 = kb_1'$, and $c_1 = kc_1'$:

$$|A| = \begin{vmatrix} ka_1' & kb_1' & kc_1' \\ a_2 & b_2 & c_2 \\ a_3 & b_3 & c_3 \end{vmatrix} = k\begin{vmatrix} a_1' & b_1' & c_1' \\ a_2 & b_2 & c_2 \\ a_3 & b_3 & c_3 \end{vmatrix} = k|B|$$

9. *If to each element of a row (or column) of a square matrix we add k times the elements of another row (or column), then the resulting matrix has the same determinant as the original matrix.*

Prove this case:

$$\begin{vmatrix} a_1 + kb_1 & b_1 & c_1 \\ a_2 + kb_2 & b_2 & c_2 \\ a_3 + kb_3 & b_3 & c_3 \end{vmatrix} = \begin{vmatrix} a_1 & b_1 & c_1 \\ a_2 & b_2 & c_2 \\ a_3 & b_3 & c_3 \end{vmatrix}$$

10. Use the property in Exercise 6(a) to evaluate $\begin{vmatrix} 3 & 5 & -5 \\ 1 & 2 & 3 \\ -4 & -8 & -12 \end{vmatrix}$.

11. Use the property in Exercise 6(b) to evaluate $\begin{vmatrix} 7 & 2 & -14 \\ 0 & 6 & 0 \\ -3 & 1 & 6 \end{vmatrix}$.

12. Use the property in Exercise 8 to explain each step.

$$\begin{vmatrix} 8 & -10 & 2 \\ 4 & 25 & -1 \\ 2 & 10 & 0 \end{vmatrix} = 2\begin{vmatrix} 4 & -10 & 2 \\ 2 & 25 & -1 \\ 1 & 10 & 0 \end{vmatrix} = 10\begin{vmatrix} 4 & -2 & 2 \\ 2 & 5 & -1 \\ 1 & 2 & 0 \end{vmatrix} = 20\begin{vmatrix} 2 & -1 & 1 \\ 2 & 5 & -1 \\ 1 & 2 & 0 \end{vmatrix}$$

$$\text{(i)} \qquad\qquad \text{(ii)} \qquad\qquad \text{(iii)}$$

13. Evaluate the determinant given on the left side of (i) in Exercise 12, and also evaluate the result given in (iii).

14. You can verify that $\begin{vmatrix} 1 & 2 & 3 \\ -4 & -1 & 5 \\ 3 & 1 & 7 \end{vmatrix} = 71$.

It follows directly from this result that $\begin{vmatrix} 1 & -4 & 3 \\ 2 & -1 & 1 \\ 3 & 5 & 7 \end{vmatrix} = 71$.

Which of the preceding properties (see Exercises 4–9) justifies this conclusion without doing any further calculations?

15. Evaluate $\begin{vmatrix} 5 & -4 & 3 \\ -6 & 6 & 2 \\ -7 & 3 & 4 \end{vmatrix}$ by first making repeated use of the property in Exercise 9 to obtain either a row or a column that has two zeros.

16. (a) Explain how two applications of the property in Exercise 7 gives $\begin{vmatrix} 1 & 2 & 3 \\ 4 & 5 & 6 \\ 7 & 8 & 9 \end{vmatrix} = \begin{vmatrix} 5 & 4 & 6 \\ 2 & 1 & 3 \\ 8 & 7 & 9 \end{vmatrix}$.

(b) Verify the equality in part **(a)** by evaluating each determinant.

17. Evaluate the determinants by inspection and without doing detailed computations.

(a) $\begin{vmatrix} -2 & 3 & 4 \\ 5 & 1 & -10 \\ 4 & 0 & -8 \end{vmatrix}$ **(b)** $\begin{vmatrix} -8 & 23 & 76 \\ 0 & 0 & -51 \\ 0 & 0 & -2 \end{vmatrix}$ **(c)** $\begin{vmatrix} 1 & 4 & -6 \\ 9 & -2 & 8 \\ 1 & 4 & -6 \end{vmatrix}$ **(d)** $\begin{vmatrix} 4 & -3 & 2 \\ 8 & -4 & -12 \\ -2 & 1 & 3 \end{vmatrix}$

18. For $A = \begin{bmatrix} a_1 & b_1 & c_1 \\ a_2 & b_2 & c_2 \\ a_3 & b_3 & c_3 \end{bmatrix}$ assume that $|A| = -10$.

Use appropriate properties of determinants to find the following.

(a) $\begin{vmatrix} a_1 & b_1 & c_1 \\ a_2 & b_2 & c_2 \\ -4a_3 & -4b_3 & -4c_3 \end{vmatrix}$ **(b)** $\begin{vmatrix} a_3 & b_3 & c_3 \\ a_1 & b_1 & c_1 \\ a_2 & b_2 & c_2 \end{vmatrix}$ **(c)** $\begin{vmatrix} a_1 & b_1 + 2c_1 & c_1 \\ a_2 & b_2 + 2c_2 & c_2 \\ a_3 & b_3 + 2c_3 & c_3 \end{vmatrix}$

(d) $\begin{vmatrix} -\frac{1}{2}a_1 & -b_1 & -2c_1 \\ -\frac{1}{2}a_2 & -b_2 & -2c_2 \\ -\frac{1}{2}a_3 & -b_3 & -2c_3 \end{vmatrix}$ **(e)** $|-2A|$ **(f)** $\begin{vmatrix} a_1 & a_2 & a_3 \\ b_1 & b_2 & b_3 \\ c_1 & c_2 & c_3 \end{vmatrix}$

Evaluate each determinant.

19. $\begin{vmatrix} 2 & 2 & -1 \\ -1 & 3 & -3 \\ 1 & 2 & 3 \end{vmatrix}$ **20.** $\begin{vmatrix} 2 & 0 & -1 \\ 3 & -2 & 1 \\ -3 & 0 & 4 \end{vmatrix}$ **21.** $\begin{vmatrix} 1 & -3 & 2 \\ -5 & 2 & 0 \\ 4 & -1 & 3 \end{vmatrix}$

22. $\begin{vmatrix} 1 & 2 & 3 \\ 4 & 5 & 6 \\ 7 & 8 & 9 \end{vmatrix}$ **23.** $\begin{vmatrix} 1 & 1 & 1 \\ -1 & 1 & 1 \\ -1 & -1 & 1 \end{vmatrix}$ **24.** $\begin{vmatrix} 1 & 1 & 4 \\ 2 & 2 & -5 \\ 3 & 3 & 6 \end{vmatrix}$

Use appropriate determinant properties in conjunction with cofactor expansions to evaluate the determinants.

25. $\begin{vmatrix} 3 & -2 & 1 & 5 \\ 0 & 4 & -3 & 1 \\ -1 & 0 & 6 & 2 \\ 1 & -5 & 0 & -4 \end{vmatrix}$ **26.** $\begin{vmatrix} 1 & 2 & 3 & 4 \\ 5 & 6 & 7 & 8 \\ 9 & 10 & 11 & 12 \\ 13 & 14 & 15 & 16 \end{vmatrix}$

27. A fourth-order determinant can be evaluated by expanding by minors along a row or column as was done for third-order determinants, and a similar pattern of signs is used in the expansion as shown at the left below. Write the expansion by minors along the first row for the determinant at the right, and evaluate the determinant.

$$\begin{vmatrix} + & - & + & - \\ - & + & - & + \\ + & - & + & - \\ - & + & - & + \end{vmatrix} \qquad \begin{vmatrix} 2 & -1 & 3 & 0 \\ 1 & 0 & 5 & -3 \\ 0 & 2 & -4 & 6 \\ -5 & 3 & 0 & 1 \end{vmatrix}$$

28. Let $A = \begin{bmatrix} -3 & 4 & 0 & 1 \\ 9 & -1 & 2 & 0 \\ 0 & 0 & -6 & 0 \\ 0 & 3 & 0 & -2 \end{bmatrix}$.

Along which row or column would you expand to evaluate $|A|$ in the most efficient way? Use this expansion to find $|A|$ and check by using a different expansion.

Solve for x.

29. $\begin{vmatrix} -1 & x & -1 \\ x & -3 & 0 \\ -3 & 5 & -1 \end{vmatrix} = 0$

30. $\begin{vmatrix} x & 5 & 2x \\ 2x & 0 & x^2 \\ 1 & -1 & 2 \end{vmatrix} = 0$

31. $\begin{vmatrix} 5-x & 0 & -2 \\ 4 & -1-x & 3 \\ 2 & 0 & 1-x \end{vmatrix} = 0$

Evaluate the following for this system:

$$3x - y + 4z = 2$$
$$-5x + 3y - 7z = 0$$
$$7x - 4y + 4z = 12$$

32. D **33.** D_x **34.** $\dfrac{D_x}{D}$ **35.** $\dfrac{D_y}{D}$

Use Cramer's rule to solve each system.

36. $\begin{aligned} x + y + z &= 2 \\ x - y + 3z &= 12 \\ 2x + 5y + 2z &= -2 \end{aligned}$

37. $\begin{aligned} x + 2y + 3z &= 5 \\ 3x - y &= -3 \\ -4x + z &= 6 \end{aligned}$

38. $\begin{aligned} x - 8y - 2z &= 12 \\ -3x + 3y + z &= -10 \\ 4x + y + 5z &= 2 \end{aligned}$

39. $\begin{aligned} 2x + y &= 5 \\ 3x - 2z &= -7 \\ -3y + 8z &= -5 \end{aligned}$

40. $\begin{aligned} 4x - 2y - z &= 1 \\ 2x + y + 2z &= 9 \\ x - 3y - z &= \tfrac{3}{2} \end{aligned}$

41. $\begin{aligned} 6x + 3y - 4z &= 5 \\ \tfrac{3}{2}x + y - 4z &= 0 \\ 3x - y + 8z &= 5 \end{aligned}$

42. Show that $\begin{vmatrix} a^2 & b^2 & c^2 \\ a & b & c \\ 1 & 1 & 1 \end{vmatrix} = (a-b)(a-c)(b-c)$.

43. Following is a proof that $x = \dfrac{D_x}{D}$ for the general system on page 596, assuming that $D \neq 0$. Give reasons for the steps that are numbered at the right. Since

$$D = \begin{vmatrix} a_1 & b_1 & c_1 \\ a_2 & b_2 & c_2 \\ a_3 & b_3 & c_3 \end{vmatrix}, \text{ then}$$

$$xD = \begin{vmatrix} a_1 x & b_1 & c_1 \\ a_2 x & b_2 & c_2 \\ a_3 x & b_3 & c_3 \end{vmatrix} \qquad \text{(i)}$$

$$xD = \begin{vmatrix} a_1 x + b_1 y & b_1 & c_1 \\ a_2 x + b_2 y & b_2 & c_2 \\ a_3 x + b_3 y & b_3 & c_3 \end{vmatrix} \qquad \text{(ii)}$$

$$xD = \begin{vmatrix} a_1 x + b_1 y + c_1 z & b_1 & c_1 \\ a_2 x + b_2 y + c_2 z & b_2 & c_2 \\ a_3 x + b_3 y + c_3 z & b_3 & c_3 \end{vmatrix} \qquad \text{(iii)}$$

$$xD = \begin{vmatrix} d_1 & b_1 & c_1 \\ d_2 & b_2 & c_2 \\ d_3 & b_3 & c_3 \end{vmatrix} \qquad \text{(iv)}$$

$$xD = D_x$$

$$x = \frac{D_x}{D} \qquad \text{(v)}$$

44. Use arguments like those in Exercise 43 to prove that $y = \dfrac{D_y}{D}$ and $z = \dfrac{D_z}{D}$ for the general system on page 596.

45. (a) For any $n \times n$ matrices A and B, it is true that $|AB| = |A||B|$. Use this result to show that if matrix A has an inverse, then $|A^{-1}| = \dfrac{1}{|A|}$. Verify this result for matrix A in Example 2 on page 577.

(b) Show that for $n \times n$ matrices, it is not true in general that $|A + B| = |A| + |B|$ by finding two 3×3 matrices A, B such that $|A| + |B| \neq |A + B|$.

Challenge   The straight line through the two points (a, b) and (c, d), where $a \neq c$, is given by this equation:

$$\begin{vmatrix} 1 & 1 & 1 \\ x & a & c \\ y & b & d \end{vmatrix} = 0$$

Show how to convert this equation into the point-slope form for this line. (*Hint:* Apply the result in Exercise 9 to the columns in the determinant.)

Written Assignment

(a) Refer to the preceding Challenge and explain why three distinct points (a, b), (c, d), and (x_1, y_1) are collinear (on the same line) provided that

$$\begin{vmatrix} 1 & 1 & 1 \\ x_1 & a & c \\ y_1 & b & d \end{vmatrix} = 0.$$

(b) Explain why $|kA| = k^n |A|$ where A is an $n \times n$ matrix and k is a real number.

Graphing Calculator Exercises   Cramer's rule can be extended to n by n systems of linear equations, for n greater than 3.

1. Write the determinant forms for the values of x, y, z, and w in the solution of this system of equations:

$$
\begin{aligned}
x + y + 2z + w &= 5 \\
2x - y + 3z - w &= 3 \\
-x + y - z - w &= -2 \\
3x + y + 2z + w &= 7
\end{aligned}
$$

2. If your graphing calculator can evaluate determinants, use it to find the values of x, y, z, and w in Exercise 1.

**9.6
SYSTEMS
OF LINEAR
INEQUALITIES**

A linear equation such as $y = 2x - 1$ can be used to identify two-dimensional regions in the plane. First observe that the graph of $y = 2x - 1$ divides the plane into two *half-planes*. These two half-planes represent the graphs for the two *linear inequalities*, $y < 2x - 1$ (below the line) and $y > 2x - 1$ (above the line). To show these graphs we use a dashed line for $y = 2x - 1$ and shade the appropriate half-plane, as in the following figures.

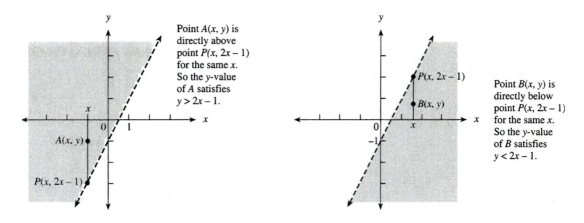

In general, the graph of $y > mx + b$ consists of all points $A(x, y)$ *above* the line $y = mx + b$. All points $B(x, y)$ *below* the line show the graph of $y < mx + b$.

In order to identify the region satisfying an inequality like $3x - y < -2$, it is useful first to graph the corresponding line $3x - y = -2$. Then there are a number of ways to proceed. Here are two methods:

1. Solve the given inequality for y; that is, $y > 3x + 2$. Therefore, the region is above the line ℓ.

2. After graphing the line, pick any convenient point, such as $(-2, 2)$, that is *not* on the line and substitute into the given inequality $3x - y < -2$. This gives the *true statement* $-8 < -2$. Therefore, $(-2, 2)$ must be on the correct side of ℓ, which may now be indicated by the shading in the following graph.

To show the graph of $3x - y \leq -2$ draw the straight line as a solid rather than as a dashed line. The graph would consist of the same shaded half-plane together with the solid line.

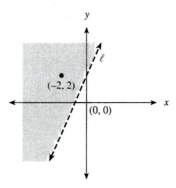

As a check, use a point on the other side, say $(0, 0)$. Substituting into the inequality $3x - y < -2$ gives the *false statement* $0 < -2$. Consequently, $(0, 0)$ is on the wrong side, and you may safely conclude that $3x - y < -2$ works for the points on the other side.

EXAMPLE 1 Locate (shade) the region that satisfies the linear inequality $2x + 3y < 1$ using methods 1 and 2.

Solution

1. Solve for y: $y < -\frac{2}{3}x + \frac{1}{3}$. Thus the required region is the half-plane below the line $y = -\frac{2}{3}x + \frac{1}{3}$.

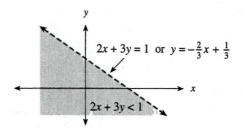

2. Substitute $(0, 0)$ into $2x + 3y < 1$ to get $0 < 1$. Hence $(0, 0)$ is on the correct side, namely below the line $2x + 3y = 1$. ∎

The methods of graphing a linear inequality can be extended to graphing systems of linear inequalities. This is demonstrated in Example 2.

EXAMPLE 2 Graph the system of linear inequalities:

$$2x + y \le 6$$
$$3x - 4y \ge 12$$

Solution Draw the lines $2x + y = 6$ and $3x - 4y = 12$. Shade the region $2x + y \le 6$ vertically and the region $3x - 4y \ge 12$ horizontally. Since the coordinates of a point $P(x, y)$ must satisfy *both* conditions, the graph consists of all points shaded in both directions, as shown. The parts of the two lines above their point of intersection are dashed since they are not included.

Observe that the unshaded region consists of the points that satisfy this system:
$$2x + y > 6$$
$$3x - 4y < 12$$
What is the system for the part shaded only vertically? (See Exercise 37.)

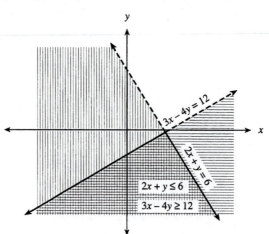

A system of linear inequalities may have more than two inequalities. Example 3 demonstrates how to graph a system of three linear inequalities.

EXAMPLE 3   Graph the system:

$$x + 3y \geq 12$$
$$-2x + y \leq 4$$
$$8x + 3y \leq 54$$

Solution Draw the lines $\ell_1: x + 3y = 12$; $\ell_2: -2x + y = 4$; and $\ell_3: 8x + 3y = 54$. Solve each inequality for y to obtain the equivalent system

$$y \geq -\frac{1}{3}x + 4$$

$$y \leq 2x + 4$$

$$y \leq -\frac{8}{3}x + 18$$

The first inequality gives the points on and above ℓ_1, and the last two inequalities give the points on and below ℓ_2 and ℓ_3. Since a point (x, y) belongs to the graph of the system when it satisfies all three inequalities, the graph is the triangular region including the sides of the triangle as shown.

As an alternative procedure, draw the three lines and use a test point from each of the 7 regions determined by the lines (six regions are outside the triangle and one is inside). When a point from a region satisfies all 3 inequalities, then that region is part of the graph.

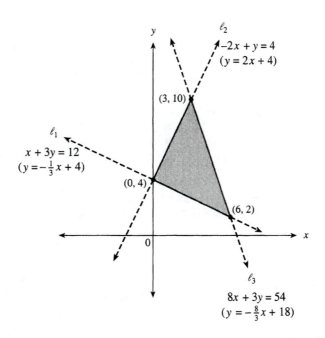

Example 4 shows how absolute values can be used to write and graph systems of linear inequalities.

EXAMPLE 4 Graph the system:

$$|x| \geq 3$$

$$|y| \leq 5$$

Such inequalities were introduced in Section 1.4.

Solution $|x| \geq 3$ is equivalent to the compound inequality $x \leq -3$ or $x \geq 3$. Thus the points in the plane that satisfy $|x| \geq 3$ are all points to the left or on the vertical line $x = -3$, as well as all points to the right or on the line $x = 3$.

Similarly, since $|y| \leq 5$ is equivalent to $-5 \leq y \leq 5$, the points for this inequality are those on or between the horizontal lines $y = \pm 5$.

Since a point is on the graph of the given system provided *both* inequalities of the system are satisfied, the graph of the given system is the shaded region below.

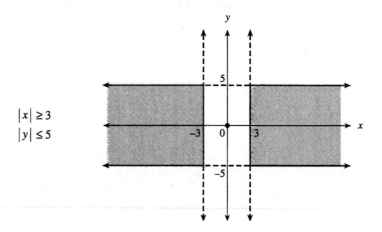

The final example demonstrates how a system of linear inequalities can be used in an applied situation. The reasoning needed here will be built upon in the next section on linear programming.

EXAMPLE 5 A gardener has 450 square feet available for planting flowers and vegetables. The gardener wants the space used for flowers to be no more than one-half the space allotted for vegetables. In addition, at least 90 square feet is to be used for flowers. Find a system of linear inequalities that satisfies all the conditions stipulated, and sketch the graph.

Solution Using x for the number of square feet for flowers and y for the number of square feet for vegetables, their sum cannot exceed 450. That is,

$$x + y \leq 450$$

Also, the condition that the space for flowers, x, is no more than $\frac{1}{2}$ the space for vegetables, y, translates into the inequality

$$x \leq \tfrac{1}{2}y \qquad (\text{or } y \geq 2x)$$

Finally, since at least 90 square feet will be for flowers, we have

$$x \geq 90$$

The resulting system is

$$x + y \leq 450$$
$$y \geq 2x$$
$$x \geq 90$$

The graph is the shaded triangle shown including the sides. The coordinates of each point in this shaded region satisfy all three of the conditions in the system.

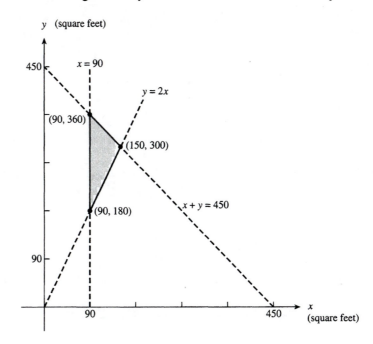

EXERCISES 9.6

Write the linear inequality whose graph is the shaded region.

1.

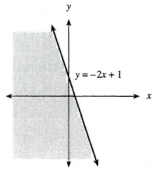

2.

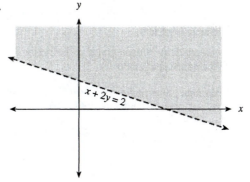

3.

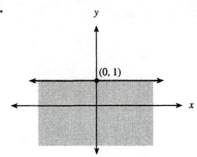

4.

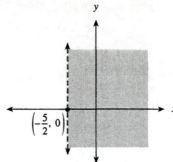

Graph the regions satisfying the given inequalities.

5. $y \geq x - 2$ **6.** $y \leq x + 1$ **7.** $y > x$ **8.** $y \geq -\frac{1}{3}x - 2$

9. $x + y \leq 4$ **10.** $3x - y < 6$ **11.** $2x + y > 4$ **12.** $x - 2y + 2 \leq 0$

13. $2y + x - 4 < 0$ **14.** $3y + x + 6 \geq 0$ **15.** $|x| \leq 3$ **16.** $|x| \geq 3$

17. $|x| < 2$ **18.** $|y| > 1$ **19.** $|y| \leq 2$ **20.** $|y| \geq 2$

Graph each system of inequalities.

21. $y \geq \quad x + 2$
$\quad\;\; y \leq -x + 1$

22. $y \geq -4x + 1$
$\quad\;\; y \leq \quad 4x + 1$

23. $y \leq -4x + 1$
$\quad\;\; y \geq \quad 4x + 1$

24. $2x - \; y + 1 \geq 0$
$\quad\;\; x - 2y + 2 \leq 0$

25. $3x + y < 6$
$\quad\;\;\;\; x > 1$

26. $2x - 3y > \quad 6$
$\quad\quad\quad\;\; y > -2$

27. $\quad x + 2y \leq 10$
$\quad 3x + 2y \leq 18$
$\quad\quad\quad\; x \geq 0$
$\quad\quad\quad\; y \geq 0$

28. $\quad x + 2y \geq 10$
$\quad 3x + 2y \leq 18$
$\quad\quad\quad\; x \geq 0$

29. $\quad x + 2y \leq 10$
$\quad 3x + 2y \geq 18$
$\quad\quad\quad\; y \geq 0$

30. $\quad x + 2y \geq 10$
$\quad 3x + 2y \geq 18$
$\quad\quad\quad\; x \geq 0$
$\quad\quad\quad\; y \geq 0$

31. $\quad x - y \leq -1$
$\quad 2x + y \geq \quad 7$
$\quad 4x - y \leq \quad 11$

32. $\quad x - y \geq -1$
$\quad 2x + y \geq \quad 7$
$\quad 4x - y \leq \quad 11$

33. $\quad x - y \geq -1$
$\quad 2x + y \leq \quad 7$
$\quad 4x - y \leq \quad 11$
$\quad\quad\quad x \geq \quad 0$
$\quad\quad\quad y \geq \quad 0$

34. $|x| > 2$
$\quad |y| > 1$

35. $|x - 2| \leq 1$
$\quad |y + 1| \geq 2$

36. $|x| \leq 2$
$\quad |y| \leq 1$

37. Find the system of linear inequalities whose graph is the region having only vertical shading in the figure for Example 2, page 604.

38. Find the system of linear inequalities whose graph is the region having only horizontal shading in the figure for Example 2, page 604.

Using the figure at the top of the next page, find the system of linear inequalities whose graph is indicated by the Roman numeral. Note, for example, that region I includes all three sides, whereas region III includes one side and excludes the other two.

39. I   **40.** II   **41.** III   **42.** IV   **43.** V   **44.** VI   **45.** VII

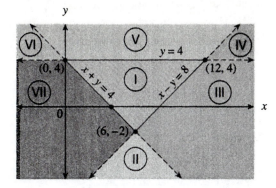

46. A shipping company will only accept a rectangular box having a square base provided that its girth (the perimeter of the base) and its height are no more than 70 inches. Also, each dimension must be at least 4 inches. Find a linear system of inequalities that satisfies the stated condition and sketch the graph.

47. A coach of a women's basketball team needs new basketballs and uniforms for the coming season. She needs at least 4 basketballs and at least 8 new uniforms. The cost of a basketball is $30 and a uniform costs $40. The school budget allows no more than $600 for these purchases. Find a linear system of inequalities that satisfies all the conditions and sketch the graph.

48. A cinema complex has two separate motion-picture theaters. Theater 1 has a 900-seat capacity and Theater 2 has 600 seats. Past attendance figures indicate that on a Saturday night each theater is at least one-third full, regardless of the films being shown. Also, new film releases are always shown in Theater 1 on Saturday nights, and on such occasions, Theater 1 will have up to one-and-one-half times as many viewers as in Theater 2. Find a system of linear inequalities that meets all of the conditions and sketch the graph.

49. A clothing store is planning to have a sale on two styles of jeans. The store pays the wholesale price of $10 for each pair of style 1, $15 per style 2, and at most $3000 has been budgeted for the store's stock of these jeans. Based on past sales, the store anticipates selling no less than 60 pairs of style 1, and they expect to sell at least one-and-one-half times as many pairs of style 1 as pairs of style 2. Find a system of linear inequalities that meets all the conditions and sketch the graph.

Carol Blazejowski, a graduate of Montclair State University, was recently elected to the Basketball Hall of Fame.

Source: Courtesy of MSU Sports Information.

 Challenge For Exercise 47, use the graph of the system to list all of the possibilities that the coach has as to how many basketballs and uniforms she can buy.

 Graphing Calculator Exercises Some graphing calculators can graph systems of inequalities, showing the shaded solution region on the graphics screen. If yours can, (see your user's manual) use it to solve Exercise 1. Otherwise, just graph the boundary line of each inequality and find the region of intersection as explained in the text. Use a suitable friendly window.

1. Solve the following system of inequalities:

$$y \leq 2x + 1$$
$$y \leq -x + 4$$
$$y \geq x + 1$$
$$y \geq -x - 2$$

2. (a) Use your TRACE facility to find the vertices of the boundary of the solution set in Exercise 1.

(b) Find the vertices of the boundary lines that are not in the solution set. Check that these do not satisfy all of the inequalities.

Critical Thinking

1. What must be true for the determinant of the coefficients of a linear system having two variables and two equations in order for the system to have a unique solution?

2. If two lines are parallel, what can you say about the determinant of the matrix of coefficients for the corresponding linear system?

3. Consider the procedure for finding $|A|$ where A is a 3×3 matrix in which the first two columns are repeated to the right. Find similar procedures by repeating parts of A to the left of $|A|$, also by repeating parts above $|A|$, and also by repeating parts below $|A|$.

4. Suppose a linear system in three variables and three equations has a nonzero determinant for the matrix of coefficients. If each of the constants c_1, c_2, c_3 is zero, what can you say about the solution of the system? Justify your answer.

5. Write the system of linear inequalities whose graph consists of all points inside and on the square whose vertices are on the coordinate axes and whose diagonals have length d.

9.7 LINEAR PROGRAMMING

In this chapter, you have learned additional procedures for finding the solution of linear systems of equations, and have graphed systems of linear inequalities. We will now apply some of this knowledge as we consider **linear programming,** a mathematical procedure for solving problems related to the logistics of decision making. Applications of this procedure can be found in numerous areas, including business and industry, agriculture, the field of nutrition, and the military.

This introduction to linear programming will primarily be a geometric approach based on the graphing of linear systems in two variables. We begin with the following system and use it to develop some of the basic concepts.

$$x \geq 0$$

$$y \geq 0$$

$$\tfrac{3}{4}x + \tfrac{1}{2}y \leq 6$$

$$\tfrac{1}{2}x + y \leq 6$$

The graph of this system is a region R consisting of all points (x, y) that satisfy *each* of the inequalities in the system. First note that the conditions $x \geq 0$ and $y \geq 0$ require the points to be in quadrant I, or on the nonnegative parts of each axis. Now graph the lines $\tfrac{3}{4}x + \tfrac{1}{2}y = 6$, $\tfrac{1}{2}x + y = 6$ and determine their point of intersection $(6, 3)$. Then, by the procedure used in Section 6.6, the (x, y) that satisfy both inequalities $\tfrac{3}{4}x + \tfrac{1}{2}y \leq 6$ and $\tfrac{1}{2}x + y \leq 6$ are found to be below or on these lines. The completed region is shown on the next page.

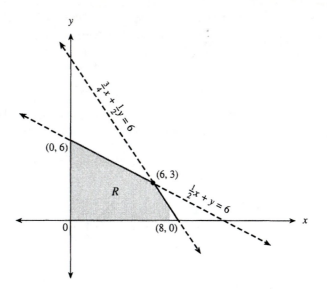

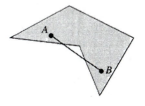

The region R is said to be a **convex set of points,** because any two points inside R can be joined by a line segment that is totally inside R. Roughly speaking, the boundary of R "bends outward." In the margin is a figure of a region that is *not* convex; note, in this figure, that points A and B cannot be connected by a line segment that is totally inside the region.

Draw the line ℓ_1: $x + y = 4$ in the same coordinate system as region R. (See the next figure.) All points (x, y) in R that are on this line have coordinates whose sum is 4. Now draw lines ℓ_2 and ℓ_3 parallel to ℓ_1 and also having equations $x + y = 2$ and $x + y = 8$, respectively. Obviously, all points in R on ℓ_2 or on ℓ_3 have coordinates whose sum is *not* 4. These three lines all have the form $x + y = k$, where k is some constant. There is an infinite number of such parallel lines, depending on the constant k. They all have slope -1. Some of these lines intersect R and others do not; we will be interested only in those that do.

Can you guess which line of this form will intersect R and have the largest possible value k? (See the following figure.)

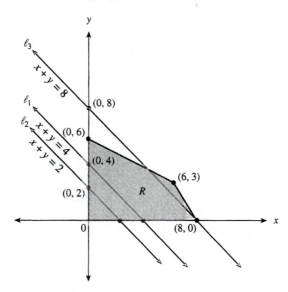

It is not difficult to see the answer, because all these lines are parallel; and the "higher" the line, the larger will be the value k. As a matter of fact, the line $x + y = k$ has k as its y-intercept. So the line we are looking for will be the line that intersects the y-axis as high up as possible, has slope -1, and still meets region R. Because of the shape of R (it is convex), this will be the line through $(6, 3)$ with slope -1, as shown in the following figure. The equation of the line through $(6, 3)$ with slope -1 is $x + y = 9$. Any other parallel line with a larger k-value will be higher and cannot intersect R; and those others that do will be lower and have a smaller k-value.

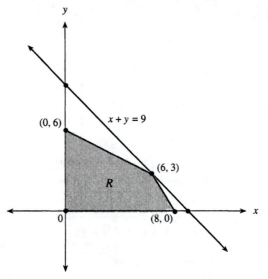

To sum up, we can say that of all the lines with form $x + y = k$ that intersect region R, the one that has the largest k-value is the line $x + y = 9$. Putting it another way, we can say that of all the points in R, the point $(6, 3)$ is the point whose coordinates give the maximum value for the quantity $k = x + y$.

Suppose that we now look for the point in R that produces the maximum value for k, where $k = 2x + 3y$. First we draw a few parallel lines, each with equation of the form $k = 2x + 3y$. The following figure shows such lines for k taking on the values 6, 10, and 18.

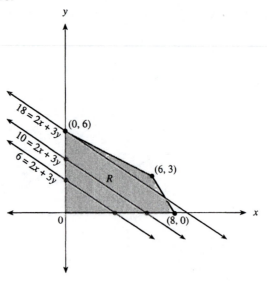

We see that the higher the line, the larger will be the value of k. Because of the convex shape of R, the highest such parallel line (intersecting R) must pass through the vertex $(6, 3)$. For this line we get $2(6) + 3(3) = 21 = k$. This is the largest value of $k = 2x + 3y$ for the (x, y) in R.

The preceding observations should be convincing evidence for the following result:

> **Whenever we have a convex-shaped region R, then the point in R that produces the maximum value of a quantity of the form $k = ax + by$ will be at a vertex of R.**

Because of this result, it is no longer necessary to draw lines through R. All that needs to be done to find the maximum of $k = ax + by$ is to graph the region R, find the coordinates of all vertices on the boundary, and see which one produces the largest value for k.

EXAMPLE 1 Maximize the quantity $k = 4x + 5y$ for (x, y) in the region S given by this system of four inequalities:

$$-x + 2y \leq 2$$
$$3x + 2y \leq 10$$
$$x + 6y \geq 6$$
$$3x + 2y \geq 6$$

Solution First graph the corresponding four lines and shade in the required region. Next find the four vertices of the region S by solving the appropriate pairs of equations. Since $k = 4x + 5y$ will be a maximum only at a vertex, the listing in the following table shows that 18 is the maximum value.

If $k > 18$, then the line given by $4x + 5y = k$ will not intersect S so that $4x + 5y = 18$ is the highest such line that does intersect S, and any parallel line $4x + 5y = k$ below $4x + 5y = 18$ will have $k < 18$.

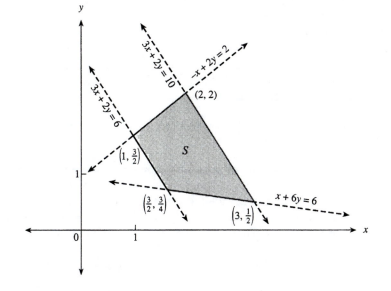

Vertex	$k = 4x + 5y$
$\left(1, \frac{3}{2}\right)$	$4(1) + 5\left(\frac{3}{2}\right) = 11\frac{1}{2}$
$\left(\frac{3}{2}, \frac{3}{4}\right)$	$4\left(\frac{3}{2}\right) + 5\left(\frac{3}{4}\right) = 9\frac{3}{4}$
$(2, 2)$	$4(2) + 5(2) = 18$
$\left(3, \frac{1}{2}\right)$	$4(3) + 5\left(\frac{1}{2}\right) = 14\frac{1}{2}$

■

Suppose we wish to maximize $k = 3x + 2y$ in Example 1. Then all points on the side from $(2, 2)$ to $\left(3, \frac{1}{2}\right)$ will give the same maximum value for k.

Instead of asking for the maximum of $k = 4x + 5y$, as in Example 1, it is also possible to find the minimum value of $k = 4x + 5y$ for the same region S. The earlier discussions explaining why a vertex of a convex region gives the maximum can easily be adjusted to show that a vertex will also give the minimum. Essentially, it becomes a matter of finding the line with correct slope, that has the lowest possible y-intercept, and that still intersects S. Because of this, the table in the solution to the preceding example shows that $9\frac{3}{4}$ is the minimum value of $k = 4x + 5y$. It occurs at the vertex $\left(\frac{3}{2}, \frac{3}{4}\right)$.

TEST YOUR UNDERSTANDING	*Find the required values for the region R, where R is the graph of the system.*

$$5x + 4y \geq 40 \qquad x + 4y \leq 40 \qquad 7x + 4y \leq 112$$

1. Maximum of $k = x + y$ **2.** Minimum of $k = x + y$

3. Maximum of $k = x + 2y$ **4.** Minimum of $k = x + 2y$

5. Maximum of $k = x + 5y$ **6.** Minimum of $k = x + 5y$

7. Maximum of $k = 2x + y$ **8.** Minimum of $k = 2x + y$

(Answers: Page 632)   **9.** Maximum of $k = 3x + 20y$ **10.** Minimum of $k = 3x + 20y$

Since a and b in $k = ax + by$ are constant, the values of k depend on the variables x and y. Therefore, $k = ax + by$ defines k to be a function of two variables.

The question in Example 1 is referred to as a *linear programming problem*. Such problems call for *optimal* (maximum or minimum) *values* of an expression of the form $ax + by = k$, which is called the *objective function*. The inequalities of the system are also called *constraints,* and the region R is referred to as the set of *feasible points* for the problem. A point in R that gives an optimal value of the objective function is referred as an *optimal point*.

The following guidelines can be helpful when solving a linear programming problem.

Step 1. Find the objective function $k = ax + by$.

Step 2. Find the system of constraints.

Step 3. Locate the set of feasible points. This is the graph of the region R for the system of constraints. Determine the coordinates of the vertices of R.

Step 4. Evaluate k at each vertex found in Step 3.

Step 5. Select the maximum or minimum values of k from the set of k-values found in Step 4.

The examples that follow are applications of linear programming that are solved using the preceding guidelines.

EXAMPLE 2   A manufacturer of calculators produces a scientific model that yields a $25 profit, and a graphics model yielding a profit of $50. To meet the daily demand, the company needs to produce at least 200 of the scientific calculators and at least 80 of the graphics calculators. The largest daily output possible is a total of

400 such calculators. How many of each model should be produced to maximize the profits?

Solution

Step 1. Let x = number of scientific calculators produced daily, and y = number of graphics calculators produced daily. Since $25x$ is the daily profit on the scientific model and $50y$ is the daily profit on the graphics models, the objective function p is given by $p = 25x + 50y$.

Step 2. The condition that the number of scientific models produced daily must be at least 200 translates into the inequality $x \geq 200$. Likewise $y \geq 80$. Also the condition that the total daily output cannot exceed 400 calculators translates into $x + y \leq 400$. We now have this system of constraints:

$$x + y \leq 400$$

$$x \geq 200$$

$$y \geq 80$$

Step 3. To sketch region R, first draw the three lines $x = 200$, $y = 80$, and $x + y = 400$. Then use the inequalities to determine the shaded region R whose vertices have the coordinates as shown.

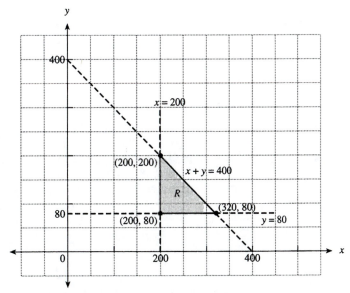

Step 4.

Vertex	$k = 25x + 50y$
(200, 80)	$25(200) + 50(80) = 9{,}000$
(200, 200)	$25(200) + 50(200) = 15{,}000$
(320, 80)	$25(320) + 50(80) = 12{,}000$

Step 5. The results in the table of values for p show that the maximum daily profit is \$15,000 when 200 scientific and 200 graphic calculators are produced. ∎

Before studying the given solution, read this problem several times. Then write the objective function and the constraints.

EXAMPLE 3 A store sells two kinds of bicycles, model A and model B. The store buys them unassembled from a wholesaler. Two employees are responsible for assembling the bicycles, and they are permitted to work no more than 6 hours each per week to do this job. Working together to assemble model A, employee I works $\frac{3}{4}$ hour and employee II works $\frac{1}{2}$ hour. Model B requires $\frac{1}{2}$ hour's work by employee I as well as 1 hour's work by employee II. There is a $110 profit on each model A sold and $96 on each model B. Because of the popularity of the sport, the store is able to sell as many bicycles as they decide to assemble. How many bicycles of each model should they assemble in order to get the maximum profit?

Solution Let x be the number of model A bicycles assembled per week and y the number of model B's per week. Then $110x$ is the profit earned for model A, and $96y$ is the profit for model B. The total profit p is given by $p = 110x + 96y$, which is the objective function.

Next we find the time each employee works. For employee I, $\frac{3}{4}x + \frac{1}{2}y$ will be the number of hours worked per week, because each model A (there are x of these) requires $\frac{3}{4}$ hour, and each model B (there are y of these) requires $\frac{1}{2}$ hour. But each employee works *no more than* 6 hours per week. Thus the total time for this worker satisfies

$$\tfrac{3}{4}x + \tfrac{1}{2}y \le 6$$

By very similar reasoning, the total time for employee II is $\frac{1}{2}x + y$, which satisfies

$$\tfrac{1}{2}x + y \le 6$$

It is also known that $x \ge 0$ and $y \ge 0$ because there cannot be a negative number of either model.

Collecting the preceding conditions, we have that x and y must satisfy this system of constraints:

$$x \ge 0$$
$$y \ge 0$$
$$\tfrac{3}{4}x + \tfrac{1}{2}y \le 6$$
$$\tfrac{1}{2}x + y \le 6$$

We want to find the (x, y) for this system so that $p = 110x + 96y$ is a maximum. The graph of the system is the same region R found earlier in the section and also shown in the following figure. The points in R are all the feasible points, since for each such point the x and y satisfy *all* the constraints of the problem. Any point (x, y) not in R is not feasible (not a possibility) because the x- and y-values do not satisfy *all* the constraints.

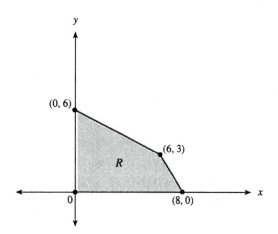

Vertex	$p = 110x + 96y$
(0, 0)	0
(0, 6)	576
(6, 3)	948
(8, 0)	880

The table in the margin shows that the vertex (6, 3) is the optimal point. Therefore, a maximum weekly profit is realized by assembling 6 bicycles of model A and 3 of model B. ∎

EXERCISES 9.7

In Exercises 1, 2, and 3, find the maximum and minimum values for each of the two objective functions for the region R.

1.

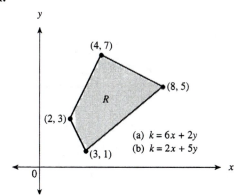

(a) $k = 6x + 2y$
(b) $k = 2x + 5y$

2.

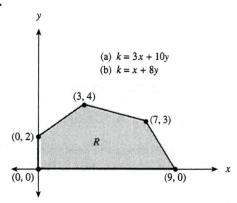

(a) $k = 3x + 10y$
(b) $k = x + 8y$

3.

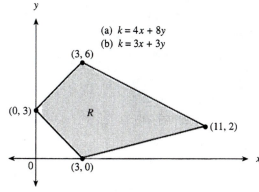

(a) $k = 4x + 8y$
(b) $k = 3x + 3y$

4. Show that the objective function in Exercise 3(a) has 60 as a maximum value for all points on the side joining the vertices (3, 6) and (11, 2). (*Hint:* Consider the equation of the line through these vertices.)

5. **(a)** Graph the region given by this system of constraints.

$$x + 3y \leq 6, \qquad x \geq 0, \qquad y \geq 0$$

(b) Find all the vertices.   **(c)** Find the maximum and minimum of $p = x + y$.

(d) Find the maximum and minimum of $q = 6x + 10y$.

(e) Find the maximum and minimum of $r = 2x + 9y$.

Follow the instructions in Exercise 5 for each of these systems.

6.	7.	8.	9.
$x + 4y \leq 72$	$x + 6y \leq 96$	$x - 2y \geq -10$	$y - x \leq 3$
$5x + 4y \leq 120$	$4x + 5y \leq 118$	$2x + 7y \leq 57$	$x + 4y \leq 22$
$x \geq 0$	$3x + y \leq 72$	$5x + 6y \leq 85$	$2x + y \leq 16$
$y \geq 0$	$x \geq 0$	$5x + 2y \leq 75$	$x - 3y \leq 1$
	$y \geq 0$	$x \geq 0$	$x + 4y \geq 8$
		$y \geq 0$	$x \geq 0$

Find the maximum and minimum values of p, q, and r for the given system of constraints.

10.		11.	
$y \geq 0$	$p = 2x + y$	$x \geq 0$	$p = x + 3y$
$8x + 7y \leq 56$	$q = 5x + 4y$	$6x + 5y \leq 60$	$q = 4x + 3y$
$8x + 3y \geq 24$	$r = 3x + 3y$	$2x + 5y \geq 20$	$r = \frac{1}{2}x + \frac{1}{4}y$

12.		13.	
$x \geq 4$	$p = x + 3y$	$x \leq 7$	$p = 2x + 2y$
$y \geq 2$	$q = 2x + 3y$	$0 \leq y \leq 5$	$q = 3x + y$
$x + 2y \leq 20$	$r = 2x + 5y$	$2x + y \geq 8$	$r = x + \frac{1}{5}y$

The regions for the systems in Exercises 14 and 15 are "open-ended"; that is, their borders do not form a closed polygon and the graph extends endlessly. These regions are still convex in the sense discussed in the text.

14. Graph the region given by the system of constraints.

$$4x - y \geq 20, \qquad 2x - 3y \geq 0, \qquad x - 4y \geq -20, \qquad y \geq 0$$

and evaluate the maximum and minimum (when they exist) of the following:
(a) $p = x + y$ **(b)** $q = 6x + 10y$ **(c)** $r = 2x + y$

15. Follow the instructions of Exercise 14 for this system of constraints.

$$5x + 2y \geq 22, \qquad 6x + 7y \geq 54, \qquad 2x + 11y \geq 44, \qquad 4x - 5y \leq 34, \qquad x \geq 0$$

16. A publisher prints and sells both hardcover and paperback copies of the same book. Two machines, M_1 and M_2, are needed jointly to manufacture these books. To produce one hardcover copy, machine M_1 works $\frac{1}{6}$ hour and machine M_2 works $\frac{1}{12}$ hour. For a paperback copy, machines M_1 and M_2 work $\frac{1}{15}$ hour and $\frac{1}{10}$ hour, respectively. Each machine may be operated no more than 12 hours per day. If the profit is $12 on a hardcover copy and $8 on a paperback copy, how many of each type should be made per day to earn the maximum profit?

17. A manufacturer produces two models of a certain product: model I and model II. There is a $5 profit on model I and an $8 profit on model II. Three machines, M_1, M_2, and M_3, are

used jointly to manufacture these models. The number of hours that each machine operates to produce 1 unit of each model is given in the table:

	Model I	Model II
Machine M_1	$1\frac{1}{2}$	1
Machine M_2	$\frac{3}{4}$	$1\frac{1}{2}$
Machine M_3	$1\frac{1}{3}$	$1\frac{1}{3}$

No machine is in operation more than 12 hours per day.

(a) If x is the number of model I made per day, and y the number of model II per day, show that x and y satisfy the following constraints.

$$x \geq 0, \qquad y \geq 0, \qquad \tfrac{3}{2}x + y \leq 12, \qquad \tfrac{3}{4}x + \tfrac{3}{2}y \leq 12, \qquad \tfrac{4}{3}x + \tfrac{4}{3}y \leq 12$$

(b) Express the daily profit p in terms of x and y.

(c) Graph the feasible region given by the constraints in part (a) and find the coordinates of the vertices.

(d) What is the maximum profit, and how many of each model are produced daily to realize it?

(e) Find the maximum profit possible for the constraints stated in part (a) if the unit profits are $8 and $5 for models I and II, respectively.

18. A farmer buys two varieties of animal feed. Type A contains 8 ounces of corn and 4 ounces of oats per pound; type B contains 6 ounces of corn and 8 ounces of oats per pound. The farmer wants to combine the two feeds so that the resulting mixture has at least 60 pounds of corn and at least 50 pounds of oats. Feed A costs him 5¢ per pound and feed B costs 6¢ per pound. How many pounds of each type should the farmer buy to minimize the cost?

19. An appliance manufacturer makes two kinds of refrigerators: model A, which earns $100 profit, and model B, which earns $120 profit. Each month the manufacturer can produce up to 600 units of model A and up to 500 units of model B. If there are only enough man-hours available to produce no more than a total of 900 refrigerators per month, how many of each kind should be produced to obtain the maximum profit?

20. Two dog foods, A and B, each contain three types of ingredients, I_1, I_2, I_3. The number of ounces of these ingredients in each pound of a dog food is given in the table.

	I_1	I_2	I_3
A	8	3	1
B	8	1	3

A mixture of the two dog foods is to be formed to contain at least 1600 ounces of I_1, at least 300 ounces of I_2, and at least 360 ounces of I_3. If a pound of dog food A costs 8¢ and a pound of dog food B costs 6¢, how many pounds of each dog food should be used so that the resulting mixture meets all the requirements at the least cost?

Written Assignment Give a geometric explanation as to why 18 is the maximum value in Example 1.

Graphing Calculator Exercises

1. Consider the objective function $P = 2x + y$. Solving this equation for y, we get $y = -2x + P$. **(a)** Graph the solution set of Graphing Calculator Exercises 1 and 2 of Section 6.6 on the same set of axes as the lines $y = -2x + 4$, $y = -2x + 4.5$, $y = -2x + 5$, $y = -2x + 5.5$, and $y = -2x + 6$. **(b)** Which is the highest line that intersects the set of feasible points? **(c)** What is the maximum value of the objective function $P = 2x + y$ that satisfies the set of inequalities?

2. Suppose that a lake is being stocked with two species of fish, species S_1 and species S_2. Both of them feed on two foods, F_1 and F_2. The first species eats 2.3 units a day of F_1 and 1.4 units a day of food F_2. The second species consumes 5.1 units a day of F_1 and 10.6 units a day of F_2. If 3500 units of F_1 and 4000 units of F_2 grow in the lake each day, and the first species weighs 1.8 pounds while the second species weighs 2.7 pounds, on the average, how should the lake be stocked so that it supports a maximum total weight of these two species of fish? (*Hint:* Your answers must be whole numbers, but rounding answers may not give feasible solutions.)

CHAPTER REVIEW EXERCISES

Section 9.1 Solving Linear Systems Using Matrices (Page 550)

1. What is the underlying strategy when solving a linear system of 3 equations in 3 variables without the use of matrices or determinants?

Solve the linear system without using matrices or determinants.

2. $\begin{aligned} x + y + z &= 2 \\ x - y + 3z &= 12 \\ 2x + 5y + 2z &= -2 \end{aligned}$

3. $\begin{aligned} 2x - 2y + z &= -7 \\ 3x + y + 2z &= -2 \\ 5x + 3y - 3z &= -7 \end{aligned}$

4. $\begin{aligned} 0.1x + 0.2y + 0.2z &= 0.2 \\ 0.3x + 0.5y + 0.1z &= -0.1 \\ 0.2x - 0.3y - 0.5z &= 0.7 \end{aligned}$

Solve the system using the augmented matrix and fundamental row operations.

5. $\begin{aligned} x + 2y + 3z &= 5 \\ -4x + z &= 6 \\ 3x - y &= -3 \end{aligned}$

6. $\begin{aligned} -3x + 3y + z &= -10 \\ 4x + y + 5z &= 2 \\ x - 8y - 2z &= 12 \end{aligned}$

7. $\begin{aligned} 2x + y &= 5 \\ -3x + 2z &= 7 \\ 3y - 8z &= 5 \end{aligned}$

8. $\begin{aligned} 5x - y - 2z &= 2 \\ 3y + 2z &= 5 \\ -5x + 4y + 4z &= -3 \end{aligned}$

9. $\begin{aligned} x + 2y - z &= 3 \\ 2x - 3y + 3z &= 0 \\ y - 2z &= 6 \end{aligned}$

10. $\begin{aligned} 2x + 3y - z &= 8 \\ x + y + z &= 5 \\ -4x - 6y + 2z &= 3 \end{aligned}$

Section 9.2 Matrix Algebra (Page 561)

For Exercises 11–25, evaluate the given matrix expression, when possible, using these matrices.

$$A = \begin{bmatrix} 3 \\ -2 \\ 1 \end{bmatrix} \qquad B = \begin{bmatrix} -5 & 6 & 4 \end{bmatrix} \qquad C = \begin{bmatrix} 4 & 1 \\ -2 & 3 \end{bmatrix} \qquad D = \begin{bmatrix} -3 & 7 \\ 1 & -1 \end{bmatrix}$$

$$E = \begin{bmatrix} 2 & 0 & -4 \\ -1 & 6 & 2 \\ 0 & 3 & -2 \end{bmatrix} \qquad F = \begin{bmatrix} 1 & 2 & 3 \\ 1 & -3 & 2 \\ 1 & 4 & -5 \end{bmatrix} \qquad G = \begin{bmatrix} 4 & 1 & 5 \\ -2 & 0 & 2 \\ 0 & 3 & -3 \\ 1 & -1 & 0 \end{bmatrix} \qquad H = \begin{bmatrix} 5 & 4 & 3 \\ 2 & 1 & 0 \end{bmatrix}$$

11. (a) $D + C$ **(b)** $B - A$ **12.** $-2F + 3E$ **13.** CD **14.** HF **15.** AB

16. BA **17.** HD **18.** GE **19.** $(HA)B$ **20.** C^3

21. $(-F)(2E)$ **22.** EI_3 **23.** $G\begin{bmatrix} 0 \\ 0 \\ 0 \end{bmatrix}$ **24.** $(C + D)H$ **25.** $H(C + D)$

26. If $A = \begin{bmatrix} -2 & 0 & 0 \\ 0 & 2 & 0 \\ 0 & 0 & \frac{1}{2} \end{bmatrix}$, find A^5.

27. If $A = \begin{bmatrix} a & b \\ c & d \end{bmatrix}$ and $EA = \begin{bmatrix} a + 2c & b + 2d \\ c & d \end{bmatrix}$, find matrix E.

28. If $A = \begin{bmatrix} a & b \\ c & d \end{bmatrix}$ and $AF = \begin{bmatrix} b & a \\ d & c \end{bmatrix}$, find F.

29. Simplify $\left(\frac{3}{2}D\right)H + D\left(-\frac{7}{2}H\right)$ and evaluate for D and H given preceding Exercise 11.

30. Two grocery stores publish the sale prices for various canned foods including beans, corn, peas, and tomatoes. The prices per can that store I charges are 48¢ for beans, 40¢ for corn, 46¢ for peas, and 64¢ for tomatoes. At store II the corresponding prices, for the same size cans as in store I, are 50¢, 37¢, 48¢, and 60¢. A shopper plans to purchase 6 cans of beans, 10 cans of corn, 8 cans of peas, and 9 cans of tomatoes.

(a) Let A be the 1×4 matrix that gives the quantity of each vegetable in alphabetical order, and let C be an appropriate cost matrix containing the prices for each store. Show both A and C.

(b) Find AC and explain what the entries in AC represent.

Section 9.3 Solving Linear Systems Using Inverses (Page 573)

Find the inverse of the given matrix when it exists.

31. $\begin{bmatrix} 0 & 4 \\ -5 & -8 \end{bmatrix}$ **32.** $\begin{bmatrix} -\frac{1}{4} & 2 \\ \frac{2}{3} & -\frac{8}{3} \end{bmatrix}$ **33.** $\begin{bmatrix} \frac{3}{4} & -\frac{6}{5} \\ -\frac{3}{8} & \frac{3}{5} \end{bmatrix}$

34. $\begin{bmatrix} 0 & 2 & -1 \\ -1 & -1 & 1 \\ 1 & -2 & 1 \end{bmatrix}$ **35.** $\begin{bmatrix} 1 & 2 & 3 \\ 4 & 5 & 6 \\ 7 & 8 & 9 \end{bmatrix}$ **36.** $\begin{bmatrix} 1 & 1 & 1 \\ 5 & 6 & 6 \\ 5 & 5 & 6 \end{bmatrix}$

37. $\begin{bmatrix} 1 & 0 & 0 & 0 \\ 0 & 0 & 0 & 1 \\ 0 & 0 & 1 & 0 \\ 0 & 1 & 0 & 0 \end{bmatrix}$ **38.** $\begin{bmatrix} 1 & 2 & 3 & 0 \\ 0 & 1 & 2 & 3 \\ 0 & 0 & 1 & 2 \\ 0 & 0 & 0 & 1 \end{bmatrix}$ **39.** $\begin{bmatrix} 1 & -1 & 0 & 1 \\ -1 & 1 & 0 & 1 \\ -1 & 1 & 1 & -1 \\ 1 & 1 & 0 & 1 \end{bmatrix}$

Solve the system making use of the inverse of the matrix of coefficients. (Use the results for Exercises 31–39 where appropriate.)

40. $-\frac{1}{4}x + 2y = 9$
$\frac{2}{3}x - \frac{8}{3}y = -8$

41. $2y - z = -1$
$-x - y + z = 9$
$x - 2y + z = -4$

42. $\begin{aligned} x + y + z &= 0 \\ 5x + 6y + 6z &= 2 \\ 5x + 5y + 6z &= 6 \end{aligned}$

43. $\begin{aligned} x - y \quad\; + w &= 5 \\ -x + y \quad\; + w &= 9 \\ -x + y + z - w &= -1 \\ x + y \quad\; + w &= 6 \end{aligned}$

Section 9.4 Introduction to Determinants (Page 584)

Evaluate the determinant.

44. $\begin{vmatrix} -7 & 4 \\ 10 & 3 \end{vmatrix}$ 　　**45.** $\begin{vmatrix} -\frac{1}{2} & \frac{2}{3} \\ 9 & -6 \end{vmatrix}$ 　　**46.** $\begin{vmatrix} \frac{4}{5} & -\frac{3}{2} \\ -8 & 15 \end{vmatrix}$

47. Evaluate and simplify: $\begin{vmatrix} x & x - 1 \\ x - 2 & x - 3 \end{vmatrix}$

Solve the system using Cramer's rule.

48. $\begin{aligned} x - 2y &= 3 \\ y - 3x &= -14 \end{aligned}$ 　　**49.** $\begin{aligned} -3x + 8y &= 16 \\ 16x - 5y &= 103 \end{aligned}$ 　　**50.** $\begin{aligned} 4x &= 7y - 6 \\ 9y &= -12x + 12 \end{aligned}$

51. Solve the system: $\begin{vmatrix} x & y \\ 7 & 3 \end{vmatrix} = -4$

$$\begin{vmatrix} -4 & 9 \\ -y & x \end{vmatrix} = 6$$

52. Prove that the value of $\begin{vmatrix} a & b \\ c & d \end{vmatrix}$ is unaffected if twice the elements in the first column are added to the corresponding elements in the second column.

Section 9.5 Higher-Order Determinants and Their Properties (Page 590)

Evaluate the given determinant using the indicated procedure.

$$\begin{vmatrix} 2 & -1 & 4 \\ 0 & 3 & -2 \\ 5 & 1 & 6 \end{vmatrix}$$

53. Expand by minors along the first column.

54. Expand by minors along the second row.

55. Use a fundamental property to obtain two zeros in column two and then expand along the second column.

56. Rewrite the first two columns.

57. Write the cofactor expansion along the first row and then evaluate this determinant: $\begin{vmatrix} -1 & 3 & 2 \\ 4 & 6 & 1 \\ 8 & 5 & -3 \end{vmatrix}$

Evaluate the determinant.

58. $\begin{vmatrix} -5 & 2 & 3 \\ 7 & 4 & 0 \\ -5 & 2 & 3 \end{vmatrix}$ 　　**59.** $\begin{vmatrix} 3 & -6 & 7 \\ 1 & -2 & 4 \\ -6 & 12 & 9 \end{vmatrix}$ 　　**60.** $\begin{vmatrix} -4 & 7 & 1 \\ 3 & 0 & 6 \\ 5 & -2 & 4 \end{vmatrix}$

61. $\begin{vmatrix} 3 & 0 & 0 \\ -3 & 3 & 0 \\ 4 & -4 & 4 \end{vmatrix}$ **62.** $\begin{vmatrix} 7 & -5 & 0 \\ 3 & 0 & 6 \\ 0 & 4 & -2 \end{vmatrix}$ **63.** $\begin{vmatrix} \frac{1}{2} & 0 & 0 \\ 0 & \frac{1}{2} & 0 \\ 0 & 0 & \frac{1}{2} \end{vmatrix}$

64. $\begin{vmatrix} -2 & 1 & 0 & 3 \\ 4 & 0 & -1 & 2 \\ -3 & 4 & 1 & 0 \\ 1 & 5 & 0 & 7 \end{vmatrix}$ **65.** $\begin{vmatrix} -3 & 4 & 0 & 2 \\ 6 & 0 & 1 & 0 \\ -9 & -2 & 5 & 6 \\ 3 & -5 & 1 & -4 \end{vmatrix}$ **66.** $\begin{vmatrix} 1 & -2 & 3 & -4 \\ 3 & -6 & 9 & -12 \\ -2 & 1 & 0 & 7 \\ 5 & -4 & 5 & -2 \end{vmatrix}$

67. Solve the system using Cramer's rule.

$$x + y + 2z = 7$$
$$3x - 2y + z = 6$$
$$2x + 5y + 3z = 11$$

68. If $\begin{vmatrix} a_1 & a_2 & a_3 \\ b_1 & b_2 & b_3 \\ c_1 & c_2 & c_3 \end{vmatrix} = 6$, find the following:

(a) $\begin{vmatrix} a_2 & a_1 & a_3 \\ b_2 & b_1 & b_3 \\ c_2 & c_1 & c_3 \end{vmatrix}$ **(b)** $\begin{vmatrix} a_1 & a_2 & \frac{1}{2}a_3 \\ b_1 & b_2 & \frac{1}{2}b_3 \\ c_1 & c_2 & \frac{1}{2}c_3 \end{vmatrix}$

(c) $\begin{vmatrix} a_1 & a_2 & a_3 \\ b_1 - 5c_1 & b_2 - 5c_2 & b_3 - 5c_2 \\ c_1 & c_2 & c_3 \end{vmatrix}$ **(d)** $\begin{vmatrix} -a_1 & -a_2 & -a_3 \\ 2c_1 & 2c_2 & 2c_3 \\ -3b_1 & -3b_2 & -3b_3 \end{vmatrix}$

69. If A is an n by n matrix, then $|3A| = 3^n|A|$. Explain why this is true.

70. If $A = [a_{ij}]$ is a 5 by 5 matrix such that each number on the main diagonal $a_{ii} = 2$, and all other numbers are zero, find $|A^{-1}|$.

Section 9.6 System of Linear Inequalities (Page 602)

Graph the system of inequalities.

71. $y \geq 2x - 2$
$\quad\ y \geq -2x - 10$

72. $-x + 2y < 0$
$\quad\ 4x + y > 9$

73. $4x - y > 6$
$\quad\ 2x + 3y < 10$

74. $x \geq 2$
$\quad\ y \leq 3$

75. $|x| \leq 3$
$\quad\ |y| \geq 1$

76. $|x - 1| \leq 4$
$\quad\ |y + 2| \leq 3$

77. $x - 2y \geq 3$
$\quad\ y - 3x \geq -14$
$\quad\quad\ x \geq 0$

78. $y \leq -2x + 14$
$\quad\ y \leq \frac{4}{5}x$
$\quad\ y \geq 0$

79. $x + 5y \leq 15$
$\quad\ x + y \leq 7$
$\quad\ x \geq 0, y \geq 0$

80. $y + 2x \geq 6$
$\quad\ -x + 3y \leq 18$
$\quad\ 8x - 3y \leq 24$

Write a system of linear inequalities for the region with the given Roman numeral. Observe that all regions are in the first quadrant and assume that all boundary lines (or line segments) are included for each region. Note: Regions I, II, and III are bounded triangular regions, whereas regions IV, V, and VI are unbounded.

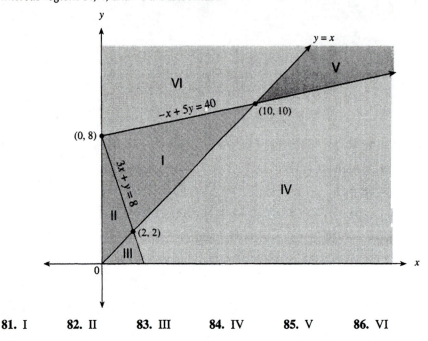

81. I **82.** II **83.** III **84.** IV **85.** V **86.** VI

87. A 600-square-foot garden is to be used for planting tomatoes and corn. The space for tomatoes is to be at most $\frac{2}{3}$ of the space used for corn, and at least 150 square feet is to be used for tomatoes. Write a system of linear inequalities that meets all conditions and sketch the graph. (Let x be the amount of space for tomatoes and let y be the amount of space for corn.)

Section 9.7 Linear Programming (Page 610)

For the region R shown, find the maximum and minimum values (where they exist) of the given objective functions.

88. (a) $k = 8x + 6y$ (b) $k = 2x + 10y$

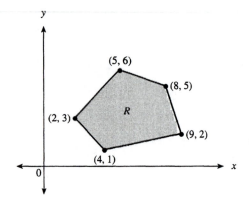

89. (a) $k = 3x + 5y$ **(b)** $k = 7x + 4y$

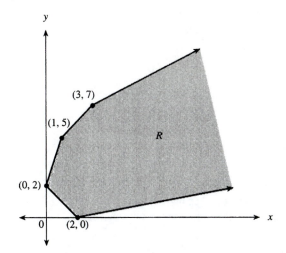

Find a set of constraints that will produce the given feasible region R.

90.

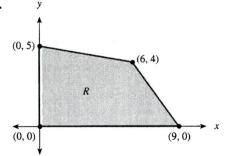

91.

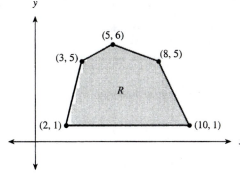

For the given set of constraints, find the maximum and minimum values (when they exist) of the objective functions p and q.

92. $x \geq 2$, $y \geq 1$, $y \leq 8 - x$

   (a) $p = 7x + 5y$ **(b)** $q = 5x + 7y$

93. $x \geq 0$, $y \geq 0$, $x + 5y \leq 30$, $5x + 4y \leq 45$

   (a) $p = 10x + 7y$ **(b)** $q = 7x + 4y$

94. $x \leq 4$, $y \leq 6$, $2x + 3y \geq 20$

   (a) $p = 100x + 500y$ **(b)** $q = 200x + 200y$

95. $0 \leq x \leq 4$, $0 \leq y \leq 6$, $2x + 3y \leq 20$

   (a) $p = 100x + 500y$ **(b)** $q = 200x + 200y$

96. The cost, in dollars, to produce a product is given by the function $C = 5x + 7y$. Find the minimum cost subject to the following constraints: $x + 2y \geq 12$, $y \geq 4x - 30$, $9y - 2x \leq 70$, $y \geq -3x + 11$.

97. A bakery sells loaves of whole wheat bread and loaves of rye bread. From past experience, the number of loaves of each type of bread sold daily is at least 10 but no more than 30. To prepare the loaves for baking, the baker needs (on the average) 3 minutes for a wheat bread, and 2 minutes for a rye bread. There are at most 2 hours available daily for the baker to prepare the loaves. How many of each type should be prepared to maximize the daily profit, if there is 90¢ profit on a wheat bread, and 80¢ on a rye bread?

CHAPTER 9: STANDARD ANSWER TEST

Use the questions to test your knowledge of the basic skills and concepts of Chapter 9. Then check your answers with those given at the back of the book.

Solve the linear system using row-equivalent matrices.

1. $-6x + 3y = 9$
$10x - 5y = -12$

2. $2x - y + 2z = -3$
$x + 4y - 3z = 18$
$-4x + 2y - 3z = 0$

3. $x + y + 2z = 3$
$3x + 2z = 2$
$-x + 2y + 2z = 4$

4. (a) Write the system in Question 1 as a matrix equation.

(b) Write the system in Question 2 as a matrix equation.

5. A nutritionist wants to combine three foods so that the resulting mixture will contain 550 units of ingredient *A*, 300 units of ingredient *B*, and 350 units of ingredient *C*. The units of each ingredient per ounce of food are given in the table. How many ounces of each food should be combined to obtain the required mixture?

	A	B	C
1 oz of food I	25 units	20 units	15 units
1 oz of food II	35 units	15 units	25 units
1 oz of food III	40 units	20 units	20 units

Evaluate whenever possible, using these matrices.

$$A = \begin{bmatrix} 2 & 0 & 3 \\ -1 & 4 & 9 \end{bmatrix} \quad B = \begin{bmatrix} 1 & 0 & -1 \\ 0 & -1 & 1 \\ -1 & 0 & 11 \end{bmatrix} \quad C = \begin{bmatrix} 1 & 2 & 3 \\ 4 & 5 & 6 \\ 7 & 8 & 9 \\ 1 & 0 & 1 \end{bmatrix}$$

$$D = \begin{bmatrix} -1 & -1 \\ 2 & 2 \\ -3 & -3 \end{bmatrix} \quad E = \begin{bmatrix} 3 & 0 \\ 1 & 1 \end{bmatrix} \quad F = \begin{bmatrix} -5 & 7 \\ 4 & -9 \end{bmatrix}$$

$$G = \begin{bmatrix} 0 & -1 & 1 \\ 6 & 2 & 3 \\ -5 & -5 & -8 \end{bmatrix} \quad H = \begin{bmatrix} -1 \\ 1 \\ 2 \end{bmatrix} \quad J = [2 \ -2 \ 4]$$

6. (a) $B + G$ **(b)** $A - B$ **7.** $5E - 2F$ **8.** AB

9. (a) m_{32} in the product $CB = M$. **(b)** $(|E|)(|F|)$

10. (a) AD **(b)** $A(EF)$ **11.** E^3 (means EEE) **12.** HJ

13. (a) The dimension of $CBDA$. **(b)** The dimension of $DBAE$.

14. Evaluate: $(A + \frac{1}{2}B)(C - 2D)$ for:

$$A = [11 \ 7 \ 2] \quad B = [-2 \ 6 \ -2] \quad C = \begin{bmatrix} -5 \\ -2 \\ 4 \end{bmatrix} \quad D = \begin{bmatrix} -8 \\ 5 \\ 1 \end{bmatrix}$$

15. Find the inverse of $A = \begin{bmatrix} 3 & 1 \\ 2 & -2 \end{bmatrix}$ and use it to solve this system: $\begin{aligned} 3x + y &= 9 \\ 2x - 2y &= 14 \end{aligned}$

16. Find the inverse of $A = \begin{bmatrix} 1 & 0 & -2 \\ 0 & 1 & 0 \\ 3 & 2 & 0 \end{bmatrix}$.

17. The matrix $A = \begin{bmatrix} -11 & 2 & 2 \\ -4 & 0 & 1 \\ 6 & -1 & -1 \end{bmatrix}$ has inverse $A^{-1} = \begin{bmatrix} 1 & 0 & 2 \\ 2 & -1 & 3 \\ 4 & 1 & 8 \end{bmatrix}$. Use this

information to solve this linear system: $\begin{aligned} -11x + 2y + 2z &= 3 \\ -4x \qquad + z &= -1 \\ 6x - y - z &= -2 \end{aligned}$

Evaluate each determinant.

18. $\begin{vmatrix} -2 & 4 & 3 \\ 6 & -1 & -4 \\ 5 & 2 & 0 \end{vmatrix}$

19. $\begin{vmatrix} 4 & 0 & -2 & 6 \\ -3 & 6 & 0 & -7 \\ -1 & 3 & -4 & 0 \\ 2 & 9 & -9 & 4 \end{vmatrix}$

Use Cramer's rule to solve the system.

20. $\begin{aligned} 4x - 2y &= 15 \\ 3x + 2y &= -8 \end{aligned}$

21. $\begin{aligned} 2x + y - z &= -3 \\ x - 2y + z &= 8 \\ 3x - y - 2z &= -1 \end{aligned}$

22. A furniture store is arranging a showroom having 1200 square feet of floor space to display sofas and easy chairs. The arrangement calls for no more than $\frac{2}{3}$ of the floor space to be used for sofas. Also the space for sofas is to be at least 1.5 times the space used for chairs, and the space for chairs is to be at least $\frac{1}{4}$ of the total space. Use x for the floor space for sofas and y for the space for chairs, and write a system of linear inequalities that satisfies all the conditions. Sketch the graph of the system.

23. Graph the system

$$x \geq 0, \qquad y \geq 0, \qquad x + 2y \leq 8, \qquad x + y \leq 5, \qquad 2x + y \leq 8$$

24. Find the maximum value of $k = 3x + 2y$ for the region in Question 23.

25. A manufacturer of television sets makes two kinds of 19-inch color TV sets: model A, which earns $80 profit, and model B, which earns $100 profit. Each month the manufacturer can produce up to 400 units of model A and up to 300 units of model B but no more than 600 sets altogether. How many of each should be produced to obtain the maximum profit?

CHAPTER 9: MULTIPLE CHOICE TEST

1. Which of the following are permissible fundamental row operations when using matrices to solve a linear system?

   I. Interchange two rows.

   II. Multiply a row by a nonzero constant.

   III. Add a multiple of a row to another row.

   (a) Only I   **(b)** Only II   **(c)** Only III   **(d)** I, II, and III   **(e)** None of the preceding

2. Which of the following is equal to the determinant $\begin{vmatrix} 2 & 5 \\ 7 & 3 \end{vmatrix}$?

   (a) $\begin{vmatrix} 5 & 7 \\ 3 & 2 \end{vmatrix}$ **(b)** $\begin{vmatrix} 7 & 3 \\ 2 & 5 \end{vmatrix}$ **(c)** $\begin{vmatrix} 2 & 5 \\ 3 & 7 \end{vmatrix}$ **(d)** $\begin{vmatrix} -3 & -5 \\ 7 & 2 \end{vmatrix}$ **(e)** None of the preceding

3. For the matrix A shown at the right, which of the following represents $9A - 2A$? $A = \begin{bmatrix} 3 & -5 \\ -2 & 6 \\ 0 & 4 \end{bmatrix}$

   (a) $\begin{bmatrix} 21 & -35 \\ -2 & 6 \\ 0 & 4 \end{bmatrix}$ **(b)** $\begin{bmatrix} 21 & -5 \\ -14 & 6 \\ 0 & 4 \end{bmatrix}$ **(c)** $\begin{bmatrix} 21 & -35 \\ -14 & 42 \\ 0 & 28 \end{bmatrix}$ **(d)** $\begin{bmatrix} 21 & -55 \\ -22 & 42 \\ 0 & 28 \end{bmatrix}$ **(e)** None of the preceding

4. Which of the following is equal to AB for matrices A and B below?

   $$A = \begin{bmatrix} -2 & 1 & 5 \\ 3 & 0 & 2 \end{bmatrix} \quad B = \begin{bmatrix} 4 & -2 \\ 3 & 4 \\ 0 & -1 \end{bmatrix}$$

   (a) $\begin{bmatrix} -5 & 3 \\ 12 & -8 \end{bmatrix}$ **(b)** $\begin{bmatrix} -5 & 12 \\ 3 & -8 \end{bmatrix}$ **(c)** $\begin{bmatrix} -6 & 4 & 5 \\ 1 & 4 & 1 \end{bmatrix}$ **(d)** AB is impossible to compute.

   (e) None of the preceding

5. Suppose a linear system of three equations in three variables is being solved by making use of row-equivalent matrices. What does it mean when we arrive at the following matrix?

   $$\begin{bmatrix} 1 & -3 & 6 & | & 4 \\ 0 & 4 & 3 & | & -1 \\ 0 & 0 & 0 & | & 0 \end{bmatrix}$$

   (a) $(4, -1, 0)$ is a solution of the original system.

   (b) The original system of equations has more than one solution.

   (c) The original system of equations has exactly one solution.

   (d) The original system of equations has no solutions.

   (e) None of the preceding

6. Which of the following is the inverse of the matrix shown at the right? $\begin{bmatrix} -1 & 2 \\ 2 & -3 \end{bmatrix}$

(a) $\begin{bmatrix} -3 & 2 \\ -1 & 1 \end{bmatrix}$ (b) $\begin{bmatrix} -1 & 2 \\ 0 & 1 \end{bmatrix}$ (c) $\begin{bmatrix} 2 & 3 \\ 1 & 2 \end{bmatrix}$ (d) $\begin{bmatrix} 3 & 2 \\ 2 & 1 \end{bmatrix}$ (e) None of the preceding

7. Matrix A has dimensions 3 by 5, matrix B has dimensions 7 by 4, and matrix C has dimensions 5 by 7. Which of the following matrix products is possible?

(a) ABC (b) BCA (c) CBA (d) ACB (e) None of the preceding

8. If $\begin{bmatrix} -2 & 4 \\ 6 & -8 \end{bmatrix}\begin{bmatrix} w & x \\ y & z \end{bmatrix} = I_2$, then $z =$

(a) $-\frac{1}{2}$ (b) 0 (c) $\frac{1}{4}$ (d) 1 (e) None of the preceding

9. Let $AX = C$ be the matrix equation of a system of linear equations, where A is the 3 by 3 matrix of coefficients, X is the 3 by 1 matrix of variables, and C is the 3 by 1 matrix of constants. If A has an inverse, which of the following is true about the solution X?

(a) $X = CA^{-1}$ (b) $X = A^{-1}C$ (c) $X = \dfrac{1}{A}C$ (d) No solution is possible. (e) None of the preceding

10. If for the system $\begin{matrix} a_1 x + b_1 y = c_1 \\ a_2 x + b_2 y = c_2 \end{matrix}$ we have $\begin{vmatrix} a_1 & b_1 \\ a_2 & b_2 \end{vmatrix} \neq 0$, then which of the following is true?

(a) The system has a unique solution.

(b) The system has no solution.

(c) The system has many solutions.

(d) The value of x is equal to the determinant $\begin{vmatrix} a_1 & c_1 \\ a_2 & c_2 \end{vmatrix}$ divided by $a_1 b_2 - a_2 b_1$.

(e) None of the preceding

11. Evaluate: $\begin{vmatrix} 1 & -2 & 3 \\ -1 & 0 & 1 \\ -2 & 3 & -1 \end{vmatrix}$

(a) 8 (b) -14 (c) -6 (d) 0 (e) None of the preceding

12. Which response below is true regarding the graph of this system?

$$2x + y \leq 6$$
$$x + y \geq 4$$
$$y \geq 0$$
$$x \geq 0$$

(a) The graph consists of all points inside and on the border of just one triangle.

(b) The graph consists of all points inside and on the border of two triangles.

(c) The graph is a four-sided figure.

(d) All points on the positive part of the y-axis are in the graph.

(e) None of the preceding

13. When Cramer's rule is used to solve for y for the system shown below, which of the following becomes the numerator of the fraction?

$$2x + y - 3z = 5$$
$$-x + 2y + z = 1$$
$$3x - y + 2z = -3$$

(a) $\begin{vmatrix} 2 & 1 & -3 \\ -1 & 2 & 1 \\ 3 & -1 & 2 \end{vmatrix}$
(b) $\begin{vmatrix} 2 & -3 & 5 \\ -1 & 1 & 1 \\ 3 & 2 & -3 \end{vmatrix}$
(c) $\begin{vmatrix} 5 & 2 & -3 \\ 1 & -1 & 1 \\ -3 & 3 & 2 \end{vmatrix}$

(d) $\begin{vmatrix} 5 & 2 & 1 \\ 1 & -1 & 2 \\ -3 & 3 & -1 \end{vmatrix}$
(e) None of the preceding

14. Which of the following is the second row of the inverse of matrix A shown below?

$$A = \begin{bmatrix} 1 & 0 & 2 \\ 0 & 2 & -1 \\ 2 & 5 & 2 \end{bmatrix}$$

(a) $0 \quad 1 \quad -\frac{1}{2}$ (b) $-2 \quad -2 \quad 1$ (c) $0 \quad -2 \quad 1$ (d) $0 \quad \frac{1}{2} \quad 0$ (e) None of the preceding

15. What is the maximum of $k = 4x + 3y$ for the region given by the system shown below?

$$x \geq 0$$
$$y \geq 0$$
$$x + 2y \leq 22$$
$$3x + 2y \leq 30$$

(a) 88 (b) 44 (c) 43 (d) 33 (e) None of the preceding

16. Which of the following is one of the three numbers in the solution of this system? $\quad 2x - 3y + z = 11$
$$3x + y - z = 9$$
$$-x + 3y - 2z = -1$$

(a) -3 (b) -2 (c) 5 (d) -8 (e) None of the preceding

17. The Auto-Rental Company buys three particular spare parts from one of two suppliers. They intend to purchase 200 units of part p_1, 350 units of part p_2, and 300 units of part p_3. Supplier A charges \$8.50 for each part p_1, \$10 for each of p_2, and \$12.50 for each of p_3. The corresponding prices charged by B are \$8, \$10.50, and \$12. If the 1 by 3 matrix P is the purchase order for the spare parts, and if C is the cost matrix containing the prices from both suppliers, which of the following is one of the elements in PC?
(a) $350 (10.50) = 3675$ (b) $200 (16.5) = 3300$
(c) 7175 (d) 8875 (e) None of the preceding

18. If $\begin{vmatrix} a_1 & a_2 & a_3 \\ b_1 & b_2 & b_3 \\ c_1 & c_2 & c_3 \end{vmatrix} = 25$, then $\begin{vmatrix} b_1 & b_2 & b_3 \\ a_1 & a_2 & a_3 \\ c_1 + 3a_1 & c_2 + 3a_2 & c_3 + 3a_3 \end{vmatrix} =$

    (a) 25   **(b)** -25 **(c)** 75   **(d)** -75 **(e)** None of the preceding

19. A is a 3 by 3 matrix such that $|A| = 2$, and A^t is the transpose of A. (The 3 rows of A become the 3 columns of A^t in the same order). Find $|A^t|$.

    (a) 2   **(b)** $\frac{1}{8}$ **(c)** $\frac{1}{2}$ **(d)** -2 **(e)** None of the preceding

20. A is an n by n matrix and I is the n by n identity matrix for the same n. If $A^2 = A$, then which of the following is $(I - A)^2$?

    (a) $I^2 + A$ **(b)** I^2A **(c)** $I - A$ **(d)** $I - 3A$ **(e)** None of the preceding

ANSWERS TO THE TEST YOUR UNDERSTANDING EXERCISES

Page 554

1. $(1, 2, -3)$ **2.** $\left(\frac{1}{2}, -2, 5\right)$

Page 564

1. Not possible   **2.** $\begin{bmatrix} 3 & 3 \\ 2 & 3 \\ 9 & 6 \end{bmatrix}$ **3.** $\begin{bmatrix} 8 & 4 \\ 1 & 9 \\ 17 & 8 \end{bmatrix}$ **4.** $\begin{bmatrix} -7 & 1 \\ 4 & -9 \\ -7 & 2 \end{bmatrix}$

5. $\begin{bmatrix} 7 & -1 \\ -4 & 9 \\ 7 & 2 \end{bmatrix}$ **6.** $\begin{bmatrix} 2 & 8 & -6 \\ 4 & 10 & 0 \\ 0 & 2 & 2 \end{bmatrix}$ **7.** Not possible   **8.** $\begin{bmatrix} -15 & -15 \\ -10 & -15 \\ -45 & -30 \end{bmatrix}$ **9.** $\begin{bmatrix} -10 & -2 \\ 2 & -12 \\ -16 & -4 \end{bmatrix}$

Page 566

1. $\begin{bmatrix} -6 & -22 \\ -59 & 11 \end{bmatrix}$ **2.** $\begin{bmatrix} 8 & 0 & 9 \\ 13 & -22 & 27 \end{bmatrix}$ **3.** Not possible   **4.** $\begin{bmatrix} 24 & 29 \\ 14 & -38 \\ -25 & 49 \end{bmatrix}$

5. Not possible   **6.** $\begin{bmatrix} 52 & -63 \\ -5 & 19 \end{bmatrix}$ **7.** $\begin{bmatrix} 25 & 2 & 27 \\ -18 & 28 & -36 \\ 3 & -26 & 18 \end{bmatrix}$ **8.** $\begin{bmatrix} 42 & -25 \\ 181 & -284 \end{bmatrix}$

Page 576

1. $\begin{bmatrix} \frac{5}{3} & 2 \\ \frac{2}{3} & 1 \end{bmatrix}$ **2.** Does not exist   **3.** $\begin{bmatrix} 0 & -1 \\ -\frac{1}{2} & \frac{3}{2} \end{bmatrix}$ **4.** $\begin{bmatrix} 4 & 3 \\ -4 & -6 \end{bmatrix}$

5. Does not exist   **6.** $\begin{bmatrix} \frac{1}{110} & \frac{1}{110} \\ \frac{1}{110} & -\frac{1}{1100} \end{bmatrix}$

Page 585

1. −2   **2.** −2   **3.** 2   **4.** −10   **5.** 5   **6.** 10

7. 20   **8.** 2   **9.** −4   **10.** 1   **11.** −4   **12.** 0

Page 592

1. −29   **2.** 0   **3.** 27   **4.** −93   **5.** −8   **6.** 0

Page 614

1. 19   **2.** 1   **3.** 26   **4.** −34   **5.** 50   **6.** −139

7. 37   **8.** 10   **9.** 200   **10.** −592

The Conic Sections

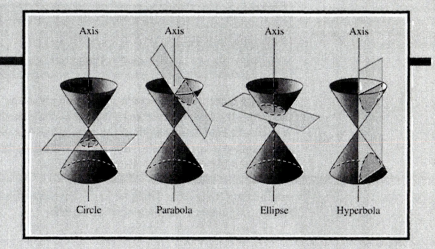

A **conic section** is a curve formed by the intersection of a plane with a double right circular cone. These curves, also called conics, are known as the **circle, ellipse, parabola,** and **hyperbola.** The figures above indicate that the inclination of the plane in relation to the vertical axis of the cone determines the nature of the curve.

The study of the conic sections dates back to the ancient Greek geometers. Euclid (about 300 B.C.) is said to have written four books on the subject, but this work has been lost. Appollonius (c. 260–190 B.C.) made major contributions to this study in his eight-volume work, *Conic Sections.* This work was purely geometric and the algebraic formulations were not introduced until the seventeenth century.

These four curves have played a vital role in mathematics and its applications from the time of the ancient Greeks until the present day. It was Johannes Kepler (1571–1630) who discovered that the planets revolve around the sun in elliptic orbits rather than circular as had been believed. (There is an introduction to elliptic orbits included in Section 10.1.)

It was the British astronomer Edmund Halley (1656–1742) who used the fact about elliptic orbits to predict that a certain comet would appear every 76 years. Named in his honor, Halley's comet's most recent visit was in 1985–1986, and will return again in the year 2062. This fact is the basis for Exercise 43 of Section 10.1.

10.1
THE ELLIPSE

This definition suggests that an ellipse can be drawn as follows. Take a loop of string and place it around two thumbtacks. Use a pencil to pull the string taut to form a triangle, and move the pencil around the loop.

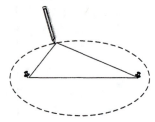

Instead of the three-dimensional approach to the conics as described in the preceding introduction, each conic will be defined as a set of points in the plane satisfying certain geometric conditions. This will directly lead to their algebraic formulation, as has been done in Section 4.1 where a circle was defined as *the set of points in the plane each of which is at a fixed distance r from a given point called the center.* The definition of the **ellipse** follows.

> ### DEFINITION OF AN ELLIPSE
>
> An **ellipse** is the set of all points in a plane such that the sum of the distances from two fixed points (called the **foci**) is a constant.

The reason that the construction shown in the margin gives an ellipse can be seen from the figure. From any position of the pencil the sum of the distances to the thumbtacks equals the length of the loop minus the distance between the thumbtacks. Since both the length of the loop and the distance between the thumbtacks are constants, their difference must also be a constant.

We first consider an ellipse whose foci, F_1 and F_2, are symmetric about the origin along the x-axis. Thus we let F_1 have coordinates $(-c, 0)$ and F_2 have coordinates $(c, 0)$, where c is some positive number.

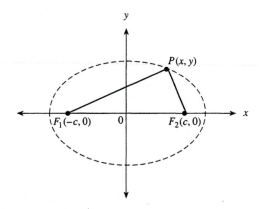

Since the sum of the distances PF_1 and PF_2 must be constant, we choose some positive number a and let this constant equal $2a$. (The form $2a$ will prove to be useful to simplify the algebraic computations.) Thus a point $P(x, y)$ is on the ellipse if and only if $PF_1 + PF_2 = 2a$. Observe that $a \geq c$. This follows from the fact that the sum of the lengths of two sides of a triangle is greater than the length of the third side. That is, $PF_1 + PF_2 > F_1F_2$, or, $2a > 2c$ and $a > c$.

Using the distance formula gives

The distance formula is given on page 248.

$$PF_1 = \sqrt{(x + c)^2 + y^2} \quad \text{and} \quad PF_2 = \sqrt{(x - c)^2 + y^2}$$

Thus

$$PF_1 + PF_2 = \sqrt{(x + c)^2 + y^2} + \sqrt{(x - c)^2 + y^2} = 2a$$

which implies the following:

$$\sqrt{(x+c)^2 + y^2} = 2a - \sqrt{(x-c)^2 + y^2}$$

Squaring both sides and collecting terms, we have

$$a\sqrt{(x-c)^2 + y^2} = a^2 - cx$$

Square both sides again and simplify:

$$(a^2 - c^2)x^2 + a^2y^2 = a^2(a^2 - c^2)$$

Since $a^2 - c^2 > 0$, we may let $b = \sqrt{a^2 - c^2}$ so that $b^2 = a^2 - c^2$. Therefore,

$$b^2x^2 + a^2y^2 = a^2b^2$$

Now divide through by a^2b^2 to obtain the following standard form:

> ### STANDARD FORM FOR THE EQUATION OF AN ELLIPSE WITH FOCI AT $(-c, 0)$ AND $(c, 0)$
>
> $$\frac{x^2}{a^2} + \frac{y^2}{b^2} = 1, \qquad \text{where } b^2 = a^2 - c^2$$

Note: $2b < 2a$ since $b = \sqrt{a^2 - c^2} < a$.

The geometric interpretations of a and b can be found from this last equation. Letting $y = 0$ produces the x-intercepts, $x = \pm a$. The points $V_1\,(-a, 0)$ and $V_2\,(a, 0)$ are called the **vertices** of the ellipse. The **major axis** of the ellipse is the chord V_1V_2, which has length $2a$. Letting $x = 0$ produces the y-intercepts, $y = \pm b$. The points $(0, -b)$ and $(0, b)$ are the endpoints of the **minor axis** that has length $2b$. The intersection of the major and minor axes is the **center** of the ellipse; in this case the center is the origin. Half of the axes are called the **semiaxes** whose lengths are a and b. Also, c is called the *focal distance,* since it is the distance from the center to the foci.

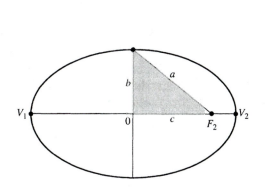

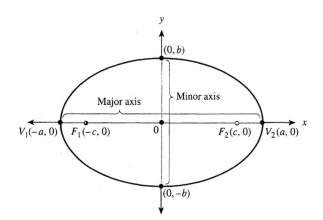

The figure at the left above shows the Pythagorean relationship of a, b, and c, namely $a^2 = b^2 + c^2$, which comes from $b^2 = a^2 - c^2$ in the derivation of the ellipse.

EXAMPLE 1   Sketch the graph of the ellipse $\dfrac{x^2}{25} + \dfrac{y^2}{9} = 1$. Find the coordinates of the foci.

Solution   For $y = 0$, $x = \pm 5$. Therefore, the vertices are $(-5, 0)$ and $(5, 0)$, which are also the endpoints of the major axis. The endpoints of the minor axis are $(0, -3)$ and $(0, 3)$ since $y = \pm 3$ when $x = 0$. Locate these points, and several others, and draw the ellipse as shown.

If $x = \pm 4$, then

$$\frac{(\pm 4)^2}{25} + \frac{y^2}{9} = 1$$

$$\frac{y^2}{9} = 1 - \frac{16}{25} = \frac{9}{25}$$

$$y^2 = \frac{81}{25}$$

$$y = \pm \frac{9}{5}$$

This gives the four points shown.

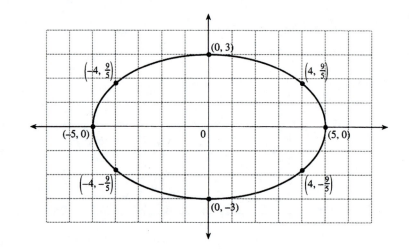

To find the foci, use $b^2 = a^2 - c^2$ with $a^2 = 25$ and $b^2 = 9$.

$$9 = 25 - c^2$$

$$c^2 = 16$$

$$c = +4 \qquad c \text{ is a positive number.}$$

The foci are located at the points $(\pm 4, 0)$.   ∎

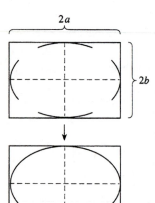

An additional aid in drawing an ellipse makes use of the rectangle with dimensions $2a$ by $2b$, as shown in the margin. The ellipse is inscribed in this rectangle in the sense that at the endpoints of both axes, the sides of the rectangle are tangent to the ellipse. Greater accuracy can be obtained by first sketching the arcs of the ellipse that show the tangency at the four points, and then connecting the arcs.

EXAMPLE 2   Write the equation of the ellipse in standard form having vertices $(\pm 10, 0)$ and foci $(\pm 6, 0)$.

Solution   Since $a = 10$ and $c = 6$, we get $b^2 = a^2 - c^2 = 100 - 36 = 64$. Now substitute into the standard form.

$$\frac{x^2}{a^2} + \frac{y^2}{b^2} = 1 \;\;\Rightarrow\;\; \frac{x^2}{100} + \frac{y^2}{64} = 1$$ ∎

When the foci of the ellipse are on the y-axis, a similar development produces the following standard form:

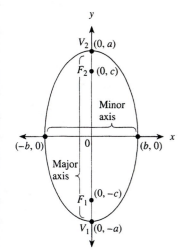

> ## STANDARD FORM FOR THE EQUATION OF AN ELLIPSE WITH FOCI AT $(0, -c)$ AND $(0, c)$
>
> $$\frac{x^2}{b^2} + \frac{y^2}{a^2} = 1, \qquad \text{where } b^2 = a^2 - c^2$$

The major axis is on the y-axis, and its endpoints are the vertices $(0, \pm a)$. The minor axis is on the x-axis and has endpoints $(\pm b, 0)$. Center: $(0, 0)$.

EXAMPLE 3 Change $25x^2 + 16y^2 = 400$ into standard form and graph the ellipse; show the foci.

Solution Divide both sides of $25x^2 + 16y^2 = 400$ by 400 to obtain the standard form $\dfrac{x^2}{16} + \dfrac{y^2}{25} = 1$. Since $a^2 = 25$, $a = 5$ and the major axis is on the y-axis with length $2a = 10$. Similarly, $b^2 = 16$ gives $b = 4$, and the minor axis has length $2b = 8$. Also, $c^2 = a^2 - b^2 = 25 - 16 = 9$, so that $c = 3$, which locates the foci at $(0, \pm 3)$.

When the equation of an ellipse is in standard form, the major axis is horizontal if the x^2-term has the largest denominator. It will be vertical if the y^2-term has the largest denominator. Another way to determine this is to locate the endpoints of the two axes. The longer of the two is the major axis.

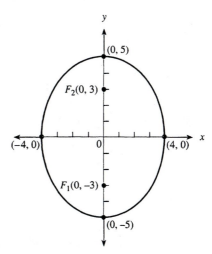

As mentioned in the introduction to this chapter, the planets of our solar system have elliptical orbits around the sun. In fact, the sun is a focus (see Exercise 42). It is also true that satellites revolving around the earth have elliptical orbits. This is illustrated in the following example.

EXAMPLE 4 A satellite follows an elliptical orbit around the earth such that the center of the earth, E, is one of the foci. The figure on the next page indicates that the furthest point that the satellite will be from the earth's surface is 2500 miles and the closest will be 1000 miles. Observe that these distances are measured along the major axis, which is assumed to be on the y-axis. Use 4000 miles as the radius of the earth and find the equation of the orbit.

Solution Since $2a = V_1V_2$ is the length of the major axis, we have

$$2a = 2500 + 8000 + 1000 = 11{,}500$$

and

$$a = 5750$$

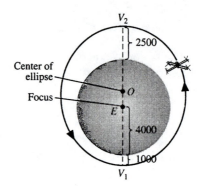

The orbital point furthest from the earth is called the **apogee**, *and the closest is the* **perigee**.

Also,

$$c = OE = OV_1 - EV_1 = a - 5000 = 5750 - 5000 = 750$$

Now find b^2.

$$b^2 = a^2 - c^2$$
$$= 5750^2 - 750^2$$
$$\approx 5701^2$$

The symbol $\approx$ stands for "is approximately equal to."

Since we have chosen the major axis to be vertical, the equation of the orbit becomes

$$\frac{x^2}{5701^2} + \frac{y^2}{5750^2} = 1 \qquad \blacksquare$$

Let us now consider the equation of an ellipse where the center is at a point (h, k), not necessarily the origin. If the major axis is horizontal, then the foci have coordinates $(h - c, k)$ and $(h + c, k)$, and it can be shown that the equation has this standard form:

Note: When $a = b$, the standard form produces $(x - h)^2 + (y - k)^2 = a^2$, which is the equation of a circle with center (h, k) and radius a. Thus a circle may be regarded as a special kind of ellipse, one for which the foci and center coincide.

STANDARD FORM FOR THE EQUATION OF AN ELLIPSE WITH CENTER (h, k) AND FOCI $(h \pm c, k)$

$$\frac{(x - h)^2}{a^2} + \frac{(y - k)^2}{b^2} = 1, \qquad \text{where } b^2 = a^2 - c^2$$

Similarly, an ellipse with center (h, k) whose major axis is vertical has this standard form:

Note that the distances a, b, and c are measured from the center (h, k), just as earlier they were measured from the origin for an ellipse having the origin as center.

STANDARD FORM FOR THE EQUATION OF AN ELLIPSE WITH CENTER (h, k) AND FOCI $(h, k \pm c)$

$$\frac{(x - h)^2}{b^2} + \frac{(y - k)^2}{a^2} = 1, \qquad \text{where } b^2 = a^2 - c^2$$

Observe that in this expanded form of an ellipse, the coefficients of x^2 and y^2 are not equal, but they have the same sign.

EXAMPLE 5 Write in standard form and graph:

$$4x^2 - 16x + 9y^2 + 18y = 11$$

Solution We follow a procedure much like that used in Section 4.1 for circles; that is, complete the square in both variables.

$$4x^2 - 16x + 9y^2 + 18y = 11$$
$$4(x^2 - 4x) + 9(y^2 + 2y) = 11$$
$$4(x^2 - 4x + 4) + 9(y^2 + 2y + 1) = 11 + 16 + 9$$
$$4(x - 2)^2 + 9(y + 1)^2 = 36$$

Divide both sides by 36:

$$\frac{(x - 2)^2}{9} + \frac{(y + 1)^2}{4} = 1$$

This is the equation of an ellipse having center at $(2, -1)$, with major axis $2a = 6$ and minor axis $2b = 4$. Since $c^2 = a^2 - b^2 = 5$, $c = \sqrt{5}$, which gives the foci $\left(2 \pm \sqrt{5}, -1\right)$.

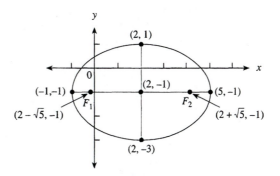

Observe that the ellipse shown above has the same size and shape as that for $\frac{x^2}{9} + \frac{y^2}{4} = 1$. However, it has been shifted 2 units to the right and 1 unit down.

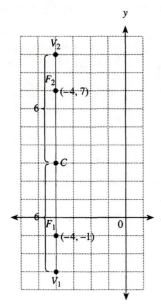

EXAMPLE 6   Write the standard form of the ellipse having major axis of length 12 and foci at $(-4, -1)$ and $(-4, 7)$.

Solution   Since the foci have the same first coordinate, the major axis is vertical. Also, since the center C of the ellipse is the midpoint of F_1F_2, C has coordinates $\left(-4, \dfrac{7 + (-1)}{2}\right) = (-4, 3)$. The major axis has length $2a = 12$ so that $a = 6$, and the focal length $c = 4$ since $2c = F_1F_2 = 7 - (-1) = 8$. Therefore,

$$b^2 = a^2 - c^2 = 36 - 16 = 20$$

and the standard form is

$$\frac{(x + 4)^2}{20} + \frac{(y - 3)^2}{36} = 1$$

∎

(Answers: Page 682)

TEST YOUR UNDERSTANDING

Find the coordinates of the center, vertices, and foci for each ellipse.

1. $\dfrac{x^2}{100} + \dfrac{y^2}{64} = 1$ **2.** $16x^2 + y^2 = 16$

Write in standard form the equation of the ellipse having the given properties.

3. Vertices: $(\pm 8, 0)$ **4.** Vertices: $(0, \pm 9)$
   Foci: $(\pm 7, 0)$ Length of minor axis: 12

5. Write the standard form of the ellipse $x^2 - 10x + 4y^2 + 16y = -37$, and give the center, vertices, and foci.

When the numbers a and b are relatively close in value, the ellipse will have a circular appearance, but the ellipse will be flatter when b is relatively small in comparison to a. This is demonstrated in the following ellipses in which $a = 4$ is held fixed, while b first equals 3.8, which is close in value to a, followed by $b = 1$, which is small in comparison to the value of a.

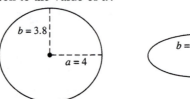   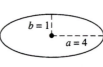

This "roundness" of an ellipse can be measured by the *eccentricity*:

The **eccentricity** e of an ellipse with semimajor axis a and focal length c, is defined as

$$e = \frac{c}{a}$$

The eccentricity of an ellipse is always between 0 and 1. That is, since $0 < c < a$, dividing by a gives $0 < \dfrac{c}{a} = e < 1$. Furthermore, $c = \sqrt{a^2 - b^2}$ gives $e = \dfrac{\sqrt{a^2 - b^2}}{a}$. From this form we see that when b is close to a, then $c = \sqrt{a^2 - b^2} \approx 0$ so that $e = \dfrac{c}{a} \approx 0$, and the ellipse is almost circular. But

This use of the letter e should not be confused with its earlier use as the base of natural logarithms.

when b is close to 0, then $c = \sqrt{a^2 - b^2} \approx a$ so that $e = \dfrac{c}{a} \approx \dfrac{a}{a} = 1$, and the ellipse is flatter. In brief,

---

Smaller Eccentricity ---- Rounder Ellipses

Larger Eccentricity ---- Flatter Ellipses

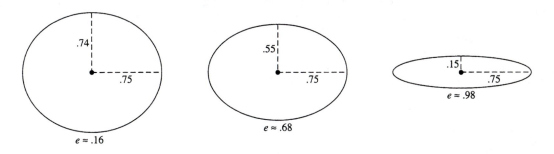

$e \approx .16$ $e \approx .68$ $e \approx .98$

EXAMPLE 7   Find the eccentricity of the ellipse $25x^2 + 16y^2 = 400$.

Solution   Divide by 400 to obtain

$$\frac{x^2}{16} + \frac{y^2}{25} = 1$$

This ellipse is shown on page 637.

Then $a = 5$, $b = 4$, and $c = \sqrt{25 - 16} = 3$. Now find the eccentricity:

$$e = \frac{c}{a} = \frac{3}{5} \qquad \blacksquare$$

EXAMPLE 8   An ellipse with center $(2, 6)$ has a vertex at $(-2, 6)$ and has eccentricity $= \frac{5}{8}$. Find the foci and write an equation of the ellipse.

Solution   Since the center and vertex have the same second coordinate, the major axis must be horizontal and $a = 2 - (-2) = 4$. Now the equation $e = \dfrac{c}{a}$ can be used to find the focal length c as follows.

$$e = \frac{c}{a} = \frac{c}{4} = \frac{5}{8} \qquad \text{and} \qquad c = 4\left(\frac{5}{8}\right) = 2.5$$

The foci are $(2 - c, 6) = (-0.5, 6)$ and $(2 + c, 6) = (4.5, 6)$. Also,

$$b^2 = a^2 - c^2 = 4^2 - (2.5)^2 = 9.75$$

The equation is

$$\frac{(x - 2)^2}{16} + \frac{(y - 6)^2}{9.75} = 1$$ ∎

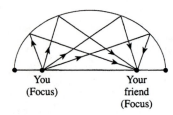

You
(Focus)

Your
friend
(Focus)

The ellipse shown in the figure at the left has an interesting reflecting property. If a source of sound (or light) is positioned at one focus, the sound (or light) waves will reflect off the ellipse and pass through the other focus. Now suppose the ceiling of a large room is shaped like part of an ellipsoid (an ellipsoid is obtained by revolving an ellipse around its major axis), and you and a friend are each standing at a focus. If you whisper so that others in the room are unable to hear you, your friend will hear you because the sound waves will bounce off the ellipsoid and pass through the other focus. The Capitol building in Washington, D.C., has such a "whispering gallery."

EXERCISES 10.1

Graph each ellipse. State the coordinates of the center, vertices, and foci.

1. $\dfrac{x^2}{25} + \dfrac{y^2}{16} = 1$ **2.** $\dfrac{x^2}{16} + \dfrac{y^2}{25} = 1$ **3.** $\dfrac{x^2}{16} + \dfrac{y^2}{9} = 1$ **4.** $\dfrac{x^2}{9} + \dfrac{y^2}{16} = 1$

5. $\dfrac{x^2}{4} + \dfrac{y^2}{36} = 1$ **6.** $\dfrac{x^2}{36} + \dfrac{y^2}{4} = 1$ **7.** $9x^2 + y^2 = 9$ **8.** $x^2 + 9y^2 = 9$

9. $4x^2 + 9y^2 = 36$ **10.** $x^2 + 4y^2 = 4$ **11.** $25x^2 + 9y^2 = 225$ **12.** $16x^2 + 9y^2 = 144$

13. $\dfrac{(x - 1)^2}{9} + \dfrac{(y + 2)^2}{4} = 1$ **14.** $\dfrac{(x + 2)^2}{16} + \dfrac{(y - 1)^2}{9} = 1$

15. $\dfrac{(x + 2)^2}{4} + \dfrac{(y - 3)^2}{9} = 1$ **16.** $\dfrac{(x - 3)^2}{9} + \dfrac{(y + 2)^2}{16} = 1$

Write an equation for the given ellipse.

17.

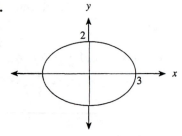

18.

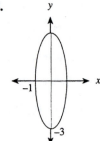

19.

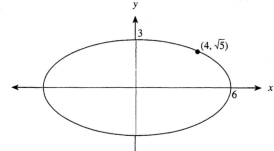

20.

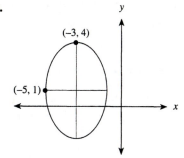

21.

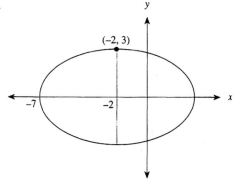

22.

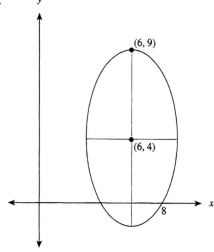

Write, in standard form, the equation of the ellipse having the given properties.

23. Center $(0, 0)$; horizontal major axis of length 10; minor axis of length 6.

24. Center $(0, 0)$; foci $(\pm 2, 0)$; vertices $(\pm 5, 0)$.

25. Center $(2, 3)$; foci $(-2, 3)$ and $(6, 3)$; minor axis of length 8.

26. Center $(2, -3)$; vertical major axis of length 12; minor axis of length 8.

27. Vertices $(0, \pm 5)$; foci $(0, \pm 3)$.

28. Center $(-5, 0)$; foci $(-5, \pm 2)$; $b = 3$.

29. Endpoints of the major and minor axes are $(-8, 0)$, $(8, 0)$, $(0, -4)$, and $(0, 4)$.

30. Endpoints of the major and minor axes are $(-1, 0)$, $(1, 0)$, $(0, -3)$, and $(0, 3)$.

31. Endpoints of the major and minor axes are $(3, 1)$, $(9, 1)$, $(6, -1)$, and $(6, 3)$.

32. Endpoints of the major and minor axes are $(-4, 3)$, $(2, 3)$, $(-1, -2)$, and $(-1, 8)$.

33. Foci $(\pm 4, 0)$; eccentricity $\frac{2}{3}$.

34. Vertices $(0, \pm 6)$; eccentricity $\frac{11}{12}$.

35. Minor axis along the x-axis; vertex $(-1, 10)$; eccentricity $\frac{4}{5}$.

36. Center $(5, -1)$; focus $(8, -1)$; eccentricity $\frac{1}{2}$.

Write each ellipse in standard form. Find the coordinates of the center, vertices, and foci.

37. $25x^2 + y^2 - 12y = -11$

38. $x^2 + 4x + 9y^2 = 5$

39. $4x^2 + 24x + 13y^2 - 26y = 3$

40. $16x^2 - 32x + 9y^2 - 72y = -16$

41. In 1957, the Russians launched the first man-made satellite, Sputnik. Its orbit around the earth was elliptical with the center of the earth as one focus. The maximum height above the earth was about 580 miles, and the minimum height was approximately 130 miles.

   (a) Assuming that the earth's radius is 4000 miles, find the equation of Sputnik's orbit. (Leave the value b^2 in unsimplified form.)

   (b) Find the value of b to the nearest mile, and rewrite the equation of the ellipse using this result.

42. The orbit of the earth around the sun is elliptical with the sun being one of the foci. The earth's maximum distance from the sun (the apogee) is approximately 94.56 million miles, and the minimum distance (the perigee) is about 91.45 million miles.

   (a) Find a and b in millions of miles. (*Hint:* Use $a + c$ and $a - c$.)

   (b) Find the eccentricity of the earth's orbit and comment on this result in relation to the answer in part (a).

   (c) The eccentricities of the other eight planets are: Mercury, 0.206; Venus, 0.007; Mars, 0.093; Jupiter, 0.048; Saturn, 0.054; Uranus, 0.046; Neptune, 0.008; Pluto, 0.248. List the planets in order of the circularity of their orbits, from the most circular to the least.

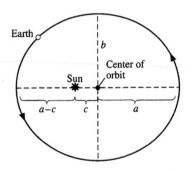

43. Halley's comet is named in honor of the British astronomer Edmund Halley (1656–1742) who successfully predicted that the comet that appeared in 1682 had a period of close to 76 years and would reappear in 1758, as it did. The most recent visit was in 1985–1986 and it will return again in 2062. The orbit of this comet around the sun is an ellipse with the sun as a focus and eccentricity $e = 0.967$. The position where the comet is closest to the sun to the position where it is furthest from the sun is approximately 3.365×10^9 miles. Approximate how close the comet gets to the sun.

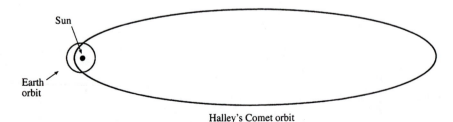

Halley's Comet orbit

44. Several years ago some manufacturers of bicycle parts designed chainrings that had an elliptical shape, rather than circular. Such rings were named *biopace chainrings*. The intent of this design was to make the pedaling more efficient, but the outcome has been questionable. Suppose a manufacturer needed a biopace chainring with axes of length $8\frac{1}{8}$ inches by $7\frac{7}{8}$ inches. Find an equation for this chainring that could be used in the production process.

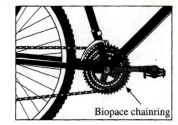

Biopace chainring

45. The underside of a bridge over a two-lane roadway is in the shape of a semiellipse. The elliptical arch spans 60 feet (the length of the major axis) and the height at the center is 20 feet.

(a) The outsides of the driving lanes are marked by lines that are 10 feet from the base of the bridge. What is the bridge clearance y above these lines? (Write the exact numerical expressions.)

(b) Find the clearance rounded to the nearest tenth of a foot.

? **Challenge** The circle in the figure has radius 3 with center at the origin. A number of horizontal chords have been drawn and the y-axis divides these chords into half-chords. On each half-chord the point P is found so that $BP = \frac{2}{3} BA$. Prove that when this is done for all possible half-chords, the set of all points P determines an ellipse.

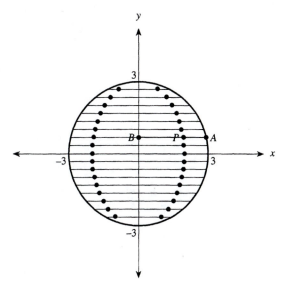

Written Assignment In calculus it can be proved that the area of an ellipse with semimajor axis a and semiminor axis b is πab. The purpose of this exercise is to find some evidence (not a proof) as to why this result is sensible. First, let's assume that if a specific ellipse has area πab, then the area of any ellipse should have this same form. Now select a specific semimajor axis such as $a = 3$ and describe why, with your choices of b, the area of the ellipse ought to be πab.

Graphing Calculator Exercises To graph an equation of an ellipse with your graphing calculator, solve for the variable y in terms of x. In Example 5, we have $4x^2 - 16x + 9y^2 + 18y = 11$ or $4(x - 2)^2 + 9(y + 1)^2 = 36$. Solving for y gives the two functions $y = \pm (1/3) \sqrt{36 - 4(x - 2)^2} - 1$. Graph these semiellipses and join them to obtain the figure in Example 5.

Use your calculator to graph the ellipses in Exercises 1–3 as the union of the graphs of two functions of x. Indicate the points where the two functions join.

1. $(x + 1)^2 + 2y^2 = 1$

2. $x^2/9 + y^2/16 = 1$

3. $3x^2 + 6x + y^2 + 4y = 20$

10.2 THE HYPERBOLA

The next conic section to be studied is the **hyperbola.** The definition of the hyperbola is similar to that of the ellipse, except that we now make use of the *difference* of the distances from the foci.

> ### DEFINITION OF A HYPERBOLA
>
> A **hyperbola** is the set of all points in the plane such that the difference of the distances from two fixed points (called the **foci**) is a constant.

The two fixed points, F_1 and F_2, are called the **foci** of the hyperbola, and its **center** is the midpoint of the segment F_1F_2. It turns out that a hyperbola consists of two congruent branches which open in opposite directions.

We begin with a hyperbola with foci on the x-axis at $F_1(-c, 0)$ and $F_2(c, 0)$, where $c > 0$. Select a number $a > 0$ so that for any point P on the right branch of the hyperbola we have $PF_1 - PF_2 = 2a$. Also, for any point P on the left branch $PF_2 - PF_1 = 2a$, as in the figure below.

The form 2a is used to simplify computations.

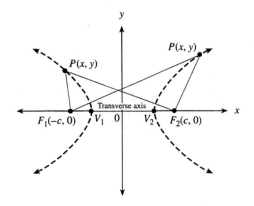

Note that $a < c$, since from triangle F_1PF_2 (P on the right branch) we have $PF_1 < PF_2 + F_1F_2$, which gives $PF_1 - PF_2 < F_1F_2$, or $2a < 2c$.

If we now follow the same type of analysis used to derive the equation of an ellipse, it can be shown that the equation of a hyperbola may be written in this standard form:

STANDARD FORM FOR THE EQUATION OF A HYPERBOLA WITH FOCI AT $(-c, 0)$ AND $(c, 0)$

$$\frac{x^2}{a^2} - \frac{y^2}{b^2} = 1, \qquad \text{where } b^2 = c^2 - a^2$$

Letting $y = 0$ gives $x = \pm a$; the points $V_1(-a, 0)$ and $V_2(a, 0)$ are the **vertices** of the hyperbola. The segment V_1V_2 is called **transverse axis** of the hyperbola and the midpoint of the transverse axis is the **center** of the hyperbola.

EXAMPLE 1 Write in standard form and identify the foci and the vertices: $16x^2 - 25y^2 = 400$.

Solution Divide through by 400 to place in standard form.

$$\frac{16x^2}{400} - \frac{25y^2}{400} = \frac{400}{400}$$

$$\frac{x^2}{25} - \frac{y^2}{16} = 1$$

Note that $a^2 = 25$ and $b^2 = 16$, so that $a = 5$ and $b = 4$. Then $c^2 = a^2 + b^2 = 25 + 16 = 41$ and $c = \sqrt{41}$. The vertices of the hyperbola are located at $(-5, 0)$ and $(5, 0)$; the foci are at $\left(-\sqrt{41}, 0\right)$ and $\left(\sqrt{41}, 0\right)$. ∎

Now return to the standard form for the equation of a hyperbola and solve for y:

$$\frac{x^2}{a^2} - \frac{y^2}{b^2} = 1$$

$$y^2 = \frac{b^2}{a^2}(x^2 - a^2)$$

$$y = \pm \frac{b}{a} \sqrt{x^2 - a^2}$$

Consequently, since $x^2 - a^2 \geq 0$, $|x| \geq a$, which means that there are no points of the hyperbola for $-a < x < a$.

An efficient way to sketch a hyperbola is first to draw the rectangle that is $2a$ units wide and $2b$ units high, as shown in the following figure. Note that the center of the hyperbola is also the center of this rectangle. Draw the diagonals of the rectangle and extend them in both directions; these are the **asymptotes**. Now sketch the hyperbola by beginning at the vertices $(\pm a, 0)$ so that the lines are asymptotes

The distance between the curve and an asymptote becomes smaller and smaller and approaches 0; the curve approaches the line.

to the curve and the branches are between the asymptotes whose equations are $y = \pm\dfrac{b}{a}x$.

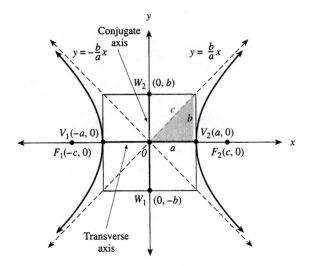

Since $b^2 = c^2 - a^2$, or $a^2 + b^2 = c^2$, it follows that b can be used as a side of a right triangle having hypotenuse c. Thus if we construct a perpendicular at V_2 of the length b, the resulting right triangle has hypotenuse c. We also see that this hypotenuse lies on the line $y = \dfrac{b}{a}x$.

The segment W_1W_2 is the **conjugate axis** of the hyperbola. Observe that the transverse and conjugate axes are perpendicular bisectors of one another, and their lengths are $2a$ and $2b$, respectively.

Greater accuracy in drawing the branches can be obtained by first sketching the arcs to show the tangency at the vertices (see below) and then extending them to show the asymptotic approaches to the asymptotes.

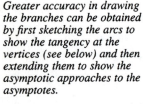

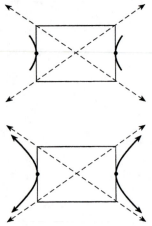

EXAMPLE 2 Sketch the graph of $\dfrac{x^2}{9} - \dfrac{y^2}{4} = 1$ and show the foci.

Solution Since $a^2 = 9$, the vertices are $(\pm 3, 0)$. Also,

$$b^2 = c^2 - a^2 \qquad \text{or} \qquad c^2 = a^2 + b^2 = 9 + 4 = 13$$

Therefore, $\left(\pm\sqrt{13}, 0\right)$ are the foci. To sketch the hyperbola, first note that $2a = 6$, $2b = 4$ and draw the 6 by 4 rectangle with center $(0, 0)$. Draw the asymptotes by extending the diagonals and sketch the two branches. The branches approach the asymptotes in the sense that as x gets larger (toward the right) the hyperbola gets closer to the asymptotes, but never touches them; similarly, for x, toward the left.

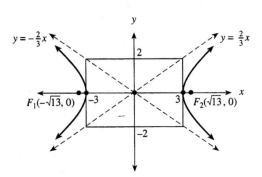

When the foci of a hyperbola are on the y-axis, the equation has this standard form and graph.

**STANDARD FORM FOR THE EQUATION OF A HYPERBOLA
WITH FOCI AT $(0, -c)$ AND $(0, c)$**

$$\frac{y^2}{a^2} - \frac{x^2}{b^2} = 1, \qquad \text{where } b^2 = c^2 - a^2$$

When the equation of a hyperbola is in standard form, the branches open vertically if the minus sign precedes the x^2-term. They will open horizontally if the minus sign precedes the y^2-term.

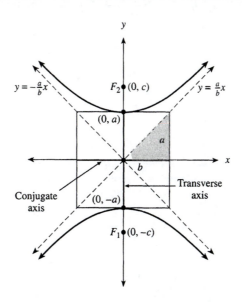

The vertices $(0, \pm a)$ are the endpoints of the vertical transverse axis that has length $2a$. The conjugate axis is horizontal and has length $2b$. The asymptotes are the lines $y = \pm\dfrac{a}{b}x$ and the branches of this hyperbola open upward and downward.

EXAMPLE 3 Write the standard form of the hyperbola with vertices $(0, \pm 2)$ and asymptotes $y = \pm\frac{1}{2}x$. Also find the foci and sketch the hyperbola.

Solution The given vertices imply that the transverse axis is vertical. Since $y = \pm\frac{1}{2}x$ are the asymptotes and $a = 2$, then $\dfrac{a}{b} = \dfrac{1}{2}$ or $b = 2a$, and $b = 2(2) = 4$. Consequently, the equation is

$$\frac{y^2}{4} - \frac{x^2}{16} = 1$$

Since $c^2 = a^2 + b^2 = 4 + 16 = 20$, $c = \sqrt{20} = 2\sqrt{5}$ so that the foci are at $(0, \pm 2\sqrt{5})$ as shown in the figure on the next page.

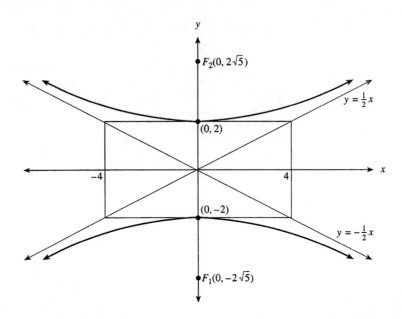

Find the coordinates of the center, vertices, and foci, and also give the equations of the asymptotes .

1. $\dfrac{x^2}{9} - \dfrac{y^2}{16} = 1$ **2.** $\dfrac{y^2}{64} - \dfrac{x^2}{36} = 1$

Write in standard form the equation of the hyperbola having the given properties.

3. Vertices: $(0, \pm 5)$ **4.** Vertices: $(\pm 2, 0)$

   Foci: $(0, \pm 7)$ Asymptotes: $y = \pm \dfrac{3}{2} x$

(Answers: Page 682)

When the center of the hyperbola is at some point (h, k) and the transverse axis is horizontal, we have the following standard form:

STANDARD FORM FOR THE EQUATION OF A HYPERBOLA WITH CENTER AT (h, k) AND FOCI $(h \pm c, k)$

$$\frac{(x - h)^2}{a^2} - \frac{(y - k)^2}{b^2} = 1, \qquad \text{where } b^2 = c^2 - a^2$$

For a hyperbola with center at (h, k) and with a vertical transverse axis, the standard form is as follows:

STANDARD FORM FOR THE EQUATION OF A HYPERBOLA WITH CENTER AT (h, k) AND FOCI $(h, k \pm c)$

$$\frac{(y - k)^2}{a^2} - \frac{(x - h)^2}{b^2} = 1, \qquad \text{where } b^2 = c^2 - a^2$$

EXAMPLE 4   Write in standard form and graph:

$$4x^2 + 16x - 9y^2 + 18y = 29$$

Observe that in this expanded form of a hyperbola the coefficients of x^2 and y^2 have opposite signs.

Solution Complete the square in x and y.

$$4(x^2 + 4x) - 9(y^2 - 2y) = 29$$

$$4(x^2 + 4x + 4) - 9(y^2 - 2y + 1) = 29 + 16 - 9$$

$$4(x + 2)^2 - 9(y - 1)^2 = 36$$

Divide both sides by 36:

$$\frac{(x + 2)^2}{9} - \frac{(y - 1)^2}{4} = 1$$

This is the standard form for a hyperbola with center at $(-2, 1)$. Since $a^2 = 9$, we have $a = 3$ and the vertices are located 3 units from the center at $(-5, 1)$ and $(1, 1)$. Since $c^2 = a^2 + b^2 = 13$, the foci are located at $\sqrt{13}$ units from the center, namely at $(-2 \pm \sqrt{13}, 1)$.

To sketch the hyperbola, first draw the 6 by 4 rectangle with center at $(-2, 1)$ as shown. Draw the asymptotes by extending the diagonals and sketch the branches of the hyperbola.

This hyperbola can also be obtained by shifting the hyperbola
$$\frac{x^2}{9} - \frac{y^2}{4} = 1 \; two$$
units left and one unit up.

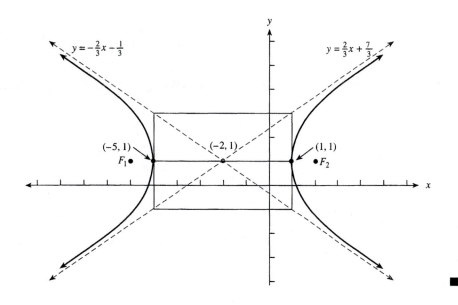

Observe that in the preceding example the asymptotes pass through the center $(-2, 1)$, their slopes are $\pm \dfrac{b}{a} = \pm \dfrac{2}{3}$, and their equations are $y - 1 = \pm \dfrac{2}{3}(x + 2)$.

The results of Example 4 are included in the following generalization:

The Asymptotes of a Hyperbola with Center (h, k) are:

$$y - k = \pm \frac{b}{a}(x - h) \quad \bigg| \quad y - k = \pm \frac{a}{b}(x - h)$$

when the transverse axis is horizontal. | when the transverse axis is vertical.

The final example demonstrates how hyperbolas can be applied to solve some location problems.

EXAMPLE 5   A park ranger is stranded in his vehicle on a wooded mountain trail in a snowstorm. The trail runs parallel to a north-south highway at a distance of 2 miles from the highway. There are two rescue vehicles on the highway that are parked 3 miles apart. The ranger sets off an explosive device (as a distress signal) and the sound reaches the northern rescue vehicle 3 seconds earlier than it reaches the other vehicle. Locate the position of the ranger relative to the rescue vehicles. (Use 1100 ft/sec as the speed of sound.)

Solution   Let the y-axis represent the highway and use $F_1(0, -1.5)$ and $F_2(0, 1.5)$ as the positions of the rescue vehicles whose distances from the ranger R are s_1 and s_2. Using time $=$ distance/rate we have $s_1/1100$ and $s_2/1100$ as the times that the sound of the explosive reaches F_1 and F_2.

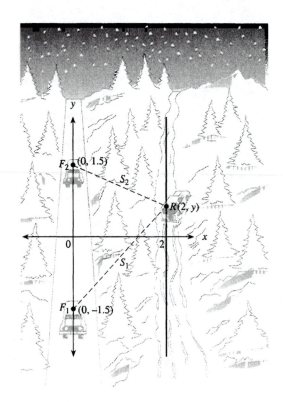

Thus

$$\frac{s_1}{1100} - \frac{s_2}{1100} = 3 \text{ seconds}$$

or, $s_1 - s_2 = 3300$ ft. Converting to miles, we obtain

$$s_1 - s_2 = \frac{3300}{5280} = \frac{5}{8} \text{ mile}$$

This shows that the difference in the distances of R from F_2 and F_1 is the constant $\frac{5}{8}$, so that the definition of a hyperbola is applicable. Then R is on a hyperbola with foci F_1, F_2 and $2a = \frac{5}{8}$ or $a = \frac{5}{16}$. Also, since $c = 1.5$

$$b^2 = \left(\frac{3}{2}\right)^2 - \left(\frac{5}{16}\right)^2 = \frac{551}{256} \text{ and } b \approx 1.47$$

Since the foci are on the y-axis and the center is $(0, 0)$, the equation of the hyperbola is

$$\frac{y^2}{a^2} - \frac{x^2}{b^2} = \frac{y^2}{\frac{25}{256}} - \frac{x^2}{\frac{551}{256}} = 1$$

Converting to decimal form, the (approximate) equation is

$$\frac{y^2}{0.0977} - \frac{x^2}{2.1523} = 1$$

Since the mountain trail is 2 miles from the highway, the x-coordinate of R is 2. Then solving the equation

$$\frac{y^2}{0.0977} - \frac{2^2}{2.1523} = 1$$

for y gives $y^2 = 0.0977 \left(1 + \frac{4}{2.1523}\right) \approx 0.2793$, and $y \approx \pm 0.53$. We use the positive value 0.53 since R is closer to F_2 than to F_1 and the coordinates of R are $(2, 0.53)$.

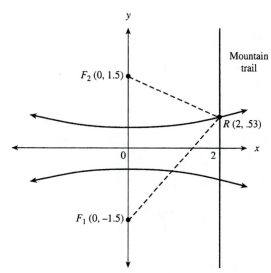

EXERCISES 10.2

Graph each hyperbola. State the coordinates of the center, vertices, and foci. Give the equations of the asymptotes.

1. $\dfrac{x^2}{25} - \dfrac{y^2}{9} = 1$

2. $\dfrac{x^2}{4} - \dfrac{y^2}{9} = 1$

3. $\dfrac{x^2}{16} - \dfrac{y^2}{25} = 1$

4. $\dfrac{y^2}{16} - \dfrac{x^2}{4} = 1$

5. $\dfrac{y^2}{36} - \dfrac{x^2}{9} = 1$

6. $\dfrac{y^2}{4} - \dfrac{x^2}{9} = 1$

7. $9x^2 - y^2 = 9$

8. $x^2 - 9y^2 = 9$

9. $4x^2 - 9y^2 = 36$

10. $9y^2 - 4x^2 = 36$

11. $25y^2 - 9x^2 = 225$

12. $9y^2 - 16x^2 = 144$

13. $\dfrac{(x-2)^2}{9} - \dfrac{(y+1)^2}{4} = 1$

14. $\dfrac{(x+3)^2}{16} - \dfrac{(y-2)^2}{9} = 1$

15. $\dfrac{(y+2)^2}{25} - \dfrac{(x-3)^2}{16} = 1$

16. $\dfrac{(y+2)^2}{16} - \dfrac{(x-1)^2}{4} = 1$

Write an equation for the given hyperbola.

17.

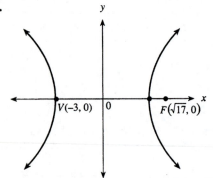

18.

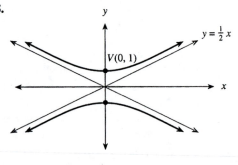

19.

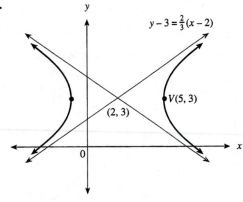

20.

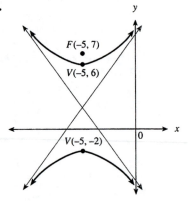

Write, in standard form, the equation of the hyperbola having the given properties.

21. Center $(0, 0)$; foci $(\pm 6, 0)$; vertices $(\pm 4, 0)$.

22. Center $(0, 0)$; foci $(0, \pm 4)$; vertices $(0, \pm 1)$.

23. Center $(-2, 3)$; vertical transverse axis of length 6; $c = 4$.

24. Center $(4, 4)$; vertex $(4, 7)$; $b = 2$.

25. Center $(0, 0)$; asymptotes $y = \pm \dfrac{1}{2} x$; vertices $(\pm 4, 0)$.

26. Asymptotes $y = \pm \dfrac{5}{12} x$; foci $(\pm 13, 0)$.

27. Asymptotes $y = \pm \dfrac{8}{15} x$; foci $(0, \pm 17)$.

Identify the center, vertices, and foci of each hyperbola.

28. $\dfrac{(x-2)^2}{25} - \dfrac{(y+1)^2}{16} = 1$

29. $\dfrac{(x+1)^2}{16} - \dfrac{(y-3)^2}{9} = 1$

30. $\dfrac{(y-3)^2}{16} - \dfrac{(x+1)^2}{4} = 1$

31. $\dfrac{(y+1)^2}{64} - \dfrac{(x-3)^2}{36} = 1$

Write in standard form and identify the center, vertices, and foci of each hyperbola.

32. $9x^2 + 36x - 16y^2 - 96y = 252$ **33.** $4x^2 - 8x - 9y^2 - 36y = 68$

34. $16y^2 + 32y - 9x^2 - 90x = 353$ **35.** $y^2 + 4y - 4x^2 + 8x = 4$

36. Let P be on the right branch of the hyperbola with foci $F_1(-c, 0)$ and $F_2(c, 0)$, and let

$PF_1 - PF_2 = 2a$ for a $< c$. Derive the equation $\dfrac{x^2}{a^2} - \dfrac{y^2}{b^2} = 1$, where $b^2 = c^2 - a^2$.

37. Write the equation of the hyperbola that has the asymptotes $y - 4 = \pm\frac{3}{5}(x+2)$ and the focus $(-2 + 2\sqrt{34}, 4)$.

38. Assume that the upper branch of the hyperbola $12y^2 + 72y - 4x^2 + 81 = 0$ is the path of a comet having the sun as the focus. Sketch the graph of this path and let 1 unit on the axes represent 93 million miles. Find the closest approach that the comet makes to the sun. (You may assume here that for a given branch of a hyperbola and its focus, the point on the branch that is closest to the focus is the vertex.)

39. (a) A small boat is anchored off shore on a lake in a dense fog. The boat sends out a loud signal on its foghorn and is received by two stations that are 2 miles apart on the shoreline. The sound, traveling at 1100 ft/sec, takes 2.7 seconds longer to reach one station than the other. Using $(0, 0)$ and $(2, 0)$ for the positions of the stations, find the possible locations (x, y) of the boat relative to the receiving stations.

(b) Assume that the shoreline is straight and that the boat was on a course parallel to and 1 mile from the shoreline. Because of the fog, the boat dropped anchor after having traveled east more than 1 mile past the station at $(0, 0)$, and then sent the signal as described in part **(a)**. Find the coordinates of the boat.

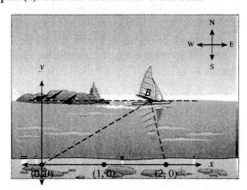

40. (a) Three search parties set out to locate a hunter lost in a wooded area. Two of the parties take positions so that they are 1 mile apart. The hunter fires his rifle and the sound reaches one party 3.6 seconds before it reaches the other. Using $(0, -0.5)$ and $(0, 0.5)$ for the positions of these two search parties, find the equation of the hyperbola whose points are the possible locations of the hunter. (Use 1100 ft/sec as the speed of sound.)

(b) Now assume that the third search party was located at point $(2, 0.5)$ when the rifle was fired, and the sound reached this point at the same time that it reached $(0, 0.5)$. Find the coordinates of the hunter if the sound reached points $(0, 0.5)$ and $(2, 0.5)$ before it reached $(0, -0.5)$.

 Challenge   Check the answer for Example 5.

 Written Assignment   The eccentricity of a hyperbola is defined by $e = \dfrac{c}{a}$ which is the same form as used for an ellipse. However, since $c > a$ for a hyperbola, it follows that $e > 1$. Describe the shape of a hyperbola when the eccentricity is close to 1, and also when the eccentricity is large. (Note: $e = \dfrac{c}{a} = \sqrt{a^2 + b^2}/a$ and consider the effect on b when e is close to 1, and when e is large.)

 Graphing Calculator Exercises   Graph each of the hyperbolas in Exercises 1 and 2 by solving for y in terms of x and then graphing each of the two functions. Graph their asymptotes on the same set of axes.

1. $x^2/9 - y^2/16 = 1$ **2.** $x^2 - 2x + 1 - y^2 - 4y = 20$

10.3 THE PARABOLA

A considerable amount of work has been done with parabolas in Chapter 2. In those earlier developments, vertical parabolas were regarded as the graphs of equations of the form $y = ax^2 + bx + c$ or, in standard form, $y = a(x - h)^2 + k$. (When the roles of the variables are interchanged, the graphs are horizontal parabolas.)

In our current work a parabola is going to be defined geometrically, as done for circles, ellipses, and hyperbolas. This definition will produce the preceding parabolic equations, so that the earlier work is consistent with the more fundamental approach here. As you will see, some new properties of parabolas will be encountered that were not discussed before.

> ### DEFINITION OF A PARABOLA
>
> A **parabola** is the set of all points in a plane equidistant from a given fixed line called the **directrix** and a given fixed point called the **focus**.

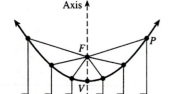

For each point P on the parabola at the left $PF = PQ$, where F is the focus and Q is the point on the directrix. The line through F and perpendicular to the directrix is called the **axis** (or **axis of symmetry**) of the parabola, and the point V, which is the intersection of the parabola with its axis, is called the **vertex**.

The parabolas we will consider will have either vertical or horizontal axes. We begin with parabolas whose axes are the y-axis.

In the figures that follow, let focus F have coordinates $(0, p)$, and let the directrix have equation $y = -p$ as indicated.

Observe that in each case the focus is within the parabola and the directrix is outside.

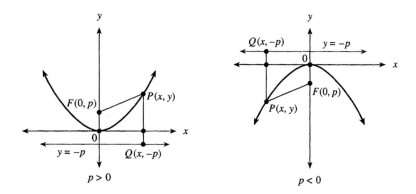

The origin, which is also the vertex, is on this parabola because it is the same distance from the focus and the directrix. In general, for any point P (x, y) on the parabola the distance formula can be used to find PF and PQ. Then set these two distances equal to each other as indicated in the definition.

$$PF = \sqrt{(x - 0)^2 + (y - p)^2} = \sqrt{x^2 + (y - p)^2}$$
$$PQ = \sqrt{(x - x)^2 + (y - (-p))^2} = \sqrt{(y + p)^2}$$
$$PF = PQ: \quad \sqrt{x^2 + (y - p)^2} = \sqrt{(y + p)^2} \qquad \text{Square each side.}$$
$$x^2 + y^2 - 2py + p^2 = y^2 + 2py + p^2$$
$$x^2 = 4py$$

In summary, we have the following:

This equation can also be written as $y = \dfrac{1}{4p}x^2$.

Letting $a = \dfrac{1}{4p}$ gives $y = ax^2$, the form we used in our earlier work with such parabolas.

> ### STANDARD FORM FOR THE EQUATION OF A PARABOLA WITH FOCUS $(0, p)$ AND DIRECTRIX $y = -p$
>
> $$x^2 = 4py$$
>
> The axis is the y-axis and the vertex is the origin.

This form for the equation of a parabola can be used to determine the coordinates of the focus and the equation of the directrix, as in Example 1.

EXAMPLE 1 Find the coordinates of the focus and the equation of the directrix for the parabola $x^2 = 4y$ and sketch the graph.

Solution Consider the general form $x^2 = 4py$, and let $4p = 4$. Thus $p = 1$ and we can locate the focus and directrix. The parabola has its focus at $(0, 1)$ and the equation of the directrix is $y = -p = -1$ as shown on the next page.

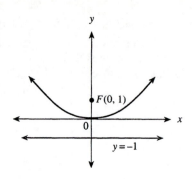

EXAMPLE 2 The focus of a parabola has coordinates $\left(0, -\frac{5}{2}\right)$ and the vertex is at the origin. Find the equation of the parabola.

Solution Both the focus and vertex are on the y-axis. Therefore, the y-axis is the axis of the parabola. Since such parabolas have focus at $(0, p)$, we get $p = -\frac{5}{2}$. Then, using $x^2 = 4py$, the equation is

$$x^2 = 4\left(-\tfrac{5}{2}\right)y = -10y$$

A parabola with vertex at the origin whose axis is the x-axis has an equation of the form $y^2 = 4px$. The derivation for this standard form is very similar to the derivation used to obtain the form $x^2 = 4py$.

STANDARD FORM FOR THE EQUATION OF A PARABOLA WITH FOCUS $(p, 0)$ AND DIRECTRIX $x = -p$

$$y^2 = 4px$$

The axis is the x-axis and the vertex is the origin.

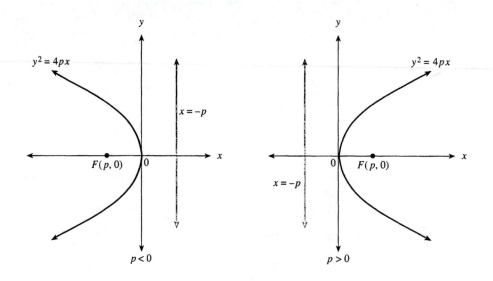

EXAMPLE 3 The directrix of a parabola is $x = -2$ and the focus is $(2, 0)$. Find the equation and graph.

Solution Since the directrix is a vertical line, the axis of the parabola is horizontal. Also, since the focus $(2, 0)$ is on the x-axis, the parabola's axis is the x-axis. Such parabolas have focus $(p, 0)$ and therefore $p = 2$. Thus, using $y^2 = 4px$, the equation is

$$y^2 = 4(2)x = 8x$$ ■

The parabola $x^2 = 4py$ has vertex $(0, 0)$ and vertical axis $x = 0$. When this parabola is translated h units horizontally and k units vertically, a congruent parabola is obtained having equation $(x - h)^2 = 4p(y - k)$. The vertex $(0, 0)$ has been translated to (h, k), and the focus, directrix, and axis have been translated as follows:

$$x^2 = 4py \longrightarrow (x - h)^2 = 4p(y - k)$$

vertex:	$(0, 0)$ $\longrightarrow$	(h, k)
focus:	$(0, p)$ $\longrightarrow$	$(h, k + p)$
directrix:	$y = -p$ $\longrightarrow$	$y = k - p$
axis:	$x = 0$ $\longrightarrow$	$x = h$

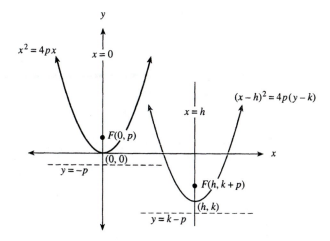

Note that the adjoining equation is based on the definition on page 656. It becomes $y - k = a(x - h)^2$, when $a = \dfrac{1}{4p}$, which is the same form mentioned in the introduction to this section.

STANDARD FORM FOR THE EQUATION OF A PARABOLA WITH FOCUS $(h, k + p)$ AND DIRECTRIX $y = k - p$

$$(x - h)^2 = 4p(y - k)$$

The axis is $x = h$ and the vertex is (h, k).

Similarly, when the parabola $y^2 = 4px$ is translated h units horizontally and k units vertically, we have

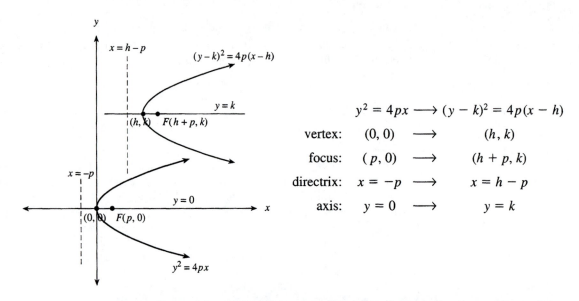

$$y^2 = 4px \longrightarrow (y - k)^2 = 4p(x - h)$$

vertex:	$(0, 0)$	$\longrightarrow$	(h, k)
focus:	$(p, 0)$	$\longrightarrow$	$(h + p, k)$
directrix:	$x = -p$	$\longrightarrow$	$x = h - p$
axis:	$y = 0$	$\longrightarrow$	$y = k$

Observe that in this and other algebraic forms of parabolas only one of the variables has a squared term and the other variable is to the first power.

STANDARD FORM FOR THE EQUATION OF A PARABOLA WITH FOCUS $(h + p, k)$ AND DIRECTRIX $x = h - p$

$$(y - k)^2 = 4p(x - h)$$

The axis is $y = k$ and the vertex is (h, k).

EXAMPLE 4 A parabola has vertex $(-2, 4)$ and focus $\left(-2, \frac{7}{2}\right)$. Write the equations of the parabola, the directrix, and the axis.

Solution Since the vertex is $(-2, 4)$, $h = -2$ and $k = 4$. Also, since the first coordinate of the vertex and focus is -2, the axis is the vertical line $x = h = -2$, or $x = -2$. The focus for such a parabola is $(h, k + p)$. Therefore,

$$k + p = 4 + p = \tfrac{7}{2}$$

$$p = -\tfrac{1}{2}$$

and the directrix is $y = k - p = 4 - \left(-\frac{1}{2}\right) = \frac{9}{2}$, or $y = \frac{9}{2}$. The equation of the parabola is

$$(x - (-2))^2 = 4\left(-\tfrac{1}{2}\right)(y - 4)$$

$$(x + 2)^2 = -2(y - 4)$$

The graph of this parabola is shown below.

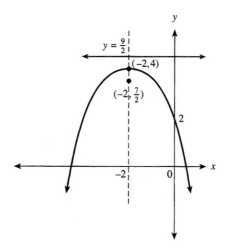

The last equation in Example 4 can be converted to $x^2 + 4x + 2y - 4 = 0$. Beginning with this form, we can get back to the standard form by completing the square in x as follows:

$$x^2 + 4x + 2y - 4 = 0$$
$$x^2 + 4x = -2y + 4$$
$$x^2 + 4x + 4 = -2y + 8$$
$$(x + 2)^2 = -2(y - 4)$$
$$(x + 2)^2 = 4\left(-\tfrac{1}{2}\right)(y - 4)$$

TEST YOUR UNDERSTANDING

Find the coordinates of the focus and vertex of the parabola. Also give the equation of the directrix and sketch the graph.

1. $\dfrac{1}{3}x^2 = y$ **2.** $-6(x + 2) = (y - 4)^2$

Write the equation of the parabola in standard form that has the given properties.

3. Directrix: $y = 3$;   Vertex: $(0, 0)$

4. Focus: $(-3, 2)$;   Vertex: $\left(-\dfrac{3}{2}, 2\right)$

5. Convert the equation of the parabola $x^2 - 6x - 8y + 1 = 0$ into standard form and identify the vertex, focus, and directrix.

(Answers: Page 682)

Parabolic reflectors are used in a wide variety of instruments, including reflecting telescopes, searchlights, microwave antennae, and solar energy devices. The surface of a parabolic reflector is obtained by revolving a parabola around its axis of symmetry.

Other forms of energy such as radio waves and sound waves have the same behavior as described here for light waves.

The following figure at the left illustrates that when light rays, parallel to the axis, strike the surface they are reflected to the focus. Conversely, the figure at the right shows that when there is a light source at the focus of a parabolic reflector the light rays will reflect off the surface, forming a beam of light parallel to the axis.

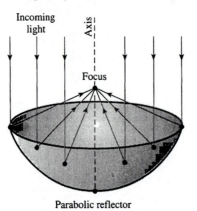

EXAMPLE 5 Suppose that the reflecting surface of a television antenna was formed by rotating the parabola $y = \frac{1}{15}x^2$ about its axis of symmetry for the interval $-5 \le x \le 5$. Assuming that the measurements are made in feet, how far from the bottom or vertex of this "dish" antenna should the receiver be placed? How deep is this antenna?

Solution The receiver must be at the focus and the distance the focus is from the vertex is the value p in the standard form $x^2 = 4py$. Since $y = \frac{1}{15}x^2$, $x^2 = 15y$. Then

$$4p = 15$$

$$p = \frac{15}{4} = 3.75$$

The receiver is 3.75 feet above the vertex along the axis.

The depth of the antenna is the y-value when $x = 5$. Thus, using 5 in $y = \frac{1}{15}x^2$, we get $y = \frac{25}{15} = \frac{5}{3}$. The anntenna is $1\frac{2}{3}$ feet deep.

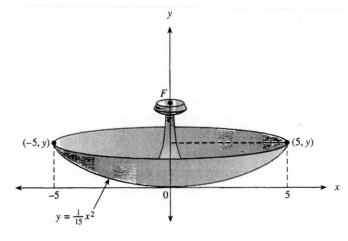

■

EXERCISES 10.3

Find the coordinates of the focus and the equation of the directrix for each of the following parabolas.

1. $x^2 = \frac{1}{4}y$

2. $y = \frac{1}{2}x^2$

3. $x^2 = \frac{1}{2}y$

4. $x^2 = -4y$

5. $y^2 = 2x$

6. $y^2 = -2x$

7. $y = -4x^2$

8. $y = -\frac{1}{8}x^2$

9. $-9(x - 2) = 6y^2$

10. $\frac{1}{2}x = \frac{1}{8}(y + 2)^2$

11. $(x - 4)^2 = -\frac{1}{3}(y + 5)$

12. $y - 1 = -\frac{3}{2}(x - 3)^2$

Write the equation of each parabola having the given properties and sketch the graph.

13. Focus $(0, -3)$; directrix $y = 3$.

14. Focus $(0, 3)$; directrix $y = -3$.

15. Directrix $y = -\frac{2}{3}$; vertex $(0, 0)$.

16. Vertex $(0, 0)$; vertical axis; $(2, -2)$ is on the parabola.

17. Focus $\left(-\frac{3}{4}, 2\right)$; directrix $x = \frac{3}{4}$.

18. Focus $(1, -3)$; vertex $(-1, -3)$.

19. Vertex $(2, 1)$; vertical axis; $(-1, 4)$ is on the parabola.

20. Directrix $x = \frac{13}{2}$; vertex $(5, 0)$.

Write an equation for the given parabola.

21.

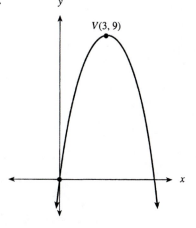

22.

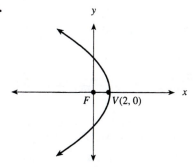

23.

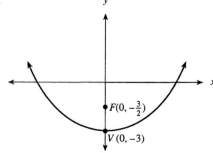

24.

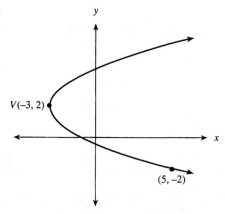

25. A parabola has vertex $(2, -5)$, focus at $(2, -3)$, and directrix $y = -7$. Write its equation in the form $(x - h)^2 = 4p(y - k)$. Then write its equation in the forms $y = a(x - h)^2 + k$ and $y = ax^2 + bx + c$.

26. A parabola has directrix $y = -1$ and focus $(3, 3)$. Write the equations of the parabola, and the axis of symmetry, and the coordinates of the vertex.

Identify the vertex, the axis of symmetry, the focus, and the directrix.

27. $(x - 2)^2 = 4(y + 5)$ 28. $\left(x + \frac{3}{2}\right)^2 = -\frac{1}{2}(y - 1)$

Write the equation of each parabola in the standard form $(x - h)^2 = 4p(y - k)$ and identify the vertex, focus, and directrix.

29. $y = \frac{1}{4}x^2 - x + 4$ 30. $y = 2x^2 - 12x + 16$ 31. $4x^2 - 4x - 28y - 83 = 0$

32. A parabola has vertex $(2, -5)$ and focus $(-1, -5)$. Write its equation in standard form. Then write its equation in the two forms $x = a(y - k)^2 + h$ and $x = ay^2 + by + c$.

33. A parabola has vertex $(-3, -2)$ and directrix $x = -\frac{9}{2}$. Find the focus and write the equations of the parabola and the axis of symmetry.

34. Find an equation of the parabola having a vertical axis of symmetry that contains the points $(0, 3)$, $(-2, 11)$, and $(3, 6)$. (*Hint:* Use $y = ax^2 + bx + c$.)

35. Find an equation of the parabola having a horizontal axis of symmetry that contains the points $(6, 1)$, $(-8, 2)$, and $(-18, -3)$. (*Hint:* Use $x = ay^2 + by + c$.)

Identify the vertex, axis of symmetry, focus, and directrix.

36. $(y + 5)^2 = \frac{4}{3}(x - 6)$ 37. $y^2 = -9(x - 4)$

38. Convert $y^2 + 6y - 2x + 9 = 0$ into the standard form $(y - k)^2 = 4p(x - h)$ and identify the vertex, axis of symmetry, focus, and directrix.

39. The reflecting surface of a radar antenna is generated by revolving the parabola $y = \frac{2}{9}x^2$ about its axis of symmetry for $-4 \le x \le 4$. Assuming the measurements are done in feet, how far from the bottom of the dish antenna is the receiver? What is the circumference of the antenna?

40. The reflecting surface of an antenna, as in Exercise 39, is generated by revolving a parabola of the form $x^2 = 4py$ about its axis of symmetry. If the antenna is 8 feet across the top (this is the length of a diameter) and $1\frac{1}{2}$ feet deep, where must the receiver be located?

41. The center cable of a suspension bridge forms a parabolic arc. The cable is suspended from the tops of the two support towers, which are 800 feet apart. The tops of the towers are 170 feet above the road and the lowest point of the cable is midway between the towers and ten feet above the road. Find the height of the cable above the road at a distance of 100 feet from a tower.

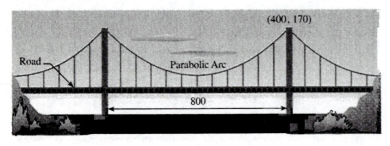

Name the conic section and, where applicable, give the coordinates of the center, vertices, and foci, the radius, and the equations of asymptotes and of the directrix.

42. $y = (x + 2)^2$ **43.** $\dfrac{x^2}{25} + \dfrac{y^2}{16} = 1$ **44.** $\dfrac{x^2}{16} + \dfrac{y^2}{25} = 1$

45. $\dfrac{x^2}{36} - \dfrac{y^2}{25} = 1$ **46.** $\dfrac{y^2}{36} - \dfrac{x^2}{25} = 1$ **47.** $x^2 + y^2 = 16$

Name the conic and graph.

48. $\dfrac{x^2}{16} + \dfrac{y^2}{9} = 1$ **49.** $16y^2 - 9x^2 = 144$ **50.** $y = 2(x + 1)^2 - 1$

51. $\dfrac{(x - 1)^2}{64} + \dfrac{(y - 2)^2}{36} = 1$ **52.** $\dfrac{(y - 1)^2}{64} - \dfrac{(x - 3)^2}{36} = 1$ **53.** $16(y - 3)^2 - 9(x + 2)^2 = -144$

54. $(x - 1)^2 + (y + 1)^2 = 25$

Identify each conic and write in standard form.

55. $x^2 + y^2 - 2x + 4y + 1 = 0$ **56.** $x^2 + y^2 + 6x - 4y + 4 = 0$

57. $x^2 + 4y^2 + 2x - 3 = 0$ **58.** $x^2 - 9y^2 - 2x - 8 = 0$

59. $9x^2 + 18x - 16y^2 + 96y = 279$ **60.** $4x^2 - 16x + y^2 + 8y = -28$

61. $y^2 + 10y = 6x - 1$ **62.** $y = 2x^2 - 4x + 5$

 Challenge The vertex of a parabola is the point on the parabola that is closest to the focus. Prove this result for the specific parabola $y^2 = 8x$. (*Hint:* Use the distance formula and compare the lengths of *VF* and *AF*, where *A* is any point on the parabola other than *V*.)

 Graphing Calculator Exercise Graph the horizontal parabola $x = y^2 - 4y + 1$. Solve for y by completing the square and taking square roots, and then graph the two functions of x on the same set of axes. Graph the axis of symmetry as well.

 Critical Thinking

1. Sketch the graph of the ellipse $\dfrac{x^2}{25} + \dfrac{y^2}{16} = 1$. Join the endpoints of the major axis to those of the minor axis. Find the equation of this new figure, said to be *analogous* to the ellipse.

2. Sketch the graph for ... of $|3x| + |2y| = 6$ where $|x| \le 2$ *and* $|y| \le 3$. Find the equation of the ellipse that is analogous to this figure.

3. The graphs of the following are said to be *conjugate hyperbolas:*

$$\frac{x^2}{25} - \frac{y^2}{16} = 1 \qquad \frac{y^2}{16} - \frac{x^2}{25} = 1$$

Graph both on the same set of axes and describe the relationship between the two graphs.

4. All of the conic sections that we have studied in this chapter may be written in the form $Ax^2 + Cy^2 + Dx + Ey + F = 0$, where A and C are not both equal to 0. What type of conic is represented if (a) $A = C \ne 0$; (b) $A = 0$ or $C = 0$; (c) $A \ne C$ and both are either positive or negative; (d) $A \ne C$ and are of opposite signs?

5. In the figure the point P is equidistant from the circle and the line ℓ. That is, $PA = PB$ where PA is perpendicular to line ℓ and B is the intersection of the line OP with the circle. Prove that the set of all such points P determines a parabola.

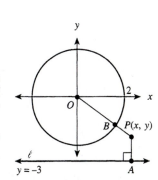

10.4 SOLVING NONLINEAR SYSTEMS

A straight line will intersect a parabola or a circle twice, or once, or not at all. Two parabolas of the form $y = ax^2 + bx + c$ can intersect at most two times; the same is true for two circles. A circle and a parabola can intersect at most four times. These diagrams illustrate some of the possibilities.

When you study calculus, you will learn how to find the areas of the regions between curves. For example, the areas of the shaded regions in the first two diagrams can be found once the coordinates of the points of intersection are known. Here we will address ourselves only to this part of the problem: finding the points of intersection.

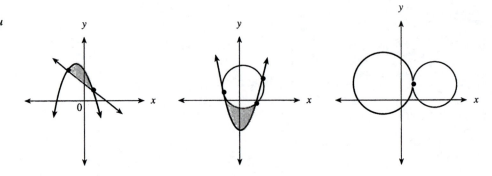

In each of the following examples, at least one of the two equations is not linear, and is referred to as a *nonlinear system*. A few such systems were solved earlier. (See page 205.) Now, with the availability of the conic sections, we are able to solve algebraically a wide variety of such systems. The underlying strategy will be the same as for linear systems; namely, first eliminate one of the two variables to obtain an equation in one unknown.

This example involves a parabola and a line.

EXAMPLE 1 Solve the system of two equations and graph:

$$y = x^2$$
$$y = -2x + 8$$

Solution Let (x, y) represent the points of intersection. Since these x- and y-values are the same in both equations, we may set the two values for y equal to each other and solve for x.

$$x^2 = -2x + 8$$
$$x^2 + 2x - 8 = 0$$
$$(x + 4)(x - 2) = 0$$
$$x = -4 \quad \text{or} \quad x = 2$$

To find the corresponding y-values, either of the original equations may be used. Using $y = -2x + 8$, we have:

For $x = -4$: $y = -2(-4) + 8 = 16$

For $x = 2$: $y = -2(2) + 8 = 4$

The solution of the system consists of the two ordered pairs $(-4, 16)$ and $(2, 4)$. The other equation can be used as a check of these results.

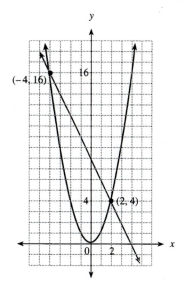

■

This example involves two parabolas.

EXAMPLE 2   Solve the system and graph:

$$y = x^2 - 2$$
$$y = -2x^2 + 6x + 7$$

Solution Set the two values for y equal to each other and solve for x.

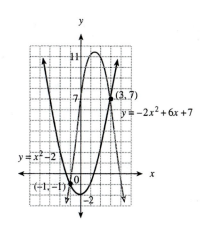

$$x^2 - 2 = -2x^2 + 6x + 7$$
$$3x^2 - 6x - 9 = 0$$
$$x^2 - 2x - 3 = 0$$
$$(x + 1)(x - 3) = 0$$
$$x = -1 \quad \text{or} \quad x = 3$$

Use $y = x^2 - 2$ to solve for y.

$$y = (-1)^2 - 2 = -1 \qquad y = 3^2 - 2 = 7$$

Check these points in the second equation of the given system.

The points of intersection are $(-1, -1)$ and $(3, 7)$.

■

This example involves an ellipse and a parabola.

EXAMPLE 3   Solve the system and graph:

$$4x^2 + 9(y - 3)^2 = 36$$

$$y - x^2 = 1$$

Solution   Solve the second equation for x^2.

$$x^2 = y - 1$$

Substitute into the first equation and solve for y.

Another way to solve this system begins by solving $y - x^2 = 1$ for y and substituting into the first equation. Try this.

$$4(y - 1) + 9(y - 3)^2 = 36$$

$$4y - 4 + 9y^2 - 54y + 81 = 36$$

$$9y^2 - 50y + 41 = 0$$

$$(y - 1)(9y - 41) = 0$$

$$y = 1 \quad \text{or} \quad y = \frac{41}{9}$$

Use $x^2 = y - 1$ to solve for x.

$$\text{For } y = 1: \quad x^2 = 1 - 1 = 0; \ x = 0$$

$$\text{For } y = \frac{41}{9}: \ x^2 = \frac{41}{9} - 1 = \frac{32}{9}; \ x = \pm\frac{4\sqrt{2}}{3}$$

Check these points in the given system.

The points of intersection are $(0, 1)$, $\left(\dfrac{4\sqrt{2}}{3}, \dfrac{41}{9}\right)$, and $\left(-\dfrac{4\sqrt{2}}{3}, \dfrac{41}{9}\right)$.

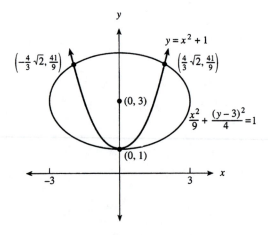

This example involves a circle and an ellipse.

EXAMPLE 4   Solve the system and graph.

$$x^2 + y^2 = 9$$

$$\frac{x^2}{25} + \frac{y^2}{16} = 1$$

Solution You should recognize the equations as those of a circle and an ellipse. First rewrite the second equation in this form:

$$16x^2 + 25y^2 = 400$$

Then solve the first equation for either x^2 or y^2, say y^2, and substitute into the second equation:

$$y^2 = 9 - x^2$$
$$16x^2 + 25(9 - x^2) = 400$$
$$16x^2 + 225 - 25x^2 = 400$$
$$-9x^2 = 175$$
$$x^2 = -\frac{175}{9}$$
$$x = \pm \sqrt{-\frac{175}{9}}$$

The attempted solution produces the square root of a negative number, which is imaginary. Thus there are no real solutions and the two curves do not intersect.

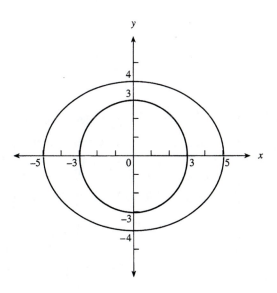

This example involves an ellipse and a hyperbola.

EXAMPLE 5 Solve the system and graph.

$$x^2 - y^2 = 1$$
$$9x^2 + y^2 = 9$$

Solution Add the two equations:

$$x^2 - y^2 = 1$$
$$\underline{9x^2 + y^2 = 9}$$
$$10x^2 \qquad = 10$$
$$x^2 = 1$$
$$x = \pm 1 \quad \text{and} \quad y = 0$$

The points of intersection are located at $(\pm 1, 0)$.

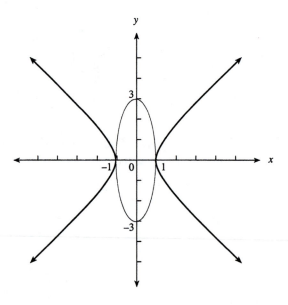

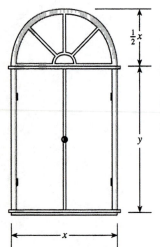

EXAMPLE 6 An architect is designing an arched window consisting of a rectangle surmounted by a semicircle. The design also requires that the total window area be 30 ft², and that the perimeter of the rectangle be 20 ft. Find the dimensions.

Solution Let the base of the rectangle be x feet and the height be y feet. Then the radius of the semicircle is $\dfrac{x}{2}$ feet and the total area is

$$\begin{pmatrix} \text{area of} \\ \text{rectangle} \end{pmatrix} \quad + \quad \begin{pmatrix} \text{area of} \\ \text{semicircle} \end{pmatrix} \quad = \quad \begin{pmatrix} \text{total} \\ \text{area} \end{pmatrix}$$

$$xy \quad + \quad \tfrac{1}{2}\pi\left(\frac{x}{2}\right)^2 \quad = \quad 30$$

Also, the perimeter is

$$2x + 2y \quad = \quad 20$$

Now solve this system:

$$xy + \frac{\pi}{8}x^2 = 30$$

$$x + y = 10$$

Substitute $y = 10 - x$ into the first equation.

$$x(10 - x) + \frac{\pi}{8}x^2 = 30$$

$$10x - x^2 + \frac{\pi}{8}x^2 = 30$$

$$x^2 - \frac{\pi}{8}x^2 - 10x + 30 = 0$$

$$\left(1 - \frac{\pi}{8}\right)x^2 - 10x + 30 = 0$$

Using the quadratic formula, we have

$$x = \frac{10 \pm \sqrt{10^2 - 4(1 - \frac{\pi}{8})30}}{2(1 - \frac{\pi}{8})}$$

$$\approx 12.5 \text{ or } 3.9$$

Since $y = 10 - 12.5 < 0$, we use $x = 3.9$ and $y = 10 - 3.9 = 6.1$. The dimensions of the rectangle are approximately 3.9 ft by 6.1 ft. ■

The graphs of systems of linear inequalities were studied in the preceding chapter. The procedures used there also apply here to the graphing of systems of nonlinear inequalities.

EXAMPLE 7   Graph the system of inequalities.

$$y \geq x^2$$

$$y \leq -2x + 8$$

Solution First draw the graphs of the equations $y = x^2$ and $y = -2x + 8$ (see Example 1). Now use test points from the numbered regions determined by the two graphs. If a test point satisfies both inequalities, then the region it comes from is part of the graph of the system. Otherwise, the region is not part of the graph. Using $(0, 4)$ from region I gives true statements:

$$4 \geq 0^2 = 0$$

$$4 \leq -2(0) + 8 = 8$$

Therefore, region I is included. Using $(3, 4)$ from region V gives

$$4 \geq 3^2 = 9$$

$$4 \leq -2(3) + 8 = 2$$

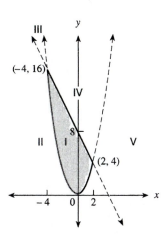

Another way to obtain this graph is to observe that $y \geq x^2$ calls for all points on and above the parabola $y = x^2$, and $y \leq -2x + 8$ calls for all points on or below the line $y = -2x + 8$.

Since $(3, 4)$ does not satisfy *both* inequalities, region V is excluded. You should verify that regions II, III, and IV are also excluded. The graph of the system is the shaded part as shown. Note that the boundaries of region I are included since the given system uses the inequality symbols $\leq$ and $\geq$. ∎

EXERCISES 10.4

Solve each system and graph.

1. $y = -x^2 - 4x + 1$
$y = 2x + 10$

2. $3x - 4y = -5$
$(x + 3)^2 + (y + 1)^2 = 25$

3. $(x + 4)^2 + (y - 1)^2 = 16$
$(x + 4)^2 + (y - 3)^2 = 4$

4. $y = x^2 - 6x + 9$
$(x - 3)^2 + (y - 9)^2 = 9$
(*Hint:* Factor $x^2 - 6x + 9$ and substitute.)

5. $y = x^2 + 6x + 6$
$y = -x^2 - 6x + 6$

6. $y = \frac{1}{3}(x - 3)^2 - 3$
$(x - 3)^2 + (y + 2)^2 = 1$
(*Hint:* Solve the first equation for $(x - 3)^2$ and substitute into the second.)

Solve each system.

7. $y = (x + 1)^2$
$y = (x - 1)^2$

8. $x^2 + y^2 = 9$
$y = x^2 - 3$

9. $y = x^2$
$y = -x^2 + 8x - 16$

10. $y = x^2$
$y = x^2 - 8x + 24$

11. $7x + 3y = 42$
$y = -3x^2 - 12x - 15$

12. $\dfrac{x^2}{16} + \dfrac{y^2}{4} = 1$
$y = x^2 - 16$

13. $y - x = 0$
$(x - 2)^2 + (y + 5)^2 = 25$

14. $y - 2x = 0$
$(x - 2)^2 + (y + 5)^2 = 25$

15. $x - 2y^2 = 0$
$x^2 - y^2 = 3$

16. $x^2 + y^2 = 25$
$2x^2 + y^2 = 34$

17. $4x^2 + y^2 = 4$
$2x - y = 2$

18. $4x^2 + y^2 = 4$
$x + y = 3$

19. $4x^2 - 9y^2 = 36$
$9x^2 + 4y^2 = 36$

20. $2x^2 - y^2 = 1$
$y^2 - x^2 = 3$

21. $2x^2 + y^2 = 11$
$x^2 - 2y^2 = -2$

22. $x^2 - 2y^2 = 8$
$3x + 4y = 4$

23. $y = -x^2 + 3$
$x^2 + \dfrac{(y - 2)^2}{9} = 1$

24. $x^2 + 4x + y^2 - 4y = -4$
$(x - 2)^2 + (y - 2)^2 = 4$

25. $(x - 1)^2 + y^2 = 1$
$x^2 + (y - 1)^2 = 1$

26. $\dfrac{y^2}{4} - (x - 2)^2 = 1$
$y = x^2 - 4x + 2$

27. $y = \frac{1}{3}(x - 3)^2 - 3$
$x^2 - 6x + y^2 + 2y = -6$

28. $x^2 - y^2 - 2y = 10$
$x^2 - 4(y + 1)^2 = 6$

Graph each system.

29. $y \geq -x^2 - 4x + 1$
$y \leq 2x + 10$
(See Exercise 1.)

30. $3x - 4y \geq -5$
$(x + 3)^2 + (y + 1)^2 \leq 25$
(See Exercise 2.)

31. $(x + 4)^2 + (y - 1)^2 \leq 16$
$(x + 4)^2 + (y - 3)^2 \geq 4$
(See Exercise 3.)

32. $y \geq x^2 - 6x + 9$
$(x - 3)^2 + (y - 9)^2 \geq 9$
(See Exercise 4.)

33. $y \geq x^2 + 6x + 6$
$y \leq -x^2 - 6x + 6$
(See Exercise 5.)

34. $y \geq \frac{1}{3}(x - 3)^2 - 3$
$(x - 3)^2 + (y + 2)^2 \geq 1$
(See Exercise 6.)

35. $9x^2 + 25y^2 \leq 225$
$x^2 + 4y^2 \geq 16$

36. $\dfrac{x^2}{16} + \dfrac{y^2}{9} \leq 1$
$\dfrac{x^2}{9} + \dfrac{y^2}{16} \leq 1$

37. $\dfrac{x^2}{16} + \dfrac{y^2}{9} \leq 1$
$\dfrac{x^2}{9} + \dfrac{y^2}{16} \geq 1$

38. $y^2 - x^2 \geq 1$
$\dfrac{x^2}{4} + \dfrac{y^2}{9} \leq 1$

39. $x^2 - y^2 \leq 1$
$\dfrac{(x - 3)^2}{4} + \dfrac{y^2}{16} \leq 1$

40. $x^2 - 4y^2 \geq 1$
$x^2 - y^2 \leq 9$

41. An architect is designing a cathedral window that consists of a rectangle surmounted by an isosceles triangle with altitude equal to one-third the height of the rectangle. The design also calls for the total window area to be 20 square feet, the perimeter of the rectangle to be 17 feet, and the height of the rectangle to be larger than the base. Find the dimensions.

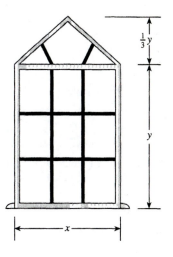

42. A design for a quilt is made up of figures consisting of two intersecting, congruent ellipses as shown. Find the four intersection points.

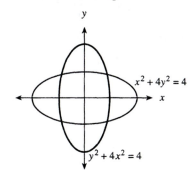

43. The pattern of the quilt is shown below. Each individual figure is a copy of the figure in Exercise 42 and uses three colors: one for the area common to both ellipses; another for the area inside the horizontal ellipse, but outside the other; and the third for the area inside the vertical ellipse, but outside the other. Write a system of inequalities for each of the three areas described.

Challenge   Solve the system and graph:

$$y = 2x - 5$$

$$y = \frac{-1}{x^2 - 2x + 1}$$

Written Assignment   Consider the following system of equations. Let r take on all values in the interval $(0, \infty)$ and describe how the various values of r affect the number of solutions for the resulting system.

$$x^2 + y^2 = r^2$$

$$\frac{x^2}{49} + \frac{y^2}{9} = 1$$

Graphing Calculator Exercises

1–4. Check each of Exercises 1–4 by finding the points of intersection with your graphing calculator. To ensure that you have found the point of intersection in each case, toggle between the curves. At a common point, the y-value should not change.

5–8. If your graphing calculator can graph inequality graphs (see your user's manual), use it to check Exercises 29–33.

CHAPTER REVIEW EXERCISES

Section 10.1   The Ellipse   (Page 634)

1. State the definition of an ellipse.

Graph the ellipse and find the coordinates of the center, vertices, and foci.

2. $\dfrac{x^2}{9} + y^2 = 1$
  3. $\dfrac{x^2}{9} + \dfrac{y^2}{49} = 1$
  4. $25x^2 + 16y^2 = 400$

5. $(x + 3)^2 + \dfrac{(y - 4)^2}{4} = 1$
  6. $\dfrac{(x - \frac{1}{2})^2}{16} + \dfrac{(y - \frac{5}{2})^2}{4} = 1$
  7. $\dfrac{(x + 5)^2}{16} + \dfrac{(y + 10)^2}{64} = 1$

Write the equation of the ellipse in standard form and find the coordinates of the center, vertices and foci.

8. $x^2 + 10x + 36y^2 = 11$
  9. $4x^2 + 24x + 9y^2 - 18y = -9$

10. $4x^2 + 5y^2 - 48x + 10y + 129 = 0$
  11. $7y^2 + 10x^2 - 100x - 42y = -243$

Write the equation of the ellipse in standard form having the given properties.

12. Center $(0, 0)$; vertex $(3, 0)$; minor axis of length 2.
  13. Vertices $(\pm 7, 0)$; foci $(\pm 6, 0)$.

14. Center $(0, 0)$; foci $(0, \pm\sqrt{5})$; major axis of length 6.
  15. Center $(4, 6)$; Vertex $(9, 6)$; $(0, 8)$ is on the ellipse.

16. Endpoints of major and minor axes are $(-5, 7)$, $(-5, -11)$ and $(-2, -2)$, $(-8, -2)$.

17. Foci $(0, \pm 6)$; eccentricity $\frac{3}{4}$.
  18. Center $(-1, 4)$; vertex $(-1, -2)$; eccentricity $\frac{1}{3}$.

19. If the distance from a point $P(x, y)$ to point $(-3, 1)$ is added to the distance from P to point $(1, 1)$ the sum is 6. Find the equation for all such points (x, y).

20. Find the eccentricity of the ellipse $8(x - 3)^2 + 24(y + 7)^2 = 192$.

21. A satellite travels around the earth in an elliptical orbit with one focus at the earth's center. The maximum and minimum distances that the satellite is from the surface of the earth are 2000 miles and 400 miles. Assume that the radius of the earth is 4000 miles, and use a rectangular system such that the origin is the center of the ellipse with major axis along the y-axis to find the equation of the orbit.

22. The shape of the base of a cylindrical tank is an ellipse whose major axis is 8 feet long. A chord perpendicular to the major axis and 1 foot from a vertex measures 2.60 feet. Find an equation of the ellipse using b rounded to two decimal places.

Section 10.2 The Hyperbola (Page 646)

23. State the definition of a hyperbola.

Graph the hyperbola. Find the coordinates of the center, foci, and vertices, and write the equations of the asymptotes.

24. $x^2 - y^2 = 1$

25. $\dfrac{x^2}{25} - \dfrac{y^2}{49} = 1$

26. $\dfrac{y^2}{4} - x^2 = 1$

27. $4y^2 - x^2 = 4$

28. $9x^2 - 4y^2 = 36$

29. $5y^2 - 3x^2 = 15$

30. $\dfrac{(x - 3)^2}{4} - (y + 4)^2 = 1$

31. $\dfrac{(y + 5)^2}{9} - \dfrac{(x + 2)^2}{16} = 1$

32. $\dfrac{(x - 6)^2}{25} - \dfrac{(y - 4)^2}{9} = -1$

33. $49(x + 1)^2 - 4(y - 5)^2 = 196$

Write the equation of the hyperbola in standard form and find the coordinates of the center, vertices, and foci. Also find the equations of the asymptotes.

34. $x^2 - 16y^2 + 96y = 160$

35. $9y^2 - 36y - 4x^2 - 8x = 4$

36. $x^2 - y^2 + 8x - 12y = 23$

37. $5y^2 - 20y - 4x^2 + 12x - 9 = 0$

Write the equation of the hyperbola in standard form having the given properties.

38. Center $(0, 0)$; vertices $(\pm 3, 0)$; foci $(\pm 5, 0)$.

39. Center $(0, 0)$; vertices $\left(0, \pm \sqrt{10}\right)$; foci $(0, \pm 5)$.

40. Center $(2, -1)$; horizontal transverse axis of length 6; conjugate axis of length 4.

41. Center $(-3, 2)$; vertical transverse axis of length 8; $c = \sqrt{20}$.

42. Asymptotes $y = \pm \frac{4}{3}x$; foci $(\pm 5, 0)$.

43. Asymptotes $y + 2 = \pm \frac{2}{3}(x - 4)$; focus $(4, -2 + \sqrt{13})$.

44. Two stations that are 2 miles apart receive a sound signal issued from a source located at point $P(x, y)$. The station located at point $(0, -1)$ gets the signal 4 seconds earlier than the station located $(0, 1)$. Use 1100 ft/sec as the speed of sound and find an equation of the hyperbola containing point P.

45. Suppose that in Exercise 44 there is a third station located at point $(3, -1)$ that received the sound signal at the same time as the station located at $(0, -1)$. Find the coordinates of P.

Section 10.3   The Parabola (Page 656)

46. State the definition of a parabola.

Find the coordinates of the focus and vertex of the parabola, and write the equation of the directrix.

47. $x^2 = 6y$ 　　　 **48.** $y = \frac{3}{4}x^2$ 　　　 **49.** $x^2 = -2y$

50. $x = -\frac{1}{6}y^2$ 　　 **51.** $y - 3 = 2x^2$ 　　 **52.** $(y - 6)^2 = -\frac{2}{3}(x + 5)$

Write the equation of the parabola in standard form and identify the vertex, focus, and directrix.

53. $y = 12x^2 - 24x + 16$ 　　　 **54.** $x = -6y^2 - 24y - \frac{47}{2}$

55. $6y^2 + 36y + 5x + 29 = 0$ 　　　 **56.** $3x^2 - 12x - 7y + 33 = 0$

Write the equation of the parabola in standard from having the given properties and sketch the graph.

57. Focus $(0, 7)$; directrix $y = 9$.

58. Focus $\left(-\frac{5}{2}, 0\right)$; directrix $x = -4$.

59. Directrix $x = \frac{3}{4}$; vertex $(0, 0)$.

60. Vertex $(0, 0)$; focus $(0, -4)$.

61. Focus $(4, -2)$; directrix $x = 6$.

62. Vertex $(-3, -4)$; directrix $x = -\frac{25}{8}$.

63. Vertex $(6, -1)$; focus $(6, 2)$.

64. Vertex $(3, 6)$; vertical axis; $(0, 0)$ is on the parabola.

65. Vertex $(-2, 6)$; horizontal axis; $(-10, 2)$ is on the parabola.

66. A parabola has a vertical axis and passes through the points $(0, 3)$, $(1, 7)$, and $(-2, -17)$. Find an equation of the parabola. (Use $y = ax^2 + bx + c$.)

67. A parabola has a horizontal axis and passes through points $(-1, 1)$, $(35, 3)$, and $(20, -2)$. Find an equation of the parabola. (Use $x = ay^2 + by + c$.)

68. Find equations for the two parabolas, one with a vertical axis and the other with a horizontal axis, each of which contains the points $(3, 0)$, $(0, 9)$, and $(5, 4)$.

69. The reflecting surface of a parabolic antenna is generated by revolving the parabola $y = \frac{1}{14}x^2$ around its axis for $-5 \le x \le 5$. Assume the measurements are in feet and find how far the receiver is from the bottom of the antenna. What is the depth of the (dish) antenna?

70. The reflecting surface of a parabolic searchlight is generated by revolving a parabola of the form $x^2 = 4py$ around its axis. If the top of the searchlight has an 8-foot diameter and if the center of the top is 3 feet above the vertex, then where is the light source located?

Name the conic and give the following information where applicable: (i) coordinates of the center; (ii) coordinates of the vertex or vertices; (iii) coordinates of the focus or foci; (iv) equations of the asymptotes; and (v) equation of the directrix.

71. $\dfrac{x^2}{9} + \dfrac{y^2}{25} = 1$

72. $4x^2 - 9y^2 = 36$

73. $y^2 = 6(x + 2)$

74. $(x - 4)^2 = \frac{3}{2}(y - 1)$

75. $\dfrac{y^2}{16} - \dfrac{(x + 3)^2}{9} = 1$

76. $25(x - 1)^2 + 4(y + 5)^2 = 100$

Convert to standard form and identify the conic.

77. $y = \frac{1}{8}x^2 + 3$

78. $x^2 + 8x - 4y^2 - 8y = -8$

79. $9x^2 - 36x + 4y^2 - 24y + 36 = 0$

80. $y = 3x^2 - 6x + 4$

Section 10.4 Solving Nonlinear Systems (Page 666)

Solve the system and graph.

81. $y = x^2 - 2x - 9$
$\quad\ y = 2x + 3$

82. $y = x^2 + 2x - 3$
$\quad\ y = -2x^2 + 2x + 9$

83. $\dfrac{(x - 2)^2}{4} + y^2 = 1$

$\quad\ \dfrac{x^2}{16} + \dfrac{y^2}{4} = 1$

84. $\dfrac{x^2}{4} + y^2 = 1$

$\quad\ y - x^2 = -1$

85. $x^2 - y^2 = 1$

$\quad\ x^2 + y^2 = 7$

86. $\dfrac{(x - 3)^2}{4} - y^2 = 1$

$\quad\ \dfrac{(x - 3)^2}{16} + \dfrac{y^2}{2} = 1$

Solve the system.

87. $y = -x^2 + 8x - 10$
$\quad\ y + 2x = 6$

88. $(x + 3)^2 + (y - 6)^2 = 25$
$\quad\ x - 3y = -26$

89. $\dfrac{x^2}{5} - \dfrac{y^2}{9} = 1$
$\quad\ 5y - 6x = 0$

90. $y = -x^2$
$\quad\ y = -x^2 + 6x - 12$

91. $(x - 4)^2 + (y - 6)^2 = 4$
$\quad\ y = -\frac{1}{2}(x - 4)^2 + 8$

92. $(x + 5)^2 + (y - 2)^2 = 16$
$\quad\ (x + 6)^2 + (y - 2)^2 = 25$

93. $3x^2 + y^2 = 12$
$\quad\ x^2 - 3y^2 = 4$

94. $y^2 - x^2 = 4$
$\quad\ x^2 + 9y^2 = 9$

95. $x^2 + 4y^2 = 4$
$\quad\ 4x^2 + y^2 = 4$

96. $x^2 + y^2 = 64$
$\quad\ 5x^2 + y^2 = 100$

97. $y = x^2 - 4x + 4$
$\quad\ (x - 2)^2 + (y - 4)^2 = 16$

98. $x^2 + y^2 = 16$
$\quad\ y = 4 - x^2$

99. $x^2 - 12x + y^2 + 4y = -15$
$\quad\ x + 7y = 17$

100. The adjoining figure represents a pattern consisting of one large square and two congruent smaller squares. If the total area is 67 square inches and the outer perimeter (this excludes the two dashed sides) is 40 inches, then what are the dimensions of the squares? (*Note:* There are two possible solutions.)

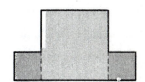

101. $y \geq x^2 - 2x - 9$

$y \leq 2x + 3$

(See Exercise 81.)

102. $y \geq x^2 + 2x - 3$

$y \leq -2x^2 + 2x + 9$

(See Exercise 82.)

103. $\dfrac{(x-2)^2}{4} + y^2 \geq 1$

$\dfrac{x^2}{16} + \dfrac{y^2}{4} \leq 1$

(See Exercise 83.)

104. $\dfrac{x^2}{4} + y^2 \leq 1$

$y - x^2 \leq -1$

(See Exercise 84.)

105. $\dfrac{x^2}{25} + \dfrac{y^2}{9} \leq 1$

$x^2 + y^2 \geq 9$

106. $\dfrac{x^2}{4} + \dfrac{y^2}{16} \leq 1$

$y^2 \geq x^2 + 1$

CHAPTER 10: STANDARD ANSWER TEST

Use the questions to test your knowledge of the basic skills and concepts of Chapter 10. Then check your answers with those given at the back of the book.

1. Find the coordinates of the center, vertices, and foci of the ellipse $\dfrac{(x-3)^2}{16} + \dfrac{y^2}{9} = 1$, and sketch the graph.

2. Find the coordinates of the center, vertices, and foci of the hyperbola

$$\frac{(y-2)^2}{4} - \frac{(x-4)^2}{9} = 1$$

   Also write the equations of the asymptotes and sketch the graph of the hyperbola.

3. Find the coordinates of the vertex and focus of the parabola $(y-5)^2 = -8(x-1)$. Also write the equation of the directrix and sketch the graph of the parabola.

Write the equation of the ellipse in standard form having the given properties.

4. Center: $(0, 0)$, horizontal major axis of length 8, and minor axis of length 6.

5. Vertices: $(-1, -1)$ and $(-1, 5)$; $c = \sqrt{5}$.

6. Center: $(-2, 4)$; Vertex: $(4, 4)$; eccentricity $= \frac{2}{3}$.

Write the equation of the hyperbola in standard form having the given properties.

7. Center: $(0, 0)$; foci: $(\pm 8, 0)$; vertices: $(\pm 6, 0)$.

8. Center: $(4, -5)$; focus: $(4, 0)$; transverse axis of length 8.

9. Horizontal transverse axis; center: $(0, 1)$; asymptotes: $y - 1 = \pm\frac{1}{3}x$; $c = \sqrt{40}$.

Write the equation of the parabola in standard form having the given properties.

10. Directrix: $y = \frac{2}{3}$; vertex: $(0, 0)$.

11. Vertex: $(-5, -3)$; focus: $(-2, -3)$.

12. Vertex: $(2, -1)$; vertical axis of symmetry; point $(5, 8)$ is on the parabola.

Name the conic and give the following information where applicable: (i) coordinates of the center; (ii) coordinates of the vertex or vertices; (iii) coordinates of the focus or foci; (iv) equations of the asymptotes; (v) equation of the directrix.

13. $y^2 - 9x^2 = 9$

14. $9(x+4)^2 + 16(y-1)^2 = 144$

15. $(x+5)^2 = -12(y-5)$

Convert to standard form and identify the conic.

16. $4x^2 - 16x - 9y^2 = 20$

17. $y = 2x^2 - 4x + 9$

18. $9x^2 - 18x + 4y^2 + 16y = 11$

19. $4x^2 + 16x - y^2 + 6y = -3$

20. Solve the system and graph: $9x^2 + 4y^2 = 36$
$$y^2 = x + 2$$

21. Solve the system: $2x^2 - 3y^2 = 15$
$$3x^2 + 2y^2 = 29$$

22. Graph the system: $\dfrac{x^2}{9} + \dfrac{y^2}{25} \le 1$
$$x^2 \ge y^2 + 4$$

(The coordinates of the intersection points are not required.)

23. The underside of a bridge over a roadway is in the shape of a semiellipse. The elliptical arch spans 50 feet (the length of the major axis) and the height at the center is 20 feet. Find the bridge clearance 10 feet from one of the bases of the bridge.

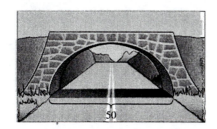

24. A satellite has an elliptical orbit around the earth with one focus at the earth's center, E. As indicated, the earth's radius is 4000 miles, the highest point that the satellite is from the surface of the earth is 800 miles and the lowest is 200 miles. Find the eccentricity of the satellite's orbit.

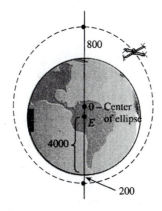

25. A vehicle moving along a path is sending sound signals, traveling 1100 ft/sec, that reach two receivers positioned so that they are 3 miles apart. Assume that the receivers are located on the x-axis at points $A(-1.5, 0)$ and $B(1.5, 0)$, and the coordinates of the vehicle are (x, y). If the signals reach A three seconds before they reach B, find an equation of the path of the vehicle.

CHAPTER 10: MULTIPLE CHOICE TEST

1. Which of the following is the equation of the ellipse with foci $(0, \pm 3)$ and major axis of length 10?

   (a) $\dfrac{x^2}{9} + \dfrac{y^2}{100} = 1$ (b) $\dfrac{x^2}{100} + \dfrac{y^2}{9} = 1$ (c) $\dfrac{x^2}{16} + \dfrac{y^2}{25} = 1$ (d) $\dfrac{x^2}{25} + \dfrac{y^2}{9} = 1$ (e) None of the preceding

2. The graph of $\dfrac{x^2}{10} - \dfrac{y^2}{15} = 1$ is

   (a) A hyperbola with asymptotes $y = \pm \frac{3}{2} x$

   (b) A hyperbola with foci $(\pm 5, 0)$

   (c) An ellipse with a vertical major axis

   (d) An ellipse with vertices $(0, \pm 5)$

   (e) None of the preceding

3. The focus and directrix of the parabola $-2y^2 = x$ are

   (a) $\left(-\frac{1}{8}, 0\right); x = \frac{1}{8}$ (b) $\left(0, \frac{1}{8}\right); y = -\frac{1}{8}$ (c) $(-2, 0); x = 2$ (d) $(0, 0); x = -2$

   (e) None of the preceding

4. The foci of the conic $\dfrac{x^2}{36} + \dfrac{y^2}{4} = 1$ are

   (a) $(\pm \sqrt{40}, 0)$ (b) $(0, \pm \sqrt{40})$ (c) $(0, \pm \sqrt{32})$ (d) $(\pm \sqrt{32}, 0)$ (e) None of the preceding

5. The asymptotes for the conic $25y^2 - 4x^2 = 100$ are

   (a) $y = \pm \frac{5}{2} x$ (b) $y = \pm \frac{2}{5} x$ (c) $y = \pm \frac{4}{25} x$ (d) $y = \pm \frac{25}{4} x$ (e) None of the preceding

6. For the ellipse $\dfrac{(x-6)^2}{9} + \dfrac{(y+7)^2}{16} = 1$, find the eccentricity e.

   (a) $\frac{3}{5}$ (b) $\frac{5}{4}$ (c) $\frac{25}{16}$ (d) $\frac{5}{3}$ (e) None of the preceding

7. Which of the following is the equation of the parabola with focus $(0, 0)$, directrix $x = 2$, and horizontal axis of symmetry?

   (a) $y^2 = -4(x-1)$ (b) $y^2 = 8(x-1)$ (c) $(x-1)^2 = 4y$ (d) $(x-1)^2 = -8y$

   (e) None of the preceding

8. Which of the following is the equation of an ellipse with center $(-3, 2)$, minor axis of length 4, and foci at $(-4, 2)$ and $(-2, 2)$?

   (a) $\dfrac{(x-3)^2}{5} + \dfrac{(y+2)^2}{4} = 1$ (b) $\dfrac{(x+3)^2}{4} + \dfrac{(y-2)^2}{5} = 1$

   (c) $\dfrac{(x+3)^2}{5} + \dfrac{(y-2)^2}{4} = 1$ (d) $\dfrac{(x-3)^2}{4} + \dfrac{(y+2)^2}{5} = 1$

   (e) None of the preceding

9. Which of the following is the equation of a hyperbola with center $(-2, 1)$, vertices at $(1, 1)$ and $(-5, 1)$, and foci located $\sqrt{13}$ units from the center?

   (a) $\dfrac{(x+2)^2}{4} - \dfrac{(y-1)^2}{9} = 1$ (b) $\dfrac{(x-2)^2}{9} - \dfrac{(y+1)^2}{4} = 1$

(c) $\dfrac{(x-2)^2}{4} + \dfrac{(y+1)^2}{9} = 1$ **(d)** $\dfrac{(x+2)^2}{9} - \dfrac{(y-1)^2}{4} = 1$

(e) None of the preceding

10. The vertices of the conic $\dfrac{(y-2)^2}{7} - \dfrac{(x+4)^2}{5} = 1$ are

(a) $(-4, 9)$ and $(-4, -5)$ **(b)** $(-4, 2 \pm \sqrt{7})$ **(c)** $(-4 \pm \sqrt{7}, 2)$ **(d)** $(-4 \pm \sqrt{5}, 2)$

(e) None of the preceding

11. A parabola has focus $(2, -2)$ and directrix $y = -5$. The vertex of this parabola is

(a) $\left(\frac{7}{2}, -2\right)$ **(b)** $\left(2, -\frac{7}{2}\right)$ **(c)** $\left(2, \frac{7}{2}\right)$ **(d)** $\left(2, -\frac{3}{2}\right)$ **(e)** None of the preceding

12. The graph of the system of two equations $y^2 - 2x^2 = 16$ and $y - x = 2$ has how many points of intersection?

(a) None **(b)** One **(c)** Two **(d)** Three **(e)** None of the preceding

13. An ellipse has vertices $(-1, -1)$ and $(5, -1)$ with minor axis of length 4. Which of the following is a focus of this ellipse?

(a) $\left(2 - \sqrt{5}, -1\right)$ **(b)** $\left(-\sqrt{5}, -1\right)$ **(c)** $(2, -1 + \sqrt{5})$ **(d)** $\left(2 + \sqrt{20}, -1\right)$

(e) None of the preceding

14. Which of the following is true for the parabola given by $x^2 - 10x + 13 = -12y$?

(a) The directrix is $y = -4$. **(b)** The focus is $(5, 4)$. **(c)** The focus is $(5, -2)$.

(d) The axis of symmetry is $y = 1$. **(e)** None of the preceding.

15. Which of the following is true for the conic $x^2 - 4y^2 + 6x + 16y - 11 = 0$?

(a) It is an ellipse with a horizontal major axis.

(b) It is an ellipse with a vertical major axis.

(c) It is a hyperbola with a vertical transverse axis.

(d) It is a hyperbola with a horizontal transverse axis.

(e) None of the preceding.

16. A hyperbola having a horizontal transverse axis has asymptotes $y - 2 = \pm\frac{3}{4}(x - 5)$. If also $a = 8$, then one focus is

(a) $(15, 2)$ **(b)** $(10, 2)$ **(c)** $(5, 12)$ **(d)** $\left(5 + \sqrt{14}, 2\right)$ **(e)** None of the preceding

17. If the distance from point $P(x, y)$ to the horizontal line $y = -4$ is equal to the distance from P to point $(3, 0)$, then the coordinates of P satisfy which of these equations?

(a) $4(x - 3)^2 = y + 2$ **(b)** $y + 2 = \frac{1}{8}(x - 3)^2$ **(c)** $x^2 - 6x - 7 = 4y$

(d) $y^2 - 8y + 16 = x - 3$ **(e)** None of the preceding

18. The number of solutions of the form (a, b) for the system of the two equations $y^2 - x^2 = 4$ and $8x^2 + 24y^2 = 192$ is which of the following?

(a) None **(b)** Two **(c)** Three **(d)** Four **(e)** None of the preceding

19. The reflecting surface of a parabolic antenna is generated by revolving the parabola $y = \frac{1}{18}x^2$ around its axis of symmetry for the interval $-6 \le x \le 6$. Assuming that the measurements are in feet, how far from the vertex (bottom) of the generating parabola should the receiver be placed?

(a) 2 ft **(b)** 3.5 ft **(c)** 4.5 ft **(d)** 6 ft **(e)** None of the preceding

20. The figure represents a pattern consisting of a large square *ABFG* and a small square *BCDE*. The total area of the pattern is 97 square inches and the total outside perimeter (excluding *BE*) is 44 inches. Which of the following systems can be used to find the dimensions *x* and *y* as shown on the figure?

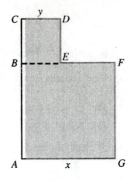

(a) $x^2 + y^2 = 97$
$4x + 4y = 44$

(b) $x^2y^2 = 97$
$x + y = 11$

(c) $(x + y)x = 97$
$3x + 4y = 44$

(d) $x^2 + y^2 = 97$
$2x + y = 22$

(e) None of the preceding

ANSWERS TO THE TEST YOUR UNDERSTANDING EXERCISES

Page 640

1. Center: $(0, 0)$
Vertices: $(\pm 10, 0)$
Foci: $(\pm 6, 0)$

2. Center: $(0, 0)$
Vertices: $(0, \pm 4)$
Foci: $\left(0, \pm\sqrt{15}\right)$

3. $\dfrac{x^2}{64} + \dfrac{y^2}{15} = 1$

4. $\dfrac{x^2}{36} + \dfrac{y^2}{81} = 1$

5. $\dfrac{(x - 5)^2}{4} + (y + 2)^2 = 1$; Center: $(5, -2)$
Vertices: $(3, -2), (7, -2)$
Foci: $\left(5 - \sqrt{3}, -2\right), \left(5 + \sqrt{3}, -2\right)$

Page 650

1. Center: $(0, 0)$
Vertices: $(\pm 3, 0)$
Foci: $(\pm 5, 0)$
Asymptotes: $y = \pm\frac{4}{3}x$

2. Center: $(0, 0)$
Vertices: $(0, \pm 8)$
Foci: $(0, \pm 10)$
Asymptotes: $y = \pm\frac{4}{3}x$

3. $\dfrac{y^2}{25} - \dfrac{x^2}{24} = 1$

4. $\dfrac{x^2}{4} - \dfrac{y^2}{9} = 1$

Page 661

1.

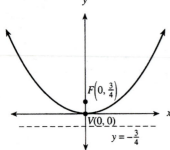

2.

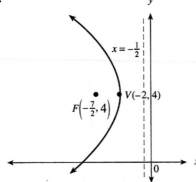

3. $x^2 = 4(-3)y$

4. $(y - 2)^2 = 4\left(-\frac{3}{2}\right)\left(x + \frac{3}{2}\right)$

5. $(x - 3)^2 = 4(2)(y + 1)$; vertex: $(3, -1)$; focus: $(3, 1)$; directrix: $y = -3$.

Sequences and Series

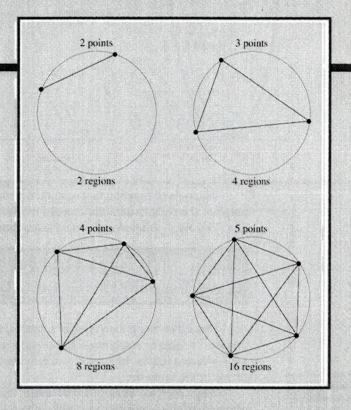

2 points

3 points

2 regions

4 regions

4 points

5 points

8 regions

16 regions

Can you decide what comes next in this sequence? 1, 4, 7, 10, . . .

If it is assumed that the pattern used to obtain a new term from the preceding term continues, then it should be clear that the next term is 13; each term after the first is obtained by adding 3 to the preceding term. This is an example of an *arithmetic sequence*.

Here is an interesting sequence whose terms are determined as follows. Take a circle and locate two points on it, then 3, then 4; continue in this manner, each time selecting one more point than before. Connect the points with lines in all possible ways, and then count the number of nonoverlapping regions formed within the circle. The numbers of such regions are the terms of a sequence, and the first four terms are 2, 4, 8, and 16 as shown.

After the first term, 2, each term can be obtained by multiplying the preceding term by 2. Does this pattern continue? Is the next term 2(16) = 32? Draw a circle with 6 points connected in all possible ways. Count the number of nonoverlapping regions. Is your answer 32? If not, what is your reaction?

11.1 SEQUENCES

The same equation can be used to define a variety of functions by changing the domain. For example, below are the graphs of three functions all of whose range values are given by the equation $y = x^2$ for the indicated domains.

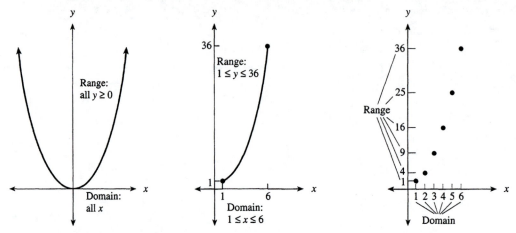

The type of function that is studied in this chapter is illustrated by the preceding graph at the right, where the domain consists of the consecutive integers 1, 2, 3, 4, 5, 6. This kind of function is called a **sequence.**

> ## DEFINITION OF A SEQUENCE
>
> A sequence is a function whose domain is a set of consecutive positive integers.

Instead of using the variable x, letters, such as n, k, i are normally used for the domain variable of a sequence. Frequently, sequences (functions) will be denoted by the lowercase letters, such as a, and the range values by a_n, which are also called the **terms** of the sequence.

a_n is read "a-sub-n" and has the same meaning as the functional notation $a(n)$, that is, "a of n."

Sequences are often given by stating their **general** or **nth terms.** Thus the general term of the sequence, previously given by $y = x^2$, becomes $a_n = n^2$.

EXAMPLE 1 Find the range values of the sequence given by $a_n = \dfrac{1}{n}$ for the domain $\{1, 2, 3, 4, 5\}$ and graph.

Solution The range values and graph are as follows.

This is an example of a finite sequence since the domain is finite. That is, the domain is a set of positive integers having a last element.

$$a_1 = \tfrac{1}{1} = 1$$

$$a_2 = \tfrac{1}{2}$$

$$a_3 = \tfrac{1}{3}$$

$$a_4 = \tfrac{1}{4}$$

$$a_5 = \tfrac{1}{5}$$

EXAMPLE 2 List the first six terms of the sequence given by $b_k = \dfrac{(-1)^k}{k^2}$.

Solution Use the given nth term and let $k = 1, 2, 3, 4, 5,$ and $6,$ respectively.

$$b_1 = \frac{(-1)^1}{1^2} = -1 \qquad\qquad b_4 = \frac{(-1)^4}{4^2} = \frac{1}{16}$$

$$b_2 = \frac{(-1)^2}{2^2} = \frac{1}{4} \qquad\qquad b_5 = \frac{(-1)^5}{5^2} = -\frac{1}{25}$$

$$b_3 = \frac{(-1)^3}{3^2} = -\frac{1}{9} \qquad\qquad b_6 = \frac{(-1)^6}{6^2} = \frac{1}{36} \qquad\blacksquare$$

TEST YOUR UNDERSTANDING

Write the first five terms of the given sequence.

1. $a_n = 2n + 1$ **2.** $a_n = -2n$ **3.** $a_n = -2n + 2$

4. $b_k = \dfrac{(-1)^k}{k}$ **5.** $b_k = \dfrac{1}{k^2}$ **6.** $b_k = \dfrac{-3}{k(k+1)}$

7. $c_n = \dfrac{3}{n(2n-1)}$ **8.** $c_n = \left(\dfrac{1}{3}\right)^n$ **9.** $c_n = 1 - (-1)^n$

(Answers: Page 736)

Sometimes a sequence is given by a verbal description. If, for example, we ask for the increasing sequence of odd integers beginning with -3, then this implies the **infinite sequence** whose first few terms are

$$-3, \quad -1, \quad 1, \ldots$$

This is an example of an infinite sequence since the domain is infinite. That is, the domain consists of all the positive integers.

A sequence can also be given by presenting a listing of its first few terms, possibly including the general term. Thus the preceding sequence can be defined as

$$-3, \quad -1, \quad 1, \quad \ldots, \quad 2n - 5, \quad \ldots$$

EXAMPLE 3 Find the tenth term of the sequence

$$-3, \quad 4, \quad \frac{5}{3}, \quad \ldots, \quad, \quad \frac{n+2}{2n-3}, \quad \ldots$$

Solution Since the first term, -3, is obtained by letting $n = 1$ in the general term $\dfrac{n+2}{2n-3}$, the tenth term is

$$a_{10} = \frac{10+2}{2(10)-3} = \frac{12}{17} \qquad\blacksquare$$

EXAMPLE 4 Write the first four terms of the sequence given by $a_n = \left(1 + \dfrac{1}{n}\right)^n$. Round off to two decimal places when appropriate.

Solution

$$a_1 = \left(1 + \tfrac{1}{1}\right)^1 = 2$$

$$a_2 = \left(1 + \tfrac{1}{2}\right)^2 = \left(\tfrac{3}{2}\right)^2 = \tfrac{9}{4} = 2.25$$

$$a_3 = \left(1 + \tfrac{1}{3}\right)^3 = \left(\tfrac{4}{3}\right)^3 = \tfrac{64}{27} = 2.37$$

$$a_4 = \left(1 + \tfrac{1}{4}\right)^4 = \left(\tfrac{5}{4}\right)^4 = \tfrac{625}{256} = 2.44$$ ∎

The terms of the sequence in Example 4 are getting successively larger. But the increase from term to term is getting smaller. That is, the differences between successive terms are decreasing:

$$a_2 - a_1 = 0.25$$

$$a_3 - a_2 = 0.12$$

$$a_4 - a_3 = 0.07$$

Use a calculator to verify these table entries to four decimal places.

n	$a_n = \left(1 + \dfrac{1}{n}\right)^n$
10	2.5937
50	2.6916
100	2.7048
500	2.7156
1000	2.7169
5000	2.7180
10,000	2.7181

If more terms of $a_n = \left(1 + \dfrac{1}{n}\right)^n$ were computed, you would see that while the terms keep on increasing, the amount by which each new term increases keeps getting smaller.

It turns out that no matter how large n is, the value of $\left(1 + \dfrac{1}{n}\right)^n$ is never more than 2.72. In fact, the larger the n that is taken, the closer $\left(1 + \dfrac{1}{n}\right)^n$ gets to the irrational value $e = 2.71828 \ldots$. This is the number that was introduced in Chapter 5 in reference to natural logarithms and exponential functions.

Sequences can also be defined **recursively,** which means that the first term is given and the nth term a_n is defined in terms of the preceding term a_{n-1}. This is illustrated in the next example.

EXAMPLE 5 Let a_n be the nth term of a sequence defined recursively by

$$a_1 = 6$$

$$a_n = 3a_{n-1} - 7 \qquad \text{for } n > 1$$

Find the first five terms of this sequence.

Solution

Observe that each term in a
recursive sequence, except
the first, is obtained from the
preceding term according to
a specified rule of formation.
Thus, unless otherwise indicated, a recursive sequence
is an infinite sequence.

$$a_1 = 6$$

$$a_2 = 3a_1 - 7 = 3 \cdot 6 - 7 = 11$$

$$a_3 = 3a_2 - 7 = 3 \cdot 11 - 7 = 26$$

$$a_4 = 3a_3 - 7 = 3 \cdot 26 - 7 = 71$$

$$a_5 = 3a_4 - 7 = 3 \cdot 71 - 7 = 206$$

■

EXERCISES 11.1

The domain of the sequence in each exercise consists of the integers 1, 2, 3, 4, 5. Write the corresponding range values.

1. $a_n = 2n - 1$ **2.** $a_n = 10 - n^2$ **3.** $a_k = (-1)^k$

4. $b_k = -\dfrac{6}{k}$ **5.** $b_i = 8(-\tfrac{1}{2})^i$ **6.** $b_i = (\tfrac{1}{2})^{i-3}$

Write the first four terms of the sequence given by the formula in each exercise.

7. $c_k = (-1)^k k^2$ **8.** $c_j = 3(\tfrac{1}{10})^{j-1}$ **9.** $c_j = 3(\tfrac{1}{10})^j$

10. $a_j = 3(\tfrac{1}{10})^{j+1}$ **11.** $a_j = 3(\tfrac{1}{10})^{2j}$ **12.** $a_n = \dfrac{(-1)^{n+1}}{n+3}$

13. $c_n = \dfrac{1}{n} - \dfrac{1}{n+1}$ **14.** $a_n = \dfrac{n^2 - 4}{n+2}$ **15.** $a_k = (2k - 10)^2$

16. $a_k = 1 + (-1)^k$ **17.** $a_n = -2 + (n-1)(3)$ **18.** $a_n = a_1 + (n-1)(d)$

19. $b_i = \dfrac{i-1}{i+1}$ **20.** $b_i = 64^{1/i}$ **21.** $b_n = \left(1 + \dfrac{1}{n}\right)^{n-1}$

22. $u_n = \dfrac{1}{2^n}$ **23.** $u_n = -2(\tfrac{3}{4})^{n-1}$ **24.** $u_k = a_1 r^{k-1}$

25. $x_k = \dfrac{k}{2^k}$ **26.** $x_n = \dfrac{(-1)^n}{n} + n$ **27.** $x_k = \dfrac{k}{k+1} - \dfrac{k+1}{k}$

28. $y_n = \left(1 + \dfrac{1}{n+1}\right)^n$ **29.** $y_n = 4$ **30.** $y_n = \dfrac{n}{(n+1)(n+2)}$

31. Find the sixth term of $1, 2, 5, \ldots, \frac{1}{2}(1 + 3^{n-1}), \ldots$

32. Find the ninth and tenth terms of $0, 4, 0, \ldots, \dfrac{2^n + (-2)^n}{n}, \ldots$

33. Find the seventh term of $a_k = 3(0.1)^{k-1}$.

34. Find the twentieth term of $a_n = (-1)^{n-1}$.

35. Find the twelfth term of $a_i = i$.

36. Find the twelfth term of $a_i = (i - 1)^2$.

37. Find the twelfth term of $a_i = (1 - i)^3$.

38. Find the hundredth term of $a_n = \dfrac{n + 1}{n^2 + 5n + 4}$.

39. Write the first four terms of the sequence of even increasing integers beginning with 4.

40. Write the first four terms of the sequence of decreasing odd integers beginning with 3.

41. Write the first five positive multiples of 5 and find the formula for the nth term.

42. Write the first five powers of 5 and find the formula for the nth term.

43. Write the first five powers of -5 and find the formula for the nth term.

44. Write the first five terms of the sequence of reciprocals of the negative integers and find the formula for the nth term.

45. The numbers 1, 3, 6, and 10 are called **triangular numbers** because they correspond to the number of dots in the triangular arrays. Find the next three triangular numbers.

46. When an investment earns **simple interest** it means the interest is earned only on the original investment. For example, if P dollars are invested in a bank that pays simple interest at the annual rate of r percent, then the interest for the first year is Pr, and the amount in the bank at the end of the year is $P + Pr$. For the second year, the interest is again Pr; the amount now would be $(P + Pr) + Pr = P + 2Pr$.
   (a) What is the amount after n years?
   (b) What is the amount in the bank if an investment of $750 has been earning simple interest for 5 years at the annual rate of 12%?
   (c) If the amount in the bank is $5395 after 12 years, what was the original investment P if it has been earning simple interest at the annual rate of $12\frac{1}{2}\%$?

47. Find the first eight terms of the sequence defined recursively by $a_1 = 12$ and $a_n = -\frac{1}{2}a_{n-1} + 2$ for $n > 1$.

48. Find the first six terms of the sequence defined recursively by $a_1 = 6$ and $a_n = \dfrac{3}{a_{n-1}}$ for $n > 1$.

49. Write the first eight terms of $a_n = \dfrac{1 + (-1)^{n+1}}{2i^{n-1}}$. (See page 73 for the powers of $i = \sqrt{-1}$).

50. (a) Write the first seven terms of $a_n = n!$ where $n!$ is read as "n factorial" and is defined by $n! = n(n - 1)(n - 2) \cdots 3 \cdot 2 \cdot 1$
   (b) If $a_n = \dfrac{(2n)!}{(n!)^2}$, then show that $\dfrac{a_{n+1}}{a_n} = 2\left(\dfrac{2n + 1}{n + 1}\right)$.

51. Write the first four terms of

$$a_n = \frac{3 \cdot 5 \cdots (2n - 1)(2n + 1)}{2 \cdot 4 \cdots (2n - 2)(2n)}$$

Challenge   Find a formula for the *n*th triangular number. (*Hint:* See Exercise 45 and consider the additional number of dots in each new figure.)

Written Assignment   Refer to the function concept and explain what it means to add two sequences and to multiply two sequences.

Graphing Calculator Exercises   When a sequence is given by means of a function of the form $a_n = f(n)$, we can study the behavior of the sequence by graphing the function. The RANGE variables should be set so that the integers n will correspond to some of the pixels on your screen, and the MODE to DOT or DISC. This may still give more points than those corresponding to integers, of course, especially if you don't want n to get too big. However, if your calculator allows the use of relational operators (see your user's manual), graphing the function $y = f(x)*(x\text{-IPart } x = 0)$ will only graph the points where x is an integer. (*Note:* IPart = "integer part of".)

Use the preceding technique as far as your calculator will permit to graph the first 10 terms of the sequence in Exercises 1–5. On your graph, distinguish points of the sequence from other points graphed, if any.

1. $a_n = 2n - 1$

2. $a_n = n^2 - n + 2$

3. $a_n = (2n - 1)/(n + 1)$

4. $a_n = (-1)^n n^2$

5. $a_n = (1 + 1/n)^n$

6. The sequence of *Fibonacci numbers* may be defined recursively by $a_n = a_{n-1} + a_{n-2}$, $a_1 = a_2 = 1$. The first 5 terms, therefore, are 1, 1, 2, 3, and 5.

   (a) Find the first 20 terms of the Fibonacci numbers.

   (b) The terms of the sequence may also be given by

   $$a_n = (1/\sqrt{5})((1 + \sqrt{5})/2)^n - (1/\sqrt{5})((1 - \sqrt{5})/2)^n.$$

   Verify this for $n = 18$, 19, and 20.

   (c) Find the sequence of quotients $q_n = a_{n+1}/a_n$ of the Fibonacci numbers through $n = 10$, rounded to four decimal places where appropriate. The quotients q_n get closer and closer to a number called the *golden mean*. What is this number to the nearest thousandth?

11.2
SUMS OF FINITE SEQUENCES

How long would it take you to add the integers from 1 to 1000? Here is a quick way. List the sequence displaying the first few and last few terms.

$$1, \quad 2, \quad 3, \quad \ldots, \quad 998, \quad 999, \quad 1000$$

Add them in pairs, the first and last, the second and second from last, and so on.

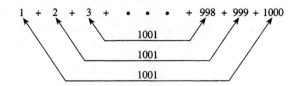

Since there are 500 such pairs to be added, the total is

$$500(1001) = 500{,}500$$

For any finite sequence we can add all of its terms and say that we have found the *sum of the sequence*. The sum of a sequence is called a **series.** For example, the sequence

$$1, \quad 3, \quad 5, \quad 7, \quad 9, \quad 11$$

can be associated with the series

$$1 + 3 + 5 + 7 + 9 + 11$$

The sum of the terms in this series can easily be found, by addition, to be 36. As another example, the sequence $a_n = \dfrac{1}{n}$ for $n = 1, 2, 3, 4, 5$ has the sum

$$1 + \frac{1}{2} + \frac{1}{3} + \frac{1}{4} + \frac{1}{5} = \frac{60 + 30 + 20 + 15 + 12}{60} = \frac{137}{60}$$

EXAMPLE 1   Find the sum of the first seven terms of $a_n = 2n$.

Solution

$$a_1 + a_2 + a_3 + a_4 + a_5 + a_6 + a_7 = 2 + 4 + 6 + 8 + 10 + 12 + 14 = 56$$

∎

Just think of sigma as a command to add.

There is a very handy notational device available for expressing the sum of a sequence. The Greek letter Σ (capital sigma) is used for this purpose. For example, we can represent the sum of the first seven terms of the sequence whose general term is a_k by the symbol $\displaystyle\sum_{k=1}^{7} a_k$; that is,

$$\sum_{k=1}^{7} a_k = a_1 + a_2 + a_3 + a_4 + a_5 + a_6 + a_7$$

Add the terms a_k for consecutive values of k, starting with $k = 1$ up to and including $k = 7$. With this symbolism, the question in Example 1 can now be stated by asking for the value of $\displaystyle\sum_{k=1}^{7} 2k$.

> ## SUMMATION NOTATION
>
> For a sequence whose general term is a_k, the sum of the first n terms can be represented by
>
> $$\sum_{k=1}^{n} a_k = a_1 + a_2 + a_3 + \cdots + a_n$$
>
> The **index of summation** is k; the summation starts with $k = 1$ and ends with $k = n$.

Observe that the domain variable k of the given sequence becomes the index of summation in this notation.

EXAMPLE 2   Find $\displaystyle\sum_{n=1}^{5} b_n$ where $b_n = \dfrac{n}{n+1}$.

Solution

$$\sum_{n=1}^{5} b_n = b_1 + b_2 + b_3 + b_4 + b_5$$

Now replace each b_n by its numerical value:

$$\sum_{n=1}^{5} b_n = \frac{1}{2} + \frac{2}{3} + \frac{3}{4} + \frac{4}{5} + \frac{5}{6}$$

$$= \frac{30 + 40 + 45 + 48 + 50}{60}$$

$$= \frac{213}{60} = \frac{71}{20}$$ ∎

EXAMPLE 3   Evaluate $\displaystyle\sum_{k=1}^{4} (2k + 1)$.

Solution It is understood here that we are to find the sum of the first four terms of the sequence whose general term is $a_k = 2k + 1$.

$$\sum_{k=1}^{4} (2k + 1) = (2 \cdot 1 + 1) + (2 \cdot 2 + 1) + (2 \cdot 3 + 1) + (2 \cdot 4 + 1)$$

$$= 3 + 5 + 7 + 9$$

$$= 24$$ ∎

In Example 3, the sigma gives the sum of the odd numbers from 3 to 9. How can this sigma be modified so that it gives the sum of the first five odd numbers?

EXAMPLE 4 Find $\displaystyle\sum_{i=1}^{5} x_i$, where $x_i = (-1)^i(i+1)$.

Solution

$$\sum_{i=1}^{5} x_i = x_1 + x_2 + x_3 + x_4 + x_5$$

$$= (-1)^1(1+1) + (-1)^2(2+1) + (-1)^3(3+1)$$
$$+ (-1)^4(4+1) + (-1)^5(5+1)$$

$$= -2 + 3 - 4 + 5 - 6$$

$$= -4 \qquad\blacksquare$$

TEST YOUR UNDERSTANDING

Evaluate each of the following.

1. $\displaystyle\sum_{k=1}^{5}(4k)$ **2.** $\displaystyle\sum_{k=1}^{5}(2k-1)$ **3.** $\displaystyle\sum_{k=1}^{5}(k^2-k)$

4. $\displaystyle\sum_{n=1}^{6}(-1)^n$ **5.** $\displaystyle\sum_{n=1}^{4}(2n^2-n)$ **6.** $\displaystyle\sum_{n=1}^{4}2\left(-\tfrac{1}{2}\right)^{n+1}$

(Answers: Page 736)

When the general term of a given series in expanded form can be found, then the series can be rewritten using the sigma notation. For example, since the series

$$3 + 6 + 9 + 12 + 15 + 18 + 21$$

Note: In some of the exercises that follow, the summation begins with values of i, n, or k other than 1. For example, see Exercises 8, 22, 24, 25 and 30.

is the sum of the first seven multiples of 3, it can be written as $\displaystyle\sum_{n=1}^{7} 3n$.

EXERCISES 11.2

Find the sum of the first five terms of the sequence given by the formula in each exercise.

1. $a_n = 3n$ **2.** $a_k = (-1)^k\dfrac{1}{k}$ **3.** $a_i = i^2$ **4.** $b_i = i^3$ **5.** $b_k = \dfrac{3}{10^k}$ **6.** $b_n = -6 + 2(n-1)$

7. Find $\displaystyle\sum_{n=1}^{8} t_n$ where $t_n = 2^n$. **8.** Find $\displaystyle\sum_{n=0}^{8} x_n$ where $x_n = \dfrac{1}{2^n}$. **9.** Find $\displaystyle\sum_{k=1}^{20} y_k$ where $y_k = 3$.

Find each of the following sums for n = 7.

10. $2 + 4 + \cdots + 2n$ **11.** $2 + 4 + \cdots + 2^n$

12. $-7 + 2 + \cdots + (9n - 16)$ **13.** $3 + \tfrac{3}{2} + \cdots + 3\left(\tfrac{1}{2}\right)^{n-1}$

Evaluate each of the following.

14. $\displaystyle\sum_{k=1}^{6} (5k)$

15. $\displaystyle 5\left(\sum_{k=1}^{6} k\right)$

16. $\displaystyle\sum_{n=1}^{4} (n^2 + n)$

17. $\displaystyle\sum_{n=1}^{4} n^2 + \sum_{n=1}^{4} n$

18. $\displaystyle\sum_{t=1}^{8} (i - 2i^2)$

19. $\displaystyle\sum_{k=1}^{4} \frac{k}{2^k}$

20. $\displaystyle\sum_{k=1}^{7} (-1)^k$

21. $\displaystyle\sum_{k=1}^{8} (-1)^k$

22. $\displaystyle\sum_{k=3}^{7} (2k - 5)$

23. $\displaystyle\sum_{j=1}^{6} [-3 + (j - 1)5]$

24. $\displaystyle\sum_{k=-3}^{3} 10^k$

25. $\displaystyle\sum_{k=-3}^{3} \frac{1}{10^k}$

26. $\displaystyle\sum_{k=1}^{5} 4(-\tfrac{1}{2})^{k-1}$

27. $\displaystyle\sum_{i=1}^{4} (-1)^i 3^i$

28. $\displaystyle\sum_{n=1}^{3} \left(\frac{n+1}{n} - \frac{n}{n+1}\right)$

29. $\displaystyle\sum_{n=1}^{3} \frac{n+1}{n} - \sum_{n=1}^{3} \frac{n}{n+1}$

30. $\displaystyle\sum_{k=0}^{8} \frac{1 + (-1)^k}{2}$

31. $\displaystyle\sum_{k=1}^{3} (0.1)^{2k}$

Rewrite each series using sigma notation.

32. $4 + 8 + 12 + 16 + 20 + 24$

33. $5 + 10 + 15 + 20 + \cdots + 50$

34. $-4 - 2 + 0 + 2 + 4 + 6 + 8$

35. $-9 - 6 - 3 + 0 + 3 + \cdots + 24$

36. Read the discussion at the beginning of this section, where we found the sum of the first 1000 positive integers, and find a formula for the sum of the first n positive integers when n is even.

37. (a) Find $\displaystyle\sum_{k=1}^{n} (2k - 1)$ for each of the following values on n: $2, 3, 4, 5, 6$.

   (b) On the basis of the results in part (a), find a formula for the first n odd positive integers.

38. The sequence $1, 1, 2, 3, 5, 8, 13, \ldots$ is called the *Fibonacci sequence.* Its first two terms are one, and each term thereafter is computed by adding the preceding two terms.

   (a) Write the next seven terms of this sequence.

   (b) Let $u_1, u_2, u_3, \ldots, u_n, \ldots$ be the Fibonacci sequence. Evaluate $S_n = \displaystyle\sum_{k=1}^{n} u_k$ for these values of n: $1, 2, 3, 4, 5, 6, 7, 8$.

   (c) Note that $u_1 = u_3 - u_2$, $u_2 = u_4 - u_3$, $u_3 = u_5 - u_4$, and so on. Use this form for the first n numbers to derive a formula for the sum of the first n Fibonacci numbers.

39. Let a_n be a sequence with $a_1 = 2$ and $a_m + a_n = a_{m+n}$, where m and n are any positive integers. Show that $a_n = 2n$ for any n.

40. Show that $\displaystyle\sum_{k=1}^{9} \log_{10} \frac{k + 1}{k} = 1$. $\left(Hint: \log_{10} \dfrac{a}{b} = \log_{10} a - \log_{10} b.\right)$

41. Prove: $\displaystyle\sum_{k=1}^{n} s_k + \sum_{k=1}^{n} t_k = \sum_{k=1}^{n} (s_k + t_k)$.

42. Prove: $\displaystyle\sum_{k=1}^{n} cs_k = c\sum_{k=1}^{n} s_k$, c a constant.

43. Prove $\displaystyle\sum_{k=1}^{n} (s_k + c) = \left(\sum_{k=1}^{n} s_k\right) + nc$, c a constant.

44. (a) Evaluate $\displaystyle\sum_{k=1}^{10} \frac{1}{k(k+1)}$ using the result $\dfrac{1}{k(k+1)} = \dfrac{1}{k} - \dfrac{1}{k+1}$.

(b) Use the identity given in part (a) to prove that $\dfrac{1}{1\cdot2} + \dfrac{1}{2\cdot3} + \dfrac{1}{3\cdot4} + \cdots + \dfrac{1}{98\cdot99} + \dfrac{1}{99\cdot100} = \dfrac{99}{100}$.

(c) Write $\displaystyle\sum_{k=1}^{n} \left(\frac{1}{k} - \frac{1}{k+1}\right)$ without the sigma notation and showing at least the first three and last three terms. Then simplify this summation. (Note: Such a sum is referred to as a *telescoping sum*. Can you see why it has this name?)

45. (a) Use the identity $\dfrac{2}{k(k+2)} = \dfrac{1}{k} - \dfrac{1}{k+2}$ to show that $\displaystyle\sum_{k=1}^{10} \frac{2}{k(k+2)} = \frac{175}{132}$.

(b) Write $\displaystyle\sum_{k=1}^{n} \left(\frac{1}{k} - \frac{1}{k+2}\right)$ without the sigma notation and showing at least the first four and last four terms. Then simplify this summation.

? **Challenge**   This question refers to matrix multiplication developed in Section 9.2. Let matrix $A=[a_{ij}]$ be m by n and $B = [b_{ij}]$ be n by p so that $AB = C = [c_{ij}]$ is m by p. The computation that produces an element c_{ij} in the product involves the ith row of A and jth column of B. Express this computation using the sigma notation.

 Written Assignment   Consider the sequence $a_k = c$ for all k and evaluate $\displaystyle\sum_{k=1}^{n} c$.

Explain how you obtained your answer.

11.3 ARITHMETIC SEQUENCES

Here are the first five terms of the sequence $a_n = 7n - 2$.

$$5, \quad 12, \quad 19, \quad 26, \quad 33$$

Do you notice any special pattern? It should not take long to observe that each term, after the first, is 7 more than the preceding term. This sequence is an example of an *arithmetic sequence*.

An arithmetic sequence is also referred to as an arithmetic progression.

> **DEFINITION OF AN ARITHMETIC SEQUENCE**
>
> A sequence is said to be *arithmetic* if each term, after the first, is obtained from the preceding term by adding a common value.

Let us consider the first four terms of two different arithmetic sequences:

$$2, \quad 4, \quad 6, \quad 8, \ldots$$

$$-\tfrac{1}{2}, \quad -1, \quad -\tfrac{3}{2}, \quad -2, \quad \ldots$$

For the first sequence, the common value (or difference) that is added to each term to get the next is 2. Thus it is easy to see that 10, 12, and 14 are the next three terms. You might guess that the nth term is $a_n = 2n$.

The second sequence has the common difference $-\tfrac{1}{2}$. This can be found by subtracting the first term from the second, or the second from the third, and so forth. The nth term is $a_n = -\tfrac{1}{2}n$.

Unlike the preceding sequences, it is not always easy to see what the nth term of the specific arithmetic sequence is. Therefore, we will now develop a general form that makes it possible to write the nth term of any such sequence.

Can you find the nth term of the arithmetic sequence 11, 2, −7, −16, . . . ?

Let a_n be the nth term of an arithmetic sequence, and let d be the **common difference.** Then the first four terms are:

$$a_1$$
$$a_2 = a_1 + d$$
$$a_3 = a_2 + d = (a_1 + d) + d = a_1 + 2d$$
$$a_4 = a_3 + d = (a_1 + 2d) + d = a_1 + 3d$$

The pattern is clear. Without further computation we see that

$$a_5 = a_1 + 4d \quad \text{and} \quad a_6 = a_1 + 5d$$

Since the coefficient of d is always 1 less than the number of the term, the nth term is given as follows.

GENERAL TERM OF AN ARITHMETIC SEQUENCE

This formula says that the nth term of an arithmetic sequence is completely identified by its first term a_1 and its common difference d. Also note that $a_n = a_{n-1} + d$ and $d = a_n - a_{n-1}$.

The nth term of an arithmetic sequence is

$$a_n = a_1 + (n - 1)d$$

where a_1 is the first term and d is the common difference.

By substituting the values $n = 1, 2, 3, 4, 5, 6$, you can check that this formula gives the preceding terms a_1 through a_6.

EXAMPLE 1   Find the nth term of the arithmetic sequence $11, 2, -7, \ldots$.

Solution Use the formula for a_n with $a_1 = 11$ and $d = a_2 - a_1 = 2 - 11 = -9$. Thus

$$a_n = a_1 + (n - 1)d$$
$$= 11 + (n - 1)(-9) = -9n + 20 \qquad \blacksquare$$

| **TEST YOUR UNDERSTANDING** | *Each of the following gives the first few terms of an arithmetic sequence. Find the nth term in each case.* |

1. 5, 10, 15, . . .   **2.** 6, 2, −2, . . .   **3.** $\frac{1}{10}$, $\frac{1}{5}$, $\frac{3}{10}$

4. −5, −13, −21, . . .   **5.** 1, 2, 3, . . .   **6.** −3, −2, −1, . . .

Find the nth term a_n of the arithmetic sequence with the given values for the first term and the common difference.

7. $a_1 = \frac{2}{3}$; $d = \frac{2}{3}$ **8.** $a_1 = 53$; $d = -12$

9. $a_1 = 0$; $d = \frac{1}{5}$ **10.** $a_1 = 2$; $d = 1$

EXAMPLE 2   The first term of an arithmetic sequence is −15 and the fifth term is 13. Find the fortieth term.

Solution   Since $a_5 = 13$ use $n = 5$ in the formula $a_n = a_1 + (n - 1)d$ in order to solve for d.

$$a_5 = a_1 + (5 - 1)d$$
$$13 = -15 + 4d$$
$$28 = 4d$$
$$7 = d$$

Then $a_{40} = -15 + (39)7 = 258$.   ∎

Adding the terms of a finite sequence may not be much work when the number of terms to be added is small. When many terms are to be added, however, the amount of time and effort needed can be overwhelming. For example, to add the first 10,000 terms of the arithmetic sequence beginning with

$$246, \quad 261, \quad 276, \ldots$$

would call for an enormous effort, unless some shortcut could be found. Fortunately, there is an easy way to find such sums. This method (in disguise) was already used in the question at the start of Section 11.2. Let us look at the general situation. Let S_n denote the sum of the first n terms of the arithmetic sequence given by $a_k = a_1 + (k - 1)d$:

*The sum of an arithmetic sequence is also called an **arithmetic series**.*

$$S_n = \sum_{k=1}^{n} [a_1 + (k - 1)d]$$

$$= a_1 + [a_1 + d] + [a_1 + 2d] + \cdots + [a_1 + (n - 1)d]$$

Put this sum in reverse order and write the two equalities together as follows:

$$S_n = \quad a_1 \quad + \quad [a_1 + d] \quad + \cdots + [a_1 + (n - 2)d] + [a_1 + (n - 1)d]$$

$$\updownarrow \qquad\qquad \updownarrow \qquad\qquad\qquad \updownarrow \qquad\qquad\qquad \updownarrow$$

$$S_n = [a_1 + (n - 1)d] + [a_1 + (n - 2)d] + \cdots + \quad [a_1 + d] \quad + \quad a_1$$

Now add to get

$$2S_n = [2a_1 + (n-1)d] + [2a_1 + (n-1)d] + \cdots + [2a_1 + (n-1)d] + [2a_1 + (n-1)d]$$

On the right-hand side of this equation there are n terms, each of the form $2a_1 + (n-1)d$. Therefore,

$$2S_n = n[2a_1 + (n-1)d]$$

Divide by 2 to solve for S_n:

$$S_n = \frac{n}{2}[2a_1 + (n-1)d]$$

Returning to the sigma notation, we can summarize our results this way:

SUM OF AN ARITHMETIC SEQUENCE

$$S_n = \sum_{k=1}^{n} [a_1 + (k-1)d] = \frac{n}{2}[2a_1 + (n-1)d]$$

EXAMPLE 3 Find S_{20} for the arithmetic sequence whose first term is $a_1 = 3$ and whose common difference is $d = 5$.

Solution Substituting $a_1 = 3$, $d = 5$, and $n = 20$ into the formula for S_n, we have

$$S_{20} = \frac{20}{2}[2(3) + (20-1)5]$$

$$= 10(6 + 95)$$

$$= 1010 \qquad \blacksquare$$

EXAMPLE 4 Find the sum of the first 10,000 terms of the arithmetic sequence beginning with 246, 261, 276, ...

Solution Since $a_1 = 246$ and $d = 15$,

$$S_{10,000} = \frac{10,000}{2}[2(246) + (10,000-1)15]$$

$$= 5000(150,477)$$

$$= 752,385,000 \qquad \blacksquare$$

EXAMPLE 5 Find the sum of the first n positive integers.

Solution First observe that the problem calls for the sum of the sequence $a_k = k$ for $k = 1, 2, \ldots, n$. This is an arithmetic sequence with $a_1 = 1$ and $d = 1$. Therefore,

$$\sum_{k=1}^{n} k = \frac{n}{2}[2(1) + (n-1)1] = \frac{n(n+1)}{2} \qquad \blacksquare$$

With the result of Example 5 we are able to check the answer for the sum of the first 1000 positive integers, found at the beginning of Section 11.2, as follows:

$$\sum_{k=1}^{1000} k = \frac{1000(1001)}{2} = 500{,}500$$

The form $a_k = a_1 + (k - 1)d$ for the general term of an arithmetic sequence easily converts to $a_k = dk + (a_1 - d)$. It is this latter form that is ordinarily used when the general term of a *specific* arithmetic sequence is given. For example, we would usually begin with the form $a_k = 3k + 5$ instead of $a_k = 8 + (k - 1)3$. The important thing to notice in the form $a_k = dk + (a_1 - d)$ is that the common difference is the coefficient of k.

EXAMPLE 6 Evaluate: $\displaystyle\sum_{k=1}^{50} (-6k + 10)$

Solution First note that $a_k = -6k + 10$ is an arithmetic sequence with $d = -6$ and with $a_1 = 4$.

$$\sum_{k=1}^{50} (-6k + 10) = \frac{50}{2}[2(4) + (50 - 1)(-6)]$$
$$= -7150 \qquad \blacksquare$$

The formula for an arithmetic series can be converted into another useful form when a_n is substituted for $a_1 + (n - 1)d$. Thus

$$S_n = \frac{n}{2}[2a_1 + (n - 1)d]$$

When this form is rewritten as

$$S_n = n\left(\frac{a_1 + a_n}{2}\right)$$

the sum can be viewed as n times the average of the first and nth (or last) terms.

$$= \frac{n}{2}[a_1 + a_1 + (n - 1)d]$$

$$= \frac{n}{2}(a_1 + a_n)$$

SUM OF AN ARITHMETIC SEQUENCE (Alternative Form)

$$S_n = \frac{n}{2}(a_1 + a_n)$$

Applying this result to Example 6, we have $a_1 = 4$, $a_{50} = -290$, and $S_{50} = \frac{50}{2}(4 + (-290)) = -7150$.

In order to decide when a positive integer N is divisible by 3, use the fact that N is divisible by 3 if the sum of its digits is divisible by 3. Thus 261 is divisible by 3 because 2 + 6 + 1 = 9 is divisible by 3.

EXAMPLE 7 Find the sum of all the multiples of 3 between 4 and 262.

Solution The first multiple of 3 after 4 is $6 = a_1$ and the one preceding 262 is $261 = a_n$. Now find n.

$$a_n = a_1 + (n - 1)d$$
$$261 = 6 + (n - 1)3$$
$$86 = n$$

Then

$$S_{86} = \frac{86}{2}[a_1 + a_{86}] = \frac{86}{2}[6 + 261] = 11{,}481 \qquad \blacksquare$$

Verify that the common difference is $d = \dfrac{b - a}{2}$.

When the average of two numbers, $(a + b)/2$, is inserted between the numbers, we get the arithmetic sequence $a, \dfrac{a + b}{2}, b$. The average $(a + b)/2$ is called the **arithmetic mean** of the two numbers a and b. This concept can be extended. That is, for two numbers there can be more than one arithmetic mean between them according to the following. If k is a positive integer and

$$a, \quad m_1, \quad m_2, \quad \ldots, \quad m_k, \quad b$$

is an arithmetic sequence, then the numbers, $m_1, m_2, \ldots, m_k$ are k arithmetic means between a and b. For example, since 2, 6, 10, 14, 18 is an arithmetic sequence, then 6, 10, and 14 are three arithmetic means between 2 and 18.

EXAMPLE 8 Insert four arithmetic means between 5 and 20.

Solution Numbers m_1, m_2, m_3, m_4 are needed so that 5, m_1, m_2, m_3, m_4, 20 is an arithmetic sequence. To find the common difference, d, use the formula $a_n = a_1 + (n - 1)d$ where $a_1 = 5$, $n = 6$, and $a_n = a_6 = 20$. Then

$$20 = 5 + (6 - 1)d$$
$$5d = 15$$
$$d = 3$$

Now begin with 5 and successively add 3 to obtain the arithmetic means 8, 11, 14, 17 so that the sequence is

$$5, \quad 8, \quad 11, \quad 14, \quad 17, \quad 20 \qquad \blacksquare$$

The final example is an application of arithmetic sequences to a banking situation.

EXAMPLE 9 Suppose you take out a short-term loan of $6000 that you agree to pay back in 12 equal monthly payments, plus 3% interest per month on the monthly balance. Each month's payment is made during the first week of the new month.

(a) Write the general term of the sequence that gives the monthly balance.

(b) How much is the interest for each of the first 3 months? Write the general term of the sequence that gives the amount of the monthly interest.

(c) What is the total interest payment for the year, and what percent is this total of the $6000 loan?

Solution

At the end of the first month your balance is
$6000 - 500(0) = 6000,$
and for $k = 12$,
$6000 - 500(11) = 500,$
which is the last payment made during the first week of the thirteenth month.

(a) The equal monthly payments on the loan are $6000 \div 12 = 500$ dollars. Since these payments are made during the first week for each month, the monthly balance is given by

$$a_k = 6000 - 500(k - 1)$$

for $k = 1, 2, \ldots, 12$.

(b) The first month's interest is 3% of $6000, or $6000(0.03) = 180$ dollars. You also pay back $500, so your second month's interest is $5500(0.03) = 165$ dollars. Similarly, the interest for the third month is $5000(.03) = 150$ dollars. In general, the monthly interest payment is

This is an arithmetic sequence with $a_1 = 180$ and $d = -15$.

$$\underbrace{\begin{pmatrix} \text{monthly} \\ \text{balance} \end{pmatrix}}_{\downarrow} \quad \underbrace{\begin{pmatrix} \text{monthly} \\ \text{rate} \end{pmatrix}}_{\downarrow}$$

$$[6000 - 500(k - 1)](.03) = 180 - 15(k - 1)$$
$$= 195 - 15k$$

(c) The total interest is the sum of the 12 interest payments. Thus

$$\sum_{k=1}^{12} (195 - 15k) = \frac{12}{2}(2 \cdot 180 + 11(-15))$$
$$= 6(195)$$
$$= 1170$$

The annual rate is $\dfrac{1170}{6000} = 0.195 = 19.5\%$. ∎

EXERCISES 11.3

Each of the following gives the first two terms of an arithmetic sequence. Write the next three terms; find the nth term; and find the sum of the first 20 terms.

1. $1, 3, \ldots$
2. $2, 4, \ldots$
3. $2, -4, \ldots$
4. $1, -3, \ldots$

5. $\frac{15}{2}, 8, \ldots$
6. $-\frac{4}{3}, -\frac{11}{3}, \ldots$
7. $\frac{2}{5}, -\frac{1}{5}, \ldots$
8. $-\frac{1}{2}, \frac{1}{4}, \ldots$

9. $50, 100, \ldots$
10. $-27, -2, \ldots$
11. $-10, 10, \ldots$
12. $225, 163, \ldots$

Find the indicated sum by using ordinary addition; also find the sum by using the formula for the sum of an arithmetic sequence.

13. $5 + 10 + 15 + 20 + 25 + 30 + 35 + 40 + 45 + 50 + 55 + 60 + 65$

14. $-33 - 25 - 17 - 9 - 1 + 7 + 15 + 23 + 31 + 39$

15. $\frac{3}{4} + 1 + \frac{5}{4} + \frac{3}{2} + \frac{7}{4} + 2 + \frac{9}{4} + \frac{5}{2} + \frac{11}{4}$

16. $128 + 71 + 14 - 43 - 100 - 157$

17. Find a_{30} for the arithmetic sequence having $a_1 = -30$ and $a_{10} = 69$.

18. Find a_{51} for the arithmetic sequence having $a_1 = 9$ and $a_8 = -19$.

Find S_{100} for the arithmetic sequence with the given values for a_1 and d.

19. $a_1 = 3; d = 3$ **20.** $a_1 = 1; d = 8$ **21.** $a_1 = -91; d = 21$ **22.** $a_1 = -7; d = -10$

23. $a_1 = \frac{1}{7}; d = 5$ **24.** $a_1 = \frac{2}{5}; d = -4$ **25.** $a_1 = 725; d = 100$ **26.** $a_1 = 0.1; d = 10$

27. Find S_{28} for the sequence $-8, 8, \ldots, 16n - 24, \ldots$.

28. Find S_{25} for the sequence $96, 100, \ldots, 4n + 92, \ldots$.

29. Find the sum of the first 50 positive multiples of 12.

30. **(a)** Find the sum of the first 100 positive even numbers.

   (b) Find the sum of the first n positive even numbers.

31. **(a)** Find the sum of the first 100 positive odd numbers.

   (b) Find the sum of the first n positive odd numbers.

Evaluate.

32. $\displaystyle\sum_{k=1}^{12} [3 + (k-1)9]$ **33.** $\displaystyle\sum_{k=1}^{9} \left[-6 + (k-1)\frac{1}{2}\right]$ **34.** $\displaystyle\sum_{k=1}^{20} (4k - 15)$ **35.** $\displaystyle\sum_{k=1}^{30} (10k - 1)$

36. $\displaystyle\sum_{k=1}^{40} \left(-\frac{1}{3}k + 2\right)$ **37.** $\displaystyle\sum_{k=1}^{49} \left(\frac{3}{4}k - \frac{1}{2}\right)$ **38.** $\displaystyle\sum_{k=1}^{20} 5k$ **39.** $\displaystyle\sum_{k=1}^{n} 5k$

40. Find u such that $7, u, 19$ is an arithmetic sequence.

41. Find u such that $-7, u, \frac{5}{2}$ is an arithmetic sequence.

42. Find the twenty-third term of the arithmetic sequence $6, -4, \ldots$.

43. Find the thirty-fifth term of the arithmetic sequence $-\frac{2}{3}, -\frac{1}{5}, \ldots$.

44. Insert three arithmetic means between 8 and 44.

45. Insert four arithmetic means between 3 and 38.

46. Insert five arithmetic means between 6 and 10.

47. Insert six arithmetic means between -8 and 48.

48. Insert seven arithmetic means between -36 and 4.

49. Insert six arithmetic means between $-\frac{1}{5}$ and -3.

50. An object is dropped from an airplane and falls 32 feet during the first second. During each successive second it falls 48 feet more than in the preceding second. How many feet does it travel during the first 10 seconds? How far does it fall during the tenth second?

51. Suppose you save $10 one week and that each week thereafter you save 50¢ more than the preceding week. How much will you have saved by the end of 1 year?

52. Suppose a 100-pound bag of grain has a small hole in the bottom that is steadily getting larger. The first minute $\frac{1}{3}$ ounce of grain leaks out, and each successive minute $\frac{1}{3}$ ounce more grain leaks out than during the preceding minute. How many pounds of grain remain in the bag after one hour?

53. A $12,000 loan is paid back in 12 equal monthly payments, made during the first week of the new month. The interest rate is 2% per month on the monthly balance. (This is the amount at the end of a month before the 1000 dollar payment for that month is made.)

(a) Find the monthly interest payment for each of the first 3 months.

(b) Write the general term of the sequences that give the monthly loan balance and the monthly interest.

(c) Find the total interest paid for the year and the annual interest rate.

54. A pyramid of blocks has 26 blocks in the bottom row and 2 fewer blocks in each successive row thereafter. How many blocks are there in the pyramid?

55. Evaluate $\displaystyle\sum_{n=6}^{20} (5n - 3)$.

Use the form $S_n = \dfrac{n}{2}(a_1 + a_n)$ to find S_{80} for the sequences in Exercises 56 and 57.

56. $a_k = \frac{1}{2}k + 10$ **57.** $a_k = 3k - 8$

58. Find the sum of all the even numbers between 33 and 427.

59. Find the sum of all odd numbers from 33 to 427.

60. Find the sum of all multiples of 4 from -100 to 56.

61. If $\displaystyle\sum_{k=1}^{30} [a_1 + (k-1)d] = -5865$ and $\displaystyle\sum_{k=1}^{20} [a_1 + (k-1)d] = -2610$, find a_1 and d.

62. Listing the first few terms of a sequence like 2, 4, 6, . . . , without stating its general term or describing just what kind of sequence it is makes it impossible to predict the next term. Show that both sequences $t_n = 2n$ and $u_n = 2n + (n-1)(n-2)(n-3)$ produce these first three terms but that their fourth terms are different.

63. Find u and v such that 3, u, v, 10 is an arithmetic sequence.

64. What is the connection between arithmetic sequences and linear functions?

65. The function f defined by $f(x) = 3x + 7$ is a linear function. Evaluate $\displaystyle\sum_{k=1}^{16} f_k$, where $f_k = f(k)$ is the arithmetic sequence associated with f.

Written Assignment Let $f(x) = x^2 - 8x + 12$ for all real numbers x, and let $a_n = a_1 + (n - 1)d$ be any arithmetic sequence. What is the maximum number of points (x, y) on the graph of $y = f(x)$ that coincide with the sequential points (n, a_n)? Explain. (Note: $a_n = dn + b$ where b is the constant $a_1 - d$.)

Critical Thinking

1. Infinite sequences may or may not have an infinite number of distinct terms. Give examples of general terms for infinite sequences having the following number of distinct terms: (a) one, (b) two, (c) three.

2. If the addend $2k - 1$ in $\sum_{k=1}^{n} (2k - 1)$ is changed to $2k + 1$, then the value of the summation will remain the same provided that the proper change is made with the index of summation. Make this adjustment by completing the equation $\sum_{k=1}^{n} (2k - 1) = \sum (2k + 1)$.

   Now complete the equality $\sum_{k=3}^{n+1} (k + 1)^2 = \sum_{k=1}^{n-1} \ldots$, in which the index of summation has been modified, by adjusting the addend.

3. The procedure used in Exercises 44 and 45 on page 694 avoids tedious computations in evaluating certain sums. Use the method that was presented in Section 3.8 to convert

   $$a_k = \frac{4}{(k + 1)(k + 3)}$$

   into the difference of two fractions and then evaluate $\sum_{k=1}^{10} a_k$.

4. In Exercises 30(b) and 31(b), the sum of the first n odd integers and the sum of the first n even integers are called for. When these two answers are added, the result represents the sum of which consecutive integers beginning with 1? Use the formula for the sum of an arithmetic sequence to verify your result.

11.4 GEOMETRIC SEQUENCES

Suppose that a ball is dropped from a height of 4 feet and bounces straight up and down, always bouncing up exactly one-half the distance it just came down. How far will the ball have traveled if you catch it after it reaches the top of the fifth bounce? The following figure will help you to answer this question. For the sake of clarity, the bounces have been separated in the figure.

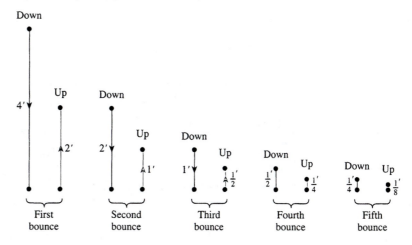

From this diagram we can determine how far the ball has traveled on each bounce. On the first bounce it goes 4 feet down and 2 feet up, for a total of 6 feet; on the second bounce the total distance is $2 + 1 = 3$ feet; and so on. These distances form the following sequence of five terms (one for each bounce):

$$6, \quad 3, \quad \frac{3}{2}, \quad \frac{3}{4}, \quad \frac{3}{8}$$

This sequence has the special property that, after the first term, each successive term can be obtained by multipying the preceding term by $\frac{1}{2}$; that is, the second term, 3, is half the first, 6, and so on. This is an example of a **geometric sequence.** Later we will develop a formula for finding the sum of such a sequence; in the meantime, we can find the total distance the ball has traveled during the five bounces by adding the first five terms as follows:

$$6 + 3 + \frac{3}{2} + \frac{3}{4} + \frac{3}{8} = \frac{48 + 24 + 12 + 6 + 3}{8} = 11\frac{5}{8}$$

*A geometric sequence is also referred to as a **geometric progression**.*

DEFINITION OF A GEOMETRIC SEQUENCE

A sequence is said to be *geometric* if each term, after the first, is obtained by multiplying the preceding term by a common value.

Here are the first four terms of a geometric sequence

$$2, \quad -4, \quad 8, \quad -16, \quad \ldots$$

By inspection, you can determine that the common multiplier for this sequence is -2. To find the nth term we will first derive a formula for the nth term of any geometric sequence.

Let a_n be the nth term of a geometric sequence, and let a_1 be its first term. The common multiplier, which is also called the **common ratio,** is denoted by r. Here are the first four terms:

$$a_1$$
$$a_2 = a_1 r$$
$$a_3 = a_2 r = (a_1 r)r = a_1 r^2$$
$$a_4 = a_3 r = (a_1 r^2)r = a_1 r^3$$

Notice that the exponent of r is 1 less than the number of the term. This observation allows us to write the nth term as follows:

GENERAL TERM OF A GEOMETRIC SEQUENCE

This formula says that the nth term of a geometric sequence is completely determined by its first term a_1 and common ratio r.

The nth term of a geometric sequence is

$$a_n = a_1 r^{n-1}$$

where a_1 is the first term and r is the common ratio.

With this result, the first four terms and the nth terms of the geometric sequence given previously are as follows.

$$2, -4, 8, -16, \ldots, 2(-2)^{n-1} \quad (r = -2)$$

Here are two more illustrations:

$$1, \; \frac{1}{3}, \; \frac{1}{9}, \; \frac{1}{27}, \; \ldots, \; 1\left(\frac{1}{3}\right)^{n-1} \quad \left(r = \tfrac{1}{3}\right)$$

$$5, \; -5, \; 5, \; -5, \; \ldots, \; 5(-1)^{n-1} \quad (r = -1)$$

You can substitute the values $n = 1, 2, 3$, and 4 into the forms for the nth terms and see that the given first four terms are obtained in each case.

EXAMPLE 1 Find the hundredth term of the geometric sequence having $r = \tfrac{1}{2}$ and $a_1 = \tfrac{1}{2}$.

Solution The nth term of this sequence is given by

$$a_n = \frac{1}{2}\left(\frac{1}{2}\right)^{n-1} = \frac{1}{2}\left(\frac{1}{2^{n-1}}\right) = \frac{1}{2^n}$$

Thus $a_{100} = \dfrac{1}{2^{100}}$. ∎

The reason r is called the common ratio of a geometric sequence is that for each n the ratio of the $(n + 1)$st term to the nth term equals r. Thus

$$\frac{a_{n+1}}{a_n} = \frac{a_1 r^n}{a_1 r^{n-1}} = r$$

Note that r can also be found using a_2 and a_3.

$$\frac{a_3}{a_2} = \frac{\frac{27}{2}}{9} = \frac{27}{2 \cdot 9} = \frac{3}{2}$$

EXAMPLE 2 Find the nth term of the geometric sequence $6, 9, \frac{27}{2}, \ldots$ and find the seventh term.

Solution First find r.

$$r = \frac{a_2}{a_1} = \frac{9}{6} = \frac{3}{2}$$

Then the nth term is

$$a_n = 6\left(\frac{3}{2}\right)^{n-1}$$

Let $n = 7$: $\quad a_7 = 6\left(\frac{3}{2}\right)^{7-1} = 6\left(\frac{3}{2}\right)^6 = (3 \cdot 2) \cdot \frac{3^6}{2^6} = \frac{3^7}{2^5} = \frac{2187}{32}$ ∎

EXAMPLE 3 Write the kth term of the geometric sequence $a_k = \left(\frac{1}{2}\right)^{2k}$ in the form $a_1 r^{k-1}$ and find the value of a_1 and r.

Solution

The first term a_1 can also be found by substituting 1 into the given formula for a_k; then find a_2 and evaluate
$$r = \frac{a_2}{a_1}.$$

$$a_k = \left(\frac{1}{2}\right)^{2k} = \left[\left(\frac{1}{2}\right)^2\right]^k = \left(\frac{1}{4}\right)^k$$

$$= \frac{1}{4}\left(\frac{1}{4}\right)^{k-1} \qquad \longleftarrow \qquad \text{This is now in the form } a_1 r^{n-1}.$$

Then $a_1 = \frac{1}{4}$ and $r = \frac{1}{4}$. ∎

TEST YOUR UNDERSTANDING

Write the first five terms of the geometric sequences with the given general term. Also write the nth term in the form $a_1 r^{n-1}$ and find r.

1. $a_n = \left(\frac{1}{2}\right)^{n-1}$ **2.** $a_n = \left(\frac{1}{2}\right)^{n+1}$ **3.** $a_n = \left(-\frac{1}{2}\right)^{n}$ **4.** $a_n = \left(-\frac{1}{3}\right)^{3n}$

Find r and the nth term of the geometric sequence with the given first two terms.

5. $\frac{1}{5}, 2$ **6.** $27, -12$

(Answers: Page 736)

EXAMPLE 4 A geometric sequence consisting of positive numbers has $a_1 = 18$ and $a_5 = \frac{32}{9}$. Find r.

Solution Use $n = 5$ in $a_n = a_1 r^{n-1}$.

$$\frac{32}{9} = 18r^4$$

$$r^4 = \frac{32}{9 \cdot 18} = \frac{16}{81}$$

$$r = \pm\sqrt[4]{\frac{16}{81}} = \pm\frac{2}{3}$$

Check this result by writing $a_1 = 18$ and finding $a_2, a_3, a_4,$ and a_5 using $a_n = a_{n-1} r$.

Since the terms are positive, $r = \frac{2}{3}$. ∎

Let us return to the original problem of this section. We found that the total distance the ball traveled was $11\frac{5}{8}$ feet. This is the sum of the first five terms of the geometric sequence whose nth term is $6\left(\frac{1}{2}\right)^{n-1}$. Adding these five terms was easy. But what about adding the first 100 terms? There is a formula for the sum of a geometric sequence that will enable us to find such answers efficiently.

The sum of a geometric sequence is called a **geometric series.** Just as with arithmetic series, there is a formula for finding such sums. To discover this formula, let $a_k = a_1 r^{k-1}$ be a geometric sequence and denote the sum of the first n terms by $S_n = \sum\limits_{k=1}^{n} a_1 r^{k-1}$. Then

$$S_n = a_1 + a_1 r + a_1 r^2 + \cdots + a_1 r^{n-2} + a_1 r^{n-1}$$

Multiplying this equation by r gives

$$rS_n = a_1 r + a_1 r^2 + \cdots + a_1 r^{n-1} + a_1 r^n$$

Now consider these two equations:

$$S_n = a_1 + a_1 r + a_1 r^2 + \cdots + a_1 r^{n-2} + a_1 r^{n-1}$$

$$\updownarrow \quad \updownarrow \qquad\qquad \updownarrow \qquad\quad \updownarrow$$

$$rS_n = \quad\; a_1 r + a_1 r^2 + \cdots + a_1 r^{n-2} + a_1 r^{n-1} + a_1 r^n$$

Subtract and factor:

$$S_n - rS_n = a_1 - a_1 r^n$$
$$(1 - r)S_n = a_1(1 - r^n)$$

Divide by $1 - r$ to solve for S_n:

Here $r \neq 1$. However, when $r = 1$, $a_k = a_1 r^{k-1} = a_1$, which is also an arithmetic sequence having $d = 0$.

$$S_n = \frac{a_1(1 - r^n)}{1 - r}$$

Returning to sigma notation, we can summarize our results this way:

SUM OF A GEOMETRIC SEQUENCE

$$S_n = \sum_{k=1}^{n} a_1 r^{k-1} = \frac{a_1(1 - r^n)}{1 - r} \qquad (r \neq 1)$$

The preceding formula can be used to verify the earlier result for the bouncing ball:

$$\sum_{k=1}^{5} 6\left(\frac{1}{2}\right)^{k-1} = \frac{6\left[1 - \left(\frac{1}{2}\right)^5\right]}{1 - \frac{1}{2}}$$

$$= \frac{6\left(1 - \frac{1}{32}\right)}{\frac{1}{2}}$$

$$= \frac{93}{8} = 11\frac{5}{8}$$

EXAMPLE 5 Find the sum of the first 100 terms of the geometric sequence given by $a_k = 6\left(\frac{1}{2}\right)^{k-1}$ and show that the answer is very close to 12.

Solution

$$S_{100} = \frac{6\left(1 - \dfrac{1}{2^{100}}\right)}{1 - \dfrac{1}{2}}$$

$$= 12\left(1 - \frac{1}{2^{100}}\right)$$

Next observe that the fraction $\dfrac{1}{2^{100}}$ is so small that $1 - \dfrac{1}{2^{100}}$ is very nearly equal to 1, and therefore S_{100} is very close to 12. ∎

EXAMPLE 6 Evaluate: $\displaystyle\sum_{k=1}^{8} 3\left(\frac{1}{10}\right)^{k+1}$

Solution

$$3\left(\frac{1}{10}\right)^{k+1} = \frac{3}{100}\left(\frac{1}{10}\right)^{k-1}$$

Another way to find a_1 and r is to write the first few terms as follows:

$$\frac{3}{100} + \frac{3}{1000} + \frac{3}{10{,}000} + \cdots$$

$$= \frac{3}{100} + \frac{3}{100}\left(\frac{1}{10}\right)$$

$$+ \frac{3}{100}\left(\frac{1}{10}\right)^2 + \cdots$$

Then $a_1 = 0.03$ and $r = 0.1$.

Then $a_1 = 0.03$, $r = 0.1$, and

$$S_8 = \frac{0.03[1 - (0.1)^8]}{1 - 0.1}$$

$$= \frac{0.03(1 - 0.00000001)}{0.9}$$

$$= 0.033333333 \qquad ∎$$

Geometric sequences have many applications, as illustrated by Examples 7 and 8. You will find others in the exercises at the end of this section.

EXAMPLE 7 Suppose that you save $128 in January and that each month thereafter you only manage to save half of what you saved the previous month. How much do you save in the tenth month, and what are your total savings after 10 months?

Solution The amounts saved each month form a geometric sequence with $a_1 = 128$ and $r = \frac{1}{2}$. Then $a_n = 128\left(\frac{1}{2}\right)^{n-1}$ and

$$a_{10} = 128\left(\frac{1}{2}\right)^9 = \frac{2^7}{2^9} = \frac{1}{4} = 0.25$$

This means that you saved 25¢ in the tenth month. Your total savings are:

$$S_{10} = \frac{128\left(1 - \dfrac{1}{2^{10}}\right)}{1 - \frac{1}{2}} \quad = 256\left(1 - \frac{1}{2^{10}}\right)$$

$$= 256 - \frac{256}{2^{10}}$$

$$= 256 - \frac{2^8}{2^{10}}$$

$$= 255.75$$

The total savings is $255.75. ■

EXAMPLE 8 A roll of wire is 625 feet long. If $\frac{1}{5}$ of the wire is cut off repeatedly, what is the general term of the sequence for the length of wire remaining? Use a calculator and the general term to determine the length of wire remaining after 7 cuts.

Solution Since $\frac{1}{5}$ is cut off, $\frac{4}{5} = 0.8$ must remain. Thus, $625(0.8) = 500$ feet remain after one cut, $625(0.8)(0.8) = 500(0.8) = 400$ feet remain after two cuts, and after n cuts are made, the length of wire remaining is $625(0.8)^n$ feet. Using a calculator, we have

$$625\,(0.8)^7 = 131.072$$

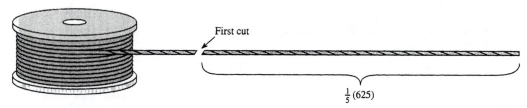

Therefore, to the nearest tenth of a foot, 131.1 feet of wire remains after 7 cuts. ■

A **geometric mean** of two real numbers a and b is a number g such that a, g, b is a geometric sequence. Letting r be the common ratio, we have $ar = g$ and $gr = b$. Solving for r and equating the results gives

$$\frac{g}{a} = \frac{b}{g} \quad \text{or} \quad g^2 = ab$$

Then $g = \pm\sqrt{ab}$, provided that $\sqrt{ab}$ is a real number. For example, the geometric means between 18 and 32 are $\pm\sqrt{18\cdot 32} = \pm\sqrt{36\cdot 16} = \pm 24$.

As done for arithmetic means, the concept of geometric means can also be extended as follows. For real numbers, a, b, and a positive integer k, if there are k numbers $g_1, g_2, \cdots, g_k$ such that

Note that -18 and -32 have the same geometric means, whereas -18 and 32 have none since $\sqrt{-18 \cdot 32}$ is not a real number.

$$a, \quad g_1, \quad g_2, \quad \ldots, \quad g_k, \quad b$$

is a geometric sequence, then the g_i's are k geometric means between a and b.

EXAMPLE 9   Insert three geometric means between 7 and 567.

Solution The question calls for three numbers g_1, g_2, g_3 so that $7, g_1, g_2, g_3, 567$ is a geometric sequence. To find the common ratio r use $a_n = a_1 r^{n-1}$ where $a_1 = 7$, $a_5 = 567$ and $n = 5$:

$$567 = 7r^4$$

$$81 = r^4$$

$$\pm 3 = r$$

There are two solutions:

$$\text{For } r = +3 \text{ we have } 7, 21, 63, 189, 567$$

$$\text{For } r = -3 \text{ we have } 7, -21, 63, -189, 567 \qquad \blacksquare$$

EXERCISES 11.4

The first three terms of a geometric sequence are given. Write the next three terms and also find the formula for the nth term.

1. $2, 4, 8, \ldots$

2. $2, -4, 8, \ldots$

3. $1, 3, 9, \ldots$

4. $2, -2, 2, \ldots$

5. $-3, 1, -\frac{1}{3}, \ldots$

6. $100, 10, 1, \ldots$

7. $-1, -5, -25, \ldots$

8. $12, -6, 3, \ldots$

9. $-6, -4, -\frac{8}{3}, \ldots$

10. $-64, 16, -4, \ldots$

11. $\frac{1}{1000}, \frac{1}{10}, 10, \ldots$

12. $\frac{27}{8}, \frac{3}{2}, \frac{2}{3}, \ldots$

Find the sum of the first six terms of the indicated sequence by using ordinary addition and also by using the formula for the sum of a geometric sequence.

13. The sequence in Exercise 1.

14. The sequence in Exercise 5.

15. The sequence in Exercise 9.

16. Find the tenth term of the geometric sequence $2, 4, 8, \ldots$

17. Find the fourteenth term of the geometric sequence $\frac{1}{8}, \frac{1}{4}, \frac{1}{2}, \ldots$

18. Find the fifteenth term of the geometric sequence $\dfrac{1}{100,000}, \dfrac{1}{10,000}, \dfrac{1}{1000}, \ldots$

19. What is the one-hundred-first term of the geometric sequence having $a_1 = 3$ and $r = -1$?

20. For the geometric sequence with $a_1 = 100$ and $r = \frac{1}{10}$, use the formula $a_n = a_1 r^{n-1}$ to find which term is equal to $\dfrac{1}{10^{10}}$.

21. Find r for the geometric sequence having $a_1 = 20$ and $a_6 = -\frac{5}{8}$.

22. Find r for geometric sequence having $a_1 = -25$ and $a_5 = -3.24$.

Evaluate.

23. $\displaystyle\sum_{k=1}^{10} 2^{k-1}$

24. $\displaystyle\sum_{j=1}^{10} 2^{j+2}$

25. $\displaystyle\sum_{k=1}^{n} 2^{k-1}$

26. $\displaystyle\sum_{k=1}^{8} 3\left(\tfrac{1}{10}\right)^{k-1}$

27. $\displaystyle\sum_{k=1}^{5} 3^{k-4}$

28. $\displaystyle\sum_{k=1}^{6} (-3)^{k-2}$

29. $\displaystyle\sum_{j=1}^{5} \left(\tfrac{2}{3}\right)^{j-2}$

30. $\displaystyle\sum_{k=1}^{8} 16\left(\tfrac{1}{2}\right)^{k+2}$

31. $\displaystyle\sum_{k=1}^{8} 16\left(-\tfrac{1}{2}\right)^{k+2}$

32. Find $u > 0$ such that $2, u, 98$ forms a geometric sequence.

33. Find $u < 0$ such that $\frac{1}{7}, u, \frac{25}{63}$ forms a geometric sequence.

34. Find a sequence whose first term is 5 that is both geometric and arithmetic. What are r and d?

35. Find the geometric means of 8 and 12.

36. Insert three geometric means between 2 and 162.

37. Insert three geometric means between 6 and 1536.

38. Insert four geometric means between 128 and 4.

39. Suppose someone offered you a job that pays 1¢ the first day, 2¢ the second day, 4¢ the third day, etc., each day earning double what was earned the preceding day. How many dollars would you earn for 30 days work?

40. Suppose that the amount you save in any given month is twice the amount you saved in the previous month. How much will you have saved at the end of 1 year if you save $1 in January? How much if you saved 25¢ in January?

41. A certain bacterial culture doubles in number every day. If there were 1000 bacteria at the end of the first day, how many will there be after 10 days? How many after n days?

42. A radioactive substance is decaying so that at the end of each month there is only one-third as much as there was at the beginning of the month. If there were 75 grams of the substance at the beginning of the year, how much is left at midyear?

43. Suppose that an automobile depreciates 10% in value each year for the first 5 years. What is it worth after 5 years if its original cost was $14,280?

44. In the compound-interest formula $A_t = P(1 + r)^t$, developed in Section 5.5, P is the initial investment, r is the annual interest rate, and t is the number of years during which the interest has been compounded annually to obtain the total value A_t. Explain how this formula may be viewed as the general term of a geometric sequence.

45. A sum of $800 is invested at 11% interest compounded annually.
   (a) What is the amount after n years?
   (b) What is the amount after 5 years?

46. How much money must be invested at the interest rate of 12%, compounded annually, so that after 3 years the amount is $1000?

47. Find the amount of money that an investment of $1500 earns at the interest rate of 8% compounded annually for 5 years.

48. (a) If $\frac{3}{5}$ of the wire in Example 8 is cut off repeatedly, what is the general form of the sequence for the length of the remaining wire?

    (b) What length remains after six cuts have been made? Give the answer to the nearest tenth of a foot.

    (c) What is the general form of the total length of wire that has been cut off after n cuts have been made?

49. (a) A set of containers are decreasing in size so that the second container is $\frac{1}{2}$ the volume of the first, the third is $\frac{1}{2}$ the volume of the second, etc. If the first container is empty and the other five are filled with water, can all the full ones be emptied into the first without the water spilling over? Explain.

    (b) Answer the question in part (a) assuming that each container, after the first, is $\frac{2}{3}$ the volume of the one preceding it.

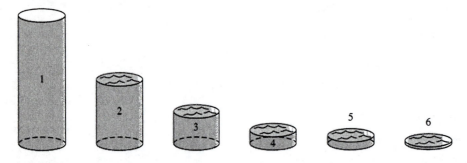

50. Find the first term and common ratio of a geometric sequence whose fourth term is $\frac{2}{9}$ and whose sixth term is $\frac{8}{81}$. (There are two possible answers.)

 Challenge Suppose you snap your fingers, wait 1 minute, and snap them again. Then you snap them after 2 more minutes, then again after 4 more minutes, again after 8 more minutes, etc., each time waiting twice as long as you waited for the preceding snap. First *guess* how many times you would snap your fingers if you continued this process for one year. Following this process, compute how long it would take you to snap your fingers (a) 10 times, (b) 15 times, and (c) 20 times.

Written Assignment

1. Use specific examples to explain the difference between an arithmetic sequence and a geometric sequence.

2. In the formula for the sum of a geometric sequence $r \neq 1$. Explain how the case $r = 1$ is somehow otherwise included in our study of sequences.

11.5 INFINITE GEOMETRIC SERIES

The ancient Greek philosopher Zeno (c. 450 B.C.) proposed four paradoxes that baffled the philosophers of that time. In one of these he argued that one could never cross a room, for example, because to do so you would first have to reach the halfway point to the wall. Thereafter, you would still have to travel one half of the remaining distance before continuing, so that one-fourth of the distance to be traversed would still remain . . . and so forth. Using w for the width of the room, the distance traveled could be indicated by this sum of an *infinite number of terms:*

$$\tfrac{1}{2}w + \tfrac{1}{4}w + \tfrac{1}{8}w + \tfrac{1}{16}w + \cdots$$

Thus, no matter how close you might get to the other side of the room, you must always travel one-half of the remaining distance before you get to the wall. Therefore, Zeno argued, you could never reach your goal!

A mathematical solution for this paradox will be given later in this section, but first we consider the meaning of *the sum of an infinite number of terms.* Your familiarity with decimal forms will be helpful in this regard.

In decimal form the fraction $\frac{3}{4}$ becomes 0.75, which means $\frac{75}{100}$. This can also be written as $\frac{7}{10} + \frac{5}{100}$. What about $\frac{1}{3}$? As a decimal we can write

$$\frac{1}{3} = 0.333 \ldots$$

where the dots mean that the 3 repeats endlessly. This decimal can be expressed as the sum of fractions whose denominators are powers of 10:

$$\frac{1}{3} = \frac{3}{10} + \frac{3}{100} + \frac{3}{1000} + \cdots$$

*The sum of an infinite sequence is an **infinite series**.*

The numbers being added here are the terms of the *infinite geometric sequence* with first term $a_1 = \frac{3}{10}$ and common ratio $r = \frac{1}{10}$. Thus the nth term is

$$a_1 r^{n-1} = \frac{3}{10}\left(\frac{1}{10}\right)^{n-1} = 3\left(\frac{1}{10}\right)\left(\frac{1}{10}\right)^{n-1}$$

$$= 3\left(\frac{1}{10}\right)^n$$

$$= \frac{3}{10^n}$$

The sum of the first n terms, S_n, is called an **nth partial sum** and is found by using the formula

$$S_n = \sum_{k=1}^{n} a_1 r^{k-1} = \frac{a_1(1 - r^n)}{1 - r}$$

Here are some cases:

$$S_1 = \frac{\frac{3}{10}\left(1 - \frac{1}{10}\right)}{1 - \frac{1}{10}} = \frac{1}{3}\left(1 - \frac{1}{10}\right) = 0.3$$

$$S_2 = \frac{\frac{3}{10}\left(1 - \frac{1}{10^2}\right)}{1 - \frac{1}{10}} = \frac{1}{3}\left(1 - \frac{1}{10^2}\right) = 0.33$$

$$S_{10} = \frac{\frac{3}{10}\left(1 - \frac{1}{10^{10}}\right)}{1 - \frac{1}{10}} = \frac{1}{3}\left(1 - \frac{1}{10^{10}}\right) = 0.\underbrace{3333333333}_{10 \text{ places}}$$

$$S_n = \frac{\frac{3}{10}\left(1 - \frac{1}{10^n}\right)}{1 - \frac{1}{10}} = \frac{1}{3}\left(1 - \frac{1}{10^n}\right) = 0.\underbrace{333 \ldots 3}_{n \text{ places}}$$

You can see that as more and more terms are added, the closer and closer the answer gets to $\frac{1}{3}$. This can be seen by studying the form for the sum of the first n terms:

$$S_n = \frac{1}{3}\left(1 - \frac{1}{10^n}\right)$$

It is clear that the bigger n is, the closer $\frac{1}{10^n}$ is to zero, the closer $1 - \frac{1}{10^n}$ is to 1, and, finally, the closer S_n is to $\frac{1}{3}$. In symbols, we may express this as follows:

$$\text{As } n \longrightarrow \infty, \frac{1}{10^n} \longrightarrow 0 \text{ and } 1 - \frac{1}{10^n} \longrightarrow 1$$

Thus, as $n \longrightarrow \infty, S_n \longrightarrow \frac{1}{3}$

Although it is true that S_n is never exactly equal to $\frac{1}{3}$, for very large n the difference between S_n and $\frac{1}{3}$ is very small. Saying this another way:

By taking n large enough, the partial sums S_n can be made as close to $\frac{1}{3}$ as we like.

This is what we mean when we say that the sum of all the terms is $\frac{1}{3}$.

$$\frac{3}{10} + \frac{3}{10^2} + \frac{3}{10^3} + \cdots + \frac{3}{10^n} + \cdots = \frac{1}{3}$$

The summation symbol, $\sum$, can also be used here after an adjustment in notation is made. Traditionally, the symbol ∞ has been used to suggest an infinite number of objects. So we use this and make the transition from the sum of a finite number of terms

$$S_n = \sum_{k=1}^{n} \frac{3}{10^k} = \frac{3}{10} + \frac{3}{10^2} + \cdots + \frac{3}{10^n} = \frac{1}{3}\left(1 - \frac{1}{10^n}\right)$$

In calculus the symbol S_∞ is replaced by $\lim\limits_{n \to \infty} S_n = \frac{1}{3}$, which is read as "the limit of S_n as n gets arbitrarily large is $\frac{1}{3}$."

to the sum of an infinite number of terms:

$$S_\infty = \sum_{k=1}^{\infty} \frac{3}{10^k} = \frac{3}{10} + \frac{3}{10^2} + \cdots + \frac{3}{10^n} + \cdots = \frac{1}{3}$$

Not every geometric sequence produces an infinite geometric series that has a finite sum. For instance, the sequence

$$2, \quad 4, \quad 8, \ldots, \quad 2^n, \ldots$$

is geometric, but the corresponding geometric series

$$2 + 4 + 8 + \cdots + 2^n + \cdots$$

cannot have a finite sum. The partial sums become larger and larger without bound.

By now you might suspect that the common ratio r determines whether or not an infinite geometric sequence can be added. This turns out to be true. To see how this works, the general case will be considered next.

Let

$$a_1, \quad a_1 r, \quad a_1 r^2, \ldots, \quad a_1 r^{n-1}, \ldots$$

be an infinite geometric sequence.

Then the sum of the first n terms, the nth partial sum, is

$$S_n = \frac{a_1(1 - r^n)}{1 - r}$$

Rewrite in this form:

$$S_n = \frac{a_1}{1 - r}(1 - r^n)$$

Use a calculator to verify the powers of $r = 0.9$ and $r = 1.1$ to the indicated decimal places.

$$
\begin{array}{ll}
(0.9)^1 & = 0.9 \\
(0.9)^{10} & = 0.35 \\
(0.9)^{20} & = 0.12 \\
(0.9)^{40} & = 0.015 \\
(0.9)^{80} & = 0.0002 \\
(0.9)^{100} & = 0.00003
\end{array}
$$

$\downarrow$

getting close to 0

$$
\begin{array}{ll}
(1.1)^1 & = 1.1 \\
(1.1)^5 & = 1.6 \\
(1.1)^{10} & = 2.6 \\
(1.1)^{20} & = 6.7 \\
(1.1)^{50} & = 117.4 \\
(1.1)^{100} & = 13780.6
\end{array}
$$

$\downarrow$

getting very large

At this point the importance of r^n becomes clear. If, as n gets larger, r^n gets very large, then the infinite geometric series will not have a finite sum. But if r^n gets arbitrarily close to zero as n gets larger, then $1 - r^n$ gets close to 1 and S_n gets closer and closer to $\dfrac{a_1}{1 - r}$.

The values of r for which r^n gets arbitrarily close to zero are precisely those values between -1 and 1; that is, $|r| < 1$. For instance, $\frac{3}{5}$, $-\frac{1}{10}$, and 0.09 are values of r for which r^n gets close to zero; and 1.01, -2, and $\frac{3}{2}$ are values for which the series does not have a finite sum.

To sum up, we have the following result:

SUM OF AN INFINITE GEOMETRIC SEQUENCE

If $|r| < 1$, then $S_\infty = \displaystyle\sum_{k=1}^{\infty} a_1 r^{k-1} = \dfrac{a_1}{1 - r}$. For other values of r the series has no finite sum.

EXAMPLE 1   Find the sum of the infinite geometric series

$$
27 + 3 + \frac{1}{3} + \cdots
$$

Solution Since $r = \frac{3}{27} = \frac{1}{9}$ and $a_1 = 27$, the preceding result gives

$$
S_\infty = 27 + 3 + \frac{1}{3} + \cdots = \frac{27}{1 - \frac{1}{9}} = \frac{27}{\frac{8}{9}} = \frac{243}{8}
$$
∎

EXAMPLE 2   Why does the infinite geometric series $\displaystyle\sum_{k=1}^{\infty} 5\left(\frac{4}{3}\right)^{k-1}$ have no finite sum?

Solution The series has no finite sum because the common ratio $r = \frac{4}{3}$ is not between -1 and 1.
∎

Another way to find a_1 is to let $k = 1$ in $\dfrac{7}{10^{k+1}}$:

$$
a_1 = \frac{7}{10^2} = \frac{7}{100}
$$

Also, r can be found by taking the ratio of the second term to the first term.

$$
r = \frac{\dfrac{7}{10^3}}{\dfrac{7}{10^2}} = \frac{1}{10}
$$

EXAMPLE 3   Find: $\displaystyle\sum_{k=1}^{\infty} \frac{7}{10^{k+1}}$

Solution Since $\dfrac{7}{10^{k+1}} = 7\left(\dfrac{1}{10^{k+1}}\right) = 7\left(\dfrac{1}{10^2}\right)\dfrac{1}{10^{k-1}} = \dfrac{7}{100}\left(\dfrac{1}{10}\right)^{k-1}$, it follows that $a_1 = \frac{7}{100}$ and $r = \frac{1}{10}$. Therefore, by the formula for the sum of an infinite geometric series we have

$$
S_\infty = \sum_{k=1}^{\infty} \frac{7}{10^{k+1}} = \frac{\frac{7}{100}}{1 - \frac{1}{10}} = \frac{7}{100 - 10} = \frac{7}{90}
$$
∎

TEST YOUR UNDERSTANDING

Find the common ratio r, and then find the sum, if it exists, of the given infinite geometric series.

1. $10 + 1 + \frac{1}{10} + \cdots$ **2.** $\frac{1}{64} + \frac{1}{16} + \frac{1}{4} + \cdots$

3. $36 - 6 + 1 - \cdots$ **4.** $-16 - 4 - 1 - \cdots$

5. $\displaystyle\sum_{k=1}^{\infty} \left(\frac{4}{3}\right)^{k-1}$ **6.** $\displaystyle\sum_{k=1}^{\infty} 3(0.01)^k$

7. $\displaystyle\sum_{i=1}^{\infty} (-1)^i 3^i$ **8.** $\displaystyle\sum_{n=1}^{\infty} 100\left(-\frac{9}{10}\right)^{n+1}$

9. $101 - 102.01 + 103.0301 - \cdots$

(Answers: Page 736)

Earlier we saw how the endless repeating decimal $0.333\ldots$ can be regarded as an infinite geometric series. The next example illustrates how such decimal fractions can be written in the rational form $\frac{a}{b}$ (the ratio of two integers) by using the formula for the sum of an infinite geometric series.

Compare the method shown in Example 4 with that developed in Exercise 57 of Section 1.1.

EXAMPLE 4 Express the repeating decimal $0.242424\ldots$ in rational form.

Solution First write

$$0.242424\ldots = \frac{24}{100} + \frac{24}{10,000} + \frac{24}{1,000,000} + \cdots$$

$$= \frac{24}{10^2} + \frac{24}{10^4} + \frac{24}{10^6} + \cdots + \frac{24}{10^{2k}} + \cdots$$

$$= \frac{24}{100} + \frac{24}{100}\left(\frac{1}{100}\right) + \frac{24}{100}\left(\frac{1}{100}\right)^2 + \cdots + \frac{24}{100}\left(\frac{1}{100}\right)^{k-1} + \cdots$$

Observe that

$$\frac{24}{10^{2k}} = 24\left(\frac{1}{10^{2k}}\right)$$
$$= 24\left(\frac{1}{10^2}\right)^k = 24\left(\frac{1}{100}\right)^k$$
$$= \frac{24}{100}\left(\frac{1}{100}\right)^{k-1}$$

Then $a_1 = \frac{24}{100}$, $r = \frac{1}{100}$, and

$$0.242424\ldots = \sum_{k=1}^{\infty} \frac{24}{100}\left(\frac{1}{100}\right)^{k-1}$$

$$= \frac{\frac{24}{100}}{1 - \frac{1}{100}}$$

$$= \frac{24}{99} = \frac{8}{33} \qquad \text{Check this result by dividing 33 into 8.} \qquad \blacksquare$$

It seems as if the horse cannot finish the race this way. But read on to see that there is really no contradiction with this interpretation.

EXAMPLE 5 A racehorse running at the constant rate of 30 miles per hour will finish a 1-mile race in 2 minutes. Now consider the race broken down into the following parts: before the racehorse can finish the 1-mile race it must first reach the halfway mark; having done that, the horse must next reach the quarter pole; then it must reach the eighth pole; and so on. That is, it must always cover half the distance remaining before it can cover the whole distance. Show that the sum of the infinite number of time intervals is also 2 minutes.

$T = \dfrac{D}{R} \left(time = \dfrac{distance}{rate} \right)$

Observe that the rate of 30 mph converts to $\frac{1}{2}$ mile per minute.

Solution For the first $\frac{1}{2}$ mile the time will be $\dfrac{\frac{1}{2}}{\frac{1}{2}} = 1$ minute; for the next $\frac{1}{4}$ mile the time will be $\dfrac{\frac{1}{4}}{\frac{1}{2}} = \frac{1}{2}$ minute; for the next $\frac{1}{8}$ mile the time will be $\dfrac{\frac{1}{8}}{\frac{1}{2}} = \frac{1}{4}$ minute; and for the nth distance, which is $\dfrac{1}{2^n}$ miles, the time will be $\dfrac{\frac{1}{2^n}}{\frac{1}{2}} = \dfrac{1}{2^{n-1}}$ minute.

Thus the total time is given by this series:

$$\sum_{k=1}^{\infty} \frac{1}{2^{k-1}} = 1 + \frac{1}{2} + \frac{1}{4} + \cdots + \frac{1}{2^{n-1}} + \cdots$$

This is an infinite geometric series having $a_1 = 1$ and $r = \frac{1}{2}$. Thus

$$\sum_{k=1}^{\infty} 1 \left(\frac{1}{2}\right)^{k-1} = \frac{1}{1 - \frac{1}{2}} = 2 \text{ minutes}$$

which is the same result as before. ■

You probably recognized the problem in Example 5 as a variation of Zeno's paradox which was used to introduce this section. In that situation, we dealt with what we now know to be an infinite geometric series

$$\frac{1}{2}w + \frac{1}{4}w + \frac{1}{8}w + \frac{1}{16}w + \cdots$$

with $a_1 = \dfrac{1}{2}w$ and $r = \dfrac{1}{2}$, which has the sum $S_\infty = \dfrac{a_1}{1-r} = \dfrac{\frac{1}{2}w}{1 - \frac{1}{2}} = w.$

This provides a mathematical solution to the paradox.

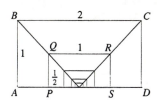

EXAMPLE 6   Rectangle *ABCD* has dimensions 1 by 2. The next rectangle *PQRS* has dimensions $\frac{1}{2}$ by 1. In like manner, each inner rectangle has dimensions half the size of the preceding rectangle. If this sequence of rectangles continues endlessly, what is the sum of the areas of all the rectangles?

Solution   The area of rectangle *ABCD* is $1 \cdot 2$, the area of rectangle *PQRS* is $\frac{1}{2} \cdot 1$, the next has area $\frac{1}{4} \cdot \frac{1}{2}$, and so on. The sum of all the areas is this infinite geometric series:

$$1 \cdot 2 + \tfrac{1}{2} \cdot 1 + \tfrac{1}{4} \cdot \tfrac{1}{2} + \tfrac{1}{8} \cdot \tfrac{1}{4} + \cdots$$

$$= 2 + \tfrac{1}{2} + \tfrac{1}{8} + \tfrac{1}{32} + \cdots$$

$$= 2 + \tfrac{1}{2} + \left(\tfrac{1}{2}\right)^3 + \left(\tfrac{1}{2}\right)^5 + \cdots$$

Since $a_1 = 2$ and $r = \frac{1}{4}$, the sum equals

$$\frac{a_1}{1 - r} = \frac{2}{1 - \frac{1}{4}} = \frac{8}{3} = 2\tfrac{2}{3} \qquad \blacksquare$$

CAUTION: Learn to Avoid These Mistakes	
WRONG	**RIGHT**
$\displaystyle\sum_{k=1}^{\infty} \left(\tfrac{1}{3}\right)^{n+1} = \dfrac{1}{1 - \frac{1}{3}}$	$\displaystyle\sum_{n=1}^{\infty} \left(\tfrac{1}{3}\right)^{n+1} = \sum_{n=1}^{\infty} \tfrac{1}{9}\left(\tfrac{1}{3}\right)^{n-1}$ $= \dfrac{\frac{1}{9}}{1 - \frac{1}{3}}$
$\displaystyle\sum_{n=1}^{\infty} \left(-\tfrac{1}{2}\right)^{n-1} = \dfrac{1}{1 - \frac{1}{2}}$	$\displaystyle\sum_{n=1}^{\infty} \left(-\tfrac{1}{2}\right)^{n-1} = \dfrac{1}{1 - \left(-\frac{1}{2}\right)}$
$\displaystyle\sum_{n=1}^{\infty} 2(1.03)^{n-1} = \dfrac{2}{1 - 1.03}$	$\displaystyle\sum_{n=1}^{\infty} 2(1.03)^{n-1}$ is not a finite sum since $r = 1.03 > 1$.

EXERCISES 11.5

Find the sum, if it exists, of each infinite geometric series.

1. $2 + 1 + \frac{1}{2} + \cdots$ **2.** $8 + 4 + 2 + \cdots$ **3.** $25 + 5 + 1 + \cdots$

4. $1 + \frac{4}{3} + \frac{16}{9} + \cdots$ **5.** $1 - \frac{1}{2} + \frac{1}{4} - \cdots$ **6.** $100 - 1 + \frac{1}{100} - \cdots$

7. $1 + 0.1 + 0.01 + \cdots$ **8.** $52 + 0.52 + 0.0052 + \cdots$ **9.** $-2 - \frac{1}{4} - \frac{1}{32} - \cdots$

10. $-729 + 81 - 9 + \cdots$

Explain what is wrong with each of these statements.

11. $\displaystyle\sum_{n=1}^{\infty} \left(\frac{1}{2}\right)^{n+1} = \frac{1}{1 - \frac{1}{2}}$ **12.** $\displaystyle\sum_{n=1}^{\infty} (-1)^{n-1} = \frac{1}{1 - (-1)} = \frac{1}{2}$

13. $\displaystyle\sum_{n=1}^{\infty} \left(-\frac{1}{3}\right)^{n-1} = \frac{1}{1 - \frac{1}{3}}$ **14.** $\displaystyle\sum_{n=1}^{\infty} 3(1.02)^{n-1} = \frac{3}{1 - 1.02}$

Decide whether or not the given infinite geometric series has a sum. If it does, find it using

$$S_\infty = \frac{a_1}{1 - r}.$$

15. $\displaystyle\sum_{k=1}^{\infty} \left(\frac{1}{3}\right)^{k-1}$ **16.** $\displaystyle\sum_{k=1}^{\infty} \left(\frac{1}{3}\right)^{k}$ **17.** $\displaystyle\sum_{k=1}^{\infty} \left(\frac{1}{3}\right)^{k+1}$ **18.** $\displaystyle\sum_{n=1}^{\infty} \frac{1}{2^{n+1}}$ **19.** $\displaystyle\sum_{n=1}^{\infty} \frac{1}{2^{n-2}}$

20. $\displaystyle\sum_{k=1}^{\infty} \left(\frac{1}{10}\right)^{k-1}$ **21.** $\displaystyle\sum_{k=1}^{\infty} 2(0.1)^{k-1}$ **22.** $\displaystyle\sum_{k=1}^{\infty} \left(-\frac{1}{2}\right)^{k-1}$ **23.** $\displaystyle\sum_{n=1}^{\infty} \left(\frac{3}{2}\right)^{n-1}$ **24.** $\displaystyle\sum_{n=1}^{\infty} \left(-\frac{1}{3}\right)^{n+2}$

25. $\displaystyle\sum_{k=1}^{\infty} (0.7)^{k-1}$ **26.** $\displaystyle\sum_{k=1}^{\infty} 5(0.7)^{k}$ **27.** $\displaystyle\sum_{k=1}^{\infty} 5(1.01)^{k}$ **28.** $\displaystyle\sum_{k=1}^{\infty} \left(\frac{1}{10}\right)^{k-4}$ **29.** $\displaystyle\sum_{k=1}^{\infty} 10\left(\frac{2}{3}\right)^{k-1}$

30. $\displaystyle\sum_{k=1}^{\infty} (-1)^{k}$ **31.** $\displaystyle\sum_{k=1}^{\infty} (0.45)^{k-1}$ **32.** $\displaystyle\sum_{k=1}^{\infty} (-0.9)^{k+1}$ **33.** $\displaystyle\sum_{n=1}^{\infty} 7\left(-\frac{3}{4}\right)^{n-1}$ **34.** $\displaystyle\sum_{k=1}^{\infty} (0.1)^{2k}$

35. $\displaystyle\sum_{k=1}^{\infty} \left(-\frac{2}{5}\right)^{2k}$

Find a rational form for each of the following repeating decimals in a manner similar to that in Example 4. Check your answers.

36. $0.444\ldots$ **37.** $0.777\ldots$ **38.** $7.777\ldots$ **39.** $0.131313\ldots$

40. $13.131313\ldots$ **41.** $0.0131313\ldots$ **42.** $0.050505\ldots$ **43.** $0.999\ldots$

44. $0.125125125\ldots$

45. Suppose that a 1-mile distance a racehorse must run is divided into an infinite number of parts, obtained by always considering $\frac{2}{3}$ of the remaining distance to be covered. Then the lengths of these parts form the sequence

$$\frac{2}{3}, \frac{2}{9}, \frac{2}{27}, \ldots, \frac{2}{3^n}, \ldots$$

(a) Find the sequence of times corresponding to these distances. (Assume that the horse is moving at a rate of $\frac{1}{2}$ mile per minute.)

(b) Show that the sum of the times in part (a) is 2 minutes.

46. A certain ball always rebounds $\frac{1}{3}$ of the distance it falls. If the ball is dropped from a height of 9 feet, how far does it travel before coming to rest? (See the similar situation at the beginning of Section 11.4.)

47. A substance initially weighing 64 grams is decaying at a rate such that after 4 hours there are only 32 grams left. In another 2 hours only 16 grams remain; in another 1 hour after that only 8 grams remain; and so on so that the time intervals and amounts remaining form geometric sequences. How long does it take altogether until nothing of the substance is left?

48. After it is set in motion, each swing in either direction of a particular pendulum is 40% as long as the preceding swing. What is the total distance that the end of the pendulum travels before coming to rest if the first swing is 30 inches long?

49. Assume that a racehorse takes 1 minute to go the first $\frac{1}{2}$ mile of a 1-mile race. After that, the horse's speed is no longer constant: for the next $\frac{1}{4}$ mile it takes $\frac{2}{5}$ minute; for the next $\frac{1}{8}$ mile it takes $\frac{4}{9}$ minute; for the next $\frac{1}{16}$ mile it takes $\frac{40}{81}$ minute; and so on, so that the time intervals form a geometric sequence. Why can't the horse finish the race?

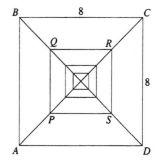

50. (a) *ABCD* is a square whose sides measure 8 units. *PQRS* is a square whose sides are $\frac{1}{2}$ the length of the sides of square *ABCD*. The next square has sides $\frac{1}{2}$ the length of those of square *PQRS*. In like manner each inner square has sides $\frac{1}{2}$ the length of the preceding square. If this sequence of squares continues endlessly, what is the sum of all the areas of the squares in the sequence?

(b) What is the sum of all the perimeters?

51. *ABC* is an isosceles right triangle with right angle at *C*. P_1 is the midpoint of the hypotenuse *AB* so that CP_1 divides triangle *ABC* into two congruent triangles. P_2 is the midpoint of *BC* so that P_1P_2 divides triangle CBP_1 into two congruent triangles. This process continues endlessly.

(a) If $AC = CB = 4$, what do you expect the sum of the areas of all the triangles labeled 1, 2, 3, . . . to be equal to?

(b) Verify the result in part (a) by using an infinite geometric series.

(c) Find the sum of all the triangles labeled with odd numbers and also find the sum of all the triangles labeled with even numbers. What is the sum of the two sums?

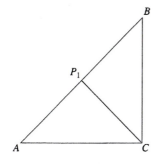
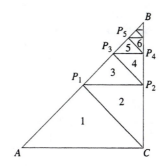

52. The largest circle has radius $A_1B = 1$. The next circle has radius $A_2B = \frac{1}{2}A_1B$, the one after that has radius $A_3B = \frac{1}{2}A_2B$, and so on. If these circles continue endlessly in this manner, what is the sum of the areas of all the circles?

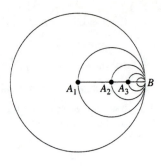

53. Triangle AB_1C_1 has a right angle at C_1. $AC_1 = 9$ and $B_1C_1 = 3$. Points $C_2, C_3, C_4, \ldots$ are chosen so that $AC_2 = \frac{2}{3}AC_1$, $AC_3 = \frac{2}{3}AC_2$, and so on. Find the sum of the areas of all the right triangles labeled AB_kC_k for $k = 1, 2, 3, \ldots$.

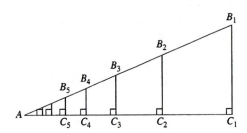

54. A ball is released on a semicircular track and travels 8 feet before it reverses direction. Then it travels 6 feet before reversing a second time. This movement with reversals continues indefinitely so that the distances traveled between reversals forms an infinite geometric sequence. Find the total distance traveled before coming to rest.

55. (a) Suppose you have $2000 in a special savings account and spend 60% on various products. Next, assume that the owners of the stores from which the products were purchased also spend 60% of what you paid them. If this process continues indefinitely, the total amount spent forms an infinite geometric series whose first term is 1200. Find the total spent on all the purchases made.

(b) Assume that at each step in part (a) the amount not spent is put into savings. Use an infinite geometric series, having first term 800, to find the total savings.

56. In the study of calculus it is shown that the number $e = 2.718281 \ldots$ (see section 5.4) is given by the infinite series $e = 1 + 1 + \dfrac{1}{2!} + \dfrac{1}{3!} + \cdots + \dfrac{1}{k!} + \cdots$

where $k! = k(k-1)(k-2)\cdots 3 \cdot 2 \cdot 1$. Approximate e by adding the first six terms of the series and compare to the decimal form of e.

Challenge Use the identity $\dfrac{1}{k(k+1)} = \dfrac{1}{k} - \dfrac{1}{(k+1)}$ to find the sum $\displaystyle\sum_{k=1}^{\infty} \dfrac{1}{k(k+1)}$.

(*Hint:* Consider the partial sum $S_n = \displaystyle\sum_{k=1}^{n} \dfrac{1}{k(k+1)}$ as n gets arbitrarily large.)

 Graphing Calculator Exercises

1. The sum of the infinite geometric sequence with the first term equal to 1 and common ratio equal to x may be written

$$1 + x + x^2 + x^3 + \cdots = 1/(1 - x)$$

as long as $|x| < 1$. (We call this a *power series* representation of $1/(1-x)$.) Graph the functions $f(x) = 1/(1-x)$, $1 + x$, $1 + x + x^2$, and $1 + x + x^2 + x^3$ on the same set of axes with $-1 \le x \le 2$. What do you notice as more terms are added to the series?

2. Add three more terms to the series in Exercise 1. As each new term is added, is your observation in Exercise 1 confirmed?

3. Repeat Exercises 1 and 2 with the series $1 - x + x^2 - x^3 + \ldots$ and the function $f(x) = 1/(1 + x)$ with $-2 \le x \le 1$.

4. Give a four-term power series approximation of the function $1/(1 + 2x)$ and confirm your guess by graphing. For what values of x do you suppose the corresponding power series representation of $1/(1 + 2x)$ would be valid?

 Critical Thinking

1. Which is larger, the sum of the first n powers of 2, starting with 2^0, or the single power 2^n? Justify your answer.

2. Observe that $\displaystyle\sum_{k=1}^{2n} [2^{k-1} + (-2)^{k-1}] = \sum_{k=1}^{2n} 2^{k-1} + \sum_{k=1}^{2n} (-2)^{k-1}$. Therefore, the series at the left can be evaluated by evaluating each series at the right and adding the results. Do this computation and use the answer to show that the series on the left can be written in the form $\displaystyle\sum_{k=1}^{n} a_1 r^{k-1}$.

3. For each x such that $|x| < 1$, $\displaystyle\sum_{k=1}^{\infty} x^{k-1}$ is an infinite geometric series having a finite sum.

   Thus we may define the function f by $f(x) = \displaystyle\sum_{k=1}^{\infty} x^{k-1}$ for $|x| < 1$. Sketch the graph of this function.

4. Decide if the series $\displaystyle\sum_{k=1}^{\infty} \dfrac{3^{k+1}}{5^{k-1}}$ is geometric having a ratio r, where $-1 < r < 1$. If it is, then find the sum. If it isn't, give a reason.

5. For which values of x does the geometric series $\displaystyle\sum_{k=1}^{\infty} (3x - 4)^{k-1}$ have a finite sum? Find the sum for such x.

11.6 MATHEMATICAL INDUCTION

Study these statements and search for a pattern.

$$1 = 1^2$$
$$1 + 3 = 2^2$$
$$1 + 3 + 5 = 3^2$$
$$1 + 3 + 5 + 7 = 4^2$$
$$1 + 3 + 5 + 7 + 9 = 5^2$$

Do you see the pattern? The last statement shows that the sum of the first five positive odd integers is 5^2. What about the sum of the first six positive odd integers? The pattern is the same:

$$1 + 3 + 5 + 7 + 9 + 11 = 6^2$$

It would be reasonable to guess that the sum of the first n positive odd integers is n^2. That is,

$$1 + 3 + 5 + \cdots + (2n - 1) = n^2$$

But a guess is not a proof. It is our objective in this section to learn how to prove a statement that involves an infinite number of cases.

Let us refer to the nth statement above as S_n. Thus S_1, S_2, S_3, S_4, S_5, and S_6 are the first six cases of S_n that we know are true.

Does the truth of the first six cases allow us to conclude that S_n is true for all positive integers n? No! We cannot assume that a few special cases guarantee an infinite number of cases. If we allowed "proving by a finite number of cases," then the following example is such a "proof" that all positive even integers are less than 100.

The first positive even integer is 2, and we know that $2 < 100$.
The second is 4, and we know that $4 < 100$.
The third is 6, and $6 < 100$.
Therefore, since $2n < 100$ for a finite number of cases, we might conclude that $2n < 100$ for all n.

This false result should convince you that in trying to prove a collection of statements S_n for all positive integers $n = 1, 2, 3, \ldots$, we need to do more than just check it out for a finite number of cases. We need to call on a type of proof known as **mathematical induction**.

Suppose that we had a long (endless) row of dominoes each 2 inches long all standing up in a straight row so that the distance between any two of them is $1\frac{1}{2}$ inches. How can you make them all fall down with the least effort?

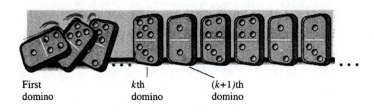

First kth $(k+1)$th
domino   domino   domino

The answer is simple. Push the first domino down toward the second. Since the first one falls, and because the space between each pair is less than the length of a domino, they will all (eventually) fall down. The first knocks down the second, the second knocks down the third, and, in general, the kth domino knocks down the $(k + 1)$st. Two things guaranteed this "chain reaction":

1. The first domino will fall.
2. If any domino falls, then so will the next.

These two conditions are guidelines in forming the **principle of mathematical induction**.

THE PRINCIPLE OF MATHEMATICAL INDUCTION

Let S_n be a statement for each positive integer n. Suppose that the following two conditions hold:

1. S_1 is true.
2. If S_k is true, then S_{k+1} is true, where k is any positive integer.

Then S_n is a true statement for all positive integers n.

Condition 1 starts the "chain reaction" and condition 2 keeps it going.

Note that we are not proving this principle; rather, it is a basic principle that we accept and use to construct proofs. It is very important to realize that in condition 2 we are not proving S_k to be true; rather, we must prove this proposition:

If S_k is true, then S_{k+1} is true.

Consequently, a proof by mathematical induction *includes* a proof of the proposition that S_k implies S_{k+1}, a proof within a proof. Within that inner proof, we are allowed to *assume* and use S_k.

At the beginning of this section, we guessed at the formula for the sum of the first n odd integers. Now, using mathematical induction, this formula is proved in Example 1.

EXAMPLE 1 Prove by mathematical induction that S_n is true for all positive integers n, where S_n is the statement

$$1 + 3 + 5 + \cdots + (2n - 1) = n^2$$

PROOF Both conditions 1 and 2 of the principle of mathematical induction must be satisfied. We begin with the first.

Proving S_1 starts the chain. The first domino has fallen.

1. S_1 is true because $1 = 1^2$.

We assume S_k to be true to see what effect it has on the next case, S_{k+1}. This is comparable to considering what happens when the kth domino falls.

2. Suppose that S_k is true, where k is a positive integer. That is, we *assume*

$$1 + 3 + 5 + \cdots + (2k - 1) = k^2$$

We want to prove that S_{k+1} follows from this. To do so note that the next odd number after $2k - 1$ is $2(k + 1) - 1 = 2k + 1$, which is added to the preceding equation:

$$1 + 3 + 5 + \cdots + (2k - 1) \qquad = k^2$$

$$\underline{\hspace{3cm} 2k + 1 = 2k + 1 \hspace{2cm}}$$

$$1 + 3 + 5 + \cdots + (2k - 1) + (2k + 1) = k^2 + 2k + 1$$

Now factor $k^2 + 2k + 1$:

$$1 + 3 + 5 + \cdots + (2k - 1) + (2k + 1) = (k + 1)^2$$

This is the statement S_{k+1}. Therefore, we have shown that if S_k is given, then S_{k+1} must follow. This, together with the fact that S_1 is true, allows us to say that S_n is true for all n by the principle of mathematical induction. ■

EXAMPLE 2 Prove that the sum of the squares of the first n consecutive positive integers is given by $\dfrac{n(n + 1)(2n + 1)}{6}$.

PROOF Let S_n be the statement

$$\underbrace{1^2 + 2^2 + 3^2 + \cdots + n^2}_{\text{(sum of first } n \text{ squares)}} = \frac{n(n + 1)(2n + 1)}{6}$$

for any positive integer n.

1. S_1 is true because $1^2 = \dfrac{1(1 + 1)(2 \cdot 1 + 1)}{6}$.

2. Suppose that S_k is true for any k. That is, we *assume*

$$1^2 + 2^2 + 3^2 + \cdots + k^2 = \frac{k(k + 1)(2k + 1)}{6}$$

We want to prove that S_{k+1} follows from this. To do so, add the next square, $(k + 1)^2$, to both sides.

(A) $\quad 1^2 + 2^2 + 3^2 + \cdots + k^2 + (k + 1)^2 = \dfrac{k(k + 1)(2k + 1)}{6} + (k + 1)^2$

Combine the right side.

$$\frac{k(k + 1)(2k + 1)}{6} + (k + 1)^2 = \frac{k(k + 1)(2k + 1) + 6(k + 1)^2}{6}$$

$$= \frac{(k + 1)[k(2k + 1) + 6(k + 1)]}{6}$$

$$= \frac{(k + 1)(2k^2 + 7k + 6)}{6}$$

$$= \frac{(k + 1)(k + 2)(2k + 3)}{6}$$

Substitute back into equation (A).

$$1^2 + 2^2 + 3^2 + \cdots + (k + 1)^2 = \frac{(k + 1)(k + 2)(2k + 3)}{6}$$

To see that this is S_{k+1}, rewrite the right side:

Observe that on the left the first k + 1 squares are added, and on the right we have the required form in terms of k + 1.

$$1^2 + 2^2 + 3^2 + \cdots + (k + 1)^2 = \frac{(k + 1)[(k + 1) + 1][2(k + 1) + 1]}{6}$$

Now both conditions of the principle of mathematical induction have been satisfied, and it follows that

$$1^2 + 2^2 + 3^3 + \cdots + n^2 = \frac{n(n + 1)(2n + 1)}{6}$$

is true for all integers $n \geq 1$. ∎

Examples 1 and 2 involved equations that were established by mathematical induction. However, this principle is also used for other types of mathematical situations. The next example demonstrates the application of this principle when an inequality is involved. Also note that in Example 3 the proof begins by establishing S_2 rather than S_1. This is an acceptable use of the principle of induction as long as we also establish part (2). In fact, if S_1 in a proof by induction is replaced by S_a, where a is any (fixed) positive integer, and part (2) proves that S_k implies S_{k+1} for all $k \geq a$, then S_n holds for all $n \geq a$.

Try a few specific cases. Use a calculator to verify these:

$(1.02)^2 > 1 + 2(0.02)$

$(1.001)^2 > 1 + 2(0.001)$

$(1.00054)^2 > 1 + 2(0.00054)$

EXAMPLE 3 Let $t > 0$ and use mathematical induction to prove that $(1 + t)^n > 1 + nt$ for all positive integers $n \geq 2$.

PROOF Let S_n be the statement $(1 + t)^n > 1 + nt$, where $t > 0$ and n is any integer where $n \geq 2$.

1. For $n = 2$, $(1 + t)^2 = 1 + 2t + t^2$. Since $t^2 > 0$, we get

$$1 + 2t + t^2 > 1 + 2t$$

because $(1 + 2t + t^2) - (1 + 2t) = t^2 > 0$.   ($a > b$ means that $a - b > 0$.)

2. Suppose that for $k \geq 2$ we have $(1 + t)^k > 1 + kt$. Multiply both sides by the positive number $(1 + t)$.

Our objective here is to prove
$(1 + t)^{k+1} > 1 + (k + 1)t.$

$$(1 + t)^k(1 + t) > (1 + kt)(1 + t)$$

Then
$$(1 + t)^{k+1} > 1 + (k + 1)t + kt^2$$
But $1 + (k + 1)t + kt^2 > 1 + (k + 1)t$. Therefore, by transitivity of $>$,

$$(1 + t)^{k+1} > 1 + (k + 1)t$$

By **(1)** and **(2)** above, the principle of mathematical induction implies that $(1 + t)^n > 1 + nt$ for all integers $n \geq 2$. ∎

EXERCISES 11.6

Use mathematical induction to prove the following statements for all positive integers n.

1. $1 + 2 + 3 + \cdots + n = \dfrac{n(n + 1)}{2}$

2. $2 + 4 + 6 + \cdots + 2n = n(n + 1)$

3. $\displaystyle\sum_{i=1}^{n} 3i = \dfrac{3n(n + 1)}{2}$

4. $1 + 4 + 7 + \cdots + (3n - 2) = \dfrac{n(3n - 1)}{2}$

5. $\dfrac{5}{3} + \dfrac{4}{3} + 1 + \cdots + \left(-\dfrac{1}{3}n + 2\right) = \dfrac{n(11 - n)}{6}$

6. $1 \cdot 2 + 2 \cdot 3 + 3 \cdot 4 + \cdots + n(n + 1) = \dfrac{n(n + 1)(n + 2)}{3}$

7. $\dfrac{1}{1 \cdot 2} + \dfrac{1}{2 \cdot 3} + \dfrac{1}{3 \cdot 4} + \cdots + \dfrac{1}{n(n + 1)} = \dfrac{n}{n + 1}$

8. $3 + 3^2 + 3^3 + \cdots + 3^n = \dfrac{3^{n+1} - 3}{2}$

9. $-2 - 4 - 6 - \cdots - 2n = -n - n^2$

10. $1 + \dfrac{1}{2} + \dfrac{1}{2^2} + \cdots + \dfrac{1}{2^{n-1}} = 2\left(1 - \dfrac{1}{2^n}\right)$

11. $1 + \frac{2}{5} + \frac{4}{25} + \cdots + \left(\frac{2}{5}\right)^{n-1} = \frac{5}{3}\left[1 - \left(\frac{2}{5}\right)^n\right]$

12. $1 - \frac{1}{3} + \frac{1}{9} - \cdots + \left(-\frac{1}{3}\right)^{n-1} = \frac{3}{4}\left[1 - \left(-\frac{1}{3}\right)^n\right]$

13. $1^3 + 2^3 + 3^3 + \cdots + n^3 = \dfrac{n^2(n + 1)^2}{4}$

14. Which of Exercises 1–13 can also be proved using the formula for the sum of an arithmetic sequence? Which ones can be proved using the formula for the sum of a geometric sequence?

For Exercises 15 and 16 use mathematical induction to prove the following for all positive integers $n \geq 1$.

15. $\displaystyle\sum_{i=1}^{n} ar^{i-1} = \dfrac{a(1 - r^n)}{1 - r}, r \neq 1$

16. $\displaystyle\sum_{i=1}^{n} [a + (i - 1)d] = \dfrac{n}{2}[2a + (n - 1)d]$

17. Use mathematical induction to prove $2^n > 4n$ for $n \geq 5$.

18. Use mathematical induction to prove $\left(\frac{3}{4}\right)^n < \frac{3}{4}$ for $n \geq 2$.

19. Use mathematical induction to prove $a^n < 1$, where $0 < a < 1$, for $n \geq 1$.

20. Let $0 < a < 1$. Use mathematical induction to prove that $a^n < a$ for all integers $n \geq 2$.

21. Let a and b be real numbers. Then use mathematical induction to prove that $(ab)^n = a^n b^n$ for all positive integers n.

22. Use mathematical induction to prove the generalized distributive property $a(b_0 + b_1 + \cdots + b_n) = ab_0 + ab_1 + \cdots + ab_n$ for all positive integers n, where a_i and b_i are real numbers. (Assume that parentheses may be inserted into or extracted from an indicated sum of real numbers).

23. Give an inductive proof of $|a_0 + a_1 + \cdots + a_n| \leq |a_0| + |a_1| + \cdots + |a_n|$ for all positive integers n, where the a_i are real numbers.

24. Use induction to prove that if $a_0 a_1 \ldots a_n = 0$, then at least one of the factors is zero for all positive integers n. (Assume that parentheses may be inserted into or extracted from an indicated product of real numbers.)

25. (a) Prove by induction that $\dfrac{a^n - b^n}{a - b} = a^{n-1} + a^{n-2}b + \cdots + ab^{n-2} + b^{n-1}$ for all integers $n \geq 2$.

$$\left(Hint: \quad \text{Consider } \frac{a^{n+1} - b^{n+1}}{a - b} = \frac{a^n a - b^n b}{a - b} = \frac{a^n a - b^n a + b^n a - b^n b}{a - b} \right)$$

(b) How does the result give the factorization of the difference of two nth powers?

26. Let u_n be the sequence such that $u_1 = 1$, $u_2 = 1$, and $u_{n+1} = u_{n-1} + u_n$ for $n \geq 2$. (This is the Fibonacci sequence; see Exercise 38, page 693.) Use mathematical induction to prove that for any positive integer n, $\sum_{i=1}^{n} u_i = u_{n+2} - 1$.

27. (a) Complete the statements

$$
\begin{aligned}
1 & = \\
1 + 2 + 1 & = \\
1 + 2 + 3 + 2 + 1 & = \\
1 + 2 + 3 + 4 + 3 + 2 + 1 & = \\
1 + 2 + 3 + 4 + 5 + 4 + 3 + 2 + 1 & =
\end{aligned}
$$

(b) Use the results in part (a) to *guess* the sum in the general case.

$$1 + 2 + 3 + \cdots + (n-1) + n + (n-1) + \cdots + 3 + 2 + 1 =$$

(c) Prove part (b) by mathematical induction.

28. (a) The left-hand column contains a number of points in a plane, no three of which are collinear. The right-hand column contains the number of distinct lines that the points determine. Complete this information for the last three cases.

Number of Points		Number of Lines	
∴	2	⁄	1
∴	3	⋈	3
∷	4	∷	
∴∙	5	∴∙	
∴∷	6	∴∷	

(b) Observe that when there are 4 points, there are $6 = \dfrac{4(3)}{2}$ lines. Write similar statements when there are 2, 3, 5, or 6 points.

(c) Conjecture (guess) what the number of distinct lines is for n points, no three of which are collinear. Prove the conjecture by mathematical induction.

29. Use mathematical induction to prove that $7^n - 1$ is divisible by 6 for $n \geq 1$. (*Hint:* A number divisible by 6 can be written in the form $6b$, where b is an integer.)

30. Use mathematical induction to prove that $7^n - 4$ is divisible by 3 for $n \geq 1$. (*Hint:* A number divisible by 3 can be written in the form $3b$, where b is an integer.)

Challenge   In Exercise 29 you were asked to prove that $7^n - 1$ is divisible by 6 for $n \geq 1$. Prove this result *without* using mathematical induction.

Written Assignment   In your own words, explain why *both* parts of the principle of mathematical induction are essential.

CHAPTER REVIEW EXERCISES

Section 11.1   Sequences   (Page 684)

1. State the definition of a sequence. What does it mean to say that a sequence is infinite?

2. The domain of a sequence consists of the integers 1, 2, 3, 4, 5. Write the corresponding range values for $a_n = 3n - 2$.

3. Write the first seven terms of the sequence given by $\left(1 - \dfrac{1}{n}\right)^n$. Round off to three decimal places where appropriate, and find the differences $a_n - a_{n-1}$ for $n = 2, 3, \ldots, 7$.

Write the first five terms of each sequence.

4. $a_n = 3n - (-1)^n$ **5.** $b_k = \dfrac{2}{k(k + 2)}$ **6.** $a_n = \left(-\dfrac{1}{2}\right)^{n-2}$ **7.** $a_n = \dfrac{(-2)^n}{n(n + 1)}$

8. Find the tenth term of the sequence $2, 1, \dfrac{4}{5}, \ldots, \dfrac{n + 1}{2n - 1}, \ldots$.

9. Let a_n be the nth term of a sequence defined by $a_1 = 5$ and $a_n = 2a_{n-1}$ for $n > 1$. Find the first five terms of this sequence.

10. Write the first five powers of -3 and write a formula for the nth term.

Section 11.2   Sums of Finite Sequences   (Page 689)

11. Explain the meaning of the symbol $\displaystyle\sum_{i=1}^{n} a_i$.

Find the sum of the first six terms:

12. $a_n = 5n$ **13.** $a_k = (-1)^k(k + 1)$ **14.** Find $\displaystyle\sum_{n=1}^{5} b_n$ where $b_n = \dfrac{n}{2n - 1}$.

Evaluate:

15. $\displaystyle\sum_{k=1}^{4} (3k - 1)$ **16.** $\displaystyle\sum_{n=1}^{5} (n^2 + n)$

17. Express the series $4 + 8 + 12 + 16 + 20 + 24$ using sigma notation.

Section 11.3 Arithmetic Sequences (Page 694)

18. State the definition of an arithmetic sequence.

19. What is the nth term of an arithmetic sequence whose first term is a_1 and whose common difference is d?

20. Find the nth term of the arithmetic sequence $10, 3, -4, \ldots$.

21. The first term of an arithmetic sequence is 12, and the fifth term is -8. Find the twentieth term.

22. State the formula for the sum of an arithmetic sequence with first term a_1 and common difference d.

23. Find S_{30} for the arithmetic sequence whose first term is $a_1 = 5$ and whose common difference is $d = -3$.

24. Find the sum of the first 1000 terms of the arithmetic sequence beginning with $257, 269, 281, \ldots$.

25. Evaluate: **(a)** $\displaystyle\sum_{k=1}^{40} (-3k + 5)$ **(b)** $\displaystyle\sum_{k=1}^{40} \left(-\frac{1}{3}k + 5\right)$

26. Find the sum of all the multiples of 5 between 9 and 297.

27. What is meant by the arithmetic mean of two numbers a_1 and a_2?

28. Insert four arithmetic means between 8 and 43.

Section 11.4 Geometric Sequences (Page 703)

29. What is meant by a geometric sequence?

30. What is the nth term of a geometric sequence whose first term is a_1 and whose common ratio is r?

31. What is the hundredth term of the geometric sequence having $r = \dfrac{1}{3}$ and $a_1 = \dfrac{1}{3}$?

Consider the geometric sequence $12, -8, \dfrac{16}{3}, \ldots$

32. Find the nth term.   **33.** Find the eighth term.

34. Write the kth term of the geometric sequence $a_k = \left(\dfrac{1}{3}\right)^{3k}$ in the form $a_1 r^{k-1}$ and find the value of a_1 and r.

35. A geometric sequence has $a_1 = 36$ and $a_5 = \dfrac{9}{4}$. Find r.

36. Write a formula for $\displaystyle\sum_{k=1}^{n} a_1 r^{k-1}$.

37. Evaluate: **(a)** $\displaystyle\sum_{k=1}^{7} 2\left(\frac{1}{10}\right)^{k+1}$ **(b)** $\displaystyle\sum_{k=1}^{7} 2\left(-\frac{1}{10}\right)^{k+1}$

38. A roll of wire is 800 feet long. If $\dfrac{1}{4}$ of the wire is cut off repeatedly, what is the general term of the sequence for the length of wire remaining? Find the length of wire remaining after 8 cuts.

39. What are the geometric means between two numbers a and b?

40. Insert three geometric means between 6 and 96.

Section 11.5 Infinite Geometric Series (Page 713)

41. What is the nth partial sum of an infinite geometric sequence? Give a specific example of one.

42. State the formula for finding the sum of an infinite geometric sequence and the conditions for the common ratio r.

Find the sum of each infinite geometric series:

43. $36 + 24 + 16 + \ldots$

44. $48 - 12 + 3 - \ldots$

45. Explain why the infinite geometric series $\displaystyle\sum_{k=1}^{\infty} 2\left(\frac{5}{3}\right)^{k-1}$ has no finite sum.

Evaluate if possible.

46. $\displaystyle\sum_{k=1}^{\infty} \frac{5}{10^{k+1}}$

47. $\displaystyle\sum_{i=1}^{\infty} (-1)^i 5^i$

48. Express the repeating decimal $0.727272\ldots$ in the form $\dfrac{a}{b}$ where a and b are integers.

49. After it is set in motion, each swing in either direction of a particular pendulum is 60% as long as the previous swing. What is the total distance that the end of the pendulum travels before coming to rest if the first swing is 40 inches long?

Section 11.6 Mathematical Induction (Page 724)

50. To prove that a statement S_n is true for each positive integer n, what two conditions must be established according to the principle of mathematical induction?

Use mathematical induction to prove that the statement is true for all positive integers n.

51. $3 + 6 + 9 + \cdots + 3n = \dfrac{3}{2} n(n + 1)$

52. $3 + 6 + 12 + \cdots + 3 \cdot 2^{n-1} = 3(2^n - 1)$

53. $\dfrac{1}{1 \cdot 3} + \dfrac{1}{3 \cdot 5} + \dfrac{1}{5 \cdot 7} + \cdots + \dfrac{1}{(2n-1)(2n+1)} = \dfrac{n}{2n+1}$

54. $1 + \dfrac{1}{3} + \dfrac{1}{3^2} + \cdots + \dfrac{1}{3^{n-1}} = \dfrac{3}{2}\left(1 - \dfrac{1}{3^n}\right)$

55. Use mathematical induction to prove that $3^n > 27n$ for all integers $n \geq 5$.

56. Use mathematical induction to prove that $n^2 + 3n$ is an even integer for every positive integer n. (Recall that a positive integer b is even, provided that $b = 2k$ for some integer k.)

CHAPTER 11: STANDARD ANSWER TEST

Use these questions to test your knowledge of the basic skills and concepts of Chapter 11. Then check your answers with those given at the back of the book.

1. Find the first four terms and the fortieth term of the sequence given by $a_n = \dfrac{n^2}{6 - 5n}$.

2. Write the hundredth term of $a_n = \dfrac{n + 2}{3n^2 + 6n}$ in simplified form.

3. Write the first four terms of the sequence given by $b_n = (-1)^n + n$.

4. Find the tenth term of $u_k = \dfrac{(1 - k)^4}{(-3)^{k - 1}}$ and simplify.

5. Find a_5 for the sequence where $a_1 = -\frac{2}{3}$ and $a_n = (-1)^n a_{n - 1} + \frac{4}{3}$ for $n > 1$.

6. Find $\displaystyle\sum_{n = 1}^{5} c_n$, where $c_n = \dfrac{(-2)^n}{n}$.

In questions 7 and 8, an arithmetic sequence has $a_1 = -3$ and $d = \frac{1}{2}$.

7. Find the forty-ninth term.

8. What is the sum of the first 20 terms?

9. Write the next three terms and the nth term of the geometric sequence $-768, 192, -48, \ldots$.

10. Find the sum of the first 50 positive multiples of 7.

11. For an arithmetic sequence $a_1 = -\frac{2}{3}$ and $a_9 = 10$, find a_{21}.

12. Find the sum of all multiples of 3 between 89 and 301.

13. Use the formula for the sum of a finite geometric sequence to show that

$$\sum_{k = 1}^{4} 8\left(\frac{1}{2}\right)^k = \frac{15}{2}$$

14. Evaluate $\displaystyle\sum_{j = 1}^{101} (4j - 50)$.   15. Evaluate $\displaystyle\sum_{k = 1}^{8} 12\left(\frac{1}{2}\right)^{k - 1}$.

16. A geometric sequence has $a_1 = -24$ and $a_6 = \frac{3}{4}$. Find r.

Decide whether each of the given infinite geometric series has a sum. Find the sum if it exists; otherwise, give a reason why there is no sum.

17. $\displaystyle\sum_{k = 1}^{\infty} 8\left(\frac{3}{4}\right)^{k + 1}$ 18. $1 + \frac{3}{2} + \frac{9}{4} + \cdots$ 19. $0.06 - 0.009 + 0.00135 - \cdots$

20. Change the repeating decimal $0.363636 \ldots$ into rational form.

21. Suppose you save \$10 one week and that each week thereafter you save 10¢ more than the week before. How much will you have saved after 1 year?

22. A promotional display of canned peaches in a supermarket is in the form of a pyramid. A sign that gives the discount price for the peaches is standing on the top row, which consists of 3 cans of peaches. The second row has 4 cans, the third row has 5, etc. If the pyramid has 20 rows, how many cans are there in the pyramid?

23. An object is moving along a straight line such that each minute it travels one-third as far as it did during the preceding minute. How far will the object have moved before coming to rest if it moves 24 feet during the first minute?

24. Prove by mathematical induction that for all $n \geq 1$

$$5 + 10 + 15 + \cdots + 5n = \frac{5n(n + 1)}{2}$$

25. Prove by mathematical induction that for all $n \geq 1$

$$\frac{1}{1 \cdot 4} + \frac{1}{2 \cdot 6} + \frac{1}{3 \cdot 8} + \cdots + \frac{1}{n(2n + 2)} = \frac{n}{2(n + 1)}$$

CHAPTER 11: MULTIPLE CHOICE TEST

1. What is the tenth term of the sequence $1, \dfrac{1}{3}, \ldots, \dfrac{n + 2}{2n^2 + 3n - 2}, \ldots$?

   (a) $\dfrac{1}{7}$ **(b)** $\dfrac{1}{19}$ **(c)** $\dfrac{3}{107}$ **(d)** $\dfrac{2}{201}$ **(e)** None of the preceding

2. List the first four terms of the sequence given by the formula $a_n = 1 + (-1)^n$.

   (a) $0, 2, 0, 2$ **(b)** $2, 0, 2, 0$ **(c)** $0, -2, 0, -2$ **(d)** $-2, 0, -2, 0$ **(e)** None of the preceding

3. Find $\displaystyle\sum_{n=1}^{4} c_n$ where $c_n = 3^{n-1}(n - 1)$.

   (a) 38   **(b)** 102   **(c)** 103   **(d)** 306   **(e)** None of the preceding

4. What is the fiftieth term of the arithmetic sequence having $a_1 = -2$ and $d = 5$?

   (a) 243   **(b)** 245   **(c)** 248   **(d)** 252   **(e)** None of the preceding

5. Find S_{20} for the arithmetic sequence whose first term is $a_1 = 2$ and whose common difference is $d = -3$.

   (a) 640   **(b)** 610   **(c)** -690 **(d)** -530 **(e)** None of the preceding

6. Find $\displaystyle\sum_{k=1}^{50}(-2k + 3)$.

   (a) 2550   **(b)** 2500   **(c)** -2300 **(d)** -2400 **(e)** None of the preceding

7. A slow leak in a water pipe develops in such a way that the first day of the leak one ounce of water drips out. Each day thereafter the amount of water lost is one-half ounce more than the day before. How many ounces of water will leak out in 60 days?

   (a) $30\frac{1}{2}$ **(b)** 915   **(c)** 945   **(d)** 960   **(e)** None of the preceding

8. What is the hundredth term of the geometric sequence having $r = \frac{1}{2}$ and $a_1 = -\frac{1}{2}$?

   (a) 2^{-100} **(b)** $-\dfrac{1}{4^{99}}$ **(c)** $-\dfrac{1}{2^{100}}$ **(d)** $-\dfrac{1}{2^{98}}$ **(e)** None of the preceding

9. Find $\displaystyle\sum_{k=1}^{100} 3\left(\frac{1}{3}\right)^k$.

   (a) $\dfrac{3}{2}\left(1 - \dfrac{1}{3^{100}}\right)$ **(b)** $1 - \dfrac{1}{3^{100}}$ **(c)** $\dfrac{2}{3}\left(1 - \dfrac{1}{3^{100}}\right)$ **(d)** $1 - \left(\dfrac{1}{3}\right)^{99}$ **(e)** None of the preceding

10. Suppose you save \$512 in January and then each month thereafter you save only half as much as you saved the preceding month. How much money will you have saved after one year?

   (a) \$1023   **(b)** \$1023.25   **(c)** \$1023.50   **(d)** 1023.75   **(e)** None of the preceding

11. Find the sum of the infinite geometric sequence $-27, -9, -3, \ldots$.

(a) $\frac{81}{4}$ (b) $-\frac{81}{4}$ (c) $\frac{81}{2}$ (d) -81 (e) None of the preceding

12. Which of the following shows the sum of the multiples of 5 from 5 through 50?

(a) $\sum_{n=1}^{10} 5n$ (b) $\sum_{n=0}^{10} 5n$ (c) $\sum_{n=1}^{10} 5$ (d) $\sum_{n=5}^{50} n$ (e) None of the preceding

13. For the numbers $\frac{5}{2}$ and $\frac{45}{2}$, which of the following statements is true concerning their arithmetic mean (A.M.) and their positive geometric mean (G.M.)?

(a) A.M. = G.M. (b) A.M. > G.M. (c) A.M. < G.M.

(d) The relationship between the A.M. and G.M. cannot be determined with the information given.

(e) None of the preceding.

14. If three geometric means are inserted between $\frac{16}{9}$ and 144, one of these means is

(a) $\frac{8}{9}$ (b) $\frac{32}{9}$ (c) 16 (d) 32 (e) None of the preceding

15. Evaluate: $\sum_{n=1}^{\infty} 100\left(\frac{7}{100}\right)^{n+1}$.

(a) $\frac{49}{93}$ (b) $\frac{490}{3}$ (c) $\frac{1000}{3}$ (d) No finite sum (e) None of the preceding

16. Evaluate: $\sum_{n=1}^{\infty} (-1)^n 2^n$.

(a) $-\frac{1}{2}$ (b) $\frac{1}{2}$ (c) 2 (d) No finite sum (e) None of the preceding

17. The first triangle in an infinite sequence of triangles has vertices $A(0, 16)$, $B(-4, 0)$, and $C(4, 0)$. Then each triangle after the first has coordinates that are one-half the coordinates of the preceding triangle. What is the sum of the areas of all the triangles?

(a) $\frac{128}{3}$ (b) $\frac{256}{3}$ (c) $\frac{512}{3}$ (d) 96 (e) 128

18. Which of the following is correct regarding the proof, by mathematical induction, of a statement S_n for all positive integers n?

I. In part (1) of the proof we prove that S_1 is true.

II. In part (2) of the proof we assume that S_k is true for all integers $k \geq 1$, and then prove that S_k implies S_{k+1}.

III. In part (2) of the proof we first prove S_k is true for all integers $k \geq 1$ and then use this to prove that S_{k+1} is true.

(a) Only I (b) Only III (c) Only I and II (d) Only I and III (e) None of the preceding

19. An office machine costing $6000 depreciates in value at the rate of 12% per year. Which expression gives the value of the machine after four years?

(a) $6000\,(0.12)^4$ (b) $6000\,(0.88)^4$ (c) $600\,(0.12)^5$ (d) $6000\,(0.88)^5$ (e) None of the preceding

20. Find $\displaystyle\sum_{n=21}^{40}(4n-11)$.

(a) 3460   (b) 2840   (c) 2220   (d) 620   (e) None of the preceding

ANSWERS TO THE TEST YOUR UNDERSTANDING EXERCISES

Page 685

1. $3, 5, 7, 9, 11$

2. $-2, -4, -6, -8, -10$

3. $0, -2, -4, -6, -8$

4. $-1, \frac{1}{2}, -\frac{1}{3}, \frac{1}{4}, -\frac{1}{5}$

5. $1, \frac{1}{4}, \frac{1}{9}, \frac{1}{16}, \frac{1}{25}$

6. $-\frac{3}{2}, -\frac{1}{2}, -\frac{1}{4}, -\frac{3}{20}, -\frac{1}{10}$

7. $3, \frac{1}{2}, \frac{1}{5}, \frac{3}{28}, \frac{1}{15}$

8. $\frac{1}{3}, \frac{1}{9}, \frac{1}{27}, \frac{1}{81}, \frac{1}{243}$

9. $2, 0, 2, 0, 2$

Page 692

1. 60   **2.** 25   **3.** 40   **4.** 0   **5.** 50   **6.** $\frac{5}{16}$

Page 696

1. $5n$ **2.** $-4n+10$ **3.** $\frac{1}{10}n$ **4.** $-8n+3$ **5.** n

6. $n-4$ **7.** $\frac{2}{3}n$ **8.** $-12n+65$ **9.** $\frac{1}{5}n-\frac{1}{5}$ **10.** $n+1$

Page 706

1. $1, \frac{1}{2}, \frac{1}{4}, \frac{1}{8}, \frac{1}{16}, 1\left(\frac{1}{2}\right)^{n-1}; r=\frac{1}{2}$

2. $\frac{1}{4}, \frac{1}{8}, \frac{1}{16}, \frac{1}{32}, \frac{1}{64}, \frac{1}{4}\left(\frac{1}{2}\right)^{n-1}; r=\frac{1}{2}$

3. $-\frac{1}{2}, \frac{1}{4}, -\frac{1}{8}, \frac{1}{16}, -\frac{1}{32}; -\frac{1}{2}\left(-\frac{1}{2}\right)^{n-1}; r=-\frac{1}{2}$

4. $-\frac{1}{27}, \frac{1}{27^2}, -\frac{1}{27^3}, \frac{1}{27^4}, -\frac{1}{27^5}, -\frac{1}{27}\left(-\frac{1}{27}\right)^{n-1}; r=-\frac{1}{27}$

5. $r=10; \frac{1}{5}(10)^{n-1}$

6. $r=-\frac{4}{9}; 27\left(-\frac{4}{9}\right)^{n-1}$

Page 717

1. $r=\frac{1}{10}; S_\infty=11\frac{1}{9}$

2. $r=4;$ no finite sum.

3. $r=-\frac{1}{6}; S_\infty=30\frac{6}{7}$

4. $r=\frac{1}{4}; S_\infty=-21\frac{1}{3}$

5. $r=\frac{4}{3};$ no finite sum.

6. $r=0.01; S_\infty=\frac{1}{33}$

7. $r=-3;$ no finite sum.

8. $r=-\frac{9}{10}; S_\infty=\frac{810}{19}$

9. $r=-1.01;$ no finite sum.

CHAPTER 12

Permutations, Combinations, Probability

Several years ago there was a great deal of interest generated by the following problem:

Suppose you are a contestant on a game show and are ushered into a room where there are three closed doors. Your host indicates that behind one of these doors is a new car. Behind each of the other two doors is a goat, for which you have no use. You are to select one door and will receive whatever is behind that door.

You then proceed to select a door, say door 1. However, before you open that door, the host opens one of the other two doors to reveal a goat. You are then asked whether you would like to stay with your original choice (of door 1) or switch to the other unopened door.

The question is: Should you remain with your original choice? Should you switch to the remaining door? Or doesn't it make any difference? The answer to this problem was vigorously debated by numerous people, including some mathematicians, yet it does not take any advanced mathematical knowledge to solve. What do you think is the correct answer? We will discuss this problem in detail later in this chapter. The answer may surprise you!

12.1 PERMUTATIONS

Suppose that there are 30 students in your class, and on the first day of class you all decide to become acquainted by shaking hands with one another. That is, every student shakes hands with every other student. How many handshakes take place? This problem suggests the types of problems we may solve using counting procedures that will be studied in this chapter. However, we can also solve this problem informally in this way:

> If every one of the 30 students shakes hands with each of the remaining 29 students, there will be a total of 30×29 handshakes in all. However, this product will give us twice the answer because it includes duplications. For example, if Julie shakes hands with Andrew, there is really no need for Andrew to shake hands with Julie! Thus the answer to the problem is $\frac{1}{2}(30 \times 29) = 435$.

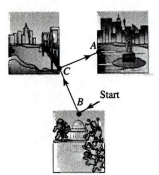

This map shows the trip where the first city is B, then C, and then A.

Convince yourself that this answer is correct by using a smaller number. Thus, in a class of five students the number of handshakes will be $\frac{1}{2}(5 \times 4) = 10$. Use five students, shake hands, and count the number of handshakes to verify this result.

As another problem, assume that you are planning a trip that will consist of visiting three cities, A, B, and C. You have your choice as to the order in which you are to visit the cities. How many different trips are possible? A trip begins with a stop at any one of the three cities, the second stop will be at any one of the remaining two cities, and the trip is completed by stopping at the remaining city. One way to answer this question is to sketch all possible trips, one of which is shown in the margin. However, it can get clumsy or tedious to find all trips using such diagrams, especially if more cities are involved.

A better way to obtain the solution is to draw a **tree diagram** that illustrates all possible routes. From the diagram we can read the six possible trips. The arrangement ABC means that the trip begins with city A, goes to B, and ends at C. The other arrangements have similar interpretations.

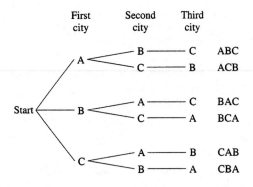

Tree diagrams can be useful in solving such counting problems when the number of possibilities is relatively small. However, when large numbers of possibilities are involved, such tree diagrams are not practical. A more efficient method is needed. There is such a method whose underlying idea can be observed using the preceding tree diagram.

After the start, the tree branches into 3 paths A, B, and C. Then, from each of these 3 points there are 2 new branches, giving 6 paths thus far, as in the figure below. In other words, using the concept of multiplication, we have 3 groups each containing 2 paths for a total of $3 \cdot 2 = 6$ possibilities for the first two cities. Then the last choice consists of only the 1 remaining city. So now we have $3 \cdot 2 = 6$ groups, each of which contains 1 path, for a final total of $(3 \cdot 2) \cdot 1 = 6$ possible trips.

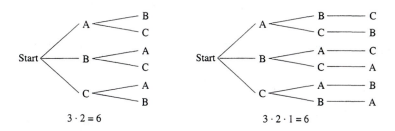

When the preceding observations are generalized, we obtain the following important principle of counting.

FUNDAMENTAL PRINCIPLE OF COUNTING

Suppose that a first task can be completed in m_1 ways, a second task in m_2 ways, and so on, until we reach the rth task that can be done in m_r ways; then the total number of ways in which these tasks can be completed together is the product

$$m_1 m_2 \cdots m_r$$

The following examples illustrate the use of this counting principle. Because of the large number of possibilities involved, these examples also demonstrate the advantage this principle has over the construction of tree diagrams.

EXAMPLE 1   A club consists of 15 boys and 20 girls. They wish to elect officers consisting of a girl as president and a boy as vice-president. They also wish to elect a treasurer and a secretary, who may be of either sex. How many sets of officers are possible?

We must assume here that no person may hold two offices at the same time.

Solution   There are 20 choices for president and 15 choices for vice-president. Thereafter, since two club members have been chosen and the remaining positions can be filled by either a boy or a girl, 33 members are left for the post of treasurer, and then 32 choices for secretary. Then by the fundamental principle of counting, the total number of choices is

$$20 \cdot 15 \cdot 33 \cdot 32 = 316{,}800 \qquad \blacksquare$$

EXAMPLE 2

(a) How many three-digit whole numbers can be formed if zero is not an acceptable digit in the hundreds place and repetitions of digits are allowed?

(b) How many if repetitions are not allowed?

Solution

(a) Imagine that you must place a digit in each of three positions, as in the display below.

____ ____ ____   Digits to Use

0 cannot be used here   0, 1, 2, 3, 4, 5, 6, 7, 8, 9

A tree diagram would begin this way:

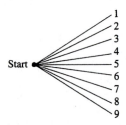

There are 9 choices for the first position and 10 for each of the others. By the fundamental principle of counting, the solution is

$$9 \cdot 10 \cdot 10 = 900$$

(b) There are 9 choices for the hundreds place since zero is not allowed here. Once a choice is made, there are still 9 choices available for the tens digit since zero is permissible here. Finally, there are only 8 choices available for the units digits. The final solution is the product $9 \cdot 9 \cdot 8 = 648$. ∎

If the tree were continued what nine numbers would follow the 4? What eight numbers would follow the path 47?

In the earlier illustration, where the six trips to the three cities were listed, the order of the elements *A*, *B*, *C* was crucial; trip *ABC* is different from trip *BAC*. We say that each of the six arrangements is a *permutation* of three objects taken three at a time. In general, for *n* elements (*n* a positive integer) and *r* a positive integer, $r \leq n$, we have this definition.

When considering permutations, the order of the elements is important. Thus 213 and 312 are two different permutations of the digits 1, 2, 3.

> ### DEFINITION OF PERMUTATION
>
> A **permutation** of *n* distinct elements taken *r* at a time is an ordered arrangement, without repetitions, of *r* of the *n* elements. The number of permutations of *n* elements taken *r* at a time is denoted by $_nP_r$.

Note that a "word" here is to be interpreted as any collection of three letters. That is, "nlm" is considered to be a word in this example.

EXAMPLE 3 How many three-letter "words" composed from the 26 letters of the alphabet are possible? No duplication of letters is permitted.

Solution Since duplication of letters is *not* permitted, once a letter is chosen, it may not be selected again. Therefore, the first letter may be any one of the 26 letters of the alphabet, the second may be any one of the remaining 25, and the third is chosen from the remaining 24. Thus the total number of different "words" is $26 \cdot 25 \cdot 24 = 15,600$. Since we have taken 3 elements out of 26, without repetitions and in all possible orders, we may say that there are 15,600 permutations; that is

$$_{26}P_3 = 26 \cdot 25 \cdot 24 = 15,600$$ ∎

Note that $_{26}P_3 = 26 \cdot 25 \cdot 24$ has three factors beginning with 26 and with each successive factor decreasing by 1. In general, for $_nP_r$ there will be r factors beginning with n, as follows.

$$_nP_r = n(n-1)(n-2)(n-3) \cdots [n - (r-1)]$$

Thus

$$_nP_r = n(n-1)(n-2)(n-3) \cdots (n - r + 1)$$

When applying this formula, begin with n and use a total of r factors successively decreasing by 1.

Illustration:

$$\underset{r = 8 \text{ factors successively decreasing by 1}}{_{20}P_8 \underbrace{= 20 \cdot 19 \cdot 18 \cdot 17 \cdot 16 \cdot 15 \cdot 14 \cdot 13}}$$

with $n = 20$

In the notation $n!$ the $!$ is not used as a typical exclamation mark. Rather, it means to multiply all the positive integers from n down to 1 as shown.

A specific application of this formula occurs when $n = r$. In this case we have the permutation of n elements taken n at a time, and the product has n factors.

$$_nP_n = n(n-1)(n-2)(n-3) \cdots 3 \cdot 2 \cdot 1$$

We may abbreviate this formula by using **factorial notation.**

$$_nP_n = n!$$

For example:

$$_3P_3 = 3! = 3 \cdot 2 \cdot 1 = 6$$
$$_4P_4 = 4! = 4 \cdot 3 \cdot 2 \cdot 1 = 24$$
$$_5P_5 = 5! = 5 \cdot 4 \cdot 3 \cdot 2 \cdot 1 = 120$$

For future consistency we find it convenient to define $0!$ as equal to 1.

$$0! = 1$$

A tree diagram would begin this way.

First coin

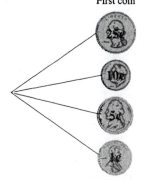

EXAMPLE 4 In how many ways can four coins (a quarter, a dime, a nickel, and a penny) be arranged in a row?

Solution This question calls for all *ordered* arrangements of 4 objects taken 4 at a time, which makes it a permutation problem with $n = 4$ and $r = 4$. Therefore, the number of such arrangements is

$$_4P_4 = 4! = 24$$

Recognizing that this is a permutation problem leads to the preceding simple computation. However, the fundamental principle of counting could be used as well.

In addition, a tree diagram or a list showing all the arrangements can also be made. For example, here are 6 of the 24 arrangements consisting of those in which the quarter is in the first position.

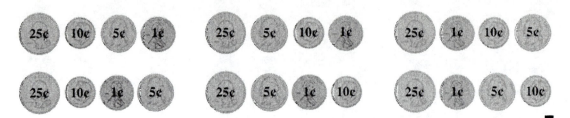

TEST YOUR UNDERSTANDING

Evaluate.

1. $_{10}P_4$ **2.** $_8P_3$ **3.** $_6P_6$ **4.** $\dfrac{10!}{8!}$ **5.** $\dfrac{12!}{9!\,3!}$

6. How many four-letter "words" from the 26 letters of the alphabet are possible? No duplication of letters is permitted.

7. Answer Exercise 6 if duplications are permitted.

8. How many different ways can the letters of the word EAT be arranged using each letter once in each arrangement? List all the possibilities.

9. Draw a tree diagram that shows all the three-digit whole numbers that can be formed using the digits 2, 5, 8 so that each of the digits is used once in each number.

10. How many four-digit whole numbers greater than 5000 can be formed using each of the digits 3, 5, 6, 8 once in each number?

(Answers: Page 778)

You have seen that $n!$ consists of n factors, beginning with n and successively decreasing to 1. At times it will be useful to display only some of the specific factors in $n!$ For example:

$$n! = n(n-1)! = n(n-1)(n-2)! = n(n-1)(n-2)(n-3)!$$

In particular,

$$5! = 5 \cdot 4! = 5 \cdot 4 \cdot 3! = 5 \cdot 4 \cdot 3 \cdot 2!$$

We can use the formula for $_nP_n$ to obtain a different form for $_nP_r$. To do so write the formula for $_nP_r$ and then multiply numerator and denominator by $(n-r)!$ as follows.

$$_nP_r = \frac{n(n-1)(n-2) \cdots [n-(r-1)]}{1} \cdot \frac{(n-r)!}{(n-r)!}$$

Now observe that, after multiplying, the numerator can be written as $n!$ to produce another useful formula for the permutation of n elements taken r at a time.

When applying these formulas there are three things to keep in mind. First, the n elements or objects must all be different. Second, no element is repeated in a permutation. And third, order of the elements is important.

PERMUTATION FORMULAS

For n distinct elements taken r at a time, where $1 \leq r \leq n$:

$$_nP_r = n(n-1)(n-2) \cdots (n-r+1)$$

$$_nP_r = \frac{n!}{(n-r)!}$$

Illustrations:

Using the first formula, $_7P_4 = 7 \cdot 6 \cdot 5 \cdot 4 = 840$

Using the second formula, $_7P_4 = \dfrac{7!}{3!} = \dfrac{7 \cdot 6 \cdot 5 \cdot 4 \cdot \cancel{3!}}{\cancel{3!}} = 840$

EXAMPLE 5 A club contains 10 members. They wish to elect officers consisting of a president, vice-president, and secretary-treasurer. How many sets of officers are possible?

Solution We may think of the three offices to be filled in terms of a first office (president), a second office (vice-president), and a third office (secretary-treasurer). Therefore it is necessary to select 3 out of 10 members and arrange them in all possible orders; we need to find the permutations of 10 elements taken 3 at a time.

$$_{10}P_3 = \frac{10!}{7!} = \frac{10 \cdot 9 \cdot 8 \cdot \cancel{7!}}{\cancel{7!}} = 720 \qquad \blacksquare$$

EXAMPLE 6 A family of 5 consisting of the parents and 3 children are going to be arranged in a row by a photographer. If the parents are to be next to each other, how many arrangements are possible?

Solution Suppose the parents occupy the first two positions and the children the last three. In this case we have

This example makes use of both permutations and the fundamental counting principle.

$$\begin{pmatrix} \text{Parents occupy} \\ \text{these two positions} \\ \text{in } _2P_2 \text{ ways} \\ \underline{2} \cdot \underline{1} = 2 \end{pmatrix} \cdot \begin{pmatrix} \text{Children occupy} \\ \text{these three positions} \\ \text{in } _3P_3 \text{ ways} \\ \underline{3} \cdot \underline{2} \cdot \underline{1} = 6 \end{pmatrix}$$

Since the parents can occupy the first two positions in 2 ways and the children occupy the remaining three positions in 6 ways, the fundamental counting principle gives $2 \cdot 6 = 12$ ways. Also there are 4 adjacent positions possible for the parents as indicated by the blue dashes in the diagram; the three red dashes are the children's positions.

Each of these 4 cases produces 12 arrangements. Therefore, the total number of arrangements is $4(12) = 48$. ∎

Permutations can also be formed using collections of objects not all of which are distinct from one another. For example, in the word ELEMENT the 3 E's are not distinguishable. How many ways can these 7 letters be arranged to form all *distinguishable permutations*?

One such permutation is TENEMEL. If the E's were momentarily made distinguishable, by using subscripts, and if the consonants remain fixed, then TENEMEL produces these $3! = 6$ permutations:

$E_1, E_2 E_3$ can be put into 3 positions in $3! = 6$ ways.

$$\text{TE}_1\text{NE}_2\text{ME}_3\text{L} \qquad \text{TE}_2\text{NE}_1\text{ME}_3\text{L} \qquad \text{TE}_3\text{NE}_1\text{ME}_2\text{L}$$

$$\text{TE}_1\text{NE}_3\text{ME}_2\text{L} \qquad \text{TE}_2\text{NE}_3\text{ME}_1\text{L} \qquad \text{TE}_3\text{NE}_2\text{ME}_1\text{L}$$

Similarly, every distinguishable permutation of the letters in ELEMENT produces $3!$ permutations of the letters in $E_1LE_2ME_3NT$, of which there are $7! = 5040$. Then letting P be the number of distinguishable permutations of the letters in ELEMENT, we have $6P = 5040$, or $P = \dfrac{5040}{6} = \dfrac{7!}{3!}$.

This type of reasoning can be extended to produce the following result:

> The number of distinguishable permutations of n objects of which n_1 are of one kind, n_2 are of another kind, . . . , n_k are of another kind is given by
>
> $$\frac{n!}{n_1! \, n_2! \cdots n_k!}$$

These are the permutations for part (a):

BKOO	OBKO
BOKO	OKBO
BOOK	OBOK
KBOO	OKOB
KOBO	OOBK
KOOB	OOKB

EXAMPLE 7 Find the number of distinguishable permutations using all of the letters in each word.

(a) BOOK **(b)** REFERRED **(c)** BEGINNING

Solution

(a) Use $n_1 = 2$, since there are two O's, and $n = 4$ to get $\dfrac{4!}{2!} = 12$.

(b) Use $n_1 = 3$, $n_2 = 3$, and $n = 8$ to get $\dfrac{8!}{3! \, 3!} = 1120$.

(c) Use $n_1 = 2$, $n_2 = 2$, and $n_3 = 3$, and $n = 9$ to get

$$\frac{9!}{2! \, 2! \, 3!} = \frac{9 \cdot 8 \cdot 7 \cdot 6 \cdot 5 \cdot 4 \cdot 3!}{2 \cdot 2 \cdot 3!}$$

$$= 9 \cdot 8 \cdot 7 \cdot 6 \cdot 5 = 15{,}120 \qquad ∎$$

EXERCISES 12.1

Evaluate.

1. $\dfrac{7!}{6!}$ **2.** $\dfrac{12!}{10!}$ **3.** $\dfrac{12!}{2!\,10!}$ **4.** $\dfrac{15!}{10!\,5!}$ **5.** $_5P_4$ **6.** $_5P_5$ **7.** $_4P_1$ **8.** $_8P_5$

9. Write $_nP_{n-3}$ in factorial notation.   **10.** Show that $_nP_n = {_nP_{n-1}}$.

11. How many ways can the manager of a baseball team select a pitcher and a catcher for a game if there are 5 pitchers and 3 catchers on the team?

12. How many different outfits can Laura wear if she is able to match any one of five blouses, four skirts, and three pairs of shoes?

Consider three-letter "words" to be formed by using the vowels a, e, i, o, and u.

13. How many different words can be formed if repetitions are not allowed?

14. How many different words can be formed if repetitions are allowed?

15. How many different words without repetitions can be formed whose middle letter is *o*?

16. How many different words without repetitions can be formed whose first letter is *e*?

17. How many different words without repetitions can be formed whose letters at the ends are *u* and *i*?

18. If repetitions are allowed, how many different words can be formed whose middle letter is *a*?

19. How many different words can be formed containing the letter *a* and two other letters?

20. How many different words can be formed containing the letters *a* and *e* so that these letters are not next to each other?

For Exercises 21–26, consider three-digit numbers to be formed using the digits 1, 2, 3, 4, 5, 6, 7, 8, 9. Also assume that repetition of digits is not allowed unless specified otherwise.

21. How many three-digit whole numbers can be formed?

22. How many three-digit whole numbers can be formed if repetition of digits is allowed?

23. How many three-digit whole numbers can be formed that are even?

24. How many three-digit whole numbers can be formed that are divisible by 5?

25. How many three-digit whole numbers can be formed that are greater than 600?

26. How many three-digit whole numbers can be formed that are less than 400 and are divisible by 5?

27. **(a)** In how many different ways can the letters of STUDY be arranged using each letter only once in each arrangement?

     (b) How many arrangements are there if the S and T are in the first two positions?

     (c) How many arrangements are there if S and T are next to each other?

     (d) How many arrangements are there if the S and T are not next to each other?

28. A class consists of 20 members. In how many different ways can the class select a set of officers consisting of a president, a vice-president, a secretary, and treasurer?

29. A baseball team consists of nine players. How many different batting orders are possible? How many are possible if the pitcher bats last?

30. (a) Each question in a multiple-choice exam has the four choices indicated by the letters *a*, *b*, *c*, and *d*. If there are eight questions, how many ways can the test be answered? (*Hint:* Use the fundamental counting principle.)

   (b) How many ways are there if no two consecutive questions can have the same answer?

31. When people are seated at a circular table, we consider only their positions relative to each other and are not concerned with the particular seat that a person occupies. How many arrangements are there for seven people to seat themselves around a circular table? (*Hint:* Consider one person's position as fixed.)

32. Review Exercise 31 and conjecture a formula for the number of different permutations of *n* distinct objects placed around a circle.

33. A license plate is formed by listing two letters of the alphabet followed by three digits. How many different license plates are possible:

   (a) If repetitions of letters and digits are not allowed?

   (b) If repetitions are allowed?

34. Write an expression that gives the number of arrangements of *n* objects taken *r* at a time if repetitions are allowed.

35. Solve for *n*: **(a)** $_nP_1 = 10$; **(b)** $_nP_2 = 12$.

36. Show that $2(_nP_{n-2}) = {_nP_{n-1}}$.

37. A pair of dice is rolled. Each die has six faces on each of which is one of the numbers 1 through 6. One possible outcome is (3, 5), where the first digit shows the outcome of one die and the second digit shows the outcome of the other die. How many different outcomes are possible? List all possible outcomes as pairs of numbers (x, y).

38. To avoid electronic detection, a ship can send coded messages to neighboring ships by displaying a sequence of signal flags having different shapes. If 12 different-shaped flags are available, how many messages can be displayed using a 4-flag sequence?

39. (a) A social security number is a sequence of nine digits. How many different social security numbers are possible?

   (b) How many are there in which no digits repeat?

   (c) How many are there in which there are some repetitions of digits?

(d) How many are there in which a digit appears exactly three times in succession and no other digits repeat?

40. A local telephone number consists of 7 digits, the first 3 of which are called the telephone exchange, such as

$$627\text{–}4195$$

(telephone exchange)

(a) For a given exchange, how many different telephone numbers are possible?

(b) Suppose a city has 73,500 telephones. What is the minimum number of exchanges needed to accommodate the city's phones?

41. A student has room for 6 books on a shelf near her study area. The books consist of a dictionary and textbooks in the areas of chemistry, English, history, mathematics, and philosophy.

(a) In how many ways can the books be arranged on the shelf?

(b) In how many ways can they be arranged if the dictionary is put into the first position?

(c) How many arrangements are possible in which the mathematics and philosophy books are next to each other?

42. Solve for n: **(a)** $_nP_6 = 15(_nP_5)$ **(b)** $_nP_6 = 90(_nP_4)$

43. Solve for r: $_{12}P_r = 8(_{12}P_{r-1})$

How many distinguishable permutations can be formed using all the letters in each word?

44. (a) SEVEN **(b)** INNING **(c)** ORDERED

45. (a) DELEGATE **(b)** COLLEGE **(c)** STATEMENTS

46. List the distinguishable permutations using all the letters in ERIE. (*Hint:* Use a tree diagram.)

?

Challenge How many ways can you choose the letters to form the word PYRAMID in the given diagram if

(a) each letter in the word PYRAMID can be any one of that particular letter listed?

(b) the letter being chosen, other than P, is below and to the immediate left or right of the preceding letter?

(c) the letters are chosen according to part (b) except that the D must be the middle D in the last row?

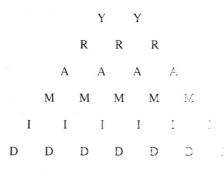

Written Assignment A permutation of n distinct objects using all of the objects is not the same as a permutation using all but one of the objects. However, algebraically, $_nP_{n-1} = {_nP_n}$. Use an effective diagram to explain why this is so for $n = 3$.

12.2
COMBINATIONS

A permutation may be regarded as an *ordered* collection of elements. For instance, a visit to cities A, B, and C in the order ABC is different from a visit in the order ACB. On the other hand, there are times when we need to consider situations where the order of the elements is *not* essential. For example, suppose that on a mathematics test you are given the choice of answering any three of five given questions denoted by Q_1, Q_2, Q_3, Q_4, Q_5.

Precalculus Examination

Answer any 3 questions.

Q_1 Find the maximum or minimum value of $f(x) = -8x^2 - 64x + 3$.

Q_2 Find the vertices, foci, and asymptotes of the hyperbola
$4x^2 - 8x - 9y^2 - 36y = 68$.

Q_3 Sketch the curve $y = f(x) = -x^3 - x^2 + 6x$ and show all intercepts.

Q_4 Solve: (a) $\log_8 (x - 6) + \log_8 (x + 6) = 2$
(b) $e^{\ln(6x^2 - 4)} = 5x$.

Q_5 From a group of 12 women and 10 men, how many committees of 3 women and 4 men can be formed?

Now suppose you decide to answer questions Q_2, Q_3, and Q_4. It doesn't make any difference in which order you answer these three questions, so this is *not* a permutation. Rather, we say that the set $\{Q_2, Q_3, Q_4\}$ is a *combination* of 3 things taken out of 5, according to the following definition in which n and r are integers with $0 \le r \le n$.

DEFINITION OF COMBINATION

Recall that a set A is a subset of a set B provided that each element in A is also in B.

A **combination** is a subset of r elements selected out of n distinct elements *without regard to order*. The number of combinations of n elements taken r at a time is denoted by either of these symbols:

$$_nC_r \quad \text{or} \quad \binom{n}{r}$$

As an example, suppose that a class of 10 members wishes to elect a committee consisting of three of its members. These three members are not to be designated as holding any special office. Then a committee consisting of members David, Ellen, and Robert is the same regardless of the order in which they are selected. In other words, using D, E, and R respectively as abbreviations for their names, the combination $\{D, E, R\}$ gives rise to the following six permutations:

DER   DRE   EDR   ERD   RDE   RED

Each combination of three members in this illustration actually gives rise to $3! = 6$ permutations. To find the number of possible committees we need to find the combinations of 10 elements taken 3 at a time. The *number* of such combinations is expressed as $_{10}C_3$ and is read as "the number of combinations of 10 elements taken 3 at a time." Since each of these combinations produces $3! = 6$ permutations, it follows that

The 120 combinations give rise to 720 permutations.

$$3!(_{10}C_3) = {}_{10}P_3 \quad \text{or} \quad {}_{10}C_3 = \frac{_{10}P_3}{3!} = \frac{10 \cdot 9 \cdot 8}{3 \cdot 2 \cdot 1} = 120$$

In general, using $_nP_r = \dfrac{n!}{(n-r)!}$, we obtain

$$_nC_r = \frac{_nP_r}{r!} = \frac{\dfrac{n!}{(n-r)!}}{r!} = \frac{n!}{r!(n-r)!}$$

*When applying these formulas there are three things to keep in mind. First, the n elements or objects must all be different. Second, no element is used more than once in a combination. And third, order of the elements is **not** important.*

COMBINATION FORMULAS

For n elements taken r at a time, where $0 \le r \le n$:

$$_nC_r = \frac{_nP_r}{r!}$$

$$_nC_r = \frac{n!}{r!(n-r)!}$$

Recall that we have defined $0! = 1$. Do you see here why this definition was made?

As noted, the symbol $\dbinom{n}{r}$ can be used in place of $_nC_r$. Thus $\dbinom{n}{0} = \dfrac{n!}{0!\,n!} = 1$; also

$\dbinom{n}{n} = \dfrac{n!}{n!\,0!} = 1$. Can you prove that $\dbinom{n}{r} = \dbinom{n}{n-r}$?

Illustrations:

$$\binom{10}{2} = \frac{10!}{2!\,8!} = \frac{10 \cdot 9 \cdot \cancel{8!}}{2 \cdot \cancel{8!}} = 45$$

$$_{10}C_9 = \frac{10!}{9!\,1!} = \frac{10 \cdot 9!}{9!} = 10$$

The question of order is the essential ingredient that determines whether a problem involves permutations or combinations. Examples 1 and 2 will illustrate this distinction.

EXAMPLE 1 Using the digits 1 through 9, how many different four-digit whole numbers can be formed if repetition of digits is not allowed?

Solution Order is important here; thus 4923 is a different number from 9432. Therefore we need to find the number of *permutations* of nine elements taken four at a time.

$$_9P_4 = \frac{9!}{5!} = 9 \cdot 8 \cdot 7 \cdot 6 = 3024$$

■

EXAMPLE 2 A student has a penny, a nickel, a dime, a quarter, and a half-dollar and wishes to leave a tip consisting of exactly three coins. How many different amounts as tips are possible?

Here are four of the 10 tips possible:

$$1 + 5 + 10 = 16$$
$$1 + 5 + 25 = 31$$
$$1 + 5 + 50 = 56$$
$$1 + 10 + 25 = 36$$

Complete this list to find the other 6 possibilities.

Solution Order is not important here; a tip of 5¢ + 10¢ + 25¢ is the same as one of 25¢ + 10¢ + 5¢. Therefore, we need to find the number of *combinations* of five things taken three at a time.

$$\binom{5}{3} = {_5}C_3 = \frac{5!}{3! \, 2!} = \frac{5 \cdot 4}{2} = 10 \qquad \blacksquare$$

The next example illustrates how the fundamental principle of counting is used in a problem involving combinations.

EXAMPLE 3 A class consists of 10 boys and 15 girls. How many committees of five can be selected if each committee is to consist of two boys and three girls?

Solution The order is not essential here since the committee members do not hold any special offices. Thus the problem involves combinations.

$$\text{To select two boys:} \qquad {_{10}}C_2 = \frac{10!}{2! \, 8!} = 45$$

$$\text{To select three girls:} \qquad {_{15}}C_3 = \frac{15!}{3! \, 12!} = 455$$

Since there are 45 pairs of boys possible, and since each of these pairs can be matched with any of the possible 455 triples of girls, the fundamental principle of counting gives

$$45 \cdot 455 = 20{,}475$$

as the total number of committees that can be formed. $\qquad \blacksquare$

TEST YOUR UNDERSTANDING

Evaluate.

1. ${_{10}}C_4$ 2. ${_5}C_5$ 3. ${_8}C_0$ 4. $\binom{12}{3}$ 5. $\binom{8}{5}$

6. Show that ${_{10}}C_3 = {_{10}}C_7$.

7. How many different ways can a committee of 4 members be selected from a group of 12 students?

8. A supermarket carries 6 brands of canned peas and 8 brands of canned corn. A shopper wants to try 2 different brands of peas and 3 different brands of corn. How many ways can the shopper select the 5 items?

9. How many lines are determined by eight points in a plane if no three points are on the same line?

10. How many triangles can be formed using five points in a plane no three of which are on the same line?

(Answers: Page 778)

In the next example an ordinary deck of playing cards consisting of 52 different cards is used. These are divided into four suits: spades, hearts, diamonds, and clubs. There are 13 cards in each suit: from 1 (ace) through 10, jack, queen, and king.

EXAMPLE 4 A "poker hand" consists of 5 cards. How many different hands can be dealt from a deck of 52 cards?

Solution The order of the 5 cards dealt is not important, so that this becomes a problem involving combinations rather than permutations. We wish to find the number of combinations of 52 elements taken 5 at a time.

Use a calculator to solve problems with extensive computations such as shown here. It may be helpful to first simplify the fraction.

$$_{52}C_5 = \frac{52!}{5!\,47!} = \frac{52 \cdot 51 \cdot 50 \cdot 49 \cdot 48 \cdot \cancel{47!}}{5 \cdot 4 \cdot 3 \cdot 2 \cdot 1 \cdot \cancel{47!}} = 2{,}598{,}960 \qquad \blacksquare$$

EXAMPLE 5 An ice cream parlor advertises that you may have your choice of five different toppings, and you may choose none, one, two, three, four, or all five toppings. How many choices are there in all?

This example illustrates how the fundamental counting principle is used in a problem involving combinations.

Solution There are several ways to approach this problem. From one point of view you may consider yourself on a cafeteria line with five stations. At each one you have two choices, to accept the topping or not to accept it. Thus, by the fundamental principal of counting, the total number of choices is $2 \cdot 2 \cdot 2 \cdot 2 \cdot 2 = 32$. From another point of view, the solution is the number of different ways that we can select none, one, two, three, four, or five elements from a total of five possibilities; that is,

Since, for example, choosing one topping and choosing two toppings are not done together, we add the number of possibilities rather than multiply.

$$\binom{5}{0} + \binom{5}{1} + \binom{5}{2} + \binom{5}{3} + \binom{5}{4} + \binom{5}{5}$$

Show that this sum is also equal to 32. $\qquad \blacksquare$

EXAMPLE 6 How many subsets, each consisting of 3 elements, does the set $S = \{a, b, c, d, e\}$ have? List these subsets.

Solution Since a subset is a combination, the number of required subsets is given by

$$_5C_3 = \frac{5!}{3!\,2!} = 10$$

The subsets are

How many times does each of the 5 elements of S appear in the 10 subsets?

$$\{a, b, c\}, \quad \{a, b, d\}, \quad \{a, b, e\}, \quad \{a, c, d\}, \quad \{a, c, e\}$$
$$\{a, d, e\}, \quad \{b, c, d\}, \quad \{b, c, e\}, \quad \{b, d, e\}, \quad \{c, d, e\} \qquad \blacksquare$$

EXAMPLE 7 A dinner party of 10 people arrives at a restaurant that has only two tables available. One table seats 6 and the other 4. If the seating arrangement at either table is not taken into account, how many ways can the 10 people divide themselves to be seated at these two tables?

Solution Each time 6 people sit at the table for 6, the remaining 4 will be at the other table. Therefore, it is necessary to find the number of ways we can select subsets of 6 out of 10 people. Thus

If the seating arrangements at each table are taken into account, how many ways can the people be seated? (See Exercise 41.)

$$\binom{10}{6} = \frac{10!}{6!\,4!} = \frac{10 \cdot 9 \cdot 8 \cdot 7 \cdot 6!}{6! \cdot 4 \cdot 3 \cdot 2 \cdot 1} = 210$$

The dinner party can split into the two tables in 210 ways.   ∎

EXERCISES 12.2

Evaluate.

1. $_5C_2$ **2.** $_{10}C_1$ **3.** $_{10}C_0$ **4.** $_4C_3$ **5.** $\binom{15}{15}$ **6.** $\binom{30}{3}$ **7.** $\binom{30}{27}$ **8.** $\binom{n}{3}$

9. A class consists of 20 members. In how many different ways can the class select **(a)** a committee of 4? **(b)** a set of 4 officers?

10. On a test a student must select 8 questions out of a total of 10. In how many different ways can this be done?

11. How many different ways can six people be split up into two teams of three each?

12. How many straight lines are determined by five points, no three of which are collinear?

13. There are 15 women on a basketball team. In how many different ways can a coach field a team of 5 players?

14. Answer Exercise 13 if two of the players can only play center and the others can play any of the remaining positions. (Assume that exactly one center is in a game at one time.)

15. How many handshakes take place when each person in a group of 20 shakes hands once with every other person?

16. Box *A* contains 8 balls and box *B* contains 10 balls. In how many different ways can 5 balls be selected from these boxes if 2 are to be taken from box *A* and 3 from box *B*?

A   B

17. A class consists of 12 women and 10 men. A committee is to be selected consisting of 3 women and 4 men. How many different committees are possible?

18. Solve for *n*: **(a)** $_nC_1 = 6$; **(b)** $_nC_2 = 6$.

19. Convert to fraction form and simplify: **(a)** $_nC_{n-1}$; **(b)** $_nC_{n-2}$.

20. Prove: $\binom{n}{r} = \binom{n}{n-r}$.   **21.** Evaluate $_nC_4$ given that $_nP_4 = 1680$.

22. Solve for *n*: $5\binom{n}{2} = 2\binom{n+2}{2}$

23. Consider this expression: $\binom{n}{0} + \binom{n}{1} + \binom{n}{2} + \cdots + \binom{n}{n-1} + \binom{n}{n}$.

Evaluate for: **(a)** $n = 2$ **(b)** $n = 3$ **(c)** $n = 4$ **(d)** $n = 5$

24. Use the results of Exercise 23 and conjecture the value of the expression for any positive integer n.

25. Explain the equality in Exercise 20 in terms of subsets.

26. Give subset interpretations of the results $_nC_0 = 1$ and $_nC_n = 1$.

27. Interpret the result in Exercise 24 in terms of subsets of a set.

28. Ten points are marked on a circle. How many different triangles do these points determine so that the vertices of each triangle are marked points on the circle?

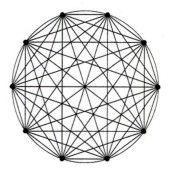

29. How many ways can 5-card hands be selected out of a deck of 52 cards so that all 5 cards are in the same suit?

30. How many ways can 5-card hands be selected out of a deck of 52 cards such that 4 of the cards have the same face value? (Four 10's and some fifth card is one such hand.)

31. How many ways can 5-card hands be selected out of a deck of 52 cards so that the 5 cards consist of a pair and three of a kind? (Two kings and three 7's is one such hand.)

32. A student wants to form a schedule consisting of 2 mathematics courses, 2 history courses, and 1 art course. The student can make these selections from 6 mathematics courses, 10 history courses, and 5 art courses. Assuming that there are no time conflicts, how many ways can the student select the 5 courses?

33. Suppose that in the U.S. Senate 25 Republicans and 19 Democrats are eligible for membership on a new committee. If this new committee is to consist of 9 senators, how many committees would be possible if the committee contained:

(a) 5 Republicans and 4 Democrats?   **(b)** 5 Democrats and 4 Republicans?

34. In Exercise 33, how many committees are possible if the chairperson of the committee is an eligible Republican senator appointed by the Vice-President, and the rest of the committee is evenly divided between Democrats and Republicans?

35. A basketball squad has 12 players consisting of 3 centers, 5 forwards, and 4 guards. How many ways can the coach field a team having 1 center, 2 forwards, and 2 guards?

36. The Green Lawn Tennis Club has scheduled a round-robin tennis tournament in which each player plays one match against every other player. If 12 players are signed up for the tournament, how many matches have been scheduled?

37. A college club has 18 members of which 10 are men and 8 are women. The chairperson of the club is one of the men. A committee of 5 club members is to be formed that must include the chairperson of the club.

(a) How many committees can be formed consisting of 2 women and 3 men?

(b) How many committees can be formed containing no less than 1 woman?

38. Solve for n: $_nC_6 = 3(_{n-2}C_4)$

39. Suppose that a 9-member committee is to vote on an amendment. How many different ways can the votes be cast so that the amendment passes by a simple majority? (A simple majority means 5 or more yes votes.)

40. In Exercise 39, how many different favorable ways can the votes be cast if the amendment needs at least a $\frac{2}{3}$ majority?

41. If the dinner tables in Example 7 of this section are round, how many seating arrangements are possible if the arrangement at each table is taken into account?

Written Assignment

(a) Use specific examples of your own to explain the distinction between a permutation and a combination of n elements taken r at a time.

(b) In general, the connection between permutations and combinations is given by the equation $r! \, ({}_nC_r) = {}_nP_r$. Explain why this equation is true without using algebra, for the case when $r = 4$. (Suggestion: Begin by considering one of the combinations, say $\{A, B, C, D\}$.)

? Challenge

Prove: $\dbinom{n}{r-1} + \dbinom{n}{r} = \dbinom{n+1}{r}$

12.3
THE BINOMIAL THEOREM

The factored form of the trinomial square $a^2 + 2ab + b^2$ is $(a + b)^2$. Turning this around, we say that the *expanded form* of $(a + b)^2$ is $a^2 + 2ab + b^2$. And if $(a + b)^2$ is multiplied by $a + b$, we get the expansion of $(a + b)^3$. Here is a list of the expansions of the first five powers of the binomial $a + b$.

You can verify these results by multiplying the expansion in each row by $a + b$ to get the expansion in the next row.

$$(a + b)^1 = a + b$$

$$(a + b)^2 = a^2 + 2ab + b^2$$

$$(a + b)^3 = a^3 + 3a^2b + 3ab^2 + b^3$$

$$(a + b)^4 = a^4 + 4a^3b + 6a^2b^2 + 4ab^3 + b^4$$

$$(a + b)^5 = a^5 + 5a^4b + 10a^3b^2 + 10a^2b^3 + 5ab^4 + b^5$$

When the coefficients are extracted from the preceding expansions, we obtain the beginning of what is known as **Pascal's triangle.** The number 1 has been placed on top of the array for completeness but is not counted as the first row. The first row of the triangle is 1 1, and the second entry in each row is also the number of that row.

HISTORICAL NOTE
Blaise Pascal (1623–1662) was a French mathematician and philosopher. Although this triangular array of numbers is named after him because of his extensive writings, a similar array of numbers has been found in Chinese writings in the early 1300s.

French stamp honoring mathematician Blaise Pascal.
Source: Dr. Marvin Lang.

```
                1
              1   1
            1   2   1
          1   3   3   1
        1  (4) + (6)  4   1
      1   5  (10)  10   5   1
```

Observe that each row begins and ends with a 1 and that each number, other than a 1, is the sum of the two numbers in the previous row, to the left and right of the chosen number. For example, as shown in the display, $10 = 4 + 6$. After the first row, this property can be used to generate all the rows of the triangle. In particular, note below how the fifth row is used to find the entries in the sixth row.

```
      1    5    10    10    5    1
        \+/  \+/  \+/  \+/  \+/
    1    6    15   20   15    6    1
```

Furthermore, the entries in the sixth row are the coefficients in the expansion of $(a + b)^6$. You can verify this by multiplying the given expansion of $(a + b)^5$ by $(a + b)$ and simplify. In general, *the entries in the nth row of the Pascal triangle are the coefficients in the expansion of $(a + b)^n$.* However, this is not a convenient way to find such coefficients for large values of n, and multiplying $a + b$ by itself repeatedly is even more laborious. So now we return to the given expansions of $(a + b)^n$ for $n = 1, 2, 3, 4, 5$ and search for another way to obtain the coefficients.

First observe that each expansion begins with a^n and ends with b^n. Moreover, each expansion has $n + 1$ terms that are all preceded by plus signs. Now look at the case for $n = 5$. Replace the first term a^5 by a^5b^0 and use a^0b^5 in place of b^5. Then

$$(a + b)^5 = a^5b^0 + 5a^4b + 10a^3b^2 + 10a^2b^3 + 5ab^4 + a^0b^5$$

In this form it becomes clear that (from left to right) the exponents of a successively decrease by 1, beginning with 5 and ending with zero. At the same time, the exponents of b increase from zero to 5. Also note that the sum of the exponents for each term is 5. Verify that similar patterns also hold for the other cases shown.

Using the preceding observations, we would *expect* the expansion of $(a + b)^6$ to have seven terms that, except for the unknown coefficients, look like this:

$$a^6 + \underline{\quad} a^5b + \underline{\quad} a^4b^2 + \underline{\quad} a^3b^3 + \underline{\quad} a^2b^4 + \underline{\quad} ab^5 + b^6$$

Our list of expansions reveals that the second coefficient, as well as the coefficient of the next to the last term, is the number n. Filling in these coefficients for the case $n = 6$ gives

$$a^6 + 6a^5b + \underline{\quad} a^4b^2 + \underline{\quad} a^3b^3 + \underline{\quad} a^2b^4 + 6ab^5 + b^6$$

To get the remaining coefficients we return to the case $n = 5$ and learn how such coefficients can be generated. Look at the second and third terms.

$$\underbrace{\textcircled{5}\, a^{\textcircled{4}} b}_{\text{2nd term}} \qquad \underbrace{10 a^3 b^{\textcircled{2}}}_{\text{3rd term}}$$

If the exponent 4 of a in the *second* term is multiplied by the coefficient 5 of the *second* term and then divided by the exponent 2 of b in the *third* term, the result is 10, the coefficient of the third term.

$$\text{coefficient of third term} = \frac{4(5)}{2} = 10$$

exponent of a in 2nd term — coefficient of 2nd term — exponent of b in 3rd term

Verify that this procedure works for the next coefficient.

On the basis of the evidence, we expect the missing coefficients for the case $n = 6$ to be obtainable in the same way. Here are the computations:

$$\text{Use } ⑥\, a^{⑤}b + \underline{} a^{4}b^{②}: \qquad \text{3rd coefficient} = \frac{5(6)}{2} = 15$$

$$\text{Use } ⑮\, a^{④}b^{2} + \underline{} a^{3}b^{③}: \qquad \text{4th coefficient} = \frac{4(15)}{3} = 20$$

$$\text{Use } ⑳\, a^{③}b^{3} + \underline{} a^{2}b^{④}: \qquad \text{5th coefficient} = \frac{3(20)}{4} = 15$$

Finally, we may write the following expansion:

$$(a + b)^6 = a^6 + 6a^5b + 15a^4b^2 + 20a^3b^3 + 15a^2b^4 + 6ab^5 + b^6$$

You can verify that this equality is correct by multiplying the expansion for $(a + b)^5$ by $a + b$. Also, the coefficients in this expansion are the entries in the sixth row of Pascal's triangle.

More labor can be saved by observing the symmetry in the expansions of $(a + b)^n$. For instance, when $n = 6$ the coefficients around the middle term are symmetric. Similarly, when $n = 5$ the coefficients around the two middle terms are symmetric.

To get an expansion of the binomial $a - b$, write $a - b = a + (-b)$ and substitute into the previous form. For example, with $n = 6$,

$$(a - b)^6 = [a + (-b)]^6 = a^6 + 6a^5(-b) + 15a^4(-b)^2 + 20a^3(-b)^3$$
$$+ 15a^2(-b)^4 + 6a(-b)^5 + (-b)^6$$

Be certain that you recognize the difference between the expansion of $(a - b)^n$ and the factorization of $a^n - b^n$ (see page 55).

$$= a^6 - 6a^5b + 15a^4b^2 - 20a^3b^3 + 15a^2b^4 - 6ab^5 + b^6$$

This result indicates that the expansion of $(a - b)^n$ is the same as the expansion of $(a + b)^n$ except that the signs alternate, beginning with plus.

EXAMPLE 1 Expand: **(a)** $(x + 2)^7$ **(b)** $(x - 2)^7$

Solution

(a) Let x and 2 play the role of a and b in $(a + b)^7$, respectively.

$$(x + 2)^7 = x^7 + 7x^62 + \underline{} x^52^2 + \underline{} x^42^3 + \underline{} x^32^4 + \underline{} x^22^5 + 7x2^6 + 2^7$$

Now find the missing coefficients:

$$\text{3rd coefficient} = \frac{6(7)}{2} = 21 = \text{6th coefficient}$$

$$\text{4th coefficient} = \frac{5(21)}{3} = 35 = \text{5th coefficient}$$

The completed expansion may now be given as follows:

$$(x + 2)^7 = x^7 + 7x^6\,2 + 21\,x^52^2 + 35x^42^3 + 35x^32^4 + 21x^22^5 + 7x2^6 + 2^7$$
$$= x^7 + 14x^6 + 84x^5 + 280x^4 + 560x^3 + 672x^2 + 448x + 128$$

Note that since
$x - 2 = x + (-2)$, *you can set* $a = x$ *and* $b = -2$. *Then each odd power of b is negative, and the signs alternate.*

(b) The expansion of $(x - 2)^7$ may be obtained from the expansion of $(x + 2)^7$ by alternating the signs. Thus,

$$(x - 2)^7 = x^7 - 14x^6 + 84x^5 - 280x^4 + 560x^3 - 672x^2 + 448x - 128 \quad \blacksquare$$

The preceding work can be used to obtain the expansion of the **general form** $(a + b)^n$. Begin by writing the variable parts of the first few terms.

$$a^n + \underline{\quad} a^{n-1}b^1 + \underline{\quad} a^{n-2}b^2 + \underline{\quad} a^{n-3}b^3 + \cdots$$

As before, to get the second coefficient multiply 1 by n and divide by 1.

$$a^n + \frac{n}{1} a^{n-1}b^1 + \underline{\quad} a^{n-2}b^2 + \underline{\quad} a^{n-3}b^3 + \cdots$$

To get the third coefficient multiply $\dfrac{n}{1}$ by $n - 1$ and divide by 2.

$$a^n + \frac{n}{1} a^{n-1}b^1 + \frac{n(n - 1)}{1 \cdot 2} a^{n-2}b^2 + \underline{\quad} a^{n-3}b^3 + \cdots$$

The next coefficient is $\dfrac{n(n - 1)}{1 \cdot 2}$ times $n - 2$ divided by 3, and we now have

$$a^n + \frac{n}{1} a^{n-1}b^1 + \frac{n(n - 1)}{1 \cdot 2} a^{n-2}b^2 + \frac{n(n - 1)(n - 2)}{1 \cdot 2 \cdot 3} a^{n-3}b^3 + \cdots$$

Proceeding in this manner and noting the symmetry of the coefficients, we obtain the following result:

BINOMIAL THEOREM

For real numbers a and b, if n is a positive integer, then

$$(a + b)^n = a^n + \frac{n}{1} a^{n-1}b + \frac{n(n - 1)}{1 \cdot 2} a^{n-2}b^2$$

$$+ \frac{n(n - 1)(n - 2)}{1 \cdot 2 \cdot 3} a^{n-3}b^3 + \cdots + \frac{n}{1} ab^{n-1} + b^n$$

The term having the factor b^r is the $(r + 1)$st term and can be written as

$$\frac{n(n - 1)(n - 2) \cdots (n - r + 1)}{r!} a^{n-r}b^r$$

When expanding a binomial you may find it easier to follow the steps that precede the statement of the binomial formula rather than substitute directly into it.

EXAMPLE 2 Expand $(2x - y)^5$ and simplify.

Solution Use $a = 2x$, $b = y$, and $n = 5$ in the binomial formula with alternating signs.

$$(2x - y)^5 = (2x)^5 - \frac{5}{1}(2x)^4y + \frac{5 \cdot 4}{1 \cdot 2}(2x)^3y^2 - \frac{5 \cdot 4 \cdot 3}{1 \cdot 2 \cdot 3}(2x)^3y^3$$

$$+ \frac{5 \cdot 4 \cdot 3 \cdot 2}{1 \cdot 2 \cdot 3 \cdot 4}(2x)y^4 - \frac{5 \cdot 4 \cdot 3 \cdot 2 \cdot 1}{1 \cdot 2 \cdot 3 \cdot 4 \cdot 5}y^5$$

$$= 32x^5 - 80x^4y + 80x^3y^2 - 40x^2y^3 + 10xy^4 - y^5 \qquad \blacksquare$$

EXAMPLE 3 Evaluate 2^7 by expanding $(1 + 1)^7$.

The coefficients in the expansion of $(a + b)^7$ were found in Example 1.

Solution Since all powers of 1 equal 1, the expansion of $(1 + 1)^7$ is the sum of the coefficients in the expansion of $(a + b)^7$. Thus

$$2^7 = (1 + 1)^7 = 1 + 7 + 21 + 35 + 35 + 21 + 7 + 1 = 128 \qquad \blacksquare$$

The term "binomial formula" refers to the equation that gives the expansion of $(a + b)^n$; the conclusion of the binomial theorem.

Another way to develop the *binomial formula* is to make use of our knowledge of combinations. Let us consider the expansion of $(a + b)^5$ again from a different point of view.

$$(a + b)^5 = \underbrace{(a + b)(a + b)(a + b)(a + b)(a + b)}_{}$$

5 factors

To expand $(a + b)^5$, consider each term as follows.

First term: Multiply all the a's together to obtain a^5.

Second term: We need to combine all terms of the form a^4b. How are these terms formed in the multiplication process that gives the expansion? One of these terms is formed by multiplying the a's in the first four factors times the b in the last factor.

$$(\textcircled{a} + b)(\textcircled{a} + b)(\textcircled{a} + b)(\textcircled{a} + b)(a + \boxed{b})$$

Product $= a^4b$

Another of these terms is formed like this:

$$(\textcircled{a} + b)(\textcircled{a} + b)(\textcircled{a} + b)(a + \boxed{b})(\textcircled{a} + b)$$

Product $= a^4b$

Now you can see that the number of such terms is the same as the number of ways we can select just one of the b's from the five factors. This can be done in five ways, which can be expressed as $_5C_1$ or as $\binom{5}{1}$, the coefficient of a^4b.

Third term: Search for all terms of the form a^3b^2. The number of ways of selecting two b's from five factors, that is, $_5C_2$ or $\binom{5}{2}$.

Fourth term: The number of terms of the form a^2b^3 is the number of ways of selecting three b's from five factors, that is $_5C_3$ or $\binom{5}{3}$.

Fifth term: The number of ways of selecting four b's from the five factors is $_5C_4$ of $\binom{5}{4}$, the coefficient of ab^4.

Sixth term: Multiply all the b's together to obtain b^5.

Thus we may write the expansion of $(a + b)^5$ in this form:

$$a^5 + \binom{5}{1}a^4b + \binom{5}{2}a^3b^2 + \binom{5}{3}a^2b^3 + \binom{5}{4}ab^4 + b^5$$

For consistency of form, we may write the coefficient of a^5 as $\binom{5}{0}$ and that of b^5 as $\binom{5}{5}$. In each case note that $\binom{5}{0} = \binom{5}{5} = 1$.

A similar argument can be made for each of the terms of the expansion of $(a + b)^n$. For example, to find the coefficient of the term that has the factor $a^{n-r}b^r$, we need to find the number of different ways of selecting r b's from n factors. This can be expressed as $_nC_r$ or $\binom{n}{r}$. Now we are ready to generalize and write this form of the **binomial formula:**

This expansion can be written in sigma notation as

$$(a + b)^n = \sum_{r=0}^{n} \binom{n}{r} a^{n-r}b^r$$

$$(a + b)^n = \binom{n}{0}a^nb^0 + \binom{n}{1}a^{n-1}b^1 + \binom{n}{2}a^{n-2}b^2 + \cdots$$

$$+ \binom{n}{r}a^{n-r}b^r + \cdots + \binom{n}{n-1}a^1b^{n-1} + \binom{n}{n}a^0b^n$$

*The numbers $\binom{n}{r}$ are referred to as **binomial coefficients**.*

Observe that the $(r + 1)$st term here and in the formula on page 757 are the same:

$$\binom{n}{r}a^{n-r}b^r = \frac{_nP_r}{r!}a^{n-r}b^r = \frac{n(n-1)(n-2)\cdots(n-r+1)}{r!}a^{n-r}b^r$$

EXAMPLE 4 Find the sixth term in the expansion of $(a + b)^8$.

Recall that the sum of the exponents in each term is equal to n.

Solution Note that for any term in the expansion the exponent r of b is one less than the number of the term. Then, since we need the sixth term, $r = 5$. Also, since the sum of the exponents is 8, the sixth term is

$$\binom{8}{5}a^3b^5 = 56a^3b^5 \qquad \blacksquare$$

Example 4 can also be solved by using the method in the next example.

EXAMPLE 5 Find the fourth term in the expansion of $(x - 2y)^{10}$.

Note that in Example 5 we may think of $(x - 2y)^{10}$ as $[x + (-2y)]^{10}$ so that we may apply the binomial formula for $(a + b)^n$.

Solution Use the general term $\binom{n}{r} a^{n-r} b^r$, which is the $(r + 1)$st term. Then $r + 1 = 4$ and $r = 3$, $n = 10$, and $n - r = 7$. Thus the fourth term is

$$\binom{10}{3} x^7 (-2y)^3 = 120 \, x^7 (-8y^3) = -960 x^7 y^3 \qquad \blacksquare$$

TEST YOUR UNDERSTANDING	*Use the binomial theorem to expand and simplify.*

1. $(x + 2)^3$ **2.** $(x - 3)^4$ **3.** $(a + 2b)^5$ **4.** $(1 - y)^6$

5. Find the fourteenth term in the expansion of $(x + y)^{16}$.

(Answers: Page 778)

6. Find the fifth term in the expansion of $(2x - y)^9$.

EXERCISES 12.3

Expand and simplify.

1. $(x + 1)^5$ **2.** $(x - 1)^6$ **3.** $(x + 1)^7$ **4.** $(x - 1)^8$

5. $(a - b)^4$ **6.** $(3x - 2)^4$ **7.** $(3x - y)^5$ **8.** $(x + y)^8$

9. $(a^2 + 1)^5$ **10.** $(2 + h)^9$ **11.** $(1 - h)^{10}$ **12.** $(-2 + x)^7$

13. $\left(\dfrac{1}{2} - a\right)^4$ **14.** $\left(\dfrac{x}{2} + \dfrac{2}{y}\right)^5$ **15.** $\left(\dfrac{1}{x} - x^2\right)^6$ **16.** $\left(2a - \dfrac{1}{a^2}\right)^6$

17. $\left(\dfrac{x}{2} + 4y\right)^6$ **18.** $(a + 2)^6$ **19.** $(2x - 3)^3$ **20.** $(3p + 2q)^4$

21. Write the entries in the seventh, eighth, ninth, and tenth rows of Pascal's triangle.

22. Use the ninth row found in Exercise 21 to expand $(x + h)^9$.

23. Use the tenth row found in Exercise 21 to expand $(x - h)^{10}$.

24. Evaluate 2^{10} by expanding $(1 + 1)^{10}$.

25. Write the first five terms in the expansion of $(x + 1)^{15}$. What are the last five terms?

26. Write the first five terms and the last five terms in the expansion of $(c + h)^{20}$.

27. Write the first four terms and the last four terms in the expansion of $(a - 1)^{30}$.

Simplify.

28. $\dfrac{(1 + h)^3 - 1}{h}$ **29.** $\dfrac{(3 + h)^4 - 81}{h}$ **30.** $\dfrac{(c + h)^3 - c^3}{h}$

31. $\dfrac{(x + h)^6 - x^6}{h}$ **32.** $\dfrac{2(x + h)^5 - 2x^5}{h}$ **33.** $\dfrac{\dfrac{1}{(2 + h)^2} - \dfrac{1}{4}}{h}$

34. Why does the sum of all the numbers in one line of Pascal's triangle equal twice the sum of the numbers in the preceding line?

35. Find the sixth term in the expansion of $(a + 2b)^{10}$.

36. Find the fifth term in the expansion of $(2x - y)^8$.

37. Find the fourth term in the expansion of $\left(\dfrac{1}{x} + \sqrt{x}\right)^7$.

38. Find the eighth term in the expansion of $\left(\dfrac{a}{2} + \dfrac{b^2}{3}\right)^{10}$.

39. Find the term that contains x^5 in the expansion of $(2x + 3y)^8$.

40. Find the term that contains y^{10} in the expansion of $(x - 2y^2)^8$.

41. Write the middle term of the expansion of $\left(3a - \dfrac{b}{2}\right)^{10}$.

42. Write the last three terms of the expansion of $(a^2 - 2b^3)^7$.

43. Evaluate $(2.1)^4$ by expanding $(2 + 0.1)^4$.   **44.** Evaluate $(1.9)^4$ by expanding $(2 - 0.1)^4$.

45. Evaluate $(3.98)^3$ by expanding an appropriate binomial.   **46.** Evaluate $(1.2)^5$ by expanding an appropriate binomial.

47. Write the rth term of $(a + b)^n$ and of $(a - b)^n$.

48. Write the term in $(a + b)^n$ that contains the factor a^r.

49. Repeat Exercise 48 for the term that contains the factor b^r.

Written Assignment   Assume that the proof of $\dbinom{n}{r-1} + \dbinom{n}{r} = \dbinom{n+1}{r}$ has been done in the Challenge on page 754. Explain how this result is connected to the Pascal triangle and the binomial theorem.

Challenge

1. For x such that $|x| < 1$, find the infinite binomial expansion of $(1 - x)^{-1}$ and $(1 + x)^{-1}$. (*Hint:* Consider infinite geometric series.)

2. Prove that the sum of all the numbers in any two successive rows of Pascal's triangle is divisible by 6.

Graphing Calculator Exercises   In Exercises 1–2, graph the three functions on the same set of axes. Which functions are equivalent? Check your answer using the binomial theorem.

1. $f(x) = (1 + x)^3$, $g(x) = 1 + 3x + 3x^2 + x^3$, $h(x) = 1 + x^3$.

2. $f(x) = (1 + x)^4$, $g(x) = 1 + 4x + 6x^2 + 4x^3 + x^4$, $h(x) = 1 + x^4$.

In Exercises 3–5, graph the function $d(x)$. On the same set of axes also graph the functions $g(x) = (f(x + h) - f(x))/h$, using the values $h = 1, 0.5, 0.25$, and 0.1 and the given function $f(x)$. Describe the relationship between the graphs in each case. Use RANGE values of $-2 \le x \le 2$ and $-10 \le y \le 10$.

3. $d(x) = 2x$, $f(x) = x^2$

4. $d(x) = 3x^2$, $f(x) = x^3$

5. $d(x) = 4x^3$, $f(x) = x^4$

6. In the language of calculus, we say that $d(x)$ is the *derivative* of $f(x)$ in Exercises 3–5. If $f(x) = x^5$, guess at its derivative $d(x)$. Confirm your guess by graphing $g(x)$, as in Exercises 3–5, for $h = 1, 0.5,$ and 0.1.

Critical Thinking

1. True or false: The permutation formula $_nP_r = n(n - 1)(n - 2) \cdots (n - r + 1)$ is based directly on the fundamental principle of counting. Justify your answer.

2. From Exercise 27, page 753, the number of subsets of a set of n elements is 2^n. Arrive at this result by making use of the fundamental principle of counting. (*Hint:* Use a tree diagram applied to a set of three elements and generalize for n elements.)

3. In order to win a certain state lottery, a player must have chosen the same six integers that are selected at random, from the integers 1 through 40, by the lottery agency. Is the number of groups of six numbers a matter of permutations or combinations? How many such groups are there?

4. What can be done to the formula $_nP_r = \dfrac{n!}{(n - r)!}$ in order to produce the formula for $_{n+1}P_r$? On the basis of the preceding observation explain why $_{n+1}P_r \geq _nP_r$.

5. For any positive integer n evaluate the expression

$$\binom{n}{0} - \binom{n}{1} + \binom{n}{2} - \binom{n}{3} + \cdots + (-1)^n\binom{n}{n}$$

On the basis of the preceding result, what can you conclude about the number of subsets formed from a set of n elements that have an even number of elements and the number of subsets that have an odd number of elements?

6. There are numerous number patterns that can be found in Pascal's triangle. For example, the first four rows give powers of 11 directly. As another example, consider the sequence formed by the sum of the numbers in each row. Also consider the sequence formed by the sums of the numbers in each diagonal, as in the following array.

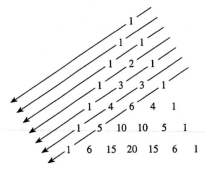

Try to discover some other patterns in this triangular array of numbers.

**12.4
AN
INTRODUCTION
TO
PROBABILITY**

Concepts of probability are encountered frequently in daily life, such as a weather forecaster's statement that there is "a 20% chance of rain." Actually, the formal study of probability started in the seventeenth century, when two famous mathematicians, Pascal and Fermat, considered the following problem that was posed to them by a gambler. Two people are involved in a game of chance and are forced to quit before either one has won. The number of points needed to win the game is known, and the number of points that each player has at the time is known. The problem was to determine how the stakes should be divided.

From this beginning mathematicians developed the theory of probability that has had far-reaching effects in many fields of endeavor. In this section we explore some basic aspects of probability, and the counting procedures studied earlier in this chapter will be useful in solving a variety of probability questions.

Let us begin by considering an *experiment* that consists of tossing two coins to determine the probability of obtaining two heads. A tree diagram is helpful in identifying all of the possible outcomes.

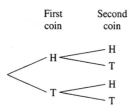

The tree diagram helps us to list the **sample space S,** which is the *set of all possible outcomes of the experiment;* the number of elements in S is denoted by $n(S)$. In this case

$$S = \{HH,\ HT,\ TH,\ TT\} \text{ and } n(S) = 4$$

Any subset E of the sample space S is called an event, and $n(E)$ is the number of elements in subset E. In this situation the event is two heads, so that $E = \{HH\}$ and $n(E) = 1$.

The four outcomes in S are *equally likely,* which means that each of the four outcomes has the same chance of happening as any of the others. Since $n(E) = 1$ and $n(S) = 4$, we expect the probability of two heads to be $\dfrac{1}{4}$. This is consistent with the formal definition of the *probability of an event* that follows, in which S is a sample space of an experiment with $n(S)$ equally likely outcomes, and E is an event.

PROBABILITY OF AN EVENT

$$P(E) = \frac{n(E)}{n(S)} = \frac{\text{number of possible outcomes of event } E}{\text{total number of outcomes in sample space } S}$$

Applying this definition to the previous experiment, we have $P(E) = \dfrac{n(E)}{n(S)} = \dfrac{1}{4}$. Sometimes we use a less formal notation in place of $P(E)$ by replacing E with the elements in E or some other type of description. Thus

$$P(E) = P(HH) = P(\text{two heads}) = \frac{1}{4}$$

This is read as "the probability of two heads is $\frac{1}{4}$."

This answer reflects what we can expect to happen *in the long run*. If we continue to toss two coins continuously, we would expect that *on the average* one out of four tosses will show two heads. Of course, there may be times when we toss double heads several times in a row; but if the experiment were to be repeated 1000 times, we would expect to have *about* 250 cases of double heads. The following table shows the probabilities for events that describe the number of possible heads in this experiment.

First Coin	Second Coin	Event	Probability
Heads	Heads	2 heads	$\dfrac{1}{4}$
Heads	Tails	1 head	$\dfrac{2}{4}$
Tails	Heads		
Tails	Tails	0 heads	$\dfrac{1}{4}$

EXAMPLE 1   A die is tossed. What is the probability of tossing an odd number?

Solution   First note that the sample space is a set of six equally likely outcomes.

$$S = \{1, 2, 3, 4, 5, 6\}$$

The event E is the subset of S consisting of the odd numbers; $E = \{1, 3, 5\}$. Thus

$$P(E) = \frac{n(E)}{n(S)} = \frac{3}{6} = \frac{1}{2}$$ ■

EXAMPLE 2   A pair of dice is tossed. What is the probability that the sum of the numbers showing will be 7?

Solution   To find the sample space for this experiment, it will be convenient to use an ordered pair of numbers to show an outcome on the two dice. It might help you to think of the two dice being of different colors, say red and blue, with the first number in the ordered pair showing the outcome of the red die and the second number showing the outcome of the blue die.

From the fundamental principle of counting, if the first die can land showing one of 6 faces, as can the second die, then the total number of possibilities is $6 \times 6 = 36$.

(1, 1)	(1, 2)	(1, 3)	(1, 4)	(1, 5)	(1, 6)
(2, 1)	(2, 2)	(2, 3)	(2, 4)	(2, 5)	(2, 6)
(3, 1)	(3, 2)	(3, 3)	(3, 4)	(3, 5)	(3, 6)
(4, 1)	(4, 2)	(4, 3)	(4, 4)	(4, 5)	(4, 6)
(5, 1)	(5, 2)	(5, 3)	(5, 4)	(5, 5)	(5, 6)
(6, 1)	(6, 2)	(6, 3)	(6, 4)	(6, 5)	(6, 6)

From the sample space you can see that there are 6 outcomes that give a sum of 7, out of a total of 36 possible outcomes. Thus $P(7) = \dfrac{6}{36} = \dfrac{1}{6}$. ∎

In Example 2 we could have used the fundamental principle of counting, as indicated in the marginal note, to obtain the number of possible outcomes, $36 = n(S)$. Then all we would need to do is find the set E of possible outcomes that produces a sum of 7, which is

$$E = \{(1, 6), \quad (2, 5), \quad (3, 4), \quad (4, 3), \quad (5, 2), \quad (6, 1)\}$$

Now return to the problem stated in the introduction to this chapter. Recall that there are three doors, and you are to choose one and will receive whatever is behind that door. Two of the doors have goats behind them, and one has a new car. After you choose a door, the host opens one of the other doors to reveal a goat. You then are given the choice of staying with your original choice or switching to the other un-opened door. What should you do? Most people claim that it does not make any difference, but it proves to be to your advantage to switch! This can best be explained by the following diagrams. First, let's assume that you do *not* switch; that is, you remain with your original choice of a door.

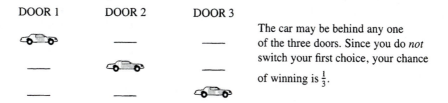

The car may be behind any one of the three doors. Since you do *not* switch your first choice, your chance of winning is $\frac{1}{3}$.

In the next diagram, let's *assume that you choose door* #1 in each case, but then switch after the host opens one of the other doors. Note what happens:

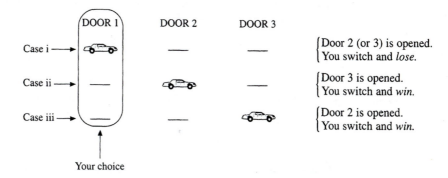

Thus, when you switch, you win 2 out of 3 times; the probability of winning is $\frac{2}{3}$ as against $\frac{1}{3}$ when you do not switch. Therefore, it is definitely to your advantage to switch doors under these conditions!

In the examples that follow we will see how to use our knowledge of combinations to solve probability problems where the listing of a sample space might prove to be quite difficult or tedious.

EXAMPLE 3   To win the jackpot of a state lottery, the six numbers chosen by a person, from the numbers 1 through 54, must be the same as the six numbers selected at random by the state lottery system. On each lottery ticket purchased, there are two separate selections for the six numbers. What is the probability of winning the jackpot with one lottery ticket?

Solution The sample space consists of all possible six-number selections out of 54 possible choices. Since the order of the six numbers chosen does not matter, the number of ways of selecting six numbers out of 54 is $_{54}C_6$. Thus

$$n(S) = {_{54}C_6} = \frac{54 \cdot 53 \cdot 52 \cdot 51 \cdot 50 \cdot 49 \cdot 48!}{6 \cdot 5 \cdot 4 \cdot 3 \cdot 2 \cdot 1 \cdot 48!}$$

$$= 25{,}827{,}165$$

The chances of winning the jackpot is about 1 in 13 million, an extremely small possibility.

Then, since there are two chances on one ticket, we have $n(E) = 2$.

$$P(\text{jackpot}) = P(E) = \frac{n(E)}{n(S)} = \frac{2}{25{,}827{,}165} \approx \frac{1}{12{,}913{,}583} \qquad \blacksquare$$

EXAMPLE 4   A die is tossed. What is the probability of tossing a 7?

We can phrase the result of Example 4 in another way: $P(7$ cannot occur$) = 1$.

Solution None of the outcomes is 7. Thus $P(7) = \frac{0}{6} = 0$. $\qquad \blacksquare$

From Example 4 we see that the probability of an event that *cannot* occur is 0. Furthermore, the probability for an event that will *always* occur is 1. For example, the probability of tossing a number less than 7 on a single toss of a die is $\frac{6}{6} = 1$ since all numbers are less than 7. This leads to the following general observation for the probability that an event E will occur:

$$0 \le P(E) \le 1$$

As an extension of this idea, we note that every event will either occur or fail to occur. That is, $P(E) + P(\text{not } E) = 1$. Therefore,

$$P(\text{not } E) = 1 - P(E)$$

EXAMPLE 5   Two cards are drawn simultaneously from a deck of playing cards. What is the probability that not both cards are spades?

Solution Two cards can be selected out of 52 in $_{52}C_2$ ways. Also, since there are 13 spades, $_{13}C_2$ is the number of ways of selecting 2 spades. Then

The advantage of the formula P (not E) = 1 − P (E) is demonstrated by this example. It is much more difficult to solve otherwise. Try to explain this solution:

P(not 2 spades)

$$= \frac{{}_{39}C_2 + {}_{39}C_1 \cdot {}_{13}C_1}{{}_{52}C_2}$$

$$= \frac{16}{17}.$$

Mutually exclusive events have no elements in common.

$$P(2 \text{ spades}) = \frac{{}_{13}C_2}{{}_{52}C_2} = \frac{1}{17}$$

Now use the formula P (not E) = $1 - P$ (E).

$$P \text{ (not 2 spades)} = 1 - P \text{ (2 spades)} = 1 - \frac{1}{17} = \frac{16}{17} \qquad \blacksquare$$

We need to be careful when adding probabilities, since we may do this only when events are *mutually exclusive,* that is, when they cannot both happen at the same time. For example, consider the probability of drawing a king or a queen when a single card is drawn from a deck of cards.

$$P \text{ (king)} = \frac{4}{52} \qquad P \text{ (queen)} = \frac{4}{52}$$

$$P \text{ (king or queen)} = \frac{4}{52} + \frac{4}{52} = \frac{8}{52} = \frac{2}{13}$$

This seems to agree with our intuition, since there are 8 cards in a deck of 52 cards that meet the necessary conditions. These conditions are mutually exclusive because a card cannot be a king and a queen at the same time. Note, however, the difference in the conditions of Example 6.

PROBABILITY FOR MUTUALLY EXCLUSIVE EVENTS

For **mutually exclusive** events E or F,

$$P(E \text{ or } F) = P(E) + P(F)$$

EXAMPLE 6   A single card is drawn from a deck of cards. What is the probability that the card is either a queen or a spade?

*In Example 6 the events are **not** mutually exclusive because it is possible for a card to be both a queen and a spade.*

Solution   The probability of drawing a queen is $\frac{4}{52}$, and the probability of drawing a spade is $\frac{13}{52}$. However, there is one card that is being counted twice in these probabilities, namely the queen of spades. Thus we account for this as follows:

$$P \text{ (queen or spade)} = \frac{13}{52} + \frac{4}{52} - \frac{1}{52} = \frac{16}{52} = \frac{4}{13} \qquad \blacksquare$$

In general, for events E and F,

$$P(E \text{ or } F) = P(E) + P(F) - P(E \text{ and } F)$$

Source: Monkmeyer Press

Sometimes a probability example can be solved using multiplication. For example, consider the probability of drawing two aces when two cards are selected from a deck of cards. In all such cases, unless stated otherwise, we shall assume that a card is drawn and *not* replaced in the deck before the second card is drawn.

Probability that the first card is an ace $= \frac{4}{52}$.

Assume that an ace was drawn. Then there are only 51 cards left in the deck, of which 3 are aces. Thus the probability that the second card is an ace $= \frac{3}{51}$.

We now make use of the following principle. Suppose that $P(E)$ represents the probability that an event E will occur, and $P(F \text{ given } E)$ is the probability that event F occurs after E has occurred. Then

$$P(E \text{ and } F) = P(E) \times P(F \text{ given } E)$$

Thus to find the probability that *both* cards are aces in the preceding illustration, we must multiply:

$$\frac{4}{52} \cdot \frac{3}{51} = \frac{1}{221} \approx 0.0045$$

This example can also be expressed through the use of combinations. Thus the total number of ways to select two cards from the deck is $_{52}C_2$. Furthermore, the number of ways of selecting two aces from the four aces in a deck of cards is $_4C_2$. Therefore,

Show that $\frac{_4C_2}{_{52}C_2} = \frac{1}{221}$.

$$\text{Probability of selecting two aces} = \frac{_4C_2}{_{52}C_2} = \frac{1}{221}$$

When each of two events can occur so that neither one affects the probability of the other, we say that the events are **independent.** For example, if E is the event that a head will come up when tossing a coin and F is the event that a 4 comes up when rolling a die, then neither outcome affects the probability of the other and therefore they are independent events. $P(E) = \frac{1}{2}$ and $P(F) = \frac{1}{6}$; the probability of both events occurring is found by multiplying. (See the note in the margin.) Thus

This can also be found by noting that there are 2 outcomes when a coin is tossed and 6 when a die is rolled. Then there are $2 \cdot 6 = 12$ ways for both events to occur together. Since only 1 of these 12 consists of heads on the coin and 4 on the die,

$$P(E \text{ and } F) = \frac{1}{2} \cdot \frac{1}{6} = \frac{1}{12}$$

In general,

$P(E \text{ and } F) = \frac{1}{12}$.

PROBABILITY FOR INDEPENDENT EVENTS

For **independent events** E and F,

$$P(E \text{ and } F) = P(E) \cdot P(F)$$

EXAMPLE 7 A single card is drawn from a deck of cards. It is then replaced and a second card is drawn. What is the probability that both cards are aces?

Solution Since the card is replaced after the first drawing, the outcome of either card cannot possibly affect the probability of the other, so that the events are independent. In each case the probability is $4/52$, and the probability that both are aces is the product of the individual probabilities, namely, $\dfrac{4}{52} \cdot \dfrac{4}{52} = \dfrac{1}{169}$. ∎

The final two illustrative examples demonstrate the use of combinations in conjunction with the fundamental counting principle to solve probability problems.

---

*In Example 8 we assume that we are to draw **exactly** two aces. A more difficult problem is to find the probability of drawing **at least** two aces. Can you solve that problem? (See Exercise 43e.)*

EXAMPLE 8 Five cards are dealt from a deck of 52 cards. What is the probability that exactly two of the cards are aces?

Solution The number of ways of selecting two aces from the four aces in the deck is $_4C_2$. However, we also need to select three additional cards from the remaining 48 cards in the deck that are *not* aces; this can be done in $_{48}C_3$ ways. Thus, by the fundamental counting principle, the total number of ways of drawing two aces (and three other cards) is the product $(_4C_2) \cdot (_{48}C_3)$.

$$\text{Probability of drawing two aces in five cards} = \frac{(_4C_2) \cdot (_{48}C_3)}{_{52}C_5}$$

Show that this ratio reduces to $\dfrac{2162}{54{,}145} \approx 0.04$. ∎

EXAMPLE 9 A class consists of 10 men and 8 women. Four members are to be selected *at random* to represent the class. What is the probability that the selection will consist of two men and two women?

*If the members are selected **at random**, then each one has an equally likely chance of being selected. For example, a random selection could involve having all 18 names placed in a hat and 4 names drawn, as in a lottery.*

Solution Number of ways of selecting two men = $_{10}C_2$.
Number of ways of selecting two women = $_8C_2$.
Number of ways of selecting four members of the class = $_{18}C_4$.
Thus the probability that the four members will consist of two men and two women is given as

$$\frac{(_{10}C_2) \cdot (_8C_2)}{_{18}C_4} = \frac{7}{17}$$ ∎

EXERCISES 12.4

Use the tree diagram, showing the results of tossing three coins, to find the probability of each event.

1. All three coins are heads.
2. Exactly one coin is heads.
3. At least one coin is heads.
4. None of the coins is heads.
5. At most one coin is heads.
6. At least two coins are heads.

```
First    Second    Third
coin      coin      coin
                  ┌ H
          ┌ H ──< 
          │       └ T
    ┌ H ──┤
    │     │       ┌ H
    │     └ T ──<
    │             └ T
  ──┤
    │             ┌ H
    │     ┌ H ──<
    │     │       └ T
    └ T ──┤
          │       ┌ H
          └ T ──<
                  └ T
```

Assume that after spinning the pointer, its chance of stopping in any one region is just as great as its chance of stopping in any of the other regions and that it does not stop on a line. Find the probabilities.

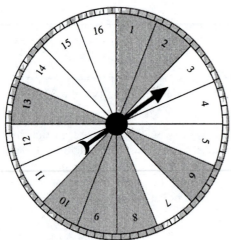

7. $P(\text{odd number})$
8. $P(\text{multiple of 3})$
9. $P(\text{white})$
10. $P(7 \text{ or } 11)$
11. $P(\text{even and red})$
12. $P(\text{no more than 7})$
13. $P(\text{prime})$
14. $P(\text{red and between 3 and 10})$

Use the sample space for the tossing of a pair of dice, given on page 764, to find the probability of each event.

15. Both dice show the same number.
16. The sum is 11.
17. The sum is 7.
18. The sum is 7 or 11.

19. The sum is 2, 3, or 12.

20. The sum is 6 or 8.

21. The sum is an odd number.

22. The sum is not 7.

Two cards are drawn from a deck of 52 playing cards without replacement. Find the probability of each event.

23. Both cards are red.

24. Both cards are spades.

25. Both cards are the ace of hearts.

26. Both cards are the same suit.

(*Hint for Exercise* 26: A successful outcome is to have both cards spades *or* both hearts *or* both diamonds *or* both clubs. The sum of these probabilities gives the solution.)

Two cards are drawn from a deck of 52 playing cards with the first card replaced before the second card is drawn. Find the probability of each event.

27. Both cards are black or both are red.

28. Both cards are hearts.

29. Both cards are the ace of hearts.

30. Both cards are picture cards.

31. Neither card is an ace.

32. Neither card is a spade.

33. The first card is an ace and the second card is a king.

34. The first card is an ace and the second card is not an ace.

A bag of marbles contains 8 red marbles and 5 green marbles. Three marbles are drawn at random at the same time. Find the probability of each event.

35. All are red.

36. All are green.

37. Two are red and one is green.

38. One is red and two are green.

39. A student takes a true-false test consisting of 10 questions by just guessing at each answer. Find the probability that the student's score will be:

(a) 100%   (b) 0%   (c) 80% or better

40. A die is tossed three times in succession. Find the probability that:

(a) All three tosses show 5.   (b) Exactly one of the tosses shows 5.

(c) At least one of the tosses shows 5.

In Exercises 41 and 42 use $P(not\ E) = 1 - P(E)$.

41. Two cards are drawn simultaneously from a deck of playing cards. Find $P($not two red cards$)$.

42. Three cards are drawn simultaneously from a deck of playing cards. Find $P($not three hearts$)$.

43. Five cards are dealt from a deck of 52 cards. Find the probability of obtaining:

(a) Four aces

(b) Four of a kind (that is, four aces or four twos or four threes, etc.).

(c) A flush (that is, all five cards of the same suit).

(d) A royal straight flush (10, jack, queen, king, ace, all in the same suit).

(e) At least two aces.

44. You are given 10 red marbles and 10 green marbles to distribute into two boxes. You will then be blindfolded and asked to select one of the boxes and then draw one marble from that box. You win $10,000 if the marble you select is red. Finally, you are allowed to distribute the 20 marbles into the two boxes in any way that you wish before you begin to make the selection. Try to determine what is the best strategy for distributing the marbles; that is, how many of each color should you place in each box?

*The **odds in favor of an event** occurring is the ratio of the probability that it will occur to the probability that it will not occur. For example, the odds in favor of tossing two heads in two tosses of a coin are* $\dfrac{1/4}{3/4} = \dfrac{1}{3}$ *or 1 to 3. The odds against this event are 3 to 1.*

Exercises 45–52 involve computation of odds.

45. A card is drawn from a deck of cards. What are the odds in favor of obtaining an ace? What are the odds against this event occurring?

46. What are the odds in favor of tossing three successive heads with a coin?

47. What are the odds against tossing a 7 or an 11 in a single throw of a pair of dice?

48. Two cards are drawn simultaneously from a deck of cards. What are the odds in favor of both cards being hearts?

49. What are the odds for getting 2 heads when tossing 3 coins?

50. Suppose you have four playing cards in your hand whose face values are 5, 6, 7, 8. What are the odds of selecting a fifth card from the remaining 48 that has a face value of either 4 or 9?

51. Show that if the odds for event E are a to b, then $P(E) = \dfrac{a}{a+b}$.

52. Suppose you are given 1 to 3 odds that event E occurs and 3 to 5 odds that event F occurs. If E and F cannot occur together, what are the odds that E or F will occur?

Assume that the probability of giving birth to a boy or girl is $\frac{1}{2}$.

53. For a family of four children, what is the probability that there will be two boys and two girls?

54. For a family of four children, what is the probability that there will be more boys than girls?

55. Several years ago a young teacher was the single winner of a lottery that paid him a total of $111,000,000, payable over a period of 20 years. In order to win, he had to select 5 numbers from the numbers 1 through 45, and then a sixth "powerball number" from a different set of numbers 1 through 45.

The numbers shown are the actual winning numbers for this lottery.

Approximately what is the probability of choosing the winning six numbers?

56. Two mathematicians at Harvard University used what is known as a *Poisson distribution* to find the probability P_x that there would be x winners, assuming that all players selected their numbers randomly. The formula is:

$$P_x = \frac{(e^{-a})(a^x)}{x!}$$

x = the number of winners; in this case $x = 1$
e = the base of the natural exponential function
a = the probability of winning times the number of players

Use the information given in Exercise 55 to find P_x if there were 68,400,000 players for this particular lottery.

57. When five cards are selected at random from a deck of playing cards, what is the probability of getting a straight, but *not* a straight flush. (A straight flush is five cards in succession in the same suit.) The lowest straight begins with 1 (an ace), and the highest ends with an ace.)

58. Sometimes probabilities are found *empirically,* that is, by experimentation. For example, conduct the following experiment. Rule a piece of paper with parallel lines that are two inches apart. Then cut a matchstick or toothpick to obtain a one-inch length. From a convenient distance, drop the object onto the paper about 50 times and estimate the probability that the object will cross a line:

$$P(E) = \frac{\text{number of times object crosses a line}}{\text{total number of tosses}}$$

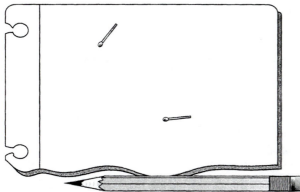

*This experiment was first conducted by Count Buffon in the 1770s and is noted in the literature as **Buffon's needle problem**. Note that the distance between the lines must be twice that of the length of the object dropped.*

Write this probability as a decimal, to four decimal places. Use a calculator to find the reciprocal of this number and see if you obtain a familiar result. Then repeat the experiment 50 more times, for a total of 100 trials, and comment on your results.

59. A pair of dice is tossed. You win if the sum of the numbers is even; you lose if the sum is odd. Is this a *fair game*? That is, will two players have an equally likely chance of winning? Explain your answer. Then answer the same question for the product of two numbers.

Written Assignment   Consider the information given for Exercises 55–56. The Poisson distribution, named after an eighteenth-century physicist, provides statisticians with a method to determine whether events are occurring *randomly*. Explain why it is very likely that most player's selections of lottery numbers are not made at random. Are there conditions when the numbers might be randomly chosen?

Challenge   Three cards are placed in a hat. One is red on each side, one is green on each side, and the third is red on one side and green on the other side. A card is drawn at random and placed on the table. Assume that the color showing is red. What is the probability that the other side is also red? Note that the answer, surprisingly, is *not* $\frac{1}{2}$. (*Hint:* Construct a sample space of all possibilities using R_1, R_2, and R_3 for the red sides and G_1, G_2, and G_3 for the green sides.)

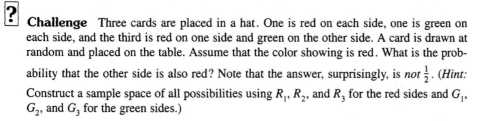

Graphing Calculator Exercises   Consider the idea of the probability of events with an infinite number of outcomes. For example, if a dart is thrown at a circular dartboard in such a way that it is sure to hit the board but it is otherwise aimed *randomly*, we might ask for the probability that it would hit the bull's eye. If the bull's eye is also a circular region having, let us say, one third the area of the entire dartboard, then our intuition tells us that the probability of hitting the bull's eye is $\frac{1}{3}$.

For the following experiments, look up in your calculator's user's manual how to (1) plot a single point on your viewing window, and (2) how to select a *random number*, rnd #, between 0 and 1. By a *random point*, we will mean coordinates of the form (rnd #, rnd #). (*Note:* A different random number is selected each time.)

1. Use the window $0 \le x \le 1$, $0 \le y \le 1$ and graph the straight line $y = x$ and 20 random points. How many would you expect there to be above the line? How many did you get? Repeat the experiment several times. Take the average. Is your average close to the expected probability?

2. Repeat the experiment in Exercise 1 with the line $y = 2x$.

3. Repeat the experiment in Exercise 1 with the curve $y = \sqrt{1 - x^2}$.

CHAPTER REVIEW EXERCISES

Section 12.1 Permutations (Page 738)

1. What is the fundamental principle of counting? Illustrate with a specific example.

2. What is the definition of the permutation of n elements taken r at a time?

3. State two different permutation formulas for $_nP_r$.

4. State the formula for the number of distinguishable permutations of n objects of which n_1 are of one kind, n_2 of another kind, . . . , n_k are of another kind.

5. Evaluate: (a) $_9P_3$ (b) $_5P_5$ (c) $\dfrac{10!}{7! \, 3!}$

6. A club has 8 male and 12 female members. They decide to elect a female as president, a male as vice president, and a combined secretary/treasurer of either sex. How many such sets of officers are possible?

7. Repeat Exercise 6 if there is to be no restriction as to the sex of an officer.

8. How many four-letter "words" can be formed from the 26 letters of the alphabet if no duplication of letters are permitted?

9. Repeat Exercise 8 if duplications are permitted.

10. In how many ways can the letters of the word CALCULUS be arranged so as to form distinguishable permutations?

Section 12.2 Combinations (Page 748)

11. What is the definition of the combination of n elements taken r at a time?

12. State two different combination formulas for $_nC_r$.

13. Evaluate: (a) $_{25}C_5$ (b) $_{25}C_0$ (c) $\dbinom{12}{3}$

14. Evaluate: $\dbinom{6}{0} + \dbinom{6}{1} + \dbinom{6}{2} + \dbinom{6}{3} + \dbinom{6}{4} + \dbinom{6}{5} + \dbinom{6}{6}$

15. How many subsets, each consisting of three elements, does the set $\{a, b, c, d, e, f\}$ have?

16. In how many ways can a club of 15 members select a committee that is to consist of three of its members?

17. Suppose you have a half-dollar, a quarter, a dime, and a nickel. How many different amounts are possible if you are to use three of these coins?

18. A deck of 10 cards are numbered from 1 through 10. How many different hands of three cards can be dealt from this deck?

19. A club has 10 members. They wish to split themselves into two committees of six and four members each. In how many different ways can this be done?

20. How many ways can 5-card hands be selected out of a deck of 52 cards so that all 5 cards are picture cards?

Section 12.3   The Binomial Theorem (Page 754)

21. List the elements in the row of Pascal's triangle that begins $1, 8, \ldots$.

22. State a formula for the expansion of $(a + b)^n$ whose coefficients have the form $\binom{n}{r}$.

23. Write an expression for the $(r + 1)$st term in the expansion of $(a + b)^n$.

Expand and simplify.

24. $(x + 3)^5$ **25.** $(x - 3)^5$ **26.** $(a + 2b)^4$ **27.** $(2a - 3b)^3$ **28.** $(2x + \frac{1}{2}y)^4$

29. Find the seventh term in the expansion of $(a + b)^{10}$.

30. Find the term in the expansion of $(2x - 3y)^8$ in which the power of y is 5.

Section 12.4   An Introduction to Probability (Page 762)

31. List a sample space for an experiment that consists of the tossing of three coins.

Use the sample space of Exercise 31 to find:

32. The probability of tossing three tails.   **33.** The probability of tossing three heads or three tails.

34. The probability of tossing exactly one head.

A pair of dice is tossed. Find the probability of obtaining:

35. A sum of 5.   **36.** A sum of 7 or 11.   **37.** A sum of 2, 3, or 12.

38. A sum greater than 9.   **39.** A pair with one number twice as great as the other.

Two cards are drawn simultaneously from a deck of playing cards. Find the probability that:

40. Neither card is an ace.   **41.** Both cards are red.   **42.** Both cards are hearts.

A single card is drawn from a deck of cards. It is then replaced and a second card is drawn. Find the probability that:

43. Both are kings.   **44.** Neither one is an ace.   **45.** The first is an ace and the second is a king.

Five cards are dealt from a deck of 52 cards. Find the probability that:

46. Exactly three of the cards are hearts.   **47.** At least three of the cards are hearts.

48. Three of the cards are jacks and the other two cards have the same face value (e.g., two sevens).

CHAPTER 12: STANDARD ANSWER TEST

Use these questions to test your knowledge of the basic skills and concepts of Chapter 12. Then check your answers with those given at the back of the book.

1. Evaluate: **(a)** $_{10}P_3$ **(b)** $_{10}C_3$.

2. How many different ways can the letters of the word TODAY be arranged so that each arrangement uses each of the letters once?

3. How many three-digit whole numbers can be formed if zero is not an acceptable hundreds digit?

   (a) Repetitions are allowed.   (b) Repetitions are not allowed.

4. In how many different ways can a class of 15 students select a committee consisting of three students?

5. How many 3-digit even numbers can be formed if repetition of digits is allowed?

6. How many different four-letter "words" can be formed from the letters m, o, n, d, a, y if no letter may be used more than once?

7. How many distinguishable permutations can be made from all of the letters in the word MINIMUM?

8. A class consists of 12 boys and 10 girls. How many committees of four can be selected if each committee is to consist of two boys and two girls?

9. Solve for n: $7\binom{n}{2} = 3\binom{n+1}{3}$

10. A student takes a five-question true-false test just by guessing at each answer.

    (a) How many different sets of answers are possible?

    (b) What is the probability of obtaining a score of 100% on the test?

    (c) What is the probability that the score will be 0%?

11. A single die is tossed. What is the probability that it will show an even number or a number less than 3?

12. A pair of dice is tossed. What is the probability that the sum will be:

    (a) 12   (b) Not 12   (c) Less than 5

13. Two cards are drawn from a deck of 52 playing cards. What is the probability that they are both aces or both kings if:

    (a) The first card is replaced before the second card is drawn?   (b) No replacements are made?

14. Five cards are dealt from a deck of 52 playing cards. What is the probability that exactly four of the cards are picture cards?

15. A box contains 20 red chips and 15 green chips. Five chips are selected from the box at random. What is the probability that exactly four of the chips will be red?

16. Four coins are tossed. Find the probability that:

    (a) All four coins will land showing heads.   (b) None of the coins will show heads.

17. Two integers from 0 through 10 are selected simultaneously and at random. What is the probability that they will both be odd?

18. Box A contains 5 green and 3 red chips. Box B contains 4 green and 6 red chips. You are to select one chip from each box. What is the probability that both chips will be red?

19. In Exercise 18, suppose that you are blindfolded and told to select just one chip from one of the boxes. What is the probability that the chip you select will be red?

20. A box contains 8 red marbles, 13 blue marbles, and 6 green marbles. If two marbles are drawn simultaneously from the box, what is the probability that neither marble is blue?

21. Expand: $(x - 2y)^5$.

22. Write the first four terms of the expansion of $\left(\dfrac{1}{2a} - b\right)^{10}$.

23. Write the seventh term in the expansion of $(3a + b)^{11}$.

24. Write the middle term in the expansion of $(2x - y)^{16}$.

25. Evaluate $(3.1)^4$ by using the expansion of $(a + b)^n$ for appropriate values of $n, a,$ and b.

CHAPTER 12: MULTIPLE CHOICE TEST

1. How many four-digit whole numbers can be formed if zero is not an acceptable digit in the thousands place and repetition of digits is not allowed?

    (a) 3360   (b) 4032   (c) 4536   (d) 5040   (e) None of the preceding

2. Which of the following are correct?

    I. $_nP_r = \dfrac{n!}{(n-r)!}$ **II.** $_nP_n = n!$ **III.** $_nC_r = \dfrac{n!}{r!(n-r)!}$

    (a) Only I   (b) Only II   (c) Only III   (d) I, II, and III   (e) None of the preceding

3. How many different ways can the five letters of the word EIGHT be arranged using each letter only once in each arrangement?

    (a) $_8P_5$ (b) $_5C_5$ (c) 5!   (d) 5^5 (e) None of the preceding

4. A group consists of 5 boys and 6 girls. How many committees of five can be selected if each committee is to consist of 2 boys and 3 girls?

    (a) 150   (b) 200   (c) 1800   (d) 2400   (e) None of the preceding

5. How many distinguishable permutations can be formed using all of the letters in the word PRESSURE?

    (a) 8   (b) 5040   (c) 6720   (d) 56   (e) None of the preceding

6. Which of the following are true?

    I. $_{10}C_3 = {_{10}}C_7$ **II.** $_{10}C_3 = \dfrac{_{10}P_3}{3!}$ **III.** $_{10}P_3 = {_{10}}P_7$

    (a) Only I   (b) Only II   (c) Only III   (d) I, II, and III   (e) None of the preceding

7. How many triangles can be formed using six points in a plane no three of which are on the same straight line?

    (a) 10   (b) 12   (c) 18   (d) 20   (e) None of the preceding

8. What is the coefficient of x^3 in the expansion of $(2 + x)^5$?

    (a) 40   (b) 20   (c) 10   (d) 80   (e) None of the preceding

9. What is the fifth term in the expansion of $(3a - b)^6$?

    (a) $-135a^2b^4$ (b) $135a^2b^4$ (c) $-540a^3b^3$ (d) $-18ab^5$ (e) None of the preceding

10. What is the middle term in the expansion of $(x - 2)^6$?

    (a) $-20x^3$ (b) $60x^4$ (c) $-160x^3$ (d) $240x^2$ (e) None of the preceding

11. Which of the following expressions represents the ninth term in the expansion of $(a + b)^n$?

    (a) $\dbinom{n}{8} a^{n-8}b^8$ (b) $\dbinom{n}{8} a^8b^{n-8}$ (c) $\dbinom{n}{9} a^{n-9}b^9$ (d) $\dbinom{n}{9} a^9b^{n-9}$ (e) None of the preceding

12. A coin is tossed three times. What is the probability that not all three tosses are the same?

    (a) $\frac{1}{8}$ (b) $\frac{3}{8}$ (c) $\frac{1}{4}$ (d) $\frac{3}{4}$ (e) None of the preceding

13. Five cards are dealt from a deck of 52 cards. Which of the following shows the probability that four aces will be dealt?

    (a) $\dfrac{_4C_4}{_{12}C_5}$ (b) $\dfrac{_{52}C_4}{_{52}C_5}$ (c) $\dfrac{(_4C_4)(_{48}C_1)}{_{52}C_5}$ (d) $\dfrac{(_4C_4)(_{52}C_1)}{_{52}C_5}$ (e) None of the preceding

14. Two cards are drawn from a deck of 52 cards with the first card replaced before the second card is drawn. What is the probability that neither card is a spade?

    (a) $\frac{9}{16}$ (b) $\frac{3}{4}$ (c) $\frac{1}{16}$ (d) $\frac{19}{34}$ (e) None of the preceding

15. A pair of dice is tossed. What is the probability that the sum of the faces showing on top is 10?

(a) $\frac{2}{9}$ (b) $\frac{1}{12}$ (c) $\frac{1}{9}$ (d) $\frac{1}{6}$ (e) None of the preceding

16. A coin is tossed first, and then a die is tossed. Which of the following would be an appropriate sample space for this experiment?

(a) $\{(H, T), (1, 2, 3, 4, 5, 6)\}$ (b) $\{(H, 1, 2, 3, 4, 5, 6), (T, 1, 2, 3, 4, 5, 6)\}$

(c) $\{(H, 1), (H, 2), (H, 3), (H, 4), (H, 5), (H, 6), (T, 1), (T, 2), (T, 3), (T, 4), (T, 5), (T, 6)\}$

(d) $\{(H, T, 1), (H, T, 2), (H, T, 3), (H, T, 4), (H, T, 5), (H, T, 6)\}$ (e) None of the preceding

17. What are the elements in the row of Pascal's triangle following the row 1, 6, 15, 20, 15, 6, 1?

(a) 1, 7, 21, 35, 21, 7, 1 (b) 1, 6, 21, 35, 21, 6, 1 (c) 1, 6, 21, 35, 35, 21, 6, 1

(d) 1, 7, 21, 35, 35, 21, 7, 1 (e) None of the preceding

18. Which of the following are correct?

I. $_nC_1 = {_n}P_1$ **II.** $_nC_n = {_n}P_n$ **III.** $n! = (n-1)(n-2)!$

(a) Only I (b) Only II (c) Only III (d) I, II, and III (e) None of the preceding

19. In the expansion of $(a + b)^n$, the term containing the factor a^r is:

(a) $\binom{n}{r}a^r b^r$ (b) $\binom{n}{n-r}a^r b^{n-r}$ (c) $\binom{n}{r}a^r b^{n-r}$ (d) $\binom{n}{n-r+1}a^r b^{n-r}$ (e) None of the preceding

20. A box contains 8 red marbles and 5 blue marbles. If four marbles are selected from the box at random, what is the probability of drawing 2 red and 2 blue marbles?

(a) $\dfrac{_{13}C_2}{_{13}C_4}$ (b) $\dfrac{_{13}C_4}{(_8C_2)(_5C_2)}$ (c) $\dfrac{(_8C_2)(_5C_2)}{_{13}C_4}$ (d) $\dfrac{(_8C_2)(_5C_2)}{_{13}C_2}$ (e) None of the preceding

ANSWERS TO THE TEST YOUR UNDERSTANDING EXERCISES

Page 742

1. 5040 **2.** 336 **3.** 720 **4.** 90 **5.** 220 **6.** 358,800 **7.** 456,976

8. 6; EAT, ETA, AET, ATE, TEA, TAE **9.**

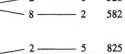

258 **10.** 18.

5 — 8	258	
8 — 5	285	
2 — 8	528	
8 — 2	582	
2 — 5	825	
5 — 2	852	

Page 750

1. 210 **2.** 1 **3.** 1 **4.** 220 **5.** 56 **6.** $_{10}C_3 = \dfrac{10!}{3!\,7!} = \dfrac{10!}{7!\,3!} = {_{10}}C_7$ **7.** 495

8. 840 **9.** 28 **10.** 10

Page 760

1. $x^3 + 6x^2 + 12x + 8$ **2.** $x^4 - 12x^3 + 54x^2 - 108x + 81$

3. $a^5 + 10a^4b + 40a^3b^2 + 80a^2b^3 + 80ab^4 + 32b^5$ **4.** $1 - 6y + 15y^2 - 20y^3 + 15y^4 - 6y^5 + y^6$

5. $560x^2y^{13}$ **6.** $4032x^5y^4$

Page 768

1. $\frac{1}{4}$ **2.** $\frac{1}{2}$ **3.** $\frac{3}{4}$ **4.** $\frac{1}{2}$ **5.** $\frac{1}{2}$ **6.** $\frac{2}{3}$ **7.** $\frac{2}{3}$ **8.** $\frac{1}{2}$ **9.** $\frac{1}{2}$ **10.** $\frac{4}{13}$ **11.** $\frac{3}{13}$ **12.** $\frac{4}{13}$

Graphing Calculator Appendix

A large portion of this book is devoted to graphs and their uses. This is because graphs provide a vital link between algebra, trigonometry, geometry, and their applications. In this appendix you will find useful information on using graphing calculators to obtain the graphs of functions and other relations. But first we consider some procedures on entering data and carrying out ordinary computations on graphing calculators.

CALCULATIONS ON GRAPHING CALCULATORS

Your user's manual should be used throughout this appendix. Thus refer to the manual to read more about replaying and editing expressions.

Your graphing calculator's viewing screen can display more than just graphs. It can also display strings of calculations somewhat as you would see them on your blackboard or in your notebook. This gives graphing calculators the potential to help you do ordinary calculations better than on nongraphing calculators. One reason for this is that you can more easily check your calculations for mistakes when you can see all the symbols and numbers at once. Moreover, if you find that you have made a mistake in entering part of an expression, you usually will not have to reenter the whole expression, symbol by symbol. A *replay key* lets you edit the expression until it is entered correctly.

Most graphing calculators follow standard algebraic usage with regard to order of operations. Operations are carried out from left to right, except that **(1)** expressions in parentheses are evaluated first; **(2)** next, calculator-defined functions, such as roots, are evaluated; **(3)** multiplications and divisions are done before additions and subtractions, in order from left to right, but after calculator-defined functions.

Note: On some calculators the symbol X is used to denote multiplication, instead of ∗. *Also* ÷ *is used instead of* / *to denote division.*

EXAMPLE 1   Place the following expression on your calculator screen:

$$2 + 10/\sqrt{25} * (4 - 1)$$

(a) What is the answer, and why?

(b) Suppose we want the 25 to be multiplied by the 4 before we take the root. How should we correct the calculation, and what is the new answer?

Solution

(a) The subtraction $4 - 1$ (= 3) is done first, since it is in parentheses. Then the square root of 25, as a calculator-defined function, is evaluated as 5. Next, this 5 is divided into the 10, and the quotient of 2 is multiplied by the difference, 3, giving a product of 6. Finally, the initial 2 and this 6 are added for a sum of 8.

(b) To edit the given expression, we press the replay key to obtain the original expression. The blinking cursor will now be at the beginning or end of the ex-

A1

pression. With the directional cursor keys, move the cursor over each parenthesis and press the delete key, removing both parentheses. Move the cursor onto the digit "2" in 25 and press the insert key. Pressing the left parenthesis key places the left parenthesis before the digit "2". Now move the cursor to the right of the 4 and insert the right parenthesis. Your expression should look like this: $2 + 10/\sqrt{(25 * 4)} - 1$ and the new answer is 2. ■

---

1. Evaluate the following expression mentally, and then try it on your graphing calculator.

$$3 - (10 + 5\sqrt{16})/6 * (-2)$$

SETTING MODES

See page 115 or your user's manual for a discussion of default settings. In general, always refer to your manual when you encounter an unfamiliar term.

Find the MODE key on your graphing calculator. Pressing it reveals a menu with several choices of settings. To evaluate certain functions properly and to obtain answers and graphs in the form you want, you must have the MODES set correctly. For the most part, the *default settings* will usually suffice. On most graphing calculators these are *Normal* (numerical display), *Float* (floating-point display instead of rounded-off answers), *Rad* (the radian measure for arguments of trigonometric functions, instead of degrees), *Rect* (rectangular coordinates), and *Connected* (connected graphs, as opposed to separated dots). However, there are times, described later, where the other settings are useful. Read your user's manual on how to change modes on your calculator, and make sure that your calculator is set to these modes until further notice.

POINTS, PIXELS, AND WINDOWS

As discussed in this text, the graph of an equation of the form $y = f(x)$ may be obtained by making a table of ordered pairs of values (x, y) that satisfy the equation. Making such tables and plotting the points by hand is limited by the number of points we are willing or realistically able to plot. In addition, unless we choose our points with care and plot a large number of them, the graph we obtain might be very misleading.

EXAMPLE 2   Use the five points corresponding to $x = 0, \pm 1$, and ± 2, to graph the equation $y = x^5 - 5x^3 + 4x$.

Solution   The y-values corresponding to these x's are all zero, giving us the points $(0, 0), (-1, 0), (1, 0), (2, 0)$, and $(-2, 0)$. If only these points were used, we might conclude that the graph is the x-axis! (Figure 1). However, plotting several additional points shows us that the graph is considerably more interesting. (Figure 2).

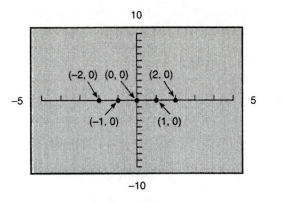

Figure 1   Five collinear points

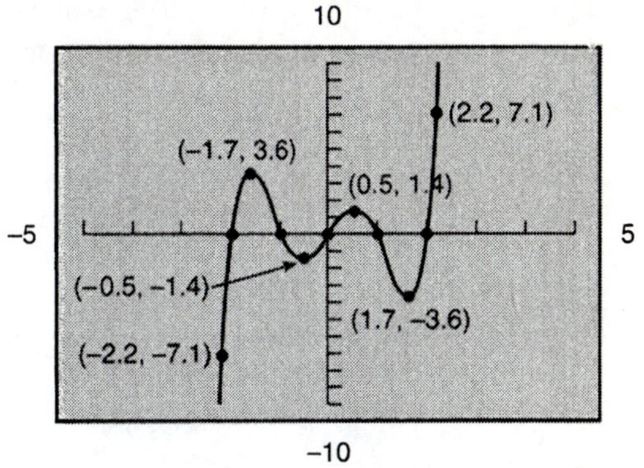

Figure 2   Coordinates rounded to tenths   ■

Graphing calculators let us plot many points in a relatively short time. (Ninety-five or more in seconds!) The calculator's *viewing screen* is a rectangular grid of thousands of small areas, called *pixels*, each of which can become visible to represent a plotted point. The range of values of the x, y and other variables is often called the *window*. Selecting an appropriate window is one of the most important steps in graphing on your graphing calculator since, just as with a real window, you will only see that part of the scenery that your window is facing.

BASIC STEPS FOR GRAPHING (X, Y) EQUATIONS ON YOUR CALCULATOR

Here is how to proceed with equations in x and y.

1. Solve the equation for y in terms of an expression in the variable x.
2. Enter the expression into your calculator for graphing, as directed in your user's manual.
3. Decide on and set an appropriate window for your graph.
4. Execute the graphing command.

On some calculators, the expression in Step 2 is entered on a special $Y=$ menu. On others, the expression is entered after "Graph $Y =$", which is obtained by pressing the Graph key. The exact way to enter the expression on your calculator is described in your user's manual.

If you don't know what window would be appropriate, it is often a good idea to start with a *standard window*: $-10 \leq X \leq 10$ and $-10 \leq Y \leq 10$. This is the *default,* or initial, window, on many calculators. On other graphing calculators the initial setting is $-4.7 \leq X \leq 4.7$ and $-3.1 \leq Y \leq 3.1$, and this too is sometimes a good place to start. This window has other advantages that we will describe later.

EXAMPLE 3   Use your graphing calculator to graph the following equation: $0.04x^4 - 1.04x^2 = y - 1$

Solution   Begin by solving for y in terms of x:

$$y = 0.04x^4 - 1.04x^2 + 1$$

The expression $0.04x^4 - 1.04x^2 + 1$ is now entered for graphing as directed in your user's manual. To view the graph, we will select one of the standard windows defined earlier. (In Example 4 the same function will be graphed using a different initial setting.) This can be done by depressing your RANGE key and changing the settings, one by one, until they match those shown in Figure 3. Completing the operation as directed in your manual yields the graph in Figure 4.

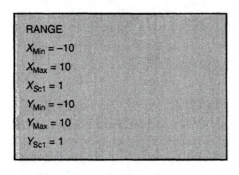

RANGE
$X_{Min} = -10$
$X_{Max} = 10$
$X_{Sc1} = 1$
$Y_{Min} = -10$
$Y_{Max} = 10$
$Y_{Sc1} = 1$

Figure 3   Standard range   **Figure 4** $y = 0.04x^4 - 1.04x^2 + 1$ ∎

TEST YOUR UNDERSTANDING

2. Graph the equation $4x - 2y + 6 = 2x - 3$ on a standard window.

(Answer: Page A12)

ZOOMING OUT

EXAMPLE 4 Graph the function in Example 3 using the Range settings $-4.7 \le X \le 4.7$, $-3.1 \le Y \le 3.1$.

Solution On some calculators this is the initial setting. Hence, it may be selected at one stroke on the Range menu. On other calculators these settings must be entered line by line. (For example, set $X\text{min} = -4.7$.) In both cases, however, we get the graph shown in Figure 5.

While this is a useful window on some calculators, here it lops off important parts of the graph, such as the x-intercepts, which usually are important. However, by pressing the Zoom key and selecting Zoom Out on the screen menu, we increase all the range settings. For example, we get $-9.4 \le X \le 9.4$, and $-6.2 \le Y \le 6.2$ if our ZOOM FACTORS are set at 2, and if the previous window is centered at $(0, 0)$; this gives the graph in Figure 6. (On some calculators, Zooming Out quadruples the Range settings. Check the default FACTORS settings on yours. Also, if your initial screen was not centered at $(0, 0)$, even ZOOM FACTORS set at 2 will not precisely double the range variables. Your new values, however, will be close to the ones given.)

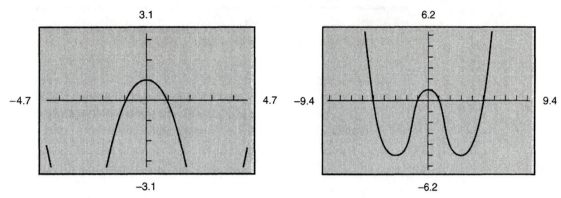

Figure 5   Truncated view
of $y = 0.04x^4 - 1.04x^2 + 1$

Figure 6   Better view of
$y = 0.04x^4 - 1.04x^2 + 1$ ■

Zooming Out is a good general technique for seeing the overall characteristics of some graphs. By multiplying the Range factors by numbers greater than one, it gives the viewer the illusion of stepping back away from the graph, permitting a sort of bird's-eye view.

TRACING A GRAPH AND FRIENDLY INTERVALS

Sometimes we want to do a point-by-point investigation of a graph that has been plotted by the graphing calculator. For example, it might be useful to see precisely where the graph crosses the x- and y-axes. While we can move around the screen with the directional cursor keys, point-by-point examination is best accomplished by means of your calculator's Trace key.

Suppose, for example, we have the settings of Figure 5. Then, pressing first the Trace key and then the directional left or right cursor key allows us to move and keep the cursor exactly on the graph. If the calculator's screen is exactly 95 pixels wide (which is true where our window defines the initial settings), notice that the x-coordinates at the bottom of the screen are increased by exactly one tenth for each step to the left.

We will say that $-4.7 \le X \le 4.7$ is a *friendly interval* for a certain calculator if each step from pixel to pixel on that calculator is a convenient number, such as one tenth, rather than a number like 2.1245789. Similarly friendly intervals on calculators that are 96 pixels wide are $-4.8 \le X \le 4.7$ and $-4.7 \le X \le 4.8$.

If both the x- and y-intervals are friendly, we will say that we have a *friendly window*. (The window in Figure 6 is friendly. Why?)

EXAMPLE 5   What friendly interval on a 96 pixel wide calculator screen will allow us to trace along a curve at steps of one unit?

Solution   Since there are 96 pixels horizontally, to include the point where $x = 0$, we might choose the interval $0 \le X \le 95$. This makes 0 correspond to the first pixel, 1 to the second, . . . , and 95 to the ninety-sixth pixel. Similarly, the interval $-95 \le X \le 0$ gives such a correspondence for negative integers. Still another suitable choice is $-48 \le X \le 47$.   ∎

To find the length of a traced step on an interval on the x- or y-axis, we take $(X\text{max} - X\text{min})/(n - 1)$ or $(Y\text{max} - Y\text{min})/(n-1)$, where $n =$ the number of pixels in the x- or y-direction, respectively.

EXAMPLE 6   If the calculator's screen is 95 pixels wide and we Trace the graph in Figure 6, what part of a unit is traced each step (or pixel) on the x-axis?

Solution

$$(X\text{max} - X\text{min})/(94) = (9.4 + 9.4)/(94) = 0.2$$

Therefore, each time we move the right or left cursor key one step, the x-coordinate shown at the bottom of the screen will increase or decrease by 0.2 units respectively.   ∎

TEST YOUR UNDERSTANDING

4. Suppose that your calculator screen is 96 pixels wide. What interval will start at $x = 0$ and give a Trace step of 0.5 on the x-axis?

(Answer: Page A12)

EXAMPLE 7   Find the x- and y-intercepts of the graph of the equation $y = 0.04x^4 - 1.04x^2 + 1$ using (a) the friendly window in Figure 6 and (b) the standard window in Figure 4.

Solution
(a) Tracing along the graph in Figure 6, and assuming that the x-interval is friendly, we see at the bottom of the window that the y-coordinates are 0 when the x-coordinates are precisely -5, -1, 1, and 5. These, then, are the x-intercepts. Tracing also shows that the x-coordinate is zero when the y-coordinate is 1. Hence, the y-intercept is 1.
(b) The standard window of Figure 4 is not friendly on any calculator. Let's suppose again that we have a screen 95 pixels wide. Then Tracing near the x-value -5 lets us get as close as -5.106382, after which we jump to -4.893617, since our step size is now about $(10 + 10)/94 = 0.2127659574$ (Figure 7). Neither value of x gives us 0 for y. We get similar approximate results with the other intercepts. To get better approximations, we first need to discuss the technique of "Zooming In."   ∎

ZOOMING IN

If we didn't already know the intercepts in Example 7, we could try to improve our accuracy by zooming in, that is, by choosing the "zoom-in" option on our Zoom menu to reduce our range settings by preselected factors. The technique for zooming in is to:

1. Trace as close as possible to the desired point, and then
2. Zoom in, noting any improvement in the coordinates on the window, and go back to the first step, if desired, with the new window.

EXAMPLE 8   Continuing Example 7(**b**), use Trace and Zoom to achieve accuracy to the hundredth's place in the x-intercept near -5.

Solution   Tracing to the x-value -5.106382, we get the graph in Figure 7. Zooming in and tracing again, we land precisely on the point $(-5, 0)$. (This is sheer good luck, as you will see if you try the same procedure to locate the x-intercept "near" $x = -1$.) It takes several Trace and Zoom-Ins to decide that the x-intercept is within 0.01 of -5.

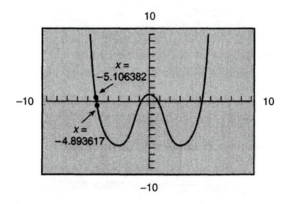

Figure 7   One pixel apart   ■

TEST YOUR UNDERSTANDING

5. Use Trace and Zoom In to find the largest positive x to the nearest hundredth for which the graph of $y = x^2 - 2x - 2$ touches the x-axis.

(Answer: Page A12)

SQUARE WINDOWS

The dimensions of your viewing screen also have an effect on the shape of the graphs shown on that screen. Most screens are wider than they are high. Hence, if there are as many scale units on the y-axis as on the x-axis, the y-units must be smaller. This introduces distortion of shapes and angles, especially noticeable when graphing circles, for example. To correct this, some calculators have a menu selection for a *square window,* that is, one in which the Range settings are adjusted so that the units on both axes are the same length. For example, on some calculators, "Square" changes $0 \le X \le 10$ and $-10 \le Y \le 10$ to $-10 \le X \le 20$ and $-10 \le Y \le 10$. So we will say that 3 to 2 is its *square ratio*.

EXAMPLE 9 Suppose that your calculator has a square ratio of 3 to 2. Graph the circle $y = \pm\sqrt{100 - x^2}$ **(a)** on a standard window and **(b)** on a square window. Compare the results.

Solution

(a) The circle, given as two semicircles, has its center at $(0, 0)$ and a radius of ten units. Therefore, a standard window of $-10 \le X \le 10$ by $-10 \le Y \le 10$ accommodates the entire circle, as in Figure 8. The elliptical shape on the screen is due to the x-units being larger than the y-units.

(b) Changing the window settings to $-15 \le X \le 15$ by $-10 \le Y \le 10$ gives us the circular-looking graph in Figure 9. Notice that if we take a square setting with a smaller y-interval, we will not obtain the entire graph.

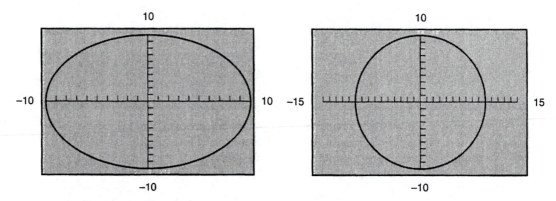

Figure 8 Different unit lengths on x and y axes   **Figure 9** Equivalent unit lengths on x and y axes

■

TEST YOUR UNDERSTANDING

6. Graph the circle $y = \pm\sqrt{400 - x^2}$ on a square window just large enough to accommodate it.

(Answer: Page A12)

Zooming out and zooming in will maintain a square window if the ZOOM FACTORS are the same for x and y. For example, if your calculator has an initial square setting of $-4.7 \le X \le 4.7$ by $-3.1 \le Y \le 3.1$ and ZOOM FACTORS of 2 for both x and for y, then zooming out twice will give a square window large enough to accommodate the circle $y = \pm\sqrt{100 - x^2}$, namely, $-18.8 \le X \le 18.8$ by $-12.4 \le Y \le 12.4$.

USING A ZOOM BOX

Your graphing calculator allows you to Zoom In on an interesting section of graph by framing the section in a rectangle (or *zoom box*) right on the screen and using the zoom box as your new window. To do this:

1. Select Box or ZBOX from your Zoom menu.
2. Move the blinking cursor to any corner P of the zoom box you wish to define, and then press ENTER (or EXE).
3. Move the cursor (**a**) horizontally to an adjacent corner Q of the desired zoom box, and (**b**) then move it vertically to the next adjacent corner R. The box should now be completely formed.
4. Press ENTER (or EXE). A new window will appear including only the material in your zoom box.

EXAMPLE 10   Graph $y = x^4/8 - x^2/4 + 3$ in a standard window (Figure 10). Then use a zoom box to Zoom In on the section that appears to be flat.

Solution   The figure on the left shows the graph in the standard window after we have performed the preceding Steps 1 and 2. The middle figure shows how to carry out Step 3. The right-hand figure shows the window and additional details obtained by Step 4.

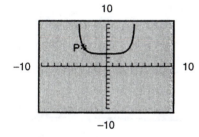
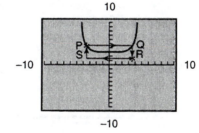
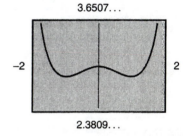

Figure 10   ■

TRIGONOMETRIC AND OTHER SPECIAL WINDOWS

Some graphing calculators have special windows for several different functions, for example, one for the sin x, another for the tan x, and still another for ln x. (Check your user's manual to see if yours does.) Others have a special Trig window selection on their Zoom menu of $-2\pi \le X \le 2\pi$ by $-3 \le Y \le 3$, which is good for viewing several standard trigonometric functions. However, you may have to modify these Range settings to get one that is best for your function.

EXAMPLE 1   (**a**) Use the window $-2\pi \le X \le 2\pi$ by $-3 \le Y \le 3$ to graph the equation $y = \sin 2x + 3$. (**b**) Then find a better window and graph the equation on it.

Solution

(a) The graph in Figure 11 is the result of graphing $y = \sin 2x + 3$ on the given window. (Make sure that your mode setting is in radians.) You may recognize that this is not typical of a sine wave. Tracing the graph shows that it rises about 1 unit higher than the present window.

(b) Choose $-2\pi \leq X \leq 2\pi$ by $0 \leq Y \leq 5$ as the new window. This gives the graph in Figure 12.

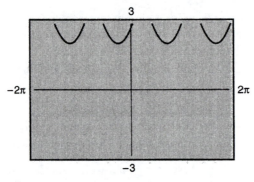

Figure 11   Truncated Graph of $y = \sin 2x + 3$

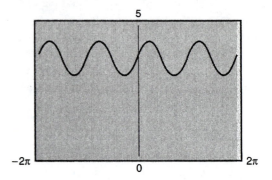

Figure 12 $y = \sin 2x + 3$ on A Better Window   ∎

TEST YOUR UNDERSTANDING

7. Graph $y = 2 \sin x - 5$ in a trigonometric window 2π radians wide and just big enough to accommodate it.

(Answer: Page A12)

PARAMETRIC AND POLAR MODES

All graphing calculators have what is called a *parametric mode*, which allows the user to graph the x- and y-coordinates independently of one another. The x- and y-coordinates are each given by the user as a function of a third variable, T. For example, you can define $X(T) = \cos T$ and $Y(T) = \sin T$. Since $\cos^2 T + \sin^2 T = 1$ for all real numbers T, the unit circle is obtained by letting $0 \leq T \leq 2\pi$. Similarly, letting $X(T) = T$ and $Y(T) = T^2$ for all T, you obtain the parabola $y = x^2$.

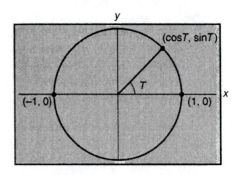

Figure 13 $(X(T), Y(T)) = (\cos T, \sin T)$

Some, but not all, graphing calculators have a mode in which one can directly graph *polar equations*, that is, equations of the form $r = f(\theta)$. If yours does, it is called the *polar mode*, where we say that we are graphing points of the form (r, θ) in *polar coordinates*. In this case, one can enter an equation such as $r = 4 \cos \theta$ and immediately obtain a graph such as that in Figure 14.

The rectangular coordinates $(0, 0)$ *can be replaced by the polar coordinates* $\left(0, \frac{\pi}{2}\right)$; $(4, 0)$ *are both polar and rectangular coordinates of the indicated point.*

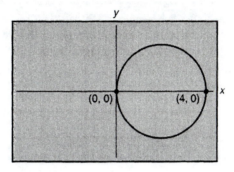

Figure 14 $r = 4 \cos \theta$

If your calculator has a parametric mode but not a polar mode, you can still graph polar equations. Given $r = f(\theta)$, put your calculator in parametric mode. Then let $\theta = T$, and enter $X(T) = r(T) \cos T$, and $Y(T) = r(T) \sin T$. Letting T range over the same values that θ does, graphs the polar equation. For example, to graph $r = 4 \cos \theta$, where $0 \leq \theta \leq 2\pi$, in parametric mode, we let $X(T) = (4 \cos T) \cos T$ and $Y = (4 \cos T) \sin T$, where $0 \leq T \leq 2\pi$. This gives the graph in Figure 14.

TEST YOUR UNDERSTANDING	**8.** Graph the circle $r = 5 \sin \theta$, $0 \leq \theta \leq 2\pi$, in either polar or parametric mode. Choose an appropriate window for X and Y when using parametric mode.

(Answer: Page A12)

SOME USEFUL SPECIAL FEATURES

Not all graphing calculators are exactly alike, but they all can do what has been described previously. Some of them also have especially nice additional features that are called for in a few graphing calculator exercises.

Inequality graphing in two dimensions, for example, is a feature on the Casio fx 7700G series that enables the user to graph and solve systems of inequalities such as $y > x^2 - 5x - 5$ and $y \leq x - 2$. The solution is represented as the shaded intersection of the graphs of each inequality. (See the Casio fx 7700G manual, particularly page xvi of "Quick Start.")

Relational operations may be used in defining functions on the TI 81–85 series. For example, expressions like $(X < 2)$, called *relational operations*, yield a value of 1 when true and 0 when false. Hence, $(5) (X < 2)$ is $5*1$ when $(X < 2)$ is true and $5*0$ when $(X < 2)$ is false. This feature enables the user to define the function differently on different intervals. The user can also qualify the functions in other ways, such as only plotting values at integers. For example, $y = (5)(X < 2) + (3X)(X \geq 2)(X < 4) + (X^2) (X \geq 4)(X - \text{IPart } X = 0)$ gives the value

5 when x is less than 2. It will give the value $3x$ for x's greater than or equal to 2 but less than 4. And it will give the value of x^2 when x is greater than or equal to 4 and x is an integer. ("IPart X" gives the integer part of X. For example, IPart $2.345 = 2$. See the TI 81–85 user's manual for details and examples.)

TEST YOUR UNDERSTANDING	**9.** **(a)** When does $(X - \text{IPart } X = 0)$ give a value of 1? Give an example. **(b)** When is $(X - \text{IPart } X = 0)$ equal to 0? Give an example.
(Answer: Page A12)	

ANSWERS TO THE TEST YOUR UNDERSTANDING EXERCISES

1. 13

2. $y = x + 4.5$

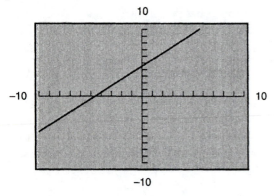

3. $y = -0.4x + 20$; x-intercept: 50; y-intercept: 20

4. $0 \le X \le 47.5$

5. 2.73

6. $y = \pm\sqrt{400 - x^2}$

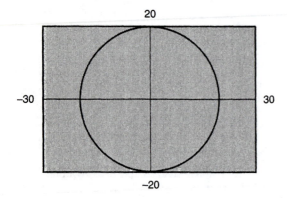

7.

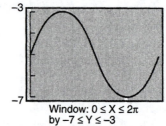

Window: $0 \le X \le 2\pi$
by $-7 \le Y \le -3$

8.

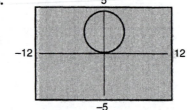

9. **(a)** When X is an integer, that is, when X equals its integer part. For example, $X = 2$.
(b) When X does not equal its integer part, that is, when X is not an integer. For example, $X = 1.5$.

Tables

TABLE I: EXPONENTIAL FUNCTIONS					
x	e^x	e^{-x}	x	e^x	e^{-x}
0.0	1.00	1.000	3.1	22.2	0.045
0.1	1.11	0.905	3.2	24.5	0.041
0.2	1.22	0.819	3.3	27.1	0.037
0.3	1.35	0.741	3.4	30.0	0.033
0.4	1.49	0.670	3.5	33.1	0.030
0.5	1.65	0.607	3.6	36.6	0.027
0.6	1.82	0.549	3.7	40.4	0.025
0.7	2.01	0.497	3.8	44.7	0.022
0.8	2.23	0.449	3.9	49.4	0.020
0.9	2.46	0.407	4.0	54.6	0.018
1.0	2.72	0.368	4.1	60.3	0.017
1.1	3.00	0.333	4.2	66.7	0.015
1.2	3.32	0.301	4.3	73.7	0.014
1.3	3.67	0.273	4.4	81.5	0.012
1.4	4.06	0.247	4.5	90.0	0.011
1.5	4.48	0.223	4.6	99.5	0.010
1.6	4.95	0.202	4.7	110	0.0091
1.7	5.47	0.183	4.8	122	0.0082
1.8	6.05	0.165	4.9	134	0.0074
1.9	6.69	0.150	5.0	148	0.0067
2.0	7.39	0.135	5.5	245	0.0041
2.1	8.17	0.122	6.0	403	0.0025
2.2	9.02	0.111	6.5	665	0.0015
2.3	9.97	0.100	7.0	1097	0.00091
2.4	11.0	0.091	7.5	1808	0.00055
2.5	12.2	0.082	8.0	2981	0.00034
2.6	13.5	0.074	8.5	4915	0.00020
2.7	14.9	0.067	9.0	8103	0.00012
2.8	16.4	0.061	9.5	13360	0.00075
2.9	18.2	0.055	10.0	22026	0.000045
3.0	20.1	0.050			

TABLE II: NATURAL LOGARITHMS (BASE e)

x	$\ln x$	x	$\ln x$	x	$\ln x$
0.0		3.4	1.224	6.8	1.917
0.1	−2.303	3.5	1.253	6.9	1.932
0.2	−1.609	3.6	1.281	7.0	1.946
0.3	−1.204	3.7	1.308	7.1	1.960
0.4	−0.916	3.8	1.335	7.2	1.974
0.5	−0.693	3.9	1.361	7.3	1.988
0.6	−0.511	4.0	1.386	7.4	2.001
0.7	−0.357	4.1	1.411	7.5	2.015
0.8	−0.223	4.2	1.435	7.6	2.028
0.9	−0.105	4.3	1.459	7.7	2.041
1.0	0.000	4.4	1.482	7.8	2.054
1.1	0.095	4.5	1.504	7.9	2.067
1.2	0.182	4.6	1.526	8.0	2.079
1.3	0.262	4.7	1.548	8.1	2.092
1.4	0.336	4.8	1.569	8.2	2.104
1.5	0.405	4.9	1.589	8.3	2.116
1.6	0.470	5.0	1.609	8.4	2.128
1.7	0.531	5.1	1.629	8.5	2.140
1.8	0.588	5.2	1.649	8.6	2.152
1.9	0.642	5.3	1.668	8.7	2.163
2.0	0.693	5.4	1.686	8.8	2.175
2.1	0.742	5.5	1.705	8.9	2.186
2.2	0.788	5.6	1.723	9.0	2.197
2.3	0.833	5.7	1.740	9.1	2.208
2.4	0.875	5.8	1.758	9.2	2.219
2.5	0.916	5.9	1.775	9.3	2.230
2.6	0.956	6.0	1.792	9.4	2.241
2.7	0.993	6.1	1.808	9.5	2.251
2.8	1.030	6.2	1.825	9.6	2.262
2.9	1.065	6.3	1.841	9.7	2.272
3.0	1.099	6.4	1.856	9.8	2.282
3.1	1.131	6.5	1.872	9.9	2.293
3.2	1.163	6.6	1.887	10.0	2.303
3.3	1.194	6.7	1.902		

TABLE III: SQUARE ROOTS AND CUBE ROOTS

N	$\sqrt{N}$	$\sqrt[3]{N}$	N	$\sqrt{N}$	$\sqrt[3]{N}$	N	$\sqrt{N}$	$\sqrt[3]{N}$	N	$\sqrt{N}$	$\sqrt[3]{N}$
1	1.000	1.000	51	7.141	3.708	101	10.050	4.657	151	12.288	5.325
2	1.414	1.260	52	7.211	3.733	102	10.100	4.672	152	12.329	5.337
3	1.732	1.442	53	7.280	3.756	103	10.149	4.688	153	12.369	5.348
4	2.000	1.587	54	7.348	3.780	104	10.198	4.703	154	12.410	5.360
5	2.236	1.710	55	7.416	3.803	105	10.247	4.718	155	12.450	5.372
6	2.449	1.817	56	7.483	3.826	106	10.296	4.733	156	12.490	5.383
7	2.646	1.913	57	7.550	3.849	107	10.344	4.747	157	12.530	5.395
8	2.828	2.000	58	7.616	3.871	108	10.392	4.762	158	12.570	5.406
9	3.000	2.080	59	7.681	3.893	109	10.440	4.777	159	12.610	5.418
10	3.162	2.154	60	7.746	3.915	110	10.488	4.791	160	12.649	5.429
11	3.317	2.224	61	7.810	3.936	111	10.536	4.806	161	12.689	5.440
12	3.464	2.289	62	7.874	3.958	112	10.583	4.820	162	12.728	5.451
13	3.606	2.351	63	7.937	3.979	113	10.630	4.835	163	12.767	5.463
14	3.742	2.410	64	8.000	4.000	114	10.677	4.849	164	12.806	5.474
15	3.873	2.466	65	8.062	4.021	115	10.724	4.863	165	12.845	5.485
16	4.000	2.520	66	8.124	4.041	116	10.770	4.877	166	12.884	5.496
17	4.123	2.571	67	8.185	4.062	117	10.817	4.891	167	12.923	5.507
18	4.243	2.621	68	8.246	4.082	118	10.863	4.905	168	12.961	5.518
19	4.359	2.668	69	8.307	4.102	119	10.909	4.919	169	13.000	5.529
20	4.472	2.714	70	8.367	4.121	120	10.954	4.932	170	13.038	5.540
21	4.583	2.759	71	8.426	4.141	121	11.000	4.946	171	13.077	5.550
22	4.690	2.802	72	8.485	4.160	122	11.045	4.960	172	13.115	5.561
23	4.796	2.844	73	8.544	4.179	123	11.091	4.973	173	13.153	5.572
24	4.899	2.884	74	8.602	4.198	124	11.136	4.987	174	13.191	5.583
25	5.000	2.924	75	8.660	4.217	125	11.180	5.000	175	13.229	5.593
26	5.099	2.962	76	8.718	4.236	126	11.225	5.013	176	13.266	5.604
27	5.196	3.000	77	8.775	4.254	127	11.269	5.027	177	13.304	5.615
28	5.292	3.037	78	8.832	4.273	128	11.314	5.040	178	13.342	5.625
29	5.385	3.072	79	8.888	4.291	129	11.358	5.053	179	13.379	5.636
30	5.477	3.107	80	8.944	4.309	130	11.402	5.066	180	13.416	5.646
31	5.568	3.141	81	9.000	4.327	131	11.446	5.079	181	13.454	5.657
32	5.657	3.175	82	9.055	4.344	132	11.489	5.092	182	13.491	5.667
33	5.745	3.208	83	9.110	4.362	133	11.533	5.104	183	13.528	5.677
34	5.831	3.240	84	9.165	4.380	134	11.576	5.117	184	13.565	5.688
35	5.916	3.271	85	9.220	4.397	135	11.619	5.130	185	13.601	5.698
36	6.000	3.302	86	9.274	4.414	136	11.662	5.143	186	13.638	5.708
37	6.083	3.332	87	9.327	4.431	137	11.705	5.155	187	13.675	5.718
38	6.164	3.362	88	9.381	4.448	138	11.747	5.168	188	13.711	5.729
39	6.245	3.391	89	9.434	4.465	139	11.790	5.180	189	13.748	5.739
40	6.325	3.420	90	9.487	4.481	140	11.832	5.192	190	13.784	5.749
41	6.403	3.448	91	9.539	4.498	141	11.874	5.205	191	13.820	5.759
42	6.481	3.476	92	9.592	4.514	142	11.916	5.217	192	13.856	5.769
43	6.557	3.503	93	9.644	4.531	143	11.958	5.229	193	13.892	5.779
44	6.633	3.530	94	9.695	4.547	144	12.000	5.241	194	13.928	5.789
45	6.708	3.557	95	9.747	4.563	145	12.042	5.254	195	13.964	5.799
46	6.782	3.583	96	9.798	4.579	146	12.083	5.266	196	14.000	5.809
47	6.856	3.609	97	9.849	4.595	147	12.124	5.278	197	14.036	5.819
48	6.928	3.634	98	9.899	4.610	148	12.166	5.290	198	14.071	5.828
49	7.000	3.659	99	9.950	4.626	149	12.207	5.301	199	14.107	5.838
50	7.071	3.684	100	10.000	4.642	150	12.247	5.313	200	14.142	5.848

TABLE IV: FOUR-PLACE COMMON LOGARITHMS (BASE 10)

N	0	1	2	3	4	5	6	7	8	9
1.0	0.0000	0.0043	0.0086	0.0128	0.0170	0.0212	0.0253	0.0294	0.0334	0.0374
1.1	0.0414	0.0453	0.0492	0.0531	0.0569	0.0607	0.0645	0.0682	0.0719	0.0755
1.2	0.0792	0.0828	0.0864	0.0899	0.0934	0.0969	0.1004	0.1038	0.1072	0.1106
1.3	0.1139	0.1173	0.1206	0.1239	0.1271	0.1303	0.1335	0.1367	0.1399	0.1430
1.4	0.1461	0.1492	0.1523	0.1553	0.1584	0.1614	0.1644	0.1673	0.1703	0.1732
1.5	0.1761	0.1790	0.1818	0.1847	0.1875	0.1903	0.1931	0.1959	0.1987	0.2014
1.6	0.2041	0.2068	0.2095	0.2122	0.2148	0.2175	0.2201	0.2227	0.2253	0.2279
1.7	0.2304	0.2330	0.2355	0.2380	0.2405	0.2430	0.2455	0.2480	0.2504	0.2529
1.8	0.2553	0.2577	0.2601	0.2625	0.2648	0.2672	0.2695	0.2718	0.2742	0.2765
1.9	0.2788	0.2810	0.2833	0.2856	0.2878	0.2900	0.2923	0.2945	0.2967	0.2989
2.0	0.3010	0.3032	0.3054	0.3075	0.3096	0.3118	0.3139	0.3160	0.3181	0.3201
2.1	0.3222	0.3243	0.3263	0.3284	0.3304	0.3324	0.3345	0.3365	0.3385	0.3404
2.2	0.3424	0.3444	0.3464	0.3483	0.3502	0.3522	0.3541	0.3560	0.3579	0.3598
2.3	0.3617	0.3636	0.3655	0.3674	0.3692	0.3711	0.3729	0.3747	0.3766	0.3784
2.4	0.3802	0.3820	0.3838	0.3856	0.3874	0.3892	0.3909	0.3927	0.3945	0.3962
2.5	0.3979	0.3997	0.4014	0.4031	0.4048	0.4065	0.4082	0.4099	0.4116	0.4133
2.6	0.4150	0.4166	0.4183	0.4200	0.4216	0.4232	0.4249	0.4265	0.4281	0.4298
2.7	0.4314	0.4330	0.4346	0.4362	0.4378	0.4393	0.4409	0.4425	0.4440	0.4456
2.8	0.4472	0.4487	0.4502	0.4518	0.4533	0.4548	0.4564	0.4579	0.4594	0.4609
2.9	0.4624	0.4639	0.4654	0.4669	0.4683	0.4698	0.4713	0.4728	0.4742	0.4757
3.0	0.4771	0.4786	0.4800	0.4814	0.4829	0.4843	0.4857	0.4871	0.4886	0.4900
3.1	0.4914	0.4928	0.4942	0.4955	0.4969	0.4983	0.4997	0.5011	0.5024	0.5038
3.2	0.5051	0.5065	0.5079	0.5092	0.5105	0.5119	0.5132	0.5145	0.5159	0.5172
3.3	0.5185	0.5198	0.5211	0.5224	0.5237	0.5250	0.5263	0.5276	0.5289	0.5302
3.4	0.5315	0.5328	0.5340	0.5353	0.5366	0.5378	0.5391	0.5403	0.5416	0.5428
3.5	0.5441	0.5453	0.5465	0.5478	0.5490	0.5502	0.5514	0.5527	0.5539	0.5551
3.6	0.5563	0.5575	0.5587	0.5599	0.5611	0.5623	0.5635	0.5647	0.5658	0.5670
3.7	0.5682	0.5694	0.5705	0.5717	0.5729	0.5740	0.5752	0.5763	0.5775	0.5786
3.8	0.5798	0.5809	0.5821	0.5832	0.5843	0.5855	0.5866	0.5877	0.5888	0.5899
3.9	0.5911	0.5922	0.5933	0.5944	0.5955	0.5966	0.5977	0.5988	0.5999	0.6010
4.0	0.6021	0.6031	0.6042	0.6053	0.6064	0.6075	0.6085	0.6096	0.6107	0.6117
4.1	0.6128	0.6138	0.6149	0.6160	0.6170	0.6180	0.6191	0.6201	0.6212	0.6222
4.2	0.6232	0.6243	0.6253	0.6263	0.6274	0.6284	0.6294	0.6304	0.6314	0.6325
4.3	0.6335	0.6345	0.6355	0.6365	0.6375	0.6385	0.6395	0.6405	0.6415	0.6425
4.4	0.6435	0.6444	0.6454	0.6464	0.6474	0.6484	0.6493	0.6503	0.6513	0.6522
4.5	0.6532	0.6542	0.6551	0.6561	0.6571	0.6580	0.6590	0.6599	0.6609	0.6618
4.6	0.6628	0.6637	0.6646	0.6656	0.6665	0.6675	0.6684	0.6693	0.6702	0.6712
4.7	0.6721	0.6730	0.6739	0.6749	0.6758	0.6767	0.6776	0.6785	0.6794	0.6803
4.8	0.6812	0.6821	0.6830	0.6839	0.6848	0.6857	0.6866	0.6875	0.6884	0.6893
4.9	0.6902	0.6911	0.6920	0.6928	0.6937	0.6946	0.6955	0.6964	0.6972	0.6981
5.0	0.6990	0.6998	0.7007	0.7016	0.7024	0.7033	0.7042	0.7050	0.7059	0.7067
5.1	0.7076	0.7084	0.7093	0.7101	0.7110	0.7118	0.7126	0.7135	0.7143	0.7152
5.2	0.7160	0.7168	0.7177	0.7185	0.7193	0.7202	0.7210	0.7218	0.7226	0.7235
5.3	0.7243	0.7251	0.7259	0.7267	0.7275	0.7284	0.7292	0.7300	0.7308	0.7316
5.4	0.7324	0.7332	0.7340	0.7348	0.7356	0.7364	0.7372	0.7380	0.7388	0.7396
N	0	1	2	3	4	5	6	7	8	9

TABLE IV: (continued)									
N	0	1	2	3	4	5	6	7	8	9
5.5	0.7404	0.7412	0.7419	0.7427	0.7435	0.7443	0.7451	0.7459	0.7466	0.7474
5.6	0.7482	0.7490	0.7497	0.7505	0.7513	0.7520	0.7528	0.7536	0.7543	0.7551
5.7	0.7559	0.7566	0.7574	0.7582	0.7589	0.7597	0.7604	0.7612	0.7619	0.7627
5.8	0.7634	0.7642	0.7649	0.7657	0.7664	0.7672	0.7679	0.7686	0.7694	0.7701
5.9	0.7709	0.7716	0.7723	0.7731	0.7738	0.7745	0.7752	0.7760	0.7767	0.7774
6.0	0.7782	0.7789	0.7796	0.7803	0.7810	0.7818	0.7825	0.7832	0.7839	0.7846
6.1	0.7853	0.7860	0.7868	0.7875	0.7882	0.7889	0.7896	0.7903	0.7910	0.7917
6.2	0.7924	0.7931	0.7938	0.7945	0.7952	0.7959	0.7966	0.7973	0.7980	0.7987
6.3	0.7993	0.8000	0.8007	0.8014	0.8021	0.8028	0.8035	0.8041	0.8048	0.8055
6.4	0.8062	0.8069	0.8075	0.8082	0.8089	0.8096	0.8102	0.8109	0.8116	0.8122
6.5	0.8129	0.8136	0.8142	0.8149	0.8156	0.8162	0.8169	0.8176	0.8182	0.8189
6.6	0.8195	0.8202	0.8209	0.8215	0.8222	0.8228	0.8235	0.8241	0.8248	0.8254
6.7	0.8261	0.8267	0.8274	0.8280	0.8287	0.8293	0.8299	0.8306	0.8312	0.8319
6.8	0.8325	0.8331	0.8338	0.8344	0.8351	0.8357	0.8363	0.8370	0.8376	0.8382
6.9	0.8388	0.8395	0.8401	0.8407	0.8414	0.8420	0.8426	0.8432	0.8439	0.8445
7.0	0.8451	0.8457	0.8463	0.8470	0.8476	0.8482	0.8488	0.8494	0.8500	0.8506
7.1	0.8513	0.8519	0.8525	0.8531	0.8537	0.8543	0.8549	0.8555	0.8561	0.8567
7.2	0.8573	0.8579	0.8585	0.8591	0.8597	0.8603	0.8609	0.8615	0.8621	0.8627
7.3	0.8633	0.8639	0.8645	0.8651	0.8657	0.8663	0.8669	0.8675	0.8681	0.8686
7.4	0.8692	0.8698	0.8704	0.8710	0.8716	0.8722	0.8727	0.8733	0.8739	0.8745
7.5	0.8751	0.8756	0.8762	0.8768	0.8774	0.8779	0.8785	0.8791	0.8797	0.8802
7.6	0.8808	0.8814	0.8820	0.8825	0.8831	0.8837	0.8842	0.8848	0.8854	0.8859
7.7	0.8865	0.8871	0.8876	0.8882	0.8887	0.8893	0.8899	0.8904	0.8910	0.8915
7.8	0.8921	0.8927	0.8932	0.8938	0.8943	0.8949	0.8954	0.8960	0.8965	0.8971
7.9	0.8976	0.8982	0.8987	0.8993	0.8998	0.9004	0.9009	0.9015	0.9020	0.9025
8.0	0.9031	0.9036	0.9042	0.9047	0.9053	0.9058	0.9063	0.9069	0.9074	0.9079
8.1	0.9085	0.9090	0.9096	0.9101	0.9106	0.9112	0.9117	0.9122	0.9128	0.9133
8.2	0.9138	0.9143	0.9149	0.9154	0.9159	0.9165	0.9170	0.9175	0.9180	0.9186
8.3	0.9191	0.9196	0.9201	0.9206	0.9212	0.9217	0.9222	0.9227	0.9232	0.9238
8.4	0.9243	0.9248	0.9253	0.9258	0.9263	0.9269	0.9274	0.9279	0.9284	0.9289
8.5	0.9294	0.9299	0.9304	0.9309	0.9315	0.9320	0.9325	0.9330	0.9335	0.9340
8.6	0.9345	0.9350	0.9355	0.9360	0.9365	0.9370	0.9375	0.9380	0.9385	0.9390
8.7	0.9395	0.9400	0.9405	0.9410	0.9415	0.9420	0.9425	0.9430	0.9435	0.9440
8 8	0.9445	0.9450	0.9455	0.9460	0.9465	0.9469	0.9474	0.9479	0.9484	0.9489
8.9	0.9494	0.9499	0.9504	0.9509	0.9513	0.9518	0.9523	0.9528	0.9533	0.9538
9.0	0.9542	0.9547	0.9552	0.9557	0.9562	0.9566	0.9571	0.9576	0.9581	0.9586
9.1	0.9590	0.9595	0.9600	0.9605	0.9609	0.9614	0.9619	0.9624	0.9628	0.9633
9.2	0.9638	0.9643	0.9647	0.9652	0.9657	0.9661	0.9666	0.9671	0.9675	0.9680
9.3	0.9685	0.9689	0.9694	0.9699	0.9703	0.9708	0.9713	0.9717	0.9722	0.9727
9.4	0.9731	0.9736	0.9741	0.9745	0.9750	0.9754	0.9759	0.9763	0.9768	0.9773
9.5	0.9777	0.9782	0.9786	0.9791	0.9795	0.9800	0.9805	0.9809	0.9814	0.9818
9.6	0.9823	0.9827	0.9832	0.9836	0.9841	0.9845	0.9850	0.9854	0.9859	0.9863
9.7	0.9868	0.9872	0.9877	0.9881	0.9886	0.9890	0.9894	0.9899	0.9903	0.9908
9 8	0.9912	0.9917	0.9921	0.9926	0.9930	0.9934	0.9939	0.9943	0.9948	0.9952
9.9	0.9956	0.9961	0.9965	0.9969	0.9974	0.9978	0.9983	0.9987	0.9991	0.9996
N	0	1	2	3	4	5	6	7	8	9

TABLE V: TRIGONOMETRIC FUNCTION VALUES—RADIANS

Radians	sin	cos	tan	cot	sec	csc
.00	.0000	1.0000	.0000		1.000	
.01	.0100	1.0000	.0100	99.997	1.000	100.00
.02	.0200	.9998	.0200	49.993	1.000	50.00
.03	.0300	.9996	.0300	33.323	1.000	33.34
.04	.0400	.9992	.0400	24.987	1.001	25.01
.05	.0500	.9988	.0500	19.983	1.001	20.01
.06	.0600	.9982	.0601	16.647	1.002	16.68
.07	.0699	.9976	.0701	14.262	1.002	14.30
.08	.0799	.9968	.0802	12.473	1.003	12.51
.09	.0899	.9960	.0902	11.081	1.004	11.13
.10	.0998	.9950	.1003	9.967	1.005	10.02
.11	.1098	.9940	.1104	9.054	1.006	9.109
.12	.1197	.9928	.1206	8.293	1.007	8.353
.13	.1296	.9916	.1307	7.649	1.009	7.714
.14	.1395	.9902	.1409	7.096	1.010	7.166
.15	.1494	.9888	.1511	6.617	1.011	6.692
.16	.1593	.9872	.1614	6.197	1.013	6.277
.17	.1692	.9856	.1717	5.826	1.015	5.911
.18	.1790	.9838	.1820	5.495	1.016	5.586
.19	.1889	.9820	.1923	5.200	1.018	5.295
.20	.1987	.9801	.2027	4.933	1.020	5.033
.21	.2085	.9780	.2131	4.692	1.022	4.797
.22	.2182	.9759	.2236	4.472	1.025	4.582
.23	.2280	.9737	.2341	4.271	1.027	4.386
.24	.2377	.9713	.2447	4.086	1.030	4.207
.25	.2474	.9689	.2553	3.916	1.032	4.042
.26	.2571	.9664	.2660	3.759	1.035	3.890
.27	.2667	.9638	.2768	3.613	1.038	3.749
.28	.2764	.9611	.2876	3.478	1.041	3.619
.29	.2860	.9582	.2984	3.351	1.044	3.497
.30	.2955	.9553	.3093	3.233	1.047	3.384
.31	.3051	.9523	.3203	3.122	1.050	3.278
.32	.3146	.9492	.3314	3.018	1.053	3.179
.33	.3240	.9460	.3425	2.920	1.057	3.086
.34	.3335	.9428	.3537	2.827	1.061	2.999
.35	.3429	.9394	.3650	2.740	1.065	2.916
.36	.3523	.9359	.3764	2.657	1.068	2.839
.37	.3616	.9323	.3879	2.578	1.073	2.765
.38	.3709	.9287	.3994	2.504	1.077	2.696
.39	.3802	.9249	.4111	2.433	1.081	2.630
.40	.3894	.9211	.4228	2.365	1.086	2.568
.41	.3986	.9171	.4346	2.301	1.090	2.509
.42	.4078	.9131	.4466	2.239	1.095	2.452
.43	.4169	.9090	.4586	2.180	1.100	2.399
.44	.4259	.9048	.4708	2.124	1.105	2.348
	sin	cos	tan	cot	sec	csc

TABLE V: TRIGONOMETRIC FUNCTION VALUES—RADIANS
(continued)

Radians	sin	cos	tan	cot	sec	csc
.45	.4350	.9004	.4831	2.070	1.111	2.299
.46	.4439	.8961	.4954	2.018	1.116	2.253
.47	.4529	.8916	.5080	1.969	1.122	2.208
.48	.4618	.8870	.5206	1.921	1.127	2.166
.49	.4706	.8823	.5334	1.875	1.133	2.125
.50	.4794	.8776	.5463	1.830	1.139	2.086
.51	.4882	.8727	.5594	1.788	1.146	2.048
.52	.4969	.8678	.5726	1.747	1.152	2.013
.53	.5055	.8628	.5859	1.707	1.159	1.978
.54	.5141	.8577	.5994	1.668	1.166	1.945
.55	.5227	.8525	.6131	1.631	1.173	1.913
.56	.5312	.8473	.6269	1.595	1.180	1.883
.57	.5396	.8419	.6410	1.560	1.188	1.853
.58	.5480	.8365	.6552	1.526	1.196	1.825
.59	.5564	.8309	.6696	1.494	1.203	1.797
.60	.5646	.8253	.6841	1.462	1.212	1.771
.61	.5729	.8196	.6989	1.431	1.220	1.746
.62	.5810	.8139	.7139	1.401	1.229	1.721
.63	.5891	.8080	.7291	1.372	1.238	1.697
.64	.5972	.8021	.7445	1.343	1.247	1.674
.65	.6052	.7961	.7602	1.315	1.256	1.652
.66	.6131	.7900	.7761	1.288	1.266	1.631
.67	.6210	.7838	.7923	1.262	1.276	1.610
.68	.6288	.7776	.8087	1.237	1.286	1.590
.69	.6365	.7712	.8253	1.212	1.297	1.571
.70	.6442	.7648	.8423	1.187	1.307	1.552
.71	.6518	.7584	.8595	1.163	1.319	1.534
.72	.6594	.7518	.8771	1.140	1.330	1.517
.73	.6669	.7452	.8949	1.117	1.342	1.500
.74	.6743	.7385	.9131	1.095	1.354	1.483
.75	.6816	.7317	.9316	1.073	1.367	1.467
.76	.6889	.7248	.9505	1.052	1.380	1.452
.77	.6961	.7179	.9697	1.031	1.393	1.437
.78	.7033	.7109	.9893	1.011	1.407	1.422
.79	.7104	.7038	1.009	.9908	1.421	1.408
.80	.7174	.6967	1.030	.9712	1.435	1.394
.81	.7243	.6895	1.050	.9520	1.450	1.381
.82	.7311	.6822	1.072	.9331	1.466	1.368
.83	.7379	.6749	1.093	.9146	1.482	1.355
.84	.7446	.6675	1.116	.8964	1.498	1.343
.85	.7513	.6600	1.138	.8785	1.515	1.331
.86	.7578	.6524	1.162	.8609	1.533	1.320
.87	.7643	.6448	1.185	.8437	1.551	1.308
.88	.7707	.6372	1.210	.8267	1.569	1.297
.89	.7771	.6294	1.235	.8100	1.589	1.287
	sin	cos	tan	cot	sec	csc

$\dfrac{\pi}{6} \rightarrow$ (at row .52)

$\dfrac{\pi}{4} \rightarrow$ (at row .79)

TABLE V: TRIGONOMETRIC FUNCTION VALUES—RADIANS (continued)						
Radians	**sin**	**cos**	**tan**	**cot**	**sec**	**csc**
.90	.7833	.6216	1.260	.7936	1.609	1.277
.91	.7895	.6137	1.286	.7774	1.629	1.267
.92	.7956	.6058	1.313	.7615	1.651	1.257
.93	.8016	.5978	1.341	.7458	1.673	1.247
.94	.8076	.5898	1.369	.7303	1.696	1.238
.95	.8134	.5817	1.398	.7151	1.719	1.229
.96	.8192	.5735	1.428	.7001	1.744	1.221
.97	.8249	.5653	1.459	.6853	1.769	1.212
.98	.8305	.5570	1.491	.6707	1.795	1.204
.99	.8360	.5487	1.524	.6563	1.823	1.196
1.00	.8415	.5403	1.557	.6421	1.851	1.188
1.01	.8468	.5319	1.592	.6281	1.880	1.181
1.02	.8521	.5234	1.628	.6142	1.911	1.174
1.03	.8573	.5148	1.665	.6005	1.942	1.166
1.04	.8624	.5062	1.704	.5870	1.975	1.160
1.05	.8674	.4976	1.743	.5736	2.010	1.153
1.06	.8724	.4889	1.784	.5604	2.046	1.146
1.07	.8772	.4801	1.827	.5473	2.083	1.140
1.08	.8820	.4713	1.871	.5344	2.122	1.134
1.09	.8866	.4625	1.917	.5216	2.162	1.128
1.10	.8912	.4536	1.965	.5090	2.205	1.122
1.11	.8957	.4447	2.014	.4964	2.249	1.116
1.12	.9001	.4357	2.066	.4840	2.295	1.111
1.13	.9044	.4267	2.120	.4718	2.344	1.106
1.14	.9086	.4176	2.176	.4596	2.395	1.101
1.15	.9128	.4085	2.234	.4475	2.448	1.096
1.16	.9168	.3993	2.296	.4356	2.504	1.091
1.17	.9208	.3902	2.360	.4237	2.563	1.086
1.18	.9246	.3809	2.427	.4120	2.625	1.082
1.19	.9284	.3717	2.498	.4003	2.691	1.077
1.20	.9320	.3624	2.572	.3888	2.760	1.073
1.21	.9356	.3530	2.650	.3773	2.833	1.069
1.22	.9391	.3436	2.733	.3659	2.910	1.065
1.23	.9425	.3342	2.820	.3546	2.992	1.061
1.24	.9458	.3248	2.912	.3434	3.079	1.057
1.25	.9490	.3153	3.010	.3323	3.171	1.054
1.26	.9521	.3058	3.113	.3212	3.270	1.050
1.27	.9551	.2963	3.224	.3102	3.375	1.047
1.28	.9580	.2867	3.341	.2993	3.488	1.044
1.29	.9608	.2771	3.467	.2884	3.609	1.041
1.30	.9636	.2675	3.602	.2776	3.738	1.038
1.31	.9662	.2579	3.747	.2669	3.878	1.035
1.32	.9687	.2482	3.903	.2562	4.029	1.032
1.33	.9711	.2385	4.072	.2456	4.193	1.030
1.34	.9735	.2288	4.256	.2350	4.372	1.027
	sin	**cos**	**tan**	**cot**	**sec**	**csc**

$\dfrac{\pi}{3} \rightarrow$ (at row 1.05)

TABLE V: TRIGONOMETRIC FUNCTION VALUES—RADIANS (continued)

Radians	sin	cos	tan	cot	sec	csc
1.35	.9757	.2190	4.455	.2245	4.566	1.025
1.36	.9779	.2092	4.673	.2140	4.779	1.023
1.37	.9799	.1994	4.913	.2035	5.014	1.021
1.38	.9819	.1896	5.177	.1931	5.273	1.018
1.39	.9837	.1798	5.471	.1828	5.561	1.017
1.40	.9854	.1700	5.798	.1725	5.883	1.015
1.41	.9871	.1601	6.165	.1622	6.246	1.013
1.42	.9887	.1502	6.581	.1519	6.657	1.011
1.43	.9901	.1403	7.055	.1417	7.126	1.010
1.44	.9915	.1304	7.602	.1315	7.667	1.009
1.45	.9927	.1205	8.238	.1214	8.299	1.007
1.46	.9939	.1106	8.989	.1113	9.044	1.006
1.47	.9949	.1006	9.887	.1011	9.938	1.005
1.48	.9959	.0907	10.983	.0910	11.029	1.004
1.49	.9967	.0807	12.350	.0810	12.390	1.003
1.50	.9975	.0707	14.101	.0709	14.137	1.003
1.51	.9982	.0608	16.428	.0609	16.458	1.002
1.52	.9987	.0508	19.670	.0508	19.695	1.001
1.53	.9992	.0408	24.498	.0408	24.519	1.001
1.54	.9995	.0308	32.461	.0308	32.477	1.000
1.55	.9998	.0208	48.078	.0208	48.089	1.000
1.56	.9999	.0108	92.620	.0108	92.626	1.000
1.57	1.0000	.0008	1255.8	.0008	1255.8	1.000
	sin	cos	tan	cot	sec	csc

$\dfrac{\pi}{2} \rightarrow$ (at row 1.57)

Table V contains the trigonometric function values for the radians, to two decimal places, from .00 to $1.57 \approx \dfrac{\pi}{2}$.

For radian values of 1.58 and larger, or -1.58 and smaller, the trigonometric function values can be found by locating the terminal side of the angle and using the reference angles. This figure can be useful in this regard.

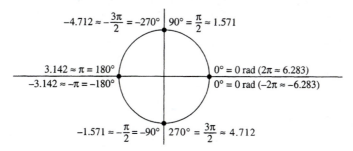

Illustrations:

The terminal side for $\theta = 2.95$ is in quadrant II, since $1.571 < 2.95 < 3.142$. The reference angle $\phi = 3.14 - 2.95 = .19$. Thus, $\tan 2.95 = -\tan .19 = -.1923$.

For $\theta = -9.05$ the terminal side is in quadrant III, since -9.05 is coterminal with $-9.05 + 6.28 = -2.77$. The reference angle $\phi = 3.14 - 2.77 = .37$. Thus, $\sin(-9.05) = -\sin(.37) = -.3616$.

Note: These results will differ slightly from the more accurate results obtained with a calculator.

TABLE VI: TRIGONOMETRIC FUNCTION VALUES—DEGREES

Degrees	sin	cos	tan	cot	sec	csc	
.0	.0000	1.0000	.0000	undef	1.0000	undef	**90.0**
.1	.0017	1.0000	.0017	572.96	1.0000	572.96	89.9
.2	.0035	1.0000	.0035	286.48	1.0000	286.48	89.8
.3	.0052	1.0000	.0052	190.98	1.0000	190.99	89.7
.4	.0070	1.0000	.0070	143.24	1.0000	143.24	89.6
.5	.0087	1.0000	.0087	114.59	1.0000	114.59	**89.5**
.6	.0105	.9999	.0105	95.489	1.0001	95.495	89.4
.7	.0122	.9999	.0122	81.847	1.0001	81.853	89.3
.8	.0140	.9999	.0140	71.615	1.0001	71.622	89.2
.9	.0157	.9999	.0157	63.657	1.0001	63.665	89.1
1.0	.0175	.9998	.0175	57.290	1.0002	57.299	**89.0**
1.1	.0192	.9998	.0192	52.081	1.0002	52.090	88.9
1.2	.0209	.9998	.0209	47.740	1.0002	47.750	88.8
1.3	.0227	.9997	.0227	44.066	1.0003	44.077	88.7
1.4	.0244	.9997	.0244	40.917	1.0003	40.930	88.6
1.5	.0262	.9997	.0262	38.188	1.0003	38.202	**88.5**
1.6	.0279	.9996	.0279	35.801	1.0004	35.815	88.4
1.7	.0297	.9996	.0297	33.694	1.0004	33.708	88.3
1.8	.0314	.9995	.0314	31.821	1.0005	31.836	88.2
1.9	.0332	.9995	.0332	30.145	1.0006	30.161	88.1
2.0	.0349	.9994	.0349	28.636	1.0006	28.654	**88.0**
2.1	.0366	.9993	.0367	27.271	1.0007	27.290	87.9
2.2	.0384	.9993	.0384	26.031	1.0007	26.050	87.8
2.3	.0401	.9992	.0402	24.898	1.0008	24.918	87.7
2.4	.0419	.9991	.0419	23.859	1.0009	23.880	87.6
2.5	.0436	.9990	.0437	22.904	1.0010	22.926	**87.5**
2.6	.0454	.9990	.0454	22.022	1.0010	22.044	87.4
2.7	.0471	.9989	.0472	21.205	1.0011	21.229	87.3
2.8	.0488	.9988	.0489	20.446	1.0012	20.471	87.2
2.9	.0506	.9987	.0507	19.740	1.0013	19.766	87.1
3.0	.0523	.9986	.0524	19.081	1.0014	19.107	**87.0**
3.1	.0541	.9985	.0542	18.464	1.0015	18.492	86.9
3.2	.0558	.9984	.0559	17.886	1.0016	17.914	86.8
3.3	.0576	.9983	.0577	17.343	1.0017	17.372	86.7
3.4	.0593	.9982	.0594	16.832	1.0018	16.862	86.6
3.5	.0610	.9981	.0612	16.350	1.0019	16.380	**86.5**
3.6	.0628	.9980	.0629	15.895	1.0020	15.926	86.4
3.7	.0645	.9979	.0647	15.464	1.0021	15.496	86.3
3.8	.0663	.9978	.0664	15.056	1.0022	15.089	86.2
3.9	.0680	.9977	.0682	14.669	1.0023	14.703	86.1
4.0	.0698	.9976	.0699	14.301	1.0024	14.336	**86.0**
4.1	.0715	.9974	.0717	13.951	1.0026	13.987	85.9
4.2	.0732	.9973	.0734	13.617	1.0027	13.654	85.8
4.3	.0750	.9972	.0752	13.300	1.0028	13.337	85.7
4.4	.0767	.9971	.0769	12.996	1.0030	13.035	85.6
4.5	.0785	.9969	.0787	12.706	1.0031	12.746	**85.5**
4.6	.0802	.9968	.0805	12.429	1.0032	12.469	85.4
4.7	.0819	.9966	.0822	12.163	1.0034	12.204	85.3
4.8	.0837	.9965	.0840	11.909	1.0035	11.951	85.2
4.9	.0854	.9963	.0857	11.665	1.0037	11.707	85.1
5.0	.0872	.9962	.0875	11.430	1.0038	11.474	**85.0**
	cos	sin	cot	tan	csc	sec	Degrees

TABLE VI: TRIGONOMETRIC FUNCTION VALUES—DEGREES
(continued)

Degrees	sin	cos	tan	cot	sec	csc	
5.0	.0872	.9962	.0875	11.430	1.0038	11.474	**85.0**
5.1	.0889	.9960	.0892	11.205	1.0040	11.249	84.9
5.2	.0906	.9959	.0910	10.988	1.0041	11.034	84.8
5.3	.0924	.9957	.0928	10.780	1.0043	10.826	84.7
5.4	.0941	.9956	.0945	10.579	1.0045	10.626	84.6
5.5	.0958	.9954	.0963	10.385	1.0046	10.433	**84.5**
5.6	.0976	.9952	.0981	10.199	1.0048	10.248	84.4
5.7	.0993	.9951	.0998	10.019	1.0050	10.069	84.3
5.8	.1011	.9949	.1016	9.8448	1.0051	9.8955	84.2
5.9	.1028	.9947	.1033	9.6768	1.0053	9.7283	84.1
6.0	.1045	.9945	.1051	9.5144	1.0055	9.5668	**84.0**
6.1	.1063	.9943	.1069	9.3572	1.0057	9.4105	83.9
6.2	.1080	.9942	.1086	9.2052	1.0059	9.2593	83.8
6.3	.1097	.9940	.1104	9.0579	1.0061	9.1129	83.7
6.4	.1115	.9938	.1122	8.9152	1.0063	8.9711	83.6
6.5	.1132	.9936	.1139	8.7769	1.0065	8.8337	**83.5**
6.6	.1149	.9934	.1157	8.6427	1.0067	8.7004	83.4
6.7	.1167	.9932	.1175	8.5126	1.0069	8.5711	83.3
6.8	.1184	.9930	.1192	8.3863	1.0071	8.4457	83.2
6.9	.1201	.9928	.1210	8.2636	1.0073	8.3238	83.1
7.0	.1219	.9925	.1228	8.1443	1.0075	8.2055	**83.0**
7.1	.1236	.9923	.1246	8.0285	1.0077	8.0905	82.9
7.2	.1253	.9921	.1263	7.9158	1.0079	7.9787	82.8
7.3	.1271	.9919	.1281	7.8062	1.0082	7.8700	82.7
7.4	.1288	.9917	.1299	7.6996	1.0084	7.7642	82.6
7.5	.1305	.9914	.1317	7.5958	1.0086	7.6613	**82.5**
7.6	.1323	.9912	.1334	7.4947	1.0089	7.5611	82.4
7.7	.1340	.9910	.1352	7.3962	1.0091	7.4635	82.3
7.8	.1357	.9907	.1370	7.3002	1.0093	7.3684	82.2
7.9	.1374	.9905	.1388	7.2066	1.0096	7.2757	82.1
8.0	.1392	.9903	.1405	7.1154	1.0098	7.1853	**82.0**
8.1	.1409	.9900	.1423	7.0264	1.0101	7.0972	81.9
8.2	.1426	.9898	.1441	6.9395	1.0103	7.0112	81.8
8.3	.1444	.9895	.1459	6.8548	1.0106	6.9273	81.7
8.4	.1461	.9893	.1477	6.7720	1.0108	6.8454	81.6
8.5	.1478	.9890	.1495	6.6912	1.0111	6.7655	**81.5**
8.6	.1495	.9888	.1512	6.6122	1.0114	6.6874	81.4
8.7	.1513	.9885	.1530	6.5350	1.0116	6.6111	81.3
8.8	.1530	.9882	.1548	6.4596	1.0119	6.5366	81.2
8.9	.1547	.9880	.1566	6.3859	1.0122	6.4637	81.1
9.0	.1564	.9877	.1584	6.3138	1.0125	6.3925	**81.0**
9.1	.1582	.9874	.1602	6.2432	1.0127	6.3228	80.9
9.2	.1599	.9871	.1620	6.1742	1.0130	6.2546	80.8
9.3	.1616	.9869	.1638	6.1066	1.0133	6.1880	80.7
9.4	.1633	.9866	.1655	6.0405	1.0136	6.1227	80.6
9.5	.1650	.9863	.1673	5.9758	1.0139	6.0589	**80.5**
9.6	.1668	.9860	.1691	5.9124	1.0142	5.9963	80.4
9.7	.1685	.9857	.1709	5.8502	1.0145	5.9351	80.3
9.8	.1702	.9854	.1727	5.7894	1.0148	5.8751	80.2
9.9	.1719	.9851	.1745	5.7297	1.0151	5.8164	80.1
10.0	.1736	.9848	.1763	5.6713	1.0154	5.7588	**80.0**
	cos	sin	cot	tan	csc	sec	Degrees

TABLE VI: TRIGONOMETRIC FUNCTION VALUES—DEGREES
(continued)

Degrees	sin	cos	tan	cot	sec	csc	
10.0	.1736	.9848	.1763	5.6713	1.0154	5.7588	**80.0**
10.1	.1754	.9845	.1781	5.6140	1.0157	5.7023	79.9
10.2	.1771	.9842	.1799	5.5578	1.0161	5.6470	79.8
10.3	.1788	.9839	.1817	5.5026	1.0164	5.5928	79.7
10.4	.1805	.9836	.1835	5.4486	1.0167	5.5396	79.6
10.5	.1822	.9833	.1853	5.3955	1.0170	5.4874	**79.5**
10.6	.1840	.9829	.1871	5.3435	1.0174	5.4362	79.4
10.7	.1857	.9826	.1890	5.2924	1.0177	5.3860	79.3
10.8	.1874	.9823	.1908	5.2422	1.0180	5.3367	79.2
10.9	.1891	.9820	.1926	5.1929	1.0184	5.2883	79.1
11.0	.1908	.9816	.1944	5.1446	1.0187	5.2408	**79.0**
11.1	.1925	.9813	.1962	5.0970	1.0191	5.1942	78.9
11.2	.1942	.9810	.1980	5.0504	1.0194	5.1484	78.8
11.3	.1959	.9806	.1998	5.0045	1.0198	5.1034	78.7
11.4	.1977	.9803	.2016	4.9594	1.0201	5.0593	78.6
11.5	.1994	.9799	.2035	4.9152	1.0205	5.0159	**78.5**
11.6	.2011	.9796	.2053	4.8716	1.0209	4.9732	78.4
11.7	.2028	.9792	.2071	4.8288	1.0212	4.9313	78.3
11.8	.2045	.9789	.2089	4.7867	1.0216	4.8901	78.2
11.9	.2062	.9785	.2107	4.7453	1.0220	4.8496	78.1
12.0	.2079	.9781	.2126	4.7046	1.0223	4.8097	**78.0**
12.1	.2096	.9778	.2144	4.6646	1.0227	4.7706	77.9
12.2	.2113	.9774	.2162	4.6252	1.0231	4.7321	77.8
12.3	.2130	.9770	.2180	4.5864	1.0235	4.6942	77.7
12.4	.2147	.9767	.2199	4.5483	1.0239	4.6569	77.6
12.5	.2164	.9763	.2217	4.5107	1.0243	4.6202	**77.5**
12.6	.2181	.9759	.2235	4.4737	1.0247	4.5841	77.4
12.7	.2198	.9755	.2254	4.4374	1.0251	4.5486	77.3
12.8	.2215	.9751	.2272	4.4015	1.0255	4.5137	77.2
12.9	.2233	.9748	.2290	4.3662	1.0259	4.4793	77.1
13.0	.2250	.9744	.2309	4.3315	1.0263	4.4454	**77.0**
13.1	.2267	.9740	.2327	4.2972	1.0267	4.4121	76.9
13.2	.2284	.9736	.2345	4.2635	1.0271	4.3792	76.8
13.3	.2300	.9732	.2364	4.2303	1.0276	4.3469	76.7
13.4	.2317	.9728	.2382	4.1976	1.0280	4.3150	76.6
13.5	.2334	.9724	.2401	4.1653	1.0284	4.2837	**76.5**
13.6	.2351	.9720	.2419	4.1335	1.0288	4.2527	76.4
13.7	.2368	.9715	.2438	4.1022	1.0293	4.2223	76.3
13.8	.2385	.9711	.2456	4.0713	1.0297	4.1923	76.2
13.9	.2402	.9707	.2475	4.0408	1.0302	4.1627	76.1
14.0	.2419	.9703	.2493	4.0108	1.0306	4.1336	**76.0**
14.1	.2436	.9699	.2512	3.9812	1.0311	4.1048	75.9
14.2	.2453	.9694	.2530	3.9520	1.0315	4.0765	75.8
14.3	.2470	.9690	.2549	3.9232	1.0320	4.0486	75.7
14.4	.2487	.9686	.2568	3.8947	1.0324	4.0211	75.6
14.5	.2504	.9681	.2586	3.8667	1.0329	3.9939	**75.5**
14.6	.2521	.9677	.2605	3.8391	1.0334	3.9672	75.4
14.7	.2538	.9673	.2623	3.8118	1.0338	3.9408	75.3
14.8	.2554	.9668	.2642	3.7848	1.0343	3.9147	75.2
14.9	.2571	.9664	.2661	3.7583	1.0348	3.8890	75.1
15.0	.2588	.9659	.2679	3.7321	1.0353	3.8637	**75.0**
	cos	sin	cot	tan	csc	sec	Degrees

TABLE VI: TRIGONOMETRIC FUNCTION VALUES—DEGREES
(continued)

Degrees	sin	cos	tan	cot	sec	csc	
15.0	.2588	.9659	.2679	3.7321	1.0353	3.8637	**75.0**
15.1	.2605	.9655	.2698	3.7062	1.0358	3.8387	74.9
15.2	.2622	.9650	.2717	3.6806	1.0363	3.8140	74.8
15.3	.2639	.9646	.2736	3.6554	1.0367	3.7897	74.7
15.4	.2656	.9641	.2754	3.6305	1.0372	3.7657	74.6
15.5	.2672	.9636	.2773	3.6059	1.0377	3.7420	**74.5**
15.6	.2689	.9632	.2792	3.5816	1.0382	3.7186	74.4
15.7	.2706	.9627	.2811	3.5576	1.0388	3.6955	74.3
15.8	.2723	.9622	.2830	3.5339	1.0393	3.6727	74.2
15.9	.2740	.9617	.2849	3.5105	1.0398	3.6502	74.1
16.0	.2756	.9613	.2867	3.4874	1.0403	3.6280	**74.0**
16.1	.2773	.9608	.2886	3.4646	1.0408	3.6060	73.9
16.2	.2790	.9603	.2905	3.4420	1.0413	3.5843	73.8
16.3	.2807	.9598	.2924	3.4197	1.0419	3.5629	73.7
16.4	.2823	.9593	.2943	3.3977	1.0424	3.5418	73.6
16.5	.2840	.9588	.2962	3.3759	1.0429	3.5209	**73.5**
16.6	.2857	.9583	.2981	3.3544	1.0435	3.5003	73.4
16.7	.2874	.9578	.3000	3.3332	1.0440	3.4799	73.3
16.8	.2890	.9573	.3019	3.3122	1.0446	3.4598	73.2
16.9	.2907	.9568	.3038	3.2914	1.0451	3.4399	73.1
17.0	.2924	.9563	.3057	3.2709	1.0457	3.4203	**73.0**
17.1	.2940	.9558	.3076	3.2506	1.0463	3.4009	72.9
17.2	.2957	.9553	.3096	3.2305	1.0468	3.3817	72.8
17.3	.2974	.9548	.3115	3.2106	1.0474	3.3628	72.7
17.4	.2990	.9542	.3134	3.1910	1.0480	3.3440	72.6
17.5	.3007	.9537	.3153	3.1716	1.0485	3.3255	**72.5**
17.6	.3024	.9532	.3172	3.1524	1.0491	3.3072	72.4
17.7	.3040	.9527	.3191	3.1334	1.0497	3.2891	72.3
17.8	.3057	.9521	.3211	3.1146	1.0503	3.2712	72.2
17.9	.3074	.9516	.3230	3.0961	1.0509	3.2536	72.1
18.0	.3090	.9511	.3249	3.0777	1.0515	3.2361	**72.0**
18.1	.3107	.9505	.3269	3.0595	1.0521	3.2188	71.9
18.2	.3123	.9500	.3288	3.0415	1.0527	3.2017	71.8
18.3	.3140	.9494	.3307	3.0237	1.0533	3.1848	71.7
18.4	.3156	.9489	.3327	3.0061	1.0539	3.1681	71.6
18.5	.3173	.9483	.3346	2.9887	1.0545	3.1515	**71.5**
18.6	.3190	.9478	.3365	2.9714	1.0551	3.1352	71.4
18.7	.3206	.9472	.3385	2.9544	1.0557	3.1190	71.3
18.8	.3223	.9466	.3404	2.9375	1.0564	3.1030	71.2
18.9	.3239	.9461	.3424	2.9208	1.0570	3.0872	71.1
19.0	.3256	.9455	.3443	2.9042	1.0576	3.0716	**71.0**
19.1	.3272	.9449	.3463	2.8878	1.0583	3.0561	70.9
19.2	.3289	.9444	.3482	2.8716	1.0589	3.0407	70.8
19.3	.3305	.9438	.3502	2.8556	1.0595	3.0256	70.7
19.4	.3322	.9432	.3522	2.8397	1.0602	3.0106	70.6
19.5	.3338	.9426	.3541	2.8239	1.0608	2.9957	**70.5**
19.6	.3355	.9421	.3561	2.8083	1.0615	2.9811	70.4
19.7	.3371	.9415	.3581	2.7929	1.0622	2.9665	70.3
19.8	.3387	.9409	.3600	2.7776	1.0628	2.9521	70.2
19.9	.3404	.9403	.3620	2.7625	1.0636	2.9379	70.1
20.0	.3420	.9397	.3640	2.7475	1.0642	2.9238	**70.0**
	cos	sin	cot	tan	csc	sec	**Degrees**

TABLE VI: TRIGONOMETRIC FUNCTION VALUES—DEGREES
(continued)

Degrees	sin	cos	tan	cot	sec	csc	
20.0	.3420	.9397	.3640	2.7475	1.0642	2.9238	**70.0**
20.1	.3437	.9391	.3659	2.7326	1.0649	2.9099	69.9
20.2	.3453	.9385	.3679	2.7179	1.0655	2.8960	69.8
20.3	.3469	.9379	.3699	2.7034	1.0662	2.8824	69.7
20.4	.3486	.9373	.3719	2.6889	1.0669	2.8688	69.6
20.5	.3502	.9367	.3739	2.6746	1.0676	2.8555	**69.5**
20.6	.3518	.9361	.3759	2.6605	1.0683	2.8422	69.4
20.7	.3535	.9354	.3779	2.6464	1.0690	2.8291	69.3
20.8	.3551	.9348	.3799	2.6325	1.0697	2.8161	69.2
20.9	.3567	.9342	.3819	2.6187	1.0704	2.8032	69.1
21.0	.3584	.9336	.3839	2.6051	1.0711	2.7904	**69.0**
21.1	.3600	.9330	.3859	2.5916	1.0719	2.7778	68.9
21.2	.3616	.9323	.3879	2.5782	1.0726	2.7653	68.8
21.3	.3633	.9317	.3899	2.5649	1.0733	2.7529	68.7
21.4	.3649	.9311	.3919	2.5517	1.0740	2.7407	68.6
21.5	.3665	.9304	.3939	2.5386	1.0748	2.7285	**68.5**
21.6	.3681	.9298	.3959	2.5257	1.0755	2.7165	68.4
21.7	.3697	.9291	.3979	2.5129	1.0763	2.7046	68.3
21.8	.3714	.9285	.4000	2.5002	1.0770	2.6927	68.2
21.9	.3730	.9278	.4020	2.4876	1.0778	2.6811	68.1
22.0	.3746	.9272	.4040	2.4751	1.0785	2.6695	**68.0**
22.1	.3762	.9265	.4061	2.4627	1.0793	2.6580	67.9
22.2	.3778	.9259	.4081	2.4504	1.0801	2.6466	67.8
22.3	.3795	.9252	.4101	2.4383	1.0808	2.6354	67.7
22.4	.3811	.9245	.4122	2.4262	1.0816	2.6242	67.6
22.5	.3827	.9239	.4142	2.4142	1.0824	2.6131	**67.5**
22.6	.3843	.9232	.4163	2.4023	1.0832	2.6022	67.4
22.7	.3859	.9225	.4183	2.3906	1.0840	2.5913	67.3
22.8	.3875	.9219	.4204	2.3789	1.0848	2.5805	67.2
22.9	.3891	.9212	.4224	2.3673	1.0856	2.5699	67.1
23.0	.3907	.9205	.4245	2.3559	1.0864	2.5593	**67.0**
23.1	.3923	.9198	.4265	2.3445	1.0872	2.5488	66.9
23.2	.3939	.9191	.4286	2.3332	1.0880	2.5384	66.8
23.3	.3955	.9184	.4307	2.3220	1.0888	2.5282	66.7
23.4	.3971	.9178	.4327	2.3109	1.0896	2.5180	66.6
23.5	.3987	.9171	.4348	2.2998	1.0904	2.5078	**66.5**
23.6	.4003	.9164	.4369	2.2889	1.0913	2.4978	66.4
23.7	.4019	.9157	.4390	2.2781	1.0921	2.4879	66.3
23.8	.4035	.9150	.4411	2.2673	1.0929	2.4780	66.2
23.9	.4051	.9143	.4431	2.2566	1.0938	2.4683	66.1
24.0	.4067	.9135	.4452	2.2460	1.0946	2.4586	**66.0**
24.1	.4083	.9128	.4473	2.2355	1.0955	2.4490	65.9
24.2	.4099	.9121	.4494	2.2251	1.0963	2.4395	65.8
24.3	.4115	.9114	.4515	2.2148	1.0972	2.4300	65.7
24.4	.4131	.9107	.4536	2.2045	1.0981	2.4207	65.6
24.5	.4147	.9100	.4557	2.1943	1.0989	2.4114	**65.5**
24.6	.4163	.9092	.4578	2.1842	1.0998	2.4022	65.4
24.7	.4179	.9085	.4599	2.1742	1.1007	2.3931	65.3
24.8	.4195	.9078	.4621	2.1642	1.1016	2.3841	65.2
24.9	.4210	.9070	.4642	2.1543	1.1025	2.3751	65.1
25.0	.4226	.9063	.4663	2.1445	1.1034	2.3662	**65.0**
	cos	sin	cot	tan	csc	sec	Degrees

TABLE VI: TRIGONOMETRIC FUNCTION VALUES—DEGREES
(continued)

Degrees	sin	cos	tan	cot	sec	csc	
25.0	.4226	.9063	.4663	2.1445	1.1034	2.3662	**65.0**
25.1	.4242	.9056	.4684	2.1348	1.1043	2.3574	64.9
25.2	.4258	.9048	.4706	2.1251	1.1052	2.3486	64.8
25.3	.4274	.9041	.4727	2.1155	1.1061	2.3400	64.7
25.4	.4289	.9033	.4748	2.1060	1.1070	2.3314	64.6
25.5	.4305	.9026	.4770	2.0965	1.1079	2.3228	**64.5**
25.6	.4321	.9018	.4791	2.0872	1.1089	2.3144	64.4
25.7	.4337	.9011	.4813	2.0778	1.1098	2.3060	64.3
25.8	.4352	.9003	.4834	2.0686	1.1107	2.2976	64.2
25.9	.4368	.8996	.4856	2.0594	1.1117	2.2894	64.1
26.0	.4384	.8988	.4877	2.0503	1.1126	2.2812	**64.0**
26.1	.4399	.8980	.4899	2.0413	1.1136	2.2730	63.9
26.2	.4415	.8973	.4921	2.0323	1.1145	2.2650	63.8
26.3	.4431	.8965	.4942	2.0233	1.1155	2.2570	63.7
26.4	.4446	.8957	.4964	2.0145	1.1164	2.2490	63.6
26.5	.4462	.8949	.4986	2.0057	1.1174	2.2412	**63.5**
26.6	.4478	.8942	.5008	1.9970	1.1184	2.2333	63.4
26.7	.4493	.8934	.5029	1.9883	1.1194	2.2256	63.3
26.8	.4509	.8926	.5051	1.9797	1.1203	2.2179	63.2
26.9	.4524	.8918	.5073	1.9711	1.1213	2.2103	63.1
27.0	.4540	.8910	.5095	1.9626	1.1223	2.2027	**63.0**
27.1	.4555	.8902	.5117	1.9542	1.1233	2.1952	62.9
27.2	.4571	.8894	.5139	1.9458	1.1243	2.1877	62.8
27.3	.4586	.8886	.5161	1.9375	1.1253	2.1803	62.7
27.4	.4602	.8878	.5184	1.9292	1.1264	2.1730	62.6
27.5	.4617	.8870	.5206	1.9210	1.1274	2.1657	**62.5**
27.6	.4633	.8862	.5228	1.9128	1.1284	2.1584	62.4
27.7	.4648	.8854	.5250	1.9047	1.1294	2.1513	62.3
27.8	.4664	.8846	.5272	1.8967	1.1305	2.1441	62.2
27.9	.4679	.8838	.5295	1.8887	1.1315	2.1371	62.1
28.0	.4695	.8829	.5317	1.8807	1.1326	2.1301	**62.0**
28.1	.4710	.8821	.5340	1.8728	1.1336	2.1231	61.9
28.2	.4726	.8813	.5362	1.8650	1.1347	2.1162	61.8
28.3	.4741	.8805	.5384	1.8572	1.1357	2.1093	61.7
28.4	.4756	.8796	.5407	1.8495	1.1368	2.1025	61.6
28.5	.4772	.8788	.5430	1.8418	1.1379	2.0957	**61.5**
28.6	.4787	.8780	.5452	1.8341	1.1390	2.0890	61.4
28.7	.4802	.8771	.5475	1.8265	1.1401	2.0824	61.3
28.8	.4818	.8763	.5498	1.8190	1.1412	2.0757	61.2
28.9	.4833	.8755	.5520	1.8115	1.1423	2.0692	61.1
29.0	.4848	.8746	.5543	1.8040	1.1434	2.0627	**61.0**
29.1	.4863	.8738	.5566	1.7966	1.1445	2.0562	60.9
29.2	.4879	.8729	.5589	1.7893	1.1456	2.0498	60.8
29.3	.4894	.8721	.5612	1.7820	1.1467	2.0434	60.7
29.4	.4909	.8712	.5635	1.7747	1.1478	2.0371	60.6
29.5	.4924	.8704	.5658	1.7675	1.1490	2.0308	**60.5**
29.6	.4939	.8695	.5681	1.7603	1.1501	2.0245	60.4
29.7	.4955	.8686	.5704	1.7532	1.1512	2.0183	60.3
29.8	.4970	.8678	.5727	1.7461	1.1524	2.0122	60.2
29.9	.4985	.8669	.5750	1.7391	1.1535	2.0061	60.1
30.0	.5000	.8660	.5774	1.7321	1.1547	2.0000	**60.0**
	cos	sin	cot	tan	csc	sec	**Degrees**

TABLE VI: TRIGONOMETRIC FUNCTION VALUES—DEGREES
(continued)

Degrees	sin	cos	tan	cot	sec	csc	
30.0	.5000	.8660	.5774	1.7321	1.1547	2.0000	**60.0**
30.1	.5015	.8652	.5797	1.7251	1.1559	1.9940	59.9
30.2	.5030	.8643	.5820	1.7182	1.1570	1.9880	59.8
30.3	.5045	.8634	.5844	1.7113	1.1582	1.9821	59.7
30.4	.5060	.8625	.5867	1.7045	1.1594	1.9762	59.6
30.5	.5075	.8616	.5890	1.6977	1.1606	1.9703	**59.5**
30.6	.5090	.8607	.5914	1.6909	1.1618	1.9645	59.4
30.7	.5105	.8599	.5938	1.6842	1.1630	1.9587	59.3
30.8	.5120	.8590	.5961	1.6775	1.1642	1.9530	59.2
30.9	.5135	.8581	.5985	1.6709	1.1654	1.9473	59.1
31.0	.5150	.8572	.6009	1.6643	1.1666	1.9416	**59.0**
31.1	.5165	.8563	.6032	1.6577	1.1679	1.9360	58.9
31.2	.5180	.8554	.6056	1.6512	1.1691	1.9304	58.8
31.3	.5195	.8545	.6080	1.6447	1.1703	1.9249	58.7
31.4	.5210	.8536	.6104	1.6383	1.1716	1.9194	58.6
31.5	.5225	.8526	.6128	1.6319	1.1728	1.9139	**58.5**
31.6	.5240	.8517	.6152	1.6255	1.1741	1.9084	58.4
31.7	.5255	.8508	.6176	1.6191	1.1753	1.9031	58.3
31.8	.5270	.8499	.6200	1.6128	1.1766	1.8977	58.2
31.9	.5284	.8490	.6224	1.6066	1.1779	1.8924	58.1
32.0	.5299	.8480	.6249	1.6003	1.1792	1.8871	**58.0**
32.1	.5314	.8471	.6273	1.5941	1.1805	1.8818	57.9
32.2	.5329	.8462	.6297	1.5880	1.1818	1.8766	57.8
32.3	.5344	.8453	.6322	1.5818	1.1831	1.8714	57.7
32.4	.5358	.8443	.6346	1.5757	1.1844	1.8663	57.6
32.5	.5373	.8434	.6371	1.5697	1.1857	1.8612	**57.5**
32.6	.5388	.8425	.6395	1.5637	1.1870	1.8561	57.4
32.7	.5402	.8415	.6420	1.5577	1.1883	1.8510	57.3
32.8	.5417	.8406	.6445	1.5517	1.1897	1.8460	57.2
32.9	.5432	.8396	.6469	1.5458	1.1910	1.8410	57.1
33.0	.5446	.8387	.6494	1.5399	1.1924	1.8361	**57.0**
33.1	.5461	.8377	.6519	1.5340	1.1937	1.8312	56.9
33.2	.5476	.8368	.6544	1.5282	1.1951	1.8263	56.8
33.3	.5490	.8358	.6569	1.5224	1.1964	1.8214	56.7
33.4	.5505	.8348	.6594	1.5166	1.1978	1.8166	56.6
33.5	.5519	.8339	.6619	1.5108	1.1992	1.8118	**56.5**
33.6	.5534	.8329	.6644	1.5051	1.2006	1.8070	56.4
33.7	.5548	.8320	.6669	1.4994	1.2020	1.8023	56.3
33.8	.5563	.8310	.6694	1.4938	1.2034	1.7976	56.2
33.9	.5577	.8300	.6720	1.4882	1.2048	1.7929	56.1
34.0	.5592	.8290	.6745	1.4826	1.2062	1.7883	**56.0**
34.1	.5606	.8281	.6771	1.4770	1.2076	1.7837	55.9
34.2	.5621	.8271	.6796	1.4715	1.2091	1.7791	55.8
34.3	.5635	.8261	.6822	1.4659	1.2105	1.7745	55.7
34.4	.5650	.8251	.6847	1.4605	1.2120	1.7700	55.6
34.5	.5664	.8241	.6873	1.4550	1.2134	1.7655	**55.5**
34.6	.5678	.8231	.6899	1.4496	1.2149	1.7610	55.4
34.7	.5693	.8221	.6924	1.4442	1.2163	1.7566	55.3
34.8	.5707	.8211	.6950	1.4388	1.2178	1.7522	55.2
34.9	.5721	.8202	.6976	1.4335	1.2193	1.7478	55.1
35.0	.5736	.8192	.7002	1.4281	1.2208	1.7434	**55.0**
	cos	sin	cot	tan	csc	sec	Degrees

TABLE VI: TRIGONOMETRIC FUNCTION VALUES—DEGREES
(continued)

Degrees	sin	cos	tan	cot	sec	csc	
35.0	.5736	.8192	.7002	1.4281	1.2208	1.7434	**55.0**
35.1	.5750	.8181	.7028	1.4229	1.2223	1.7391	54.9
35.2	.5764	.8171	.7054	1.4176	1.2238	1.7348	54.8
35.3	.5779	.8161	.7080	1.4124	1.2253	1.7305	54.7
35.4	.5793	.8151	.7107	1.4071	1.2268	1.7263	54.6
35.5	.5807	.8141	.7133	1.4019	1.2283	1.7221	**54.5**
35.6	.5821	.8131	.7159	1.3968	1.2299	1.7179	54.4
35.7	.5835	.8121	.7186	1.3916	1.2314	1.7137	54.3
35.8	.5850	.8111	.7212	1.3865	1.2329	1.7095	54.2
35.9	.5864	.8100	.7239	1.3814	1.2345	1.7054	54.1
36.0	.5878	.8090	.7265	1.3764	1.2361	1.7013	**54.0**
36.1	.5892	.8080	.7292	1.3713	1.2376	1.6972	53.9
36.2	.5906	.8070	.7319	1.3663	1.2392	1.6932	53.8
36.3	.5920	.8059	.7346	1.3613	1.2408	1.6892	53.7
36.4	.5934	.8049	.7373	1.3564	1.2424	1.6852	53.6
36.5	.5948	.8039	.7400	1.3514	1.2440	1.6812	**53.5**
36.6	.5962	.8028	.7427	1.3465	1.2456	1.6772	53.4
36.7	.5976	.8018	.7454	1.3416	1.2472	1.6733	53.3
36.8	.5990	.8007	.7481	1.3367	1.2489	1.6694	53.2
36.9	.6004	.7997	.7508	1.3319	1.2505	1.6655	53.1
37.0	.6018	.7986	.7536	1.3270	1.2521	1.6616	**53.0**
37.1	.6032	.7976	.7563	1.3222	1.2538	1.6578	52.9
37.2	.6046	.7965	.7590	1.3175	1.2554	1.6540	52.8
37.3	.6060	.7955	.7618	1.3127	1.2571	1.6502	52.7
37.4	.6074	.7944	.7646	1.3079	1.2588	1.6464	52.6
37.5	.6088	.7934	.7673	1.3032	1.2605	1.6427	**52.5**
37.6	.6101	.7923	.7701	1.2985	1.2622	1.6390	52.4
37.7	.6115	.7912	.7729	1.2938	1.2639	1.6353	52.3
37.8	.6129	.7902	.7757	1.2892	1.2656	1.6316	52.2
37.9	.6143	.7891	.7785	1.2846	1.2673	1.6279	52.1
38.0	.6157	.7880	.7813	1.2799	1.2690	1.6243	**52.0**
38.1	.6170	.7869	.7841	1.2753	1.2708	1.6207	51.9
38.2	.6184	.7859	.7869	1.2708	1.2725	1.6171	51.8
38.3	.6198	.7848	.7898	1.2662	1.2742	1.6135	51.7
38.4	.6211	.7837	.7926	1.2617	1.2760	1.6099	51.6
38.5	.6225	.7826	.7954	1.2572	1.2778	1.6064	**51.5**
38.6	.6239	.7815	.7983	1.2527	1.2796	1.6029	51.4
38.7	.6252	.7804	.8012	1.2482	1.2813	1.5994	51.3
38.8	.6266	.7793	.8040	1.2437	1.2831	1.5959	51.2
38.9	.6280	.7782	.8069	1.2393	1.2849	1.5925	51.1
39.0	.6293	.7771	.8098	1.2349	1.2868	1.5890	**51.0**
39.1	.6307	.7760	.8127	1.2305	1.2886	1.5856	50.9
39.2	.6320	.7749	.8156	1.2261	1.2904	1.5822	50.8
39.3	.6334	.7738	.8185	1.2218	1.2923	1.5788	50.7
39.4	.6347	.7727	.8214	1.2174	1.2941	1.5755	50.6
39.5	.6361	.7716	.8243	1.2131	1.2960	1.5721	**50.5**
39.6	.6374	.7705	.8273	1.2088	1.2978	1.5688	50.4
39.7	.6388	.7694	.8302	1.2045	1.2997	1.5655	50.3
39.8	.6401	.7683	.8332	1.2002	1.3016	1.5622	50.2
39.9	.6414	.7672	.8361	1.1960	1.3035	1.5590	50.1
40.0	.6428	.7660	.8391	1.1918	1.3054	1.5557	**50.0**
	cos	sin	cot	tan	csc	sec	Degrees

TABLE VI: TRIGONOMETRIC FUNCTION VALUES—DEGREES
(continued)

Degrees	sin	cos	tan	cot	sec	csc	
40.0	.6428	.7660	.8391	1.1918	1.3054	1.5557	**50.0**
40.1	.6441	.7649	.8421	1.1875	1.3073	1.5525	49.9
40.2	.6455	.7638	.8451	1.1833	1.3093	1.5493	49.8
40.3	.6468	.7627	.8481	1.1792	1.3112	1.5461	49.7
40.4	.6481	.7615	.8511	1.1750	1.3131	1.5429	49.6
40.5	.6494	.7604	.8541	1.1708	1.3151	1.5398	**49.5**
40.6	.6508	.7593	.8571	1.1667	1.3171	1.5366	49.4
40.7	.6521	.7581	.8601	1.1626	1.3190	1.5335	49.3
40.8	.6534	.7570	.8632	1.1585	1.3210	1.5304	49.2
40.9	.6547	.7559	.8662	1.1544	1.3230	1.5273	49.1
41.0	.6561	.7547	.8693	1.1504	1.3250	1.5243	**49.0**
41.1	.6574	.7536	.8724	1.1463	1.3270	1.5212	48.9
41.2	.6587	.7524	.8754	1.1423	1.3291	1.5182	48.8
41.3	.6600	.7513	.8785	1.1383	1.3311	1.5151	48.7
41.4	.6613	.7501	.8816	1.1343	1.3331	1.5121	48.6
41.5	.6626	.7490	.8847	1.1303	1.3352	1.5092	**48.5**
41.6	.6639	.7478	.8878	1.1263	1.3373	1.5062	48.4
41.7	.6652	.7466	.8910	1.1224	1.3393	1.5032	48.3
41.8	.6665	.7455	.8941	1.1184	1.3414	1.5003	48.2
41.9	.6678	.7443	.8972	1.1145	1.3435	1.4974	48.1
42.0	.6691	.7431	.9004	1.1106	1.3456	1.4945	**48.0**
42.1	.6704	.7420	.9036	1.1067	1.3478	1.4916	47.9
42.2	.6717	.7408	.9067	1.1028	1.3499	1.4887	47.8
42.3	.6730	.7396	.9099	1.0990	1.3520	1.4859	47.7
42.4	.6743	.7385	.9131	1.0951	1.3542	1.4830	47.6
42.5	.6756	.7373	.9163	1.0913	1.3563	1.4802	**47.5**
42.6	.6769	.7361	.9195	1.0875	1.3585	1.4774	47.4
42.7	.6782	.7349	.9228	1.0837	1.3607	1.4746	47.3
42.8	.6794	.7337	.9260	1.0799	1.3629	1.4718	47.2
42.9	.6807	.7325	.9293	1.0761	1.3651	1.4690	47.1
43.0	.6820	.7314	.9325	1.0724	1.3673	1.4663	**47.0**
43.1	.6833	.7302	.9358	1.0686	1.3696	1.4635	46.9
43.2	.6845	.7290	.9391	1.0649	1.3718	1.4608	46.8
43.3	.6858	.7278	.9424	1.0612	1.3741	1.4581	46.7
43.4	.6871	.7266	.9457	1.0575	1.3763	1.4554	46.6
43.5	.6884	.7254	.9490	1.0538	1.3786	1.4527	**46.5**
43.6	.6896	.7242	.9523	1.0501	1.3809	1.4501	46.4
43.7	.6909	.7230	.9556	1.0464	1.3832	1.4474	46.3
43.8	.6921	.7218	.9590	1.0428	1.3855	1.4448	46.2
43.9	.6934	.7206	.9623	1.0392	1.3878	1.4422	46.1
44.0	.6947	.7193	.9657	1.0355	1.3902	1.4396	**46.0**
44.1	.6959	.7181	.9691	1.0319	1.3925	1.4370	45.9
44.2	.6972	.7169	.9725	1.0283	1.3949	1.4344	45.8
44.3	.6984	.7157	.9759	1.0247	1.3973	1.4318	45.7
44.4	.6997	.7145	.9793	1.0212	1.3996	1.4293	45.6
44.5	.7009	.7133	.9827	1.0176	1.4020	1.4267	**45.5**
44.6	.7022	.7120	.9861	1.0141	1.4044	1.4242	45.4
44.7	.7034	.7108	.9896	1.0105	1.4069	1.4217	45.3
44.8	.7046	.7096	.9930	1.0070	1.4093	1.4192	45.2
44.9	.7059	.7083	.9965	1.0035	1.4118	1.4167	45.1
45.0	.7071	.7071	1.0000	1.0000	1.4142	1.4142	**45.0**
	cos	sin	cot	tan	csc	sec	Degrees

Answers to Odd-Numbered Exercises and Chapter Tests

CHAPTER 1 FUNDAMENTALS OF ALGEBRA

1.1 Real Numbers and Their Properties (page 6)

1. (c), (d), (f) **3.** (e), (f) **5.** (a), (b), (c), (d), (f) **7.** (b), (c), (d), (f) **9.** (e), (f) **11.** $\{1, 2, 3, 4\}$
13. $\{3, 4, 5, 6\}$ **15.** $\{-2, -1\}$ **17.** $\{\ldots, -3, -2, -1, 0\}$ **19.** There are none. **21.** True
23. False; $7 \div 3$ is not an integer. **25.** False; $8 - 2 \neq 2 - 8$ **27.** True **29.** True
31. Closure property for addition **33.** Inverse property for addition **35.** Commutative property for multiplication
37. Commutative property for addition **39.** Identity property for addition **41.** Inverse property for addition
43. Multiplication property of zero **45.** Distributive property **47.** 3 **49.** 5 **51.** 7
53. If $ab = 0$, then at least one of a or b is zero. Since $5 \neq 0$, $x - 2 = 0$ which implies $x = 2$.
55. No; as a counterexample $2^3 \neq 3^2$. **57. (a)** $\dfrac{5}{11}$; **(b)** $\dfrac{37}{99}$; **(c)** $\dfrac{26}{111}$
59. Let $\frac{0}{0} = x$, where x is some number. Then the definition of division gives $0 \cdot x = 0$. Since any number x will work, the answer to $\frac{0}{0}$ is not unique; therefore, $\frac{0}{0}$ is undefined.
61. The results get closer and closer to $\sqrt{2}$.
63. At the point with coordinate 2, construct a 1-unit segment perpendicular to the number line. Connect the endpoint of this segment to the point labeled 0. This segment will be the length $\sqrt{5}$ and can then be transferred to the number line.

1.2 Introduction to Equations and Problem Solving (page 13)

1. $x = 4$ **3.** $x = -4$ **5.** $x = -4$ **7.** $x = -8$ **9.** $x = \frac{9}{2}$ **11.** $x = -9$ **13.** $x = \frac{20}{3}$ **15.** $x = 15$
17. $x = -10$ **19.** $x = -\frac{22}{3}$ **21.** $x = 5$ **23.** $x = 3$ **25.** $x = 90$ **27.** $h = \dfrac{2A}{b_1 + b_2}$ **29.** $h = \dfrac{V}{\pi r^2}$
31. $P = \dfrac{I}{rt}$; \$2000 **33.** $R = \dfrac{V + \frac{1}{3}\pi h^3}{\pi h^2} = \dfrac{3V + \pi h^3}{3\pi h^2}$ **35.** $w = 12, l = 16$ **37.** 10 at 10¢, 13 at 30¢, 5 at 25¢
39. $3\frac{1}{2}$ hours **41.** Let $x + (x + 2) + (x + 4) = 180$. Then $3x = 174$ and $x = 58$. Therefore, the integers must be even.
43. $w = 5, l = 14$ **45.** 77, 79, 81 **47.** 36 min. **49.** \$7000 at 9%; \$9700 at 12% **51.** \$8500

1.3 Statements of Inequality and Their Graphs (page 22)

1. False; $1 < 2$ and 1 is not negative. **3.** True **5.** False; $\frac{1}{2} > 0$ but $\frac{1}{4} < \frac{1}{2}$. **7.** True **9.** $[-5, 2]$
11. $[-6, 0)$ **13.** $(-10, 10)$ **15.** $(-\infty, 5)$ **17.** $[-2, \infty)$ **19.** $(-\infty, -1]$

21. $\xleftarrow{\hspace{1cm}}\overset{\hspace{0.3cm}-3\hspace{0.6cm}-1\hspace{0.4cm}0}{\circ\!-\!\!+\!\!-\!\circ\!-\!\!+}\xrightarrow{\hspace{0.3cm}}$ **23.** $\xleftarrow{\hspace{1cm}}\overset{\hspace{0.3cm}-3\hspace{0.6cm}-1\hspace{0.4cm}0}{\bullet\!\!\blacksquare\!\!\circ\!\!+\!\!+}\xrightarrow{\hspace{0.3cm}}$ **25.** $\xleftarrow{\hspace{0.8cm}}\overset{\hspace{0.3cm}0\hspace{1cm}5}{\bullet\!\!+\!\!+\!\!+\!\!+\!\!\bullet}\xrightarrow{\hspace{0.3cm}}$ **27.** $\xleftarrow{\hspace{2cm}}\overset{\hspace{0.3cm}0}{\bullet}$

29. $-1 \le x < 3$; $[-1, 3)$ **31.** $-1 < x < 3$; $(-1, 3)$ **33.** $x < 1$; $(-\infty, 1)$ **35.** $\{x \mid x > 12\}$ **37.** $\{x \mid x \ge 4\}$
39. $\{x \mid x < 10\}$ **41.** $\{x \mid x > -7\}$ **43.** $\{x \mid x > -10\}$ **45.** $\{x \mid x \le 5\}$ **47.** $\{x \mid x \le -\frac{1}{4}\}$ **49.** $\{x \mid x > 2\}$
51. $\{x \mid x < -3\}$ **53.** $\{x \mid x > 32\}$ **55.** $\{x \mid x > -65\}$ **57.** $\{x \mid x < 0\}$ **59.** $\{x \mid x \ne 1\}$ **61.** $\{x \mid x \ge 1\}$
63. $\{x \mid x \ge -\frac{13}{2}\}$ **65.** $\{x \mid x \le 3\}$ **67.** $\{x \mid x < 4\}$ **69.** $\{x \mid x \le -2\}$
71. $\{x \mid x > -4\}$ **73.** $(10, 25), (11, 28), (12, 31), (13, 34), (14, 37)$

75. At least 83% but less than 93% **77.** Between $77\,°F$ and $86\,°F$ **79.** (a) $8 \le g \le 68$; (b) $2 \le h \le 16$
81. Two earn between \$74 and \$88, and the third earns between \$62 and \$76.

1.4 Absolute Value (page 28)

1. False **3.** False **5.** True **7.** True **9.** False **11.** (a) 6 (b) -5 (c) -6 (d) $\frac{3}{2}$ **13.** $x = \pm\frac{1}{2}$

15. $x = -2, x = 4$ **17.** $x = -2, x = 5$ **19.** $x = 1, x = 7$ **21.** $x = -\frac{20}{3}, x = 4$ **23.** $x > 0$

25. $x = -4$ or $x = 2$: **27.** $x \le -2$ or $x \ge 4$:

29. $-5 \le x \le 1$: **31.** $x = -5$ or $x = 5$:

33. $x \le -5$ or $x \ge 5$: **35.** $2 \le x \le 8$:

37. $2.9 < x < 3.1$: **39.** $-3 < x < 4$:

41. $2 < x < 6$: **43.** $3 \le x \le 5$:

45. All $x \ne 3$:

47. (a) $|5 - 2| \ge ||5| - |2||$; $3 \ge 3$ (b) $|3 + 4| \le |3| + |4|$; $7 \le 7$
　　$|5 - (-2)| \ge ||5| - |-2||$; $7 \ge 3$ 　$|3 + (-4)| \le |3| + |-4|$; $1 \le 7$
　　$|-5 - 2| \ge ||-5| - |2||$; $7 \ge 3$ 　$|-3 + 4| \le |-3| + |4|$; $1 \le 7$
　　$|-5 - (-2)| \ge ||-5| - |-2||$; $3 \ge 3$ 　$|-3 + (-4)| \le |-3| + |-4|$; $7 \le 7$

1.5 Integral Exponents (page 36)

1. False; 3^6 **3.** False; 2^7 **5.** True **7.** True **9.** False; $2 \cdot 3^4$ **11.** False; 1 **13.** False; 2^3
15. False; 2^{-8} **17.** 7 **19.** $\frac{5}{3}$ **21.** 1 **23.** 4 **25.** 24 **27.** 16 **29.** $\frac{15}{4}$ **31.** 2×10^8
33. 3.7×10^{-4} **35.** 5.55×10^{-8} **37.** 0.000789 **39.** 225,000 **41.** $\dfrac{1}{x^6}$ **43.** x^{12} **45.** $72a^5$ **47.** $16a^8$
49. $x^6 y^2$ **51.** $\dfrac{x^8}{y^6}$ **53.** $\dfrac{y^7}{x^5}$ **55.** $5x$ **57.** $\dfrac{b^2}{9}$ **59.** $(a + b)^6$ **61.** $-3x^3 y^3$ **63.** $\dfrac{1}{(a + 3b)^{22}}$
65. $\dfrac{1}{x^2} + \dfrac{1}{y^2}$ **67.** $x^2 y^3$ **69.** $\dfrac{10}{(4 - 5x)^3}$ **71.** 9 **73.** 8 **75.** -3 **77.** 5 grams; $640\left(\frac{1}{2}\right)^n$ grams
79. 32 feet; $243\left(\frac{2}{3}\right)^n$ feet **81.** \$1331 **83.** $3^4 \ne 4^3$; $(2^3)^4 = 2^{12} \ne 2^{81} = 2^{(3^4)}$ **85.** 500 seconds

1.6　Radicals and Rational Exponents　(page 45)

1. $\frac{1}{9}$　　**3.** $\frac{1}{16}$　　**5.** 25　　**7.** -3　　**9.** $\frac{1}{2}$　　**11.** 9　　**13.** 50　　**15.** -2　　**17.** 13　　**19.** $\sqrt[3]{35}/6$

21. $\frac{148}{135}$　　**23.** $-\frac{11}{4}$　　**25.** $4\sqrt{2}$　　**27.** $6\sqrt{2}$　　**29.** $17\sqrt{5}$　　**31.** $10\sqrt{2}$　　**33.** $6\sqrt[3]{2}$　　**35.** $14\sqrt{2}$　　**37.** $|x|\sqrt{2}$

39. $5\sqrt{10}$　　**41.** $7\sqrt{5}$　　**43.** $12|x|$　　**45.** $|x|\sqrt{y} + 12|x|\sqrt{2y}$　　**47.** $4x\sqrt{2}$　　**49.** $\sqrt{2}/6$

51. $8\sqrt{3}/|x|$　　**53.** $4\sqrt[3]{4}$　　**55.** 1　　**57.** $16a^4b^6$　　**59.** a^2/b　　**61.** $\dfrac{3x}{(3x^2 + 2)^{1/2}}$　　**63.** $\dfrac{x + 2}{(x^2 + 4x)^{1/2}}$

65. (a) 25 cm　(b) 18.9 cm　　**67.** $\dfrac{\sqrt[3]{4x}}{x}$　　**69.** $\dfrac{1}{xy^n}$　　**71.** x

73. Let $x = \sqrt[n]{a}$, $y = \sqrt[n]{b}$. Then $x^n = a$ and $y^n = b$. Now we get $\dfrac{a}{b} = \dfrac{x^n}{y^n} = \left(\dfrac{x}{y}\right)^n$. Thus, by definition, $\sqrt[n]{\dfrac{a}{b}} = \dfrac{x}{y}$.

But $x = \sqrt[n]{a}$ and $y = \sqrt[n]{b}$. Therefore, $\sqrt[n]{\dfrac{a}{b}} = \dfrac{\sqrt[n]{a}}{\sqrt[n]{b}}$.　　**75.** $|xy| = \sqrt{(xy)^2} = \sqrt{x^2y^2} = \sqrt{x^2}\sqrt{y^2} = |x||y|$

1.7　Fundamental Operations with Polynomials　(page 51)

1. $8x^2 - 2x + 7$　　**3.** $4x^3 - 9x^2 + 16x + 3$　　**5.** $2x^3 + x^2 + 11x - 28$　　**7.** $x^3 - 7x^2 - 10x + 8$

9. $4x^3 - x^2 - 5x - 22$　　**11.** $3x + 1$　　**13.** $6y + 6$　　**15.** $x^3 + 2x^2 + x$　　**17.** $7y + 8$　　**19.** $3x - 3xy + 2x^2y^2$

21. $-20x^4 + 4x^3 + 2x^2$　　**23.** $x^2 + 2x + 1$　　**25.** $4x^2 + 26x - 14$　　**27.** $-6x^2 - 3x + 18$　　**29.** $-6x^2 + 3x + 18$

31. $\frac{4}{9}x^2 + 8x + 36$　　**33.** $-63 + 55x - 12x^2$　　**35.** $\frac{1}{25}x^2 - \frac{1}{10}x + \frac{1}{16}$　　**37.** $x - 100$　　**39.** $x - 2$

41. $x^4 - 2x^3 + 2x^2 - 31x - 36$　　**43.** $x^3 - 8$　　**45.** $x^5 - 32$　　**47.** $x^{4n} - x^{2n} - 2$　　**49.** $3x - 6x^2 + 3x^3$

51. $-6x^3 + 19x^2 - x - 6$　　**53.** $5x^4 - 12x^2 + 2x + 4$　　**55.** $6x^5 + 15x^4 - 12x^3 - 27x^2 + 10x + 15$

57. $x^3 - 3x^2 + 3x - 1$　　**59.** $a^4 + 4a^3b + 6a^2b^2 + 4ab^3 + b^4$　　**61.** $8x^3 + 36x^2 + 54x + 27$

63. $\frac{1}{27}x^3 + x^2 + 9x + 27$　　**65.** $6(\sqrt{5} + \sqrt{3})$　　**67.** $-2(\sqrt{2} + 3)$　　**69.** $\dfrac{x + 2\sqrt{xy} + y}{x - y}$　　**71.** $\dfrac{-4}{5 - 3\sqrt{5}}$

73. $\dfrac{x - y}{x - 2\sqrt{xy} + y}$　　**75.** Multiply numerator and denominator of the first fraction by $\sqrt{4 + h} + 2$ and simplify.

1.8　Factoring Polynomials　(page 59)

1. $(2x + 3)(2x - 3)$　　**3.** $(a + 11b)(a - 11b)$　　**5.** $(x + 4)(x^2 - 4x + 16)$　　**7.** $(5x - 4)(25x^2 + 20x + 16)$

9. $(2x + 7y)(4x^2 - 14xy + 49y^2)$　　**11.** $(\sqrt{3} + 2x)(\sqrt{3} - 2x)$　　**13.** $(\sqrt{x} + 6)(\sqrt{x} - 6)$

15. $(2\sqrt{2} + \sqrt{3x})(2\sqrt{2} - \sqrt{3x})$　　**17.** $(\sqrt[3]{7} + a)[\sqrt[3]{49} - a\sqrt[3]{7} + a^2]$

19. $(3\sqrt[3]{x} + 1)(9\sqrt[3]{x^2} - 3\sqrt[3]{x} + 1)$　　**21.** $(\sqrt[3]{3x} - \sqrt[3]{4})(\sqrt[3]{9x^2} + \sqrt[3]{12x} + \sqrt[3]{16})$　　**23.** $(x - 1)(x + y)$

25. $(x - 1)(y - 1)$　　**27.** $(2 - y^2)(1 + x)$　　**29.** $7(x + h)(x^2 - hx + h^2)$　　**31.** $(a^4 + b^4)(a^2 + b^2)(a + b)(a - b)$

33. $(a - 2)(a^4 + 2a^3 + 4a^2 + 8a + 16)$　　**35.** $(x - y)(a + b)(a^2 - ab + b^2)$　　**37.** $(x^2 + 4y^2)(x + 2y)(x - 2y)^2$

39. $(5a - 1)(4a - 1)$　　**41.** $(3x + 1)^2$　　**43.** $(7x + 1)(2x + 5)$　　**45.** $(8x - 1)(x - 1)$　　**47.** $2(2x - 3)(2x - 1)$

49. $(4a - 3)(3a - 4)$　　**51.** $(2x - 1)(2x + 3)$　　**53.** $(8a + 3b)(3a + 2b)$　　**55.** Not factorable

57. $2(3x - 5)(x + 2)$　　**59.** $2(b + 2)(b + 4)$　　**61.** $ab(a - b)^2$　　**63.** $8(2x - 1)(x - 1)$　　**65.** $25(a + b)^2$

67. $(a - 1)^2(a^2 + a + 1)^2$　　**69.** $3xy(x^2 + 2y)(2x^2 - 5y)$　　**71.** $(x + 2)^2(8x + 7)$

73. $x(x + 1)^2(x^2 - x + 1)^2(11x^3 - 9x + 2)$ or, in complete factored form, $x(x + 1)^3(x^2 - x + 1)^2(11x^2 - 11x + 2)$.

75. (a) $(x + 2)(x^4 - 2x^3 + 4x^2 - 8x + 16)$; (b) $x(2x + y)(64x^6 - 32x^5y + 16x^4y^2 - 8x^3y^3 + 4x^2y^4 - 2xy^5 + y^6)$

77. $(x^2 - x + 1)(x^2 + x + 1)$　　**79.** $20(2x + 1)^9$

81. $\sqrt{1 + y^2} = \sqrt{1 + x^2(x^2 + 2)} = \sqrt{x^4 + 2x^2 + 1} = \sqrt{(x^2 + 1)^2} = x^2 + 1$

83. $\pi R^2 - 9\pi a^2 = \pi(R - 3a)(R + 3a)$; $\pi(15.7 - 9.3)(15.7 + 9.3) = \pi(6.4)(25) = 160\pi$

1.9 Fundamental Operations with Rational Expressions   (page 67)

1. False; $\frac{1}{21}$ **3.** False; $\frac{ax}{2} - \frac{5b}{6}$ **5.** True   **7.** $\frac{2x}{3z}$ **9.** $3x + 1$ **11.** $3x^2 + 2x + 1$ **13.** $a + 1 - ab$

15. $3 - 4a^2x^4$ **17.** $-x$ **19.** $\frac{n + 1}{n^2 + 1}$ **21.** $\frac{3(x - 1)}{2(x + 1)}$ **23.** $\frac{2x + 3}{2x - 3}$ **25.** $\frac{a - 4b}{a^2 - 4ab + 16b^2}$ **27.** $\frac{2y}{x}$

29. $\frac{2}{a^2}$ **31.** $\frac{x}{y}$ **33.** $\frac{-3a - 8b}{6}$ **35.** $\frac{x^2 + 1}{3(x + 1)}$ **37.** -1 **39.** $\frac{2 - xy}{x}$ **41.** $\frac{5y^2 - y}{y^2 - 1}$ **43.** $\frac{3x}{x + 1}$

45. $\frac{8 - x}{x^2 - 4}$ **47.** $\frac{3(x + 9)}{(x - 3)(x + 3)^2}$ **49.** $\frac{2}{x + 5}$ **51.** $x - 3$ **53.** $\frac{2(n^2 + 4)}{n + 2}$ **55.** $x^2 - 5x + 2; r = 0$

57. $x^2; r = 7$ **59.** $x^2 + 7x + 7; r = 22$ **61.** $4x + 3; r = 7x - 7$ **63.** $x + 2; r = 0$ **65.** $\frac{1}{2(x + 2)}$

67. $-\dfrac{1}{4(4 + h)}$ **69.** $-\dfrac{1}{3(x + 3)}$ **71.** $\dfrac{4 - x}{16x^2}$ **73.** $\dfrac{xy}{y + x}$ **75.** $\dfrac{-4x}{(1 + x^2)^2}$ **77.** $-\dfrac{4}{x^2}$ **79.** $-\dfrac{1}{ab}$

81. $\dfrac{y^2 - x^2}{x^3y^3}$

83. (a) $\dfrac{\dfrac{AD}{B} + C}{D} = \dfrac{AD + BC}{BD} = \dfrac{AD}{BD} + \dfrac{BC}{BD} = \dfrac{A}{B} + \dfrac{C}{D}$

(b) $\left[\dfrac{\left(\dfrac{AB}{D} + C\right)D}{F} + E\right] \cdot F = \left[\dfrac{AB + CD}{F} + E\right] \cdot F = AB + CD + EF$

85. 48 mph; $\dfrac{2s_1s_2}{s_1 + s_2}$ **87.** $\dfrac{2xy^2 - 2x^2y\left(\dfrac{x^2}{y^2}\right)}{y^4} = \dfrac{2xy^2 - \dfrac{2x^4}{y}}{y^4} = \dfrac{2xy^3 - 2x^4}{y^5} = \dfrac{2x(x^3 + 8) - 2x^4}{y^5} = \dfrac{16x}{y^5}$

89. $\sqrt{1 + y^2} = \sqrt{1 + \dfrac{x^4}{64} - \dfrac{1}{2} + \dfrac{4}{x^4}} = \sqrt{\dfrac{x^4}{64} + \dfrac{1}{2} + \dfrac{4}{x^4}} = \sqrt{\left(\dfrac{x^2}{8} + \dfrac{2}{x^2}\right)^2} = \dfrac{x^2}{8} + \dfrac{2}{x^2}$

1.10 Introduction to Complex Numbers   (page 74)

1. True   **3.** True   **5.** False   **7.** $5 + 2i$ **9.** $-5 + 0i$ **11.** $4i$ **13.** $12i$ **15.** $\frac{3}{4}i$ **17.** $-\sqrt{5}i$

19. -27 **21.** $-\sqrt{6}$ **23.** $15i$ **25.** $12i$ **27.** $5i\sqrt{2}$ **29.** $(3 - \sqrt{3})i$ **31.** $10 + 7i$ **33.** $5 - 3i$

35. $10 + 2i$ **37.** $-10 + 6i$ **39.** $0 + 13i$ **41.** $23 + 14i$ **43.** $5 - 3i$ **45.** $\frac{13}{5} + \frac{1}{5}i$ **47.** $\frac{4}{5} - \frac{3}{5}i$

49. $-3i$ **51.** $-2i$ **53.** 4 **55.** $\frac{3}{13} - \frac{2}{13}i$ **57.** $\sqrt{(-4)(-9)} = \sqrt{36} = 6; \sqrt{-4}\sqrt{-9} = (2i)(3i) = 6i^2 = -6$

59. $\dfrac{ac + bd}{c^2 + d^2} + \dfrac{bc - ad}{c^2 + d^2}i$

61. $[(3 + i)(3 - i)](4 + 3i) = (9 - i^2)(4 + 3i) = 10(4 + 3i) = 40 + 30i$
$(3 + i)[(3 - i)(4 + 3i)] = (3 + i)(15 + 5i) = 40 + 30i$

63. $4 - 3i$ **65.** $\frac{53}{13} - \frac{21}{13}i$ **67.** $(x + i)(x - i)$ **69.** $3(x + 5i)(x - 5i)$ **71.** -1

CHAPTER 1   REVIEW EXERCISES   (page 76)

1. See page 2   **2.** See page 2   **3.** (a) real number, rational number, integer; (b) real number, irrational number; (c) real number, rational number; (d) real number, rational number, integer, whole number, natural number; (e) same as (d)   **4.** (a) distributive property; (b) commutative property for multiplication; (c) associative property for addition; (d) inverse property for addition; (e) closure property for addition

5. (a) $5 - 7 \neq 7 - 5$; (b) $8 - (5 - 2) \neq (8 - 5) - 2$ **6.** See page 4   **7.** 0, 1   **8.** See page 5   **9.** $n = 9$

10. Any real number   **11.** $n = 10$ **12.** $n = 8$ **13.** See page 9   **14.** See page 9

15. (a) $x = \frac{5}{2}$; (b) $x = -3$ **16.** $r = \dfrac{C}{2\pi}$; 14 cm   **17.** 5 cm by 11 cm   **18.** \$1125   **19.** $4\frac{1}{2}\%$

20. 3 hours   **21.** See page 16   **22.** See page 17   **23.** See page 18   **24.** $n < 6$ **25.** $\{x \mid x \leq 4\}$

26. (a) $x < -1$; (b) $x > 2$

27. (a)   (b)   (c)

(d)   **28.** (a) $x \geq 1$:   (b) $x < 2$:   **29.** See page 21

30. See page 21   **31.** See page 16   **32.** (3, 11), (4, 14), (5, 17), (6, 20), (7, 23)   **33.** See page 25

34. $a < -k$ or $a > k$ **35.** $-k < a < k$ **36.** $\dfrac{x_1 + x_2}{2}$ **37.** $x > 2$

38. (a) $x = -3$ or $x = 7$; (b) $-1 \leq x \leq 2$

39. (a)   (b)

40. The distance between points a and b. If $a = 2$ and $b = 7$, $|a - b| = 5$; if $a = -3$ and $b = 8$, $|a - b| = 11$

41. True if x and y are both positive or both negative; false if one is positive and the other is negative.

42. Yes; See Ex. 75, page 47.   **43.** $-5^2 = -25$; $(-5)^2 = 25$ **44.** See page 32

45. (a) $3x^6$; (b) -32; (c) 8; (d) $\dfrac{x^{10}y}{2}$ **46.** $\dfrac{a^{15}}{b^{15}}$ **47.** $x = 7$ **48.** $x = 4$ **49.** $x = 7$

50. (a) 4.2×10^8; (b) 2.3×10^{-8} **51.** (a) 325,000; (b) 0.0000025   **52.** 15 grams; $(960)\left(\dfrac{1}{2}\right)^n$

53. See page 39   **54.** (a) False; (b) false   **55.** (a) True; (b) false   **56.** (a) 5; (b) 4 (c) -2

57. (a) $\sqrt{15ab}$; (b) $-3x$ (c) $-2x$ **58.** (a) $\dfrac{27}{8}$; (b) $-\dfrac{1}{16}$ **59.** $\dfrac{16}{a^2 b^4}$

60. (a) $5\sqrt{3}$; (b) $2\sqrt[3]{3}$; (c) $-5\sqrt[3]{2}$ **61.** $-4x\sqrt[3]{2}$ **62.** (a) $5\sqrt{5}$; (b) $3\sqrt[3]{4}$ **63.** $-12\sqrt{3}$

64. $|a|$ **65.** $\sqrt[mn]{a}$; 2   **66.** No; $\sqrt{9 + 16} \neq \sqrt{9} + \sqrt{16}$ **67.** See page 48   **68.** See page 47

69. $3x^3 - 7x^2 - 5x - 4$ **70.** $5a^3 - 9a^2b + 4b$ **71.** $6x^5 - 4x^4 + 2x^3 - 10x^2$ **72.** $acx^2 + (bc + ad)x + bd$

73. $6x^2 - 7x - 5$ **74.** $16x^2 - 25$ **75.** $2x^5 - 6x^4 - x^3 + 16x^2 - 13x + 2$ **76.** (a) $4a^2 - 4ab + b^2$

(b) $a^3 - 6a^2b + 12ab^2 - 8b^3$ **77.** $2(\sqrt{5} + \sqrt{2})$ **78.** See page 54   **79.** $5x^2(x^4 + 5x^2 - 3)$

80. $3(x - 5)(x + 5)$ **81.** $(3 - 2x)(9 + 6x + 4x^2)$ **82.** $(x - 4)(x + 2)$ **83.** $(a^2 + b^2)(a + b)(a - b)$

84. $2(a - 2)(a^4 + 2a^3 + 4a^2 + 8a + 16)$ **85.** $(3 - ax)(5 - x)$ or $(ax - 3)(x - 5)$ **86.** $(x + y)(x + y)$

87. $(x - y)(x - y)$ **88.** $(8x - 3)(x + 4)$ **89.** $(3x - 1)(2x + 1)$ **90.** $x(2x + 1)(2x + 1)$

91. $(x - \sqrt{3})(x + \sqrt{3})$ **92.** $(\sqrt[3]{x} - 2)(\sqrt[3]{x^2} + 2\sqrt[3]{x} + 4)$ **93.** See page 55   **94.** See page 56

95. See page 61   **96.** $\dfrac{x - 2}{x}$ **97.** -1 **98.** Cannot be reduced   **99.** $-1(x + 1)(x + 2)$ **100.** $\dfrac{4x + 1}{x - 2}$

101. $\dfrac{x + 4}{x(x - 1)(x + 1)}$ **102.** $\dfrac{4x^3 - 7x^2 + 9x - 9}{6x^2(x - 1)}$ **103.** $x^2 + x - 2$ **104.** $2x^2 - 3x + 1$ **105.** $-\dfrac{1}{2(2 + h)}$

106. $\dfrac{xy}{y-x}$ **107.** See page 71   **108.** See page 73   **109. (a)** $9i$; **(b)** -18 **110.** $-9+19i$ **111.** $\dfrac{8+i}{5}$

112. $-56+80i$ **113.** $7-8i$ **114. (a)** 1; **(b)** -1; **(c)** i; **(d)** $-i$ **115.** $5i$

116. Real part: $\dfrac{1}{5}$; imaginary part: $-\dfrac{2}{5}$ **117.** See page 72   **118.** $\dfrac{8}{5}+\dfrac{1}{5}i$

CHAPTER 1   STANDARD ANSWER TEST   (page 80)

1. (a) False; **(b)** true; **(c)** false; **(d)** true; **(e)** false; **(f)** false; **(g)** true; **(h)** false

2. 3   **3.** $x=-3$ **4.** Width = 7 inches; length = 19 inches   **5.** 2:27 P.M.

6. (a) 3.75×10^8; **(b)** 3.18×10^{-5} **7.** $x<-2$ **8.** $-3<x<-1$

9. (a) $x<\dfrac{2}{9}$:   **(b)** $x\le-1$ or $x\ge2$:

10. (a) False; **(b)** true; **(c)** false; **(d)** false; **(e)** true; **(f)** false

11. (a) $\dfrac{4x^8}{y^7}$; **(b)** $\dfrac{4y^7}{3x^6}$ **12. (a)** $4\sqrt{3}$; **(b)** $3\sqrt{5}$; **(c)** $-3x^2\sqrt[3]{3}$ **13. (a)** $10\sqrt{2}$; **(b)** $6\sqrt{3}$

14. $-|x|\sqrt{3}$ **15.** $5x^4+12x^3-2x-3$ **16.** $3x^2-x+2$ **17.** $(4-3b)(16+12b+9b^2)$

18. $(2x-3)(3x+1)$ **19.** $(x-3y)(2x+y^2)$ **20.** $\dfrac{2(x+3)}{(x+2)(x+5)}$ **21.** x **22.** $-\dfrac{x+7}{49x^2}$

23. $\dfrac{3x^2-18x+25}{(x-3)(x+3)(2x-5)}$ **24.** $43+23i$ **25.** $-\frac{13}{41}+\frac{47}{41}i$

CHAPTER 1   MULTIPLE CHOICE TEST   (page 81)

1. (c)   **2.** (c)   **3.** (d)   **4.** (b)   **5.** (a)   **6.** (d)   **7.** (b)   **8.** (c)   **9.** (b)   **10.** (a)   **11.** (c)   **12.** (b)
13. (d)   **14.** (e)   **15.** (d)   **16.** (a)   **17.** (b)   **18.** (c)   **19.** (c)   **20.** (d)

CHAPTER 2   LINEAR AND QUADRATIC FUNCTIONS WITH APPLICATIONS

2.1   Introduction to the Function Concept   (page 92)

1. True   **3.** False; -3 **5.** True   **7.** False; $-x^2+16$ **9.** True   **11.** Function: all reals

13. Function: all $x>0$ **15.** Not a function   **17.** Function: all $x\ne-1$ **19.** Not a function

21. (a) -3; **(b)** -1; **(c)** 0   **23. (a)** 1; **(b)** 0; **(c)** $\frac{1}{4}$ **25. (a)** -2; **(b)** -1; **(c)** $-\frac{7}{8}$

27. (a) 2; **(b)** 0; **(c)** $\frac{5}{16}$ **29. (a)** $-\frac{1}{2}$; **(b)** -1; **(c)** -2 **31. (a)** -1; **(b)** does not exist; **(c)** $\sqrt[3]{2}$

33. (a) 81; **(b)** 5; **(c)** $\frac{25}{36}$; **(d)** $\frac{1}{36}$ **35.** $3h(2) = 24 \neq 48 = h(6)$

37. (a) -11; **(b)** -5; **(c)** not defined; **(d)** not defined; **(e)** -9 **39.** $x + 3$ **41.** $-\dfrac{1}{3x}$ **43.** 2

45. 1   **47.** $-4 - h$ **49.** $-\dfrac{4 + h}{4(2 + h)^2}$

51. $h(t) = -16t^2 + 128t$; $h(0) = h(8) = 0$; the object is at ground level at $t = 0$ and $t = 8$.

53. (a) $h(x) = \frac{5}{2}x$; **(b)** $A(x) = \frac{25}{4}x$ **55.** $s(x) = \frac{1}{3}x$ **57.** $V(x) = x(50 - 2x)^2$ **59.** $V(x) = \frac{4}{3}x(30 - x^2)$

2.2 Graphing Lines in the Rectangular Coordinate System (page 102)

1.

x	−3	−2	−1	0	1	2
y	−5	−4	−3	−2	−1	0

3.

x	−2	−1	0	1	2
y	−8	−6	−4	−2	0

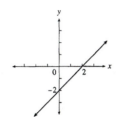

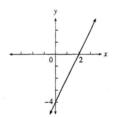

5. 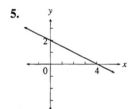   **7.** 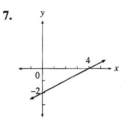   **9.** 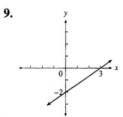   **11.**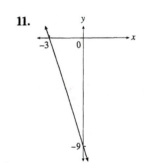

13. 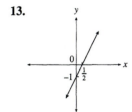   **15.** 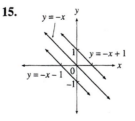   **17.** 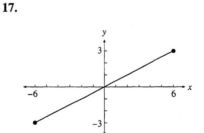   **19.**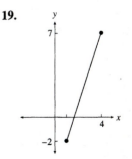

21. Each quotient gives the same slope, $-\frac{1}{2}$. **(a)** $\dfrac{3-1}{-2-2}$; **(b)** $\dfrac{2-0}{0-4}$; **(c)** $\dfrac{1-0}{2-4}$; **(d)** $\dfrac{3-(-1)}{-2-6}$;

(e) $\dfrac{2-(-1)}{0-6}$; **(f)** $\dfrac{1-(-1)}{2-6}$ **23.** $\frac{1}{9}$ **25.** 0   **27.** $-\dfrac{17}{28}$

29. 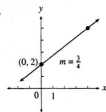   **31.** 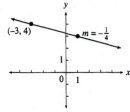   **33.**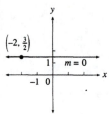

35. 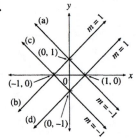   **37.** Both have the same slope of $-\frac{3}{2}$.

39. The slopes of PQ and RS are each $-\frac{5}{12}$; the slopes of PS and QR are each $\frac{12}{5}$. Thus the sides of the figure are

perpendicular to each other and form right angles. Also, the diagonals are perpendicular because the slope of PR is $\frac{7}{17}$

and the slope of QS is the negative reciprocal, $-\frac{17}{7}$.

41. Horizontal lines have slope 0, which does not have a reciprocal. Also, vertical lines have no slope, and therefore a slope comparison cannot be made.

43. $\frac{49}{2}$

45. **(a)** Slope $\ell_1 = m_1 = \dfrac{DA}{CD} = \dfrac{DA}{1} = DA$; **(b)** slope $\ell_2 = m_2 = \dfrac{DB}{CD} = \dfrac{DB}{1} = DB$; $m_2 < 0$;

(c) $\dfrac{DA}{CD} = \dfrac{CD}{BD}$ or $\dfrac{m_1}{1} = \dfrac{1}{-m_2}$, since $BD = -DB = -m_2$.

47. \$135; 1625   **49.** $A(x) = \dfrac{5}{4}x^2$ **51.** $A(x) = \dfrac{x^2}{16} + \dfrac{(50-x)^2}{4\pi}$

2.3 Algebraic Forms of Linear Functions   (page 113)

1. $y = 2x + 3$ **3.** $y = x + 1$ **5.** $y = 5$ **7.** $y = \frac{1}{2}x + 3$ **9.** $y = \frac{1}{4}x - 2$ **11.** $y - 3 = x - 2$

13. $y - 3 = 4(x + 2)$ **15.** $y - 5 = 0$ **17.** $y - 1 = \frac{1}{2}(x - 2)$ **19.** $y = 5x$ **21.** $y + \sqrt{2} = 10(x - \sqrt{2})$

23. $y = -3x + 4$; $m = -3$; $b = 4$ **25.** $y = 2x - \frac{1}{3}$; $m = 2$; $b = -\frac{1}{3}$ **27.** $y = \frac{5}{3}$; $m = 0$; $b = \frac{5}{3}$

29. $y = \frac{4}{3}x - \frac{7}{3}$; $m = \frac{4}{3}$; $b = -\frac{7}{3}$ **31.** $y = \frac{1}{2}x - 2$; $m = \frac{1}{2}$; $b = -2$ **33.** $x + y = 5$ **35.** $x - y = 3$

37. $2x + y = -15$ **39.** $y = 27$ **41.** $x = 5, y = -7$ **43.** $y = -\frac{2}{3}x - \frac{1}{3}$ **45.** $y = -\frac{1}{3}x + \frac{25}{3}$

47. $y = -\frac{1}{2}x - \frac{3}{2}$ **49.** $y = 3x + 2$; $y = -\frac{1}{2}x + 4$; $y = -\frac{5}{3}x + \frac{14}{3}$ **51.** They are negative reciprocals.

53.

55.
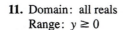

57. **(a)** If $c = 0$ then $(0, 0)$ fits the equation and the line would then pass through the origin.

   (b) $\dfrac{c}{b} = y$-intercept; $\dfrac{c}{a} = x$-intercept

   (c) $ax + by = c$; $\dfrac{ax}{c} + \dfrac{by}{c} = 1$; $\dfrac{x}{\left(\dfrac{c}{a}\right)} + \dfrac{y}{\left(\dfrac{c}{b}\right)} = 1$; $\dfrac{x}{q} + \dfrac{y}{p} = 1$

   (d) $\dfrac{x}{\frac{3}{2}} + \dfrac{y}{-5} = 1$ or $10x - 3y = 15$ **(e)** $m = \dfrac{0 - (-5)}{\frac{3}{2} - (0)} = \dfrac{5}{\frac{3}{2}} = \dfrac{10}{3}$; $y = \dfrac{10}{3}x - 5$

2.4 Piecewise Linear Functions (page 119)

1. 99   **3.** -100 **5.** -4 **7.** 1

9. Domain: all reals
  Range: $y \geq 0$

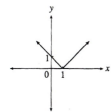

11. Domain: all reals
  Range: $y \geq 0$

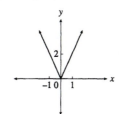

13. Domain: all reals
  Range: $y \geq 0$

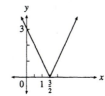

15. Domain: $x \geq -1$
  Range: $y \leq 3$

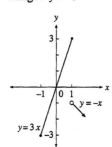

17. Domain: $-2 < x \leq 3$
  Range: $-2 < y \leq 4$

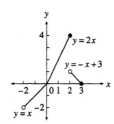

19.

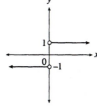

21.

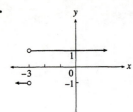

23.

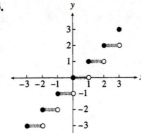

25.

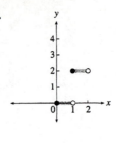

27.

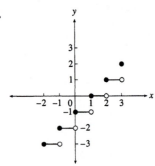

29.

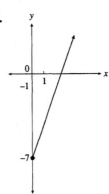

31.

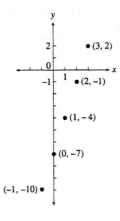

33. $x = -3$; not a function   **35.** $y = x + 2$
Domain: $x \geq -2$
Range: $y \geq 0$

37. $y = -1$ for $-1 \leq x < 0$; $y = 3$ for $0 \leq x < 3$; $y = 1$ for $3 \leq x \leq 4$; Domain: $-1 \leq x \leq 4$; Range: $\{-1, 1, 3\}$

39.

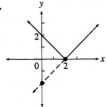

41. $P(x) = \frac{1}{10}(8 + [\![x]\!])$ for $0 < x < 10$

2.5 Introduction to Systems of Equations (page 128)

1. $(2, 2)$ **3.** $\left(4, \frac{1}{2}\right)$ **5.** $\left(\frac{21}{88}, -\frac{9}{22}\right)$ **7.** $(5, 1)$ **9.** $(8, 5)$ **11.** $(3, -11)$ **13.** $(1, 0)$ **15.** $\left(1, \frac{1}{2}\right)$

17. $(1, 0)$ **19.** $\left(\frac{17}{18}, \frac{7}{18}\right)$ **21.** Consistent   **23.** Dependent   **25.** Inconsistent   **27.** $\left(-\frac{3}{2}c, \frac{9}{2}c\right)$

29. $(-20, -20)$ **31.** $a = 4, b = -7$ **33.** 40 field goals; 16 free throws   **35.** $\ell = 21$ cm, $w = 9$ cm

37. 8 lb of potatoes and $1\frac{1}{2}$ lb of string beans   **39.** $3200 room and board; $5200 tuition

41. 14 quarters; 8 nickels   **43.** 344   **45.** $2\frac{3}{4}$ hours at 5 kph; $1\frac{3}{4}$ hours at 3 kph; total time $= 4\frac{1}{2}$ hours

47. 550 at 25¢ and 360 at 45¢   **49.** $2800 at 8%; $3200 at $7\frac{1}{2}$%   **51.** 16 milliliters of each

53. 6 miles by car; 72 miles by train   **55.** An infinite number of answers

57. The clerk was wrong because if there were common unit prices, say x = cost per orange and y = cost per tangerine, then $6x + 12y = 2.34$, $2x + 4y = .77$, which is an inconsistent system. (*Note*: The smaller bag is a better buy since 2 (oranges) + 4 (tangerines) = $\frac{1}{3}$ (6 oranges + 12 tangerines) = $\frac{1}{3}$(2.34) = 0.78, and 77¢ is cheaper.)

59. $120   **61.** 90   **63.** (90, 1800)   **65.** 115

2.6 Graphing Quadratic Functions   (page 139)

1.
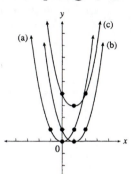

3.
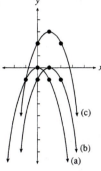

5.
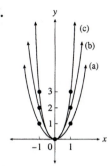

7.

9.
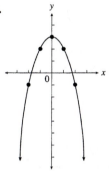

11.
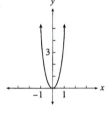

13.
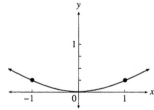

15.
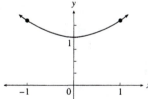

17.
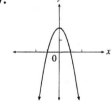

19. Decreasing on $(-\infty, 1]$
Increasing on $[1, \infty)$
Concave up

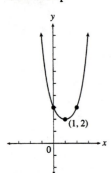

21. Increasing on $(-\infty, -1]$
Decreasing on $[-1, \infty)$
Concave down

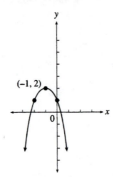

23. Decreasing on $(-\infty, 3]$
Increasing on $[3, \infty)$
Concave up

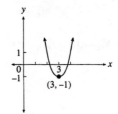

25. (a) $(3, 5)$; (b) $x = 3$; (c) set of real numbers; (d) $y \geq 5$

27. (a) $(3, 5)$; (b) $x = 3$; (c) set of real numbers; (d) $y \leq 5$

29. (a) $(-1, -3)$; (b) $x = -1$; (c) set of real numbers; (d) $y \geq -3$

31. (a) $(1, 2)$; (b) $x = 1$; (c) set of real numbers; (d) $y \leq 2$

33. (a) $(-2, -4)$; (b) $x = -2$; (c) set of real numbers; (d) $y \geq -4$

35.

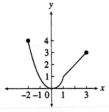

37. Not a function because for $x > 0$ there are two corresponding y-values.

39. The graph of $x^2 - 4 > 0$ consists of all points on the number line where $x < -2$ or where $x > 2$.
On the coordinate plane, the graph of $y = x^2 - 4$ is positive (above the x-axis) for $x < -2$ or $x > 2$.

41. $a = -2$ **43.** $k = 3$

45.

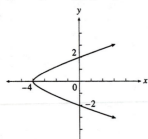

47.

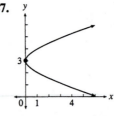

49.

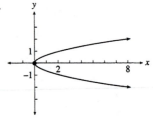

51.

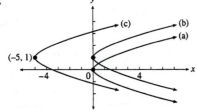

53. $y = (x - 4)^2 + 2$ **55.** $x = (y - 2)^2 - 5$ **57.** $y = \frac{1}{2}(x - 3)^2 - 2$ **59.** Yes   **61.** Yes   **63.** No

65. No   **67.**   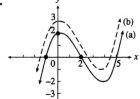   **69.** $A(x) = -2x^3 + 18x$; domain: $-3 < x < 3$

2.7 The Quadratic Formula (page 152)

1. $(x + 1)^2 - 6$ **3.** $-(x + 3)^2 + 11$ **5.** $-\left(x - \frac{3}{2}\right)^2 - \frac{7}{4}$ **7.** $\left(x + \frac{5}{2}\right)^2 - \frac{33}{4}$ **9.** $3(x - 1)^2 + 2$

11. $-3(x + 1)^2 + 8$ **13.** $y = (x + 1)^2 - 2$; $(-1, -2)$; $x = -1$; -1 **15.** $y = -(x - 2)^2 + 3$; $(2, 3)$; $x = 2$; -1

17. $y = 3(x + 1)^2 - 6$; $(-1, -6)$; $x = -1$; -3 **19.** $x = 2, x = 3$ **21.** $x = 0, x = 10$ **23.** $x = -\frac{1}{5}, x = \frac{3}{2}$

25. $x = 4$ **27.** $x = 0, x = 5$ **29.** $x = 2 \pm \sqrt{3}$ **31.** $x = -2; x = 5$

33. $x = 1 - \sqrt{5} = -1.24$; $x = 1 + \sqrt{5} = 3.24$ **35.** $x = -2$; $x = -\frac{1}{3}$ **37.** $x = 3 \pm \sqrt{3}$; $4.73, 1.27$

39. $x = 3 \pm i\sqrt{5}$ **41.** $x = \dfrac{3 \pm \sqrt{17}}{4}$; 1.78; -0.28 **43.** 3; $-\frac{1}{2}$ **45.** None   **47.** $\dfrac{-1 \pm \sqrt{13}}{6}$ **49.** (a)

51. (b)   **53.** (c)   **55.** Once; **(a)** $(2, 0)$; **(b)** 4; **(c)** 2 **57.** Once; **(a)** $\left(\frac{1}{3}, 0\right)$; **(b)** 1; **(c)** $\frac{1}{3}$

59. None; **(a)** $(1, 1)$; **(b)** 3; **(c)** none   **61.** -6 or 6 **63.** $\pm 2\sqrt{7}$ **65.** $k > -4$

67. $k > -\frac{1}{4}$ **69.** $t > 9$ **71.** $t < -\frac{1}{4}$

73. Sum: $\dfrac{-b + \sqrt{b^2 - 4ac}}{2a} + \dfrac{-b - \sqrt{b^2 - 4ac}}{2a} = \dfrac{-2b}{2a} = -\dfrac{b}{a}$

Product: $\dfrac{-b + \sqrt{b^2 - 4ac}}{2a} \cdot \dfrac{-b - \sqrt{b^2 - 4ac}}{2a} = \dfrac{b^2 - (b^2 - 4ac)}{4a^2} = \dfrac{4ac}{4a^2} = \dfrac{c}{a}$

75. $-1 < x < 3$ **77.** $x \leq -5$ or $x \geq 2$ **79.** No solution.

81.   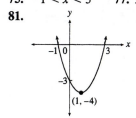   **83.**   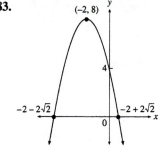   **85.**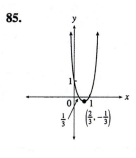

87. $x = \pm\sqrt{7}, x = \pm\dfrac{\sqrt{2}}{2}i$ **89.** $x = -3, x = -2, x = 2$ **91.** $x = -\frac{3}{2}, x = \frac{1}{4}$

93. Decreasing on $(-\infty, -3]$ and on $[0, 3]$

Increasing on $[-3, 0]$ and on $[3, \infty)$
Concave up on $(-\infty, -3)$ and on $(3, \infty)$
Concave down on $(-3, 3)$
Range: $y \geq 0$

95. Decreasing on $(-\infty, -2]$ and on $\left[\frac{1}{2}, 3\right]$

Increasing on $\left[-2, \frac{1}{2}\right]$ and on $[3, \infty)$
Concave up on $(-\infty, -2)$ and on $(3, \infty)$
Concave down on $(-2, 3)$
Range: $y \geq 0$

97.

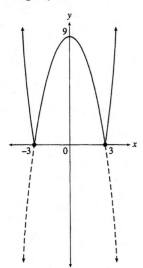

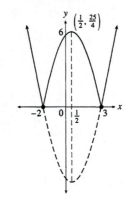

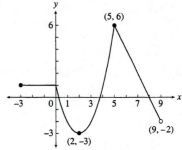

99. $a > 0$; $k = 0$ **101.** $a > 0$; $k = 2$

2.8 Applications of Quadratic Functions   (page 161)

1. 14, 15   **3.** 5, 8   **5.** 2 feet   **7.** Width = 3 cm; length = 5 cm   **9.** $-7 + \sqrt{62}$ meters   **11.** $n = 15$
13. 20 mph; 1.5 hours downstream; 2.25 hours upstream   **15.** Maximum = 7 at $x = 5$

17. Minimum = 36 at $x = 2$ **19.** Minimum = 0 at $x = \frac{7}{2}$ **21.** Maximum = $\frac{1}{9}$ at $x = -\frac{1}{3}$ **23.** 60; $600

25. 6, 6   **27.** -11 and 11; product = -121 **29.** $x = 25$ feet, $y = 50$ feet for each rectangle   **31.** 4 seconds
33. 2 seconds and 6 seconds. It will reach a height of 192 feet on the way up and again on the way down.   **35.** $9.75

37. $(1, 2)$ **39.** $\left(\frac{5}{2}, \frac{7}{2}\right)$; $\frac{35}{2}$ square units   **41.** $\left(\frac{3}{2}, \frac{9}{2}\right)$; minimum value = $\frac{9}{2}$

43. $\left(-\frac{1}{3}, \frac{19}{18}\right)$; maximum value = $\frac{19}{18}$ **45.** $(0, 9)$; maximum value = 9

47. (a) 116 feet; (b) 197 feet; (c) 262 feet   **49.** $15

CHAPTER 2   REVIEW EXERCISES   (page 165)

1. See page 86   **2.** Function: all reals   **3.** Function: all reals   **4.** Function: all $x \geq 5$ **5.** Not a function
6. Function: all $x < 1$ **7.** Not a function   **8.** (a) 9; (b) 3; (c) 5   **9.** 32; 52; $2f(3) \neq f(6)$

10. (a) 18; (b) 0; (c) $4x^2 - 10x$; (d) $8x^2 - 10x$ **11.** $x + 2$ **12.** $-2 - h$ **13.** (a) $\dfrac{2}{3x - 1}$; (b) $\dfrac{6}{x - 1}$

14. $h = \dfrac{8x}{x + 5}$; $A = \dfrac{4x^2}{x + 5}$ **15.** See page 98   **16.** 0; undefined

17. The slopes are negative reciprocals of one another.

18.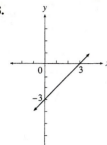

19.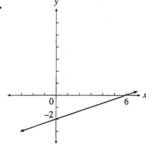

20. $-\frac{5}{2}$ **21.**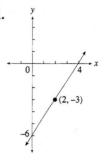

22. See page 101   **23.** (20, 1000); demand = supply   **24.** $y = mx + b$ **25.** $y - y_1 = m(x - x_1)$

26. A horizontal line, domain = all reals; range = the constant value of the function

27. 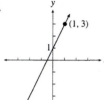   **28.** $y - 2 = -\frac{3}{4}x$ **29. (a)** $y = 2$; **(b)** $x = -3$ **30.** $y + 3 = -2(x - 1)$

31. (a) $y - 2 = 3(x + 3)$ or $y - 5 = 3(x + 2)$; **(b)** $y = 3x + 11$; **(c)** $3x - y = -11$

32. $\frac{3}{2}$; 3   **33.** $y - 1 = -\frac{2}{3}(x + 3)$; $y - 1 = \frac{3}{2}(x + 3)$

34.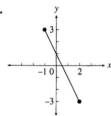

35.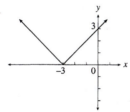

36.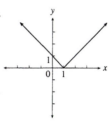

37.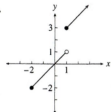

38. (a) 4; **(b)** -5; **(c)** 3; **(d)** -1 **39.**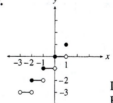

Domain: $-3 < x \le 1$
Range: $[-3, -2, -1, 0, 1]$

40.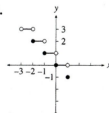

Domain: $-3 < x \le 1$
Range: $[-1, 0, 1, 2, 3]$

41. $x = 2, y = 3$ **42.** $x = -1, y = 5$ **43.** $s = 5, t = -1$

44. $x = -2, y = \frac{1}{2}$ **45.** $x = -\frac{3}{2}, y = \frac{1}{4}$ **46.** $x = 4, y = -6$

47. Dependent   **48.** Inconsistent   **49.** Consistent

50. 8, 12   **51. (a)** $240 + 6x$; **(b)** $10x$; **(c)** (60, 600)

52. Cost = $18, profit = $13   **53.** See page 132   **54.** See page 132

55.

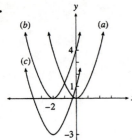

56.

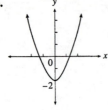

57.

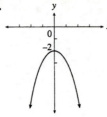

58.

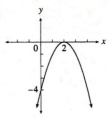

59.

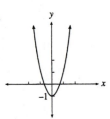

60.

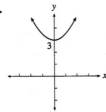

61.

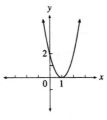

62. Decreasing on $(-\infty, 2]$; increasing on $[2, \infty)$; concave up; vertex at $(2, 1)$; axis: $x = 2$

63. Decreasing on $[-2, \infty)$; increasing on $(-\infty, -2]$; concave down; vertex at $(-2, 3)$; axis: $x = -2$

64. Decreasing on $(-\infty, 1]$; increasing on $[1, \infty)$; concave up; vertex at $(1, -3)$; axis: $x = 1$

65. (a) (h, k); **(b)** $x = h$ **66.** See page 139   **67.** Vertex: $(3, 2)$; axis of symmetry: $y = 2$

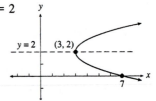

68. $y = (x + 2)^2 - 7$ **69.** $y = -2(x + 3)^2 + 29$ **70.** $y = \frac{1}{2}(x + 3)^2 - \frac{15}{2}$ **71.** See page 149

72. $x = \dfrac{-2 \pm \sqrt{10}}{3}$ **73.** $x = -2, x = -5$ **74.** $x = 5$ **75.** $x = -2 \pm \sqrt{11}$ **76.** $x = 2 \pm i\sqrt{3}$

77. $x = -1, x = \frac{3}{2}$ **78.** $x = \dfrac{-1 \pm i\sqrt{35}}{6}$

79. $a < 0$ and $k > 0$; domain: all x; range: $y \le k$ **80.** See page 151   **81.** $x = -2, x = \frac{3}{2}$

82. Two imaginary numbers   **83.** One rational number   **84.** Two irrational numbers

85.

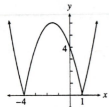

86. $x < -\frac{5}{3}$ or $x > \frac{1}{2}$

87. Minimum value of 0 at $x = 1$ **88.** Maximum value of -5 at $x = -1$

89. Minimum value of -1 at $x = -2$ **90.** Minimum value of -1 at $x = 3$

91. Minimum value of 0 at $x = -2$ **92.** Maximum value of $\frac{19}{2}$ at $x = -\frac{3}{2}$

93. 20, 20   **94.** 30 m by 60 m   **95. (a)** 96 feet; **(b)** 1 second; **(c)** $1 + \sqrt{6}$ seconds

96. (a) Maximum value of k; **(b)** minimum value of k

CHAPTER 2   STANDARD ANSWER TEST   (page 169)

1. All $x \neq -2$ **2. (a)** $\dfrac{3}{2 + x}$; **(b)** $\dfrac{3}{2} + \dfrac{3}{x} = \dfrac{3x + 6}{2x}$ **3.** $x + 8$ **4.**

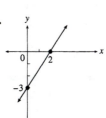

5. (a) $-\dfrac{7}{3}$ **(b)** 0 **6.** Domain: all x **7. (a)** $x = 3$; **(b)** $y = -2$ **8.** $y = \frac{1}{2}x - 3$
Range: all y

9. $y = \frac{2}{3}x - \frac{5}{3}$; $\frac{2}{3}$; $-\frac{5}{3}$ **10.** $y = -\frac{9}{5}x + \frac{2}{5}$ **11.** $y - 8 = \frac{5}{2}(x - 2)$ **12.**

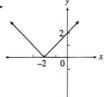

13. 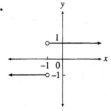   **14.** 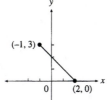   **15.** 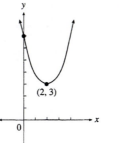   **16.**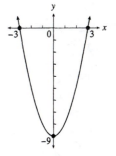

17. (a) $y = -5(x - 2)^2 + 19$; **(b)** $(2, 19)$; **(c)** $x = 2$; **(d)** domain: all real numbers; range: all $y \leq 19$

18. $x = -\frac{1}{3}, x = 3$ **19.** $2 - \sqrt{11}, 2 + \sqrt{11}$ **20.** $9, 11$

21. (a) 5, two irrational numbers; **(b)** 169, two rational numbers   **22.** $x = 5, y = -2$
23. Maximum $= 20$ at $x = -6$ **24.** 6 feet   **25.** 64 feet

CHAPTER 2   MULTIPLE CHOICE TEST   (page 170)

1. (b)   **2.** (d)   **3.** (d)   **4.** (a)   **5.** (c)   **6.** (b)   **7.** (b)   **8.** (d)   **9.** (e)   **10.** (a)   **11.** (b)
12. (d)   **13.** (c)   **14.** (e)   **15.** (b)   **16.** (a)   **17.** (d)   **18.** (c)   **19.** (a)   **20.** (c)

CHAPTER 3   POLYNOMIAL AND RATIONAL FUNCTIONS

3.1   Hints for Graphing   (page 179)

1.

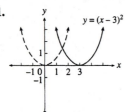

For $y = (x - 3)^2$:
Domain: all reals
Range: all $y \geq 0$
Decreasing on $(-\infty, 3)$
Increasing on $(3, \infty)$
Concave up on $(-\infty, \infty)$

3.

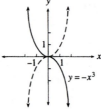

For $y = -x^3$:
Domain: all reals
Range: all reals
Decreasing on $(-\infty, \infty)$
Concave up on $(-\infty, 0)$
Concave down on $(0, \infty)$

5.

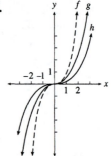

For $h(x) = \frac{1}{4} x^3$:
Domain: all reals
Range: all reals
Increasing on $(-\infty, \infty)$
Concave down on $(\infty, 0)$
Concave up on $(0, \infty)$

7.

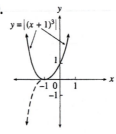

For $h(x) = (x - 2)^4 - 2$
Domain: all reals
Range: all $y \geq -2$
Decreasing on $(-\infty, 2)$
Increasing on $(2, \infty)$
Concave up on $(-\infty, \infty)$

9.

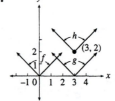

$y = |(x + 1)^3|$

11.

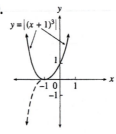

13. Translate the graph of $y = x^3$ one unit to the right and 2 units up.

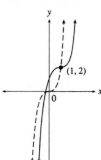

15. First sketch $y = 2x^3$ by multiplying the ordinates of $y = x^3$ by 2. Then translate $y = 2x^3$ three units left, and 3 units down.

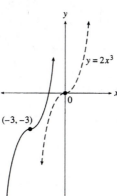

17. Reflect the graph of $y = x^3$ in the x-axis. Then translate 1 unit to the left and 1 unit down.

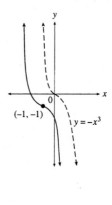

19. $y = (x - 3)^4 + 2$ **21.** $y = \left| x + \frac{3}{4} \right|$ **23.** $y = \left| x^3 - 1 \right|$ **25.**

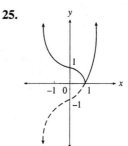

27.

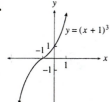

29.

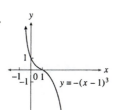

31. $x^2 + 3x + 9$ **33.** $4 + 6h + 4h^2 + h^3$

3.2 Graphing Some Special Rational Functions (page 188)

1.

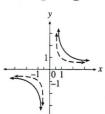

3.

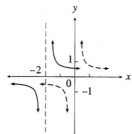

5.

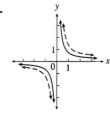

7.

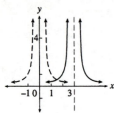

9. Asymptotes: $x = 0$, $y = 2$

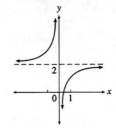

Domain: all $x \neq 0$
Range: all $y \neq 2$
Increasing and
   concave up on $(-\infty, 0)$
Increasing and
   concave down on $(0, \infty)$

11. Asymptotes: $x = -4$, $y = -2$

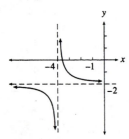

Domain: all $x \neq -4$
Range: all $y \neq -2$
Decreasing and
   concave down on $(-\infty, -4)$
Decreasing and
   concave up on $(-4, \infty)$

13. Asymptotes: $x = 2$, $y = 1$

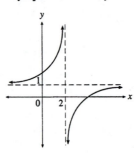

Domain: all $x \neq 2$
Range: all $y \neq 1$
Increasing and
   concave up on $(-\infty, 2)$
Increasing and
   concave down on $(2, \infty)$

15. Asymptotes: $x = -1$, $y = -2$

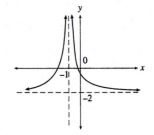

Domain: all $x \neq -1$
Range: all $y > -2$
Increasing and
   concave up on $(-\infty, -1)$
Decreasing and
   concave up on $(-1, \infty)$

17. Asymptotes: $x = 2$, $y = 0$

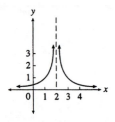

Domain: all $x \neq 2$
Range: all $y > 0$
Increasing and
   concave up on $(-\infty, 2)$
Decreasing and
   concave up on $(2, \infty)$

19. Asymptotes: $x = 0$, $y = 0$

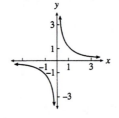

21. Asymptotes: $x = 1$, $y = 0$

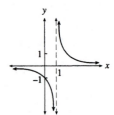

23. $f(x) = x + 3$, $x \neq 3$

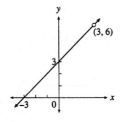

25. $f(x) = x + 2$, $x \neq 3$

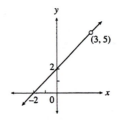

27. $f(x) = \dfrac{1}{x - 1}$, $x \neq -1$
Asymptotes: $x = 1$, $y = 0$

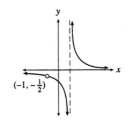

29. $f(x) = x^2 + 2x + 4, x \neq 2$

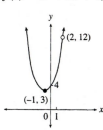

31.

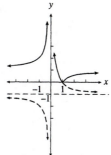

33. $-\dfrac{1}{3x}$

3.3 Polynomial and Rational Functions (page 196)

1. $f(x) < 0$ on $(-\infty, 1)$ and $(2, 3)$; $f(x) > 0$ on $(1, 2)$ and $(3, \infty)$

3. $f(x) < 0$ on $(-\infty, -4)$ and $\left(\frac{1}{3}, 2\right)$; $f(x) > 0$ on $(-4, 0)$, $\left(0, \frac{1}{3}\right)$, and $(2, \infty)$

5. $f(x) < 0$ on $(-1, 4)$; $f(x) > 0$ on $(-\infty, -1)$ and $(4, \infty)$

7. $f(x) < 0$ on $(-\infty, -1)$ and $\left(\frac{1}{5}, 10\right)$; $f(x) > 0$ on $\left(-1, \frac{1}{5}\right)$ and $(10, \infty)$

9.

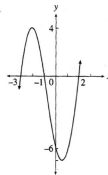

11.

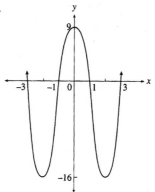

13.

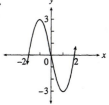

15.

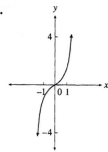

17.

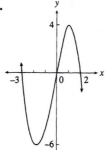

19.

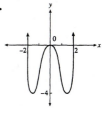

21.

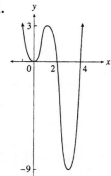

23. Asymptotes: $x = -2$; $x = 1$; $y = 0$

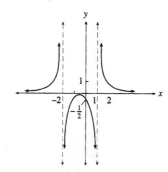

25. Asymptotes: $x = -2, x = 2$; $y = 0$

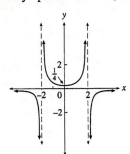

27. Asymptotes: $x = -1$; $x = 1$; $y = 0$

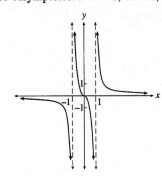

29. Asymptote: $y = 0$

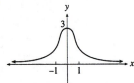

31. Asymptotes: $x = 2$; $y = 1$

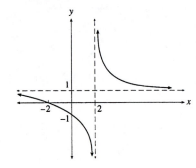

33. Asymptotes: $x = -3$; $y = 1$

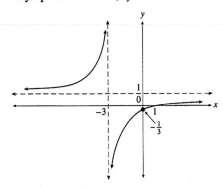

35. Asymptotes: $x = -4$; $x = 3$; $y = 1$

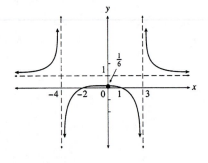

37. Horizontal asymptote: $y = 0$ **39.** Horizontal asymptote: $y = 2$ **41.** $y = x$ **43.** $y = x - 2$ **45.** $y = x - 1$

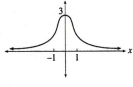

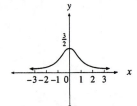

47. $g(x) = 6(x - 2)(x - 1)$; $g(x) > 0$ on $(-\infty, 1)$ and $(2, \infty)$; $g(x) < 0$ on $(1, 2)$

49. $g(x) = -\dfrac{x + 4}{x^3}$; $g(x) > 0$ on $(-4, 0)$; $g(x) < 0$ on $(-\infty, -4)$ and $(0, \infty)$

51. $g(x) = \dfrac{14(x - 5)}{(x + 2)^3}$; $g(x) > 0$ on $(-\infty, -2)$ and $(5, \infty)$; $g(x) < 0$ on $(-2, 5)$

53. $y = \dfrac{2x^2}{3 - 5x^2} = \dfrac{\dfrac{2x^2}{x^2}}{\dfrac{3}{x^2} - \dfrac{5x^2}{x^2}} = \dfrac{2}{\dfrac{3}{x^2} - 5};$ as $x \to 0,\ \dfrac{3}{x^2} \to 0$ and $y \to -\dfrac{2}{5}$

55. (a)
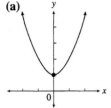
No x-intercepts
1 turning point

(b)
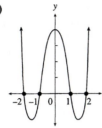
4 x-intercepts
3 turning points

(c)
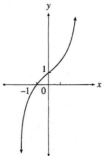
1 x-intercept
No turning points

(d)
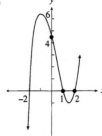
3 x-intercepts
2 turning points

3.4 Equations and Inequalities with Fractions (page 206)

1. $x = 20$ **3.** $x = 4$ **5.** $x = \frac{17}{4}$ **7.** $x = 2$ **9.** $x = 12$ **11.** No solutions   **13.** $x = -\frac{5}{2};\ x = 2$

15. $x = 1;\ x = 2$ **17.** $x = 20$ **19.** $x = -3;\ x = 3$ **21.** $-\frac{1}{2},\ 3;$ domain: all $x \neq -1$ or 0

23. $\frac{10}{7},\ 5;$ domain: all $x \neq -5$ or 0 **25.** $-5,\ 7;$ domain: all $x \neq -\frac{1}{2}$ or 2 **27.** $m = \dfrac{2gK}{v^2}$ **29.** $r_1 = \dfrac{S}{\pi s} - r_2$

31. $s = a + (n - 1)d$ **33.** $m = \dfrac{fp}{p - f}$ **35.** $x \leq 30$ **37.** $x < -1$ **39.** $x > 10$ **41.** $x < -3,\ x \geq 2$

43. $-3 \leq x < -1,\ x \geq 6$ **45.** $1 < x < 2$ **47.** 3   **49.** $1\frac{5}{7}$ hours   **51.** $\frac{2}{9},\ \frac{4}{9}$ **53.** 60 feet   **55.** 19   **57.** $\frac{2}{3}$

59. $\dfrac{a}{b} = \dfrac{c}{d}$ implies $ad = bc$. Then, $ad + bd = bc + bd;\ d(a + b) = b(c + d);$ therefore, $\dfrac{a + b}{b} = \dfrac{c + d}{d}.$

61. 110 paperbacks and 55 hardcover copies   **63.** 8000

65. Use $\dfrac{4(4) + 4(3) + 3(3) + 1(2) + x(3)}{15} = 3.4;\ x = 4;$ grade must be A.   **67.** 5000 ohms   **69.** 10 cm

71. $\left(\frac{1}{2}, 2\right), \left(3, \frac{1}{3}\right)$ **73.** $A(x) = \dfrac{x^2}{2(x - 3)};$ 48.05

3.5 Variation (page 215)

1. $P = 4s;\ k = 4$ **3.** $A = 5\ell;\ k = 5$ **5.** $z = kxy^3$ **7.** $z = \dfrac{kx}{y^3}$ **9.** $w = \dfrac{kx^2}{yz}$ **11.** $\frac{1}{2}$ **13.** $\frac{1}{5}$ **15.** $\frac{2}{5}$

17. $\frac{243}{8}$ **19.** 27 feet   **21.** 4 pounds   **23.** 75 pounds per square inch   **25.** 784 feet   **27.** $\dfrac{32\pi}{3}$ cubic inches

29. 5.12 ohms; 512 ohms   **31. (a)** $s(y) = ky\left(\frac{25}{4} - y^2\right);$ **(b)** $s(x) = \frac{1}{2}kx^2\sqrt{25 - 4x^2};$ **(c)** $5.52k$

33. 146 pounds   **35.** $x = k_1 z,\ y = k_2 z;\ x + y = k_1 z + k_2 z = (k_1 + k_2)z = kz$

37. $x = k_1 z,\ y = k_2 z;\ xy = k_1 k_2 z^2 = kz^2;$ the product varies directly as the square of z.   **39.** 5 foot-candles

41. $A = \frac{1}{2}xy = \frac{1}{2}x\,(mx) = \frac{1}{2}mx^2$

3.6 Synthetic Division, the Remainder, and Factor Theorem (page 222)

1. $x^2 + x - 2;\ r = 0$ **3.** $2x^2 - x - 2;\ r = 9$ **5.** $x^2 + 7x + 7;\ r = 22$ **7.** $x^3 - 5x^2 + 17x - 36;\ r = 73$

9. $2x^3 + 2x^2 - x + 3$; $r = 1$ **11.** $x^2 + 3x + 9$; $r = 0$ **13.** $x^2 - 3x + 9$; $r = 0$

15. $x^3 + 2x^2 + 4x + 8$; $r = 0$ **17.** $x^3 - 2x^2 + 4x - 8$; $r = 32$ **19.** $x^3 + \frac{1}{2}x^2 + \frac{5}{6}x + \frac{7}{12}$; $r = \frac{47}{60}$

21. $3x^2 - x - 2$; $r = 2$ **23.** $3x^2 - x + 2$; $r = 2$ **25.** 8   **27.** 59   **29.** 0   **31.** 1   **33.** 67   **35.** -4

For Exercises 37–47, use synthetic division to obtain $p(c) = 0$, showing $x - c$ to be a factor of $p(x)$. The remaining factors of $p(x)$ are obtained by factoring the quotient obtained in the synthetic division.

37. $(x + 1)(x + 2)(x + 3)$; $x = -1, x = -2, x = -3$ **39.** $(x - 2)(x + 3)(x + 4)$; $x = 2, x = -3, x = -4$

41. $-(x + 2)(x - 3)(x + 1)$; $x = -2, x = 3, x = -1$ **43.** $(x - 5)(3x + 4)(2x - 1)$; $x = 5, x = -\frac{4}{3}, x = \frac{1}{2}$

45. $(x + 2)^2(x + 1)(x - 1)$; $x = -2$ (double root), $x = -1, x = 1$

47. $x^2(x + 3)^2(x + 1)(x - 1)$; $x = 0$ (a double root), $x = -3$ (a double root), $x = -1, x = 1$

49. $3x^3 - 9x^2 - 12x + 36$ **51.** 2, 3   **53.** 2   **55.** $(x + 2)(x + 1)(x - 1)(x^2 - x + 3)$

57. $f(c) = 2c^4 + c^2 + 20$ which is greater than 0 for all c.

59. If $p(x) = x^n + 1, p(-1) = (-1)^n + 1$. For any positive odd integer n, $(-1)^n = -1$. Thus $-1 + 1 = 0$ and $x + 1$ is a factor of $x^n + 1$.

3.7 Solving Polynomial Equations (page 232)

1. -3 (a double root), 5   **3.** $-\frac{2}{3}, -5, 5$ **5.** -1 (a triple root), 0   **7.** $-5, -\sqrt{3}, -1, \sqrt{3}$ **9.** $-3, 3$

11. $-\sqrt{3}, 0, \sqrt{3}, 2, 3$ **13.** $-1, \dfrac{1 - \sqrt{7}}{2}, \dfrac{1 + \sqrt{7}}{2}$ **15.** $\frac{2}{3}, 4$ **17.** $-(x + 4)^2(x - 5)$

19. $3(x + 2)(2x - 1)(x^2 + 1)$

21. By the rational root theorem, the only possible rational roots are $\pm 1, \pm\frac{1}{2}, \pm 2, \pm 4, \pm 8$. Using synthetic division, we see that none of these are roots.

23. $-1, 2, 1 - 2i, 1 + 2i$ **25.** $2, -\dfrac{1}{6} \pm \dfrac{\sqrt{47}}{6}i$ **27.** 1 (a triple root), $3 + i, 3 - i$

	Number of Positive Zeros	Number of Negative Zeros	Number of Imaginary Zeros	Total Number of Zeros
29.	2	1	0	3
	0	1	2	3
31.	3	2	0	5
	3	0	2	5
	1	2	2	5
	1	0	4	5

33. $(-2, -27)$; $(2, 1)$; $(3, 8)$ **35.** $(-3, -161)$; $(5, 335)$; $\left(\frac{1}{3}, \frac{253}{27}\right)$

37. $\overline{z + w} = \overline{(-6 + 8i) + \left(\frac{1}{2} + \frac{1}{2}i\right)} = \overline{-\frac{11}{2} + \frac{17}{2}i} = -\frac{11}{2} - \frac{17}{2}i$

$\overline{z} + \overline{w} = \overline{-6 + 8i} + \overline{\frac{1}{2} + \frac{1}{2}i} = (-6 - 8i) + \left(\frac{1}{2} - \frac{1}{2}i\right) = -\frac{11}{2} - \frac{17}{2}i$

39. $\overline{zw} = \overline{(-6 + 8i)\left(\frac{1}{2} + \frac{1}{2}i\right)} = \overline{-7 + i} = -7 - i$; $\overline{z}\,\overline{w} = \overline{(-6 + 8i)}\,\overline{\left(\frac{1}{2} + \frac{1}{2}i\right)} = (-6 - 8i)\left(\frac{1}{2} - \frac{1}{2}i\right) = -7 - i$

41. $\overline{z + w} = \overline{(a + bi) + (c + di)} = \overline{(a + c) + (b + d)i} = (a + c) - (b + d)i$

$\overline{z} + \overline{w} = \overline{a + bi} + \overline{c + di} = (a - bi) + (c - di) = (a + c) - (b + d)i$

43. $\overline{\left(\dfrac{z}{w}\right)} = \overline{\left(\dfrac{a+bi}{c+di}\right)} = \overline{\dfrac{ac+bd}{c^2+d^2} + \dfrac{bc-ad}{c^2+d^2}i} = \dfrac{ac+bd}{c^2+d^2} - \dfrac{bc-ad}{c^2+d^2}i$

$\dfrac{\overline{z}}{\overline{w}} = \dfrac{\overline{a+bi}}{\overline{c+di}} = \dfrac{a-bi}{c-di} = \dfrac{a-bi}{c-di} \cdot \dfrac{c+di}{c+di} = \dfrac{ac+bd}{c^2+d^2} - \dfrac{bc-ad}{c^2+d^2}i$

45. The following proof makes repeated use of the properties in Exercises 41 and 44. Also used is the observation that the conjugate of a real number is itself.

Since $0 = p(z)$, $\overline{0} = \overline{p(z)}$. Then, since $\overline{0} = 0$, we have

$$0 = \overline{p(z)} = \overline{a_n z^n + a_{n-1}z^{n-1} + \cdots + a_1 z + a_0}$$

$$= \overline{a_n z^n} + \overline{a_{n-1}z^{n-1}} + \cdots + \overline{a_1 z} + \overline{a_0} \qquad \text{(Ex. 41)}$$

$$= \overline{a_n}\,\overline{z^n} + \overline{a_{n-1}}\,\overline{z^{n-1}} + \cdots + \overline{a_1}\,\overline{z} + \overline{a_0} \qquad \text{(Ex. 44)}$$

$$= a_n \overline{z^n} + a_{n-1}\overline{z^{n-1}} + \cdots + a_1\overline{z} + a_0 \qquad \text{(The } a_i \text{ are real numbers.)}$$

$$= a_n\left(\overline{z}\right)^n + a_{n-1}\left(\overline{z}\right)^{n-1} + \cdots + a_1\left(\overline{z}\right) + a_0 \qquad \text{(Ex. 44)}$$

$$= p\left(\overline{z}\right)$$

Thus $p\left(\overline{z}\right) = 0$

47. $f(1) = -3$ and $f(2) = 21$; thus there exists a c in $(1, 2)$ such that $f(c) = 0$.

49. By the rational root theorem, the only possible rational roots of $x^2 - 5 = 0$ are ± 1 and ± 5. Since none of these are roots, and $\sqrt{5}$ is a root $\left((\sqrt{5})^2 - 5 = 0\right)$, it follows that $\sqrt{5}$ is an irrational number.

3.8 Decomposing Rational Functions (page 239)

1. $\dfrac{1}{x+1} + \dfrac{1}{x-1}$ **3.** $\dfrac{2}{x-3} - \dfrac{1}{x+2}$ **5.** $\dfrac{6}{x-4} + \dfrac{3}{x-2} - \dfrac{4}{x+1}$ **7.** $\dfrac{3}{x-2} + \dfrac{3}{(x-2)^2}$

9. $\dfrac{6}{5x+2} - \dfrac{3}{3x-4}$ **11.** $-\dfrac{1}{x} + \dfrac{2}{x+2} - \dfrac{1}{(x+2)^2}$ **13.** $x + \dfrac{1}{x-1} - \dfrac{1}{x+1}$

15. $3x^2 + 1 + \dfrac{1}{2x-1} - \dfrac{3}{(2x-1)^2}$ **17.** $10x - 5 + \dfrac{6}{x-3} + \dfrac{14}{x+2}$ **19.** $\dfrac{4}{x+7} - \dfrac{2}{x-5} + \dfrac{3}{x+2}$

21. $\dfrac{x-3}{x^2+3} - \dfrac{1}{x-1}$ **23.** $\dfrac{3}{x-1} + \dfrac{x-2}{x^2+x+1}$

CHAPTER 3   REVIEW EXERCISES (page 240)

1.

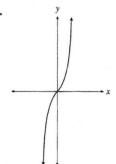

Domain: all x
Range: all x

2.

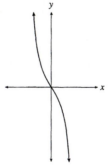

Domain: all x
Range: all y

3.

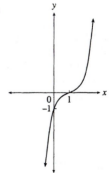

Domain: all x
Range: all y

4.

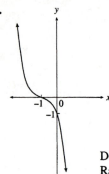

Domain: all x
Range: all y

5.

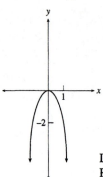

Domain: all x
Range: $y \leq 0$

6.

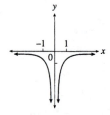

Domain: all x
Range: $y \leq 0$

7. $f(-x) = -f(x)$ **8.** $f(x) = f(-x)$ **9.** See page 178   **10.** See page 178   **11.** A ratio of polynomials

12. See page 183

13.

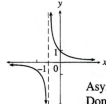

Asymptotes: $x = -1$, $y = 0$
Domain: $x \neq -1$
Range: $y \neq 0$
Concave down: $(-\infty, -1)$
Increasing up: $(-1, \infty)$

14.

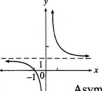

Asymptotes: $x = 0$, $y = 1$
Domain: $x \neq 0$
Range: $y \neq 1$
Concave down: $(-\infty, 0)$
Increasing up: $(0, \infty)$

15.

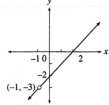

Asymptotes: $x = 0$, $y = 0$
Domain: $x \neq 0$
Range: $y < 0$
Concave down: $(-\infty, 0)$, $(0, \infty)$

16.

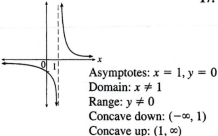

Asymptotes: $x = 1$, $y = 0$
Domain: $x \neq 1$
Range: $y \neq 0$
Concave down: $(-\infty, 1)$
Concave up: $(1, \infty)$

17.

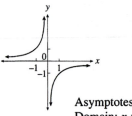

Asymptotes: $x = 0$, $y = 0$
Domain: $x \neq 0$
Range: $y \neq 0$
Concave up: $(-\infty, 0)$
Concave down: $(0, \infty)$

18.

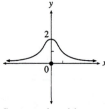

Domain: $x \neq -1$
Range: $y \neq -3$

19. See page 187   **20.** See page 188

21. $f(x) < 0$ on $(-5, 2)$; $f(x) > 0$ on $(-\infty, -5)$ and $(2, \infty)$

22. $f(x) < 0$ on $(-\infty, -2)$ and $(1, 2)$; $f(x) > 0$ on $(-2, 1)$ and $(2, \infty)$

23. Horizontal asymptote: $y = \frac{2}{3}$;   vertical asymptotes: $x = \pm\dfrac{\sqrt{3}}{3}$

24. Vertical asymptote: $x = 2$; no horizontal asymptote   **25.** Horizontal asymptote: $y = 0$;   vertical asymptote: $x = 2$

26. Horizontal asymptote: $y = 0$;   vertical asymptotes: $x = -1$

27.

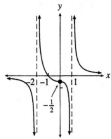

Asymptotes: $x = 1$, $x = -2$, $y = 0$

28.

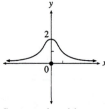

Symmetric with respect to the y-axis
Asymptote: $y = 0$

29. $y = x - 2$ **30.** $y = x - 7$ **31.** $x = 2$ **32.** $x = -\frac{1}{5}, x = 3$ **33.** See page 201 **34.** $x = 8$

35. $x = -4, x = 1$; domain: $x \neq 0$ **36.** $x = 5$; domain: $x \neq -5$ **37.** $c = \dfrac{ab - ra - rb}{r}$

38. $x \leq -\frac{1}{2}$ or $x > 2$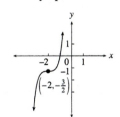

39. $x \leq \frac{2}{3}$ or $x > 2$ **40.** $(-\frac{3}{2}, 4), (-2, 3)$ **41.** 90 feet **42.** $2\frac{2}{9}$ hours **43.** See pages 210, 212

44. $A = ke^2$; $k = 6$ **45.** $A = kbh$; $k = \frac{1}{2}$ **46.** $y = 24$ **47.** $y = 3$ **48.** $z = 6$

49. $P = 60$ pounds per square inch **50.** The force will be 81 times as great. **51.** $x^2 + 2x - 5$; $r = 0$
52. $x^2 + 3x - 2$; $r = 0$ **53.** $2x^2 - 5x + 15$; $r = -23$ **54.** $3x^2 + 7x + 15$; $r = 28$
55. $x^3 + 2$; $r = -2$ **56.** $2x^3 - 6x^2 + 23x - 76$; $r = 236$ **57.** See page 220 **58.** See page 221
59. $p(3) = 1$ **60.** $f(3) = 27 - 27 - 3 + 3 = 0$ **61.** $f(-5) = -125 + 50 + 65 + 10 = 0$; $(x - 1)(x - 2)(x + 5)$

62. See page 225 **63.** $\pm 1, \pm\frac{1}{2}, \pm\frac{1}{3}, \pm\frac{1}{6}, \pm 2, \pm\frac{2}{3}, \pm 4, \pm\frac{4}{3}, \pm 8, \pm\frac{8}{3}$

64. $(2x + 1)(3x - 2)(x + 4)$ **65.** -1 (double root), $\pm\sqrt{2}$ **66.** -2 (double root), $\dfrac{1 \pm i\sqrt{3}}{2}$ **67.** See page 229
68. $\overline{z + w} = \overline{(2 - 3i) + (-3 + 4i)} = \overline{-1 + i} = -1 - i$;
$\bar{z} + \bar{w} = \overline{2 - 3i} + \overline{-3 + 4i} = 2 + 3i - 3 - 4i = -1 - i$

69. $\overline{zw} = \overline{(2 - 3i)(-3 + 4i)} = \overline{6 + 17i} = 6 - 17i$;
$\bar{z} \cdot \bar{w} = \overline{(2 - 3i)} \cdot \overline{(-3 + 4i)} = (2 + 3i)(-3 - 4i) = 6 - 17i$

70. 1 positive root, 4 negative roots; or 1 positive root, 2 negative roots, 2 imaginary roots; or 1 positive, 4 imaginary roots

71. $\dfrac{2}{x + 3} + \dfrac{3}{x - 1}$ **72.** $-\dfrac{1}{3(x + 1)} + \dfrac{10}{3(x + 4)}$ **73.** $\dfrac{1}{x + 1} - \dfrac{2}{x + 2} + \dfrac{4}{x - 3}$

74. $\dfrac{2}{x - 1} + \dfrac{5}{(x - 1)^2}$ **75.** $2x + 5 + \dfrac{59}{4(x - 3)} - \dfrac{7}{4(x + 1)}$

CHAPTER 3 STANDARD ANSWER TEST (page 243)

1. No asymptotes

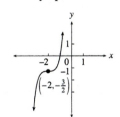

2. No asymptotes

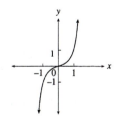

3. Asymptotes: $x = 2, y = 0$

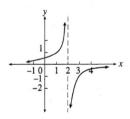

4. Asymptotes: $x = -1$, $x = 2$, $y = 0$

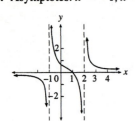

5.

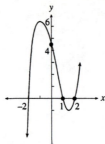

6. $f(x) < 0$ on both $(-\infty, -3)$ and $(0, 2)$;
$f(x) > 0$ on both $(-3, 0)$ and $(2, \infty)$

7.

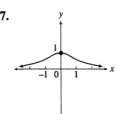

8.

Symmetric with respect
to the y-axis.
Asymptote: $y = 0$

9. $a = \dfrac{S}{n} - \dfrac{1}{2}(n - 1)d$ **10.** $-\dfrac{3}{2}$; 2   **11.** $x < -\dfrac{14}{3}$ **12.** 1   **13.** $5\dfrac{5}{7}$ feet   **14.** 90 minutes   **15.** $z = 2$

16. $V = 288\pi$ cubic inches

17.
$$-3\ \underline{\big|\ 2 + 5 + 0 - 1 - 21 + 7}$$
$$\ -6 + 3 - 9 + 30 - 27$$
$$\ \overline{2 - 1 + 3 - 10 + 9\ \big|-20}$$
quotient: $2x^4 - x^3 + 3x^2 - 10x + 9$
remainder: -20

18. When $p(x)$ is divided by $x - \dfrac{1}{3}$, the remainder is $\dfrac{2}{3}$. Then, by the remainder theorem, we have $p(\tfrac{1}{3}) = \tfrac{2}{3}$.

19. Since $p(\tfrac{1}{3}) \neq 0$, the factor theorem says that $x - \dfrac{1}{3}$ is not a factor of $p(x)$.

20.
$$2\ \underline{\big|\ 1 - 4 + 7 - 12 + 12}$$
$$\ +2 - 4 + 6 - 12$$
$$\ \overline{1 - 2 + 3 - 6\ \big|+\ 0} = r$$

Since $r = 0$, $x - 2$ is a factor of $p(x)$, we get
$$p(x) = (x - 2)(x^3 - 2x^2 + 3x - 6) = (x - 2)[x^2(x - 2) + 3(x - 2)] = (x - 2)(x^2 + 3)(x - 2)$$
$$= (x - 2)^2(x^2 + 3)$$

21. $(x + 3)^2(x^2 - x + 1)$ **22.** -1 (a double root), -3, 2   **23.** $-1, 3, \pm i$ **24.** $\dfrac{2}{x + 5} - \dfrac{1}{x - 5}$

25. $-\dfrac{1}{x - 1} + \dfrac{5}{x - 3} + \dfrac{2}{x + 2}$

CHAPTER 3   MULTIPLE CHOICE TEST   (page 244)

1. (d)   **2.** (b)   **3.** (a)   **4.** (c)   **5.** (b)   **6.** (e)   **7.** (d)   **8.** (a)   **9.** (b)   **10.** (a)
11. (d)   **12.** (c)   **13.** (b)   **14.** (c)   **15.** (a)   **16.** (a)   **17.** (c)   **18.** (d)   **19.** (b)   **20.** (b)

CHAPTER 4   CIRCLES, ADDITIONAL CURVES, AND THE ALGEBRA OF FUNCTIONS

4.1   Circles   (page 254)

1.

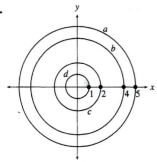

3. $(x - 2)^2 + y^2 = 25$; $C(2, 0), r = 5$ **5.** $(x - 1)^2 + (y - 3)^2 = 1$; $C(1, 3), r = 1$

7. $(x - 2)^2 + (y - 5)^2 = 1$; $C(2, 5), r = 1$ **9.** $(x - 4)^2 + (y - 0)^2 = 2$; $C(4, 0), r = \sqrt{2}$

11. $(x - 10)^2 + (y + 10)^2 = 100$; $C(10, -10), r = 10$ **13.** $\left(x + \frac{3}{4}\right)^2 + (y - 1)^2 = 9$; $C\left(-\frac{3}{4}, 1\right), r = 3$

15. $(x - 2)^2 + y^2 = 4$ **17.** $(x + 3)^2 + (y - 3)^2 = 7$

19.

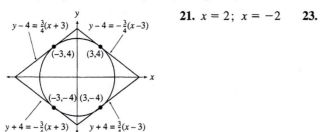

21. $x = 2$; $x = -2$ **23.**

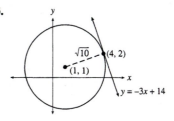

25.

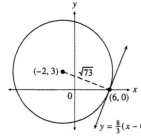

27.

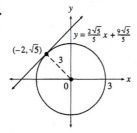

29. $\left(\frac{1}{2}, -2\right)$ **31.** $\left(-\frac{5}{2}, \frac{1}{2}\right)$ **33.** $y = -\frac{1}{2}x + 10$ **35.** $5x + 12y = 26$

37. **(a)** $y(x) = \sqrt{25 + 24x - x^2} - 5$; **(b)** $y(7) = 7$ **39.** $\left(\frac{2}{5}, \frac{16}{5}\right)$ **41.** $A(x) = 2x\sqrt{144 - x^2}$

43. $A(h) = h\left(10 + \sqrt{25 - h^2}\right)$

45. By Pythagorean theorem $AB = 5$.

The radii on BO, OA, and AB are $2, \frac{3}{2}$, and $\frac{5}{2}$, respectively. The sum of the areas of the semicircles on the legs is

$\frac{1}{2}\pi (2)^2 + \frac{1}{2}\pi \left(\frac{3}{2}\right)^2 = 2\pi + \frac{9}{8}\pi = \frac{25}{8}\pi$. The area of the semicircle on AB: $\frac{1}{2}\pi \left(\frac{5}{2}\right)^2 = \frac{25}{8}\pi$.

47. $x^2 + (y - 12)^2 = 64$ (in inches) or $x^2 + (y - 1)^2 = \frac{4}{9}$ (in feet)   **49.** $\left(\frac{1}{2}, \pm \frac{\sqrt{15}}{2} \right)$

4.2 Graphing Radical Functions (page 264)

1.

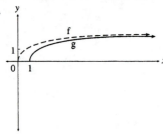

3.

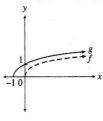

5.

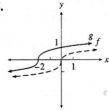

7.

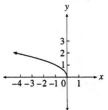

9. Domain $x \geq -2$
Increasing for $x \geq -2$; concave
down for $x > -2$.

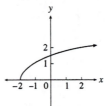

11. Domain: $x \geq 3$
Increasing for $x \geq 3$; concave
down for $x > 3$.

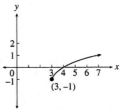

13. Domain $x \leq 0$
Decreasing for $x \leq 0$; concave
down for $x < 0$.

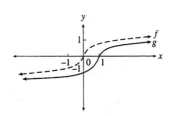

15. Domain: all real x
Increasing for all x; concave up
for $x < 0$; concave down for
$x > 0$.

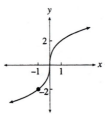

17. Domain: all real x
Decreasing for all x;
concave down for $x < 0$;
concave up for $x > 0$.

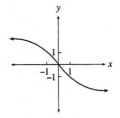

19. Domain: $x > 0$
Asymptotes: $x = 0$, $y = -1$
Decreasing and concave up
for $x > 0$.

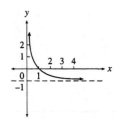

21. **(a)** $f(-x) = \dfrac{1}{\sqrt[3]{-x}} = \dfrac{1}{-\sqrt[3]{x}} = -\dfrac{1}{\sqrt[3]{x}} = -f(x)$

(b) all $x \neq 0$

(c)

x	$\frac{1}{27}$	$\frac{1}{8}$	1	8
y	3	2	1	$\frac{1}{2}$

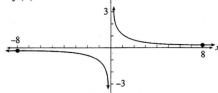

(d) $x = 0$; $y = 0$

23. $y = \sqrt[4]{x}$ is equivalent to $y^4 = x$ for $x \geq 0$.

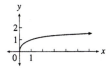

25.

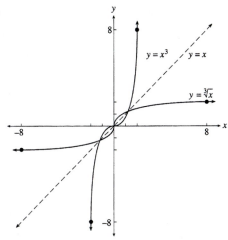

27.

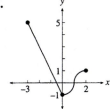

29. $\dfrac{\sqrt{4+h}-2}{h} = \dfrac{(\sqrt{4+h}-2)(\sqrt{4+h}+2)}{h(\sqrt{4+h}+2)} = \dfrac{4+h-4}{h(\sqrt{4+h}+2)} = \dfrac{1}{\sqrt{4+h}+2}$

31. **(a)** $d(x) = \sqrt{25+x^2} + \sqrt{100+(20-x)^2}$; **(b)** $t(x) = \dfrac{\sqrt{25+x^2}}{12} + \dfrac{\sqrt{100+(20-x)^2}}{10}$; **(c)** 2.4 hours

33. $d(x) = \sqrt{x^2 + \frac{1}{x}}$

4.3 Radical Equations and Graphs (page 273)

1. 17; domain: $x \geq 1$ **3.** $-2, 0$; domain: $x \leq -2$ or $x \geq 0$ **5.** $-1, 6$; domain: $x \leq -1$ or $x \geq 6$ **7.** $x = 10$

9. $x = 5$ **11.** $x = \pm 10$ **13.** No solution   **15.** $x = 2$ **17.** $x = \frac{2}{3}$ **19.** $x = \frac{5}{16}$ **21.** $x = 9$

23. $x = 2$ **25.** $x = 2$ **27.** $x = 4$ **29.** $x = 4$ **31.** $x = 0, x = 8$ **33.** $x = \frac{27}{8}$ **35.** $x = 1$

37. $x = 9$ **39.** $x = 4$ **41.** **(a)** All $x \geq 2$; **(b)** $f(x) > 0$ for all $x > 2$; **(c)** $x = 2$

43. **(a)** All real numbers; **(b)** $f(x) > 0$ for all $x \neq 2$; **(c)** $x = 2$

45. **(a)** All $x > 0$; **(b)** $f(x) > 0$ on $(0, \infty)$; **(c)** none

47. **(a)** All $x \neq 0$; **(b)** $f(x) < 0$ on both $(-\infty, -4), (0, 2)$; $f(x) > 0$ on both $(-4, 0), (2, \infty)$; **(c)** $x = -4, 2$

49. $x^{1/2} + (x - 4)\frac{1}{2}x^{-1/2} = x^{1/2} + \frac{x - 4}{2x^{1/2}} = \frac{2x + x - 4}{2\sqrt{x}} = \frac{3x - 4}{2\sqrt{x}}$

51. $\frac{1}{2}(4 - x^2)^{-1/2}(-2x) = \frac{-2x}{2(4 - x^2)^{1/2}} = -\frac{x}{\sqrt{4 - x^2}}$

53. $\frac{x}{3(x - 1)^{2/3}} + (x - 1)^{1/3} = \frac{x + 3(x - 1)}{3(x - 1)^{2/3}} = \frac{4x - 3}{3(\sqrt[3]{x - 1})^2}$

55. $f(x) = \frac{x - 1}{\sqrt{x}}$; $f(x) < 0$ on $(0, 1)$, $f(x) > 0$ on $(1, \infty)$

57. $f(x) = \frac{x(5x - 8)}{2\sqrt{x - 2}}$; $f(x) > 0$ on $(2, \infty)$

59. Domain: $x \leq -3$ or $x \geq 3$ **61.** Domain: $-3 \leq x \leq 3$ **63.** Domain: $x \geq 0$

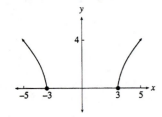

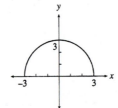

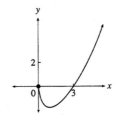

(Note: This is the upper half of the circle $x^2 + y^2 = 9$.)

65. $A = 4\pi r^2$; 16π **67.** $h = \sqrt{s^2 - r^2}$; 14.72 cm   **69.** $S = \frac{10}{h} + 2\sqrt{5\pi h}$ **71.** $(9, 3)$

73.

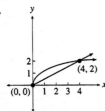

75. $y = \frac{5}{3}\sqrt{9 - x^2}$ **77.** $f = \frac{y}{\sqrt{xy} - x}$

4.4 Combining and Decomposing Functions   (page 284)

1. **(a)** $-1, 5, 4$; **(b)** $5x - 1$, all reals; **(c)** 4   **3.** **(a)** $-2, \frac{7}{2}, -7$; **(b)** $6x^2 - 5x - 6$, all reals; **(c)** -7

5. **(a)** $2, 1$; **(b)** $6x + 1$, all reals; **(c)** 1   **7.** **(a)** $x^2 + \sqrt{x}$, $x \geq 0$; **(b)** $x^{3/2}$, $x > 0$; **(c)** x, $x \geq 0$

9. **(a)** $x^3 - 1 + \frac{1}{x}$, all $x \neq 0$; **(b)** $x^4 - x$, all $x \neq 0$; **(c)** $\frac{1}{x^3} - 1$, all $x \neq 0$

11. **(a)** $x^2 + 6x + 8 + \sqrt{x - 2}$, $x \geq 2$; **(b)** $\frac{x^2 + 6x + 8}{\sqrt{x - 2}}$, $x > 2$; **(c)** $x + 6 + 6\sqrt{x - 2}$, $x \geq 2$

13. **(a)** $6x - 6$, all reals; **(b)** $-8x^2 + 22x - 5$, all reals; **(c)** $-8x + 19$, all reals

15. **(a)** $\frac{1}{2x} - 2x^2 + 1$, all $x \neq 0$; **(b)** $x - \frac{1}{2x}$, all $x \neq 0$; **(c)** $\frac{1}{4x^2 - 2}$, all $x \neq \pm\frac{1}{\sqrt{2}}$

17. $(f \circ g)(x) = (x - 1)^2$; $(g \circ f)(x) = x^2 - 1$ **19.** $(f \circ g)(x) = \dfrac{x + 3}{3 - x}$; $(g \circ f)(x) = 4 - \dfrac{6}{x}$

21. $(f \circ g)(x) = x^2$; $(g \circ f)(x) = x^2 + 2x$ **23.** $(f \circ g)(x) = 2$; $(g \circ f)(x) = 4$

25. (a) $\dfrac{1}{2\sqrt[3]{x} - 1}$; (b) $\dfrac{2}{\sqrt[3]{x}} - 1$; (c) $\dfrac{1}{\sqrt[3]{2x - 1}}$ **27.** $(f \circ f)(x) = x$; $(f \circ f \circ f)(x) = \dfrac{1}{x}$

(Other answers are possible for Exercises 29–51.)

29. $g(x) = 3x + 1$; $f(x) = x^2$ **31.** $g(x) = 1 - 4x$; $f(x) = \sqrt{x}$ **33.** $g(x) = \dfrac{x + 1}{x - 1}$; $f(x) = x^2$

35. $g(x) = 3x^2 - 1$; $f(x) = x^{-3}$ **37.** $g(x) = \dfrac{x}{x - 1}$; $f(x) = \sqrt{x}$ **39.** $g(x) = (x^2 - x - 1)^3$; $f(x) = \sqrt{x}$

41. $g(x) = 4 - x^2$; $f(x) = \dfrac{2}{\sqrt{x}}$ **43.** $f(x) = 2x + 1$; $g(x) = x^{1/2}$; $h(x) = x^3$

45. $f(x) = \dfrac{x}{x + 1}$; $g(x) = x^5$; $h(x) = x^{1/2}$ **47.** $f(x) = x^2 - 9$; $g(x) = x^2$; $h(x) = x^{1/3}$

49. $f(x) = x^2 - 4x + 7$; $g(x) = x^3$; $h(x) = -\sqrt{x}$ **51.** $f(x) = 2x - 11$; $g(x) = 1 + \sqrt{x}$; $h(x) = x^2$

53. $f(x) = (x + 1)^2$ **55.** $g(x) = \sqrt[3]{x}$ **57.** $x^2 - 2x$

59. $r(V) = \sqrt[3]{\dfrac{3V}{4\pi}}$; $(r \circ V)(t) = r(V(t)) = r(50t) = \sqrt[3]{\dfrac{150t}{4\pi}}$ = the length of the radius in feet after t seconds.

$(r \circ V)(10)$ is approximately 4.9 feet.

4.5 Inverse Functions   (page 295)

1.
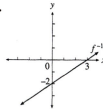
    3. Not one-to-one   **5.**
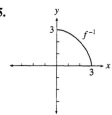

7. Not one-to-one   **9.** One-to-one   **11.** One-to-one   **13.** $\frac{3}{5}(x_1 - 4) + 6 = \frac{3}{5}(x_2 - 4) + 6$ leads to $x_1 = x_2$

15. $(f \circ g)(x) = \frac{1}{3}(3x + 9) - 3 = x$ **17.** $(f \circ g)(x) = \left(\sqrt[3]{x - 1} + 1\right)^3 = x$ **19.** $(f \circ g)(x) = \dfrac{1}{\dfrac{1}{x} + 1 - 1} = x$

$(g \circ f)(x) = 3\left(\frac{1}{3}x - 3\right) + 9 = x$ $(g \circ f)(x) = \sqrt[3]{(x + 1)^3} - 1 = x$ $(g \circ f)(x) = \dfrac{1}{\dfrac{1}{x - 1}} + 1 = x$

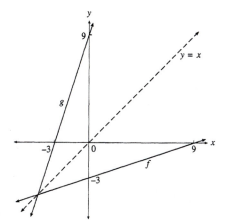

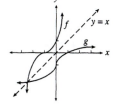

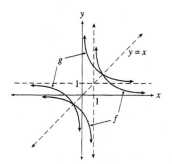

21. $g(x) = \sqrt[3]{x} + 5$ **23.** $g(x) = \frac{3}{2}x + \frac{3}{2}$ **25.** $g(x) = \sqrt[5]{x} + 1$ **27.** $g(x) = x^{5/3}$

29. $f^{-1}(x) = 2 + \dfrac{2}{x}$; $f(f^{-1}(x)) = f\left(2 + \dfrac{2}{x}\right) = \dfrac{2}{\left(2 + \dfrac{2}{x}\right) - 2} = \dfrac{2}{\dfrac{2}{x}} = x$

$f^{-1}(f(x)) = f^{-1}\left(\dfrac{2}{x-2}\right) = 2 + \dfrac{2}{\dfrac{2}{x-2}} = 2 + (x-2) = x$

31. $f^{-1}(x) = \dfrac{3}{x} - 2$; $f(f^{-1}(x)) = f\left(\dfrac{3}{x} - 2\right) = \dfrac{3}{\left(\dfrac{3}{x} - 2\right) + 2} = \dfrac{3}{\dfrac{3}{x}} = x$

$f^{-1}(f(x)) = f^{-1}\left(\dfrac{3}{x+2}\right) = \dfrac{3}{\dfrac{3}{x+2}} - 2 = (x+2) - 2 = x$

33. $f^{-1}(x) = x^{-1/5}$; $f(f^{-1}(x)) = f(x^{-1/5}) = (x^{-1/5})^{-5} = x$; $f^{-1}(f(x)) = f^{-1}(x^{-5}) = (x^{-5})^{-1/5} = x$

35. $(f \circ f)(x) = f(f(x)) = f\left(\dfrac{1}{x}\right) = \dfrac{1}{\dfrac{1}{x}} = x$ **37.** $(f \circ f)(x) = f(f(x)) = f\left(\dfrac{x}{x-1}\right) = \dfrac{\dfrac{x}{x-1}}{\dfrac{x}{x-1} - 1} = \dfrac{x}{x - (x-1)} = x$

39. $g(x) = \sqrt{x} - 1$ **41.** $g(x) = \dfrac{1}{x^2}$

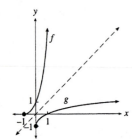

Domain of g: $x \geq 0$
Range of g: $y \geq -1$

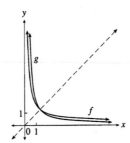

Domain of g: $x > 0$
Range of g: $y > 0$

43. $g(x) = \sqrt{x + 4}$ **45.** $g(x) = \sqrt{x + 4} + 2$ **47.** $y = mx + k$, where $m = -1$

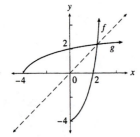

Domain of g: $x \geq -4$
Range of g: $y \geq 0$

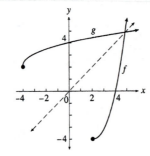

Domain of g: $x \geq -4$
Range of g: $y \geq 2$

CHAPTER 4   REVIEW EXERCISES   (page 297)

1. See page 248   **2.** $\left(\dfrac{x_1 + x_2}{2}, \dfrac{y_1 + y_2}{2}\right)$ **3.** (a) $x^2 + y^2 = r^2$; (b) $(x - h)^2 + (y - k)^2 = r^2$

4. $2\sqrt{29}$ **5.** $(1, 1)$ **6.** $(x - 2)^2 + (y + 5)^2 = 3$ **7.** $(-3, 1)$; 3   **8.** $(2, -1)$; 2   **9.** $y = -x + 3$

10. $(-1, 1)$; $\sqrt{20}$; $(x + 1)^2 + (y - 1)^2 = 20$

11.

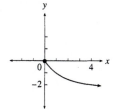

Domain: $x \geq 0$

12.

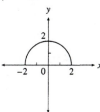

Domain: $-2 \leq x \leq 2$

13.

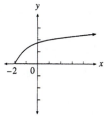

Domain: $x \geq -2$

14.

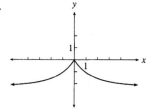

Domain: all x

15.

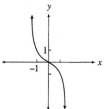

Domain: all x

16.

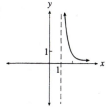

Domain: $x > 1$

17. $x = 1$, $y = 0$ **18.**

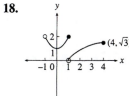

19. $x = 8$ **20.** $x = 9$ **21.** $x = 9$ **22.** $x = 12$

23.

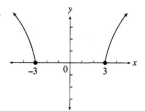

Domain: $x \leq -3$ or $x \geq 3$

24.

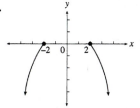

Domain: $x \leq -2$ or $x \geq 2$

25.

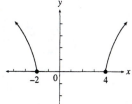

Domain: $x \leq -2$ or $x \geq 4$

26.

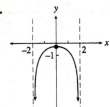

Domain: $-2 < x < 2$

27. $x > -1$; $x = 0$; $f(x) < 0$ on $(-1, 0)$ and $f(x) > 0$ on $(0, \infty)$

28. $(0, 0), (4, 2)$

29. $x = -27, x = 8$ **30.** $x = -\dfrac{1}{2}, x = 0$ **31.** $\dfrac{x + 5x - 6}{(5x - 6)^{4/5}} = \dfrac{6(x - 1)}{(5x - 4)^{4/5}}$

32. $\dfrac{3(x - 3)}{2\sqrt{x}}$; $f(x) < 0$ on $(0, 3)$, $f(x) > 0$ on $(3, \infty)$ **33.** $\sqrt{x + 1} + \dfrac{1}{x + 1}$; $x > -1$

34. $\sqrt{x + 1} - \dfrac{1}{x + 1}$; $x > -1$ **35.** $\dfrac{\sqrt{x + 1}}{x + 1}$; $x > -1$ **36.** $(x + 1)\sqrt{x + 1}$; $x > -1$ **37.** $\sqrt{2}$

38. $\frac{1}{3}$ **39.** $\sqrt{\dfrac{x + 2}{x + 1}}$ **40.** $\dfrac{1}{\sqrt{x + 1} + 1}$ **41.** $\dfrac{1}{(\sqrt{x} - 1)^2} = \dfrac{1}{x - 2\sqrt{x} + 1}$

42. Let $h(x) = g(f(x))$ where $f(x) = \sqrt{x^2 + x}$ and $g(x) = x^3$

43. Consider $t(g(f(x)))$ where $f(x) = x^2 + x$, $g(x) = x^3$ and $t(x) = \sqrt{x}$

44. Let $f(x) = x^3$ and $g(x) = \dfrac{1}{2x - 3}$

45. $d(y) = \frac{1}{4}\sqrt{16y^2 + 1}$, $y(t) = 45t$; $(d \circ y)(t) = \frac{1}{4}\sqrt{32400t^2 + 1}$ is the distance in miles that the car is from point P after t minutes. At $t = 3$, $d \approx 135$ miles.

46. See page 288   **47.** See page 288   **48.** See page 290

49. (a) Not one-to-one (b) Not one-to-one (c) One-to-one

50.

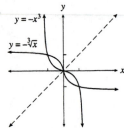

51.

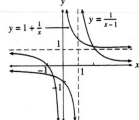

52. (a) $f^{-1}(x) = \dfrac{x}{2} - \dfrac{5}{2}$; (b) $\dfrac{1}{f(x)} = \dfrac{1}{2x + 5}$

53. $f(g(x)) = f\left(\dfrac{1}{2} - \dfrac{x}{2}\right) = 1 - 2\left(\dfrac{1}{2} - \dfrac{x}{2}\right) = 1 - 1 + x = x;$

$g(f(x)) = g(1 - 2x) = \dfrac{1}{2} - \dfrac{1 - 2x}{2} = \dfrac{1}{2} - \dfrac{1}{2} + \dfrac{2x}{2} = x$

54. $y = g(x) = \sqrt[3]{x} - 2$; domain: all x; range: all y **55.** $y = g(x) = (x - 1)^{3/2}$; domain: $x \geq 1$; range: $y \geq 0$

56.

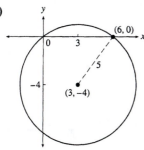

CHAPTER 4   STANDARD ANSWER TEST   (page 300)

1. (a)

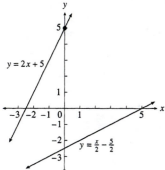

(b) $y = -\dfrac{3}{4}x + \dfrac{9}{2}$ **2.** $C\left(-\dfrac{1}{2}, 7\right); r = 5$

3. Domain: all reals

4. Domain: $x > 0$
Asymptotes: $y = 2$, $x = 0$

5. Domain: all reals

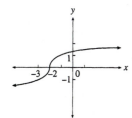

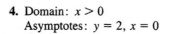

6. Domain: $x > 3$
Asymptotes: $x = 3,\ y = 0$

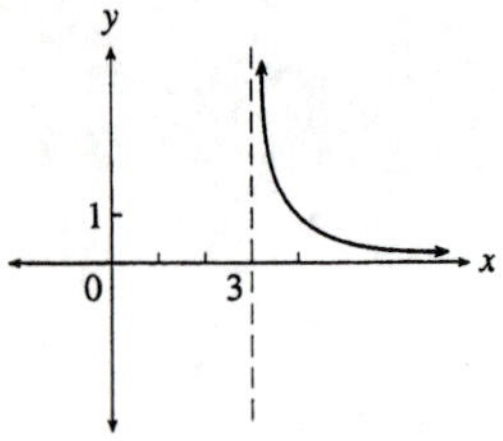

7.

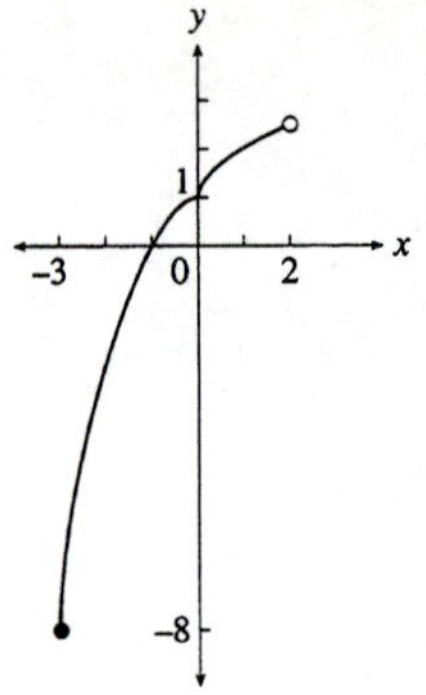

8. $\frac{2}{9}$ **9.** $-\frac{64}{27}, \frac{1}{8}$ **10.** $x = 16$

11. $(x+1)^{1/2}(2x) + (x+1)^{-1/2}\left(\frac{x^2}{2}\right) = 2x(x+1)^{1/2} + \frac{x^2}{2(x+1)^{1/2}} = \frac{4x(x+1)+x^2}{2\sqrt{x+1}} = \frac{5x^2+4x}{2\sqrt{x+1}} = \frac{x(5x+4)}{2\sqrt{x+1}}$

12. $f(x) < 0$ on $(2, 4)$; $f(x) > 0$ on $(-\infty, 2)$ and $(4, \infty)$ **13.** $f(x) < 0$ on $(0, 4)$; $f(x) > 0$ on $(4, \infty)$

14. $-2, 1$; domain: $x \le -2$ or $x \ge 1$

15. (a) $\frac{1}{x^2-1} + \sqrt{x+2}$; all $x \ge -2$ and $x \ne \pm 1$; **(b)** $\frac{1}{x^2-1} - \sqrt{x+2}$; all $x \ge -2$ and $x \ne \pm 1$

16. (a) $\frac{1}{(x^2-1)\sqrt{x+2}}$; all $x > -2$ and $x \ne \pm 1$; **(b)** $\frac{\sqrt{x+2}}{x^2-1}$; all $x \ge -2$ and $x \ne \pm 1$

17. Shift 4 units to the right and reflect in the x-axis.

18. (a) $\frac{1}{17}$; 4; $\frac{4}{17}$; **(b)** $\frac{1}{37}$; $\frac{2}{\sqrt{82}}$

19. $(f \circ g)(x) = \frac{1}{1-x}$; all $x \ge 0$ and $x \ne 1$; $(g \circ f)(x) = \frac{1}{\sqrt{1-x^2}}$; $-1 < x < 1$

(Other answers are possible for Exercises 20 and 21.)

20. $g(x) = x - 2$; $f(x) = x^{2/3}$ **21.** $h(x) = 2x - 1$; $g(x) = x^{3/2}$; $f(x) = \frac{1}{x}$ **22. (a)** $f^{-1}(x) = \frac{x+2}{3}$;

(b)

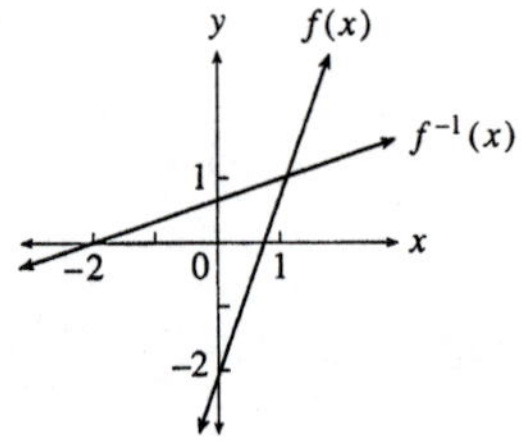

23. $g(x) = (x+1)^3$; $(f \circ g)(x) = f(g(x)) = f((x+1)^3) = \sqrt[3]{(x+1)^3 - 1} = x$

24. All real numbers; $x = -3, x = 0$ **25.** $(2, 2), (10, 6)$

CHAPTER 4 MULTIPLE CHOICE TEST (page 301)

1. (d) **2.** (b) **3.** (c) **4.** (d) **5.** (a) **6.** (e) **7.** (c) **8.** (b) **9.** (c) **10.** (d) **11.** (e)
12. (c) **13.** (d) **14.** (a) **15.** (d) **16.** (b) **17.** (b) **18.** (c) **19.** (c) **20.** (c)

CHAPTER 5   EXPONENTIAL AND LOGARITHMIC FUNCTIONS

5.1 Exponential Functions and Equations   (page 312)

1.

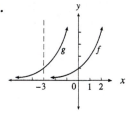

Horizontal asymptote: $y = 0$

3.

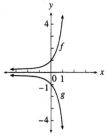

Horizontal asymptote: $y = 0$

5.

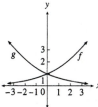

Horizontal asymptote: $y = 0$

7.

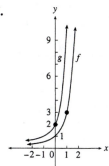

Horizontal asymptote: $y = 0$

9.

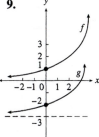

Horizontal asymptotes: $y = 0, y = -3$

11.

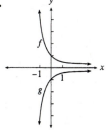

Horizontal asymptote: $y = 0$

13.

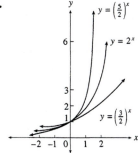

15.

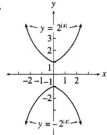

17. **(a)** $y = 2^{x-1}$ **(b)** $y = 2^{x+2} + 1$ **19.** $x = 6$ **21.** $x = \pm 3$ **23.** $x = 1$ **25.** $x = -2, x = 1$ **27.** $x = -5$

29. $x = \frac{1}{2}$ **31.** $x = \frac{3}{2}$ **33.** $x = -\frac{1}{2}$ **35.** $x = \frac{2}{3}$ **37.** $x = \frac{5}{4}$ **39.** $x = -1, x = 2$

41.

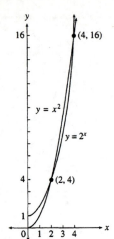

43. $x = \frac{1}{2}$

45. (a)

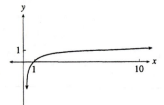

x	1.4	1.41	1.414	1.4142	1.41421
3^x	4.6555	4.7070	4.7277	4.7287	4.7288

estimate: 4.729; calculator: 4.728804…

(b)

x	1.7	1.73	1.732	1.7320	1.73205
3^x	6.4730	6.6899	6.7046	6.7046	6.7050

estimate: 6.705; calculator: 6.704991…

(c)

x	2.2	2.23	2.236	2.2360	2.23606
2^x	4.5948	4.6913	4.7109	4.7109	4.7111

estimate: 4.711; calculator: 4.711113…

(d)

x	3.1	3.14	3.141	3.1415	3.14159
4^x	73.5167	77.7085	77.8163	77.8702	77.8799

estimate: 77.880; calculator: 77.88023…

47. $a = 8$; $y = 8 \cdot (2^x) = 2^{x+3}$

5.2 Logarithmic Functions (page 320)

1. $g(x) = \log_4 x$

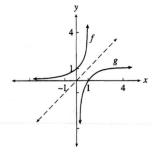

3. $g(x) = \log_{1/3} x$

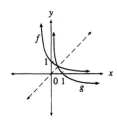

5. Shift 2 units left; $x > -2$; $x = -2$ **7.** Shift 2 units upward; $x > 0$; $x = 0$.

9. Domain: all $x > 0$

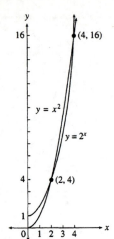

11. Domain: all $x > 0$

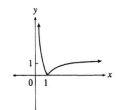

13. Domain: all $x \neq 0$

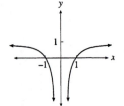

15. (a) $y = \log_2(x + 2)$ **(b)** $y = \log_2 x - 2$ **17.** $\log_2 256 = 8$ **19.** $\log_{1/3} 3 = -1$

21. $\log_{17} 1 = 0$ **23.** $10^{-4} = 0.0001$ **25.** $(\sqrt{2})^2 = 2$ **27.** $12^{-3} = \frac{1}{1728}$ **29.** 4   **31.** 4   **33.** 3   **35.** -3

37. 4   **39.** -3 **41.** $\frac{1}{216}$ **43.** 5   **45.** 4   **47.** $\frac{1}{3}$ **49.** $\frac{2}{3}$ **51.** 9   **53.** -2 **55.** $-\frac{1}{3}$ **57.** 9

59. 6   **61.** -1

63. $g(x) = 3^x - 3, (f \circ g)(x) = \log_3(3^x - 3 + 3) = \log_3 3^x = x; \; (g \circ f)(x) = 3^{\log_3(x+3)} - 3 = (x + 3) - 3 = x$

65.

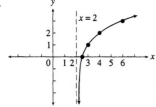

67. $f(g(x)) = f(2^{x-1} - 2) = \log_2[2(2^{x-1} - 2) + 4] = \log_2(2^x - 4 + 4) = \log_2 2^x = x;$

$g(f(x)) = g[\log_2(2x + 4)] = 2^{\log_2(2x+4)-1} - 2 = 2^{\log_2(2x+4)}(2^{-1}) - 2 = (2x + 4)(2^{-1}) - 2 = x + 2 - 2 = x$

5.3 The Laws of Logarithms (page 330)

1. $\log_b 3 + \log_b x - \log_b(x + 1)$ **3.** $\frac{1}{2}\log_b(x^2 - 1) - \log_b x = \frac{1}{2}\log_b(x + 1) + \frac{1}{2}\log_b(x - 1) - \log_b x$ **5.** $-2 \log_b x$

7. $\log_b(2x - 5) - 3\log_b x$ **9.** $\log_b \dfrac{x + 1}{x + 2}$ **11.** $\log_b \sqrt{\dfrac{x^2 - 1}{x^2 + 1}}$ **13.** $\log_b \dfrac{x^3}{2(x + 5)}$ **15.** $\log_b(x - 3)$

17. $\log_b\left(\dfrac{x^2 + 2x + 4}{x - 2}\right)^{1/3}$ **19.** $\log_b 27 + \log_b 3 = \log_b 81$ (Law 1)   **21.** $-2 \log_b \frac{4}{9} = \log_b\left(\frac{4}{9}\right)^{-2}$ (Law 3)

$\log_b 243 - \log_b 3 = \log_b 81$ (Law 2) $\quad\quad\quad\quad = \log_b \frac{81}{16}$

23. (a) 0.6020; **(b)** 0.9030; **(c)** -0.3010 **25. (a)** 1.6811; **(b)** -0.1761; **(c)** 2.0970

27. (a) 0.2330; **(b)** 1.9515; **(c)** 1.4771   **29.** 3.4534   **31.** 2   **33.** 64   **35.** x^3 **37.** 20   **39.** $\frac{1}{20}$

41. 17   **43.** 8   **45.** 2   **47.** 7   **49.** 1.01   **51.** 5

53. Let $r = \log_b M$ and $s = \log_b N$. Then $b^r = M$ and $b^s = N$. Divide:

$\dfrac{M}{N} = \dfrac{b^r}{b^s} = b^{r-s}$. Convert to log form and substitute: $\log_b \dfrac{M}{N} = r - s = \log_b M - \log_b N$

55. $-1; \; 0$ **57.** 8   **59.** 10^{-6} watts per sq cm   **61.** 100 times as great

63.

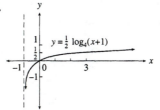

Domain: $x > 0$

Asymptote: $x = 0$

65.

Domain: $x > -1$

Asymptote: $x = -1$

67.

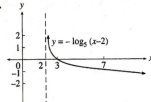

Domain: $x > 2$

Asymptote: $x = 2$

69.

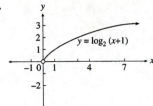

Domain: $x > 0$

No asymptote, no intercepts

5.4 The Natural Exponential and Logarithmic Functions (page 340)

1.

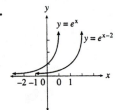

3.

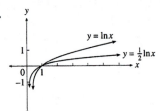

5.

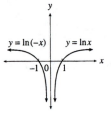

7.

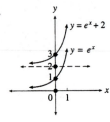

9.

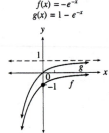

11.

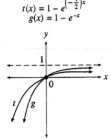

13. Since $f(x) = 1 + \ln x$, shift 1 unit upward.   **15.** Since $f(x) = \frac{1}{2} \ln x$, multiply the ordinates by $\frac{1}{2}$.

17. Since $f(x) = \ln(x - 1)$, shift 1 unit to the right.   **19.** All $x > -2$; $x = -1$ **21.** All $x > \frac{1}{2}$; $x = 1$

23. All $x > 1$ except $x = 2$; none   **25.** $\ln 5 + \ln x - \ln(x^2 - 4) = \ln 5 + \ln x - \ln(x + 2) - \ln(x - 2)$

27. $\ln(x - 1) + 2 \ln(x + 3) - \frac{1}{2} \ln(x^2 + 2)$ **29.** $\frac{3}{2} \ln x + \frac{1}{2} \ln(x + 1)$ **31.** $\ln \sqrt{x}(x^2 + 5)$ **33.** $\ln(x^2 - 1)^3$

35. $\ln \dfrac{\sqrt{x}}{(x - 1)^2 \sqrt[3]{x^2 + 1}}$ **37.** $\sqrt{x}$ **39.** $\dfrac{1}{x^2}$ **41.** 1   **43.** $x = -100 \ln 27$ **45.** $x = \frac{1}{3}$ **47.** $x = 0$

49. $x = 8$ **51.** $x = 3$ **53.** $x = 8$ **55.** $x = \dfrac{e^2}{1 + e^2}$

57.

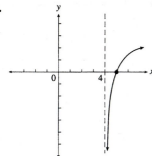

59.

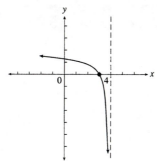

61.

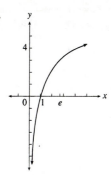

Domain: all $x > 4$
x-intercept: 5
Asymptote: $x = 4$

Domain: all $x < 4$
x-intercept: 3; y-intercept: 1.4
Asymptote: $x = 4$

Domain: all $x > 0$
x-intercept: 1
Asymptote: $x = 0$

(Other answers are possible for Exercises 63–71.)

63. Let $g(x) = -x^2 + x$ and $f(x) = e^x$. Then $(f \circ g)(x) = f(g(x)) = f(-x^2 + x) = e^{-x^2+x} = h(x)$.

65. Let $g(x) = \dfrac{x}{x + 1}$ and $f(x) = \ln x$. Then $(f \circ g)(x) = f(g(x)) = f\left(\dfrac{x}{x+1}\right) = \ln \dfrac{x}{x+1} = h(x)$.

67. Let $g(x) = \ln x$ and $f(x) = \sqrt[3]{x}$. Then $(f \circ g)(x) = f(g(x)) = f(\ln x) = \sqrt[3]{\ln x} = h(x)$.

69. Let $h(x) = 3x - 1$, $g(x) = x^2$, and $f(x) = e^x$. Then $(f \circ g \circ h)(x) = f(g(h(x))) = f(g(3x - 1)) = f((3x - 1)^2) = e^{(3x-1)^2} = F(x)$

71. Let $h(x) = e^x + 1$, $g(x) = \sqrt{x}$, and $f(x) = \ln x$. Then $(f \circ g \circ h)(x) = f(g(h(x))) = f(g(e^x + 1)) = f(\sqrt{e^x + 1}) = \ln\sqrt{e^x + 1} = F(x)$.

73. All real numbers; $x = \frac{1}{2}$ **75.** All $x > 0$; $x = \dfrac{1}{e}$

77. $\ln\left(\dfrac{x}{4} - \dfrac{\sqrt{x^2 - 4}}{4}\right) = \ln\left(\dfrac{x - \sqrt{x^2 - 4}}{4}\right) = \ln\left(\dfrac{x - \sqrt{x^2 - 4}}{4} \cdot \dfrac{x + \sqrt{x^2 - 4}}{x + \sqrt{x^2 - 4}}\right)$

$= \ln\left(\dfrac{x^2 - (x^2 - 4)}{4(x + \sqrt{x^2 - 4})}\right) = \ln\left(\dfrac{1}{x + \sqrt{x^2 - 4}}\right) = -\ln(x + \sqrt{x^2 - 4})$

79. $x = \ln(y + \sqrt{y^2 + 1})$ **81.** $\dfrac{\ln(2 + h) - \ln 2}{h} = \dfrac{\ln\left(\dfrac{2 + h}{2}\right)}{h} = \dfrac{1}{h}\ln\left(1 + \dfrac{h}{2}\right)$

83. $\dfrac{\ln e^{3x} - \ln e^6}{x - 2} = \dfrac{3x \ln e - 6 \ln e}{x - 2} = \dfrac{3x - 6}{x - 2} = 3$ **85.** $x > \ln 500$

5.5 Applications: Exponential Growth and Decay (page 349)

1. 2.585   **3.** -1.292 **5.** 1.232   **7.** -0.5 **9.** 5.492   **11.** -6.966 **13.** 2.107   **15.** 0.405

17. 2008.55   **19.** 27.32   **21.** $\frac{1}{2} \ln 100$ **23.** $\frac{1}{4} \ln \frac{1}{3}$ **25.** 667,000   **27.** 1.83 days   **29.** 17.33 years

31. **(a)** $\frac{1}{5} \ln \frac{4}{5}$; **(b)** 6.4 grams; **(c)** 15.5 years   **33.** 93.2 seconds   **35.** 18.47 years   **37.** 4200 years

39. 115,000 years   **41.** Approximately 3 years   **43.** 5.43 hours
45. **(a)** \$15,605; **(b)** \$15,657; **(c)** \$15,677; **(d)** \$15,682; **(e)** \$15,683
47. **(a)** \$13,655; **(b)** \$13,686; **(c)** \$13,699; **(d)** \$13,702; **(e)** \$13,703
49. 7.7 years; 5.8 years   **51.** 13.86%   **53.** 8.69 years   **55.** \$2914   **57.** \$11,009   **59.** $y = Ae^{kx}$ **61.** 4.64%

CHAPTER 5 REVIEW EXERCISES (page 352)

1. See page 311

2.

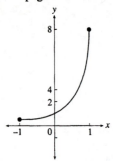

3.

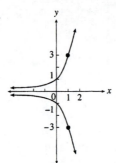

4. Shift the graph of $y = 5^x$ to the left 3 units to obtain $y = 5^{x+3}$ and up 3 units to obtain $y = 5^x + 3$.

5.

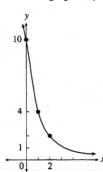

6. $x_1 = x_2$

7. $x = \frac{4}{3}$ **8.** $x = \sqrt[3]{3}$

9. $x = -5$ **10.** $x = -3$

11. The graph is a straight line parallel to and one unit above the x-axis.

12. (a), (c), and (d)

13.

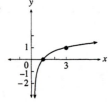

14.

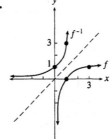

15.

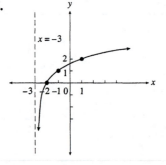

Domain: all $x > -3$

16.

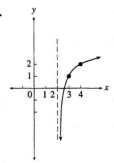

17. (a) $3^4 = 81$; (b) $2^5 = 32$ **18.** (a) $\log_{0.1}\left(\dfrac{1}{1000}\right) = 3$; (b) $\log_{1/64}16 = -\dfrac{2}{3}$ **19.** $y = 6$ **20.** $x = \dfrac{1}{32}$

21. $x = \dfrac{1}{25}$ **22.** See page 323   **23.** $\log_b A + \log_b B - 2\log_b C$ **24.** $\log_b \dfrac{x^3}{\sqrt{x+2}}$ **25.** See page 324

26. See page 325   **27.** 6.2288   **28.** 3.0478   **29.** 2.7799   **30.** $x = 10$ **31.** $x = 11$ **32.** $x = 1$

33. $x = 97$

34. $\log_2 8(x+2) = \log_2 8 + \log_2(x+2) = 3 + \log_2(x+2)$; shift the graph of $y = \log_2 x$ to the left 2 units, and then up 3 units; asymptote: $x = -2$; x-intercept $= -\dfrac{15}{8}$

35. $y = f(x) = \log \dfrac{x^2 - 2x}{x - 2} = \log x$; $x > 2$.

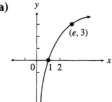

36. See page 333   **37.** See page 335   **38.** See page 336

39. See page 336   **40.** Shift the graph of $y = \ln x$ to the right 2 units.

41. $\ln 2 + 2\ln x - \ln(x - 4)$ **42.** $2\ln x + \dfrac{1}{2}\ln(x - 1) - \ln(x + 5) - \ln(x - 5)$

43. $\ln 3(x + 1)$; $x > 1$ **44.** $\ln \dfrac{x(x^2 + 4)^2}{\sqrt{x + 7}}$ **45.** $x > \dfrac{2}{3}$; $x = \dfrac{2}{3}$ **46.** (a) x^2; (b) x^3 **47.** (a) $x = \dfrac{3}{7}$; (b) $x = 2e$

48. $h(x) = (f \circ g)(x)$, where $f(x) = \ln x$ and $g(x) = 3x - 2$ **49.** $x = \dfrac{e}{e - 1}$ **50.** $x = 0$, $x = \dfrac{1}{2}$; domain all x

51. i − (d); ii − (e); iii − (f); iv − (a); v − (b); vi − (c)

52. (a)   (b)

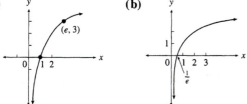

53. $x = 7.044$ **54.** $x = -6.322$ **55.** $x = 2.092$ **56.** 3004.17   **57.** 149.41   **58.** 72.3 seconds
59. 144.6 seconds   **60.** See page 347   **61.** See page 348   **62.** $6326.60   **63.** $6341.21   **64.** $6356.25
65. 8.664 years   **66.** See page 344

CHAPTER 5   STANDARD ANSWER TEST   (page 355)

1. (i) a; (ii) c; (iii) e; (iv) h; (v) d; (vi) b **2.** (a) $5^3 = 125$; (b) $\log_{16}8 = \dfrac{3}{4}$ **3.** (a) $x = \dfrac{1}{2}$; (b) $x = -2$, $x = 3$

4. $x = -7$ **5.** 10   **6.** (a) $b = \dfrac{2}{3}$; (b) $b = -2$

7. (a) Domain: all real x; range: $y > 0$; (b) increasing for all x; (c) concave up for all x;
(d) y-intercept $= 1$; no x-intercept; (e) horizontal asymptote: $y = 0$

8.

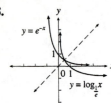

9. All real x; $y = -4$ **10.** All $x > -4$; $x = -4$

11. (a) $\log_b x + 10 \log_b(x^2 + 1)$; **(b)** $3 \ln x - \ln(x + 1) - \frac{1}{2} \ln(x^2 + 2)$

12. (a) $\log_7 10x^2$; **(b)** $\ln \dfrac{\sqrt{x}}{(x + 2)^2}$

13. 6.2288

14.

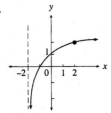

Domain: $x > -2$
x-intercept: -1, y-intercept: 1
Asymptote: $x = -2$

15.

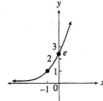

Domain: all x
y-intercept: e
Asymptote: $y = 0$

16. $x = 5$ **17.** $x = \frac{10}{3}$ **18.** Shift up 2 units and to the right 1 unit   **19.** 0.5805   **20.** 17.3 years

21. $2000(1.02)^{24}$ **22.** 8.75 years   **23.** \$12,083   **24.** $k = \frac{1}{3} \ln \frac{3}{2} = 0.1352$ **25.** $x = \dfrac{e - 2}{e + 2}$

CHAPTER 5 MULTIPLE CHOICE TEST (page 357)

1. (d)   **2.** (a)   **3.** (c)   **4.** (e)   **5.** (a)   **6.** (c)   **7.** (e)   **8.** (d)   **9.** (b)   **10.** (d)   **11.** (c)   **12.** (b)
13. (a)   **14.** (a)   **15.** (c)   **16.** (b)   **17.** (c)   **18.** (b)   **19.** (c)   **20.** (d)

CHAPTER 6 THE TRIGONOMETRIC FUNCTIONS

6.1 Angle Measure (page 367)

1. $\dfrac{\pi}{4}$ **3.** $\dfrac{\pi}{2}$ **5.** $\dfrac{3\pi}{2}$ **7.** $\dfrac{5\pi}{6}$ **9.** $\dfrac{5\pi}{4}$ **11.** $\dfrac{7\pi}{6}$ **13.** $\dfrac{11\pi}{6}$ **15.** $\dfrac{5\pi}{12}$ **17.** $\dfrac{5\pi}{9}$; 1.75

19. $\dfrac{17\pi}{9}$; 5.93   **21.** 180°   **23.** 360°   **25.** 100°   **27.** 120°   **29.** 225°   **31.** 300°   **33.** 50°

35. 12°; 12.0°   **37.** $\left(\dfrac{540}{\pi}\right)^{\circ}$; 171.9°   **39.** $s = 2\pi$ **41.** $\theta = \dfrac{\pi}{2}$ **43.** $\theta = \dfrac{4\pi}{3}$ **45.** 12π square centimeters

47. 54π square centimeters   **49.** 126π square centimeters   **51.** 8 square inches   **53.** $\dfrac{\pi}{2}$; 1.6   **55.** 11°

57. 150.8 square inches   **59.** $\dfrac{1225\pi}{6}$ square feet   **61. (a)** 15,439 mph; 22,644 ft/sec   **(b)** $\dfrac{8\pi}{7}$ rad/hr

63. (a) $\dfrac{\pi}{30}$ rad/min   **(b)** 2π rad/hr   **65. (a)** 440π ft/min   **(b)** approximately 3.8 min   **67.** 1.4 rad

6.2 Trigonometric Functions   (page 379)

1. $\dfrac{\sqrt{2}}{2}$ **3.** $\dfrac{1}{2}$ **5.** $\dfrac{1}{\sqrt{3}}$ **7.** II; $\dfrac{\pi}{3}$ **9.** III; $\dfrac{\pi}{4}$ **11.** IV; $\dfrac{\pi}{3}$ **13.** II; $\dfrac{5\pi}{6}$ **15.** IV; $\dfrac{5\pi}{3}$

17. IV; $\dfrac{7\pi}{4}$

19. $\sin\dfrac{2\pi}{3} = \dfrac{\sqrt{3}}{2}$ $\qquad$ $\csc\dfrac{2\pi}{3} = \dfrac{2}{\sqrt{3}}$

$\cos\dfrac{2\pi}{3} = -\dfrac{1}{2}$ $\qquad$ $\sec\dfrac{2\pi}{3} = -2$

$\tan\dfrac{2\pi}{3} = -\sqrt{3}$ $\qquad$ $\cot\dfrac{2\pi}{3} = -\dfrac{1}{\sqrt{3}}$

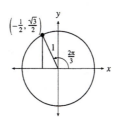

reference angle $= \frac{\pi}{3}$

21. $\sin\left(-\dfrac{7\pi}{4}\right) = \dfrac{1}{\sqrt{2}}$ $\quad$ $\csc\left(-\dfrac{7\pi}{4}\right) = \sqrt{2}$

$\cos\left(-\dfrac{7\pi}{4}\right) = \dfrac{1}{\sqrt{2}}$ $\quad$ $\sec\left(-\dfrac{7\pi}{4}\right) = \sqrt{2}$

$\tan\left(-\dfrac{7\pi}{4}\right) = 1$ $\qquad$ $\cot\left(-\dfrac{7\pi}{4}\right) = 1$

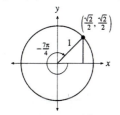

reference angle $= \frac{\pi}{4}$

23. $\sin\left(-\dfrac{7\pi}{6}\right) = \dfrac{1}{2}$ $\qquad$ $\csc\left(-\dfrac{7\pi}{6}\right) = 2$

$\cos\left(-\dfrac{7\pi}{6}\right) = -\dfrac{\sqrt{3}}{2}$ $\quad$ $\sec\left(-\dfrac{7\pi}{6}\right) = -\dfrac{2}{\sqrt{3}}$

$\tan\left(-\dfrac{7\pi}{6}\right) = -\dfrac{1}{\sqrt{3}}$ $\quad$ $\cot\left(-\dfrac{7\pi}{6}\right) = -\sqrt{3}$

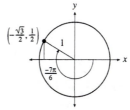

reference angle $= \frac{\pi}{6}$

25. $(0, -1)$; $\sin\theta = -1$; $\cos\theta = 0$; $\tan\theta$ is undefined; $\cot\theta = 0$, $\sec\theta$ is undefined; $\csc\theta = -1$.
27. $(0, 1)$; $\sin\theta = 1$; $\cos\theta = 0$; $\tan\theta$ is undefined; $\cot\theta = 0$; $\sec\theta$ is undefined; $\csc\theta = 1$.

29. -1 $\quad$ **31.** $-\dfrac{\sqrt{3}}{2}$ $\quad$ **33.** Undefined $\quad$ **35.** 0 $\quad$ **37.** Undefined $\quad$ **39.** 0 $\quad$ **41.** 0.9737 $\quad$ **43.** 0.7151

45. $-\dfrac{1}{\sqrt{2}}$ $\quad$ **47.** 10.983 $\quad$ **49.** 0.8391 $\quad$ **51.** -0.9877 $\quad$ **53.** -0.5774 $\quad$ **55.** 0.3007

	$\sin\theta$	$\cos\theta$	$\tan\theta$	$\cot\theta$	$\sec\theta$	$\csc\theta$
57.	0.4894	0.8721	0.5612	1.7820	1.1467	2.0434
59.	0.4571	-0.8894	-0.5139	-1.9458	-1.1243	2.1877

61. $P_1\left(\dfrac{\sqrt{3}}{2}, \dfrac{1}{2}\right)$; $P_2\left(\dfrac{\sqrt{2}}{2}, \dfrac{\sqrt{2}}{2}\right)$; $P_3\left(\dfrac{1}{2}, \dfrac{\sqrt{3}}{2}\right)$; $P_4\left(-\dfrac{\sqrt{3}}{2}, -\dfrac{1}{2}\right)$; $P_5\left(-\dfrac{\sqrt{2}}{2}, -\dfrac{\sqrt{2}}{2}\right)$; $P_6\left(-\dfrac{1}{2}, -\dfrac{\sqrt{3}}{2}\right)$

63. (a) $\left(\dfrac{2}{3}\right)^2 + \left(\dfrac{\sqrt{5}}{3}\right)^2 = \dfrac{4}{9} + \dfrac{5}{9} = \dfrac{9}{9} = 1$

(c) $\sin \theta = \dfrac{\sqrt{5}}{3}$; $\cos \theta = \dfrac{2}{3}$; $\tan \theta = \dfrac{\sqrt{5}}{2}$;

(b)

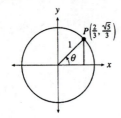

$\cot \theta = \dfrac{2}{\sqrt{5}}$; $\sec \theta = \dfrac{3}{2}$; $\csc \theta = \dfrac{3}{\sqrt{5}}$

65. $y = -\dfrac{\sqrt{13}}{4} = -0.9014$; -2.0817 **67.** $\dfrac{\pi}{2}$ **69.** 0; π **71.** $\dfrac{\pi}{6}$; $\dfrac{5\pi}{6}$ **73.** $\dfrac{3\pi}{4}$; $\dfrac{5\pi}{4}$

75. III; $\cos \theta = -\dfrac{\sqrt{2}}{2}$; $\cot \theta = 1$; $\sec \theta = -\sqrt{2}$; $\csc \theta = -\sqrt{2}$

77. IV; $\cos \theta = \dfrac{\sqrt{3}}{2}$; $\tan \theta = -\dfrac{\sqrt{3}}{3}$; $\sec \theta = \dfrac{2\sqrt{3}}{3}$; $\csc \theta = -2$ **79.** $-\dfrac{2\sqrt{2}}{3}$ **81.** $\dfrac{4\sqrt{15}}{15}$

83. $AB = \dfrac{AB}{1} = \dfrac{AB}{OA} = \dfrac{y}{x} = \tan \theta$; OB

6.3 Graphing the Sine and Cosine Functions (page 391)

1.

x	$-\pi$	$-\dfrac{5\pi}{6}$	$-\dfrac{2\pi}{3}$	$-\dfrac{\pi}{2}$	$-\dfrac{\pi}{3}$	$-\dfrac{\pi}{6}$	0
$y = \sin x$	0	$-\dfrac{1}{2}$	$-\dfrac{\sqrt{3}}{2}$	-1	$-\dfrac{\sqrt{3}}{2}$	$-\dfrac{1}{2}$	0

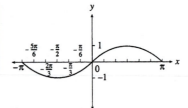

3. Shift the graph of f to the right $\dfrac{\pi}{2}$ units.

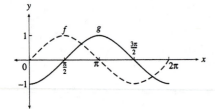

5. Shift the graph of f to the right $\dfrac{\pi}{3}$ units.

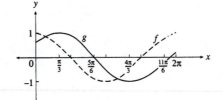

7. Shift the graph f to the left π units and multiply the ordinates by 2.

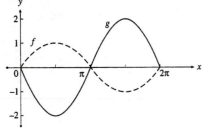

9.

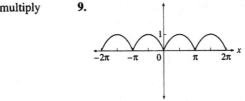

11.

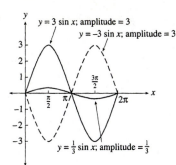

$y = 3 \sin x$; amplitude = 3

$y = -3 \sin x$; amplitude = 3

$y = \frac{1}{3} \sin x$; amplitude = $\frac{1}{3}$

13. Amplitude = 1; period = π.

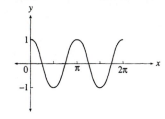

15. Amplitude = $\dfrac{3}{2}$; period = $\dfrac{\pi}{2}$.

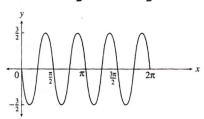

17. Amplitude = 1; period = 4π.

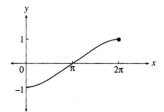

19. Period = 8π

21. Period = 2

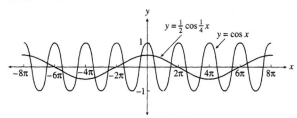

$y = \frac{1}{2} \cos \frac{1}{4} x$ $y = \cos x$

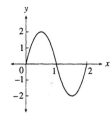

23. $a = 1, b = 3$; $y = \sin 3x$ **25.** $a = -3, b = \frac{1}{2}$; $y = -3 \sin \frac{1}{2} x$

	Amplitude	Period	Phase Shift
27.	5	$\frac{2\pi}{3}$	$\frac{\pi}{6}$
29.	2	π	$-\frac{\pi}{2}$
31.	$\frac{3}{2}$	8π	4

33. Amplitude = 1; period = $\dfrac{\pi}{2}$; phase shift = $\dfrac{\pi}{4}$.

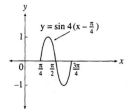

$y = \sin 4\left(x - \frac{\pi}{4}\right)$

35. Amplitude = 2; period = π; phase shift = $\dfrac{\pi}{4}$.

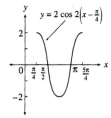

$y = 2 \cos 2\left(x - \frac{\pi}{4}\right)$

37. Amplitude $= \dfrac{5}{2}$; period $= 4\pi$; phase shift $= -\dfrac{\pi}{2}$.

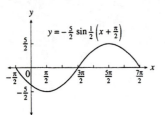

39. Shift the graph of $y = \sin\frac{1}{2}(x + \pi)$ three units upward.

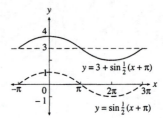

41.

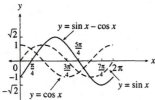

43.

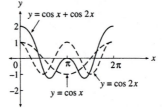

45.

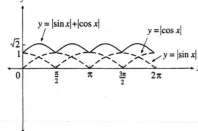

47. $(f \circ g)(x) = 2\cos x + 5$; $(g \circ f)(x) = \cos(2x + 5)$ **49.** $g(x) = 5x^2$; $f(x) = \cos x$

51. $h(x) = 1 - 2x$; $g(x) = \sqrt[3]{x}$; $f(x) = \cos x$ **53.** $f(x + 2p) = f((x + p) + p) = f(x + p) = f(x)$

6.4 Graphing Other Trigonometric Functions (page 401)

1.

x	-1.4	-1.3	$-\frac{\pi}{3}$	$-\frac{\pi}{4}$	$-\frac{\pi}{6}$	0
$y = \tan x$	-5.8	-3.6	$-\sqrt{3}$	-1	$-\frac{\sqrt{3}}{3}$	0

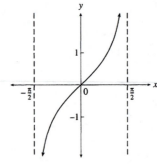

3. (a) $\cot(-x) = \dfrac{\cos(-x)}{\sin(-x)} = \dfrac{\cos x}{-\sin x} = -\dfrac{\cos x}{\sin x} = -\cot x$

(b) $\csc(-x) = \dfrac{1}{\sin(-x)} = \dfrac{1}{-\sin x} = -\dfrac{1}{\sin x} = -\csc x$

(c) $\csc(x + 2\pi) = \dfrac{1}{\sin(x + 2\pi)} = \dfrac{1}{\sin x} = \csc x$

5. Shift the graph of g to the left $\dfrac{\pi}{2}$ units.

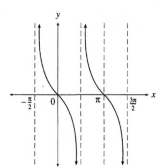

7. Shift the graph of g to the right $\dfrac{\pi}{3}$ units.

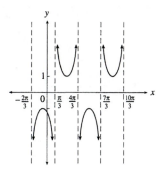

9. Period $= \dfrac{\pi}{3}$

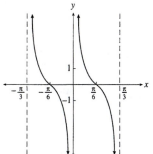

11. Period $= 2\pi$

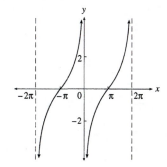

13. Period $= \dfrac{\pi}{2}$

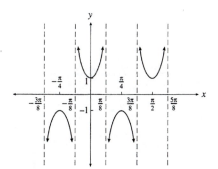

15. Period $= 6\pi$

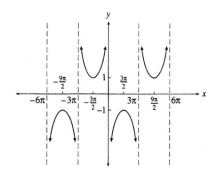

17. Period $= \dfrac{4\pi}{3}$

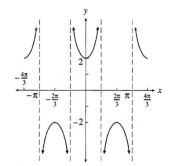

19. Period $= \dfrac{\pi}{2}$;   phase shift $= -\dfrac{\pi}{4}$.

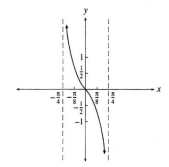

21. Period $= 4\pi$; phase shift $= \dfrac{\pi}{2}$.

23. $a = 1, b = \dfrac{2}{3}, h = -\dfrac{\pi}{4}$ **25.** $a = \dfrac{1}{2}, b = 3, h = \dfrac{\pi}{6}$

27.

x	0.5	0.1	0.01	0.001	0.0001	0.00001
csc x	2	10	100	1000	10,000	100,000

csc $x \longrightarrow \infty$ as $x \longrightarrow 0^+$

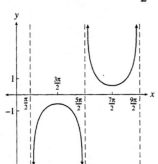

29.

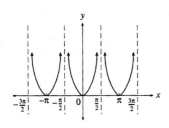

31. $(f \circ g)(x) = \tan^2 x$; $(g \circ f)(x) = \tan x^2$

33. $(f \circ g \circ h)(x) = e^{\sqrt{\sec x}}$; $(g \circ f \circ h)(x) = \sqrt{e^{\sec x}}$; $(h \circ g \circ f)(x) = \sec \sqrt{e^x}$

35. $h(x) = 2x + 1, g(x) = \tan x$; $f(x) = \sqrt{x}$ **37.** $0 < QP = \sin \theta < \overset{\frown}{AP} = \theta$; divide by θ to get $0 < \dfrac{\sin \theta}{\theta} < 1$

39.

θ	1	0.5	0.25	0.1	0.01
$\dfrac{\sin \theta}{\theta}$	0.841471	0.958851	0.989616	0.998334	0.999983

$\dfrac{\sin \theta}{\theta}$ appears to be getting close to 1 as θ approaches 0.

6.5 Inverse Trigonometric Functions (page 414)

1. 0   **3.** $-\dfrac{\pi}{2}$ **5.** $-\dfrac{\pi}{4}$ **7.** $\dfrac{3\pi}{4}$ **9.** 1.56   **11.** 0.59   **13.** 2.23   **15.** Undefined   **17.** 0.57

19. $\dfrac{x}{2}$ **21.** 0.7840   **23.** 0.0830   **25.** 0   **27.** $\dfrac{\sqrt{3}}{2}$ **29.** Undefined   **31.** 0

33. Sin and arcsin are inverse functions.

35. Domain: $-1 \le x \le 1$.
Range: $-\pi \le y \le \pi$.

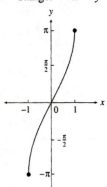

37. Domain: all real x
Range: $2 - \dfrac{\pi}{2} < y < 2 + \dfrac{\pi}{2}$.

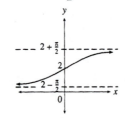

39. Domain: $-\dfrac{1}{2} \le x \le \dfrac{1}{2}$.
Range: $-\dfrac{\pi}{4} \le y \le \dfrac{\pi}{4}$.

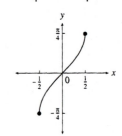

41. Domain: $-1 \leq x \leq 1$.   **43.** -0.6364 **45.** 0.9955 **47.** 0.7778 **49.** $-\frac{3}{2}$ **51.** -2
Range: $-1 \leq y \leq 1$.

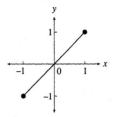

53. Let $y = \arcsin x$. Then $x = \sin y$; $-x = -\sin y = \sin(-y)$; $-y = \arcsin(-x)$; $-\arcsin x = \arcsin(-x)$.

55. $\theta = \arctan\dfrac{6}{7} - \arctan\dfrac{1}{2} = 0.24$

57. Let $\theta = \arcsin x$, $-1 \leq x \leq 1$. Then $\sin \theta = x$, where $-\dfrac{\pi}{2} \leq x \leq \dfrac{\pi}{2}$, and $\sin^2\theta = x^2$. From Exercise 82, page 381,

$\sin^2\theta + \cos^2\theta = 1$ so that $\cos^2\theta = 1 - \sin^2\theta = 1 - x^2$, which gives $\cos \theta = \pm\sqrt{1 - x^2}$. But $\cos \theta \geq 0$, since

$-\dfrac{\pi}{2} \leq \theta \leq \dfrac{\pi}{2}$. Therefore, $\cos \theta = \sqrt{1 - x^2}$ and replacing θ by $\arcsin x$ gives $\cos(\arcsin x) = \sqrt{1 - x^2}$.

59. $(f \circ g)(x) = \arcsin(3x + 2)$; $(g \circ f)(x) = 3 \arcsin x + 2$ **61.** $g(x) = \arccos x$; $f(x) = \ln x$

CHAPTER 6   REVIEW EXERCISES (page 416)

1. See page 362.   **2.** (a) $\dfrac{23\pi}{36}$ (b) $\dfrac{9\pi}{4}$ (c) $\dfrac{\pi}{120}$ **3.** (a) $144°$ (b) $510°$ (c) $\left(\dfrac{36}{\pi}\right)°$ **4.** 7π cm   **5.** $300°$

6. $A = \frac{1}{2}r^2\theta$ **7.** $\dfrac{100\pi}{3}$ cm^2 **8.** $120°$ **9.** 10.9 in.   **10.** 300 ft/sec

11. (a) $17{,}593$ mph (b) $\dfrac{4\pi}{3}$ rad/hr   **12.** $(4 - \pi)$ cm^2

	$p(x, y)$	$\sin \theta$	$\cos \theta$	$\tan \theta$	$\cot \theta$	$\sec \theta$	$\csc \theta$
13.	$\left(-\frac{1}{\sqrt{2}}, \frac{1}{\sqrt{2}}\right)$	$\frac{1}{\sqrt{2}}$	$-\frac{1}{\sqrt{2}}$	-1	-1	$-\sqrt{2}$	$\sqrt{2}$
14.	$\left(\frac{\sqrt{3}}{2}, -\frac{1}{2}\right)$	$-\frac{1}{2}$	$\frac{\sqrt{3}}{2}$	$-\frac{1}{\sqrt{3}}$	$-\sqrt{3}$	$\frac{2}{\sqrt{3}}$	-2
15.	$\left(-\frac{1}{2}, -\frac{\sqrt{3}}{2}\right)$	$-\frac{\sqrt{3}}{2}$	$-\frac{1}{2}$	$\sqrt{3}$	$\frac{1}{\sqrt{3}}$	-2	$-\frac{2}{\sqrt{3}}$
16.	$\left(\frac{1}{2}, -\frac{\sqrt{3}}{2}\right)$	$-\frac{\sqrt{3}}{2}$	$\frac{1}{2}$	$-\sqrt{3}$	$-\frac{1}{\sqrt{3}}$	2	$-\frac{2}{\sqrt{3}}$

	sin θ	cos θ	tan θ	cot θ	sec θ	csc θ
17.	0	1	0	undefined	1	undefined
18.	−1	0	undefined	0	undefined	−1
19.	1	0	undefined	0	undefined	1
20.	0	−1	0	undefined	−1	undefined
21.	0.9659	0.2588	3.7321	0.2679	3.8637	1.0353
22.	0.8387	−0.5446	−1.5399	−0.6494	−1.8361	1.1924
23.	−0.2419	0.9703	−0.2493	−4.0108	1.0306	−4.1336
24.	0.3420	−0.9397	−0.3640	−2.7475	−1.0642	2.9238

25. −0.5736 **26.** 0.0872 **27.** $\sqrt{3}$ **28.** 0.7212 **29.** −1.0086 **30.** 1.0111 **31.** 0.3249 **32.** −1

33. −1 **34.** Undefined **35.** Undefined **36.** $\sqrt{2}$ **37.** −2 **38.** $\dfrac{1}{\sqrt{2}}$ **39.** −0.3420 **40.** 0.9260

41. $\dfrac{3\pi}{2}$ **42.** $\dfrac{\pi}{6}, \dfrac{11\pi}{6}$ **43.** $\dfrac{3\pi}{4}, \dfrac{5\pi}{4}$ **44.** $\dfrac{5\pi}{6}, \dfrac{11\pi}{6}$ **45.** −90°, −270° **46.** −120°, −240°

47. −45°, −135° **48.** $-\dfrac{4}{\sqrt{7}}$ **49.** $-\dfrac{3}{\sqrt{7}}$ **50.** $\dfrac{1}{2}$ **51.** $-\dfrac{\sqrt{55}}{8}$

52. Domain: all real numbers; range: $-1 \le y \le 1$; symmetric through the origin (an odd function).

53. $\cos \theta = \sin\left(\theta + \dfrac{\pi}{2}\right)$

54.

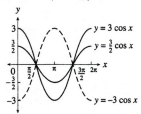

55.

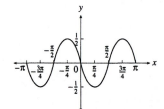

56.

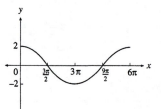

57. $a = -\dfrac{3}{2}, b = 3;\ y = -\dfrac{3}{2}\sin 3x$

58. $a = -\dfrac{1}{2}, b = 2;\ y = -\dfrac{1}{2}\cos 2x$

59.

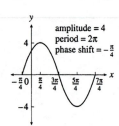

60.

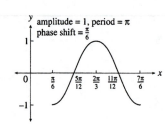

	Amplitude	Period	Phase Shift
61.	4	$\frac{\pi}{2}$	$\frac{\pi}{2}$
62.	3	3π	0
63.	$\frac{2}{3}$	π	$\frac{\pi}{4}$
64.	$\frac{5}{2}$	4π	$-\frac{\pi}{2}$
65.	$\frac{1}{3}$	$\frac{2\pi}{3}$	$\frac{\pi}{3}$

66. See page 397.   **67.** See page 400.

68. $\csc x = \dfrac{1}{\sin x}$; take reciprocals of the ordinates

of $y = \sin x$ to locate the points on the curve
$y = \csc x$ for $x \neq k\pi$.

69. See page 398.   **70.** See page 400.

71. Decreasing on the intervals $\left(0, \dfrac{\pi}{2}\right]$ and $\left[\dfrac{3\pi}{2}, 2\pi\right)$;

increasing on the intervals $\left[\dfrac{\pi}{2}, \pi\right)$ and $\left(\pi, \dfrac{3\pi}{2}\right]$ concave up
on $(0, \pi)$; concave down on $(\pi, 2\pi)$.

72. $\sec(-x) = \dfrac{1}{\cos(-x)} = \dfrac{1}{\cos x} = \sec x$

73. Period $= \dfrac{\pi}{2}$

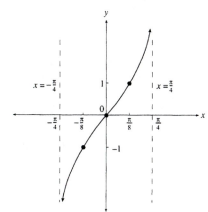

74. Period $= \dfrac{\pi}{2}$, phase shift $= \dfrac{\pi}{4}$.

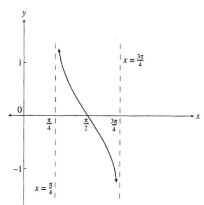

75. Period $= 4\pi$, phase shift $= \dfrac{\pi}{4}$.

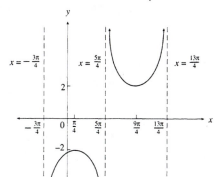

76. Period $= 2\pi$; Phase shift $= \dfrac{\pi}{2}$; Asymptote: $x = \dfrac{3\pi}{2}$

77. Period $= \dfrac{\pi}{2}$; Phase shift $= \dfrac{\pi}{8}$; Asymptote: $x = \dfrac{\pi}{8}$

78. $a = -1$, $b = 2$, $h = 0$

79. Reflect the negative parts of $y = \sec x$ through the x-axis to obtain
$y = |\sec x|$. Then reflect all of $y = |\sec x|$ through the x-axis
to obtain $y = -|\sec x|$. (Note: The same result is obtained by
reflecting the positive parts of $y = \sec x$ through the x-axis.)

80.

	Domain	Range
(a)	$-1 \leq x \leq 1$	$-\frac{\pi}{2} \leq y \leq \frac{\pi}{2}$
(b)	$-1 \leq x \leq 1$	$0 \leq y \leq \pi$
(c)	all real numbers	$-\frac{\pi}{2} < y < \frac{\pi}{2}$

81. $-\dfrac{\pi}{4}$ **82.** $\dfrac{\pi}{3}$

83. Undefined   **84.** 0.3627   **85.** $\dfrac{5}{12}$ **86.** 0.41 rad   **87.** 1.21 rad   **88.** Undefined   **89.** 0

90. Domain: $-2 \le x \le 2$; range: $-\dfrac{3\pi}{2} \le y \le \dfrac{3\pi}{2}$

91. **(a)** -0.2727 **(b)** -0.2606 **92.** **(a)** 0.875 **(b)** 0.3780   **93.** -0.3 **94.** 0.18 rad

95. Let $\theta = \arccos x$, $-1 \le x \le 1$. Then $\cos \theta = x$, where $0 \le \theta \le \pi$, and $\cos^2\theta = x^2$. Since $\sin^2\theta + \cos^2\theta = 1$, $\sin^2\theta = 1 - \cos^2\theta = 1 - x^2$, which gives $\sin \theta = \pm\sqrt{1 - x^2}$. But $\sin \theta \ge 0$ since $0 \le \theta \le \pi$. Therefore, $\sin \theta = \sqrt{1 - x^2}$ and substituting $\arccos x$ for θ gives $\sin(\arccos x) = \sqrt{1 - x^2}$.

CHAPTER 6   STANDARD ANSWER TEST   (page 419)

1. **(a)** $75°$; **(b)** $\left(\dfrac{270}{\pi}\right)°$; **(c)** $\dfrac{7\pi}{4}$; **(d)** $\dfrac{\pi}{9}$ **2.** $2\sqrt{3} - \pi$ square units

3. **(a)** 492π rad/min; **(b)** 533π ft/min; **(c)** $\dfrac{533\pi}{176}$ miles   **4.** $\dfrac{20\pi}{3}$ **5.** **(a)** $\dfrac{\sqrt{3}}{2}$; **(b)** 0; **(c)** $-\dfrac{1}{\sqrt{2}}$

6. **(a)** $\dfrac{\pi}{4}$; **(b)** $\dfrac{\pi}{6}$ **7.** **(a)** $-\dfrac{1}{\sqrt{3}}$; **(b)** -1; **(c)** 2   **8.** -2

9. $\left(-\dfrac{2}{5}\right)^2 + \left(\dfrac{\sqrt{21}}{5}\right)^2 = \dfrac{4}{25} + \dfrac{21}{25} = 1;$

$\sin \theta = \dfrac{\sqrt{21}}{5}$ $\csc \theta = \dfrac{5}{\sqrt{21}}$

$\cos \theta = -\dfrac{2}{5}$ $\sec \theta = -\dfrac{5}{2}$

$\tan \theta = -\dfrac{\sqrt{21}}{2}$ $\cot \theta = -\dfrac{2}{\sqrt{21}}$

10. **(a)** Domain: All reals
Range: $-1 \le y \le 1$

(b) Increasing on $\left[0, \dfrac{\pi}{2}\right]$ and $\left[\dfrac{3\pi}{2}, 2\pi\right]$,

Decreasing on $\left[\dfrac{\pi}{2}, \dfrac{3\pi}{2}\right]$,

Concave down on $(0, \pi)$,
Concave up on $(\pi, 2\pi)$.

11. Amplitude = 2;
period = π.

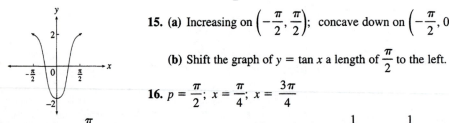

12. Period = $\dfrac{2\pi}{3}$; amplitude = $\dfrac{1}{2}$; phase shift = $\dfrac{\pi}{6}$.

13. Period = π; amplitude = 2; phase shift = $-\dfrac{\pi}{2}$.   **14.** $a = \frac{1}{2}, b = 3, y = \frac{1}{2}\cos 3x$

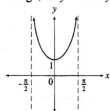

15. **(a)** Increasing on $\left(-\dfrac{\pi}{2}, \dfrac{\pi}{2}\right)$; concave down on $\left(-\dfrac{\pi}{2}, 0\right)$ and concave up on $\left(0, \dfrac{\pi}{2}\right)$.

(b) Shift the graph of $y = \tan x$ a length of $\dfrac{\pi}{2}$ to the left.

16. $p = \dfrac{\pi}{2}$; $x = \dfrac{\pi}{4}$; $x = \dfrac{3\pi}{4}$

17. Domain: all $x \ne \dfrac{\pi}{2} + k\pi$

Range; all $y \ge 1$ or $y \le -1$.

18. $\csc(-x) = \dfrac{1}{\sin(-x)} = \dfrac{1}{-\sin x}$

$= -\dfrac{1}{\sin x} = -\csc x$

19. $h(x) = 2x, g(x) = \csc x, f(x) = x^3$

20. **(a)** Domain: $-1 \le x \le 1$

Range: $-\dfrac{\pi}{2} \le y \le \dfrac{\pi}{2}$

(b) Domain: $-1 \le x \le 1$
Range: $0 \le y \le \pi$

21. (a) Domain: all reals

Range: $-\dfrac{\pi}{2} < y < \dfrac{\pi}{2}$.

(b) Symmetric through the origin; $y = \pm\dfrac{\pi}{2}$.

22. (a) $\dfrac{\pi}{4}$; **(b)** π; **(c)** $-\dfrac{\pi}{6}$ **23. (a)** .57; **(b)** 1.76; **(c)** $-.23$

24. (a) $\dfrac{1}{\sqrt{2}}$; **(b)** 0.5; **(c)** x^2 **25.** $x = -\dfrac{3}{8}$

CHAPTER 6 MULTIPLE CHOICE (page 421)

1. (a) **2.** (c) **3.** (c) **4.** (a) **5.** (d) **6.** (b) **7.** (d) **8.** (a) **9.** (d) **10.** (a) **11.** (d)
12. (b) **13.** (e) **14.** (c) **15.** (b) **16.** (d) **17.** (b) **18.** (c) **19.** (b) **20.** (c)

CHAPTER 7   RIGHT TRIANGLE TRIGONOMETRY, IDENTITIES, AND EQUATIONS

7.1 Trigonometry of Acute Angles (page 431)

	$\sin\theta$	$\cos\theta$	$\tan\theta$	$\cot\theta$	$\sec\theta$	$\csc\theta$
1.	$\dfrac{8}{17}$	$\dfrac{15}{17}$	$\dfrac{8}{15}$	$\dfrac{15}{8}$	$\dfrac{17}{15}$	$\dfrac{17}{8}$
3.	$\dfrac{\sqrt{3}}{2}$	$\dfrac{1}{2}$	$\sqrt{3}$	$\dfrac{1}{\sqrt{3}}$	2	$\dfrac{2}{\sqrt{3}}$
5.	$\dfrac{12}{13}$	$\dfrac{5}{13}$	$\dfrac{12}{5}$	$\dfrac{5}{12}$	$\dfrac{13}{5}$	$\dfrac{13}{12}$

7.

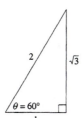

9.

11.

	sin A	cos A	tan A	cot A	sec A	csc A
13.	$\frac{3}{5}$	$\frac{4}{5}$	$\frac{3}{4}$	$\frac{4}{3}$	$\frac{5}{4}$	$\frac{5}{3}$
15.	$\frac{4}{\sqrt{41}}$	$\frac{5}{\sqrt{14}}$	$\frac{4}{5}$	$\frac{5}{4}$	$\frac{\sqrt{41}}{5}$	$\frac{\sqrt{41}}{4}$
17.	$\frac{1}{\sqrt{5}}$	$\frac{2}{\sqrt{5}}$	$\frac{1}{2}$	2	$\frac{\sqrt{5}}{2}$	$\sqrt{5}$
19.	$\frac{3}{4}$	$\frac{\sqrt{7}}{4}$	$\frac{3}{\sqrt{7}}$	$\frac{\sqrt{7}}{3}$	$\frac{4}{\sqrt{7}}$	$\frac{4}{3}$
21.	$\frac{\sqrt{21}}{5}$	$\frac{2}{5}$	$\frac{\sqrt{21}}{2}$	$\frac{2}{\sqrt{21}}$	$\frac{5}{2}$	$\frac{5}{\sqrt{21}}$
23.	$\frac{21}{29}$	$\frac{20}{29}$	$\frac{21}{20}$	$\frac{20}{21}$	$\frac{29}{20}$	$\frac{29}{21}$
	cos B	**sin B**	**cot B**	**tan B**	**csc B**	**sec B**

25. 1 **27.** $\dfrac{x^2}{z^2}$ **29.** 1 **31.** 1

33. $\cos A = \dfrac{\sqrt{7}}{4}$; $\tan A = \dfrac{3}{\sqrt{7}}$; $\cot A = \dfrac{\sqrt{7}}{3}$; $\sec A = \dfrac{4}{\sqrt{7}}$; $\csc A = \dfrac{4}{3}$

35. $\sin A = \dfrac{9}{41}$; $\cos A = \dfrac{40}{41}$; $\cot A = \dfrac{40}{9}$; $\sec A = \dfrac{41}{40}$; $\csc A = \dfrac{41}{9}$

37. $\sin B = \dfrac{3}{\sqrt{11}}$; $\cos B = \dfrac{\sqrt{2}}{\sqrt{11}}$; $\tan B = \dfrac{3}{\sqrt{2}}$; $\cot B = \dfrac{\sqrt{2}}{3}$; $\csc B = \dfrac{\sqrt{11}}{3}$

39. $\dfrac{5}{13}$ **41.** $\dfrac{3}{\sqrt{7}}$ **43.** $(\sin A)(\cos A) = \left(\sqrt{1-x^2}\right)(x) = x\sqrt{1-x^2}$

45. $\dfrac{4\sin^2 A}{\cos A} = \dfrac{4\left(\dfrac{x}{4}\right)^2}{\dfrac{\sqrt{16-x^2}}{4}} = \dfrac{x^2}{\sqrt{16-x^2}}$ **47.** $(\sin\theta)(\tan\theta) = \left(\dfrac{x}{2}\right)\left(\dfrac{x}{\sqrt{4-x^2}}\right) = \dfrac{x^2}{2\sqrt{4-x^2}}$

49. $(\tan^2\theta)(\cos^2\theta) = \left(\dfrac{x}{4}\right)^2\left(\dfrac{4}{\sqrt{x^2+16}}\right)^2 = \dfrac{x^2}{x^2+16}$ **51.** $A(s) = \dfrac{\sqrt{3}}{4}s^2$ **53.** $A(x) = 2(9-x^2)$

55. $A(x) = \dfrac{\sqrt{3}}{36}x^2 + \dfrac{(100-x)^2}{4\pi}$; domain: $0 < x < 100$

57. (a) $\sin\theta = x$; $\csc\theta = \dfrac{1}{x}$; $\cos\theta = \sqrt{1-x^2}$; $\sec\theta = \dfrac{1}{\sqrt{1-x^2}}$; $\tan\theta = \dfrac{x}{\sqrt{1-x^2}}$; $\cot\theta = \dfrac{\sqrt{1-x^2}}{x}$

(b) $\sin\theta = \dfrac{x}{\sqrt{1+x^2}}$; $\csc\theta = \dfrac{\sqrt{1+x^2}}{x}$; $\cos\theta = \dfrac{1}{\sqrt{1+x^2}}$; $\sec\theta = \sqrt{1+x^2}$; $\tan\theta = x$; $\cot\theta = \dfrac{1}{x}$

7.2 Right Triangle Trigonometry (page 439)

1. $x = \dfrac{25}{\sqrt{3}}$ **3.** $x = 15\sqrt{2}$ **5.** $x = 20\sqrt{3}$ **7.** $x = \dfrac{8}{\sqrt{3}}$ **9.** $\theta = \dfrac{\pi}{4}$ **11.** 81.0°   **13.** 65.3°

15. 6.3°   **17.** 12.7   **19.** 18.2   **21.** 71.6°   **23.** 21.1°   **25.** 35 meters   **27.** 275 feet

29. 1915 feet   **31.** 79.8°   **33.** 12.3   **35.** 1028 feet   **37.** 145 cm

7.3 Identities (page 447)

1. $\sin\theta$ **3.** 1   **5.** $\cos\theta$ **7.** 1   **9.** -1 **11.** $(\sin\theta)(\cot\theta)(\sec\theta) = (\sin\theta)\left(\dfrac{\cos\theta}{\sin\theta}\right)\left(\dfrac{1}{\cos\theta}\right) = 1$

13. $\cos^2\theta\,(\tan^2\theta + 1) = \cos^2\theta\left(\dfrac{\sin^2\theta}{\cos^2\theta} + 1\right) = \sin^2\theta + \cos^2\theta = 1$

15. All θ not coterminal with $\dfrac{\pi}{2}, \dfrac{3\pi}{2}$.   **17.** All θ not a quadrantal angle.   **19.** Same as Exercise 17.

21. $\dfrac{\sin^2\theta}{\cos^2\theta}$, θ not a quadrantal angle.   **23.** $\dfrac{\sin\theta - 1}{\cos\theta}$; θ not a quadrantal angle.

25. $\dfrac{1}{\sin\theta \cos\theta}$; θ not a quadrantal angle and θ not coterminal with $\dfrac{3\pi}{4}, \dfrac{7\pi}{4}$.

27. $1 - 2\sin^2\theta$ **29.** $\cos^2\theta + \cos\theta - 1$ **31.** $\dfrac{\cos\theta}{\cot\theta} = (\cos\theta)\dfrac{\sin\theta}{\cos\theta} = \sin\theta$; θ not a quadrantal angle.

33. $(\tan\theta - 1)^2 = \tan^2\theta - 2\tan\theta + 1 = \sec^2\theta - 2\tan\theta$; θ not coterminal with $\dfrac{\pi}{2}, \dfrac{3\pi}{2}$.

35.

$\sec\theta - \cos\theta$	$\sin\theta \tan\theta$
$\dfrac{1}{\cos\theta} - \cos\theta$	$\sin\theta\,\dfrac{\sin\theta}{\cos\theta}$
$\dfrac{1 - \cos^2\theta}{\cos\theta}$	$\dfrac{\sin^2\theta}{\cos\theta}$
$\dfrac{\sin^2\theta}{\cos\theta}$	

θ not coterminal with $\dfrac{\pi}{2}$ or $\dfrac{3\pi}{2}$

37.

$\dfrac{\cot\theta - 1}{1 - \tan\theta}$	$\dfrac{\csc\theta}{\sec\theta}$
$\dfrac{\dfrac{\cos\theta}{\sin\theta} - 1}{1 - \dfrac{\sin\theta}{\cos\theta}}$	$\dfrac{\dfrac{1}{\sin\theta}}{\dfrac{1}{\cos\theta}}$
$\dfrac{\cos^2\theta - \sin\theta\cos\theta}{\sin\theta\cos\theta - \sin^2\theta}$	$\dfrac{\cos\theta}{\sin\theta}$
$\dfrac{\cos\theta\,(\cos\theta - \sin\theta)}{\sin\theta\,(\cos\theta - \sin\theta)}$	

θ not a quadrantal angle nor coterminal with $\dfrac{\pi}{4}, \dfrac{5\pi}{4}$.

39. $\tan\theta + \cot\theta = \dfrac{\sin\theta}{\cos\theta} + \dfrac{\cos\theta}{\sin\theta} = \dfrac{\sin^2\theta + \cos^2\theta}{\sin\theta\cos\theta} = \dfrac{1}{\sin\theta\cos\theta}$

41. $(\sec \theta + \tan \theta)(1 - \sin \theta) = \left(\dfrac{1}{\cos \theta} + \dfrac{\sin \theta}{\cos \theta}\right)(1 - \sin \theta) = \left(\dfrac{1 + \sin \theta}{\cos \theta}\right)(1 - \sin \theta) = \dfrac{1 - \sin^2 \theta}{\cos \theta} = \dfrac{\cos^2 \theta}{\cos \theta} = \cos \theta$

43. $(\csc^2 \theta - 1) \sin^2 \theta = \cot^2 \theta \sin^2 \theta = \dfrac{\cos^2 \theta}{\sin^2 \theta} \sin^2 \theta = \cos^2 \theta$

45. $\tan^2 \theta - \sin^2 \theta = \dfrac{\sin^2 \theta}{\cos^2 \theta} - \sin^2 \theta = \sin^2 \theta \left(\dfrac{1}{\cos^2 \theta} - 1\right) = \sin^2 \theta (\sec^2 \theta - 1) = \sin^2 \theta \tan^2 \theta$

47. $\dfrac{1 + \sec \theta}{\csc \theta} = \dfrac{1 + \dfrac{1}{\cos \theta}}{\dfrac{1}{\sin \theta}} = \sin \theta + \dfrac{\sin \theta}{\cos \theta} = \sin \theta + \tan \theta$

49. $\dfrac{1}{1 + \cos \theta} + \dfrac{1}{1 - \cos \theta} = \dfrac{1 - \cos \theta + 1 + \cos \theta}{(1 + \cos \theta)(1 - \cos \theta)} = \dfrac{2}{1 - \cos^2 \theta} = \dfrac{2}{\sin^2 \theta} = 2 \csc^2 \theta$

51. $\dfrac{1 + \tan^2 \theta}{1 + \cot^2 \theta} = \dfrac{\sec^2 \theta}{\csc^2 \theta} = \dfrac{\dfrac{1}{\cos^2 \theta}}{\dfrac{1}{\sin^2 \theta}} = \dfrac{\sin^2 \theta}{\cos^2 \theta} = \tan^2 \theta = \sec^2 \theta - 1$

53. $\dfrac{\sec \theta + \csc \theta}{\sec \theta - \csc \theta} = \dfrac{\dfrac{1}{\cos \theta} + \dfrac{1}{\sin \theta}}{\dfrac{1}{\cos \theta} - \dfrac{1}{\sin \theta}} = \dfrac{\sin \theta + \cos \theta}{\sin \theta - \cos \theta}$

55. $(\csc \theta - \cot \theta)^2 = \left(\dfrac{1}{\sin \theta} - \dfrac{\cos \theta}{\sin \theta}\right)^2 = \dfrac{(1 - \cos \theta)^2}{\sin^2 \theta} = \dfrac{(1 - \cos \theta)^2}{1 - \cos^2 \theta} = \dfrac{(1 - \cos \theta)^2}{(1 - \cos \theta)(1 + \cos \theta)} = \dfrac{1 - \cos \theta}{1 + \cos \theta}$

57. $(\sin^2 \theta + \cos^2 \theta)^5 = 1^5 = 1$ **59.** $\dfrac{1}{\cos^2 \theta} + \dfrac{1}{\sin^2 \theta} = \dfrac{\sin^2 \theta + \cos^2 \theta}{\cos^2 \theta \sin^2 \theta} = \dfrac{1}{(1 - \sin^2 \theta) \sin^2 \theta} = \dfrac{1}{\sin^2 \theta - \sin^4 \theta}$

61. $\dfrac{\tan^2 \theta + 1}{\tan^2 \theta} = 1 + \dfrac{1}{\tan^2 \theta} = 1 + \cot^2 \theta = \csc^2 \theta$

63. $\dfrac{\tan \theta}{\sec \theta - 1} = \dfrac{\tan \theta(\sec \theta + 1)}{(\sec \theta - 1)(\sec \theta + 1)} = \dfrac{\tan \theta(\sec \theta + 1)}{\sec^2 \theta - 1} = \dfrac{\tan \theta(\sec \theta + 1)}{\tan^2 \theta} = \dfrac{\sec \theta + 1}{\tan \theta}$

65. $\sec^3 \theta + \dfrac{\sin^2 \theta}{\cos^3 \theta} = \dfrac{1}{\cos^3 \theta} + \dfrac{\sin^2 \theta}{\cos^3 \theta} = \dfrac{1 + \sin^2 \theta}{\cos^3 \theta} = \dfrac{(\cos^2 \theta + \sin^2 \theta) + \sin^2 \theta}{\cos^3 \theta} = \dfrac{\cos^2 \theta + 2 \sin^2 \theta}{\cos^3 \theta}$

67. $\dfrac{1}{(\csc \theta - \sec \theta)^2} = \dfrac{1}{\left(\dfrac{1}{\sin \theta} - \dfrac{1}{\cos \theta}\right)^2} = \dfrac{1}{\left(\dfrac{\cos \theta - \sin \theta}{\sin \theta \cos \theta}\right)^2} = \dfrac{\sin^2 \theta \cos^2 \theta}{(\cos \theta - \sin \theta)^2} = \dfrac{\sin^2 \theta \cos^2 \theta}{\cos^2 \theta - 2 \cos \theta \sin \theta + \sin^2 \theta} =$

$\dfrac{\sin^2 \theta \cos^2 \theta}{1 - 2 \cos \theta \sin \theta} = \dfrac{\sin^2 \theta}{\dfrac{1}{\cos^2 \theta} - 2\dfrac{\sin \theta}{\cos \theta}} = \dfrac{\sin^2 \theta}{\sec^2 \theta - 2 \tan \theta}$

69. $\dfrac{\sec\theta}{\tan\theta-\sin\theta}=\dfrac{\dfrac{1}{\cos\theta}}{\dfrac{\sin\theta}{\cos\theta}-\sin\theta}=\dfrac{1}{\sin\theta-\sin\theta\cos\theta}=\dfrac{1}{\sin\theta(1-\cos\theta)}=\dfrac{1+\cos\theta}{\sin\theta(1-\cos^2\theta)}$

$\qquad=\dfrac{1+\cos\theta}{\sin\theta\sin^2\theta}=\dfrac{1+\cos\theta}{\sin^3\theta}$

71. $-\ln|\cos x|=\ln|\cos x|^{-1}=\ln\dfrac{1}{|\cos x|}=\ln\left|\dfrac{1}{\cos x}\right|=\ln|\sec x|$

73. $-\ln|\sec x-\tan x|=\ln|\sec x-\tan x|^{-1}=\ln\dfrac{1}{|\sec x-\tan x|}=\ln\left|\dfrac{1}{\sec x-\tan x}\right|$

$\qquad=\ln\left|\dfrac{\sec x+\tan x}{\sec^2 x-\tan^2 x}\right|=\ln|\sec x+\tan x|$

75. $\tan\left(-\dfrac{\pi}{4}\right)=-1\neq 1=\tan\dfrac{\pi}{4}$ **77.** $\sin\dfrac{\pi}{2}=1\neq-1=-\sqrt{1-\cos^2\dfrac{\pi}{2}}$

79. $\dfrac{\cot\dfrac{\pi}{3}-1}{1-\tan\dfrac{\pi}{3}}=\dfrac{\dfrac{1}{\sqrt3}-1}{1-\sqrt3}=\dfrac{1}{\sqrt3}\neq\sqrt3=\dfrac{2}{\dfrac{2}{\sqrt3}}=\dfrac{\sec\dfrac{\pi}{3}}{\csc\dfrac{\pi}{3}}$

81. **(a)** $\ln(\sec\theta+\tan\theta)=\ln\left(\dfrac{\sqrt{9+x^2}}{3}+\dfrac{x}{3}\right)=\ln\left(\dfrac{\sqrt{9+x^2}+x}{3}\right)$ **(b)** $\ln 3$

7.4 The Addition and Subtraction Formulas (page 458)

1. $\cos 60°=\frac12$ **3.** $\tan(-30°)=-\dfrac{1}{\sqrt3}$

5. $\sin 75°=\sin(45°+30°)=\sin 45°\cos 30°+\cos 45°\sin 30°=\frac14(\sqrt6+\sqrt2)$

$\qquad\cos 75°=\cos(45°+30°)=\cos 45°\cos 30°-\sin 45°\sin 30°=\frac14(\sqrt6-\sqrt2)$

$\qquad\tan 75°=\tan(45°+30°)=\dfrac{\tan 45°+\tan 30°}{1-\tan 45°\tan 30°}=\dfrac{\sqrt3+1}{\sqrt3-1}=2+\sqrt3$

7. $\sin\dfrac{\pi}{12}=\sin\left(\dfrac{\pi}{3}-\dfrac{\pi}{4}\right)=\sin\dfrac{\pi}{3}\cos\dfrac{\pi}{4}-\cos\dfrac{\pi}{3}\sin\dfrac{\pi}{4}=\dfrac14(\sqrt6-\sqrt2)$

$\qquad\cos\dfrac{\pi}{12}=\cos\left(\dfrac{\pi}{3}-\dfrac{\pi}{4}\right)=\cos\dfrac{\pi}{3}\cos\dfrac{\pi}{4}+\sin\dfrac{\pi}{3}\sin\dfrac{\pi}{4}=\dfrac14(\sqrt2+\sqrt6)$

$\qquad\tan\dfrac{\pi}{12}=\tan\left(\dfrac{\pi}{3}-\dfrac{\pi}{4}\right)=\dfrac{\tan\dfrac{\pi}{3}-\tan\dfrac{\pi}{4}}{1+\tan\dfrac{\pi}{3}\tan\dfrac{\pi}{3}}=\dfrac{\sqrt3-1}{1+\sqrt3}=2-\sqrt3$

9. $\sin 165° = \sin(135° + 30°) = \sin 135 \cos 30° + \cos 135° \sin 30° = \frac{1}{4}(\sqrt{6} - \sqrt{2})$

$\cos 165° = \cos(135° + 30°) = \cos 135° \cos 30° - \sin 135° \sin 30° = -\frac{1}{4}(\sqrt{6} + \sqrt{2})$

$\tan 165° = \tan(135° + 30°) = \dfrac{\tan 135° + \tan 30°}{1 - \tan 135° \tan 30°} = \dfrac{-\sqrt{3} + 1}{\sqrt{3} + 1} = \sqrt{3} - 2$

11. $\sin\dfrac{17\pi}{12} = \sin\left(\dfrac{7\pi}{6} + \dfrac{\pi}{4}\right) = \sin\dfrac{7\pi}{6}\cos\dfrac{\pi}{4} + \cos\dfrac{7\pi}{6}\sin\dfrac{\pi}{4} = -\dfrac{1}{4}(\sqrt{2} + \sqrt{6})$

$\cos\dfrac{17\pi}{12} = \cos\left(\dfrac{7\pi}{6} + \dfrac{\pi}{4}\right) = \cos\dfrac{7\pi}{6}\cos\dfrac{\pi}{4} - \sin\dfrac{7\pi}{6}\sin\dfrac{\pi}{4} = \dfrac{1}{4}(\sqrt{2} - \sqrt{6})$

$\tan\dfrac{17\pi}{12} = \tan\left(\dfrac{7\pi}{6} + \dfrac{\pi}{4}\right) = \dfrac{\tan\dfrac{7\pi}{6} + \tan\dfrac{\pi}{4}}{1 - \tan\dfrac{7\pi}{6}\tan\dfrac{\pi}{4}} = \dfrac{\sqrt{3} + 1}{\sqrt{3} - 1} = 2 + \sqrt{3}$

13. $\sin(\pi - \theta) = \sin\pi\cos\theta - \cos\pi\sin\theta = -(-1)\sin\theta = \sin\theta$

15. $\tan(\pi - \theta) = \dfrac{\tan\pi - \tan\theta}{1 + \tan\pi\tan\theta} = \dfrac{0 - \tan\theta}{1 + 0} = -\tan\theta$

17. $\sin\left(\theta + \dfrac{7\pi}{2}\right) = \sin\theta\cos\dfrac{7\pi}{2} + \cos\theta\sin\dfrac{7\pi}{2} = (\sin\theta)0 + (\cos\theta)(-1) = -\cos\theta$

19. $\sin(\alpha - \beta) = \frac{16}{65};\ \tan(\alpha + \beta) = \frac{56}{33}$ **21.** $\sin(\alpha + \beta) = \frac{2}{15}(\sqrt{42} - 1);\ \cos(\alpha + \beta) = -\frac{1}{15}(\sqrt{21} + 4\sqrt{2})$

23. $\sin(\alpha - \beta) = \frac{152}{377},\ \tan(\alpha + \beta) = -\frac{352}{135}$

25. **(a)** $\cos 191° = \cos(11° + 180°) = -\cos 11° = -0.9816$

    (b) $\sin 132° = \sin(90° + 42°) = \cos 42° = 0.7431$

    (c) $\tan 173° = \tan(180° - 7°) = -\tan 7° = -0.1228$

    (d) $\cos 102° = \cos(90° + 12°) = -\sin 12° = -0.2079$

27. $\cos\left(\theta - \dfrac{\pi}{4}\right) = \cos\theta\cos\dfrac{\pi}{4} + \sin\theta\sin\dfrac{\pi}{4} = \dfrac{\sqrt{2}}{2}(\cos\theta + \sin\theta)$

29. $\cos(\theta + 30°) + \cos(\theta - 30°) = \cos\theta\cos 30° - \sin\theta\sin 30° + \cos\theta\cos 30° + \sin\theta\sin 30° = 2\cos\theta\cos 30°$

$$= \sqrt{3}\cos\theta$$

31. $\dfrac{\sin\left(\theta + \dfrac{\pi}{2}\right)}{\cos\left(\theta + \dfrac{\pi}{2}\right)} = \dfrac{\cos\theta}{-\sin\theta} = -\cot\theta$ **33.** $\csc(\pi - \theta) = \dfrac{1}{\sin(\pi - \theta)} = \dfrac{1}{\sin\theta} = \csc\theta$

35. $\sin(\alpha + \beta) - \sin(\alpha - \beta) = \sin\alpha\cos\beta + \cos\alpha\sin\beta - \sin\alpha\cos\beta + \cos\alpha\sin\beta = 2\cos\alpha\sin\beta$

37. $\dfrac{\sin(\alpha + \beta)}{\sin(\alpha - \beta)} = \dfrac{\sin\alpha\cos\beta + \cos\alpha\sin\beta}{\sin\alpha\cos\beta - \cos\alpha\sin\beta} = \dfrac{\dfrac{\sin\alpha\cos\beta}{\cos\alpha\cos\beta} + \dfrac{\cos\alpha\sin\beta}{\cos\alpha\cos\beta}}{\dfrac{\sin\alpha\cos\beta}{\cos\alpha\cos\beta} - \dfrac{\cos\alpha\sin\beta}{\cos\alpha\cos\beta}} = \dfrac{\tan\alpha + \tan\beta}{\tan\alpha - \tan\beta}$

39. $2 \sin\left(\dfrac{\pi}{4} - \theta\right) \sin\left(\dfrac{\pi}{4} + \theta\right) = 2\left[\sin\dfrac{\pi}{4}\cos\theta - \cos\dfrac{\pi}{4}\sin\theta\right]\left[\sin\dfrac{\pi}{4}\cos\theta + \cos\dfrac{\pi}{4}\sin\theta\right]$

$$= 2\left[\dfrac{\cos\theta}{\sqrt{2}} - \dfrac{\sin\theta}{\sqrt{2}}\right]\left[\dfrac{\cos\theta}{\sqrt{2}} + \dfrac{\sin\theta}{\sqrt{2}}\right] = 2\left[\dfrac{\cos^2\theta}{2} - \dfrac{\sin^2\theta}{2}\right] = \cos^2\theta - \sin^2\theta$$

41. $\cos\left(\dfrac{\pi}{2} - x\right) = \sin x;$

$\cos\left[\dfrac{\pi}{2} - \left(\dfrac{\pi}{2} - \theta\right)\right] = \sin\left(\dfrac{\pi}{2} - \theta\right);$

$\cos\theta = \sin\left(\dfrac{\pi}{2} - \theta\right)$

43. $\sin(\alpha - \beta) = \sin[\alpha + (-\beta)]$
$= \sin\alpha\cos(-\beta) + \cos\alpha\sin(-\beta)$
$= \sin\alpha\cos\beta - \cos\alpha\sin\beta$

45. $\cot(\alpha + \beta) = \dfrac{\cos(\alpha + \beta)}{\sin(\alpha + \beta)} = \dfrac{\cos\alpha\cos\beta - \sin\alpha\sin\beta}{\sin\alpha\cos\beta + \cos\alpha\sin\beta} = \dfrac{\dfrac{\cos\alpha\cos\beta}{\sin\alpha\sin\beta} - \dfrac{\sin\alpha\sin\beta}{\sin\alpha\sin\beta}}{\dfrac{\sin\alpha\cos\beta}{\sin\alpha\sin\beta} + \dfrac{\cos\alpha\sin\beta}{\sin\alpha\sin\beta}} = \dfrac{\cot\alpha\cot\beta - 1}{\cot\beta + \cot\alpha}$

47. Since $\sin\theta = -\cos\left(\theta + \dfrac{\pi}{2}\right)$, shift $y = \cos\theta$ by $\dfrac{\pi}{2}$ units to the left and reflect through the θ-axis.

49. $\sin\left(\theta + \dfrac{\pi}{2}k\right) = \sin\theta\cos\dfrac{\pi}{2}k + \cos\theta\sin\dfrac{\pi}{2}k$. If k is even, then $\sin\dfrac{\pi}{2}k = 0$, $\cos\dfrac{\pi}{2}k = \pm 1$, giving

$\sin\left(\theta + \dfrac{\pi}{2}k\right) = \pm\sin\theta$. If k is odd, then $\cos\dfrac{\pi}{2}k = 0$, $\sin\dfrac{\pi}{2}k = \pm 1$, giving $\sin\left(\theta + \dfrac{\pi}{2}k\right) = \pm\cos\theta$.

51. $y = 3\sqrt{2}\sin\left(x - \dfrac{\pi}{4}\right)$ **53.** $f(x) = 4\sin\left(x + \dfrac{\pi}{3}\right)$

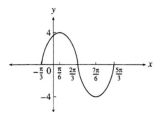

55. $\dfrac{12}{37}$ **57.** $-\dfrac{7}{11}$ **59.** $\dfrac{220}{221}$

61. (a) $\dfrac{f(x + h) - f(x)}{h} = \dfrac{\sin(x + h) - \sin x}{h} = \dfrac{\sin x\cos h + \cos x\sin h - \sin x}{h}$

$$= \dfrac{\sin x(\cos h - 1)}{h} + \dfrac{\cos x\sin h}{h} = \left(\dfrac{\cos h - 1}{h}\right)\sin x + \left(\dfrac{\sin h}{h}\right)\cos x$$

(b) $\dfrac{g(x + h) - g(x)}{h} = \dfrac{\cos(x + h) - \cos x}{h} = \dfrac{\cos x\cos h - \sin x\sin h - \cos x}{h}$

$$= \dfrac{\cos x(\cos h - 1)}{h} - \dfrac{\sin x\sin h}{h} = \left(\dfrac{\cos h - 1}{h}\right)\cos x - \left(\dfrac{\sin h}{h}\right)\sin x$$

63. $24.2°$ **65.** $121.8°$

7.5 The Double- and Half-Angle Formulas (page 468)

1. $\cos \dfrac{7\pi}{6} = -\dfrac{\sqrt{3}}{2}$ **3.** $3 \tan 150° = -\sqrt{3}$

5. $\cos 75° = \cos \dfrac{150°}{2} = \sqrt{\dfrac{1 + \cos 150°}{2}} = \sqrt{\dfrac{1 - \dfrac{\sqrt{3}}{2}}{2}} = \dfrac{1}{2}\sqrt{2 - \sqrt{3}}$

7. $\tan \dfrac{3\pi}{8} = \dfrac{1 - \cos \dfrac{3\pi}{4}}{\sin \dfrac{3\pi}{4}} = \dfrac{1 + \dfrac{1}{\sqrt{2}}}{\dfrac{1}{\sqrt{2}}} = \sqrt{2} + 1$

9. (a) $\tan 15° = \tan \dfrac{30°}{2} = \dfrac{1 - \cos 30°}{\sin 30°} = \dfrac{1 - \dfrac{\sqrt{3}}{2}}{\dfrac{1}{2}} = 2 - \sqrt{3}$

   (b) $\tan 15° = 0.2679$ and $2 - \sqrt{3} = 0.2679$, to four decimal places.

11. (a) $\tan 22.5° = \dfrac{1 - \cos 45°}{\sin 45°} = \dfrac{1 - \dfrac{1}{\sqrt{2}}}{\dfrac{1}{\sqrt{2}}} = \sqrt{2} - 1$

   (b) $\tan 22.5° = 0.4142$ and $\sqrt{2} - 1 = 0.4142$, to four decimal places.

13. (a) $-\dfrac{240}{289}$; **(b)** $-\dfrac{161}{289}$; **(c)** $\dfrac{240}{161}$ **15.** $-\dfrac{12}{13}$

17. $9 \sin 2\theta = 18 \sin \theta \cos \theta = 18\left(\dfrac{x}{3}\right)\left(\dfrac{\sqrt{9 - x^2}}{3}\right) = 2x\sqrt{9 - x^2}$

19. $\dfrac{b}{10} = \sin 15° = \sin \dfrac{30°}{2} = \sqrt{\dfrac{1 - \cos 30°}{2}} = \dfrac{\sqrt{2 - \sqrt{3}}}{2}$; $b = 5\sqrt{2 - \sqrt{3}}$

21. Use $\tan 2\theta = \dfrac{2 \tan \theta}{1 - \tan^2 \theta}$, where $\tan 2\theta = \dfrac{h}{50}$ and $\tan \theta = \dfrac{h}{150}$ to find that $h = 50\sqrt{3}$.

23. $9\sqrt{7}$ **25.** $\cot x \sin 2x = \dfrac{\cos x}{\sin x} \cdot 2 \sin x \cos x = 2 \cos^2 x$ **27.** $\sec^2 \dfrac{x}{2} = \dfrac{1}{\cos^2 \dfrac{x}{2}} = \dfrac{1}{\dfrac{1 + \cos x}{2}} = \dfrac{2}{1 + \cos x}$

29. $\cos^2 \dfrac{x}{2} = \dfrac{1 + \cos x}{2} = \dfrac{\tan x + \tan x \cos x}{2 \tan x} = \dfrac{\tan x + \sin x}{2 \tan x}$

31. $\cot \dfrac{x}{2} = \dfrac{1}{\tan \dfrac{x}{2}} = \dfrac{1}{\dfrac{\sin x}{1 + \cos x}} = \dfrac{1 + \cos x}{\sin x} = \dfrac{1}{\sin x} + \dfrac{\cos x}{\sin x} = \csc x + \cot x$

33. $(\sin x + \cos x)^2 = \sin^2 x + 2 \sin x \cos x + \cos^2 x = 1 + 2 \sin x \cos x = 1 + \sin 2x$

35. $\tan 3x = \tan(2x + x) = \dfrac{\tan 2x + \tan x}{1 - \tan 2x \tan x} = \dfrac{\dfrac{2\tan x}{1 - \tan^2 x} + \tan x}{1 - \dfrac{2\tan x}{1 - \tan^2 x} \cdot \tan x} = \dfrac{2\tan x + \tan x - \tan^3 x}{1 - \tan^2 x - 2\tan^2 x} = \dfrac{3\tan x - \tan^3 x}{1 - 3\tan^2 x}$

37. $\cot 2x = \dfrac{1}{\tan 2x} = \dfrac{1}{\dfrac{2\tan x}{1 - \tan^2 x}} = \dfrac{1 - \tan^2 x}{2\tan x} = \dfrac{\dfrac{1}{\tan^2 x} - \dfrac{\tan^2 x}{\tan^2 x}}{\dfrac{2\tan x}{\tan^2 x}} = \dfrac{\cot^2 x - 1}{2\cot x}$

39. $\csc 2x - \cot 2x = \dfrac{1}{\sin 2x} - \dfrac{\cos 2x}{\sin 2x} = \dfrac{1 - \cos 2x}{\sin 2x} = \tan\dfrac{2x}{2} = \tan x$

41. $\cos^4 x - \sin^4 x = (\cos^2 x - \sin^2 x)(\cos^2 x + \sin^2 x) = (\cos^2 x - \sin^2 x) = \cos 2x$

43. $\cos 4x = \cos 2(2x) = 2\cos^2 2x - 1 = 2(2\cos^2 x - 1)^2 - 1 = 2(4\cos^4 x - 4\cos^2 x + 1) - 1$
$\qquad = 8\cos^4 x - 8\cos^2 x + 1$

45. $\cos^4 x = (\cos^2 x)^2 = \left(\dfrac{1 + \cos 2x}{2}\right)^2 = \dfrac{1}{4}(1 + 2\cos 2x + \cos^2 2x)$

$\qquad = \dfrac{1}{4}\left(1 + 2\cos 2x + \dfrac{1 + \cos 4x}{2}\right) = \dfrac{1}{8}\cos 4x + \dfrac{1}{2}\cos 2x + \dfrac{3}{8}$

47. $\sin^2 x \cos^2 x = \left(\dfrac{1 - \cos 2x}{2}\right)\left(\dfrac{1 + \cos 2x}{2}\right) = \dfrac{1}{4}(1 - \cos^2 2x) = \dfrac{1}{4}\left(1 - \dfrac{1 + \cos 4x}{2}\right)$

$\qquad = \dfrac{1}{4}\left(\dfrac{1}{2} - \dfrac{1}{2}\cos 4x\right) = \dfrac{1}{8}(1 - \cos 4x)$

49. Using a figure as in Exercise 50, page 459, we have $c = \sqrt{a^2 + b^2}$, $\sin\theta = \dfrac{b}{c}$, and $\cos\theta = \dfrac{a}{c}$. Then $a = c\cos\theta$,

$b = c\sin\theta$, and $a\sin kx + b\cos kx = c\cos\theta\sin kx + c\sin\theta\cos kx = c(\sin kx\cos\theta + \cos kx\sin\theta) = c\sin(kx + \theta)$

51. (a) $\quad\cos\alpha\cos\beta - \sin\alpha\sin\beta = \cos(\alpha + \beta)$

$\qquad\dfrac{\cos\alpha\cos\beta + \sin\alpha\sin\beta = \cos(\alpha - \beta)}{}$

$\qquad 2\cos\alpha\cos\beta = \cos(\alpha + \beta) + \cos(\alpha - \beta)$

$\qquad\cos\alpha\cos\beta = \tfrac{1}{2}[\cos(\alpha + \beta) + \cos(\alpha - \beta)]$

(b) $\quad\cos\alpha\cos\beta + \sin\alpha\sin\beta = \cos(\alpha - \beta)$

$\qquad\dfrac{\cos\alpha\cos\beta - \sin\alpha\sin\beta = \cos(\alpha + \beta)}{}$

$\qquad 2\sin\alpha\sin\beta = \cos(\alpha - \beta) - \cos(\alpha + \beta)$

$\qquad\sin\alpha\sin\beta = \tfrac{1}{2}[\cos(\alpha - \beta) - \cos(\alpha + \beta)]$

$\qquad\sin\alpha\cos\beta + \cos\alpha\sin\beta = \sin(\alpha + \beta)$

$\qquad\dfrac{\sin\alpha\cos\beta - \cos\alpha\sin\beta = \sin(\alpha - \beta)}{}$

$\qquad 2\sin\alpha\cos\beta = \sin(\alpha + \beta) + \sin(\alpha - \beta)$

$\qquad\sin\alpha\cos\beta = \tfrac{1}{2}[\sin(\alpha + \beta) + \sin(\alpha - \beta)]$

$\qquad\sin\alpha\cos\beta + \cos\alpha\sin\beta = \sin(\alpha + \beta)$

$\qquad\dfrac{\sin\alpha\cos\beta - \cos\alpha\sin\beta = \sin(\alpha - \beta)}{}$

$\qquad 2\cos\alpha\sin\beta = \sin(\alpha + \beta) - \sin(\alpha - \beta)$

$\qquad\cos\alpha\sin\beta = \tfrac{1}{2}[\sin(\alpha + \beta) - \sin(\alpha - \beta)]$

53. $2\cos\dfrac{u + v}{2}\cos\dfrac{u - v}{2} = \cos\left(\dfrac{u + v}{2} - \dfrac{u - v}{2}\right) + \cos\left(\dfrac{u + v}{2} + \dfrac{u - v}{2}\right) = \cos v + \cos u$

Similarly for the other formulas.

55. $\dfrac{7}{9}$ **57.** $\dfrac{40}{9}$

59. $\sin(2\arctan x) = 2\sin(\arctan x)\cdot\cos(\arctan x) = 2\left(\dfrac{x}{\sqrt{1+x^2}}\right)\left(\dfrac{1}{\sqrt{1+x^2}}\right) = \dfrac{2x}{1+x^2}$

61. $\tan(2\arctan x) = \dfrac{2\tan(\arctan x)}{1-\tan^2(\arctan x)} = \dfrac{2x}{1-x^2}$

7.6 Trigonometric Equations (page 477)

In the following answers k represents any integer.

1. $2k\pi$ **3.** $\dfrac{\pi}{6}+2k\pi;\ \dfrac{5\pi}{6}+2k\pi$ **5.** $\pi+2k\pi$ **7.** $\dfrac{7\pi}{12}+k\pi;\ \dfrac{11\pi}{12}+k\pi$ **9.** No solutions

11. $\dfrac{\pi}{6}+\dfrac{k\pi}{2};\ \dfrac{2\pi}{3}+\dfrac{k\pi}{2}$ **13.** $\pi+2k\pi-1$ **15.** $\dfrac{2k\pi}{3}$ **17.** $60°;\ 120°$ **19.** $45°;\ 315°$

21. $90°;\ 270°$ **23.** $60°;\ 300°$ **25.** $30°;\ 150°;\ 270°$ **27.** $45°;\ 225°$ **29.** $0;\ \dfrac{7\pi}{6};\ \dfrac{11\pi}{6}$ **31.** $\dfrac{\pi}{3};\ \dfrac{4\pi}{3}$

33. $\dfrac{\pi}{2};\ \dfrac{3\pi}{2};\ \dfrac{\pi}{6};\ \dfrac{11\pi}{6}$ **35.** $\dfrac{\pi}{2}$ **37.** $\dfrac{\pi}{6};\ \dfrac{5\pi}{6};\ \dfrac{\pi}{2};\ \dfrac{3\pi}{2}$ **39.** No solutions   **41.** $\dfrac{\pi}{3};\ \dfrac{5\pi}{3}$

43. $0;\ \dfrac{\pi}{4};\ \dfrac{3\pi}{4};\ \dfrac{5\pi}{4};\ \dfrac{7\pi}{4};\ \pi$ **45.** $\dfrac{\pi}{4};\ \dfrac{\pi}{2};\ \dfrac{3\pi}{4};\ \dfrac{5\pi}{4};\ \dfrac{3\pi}{2};\ \dfrac{7\pi}{4}$ **47.** $\dfrac{\pi}{4};\ \dfrac{5\pi}{4}$ **49.** $\dfrac{\pi}{3};\ \dfrac{5\pi}{3}$

51. $0;\ \dfrac{\pi}{6};\ \dfrac{5\pi}{6};\ \pi$ **53.** $\dfrac{\pi}{6};\ \dfrac{5\pi}{6};\ \dfrac{7\pi}{6};\ \dfrac{11\pi}{6}$ **55.** $0;\ \dfrac{\pi}{3};\ \dfrac{\pi}{2};\ \dfrac{2\pi}{3};\ \pi;\ \dfrac{4\pi}{3};\ \dfrac{3\pi}{2};\ \dfrac{5\pi}{3}$ **57.** $0;\ \dfrac{\pi}{2}$

59. $62.2°;\ 297.8°$ **61.** $111.4°;\ 248.6°$ **63.** $33.7°;\ 213.7°$ **65.** $69.3°;\ 290.7°;\ 110.7°;\ 249.3°$

67. (a) $\cos\dfrac{\pi}{4} = \dfrac{\sqrt{2}}{2} = .7071\ldots;\ \cos\alpha = \dfrac{28}{35} = .8 > .7071\ldots$ Then $\alpha < \dfrac{\pi}{4}$ since the cosine is decreasing

for $0 \le x \le \dfrac{\pi}{2}$.

(b) $\cos 2\alpha = 2\cos^2\alpha - 1 = 2\left(\dfrac{4}{5}\right)^2 - 1 = \dfrac{7}{25} = \dfrac{28}{100} = \cos(\alpha+\beta)$. Then, since the cosine is one-to-one

for $0 \le x \le \dfrac{\pi}{2}$, $2\alpha = \alpha + \beta$, or $\alpha = \beta$.

69. $\left(\dfrac{3\pi}{4}, \dfrac{\sqrt{2}}{2}\right), \left(\dfrac{7\pi}{4}, -\dfrac{\sqrt{2}}{2}\right)$

CHAPTER 7 REVIEW EXERCISES (page 479)

1. See page 427.

2. $\sin B = \dfrac{b}{c}, \cos B = \dfrac{a}{c}, \tan B = \dfrac{b}{a}, \cot B = \dfrac{a}{b}, \sec B = \dfrac{c}{a}, \csc B = \dfrac{c}{b}$

3. $\dfrac{2}{\sqrt{13}}$ **4.** $\sin A = \dfrac{\sqrt{5}}{3}, \cos A = \dfrac{2}{3}, \tan A = \dfrac{\sqrt{5}}{2}, \cot A = \dfrac{2}{\sqrt{5}}, \sec A = \dfrac{3}{2}, \csc A = \dfrac{3}{\sqrt{5}}$

5. $\sin B = \dfrac{\sqrt{x^2-1}}{x}, \cos B = \dfrac{1}{x}, \tan B = \sqrt{x^2-1}, \cot B = \dfrac{1}{\sqrt{x^2-1}}, \sec B = x, \csc B = \dfrac{x}{\sqrt{x^2-1}}$

6.

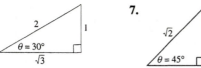

7.

8. Same as Exercise 6.   **9.** Same as Exercise 6.

10. Side opposite $\theta = \sqrt{25 - x^2}$; $\sin^2 \theta + \cos^2 \theta = \left(\dfrac{\sqrt{25 - x^2}}{5}\right)^2 + \left(\dfrac{x}{5}\right)^2 = \dfrac{25 - x^2 + x^2}{25} = 1$

11. $\tan^2 \theta + 1 = \left(\dfrac{\sqrt{25 - x^2}}{x}\right)^2 + 1 = \dfrac{25 - x^2}{x^2} + 1 = \dfrac{25 - x^2 + x^2}{x^2} = \left(\dfrac{5}{x}\right)^2 = \sec^2 \theta$

12. $\cos \theta \tan^2 \theta = \dfrac{3}{x}\left(\dfrac{\sqrt{x^2 - 9}}{3}\right)^2 = \dfrac{x^2 - 9}{3x}$

13. $x = 8\sqrt{3}$ **14.** $\dfrac{10}{\sqrt{3}}$ **15.** $\theta = \dfrac{\pi}{3}$ **16.** $x = 7\sqrt{2}$

17. 0.9563   **18.** 0.1736   **19.** 0.8421   **20.** 1.7837   **21.** 13.6°   **22.** 58.2°   **23.** 53.4°   **24.** 62.9°

25. 76.9°   **26.** 36.5°   **27.** 15.2   **28.** 19.3°   **29.** 29 m   **30.** 65.7 m   **31.** 5.3   **32.** 1

33. -1 **34.** 2   **35.** 1   **36.** $\dfrac{\sin \theta}{\cos^2 \theta}$; θ not a quadrantal angle.

37. $\dfrac{1}{\sin \theta - \cos \theta}$; θ not a quadrantal angle and θ not coterminal with $\dfrac{\pi}{4}, \dfrac{3\pi}{4}, \dfrac{5\pi}{4},$ or $\dfrac{7\pi}{4}$.

38. **(a)** See page 443.   **(b)** $\sin^2 \theta + \cos^2 \theta = 1$
$$1 + \dfrac{\cos^2 \theta}{\sin^2 \theta} = \dfrac{1}{\sin^2 \theta}$$
$$1 + \cot^2 \theta = \csc^2 \theta$$

39. $\sin^2 \theta - 2 \cos^2 \theta = \sin^2 \theta - 2(1 - \sin^2 \theta) = 3 \sin^2 \theta - 2$

40. $(\cot \theta - 1)^2 = \cot^2 \theta - 2 \cot \theta + 1 = (\cot^2 \theta + 1) - 2 \cot \theta = \csc^2 \theta - 2 \cot \theta$

41. $(1 + \cos \theta)(1 - \cos \theta) = 1 - \cos^2 \theta = \sin^2 \theta = \dfrac{1}{\csc^2 \theta} = \dfrac{1}{1 + \cot^2 \theta}$

42. $\dfrac{\tan^2 \theta + 1}{\cot^2 \theta + 1} = \dfrac{\sec^2 \theta}{\csc^2 \theta} = \dfrac{\sin^2 \theta}{\cos^2 \theta} = \tan^2 \theta$

43. $\dfrac{\cot \theta \cos \theta}{\csc \theta - 1} = \dfrac{\dfrac{\cos \theta}{\sin \theta} \cdot \cos \theta}{\dfrac{1}{\sin \theta} - 1} = \dfrac{\cos^2 \theta}{1 - \sin \theta} = \dfrac{1 - \sin^2 \theta}{1 - \sin \theta} = \dfrac{(1 - \sin \theta)(1 + \sin \theta)}{1 - \sin \theta} = 1 + \sin \theta$

44. $\dfrac{\cos \theta}{1 + \sin \theta} = \dfrac{\cos \theta (1 - \sin \theta)}{(1 + \sin \theta)(1 - \sin \theta)} = \dfrac{\cos \theta (1 - \sin \theta)}{1 - \sin^2 \theta} = \dfrac{\cos \theta (1 - \sin \theta)}{\cos^2 \theta} = \dfrac{1 - \sin \theta}{\cos \theta}$

45. $\dfrac{\cot \theta}{1 + 3 \cot^2 \theta} = \dfrac{\dfrac{\cos \theta}{\sin \theta}}{1 + 3\dfrac{\cos^2 \theta}{\sin^2 \theta}} = \dfrac{\cos \theta \sin \theta}{\sin^2 \theta + 3 \cos^2 \theta} = \dfrac{\cos \theta \sin \theta}{\sin^2 \theta + \cos^2 \theta + 2\cos^2 \theta} = \dfrac{\cos \theta \sin \theta}{1 + 2\cos^2 \theta}$

$$= \dfrac{\sin \theta}{\dfrac{1}{\cos \theta} + 2\cos \theta} = \dfrac{\sin \theta}{\sec \theta + 2\cos \theta}$$

46. $\dfrac{1 + \sin \theta}{\cos \theta} + \dfrac{\cos \theta}{1 + \sin \theta} = \dfrac{(1 + \sin \theta)^2 + \cos^2 \theta}{\cos \theta \,(1 + \sin \theta)} = \dfrac{1 + 2\sin \theta + \sin^2 \theta + \cos^2 \theta}{\cos \theta \,(1 + \sin \theta)} = \dfrac{2 + 2\sin \theta}{\cos \theta \,(1 + \sin \theta)}$

$$= \dfrac{2(1 + \sin \theta)}{\cos \theta \,(1 + \sin \theta)} = 2 \sec \theta$$

47. $\dfrac{1}{\sec \theta + \tan \theta} = \dfrac{1}{\sec \theta + \tan \theta} \cdot \dfrac{\sec \theta - \tan \theta}{\sec \theta - \tan \theta} = \dfrac{\sec \theta - \tan \theta}{\sec^2 \theta - \tan^2 \theta} = \dfrac{\sec \theta - \tan \theta}{1} = \sec \theta - \tan \theta$

48. Let $\theta = 30°$. Then $\dfrac{1 + \sin 30°}{1 - \sin 30°} = \dfrac{1.5}{0.5} = 3$; $(\sec 30° - \tan 30°)^2 = \left(\dfrac{2}{\sqrt{3}} - \dfrac{1}{\sqrt{3}}\right)^2 = \left(\dfrac{1}{\sqrt{3}}\right)^2 = \dfrac{1}{3}$; $3 \neq \dfrac{1}{3}$

49. $\sec \pi = -1$; $\sqrt{1 + \tan^2 \pi} = \sqrt{1 + 0} = 1$; $-1 \neq 1$ **50.** $\sin 6°$ **51.** $\cos 12°$ **52.** $\dfrac{\sqrt{3} - 1}{\sqrt{3} + 1}$

53. $\dfrac{1}{4}(\sqrt{2} - \sqrt{6})$ **54.** $-\dfrac{33}{65}$ **55.** $\dfrac{63}{16}$ **56.** $\dfrac{1}{12}(2\sqrt{15} + \sqrt{5})$

57. $\cos\left(\dfrac{\pi}{6} + \theta\right) = \cos \dfrac{\pi}{6} \cos \theta - \sin \dfrac{\pi}{6} \sin \theta = \dfrac{\sqrt{3}}{2} \cos \theta - \dfrac{1}{2} \sin \theta = \dfrac{1}{2}(\sqrt{3} \cos \theta - \sin \theta)$

58. $\dfrac{\cos (\alpha + \beta)}{\sin (\alpha + \beta)} = \dfrac{\cos \alpha \cos \beta - \sin \alpha \sin \beta}{\sin \alpha \cos \beta + \cos \alpha \sin \beta}$

$$= \dfrac{1 - \tan \alpha \tan \beta}{\tan \alpha + \tan \beta} \qquad \text{Dividing the numerator and the denominator by } \cos \alpha \cos \beta.$$

59. $\sin \alpha \cos (\alpha - \beta) - \cos \alpha \sin (\alpha - \beta) = \sin \alpha (\cos \alpha \cos \beta + \sin \alpha \sin \beta) - \cos \alpha (\sin \alpha \cos \beta - \cos \alpha \sin \beta)$
$= \sin \alpha \cos \alpha \cos \beta + \sin^2 \alpha \sin \beta - \cos \alpha \sin \alpha \cos \beta + \cos^2 \alpha \sin \beta = (\sin^2 \alpha + \cos^2 \alpha) \sin \beta = 1 (\sin \beta) = \sin \beta$

60. $y = 4\sqrt{2} \sin\left(x + \dfrac{3\pi}{4}\right)$ **61.** $\dfrac{9}{19}$ **62.** $\cos \dfrac{\pi}{6} = \dfrac{\sqrt{3}}{2}$ **63.** $\dfrac{1}{4} \sin 30° = \dfrac{1}{8}$ **64.** $\dfrac{1 - \cos 45°}{\sin 45°} = \sqrt{2} - 1$

65. $\sqrt{\dfrac{1 + \cos \dfrac{\pi}{4}}{2}} = \dfrac{1}{2}\sqrt{2 + \sqrt{2}}$ **66.** $4 \tan 210° = \dfrac{4}{\sqrt{3}}$ **67.** $\sin 2\theta = -\dfrac{24}{25}$, $\cos 2\theta = -\dfrac{7}{25}$, $\tan 2\theta = \dfrac{24}{7}$

68. $4\sqrt{2 + \sqrt{2}}$ **69.** (a) $\sin 2\theta = 2x\sqrt{1 - x^2}$, $\cos 2\theta = 1 - 2x^2$; (b) $\sin 4\theta = 4x(1 - 2x^2)\sqrt{1 - x^2}$ **70.** $4\sqrt{2}$

71. $\dfrac{1 + \cos 2x}{\sin 2x} = \dfrac{1 + 2\cos^2 x - 1}{\sin 2x} = \dfrac{2\cos^2 x}{2\sin x \cos x} = \dfrac{\cos x}{\sin x} = \cot x$

72. $\sin^4 x = (\sin^2 x)^2 = \left(\dfrac{1 - \cos 2x}{2}\right)^2 = \dfrac{1}{4}(1 - 2\cos 2x + \cos^2 2x) = \dfrac{1}{4} - \dfrac{1}{2}\cos 2x + \dfrac{1}{4}\cos^2 2x$

73. $\sin 4x = 2 \sin 2x \cos 2x = 2 (2 \sin x \cos x)(1 - 2\sin^2 x) = 4 \sin x \cos x - 8 \sin^3 x \cos x$

74. $x = \begin{cases} \dfrac{\pi}{6} + 2k\pi \\ \dfrac{5\pi}{6} + 2k\pi \end{cases}$ **75.** $x = \begin{cases} \dfrac{\pi}{6} + 2k\pi \\ \dfrac{11\pi}{6} + 2k\pi \end{cases}$ **76.** $x = \begin{cases} \dfrac{\pi}{6} + k\pi \\ \dfrac{5\pi}{6} + k\pi \end{cases}$ **77.** $x = \begin{cases} \dfrac{\pi}{4} + \dfrac{\pi}{3}k \\ \dfrac{7\pi}{12} + \dfrac{\pi}{3}k \end{cases}$

78. $150°, 210°$ **79.** No solutions   **80.** $90°$ **81.** $135°, 315°$ **82.** $0, \dfrac{\pi}{3}, \dfrac{2\pi}{3}, \pi$ **83.** $0, \dfrac{2\pi}{3}, \dfrac{4\pi}{3}$

84. $\dfrac{\pi}{6}, \dfrac{5\pi}{6}$ **85.** $\dfrac{\pi}{3}, \dfrac{2\pi}{3}, \dfrac{4\pi}{3}, \dfrac{5\pi}{3}$ **86.** $\dfrac{\pi}{4}, \dfrac{\pi}{3}, \dfrac{2\pi}{3}, \dfrac{3\pi}{4}, \dfrac{5\pi}{4}, \dfrac{4\pi}{3}, \dfrac{5\pi}{3}, \dfrac{7\pi}{4}$ **87.** $53.1°, 306.9°$

88. $56.3°, 161.6°, 236.3°\ 341.6°$

CHAPTER 7 STANDARD ANSWER TEST (page 484)

1. (a) $\frac{15}{8}$; (b) $\frac{8}{17}$; (c) $\frac{17}{8}$ **2.** $\sin A = \dfrac{\sqrt{5}}{3}$, $\cos A = \dfrac{2}{3}$; $\tan A = \dfrac{\sqrt{5}}{2}$.

3. $AB = \sqrt{9 + x^2}$; $3(\sin^2 A)(\sec A) = 3\left(\dfrac{x^2}{x^2 + 9}\right)\left(\dfrac{\sqrt{x^2 + 9}}{3}\right) = \dfrac{x^2}{\sqrt{x^2 + 9}}$ **4.** $20\sqrt{2}$ **5.** 373 feet   **6.** 132.3 m

7. $\cot^2 \theta - \cos^2 \theta = \dfrac{\cos^2 \theta}{\sin^2 \theta} - \cos^2 \theta = \cos^2 \theta\left(\dfrac{1}{\sin^2 \theta} - 1\right) = \cos^2 \theta\left(\dfrac{1 - \sin^2 \theta}{\sin^2 \theta}\right) = \cos^2 \theta\left(\dfrac{\cos^2 \theta}{\sin^2 \theta}\right) = \cos^2 \theta \cot^2 \theta$

8. $(\sec \theta - \tan \theta)^2 = \left(\dfrac{1}{\cos \theta} - \dfrac{\sin \theta}{\cos \theta}\right)^2 = \dfrac{(1 - \sin \theta)^2}{\cos^2 \theta} = \dfrac{(1 - \sin \theta)^2}{1 - \sin^2 \theta} = \dfrac{(1 - \sin \theta)^2}{(1 - \sin \theta)(1 + \sin \theta)} = \dfrac{1 - \sin \theta}{1 + \sin \theta}$

9. $\dfrac{\tan \theta - \sin \theta}{\csc \theta - \cot \theta} = \dfrac{\dfrac{\sin \theta}{\cos \theta} - \sin \theta}{\dfrac{1}{\sin \theta} - \dfrac{\cos \theta}{\sin \theta}} = \dfrac{\sin^2 \theta - \sin^2 \theta \cos \theta}{\cos \theta - \cos^2 \theta} = \dfrac{\sin^2 \theta(1 - \cos \theta)}{\cos \theta(1 - \cos \theta)} = \dfrac{\sin^2 \theta}{\cos \theta} = \dfrac{1 - \cos^2 \theta}{\cos \theta}$

$= \dfrac{1}{\cos \theta} - \cos \theta = \sec \theta - \cos \theta$

10. $-\frac{1}{4}(\sqrt{6} + \sqrt{2})$ **11.** $-\frac{1}{2}\sqrt{2 - \sqrt{3}}$ **12.** $\dfrac{2\sqrt{8}}{9} = \dfrac{4\sqrt{2}}{9}$ **13.** $\sqrt{3}$

14. $\sin\left(\dfrac{\pi}{4} - \theta\right) = \sin \dfrac{\pi}{4} \cos \theta - \cos \dfrac{\pi}{4} \sin \theta = \dfrac{\sqrt{2}}{2} \cos \theta - \dfrac{\sqrt{2}}{2} \sin \theta = \dfrac{\sqrt{2}}{2}(\cos \theta - \sin \theta)$

15. $\csc 4x = \dfrac{1}{\sin 4x} = \dfrac{1}{2 \sin 2x \cos 2x} = \dfrac{1}{4 \sin x \cos x \cos 2x} = \dfrac{1}{4} \csc x \sec x \sec 2x$

16. $\sin^2 \dfrac{x}{2} = \dfrac{1 - \cos x}{2} = \dfrac{\tan x - \tan x \cos x}{2 \tan x} = \dfrac{\tan x - \sin x}{2 \tan x}$ **17.** $\frac{23}{21}$ **18.** $\frac{1}{2}$ **19.** $\dfrac{\pi}{6}; \dfrac{5\pi}{6}; \dfrac{3\pi}{2}$

20. $60°, 120°, 240°, 300°$ **21.** $\dfrac{\pi}{4} + k\pi, k$ any integer   **22.** $1.16; 5.12$ **23.** $\frac{56}{65}$ **24.** $8(2 + \sqrt{3})$ **25.** 18

CHAPTER 7 MULTIPLE CHOICE TEST (page 485)

1. (b)   **2.** (c)   **3.** (b)   **4.** (d)   **5.** (c)   **6.** (b)   **7.** (a)   **8.** (a)   **9.** (c)   **10.** (d)   **11.** (c)

12. (b)   **13.** (a)   **14.** (e)   **15.** (d)   **16.** (b)   **17.** (d)   **18.** (c)   **19.** (a)   **20.** (b)

CHAPTER 8 ADDITIONAL APPLICATIONS OF TRIGONOMETRY

8.1 The Law of Cosines (page 495)

1. $a = \sqrt{93}$ **3.** $A = 30°$ **5.** $a = 6.1$; $B = 26.3°$ **7.** $C = 90°$; $A = 67.4°$ **9.** $B = 37°$; $C = 9.2°$
11. $c = 11.2$; $A = 125.7°$ **13.** $a = 5.0$; $B = 17.3°$ **15.** 151 meters **17.** 4.8 miles **19.** 22.6 centimeters
21. 42.7 **23.** 111 feet **25.** 33.6 miles **27.** 112.5 **29.** 34.4 **31.** 24 **33.** 25 **35.** 9.1 sq ft
37. 435

39. (a) $\dfrac{\cos A}{a} + \dfrac{\cos B}{b} + \dfrac{\cos C}{c} = \dfrac{1}{a} \cdot \dfrac{b^2 + c^2 - a^2}{2bc} + \dfrac{1}{b} \cdot \dfrac{a^2 + c^2 - b^2}{2ac} + \dfrac{1}{c} \cdot \dfrac{a^2 + b^2 - c^2}{2ab} =$

$\dfrac{(b^2 + c^2 - a^2) + (a^2 + c^2 - b^2) + (a^2 + b^2 - c^2)}{2abc} = \dfrac{a^2 + b^2 + c^2}{2abc}$

(b) Multiply the result in part (a) by $2abc$ and factor out the 2.

8.2 The Law of Sines (page 507)

1. $a = c = 18\sqrt{3}$ **3.** $C = \dfrac{\pi}{6}$ **5.** $B = 100°$; $a = 5.1$; $c = 10.0$ **7.** $C = 67.8°$; $a = 5.8$; $c = 6.0$

9. No solution **11.** 1 triangle **13.** No solution **15.** $B = 86.7°$; $c = 9.7$ or $B = 93.3°$; $c = 8.3$
17. No solution **19.** $C = 27.6°$ or $C = 152.4°$ **21.** 133.8° **23.** 1.6 miles **25.** 454 feet
27. 11.7 centimeters **29.** 16 **31.** 835 meters, to the nearest meter **33.** 26 pounds; 23° **35.** 42 pounds; 63°
37. 197 mph; N22°E **39.** 157 pounds **41.** 1150 pounds
43. $K = \dfrac{1}{2}\, bc \sin A = \dfrac{1}{2}\left(\dfrac{a \sin B}{\sin A}\right)\left(\dfrac{a \sin C}{\sin A}\right) \sin A = \dfrac{a^2 \sin B \sin C}{2 \sin A}$

8.3 Trigonometric Form of Complex Numbers (page 516)

1.

3.

5.

7.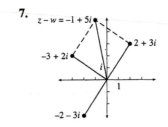

9.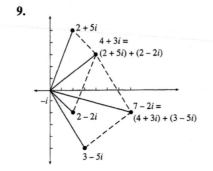

11.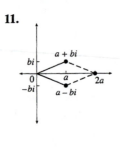

13.

$\sqrt{2}\left(\cos\frac{\pi}{4} + i\sin\frac{\pi}{4}\right)$

15.

$2\left(\cos\frac{2\pi}{3} + i\sin\frac{2\pi}{3}\right)$

17.

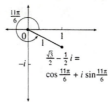

$\frac{\sqrt{3}}{2} - \frac{1}{2}i =$
$\cos\frac{11\pi}{6} + i\sin\frac{11\pi}{6}$

19.

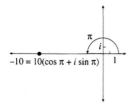

$-10 = 10(\cos\pi + i\sin\pi)$

21.

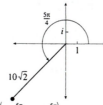

$-10 - 10i = 10\sqrt{2}\left(\cos\frac{5\pi}{4} + i\sin\frac{5\pi}{4}\right)$

23. $-\dfrac{3}{2} + \dfrac{3\sqrt{3}}{2}i$ **25.** $-5i$ **27.** $\dfrac{\sqrt{2}}{2} + \dfrac{\sqrt{6}}{2}i$ **29.** $2\sqrt{2} - 2\sqrt{2}i$ **31.** $.9848 - .1736i$

33. $zw = 5(\cos 150° + i\sin 150°) = -\dfrac{5\sqrt{3}}{2} + \dfrac{5}{2}i$

$\dfrac{z}{w} = \dfrac{1}{20}(\cos 50° + i\sin 50°) = .0321 + .0383i$

$\dfrac{w}{z} = 20(\cos 310° + i\sin 310°) = 12.8558 - 15.3209i$

35. $zw = 8\sqrt{2}(\cos\pi + i\sin\pi) = -8\sqrt{2}$

$\dfrac{z}{w} = \dfrac{\sqrt{2}}{8}\left(\cos\dfrac{3\pi}{2} + i\sin\dfrac{3\pi}{2}\right) = -\dfrac{\sqrt{2}}{8}i$

$\dfrac{w}{z} = 4\sqrt{2}\left(\cos\dfrac{\pi}{2} + i\sin\dfrac{\pi}{2}\right) = 4\sqrt{2}i$

37. $zw = \left[7\left(\cos\dfrac{\pi}{2} + i\sin\dfrac{\pi}{2}\right)\right]\left[2\left(\cos\dfrac{\pi}{6} + i\sin\dfrac{\pi}{6}\right)\right] = 14\left(\cos\dfrac{2\pi}{3} + i\sin\dfrac{2\pi}{3}\right) = -7 + 7\sqrt{3}i$

$\dfrac{z}{w} = \dfrac{7}{2}\left(\cos\dfrac{\pi}{3} + i\sin\dfrac{\pi}{3}\right) = \dfrac{7}{4} + \dfrac{7\sqrt{3}}{4}i$

39. $zw = \left[4\left(\cos\dfrac{2\pi}{3} + i\sin\dfrac{2\pi}{3}\right)\right]\left[4\left(\cos\dfrac{7\pi}{6} + i\sin\dfrac{7\pi}{6}\right)\right] = 16\left(\cos\dfrac{11\pi}{6} + i\sin\dfrac{11\pi}{6}\right) = 8\sqrt{3} - 8i$

$\dfrac{z}{w} = \cos\dfrac{3\pi}{2} + i\sin\dfrac{3\pi}{2} = -i$

41. $zw = \left[\sqrt{2}\left(\cos\dfrac{3\pi}{4} + i\sin\dfrac{3\pi}{4}\right)\right]\left[\sqrt{2}\left(\cos\dfrac{7\pi}{4} + i\sin\dfrac{7\pi}{4}\right)\right] = 2\left(\cos\dfrac{\pi}{2} + i\sin\dfrac{\pi}{2}\right) = 2i$

$\dfrac{z}{w} = \cos\pi + i\sin\pi = -1$

43. $\dfrac{\sqrt{2}}{2}(\cos 90° + i\sin 90°) = \dfrac{\sqrt{2}}{2}i$ **45.** $\dfrac{3}{2}(\cos 199° + i\sin 199°) = -1.4183 - .4884i$

47. $|\bar{z}| = |\overline{a + bi}| = |a - bi| = \sqrt{a^2 + (-b)^2} = \sqrt{a^2 + b^2} = |z|$

49. $|zw| = |(a + bi)(c + di)| = |(ac - bd) + (bc + ad)i| = \sqrt{(ac - bd)^2 + (bc + ad)^2}$

$= \sqrt{a^2c^2 + b^2d^2 + b^2c^2 + a^2d^2}$

$|z||w| = |a + bi||c + di| = \sqrt{a^2 + b^2}\sqrt{c^2 + d^2} = \sqrt{(a^2 + b^2)(c^2 + d^2)} = \sqrt{a^2c^2 + b^2d^2 + b^2c^2 + a^2d^2}$

51. Circle: $(x - 1)^2 + y^2 = 4$ **53.** Circle: $(x - 2)^2 + (y - 3)^2 = 1$ **55.** Ellipse: $\dfrac{x^2}{16} + \dfrac{y^2}{12} = 1$

8.4 De Moivre's Theorem (page 523)

1. $\cos 60° + i \sin 60° = \dfrac{1}{2} + \dfrac{\sqrt{3}}{2}i$ **3.** $\cos 320° + i \sin 320° = .7660 - .6428i$

5. $\dfrac{1}{64}\left(\cos \dfrac{3\pi}{4} + i \sin \dfrac{3\pi}{4}\right) = -\dfrac{\sqrt{2}}{128} + \dfrac{\sqrt{2}}{128}i$ **7.** $4(\cos 5\pi + i \sin 5\pi) = -4$

9. $216\left(\cos \dfrac{9\pi}{2} + i \sin \dfrac{9\pi}{2}\right) = 216i$ **11.** $\cos \dfrac{35\pi}{2} + i \sin \dfrac{35\pi}{2} = -i$ **13.** $\cos 10\pi + i \sin 10\pi = 1$

15. $3\left(\cos \dfrac{\pi}{45} + i \sin \dfrac{\pi}{45}\right),\ 3\left(\cos \dfrac{31\pi}{45} + i \sin \dfrac{31\pi}{45}\right),\ 3\left(\cos \dfrac{61\pi}{45} + i \sin \dfrac{61\pi}{45}\right)$

17. $2\left(\cos \dfrac{\pi}{40} + i \sin \dfrac{\pi}{40}\right),\ 2\left(\cos \dfrac{17\pi}{40} + i \sin \dfrac{17\pi}{40}\right), 2\left(\cos \dfrac{33\pi}{40} + i \sin \dfrac{33\pi}{40}\right), 2\left(\cos \dfrac{49\pi}{40} + i \sin \dfrac{49\pi}{40}\right),$

$2\left(\cos \dfrac{65\pi}{40} + i \sin \dfrac{65\pi}{40}\right)$

19.

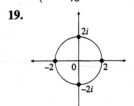

21.

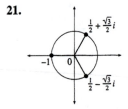

23.

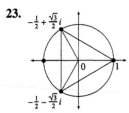

25.

cos 72° + i sin 72°

cos 144° + i sin 144°

cos 216° + i sin 216°

cos 288° + i sin 288°

1 = cos 0° + i sin 0°

27. $\dfrac{3}{2} + \dfrac{3\sqrt{3}}{2}i, -3, \dfrac{3}{2} - \dfrac{3\sqrt{3}}{2}i$ **29.** $2i, -\sqrt{3} - i, \sqrt{3} - i$

31. $2^{1/6}\left(\cos \dfrac{5\pi}{12} + i \sin \dfrac{5\pi}{12}\right), 2^{1/6}\left(\cos \dfrac{13\pi}{12} + i \sin \dfrac{13\pi}{12}\right), 2^{1/6}\left(\cos \dfrac{21\pi}{12} + i \sin \dfrac{21\pi}{12}\right)$

33. **(a)** For $k = 0$, the formula gives $r^{1/n}\left(\cos\dfrac{\theta}{n} + i\sin\dfrac{\theta}{n}\right)$

For $k = n$, the formula gives

$$r^{1/n}\left(\cos\frac{\theta + 2n\pi}{n} + i\sin\frac{\pi + 2n\pi}{n}\right) = r^{1/n}\left[\cos\left(\frac{\theta}{n} + 2\pi\right) + i\sin\left(\frac{\theta}{n} + 2\pi\right)\right] = r^{1/n}\left(\cos\frac{\theta}{n} + i\sin\frac{\theta}{n}\right)$$

(b) For $k = 1$, the formula gives $r^{1/n}\left(\cos\dfrac{\theta + 2\pi}{n} + i\sin\dfrac{\theta + 2\pi}{n}\right)$

For $k = n + 1$, the formula gives

$$r^{1/n}\left(\cos\frac{\theta + 2(n + 1)\pi}{n} + i\sin\frac{\theta + 2(n + 1)\pi}{n}\right) = r^{1/n}\left[\cos\left(\frac{\theta + 2\pi}{n} + 2\pi\right) + i\sin\left(\frac{\theta + 2\pi}{n} + 2\pi\right)\right]$$

$$= r^{1/n}\left(\cos\frac{\theta + 2\pi}{n} + i\sin\frac{\theta + 2\pi}{n}\right)$$

(c) Use k in the formula to get $r^{1/n}\left(\cos\dfrac{\theta + 2k\pi}{n} + i\sin\dfrac{\theta + 2k\pi}{n}\right)$

Use $k + n$ in the formula to get

$$r^{1/n}\left(\cos\frac{\theta + 2(n + k)\pi}{n} + i\sin\frac{\theta + 2(n + k)\pi}{n}\right) = r^{1/n}\left[\cos\left(\frac{\theta + 2k\pi}{n} + 2\pi\right) + i\sin\left(\frac{\theta + 2k\pi}{n} + 2\pi\right)\right]$$

$$= r^{1/n}\left(\cos\frac{\theta + 2k\pi}{n} + i\sin\frac{\theta + 2k\pi}{n}\right)$$

35. **(a)** $[r(\cos\theta + i\sin\theta)]^{-n} = r^{-n}(\cos\theta + i\sin\theta)^{-n} = r^{-n}[(\cos\theta + i\sin\theta)^{-1}]^{n}$

$\qquad\qquad\qquad\qquad\qquad = r^{-n}[(\cos\theta - i\sin\theta)]^{n}$　　　　　by Exercise 34

$\qquad\qquad\qquad\qquad\qquad = r^{-n}[\cos(-\theta) + i\sin(-\theta)]^{n}$

$\qquad\qquad\qquad\qquad\qquad = r^{-n}[\cos(-n\theta) + i\sin(-n\theta)]$　　　(since $n > 0$)

$\qquad\qquad\qquad\qquad\qquad = r^{-n}(\cos n\theta - i\sin n\theta)$

(b) Assuming the rule $z^{0} = 1$, the formula holds for $n = 0$. The formula is given in Section 8.4 for positive integers n. If in part (a) we let $m = -n$, where m is a negative integer, then

$[r(\cos\theta + i\sin\theta)]^{m} = [r(\cos\theta + i\sin\theta)]^{-n} = r^{-n}(\cos n\theta - i\sin n\theta)$　　　(by part (a))

$\qquad\qquad\qquad\qquad\qquad = r^{-n}[\cos(-n\theta) + i\sin(-n\theta)]$

$\qquad\qquad\qquad\qquad\qquad = r^{m}[\cos(m\theta) + i\sin(m\theta)]$

which shows that the formula holds for negative integers.

37. $1 + z + z^{2} = 1 + \cos\dfrac{2\pi}{3} + i\sin\dfrac{2\pi}{3} + \cos\dfrac{4\pi}{3} + i\sin\dfrac{4\pi}{3} = 1 - \dfrac{1}{2} + i\dfrac{\sqrt{3}}{2} - \dfrac{1}{2} - i\dfrac{\sqrt{3}}{2} = 0$

8.5 Polar Coordinates (page 529)

1.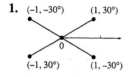
$(-1, -30°)$　　$(1, 30°)$

0

$(-1, 30°)$　　$(1, -30°)$

3.
$(-4, 0°)$　　0　　$(4, 0°)$

5.
$\left(\frac{1}{2}, \pi\right)$　0　$\left(-\frac{1}{2}, \pi\right)$

$\left(\frac{1}{2}, -\pi\right)$　　$\left(-\frac{1}{2}, -\pi\right)$

7. **(a)** $(1, 405°)$;　**(b)** $(1, -315°)$;　**(c)** $(-1, 225°)$;　**(d)** $(-1, -135°)$

9. **(a)** $\left(3, \dfrac{19\pi}{6}\right)$;　**(b)** $\left(3, -\dfrac{5\pi}{6}\right)$;　**(c)** $\left(-3, \dfrac{\pi}{6}\right)$;　**(d)** $\left(-3, -\dfrac{11\pi}{6}\right)$　　**11.** $(0, -7)$　　**13.** $\left(0, -\dfrac{3}{2}\right)$

15. $\left(\dfrac{2\sqrt{2}}{5}, \dfrac{2\sqrt{2}}{5}\right)$ **17.** $(2, 0)$ **19.** $(2, \pi)$ **21.** $\left(6\sqrt{2}, \dfrac{5\pi}{4}\right)$ **23.** $\left(4, \dfrac{\pi}{4}\right)$ **25.** $\left(2, \dfrac{11\pi}{6}\right)$ **27.** $(-7, \pi)$

29. $\left(-2, \dfrac{11\pi}{6}\right)$ **31.** $\left(-\sqrt{6}, \dfrac{5\pi}{4}\right)$ **33.** $(10, -90°)$ **35.** $r \cos\theta = 0 \cos\theta = 0 = x$; $r \sin\theta = 0 \sin\theta = 0 = y$

37. Using the polar coordinates $(-r, \theta + \pi)$, $x = -r\cos(\theta + \pi)$ and $y = -r\sin(\theta + \pi)$ since $-r > 0$. Then
$$x = -r(\cos\theta \cos\pi - \sin\theta \sin\pi) = -r(-\cos\theta) = r\cos\theta$$
$$y = -r(\sin\theta \cos\pi + \cos\theta \sin\pi) = -r(-\sin\theta) = r\sin\theta$$

8.6   Graphing Polar Equations   (page 538)

1.
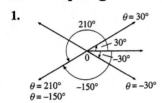

3.
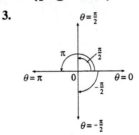

5.
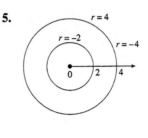

7. $\theta = -6°$ **9.** $\theta = \dfrac{\pi}{6}$

11. $x^2 + \left(y - \dfrac{1}{2}\right)^2 = \dfrac{1}{4}$

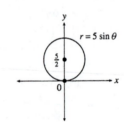

13. $x^2 + \left(y - \dfrac{5}{2}\right)^2 = \dfrac{25}{4}$

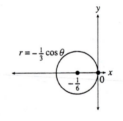

15. $\left(x + \dfrac{1}{6}\right)^2 + y^2 = \dfrac{1}{36}$

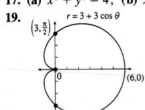

17. (a) $x^2 + y^2 = 4$;   (b) $x^2 + y^2 = 4$;   (c) $x^2 + y^2 = 5$

19.
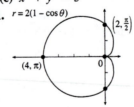

21.
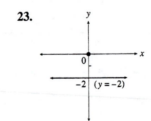

23.
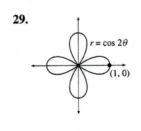

25.
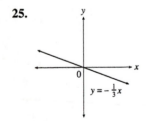

27.
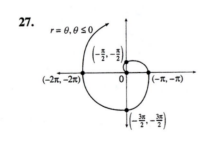

29.
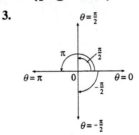

31.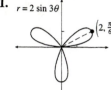
$r = 2 \sin 3\theta$
$\left(2, \frac{\pi}{6}\right)$

33.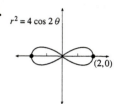
$r^2 = 4 \cos 2\theta$
$(2,0)$

35.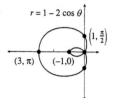
$r = 1 - 2 \cos \theta$
$\left(1, \frac{\pi}{2}\right)$
$(3, \pi)$ $(-1,0)$

37. $y^2 = 16 - 8x$ (parabola)   **39.** $x = -5$ (vertical line)   **41.** $(x - 3)^2 + (y - 4)^2 = 25$ (circle)
43. $2x + 3y = 3$ (line)   **45.** $x + \sqrt{3}y = 2\sqrt{3}$ (line)

47. Let $P(r, \theta)$ be a point on the line $x = c \ne 0$ and draw OP. From the reference triangle, $\cos \theta = \dfrac{x}{r} = \dfrac{c}{r}$, so that $r \cos \theta = c$. If $c = 0$, then (r, θ) is on ray $\dfrac{\pi}{2}$ so that $r \cos \dfrac{\pi}{2} = 0$.

49. $\left(\dfrac{3}{2}, \dfrac{\pi}{3}\right), \left(\dfrac{3}{2}, \dfrac{5\pi}{3}\right)$ **51.** $\left(0, \dfrac{\pi}{2}\right), \left(\dfrac{\sqrt{3}}{2}, \dfrac{\pi}{6}\right), \left(-\dfrac{\sqrt{3}}{2}, \dfrac{5\pi}{6}\right)$ **53.** $\left(1, -\dfrac{\pi}{3}\right)$

CHAPTER 8   REVIEW EXERCISES   (page 541)

1. See page 491.   **2.** See page 492.   **3.** $c = 11$ **4.** $A = 25°, B = 45°$ **5.** 136 meters   **6.** $\sqrt{21}$
7. See page 493.   **8.** 57 cm^2 **9.** See page 495.   **10.** $\sqrt{1836} \approx 43$ cm^2 **11.** See page 500.
12. See pages 499–500.   **13.** $5\sqrt{6}$ **14.** $A = 73°, b = 7.6, c = 5.6$ **15.** $c = 182$ **16.** No solution possible
17. $B = 74.6°, C = 65.4°, c = 11.3$ or $B = 105.4°, C = 34.6°, c = 7.1$ **18.** 20.2 mph; S 8.5° W
19. 300 mph, · N 44° W   **20.** 175 lbs   **21.** See page 511.   **22.** See page 511.   **23.** See page 513.
24. (a) $6 + i$; (b) 4; (c) $-2 + 5i$ **25.** See page 513.

26. (a) $2\left(\cos \dfrac{5\pi}{6} + i \sin \dfrac{5\pi}{6}\right)$; (b) $\sqrt{2}\left(\cos \dfrac{7\pi}{4} + i \sin \dfrac{7\pi}{4}\right)$ **27.** $1.5321 + 1.2856\,i$ **28.** See page 514.

29. See page 515.   **30.** $(2 + 2i)(-1 - i) = -4i$; $z_1 = 2\sqrt{2}\left(\cos \dfrac{\pi}{4} + i \sin \dfrac{\pi}{4}\right)$, $z_2 = \sqrt{2}\left(\cos \dfrac{5\pi}{4} + i \sin \dfrac{5\pi}{4}\right)$,

$$z_1 z_2 = 4\left[\cos\left(\dfrac{\pi}{4} + \dfrac{5\pi}{4}\right) + i \sin\left(\dfrac{\pi}{4} + \dfrac{5\pi}{4}\right)\right] = 4\left(\cos \dfrac{3\pi}{2} + i \sin \dfrac{3\pi}{2}\right) = -4i$$

31. $\dfrac{2 + 2i}{-1 - i} \cdot \dfrac{-1 + i}{-1 + i} = -2$; $\dfrac{z_1}{z_2} = \dfrac{2\sqrt{2}}{\sqrt{2}}\left[\cos\left(\dfrac{\pi}{4} - \dfrac{5\pi}{4}\right) + i \sin\left(\dfrac{\pi}{4} - \dfrac{5\pi}{4}\right)\right] = 2[\cos(-\pi) + i \sin(-\pi)] = -2$

32. See page 519.   **33.** $\left(\sqrt{2}\right)^6\left[\cos\left(6 \cdot \dfrac{\pi}{4}\right) + i \sin\left(6 \cdot \dfrac{\pi}{4}\right)\right] = 8\left(\cos \dfrac{3\pi}{2} + i \sin \dfrac{3\pi}{2}\right) = -8i$ **34.** See page 519.

35. See page 521.   **36.** $\dfrac{3\sqrt{3}}{2} + \dfrac{3}{2}i, \dfrac{-3\sqrt{3}}{2} + \dfrac{3}{2}i, -3i$
37. $2\sqrt{2} + 2\sqrt{2}i, -2\sqrt{2} + 2\sqrt{2}i, -2\sqrt{2} - 2\sqrt{2}i, 2\sqrt{2} - 2\sqrt{2}i$ **38.** See page 522.   **39.** See page 522.

40.

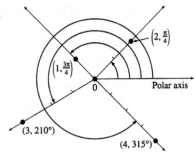

41.

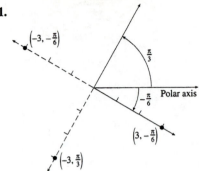

42. (a) 4; **(b)** 4; **(c)** −4; **(d)** −4   **43. (a)** $\left(5, -\dfrac{7\pi}{4}\right)$; **(b)** $\left(-5, \dfrac{5\pi}{4}\right)$ **44.** $(\sqrt{2}, \sqrt{2})$; $\left(\dfrac{\sqrt{3}}{4}, -\dfrac{1}{4}\right)$

45. $\left(2, \dfrac{7\pi}{6}\right)$; $\left(2\sqrt{2}, \dfrac{3\pi}{4}\right)$ **46.** See page 531.   **47.** See page 532.   **48.** $\theta = 120°$ **49.** $r = -6\cos\theta$

50.

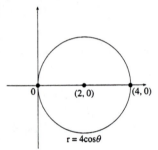

51.

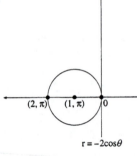

52.

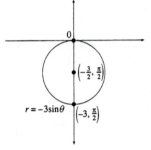

53. Twice   **54.** $x^2 + (y - a)^2 = a^2$ **55.** $x = -\dfrac{1}{2}y^2 + \dfrac{1}{2}$

56. $r = 4(1 + \sin\theta)$ **57.** $r = -\sin 2\theta$

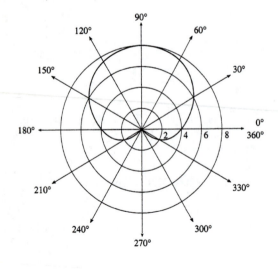

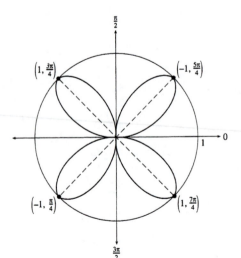

58. $r = -\theta, \theta \geq 0$

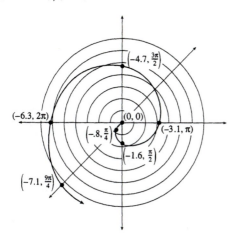

CHAPTER 8   STANDARD ANSWER TEST   (page 543)

1. (a) $4\sqrt{7}, \dfrac{\pi}{3}$; **(b)** $\sin B = \dfrac{8\sin 60°}{4\sqrt{7}} = \dfrac{\sqrt{21}}{7}$ **2. (a)** None **(b)** Two   **3.** $60\sqrt{3}$ cm²   **4.** $\dfrac{20\sqrt{6}}{3}$

5. 105 cm²   **6.** 31 cm²   **7.** $\dfrac{24}{\sqrt{3}}$ or $8\sqrt{3}$ **8.** $A = 37.4°$, $B = 118.8°$

9. Two solutions: $B = 53.1°$, $C = 96.9°$, $c = 19.9$; $B = 126.9°$, $C = 23.1°$, $c = 7.9$ **10.** 313 pounds
11. 15 mph   **12.** N 22° W   **13.** $-5 + 5i$

14. $z = \dfrac{1}{2}\left(\cos\dfrac{\pi}{6} + i\sin\dfrac{\pi}{6}\right)$, $w = 2\left(\cos\dfrac{2\pi}{3} + i\sin\dfrac{2\pi}{3}\right)$, $\dfrac{z}{w} = \dfrac{1}{4}\left(\cos\dfrac{3\pi}{2} + i\sin\dfrac{3\pi}{2}\right) = -\dfrac{1}{4}i$

15. $\dfrac{1}{32}\left(\cos\dfrac{5\pi}{6} + i\sin\dfrac{5\pi}{6}\right) = -\dfrac{\sqrt{3}}{64} + \dfrac{1}{64}i$

16.

$-1 + \sqrt{3}i$

2

$-1 - \sqrt{3}i$

17. $2\left(\cos\dfrac{3\pi}{16} + i\sin\dfrac{3\pi}{16}\right)$, $2\left(\cos\dfrac{11\pi}{16} + i\sin\dfrac{11\pi}{16}\right)$, $2\left(\cos\dfrac{19\pi}{16} + i\sin\dfrac{19\pi}{16}\right)$, $2\left(\cos\dfrac{27\pi}{16} + i\sin\dfrac{27\pi}{16}\right)$

18. (a) $(-3, -300°)$; **(b)** $(3, 240°)$.   **19. (a)** $\left(\dfrac{\sqrt{3}}{6}, -\dfrac{1}{6}\right)$; **(b)** $\left(14, \dfrac{3\pi}{4}\right)$ with $r > 0$ and $0 < \theta < 2\pi$.

20. $r = -6\sin\theta$

21. $x^2 + (y - 1)^2 = 1$; a circle with radius 1 and center $(0, 1)$ in rectangular coordinates.

22. $\left(x + \frac{3}{2}\right)^2 + (y - 2)^2 = \frac{25}{4}$; a circle with radius $\frac{5}{2}$ and center $\left(-\frac{3}{2}, 2\right)$ in rectangular coordinates.

23.

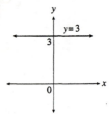

24.

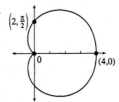

25.

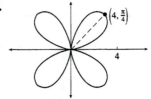

CHAPTER 8   MULTIPLE CHOICE TEST   (page 545)

1. (e)   **2.** (b)   **3.** (d)   **4.** (a)   **5.** (b)   **6.** (b)   **7.** (c)   **8.** (d)   **9.** (a)   **10.** (c)   **11.** (d)

12. (c)   **13.** (a)   **14.** (b)   **15.** (d)   **16.** (c)   **17.** (a)   **18.** (b)   **19.** (c)   **20.** (a)

CHAPTER 9   LINEAR SYSTEMS, MATRICES, AND DETERMINANTS

9.1 Solving Linear Systems Using Matrices (page 558)

1. $(-4, -1)$ **3.** $(0, 9)$ **5.** $\left(\frac{1}{8}, \frac{1}{2}\right)$ **7.** $(1, 0)$ **9.** $\left(2 - \frac{3}{2}c, c\right)$ for all c, or $\left(d, \frac{4}{3} - \frac{2}{3}d\right)$ for all d

11. No solutions   **13.** $(5, 2, -1)$ **15.** $(3, 4, 2)$ **17.** No solutions

19. $\left(\frac{3}{11} + \frac{5}{11}c, -\frac{4}{11} + \frac{8}{11}c, c\right)$ for all c, or $\left(d, -\frac{4}{5} + \frac{8}{5}d, -\frac{3}{5} + \frac{11}{5}d\right)$ for all d, or $\left(\frac{1}{2} + \frac{5}{8}e, e, \frac{1}{2} + \frac{11}{8}e\right)$ for all e

21. $(-1, 3, 4, -2)$ **23.** No solutions   **25.** 25 ones, 30 fives, 40 tens   **27.** $40°, 50°, 90°$

29. 15 pounds peanuts, 10 pounds pecans, 25 pounds Brazil nuts   **31.** 10 grams of A, 15 grams of B, 20 grams of C

9.2 Matrix Algebra (page 569)

1. $\begin{bmatrix} 6 & 1 \\ 0 & 8 \end{bmatrix}$ **3.** Does not exist   **5.** $\begin{bmatrix} 3 & -1 & 4 \\ -7 & 4 & 2 \\ -2 & -3 & 5 \end{bmatrix}$ **7.** $[-7]$ **9.** $\begin{bmatrix} 2 & 6 \\ 9 & 10 \end{bmatrix}$ **11.** $\begin{bmatrix} 7 \\ -11 \end{bmatrix}$

13. $\begin{bmatrix} 21 & -11 & 2 \\ 17 & -17 & 19 \end{bmatrix}$ **15.** Does not exist   **17.** $\begin{bmatrix} 2 & -5 \\ 4 & 3 \\ 4 & 2 \end{bmatrix}$ **19.** $\begin{bmatrix} 1 & 2 & -15 & 4 \\ 27 & -6 & -15 & 8 \end{bmatrix}$ **21.** $\begin{bmatrix} 0 \\ 0 \\ 0 \end{bmatrix}$ **23.** H

25. $\begin{bmatrix} -10 & 6 & 6 \\ 0 & 12 & -8 \end{bmatrix}$ **27.** $\begin{bmatrix} 10 & 16 & -24 & 0 \\ 10 & -8 & -18 & 10 \end{bmatrix}$

29. $B + C = \begin{bmatrix} 0+3 & 6+7 \\ 5-1 & -4+(-2) \end{bmatrix} = \begin{bmatrix} 3 & 13 \\ 4 & -6 \end{bmatrix}$; $C + B = \begin{bmatrix} 3+0 & 7+6 \\ -1+5 & -2+(-4) \end{bmatrix} = \begin{bmatrix} 3 & 13 \\ 4 & -6 \end{bmatrix}$

31. $A + (B + C) = A + \begin{bmatrix} 3 & 13 \\ 4 & -6 \end{bmatrix} = \begin{bmatrix} 5 & 14 \\ 0 & -5 \end{bmatrix}$; $(A + B) + C = \begin{bmatrix} 2 & 7 \\ 1 & -3 \end{bmatrix} + C = \begin{bmatrix} 5 & 14 \\ 0 & -5 \end{bmatrix}$

33. $(A + B)C = \begin{bmatrix} 2 & 7 \\ 1 & -3 \end{bmatrix} C = \begin{bmatrix} -1 & 0 \\ 6 & 13 \end{bmatrix}$; $AC + BC = \begin{bmatrix} 5 & 12 \\ -13 & -30 \end{bmatrix} + \begin{bmatrix} -6 & -12 \\ 19 & 43 \end{bmatrix} = \begin{bmatrix} -1 & 0 \\ 6 & 13 \end{bmatrix}$

35. $(AB)C = \begin{bmatrix} -5 & 4 & 2 \\ 3 & 21 & 33 \\ 8 & 4 & 12 \end{bmatrix} C = \begin{bmatrix} -31 \\ 6 \\ 44 \end{bmatrix}$; $A(BC) = A \begin{bmatrix} -9 \\ 11 \end{bmatrix} = \begin{bmatrix} -31 \\ 6 \\ 44 \end{bmatrix}$

37. $A(2B) + (2A)C = (2A)B + (2A)C = 2A(B + C) = [20 \quad -2]$

39. (a) $\begin{bmatrix} d & e & f \\ a & b & c \\ g & h & i \end{bmatrix}$ (b) $F = \begin{bmatrix} 0 & 0 & 1 \\ 0 & 1 & 0 \\ 1 & 0 & 0 \end{bmatrix}$ **41.** $E = \begin{bmatrix} 1 & 0 & 0 \\ 0 & 1 & 0 \\ 3 & 0 & 1 \end{bmatrix}$ $F = \begin{bmatrix} 1 & 0 & 0 \\ 0 & 1 & 0 \\ 0 & -1 & 1 \end{bmatrix}$

43. (a) $x = -2, y = -3$ (b) If there were such x, y, then $\begin{bmatrix} 2 & 2 \\ x & x \end{bmatrix} = \begin{bmatrix} 0 & 0 \\ 0 & 0 \end{bmatrix}$, which cannot be true since $2 \neq 0$.

45. (a) $\begin{bmatrix} x^2 & 0 & 0 \\ 0 & y^2 & 0 \\ 0 & 0 & z^2 \end{bmatrix}$ (b) $\begin{bmatrix} x^3 & 0 & 0 \\ 0 & y^3 & 0 \\ 0 & 0 & z^3 \end{bmatrix}$ (c) $\begin{bmatrix} x^n & 0 & 0 \\ 0 & y^n & 0 \\ 0 & 0 & z^n \end{bmatrix}$

47. (a) A

(b) $(A + B)^t = \begin{bmatrix} a_1 + b_1 & a_2 + b_2 & a_3 + b_3 \\ a_4 + b_4 & a_5 + b_5 & a_6 + b_6 \end{bmatrix}^t = \begin{bmatrix} a_1 + b_1 & a_4 + b_4 \\ a_2 + b_2 & a_5 + b_5 \\ a_3 + b_3 & a_6 + b_6 \end{bmatrix} = \begin{bmatrix} a_1 & a_4 \\ a_2 & a_5 \\ a_3 & a_6 \end{bmatrix} + \begin{bmatrix} b_1 & b_4 \\ b_2 & b_5 \\ b_3 & b_6 \end{bmatrix} = A^t + B^t$

49. (a)

$$M = \begin{bmatrix} 21 & 15 & 34 \\ 19 & 25 & 28 \end{bmatrix} \begin{matrix} \text{Store I} \\ \text{Store II} \end{matrix}$$

$$J = \begin{bmatrix} 28 & 18 & 27 \\ 25 & 17 & 28 \end{bmatrix} \begin{matrix} \text{Store I} \\ \text{Store II} \end{matrix}$$

bic. row. tread.

$$M + J = \begin{bmatrix} 49 & 33 & 61 \\ 44 & 42 & 56 \end{bmatrix}$$ Gives the total sales per item for May and June for each store.

$$M - J = \begin{bmatrix} -7 & -3 & 7 \\ -6 & 8 & 0 \end{bmatrix}$$ A positive entry means that the store sold more of that item in May than in June. If negative, more were sold in June than in May, and if 0 the sales were the same for the two months.

(b) July sales $= S - (M + J)$

$$= \begin{bmatrix} 18 & 12 & 9 \\ 15 & 20 & 7 \end{bmatrix} \begin{matrix} \text{Store I} \\ \text{Store II} \end{matrix}$$

Number of units   Total units

    a b c m p w

51. (a) $P = [0 \quad 1200 \quad 200]$; $\quad PR = [4200 \quad 7200 \quad 4400]$

    (b) $(PR)C = P(RC) = [17{,}480 \quad 17{,}530]$

       $\$17{,}530 - \$17{,}480 = \$50$; it is more economical to buy from supplier S_1.

9.3 Solving Linear Systems Using Inverses (page 581)

1. $\begin{bmatrix} 0 & \frac{1}{2} \\ -1 & 2 \end{bmatrix}$ **3.** Does not exist   **5.** $\begin{bmatrix} -\frac{4}{7} & -\frac{5}{7} \\ -\frac{3}{7} & -\frac{2}{7} \end{bmatrix}$ **7.** $\begin{bmatrix} -1 & 2 \\ 2 & -1 \end{bmatrix}$ **9.** $\begin{bmatrix} 0 & \frac{1}{b} \\ \frac{1}{a} & 0 \end{bmatrix}$ **11.** $A^{-1} = A$

13. $\begin{bmatrix} -\frac{1}{2} & \frac{3}{4} & \frac{1}{4} \\ -1 & \frac{1}{2} & \frac{1}{2} \\ \frac{3}{4} & -\frac{3}{8} & -\frac{1}{8} \end{bmatrix}$ **15.** Does not exist   **17.** $\begin{bmatrix} 1 & 0 & 2 \\ 2 & -1 & 3 \\ 4 & 1 & 8 \end{bmatrix}$ **19.** $\begin{bmatrix} -5 & 0 & \frac{5}{2} & 3 \\ 2 & 0 & -\frac{1}{2} & -1 \\ -\frac{3}{2} & \frac{1}{2} & \frac{3}{4} & 1 \\ 2 & 0 & -1 & -1 \end{bmatrix}$

21. $3x + y = 9$ **23.** $(6, 1)$ **25.** $(18, -21)$ **27.** $(-8, -8, 6)$
      $2x - 2y = 14$

29. $\left(-\frac{1}{2}, -\frac{1}{2}, -2\right)$ **31.** $(13, 7, -6)$ **33.** $(14, -4, 2, -4)$

35. $(AB)^{-1} = \begin{bmatrix} 1 & 2 \\ -4 & 8 \end{bmatrix}^{-1} = \begin{bmatrix} \frac{1}{2} & -\frac{1}{8} \\ \frac{1}{4} & \frac{1}{16} \end{bmatrix}$; $B^{-1}A^{-1} = \begin{bmatrix} 1 & 1 \\ \frac{1}{2} & 0 \end{bmatrix}\begin{bmatrix} \frac{1}{2} & \frac{1}{8} \\ 0 & -\frac{1}{4} \end{bmatrix} = \begin{bmatrix} \frac{1}{2} & -\frac{1}{8} \\ \frac{1}{4} & \frac{1}{16} \end{bmatrix}$

37. (a) $AB = I$
       $C(AB) = CI$
       $(CA)B = C$
       $IB = C$
       $B = C$

   (b) The inverse of a matrix is unique.

39. (a) $A^{-1} = \begin{bmatrix} \frac{1}{2} & \frac{1}{8} \\ 0 & -\frac{1}{4} \end{bmatrix}$; $(A^{-1})^2 = \begin{bmatrix} \frac{1}{4} & \frac{1}{32} \\ 0 & \frac{1}{16} \end{bmatrix}$

       $A^2 = \begin{bmatrix} 4 & -2 \\ 0 & 16 \end{bmatrix}$; $(A^2)^{-1} = \begin{bmatrix} \frac{1}{4} & \frac{1}{32} \\ 0 & \frac{1}{16} \end{bmatrix}$

   (b) $(A^2)^{-1} = (AA)^{-1} = A^{-1}A^{-1}$ by Ex. 38(b)
                   $= (A^{-1})^2$

   (c) $(A^n)^{-1} = (A^{-1})^n$

9.4 Introduction to Determinants (page 588)

1. 17   **3.** 94   **5.** -60 **7.** 0   **9.** $(-1, 2)$ **11.** $(3, 2)$ **13.** $(1, -1)$ **15.** $(12, 24)$ **17.** $(5, 5)$

19. $(-5, -7)$ **21.** $\left(\frac{6}{5}, -\frac{3}{10}\right)$ **23.** Inconsistent   **25.** Dependent   **27.** Inconsistent   **29.** 6   **31.** $-\frac{5}{4}$

33. $\left(\frac{2}{3}, -\frac{7}{6}\right)$ **35.** $\begin{vmatrix} a_1 & b_1 \\ a_2 & b_2 \end{vmatrix} = a_1 b_2 - a_2 b_1 = a_1 b_2 - b_1 a_2 = \begin{vmatrix} a_1 & a_2 \\ b_1 & b_2 \end{vmatrix}$

37. Let $b_1 = ka_1, b_2 = ka_2$; then $\begin{vmatrix} a_1 & b_1 \\ a_2 & b_2 \end{vmatrix} = \begin{vmatrix} a_1 & ka_1 \\ a_2 & ka_2 \end{vmatrix} = \begin{vmatrix} a_1 & a_2 \\ ka_1 & ka_2 \end{vmatrix}$ (by Exercise 35)
                                     $= 0$ (by Exercise 36)

39. $\begin{vmatrix} 27 & 3 \\ 105 & -75 \end{vmatrix} = 3\begin{vmatrix} 9 & 1 \\ 105 & -75 \end{vmatrix} = 45\begin{vmatrix} 9 & 1 \\ 7 & -5 \end{vmatrix} = -2340$

     $\begin{vmatrix} 27 & 3 \\ 105 & -75 \end{vmatrix} = 3\begin{vmatrix} 9 & 1 \\ 105 & -75 \end{vmatrix} = 9\begin{vmatrix} 3 & 1 \\ 35 & -75 \end{vmatrix} = 45\begin{vmatrix} 3 & 1 \\ 7 & -15 \end{vmatrix} = -2340$

41. $\begin{vmatrix} a_1 + kb_1 & b_1 \\ a_2 + kb_2 & b_2 \end{vmatrix} = \begin{vmatrix} a_1 & b_1 \\ a_2 & b_2 \end{vmatrix} + \begin{vmatrix} kb_1 & b_1 \\ kb_2 & b_2 \end{vmatrix}$　　　(by Exercise 40)

$$= \begin{vmatrix} a_1 & b_1 \\ a_2 & b_2 \end{vmatrix} + 0 \qquad \text{(by Exercise 37)}$$

$$= \begin{vmatrix} a_1 & b_1 \\ a_2 & b_2 \end{vmatrix}$$

43. (Sample solution)

$$\begin{vmatrix} 12 & -42 \\ -6 & 27 \end{vmatrix} = 6 \begin{vmatrix} 2 & -42 \\ -1 & 27 \end{vmatrix} = 12 \begin{vmatrix} 1 & -21 \\ -1 & 27 \end{vmatrix} = 12 \begin{vmatrix} 1 & -21 \\ 0 & 6 \end{vmatrix} = 12(6) = 72$$

　　　　　Exercise 38　　　Exercise 38　　　Exercise 41

45. Walked at 4 mph, rode at 10 mph

9.5　Higher-Order Determinants and Their Properties　(page 597)

1. (a) $6 \begin{vmatrix} -9 & 4 \\ 5 & 1 \end{vmatrix} + (-3) \begin{vmatrix} -2 & -1 \\ -9 & 4 \end{vmatrix} = -123$

(b) $-3 \begin{vmatrix} -2 & -1 \\ -9 & 4 \end{vmatrix} - 5 \begin{vmatrix} 6 & -1 \\ 0 & 4 \end{vmatrix} + 1 \begin{vmatrix} 6 & -2 \\ 0 & -9 \end{vmatrix} = -123$

(c) $\begin{vmatrix} 6 & -2 & -1 \\ 0 & -9 & 4 \\ -3 & 5 & 1 \end{vmatrix} \begin{matrix} 6 & -2 \\ 0 & -9 \\ -3 & 5 \end{matrix} = -54 + 24 + 0 - (-27) - 120 - 0 = -123$

3. (a) $(-1)^{1+1}(-5) \begin{vmatrix} 7 & 4 \\ -6 & -2 \end{vmatrix} + (-1)^{1+2}(-2) \begin{vmatrix} -3 & 4 \\ 1 & -2 \end{vmatrix} + (-1)^{1+3} \begin{vmatrix} -3 & 7 \\ 1 & -6 \end{vmatrix} = -35$

(b) $(-1)^{2+1}(-3) \begin{vmatrix} -2 & 1 \\ -6 & -2 \end{vmatrix} + (-1)^{2+2}(7) \begin{vmatrix} -5 & 1 \\ 1 & -2 \end{vmatrix} + (-1)^{2+3}(4) \begin{vmatrix} -5 & -2 \\ 1 & -6 \end{vmatrix} = -35$

(c) $(-1)^{1+3}(1) \begin{vmatrix} -3 & 7 \\ 1 & -6 \end{vmatrix} + (-1)^{2+3}(4) \begin{vmatrix} -5 & -2 \\ 1 & -6 \end{vmatrix} + (-1)^{3+3}(-2) \begin{vmatrix} -5 & -2 \\ -3 & 7 \end{vmatrix} = -35$

5. $\begin{vmatrix} a_1 & b_1 & c_1 \\ a_2 & b_2 & c_2 \\ a_3 & b_3 & c_3 \end{vmatrix} = a_1 b_2 c_3 + a_2 b_3 c_1 + a_3 b_1 c_2 - a_1 b_3 c_2 - a_2 b_1 c_3 - a_3 b_2 c_1 = \begin{vmatrix} a_1 & a_2 & a_3 \\ b_1 & b_2 & b_3 \\ c_1 & c_2 & c_3 \end{vmatrix}$

7. $\begin{vmatrix} c_1 & b_1 & a_1 \\ c_2 & b_2 & a_2 \\ c_3 & b_3 & a_3 \end{vmatrix} = c_1 b_2 a_3 + c_2 b_3 a_1 + c_3 b_1 a_2 - c_1 b_3 a_2 - c_2 b_1 a_3 - c_3 b_2 a_1$

$$= a_3 b_2 c_1 + a_1 b_3 c_2 + a_2 b_1 c_3 - a_2 b_3 c_1 - a_3 b_1 c_2 - a_1 b_2 c_3$$

$$= a_1 b_3 c_2 + a_2 b_1 c_3 + a_3 b_2 c_1 - a_1 b_2 c_3 - a_2 b_3 c_1 - a_3 b_1 c_2$$

$$= -(a_1 b_2 c_3 + a_2 b_3 c_1 + a_3 b_1 c_2 - a_1 b_3 c_2 - a_2 b_1 c_3 - a_3 b_2 c_1)$$

$$= - \begin{vmatrix} a_1 & b_1 & c_1 \\ a_2 & b_2 & c_2 \\ a_3 & b_3 & c_3 \end{vmatrix}$$

9.
$$\begin{vmatrix} a_1 + kb_1 & b_1 & c_1 \\ a_2 + kb_2 & b_2 & c_2 \\ a_3 + kb_3 & b_3 & c_3 \end{vmatrix} = (a_1 + kb_1)\begin{vmatrix} b_2 & c_2 \\ b_3 & c_3 \end{vmatrix} - (a_2 + kb_2)\begin{vmatrix} b_1 & c_1 \\ b_3 & c_3 \end{vmatrix} + (a_3 + kb_3)\begin{vmatrix} b_1 & c_1 \\ b_2 & c_2 \end{vmatrix}$$

$$= \left(a_1\begin{vmatrix} b_2 & c_2 \\ b_3 & c_3 \end{vmatrix} - a_2\begin{vmatrix} b_1 & c_1 \\ b_3 & c_3 \end{vmatrix} + a_3\begin{vmatrix} b_1 & c_1 \\ b_2 & c_2 \end{vmatrix} \right)$$

$$+ kb_1\begin{vmatrix} b_2 & c_2 \\ b_3 & c_3 \end{vmatrix} - kb_2\begin{vmatrix} b_1 & c_1 \\ b_3 & c_3 \end{vmatrix} + kb_3\begin{vmatrix} b_1 & c_1 \\ b_2 & c_2 \end{vmatrix}$$

$$= \begin{vmatrix} a_1 & b_1 & c_1 \\ a_2 & b_2 & c_2 \\ a_3 & b_3 & c_3 \end{vmatrix} + k(b_1b_2c_3 - b_1b_3c_2 - b_1b_2c_3 + b_2b_3c_1 + b_1b_3c_2 - b_2b_3c_1)$$

$$= \begin{vmatrix} a_1 & b_1 & c_1 \\ a_2 & b_2 & c_2 \\ a_3 & b_3 & c_3 \end{vmatrix}$$

11. 0, since third column is -2 times first column.   **13.** $80 = 20(4)$

15. (Sample Solution)

$$\begin{vmatrix} 5 & -4 & 3 \\ -6 & 6 & 2 \\ -7 & 3 & 4 \end{vmatrix} = \begin{vmatrix} 14 & -4 & 3 \\ 0 & 6 & 2 \\ 5 & 3 & 4 \end{vmatrix} \quad \begin{array}{l} 3 \text{ times third column} \\ \text{added to first column} \\ \text{(Exercise 9)} \end{array}$$

$$= \begin{vmatrix} 14 & -13 & 3 \\ 0 & 0 & 2 \\ 5 & -9 & 4 \end{vmatrix} \quad \begin{array}{l} -3 \text{ times third column} \\ \text{added to second column} \\ \text{(Exercise 9)} \end{array}$$

$$= -2\begin{vmatrix} 14 & -13 \\ 5 & -9 \end{vmatrix} \quad \text{Expansion by minors along row 2}$$

$$= 2\begin{vmatrix} 14 & 13 \\ 5 & 9 \end{vmatrix} \quad \text{(Exercise 8)}$$

$$= 2(126 - 65)$$

$$= 122$$

17. (a) The determinant is 0 by the property of Exercise 6 (b) applied to columns one and three.
  (b) The determinant is 0 since each of the six terms in the simplified form on page 591 has a zero factor.
  (c) The determinant is 0 by the property in Exercise 6 (a) applied to rows one and three.
  (d) The determinant is 0 by the property in Exercise 6 (a) applied to rows two and three.

19. 35   **21.** -45 **23.** 4   **25.** -156

27. $2\begin{vmatrix} 0 & 5 & -3 \\ 2 & -4 & 6 \\ 3 & 0 & 1 \end{vmatrix} -(-1)\begin{vmatrix} 1 & 5 & -3 \\ 0 & -4 & 6 \\ -5 & 0 & 1 \end{vmatrix} +3\begin{vmatrix} 1 & 0 & -3 \\ 0 & 2 & 6 \\ -5 & 3 & 1 \end{vmatrix} -0\begin{vmatrix} 1 & 0 & 5 \\ 0 & 2 & -4 \\ -5 & 3 & 0 \end{vmatrix} = -144$

29. $x = 2, x = 3$ **31.** $x = -1, x = 3$ (3 is a double root)   **33.** -92 **35.** 2

37. $(-1, 0, 2)$ **39.** $(-1, 7, 2)$ **41.** $\left(\frac{2}{3}, 1, \frac{1}{2}\right)$

43. (i) The property in Exercise 8
  (ii) y times the second column added to the first (Exercise 9)
  (iii) z times the third column added to the first (Exercise 9)
  (vi) Substituting for the given values of d_1, d_2, d_3 in the general system on page 596
  (v) Divide by D.

45. (a) $|A^{-1}||A| = |A^{-1}A|$ by the given property

$$= |I|$$
$$= 1$$

Then,

$$|A^{-1}| = \frac{1}{|A|} \text{ by dividing by } |A|$$

For A in Example 2, page 577, $|A| = -8$, and using the result A^{-1}, $|A^{-1}| = -\frac{1}{8} = \frac{1}{-8} = \frac{1}{|A|}$

(b) Sample solution:

$$\text{Let } A = \begin{vmatrix} 1 & 0 & 0 \\ 0 & 1 & 0 \\ 0 & 0 & 0 \end{vmatrix} \quad \text{and } B = \begin{vmatrix} 0 & 0 & 0 \\ 0 & 0 & 0 \\ 0 & 0 & 1 \end{vmatrix}$$

$$\text{Then, } |A| + |B| = 0 + 0 = 0 \neq 1 = \begin{vmatrix} 1 & 0 & 0 \\ 0 & 1 & 0 \\ 0 & 0 & 1 \end{vmatrix} = |A + B|$$

9.6 Systems of Linear Inequalities (page 607)

1. $y \leq -2x + 1$ **3.** $y \leq 1$

5.
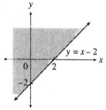

7.
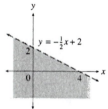

9.
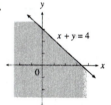

11.
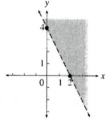

13.
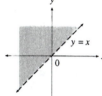

15.

17.
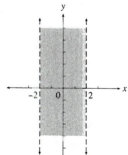

19.
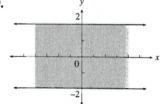

21.
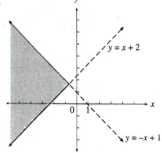

23.

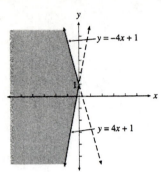

25.

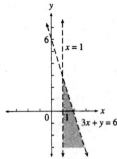

27.

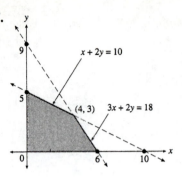

29.

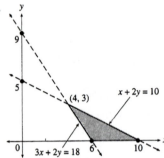

31.

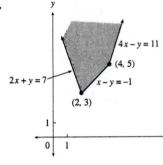

33.

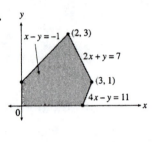

35.

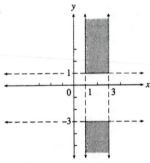

37. $2x + y < 6$
$3x - 4y \le 12$

39. $\quad y \le 4$
$x + y \ge 4$
$x - y \le 8$

41. $\quad y < 4$
$x + y > 4$
$x - y \ge 8$

43. $\quad y \ge 4$
$x + y > 4$
$x - y < 8$

45. $\quad y < 4$
$x + y \le 4$
$x - y < 8$

47. Let x be the number of basketballs and y the number of
uniforms. Then, $x \ge 4$, $y \ge 8$, $30x + 40y \le 600$
(or $3x + 4y < 60$)

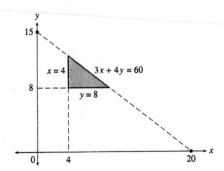

49. Let x be the number of style 1 sold and y the number
of the style 2 sold. Then
$x \ge 60$, $x \ge \frac{3}{2}y$ (or $y \le \frac{2}{3}x$)
$10x + 15y \le 3000$ (or $2x + 3y \le 600$)

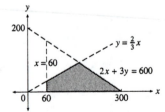

9.7 Linear Programming (page 617)

	Maximum Value	Minimum Value
1. (a)	58	18
(b)	43	11
3. (a)	60	12
(b)	39	9

5. (a)

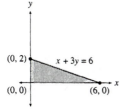

(b) (0, 0), (6, 0), (0, 2);
(c) maximum = 6, minimum = 0;
(d) maximum = 36, minimum = 0;
(e) maximum = 18, minimum = 0

7. (a)

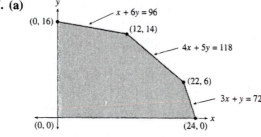

(b) (0, 0), (24, 0), (22, 6), (12, 14), (0, 16);
(c) maximum = 28, minimum = 0;
(d) maximum = 212, minimum = 0;
(e) maximum = 150, minimum = 0

9. (a)

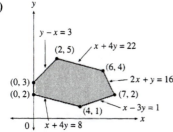

(b) (0, 2), (4, 1), (7, 2), (6, 4), (2, 5), (0, 3); **(c)** maximum = 10, minimum = 2;
(d) maximum = 76, minimum = 20; **(e)** maximum = 49, minimum = 17

11. For $p = x + 3y$, maximum = 36 at (0, 12); minimum = 10 at (10, 0)
For $q = 4x + 3y$, maximum = 40 at (10, 0); minimum = 12 at (0, 4)
For $r = \frac{1}{2}x + \frac{1}{4}y$, maximum = 5 at (10, 0); minimum = 1 at (0, 4)

13. For $p = 2x + 2y$, maximum = 24 at (7, 5); minimum = 8 at (4, 0)
For $q = 3x + y$, maximum = 26 at (7, 5); minimum = 9.5 at $(\frac{3}{2}, 5)$
For $r = x + \frac{1}{5}y$, maximum = 8 at (7, 5); minimum = $\frac{5}{2}$ at $(\frac{3}{2}, 5)$

15.

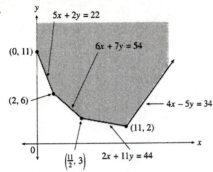

(a) No maximum; minimum = 8
(b) No maximum; minimum = 63
(c) No maximum; minimum = 10

17. (a) $x \geq 0$, $y \geq 0$ because a negative number of either model is not possible; $\frac{3}{2}x + y$ is the amount of time that machine M_1 works per day, and $\frac{3}{2}x + y \leq 12$ says that M_1 works at most 12 hours daily. The remaining inequalities are the constraints for machines M_2 and M_3; the explanations are similar, as for M_1.
(b) $p = 5x + 8y$
(c)

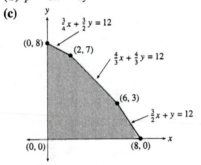

(d) maximum = \$66 when $x = 2$ and $y = 7$; **(e)** maximum = \$64 when $x = 8$ and $y = 0$
19. 400 of model A and 500 of model B. Profit \$100,000.

CHAPTER 9   REVIEW EXERCISES   (page 620)

1. Eliminate one of the variables resulting in a system of two equations in two variables that we know how to solve.
2. $(1, -2, 3)$ **3.** $(-2, 2, 1)$ **4.** $(4, -3, 2)$ **5.** $(-1, 0, 2)$ **6.** $(2, -1, -1)$ **7.** $(-1, 7, 2)$
8. Inconsistent; no solution   **9.** $(3, -2, -4)$ **10.** Inconsistent; no solution

11. (a) $\begin{bmatrix} 1 & 8 \\ -1 & 2 \end{bmatrix}$;   **(b)** undefined   **12.** $\begin{bmatrix} 4 & -4 & -18 \\ -5 & 24 & 2 \\ -2 & 1 & 4 \end{bmatrix}$ **13.** $\begin{bmatrix} -11 & 27 \\ 9 & -17 \end{bmatrix}$ **14.** $\begin{bmatrix} 12 & 10 & 8 \\ 3 & 1 & 8 \end{bmatrix}$

15. $\begin{bmatrix} -15 & 18 & 12 \\ 10 & -12 & -8 \\ -5 & 6 & 4 \end{bmatrix}$ **16.** $[-23]$ **17.** Undefined   **18.** $\begin{bmatrix} 7 & 21 & -24 \\ -4 & 6 & 4 \\ -3 & 9 & 12 \\ 3 & -6 & -6 \end{bmatrix}$ **19.** $\begin{bmatrix} -50 & 60 & 40 \\ -20 & 24 & 16 \end{bmatrix}$

20. $\begin{bmatrix} 42 & 35 \\ -70 & 7 \end{bmatrix}$ **21.** $\begin{bmatrix} 0 & -42 & 12 \\ -10 & 24 & 28 \\ 4 & -18 & -28 \end{bmatrix}$ **22.** E **23.** $\begin{bmatrix} 0 \\ 0 \\ 0 \\ 0 \end{bmatrix}$ **24.** $\begin{bmatrix} 21 & 12 & 3 \\ -1 & -2 & -3 \end{bmatrix}$ **25.** Undefined

26. $\begin{bmatrix} -32 & 0 & 0 \\ 0 & 32 & 0 \\ 0 & 0 & \frac{1}{32} \end{bmatrix}$ **27.** $\begin{bmatrix} 1 & 2 \\ 0 & 1 \end{bmatrix}$ **28.** $\begin{bmatrix} 0 & 1 \\ 1 & 0 \end{bmatrix}$ **29.** $\begin{bmatrix} 2 & 10 & 18 \\ -6 & -6 & -6 \end{bmatrix}$

30. (a) $A = [6 \quad 10 \quad 8 \quad 9]$; **(b)** $[16.32 \quad 15.94]$; the total cost in store 1 is \$16.32 and \$15.94 in store 2;

$$C = \begin{bmatrix} .48 & .50 \\ .40 & .37 \\ .46 & .48 \\ .64 & .60 \end{bmatrix}$$

31. $\begin{bmatrix} -\frac{2}{5} & -\frac{1}{5} \\ \frac{1}{4} & 0 \end{bmatrix}$ **32.** $\begin{bmatrix} 4 & 3 \\ 1 & \frac{3}{8} \end{bmatrix}$ **33.** Does not exist   **34.** $\begin{bmatrix} 1 & 0 & 1 \\ 2 & 1 & 1 \\ 3 & 2 & 2 \end{bmatrix}$

35. Does not exist   **36.** $\begin{bmatrix} 6 & -1 & 0 \\ 0 & 1 & -1 \\ -5 & 0 & 1 \end{bmatrix}$ **37.** Itself   **38.** $\begin{bmatrix} 1 & -2 & 1 & 4 \\ 0 & 1 & -2 & 1 \\ 0 & 0 & 1 & -2 \\ 0 & 0 & 0 & 1 \end{bmatrix}$ **39.** $\begin{bmatrix} 0 & -\frac{1}{2} & 0 & \frac{1}{2} \\ -\frac{1}{2} & 0 & 0 & \frac{1}{2} \\ 1 & 0 & 1 & 0 \\ \frac{1}{2} & \frac{1}{2} & 0 & 0 \end{bmatrix}$

40. $(12, 6)$ **41.** $(-5, 3, 7)$ **42.** $(-2, -4, 6)$ **43.** $(-\frac{3}{2}, \frac{1}{2}, 4, 7)$ **44.** -61 **45.** -3 **46.** 0 **47.** -2

48. $(5, 1)$ **49.** $(8, 5)$ **50.** $(\frac{1}{4}, 1)$ **51.** $(-6, -2)$

52. $\begin{vmatrix} a & b + 2a \\ c & d + 2c \end{vmatrix} = ad + 2ac - (bc + 2ac)$

$$= ad - bc = \begin{vmatrix} a & b \\ c & d \end{vmatrix}$$

53. $2 \begin{vmatrix} 3 & -2 \\ 1 & 6 \end{vmatrix} - 0 + 5 \begin{vmatrix} -1 & 4 \\ 3 & -2 \end{vmatrix} = -10$ **54.** $0 + 3 \begin{vmatrix} 2 & 4 \\ 5 & 6 \end{vmatrix} - (-2) \begin{vmatrix} 2 & -1 \\ 5 & 1 \end{vmatrix} = -10$

55. Add row three to row one and -3 times row three to row two giving

$$\begin{vmatrix} 7 & 0 & 10 \\ -15 & 0 & -20 \\ 5 & 1 & 6 \end{vmatrix} = -1 \begin{vmatrix} 7 & 10 \\ -15 & -20 \end{vmatrix} = -10$$

56. $\begin{vmatrix} 2 & -1 & 4 \\ 0 & 3 & -2 \\ 5 & 1 & 6 \end{vmatrix} \begin{matrix} 2 & -1 \\ 0 & 3 \\ 5 & 1 \end{matrix} = 36 + 10 + 0 - (60 - 4 - 0) = -10$

57. $(-1)^{1+1}(-1) \begin{vmatrix} 6 & 1 \\ 5 & -3 \end{vmatrix} + (-1)^{1+2}(3) \begin{vmatrix} 4 & 1 \\ 8 & -3 \end{vmatrix} + (-1)^{1+3}(2) \begin{vmatrix} 4 & 6 \\ 8 & 5 \end{vmatrix} = 27$

58. 0 **59.** 0 **60.** 72 **61.** 36 **62.** -198 **63.** $\frac{1}{8}$ **64.** -38 **65.** 432 **66.** 0

67. $x = \dfrac{D_x}{D} = \dfrac{20}{20} = 1, \; y = \dfrac{D_y}{D} = \dfrac{0}{20} = 0, \; z = \dfrac{D_z}{D} = \dfrac{60}{20} = 3$

68. (a) -6 by Exercise 7, page 598; **(b)** 3 by Exercise 8, page 598; **(c)** 6 by Exercise 9, page 598;
    (d) -36 by Exercises 7 and 8, page 598.

69. Multiplying matrix A by the scalar 3 gives the matrix each of whose elements are 3 times each of the elements of A. Then, n applications of Exercise 8, page 598, gives the stated result.

70. The inverse of A is the 5 by 5 matrix such that each number on the main diagonal is $\frac{1}{2}$, and all other entries are 0. Then, $|A^{-1}| = \frac{1}{32}$.

71.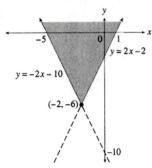

72.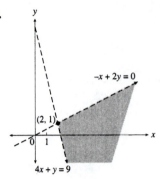

73.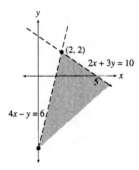

74.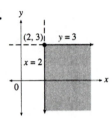

75.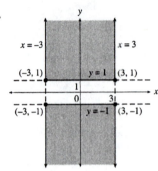

76.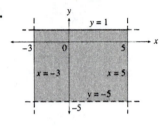

77.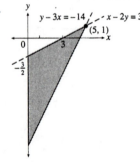

78.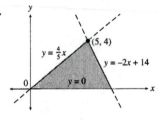

79.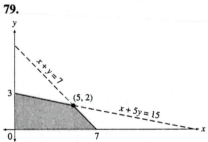

80.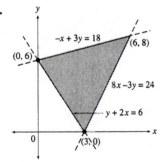

81. $y \le \frac{1}{5}x + 8$
$y \ge x$
$y \ge -3x + 8$

82. $y \le -3x + 8$
$y \ge x$
$x \ge 0$

83. $y \le -3x + 8$
$y \le x$
$y \ge 0$

84. $y \ge -3x + 8$
$y \le x$
$y \le \frac{1}{5}x + 8$
$y \ge 0$

85. $y \ge \frac{1}{5}x + 8$
$y \le x$

86. $y \ge \frac{1}{5}x + 8$
$y \ge x$
$x \ge 0$

87.

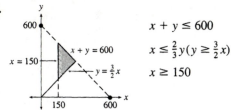

$$x + y \leq 600$$
$$x \leq \tfrac{2}{3}y\,(y \geq \tfrac{3}{2}x)$$
$$x \geq 150$$

	Maximum	Minimum
88. (a)	94	34
(b)	70	18
89. (a)	none	6
(b)	none	8

90. $x + 6y \leq 30$
$4x + 3y \leq 36$
$x \geq 0,\, y \geq 0$

91. $4x -\ y \geq\ \ 7$
$x - 2y \geq -7$
$x + 3y \leq\ 23$
$2x +\ \ y \leq\ 21$
$y \geq\ \ 1$

	Maximum	Minimum
92. (a)	54	19
(b)	52	17
93. (a)	90	0
(b)	63	0
94. (a)	3400	2400
(b)	2000	1400
95. (a)	3100	0
(b)	1600	0

96. $45 when $x = 2$ and $y = 5$
97. $42 for 20 wheat and 30 rye

CHAPTER 9   STANDARD ANSWER TEST   (page 626)

1. No solutions   **2.** $(4, -1, -6)$

3. $\left(-\tfrac{2}{3}c + \tfrac{2}{3}, -\tfrac{4}{3}c + \tfrac{7}{3}, c\right)$ for all c, or $\left(d, 1 + 2d, 1 - \tfrac{3}{2}d\right)$ for all d, or $\left(-\tfrac{1}{2} + \tfrac{1}{2}e, e, \tfrac{7}{4} - \tfrac{3}{4}e\right)$ for all e.

4. (a) $\begin{bmatrix} -6 & 3 \\ 10 & -5 \end{bmatrix} \begin{bmatrix} x \\ y \end{bmatrix} = \begin{bmatrix} 9 \\ -12 \end{bmatrix}$; **(b)** $\begin{bmatrix} 2 & -1 & 2 \\ 1 & 4 & -3 \\ -4 & 2 & -3 \end{bmatrix} \begin{bmatrix} x \\ y \\ z \end{bmatrix} = \begin{bmatrix} -3 \\ 18 \\ 0 \end{bmatrix}$

5. 6 oz. of I; 8 oz. of II; 3 oz. of III.

6. (a) $\begin{bmatrix} 1 & -1 & 0 \\ 6 & 1 & 4 \\ -6 & -5 & 3 \end{bmatrix}$; **(b)** Does not exist   **7.** $\begin{bmatrix} 25 & -14 \\ -3 & 23 \end{bmatrix}$ **8.** $\begin{bmatrix} -1 & 0 & 31 \\ -10 & -4 & 104 \end{bmatrix}$ **9. (a)** -8; **(b)** 51

10. (a) $\begin{bmatrix} -11 & -11 \\ -18 & -18 \end{bmatrix}$; **(b)** Does not exist   **11.** $\begin{bmatrix} 27 & 0 \\ 13 & 1 \end{bmatrix}$ **12.** $\begin{bmatrix} -2 & 2 & -4 \\ 2 & -2 & 4 \\ 4 & -4 & 8 \end{bmatrix}$

13. (a) 4 by 3; **(b)** Does not exist   **14.** $[-8]$ **15.** $\begin{bmatrix} \frac{1}{4} & \frac{1}{8} \\ \frac{1}{4} & -\frac{3}{8} \end{bmatrix}$; $(4, -3)$ **16.** $\begin{bmatrix} 0 & -\frac{2}{3} & \frac{1}{3} \\ 0 & 1 & 0 \\ -\frac{1}{2} & -\frac{1}{3} & \frac{1}{6} \end{bmatrix}$

17. $(-1, 1, -5)$ **18.** -45 **19.** -48 **20.** $D = 14, D_x = 14, D_y = -77$; $x = \dfrac{D_x}{D} = 1, y = \dfrac{D_y}{D} = -\dfrac{11}{2}$

21. $D = 10, D_x = 10, D_y = -20, D_z = 30$; $x = \dfrac{D_x}{D} = 1, y = \dfrac{D_y}{D} = -2, z = \dfrac{D_z}{D} = 3$

22. Let x = square feet used for sofas and y = square feet used for easy chairs. Then,

$$0 \le x \le 800, \qquad y \ge 300, \qquad x + y \le 1200, \qquad x \ge \tfrac{3}{2}y \ (y \le \tfrac{2}{3}x)$$

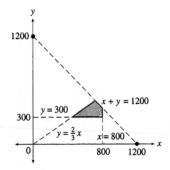

23.

y
$(0, 4)$ $x + 2y = 8$
$(2, 3)$
$x + y = 5$
$(3, 2)$
$2x + y = 8$
$(0, 0)$ $(4, 0)$ x

24. 13   **25.** 300 of model A
300 of model B

CHAPTER 9   MULTIPLE CHOICE TEST   (page 628)

1. (d)   **2.** (e)   **3.** (c)   **4.** (a)   **5.** (b)   **6.** (d)   **7.** (d)   **8.** (c)   **9.** (b)   **10.** (a)   **11.** (c)
12. (a)   **13.** (e)   **14.** (b)   **15.** (c)   **16.** (d)   **17.** (d)   **18.** (b)   **19.** (a)   **20.** (c)

CHAPTER 10   THE CONIC SECTIONS

10.1   The Ellipse   (page 642)

1.

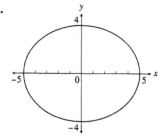

Center: $(0, 0)$
Vertices: $(\pm 5, 0)$
Foci: $(\pm 3, 0)$

3.

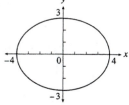

Center: $(0, 0)$
Vertices: $(\pm 4, 0)$
Foci: $(\pm\sqrt{7}, 0)$

5.

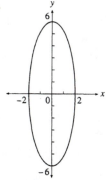

Center: $(0, 0)$
Vertices: $(0, \pm 6)$
Foci: $(0, \pm 4\sqrt{2})$

7.

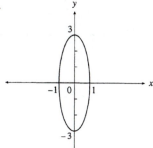

Center: $(0, 0)$
Vertices: $(0, \pm 3)$
Foci: $(0, \pm 2\sqrt{2})$

9.

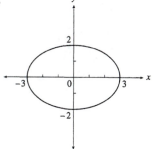

Center: $(0, 0)$
Vertices: $(\pm 3, 0)$
Foci: $(\pm\sqrt{5}, 0)$

11.

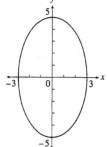

Center: $(0, 0)$
Vertices: $(0, \pm 5)$
Foci: $(0, \pm 4)$

13.

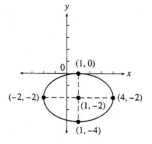

Center: $(1, -2)$
Vertices: $(-2, -2), (4, -2)$
Foci: $(1 \pm \sqrt{5}, -2)$

15.

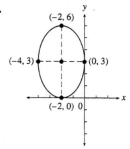

Center: $(-2, 3)$
Vertices: $(-2, 6), (-2, 0)$
Foci: $(-2, 3 \pm \sqrt{5})$

17. $\dfrac{x^2}{9} + \dfrac{y^2}{4} = 1$ **19.** $\dfrac{x^2}{36} + \dfrac{y^2}{9} = 1$ **21.** $\dfrac{(x+2)^2}{25} + \dfrac{(y^2)}{9} = 1$ **23.** $\dfrac{x^2}{25} + \dfrac{y^2}{9} = 1$

25. $\dfrac{(x-2)^2}{32} + \dfrac{(y-3)^2}{16} = 1$ **27.** $\dfrac{x^2}{16} + \dfrac{y^2}{25} = 1$ **29.** $\dfrac{x^2}{64} + \dfrac{y^2}{16} = 1$ **31.** $\dfrac{(x-6)^2}{9} + \dfrac{(y-1)^2}{4} = 1$

33. $\dfrac{x^2}{36} + \dfrac{y^2}{20} = 1$ **35.** $\dfrac{(x+1)^2}{36} + \dfrac{y^2}{100} = 1$

37. $x^2 + \dfrac{(y-6)^2}{25} = 1$; center: $(0, 6)$; vertices: $(0, 1)$, $(0, 11)$; foci: $(0, 6 \pm 2\sqrt{6})$

39. $\dfrac{(x+3)^2}{13} + \dfrac{(y-1)^2}{4} = 1$; center: $(-3, 1)$; vertices: $(-3 \pm \sqrt{13}, 1)$; foci: $(-6, 1)$, $(0, 1)$

41. (a) $\dfrac{x^2}{b^2} + \dfrac{y^2}{(4355)^2} = 1$ where $b^2 = (4355)^2 - (225)^2$; **(b)** 4349 miles; $\dfrac{x^2}{(4349)^2} + \dfrac{y^2}{(4355)^2} = 1$

43. Approximately 56 million miles.   **45. (a)** $y = \frac{2}{3}\sqrt{900 - 400} = \frac{20}{3}\sqrt{5}$ **(b)** 14.9 ft

10.2 The Hyperbola (page 654)

1.
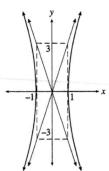

Center: $(0, 0)$
Vertices: $(\pm 5, 0)$
Foci: $(\pm\sqrt{34}, 0)$
Asymptotes: $y = \pm\frac{3}{5}x$

3.
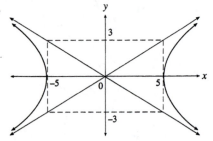

Center: $(0, 0)$
Vertices: $(\pm 4, 0)$
Foci: $(\pm\sqrt{41}, 0)$
Asymptotes: $y = \pm\frac{5}{4}x$

5.
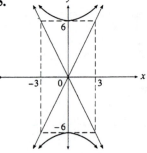

Center: $(0, 0)$
Vertices: $(0, \pm 6)$
Foci: $(0, \pm 3\sqrt{5})$
Asymptotes: $y = \pm 2x$

7.
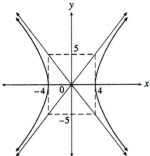

Center: $(0, 0)$
Vertices: $(\pm 1, 0)$
Foci: $(\pm\sqrt{10}, 0)$
Asymptotes: $y = \pm 3x$

9.
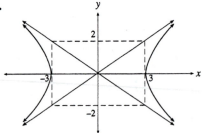

Center: $(0, 0)$
Vertices: $(\pm 3, 0)$
Foci: $(\pm\sqrt{13}, 0)$
Asymptotes: $y = \pm\frac{2}{3}x$

11.
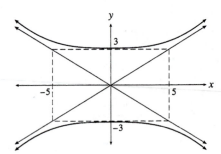

Center: $(0, 0)$
Vertices: $(0, \pm 3)$
Foci: $(0, \pm\sqrt{34})$
Asymptotes: $y = \pm\frac{3}{5}x$

13.

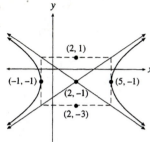

Center: $(2, -1)$
Vertices: $(-1, -1), (5, -1)$
Foci: $\left(2 \pm \sqrt{13}, -1\right)$
Asymptotes: $y + 1 = \pm\frac{2}{3}(x - 2)$

15.

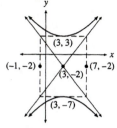

Center: $(3, -2)$
Vertices: $(3, 3), (3, -7)$
Foci: $\left(3, -2 \pm \sqrt{41}\right)$
Asymptotes: $y + 2 = \pm\frac{5}{4}(x - 3)$

17. $\dfrac{x^2}{9} - \dfrac{y^2}{8} = 1$ **19.** $\dfrac{(x - 2)^2}{9} - \dfrac{(y - 3)^2}{4} = 1$ **21.** $\dfrac{x^2}{16} - \dfrac{y^2}{20} = 1$

23. $\dfrac{(y - 3)^2}{9} - \dfrac{(x + 2)^2}{7} = 1$ **25.** $\dfrac{x^2}{16} - \dfrac{y^2}{4} = 1$ **27.** $\dfrac{y^2}{64} - \dfrac{x^2}{225} = 1$

29. Center: $(-1, 3)$
Vertices: $(-5, 3), (3, 3)$
Foci: $(-6, 3), (4, 3)$

31. Center: $(3, -1)$
Vertices: $(3, -9), (3, 7)$
Foci: $(3, -11), (3, 9)$

33. $\dfrac{(x - 1)^2}{9} - \dfrac{(y + 2)^2}{4} = 1$
Center: $(1, -2)$
Vertices: $(-2, -2), (4, -2)$
Foci: $\left(1 \pm \sqrt{13}, -2\right)$

35. $\dfrac{(y + 2)^2}{4} - \dfrac{(x - 1)^2}{1} = 1$ **37.** $\dfrac{(x + 2)^2}{100} - \dfrac{(y - 4)^2}{36} = 1$
Center: $(1, -2)$
Vertices: $(1, -4), (1, 0)$
Foci: $\left(1, -2 \pm \sqrt{5}\right)$

39. (a) The possible locations (x, y) are on the hyperbola $\dfrac{1024(x - 1)^2}{81} - \dfrac{1024y^2}{943} = 1$
(b) $(1.406, 1)$

10.3 The Parabola (page 663)

1. Focus: $\left(0, \frac{1}{16}\right)$
Directrix: $y = -\frac{1}{16}$

3. Focus: $\left(0, \frac{1}{8}\right)$
Directrix: $y = -\frac{1}{8}$

5. Focus: $\left(\frac{1}{2}, 0\right)$
Directrix: $x = -\frac{1}{2}$

7. Focus: $\left(0, -\frac{1}{16}\right)$
Directrix: $y = \frac{1}{16}$

9. Focus: $\left(\frac{13}{8}, 0\right)$
Directrix: $x = \frac{19}{8}$

11. Focus: $\left(4, -\frac{61}{12}\right)$
Directrix: $y = -\frac{59}{12}$

13. $x^2 = -12y$

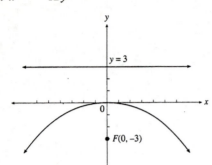

15. $x^2 = \frac{8}{3}y$

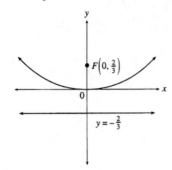

17. $(y-2)^2 = -3x$

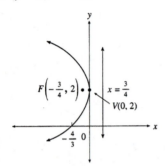

19. $(x-2)^2 = 3(y-1)$

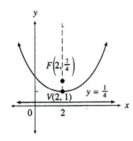

21. $x^2 = 6(y+3)$ **23.** $(x-3)^2 = -(y-9)$ **25.** $(x-2)^2 = 8(y+5)$; $y = \frac{1}{8}(x-2)^2 - 5$; $y = \frac{1}{8}x^2 - \frac{1}{2}x - \frac{9}{2}$

27. Vertex: $(2, -5)$
Axis: $x = 2$
Focus: $(2, -4)$
Directrix: $y = -6$

29. $(x-2)^2 = 4(1)(y-3)$
Vertex: $(2, 3)$
Focus: $(2, 4)$
Directrix: $y = 2$

31. $\left(x - \frac{1}{2}\right)^2 = 7(y+3)$

Vertex: $\left(\frac{1}{2}, -3\right)$

Focus: $\left(\frac{1}{2}, -\frac{5}{4}\right)$

Directrix: $y = -\frac{19}{4}$

33. $(y+2)^2 = 6(x+3)$
Focus: $\left(-\frac{3}{2}, -2\right)$
Axis: $y = -2$

35. $x = -4y^2 - 2y + 12$

37. Vertex: $(4, 0)$
Axis: $y = 0$
Focus: $\left(\frac{7}{4}, 0\right)$

Directrix: $x = \frac{25}{4}$

39. $\frac{9}{8}$ ft; 8π ft   **41.** 100 feet

43. Ellipse;
Center: $(0, 0)$
Vertices: $(\pm 5, 0)$
Foci: $(\pm 3, 0)$

45. Hyperbola
Center: $(0, 0)$
Vertices: $(\pm 6, 0)$
Foci: $\left(\pm\sqrt{61}, 0\right)$

Asymptotes: $y = \pm\frac{5}{6}x$

47. Circle
Center: $(0, 0)$
Radius: 4

49. Hyperbola:

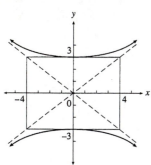

51. Ellipse:

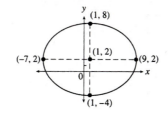

53. Hyperbola:

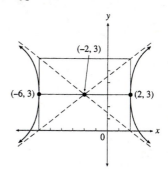

55. $(x - 1)^2 + (y + 2)^2 = 4$; circle **57.** $\dfrac{(x + 1)^2}{4} + \dfrac{y^2}{1} = 1$; ellipse

59. $\dfrac{(x + 1)^2}{16} - \dfrac{(y - 3)^2}{9} = 1$; hyperbola **61.** $(y + 5)^2 = 6(x + 4)$; parabola

10.4 Solving Nonlinear Systems (page 672)

1.

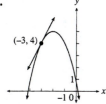

3.

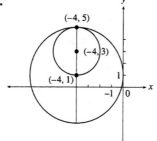

5.

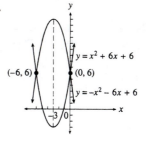

7. $(0, 1)$ **9.** No solutions **11.** No solutions **13.** $(-1, -1), (-2, -2)$ **15.** $(2, 1), (2, -1)$
17. $(1, 0), (0, -2)$ **19.** No solutions **21.** $\left(2, \sqrt{3}\right), \left(2, -\sqrt{3}\right), \left(-2, \sqrt{3}\right), \left(-2, -\sqrt{3}\right)$
23. $(\pm 1, 2)$ **25.** $(0, 0), (1, 1)$ **27.** $(3, -3), \left(3 + \sqrt{3}, -2\right); \left(3 - \sqrt{3}, -2\right)$
29.

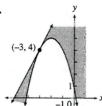

31.

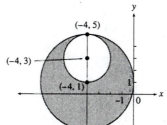

33.

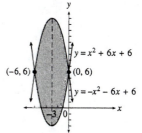

35.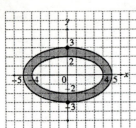

37.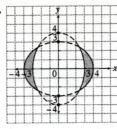

39.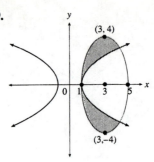

41. To the nearest tenth of a foot, the dimensions are: base of rectangle: 3.3 feet; height of rectangle: 5.2 feet; altitude of triangle: 1.7 feet.

43. Area common to both ellipses: $x^2 + 4y^2 \le 4$; $y^2 + 4x^2 \le 4$
Area inside the horizontal ellipse and outside the vertical: $x^2 + 4y^2 \le 4$; $y^2 + 4x^2 \ge 4$
Area inside the vertical ellipse and outside the horizontal: $x^2 + 4y^2 \ge 4$; $y^2 + 4x^2 \le 4$

CHAPTER 10 REVIEW EXERCISES (page 674)

1. An ellipse is the set of all points in a plane such that the sum of the distances from two points, called the foci, is a constant.

2.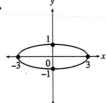

3.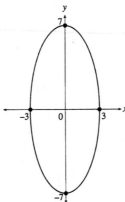

4.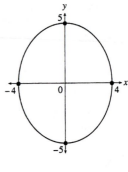

Center: (0, 0)
Vertices: (±3, 0)
Foci: $\left(\pm 2\sqrt{2}, 0\right)$

Center: (0, 0)
Vertices: (0, ±7)
Foci: $\left(0, \pm 2\sqrt{10}\right)$

Center: (0, 0)
Vertices: (0, ±5)
Foci: (0, ±3)

5.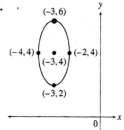

Center: $(-3, 4)$
Vertices: $(-3, 2), (-3, 6)$
Foci: $\left(-3, 4 \pm \sqrt{3}\right)$

6.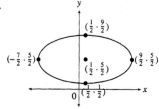

Center: $\left(\frac{1}{2}, \frac{5}{2}\right)$
Vertices: $\left(-\frac{7}{2}, \frac{5}{2}\right), \left(\frac{9}{2}, \frac{5}{2}\right)$
Foci: $\left(\frac{1}{2} \pm 2\sqrt{3}, \frac{5}{2}\right)$

7.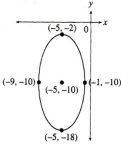

Center: $(-5, -10)$
Vertices: $(-5, -18), (-5, -2)$
Foci: $\left(-5, -10 \pm 4\sqrt{3}\right)$

8. $\dfrac{(x+5)^2}{36} + y^2 = 1$

Center: $(-5, 0)$
Vertices: $(-11, 0), (1, 0)$
Foci: $\left(-5 \pm \sqrt{35}, 0\right)$

9. $\dfrac{(x+3)^2}{9} + \dfrac{(y-1)^2}{4} = 1$

Center: $(-3, 1)$
Vertices: $(-6, 1), (0, 1)$
Foci: $\left(-3 \pm \sqrt{5}, 1\right)$

10. $\dfrac{(x-6)^2}{5} + \dfrac{(y+1)^2}{4} = 1$

Center: $(6, -1)$
Vertices: $\left(6 \pm \sqrt{5}, -1\right)$
Foci: $(5, -1), (7, -1)$

11. $\dfrac{(x-5)^2}{7} + \dfrac{(y-3)^2}{10} = 1$

Center: $(5, 3)$
Vertices: $\left(5, 3 \pm \sqrt{10}\right)$
Foci: $\left(5, 3 \pm \sqrt{3}\right)$

12. $\dfrac{x^2}{9} + y^2 = 1$

13. $\dfrac{x^2}{49} + \dfrac{y^2}{13} = 1$

14. $\dfrac{x^2}{4} + \dfrac{y^2}{9} = 1$

15. $\dfrac{(x-4)^2}{25} + \dfrac{9(y-6)^2}{100} = 1$

16. $\dfrac{(x+5)^2}{9} + \dfrac{(y+2)^2}{81} = 1$

17. $\dfrac{x^2}{28} + \dfrac{y^2}{64} = 1$

18. $\dfrac{(x+1)^2}{32} + \dfrac{(y-4)^2}{36} = 1$

19. $\dfrac{(x+1)^2}{9} + \dfrac{(y-1)^2}{5} = 1$

20. $\dfrac{2}{\sqrt{6}}$

21. $\dfrac{x^2}{5200^2 - 800^2} + \dfrac{y^2}{(5200)^2} = 1.$ (*Note:* $5200^2 - 800^2 \approx 5138^2$)

22. $\dfrac{x^2}{4^2} + \dfrac{y^2}{(1.97)^2} = 1$

23. A hyperbola is the set of all points in the plane such that the difference of the distances from two fixed points, called the foci, is a constant.

24.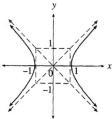

Center: $(0, 0)$
Vertices: $(\pm 1, 0)$
Foci: $\left(\pm\sqrt{2}, 0\right)$
Asymptotes: $y = \pm x$

25.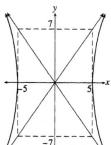

Center: $(0, 0)$
Vertices: $(\pm 5, 0)$
Foci: $\left(\pm\sqrt{74}, 0\right)$
Asymptotes: $y = \pm\frac{7}{5}x$

26.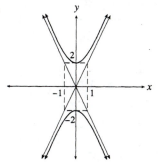

Center: $(0, 0)$
Vertices: $(0, \pm 2)$
Foci: $\left(0, \pm\sqrt{5}\right)$
Asymptotes: $y = \pm 2x$

27.

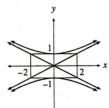

Center: $(0, 0)$
Vertices: $(0, \pm 1)$
Foci: $\left(0, \pm \sqrt{5}\right)$
Asymptotes: $y = \pm\frac{1}{2}x$

28.

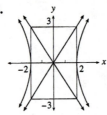

Center: $(0, 0)$
Vertices: $(\pm 2, 0)$
Foci: $\left(\pm\sqrt{13}, 0\right)$
Asymptotes: $y = \pm\frac{3}{2}x$

29.

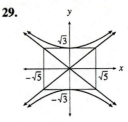

Center: $(0, 0)$
Vertices: $\left(0, \pm\sqrt{3}\right)$
Foci: $\left(0, \pm\sqrt{8}\right)$
Asymptotes: $y = \pm\sqrt{\frac{3}{5}}x$

30.

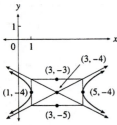

Center: $(3, -4)$
Vertices: $(1, -4), (5, -4)$
Foci: $\left(3 \pm\sqrt{5}, -4\right)$
Asymptotes: $y + 4 = \pm\frac{1}{2}(x - 3)$

31.

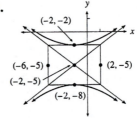

Center: $(-2, -5)$
Vertices: $(-2, -8), (-2, -2)$
Foci: $(-2, -10), (-2, 0)$
Asymptotes: $y + 5 = \pm\frac{3}{4}(x + 2)$

32.

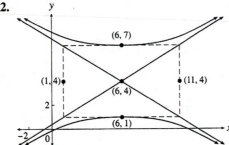

Center: $(6, 4)$
Vertices: $(6, 1), (6, 7)$
Foci: $\left(6, 4 \pm\sqrt{34}\right)$
Asymptotes: $y - 4 = \pm\frac{3}{5}(x - 6)$

33.

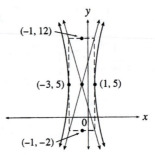

Center: $(-1, 5)$
Vertices: $(-3, 5), (1, 5)$
Foci: $\left(-1 \pm\sqrt{53}, 5\right)$
Asymptotes: $y - 5 = \pm\frac{7}{2}(x + 1)$

34. $\dfrac{x^2}{16} - (y - 3)^2 = 1$

Center: $(0, 3)$
Vertices: $(\pm 4, 3)$
Foci: $\left(\pm\sqrt{17}, 3\right)$
Asymptotes: $y - 3 = \pm\frac{1}{4}x$

35. $\dfrac{(y - 2)^2}{4} - \dfrac{(x + 1)^2}{9} = 1$

Center: $(-1, 2)$
Vertices: $(-1, 0), (-1, 4)$
Foci: $\left(-1, 2 \pm\sqrt{13}\right)$
Asymptotes: $y - 2 = \pm\frac{2}{3}(x + 1)$

36. $\dfrac{(x+4)^2}{3} - \dfrac{(y+6)^2}{3} = 1$

Center: $(-4, -6)$

Vertices: $\left(-4 \pm \sqrt{3}, -6\right)$

Foci: $\left(-4 \pm \sqrt{6}, -6\right)$

Asymptotes: $y + 6 = \pm(x + 4)$

37. $\dfrac{(y-2)^2}{4} - \dfrac{\left(x - \frac{3}{2}\right)^2}{5} = 1$

Center: $\left(\frac{3}{2}, 2\right)$

Vertices: $\left(\frac{3}{2}, 0\right), \left(\frac{3}{2}, 4\right)$

Foci: $\left(\frac{3}{2}, -1\right), \left(\frac{3}{2}, 5\right)$

Asymptotes: $y - 2 = \pm\dfrac{2}{\sqrt{5}}\left(x - \frac{3}{2}\right)$

38. $\dfrac{x^2}{9} - \dfrac{y^2}{16} = 1$ **39.** $\dfrac{y^2}{10} - \dfrac{x^2}{15} = 1$ **40.** $\dfrac{(x-2)^2}{9} - \dfrac{(y+1)^2}{4} = 1$

41. $\dfrac{(y-2)^2}{16} - \dfrac{(x+3)^2}{4} = 1$ **42.** $\dfrac{x^2}{9} - \dfrac{y^2}{16} = 1$ **43.** $\dfrac{(y+2)^2}{4} - \dfrac{(x-4)^2}{9} = 1$

44. $\dfrac{y^2}{\frac{25}{144}} - \dfrac{x^2}{\frac{119}{144}} = 1$ (or, using decimal approximations: $\dfrac{y^2}{0.174} - \dfrac{x^2}{0.826} = 1$)

45. $(1.5, -0.804)$

46. A parabola is the set of all points in a plane equidistant from a fixed line called the directrix and a given fixed point called the focus.

	47.	**48.**	**49.**	**50.**	**51.**	**52.**
Vertex	$(0, 0)$	$(0, 0)$	$(0, 0)$	$(0, 0)$	$(0, 3)$	$(-5, 6)$
Focus	$\left(0, \frac{3}{2}\right)$	$\left(0, \frac{1}{3}\right)$	$\left(0, -\frac{1}{2}\right)$	$\left(-\frac{3}{2}, 0\right)$	$\left(0, \frac{25}{8}\right)$	$\left(-\frac{31}{6}, 6\right)$
Directrix	$y = -\frac{3}{2}$	$y = -\frac{1}{3}$	$y = \frac{1}{2}$	$x = \frac{3}{2}$	$y = \frac{23}{8}$	$x = -\frac{29}{6}$

53. $(x-1)^2 = \frac{1}{12}(y-4)$

Vertex: $(1, 4)$

Focus: $\left(1, \frac{193}{48}\right)$

Directrix: $y = \frac{191}{48}$

54. $(y+2)^2 = -\frac{1}{6}\left(x - \frac{1}{2}\right)$

Vertex: $\left(\frac{1}{2}, -2\right)$

Focus: $\left(\frac{11}{24}, -2\right)$

Directrix: $x = \frac{13}{24}$

55. $(y+3)^2 = -\frac{5}{6}(x-5)$

Vertex: $(5, -3)$

Focus: $\left(\frac{115}{24}, -3\right)$

Directrix: $x = \frac{125}{24}$

56. $(x-2)^2 = \frac{7}{3}(y-3)$

Vertex: $(2, 3)$

Focus: $\left(2, \frac{43}{12}\right)$

Directrix: $y = \frac{29}{12}$

57. $x^2 = -4(y - 8)$

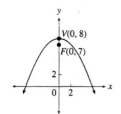

58. $y^2 = 3\left(x + \frac{13}{4}\right)$

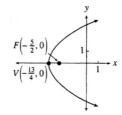

59. $y^2 = -3x$

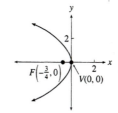

60. $x^2 = -16y$

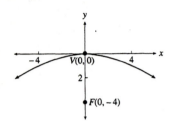

61. $(y + 2)^2 = -4(x - 5)$

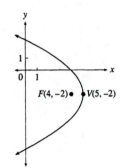

62. $(y + 4)^2 = \frac{1}{2}(x + 3)$

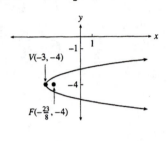

63. $(x - 6)^2 = 12(y + 1)$

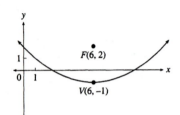

64. $(x - 3)^2 = -\frac{3}{2}(y - 6)$

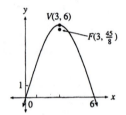

65. $(y - 6)^2 = -2(x + 2)$

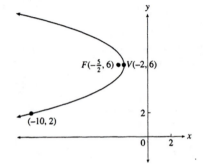

66. $y = -2x^2 + 6x + 3$ **67.** $x = 5y^2 - 2y - 4$ **68.** $y = x^2 - 6x + 9$; $x = -\frac{1}{6}y^2 + \frac{7}{6}y + 3$

69. The receiver is 3.5 feet from the vertex (bottom). The depth is $1\frac{11}{14}$ feet.

70. $\frac{4}{3}$ feet above the vertex.   **71.** Ellipse: center $(0, 0)$; vertices $(0, \pm 5)$; foci $(0, \pm 4)$

72. Hyperbola: center $(0, 0)$; vertices $(\pm 3, 0)$; foci $\left(\pm\sqrt{13}, 0\right)$; asymptotes $y = \pm\frac{2}{3}x$

73. Parabola: vertex $(-2, 0)$; focus $\left(-\frac{1}{2}, 0\right)$; directrix $x = -\frac{7}{2}$

74. Parabola: vertex $(4, 1)$; focus $(4, \frac{11}{8})$; directrix $y = \frac{5}{8}$

75. Hyperbola: center $(-3, 0)$; vertices $(-3, \pm 4)$; foci $(-3, \pm 5)$; asymptotes $y = \pm\frac{4}{3}(x + 3)$

76. Ellipse: center $(1, -5)$; vertices $(1, -10)$, $(1, 0)$; foci $\left(1, -5 \pm \sqrt{21}\right)$

77. Parabola: $x^2 = 8(y - 3)$ **78.** Hyperbola: $\dfrac{(x + 4)^2}{4} - (y + 1)^2 = 1$ **79.** Ellipse: $\dfrac{(x - 2)^2}{4} + \dfrac{(y - 3)^2}{9} = 1$

80. Parabola: $(x - 1)^2 = \frac{1}{3}(y - 1)$

81.

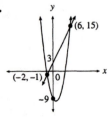

82.

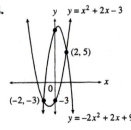

83.

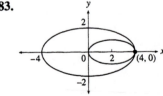

84.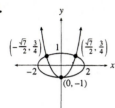

85.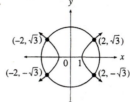

86.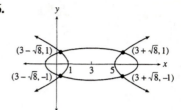

87. $(2, 2), (8, -10)$ **88.** $(-8, 6), (1, 9)$ **89.** $(5, 6), (-5, -6)$ **90.** $(2, -4)$ **91.** $(4, 8), (2, 6), (6, 6)$

92. $(-1, 2)$ **93.** $(\pm 2, 0)$ **94.** No solution   **95.** $\left(\pm \sqrt{\frac{4}{5}}, \pm \sqrt{\frac{4}{5}} \right)$ **96.** $\left(\pm 3, \pm \sqrt{55} \right)$

97. $(2, 0), \left(2 \pm \sqrt{7}, 7 \right)$ **98.** $(0, 4), \left(\pm \sqrt{7}, -3 \right)$ **99.** $(3, 2), (10, 1)$

100. Small square: 3 inch sides and large square: 7 inch sides; also, small square $3\frac{2}{3}$ inch sides and large square: $6\frac{1}{3}$ inch sides.

101.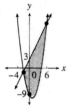

102.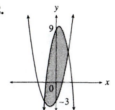

103.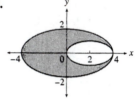

104.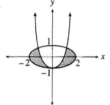

105.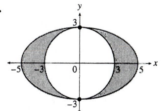

106.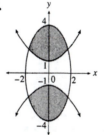

CHAPTER 10   STANDARD ANSWER TEST (page 678)

1.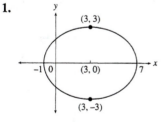

Center: $(3, 0)$
Vertices: $(-1, 0), (7, 0)$
Foci: $\left(3 \pm \sqrt{7}, 0 \right)$

2.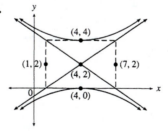

Center: $(4, 2)$
Vertices: $(4, 0), (4, 4)$
Foci: $\left(4, 2 \pm \sqrt{13} \right)$
Asymptotes: $y - 2 = \pm\frac{2}{3}(x - 4)$

3.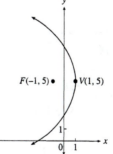

Vertex: $(1, 5)$
Focus: $(-1, 5)$
Directrix: $x = 3$

4. $\dfrac{x^2}{16} + \dfrac{y^2}{9} = 1$ **5.** $\dfrac{(x+1)^2}{4} + \dfrac{(y-2)^2}{9} = 1$ **6.** $\dfrac{(x+2)^2}{36} + \dfrac{(y-4)^2}{20} = 1$ **7.** $\dfrac{x^2}{36} - \dfrac{y^2}{28} = 1$

8. $\dfrac{(y+5)^2}{16} - \dfrac{(x-4)^2}{9} = 1$ **9.** $\dfrac{x^2}{36} - \dfrac{(y-1)^2}{4} = 1$ **10.** $x^2 = -\dfrac{8}{3}y$ **11.** $(y+3)^2 = 12(x+5)$

12. $(x-2)^2 = y + 1$ **13.** Hyperbola: center $(0, 0)$; vertices $(0, \pm 3)$, foci $\left(0, \pm \sqrt{10}\right)$; asymptotes: $y = \pm 3x$

14. Ellipse: center $(-4, 1)$; vertices $(-8, 1)$, $(0, 1)$; foci $\left(-4 \pm \sqrt{7}, 1\right)$

15. Parabola: vertex $(-5, 5)$; focus $(-5, 2)$; directrix $y = 8$

16. Hyperbola: $\dfrac{(x-2)^2}{9} - \dfrac{y^2}{4} = 1$ **17.** Parabola: $(x-1)^2 = \frac{1}{2}(y-7)$

18. Ellipse: $\dfrac{(x-1)^2}{4} + \dfrac{(y+2)^2}{9} = 1$ **19.** Hyperbola: $(x+2)^2 - \dfrac{(y-3)^2}{4} = 1$

20.

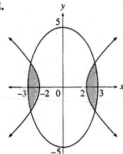

21. $(\pm 3, \pm 1)$

22.

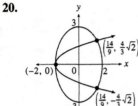

23. 16 feet

24. $\dfrac{1}{15}$

25. $\dfrac{x^2}{\frac{25}{256}} - \dfrac{y^2}{\frac{551}{256}} = 1$

or $\dfrac{256}{25}x^2 - \dfrac{256}{551}y^2 = 1$

CHAPTER 10   MULTIPLE CHOICE TEST   (page 680)

1. (c)   **2.** (b)   **3.** (a)   **4.** (d)   **5.** (b)   **6.** (e)   **7.** (a)   **8.** (c)   **9.** (d)   **10.** (b)   **11.** (b)

12. (a)   **13.** (a)   **14.** (c)   **15.** (d)   **16.** (a)   **17.** (b)   **18.** (d)   **19.** (c)   **20.** (d)

CHAPTER 11 SEQUENCES AND SERIES

11.1   Sequences   (page 687)

1. $1, 3, 5, 7, 9$ **3.** $-1, 1, -1, 1, -1$ **5.** $-4, 2, -1, \frac{1}{2}, -\frac{1}{4}$ **7.** $-1, 4, -9, 16$ **9.** $\frac{3}{10}, \frac{3}{100}, \frac{3}{1000}, \frac{3}{10,000}$

11. $\frac{3}{100}, \frac{3}{10,000}, \frac{3}{1,000,000}, \frac{3}{100,000,000}$ **13.** $\frac{1}{2}, \frac{1}{6}, \frac{1}{12}, \frac{1}{20}$ **15.** $64, 36, 16, 4$ **17.** $-2, 1, 4, 7$ **19.** $0, \frac{1}{3}, \frac{1}{2}, \frac{3}{5}$

21. $1, \frac{3}{2}, \frac{16}{9}, \frac{125}{64}$ **23.** $-2, -\frac{3}{2}, -\frac{9}{8}, -\frac{27}{32}$ **25.** $\frac{1}{2}, \frac{1}{2}, \frac{3}{8}, \frac{1}{4}$ **27.** $-\frac{3}{2}, -\frac{5}{6}, -\frac{7}{12}, -\frac{9}{20}$ **29.** $4, 4, 4, 4$ **31.** 122

33. 0.000003   **35.** 12   **37.** -1331 **39.** 4, 6, 8, 10   **41.** 5, 10, 15, 20, 25; $s_n = 5n$

43. $-5, 25, -125, 625, -3125$; $s_n = (-5)^n$ **45.** 15, 21, 28   **47.** 12, -4, 4, 0, 2, 1, $\frac{3}{2}, \frac{5}{4}$

49. 1, 0, -1, 0, 1, 0, -1, 0   **51.** $\frac{3}{2}, \frac{15}{8}, \frac{35}{16}, \frac{315}{128}$

11.2 Sums of Finite Sequences (page 692)

1. 45   **3.** 55   **5.** 0.33333   **7.** 510   **9.** 60   **11.** 254   **13.** $\frac{381}{64}$ **15.** 105   **17.** 40   **19.** $\frac{13}{8}$ **21.** 0

23. 57   **25.** 1111.111   **27.** 60   **29.** $\frac{35}{12}$ **31.** 0.010101   **33.** $\sum\limits_{k=1}^{10} 5k$ **35.** $\sum\limits_{k=-3}^{8} 3k$

37. (a) 4, 9, 16, 25, 36; (b) n^2

39. $a_n = a_1 + a_{n-1} = a_1 + (a_1 + a_{n-2}) = \cdots = \overbrace{a_1 + a_1 + a_1 + \cdots + a_1}^{n \text{ terms}} = 2 + 2 + 2 + \cdots + 2 = 2n$

41. $\sum\limits_{k=1}^{n} s_k + \sum\limits_{k=1}^{n} t_k = (s_1 + s_2 + \cdots + s_n) + (t_1 + t_2 + \cdots + t_n) = (s_1 + t_1) + (s_2 + t_2) + \cdots + (s_n + t_n)$

$$= \sum\limits_{k=1}^{n} (s_k + t_k)$$

43. $\sum\limits_{k=1}^{n} (s_k + c) = (s_1 + c) + (s_2 + c) + \cdots + (s_n + c) = (s_1 + s_2 + \cdots + s_n) + (c + c + \cdots + c) = \left(\sum\limits_{k=1}^{n} s_k\right) + nc$

45. (a) $\sum\limits_{k=1}^{10} \dfrac{2}{k(k+2)} = \sum\limits_{k=1}^{10}\left(\dfrac{1}{k} - \dfrac{1}{k+2}\right) = (1 - \frac{1}{3}) + (\frac{1}{2} - \frac{1}{4}) + (\frac{1}{3} - \frac{1}{5}) + (\frac{1}{4} - \frac{1}{6}) + (\frac{1}{5} - \frac{1}{7}) + (\frac{1}{6} - \frac{1}{8}) +$

$(\frac{1}{7} - \frac{1}{9}) + (\frac{1}{8} - \frac{1}{10}) + (\frac{1}{9} - \frac{1}{11}) + (\frac{1}{10} - \frac{1}{12}) = 1 + \dfrac{1}{2} - \dfrac{1}{11} - \dfrac{1}{12} = \dfrac{175}{132}$

(b) $(1 - \frac{1}{3}) + (\frac{1}{2} - \frac{1}{4}) + (\frac{1}{3} - \frac{1}{5}) + (\frac{1}{4} - \frac{1}{6}) + \cdots + \left(\dfrac{1}{n-3} - \dfrac{1}{n-1}\right) + \left(\dfrac{1}{n-2} - \dfrac{1}{n}\right) +$

$\left(\dfrac{1}{n-1} - \dfrac{1}{n+1}\right) + \left(\dfrac{1}{n} - \dfrac{1}{n+2}\right) = 1 + \dfrac{1}{2} - \dfrac{1}{n+1} - \dfrac{1}{n+2} = \dfrac{n(3n+5)}{2(n+1)(n+2)}$

11.3 Arithmetic Sequences (page 700)

1. 5, 7, 9; $2n - 1$; 400   **3.** $-10, -16, -22$; $-6n + 8$; -1100 **5.** $\frac{17}{2}, 9, \frac{19}{2}$; $\frac{1}{2}n + 7$; 245

7. $-\frac{4}{5}, -\frac{7}{5}, -2$; $-\frac{3}{5}n + 1$; -106 **9.** 150, 200, 250; $50n$; 10,500   **11.** 30, 50, 70; $20n - 30$; 3600

13. 455   **15.** $\frac{63}{4}$ **17.** 289   **19.** 15,150   **21.** 94,850   **23.** $\dfrac{173,350}{7}$ **25.** 567,500   **27.** 5824

29. 15,300   **31.** (a) 10,000; (b) n^2 **33.** -36 **35.** 4620   **37.** $\frac{3577}{4}$ **39.** $\frac{5}{2}n(n+1)$ **41.** $-\frac{9}{4}$

43. $\frac{228}{15} = \frac{76}{5}$ **45.** 3, 10, 17, 24, 31, 38   **47.** $-8, 0, 8, 16, 24, 32, 40, 48$

49. $-\frac{1}{5}, -\frac{3}{5}, -1, -\frac{7}{5}, -\frac{9}{5}, -\frac{11}{5}, -\frac{13}{5}, -3$ **51.** $1183

53. (a) $240, $220, $200;

(b) monthly loan balance $= 12,000 - 1000(k-1)$, monthly interest $= [12,000 - 1000(k-1)](0.02) = 260 - 20k$;

(c) $1560, 13%

55. 930   **57.** 9080   **59.** 45,540   **61.** $a_1 = -7$; $d = -13$ **63.** $u = \frac{16}{3}$; $v = \frac{23}{3}$ **65.** 520

11.4 Geometric Sequences (page 710)

1. $16, 32, 64$; 2^n **3.** $27, 81, 243$; 3^{n-1} **5.** $\frac{1}{9}, -\frac{1}{27}, \frac{1}{81}$; $-3\left(-\frac{1}{3}\right)^{n-1}$ **7.** $-125, -625, -3125$; -5^{n-1}

9. $-\frac{16}{9}, -\frac{32}{27}, -\frac{64}{81}$; $-6\left(\frac{2}{3}\right)^{n-1}$ **11.** $1000, 100{,}000, 10{,}000{,}000$; $\frac{1}{1000}(100)^{n-1}$ **13.** 126 **15.** $-\frac{1330}{81}$

17. 1024 **19.** 3 **21.** $-\frac{1}{2}$ **23.** 1023 **25.** $2^n - 1$ **27.** $\frac{121}{27}$ **29.** $\frac{211}{54}$ **31.** $-\frac{85}{64}$ **33.** $-\frac{5}{21}$

35. $\pm 4\sqrt{6}$ **37.** $6, 24, 96, 384, 1536$ or $6, -24, 96, -384, 1536$ **39.** $\$10{,}737{,}418$ **41.** $512{,}000$; $1000(2^{n-1})$

43. $\$8432.20$ **45.** (a) $\$800(1.11)^n$; (b) $\$1348.05$ **47.** $\$703.99$

49. (a) If the volume of the first container is V, then $\frac{1}{2}V + \left(\frac{1}{2}\right)^2 V + \left(\frac{1}{2}\right)^3 V + \left(\frac{1}{2}\right)^4 V + \left(\frac{1}{2}\right)^5 V = \frac{31}{32}V$ is the sum of

the volumes of the other five. Since $\frac{31}{32}V < V$, the answer is yes.

(b) $\sum_{k=1}^{5} \left(\frac{2}{3}\right)^k V = \frac{422}{243}V > V$; therefore, no.

11.5 Infinite Geometric Series (page 720)

1. 4 **3.** $\frac{125}{4}$ **5.** $\frac{2}{3}$ **7.** $\frac{10}{9}$ **9.** $-\frac{16}{7}$ **11.** The numerator at the right should be $a_1 = \frac{1}{4}$, not 1.

13. The denominator at the right should be $1 - \left(-\frac{1}{3}\right)$ since $r = -\frac{1}{3}$, not $\frac{1}{3}$ **15.** $\frac{3}{2}$ **17.** $\frac{1}{6}$ **19.** 4 **21.** $\frac{20}{9}$

23. No finite sum **25.** $\frac{10}{3}$ **27.** No finite sum **29.** 30 **31.** $\frac{20}{11}$ **33.** 4 **35.** $\frac{4}{21}$ **37.** $\frac{7}{9}$ **39.** $\frac{13}{99}$

41. $\frac{13}{990}$ **43.** 1 **45.** (a) $\frac{4}{3}, \frac{4}{9}, \frac{4}{27}, \cdots, \frac{4}{3^n}, \cdots$; (b) $\sum_{n=1}^{\infty} \frac{4}{3^n} = \frac{\frac{4}{3}}{1 - \frac{1}{3}} = 2$ **47.** 8 hours

49. The time for the last $\frac{1}{2}$ mile would have to be $\sum_{n=1}^{\infty} \frac{2}{5}\left(\frac{10}{9}\right)^{n-1}$, which is not a finite sum since $\frac{10}{9} > 1$.

51. (a) $\frac{1}{2}(AC)(CB) = \frac{1}{2}(4)(4) = 8$

(b) $4 + 2 + 1 + \frac{1}{2} + \cdots = \frac{4}{1 - \frac{1}{2}} = 8$

(c) For the odd-numbered triangles: $4 + 1 + \frac{1}{4} + \cdots = \frac{4}{1 - \frac{1}{4}} = \frac{16}{3}$

For the even-numbered triangles: $2 + \frac{1}{2} + \frac{1}{8} + \cdots = \frac{2}{1 - \frac{1}{4}} = \frac{8}{3}$; $\frac{16}{3} + \frac{8}{3} = 8$

53. $\frac{1}{2}(9)(3) + \frac{1}{2}(6)(2) + \frac{1}{2}(4)\left(\frac{4}{3}\right) + \cdots = \frac{27}{2} + 6 + \frac{8}{3} + \cdots = \frac{\frac{27}{2}}{1 - \frac{4}{9}} = \frac{243}{10}$ **55.** (a) $\$3000$; (b) $\$2000$

11.6 Mathematical Induction (page 728)

For these exercises, S_n represents the given statement where n is an integer ≥ 1 ($n \geq 2$ when appropriate). The second part of each proof begins with the hypothesis S_k, where k is an arbitrary positive integer.

1. Since $1 = \frac{1(1 + 1)}{2}$, S_1 is true. Assume S_k and add $k + 1$ to obtain $1 + 2 + 3 + \cdots + k + (k + 1) =$

$\frac{k(k + 1)}{2} + (k + 1) = \frac{k^2 + 3k + 2}{2} = \frac{(k + 1)(k + 2)}{2} = \frac{(k + 1)[(k + 1) + 1]}{2}$. Therefore, S_{k+1} holds. Since S_1 is

true and S_k implies S_{k+1}, the principle of mathematical induction makes S_n true for all integers $n \geq 1$. *Note:* The preceding sentence is an appropriate final statement for the remaining proofs. For the sake of brevity, however, it will not be repeated.

3. Since $\sum\limits_{i=1}^{1} 3i = 3 = \dfrac{3(1+1)}{2}$, S_1 is true. Assume S_k and add $3(k+1)$ to obtain the following.

$\sum\limits_{i=1}^{k+1} 3i = \left(\sum\limits_{i=1}^{k} 3i\right) + 3(k+1) = \dfrac{3k(k+1)}{2} + 3(k+1); \ \sum\limits_{i=1}^{k+1} 3i = \dfrac{3(k^2+3k+2)}{2} = \dfrac{3(k+1)(k+2)}{2} =$

$\dfrac{3(k+1)[(k+1)+1]}{2}$. Therefore, S_{k+1} holds.

5. Since $\dfrac{5}{3} = \dfrac{1(11-1)}{6}$, S_1 is true. Assume S_k and add $-\dfrac{1}{3}(k+1)+2$ to obtain $\dfrac{5}{3} + \dfrac{4}{3} + 1 + \cdots +$

$\left(-\dfrac{1}{3}k+2\right) + \left[-\dfrac{1}{3}(k+1)+2\right] = \dfrac{k(11-k)}{6} + \left[-\dfrac{1}{3}(k+1)+2\right] = \dfrac{10+9k-k^2}{6} = \dfrac{(k+1)(10-k)}{6} =$

$\dfrac{(k+1)[11-(k+1)]}{6}$. Therefore, S_{k+1} holds.

7. Since $\dfrac{1}{1\cdot 2} = \dfrac{1}{1+1}$, S_1 is true. Assume S_k and add $\dfrac{1}{(k+1)[(k+1)+1]}$ to obtain $\dfrac{1}{1\cdot 2} + \dfrac{1}{2\cdot 3} + \cdots +$

$\dfrac{1}{k(k+1)} + \dfrac{1}{(k+1)(k+2)} = \dfrac{k}{k+1} + \dfrac{1}{(k+1)(k+2)} = \dfrac{k^2+2k+1}{(k+1)(k+2)} = \dfrac{(k+1)^2}{(k+1)(k+2)} = \dfrac{k+1}{k+2} =$

$\dfrac{k+1}{(k+1)+1}$. Therefore, S_{k+1} holds.

9. S_1 is true since $-2 = -1 - (1^2)$. Assume S_k and add $-2(k+1)$ to obtain $-2 - 4 - 6 - \cdots -2k - 2(k+1) =$ $-k - k^2 - 2(k+1) = -(k+1) - (k^2+2k+1) = -(k+1) - (k+1)^2$. Therefore, S_{k+1} holds.

11. S_1 is true since $1 = \dfrac{5}{3}\left[1 - \left(\dfrac{2}{5}\right)^1\right]$. Assume S_k and add $\left(\dfrac{2}{5}\right)^k$ to obtain $1 + \dfrac{2}{5} + \dfrac{4}{25} + \cdots + \left(\dfrac{2}{5}\right)^{k-1} + \left(\dfrac{2}{5}\right)^k =$

$\dfrac{5}{3}\left[1 - \left(\dfrac{2}{5}\right)^k\right] + \left(\dfrac{2}{5}\right)^k = \dfrac{5}{3}\left[1 - \left(\dfrac{2}{5}\right)^k + \dfrac{3}{5}\left(\dfrac{2}{5}\right)^k\right] = \dfrac{5}{3}\left[1 - \left(\dfrac{2}{5}\right)^k\left(1 - \dfrac{3}{5}\right)\right] = \dfrac{5}{3}\left[1 - \left(\dfrac{2}{5}\right)^k\left(\dfrac{2}{5}\right)\right] = \dfrac{5}{3}\left[1 - \left(\dfrac{2}{5}\right)^{k+1}\right]$.

Therefore, S_{k+1} holds.

13. S_1 is true since $1^3 = 1 = \dfrac{1^2(1+1)^2}{4}$. Assume S_k and add $(k+1)^3$ to obtain $1^3 + 2^3 + 3^3 + \cdots + k^3 + (k+1)^3 =$

$\dfrac{k^2(k+1)^2}{4} + (k+1)^3 = \dfrac{k^2(k+1)^2 + 4(k+1)^3}{4} = \dfrac{(k+1)^2[k^2+4k+4]}{4} = \dfrac{(k+1)^2[k+2]^2}{4} =$

$\dfrac{(k+1)^2[(k+1)+1]^2}{4}$. Therefore, S_{k+1} holds.

15. S_1 is true since $\sum\limits_{i=1}^{1} ar^{i-1} = a = \dfrac{a(1-r^1)}{1-r}$. Assume S_k and add $ar^{(k+1)-1}$ to obtain $\sum\limits_{i=1}^{k} ar^{i-1} + ar^k =$

$\dfrac{a(1-r^k)}{1-r} + ar^k$. Then, $\sum\limits_{i=1}^{k+1} ar^{i-1} = \dfrac{a(1-r^k)}{1-r} + ar^k = \dfrac{a(1-r^k) + ar^k - ar^{k+1}}{1-r} = \dfrac{a - ar^{k+1}}{1-r}$

$= \dfrac{a(1-r^{k+1})}{1-r}$. Therefore, S_{k+1} holds.

17. $2^5 = 32 > 20 = 4\cdot 5$. So S_5 is true. Assume $2^k > 4k$. Then multiply by 2 to get $2^{k+1} > 8k = 4k + 4k \geq 4k + 20 > 4k + 4 = 4(k+1)$. Therefore, $2^{k+1} > 4(k+1)$ and S_{k+1} holds.

19. For $n = 1$, $a^1 = a < 1$ since $0 < a < 1$ (given). Thus S_1 is true. Assume $a^k < 1$. Then, since $a > 0$, $a^{k+1} < a$. But $a < 1$. Therefore, $a^{k+1} < 1$ and S_{k+1} holds.

21. S_1 is true since $(ab)^1 = a^1 b^1$. Assume that $(ab)^k = a^k b^k$ and multiply by ab to obtain $(ab)^k(ab) = (a^k b^k)ab$; $(ab)^{k+1} = (a^k a)(b^k b)$; $(ab)^{k+1} = a^{k+1} b^{k+1}$. Therefore, S_{k+1} holds.

23. S_1 is true since $|a_0 + a_1| \le |a_0| + |a_1|$. Assume S_k. Then

$$|a_0 + a_1 + \cdots + a_k + a_{k+1}| = |(a_0 + a_1 + \cdots + a_k) + a_{k+1}|$$
$$\le |a_0 + a_1 + \cdots + a_k| + |a_{k+1}| \qquad \text{(by } S_1)$$
$$\le (|a_0| + |a_1| + \cdots + |a_k|) + |a_{k+1}| \qquad \text{(by } S_k)$$
$$= |a_0| + |a_1| + \cdots + |a_{k+1}|$$

Therefore, S_{k+1} holds.

25. (a) Since $\dfrac{a^2 - b^2}{a - b} = a + b = a^{2-1} + b^{2-1}$, S_2 is true. Assume S_k. Then

$$\frac{a^{k+1} - b^{k+1}}{a - b} = \frac{a^k a - b^k b}{a - b}$$
$$= \frac{a^k a - b^k a + b^k a - b^k b}{a - b}$$
$$= \frac{a(a^k - b^k) + b^k(a - b)}{a - b} = \frac{a(a^k - b^k)}{a - b} + b^k$$
$$= a[a^{k-1} + a^{k-2}b + \cdots + ab^{k-2} + b^{k-1}] + b^k \qquad \text{(by } S_k)$$
$$= a^k + a^{k-1}b + \cdots + a^2 b^{k-1} + ab^{k-1} + b^k$$

Therefore, S_{k+1} holds.

(b) Since $\dfrac{a^n - b^n}{a - b} = a^{n-1} + a^{n-2}b + \cdots + ab^{n-2} + b^{n-1}$, multiplying by $a - b$ gives $a^n - b^n = (a - b)(a^{n-1} + a^{n-2}b + \cdots + ab^{n-2} + b^{n-1})$.

27. (a) 1; 4; 9; 16; 25; **(b)** n^2 **(c)** S_1 is true since $1 = 1^2$. Assume S_k and add $k + (k + 1)$ to obtain $1 + 2 + 3 + \cdots + (k - 1) + k + (k + 1) + k + (k - 1) + \cdots + 3 + 2 + 1 = k^2 + k + (k + 1) = k^2 + 2k + 1 = (k + 1)^2$. Therefore, S_{k+1} holds.

29. For $n = 1$, $7^1 - 1 = 6$ which is divisible by 6. So S_1 holds. Assume S_k. Then $7^k - 1 = 6b$ and $7^k = 6b + 1$. Multiply by 7 to get $7^{k+1} = 42b + 7$ and $7^{k+1} - 1 = 42b + 6 = 6(7b + 1)$ which is divisible by 6. Thus S_{k+1} holds.

CHAPTER 11   REVIEW EXERCISES   (page 730)

1. See pages 684, 685   **2.** 1, 4, 7, 10, 13

3. 0, 0.250, 0.296, 0.316, 0.328. 0.335, 0.340; differences: 0.250, 0.046, 0.020, 0.012, 0.007, 0.005

4. 4, 5, 10, 11, 16   **5.** $\frac{2}{3}, \frac{1}{4}, \frac{2}{15}, \frac{1}{12}, \frac{2}{35}$ **6.** $-2, 1, -\frac{1}{2}, \frac{1}{4}, -\frac{1}{8}$ **7.** $-1, \frac{2}{3}, -\frac{3}{2}, \frac{4}{5}, -\frac{16}{15}$ **8.** $\frac{11}{19}$

9. 5, 10, 20, 40, 80   **10.** $-3, 9, -27, 81, -243;\ (-3)^n$ **11.** See page 691   **12.** 105   **13.** 3   **14.** $\frac{3207}{945}$

15. 26   **16.** 70   **17.** $\displaystyle\sum_{k=1}^{6} 4k$ **18.** See page 694   **19.** See page 695   **20.** $-7n + 17$ **21.** -83

22. See page 697   **23.** -1155 **24.** 6,251,000   **25. (a)** -2260; **(b)** $-\frac{220}{3}$ **26.** 8845   **27.** See page 699

28. 8, 15, 22, 29, 36, 43   **29.** See page 704   **30.** See page 704   **31.** $\dfrac{1}{3^{100}}$ **32.** $12\left(-\frac{2}{3}\right)^{n-1}$ **33.** $-\frac{512}{729}$

34. $\frac{1}{27}\left(\frac{1}{27}\right)^{k-1}$; $a_1 = \frac{1}{27}, r = \frac{1}{27}$ **35.** $\pm\frac{1}{2}$ **36.** See page 707   **37. (a)** 0.02222222; **(b)** 0.01818182

38. $800(0.75)^n$; 80.1 feet   **39.** See page 709   **40.** 6, 12, 24, 48, 96 or 6, -12, 24, -48, 96   **41.** See page 713

42. See page 716   **43.** 108   **44.** $\frac{192}{5}$ **45.** $|r| = \frac{5}{3} > 1$ **46.** $\frac{1}{18}$ **47.** No finite sum; $|r| > 1$

48. $\frac{8}{11}$ **49.** 100 in.   **50.** See page 725

51. S_1 is true since $3 = \frac{3}{2}(1)(1 + 1)$. Assume S_k and add $3(k + 1)$: $3 + 6 + 9 + \cdots + 3k + 3(k + 1) =$
$\frac{3}{2}k(k + 1) + 3(k + 1) = 3(k + 1)\left(\frac{1}{2}k + 1\right) = \frac{3}{2}(k + 1)(k + 2)$. Therefore, S_{k+1} holds.

52. S_1 is true since $3 = 3(2^1 - 1)$. Assume S_k and add $3(2^{k+1-1})$: $3 + 6 + 12 + \cdots + 3 \cdot 2^{k-1} + 3 \cdot 2^k =$
$3(2^k - 1) + 3 \cdot 2^k = 3 \cdot 2^k - 3 + 3 \cdot 2^k = 2 \cdot 3 \cdot 2^k - 3 = 3(2^{k+1} - 1)$. Therefore, S_{k+1} holds.

53. S_1 is true since $\dfrac{1}{1 \cdot 3} = \dfrac{1}{2 \cdot 1 + 1}$. Assume S_k and add $\dfrac{1}{[2(k + 1) - 1][2(k + 1) + 1]} = \dfrac{1}{(2k + 1)(2k + 3)}$:
$\dfrac{1}{1 \cdot 3} + \dfrac{1}{3 \cdot 5} + \dfrac{1}{5 \cdot 7} + \cdots + \dfrac{1}{(2k - 1)(2k + 1)} + \dfrac{1}{(2k + 1)(2k + 3)} = \dfrac{k}{2k + 1} + \dfrac{1}{(2k + 1)(2k + 3)}$
$= \dfrac{k(2k + 3) + 1}{(2k + 1)(2k + 3)} = \dfrac{2k^2 + 3k + 1}{(2k + 1)(2k + 3)} = \dfrac{(2k + 1)(k + 1)}{(2k + 1)(2k + 3)} = \dfrac{k + 1}{2(k + 1) + 1}$. Therefore, S_{k+1} holds.

54. S_1 is true since $\dfrac{3}{2}\left(1 - \dfrac{1}{3^1}\right) = \dfrac{3}{2} - \dfrac{1}{2} = 1$. Assume S_k and add $\dfrac{1}{3^k}$: $1 + \dfrac{1}{3} + \dfrac{1}{3^2} + \cdots + \dfrac{1}{3^{k-1}} + \dfrac{1}{3^k} =$
$\dfrac{3}{2}\left(1 - \dfrac{1}{3^k}\right) + \dfrac{1}{3^k} = \dfrac{3}{2}(1) - \dfrac{3}{2}\left(\dfrac{1}{3^k}\right) + \left(\dfrac{3}{2}\right)\left(\dfrac{2}{3}\right)\left(\dfrac{1}{3^k}\right) = \dfrac{3}{2}\left[1 - \left(\dfrac{1}{3^k} - \dfrac{2}{3 \cdot 3^k}\right)\right] = \dfrac{3}{2}\left(1 - \dfrac{3 - 2}{3 \cdot 3^k}\right) =$
$\dfrac{3}{2}\left(1 - \dfrac{1}{3^{k+1}}\right)$. Therefore, S_{k+1} holds.

55. S_5 is true since $3^5 > 27(5)$. Assume $3^k > 27k$. Multiply by 3 to get $3^{k+1} > 81k = 27k + 54k > 27k + 27 = 27(k + 1)$. Therefore, S_{k+1} holds.

56. S_1 holds since $1^2 + 3(1) = 4$, an even integer. Assume $k^2 + 3k$ is even. Then $(k + 1)^2 + 3(k + 1) = k^2 + 2k + 1 + 3k + 3 = (k^2 + 3k) + (2k + 4)$, which is the sum of two even integers and thus is even. Therefore, S_{k+1} holds.

CHAPTER 11   STANDARD ANSWER TEST   (page 732)

1. $1, -1, -1, -\frac{8}{7}; -\frac{800}{97}$ **2.** $\frac{1}{300}$ **3.** $0, 3, 2, 5$ **4.** $-\frac{1}{3}$ **5.** $-\frac{2}{3}$ **6.** $-\frac{76}{15}$ **7.** 21 **8.** 35

9. $12, -3, \frac{3}{4}; -768\left(-\frac{1}{4}\right)^{n-1}$ **10.** 8925 **11.** 26 **12.** $13,845$ **13.** $\displaystyle\sum_{k=1}^{4} 8\left(\dfrac{1}{2}\right)^k = \dfrac{4\left(1 - \dfrac{1}{2^4}\right)}{1 - \dfrac{1}{2}} = \dfrac{15}{2}$

14. $15,554$ **15.** $24\left(1 - \dfrac{1}{2^8}\right) = \dfrac{765}{32}$ **16.** $-\frac{1}{2}$ **17.** 18 **18.** No finite sum since $r = \frac{3}{2} > 1$ **19.** $\frac{6}{115}$

20. $\frac{4}{11}$ **21.** \$652.60   **22.** 250 **23.** 36 feet   **24.** S_1 is true since $5 = \dfrac{5 \cdot 1(1 + 1)}{2}$. Assume S_k and add $5(k + 1)$:
$5 + 10 + \cdots + 5k + 5(k + 1) = \dfrac{5k(k + 1)}{2} + 5(k + 1) = \dfrac{5k(k + 1) + 10(k + 1)}{2} = \dfrac{5(k + 1)(k + 2)}{2} =$
$\dfrac{5(k + 1)[(k + 1) + 1]}{2}$. Therefore, S_{k+1} holds.

25. S_1 is true since $\dfrac{1}{1 \cdot 4} = \dfrac{1}{2(1 + 1)} = \dfrac{1}{4}$. Assume S_k and add $\dfrac{1}{(k + 1)(2k + 2 + 2)}$: $\dfrac{1}{1 \cdot 4} + \dfrac{1}{2 \cdot 6} + \cdots +$
$\dfrac{1}{k(2k + 2)} + \dfrac{1}{(k + 1)(2k + 4)} = \dfrac{k}{2(k + 1)} + \dfrac{1}{2(k + 1)(k + 2)} = \dfrac{k(k + 2) + 1}{2(k + 1)(k + 2)} = \dfrac{k^2 + 2k + 1}{2(k + 1)(k + 2)} =$
$\dfrac{k + 1}{2(k + 2)} = \dfrac{k + 1}{2[(k + 1) + 1]}$. Therefore, S_{k+1} holds.

CHAPTER 11   MULTIPLE CHOICE TEST   (page 734)

1. (b)   **2.** (a)   **3.** (b)   **4.** (a)   **5.** (d)   **6.** (d)   **7.** (c)   **8.** (c)   **9.** (a)   **10.** (d)   **11.** (e)
12. (a)   **13.** (b)   **14.** (c)   **15.** (a)   **16.** (d)   **17.** (b)   **18.** (c)   **19.** (b)   **20.** (c)

CHAPTER 12   PERMUTATIONS, COMBINATIONS, AND PROBABILITY

12.1   Permutations   (page 745)

1. 7   **3.** 66   **5.** 120   **7.** 4   **9.** $\dfrac{n!}{[n-(n-3)]!} = \dfrac{n!}{3!}$ **11.** 15   **13.** 60   **15.** 12   **17.** 6   **19.** 36

21. 504   **23.** 224   **25.** 224   **27.** (a) 120; (b) 12; (c) 48; (d) 72   **29.** 362,880; 40,320   **31.** 720

33. 468,000; 676,000   **35.** (a) 10 (b) 4

37. 36; (1, 1)   (2, 1)   (3, 1)   (4, 1)   (5, 1)   (6, 1)
           (1, 2)   (2, 2)   (3, 2)   (4, 2)   (5, 2)   (6, 2)
           (1, 3)   (2, 3)   (3, 3)   (4, 3)   (5, 3)   (6, 3)
           (1, 4)   (2, 4)   (3, 4)   (4, 4)   (5, 4)   (6, 4)
           (1, 5)   (2, 5)   (3, 5)   (4, 5)   (5, 5)   (6, 5)
           (1, 6)   (2, 6)   (3, 6)   (4, 6)   (5, 6)   (6, 6)

39. (a) 1,000,000,000; (b) 3,628,800; (c) 996,371,200; (d) $10 \cdot 7(9 \cdot 8 \cdot 7 \cdot 6 \cdot 5 \cdot 4) = 4,233,600$

41. (a) 720; (b) 120; (c) 240   **43.** 5   **45.** (a) 6720; (b) 1260; (c) 151,200

12.2   Combinations   (page 752)

1. 10   **3.** 1   **5.** 1   **7.** 4060   **9.** (a) 4845; (b) 116,280   **11.** 20   **13.** 3003   **15.** 190   **17.** 46,200

19. (a) n; (b) $\dfrac{n(n-1)}{2}$ **21.** 70   **23.** (a) 2^2 (b) 2^3 (c) 2^4 (d) 2^5

25. Each time a subset of r elements is chosen out of n elements there are $n - r$ elements left over. Similarly, when $n - r$ elements are chosen out of n elements there are $n - (n - r) = r$ elements left over. Therefore, there must be the same number of subsets of size r, $\binom{n}{r}$, as there are subsets of size $n - r$, $\binom{n}{n-r}$.

27. A set of n elements has a total of 2^n subsets of all possible sizes, including the empty set and the set itself.

29. $4\binom{13}{5} = 5148$ **31.** $13\binom{4}{2}12\binom{4}{3} = 3744$ **33.** (a) 205,931,880; (b) 147,094,200   **35.** 180

37. (a) 1008; (b) 2254   **39.** 256   **41.** $210 \times 5! \times 3! = 151,200$

12.3   The Binomial Theorem   (page 760)

1. $x^5 + 5x^4 + 10x^3 + 10x^2 + 5x + 1$ **3.** $x^7 + 7x^6 + 21x^5 + 35x^4 + 35x^3 + 21x^2 + 7x + 1$

5. $a^4 - 4a^3b + 6a^2b^2 - 4ab^3 + b^4$ **7.** $243x^5 - 405x^4y + 270x^3y^2 - 90x^2y^3 + 15xy^4 - y^5$

9. $a^{10} + 5a^8 + 10a^6 + 10a^4 + 5a^2 + 1$

11. $1 - 10h + 45h^2 - 120h^3 + 210h^4 - 252h^5 + 210h^6 - 120h^7 + 45h^8 - 10h^9 + h^{10}$

13. $\frac{1}{16} - \frac{1}{2}a + \frac{3}{2}a^2 - 2a^3 + a^4$ **15.** $\dfrac{1}{x^6} - \dfrac{6}{x^3} + 15 - 20x^3 + 15x^6 - 6x^9 + x^{12}$

17. $\frac{1}{64}x^6 + \frac{3}{4}x^5y + 15x^4y^2 + 160x^3y^3 + 960x^2y^4 + 3072xy^5 + 4096y^6$ **19.** $8x^3 - 36x^2 + 54x - 27$

21.
```
        1   7   21   35   35   21   7   1
     1   8   28   56   70   56   28   8   1
   1   9   36   84  126  126   84   36   9   1
 1  10  45  120  210  252  210  120  45  10   1
```

23. $x^{10} - 10x^9h + 45x^8h^2 - 120x^7h^3 + 210x^6h^4 - 252x^5h^5 + 210x^4h^6 - 120x^3h^7 + 45x^2h^8 - 10xh^9 + h^{10}$

25. $x^{15} + 15x^{14} + 105x^{13} + 455x^{12} + 1365x^{11} + \cdots + 1365x^4 + 455x^3 + 105x^2 + 15x + 1$

27. $a^{30} - 30a^{29} + 435a^{28} - 4060a^{27} + \cdots - 4060a^3 + 435a^2 - 30a + 1$

29. $108 + 54h + 12h^2 + h^3$ **31.** $6x^5 + 15x^4h + 20x^3h^2 + 15x^2h^3 + 6xh^4 + h^5$ **33.** $-\dfrac{4+h}{4(2+h)^2}$

35. $8064a^5b^5$ **37.** $35x^{-5/2}$ **39.** $48,384x^5y^3$ **41.** $-\dfrac{15309}{8}a^5b^5$

43. $(2 + 0.1)^4 = 2^4 + 4(2^3)(.1) + 6(2^2)(.1)^2 + 4(2)(.1)^3 + (.1)^4 = 19.4481$

45. $(4 - 0.02)^3 = 4^3 - 3(4)^2(0.02) + 3(4)(0.02)^2 - (0.02)^3 = 63.044792$

47. $\dbinom{n}{r-1}a^{n-r+1}b^{r-1}$; $\dbinom{n}{r-1}a^{n-r+1}(-b)^{r-1}$ **49.** $\dbinom{n}{r}a^{n-r}b^r$

12.4   An Introduction to Probability   (page 770)

1. $\frac{1}{8}$ **3.** $\frac{7}{8}$ **5.** $\frac{1}{2}$ **7.** $\frac{1}{2}$ **9.** $\frac{9}{16}$ **11.** $\frac{1}{4}$ **13.** $\frac{3}{8}$ **15.** $\frac{1}{6}$ **17.** $\frac{1}{6}$ **19.** $\frac{1}{9}$ **21.** $\frac{1}{2}$ **23.** $\frac{25}{102}$ **25.** 0

27. $\frac{1}{2}$ **29.** $\frac{1}{2704}$ **31.** $\frac{144}{169}$ **33.** $\frac{1}{169}$ **35.** $\frac{28}{143}$ **37.** $\frac{70}{143}$ **39.** (a) $\frac{1}{1024}$; (b) $\frac{1}{1024}$; (c) $\frac{7}{128}$ **41.** $\frac{77}{102}$

43. (a) $\dfrac{48}{\binom{52}{5}} = 0.0000185$;   (b) $\dfrac{13 \cdot 48}{\binom{52}{5}} = 0.0002401$;   (c) $\dfrac{4\binom{13}{5}}{\binom{52}{5}} = 0.0019808$;   (d) $\dfrac{4}{\binom{52}{5}} = 0.0000015$;

(e) $\dfrac{\binom{4}{2}\binom{48}{3} + \binom{4}{3}\binom{48}{2} + \binom{4}{4}\binom{48}{1}}{\binom{52}{5}} = 0.0416844$

45. 1 to 12; 12 to 1   **47.** 7 to 2   **49.** 3 to 5

51. Let $x = P(E)$. Then $P(\text{not } E) = 1 - x$ and we have the odds for $E = \dfrac{a}{b} = \dfrac{x}{1-x}$. Then,

$$a(1 - x) = bx$$
$$a - ax = bx$$
$$a = (a + b)x$$
$$\frac{a}{a+b} = x \text{ or } P(E) = \frac{a}{a+b}$$

53. $\frac{3}{8}$ **55.** $\dfrac{1}{\binom{45}{5}} \cdot \dfrac{1}{45} \approx \dfrac{1}{55,000,000}$ **57.** $\dfrac{10 \cdot 4^5 - 40}{\binom{52}{5}} = 0.0039$

59. It is a fair game using sums; $P(\text{odd sum}) = P(\text{even sum}) = \frac{1}{2}$. For products it is not a fair game; $P(\text{odd product}) = \frac{1}{4}$ and $P(\text{even product}) = \frac{3}{4}$.

CHAPTER 12   REVIEW EXERCISES   (page 774)

1. See page 739   **2.** See page 740   **3.** See page 743   **4.** See page 744   **5. (a)** 504; **(b)** 120; **(c)** 120

6. 1728   **7.** 6840   **8.** 358,800   **9.** 456,976   **10.** 5040   **11.** See page 748   **12.** See page 749

13. (a) 53,130; **(b)** 1; **(c)** 220   **14.** 64   **15.** 20   **16.** 455   **17.** 4   **18.** 120   **19.** 210   **20.** 792

21. 1, 8, 28, 56, 70, 56, 28, 8, 1   **22.** See page 759   **23.** See page 757

24. $x^5 + 15x^4 + 90x^3 + 270x^2 + 405x + 243$ **25.** $x^5 - 15x^4 + 90x^3 - 270x^2 + 405x - 243$

26. $a^4 + 8a^3b + 24a^2b^2 + 32ab^3 + 16b^4$ **27.** $8a^3 - 36a^2b + 54ab^2 - 27b^3$

28. $16x^4 + 16x^3y + 6x^2y^2 + xy^3 + \dfrac{1}{16}y^4$ **29.** $210a^4b^6$ **30.** $-108{,}864x^3y^5$

31. $\{HHH, HHT, HTH, THH, HTT, THT, TTH, TTT\}$

32. $\frac{1}{8}$ **33.** $\frac{1}{4}$ **34.** $\frac{3}{8}$ **35.** $\frac{1}{9}$ **36.** $\frac{2}{9}$ **37.** $\frac{1}{9}$ **38.** $\frac{1}{6}$ **39.** $\frac{1}{6}$ **40.** $\frac{188}{221}$ **41.** $\frac{25}{102}$ **42.** $\frac{1}{17}$ **43.** $\frac{1}{169}$

44. $\frac{144}{169}$ **45.** $\frac{1}{169}$ **46.** $\dfrac{\binom{13}{3}\binom{39}{2}}{\binom{52}{5}} = 0.0815426$ **47.** $\dfrac{\binom{13}{3}\binom{39}{2} + \binom{13}{4}\binom{39}{1} + \binom{13}{5}\binom{39}{0}}{\binom{52}{5}} = 0.0927671$

48. $\dfrac{\binom{4}{3}(12)\binom{4}{2}}{\binom{52}{5}} = 0.0001108$

CHAPTER 12   STANDARD ANSWER TEST   (page 775)

1. (a) 720; **(b)** 120   **2.** 120   **3. (a)** 900; **(b)** 648   **4.** 455   **5.** 450   **6.** 360   **7.** 420   **8.** 2970

9. 6   **10. (a)** 32; **(b)** $\frac{1}{32}$; **(c)** $\frac{1}{32}$ **11.** $\frac{2}{3}$ **12. (a)** $\frac{1}{36}$; **(b)** $\frac{35}{36}$; **(c)** $\frac{1}{6}$ **13. (a)** $\frac{2}{169}$; **(b)** $\frac{2}{221}$

14. $\dfrac{\binom{12}{4}\cdot 40}{\binom{52}{5}} = 0.0076184$ **15.** $\dfrac{\binom{20}{4}\cdot 15}{\binom{35}{5}} = 0.2238689$ **16. (a)** $\frac{1}{16}$; **(b)** $\frac{1}{16}$ **17.** $\frac{2}{11}$ **18.** $\frac{9}{40}$ **19.** $\frac{39}{80}$

20. $\frac{7}{27}$ **21.** $x^5 - 10x^4y + 40x^3y^2 - 80x^2y^3 + 80xy^4 - 32y^5$ **22.** $\dfrac{1}{1024a^{10}} - \dfrac{5b}{256a^9} + \dfrac{45b^2}{256a^8} - \dfrac{15b^3}{16a^7}$

23. $\binom{11}{6}(3a)^5b^6 = 112{,}266a^5b^6$ **24.** $\binom{16}{8}(2x)^8(-y)^8 = 3{,}294{,}720x^8y^8$ **25.** 92.3521

CHAPTER 12   MULTIPLE CHOICE TEST   (page 777)

1. (c)   **2.** (d)   **3.** (c)   **4.** (b)   **5.** (b)   **6.** (e)   **7.** (d)   **8.** (a)   **9.** (b)   **10.** (c)   **11.** (a)

12. (d)   **13.** (c)   **14.** (a)   **15.** (b)   **16.** (c)   **17.** (d)   **18.** (a)   **19.** (b)   **20.** (c)

Index

BASIC CURVES

Line

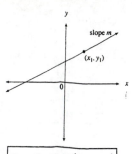

$$y - y_1 = m(x - x_1)$$

Absolute Value

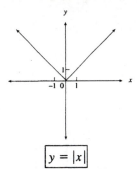

$$y = |x|$$

Greatest Integer

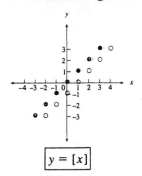

$$y = [x]$$

Parabola

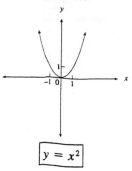

$$y = x^2$$

Cubic

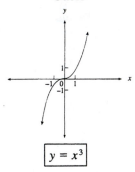

$$y = x^3$$

Reciprocal of x

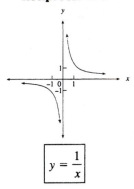

$$y = \frac{1}{x}$$

Reciprocal of x^2

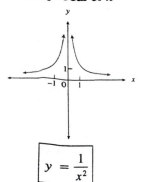

$$y = \frac{1}{x^2}$$

Square Root

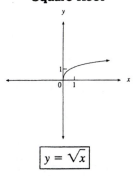

$$y = \sqrt{x}$$

Cube Root

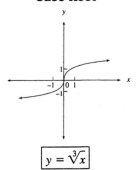

$$y = \sqrt[3]{x}$$

Natural Exponential

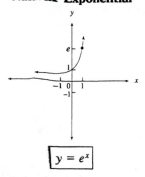

$$y = e^x$$

Natural Logarithm

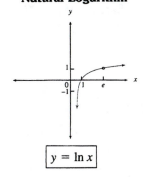

$$y = \ln x$$

Circle

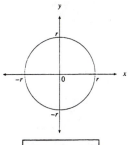

$$x^2 + y^2 = r^2$$

continued on other side

BASIC CURVES

Ellipse

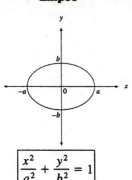

$$\frac{x^2}{a^2} + \frac{y^2}{b^2} = 1$$

Hyperbola

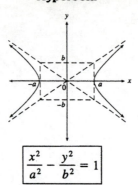

$$\frac{x^2}{a^2} - \frac{y^2}{b^2} = 1$$

Sine

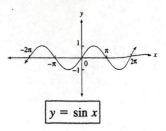

$$y = \sin x$$

Cosine

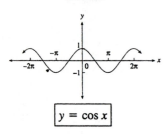

$$y = \cos x$$

Tangent

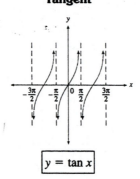

$$y = \tan x$$

Inverse Sine

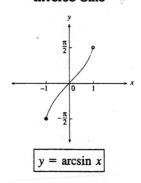

$$y = \arcsin x$$

Inverse Cosine

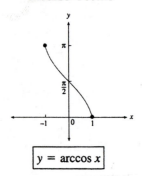

$$y = \arccos x$$

Inverse Tangent

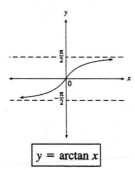

$$y = \arctan x$$

Vertical Translation of $y = f(x)$

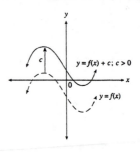

Horizontal Translation of $y = f(x)$

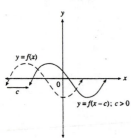

Reflection of $y = f(x)$ in x-axis

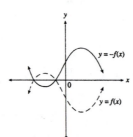

Absolute Value Reflection of $y = f(x)$

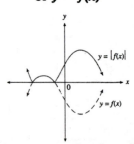

continued on other side

GEOMETRIC FORMULAS

Triangle

$\text{Area} = \frac{1}{2}bh$

Similar Triangles

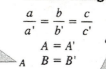 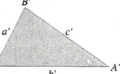

$$\frac{a}{a'} = \frac{b}{b'} = \frac{c}{c'}$$

$A = A'$
$B = B'$
$C = C'$

Right Triangle

Pythagorean theorem
$a^2 + b^2 = c^2$

Equilateral Triangle

$h = \dfrac{\sqrt{3}}{2}a$

$\text{Area} = \dfrac{\sqrt{3}}{4}a^2$

Square

$\text{Area} = s^2$
$\text{Perimeter} = 4s$

Rectangle

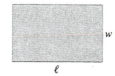

$\text{Area} = \ell w$
$\text{Perimeter} = 2\ell + 2w$

Parallelogram

$\text{Area} = bh$
$\text{Perimeter} = 2a + 2b$

Trapezoid

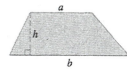

$\text{Area} = \frac{1}{2}h\,(a + b)$

Circle

$\text{Area} = \pi r^2$
$\text{Circumference} = 2\pi r$

Circular Sector

$\text{Area} = \frac{1}{2}r^2\theta$
$\text{Arc length } s = r\theta$

Sphere

$\text{Volume} = \frac{4}{3}\pi r^3$
$\text{Surface Area} = 4\pi r^2$

Right Circular Cylinder

$\text{Volume} = \pi r^2 h$
$\text{Lateral surface area} = 2\pi rh$

Right Circular Cone

$\text{Volume} = \frac{1}{3}\pi r^2 h$
$\text{Lateral surface area} = \pi r\sqrt{r^2 + h^2}$

Rectangular Box

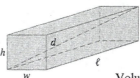

$\text{Volume} = \ell wh$
$\text{Surface Area} = 2\ell w + 2\ell h + 2wh$
$\text{Diagonal: } d^2 = \ell^2 + w^2 + h^2$